COMPANION ENCYCLOPEDIA
OF GEOGRAPHY

COMPANION ENCYCLOPEDIA OF GEOGRAPHY
the environment and humankind

EDITED BY

*IAN DOUGLAS,
RICHARD HUGGETT
and MIKE ROBINSON*

London and New York

First published 1996
by Routledge
11 New Fetter Lane, London EC4P 4EE
29 West 35th Street, New York, NY 10001

First published in paperback 2002

Routledge is an imprint of the Taylor & Francis Group

© 1996, 2002 Routledge

Typeset in Ehrhardt by Solidus (Bristol) Limited
Printed and bound in Great Britain by
TJ International Ltd, Padstow, Cornwall

British Library Cataloguing in Publication Data

A catalogue record for this book is available from the British Library

Library of Congress Cataloging in Publication Data

Companion encyclopedia of geography : the environment and humankind /
edited by Ian Douglas, Richard Huggett, and Mike Robinson.
p. cm.
Includes bibliographical references and index.
(hb: alk. paper)
1. Geography. I. Douglas, Ian. II. Huggett, Richard J.
III. Robinson, M.E. (Michael Ernest)
G116.C64S 1996
910—dc20   96—6097

ISBN 0-415-07417-7 (Hbk)
ISBN 0-415-27750-7 (Pbk)

CONTENTS

Preface   ix
The Contributors   xiii
General Introduction
Richard Huggett and Mike Robinson   1

PART I: A DIFFERENTIATED WORLD
   Introduction
   Richard Huggett and Mike Robinson   11
1   Planet Earth
   Cliff D. Ollier   15
2   The ever-changing climate
   Andrew Goudie   44
3   The biosphere
   Alfred G. Fischer   67
4   Human evolution
   Bernard Wood   86
5   The geography of language
   William C. Brice   107
6   Religion: nature and origins
   E. Geoffrey Parrinder   120
7   The modification of the earth by humans in pre-industrial times
   I. G. Simmons   137

**PART II: A WORLD TRANSFORMED BY THE GROWTH OF
A GLOBAL ECONOMY**
   Introduction
   Richard Huggett and Mike Robinson   159
8   European settlement, 1450–1750
   Len Guelke and Jeanne Kay   162

9　European expansion and land cover transformation
　　Michael Williams 　　182

10　The origins of the capitalist world economy
　　John Langton 　　206

11　Industrialization and world agriculture
　　Brian W. Ilbery and Ian R. Bowler 　　228

12　Changes in global demography
　　John I. Clarke 　　249

13　Origins of modern environmentalism
　　J. M. Powell 　　274

14　The saviour city: beneficial effects of urbanization in England and Wales
　　Brian T. Robson 　　293

15　From a 'cultural' world to a 'political' one
　　Paul Claval 　　310

PART III: THE GLOBAL SCALE OF HABITAT MODIFICATION

　　Introduction
　　Richard Huggett and Mike Robinson 　　329

16　Unity and division in global political geography
　　Peter Taylor 　　332

17　The geography of conflicts and the prospects for peace
　　John O'Loughlin 　　353

18　A new 'geo-economy': patterns, processes, problems
　　Peter Dicken 　　370

19　Third World urbanization
　　Alan Gilbert 　　391

20　From riches to rags: the international debt crisis
　　Stuart Corbridge 　　408

21　Monitoring, modelling and mothering the environment: the impact of science and technology since the Second World War
　　Richard Huggett 　　430

22　Environmentalism on the move
　　Timothy O'Riordan 　　449

PART IV: A WORLD OF QUESTIONS

　　Introduction
　　Richard Huggett and Mike Robinson 　　479

23　Climatic variation and global change
　　F. Kenneth Hare 　　482

24　Ocean uses, environment and management
　　Alastair D. Couper 　　508

25 Water: confronting the critical dilemma
    Peter Crabb   526
26 Surface instability and human modification in geomorphic systems
    Jonathan D. Phillips and William H. Renwick   553
27 The tropical rain forest
    John R. Flenley   573
28 Humanity's resources
    I. G. Simmons   599
29 Environmental hazards
    John Whittow   620
30 The sustainability of sustenance: land and agricultural production
    in the Third World
    William C. Clarke   651
31 Famines and surplus in world food production
    David Grigg   677
32 The nature of Third World cities
    David Drakakis-Smith   702
33 Western cities and their problems
    David T. Herbert   730
34 Changing countrysides
    Hugh Clout   752
35 The quality of life: human welfare and social justice
    David M. Smith   772

PART V: CHANGING WORLDS, CHANGING GEOGRAPHIES
    Introduction
    Richard Huggett and Mike Robinson   793
36 The expansion and fragmentation of geography in higher
    education
    R. J. Johnston   794
37 Achievements of spatial science
    Arild Holt-Jensen   818
38 Geography and humanism in the late twentieth century
    Anne Buttimer   837
39 Structural themes in geographical discourse
    Richard Peet   860
40 Challenging the boundaries: survival and change in a gendered
    world
    Janice Monk   888
41 Place
    Edward Relph   906

PART VI: GEOGRAPHICAL FUTURES

Introduction
Richard Huggett and Mike Robinson   925

42   Concern for geography: a case for equal emphasis of the geographical traditions
Adetoye Faniran   926

43   Home and world, cosmopolitanism and ethnicity: key concepts in contemporary human geography
Yi-Fu Tuan   939

44   Palaeoenvironmental narrative and scenario science
Frank Oldfield   952

45   Geographical futures: some personal speculations
Peter Haggett   965

Index   974

PREFACE

This encyclopedia sprung from a meeting in 1990 between Jonathan Price, then at Routledge, and Ian Douglas. Ian invited us to form an editorial team and we immediately set to work planning the structure of the volume. Routledge gave us *carte blanche* to express a personal view of geography. We decided to present the view that geography is, and at root always has been (despite excursions into spatial science and other exotic themas), about the interdependence of people and their environment, and about the evolving intercourse between humans and their earthly, and to a lesser extent celestial, habitat. Some geographers would doubtless choose to distance themselves from this view of geography, and would argue that the heart of geography lies elsewhere. That, it seems to us, is a healthy state of affairs. It reflects the rich diversity of the subject, the wide range of approaches it embraces, and the many and varied interests of its practitioners. However, we would claim that a large and growing number of geographers do focus on the people-and-environment theme; and that, in doing so, they bring different approaches to bear on the same fundamental issue of interrelationships between the human species and its habitat.

The contributors to the encyclopedia were alerted to, and asked to write within the spirit of, the editors' view of geography. They were also encouraged, however, to employ a style more personal and discursive than is usual in encyclopedic works. Happily, we can report that our grand theme is evident in most of the chapters: many contributors take a global view of their subject and stress human–environment interdependencies.

The chapters are grouped into six parts. Part I focuses on the evolution of the earth through geological, and up to historical, time. It explains how earth became more differentiated by physical and biological transformations, how stores of mineral resources were built up in the process, how it was changed by the life-forms that it supported, and how in turn the evolving life-forms responded to the changing physical environment of the planet. It then considers the origins of the human species – whose activities eventually led to the first great transformation of the planet's surface – and the evolution of their agriculture, languages, cultures, and religions.

Part II considers the second great transformation in the human occupancy of the planet – the unprecedented changes in the ecosphere occurring in the last three centuries brought about by a prolonged phase of rapid population growth and the swift rise of an industrial society powered by fossil fuels. The changes have been so multifarious and so far reaching that almost all the ecosphere now bears the hallmarks of modern human activities.

Part III examines the legacy of the momenta created during the period 1939 to 1946. After the Second World War, habitat modification switched into top gear, and environmental pollution took on a new, far more sinister aspect with the invention and use of unnatural products, often toxic in minute quantities, which, not being readily broken down in natural ecosystems, are long-lived. This part considers too the increased politicization of human habitats, the search for new social orders, and the response of geography as an academic discipline to the emergence of novel and pressing social and environmental problems.

Part IV considers earth as the twentieth century draws to a close. It examines the macro-environments of a habitat facing the threat of new and unfriendly changes, and it addresses the political, social, and cultural worlds that have emerged over the last few decades. It also reflects the growing disillusionment with the consequences of development and with the persistence of inequalities in the distribution of human well-being.

Part V examines some of the responses of the geographical discipline, in particular to the rapidly changing world of the post-war period. And last, in Part VI, four eminent geographers muse on the future of the discipline as they see it.

Putting all this together has been a big job. Even so, we are keenly aware that it is incomplete and that other people might identify topics which we ought to have tackled, and didn't. We are aware, too, that the complexion of the encyclopedia's contributors is more Anglo-American than our 'global' theme deserves. Nevertheless, we want to pay tribute to them: their knowledge, experience and skill have made our work so much easier and it is a privilege to have shared with them in this project. We hope that it approaches their expectations for it.

We want to thank other people, too. At Routledge we found helpful and congenial editors in Jonathan Price, Seth Denbo, and Colville Wemyss. In our own department we have had the invaluable skills of our cartographers Nick Scarle and Graham Bowden. It would be invidious to single out the support of individual academic colleagues, but suffice it to say that Peter Dicken has created an environment where that support is ever-present.

The last words that we will write in this project – the words that, thankfully, signal its end – are a simple record of our indebtedness to family. We dedicate it to our children in the hope that their generation will be more worthy of their world than ours has been.

Richard Huggett and Mike Robinson,
Manchester, September 1995

THE CONTRIBUTORS

IAN R. BOWLER is a Senior Lecturer in the Department of Geography at the University of Leicester, England. He holds BA and PhD degrees from the University of Liverpool, and has specialist research interests in agricultural policy analysis, the Common Agricultural Policy of the European Union and the theory of agricultural change. He is the recipient of research grants from the European Commision, the UK Ministry of Agriculture, Fisheries and Food, and local authorities. Formerly Chair of the Rural Geography Study Group of the Institute of British Geographers, he is currently Chair of the International Geographical Union's Study Group on The Sustainability of Rural Systems. In addition to numerous refereed articles, he is the author or editor of six books, including the edited volume *The Geography of Agriculture in Developed Market Economies* (Longman, 1992).

WILLIAM C. BRICE is Emeritus Professor of Geography of the University of Manchester, England. He served with the Indian Army, during which time he learned Urdu; he is also conversant with Turkish and Arabic. For twenty-five years he was the editor of *Kadmos*, a periodical concerned with early and pre-Greek scripts in the Aegean. He is the author of *South-West Asia* (1966) and editor of *An Historical Atlas of Islam* (1981).

ANNE BUTTIMER is Professor and Head of the Department of Geography at University College Dublin, Ireland. A graduate of University College Cork, she received her PhD at the University of Washington in Seattle and has held research and teaching positions in Belgium, Canada, France, Scotland, Sweden and the USA. She is the author of ten books and over 100 articles on subjects ranging from social space and urban planning to the history of ideas and environmental policy; some of her work has been published in translation in Dutch, French, German, Japanese, Portuguese, Russian, Spanish and Swedish. She has received over a dozen awards and honours, among them the Ellen Churchill Semple Award and the Association of American Geographers' Honors Award. She co-directed an international dialogue project (1978–88) and directed a Swedish-Canadian research exchange on the human use of woodlands (1989–91) and coordinated an EU-sponsored research network on landscape, life and sustainable development with partner teams in Germany, the Netherlands and Sweden (1992–95). She is Secretary of the International Geographical Union Commission on the History of Geography and member of Academia Europaea.

JOHN I. CLARKE is Emeritus Professor of Geography of the University of Durham, England; he was formerly Pro-Vice-Chancellor and Sub-Warden of the University. He is Chairman of the North Durham Health Authority and of the Commission on Population and Environment of the International Union for the Scientific Study of Population. He is a Vice-President and Victoria Medallist of the Royal Geographical Society, and was formerly Chairman of the International Geographical Union Commission on Population Geography. He is the author or editor of over twenty books on population geography and population and the environment, especially with reference to Africa and the Middle East.

WILLIAM C. CLARKE is a fellow at the Institute of Pacific Studies at the University of the South Pacific in Suva, Fiji. After an undergraduate degree in anthropology, he received his PhD from the Geography Department at the University of California at Berkeley. He has pursued research interests in tropical subsistence agriculture, ethnobotany, tourism, soil erosion and conservation issues, while teaching or holding research positions at California State University at Hayward, the University of Hawaii, the Australian National University, the University of Papua New Guinea, Monash University, the University of the South Pacific and the Macmillan Brown Centre for Pacific Studies at the University of Canterbury. His publications include *Place and People: An Ecology of a New Guinean Community* (University of California Press and Australian National University Press, 1971) and *Agroforestry in the Pacific Islands: Systems for Sustainability* (ed. with R.R. Thaman, United Nations University Press, 1993).

PAUL CLAVAL has been Professor of Geography at the Université de Paris-Sorbonne, France, since 1973. From 1949 to 1955 he studied Geography at the Université de Toulouse, and then taught at secondary schools in Bordeaux and Montpellier until 1960. From 1960 to 1972 he was Lecturer and then Reader in geography at the Université de Besançon. He has held visiting professorships at the Université de Sherbrooke, the Université Laval and the Université de Montréal in Canada, at the University of Minnesota, USA, at the Universidad federal de Bahia, Brazil, and at the National Taiwan Normal University. In 1982 he held a French National Fellowship in New Zealand. His main research interests cover the history of geographical thought and its relations to other social sciences. His publications include *Geography since the Second World War:*

An International Survey (ed. with R.J. Johnston, Croom Helm, 1984) and *Géographie culturelle* (Nathan, 1995).

HUGH CLOUT is Professor of Geography and Dean of Social and Historical Studies at University College London, England. He trained as an historical geographer and holds doctorates from the universities of London and Paris. His main research interests are: the rural geography of Western Europe, past and present; regional development in Europe, especially France; and the geography of London. His current research focuses on rural reconstruction in northern France after the First World War. His recent books include *The Land of France 1815–1914* (1982), *A Rural Policy for the EEC?* (1984), *Regional Development in Western Europe* (3rd edn, 1987), *Times London History Atlas* (1991) and *Western Europe: Geographical Perspectives* (3rd edn, 1993).

STUART CORBRIDGE is Lecturer in South Asian Geography at the University of Cambridge, England, and a Fellow of Sidney Sussex College. He has taught previously at London University and Syracuse University, New York state, and has been a Visiting Professor at Jawaharlal Nehru University, New Delhi. His main funded research interest concerns joint forest management in Bihar, India. He is the author of *Capitalist World Development* (Macmillan, 1986), *Debt and Development* (Blackwell, 1993), *Mastering Space: Hegemony, Territory and International Political Economy* (with John Agnew, Routledge, 1995) and *State, Tribe and Region: The Struggle for India's Jharkhand, 1800–1995* (Cambridge University Press, forthcoming). He has edited the volumes *World Economy* (Oxford University Press, 1993) and *Development Studies: A Reader* (Edward Arnold, 1995), and co-edited (with Nigel Thrift and Ron Martin) *Money, Power and Space* (Blackwell, 1994).

ALASTAIR D. COUPER is Professor of Maritime Studies at the University of Wales, Cardiff. He is a graduate in geography from the University of Aberdeen, and gained his PhD from the Australian National University. He has spent some years in the shipping industry and with the United Nations. He is on the Executive Board of the Law of the Sea Institute (USA) and is a member of the Board of Trustees of the National Maritime Museum, Greenwich (UK). He has published widely and is the author of several official reports.

PETER CRABB is Senior Lecturer in Geography in the University College, University of New South Wales, Australian Defence Force Academy, Canberra. He obtained his BSc from the University of Glasgow, his CertEd from the University of London, his MA from the University of Adelaide and his PhD from the University of Hull. He has held appointments at the University of Adelaide, the University of Hull, Memorial University of Newfoundland and Macquarie University. In 1989 he was awarded the Northern Telecom Five Continents Award in Canadian Studies. His main area of interest is in water resources management, especially in Australia and Canada, with a particular concern for large inter-jurisdictional river basins. His numerous publications include: 'Whither the Murray? Politics and the management of Australia's water resources', *Search* 15(1–2) (1984); *Australia's Water Resources: Their Use and Management* (Longman Cheshire, 1986); 'Managing the Murray-Darling Basin', *Australian Geographer* 19 (1988); and *The Murray-Darling Basin: A Resource at Risk* (Longman Cheshire, 1993).

PETER DICKEN is Professor of Geography at the University of Manchester, England. He has held visiting appointments at universities in the USA, Canada, Australia and Hong Kong. He acts as a consultant to the UNCTAD Programme on Transnational Corporations and was an adviser to the Commission on Global Governance. His research focuses on global economic change, the spatial behaviour and strategies of transnational corporations, foreign direct investment (notably Japanese investment) and economic change in East and South-East Asia. He is one of the editors of *Progress in Human Geography*, and he is the author of several books, including *Global Shift: The Internationalization of Economic Activity* (1992), and of some seventy academic papers.

IAN DOUGLAS is Professor of Physical Geography at the University of Manchester, England. He gained his BA and BLitt at Balliol College, University of Oxford, and his PhD at the Research School of Pacific Studies, Australian National University. From 1966 to 1971 he was a lecturer in Geography at the University of Hull, and from 1971 to 1978 he was Professor of Geography at the University of New England, Australia. His books *Humid Landforms* (ANU Press, 1977) and *The Urban Environment* (Arnold, 1983) cover his twin interests in the humid tropics and the urban environment; these interests grew out of his periods of work in Australia and Malaysia. His other publications include *Environmental Change and Tropical Geomorphology* (ed. with T. Spencer, Allen & Unwin, 1985) and papers which have appeared in *Annales de Géographie, Land Degradation and Rehabilitation, Publications of the International Association of Hydrological Sciences, Singapore Journal of Tropical Geography, Transactions of the Institute of British Geographers* and the *Zeitschrift für Geomorphologie*.

DAVID DRAKAKIS-SMITH is Professor of Economic Geography at the University of Liverpool, England. His previous posts have taken him to Hong Kong and Australia, and have enabled him to develop a long-standing interest in Pacific Asia. In recent years this has been supplemented by research and teaching in Southern Africa. His primary interest is in the impact that rapid urban growth has had on the economic and social circumstances of the urban poor and how they cope with this. He has extensive research experience in South-West and South-East Asia, the Pacific and Southern Africa. He is the author of *The Third World City* (Routledge, 1987), *Pacific Asia* (Routledge, 1993) and the co-editor of the Routledge *Introductions to Development* series.

ADETOYE FANIRAN is Professor of Geography at the University of Ibadan, Nigeria. He obtained his BA from the University of London in 1964 and his PhD from the University of Sydney. He joined the faculty of the University of

Ibadan in 1968. He was Visiting Assistant Professor at the University of Georgia in 1971, Fulbright-Hayes Scholar at the University of Wisconsin, Madison, in 1976, and Visiting Professorial Fellow at the Australian National University, Canberra, in 1986. His main academic interests are geomorphology, environment, resources and development. His publications include *A New Approach to Practical Work in Geography* (with H.I. Ajaegbu, Heinemann, 1973), *Essentials of Soil Study* (with O.O. Areola, Heinemann, 1978), *Man's Physical Environment* (with O. Ojo, Heinemann, 1980), *Humid Tropical Geographology* (with L.K. Jeje, Longman, 1983), *African Landform* (Heinemann, 1986) and *General Geomorphology* (with L.K. Jeje, Ibadan University Press, 1995).

ALFRED G. FISCHER is Professor Emeritus of Geology of the University of Southern California, Los Angeles, USA. He obtained his BA and MA from the University of Wisconsin, Madison, and in 1941 took up an appointment at the Virginia Polytechnic Institute, Blacksburg. From 1943 to 1946 he worked for the Stanolind Oil and Gas Co., Kansas, Florida, and part-time from 1946 to 1947 for the Florida Geological Survey. From 1948 to 1949 he held an appointment at the University of Rochester, and from 1949 to 1951 at the University of Kansas. He worked for the International Petroleum Co. Ltd (Esso) in Peru for five years before being appointed to the faculty of Princeton University in 1956, where he was Blair Professor from 1970 to 1984. He moved to the University of Southern California in 1984, becoming Emeritus Professor in 1991. He held an NSF Senior Postdoctoral Fellowship in Innsbruck (1962–3) and a Guggenheim Fellowship (1969–70); he has held visiting appointments at the University of Texas, Scripps Institute of Oceanography, the University of Tübingen, the Technical University of Berlin and the University of Naples. He has received numerous honours, including being admitted to the US National Academy of Sciences in 1994. He has served as an officer of several geological societies, and on various committees for the National Science Foundation, the National Research Council, NASA and Joint Oceanographic Institutions. He has published numerous papers and co-authored several books; he is an editorial consultant to Princeton University Press.

JOHN R. FLENLEY has been Professor of Geography at Massey University, New Zealand, since 1989. He graduated in botany from Cambridge University; his first visit to the rain forest was as a member of the Cambridge Botanical Expedition to Ethiopia in 1957. He completed his PhD at the Australian National University on the history of the New Guinea rain forests; this involved living for one year in the New Guinea highlands. He held appointments as Lecturer, Senior Lecturer and Reader at the University of Hull, England, and undertook extended fieldwork in Indonesian rain forests, studying their history. He was leader of biogeographical expeditions to Krakatau (Indonesia), Easter Island, Tahiti and the Cook Islands. His publications include *The Equatorial Rain Forest: A Geological History* (Butterworths, 1979) and *Easter Island, Earth Island* (with Paul Bahn, Thames and Hudson, 1992).

xvii

ALAN GILBERT is Professor of Geography at University College London, England. He obtained his BSocSci from Birmingham University and his PhD from the London School of Economics. From 1970 to 1990 he was Lecturer and then Reader at University College London and the Institute of Latin American Studies, London. He has been an adviser on employment, housing, urbanization and regional development in developing countries to the Inter-American Development Bank, the United Nations Centre for Human Settlements, UNESCO and the United Nations University; he authored a report on the Colombian economy for *Business Monitor International*. In 1984 he was the recipient of the Gill Memorial Award from the Royal Geographical Society; from 1985 to 1987 he was President of the Society of Latin American Studies. His books include *Latin American Development: A Geographical Perspective* (Penguin, 1974), *Cities, Poverty and Development: Urbanization in the Third World* (with J. Gugler, Oxford University Press, 1982), *Housing, the State and the Poor: Policy and Practice in Three Latin American Cities* (with P.M. Ward, Cambridge University Press, 1985), *The Political Economy of Land: Urban Development in an Oil Economy* (with P. Healey, Gower, 1985), *Latin America* (Routledge, 1990), *Landlord and Tenant: Housing the Poor in Urban Mexico* (with A. Varley, Routledge, 1991), *In Search of a Home: Rental and Shared Housing in Latin America* (with O.O. Camacho, R. Coulomb and A. Necochea, UCL Press and University of Arizona Press, 1993) and *The Latin American City* (Latin American Bureau and Monthly Review Press, 1994).

ANDREW GOUDIE has been Professor of Geography at the University of Oxford, England, since 1984, and is currently a Pro-Vice-Chancellor. He has been instrumental in setting up the new Environmental Change Unit, of which he has been Acting Director. He has been President of the Geographical Association and a Vice-President of the Royal Geographical Society. He is the author or editor of several books, including *Geomorphological Techniques* (editor, 2nd edn, Routledge, 1990), *Environmental Change* (3rd edn, Oxford University Press, 1992), *Desert Geomorphology* (with R.U. Cooke and A. Warren, UCL Press, 1992), *The Human Impact: On the Natural Environment* (4th edn, Blackwell, 1994) and *Atlas of the Environment of the British Isles* (with Denys Brunsden, Oxford University Press, 1995); he is co-editor of *The Encyclopedic Dictionary of Physical Geography* (2nd edn, Blackwell, 1994) and of *The Longman Encyclopedia* (Longman, 1989).

DAVID GRIGG is Professor of Geography at the University of Sheffield, England. Having been educated at Carre's Grammar School, Sleaford, and St John's College, Cambridge, in 1959 he was appointed Assistant Lecturer in Geography at the University of Sheffield. His interests have been in agricultural geography and the historical geography of agriculture; his current principal interest is in the geography of food consumption. His books include *The Harsh Lands, The World Food Problem, Population Growth and Agricultural Change* and *The Transformation of Agriculture in the West*.

LEN GUELKE is Professor of Geography at the University of Waterloo, Ontario, Canada. He is the author of *Historical Understanding in Geography: An Idealist Approach* (Cambridge and New York, 1982) and numerous articles and papers on early South African historical geography.

PETER HAGGETT has been Professor of Urban and Regional Geography at the University of Bristol, England, since 1966. His research interests over the last twenty years have been on the application of geographical models to epidemiological data. His earlier books, including *Locational Analysis in Human Geography* (1965) and *Network Analysis in Geography* (1969), were mainly on spatial analysis; these were followed by general reviews of the structure of geography, including *Geography: A Modern Synthesis* (1972) and *The Geographer's Art* (1990).

F. KENNETH HARE is former Chancellor of Trent University (1988–95) and University Professor Emeritus in Geography of the University of Toronto, Canada. He was educated at King's College London, the London School of Economics and the Université de Montréal. His academic career included appointments as Dean of Arts and Sciences at McGill University, Montreal; Master of Birkbeck College, London; President of the University of British Columbia; Director of the Institute for Environmental Studies at the University of Toronto and Provost of Trinity College at the University of Toronto. He has served with a number of official bodies, foundations, institutions and inquiries in Canada and abroad; he recently served as Chairman of the Technical Advisory Panel on Nuclear Safety, Ontario Hydro, and is a member of the Research and Development Advisory Panel of AECL Research, Canada's research agency in nuclear power technology. He is a Companion of the Order of Canada and in 1989 he received the Order of Ontario; he is a Fellow of the Royal Society of Canada. In 1989, in Geneva, he received the International Meteorological Organization Prize from the World Meteorological Organization, the second Canadian to be so awarded in thirty-four years. His fundamental interest has been and remains the global climate and its stability.

DAVID T. HERBERT is Professor of Geography and Pro Vice Chancellor at the University of Wales, Swansea. In 1962 he was appointed Lecturer in Geography at the University of Keele, and in 1965 moved to the University of Swansea where he was Dean of the Faculty of Economic and Social Studies from 1981 to 1984 and Vice-Principal from 1986 to 1989. He has held visiting appointments in many universities in Canada, the USA, Belgium, Sudan and Poland; his most recent visiting professorship was at the University of Calgary. He has written extensively on the social geography of the city, social problems, urban crime and heritage tourism. His books, sometimes co-authored, include *The Geography of Urban Crime* (Longman, 1982), *The Geography of Crime* (ed. with D.J. Evans, Routledge, 1990), *Crime, Policing and Place* (ed. with D.J. Evans and N.R. Fyfe, Routledge, 1992), *Communities within Cities* (with W.K.D. Davies, Belhaven, 1993) and *Heritage, Tourism and Society* (ed., Cassell/Mansell, 1995).

ARILD HOLT-JENSEN has been Professor of Geography at the University of Bergen, Norway, since 1991. He obtained his MA at the University of Oslo in 1963; he then held an appointment as assistant lecturer at the University of Århus, Denmark, until 1965, when he was appointed lecturer at the University of Bergen. In 1986 he was Visiting Professor at the University of Washington, Seattle; he has been a guest lecturer in Israel, Germany, the UK, the USA, Denmark and Sweden. He was Chairman of the Norwegian Association of Human Geographers from 1981 to 1983 and from 1989 to 1990. His academic interests are development planning and settlement change, and the history of geographic thought. He has written a number of articles and research publications on regional and local planning, settlement change and cultural landscape development in Norway. His books published in English are *The Norwegian Wilderness: National Parks and Related Reserves* (1978) and *Geography: History and Concepts* (1981; revised edn 1988).

RICHARD J. HUGGETT is a Reader in Geography at the University of Manchester, England. He studied geography at University College London, both as an undergraduate and postgraduate. After a brief spell as a geography teacher at the Haberdashers' Aske's School, Elstree, he moved to his current post. His research interests include catastrophism, neodiluvialism, geoecology, mathematical modelling in the environmental and physical geographical sciences, and the history of ideas in the environmental and physical geographical sciences. His publications include *Systems Analysis in Geography* (Clarendon, 1980), *Earth Surface Systems* (Springer, 1985), *Cataclysms and Earth History: the Development of Diluvialism* (Clarendon, 1989), *Catastrophism: Systems of Earth History* (Edward Arnold, 1990), *Climate, Earth Processes and Earth History* (Springer, 1991), *Modelling the Human Impact on Nature: Systems Analysis of Environmental Problems* (Oxford University Press, 1993) and *Geoecology: An Evolutionary Approach* (Routledge, 1995).

BRIAN W. ILBERY is Professor of Human Geography at Coventry University, England. He holds BA and PhD degrees from the University of Wales, Swansea, and has research interests in farm business behaviour, policy analysis of the Common Agricultural Policy of the European Union, and agricultural change in the UK. He has had over seventy refereed papers published and is the author or editor of six books, the most recent being *Agricultural Change in Great Britain* (Oxford University Press, 1992). He is the recipient of research grants from the European Commission, the UK Ministry of Agriculture, Fisheries and Food, Economic and Social Research Council, and Overseas Development Administration. He is currently Chair of the Rural Geography Study Group of the Institute of British Geographers.

R.J. JOHNSTON is Professor of Geography at the University of Bristol, having been Vice-Chancellor of the University of Essex, England, between 1992 and 1995. He obtained his BA and MA degrees from the University of Manchester and his PhD from Monash University. After spending eleven years working at Monash University and the University of Canterbury, in 1974 he became Professor of Geography at the University of Sheffield, where he stayed until 1992. He was President of the Institute of British Geographers in 1990 and has been honoured for his research publications in urban and political geography by both the Royal Geographical Society and the Association of American Geographers.

JEANNE KAY is Professor of Geography and Dean of the Faculty of Environmental Studies at the University of Waterloo, Ontario, Canada. Her research interests include environmental and women's history and religious beliefs about nature.

JOHN LANGTON has been Lecturer in Geography and Fellow of St John's College, University of Oxford, England, since 1980. He obtained his BA and PhD degrees from the University College of Wales, Aberystwyth. He was an Assistant Lecturer in Geography, first at the University of Manchester (1966–8) and then at the University of Cambridge (1968–73) where he became a Fellow of St John's College, Cambridge, in 1970. From 1973 to 1980 he was Lecturer in Geography at the University of Liverpool. His publications have been principally on the economic and regional geography of England in the seventeenth and eighteenth centuries, geographical conceptualization, and the geography of peasant societies in pre-industrial Europe, particularly nineteenth-century Sweden. These include: *Geographical Interpretations of Historical Sources* (with A.R.H. Baker and J.D. Hamshere, 1970); *Geographical Change and Industrial Revolution* (1979); *Countryside and Town in Industrialisation* (with G. Hoppe, 1979); *Town and Country in the Development of Early Modern Western Europe* (with G. Hoppe, 1983); *Atlas of Industrializing Britain* (with R.J. Morris, 1986); *Peasantry and Progress* (with C.G. Clarke, 1990); *Flows of Labour in the Early Phase of Capitalist Development* (with G. Hoppe, 1992); and *Peasantry to Capitalism* (with G. Hoppe, 1994).

JANICE MONK is Executive Director of the Southwest Institute for Research on Women and Adjunct Professor of Geography at the University of Arizona, Tucson, USA. Her research interests lie in social and cultural geography, particularly in feminist research, and in geographic education. She has published many articles and book-chapters, and co-authored or co-edited six books or special journal issues, including *Teaching Geography in Higher Education*, *The Desert Is No Lady: Southwestern Landscapes in Women's Writing and Art* and *Full Circles: Geographies of Women over the Life Course*. She is currently Vice-Chair of the International Geographical Union Commission on Gender and Geography; she served as Scientific Program Chair for the 1992

International Geographical Congress. She was awarded Honors by the Association of American Geographers in 1992 for her contributions to gender studies.

FRANK OLDFIELD was John Rankin Professor of Geography at the University of Liverpool, England, from 1975–96. He obtained his BA and MA degrees from the University of Liverpool, and his PhD from the University of Leicester. After six years at the University as Assistant Lecturer and then Lecturer, in 1964 he became Lecturer in Environmental Sciences at the University of Lancaster. In 1967 he moved to the New University of Ulster as Professor of Geography and Dean of the School of Biological and Environmental Studies. In 1973 he was Deputy and Acting Vice-Chancellor of the University of Papua New Guinea. He then returned to the University of Lancaster as Director of the School of Independent Studies, with a Personal Chair in Geography, where he remained until taking up his Liverpool appointment in 1975. He has held a Leverhulme Fellowship (1977–8) and a Sir Frederick McMaster Fellowship (at the CSIRO Division of Land and Water Resources, Canberra, 1986); in 1992, he was the recipient of the Linton Award from the British Geomorphology Research Group and in 1995, the Murchisan award from the Royal Geographical Society and Institute of British Geographers. In 1981 he was Distinguished Visiting Professor at the Quaternary Research Center, University of Washington, Seattle; from 1987 he has been Advisory Professor of Geography at the East China Normal University in Shanghai. In 1982 and 1986 he acted as Research Adviser to the International Atomic Energy Agency in Vienna. Since 1994 he has been President of the Quaternary Research Association. He has produced over 140 publications on palaeoecology, environmental magnetism, radio-isotope geochronology and environmental change. In 1996 he took up the appointment as Executive Director of the IGBP Past Global Changes (PAGES) Project Office in Bern, Switzerland.

CLIFF D. OLLIER studied geography and geology at Bristol University, gaining a DSc in 1975. He was a demonstrator in economic geology and structural geology at Bristol, and then studied soil science at Rothamsted Agricultural Station before working as a Soil Survey Officer in Uganda for three years. Later posts included Lecturer in Geomorphology at Melbourne University, Head of Geology at the University of Papua New Guinea and Professor of Geography at the University of New England, Australia. He has written over 300 publications, including books on *Weathering*, *Tectonics and Landforms*, *Volcanoes* and *Ancient Landforms*, some translated into Japanese, Russian, Italian and Polish. He has officially retired, but is still working on aspects of weathering and tectonic geomorphology.

JOHN O'LOUGHLIN is Professor of Geography and Professional Staff Member in the Program on Political and Economic Change at the Institute of Behavioral Science, University of Colorado, Boulder, USA. His main research interests are in the political geography of international relations, foreign minorities in West European cities and international political economy. He is Editor for the Americas for *Political Geography*. His recent publications include *The New Political Geography of Eastern Europe* (co-editor, Belhaven, 1993), *Dictionary of Geopolitics* (editor, Greenwood, 1994), *Social Polarization in Post-Industrial Metropolises* (co-editor, Walter de Gruyter, 1996) and *War and its Consequences: Lessons from the Persian Gulf Conflict* (co-editor, HarperCollins, 1994).

TIMOTHY O'RIORDAN is a Professor of Environmental Science at the University of East Anglia, England, and Associate Director of the Centre for Social and Economic Research on the Global Environment. He has taught in Canada, the USA and New Zealand as well as in England. His special research interests cover environmental policy analysis, environmental impact assessment, international environmental governance and the interconnections between the natural and social sciences as they relate to applied environmental problem-solving. He chairs the Environmental Science and Society Programme of the European Science Foundation and also the Environmental Research Working Group of the UK Economic and Social Research Council. He is a Member of the Broads Authority, a National Park organization in England, and chairs its Environment Committee. He has written over 100 articles and authored or co-authored over a dozen books. He has completed an international research project into the response of various European nations to the requirements imposed by the United Nations Framework Convention on Climate Change and is now involved in the relationships between democracy, locality and sustainability.

E. GEOFFREY PARRINDER is Emeritus Professor of the Comparative Study of Religions of the University of London, England, and Fellow of King's College London. He holds MA, DD and PhD degrees from the University of London, and an honorary DLitt from the University of Lancaster. From 1949 to 1958 he was Lecturer and Senior Lecturer at University College, Ibadan, Nigeria. At King's College London from 1958, he was Reader (until 1970) and then Professor (until 1977) in Comparative Study of Religions. In 1964 he was Charles Strong Lecturer in Australia; from 1966 to 1969, Wilde Lecturer in Natural and Comparative Religion at Oxford; and in 1973 Teape Lecturer in Delhi and Madras. From 1977 to 1978 he was Visiting Professor at the International Christian University of Tokyo, and from 1978 to 1982 Visiting Lecturer at the University of Surrey. He is the author of over forty books on world religions, including *African Traditional Religion* (3rd edn, Greenwood, 1970), *Religion in an African City* (Greenwood, 1973), *African Mythology* (repr. Bedrick Peter Books, 1991), *Mysticism in the World's Religions* (Oxford University Press, 1977), *Sex in the World's Religions* (Oxford University Press, 1980), *Jesus in the Qur'an* (Oneworld, 1995), *Avatar and Incarnation* (Oxford University Press, 1983), *Son of Joseph* (T. & T. Clark, 1993), *Dictionary of Non-Christian Religions* (2nd edn, Stanley Thornes, 1981) and *Dictionary of Religious and Spiritual Quotations* (ed., Routledge, 1989).

RICHARD PEET teaches at Clark University, Worcester, Massachusetts, USA. Coming from a working-class upbringing in the north of England, he obtained his BSc at the London School of Economics, his MA from the University of British Columbia and his PhD from the University of California at Berkeley. His research interests include philosophy, social theory, development and the interplay between consciousness and rationality. Editor of *Antipode: A Radical Journal of Geography* from 1970 to 1985, he is now co-editor of *Economic Geography*. He is the author of four books and dozens of articles, most of which are written in a cantankerous, argumentative style.

JONATHAN D. PHILLIPS is Professor of Geography at East Carolina University in Greenville, North Carolina, USA. He received his BA in communications and environmental studies from Virginia Polytechnic Institute, and his PhD in geography from Rutgers University. His academic interests are in fluvial, coastal and soil geomorphology, and surface hydrology; these interests are currently focused on coastal plain environments and on a nonlinear dynamical systems approach to earth surface systems. He has published more than eighty refereed research articles in geomorphology, hydrology, pedology and the environmental sciences, and he co-edited the book *Geomorphic Systems* (1992).

J.M. POWELL holds a Personal Chair in Geography at Monash University, Australia, since 1977. A former editor of *Australian Geographical Studies* and past President of the Institute of Australian Geographers, his major interests focus on the historiography of geography and historical studies of the interpretation and management of natural resources in the modern era. Elected a Fellow of the Academy of Social Sciences in Australia in 1985, he was the recipient of the Royal Society of Victoria's Research Medal for 1988. Of over 200 publications, his most recent books are *An Historical Geography of Modern Australia* (Cambridge University Press, 1988, 1991), *Watering the Garden State* (Allen & Unwin, 1989), *Plains of Promise, Rivers of Destiny* (Boolarong, 1991) and 'MDB'. *The Emergence of Bioregionalism in the Murray–Darling Basin* (Murray Darling Basin Commission, 1993).

EDWARD RELPH teaches geography at Scarborough College, University of Toronto, Canada. He is the author of *Place and Placelessness, The Modern Urban Landscape, The Toronto Guide* and numerous articles on place, phenomenology and landscapes.

WILLIAM H. RENWICK is Associate Professor of Geography at Miami University, Oxford, Ohio, USA. He received his BA in geography from Rhode Island College, and his MA and PhD from Clark University. His academic interests are fluvial and hillslope geomorphology, environmental geography and geographic information systems; recent research has focused on relationships between soil erosion and fluvial sediment yields. He has authored or co-authored a number of books and articles in geomorphology, environmental geography and environmental geology, including co-editing *Geomorphic Systems* (1992) and co-authoring the widely used text *Exploitation, Conservation, Preservation: The Geography of Natural Resource Use.*

MIKE ROBINSON has been a Lecturer in Geography at the University of Manchester, England, since 1970. He gained his BA from the University of Leicester in 1963 and his PhD from the Australian National University in 1967. He has been at the University of Manchester since 1967: to 1968 as a Demonstrator in Geography, from 1968 to 1970 as an Assistant Lecturer, and from 1970 as a Lecturer. From 1985 to 1987 he held a visiting appointment as a part-time Lecturer in Geography at De La Salle College. In 1993, with Dr D.W. Shimwell, he was responsible for the establishment of the Palaeoecological Research Unit at the University of Manchester. His research activity has fallen into four overlapping phases which have coincided with gradually developing curiosities; these are historical geography, cognitive mapping and environmental perception, applied geography, and landscape history. His publications include *The New South Wales Wheat Frontier 1851–1911* (Australian University Press, 1976), *Ceremony and Symbolism in the Japanese Home* (with M. Jeremy, Manchester University Press, 1989), many papers and reviews in academic and professional journals, several official reports, and contributions to edited collections.

BRIAN T. ROBSON is Pro-Vice-Chancellor at the University of Manchester, England, where he has been Professor of Geography since 1977. Prior to that, he taught at the University of Cambridge. He is involved in national research issues as vice-chair of the Economic and Social Research Council's Research Programmes Board and, recently, as President of the Institute of British Geographers; in Manchester and the North West of England, he plays an active role in voluntary-sector bodies. He is Director of Manchester University's Centre for Urban Policy Studies (CUPS) and, through the Centre, has undertaken a number of research projects for UK central government departments; recent publications from this work include *Assessing the Impact of Urban Policy* (Robson *et al.*, Department of the Environment, HMSO, 1994), *Relative Deprivation in Northern Ireland* (Robson *et al.*, Northern Ireland Office, 1994), *Index of Local Conditions, 1991* (Department of the Environment, HMSO, 1995) and *The Economic and Social Impact of Greater Manchester's Universities* (Robson *et al.*, Manchester University, 1995). He is the author of numerous academic books and articles; his two most recent books are *Managing the City* (1987) and *Those Inner Cities* (1988).

I.G. SIMMONS is Professor of Geography at the University of Durham, England. He holds the degrees of BSc and PhD from the University of London and DLitt from the University of Durham. He has been elected to Fellowship of the Society of Antiquaries of London and of the Academia Europaea. His detailed research is in the impact of the later Mesolithic culture on the ecosystems of upland England, and he has written more generally on environmental change at human hands in *Changing the Face of the Earth* (Blackwell, 1996), *Earth, Air and Water* (Edward Arnold, 1991) and *Environmental History* (Blackwell, 1993). A growing interest in environmental thought is signalled by *Interpreting Nature* (Routledge, 1993).

DAVID M. SMITH is Professor of Geography at Queen Mary and Westfield College, University of London, England. After gaining his PhD at the University of Nottingham, he worked in planning for two years before joining the staff of the School of Geography at the University of Manchester in 1963. In 1966 he took up a post at Southern Illinois University, followed by one at the University of Florida which he left in 1972. After a year in South Africa (at the Universities of Natal and of the Witwatersrand) and six months in Australia (at the University of New England), he was appointed to his present chair. While based in London he has made frequent return visits to the USA and South Africa for research purposes, and added Eastern Europe to his regional interests. He is author or editor of more than twenty books, including *Industrial Location: An Economic Geographical Analysis* (Wiley, 1971; revised 1981), *Human Geography: A Welfare Approach* (Edward Arnold, 1977), *Urban Inequality under Socialism: Case Studies from Eastern Europe and the Soviet Union* (Cambridge University Press, 1989), *The Apartheid City and Beyond: Urbanization and Social Change in South Africa* (ed., Routledge, 1992) and *Geography and Social Justice* (Blackwell, 1994).

PETER TAYLOR is Professor of Geography at Loughborough University, England. He is the author of many books and articles on world-systems analysis and political geography including *Political Geography: World-Economy, Nation-State and Locality* (3rd edn, 1993), *Britain and the Cold War: 1945 as Geopolitical Transition* (1990) and *The Way the Modern World Works: World Hegemony to World Impasse* (1996). He is editor and co-editor respectively of the journals *Political Geography* and *Review of International Political Economy* as well as joint editor of three volumes of essays on global issues: *World in Crisis: Geographical Perspectives* (2nd edn, 1989), *World Cities in a World-System* (1995) and *Geographies of Global Change: Remapping the World in the Late Twentieth Century* (1995).

YI-FU TUAN has been J.K. Wright and Vilas Professor at the University of Wisconsin-Madison, USA, since 1983. He read geography at Oxford between 1948 and 1951, and then went to Berkeley where he earned his PhD in 1957. He has taught at Indiana University, the University of New Mexico, the University of Toronto and the University of Minnesota. His publications include *Pediments in Southeastern Arizona* (1959), *Hydrological Cycle and the Wisdom of God* (1968), *China* (1969), *Topophilia* (1974), *Space and Place* (1977), *Landscapes of Fear* (1980), *Segmented Worlds and Self* (1982), *Dominance and Affection* (1984), *The Good Life* (1986), *Morality and Imagination* (1989) and *Passing Strange and*

Wonderful (1993). His current book-length manuscript is entitled *Cosmos and Hearth: A Cosmopolite's Viewpoint.*

JOHN B. WHITTOW was Senior Lecturer in Geography at Reading University, England, from 1973 until retirement in 1994. He was educated at Reading University, gaining his BA in 1952, his DipEd in 1953 and his PhD in 1957. From 1954 to 1957 he taught at Magee University College, Londonderry, Northern Ireland, and at Makerere University College, Uganda between 1957 and 1960. After a year teaching at the University of California at Los Angeles, in 1961 he joined the faculty of the University of Reading. He has held visiting appointments at the University of New England, Australia (1966), and Oxford University (1968). His other appointments include: editorial board, *Regional Studies*, 1975–9; Committee on Environmental Impact Analysis, Royal Town Planning Institute, 1979; Chairman, Landscape Research Group, 1973–6; Working Party on Investigation of Historic Landscapes, 1979–83; and Steering Group on Natural Disasters in Megacities, Institution of Civil Engineers, 1992–5 (part of the International Decade of Natural Disaster Reduction programme). He has published twelve books, including *Disasters: The Anatomy of Environmental Hazards* (Penguin, 1980) and *The Penguin Dictionary of Physical Geography* (Penguin, 1984), and sixty-eight refereed articles.

MICHAEL WILLIAMS is a Reader in Geography at Oxford University, England, and Sir Walter Raleigh Fellow of Oriel College. He is Director of the MSc course in Environmental Change and Management at the Environmental Change Unit, Oxford. He has previously lived and taught in Australia and the USA. In 1989 he was elected Fellow of the British Academy and Honorary Fellow of the American Forest History Society. He has a long-standing interest in initial settlement and landscape evolution in Britain, Australia and the USA; he is particularly interested in the processes of wetland draining and forest clearing, and is currently involved in assessing global transformations of land use and land use cover over time. He is the author of over eighty articles and book chapters; his books include *The Draining of the Somerset Levels* (Cambridge University Press, 1970), *The Making of the South Australian Landscape* (Academic Press, 1974), *Australian Space, Australian Time* (with J.M. Powell, Oxford University Press, 1975), *The Changing Rural Landscape of South Australia* (Heinemann, 1977), *Americans and their Forests* (Cambridge University Press, 1989), *Wetlands: A Threatened Landscape* (Blackwell, 1991), *Planet Management* (Andromeda/OUP, 1992). He is currently completing a book entitled *Deforesting the Earth* (Chicago University Press). He was editor of the *Transactions of the Institute of British Geographers* from 1983 to 1987 and is currently joint editor of *Progress in Human Geography* and *Global Environmental Change.*

BERNARD WOOD is Derby Professor of Anatomy and Chairman of the Department of Human Anatomy and Cell Biology at The University of

Liverpool, England. He leads a Hominid Palaeontology Research Group which has an international reputation. His publications include over 100 papers and book chapters. Books written or edited include the monograph *Hominid Cranial Remains* (Koobi Fora Research Project, Volume 4; Clarendon, 1991) which is both a detailed study of these remains from Koobi Fora and a reassessment of early hominid evolutionary history. His research interests centre on reconstructing the evolutionary history of early hominids, including the investigation of cranio-dental remains for evidence of form–function relationships and adaptations.

GENERAL INTRODUCTION

Richard Huggett and Mike Robinson

The distinctiveness of the geographical discipline has always lain in the extent to which it accommodates a concern for the physical as well as for the human environment and in its early vision of an integrated and holistic world: a habitat; a place of organic life. In practice, and in relatively recent times, this view has been severely strained by the widening gap between physical and human geography, by the tensions created between competing explanatory paradigms – especially in human geography – and by a multiplicity of steadily narrowing specialisms. Superficially, the outcome may seem to threaten the academic integrity and coherence of the discipline. Yet this coherence is arguably more necessary now than at any time in the past, as habitat pressures strain the weakening seams of a stretched earth. This encyclopedia, therefore, offers a view of geography from an earlier, wider, and sometimes bolder perspective by bringing together a wide variety of contributions from scholars of both the physical and human environments, and by focusing their expertise on the large theme of the evolution of the earth as a habitat. In the process, they illustrate not only the essential character of the geographical tradition but the centrality of its holistic vision to the most fundamental problems of habitat change and survival.

THE HOLISTIC TRADITION

Speculation on the interdependencies between natural phenomena and on the essential unity of all living things is probably as old as the human species. We know that by Classical times Herodotus and Plato thought that all life on earth acts in concert and maintains a stable condition. Plato envisaged a balance of nature in which the organisms are seen to be parts of an integrated whole, in the same way that organs or cells are integrated into a functioning organism itself. In his *Timaeus*, Plato wrote of how the creator made

> this world a single complete whole, consisting of parts that are wholes, and subject neither to age nor to disease. The shape he gave it was suitable to its nature. A suitable shape for a living being that was to contain within itself all living beings would be a

1

figure that contains all possible figures within itself. Therefore he turned it into a rounded spherical shape ... And he put soul in the centre and diffused it through the whole and enclosed the body in it.

(Plato 1971 edn: 45–6)

The holistic unity of Nature was a theme that re-emerged in the medieval period and through the Renaissance. The idea of holism, in which Nature is seen as an indivisible unity, has waxed and waned with the relentlessness of lunar tides throughout the modern period. Holistic views were fashionable in the late eighteenth and early nineteenth centuries. Johann Reinhold Forster, in his *Observations made during a Voyage round the World* (1778), presented the natural world as a unified and unifying whole, and attempted to weave into a coherent pattern the physical geography and climate of places with their plant life, and animal life, and human occupants (including agricultural practices, local manufactures, and customs). Gilbert White, author of the celebrated *Natural History of Selbourne* (1789), studied Nature as an interdependent whole rather than as a series of individual parts (Worster 1994: 20). James Hutton saw the world as an organic whole, floating the interesting notion, not without its precursors, that the rock cycle is comparable to the life cycle of an organism: the circulation of blood, respiration, and digestion in animals and plants having their equivalents in terrestrial processes. The idea of an organic planet was embraced and elaborated by several German philosophers including Friedrich Wilhelm Joseph Schelling and Georg Wilhelm Friedrich Hegel (Marshall 1992: 289–94).

Jean-Baptiste Pierre Antoine de Monet, Chevalier de Lamarck presented a unified system of Nature, in which life and its physical environment constantly interacted. He believed that the study of the earth should include considerations of the atmosphere (meteorology), the external crust (hydrogeology), and living organisms (biology). In particular, he maintained that the science of life (biology) could only be appreciated fully by taking into account the earth's crust and the atmosphere: living phenomena, for him, did not stand in isolation; they are part of a larger whole which we call 'Nature'. Only by recognizing the constant interaction between the living and non-living worlds, therefore, could sense be made of living things (see Jordanova 1984: 45). Alexander von Humboldt, famed for his concept of climatic zonality and its influence on vegetation, possessed a grand, holistic vision of Nature. Early acquaintance with Johann Wolfgang von Goethe, the great Romantic philosopher and poet, and a knowledge of the philosophical ideals of Immanuel Kant's universal science no doubt prompted him to think in this way. The following extracts from Humboldt's last book, *Cosmos*, a remarkable tome that describes his grand vision of the entire universe, capture the quintessence of his thinking:

In considering the study of physical phenomena, not merely in its bearings on the material wants of life, but in its general influence on the intellectual advancement of mankind, we find its noblest and most important result to be a knowledge of the chain of connection, by which all natural forces are linked together, and made mutually

2

dependent upon each other; and it is the perception of these results that exalts our views and ennobles our enjoyments.

(Von Humboldt 1849: 1)

Nature considered *rationally*, that is to say, submitted to the process of thought, is a unity in diversity of phenomena; a harmony, blending together all created things, however dissimilar in form and attributes; one great whole (το παν) animated by the breath of life.

(Von Humboldt 1849: 2–3, emphasis in original)

During the nineteenth century, Humboldtian holism was a common theme in biological, geographical, and geological discourse. Mary Somerville (1834) emphasized the connections of the physical sciences and sought to integrate the diverse elements of the organic and inorganic worlds into an ordered whole. Karl Ritter expressed the *Zusammenhang*, or 'hanging-togetherness', of all things. To Ritter, the earth was not a dead, inorganic planet, but one great organism with animate and inanimate components (Ritter 1866). A similar notion of the world as an organism was also suggested by Ritter's pupil, Arnold Henri Guyot (1850).

In the late nineteenth and early twentieth centuries, holism was revived by eminent thinkers in a range of disciplines: Henri Bergson in biology, Herbert Spencer in sociology and ecology, Friedrich Ratzel in geography, Frederic E. Clements and William Morton Wheeler in ecology, Alfred North Whitehead in biology and ecology, Lewis Mumford and Robert Park in sociology, members of the Gestalt movement in psychology, and Jan Christian Smuts in just about all fields. In his *Politische Geographie* (1897), Ratzel wrote of the earth as an organism and the union of people and the land. Paul Vidal de la Blache followed Ritter in regarding the earth as a unitary whole in which harmonious relationships appear at all scales from local to global. Jean Bruhnes and Lucien Febvre also had organismic proclivities. In the United States, Charles Redway Dryer (1920) wrote of the earth as an organism, and suggested that geography is its anatomy, physiology, and psychology. Several geographers used the organic analogy as an analytical tool. Andrew John Herbertson (1905, 1913) used the term macro-organism for the complex of physical and organic elements of the earth's surface. He compared the soil with flesh, vegetation to skin with its parasitic fauna, and water to blood, stirred daily and seasonally by solar energy and humans with nerve cells. Later work on regional methodology, including that of John Frederick Unstead, pursued this organismic conception of regions, and in cultural geography both Carl Ortwin Sauer and Wilbur Zelinsky developed a belief in the superorganic and holistic nature of culture. To Sauer, 'Culture is the agent, the natural area is the medium, the cultural landscape is the result' (see Duncan 1980).

Spencer's organicism was a fount of inspiration for Frederic E. Clements (1916), a botanist from Nebraska. Clements stressed the overriding role of climate in determining the course of vegetational change through what he called 'seres', and the establishment of stable, self-perpetuating vegetation types –

3

climatic climax formations. His concepts of succession and climatic climax formations were unashamedly holistic. To him, the climax formation, the basic unit of vegetation was an organic entity that, like an organism, grows, matures, and dies. The climax formation is the adult organism; succession is the reproductive process. Rival formations struggle to survive, those best adapted to the prevailing climate winning through. The notion of organic succession was also reflected in the thinking of human geographers, most clearly, perhaps, in Derwent Whittlesey's (1929) concept of 'sequent occupance'.

The holistic arguments of late nineteenth- and early twentieth-century geographers shaped the present status of geography as an independent discipline. However, during the 1920s and 1930s the popularity of the organic analogy declined. The so-called quantitative revolution, which started up in the mid-1940s and was in top gear by the late 1950s and early 1960s, then opened up the injurious rift between the human and physical branches of geography and led to the fragmentation of geographical knowledge. A 'quantitative' geography emerged to provide a more general, nomothetic, process-based form of exploration, a sort of space science using methods of locational analysis. Much of this work was based upon theories borrowed from other disciplines, and particularly economics: a large portion of locational theory was drawn from the ideas of Johann Heinrich von Thünen, Alfred Weber, and August Lösch. To cut an exceedingly long story very short, locational analysis, with its avowed focus on geographical pattern and process, though it led to several important discoveries, failed to produce an entirely satisfactory methodology for investigating interrelationships between humans and their environment, despite valiant attempts to do so by Richard Chorley and Peter Haggett. It lacked a philosophical base for including humanistic elements into the rather clinical, mechanistic theories of spatial systems. A result of this deficiency in 'quantitative' models has been an increasing focus on humanism and probabilism in human geography, and more recently the emergence of post-modern geographies which resist the tendency to 'closure' which is inherent in meta-theory. Physical geography, by contrast, has continued the exploration of mechanistic models, though inevitably of an increasingly sophisticated kind.

The editors of this encyclopedia believe that the physical–human cleavage within geography can be bridged by a return to geography's holistic roots. To succeed, this return must be made cautiously for, as David Stoddart (1986: 242) noted, the organic analogy as employed by early twentieth-century geographers produced little in the way of substantive results or new lines of investigation. He suggested that this was because the organic theme was superficially satisfying, and appeared to provide deep insights, but it posed no questions and therefore furnished no answers. An alternative explanation is that the early espousers of the organic analogy in geography did not have the wherewithal to analyse problems holistically – systems methods designed to deal with wholes had not been invented. Methods for studying the interdependence of geographical systems are now available and have been used to provide fresh insights about geographical and environmental problems. A new holism has emerged in the geographical and environmental sciences in which the planet Earth is seen as a unitary system, a set of interrelated components that behave holistically. This should allow the reunion of physical and human geographers, at least those who have not strayed too far from their roots, to meet the challenges of big questions posed by the modern world.

THE NEW HOLISM

In the new holism, system dynamics are defined, not by some loose analogy with organic processes, nor by some nice regulation of a stable or climax state, but by the individualistic behaviour of system components under conditions of instability, well away from equilibrium and near the edge of chaos. From these richly complex interactions emerges an evolving order – a hierarchy of spatial systems whose properties are holistic and cannot be reduced to the laws of physics and chemistry. This immensely potent idea may be traced back to Ludwig von Bertalanffy and the foundations of the general systems approach. In the mid-1920s, Bertalanffy had advocated an organismic conception of biology. Later, he explored the implications of viewing organisms as open systems, and couched the dynamics of biological systems in terms of simultaneous differential equations. To an extent, this built on the work of Alfred Lotka in the field of population ecology, and it was bolstered by Ilya Prigogine's work on open systems. An outcome of Bertalanffy's investigations was the emergence in the early 1950s of a general systems theory, at the core of which was the idea of open systems. The systems approach was soon applied throughout the earth and life sciences, including ecology, geography, geomorphology, meteorology, and pedology. It eventually led to the bold-faced reappearance of holism in two related guises: in the Gaia hypothesis and in the concept of the ecosphere.

The Gaia hypothesis is the latest recasting of the ancient, holistic belief that there exists interconnectedness and harmony among the phenomena of Nature. It was first suggested by the atmospheric chemist James Lovelock, supported by microbiologist Lynn Margulis, and named by novelist William Golding. At least two versions of the Gaia hypothesis have evolved: weak Gaia and strong Gaia (Kirchner 1991). Weak Gaia is the assertion that life wields a substantial influence over some features of the abiotic world, notably the temperature and composition of the atmosphere. In other words, it makes the simple proposal that the earth's climate and surface environment are actively regulated by animals, plants, and micro-organisms. Strong Gaia is the unashamedly teleological idea that the earth is a superorganism which controls the terrestrial environment to suit its own ends, whatever they might be. Lovelock seems to favour strong Gaia. He believes that it is useful to regard the earth, not as an inanimate globe of rock, liquid, and gas, driven by geological processes, but as a biological superorganism, a single life-form, a living planetary body that adjusts and regulates the conditions in its surroundings to suit its needs. For

Lovelock, Gaia and the biosphere are different things:

> The name of the living planet, Gaia, is not a synonym for the biosphere. The biosphere is defined as that part of the Earth where living things normally exist. Still less is Gaia the same as the biota, which is simply the collection of all individual living organisms. The biosphere and the biota taken together form part but not all of Gaia. Just as the shell is part of a snail, so the rocks, the air, and the oceans are part of Gaia.
>
> (Lovelock 1988: 10)

Lynn Margulis, on the other hand, appears to prefer a weak version of Gaia. She sees it as a hypothesis about the earth, its surface sediments, and its atmosphere, that involves the interaction of the biota with surficial materials creating anomalies of temperature, chemical composition, and alkalinity (Margulis and Hinkle 1991: 11). Margulis chooses to restrict Gaia to the surface features of the earth, simply because they can be observed: 'there is no way of testing whether the planet's organismic behaviour is limited to the surface or if the biosphere is just the skin of a global creature, alive from its molten heart goes glub, glub, glub, glub, on out' (Joseph 1990: 174). However, Herbert R. Shaw (1994: 246) concludes that 'the magmatic system is the medium and the mechanism by which the evolution of the biosphere has maintained a communication with the earth's interior via plate-tectonic processes'.

The ecosphere is the global ecosystem. It is the totality of living organisms and the inorganic environment which sustains them (Cole 1958; see Huggett 1995). By this definition, it is the same thing as Lovelock's Gaia without any vitalistic overtones. The ecosphere is a holistic concept, embodying the notion that 'everything is connected to everything else' (Commoner 1972). This is a concept that has found widespread favour with ecologists and geographers who advocate a unified approach to the study of life, soils, and the environment. For instance, Walter and Breckle (1985: i) openly encourage 'an integrative or holistic approach, by treating the many results of analytical–ecophysiological research not in isolation, but as parts of a whole', and bewail the fact that, although there is an abundance of detailed information amassed by specialists on narrowly defined problems, almost no attempt has been made at a synthesis, or what they call 'a presentation of relationships on a grand scale, with the whole as the starting point'. Likewise, the newly emerged global ecologists with their infant science of global ecology consider the earth as a single system with global dimensions (e.g. Rambler *et al.* 1989). Surprisingly perhaps, the philosophy of the ambitious and far-sighted International Geosphere–Biosphere Programme is also broadly holistic. And it is surprising because, as Friedman (1985: 21) writes, 'Scientists who emphasize the holistic nature of a geosphere–biosphere study program are often put down by colleagues as generalists who "Know nothing about everything".' They also run the risk of being confounded with one of those dreaded Gaians. Holism is advocated, however, because,

> The real connections that link the geosphere and biosphere to each other are subtle, complex, and often synergistic; their study transcends the bounds of specialized, scientific disciplines and the scope of limited, national scientific endeavours. . . . The

concept of an International Geosphere-Biosphere Program (IGBP) . . . calls for a . . . bold, 'holistic' venture in organized research – the study of whole systems of interdisciplinary science in an effort to understand global change in the terrestrial environment and its living systems.

(Friedman 1985: 20–1)

HABITAT EARTH

The basic theme of the encyclopedia is the earth as a habitat. It was chosen to draw attention to earth as the place where life exists, as the home of living things. The concept of the ecosphere provides a fitting framework within which to study this theme. The hope is that by returning to geography's holistic roots, albeit in the modern context, the gap between physical and human geographers will be bridged, the tensions between competing explanatory paradigms will be eased, and the multiplicity of narrow specialisms will be balanced by a more general approach. At the very least the range and variety of material that the encyclopedia contains may echo a message that geography, for all its apparent discontinuities, has the potential to lie at the heart of fundamental problems of habitat change and human impact, and should have much to say about the vital issue of planetary management. The structure of this material is chronological for two basic reasons: first, because chronology brings into sharp focus the earth as an evolving habitat for plants, animals, and people; and second, because it reinforces the indivisibility of space-with-time in the geographical perspective of earth as a changing habitat for all life forms.

REFERENCES

Clements, F.E. (1916) *Plant Succession: An Analysis of the Development of Vegetation*, Washington: Carnegie Institute, Publication No. 242.
Cole, L.C. (1958) 'The ecosphere', *Scientific American* 198, 83–96.
Commoner, B. (1972) *The Closing Circle: Confronting the Environmental Crisis*, London: Jonathan Cape.
Dryer, C.R. (1920) 'Genetic geography: the development of the geographic sense and concept', *Annals of the Association of American Geographers* 10, 3–16.
Duncan, J.S. (1980) 'The superorganic in American cultural geography', *Annals of the Association of American Geographers* 70, 181–98.
Forster, J.R. (1778) *Observations made during a Voyage round the World, on Physical Geography, Natural History, and Ethnic Philosophy. Especially on: 1. The Earth and Its Strata; 2. Water and the Ocean; 3. The Atmosphere; 4. The Changes of the Globe; 5. Organic Bodies; and 6. The Human Species*, London: printed for G. Robinson.
Friedman, H. (1985) 'The science of global change – an overview', pp. 20–52, in T.F. Malone and J.G. Roederer (eds) *Global Change*, Cambridge: published on behalf of the ICUS Press by Cambridge University Press.
Guyot, A.H. (1850) *The Earth and Man: Lectures on Comparative Physical Geography, in its Relation to the History of Mankind* (translated by Cornelius Conway Fulton), London: Richard Bentley.
Herbertson, A.J. (1905) 'The major natural regions of the world: an essay in systematic geography', *Geographical Journal* 25, 300–12.

Herbertson, A.J. (1913) 'The higher units: a geographical essay', *Scientia*, 14, 199–212; reprinted in *Geography* 50, 332–42.

Huggett, R.J. (1995) *Geoecology: An Evolutionary Approach*, London: Routledge.

Humboldt, A. von (1849) *Cosmos: A Sketch of a Physical Description of the Universe*, Vol. I (translated from the German by E.C. Otté), London: Henry G. Bohn.

Jordanova, L.J. (1984) *Lamarck*, Oxford: Oxford University Press.

Joseph, L.E. (1990) *Gaia: The Growth of An Idea*, Harmondsworth, Middlesex: Arkana.

Kirchner, J.W. (1991) 'The Gaia hypotheses: are they testable? Are they useful?', pp. 38–46, in S.H. Schneider and P.J. Boston (eds) *Scientists on Gaia*, Cambridge, Mass. and London, England: The MIT Press.

Lovelock, J.E. (1988) *The Ages of Gaia: A Biography of Our Living Earth*, Oxford: Oxford University Press.

Margulis, L. and Hinkle, G. (1991) 'The biota and Gaia: 150 years' support for environmental sciences', pp. 11–18, in S.H. Schneider and P.J. Boston (eds) *Scientists on Gaia*, Cambridge, Mass. and London, England: The MIT Press.

Marshall, P. (1992) *Nature's Web: An Exploration of Ecological Thinking*, London: Simon & Schuster.

Plato (1971 edn) *Timaeus and Critias* (translated with an introduction and appendix on *Atlantis* by Desmond Lee), Harmondsworth, Middlesex: Penguin Books.

Rambler, M.B., Margulis, L. and Fester, R. (eds) (1989) *Global Ecology: Towards a Science of the Biosphere*, Boston: Academic Press.

Ratzel, F. (1897) *Politische Geographie*, München: R. Oldenbourg.

Ritter, K. (1866) *The Comparative Geography of Palestine and the Sinaitic Peninsula*, 4 vols (translated and adapted for the use of Biblical students by W.L. Cage), Edinburgh: William Blackwood.

Shaw, H.R. (1994) *Craters, Cosmos, and Chronicles: A New Theory of Earth*, Stanford, Calif.: Stanford University Press.

Somerville, M. (1834) *On the Connexion of the Physical Sciences* (1st edn), London: John Murray.

Stoddart, D.R. (1986) *On Geography and Its History*, Oxford: Basil Blackwell.

Walter, H. and Breckle, S.-W. (1985) *Ecological Systems of the Geobiosphere. Vol. 1. Ecological Principles in Global Perspective* (translated by Sheila Gruber), Berlin: Springer.

White, G. (1789) *The Natural History of Selbourne with Observations on Various Parts of Nature, and the Naturalist's Calendar*, London: B. White & Son.

Whittlesey, D. (1929) 'Sequent occupance', *Annals of the Association of American Geographers* 19, 162–5.

Worster, D. (1994) *Nature's Economy: A History of Ecological Ideas* (2nd edn), Cambridge: Cambridge University Press.

8

I A DIFFERENTIATED WORLD

INTRODUCTION

Richard Huggett and Mike Robinson

Over geological time the earth has changed physically. A product of these changes was the creation of conditions in which life became possible. Once established, the burgeoning life-forms which the earth supported became a part of the substantive matrix of earth evolution. Some of the latest-developed of these opportunistic life-forms were small, bipedal apes. From the forests and savannahs of Africa, some of the hominid ancestors of these apes evolved into people. As people, and like the other life-forms with which they shared the planet, they responded in different ways to the environment of which they had become a part. And out of these responses, empowered by the development of language and bonded by the crystallization of ritual into religion, their influence upon it began to enlarge. Human groups became culturally differentiated: some became sedentary as the realization of agricultural systems and incipient urbanization signalled the possibilities of societal enlargement and the extension of socio-political ambition through the control of other people and of earth space. It was some of these groups that were to be primarily responsible for 'The Great Transformation'.

'The Great Transformation' is a topical theme in geography. The argument runs that the ecosphere has suffered changes of such magnitude and variety at the hands of the human species that its fundamental appearance and nature have been, or are about to be, radically altered. It would be wrong to belittle this view; indeed, much evidence in its favour will emerge later in the encyclopedia. None the less, it is right from the full perspective of geological time to note that the Great Transformation is but one of the changes, though possibly one of the most radical, that have occurred in the ecosphere during its history. All living beings, and not just the human species, actively change their environment. Since life first evolved, the living and non-living worlds have interacted. The course of biological evolution has not been steered solely by environmental change: life itself changes its environment and to a large extent determines the conditions in the air, sea, and land surface where it thrives. Many of earth's materials, which are presently identified as 'resources', are relics from past episodes of major

11

change in the ecosphere, some of which were probably induced by the ecosphere itself. Perhaps the earliest major change in the state of the ecosphere – the shift from a reducing to an oxidizing atmosphere – was associated with a rusting of the oceans and the laying down of the banded-iron formations that provide us now with large reserves of iron ore. Other major changes in the ecosphere can be attributed to cosmic and geological processes, particularly to bombardment by asteroids and comets and to volcanic supereruptions or bouts of protracted volcanism. Still others accompanied the colonization of land by life. Vegetation has been especially important, not only through its effects on the atmosphere but also, for example, through the evolution and spread of grasses which have radically altered the sediment budgets of landscapes.

The ecospheric transformation that is seen as 'great', however, owes its status to our perception of the impact of people. It began about two million years ago and seems to have been connected with the greater intelligence associated with a larger brain. Bigger brains had at least two effects. First, they led to the evolution of social groups (which was also encouraged by food sharing and the need to be mobile in the somewhat unreliable seasonal savannah environment). Second, they allowed the development of tool-making (probably promoted by the need to butcher large animals and as a way of unearthing the below-ground, edible parts of plants such as rhizomes, tubers, and roots). Increased intelligence and social co-operation may also have played a role in the consolidation of sounds into speech and language.

As far as we can judge, too, the rise of cultures was associated with the growth of systems of belief that came to be called 'religion'. They represented, as it were, the attempts of people to give order and meaning to realms of human imagination which were beyond the possibility of perception. Gradually, and particularly with the emergence of written records, these differentiated systems of belief developed a coherence and integrity of their own.

These developments in human evolution gradually allowed a tool-making species of hunter-gatherers to acquire the ability to overcome the severe restrictions placed on organisms living at the top of an ecological food chain. The population of any species is limited by the carrying capacity of the environment – that is, the number of individuals which a natural ecosystem can support. The human species succeeded in circumventing the limits imposed by carrying capacity in many ways, of which two were dominant. The first way was by spreading around the world: early hominids probably migrated from African homelands, entering the southerly temperate lands of Mediterranean Europe and South-East Asia about a million years ago, and more northern temperate lands of Eurasia about 700,000 to 300,000 years ago. A second wave of dispersal, involving humans whose anatomy was essentially the same as that of modern members of the species, occurred much later. In this wave, Australasia was colonized (probably around 40,000 years ago), and so were the Americas (probably around 20,000 years ago).

The second means by which humans surmounted the constraints imposed by environmental carrying capacity was by inventing multifarious, well-organized strategies for procuring, and eventually for *creating* food. These methods of obtaining food became increasingly calculated and sophisticated: animals which had been purposely hunted became purposely herded, and plants which had been deliberately gathered were deliberately manipulated, using controlled fires and other techniques during the late Palaeolithic and Mesolithic periods. By about 10,000 BC, when the population is estimated to have been around 4 million, the manipulation of plants had progressed, in some places, to their deliberate cultivation. Current thinking is that agriculture evolved slowly: it was not, as conventional wisdom once held, the result of a revolution, but rather of a series of improved ways of exploiting natural ecosystems for food. It probably evolved independently in different areas: in the Near East, for example, in China, and the Americas. Because it provided a fairly reliable food source, three consequences stemmed from the development of agriculture. First, it enabled more people to be supported, driving the world population from 4 million at about 10,000 BC to the present 5,000 million. Second, it encouraged human societies into greater internal heterogeneity: since only a part of the potential labour force needed to engage in food production, other groups in society were freed for other productive and cultural engagement. Third, because the more intensive agricultural practices forced people to establish permanent fields, it probably contributed to the establishment of permanent settlements and to the potential for enlarged, and more effective, political control.

The evolution of agriculture and of settled communities set most of human society on a path that it has followed ever since. It was the first great environmental transition in the geography and history of the human species. Because of it, far more people could be fed than natural ecosystems alone would allow. Even so, population levels commonly overshot the capacity of agriculture systems which, though generally more reliable than natural ecosystems, remained subject to the vicissitudes of weather and climate. In many places this condition still obtains and sections of human society have been dogged by malnutrition and famine ever since.

It is undeniable that the human species has had, and continues to have, a profound impact on the ecosphere. Paradoxically, there is evidence that the line leading to modern humans may well have evolved in response to a change in the ecosphere, to a shift of climate. Whether early humans had more than a small impact on their environment is difficult to say, but it seems likely that, during their early tenure of the earth, the human species did little damage to it. It might reasonably be supposed that hunter–gatherer tribes altered local fauna and flora by fire, though fires lit by natural causes would not have been unknown in some of the regions they inhabited. The first signs of the human species having a substantial impact on the ecosphere come from Pleistocene times. Pleistocene hunters might have precipitated many extirpations and extinctions of the megafauna in the western hemisphere: a possibility which remains the subject of a long and lively debate.

After the evolution of agriculture, the need to support growing populations led to more and more land being brought under cultivation and substantial environmental change inevitably ensued. Without doubt, the domestication of animals and plants played a cardinal role in the evolution of human communities and in environmental adaptations during the Holocene epoch, for it provided the means whereby the human species could control its environment. Besides spreading domesticated varieties of plants and animals, early farmers also cut forests, sometimes leaving a record of their activities in the spectra of pollen grains surviving in anaerobic deposits with low levels of mechanical attrition. Early civilizations also irrigated deserts, so changing the vegetation, soils, and local climates, and out of their co-operative skills and mutual needs created urban environments: places of functional differentiation and diverse but shared living. With the arrival of metal-using cultures, social efficiency was enlarged. The improvement of tools – and, it must be said, of weapons – made it possible for societies to exert finer, more selective, and more rapid control of their environments and to look for the accretion of territories in the waging of more efficient conflict.

1

PLANET EARTH

Cliff D. Ollier

This chapter provides an outline of the origin and evolution of the earth, its oceans and its atmosphere. It is by no means a 'physical basis of geography' – a description of an inorganic earth on which life in general, humans, and geographers in particular can operate – for it will be shown that life itself plays a vital part in the evolution of the earth as we know it. Planet Earth is an interactive system. Neither is it a simple outline of facts, for there is great controversy about many aspects of the development of our earth, and an intelligent appreciation of the earth as it is (quite apart from speculation on what may happen to it in the foreseeable future), demands an understanding of possibilities which are not apparent at first sight.

EARTH MATERIALS

The world of matter consists of atoms of ninety-one naturally occurring elements such as carbon, silicon and oxygen. These may combine together to form molecules such as carbon dioxide and silicon dioxide. Most of the earth's materials have their molecules arranged in a regular three-dimensional pattern to form minerals such as quartz (silicon dioxide) or calcite (calcium carbonate). Rocks are made up of a number of minerals: marble, for example, consists of just calcium carbonate; granite consists of quartz, feldspar and mica. With so many elements the number of possible minerals and rocks is enormous, but fortunately the reality is simpler. A small number of elements account for the bulk of the earth, and are represented by a few dominant minerals, and a fairly small number of common rocks. These are shown in Tables 1.1, 1.2 and 1.3.

The most impressive feature of the tables is the overwhelming dominance of oxygen, making up almost 94 per cent by volume of the crust and virtually 100 per cent of the water of the hydrosphere. The atmosphere contains only about 20 per cent oxygen. The solid rocks can be regarded as a mass of large oxygen atoms, glued together by tiny atoms of other elements that occupy the space between oxygen atoms.

Table 1.1 Abundances of elements, minerals, and rocks in the earth's crust

Elements in crust	Volume (%)	Weight (%)	Minerals in crust	Volume (%)	Rocks in crust	Volume (%)
Oxygen	93.8	46.6	Feldspar	51	Basalt	42
Potassium	1.8	2.6	Quartz	12	Granite	22
Sodium	1.3	2.8	Pyroxene and amphibole	16	Metamorphic	27
Calcium	1.0	3.6	Clay and mica	10	Shale	4
Silicon	0.9	27.7	Olivine	3	Limestone	2
Aluminium	0.5	8.1	Carbonates	2	Sandstone	2
Iron	0.4	5.0	Others	6	Other	1
Magnesium	0.3	2.1				
Total	100.0	98.5	Total	100	Total	100

Table 1.2 Abundance of minerals and rocks at the earth's surface

Minerals at land surface	Area (%)	Rocks at land surface	Area (%)
Feldspar	30	Shale	52
Quartz	28	Sandstone	15
Clay and mica	18	Granite	15
Carbonates	9	Limestone	7
Iron oxides	4	Basalt	3
Pyroxene and amphibole	1	Others	8
Others	10		
Total	100	Total	100

Table 1.3 Characteristics of oceanic crust and continental crust

	Oceanic crust	Continental crust
Thickness (km)	7	30 to 50 (thickest under mountains)
Seismic P-wave (km/s)	7	6 (higher in lower crust)
Density (g/cm^3)	3.0	2.7
Probable composition	Basalt (gabbro in lower crust)	Granite and metamorphic (with sedimentary rock cover)

Of the rocks two are of outstanding importance – basalt and granite. Both are igneous rocks; that is, they form from magma or molten rock. This is obvious in basalt which is erupted as liquid lava from many volcanoes, and cools to a dark, fine-grained rock (grain size indicates rate of cooling). Granite forms at some depth in the earth so only by inference is it known to cool from a magma. It is a coarse-grained rock, generally light in colour. In general the continents consist of granite. The sea-floor consists of basalt, but it is thought that basalt extends beneath the continents too.

THE ROCK CYCLE

The earth evolves through a cycle of geological events, driven by its internal and external heat engines. Rocks at the surface are weathered, and loosened particles of rock and mineral are eroded off their parent bedrock. This debris is transported by rivers, ice or wind, and eventually deposited as layers of sediment (strata) in the sea. The strata may be hardened to form sedimentary rocks. Eventual deep burial may metamorphose sedimentary rocks into metamorphic rocks, and further alteration may eventually turn metamorphic rocks into

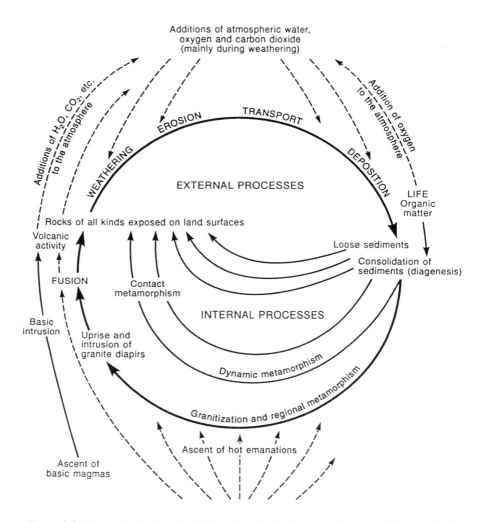

Figure 1.1 The geological cycle. Notice that basalt is not really part of the cycle, but comes from the 'side', even though it covers over three-quarters of the earth's surface. After Holmes (1965).

granite. With erosion of overlying rocks, and uplift, the granite may be exposed at the ground surface, where it will weather, and so the cycle may be repeated.

This simplified account of the geological cycle ignores features such as chemical and biological sediments (e.g. rock salt and coal), but it also ignores volcanic activity. Note that as shown in Holmes's (1965) classical depiction of the cycle (Figure 1.1), basalt comes into the cycle from the side, as an

18

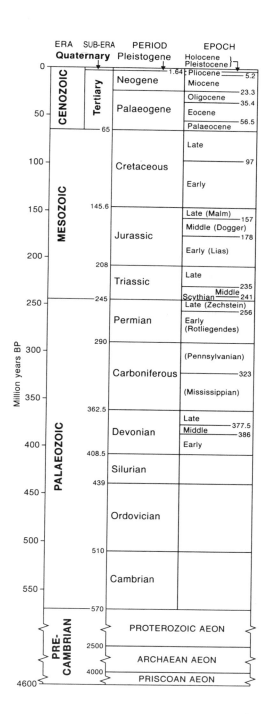

Figure 1.2 The geological time-scale. The Cenozoic is sometimes called the Cainozoic. The units are sometimes split into Lower, Middle, and Upper or lower, middle, and upper. Capital letters are used for formal stratigraphic usage, lower case letters for informal usage. The same applies to early, middle, and late. The time-scale follows Harland et al. (1991).

unremarked addition to the cycle. To some extent this is true, because a basalt erupted at the earth's surface will weather, erode, and so forth. But there is no accepted process that can produce new basalt from cycling the earth's materials. If basalt were a minor component this fact might be ignored, like salt and coal. But basalt covers three-quarters of the earth's surface. As will be explained later, most basalt is produced at mid-ocean ridges, and a very different kind of cycle is invoked to account for it.

Another kind of cycle is the geochemical cycle, in which individual elements or compounds are cycled through various reservoirs at various flux rates. Thus sodium moves from the land reservoir to the ocean reservoir, and back again. Carbon has a more complex cycle, occurring in the components of the ecosphere – atmosphere, hydrosphere, animals and plants, soils and sediments. It is not always known how materials are cycled, but if there is no appreciable build-up in geological time, then something is maintaining reasonably constant flux rates.

STRATIGRAPHY AND GEOLOGICAL TIME

A layer of sediment deposited on the sea-floor may become a rock – sand is converted into sandstone, clay into shale, and coral and shells into limestone. A sandstone deposited perhaps a million years ago may come to be overlain by shells today (which are therefore younger), and may overlie a shale which is even older. The principle of superposition allows geologists to work out the relative ages of strata. Strata can also be recognized by contained fossils. A modern beach sand will contain remains of living species. An older beach sand will contain some extinct species. The older the strata the more different the fossils become, and by knowing the type of fossils contained, strata in different places can be correlated. Stratigraphers have built up a picture of the evolution of strata and the evolution of life back to about 560 million years ago, when shelly fossils first became abundant. Strata have been named and dated, giving the geological column shown in Figure 1.2. Rocks older than the Cambrian – the Precambrian – are devoid of common fossils, and have been subdivided and dated by different means. Volcanic products can be dated when they are interbedded with fossiliferous strata. Intrusions like granite must be younger than the rocks they intrude. Many physical methods based on radioactive decay, magnetism, and other properties may be used to date rocks on an absolute time-scale.

EARTH STRUCTURE

The surface of the earth cannot be properly appreciated without some understanding of the earth's structure. Most of our knowledge comes from geophysical studies of earthquake waves, heat flow, and magnetism, and from the study of materials erupted by volcanoes.

It appears that the earth has a core which is 'liquid' in the sense that it cannot transmit compressive waves. Outside this is a mantle, and nearer to the surface

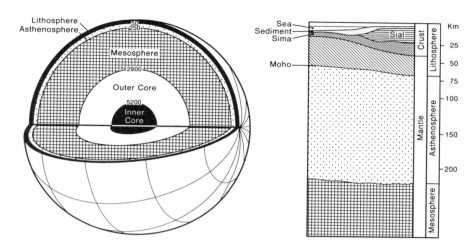

Figure 1.3 The layered earth. On the left the whole earth is simplified. Continents do not make a continuous layer. Details of a continental edge are shown on the right.

there is a crust which includes both the continents and the ocean floor. The crust and the mantle are separated by the Mohorovičić Discontinuity (the Moho), of seismic significance. A more important difference is the separation of the lithosphere (which consists of the crust and upper mantle) from the mesosphere (lower mantle) by an asthenosphere (Figure 1.3). The asthenosphere is plastic and can flow. This enables the upper, more rigid crust to move over the yielding asthenosphere.

One can envisage an ideal world built of concentric spheres: from the centre out – the core, mantle, crust, hydrosphere, and atmosphere. The only major deviation from this ideal scheme is the crust, which consists of continental crust and oceanic crust. Continental crust covers only about a quarter of the earth's surface and the hydrosphere, being liquid, is obliged to fill the gaps between continental fragments, giving the oceans that we know today.

The properties of the continental and oceanic crust are shown in Table 1.3. Essentially the continents consist of granite (sial) about 20 km thick, with a thin cover of sedimentary rocks; the sea-floor consists of a layer of basaltic material (sima) about 7 km thick. The sima layer is thought to extend under the continents.

The interior of the earth is hot, and geothermal heat flows to the surface. The energy provided by the Sun is 5,000 times greater. Any primeval heat of the earth would have been lost long ago, and modern heat results from radioactive decay. Radioactive elements are concentrated in continental rocks, so it is surprising that the heat from continents and oceans is roughly the same. Some

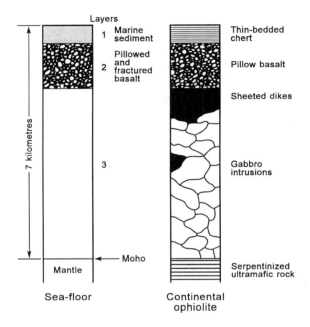

Figure 1.4 Sea-floor crust and ophiolite layers found on land. Layer 3 of the sea-floor is detected seismically, but is not directly observed by drilling. By analogy with ophiolite, it is thought to consist of gabbro and dykes.

think that oceanic heat must come from the mantle, but as there is mantle under the continents too, others prefer a dynamic explanation with convection cells.

Seismic surveys show the ocean crust to be divisible into three layers, and drilling has revealed the rock types of the top two. Where the sea-floor has been thrust onto the continent distinctive rocks called ophiolites are found, which also have three layers: an upper layer with marine sediments, pillow lavas, and gabbro intrusions (Figure 1.4). Ophiolites are presumed to reveal the structure of the sea-floor. Their presence on land, usually metamorphosed, often marks the edge of thrust sheets or terranes described later.

The really remarkable thing about the sea-floor rocks is their age. No sea-floor is older than 200 million years, whereas continental rocks date back to 3,800 million years.

EXPLAINING EARTH STRUCTURE

Until the mid-1960s the dominant theory of the earth was 'fixist': continents and oceans were permanent features of the earth in their present position. A minority of geologists upheld Wegener's concept of continental drift to account

Area of overlap

Figure 1.5 The reassembly of continents around the Atlantic. After Bullard *et al.* (1965).

for the strange distribution of plants and animals and detailed matching of rock sequences on different continents. Reconstruction of a closed Atlantic (Figure 1.5) showed such a convincing fit that at least the opening of the Atlantic, accompanied by drift, could hardly be denied. It was once thought that the continents were formerly assembled into two supercontinents – Gondwanaland in the south and Laurasia in the north, but Figure 1.5 suggests that they were assembled as a single supercontinent, Pangaea.

By 1970 the new paradigm of plate tectonics had replaced simple drift. It was

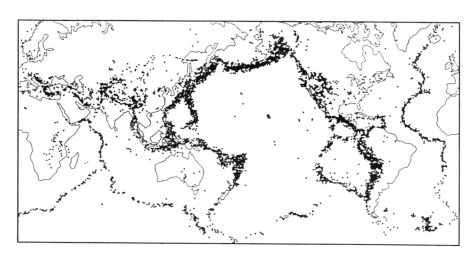

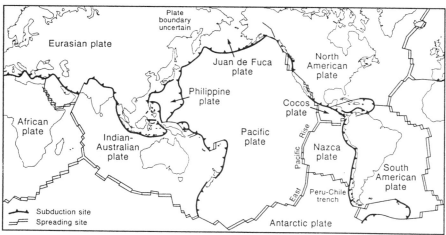

Figure 1.6 Tectonic plates and earthquakes. The top map shows the distribution of earthquakes. The bottom map depicts the 'plates' of plate tectonics, each bounded by lines of spreading and subduction.

work on rock magnetism that revolutionized tectonic ideas. The earth is known to be a magnet, with the magnetic axis aligned almost parallel to the axis of spin, so a compass needle points to the poles. It was found that at certain times in the past the polarity of the magnet was reversed, so the north magnetic pole became the south magnetic pole and vice versa.

It had been known for a long time that the distribution of earthquakes and volcanoes formed lines 'where the action is' between areas of relatively little activity (Figure 1.6). The pattern became explicable when it was discovered that rock magnetism mapped on the ocean floor revealed stripes of normal and reversed polarity that were symmetrical about the mid-ocean ridges. The ridges are spreading sites where the new sea-floor is intruded by the injection of basalt, sometimes with obvious volcanic action as in Iceland (an exposed part of the Mid-Atlantic Ridge). The new rock takes on the earth's magnetism when it cools. Older sea-floor is pushed aside, so the oceans are symmetrical, like two magnetic tape recorders (Figure 1.7). These studies also tell us that the earth is a magnet, and that repeatedly through geological time the magnetic poles have reversed (Figure 1.8).

The Atlantic Ocean is therefore growing by sea-floor spreading, with America moving away from Europe and Africa. But the Pacific Ocean is also spreading, as indeed are all the world's oceans. The rate of spreading is a few centimetres per year, about the rate of growth of finger nails, which should be easily observable in geological time, and even by direct measurement. If the earth is not expanding (and some geologists contend that it is), the newly created crust must be compensated somewhere by removal of crust. This is thought to

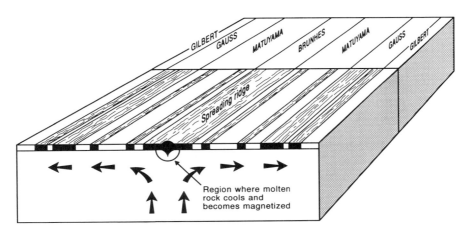

Figure 1.7 Symmetry of magnetic reversals across a sea-floor spreading site. The names refer to magnetic epochs.

25

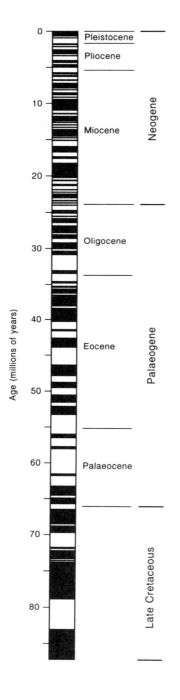

Figure 1.8 The geomagnetic polarity time-scale from the Late Cretaceous. After Cande and Kent (1992).

be by subduction, where a downgoing plate is subducted beneath an overriding plate. Subduction occurs in deep-sea trenches along continents and island arcs, and sometimes between continents. Subduction is thought to be indicated by a dipping zone of earthquakes (the Benioff Zone), and is used to explain the formation of deep trenches, the uplift of mountains, the intrusion of granites, volcanic activity, and many other features. Most volcanoes associated with subduction sites consist not of basalt but of andesite (richer in silica), which might be formed by partial melting of basalt, or by melting of both basalt and sediments of the downgoing slab.

The plate tectonic system is deemed to be driven by convection currents, and has its own cycle. Sea-floor is thrust under a continent. Convection currents return material at depth to the mid-ocean ridge, where new sea-floor is created by the intrusion of new basalt. The sea-floor moves towards the subduction site, and so the cycle goes on.

In the plate tectonic cycle, basaltic sea-floor is subducted, together with sediments of the deep-sea trench and perhaps other rocks. There is partial melting and production of granite and perhaps andesitic volcanoes, before the material returns to the mid-ocean ridge. It is quite remarkable that the random admixture of sediments and random melting to produce magma near the continental margin should leave, as a residue to be transported back to the spreading site, a perfect basalt. Mid-ocean basalts have a remarkably consistent composition. Furthermore, they emit gases such as helium which appear to be arriving at the earth's surface for the first time. Indeed the most obvious conclusion would be that the mid-ocean ridges are producing new basalt, and only the need to fit into the global cycle leads to the conclusion that it is recycled.

Plate tectonic cycling accounts for the fact that no sea-floor is older than 200 million years: any older sea-floor has been subducted. There is much debate about the thickness of the convection cells. Some continental margins have subduction sites (the western sides of the Americas), but some do not (Africa, Australia, the eastern sides of the Americas). Continental margins without subduction are called passive margins, as opposed to the active margins with subduction. Earthquakes and volcanoes are concentrated on active margins, but mountains are found on both active and passive margins.

A complication of plate tectonics theory is the concept of exotic terranes. If a plate breaks up it may produce fragments called 'terranes'. These may drift until they come into contact with a continent, when instead of being subducted they become attached to the continent and sheared along it. These are known as 'exotic terranes'. Some consider that virtually all the western side of North America consists of exotic terranes (Figure 1.9). Similar scenarios have been devised for many other places around the Pacific, and even elsewhere. It must be stressed that if exotic terranes have moved around, the excellent fit of continents, around the Atlantic in particular, should have been destroyed.

It is only right to point out that there are several problems with plate

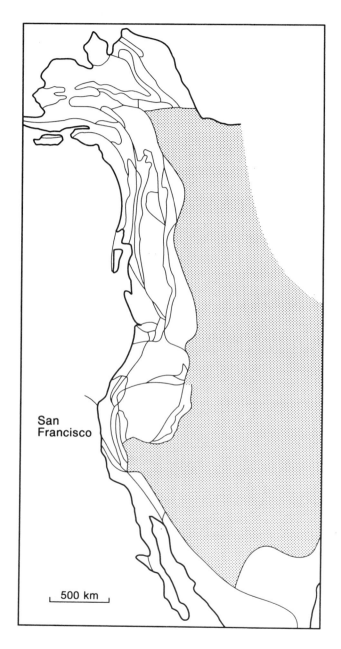

San
Francisco

500 km

Figure 1.9 Exotic terranes of North America. After Coney *et al.* (1980).

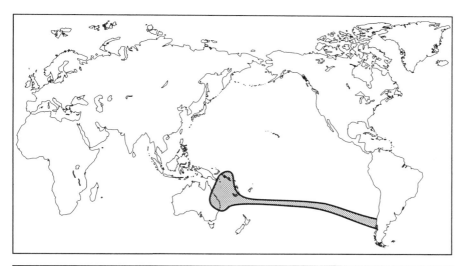

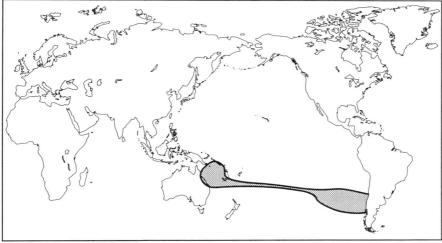

Figure 1.10 Cross-Pacific biological distributions. *Araucaria* (above) and *Gevuina* (below) produce nuts that would be difficult to disperse by ocean currents. After van Steenis (1963).

tectonics, including the fact that spreading sites are about three times as long as subduction sites, so subduction would have to be three times faster than spreading, which is not observed: island arcs exhibit tensional faults, and the areas behind arcs exhibit back-arc spreading instead of compression.

A minority of workers believe that the earth actually is expanding (Owen 1981; Carey 1976). Several have found that the continents that match so perfectly across the Atlantic can also be fitted across the Pacific on a smaller globe. Numerous biological and geological links of the kind used to suggest continental drift between Africa and America can be found across the Pacific (Figure 1.10). The discrepancy of age between the oceans and the continents ceases to be a problem, as does the creation of apparently new basalt at spreading sites.

TOPOGRAPHIC FEATURES OF THE EARTH

The most obvious division of the earth's topography is into continents and oceans, which reflects the distribution of two major rock types, granite and basalt (sial and sima). It is not an accidental distribution of water on a solid but irregular earth. Ocean waters extend onto continental rocks at continental shelves, and the true edges of the continents are the steeper continental slopes. The actual shorelines are rather accidental, depending on the height of sea-level on the sloping shelves.

Continental rocks cover about 30 per cent of the earth's surface, oceanic rocks about 70 per cent. A major unexplained feature of the earth is the antipodal relationship of continents and oceans (Figure 1.12).

Major topographic features of continents are mountain ranges, plateaux, and lowland basins or plains. At a smaller scale topography is made of valleys, hillsides, and minor landscape features such as dunes and moraines. Although some landscape features are constructed (such as lava plains), the great bulk are created by the erosion of rocks. Most of the earth's erosion is the work of running water, but in some areas wind and ice are important. The Ice Age of the last few million years has left a clear mark in polar and mountainous regions. Eroded material is transported, deposited, turned into rock, and eventually uplifted so that the rock cycle can start again. A great many measurements of erosion rates have been made, so the cycle can be quantified. A very important conclusion of this work is that the whole area of the continents could be reduced to sea-level in less than 30 million years. Since this clearly did not happen, there must be some sort of law of conservation of continents. To some extent this is the area of tectonics (earth movements) that includes uplift of plains to make plateaux, which can be eroded, and so lead back to the geological cycle.

The major features of the ocean floor include ocean depths, mid-ocean ridges, and deep-sea trenches. Marine sediments are mainly significant near continental margins. Volcanoes are widespread in the ocean, often in significant lines or chains, and in tropical regions coral reefs are important.

EXPLAINING TOPOGRAPHY

The earth is eroded by rivers, ice and wind into a landscape of erosional hills, valleys and plains, together with depositional features such as alluvial plains and sand dunes. Most of the world's landscapes are erosional, but the landscape may

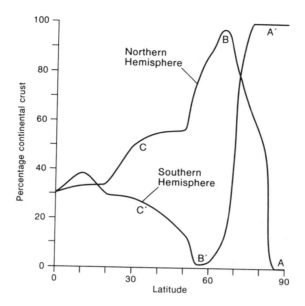

Figure 1.11 Antipodal relationships of continents and oceans. Percentage of continental crust is plotted against latitude. Seas shallower than 2,000 m are plotted as continental areas. AA': south polar continent, north polar ocean. BB': northern land girdle, southern ocean. CC': lands taper southwards, oceans taper northwards. Since the oceans cover about twice the area of the continents, some antipodal relationships might arise by chance, but 95 per cent of land is antipodal to oceans – much better than chance. After Carey (1963).

be modified by volcanic activity, and tectonic activity such as faulting. Without tectonic uplift, erosion could eventually produce a flat erosion surface, almost a plain (a peneplain). Many such erosion surfaces are found in the world (Figure 1.12).

Tectonic uplift may lead to renewed erosion and the formation of a new set of landforms as rivers cut down to a new sea-level. Alternatively, sea-level may change for some other reason, and a fall would lead to erosion on land, a rise would lead to accumulation of sediment. The level of the sea has now been traced through Phanerozoic time, and can be depicted in Vail curves (Figure 1.13), named after a pioneer in the field. Many varied scenarios have been derived to explain the course of landscape evolution, involving such features as equilibrium or cyclical theories, the effects of changing climates, and the nature of the bedrock.

Major landforms are often related to ideas about earth movement. Mountains were originally related to crustal compression on a shrinking earth, but nowadays are usually explained by compression associated with subduction at a

Figure 1.12 Erosion surfaces. (*a*) The New England Plateau, New South Wales, Australia. The erosion surface is cut across granites and steeply folded Palaeozoic rocks exposed in a gorge. Photograph by C.D. Ollier. (*b*) The African erosion surface in Uganda. In the background is a residual hill with a less distinct trace of an older erosion surface across the summit. Photograph by C.D. Ollier.

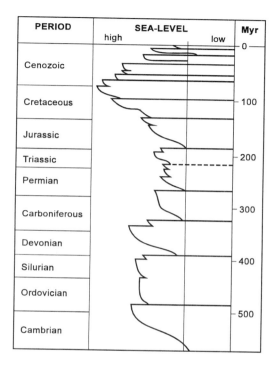

Figure 1.13 Vail curves for the Phanerozoic, greatly simplified. Note the great increase in complexity in the late Cenozoic, the very high levels in the Cretaceous flood, and the nature of the curve with very rapid falls in sea-level followed by gradual rise (the cause of which is not known).

plate boundary. There are problems with compression to form folds, because many fold belts have been shown to exhibit 'thin-skinned tectonics' (Figure 1.14) in which deformation is confined to a relatively thin but very wide belt while underlying rock is little deformed. This does not seem to accord with simple compression or with subduction. Some workers believe that thrusts and folds are related to gravity sliding, a sort of gigantic landsliding, which accounts for the intense folding of an upper layer while rock beneath the glide plane is unaffected. But compression is not necessary, and many mountains are formed in unfolded, horizontal rocks, or in granites, and many mountains are not on active margins. Other explanations for mountain formation involve vertical uplift, sometimes associated with intrusion of granite at depth. Uplift may be fairly local to give fault block mountains, or over a wide region such as eastern Australia (cymatogeny or epeirogenic uplift). Many mountains are undoubtedly formed by erosion of earlier plateaux, and even the Andes, a stronghold of plate tectonic explanations, are dominated by the altiplano (high plains).

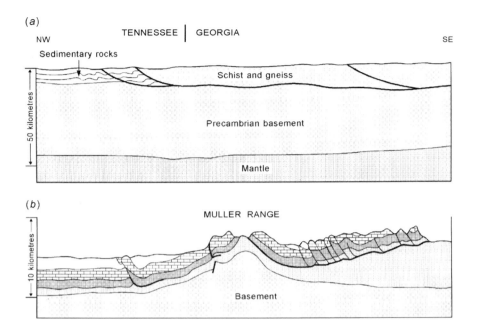

Figure 1.14 Thin-skinned tectonics. (a) The top diagram shows part of the southern Appalachians. The upper part of the crust has moved a long distance north-west above a master fault. After Cook (1983). (b) The lower diagram shows the Muller Range, Papua New Guinea. The basement is not folded, nor is the lower part of the Jurassic strata (vertical shading), but upper strata are deformed into folds and faults. After Jenkins (1974), who interprets the section as a result of gravity tectonics.

Much of the landscape is of great age, and some areas have never been under the sea since the Precambrian. Large areas of Africa, Australia and other continents retain much of their Mesozoic landscape, and deep Cretaceous weathering has been found not only there but in Sweden, Scotland and the United States. Some major river systems are older than the break-up of Gondwanaland, so landscape evolution must be seen on the same time-scale as global tectonics and biological evolution. The geomorphology on a super-continent would have been different from that of today's small continents, with greater average distance to sea, longer rivers, lower river gradients, and other changes. When supercontinents broke up new continental margins were formed, providing new base levels for erosion, and the setting for new features such as the Great Escarpments common on passive margins, and break-up uncon-formities on which offshore sediments were deposited.

The study of old landscapes helps in the reconstruction of the past. For example, Australia was once connected with Antarctica, and started to drift north about 55 million years ago. At that time, despite its high original latitude, it had a warm and wet climate, indicating that the climatic belts of today are not permanent. When Antarctica was isolated the circum-polar current came into existence, and a cold cell was set up over Antarctica, with eventual growth of an ice cap. Colder and drier conditions appeared in Australia in the mid-Tertiary. In the world as a whole there were ice ages in the late Precambrian, the Permo-Carboniferous, and the Tertiary–Quaternary. (An old idea of four major glaciations in the Quaternary must be discarded: there have been about twenty-seven major glaciations in the past 3.5 million years.) The Triassic was an especially arid period in much of the world, and many evaporite deposits were formed. The Cretaceous was a very mild period, warm and moist, with high sea-levels, deep weathering, and perhaps one of the flattest landscapes the world has known. It was also a period of magnetic quiet, suggesting that details of the earth's surface may be closely linked to global tectonics.

EXTRATERRESTRIAL EVENTS

There is no doubt that meteorites and other bits of debris in space occasionally hit the earth, making impact craters with distinctive features. Earth has suffered a bombardment from space throughout its history. The quantitative effect of these events remains debatable.

Some think large impacts in early earth history might have stocked the earth with water (Chyba 1987), and organic molecules (Chyba *et al.* 1990). Others (e.g. Alvarez *et al.* 1980) hold meteorite impact responsible for events such as the extinction of the dinosaurs, a perennial source of popular appeal, whereas others are internalists, believing that most events on earth can be explained by terrestrial mechanisms (e.g. Hallam 1987). Either way the important feature is the ability of the earth to survive and adjust. The homeostasis built into the earth system is so resilient that it is likely to survive planetary impact, or anything that humans can do to it.

THE ORIGIN OF THE EARTH

Most cosmologists today believe that the earth formed about 4,500 million years ago by accretion of cosmic dust. 'The currently favoured accretion scenario is a gravitationally interacting and collisional "swarm" of solid bodies in Keplerian orbits, with the stochastic growth of progressively larger and few bodies occurring at the expense of smaller planetesimals' (Stevenson 1983). This formed an early earth, which would be highly active. Primitive heat would have been lost very early in the earth's history. Earth heat since then comes from radioactive decay. Earth temperature was probably high in early times because there were many radioactive elements with short half-lives in the early history of the earth which have since disappeared. This early stage is called the Hadean.

Figure 1.15 The mesa landscape of the Gran Sabana, Venezuela, on flat-lying unmetamorphosed lower Proterozoic sandstone (about 1,700 million years old). The plateau is at 2,800 m, about 1,000 m above the general level of the surrounding terrain. Photograph by J.C. Dohrenwend.

Various turbulent effects would lead eventually to a layered solid earth of the sort we know today. In Archaean times the world contained areas of granite, and also greenstones, which are metamorphosed basalt. By the Proterozoic (the upper Precambrian unit) sedimentary rocks were common, especially sandstones and shales, so the rock cycle that we know today was already in existence. Only the biogenic limestones were rare or absent. There are still many places in the world where Proterozoic sandstones remain horizontal despite their great age (Figure 1.15).

THE ORIGIN OF THE ATMOSPHERE

However the earth might have originated, the early earth had no atmosphere or hydrosphere. If it originated at high temperatures, gases (which have an enhanced escape velocity at high temperature) would be lost to space. If it started as a cold body, by accretion of cold rock fragments, it would have no atmosphere until volatiles escaped from the rock (by early volcanic outgassing), and no oceans until water in the atmosphere condensed. So it is accepted that the atmosphere was created by outgassing. Extraterrestrial gains, as from ice meteorites, have probably been quantitatively trivial. Losses to space occur, but are now mainly restricted to helium (which must be replenished by outgassing) and some hydrogen.

Modern volcanoes erupt gases including water vapour (70 per cent), carbon dioxide (15 per cent) and nitrogen (about 5 per cent). There is no oxygen, because any that was erupted would instantly react with hydrogen to form water, or with carbon to make carbon dioxide or carbon monoxide. The earth's atmosphere is very different from that of other planets, and from the gas mixture erupted by volcanoes.

The earth's atmosphere is simple in composition, being made up almost entirely of three elements: nitrogen, oxygen and argon (Table 1.4). In a simple physical system the gases are in equilibrium with the solid earth at appropriate temperature. If a sample of a planetary atmosphere could be heated in the presence of planetary rock and allowed to cool, it would finish with the same solid and gaseous composition that it had originally. This is the situation found on Venus and Mars, where the atmospheres are in equilibrium. The situation on earth is quite different (Table 1.4) and the gases in the earth's atmosphere are in a persistent state of disequilibrium.

Most modern workers agree that the present composition of the earth's atmosphere is due to changes brought about by living things. The early atmosphere consisted of carbon dioxide, methane and nitrogen. It was a reducing atmosphere. Life began in this atmosphere, but eventually evolved to the stage where photosynthesis appeared, in which process carbon dioxide was converted to oxygen. This waste product accumulated until it dominated the surface of the planet, and the environment for later living things was an oxidizing environment.

The early reducing atmosphere of the earth is indicated by uraninite ores, which could only have formed in a reducing atmosphere, and by the widespread banded iron formations, the creation of which required iron to be transported in solution to the sea. This could only happen in reducing conditions. Higher

Table 1.4 Planetary atmospheres: composition and conditions

Gas	Venus	Mars	Earth without life	Earth as it is
Carbon dioxide (%)	96.5	95	98	0.03
Nitrogen (%)	3.5	2.7	1.9	79
Oxygen (%)	trace	0.13	0.0	21
Argon	70 ppm	1.6%	0.1%	1%
Methane (ppm)	0.0	0.0	0.0	1.7
Surface temperatures (°C)	459	−53	240 to 340	13
Total pressure (bars)	90	0.0064	60	1.0

Source: After Lovelock (1988).

levels of oxygen had already been attained about 2,300 million years ago, when red beds rich in oxidized iron were deposited. Later changes to atmospheric composition have been relatively minor. One important follow up of oxygen production was the eventual formation of an ozone shield, which cut down incoming ultraviolet radiation and increased the habitability of the planet.

The improbable atmosphere of today has to be maintained by recycling, mostly by organisms. Almost 1,000 million tonnes of methane must be produced annually to maintain methane at its constant level, and oxygen used in oxidizing the methane must be replaced – 2,000 million tonnes annually (Lovelock 1988).

THE ORIGIN OF THE HYDROSPHERE

Like the atmosphere, the hydrosphere must have originated by outgassing from rocks, by volcanic eruptions. Present-day volcanoes erupt volatiles which are 70 per cent water, and at the present production rate could have provided all the world's water in about 55 million years. There is no problem in producing the water, but why is there not more? Perhaps water is recycled in the plate tectonic convection cycle. Perhaps some is lost to space.

Opinions about the timing of the accumulation of water on the earth vary widely. Some think it was nearly all formed early in the earth's history, and has since been merely recycled (Fanale 1971); most favour accumulation through geological time (e.g. Rubey 1951; Anderson 1975); and some believe most water was formed as modern ocean floors were created during the past 200 million years (Carey 1988).

Although the earliest, Hadean phase of the earth might have had no oceans, the oldest known rocks are sedimentary. They were deposited in water, which implies at least the existence of lakes or shallow seas, rain, weathering, and water erosion and transport. Oceans may not yet have existed, and indeed most of the sedimentary rocks preserved on earth are shallow water sediments. Even where the sedimentary rocks are thousands of metres thick they are usually shallow water sediments deposited in a basin that sank during their accumulation. There is no evidence of deep oceans in early earth history.

Water is a remarkable substance with a whole range of unusual properties (Table 1.5). Many of these properties are vital to the preservation of life as we know it, but on earth there must have been formation of water by inorganic means, as life appears to have originated in water.

The sea is salty, and at present rather alkaline (pH 8). It may not always have been so. Early workers tried to work out the age of the oceans by dividing its salt content by the amount of salt added annually by rivers. This calculation does not work, because much of the salt is recycled, and there are additional inputs from submarine eruptions. Furthermore, the components of the salts come from different sources. The sodium comes from weathering of continental rocks, but the chlorine comes from volcanic gas. Different components of the sea are cycled at different rates, with different residence time in the ocean (Table 1.6). On top

Table 1.5 Some properties of water

Property	Comparison with substances	Importance in physical–biological environment
Heat capacity	Highest of all solids and liquids except liquid NH_3	Prevents extreme ranges in temperature. Heat transfer by water movements is very large
Latent heat of fusion	Highest except NH_3	Thermostatic effect at freezing point owing to absorption or release of latent heat
Latent heat of evaporation	Highest of all substances	Important in heat and water transfer of atmosphere. Valley glaciers are at freezing point
Thermal expansion	Temperature of maximum density decreases with increasing salinity. For pure water it is at 4°C	Fresh water and dilute sea water have their maximum density at temperatures above the freezing point. Important in controlling temperature distribution and vertical circulation in lakes
Surface tension	Highest of all liquids except mercury	Important in physiology of the cell. Controls surface phenomena, drop formation and capillarity
Dissolving power	Dissolves more substances and in greater quantities than any other	Important in both physical and biological phenomena
Dielectric constant	Pure water has the highest of all liquids	Important in behaviour of inorganic dissolved substances because of resulting high dissociation
Electrolytic dissociation	Very small	A neutral substance, yet contains both H^+ and OH^- ions
Transparency	Relatively great	Important for life in water
Conduction of heat	Highest of all liquids	Important in living cells
Absorption of radiant energy	Large in infra-red and ultra-violet	Responsible for about 75 per cent of greenhouse effect controlling temperature of atmosphere
Solid (ice) less dense than liquid	Rare feature	Important in partly frozen lakes and seas, and in weathering
Catalysis		Important catalyst in many reactions
Polar molecule	Rare in liquids	Important in many reactions and weathering

39

Table 1.6 Elements in the oceans

Element	Concentration (moles/litre)	Residence time (million years)
Sodium	0.47	56.0
Magnesium	0.05	11.0
Calcium	0.01	0.9
Potassium	0.01	5.5
Chlorine	0.53	350.0

Source: After Whitfield (1981).

of this there are interactions with the atmosphere and living things. This is especially important with carbon dioxide, which reacts with the atmosphere through partial pressure (for one part in the atmosphere there must be forty parts in water to maintain equilibrium), and carbon dioxide is removed by organisms to form the carbonate of shell and coral. This last feature is only of real significance in the Phanerozoic, as there are few carbonate sediments in the Precambrian. Nevertheless, the invariance of mineral sequences in evaporites, the small variation in the bromine/chlorine ratio of halites, and the composition of fluid inclusions in halites suggest that the concentration of major ions in sea water has not varied greatly through the Phanerozoic (Holland 1984).

THE ORIGIN OF LIFE

A brief summary of the relevant time-scale helps to put life into perspective:

Origin of Earth	4,500 million years
Origin of Life	3,800+ million years
First abundant fossils	570 million years

Life had already arisen by the time of the oldest-known sediments, 3,800 million years ago (Cloud 1983). The earliest life was something like bacteria. Bacteria and blue-green algae (cyanobacteria) are classified in the Kingdom Monera because they have simpler structure than other forms of life. They are said to be prokaryotic, and lack a nucleus. They have ring-shaped chromosomes and reproduce by fission and budding. All other organisms are eukaryotic, and have chromosomes that are not ring-shaped enclosed in a nucleus. The eukaryotes may have evolved by the assemblage of several prokaryotes (a sort of endosymbiosis). It may be that the rise of the eukaryotes is a response to the new conditions imposed by an oxygen-containing atmosphere. Green algae are thought to be ancestral to plants because they have similar chlorophyll, which traps sunlight energy and converts carbon dioxide into oxygen. Green algae showing cell division have been found in the Bitter Springs Chert of central

Australia, which is 900 million years old. Evolution may have advanced beyond the prokaryotes, but they still make up over half of the earth's biomass.

The first life must have been anaerobic, as no free oxygen existed at that time. Sulphate in sediments aged 3,400 million years shows that some oxygen was coming from somewhere – possibly bacteria. But the amount of free oxygen remained vanishingly small until about 2,300 million years ago, as indicated by widespread ores such as uraninites and banded iron formations, which could only be formed in a reducing environment. Red beds about 2,300 million years old indicate higher levels of oxygen.

Of course life is vitally concerned with carbon chemistry, organic chemistry. Photosynthesis converting carbon dioxide to water traps sunlight energy as an energy source for most life. Nevertheless the largest reservoir of carbon on earth is in organic deposits of limestone, coal, oil shale, etc. which amounts to an estimated 10^{15} tonnes. It should be noted here that fossil fuels are not restricted to the Phanerozoic, and in Siberia Precambrian oil is being recovered from Precambrian traps. It is also worth remembering that some hydrocarbons might be entirely inorganic. The carbon in sedimentary deposits cannot be derived from basement rocks, since the latter contain only a very small fraction of all carbon in rocks. If basement rocks such as granite cannot be a source of sedimentary rock carbon, where is the carbon from if not from below the crust of the earth? Some, such as Chyba *et al.* (1990) think organic molecules come from space. Recycling of these reserves in nature, or at an accelerated rate by humans, is a matter of planetary concern. The study of cycling of elements on earth is assisted by the use of stable isotopes, which can often distinguish organic and inorganic sources but is beyond this outline.

It is thought that the earth's climate has been ameliorated by living organisms, which prevented a runaway greenhouse as on Venus or a deep freeze as on Mars. The ecological tolerance of organisms is the sensor in the earth's thermostat (Hsü 1992).

The evolution and continuance of an oxidative environment was a highly improbable event and could not have happened without a kinetic lag in the thermodynamic balance. Even today the thermodynamic balance would eliminate all resident oxygen in the atmosphere in about 3 million years (Cloud 1983).

These facts lead to the most amazing feature of earth. The physical universe is essentially running down, with organized matter breaking down to disorganized matter. It is said that entropy is increasing. Entropy expresses the tendency to run down: the greater the order, the lower the entropy. But life puts more order into matter, it evolves into ever-greater complexity – it is an aberrant phenomenon that decreases entropy. In doing so it has a profound effect on the inorganic world we live in, even to the extent of changing the composition of the atmosphere. Indeed analysis of atmospheres can show whether a planet has life or not. An alien studying the atmospheres of the solar planets would conclude that there is no life on Mars or Venus, but that the earth undoubtedly had life.

41

Without life we would not have the ocean and atmosphere of today, and the physical evolution of the planet would have been entirely different. To paraphrase Hsü (1992) 'The earth is not just another heavenly body: ours is a very special planet, and we have no evidence that there is another like it in the whole universe.'

REFERENCES

Alvarez, L.W., Alvarez, W., Asaro, F. and Michel, H.V. (1980) 'Extraterrestrial cause for the Cretaceous–Tertiary extinction', *Science* 208, 1095–108.

Anderson, A.T. (1975) 'Some basaltic and andesitic gases', *Rev. Geophys. Space Phys.* 13, 37–55.

Bullard, E., Everett, J.E. and Smith, A.G. (1965) 'The fit of the continents around the Atlantic', *Phil. Trans. R. Soc., A.* 258, 41–51.

Cande, S.C. and Kent, D.V. (1992) 'A new geomagnetic polarity time scale for the Late Cretaceous and Cenozoic', *J. Geophys. Res.* 97, 13917–51.

Carey, S.E. (1963) 'The asymmetry of the earth', *Aust. J. Sci.* 25, 369–84; 479–88.

Carey, S.W. (1976) *The Expanding Earth*, Amsterdam: Elsevier.

Carey, S.W. (1988) *Theories of the Earth and Universe: A History of Dogma in the Earth Sciences*, Stanford, Calif.: Stanford University Press.

Chyba, C.F. (1987) 'The cometary contribution to the oceans of primitive Earth', *Nature* 330, 632–5.

Chyba, C.F., Thomas, P.J., Brookshaw, L. and Sagan, C. (1990) 'Cometary delivery of organic molecules to the early Earth', *Science* 249, 366–73.

Cloud, P. (1983) 'Early biogeologic history: the emergence of a paradigm', pp. 14–31 in J.W. Schopf, *Earth's Earliest Biosphere – Models in Palaeobiology*, Princeton, N.J.: Princeton University Press.

Coney, P.J. Jones, D.L. and Monger, J.W.H. (1980) 'Cordilleran suspect terranes', *Nature* 228, p. 329.

Cook, F.A. (1983) 'Some consequences of palinspastic reconstruction in the southern Appalachians', *Geology* 11, 86–9.

Fanale, F.P. (1971) 'A case for catastrophic early degassing of the earth', *Chemical Geology* 8, 79–105.

Hallam, A. (1987) 'End-Cretaceous mass extinction event: argument for terrestrial causation', *Science* 238, 1237–42.

Harland, W.B., Cox, A.V., Llewellyn, P.G., Pickton, C.A.G., Smith, A.G. and Walters, R. (1991) *A Geologic Time Scale*, Cambridge: Cambridge University Press.

Holland, H.D. (1984) *The Chemical Evolution of the Atmosphere and Oceans*, Princeton, N.J.: Princeton University Press.

Holmes, A. (1965) *Physical Geology*, Edinburgh: Nelson.

Hsü, K.J. (1992) 'Is Gaia endothermic?', *Geol. Mag.* 129, 129–41.

Jenkins, D.A.L. (1974) 'Detachment tectonics in western Papua New Guinea', *Bull. Geol. Soc. Am.* 85, 533–48.

Lovelock, J. (1988) *The Ages of Gaia: A Biography of Our Living Earth*, Oxford: Oxford University Press.

Owen, H.G. (1981) 'Constant dimensions or an expanding earth?', pp. 178–92 in L.R.M. Cox (ed.) *The Evolving Earth*, Cambridge: Cambridge University Press.

Rubey, W.W. (1951) 'Geologic history of sea water', *Bull. Geol. Soc. Amer.* 62, 1111–47.

Stevenson, D.J. (1983) 'The nature of the earth prior to the oldest known rock record: the Hadean Earth', pp. 32–40 in J.W. Schopf, *Earth's Earliest Biosphere – Models in Palaeobiology*, Princeton, N.J.: Princeton University Press.

Van Steenis, C.G.G.J. (1963) *Pacific Plant Areas*, Vol. 1, Monograph 8, Manila: National Institute of Science and Technology.

Whitfield, W. (1981) 'The world ocean. Mechanism or machination?', *Interdisciplinary Science Reviews* 6, 12–35.

FURTHER READING

Carey, S.W. (1988) *Theories of the Earth and Universe*, Stanford, Calif.: Stanford University Press.

Lovelock, J.E. (1988) *The Ages of Gaia: A Biography of Our Living Earth*, Oxford: Oxford University Press.

Ollier, C.D. (1981) *Tectonics and Landforms* (Geomorphology Texts 6), Harlow, Essex: Longman.

Francis, P. (1993) *Volcanoes: A Planetary Perspective*, Oxford: Clarendon Press.

Schopf, J.W. (1983) *Earth's Earliest Biosphere – Models in Palaeobiology*, Princeton, N.J.: Princeton University Press.

Summerfield, M.A. (1991) *Global Geomorphology. An Introduction to the Study of Landforms*, Harlow, Essex: Longman.

2

THE EVER-CHANGING CLIMATE

Andrew Goudie

INTRODUCTION

The world's climate is forever changing and such changes impact in a multitude of ways upon humans, organisms and landscapes. The changes range from the minor fluctuations within the period of instrumental record (with durations of a few years or decades) to the major geological periods (with durations of many millions of years). The shorter-term changes (Table 2.1) include such phenomena as the Sahel drought and the warm years of the last two decades. The changes with durations of hundreds of years include such events as 'The Little Ice Age', which caused alpine glaciers to expand and upland farming to contract between about AD 1500 and 1850. Within the Pleistocene epoch of the last two or so million years there have been larger duration fluctuations – cold glacial and warm interglacial cycles – which lasted for 10,000 to 100,000 years. At an even larger duration scale ice ages such as those of the Pleistocene appear to have been separated by about 250 million years (Figure 2.1).

General reviews of climatic change are provided by Goudie (1992), Crowley and North (1991) and Williams *et al.* (1993), while Bradley (1985) and Lowe and Walker (1984) provide good surveys of the techniques used to date and reconstruct past environments.

THE CENOZOIC CLIMATE DECLINE

During the Tertiary era, which started at the end of the Cretaceous about 65 million years ago, the world's climate underwent one of the longer duration changes – the so-called Cenozoic climate decline (Figure 2.2). Temperatures showed a general tendency, though not steady or uninterrupted, to fall in many parts of the world. Thus in the North Atlantic region in the early Tertiary conditions favoured a widespread, tropical, moist forest-type of vegetation. At the end of the Eocene there was a climatic deterioration so that in the Oligocene

Table 2.1 Orders of climatic variation

Time-scale unit	Duration (years)	Typical phenomena	Principal bases of evidence
1 Minor fluctuations within the instrumental record	10	Minor fluctuations which give the impression of operating over intervals of the order of 25–100 years, with somewhat irregular length and amplitude	Instrumental; behaviour of glaciers; records of river-flow and lake levels; non-instrumental diaries: crop yields, tree-rings (also for dating)
2 Post-glacial and historic	10^2	Variations over intervals of the order of 250–1,000 years, e.g. the sub-Atlantic recession and others affecting vegetation in Europe and N. America	Earlier records of extremes: fossil tree-rings; archaeological finds; lake-levels; varves and lake sediments; oceanic core-samples; pollen analysis; radio-carbon dating; ice-cores
3 Glacial	10^4	The phases within an ice age, e.g. the duration of the Würm was of the order of 50×10^3 years	Fauna and flora characteristic of interglacial deposits; pollen analysis; variation in height of snowline and extent of frozen ground; oceanic core-samples (dating through latter)
4 Minor geological	10^6	Duration of ice ages as a whole, periods of evolution of species	Geological evidence: character of deposits; fossil fauna and flora; dating largely through radioactivity of rocks
5 Major geological	10^8	Ice ages at intervals of 2.5×10^8 years	

Source: After Goudie (1992).

the climate of Britain may have been more comparable to that of a region like the south-eastern USA, though at times there may have been dry intervals that produced salt crusts, desert soils and desert dunes in the neighbouring Paris Basin.

The warmth of the first half of the Tertiary (the Palaeogene) in Britain had both local and global causes. At a local scale, Britain was at a lower latitude than today, being 10–12° further south. At a global scale the oceans and continents had a very different form, affecting the patterns of ocean currents and monsoon circulations, but there may also have been much more elevated atmospheric carbon dioxide conditions (creating a greenhouse effect) and a marked reduction

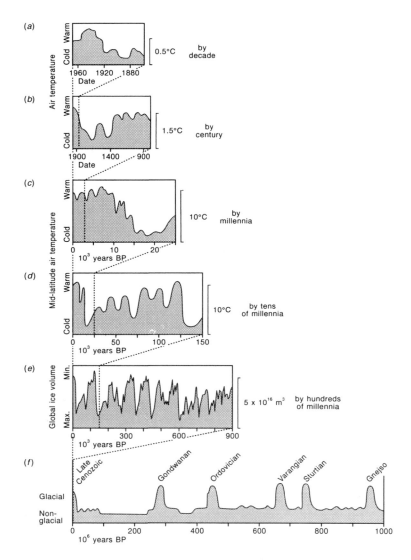

Figure 2.1 The different scales of climatic change. (a) Changes in the five-year average surface temperatures over the region 0–80°N. (b) Winter severity index for eastern Europe. (c) Generalized northern hemisphere air temperature trends, based on fluctuations in alpine glaciers, changes in tree lines, marginal fluctuations in continental glaciers, and shifts in vegetation patterns recorded in pollen spectra. (d) Generalized northern hemisphere air temperature trends, based on mid-latitude sea-surface temperature, pollen records, and on world-wide sea-level records. (e) Fluctuations in global ice-volume recorded by changes in isotopic composition of fossil plankton in a deep-sea core. (f) The occurrence of ice ages in geological time. After Goudie (1992).

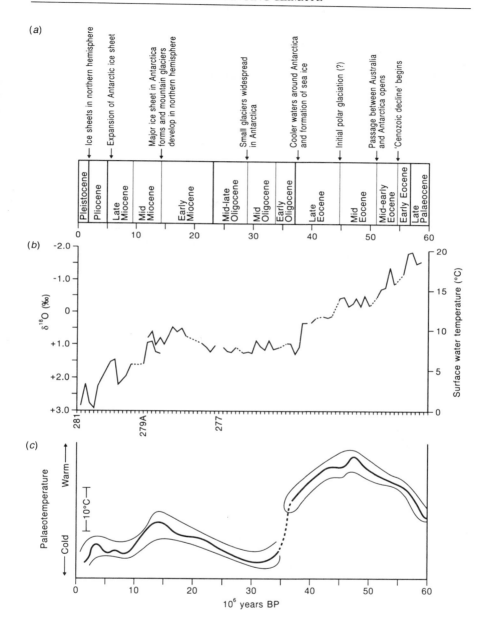

Figure 2.2 The Cenozoic climate decline. (a) A generalized outline of significant events in the Cenozoic climate decline. (b) Oxygen isotopic data and palaeotemperature indicated for planktonic foraminifera at three subantarctic sites (277, 279A, and 281). (c) Temperature changes calculated from oxygen isotope values of shells in the North Sea. After Goudie (1992).

in the angle of tilt of the earth's axis which would have affected the effect of incoming solar radiation. The long phase of warmth in the early and mid-Tertiary led to what the German geomorphologist Julius Büdel (1982) described as a palaeotropicoid earth and its impact may well still be evident in some mid- and high-latitude landscapes where the remnants of tropical planation and deep weathering occur.

By Pliocene times, the degree of cooling was such that a more temperate flora was present in the North Atlantic region, and at 2.4 million years ago glaciers started to develop in mid-latitude areas and many of the world's deserts came into being.

THE QUATERNARY

In the Quaternary period (which comprises the Pleistocene and the Holocene) the gradual and uneven progression towards cooler conditions which had characterized the earth during the Tertiary gave way to extraordinary climatic instability. Temperatures oscillated wildly from values similar to, or slightly higher than, today in interglacials to levels that were sufficiently cold to treble the volume of ice sheets on land during the glacials. Not only was the degree of change remarkable but so also, according to evidence from the sedimentary record retrieved from deep-sea cores, was the frequency of change. In all there have been about seventeen glacial/interglacial cycles in the last 1.6 million years. The cycles tend to be characterized by a gradual build up of ice volume (over a period of $c.$ 90,000 years), followed by a dramatic glacial 'termination' in only about 8,000 years. Furthermore, over the three or so millions of years during which humans have inhabited the earth, conditions such as those we experience today have been relatively short-lived and atypical of the Quaternary as a whole (Bowen 1978). Figure 2.3 illustrates the changes that have taken place over the last 850,000 years.

The last glacial cycle reached its peak about 18,000 years ago, with ice sheets extending over Scandinavia to the north German plain, over most of Britain (except the south), and over North America to 39°N (Figure 2.4). To the south of the Scandinavian ice sheet was a tundra steppe underlain by permafrost, and forest was relatively sparse to the north of the Mediterranean. In low latitudes sand deserts were considerably expanded in comparison with today.

Ice covered nearly one-third of the land area of the earth, but the additional ice-covered area in the last glacial was almost all in the Northern Hemisphere, with no more than about 3 per cent in the Southern. None the less, substantial ice cover developed over Patagonia and New Zealand. The thickness of the now-vanished ice sheets may have exceeded 4 km, with typical depths of 2 to 3 km. The total ice-covered area at a typical glacial maximum was 40×10^6 km^2, compared with the present 15×10^6 km^2.

Highly important changes also took place in the state of the oceans. During the present interglacial conditions of the Holocene, the north-eastern Atlantic

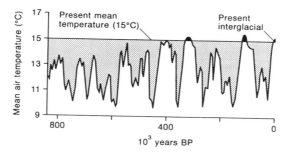

Figure 2.3 Temperature of the earth for the last 850,000 years as inferred from ice volume derived by oxygen isotope measurements from ice cores. After Gates (1993).

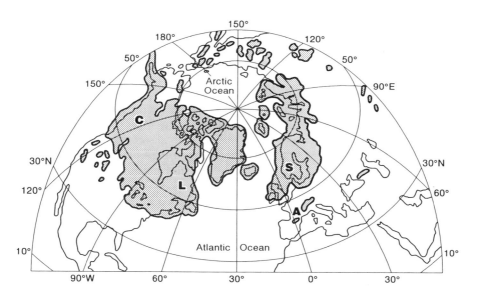

Figure 2.4 The possible maximum extent of glaciation in the Pleistocene in the northern hemisphere. C = Cordilleran ice; L = Laurentide ice; S = Scandinavian ice; A = Alpine ice. After Goudie (1992).

is at least seasonally ice-free as far north as 78°N in the Norwegian Sea. This condition reflects the advection of warm water into this region by the Gulf Stream (North Atlantic Current). During the Last Glacial Maximum, however, the oceanic polar front probably lay at about 45°N, and north of this latitude the ocean was mainly covered by sea ice during the winter.

49

The degree of temperature change that occurred over land was substantial. It was particularly great in the vicinity of the great ice sheets. The presence of permafrost in southern Britain suggests a temperature depression of the order of 15°C. Mid-latitude areas probably witnessed a lesser decline – perhaps 5–8°C was the norm – though in areas subject to maritime air masses temperatures they were more likely to have been depressed by 4–5°C.

The transfer of large volumes of water from the oceans to the ice caps caused a global fall in sea-levels. The degree of this world-wide (or eustatic) change at the time of glacial maxima is the cause of some debate, but may have been of the order of 120 to 150 m. Many areas that are at present submerged continental shelf (e.g. the North Sea) became dry land, islands became connected to each other or to the mainland, and land-bridges were created where now there are straits. When deglaciation set in the sea advanced rapidly horizontally and vertically, flooding old river systems that had cut down to the low glacial base levels.

In addition to their global effect, the ice sheets had a more localized effect on sea-levels through the process of glacio-isostasy. The great volumes of ice depressed the Earth's crust beneath them and caused some upward displacement in a surrounding zone called a peripheral prebulge. When the ice load was removed upon deglaciation the reverse process occurred, and parts of Laurentia have 'popped up' by more than 300 m during the Holocene.

The cold glacials had a multitude of impacts on the landscape that are still visible today. The ice sheets caused considerable erosion and excavation producing characteristic landform assemblages with cirques, arêtes, U-shaped valleys, roches moutonnées and other forms. They also transformed drainage patterns as the lacustrine landscapes of the Laurentian Shield and Scandinavia testify. Elsewhere they deposited boulder clay and outwash gravels, some as sheets and some as distinctive landforms (kames, eskers, etc.). Beyond the glacial limit fine particles blown from outwash plains settled to produce great belts of loess in areas like Central Europe, Tajikistan, China, New Zealand, and the Mississippi valley of the USA. Tundra conditions, with underlying permafrost, created great slope instability and drainage incision, the evidence for which is still very apparent along the escarpments of southern Britain.

Stadials and interstadials

Each glacial cycle had some complexity of form with phases of intense glacial activity and advance, called stadials, being separated by periods of slightly greater warmth (interstadials) when glacial retreat occurred. During the last glacial cycle there were various interstadials, including a particularly marked one during the period 50,000–23,000 BP, and some rather shorter ones nearer the beginning of the cycle. In England, the Chelford interstadial, which may perhaps have occurred 60,000 years ago, saw the establishment of a boreal forest where previously there had been tundra but where in a full interglacial there would have been a mixed deciduous forest.

The most extreme stadial of the last glaciation occurred round about 20,000–18,000 years BP, and in Britain is known as the Dimlington stadial. Shortly thereafter glaciers began to retreat rapidly only to advance briefly in the Younger Dryas stadial round about 11,000 years ago. This event, also called the Loch Lomond Readvance in Scotland, saw the development of cirque glaciers in the British uplands. It ended abruptly round about 10,700 years ago, whereupon the world entered the interglacial conditions of the Holocene.

Interglacials

In general terms the Quaternary interglacials were short-lived but appear to have been essentially similar in their climate, fauna, flora and landforms to the Holocene interglacial in which we live today. One of their most important characteristics was that they witnessed the rapid retreat and decay of the great ice sheets and saw the replacement of tundra conditions by forest over the now temperate lands of the northern hemisphere. At their peak they may have been a degree or two warmer than now. In recent years considerable information on conditions in the last interglacial (the Eemian) has been obtained from ice cores extracted from the polar ice caps. These ice cores provide a detailed archive of past climatic conditions derived from examination of their chemical, gaseous and particulate contents. In particular their stable isotopic composition provides a means of calculating past atmospheric temperatures. The Greenland Ice Core Project managed to drill through 3,029 m of ice under the summit of the Greenland Ice Sheet, and the core dates back to around 250,000 years ago. During the Eemian there may have been some very rapid, indeed abrupt, climate changes (GRIP 1993).

The general sequence of vegetational development during an interglacial has been described for north-west Europe by Birks (1986). The first, or *cryocratic* phase, represents cold glacial conditions, with sparse assemblages of pioneer plants growing on base-rich skeletal mineral soils under dry, continental conditions. In the second, or *protocratic* phase, there is the onset of interglacial conditions. Rising temperatures allow base-loving, shade-intolerant herbs, shrubs and trees to migrate and expand quickly to form widespread species-rich grasslands, scrub and open woodlands, which grow on unleached, fertile soils with a still low humus content. In the third, or *mesocratic* phase, temperate deciduous forest and fertile, brown–earth soils develop under warm conditions, allowing the expansion of shade-giving forest genera such as *Quercus*, *Ulmus*, *Fraxinus* and *Corylus*, followed by slower immigrants such as *Fagus* and *Carpinus*. In the fourth and last retrogressive phase, the *telocratic* phase, soil deterioration and climatic decline lead to the development of open conifer-dominated woods, ericaceous heaths and bogs growing on less fertile, humus-rich podzols and peats.

The ice ages in the tropics

The events which led to the expansions and contractions of the great ice sheets in middle and high latitudes also led to major environmental changes in lower latitudes. Periods of greater moisture (pluvials) were interspersed with periods of less moisture (interpluvials). The evidence for such changes is particularly evident on the margins of great deserts, where dry phases saw the development and advance of great sand seas, whereas in moister phases the dunes were

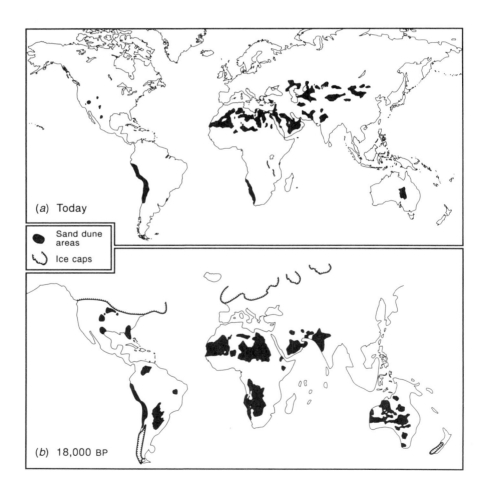

Figure 2.5 The distribution of active sand–dune areas (*ergs*): (a) today; (b) at the last glacial maximum about 18,000 years ago. After Goudie (1992).

Table 2.2 Some sources of palaeoenvironmental information for currently dry areas

Evidence	*Inference*
Fossil dune systems	Past aridity
Breaching of dunes by river systems	Increased humidity
Discordant dune trends	Changed wind direction
Lake shorelines	Balance of hydrological inputs and outputs
Lake-floor sediments	Degree of water salinity, etc.
Lunette sediments	Hydrological status of lake basin
Spring deposits and tufas	Groundwater activity
Duricrusts (lateritic) and related palaeosols	Intense chemical weathering under humid conditions
Old drainage lines	Integrated hydrological network
Fluvial sediments in ocean cores	Quantity of river flow
Aeolian dust in ocean cores	Degree of aeolian deflation
Macro-plant remains (including charcoal), e.g. in pack rat middens	Nature of vegetation cover
Pollen analysis of terrestrial sediments	Nature of vegetation cover
Pollen analysis of marine sediments	Nature of vegetation cover
Fluvial aggradation and siltation	Desiccation
Colluvial deposition	Reduced vegetation cover and stream flushing
Faunal remains	Biomes
Karstic phenomena	Increased hydrological activity
Isotopic composition of groundwaters and speleothems	Palaeotemperatures and recharge rates
Distribution of archaeological sites	Availability of water
Drought and famine record	Aridity
'Frost' screes	Palaeotemperature
Loess profiles and palaeosols	Aridity and stability

stabilized by vegetation and large lakes filled with water in areas that had previously been salty wastes.

Some of the lakes that developed in pluvial phases were enormous. One of the greatest concentrations of pluvial lakes developed in the Basin and Range Province of the American Southwest. Between 100 and 120 depressions, formed by high-angle faulting, were occupied wholly or in part by pluvial lakes during various phases of the Pleistocene. The largest of these was Lake Bonneville, which at its maximum stage had water to a depth of 325 m, was about 500 km long, and covered an area comparable to present-day Lake Michigan. In Eurasia, the Aral and Caspian seas were greatly expanded, uniting to inundate an area of 1,100,000 km^2, and many basins in the Middle East, including the Dead Sea, were occupied by large bodies of water. In northern Africa, Lake Chad expanded over large tracts of the Sahara, while in the Kalahari the Makgadikgadi pans, now disconnected salt pans, it inundated 120,000 km^2.

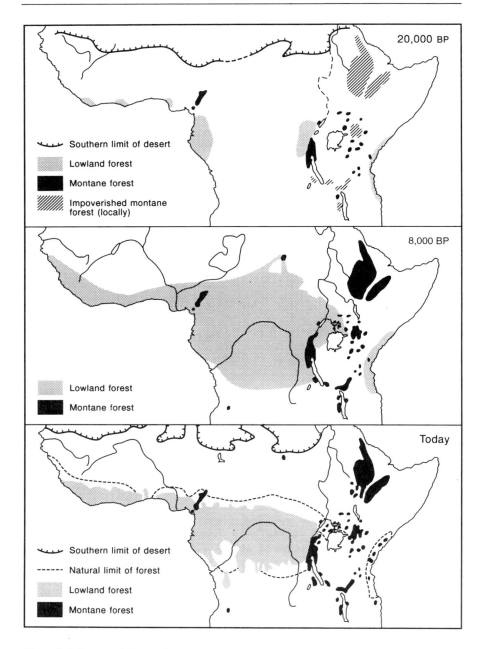

Figure 2.6 Proposed forest distribution in Africa 20,000 and 8,000 years ago compared to today. Reconstructions based on modern biogeographical patterns, plus palynological and geomorphological evidence. After Hamilton (1976).

By contrast in interpluvials, large dune fields expanded (Figure 2.5). Relict forms occur in areas where there is now a well-developed vegetation cover and annual precipitation totals of around 800 mm. The dunes probably formed when vegetation cover was much less capable of inhibiting sand movement under annual precipitation totals that were less than 100 to 300 mm.

Besides the evidence provided by fossil lakes and dunes, there is a multitude

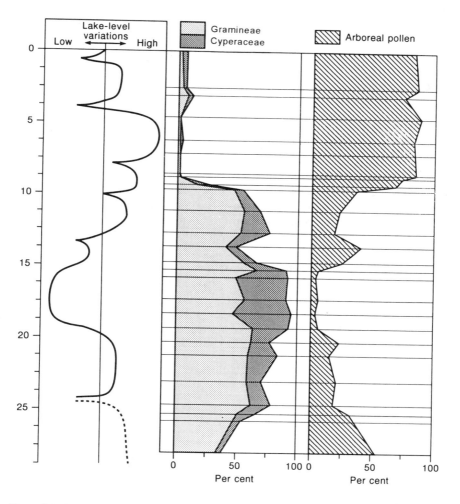

Figure 2.7 Changes in lake level and pollen at Lake Bosumtwi in Ghana during and since the last glacial maximum. Note the low arboreal pollen content in pre-Holocene times, and the very marked expansion of arboreal pollen after about 10,000 years ago. After Maley (1989).

of other lines of information that can be used to identify the former existence of conditions either more humid or more arid than the present. These are listed in Table 2.2.

The dating of pluvials and interpluvials has advanced greatly in recent years. It now appears that areas like the south-west USA had high stands of lake level at the time of the Last Glacial Maximum (18,000 years ago), partly because of reduced rates of evapotranspirational moisture loss brought about by lower temperatures, but also because of the deflection of westerly storms into that area. By contrast lower latitude areas (e.g. Lake Chad, the East African rift valley lakes, north-west India) appear to have been dry at the Last Glacial Maximum, and to have achieved some of their moistest conditions in the early to middle Holocene (round about 8,000 to 9,000 years ago).

Temperature depression also played a role in changing environmental conditions in low latitudes in the Pleistocene. The substantial degree of change in the altitudinal zonation of vegetation on tropical mountains during cold phases can be demonstrated by detailed pollen analysis from lakes and swamps from the highlands of tropical regions. Vegetation zones may have moved through as much as 1,700 m of altitude.

A combination of temperature and precipitation change had a dramatic impact on the nature and extent of rain forest in Africa (Figure 2.6) and South America. This is brought out by a consideration of pollen analyses undertaken in Lake Bosumtwi (Figure 2.7) in southern Ghana (Maley 1989). At the time of the Last Glacial Maximum, between 20,000 and 15,000 years ago, the lake had a very low level and arboreal (tree) pollen percentages reached minimum values of between 4 and 5 per cent. Trees were in effect replaced at that time by herbaceous plants, Gramineae and Cyperaceae. This compares with the situation since *c.*8,500 BP, when arboreal pollen percentages have oscillated between 75 and 85 per cent. The disruption of the tropical rain forests may have had a considerable impact on their diversity and endemism (Colinvaux 1987).

The Holocene

The ending of the Last Glacial period of the Pleistocene was not the end of substantial environmental change. Indeed, as the Holocene progressed the impact of climatic change was augmented as a cause of environmental fluctuation by the increasing role of human activities (Roberts 1989).

The warming of climate in post-glacial times set off the successive return of species of trees with different tolerances of cold and different powers of colonization (Birks 1990) to lands that had been under ice or dominated by open tundra. If we consider Europe, pollen analysis suggests that at 12,000 BP *Pinus* (pine) was mainly in southern and eastern parts of the continent, but by 6,000 BP it was abundant in northern, central and Mediterranean Europe but absent from much of the western European lowlands. *Quercus* (oak) spread progressively northwards from southern Europe and reached its maximum range limits by

6,000 BP, as did *Ulmus* (elm), *Corylus* (hazel) and *Tilia* (lime). However, not all forest trees had reached these limits by 6,000 BP. For instance, *Fagus* (beech) had a rather small range in southern and central Europe by 6,000 BP, which is in contrast to its extensive range in western Europe today. Similarly, *Carpinus* (hornbeam), which is today widespread in lowland Europe was, at 6,000 BP, very largely confined to Italy, Romania, Poland, and parts of the Balkans.

For comparison, Bernabo and Webb (1977) have produced an interesting summary of vegetational changes in the north-east of North America. As Figure 2.8 shows, there have been major changes in the relative importance of spruce, pine, oak and herb pollen (characteristic of temperate grasslands). The largest changes occurred in the early Holocene between 11,000 and 7,000 BP. Especially

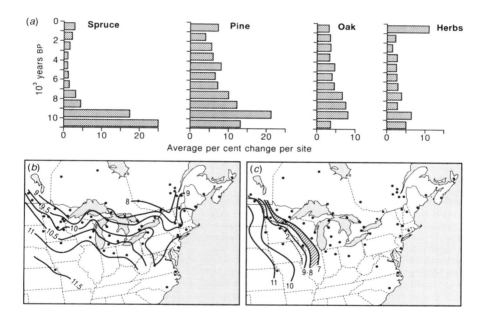

Figure 2.8 Holocene vegetation change in eastern North America. (a) Graph depicting the average percentage change per site, between each 1,000-year level from 11,000 BP to the present, for spruce, pine, oak, and herb pollen. The figure shows important shifts in the amount of change these major pollen groups underwent during the Holocene. Values were obtained by summing the total changes (regardless of signs) seen from all mapped sites and then dividing by the number of sites. (b) Isochrones plotting the time, in thousands of years BP, when spruce pollen declined to below 15 per cent. (c) Isochrones in thousands of years BP illustrating the movements of the prairie–forest ecotone. The position of the prairie border is based upon isopoll maps for herb pollen. Shaded areas show the region over which the prairie retreated after reaching its maximum post-glacial extent at 7,000 BP. After Bernabo and Webb (1977).

notable was the decline of the spruce between 11,000 and 8,000 BP as it gradually moved northwards. Another important feature of the area's Holocene vegetational history has been the fluctuating position of the boundary between prairie and forest. The signs of prairie development in the western portion of the Midwest are visible in the pollen record over 11,000 years ago, as the vast region formerly occupied by the Late Glacial boreal forest began to shrink. The largest eastward shift of the prairie took place between 10,000 and 9,000 BP, and it reached its maximum eastward extent in 8,000 BP. The rate at which plants were able to migrate in response to Holocene warming in the eastern part of North America ranges from 100 metres per year for Chestnut to 400 metres per year for Jack and Red Pines (Gates 1993).

Some portions of the Holocene may have been slightly warmer than the present and terms like 'climatic optimum', 'altithermal', or 'hypsithermal' have been used to denote the existence of a possible phase of mid-Holocene warmth, when conditions may have been 1–2°C warmer than now. There may also have been a 'Little Climatic Optimum' between AD 750 and AD 1300. However, there have also been times which have been rather colder than today, as is made evident by phases of glacial readvance (neoglaciation) in alpine valleys. The latest of these neoglaciations was 'the Little Ice Age' which peaked around AD 1700 and ended towards the end of the nineteenth century (Grove 1988).

Fluctuations of climate also occurred in lower latitudes, and of especial importance for vegetation and human activities was the mid-Holocene pluvial, which transformed the Sahara. A good demonstration of this is provided by the pollen analysis undertaken at a site called Oyo in the eastern Sahara (Ritchie *et al.* 1985). Pollen spectra at that site, dating from 8,500 BP until around 6,000 BP (Figure 2.9), show that there were strong Sudanian savannah elements with tropical affinities in an area which is now hyper-arid. After 6,000 BP the lake at Oyo became shallower, and acacia-thorn and then scrub-grassland replaced the sub-humid savannah vegetation. At around 4,500 BP the lake appears to have desiccated fully and vegetation disappeared except in a few wadis and oases. Thus, the hyper-arid belt more or less disappeared for one or two millennia before 7,000 BP. The northern limit of the Sahel shifted about 1,000 km to the north between 18,000 and 8,000 BP and about 600 km to the south between 6,000 BP and the present.

CLIMATIC CHANGES DURING THE PERIOD OF INSTRUMENTAL RECORD

Over the last two hundred years the organized growth of instrumental observations has provided direct information on climatic fluctuations. The extent of changes in climate over the last one hundred years or so is greater than was formerly believed: both temperature and rainfall have shown trends which have led periodically to great fluctuations in glaciers, lakes, and river discharges.

In many parts of the world a warming trend occurred in the late nineteenth

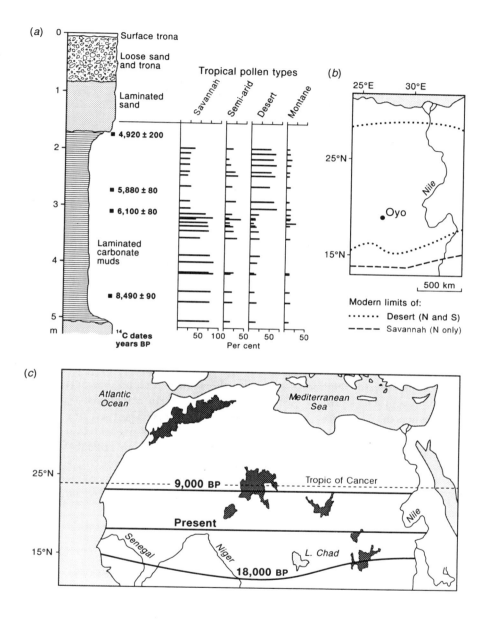

Figure 2.9 (a) Pollen diagram from Oyo in the eastern Sahara. (b) The location of Oyo. (c) The changing position of the Sahara–Sahel limit. After Goudie (1992), Ritchie *et al.* (1985), and Petit-Maire (1989).

59

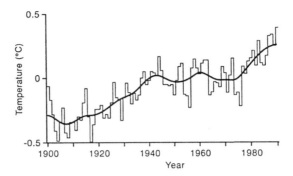

Figure 2.10 Global average surface air temperature relative to the 1950–79 average. After Jones *et al.* (1988).

century and the first decades of the twentieth century (Figure 2.10) which effectively brought to an end the Little Ice Age. From 1940 to 1960 many parts of the world saw a reversal in this trend and underwent a brief cooling phase, but in the last three decades there is mounting evidence that warming is once again a rather general feature of the world's climate. The 1980s formed a particularly warm decade.

Analyses have also been made of precipitation change patterns. Different patterns occur in different areas. In Europe, for example, annual precipitation totals have increased steadily since the middle of the nineteenth century with most of the upward trend being evident in the winter precipitation. In the USA precipitation totals showed a decline from around 1880, reaching a low in the 1930s when the Dust Bowl occurred, and generally increased thereafter. From a human point of view one of the most important spasms of precipitation change has been the persistent drought that has afflicted the Sahel zone of sub-Saharan Africa since the mid-1960s. One consequence of this long drought, aided by deteriorating land cover conditions resulting from humanly induced pressures, has been an acceleration in dust storm frequencies in the area by four to six times.

THE CAUSES OF PAST CLIMATIC CHANGES

The climatic changes that have been described have created a great deal of discussion about their causes. An indication of the great complexity of the factors that need to be considered is given in Figure 2.11. First, the quality and quantity of outputs of solar radiation may change. Second, the receipt of such radiation in the earth's atmosphere will be affected by the position and configuration of the earth. Third, once the incoming radiation reaches the atmosphere its passage to the earth's surface is controlled by the gases, moisture and particulate matter that are present. Fourth, at the earth's surface the

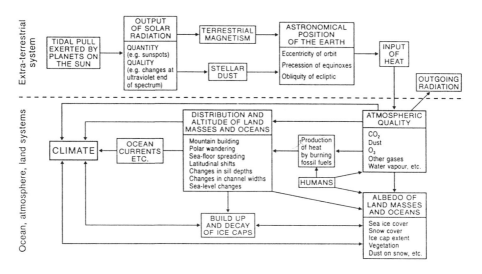

Figure 2.11 A schematic representation of some possible influences causing climatic change. After Goudie (1992).

incoming radiation may be absorbed or reflected according to the nature of the surface (i.e. such phenomena as its reflectivity or albedo, and its moisture content). Fifth, the effects of the received radiation on climate also depend on the distribution and altitude of the land masses and oceans, and this requires a consideration of the role of global tectonic history as discussed in Chapter 1. The situation is rendered complex, as the flow diagram suggests, because of the existence of various feedback loops within and between the ocean, atmosphere and land systems. It also needs to be remembered that the potential causative factors operate over a wide range of different time-scales (Figure 2.12), so that some factors may well be more appropriate than others to account for a climatic fluctuation or change of a particular span of time.

Thus the controls of very long-term climatic changes, such as the Cenozoic climatic decline, may be sought in global tectonic changes. The drift of the continents would change the albedo in different latitudes, open and close up the routes followed by ocean currents, move land masses into different configurations and cause mountain chains to be created. For example, the late Tertiary and Quaternary uplift of the mountains and plateaux of High Asia and the western USA would affect climate by creating high altitude reflective surfaces and by modifying the configurations of the major waves in the general atmospheric circulation.

On the other hand the repeated 100,000-year glacial/deglacial cycles of the Pleistocene may perhaps be best explained by a mix of mechanisms in which changes in the earth's orbital geometry are especially significant. Following the

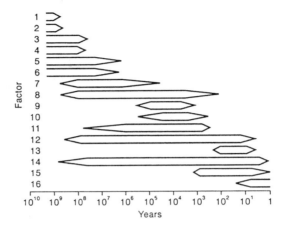

Figure 2.12 Potential causative factors in climatic change and the probable range of time-scales of change attributable to each. (1) Evolution of Sun. (2) Gravitational waves in the Universe. (3) Galactic dust. (4) Mass composition of air (except carbon dioxide, water vapour, and ozone). (5) Polar wandering. (6) Continental uplift. (7) Orogeny and continental uplift. (8) Carbon dioxide in the air. (9) Earth–orbital element. (10) Air–sea–ice–ice-cap feedback. (11) Abyssal ocean circulation. (12) Solar variability. (13) Carbon dioxide in the air from fossil fuel burning. (14) Volcanic dust in the stratosphere. (15) Ocean–atmosphere autovariation. (16) Atmosphere autovariation. After Mitchell *et al.* (1968).

so-called Milankovitch hypothesis, the position and configuration of the earth as a planet in relation to the sun changes, causing variation in the receipt of incoming radiation. Three types of change have been identified and occur in a cyclic manner: changes in the eccentricity of the earth's orbit (a 100,000-year cycle), the precession of the equinoxes (with periodicities of 24,000 and 19,000 years) and changes in the obliquity of the ecliptic (with a periodicity of 41,000 years). The great attraction of this mechanism, which has been called the pacemaker of the Ice Ages, is that the periodicity of these orbital fluctuations seem to be largely comparable with the periodicity of the ice advances and retreats of the Pleistocene (Figure 2.13). The effects of the Milankovitch signal may be magnified by various feedbacks, and this is the subject of considerable ongoing research.

When we consider the shorter-term fluctuations of the scale of the neoglacial advances of the Holocene other mechanisms may become more significant, including changes in solar activity and changes in volcanic activity. The latter mechanism is based on the idea that the presence of elevated dust levels in the atmosphere could increase the backscattering of incoming radiation and so encourage cooling. In addition volcanic dust might reduce sunshine totals further by promoting cloudiness, for dust particles, by acting as nuclei, can

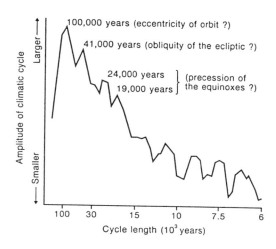

Figure 2.13 The spectrum of climatic variation over the past half-million years. This graph shows the relative importance of different climatic cycles in the isotopic record of the Indian Ocean cores and relates them to the earth's orbital variations. After Imbrie and Imbrie (1979).

promote the formation of ice crystals in sub-freezing air saturated with water vapour. At the present time, runs of cold years tend to follow severe eruptions (e.g. Krakatoa, Katmai, Pinatubo), while various workers have sought to implicate volcanism in explaining the major neoglacials of the Holocene and the Younger Dryas event.

HUMAN IMPACTS ON CLIMATE – GLOBAL WARMING?

The miscellaneous hypotheses discussed so far have been applied with varying degrees of success to a variety of different time-spans. When, however, one contemplates the present and near future, the role of man takes on a position of probable importance.

Certain mechanisms of man-induced climatic changes on a global scale can be recognized (Table 2.3). Of these, the one that currently receives the greatest attention is the role of gas emissions in creating the so-called 'greenhouse effect'. Certain gases absorb outgoing infra-red radiation, thereby reducing heat loss from the earth-atmosphere system. The metabolism of increasing numbers of domesticated cattle releases methane, a very effective greenhouse gas, as does paddy cultivation. Various industrial processes (e.g. plastic production and refrigeration) have released CFCs, which molecule for molecule are diabolically efficient greenhouse gases. Nitrous oxide is produced from vehicular exhausts and other sources. Most important of all, however, is the release of carbon

63

Table 2.3 Possible ways in which human activities may cause climatic change

Cause	Effects
Gas emissions	Global warming and
CO_2 – industrial	precipitation increase
– transport	
– deforestation, agriculture	
Methane	
Chlorofluorocarbons (CFCs)	
Nitrous oxides	
Water vapour	
Krypton 85	
Miscellaneous trace gases	
Aerosol generation	
Dust particles (from industry, agriculture)	General cooling, possible
Sulphates, etc.	reduction in precipitation
Thermal pollution	Temperate increase; accelerated convection
Albedo change	
Dust addition to ice caps	Warming of ice
Deforestation, overgrazing	Cooling
Extension of irrigation	Increase of precipitation
Alteration of ocean currents by constructing straits	Change in salinity, ocean currents, and local temperatures
Diversion of fresh water into oceans	

dioxide as a result of the burning of fossil fuels (e.g. for power generation and transport) and as a result of forest clearance and the resulting oxidation of litter and soil organic matter.

These four gases have shown an upward trajectory in recent decades with, for example, carbon dioxide levels rising from *c.* 270–280 ppm before the Industrial Revolution to nearly 360 ppm today.

There is a great deal of ongoing research activity aimed at establishing the degree of climatic change that will result in coming decades because of these anthropogenic greenhouse loadings. It is possible that in the next fifty years temperatures will rise on average by 1.5–3.5°C, with particularly large increases in certain parts of high latitudes. The amount and pattern of precipitation will also change. Globally the quantities will go up (probably by around 10 per cent), but some areas may become considerably moister (e.g. high latitudes, some monsoonal areas) while others (e.g. mid-latitude steppes and deserts) may become drier. However, there is still considerable uncertainty on this matter and some variation in the scenarios produced by different methods (Houghton *et al.* 1990).

The speed and degree of change is such that it will have great implications for a whole range of natural phenomena. Tundra and permafrost will migrate northwards in the northern hemisphere, for example, and the boreal forests will probably suffer a great reduction in range. In addition, the accelerated melting of ice caps will cause sea-level rise to take place three to six times faster than at present, with great implications for coastal wetlands and other ecosystems. Hurricane intensity, frequency and extent may increase in the tropics, while in some semi-arid areas any trend towards aridity might cause sand dune reactivation and great dust-storm activity (Goudie 1993).

REFERENCES

Bernabo, J.C. and Webb, T. III (1977) 'Changing patterns in the Holocene pollen record of north-eastern North America: a mapped summary', *Quaternary Research* 8, 64–95.

Birks, H.J.B. (1986) 'Quaternary biotic changes in terrestrial and lacustrine environments, with particular reference to north-west Europe', pp. 3–65 in B.E. Berglund (ed.) *Handbook of Holocene Palaeoecology and Palaeohydrology*, Chichester: Wiley.

Birks, H.J.B. (1990) 'Changes in vegetation and climate during the Holocene of Europe', in M.M. Boer and R.S. de Groot (eds) *Landscape-Ecological Impact of Climatic Change*, Amsterdam: IOS Press.

Bowen, D.Q. (1978) *Quaternary Geology*, Oxford: Pergamon.

Bradley, R.S. (1985) *Quaternary Palaeoclimatology*, Winchester, Mass.: Allen & Unwin.

Büdel, J. (1982) *Climatic Geomorphology*, Princeton, N.J.: Princeton University Press.

Colinvaux, P.A. (1987) 'Amazon diversity in light of the paleoecological record', *Quaternary Science Reviews* 6, 93–114.

Crowley, T.J. and North, G.R. (1991) *Paleoclimatology*, New York: Oxford University Press.

Gates, D.M. (1993) *Climate Change and its Biological Consequences*, Sunderland, Mass.: Sinauer.

Goudie, A.S. (1992) *Environmental Change* (3rd edn), Oxford: Oxford University Press.

Goudie, A.S. (1993) *The Human Impact on the Environment* (4th edn), Oxford: Blackwell.

GRIP (Greenland ice-core project) Members (1993) 'Climate instability during the last interglacial recorded in the GRIP ice core', *Nature* 364, 203–7.

Grove, J.M. (1988) *The Little Ice Age*, London: Methuen.

Hamilton, A.C. (1976) 'The significance of patterns of distribution shown by forest plants and animals in tropical Africa for the reconstruction of upper Pleistocene palaeoenvironments: a review', *Palaeoecology of Africa* 9, 63–97.

Houghton, J.T., Jenkins, G.J. and Ephraums, J.J. (eds) (1990) *Climate Change: The IPCC Scientific Assessment*, Cambridge: Cambridge University Press.

Imbrie, J. and Imbrie, K.P. (1979) *Ice Ages: Solving the Mystery*, London: Macmillan.

Jones, P.D., Wigley, T.M.L., Folland, C.K., Parker, D.E., Angell, J.K., Lebedeff, S. and Hansen, J.E. (1988) 'Evidence for global warming in the last decade', *Nature* 228, 790.

Lowe, J.J. and Walker, M.J.C. (1984) *Reconstructing Quaternary Environments*, London: Longman.

Maley, J. (1989) 'Late Quaternary climatic changes in the African rainforest: forest refugia and the major role of sea surface temperature variations', pp. 585–616 in M. Leinen and M. Sarnthein (eds) *Palaeoclimatology and Palaeometeorology*, Dordrecht: Reidel.

Mitchell, J.M. (ed.) (1968) *Causes of Climatic Change*, Meteorological Monographs 8, Boston, Mass.: American Meteorological Society.

Petit-Maire, N. (1989) 'Interglacial environments in presently hyperarid Sahara: palaeoclimatic implications', pp. 637–61 in M. Leinen and M. Sarnthein (eds) *Palaeoclimatology and palaeometeorology*, Dordrecht: Reidel.

Ritchie, J.C., Eyles, C.H. and Haynes, C.V. (1985) 'Sediment and pollen evidence for an early to mid-Holocene humid period in the Eastern Sahara', *Nature* 314, 352–5.

Roberts, N. (1989) *The Holocene: An Environmental History*, Oxford: Blackwell.

Williams, M.A.J., Dunkerley, D.L., De Deckker, P., Kershaw, A.P. and Stokes, T. (1993) *Quaternary Environments*, London: Arnold.

FURTHER READING

Bradley, R.S. (1985) *Quaternary Palaeoclimatology*, Winchester, Mass.: Allen & Unwin.

Goudie, A.S. (1992) *Environmental Change* (3rd edn), Oxford: Oxford University Press.

Williams, M.A.J., Dunkerley, D.L., De Deckker, P., Kershaw A.P. and Stokes, T. (1993) *Quaternary Environments*, London: Arnold.

Wright, H.E., Kutzbach, J.E., Webb, T.E., Muddiman, W.F., Street-Perrott, F.A. and Bartlein, P.J. (eds) (1993) *Global Climates Since the Last Glacial Maximum*, Minneapolis: University of Minnesota Press.

3

THE BIOSPHERE

Alfred G. Fischer

Earth is the watery planet, by virtue of having surficial pressure and temperature conditions in that narrow range in which ice, water, and water vapour can coexist. Life as we know it is an aqueous phenomenon, and the earth is the only solar planet on which life can now sustain itself.

Earth is also an active planet, stirred by thermally driven convection and engaged in chemical reactions (Chapter 1). The surficial region where lithosphere, hydrosphere and atmosphere interact (Penman 1970) is the abode of life, the biosphere in the sense of Vernadsky and of subsequent ecologists (Hutchinson 1970; Lapo 1987), though some geochemists restrict the term to the living fraction. Furthermore, life has modified that setting profoundly, mainly through the generation of free oxygen, unique to the earth (Cloud and Gibor 1970; Holland 1984; Lovelock 1979), which element has in turn had a powerful feedback to organic evolution.

NATURE AND ORIGIN OF LIFE

Life consists of individual organisms – complex structures of assembled macromolecules composed mainly of carbon and hydrogen in a watery matrix. Each organism is engaged in chemical reactions in which external sources of energy are used for growth, for maintenance and repair, and for reproduction. Not understood is the basic driving force for all this, a quality which manifests itself by the will of the organism to maintain these functions – qualities which in the higher animals, above all in the human species, involve consciousness, thought and imagination.

Carbon–hydrogen compounds, including amino acids, occur elsewhere in the Universe, and can be synthesized in the laboratory. But life, the ability to grow and to reproduce, has not been synthesized. Did it arrive from outer space, on a meteor, or did it develop in that primordial hydrosphere of 3.7 billion years ago in which the first traces of microbial fossils are found (Schopf 1983)? It would seem that all life goes back to one initial cluster of macromolecules that

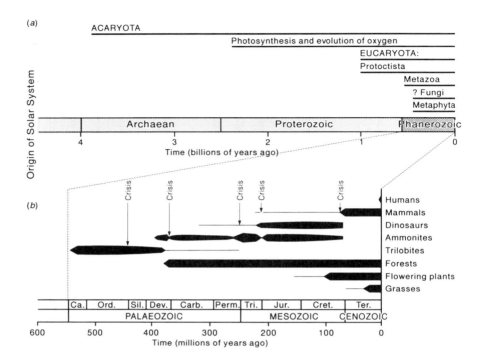

Figure 3.1 Geological time and the distribution of organisms. (a) Major division of time, and distribution of superkingdoms and kingdoms. (b) Divisions of Phanerozoic time, distribution of token taxa and communities, and timing of the five major biotic crises.

grew and reproduced either by a lucky combination or by divine providence. Green photosynthetic organisms fuelled by solar energy now furnish the basis for most life, but their complexities suggest that they were not the first organisms. Rather, as suggested in the 1920s by Oparin and by Haldane, the primordial organism was probably a decomposer. The early, anaerobic ocean was a thin broth in which many kinds of non-living carbon–hydrogen molecules drifted about, uniting and splitting, until some fortunate combination began to grow and to divide in a regular manner. The compounds absorbed came to be used in two ways: some were decomposed by fermentation, yielding the energy necessary to maintain life, and others were used as building blocks for growth, which in turn led to reproduction. This is the manner in which most present-day anaerobic eubacteria function. Alternatively the energy may have been obtained from inorganic exothermic reactions such as the oxidation of highly reduced compounds such as hydrogen sulphide in the manner of present chemolithotrophic bacteria.

68

DEVELOPMENT OF LIFE AND ITS BRANCHES

Whatever their origin, the early microbes evolved. Some of their descendants (Kingdom Eubacteria) derive both energy and materials by decomposition of organic matter, whilst others (mostly of the Kingdom Archaebacteria, see Figure 3.1) derive their energy from oxidation of ferrous iron salts, divalent manganese, reduced forms of sulphur, and other chemicals. Several bacterial stocks found yet another source of energy in solar radiation (Oort 1970; Woodwell 1970), which they harnessed by the process of photosynthesis. The most successful ones do this in a manner that emits free oxygen (Cloud and Gibor 1970). Oxygen is a systemic poison, and as it accumulated in parts of the hydrosphere and in the atmosphere it excluded the original strains of organisms, until some evolved enzymatic protection against oxygen. That in turn enabled them to use oxygen in aerobic respiration, which yields far more energy than does fermentation. From that time on fermentative organisms have played a secondary role to aerobic ones (Woodwell 1970).

Archaebacteria and Eubacteria are combined into a superphylum Acaryota ('without nucleus'), also known as Monera. While remaining microscopic, the monerans show a remarkable range of physical and chemical specializations. Some function at temperatures below 0°C and others above 100°C. Bacteria have learned to break down nearly all naturally occurring organic compounds, and have probably developed species drawing energy from every common exothermic chemical reaction in the biosphere.

In all organisms the genetic code that directs growth and reproduction is carried in certain macromolecular structures – the genes. In the Eubacteria and Archaebacteria, these are scattered throughout the cell, and cell division distributes them evenly to the daughters, leading to lineages (clones) of genetically identical individuals. It is only accident – the alteration of a gene by radiation or other trauma – that introduces heritable variations (mutations), and these in turn can spread to other members of the clone only by leaking out of one individual into another.

Almost three billion years after its origin, discovery of sexual reproduction introduced a major change in life. While asexual reproduction is subject to mutations, these may start new clones, but do not readily spread to others. In sexual reproduction, two individuals fuse into a zygote which shares genetic information from both parents and stores genetic information in duplicate. This provides a mechanism for retaining an enormous variety of mutated genes (alleles) within a species, and for making each sexually generated individual a genetic experiment in which new gene combinations are tried out, for better or for worse. The experiment is preserved if the individual reproduces, and is lost if death intervenes. Compared to the clone reproduction of the monerans, it presents a better basis for Darwin's evolutionary model: the natural selection of a variant fittest to cope with the environment that represents its niche. On the one hand, such an individual may expand that niche; on the other, this variability offers insurance against extinction by environmental change.

Basic to this new system was the segregation of genes into paired strips (chromosomes) housed in a special structure, the nucleus, a structure that characterizes the other great Superkingdom, the Eucaryota (Figure 3.1). This development, about a billion years ago, led to the enormous diversity of organisms seen today. All of the monerans are microbes that become visible only when aggregated into large colonies. Many eucaryotes are also unicellular and small, but all of the familiar multi-celled organisms are eucaryotes. Many of these have the option of either reproducing by asexual cloning (by which great numbers of identical individuals can be produced at low cost), or sexually to generate variable descendants.

It appears likely that not only the nuclei but also other structures within the cells of eucaryotes, such as the plastids which carry the photosynthetic pigments and the mitochondria which control metabolism, began as bacterial parasites, and that the association became a permanent symbiosis in which one of the partners performs the normal business of life, the other the task of carrying on the genetic code.

Four groups (kingdoms) are recognized within these eucaryotes: the Protoctista, the Fungi, the Metazoa and the Metaphyta (Margulis and Schwartz 1988; Figure 3.1). The protoctista are largely aquatic organisms. Photosynthetic ones are spoken of as algae, non-photosynthetic ones as protozoa, and some combine these lifestyles. Most protoctists are single-celled individuals or colonies of similar cells, but various algal groups have developed multicellular individuals of large form and complex structure, such as the kelps of rocky coasts.

The fungi are eucaryotic decomposers. Like bacteria, they secrete enzymes to break down organic matter around them. They too vary in size, from microbial yeasts to great fungi whose fruiting bodies (mushrooms) may be only small appendages of masses of filaments (hyphae) that may extend for tens of metres in the soil and may be thousands of years old.

The metazoans are the multicellular animals, from sponges and corals to insects and ourselves; these comprise the most diverse kingdom, having millions of species.

The metaphytes are the higher plants which, in adaptation to life on land, developed various structures such as roots for gathering water, conducting pipes (vessels) for distributing it, leaves to carry on photosynthesis, and complex reproductive structures that culminate in the development of flowers and seeds. They include the largest of living organisms, and some may live for thousands of years.

Each of these six kingdoms is classified into a hierarchy of smaller categories (taxa): phyla, classes, orders, families, genera, species, and subspecies or races according to the Linnaean system. Some millions of living species have been described, and the true number probably exceeds ten million, mostly metazoans.

FUNCTIONAL DIFFERENTIATION AND ASSOCIATION

Organisms interact in complex communities, in which some act as the basic producers of organic compounds, while others break these down, either as consumers or as decomposers, and provide a constant recycling of nutrients.

Producers

Producer organisms are able to build organic molecules of carbon and hydrogen derived from water and carbon dioxide (Bolin 1970). Most do this by photosynthesis, and are therefore confined to the sunlit parts of the earth – the surface of the lands and the upper waters of the hydrosphere. About half of the world's production is carried on by photosynthetic protoctists (mainly the phytoplankton of the oceans), about half by terrestrial metaphytes. A very small part is contributed by the chemolithotrophic bacteria utilizing chemical reactions around springs on the sea floor or in soils.

Consumers and decomposers

The consumers are the 'animals' that eat particles of organic matter, as grazers, predators or scavengers. They include the generally microscopic and unicellular protozoan protoctists and the multicellular and generally macroscopic metazoans. The decomposers are those organisms that secrete enzymes to digest organic matter in their surroundings, and soak up the dissolved products. They include all of the monerans and, amongst eucaryotes, the fungi, but some animals such as tapeworms and some metaphytes such as the indian pipe also obtain their food by diffusion from the outside.

Symbiosis

Some species may be intimately associated with each other. If one such derives benefit while harming the other, the relationship is termed parasitic. If not, it is commensal, and if it is essential for one or both it is termed symbiotic. Many animals, such as the reef-building corals or the large foraminifera or the giant clam *Tridacna*, house photosynthetic protoctists within their tissues, supplying them with nutrients and depending upon them for oxygen, for food (such as sugars leaking out of them) and for removal of carbon dioxide and nitrogenous waste. Terrestrial communities are also rife with symbiosis. Lichens are symbioses between fungi and cyanobacteria. Leguminous plants have symbiont nitrate-fixing bacteria, and many species of yuccas and figs are dependent upon symbiont insects which are essential to fertilization, at the price of eating part of the fruit. Termites and cows digest cellulose by means of bacterial symbionts. The ultimate case of symbiosis is that of complete integration, as described above for the origin of eucaryotic organisms.

DIVERSITY GRADIENTS

Communities are shaped by the interaction of two basic processes. One is the tendency to increase the number of coexisting species: a few unspecialized generalistic species give way to a larger number of more highly specialized ones, either locally evolved or entering by immigration. The other is the loss of species as environmental change exceeds their tolerances.

The richness of biotic communities is strongly related to latitude (Fischer 1980). Tens of marine snail or bivalve species characterize the American coasts in high latitudes, hundreds in the tropics. The world's richest communities – reefs and rain forests – are essentially limited to the tropics, and show greatest diversity close to the equator. Evolution in the tropics may run faster because of a more rapid succession of generations. Also selection there for fitness may be more subtle and complex, being more the result of interaction with organisms (competition, predation, parasitism, symbiosis) rather than with the blunt axe of weather that plays a larger role in the high latitudes. In the high latitudes the long winter night and extreme cold offer serious obstacles to life, and long-term changes such as glacial cycles drive regions to, or beyond, the limits of organic survival. Whether the tropics were ever similarly stressed by excessive temperatures remains to be discovered.

HABITATS

Life certainly originated in water, most likely the ocean. It spread to other parts of the surficial earth as it learned to armour itself against radiation, desiccation, temperature changes, and so forth, to utilize more and more of the available energy – chemical, solar, physical (hydroelectric power), atomic fission, and, in the future, atomic fusion here on earth. It has already paid transient visits to outer space and to the moon. The major biospheric realms are the oceanic, the limnic, the terrestrial and the subterranean.

Oceanic realm

The oceans occupy the bulk of the biosphere. Not so static as formerly believed (see Chapter 1), they change shape and size over ten million to a hundred million years. The oceans are fundamentally bipartite. At the top is a thin (100–200 m) surficial 'mixed layer' that is wave-stirred and kept in chemical balance with the atmosphere. It is partly sunlit and inhabited by producer organisms (as well as by consumers and decomposers). Below this lies the vast mass of deeper waters, dark and mostly very cold. While life extends to the deepest bottoms, it is limited to consumers and decomposers dependent on organic matter that drifts down to them from above, with the exception of local communities sustained by chemolithotrophic bacteria, such as the communities round submarine springs.

The productivity of the oceans as a whole (Thorson 1971) is mainly carried

72

on by the phytoplanktonic, microscopic protoctists (including diatoms and dinoflagellates) and a minor admixture of photosynthetic monerans, grasses and rooted algae (such as kelps) that are largely restricted to the shallowest of waters. The abundance of the phytoplankton is limited by the available nutrients, chiefly phosphate and iron, ultimately supplied by streams from the continents. Much of the organic matter thus formed is eaten by animals and decomposed by bacteria, releasing nutrients back into the mixed layer; but part of it, along with dead animals, sinks into the deeper waters, where its breakdown and the release of nutrients continues. The net result is that the intermediate and deep waters become enriched in nutrients, and that zones of upwelling along the west coasts of continents come to be particularly fertile: the belts in which organic productivity is at a maximum, and fish populations tend to be high.

In the open ocean, life is dominated by fishes and squids, and these are most diverse in the thermocline, the region below the mixed layer in which temperatures decrease rapidly with depth. Many of the animals here migrate diurnally, seeking the dark depths during daylight and rising at night to feed in the productive mixed layer. Here, where availability of energy (light) is greatest, many consumers have photosynthetic protoctist symbionts. The biotically most diverse communities are those developed in the shallow waters of the nutrient-poor 'blue water' belts – the coral reef communities of the central oceans and the eastern edges of continents especially in the tropics.

The circulation of the deep waters is presently largely driven by the generation of very cold and therefore dense waters in the polar regions. Sinking into the great depths these waters gradually become lighter due to admixture of warmer waters and to geothermal heating from the earth's interior, and thus eventually return to the surface in a cycle of about 1,000 years. The mean temperature of the oceans is only about 3°C.

The oxygen of the deep water is brought from the surface in these descending waters, and animal respiration and aerobic decomposers draw on it, thus reducing oxygen levels as water masses age. Oxygen scarcity develops seasonally in certain areas, within the oxygen minimum zone, directly under the mixed layer, and oxygen is permanently lacking in marginal basins such as the Black Sea, isolated from general circulation. The ocean was not always so well aerated. Older marine sediments contain many black shales devoid of signs of bottom animals, testifying to bottoms from which animals were excluded by lack of oxygen.

Oxygen isotope ratios in fossil skeletons retrieved from the deep-sea floor have shown that the deep waters of the Cretaceous and early Tertiary seas were much warmer, ranging up to 15°C (Douglas and Savin 1975). The circulation rates and patterns of greenhouse climates with warmer polar regions are subjects of current research. The spread of anaerobic waters (reflected in black shales) may reflect a number of factors, among them the lowered gas-carrying capacity of warmer water. Ocean circulation may at times have functioned in a reverse mode, with bottom waters derived from warm saline waters of low latitude origin. It is thus certain that oceanic behaviour and the functioning of marine communities change with time.

Limnic and terrestrial realms

Limnic and terrestrial systems (Whittaker 1970) are linked to continents. Without the buffering action of water, the terrestrial biosphere heats up quickly in sunlight and cools off drastically at night – a factor exacerbated in the polar regions where the day reaches the length of the year. These gradients are reflected primarily in vegetation which, beyond the bare ice, is tundra in the highest latitudes, passing equatorwards into the low spruce–birch–willow woods of the taiga, and onwards into the temperate forests. Where moisture permits, these pass on into equatorial rain forests, but over large parts of the continents scarcity of rainfall limits vegetation to grasslands, brushy vegetation (chaparral or maquis), steppe, or desert. This gradient is mirrored by the altitudinal succession of plant communities on mountains.

The land accounts for more than 90 per cent of the world's species: this is where metazoans and metaphytes have flourished most, and where geographical isolation has provided a stage particularly favourable to the development of multiple species. Over half are insects, and amongst these the most diversified group is that of the beetles. The richest communities are those of the tropical rain forest, and they remain the least well known.

Continents are at the mercy of three ever-changing processes: plate tectonics, the tectonic–geomorphic cycle (Chapter 1), and climatic change (Chapter 2). The break-up of Pangaea in Triassic and Jurassic times left Australia with monotremes and marsupials as the only mammals. South America, separated from the other continents in early Cretaceous time, was provided with marsupials and placentals, of which the former became mostly predators and the latter herbivores. In Africa, Eurasia and North America, which remained episodically connected, placentals won out altogether. Reconnection with South America in Pliocene time replaced most of the South American families with placentals from the African–Eurasian–North American realm, but some South American groups (e.g. opossum) have survived and have invaded North America. In Australia the indigenous marsupials have been partly replaced by placental immigrants.

Both limnic and terrestrial realms are strongly influenced by climatic change (Chapter 2), in which boundaries of moist and dry belts can shift widely in response to the rise of mountain belts. Astronomically induced (Croll–Milankovitch) changes in atmospheric circulation and heat transport drove the Pleistocene ice sheets (Imbrie and Imbrie 1979) which time and again depopulated large parts of North America and Eurasia, forced tundra and taiga southwards, and compressed the belt of temperate deciduous forests. Tree lines moved up and down in tropical mountain ranges (Hooghiemstra 1984). Glacial times expanded deserts, while in transitions lakes spread over parts of the Sahara and of the American intermontane west. These changes forced migrations and extinctions: loss of west European trees caught between ice sheets and deserts; loss of mammals (Martin and Klein 1984), and engendered extensive hybridization of plants (Anderson 1952). We are now in an interglacial age, and there is every reason to expect a progressive return of the ice, within the next 100,000 years, though the onset may be delayed by the greenhouse now being induced through the burning of fossil fuels.

Most of the limnic (freshwater) realm resides in a few very large lakes such as Lake Baikal and the Great Lakes of North America, but areally much of it takes the form of streams. If the oceans are characterized by volume, uniformity and geologic persistence, the limnic world is characterized by attenuation and isolation.

Nearly all phyla are represented in the sea, and only a few on land. The limnic realm is intermediate, having been episodically invaded from both directions. Arthropods such as horseshoe crabs had colonized fresh water in or before Devonian time, as had fishes. Freshwater bivalves and snails go back to the Carboniferous. Anadromous fishes may either come into fresh waters to breed (salmon) or go to sea to do so (eels). Extensive speciation may occur in the headwaters of river systems, where each branch may harbour endemic molluscs and little fishes. In glaciated areas, however, extensive migration through ice-front lakes has left a rather uniform limnic fauna. Very large river systems contain endemic relicts of almost vanished ancient fish groups such as the paddle fish (Yangtse, Mississippi), the bowfin (Mississippi) and large aquatic salamanders (Mississippi). The duration of most lakes is between a thousand and ten thousand years, insufficient to evolve endemic species, hence their faunas are those of the associated streams. Large lakes such as Baikal and Tanganyika, mainly developed in large rifts, may last for millions of years and develop endemic communities.

Subterranean life includes organisms in the vadose zone (above the water table), where bacteria and fungi flourish along with an assemblage of metazoans ranging from insects (especially collembolids and larvae of beetles) to amphibians, reptiles and mammals. In cavernous limestone regions blind white crayfishes, tailed amphibians, and fishes may live in the groundwater. Where there are sources of food, bacteria adapted to high pressures and temperatures exist in porous rocks, to depths of thousands of metres, feeding on petroleum by means of oxygen obtained by sulphate reduction, and generating the sulphur deposits associated with salt domes. Such groups of organisms may have persisted for millions of years without contact with others, but we know very little about them.

FOSSIL RECORD AND EVOLUTIONARY HISTORY

The accumulation of sedimentary strata and of a stratigraphic record full of historical information have been discussed in Chapter 1. Study of this record (stratigraphy) in the nineteenth century led to the recognition of successive changes in life, which served to define chapters (Systems or Periods, Chapter 1; also Figure 3.1) and their subdivisions, now radiometrically dated.

The earliest fossils (Figure 3.1) are moneran microbes of the order of 3.7 billion years old: microscopic spherules of carbon or filamentous strands of such cells, preserved in siliceous rocks (chert) (Schopf 1983). By late Archaean time microbial mats engaged in photosynthesis precipitated carbonate and built laminated algal mat limestones or mound-like columnar or branching structures termed stromatolites, of decimetre to metre dimensions (Schopf and Klein 1992). Through time these became progressively restricted and are now largely restricted to lakes and the tropical intertidal zone. Biotic oxidation of elemental nitrogen to the nitrate state, essential to all of life, was presumably developed at this time as well (Delwiche 1970).

In Proterozoic time the atmosphere had become moderately enriched in oxygen, as were parts of the hydrosphere (Holland 1984). Nitrate- and sulphate-reducing bacteria began to function (Delwiche 1970) allowing aerobic decomposition within sediments. A highly varied microbial community is preserved in the two-billion-year-old Gunflint Chert of Ontario. Sexual reproduction and the eucaryotes developed about one billion years ago. The Ediacara metazoans, soft-bodied and variously interpreted as belonging to present-day phyla or as representing a separate phylum or even kingdom (Vendozoa), appeared near the end of this era (McMenamin and McMenamin 1990).

By Cambrian time (Milne *et al.* 1985) atmospheric oxygen levels had reached at least 10 per cent of present ones, and much of the ocean had become aerobic. Skeletons appeared in several metazoan phyla, presumably serving for support of tissues, for defence against the first predators, and for tools and weapons such as claws. Most of the metazoan phyla and many of the modern classes made their appearance during that period, but faunas were dominated by the archaeocyathids (probably a group of sponges, confined to the Cambrian) and the arthropod class Trilobita. Ordovician time brought nearly all remaining classes and most of the orders. Faunas became dominated by bryozoans, brachiopods, stemmed echinoderms, and cephalopods, while trilobites reached an acme in diversity. Jawless armoured fishes appeared. By Silurian time corals and sponges were constructing large reefs, metaphytes had begun to evolve and to colonize the lands, and jawed fishes appeared. Most metaphyte classes made their appearance in the Devonian, and the first forests were dominated by lycopods and tree ferns, and inhabited by the first insects and amphibians. Sharks and bony fishes appeared, but the largest fishes were the Arthrodires, becoming extinct at the end of the period. Carboniferous time brought the invention of the amniotic egg, which, supplied with its own water, freed tetrapods from reproducing in water, and thus initiated the reptiles. In Permian time gymnosperms began to dominate forests, and reptiles became the dominant herbivores and predators. A huge extinction in late Permian time brought to an end the Palaeozoic corals and

76

trilobites, as well as many lineages of bryozoans, brachiopods, and other invertebrate groups.

Triassic faunas were of more modern stripe, dominated by gastropods and bivalves as well as by the now extinct ammonites. The tiny pelagic algae termed coccolithophores originated at this time. The rootstocks of mammals probably go back to this time, evolving simultaneously with dinosaurs. Giant reptiles (ichthyosaurs) came to dominate the seas, and the first flying reptiles (pterosaurs) appeared. Jurassic seas experienced a progressive development of microscopic calcareous plankton, from photosynthetic coccolithophores to planktonic foraminifera. By late Jurassic time this plankton was deviating perhaps half of the world's carbonate deposition from shallow water to the deep-sea floor, as it still does. Ichthyosaurs at sea were challenged by the rise of plesiosaurs, while birds represent a second aerial lineage of reptiles. Dinosaurs grew to huge size, roaming lands dominated by cycads and by great forests of araucarias and redwoods. Thermoregulation was almost certainly employed by Jurassic mammals and dinosaurs, as well as by birds and pterosaurs. Mammals imply the development of milk glands. The most notable development of Cretaceous time was that of flowering plants (angiosperms). Animals had earlier been induced to serve plants as agents of seed dispersal, but flowers now came to attract insects and other animals into the role of carrying pollen to flowers. The aimed delivery of pollen, as contrasted to the diffuse dispersion by wind, permitted flower species to maintain themselves in greatly diluted stands, thus forming the basis for the diversity of present plant communities.

At the end of Cretaceous time, the dominance of dinosaurs on land, of marine reptile groups at sea and of pterosaurs in the air was terminated by a severe biotic crisis which also caused the extinction (ammonites) or near extinction (planktonic foraminifera, coccolithophores) of many other metazoan and protoctist groups. The demise of dinosaurs opened the way to explosive Cenozoic evolution in mammals. These also invaded the waters as cetaceans and pinnipeds, and the air as bats. Flowering plants continued to diversify and dominate. Grasses, growing at the base rather than at the leaf tip, developed in the early Tertiary. Their ability to withstand cropping down to the ground had led, by Miocene time, to extensive grasslands and grazing mammals such as cattle, sheep, antelopes and horses. Primate monkeys developed in Tertiary time, giving rise in Miocene time to great apes with their highly developed brains. In Pliocene time, an African branch of these, the hominids, largely gave up arboreal life in adaptation to life in the savannah, and developed from the awkward 'knuckle walking' of the great apes into erect-postured *Homo* striding on hind legs. Stone tools and the use of fire came to be parts of a rudimentary culture. About 35,000 years ago, *Homo sapiens* appeared as a hunter and gatherer, distinguished from his ancestors not only by a somewhat different cranial structure but also by a greater care and artistry in the manufacture of stone tools, and by the development of art (rock paintings and sculptures) which remain appealing to this day. This was followed by domestication of animals, agriculture, industry and the development of cities and states, along with development of religions, arts and sciences.

BIOSPHERIC INSTABILITIES

In geological perspective, the organic evolution of the biosphere has occurred in settings involved in constant change (Huggett 1990). Normal geological change, involving uplift, erosion, subsidence, vulcanism and so on, occurs on a local to regional scale and has long been recognized, but the role of global changes has only become apparent in the last few decades. These include the slow increase in solar radiation, and its climatic compensation by a progressive (though irregular) transfer of atmospheric carbon dioxide into the crust as organic carbon and as carbonate (Bolin 1970; Holland 1984; Lasaga *et al.* 1985).

Biospheric and environmental fluctuations are summarized in Figure 3.2. The Wilson cycle of continental dispersal and re-aggregation, driven by cyclic mantle convection, has a period of about 400 million years (Fischer 1984). It is accompanied (Figure 3.3) by changes in mean sea-level and igneous activity generating carbon dioxide (highest at times of dispersal). These factors appear to drive climates from greenhouse conditions with low latitudinal temperature gradients (early Palaeozoic, late Mesozoic) associated with relatively high carbon dioxide levels, to the steep-gradient icehouse states with large and continued glaciations, although ice sheets also appeared at other times. Biotic diversity appears to have fluctuated in harmony with swings in environmental factors: Sepkoski's studies (1994) show an overall increase of marine metazoans, as measured at the family level (Figure 3.3) or the generic level, but superimposed on this is a great drop in diversity associated with the aggregation of Pangaea in late Palaeozoic times. To a lesser degree

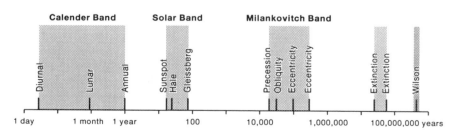

Figure 3.2 Spectral distribution of periodic biospheric disturbances, time-scale loga-rithmic to base 10. After Fischer and Bottjer (1991).

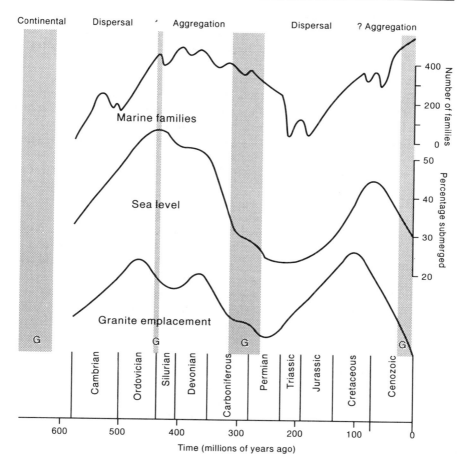

Figure 3.3 The Wilson cycle of continental dispersal and aggregation and associated phenomena: granite emplacement, sea-level, organic diversity (marine metazoan families), and icehouse states (glaciations), G. After Fischer (1984).

this pattern is shared by land plants (Milne *et al.* 1985).

Oscillations engendered by variations in the earth's orbit (Chapter 2) include the cycle of orbital eccentricity which has components of about 400,000 and 100,000 years, the cycle in the tilt of the earth's rotational axis with a period of about 41,000 years, and the precessional cycle with components of about 23,000 and 19,000 years. These cycles affect the latitudinal and seasonal distribution of solar energy. They drove the advances and retreats of ice sheets in the last two million years (Imbrie and Imbrie 1979) but also affected climates in non-glacial times. They drove climates from arid to humid and back, great lakes alternating with dry playas. They affected the oceanic plankton

79

(abundance of coccolithophorids) reflected in the rate of calcium carbonate accumulation, as well as the oxygen content of the oxygen minimum zone (thermocline) reflected in carbon burial and in the faunas of the affected bottoms (Fischer and Bottjer 1991).

Of different origin are biotic crises or mass extinctions (Figure 3.1) first recognized by Cuvier 150 years ago and recently summarized by Sepkoski (1994). They recur at intervals of some tens of millions of years. The best known of these is the Cretaceous–Tertiary crisis which 65 million years ago brought an end to dinosaurs and pterosaurs on land and to ammonites, belemnites and many other groups of organisms at sea (Raup 1986). Amongst many possible causes, that of collision with a comet or asteroid had been considered but not widely championed. Then, in 1980 and succeeding years Alvarez and his colleagues found that wherever sedimentary sequences record the events of this time, the boundary layer is enriched in the rare platinum-group element iridium, diagnostic of planetary cores and meteorites. Accordingly, Alvarez et al. (1980) postulated collision with an iron-nickel asteroid or comet which vaporized on impact and was deposited in fine dust over the world. The impact would have generated a giant earthquake, a crater with >100 km diameter and a transient depth of >10 km, a wide distribution of impact-generated glass (tektites *sensu lato*), a solar blackout caused by atmospheric turbidity, an initial heat wave that may have started global wildfires, a following Arctic chill, and the settling out of atmospheric dust containing chemical signatures of the bolide. Calculations by Prinn and Fegley (1987) show that the atmospheric passages of bolide and ejecta would oxidize much nitrogen to the NO and NO_2 state, and that the resulting rain of nitrous and nitric acids would have caused further devastation of life on land and in the mixed layer of the oceans.

Evidence for such events was found in charcoal enrichment in the boundary layer, and widespread relics of glass spherules and shocked quartz grains within a 1,500 km radius of Yucatán, and led up to the recognition of a buried, 165-km-wide crater in the Yucatán peninsula of Mexico (Hildebrand et al. 1991). Melt rock in this Chicxulub crater is dated at 65.5 million years, and grades up into glassy breccias (suevite) such as is known from other impact sites. Presence of much anhydrite at the impact site implies generation of sulphuric acid. Increasing numbers of geologists and palaeontologists are therefore ready to accept this bolide strike as a central factor in the Cretaceous–Tertiary crisis.

Are the other crises to be ascribed to the same cause? Numerous craters have been recognized, but few of them are adequately dated. Some small iridium anomalies have been found and some shocked quartz has been reported, but no other case of comparable strength has been made to date. That suggests a variety of causes, striking randomly in time. However, there is some evidence that crises are periodic (Raup 1986), with a 26-million-year period in the Mesozoic and Cenozoic. This suggests a common cause, and therefore impacts. But what might cause such impacts to run on schedules? The most likely impactors are comets from the postulated Oort cometary cloud, from which some periodic disturbance may deflect them into planet-crossing orbits. What disturbs them? Existence of a solar twin star (Nemesis), now so far away as to masquerade as a brown dwarf, no longer appears likely. The approximate 30-million-year oscillation of the solar system through the mid-galactic plane appears to have the wrong timing, and we are thus left without a good suspect. Further large collisions and crises are to be expected, but probably not for some millions of years.

THE EFFECT OF HUMANS

Every species interacts with its environment, and the effects of some species – such as that of the initial oxygen-producing photosynthesizer – have been turning points of biospheric history. The human species has achieved such a crucial position by means of its unprecedented power to exploit both the inorganic and the organic resources of the biosphere for its own benefit.

Predation and deforestation

Extinction of many Pleistocene mammals (e.g. mammoth, mastodon, and horse in America) and birds, chiefly in the last 30,000 years, was probably brought about by human hunting (Martin and Klein 1984). This extinction has continued into our day, but whereas the loss of large grazers and top carnivores is apparent, it is probably less insidious than some of the following.

Progressive deforestation commenced with the development of agriculture, and grew with human populations and the added demands for wood for building, fuel and paper. It has been greatly aided by mechanization. Much of southern Europe was cleared in antiquity, much of central Europe in the Middle Ages. Much of North America was cleared more recently, and deforestation is sweeping the tropics. There can be little doubt that this is resulting in losses of many species, and while most of these are inconspicuous and many even never named, they represent loss to the sustaining infrastructure of the community which, in the end, is likely to be more degrading to the community than losses to predation.

In the long run, the most serious consequences of these events lie in the areal restriction of the natural communities, which are being reduced to islands surrounded by agriculture. Inexorable climatic change will demand migration of entire communities, or their degradation to those species that can adapt to new climates. How will these communities be able to roll with the climatic punches when hemmed in by agriculture or agriculturally degraded soils?

Acceleration of geological cycles

Agriculture and some forms of mining speed natural rates of erosion, and have led to widespread degradation or loss of soils. Heavy use of mineral fertilizers and of synthetically fixed nitrogen salts has loaded streams with nutrients. And, as some land areas lose their soil cover and mineral reserves, aqueous communities are being stressed by the effluents – in some cases mud – and more generally with higher nutrient levels inimical to the blue-water communities such as coral reefs.

A particularly widespread effect of this sort has been the acceleration of the sulphur cycle as sulphur compounds released by combustion of coal and petroleum are added to the atmosphere (Singer 1970). Nitrous oxides emitted by engines are also added. The resulting acid rain is lowering the pH of many freshwater bodies, degrading their communities, and is impinging on forest health. Weathering is being intensified, and there is some concern that it may liberate toxic quantities of aluminium.

Spread of synthetic toxins

Related to this has been the spread of anthropogenic toxins, from pesticides such as DDT to industrial wastes such as PCBs, mercury compounds and radioactive wastes. One of the insidious features of these toxins is their tendency to be progressively concentrated as they move up the food chain. They attain highest levels in the top members (such as radiostrontium in cows and mercury and PCBs in fish) which become poisonous for human consumption and may become threatened by extinction.

The synthetic industrial compounds known as fluorocarbons have been accumulating in the atmosphere, where they react with ozone, O_3. While ozone itself is toxic, its enrichment in the stratosphere protects organisms from damage by ultraviolet radiation. As the ozone shield is weakened, receipt of ultraviolet rays will rise and with it radiation damage such as skin cancer in organisms exposed to the sun.

Global warming

It is estimated that within a generation the burning of fossil fuels will have doubled the carbon dioxide content of the atmosphere; that the mean temperature of the atmosphere will thereby rise an average 4–5°C (Budyko *et al.* 1985); and that heating will be greater in the high latitudes than in the tropics. But the complexities of the climate are such that specific predictions for regions are not reliable. Human populations over much of the world may have to adapt to new crops or will have to move. Natural plant communities will be degraded by loss of sensitive species unless they migrate, but how can those communities migrate where they have come to be island relicts on largely farmed continents? From the geological perspective, such a rise will merely restore earth to a more 'normal' state, and may postpone the inevitable advent of another ice age, but the human impact will change climate rapidly, necessitating rapid biotic response, both by humans and by natural communities, in a world in which agriculture and national boundaries tend to inhibit migration of both natural and cultural communities.

SYNOPSIS AND OUTLOOK

The biosphere has a duration limited by the existence of liquid water. In the beginning, earth was too hot for life, and in the end the sun, at the close of its helium cycle, will engulf life in a fireball. This window for life would have been much more transient had the increasing rate of heating by the sun not been counteracted by the progressive loss of atmospheric heat retention through loss of carbon dioxide.

Through the last 3.7 billion years organisms have come to function in increasingly complex interactive communities, the peaks of which are seen in the coral reefs at sea and in the tropical rain forests on land. Also, the progressive increase of complexity in some lineages has developed one species, *Homo sapiens*, capable of visualizing its role in this drama. Having attained unprecedented capacities for utilizing sources of energy for his own good, the human species has come to threaten the biotas of the biosphere in an unprecedented way. Human civilization demands equal access to resources by all people, yet even present populations, many at starvation level, make demands that shrink natural communities into islands which cannot possibly maintain themselves in the face of impending geological change. Maintenance of civilization will only succeed with restricted populations and with an ethic that, despite many ethnic roots, has been largely lost: an ethic centred on the preservation of earth's biotic diversity as a treasure to be passed on to the untold future.

REFERENCES

Alvarez, L.W., Alvarez, W., Asaro. F. and Michel, H. (1980) 'Extraterrestrial cause for the Cretaceous–Tertiary extinction', *Science* 208, 1095–108.

Anderson, Edgar (1952) *Plants, Man, and Life*, Berkeley, Calif.: University of California Press.

Bolin, B. (1970) 'The carbon cycle', *Scientific American* 223, 124–32.

Brown, L.R. (1970) 'Human food production as a process in the biosphere', *Scientific American* 223, 160–70.

Budyko, M.I., Ronov, A.B. and Yanshin, A.L. (1985) *History of the Earth's Atmosphere*, Berlin, Heidelberg, New York: Springer.

Cloud, P. and Gibor, A. (1970) 'The oxygen cycle', *Scientific American* 223, 110–23.

Deevey, E.S., Jr (1970) 'Mineral cycles', *Scientific American* 223, 148–58.

Delwiche, C.C. (1970) 'The nitrogen cycle', *Scientific American* 223, 136–46.

Douglas, R.E. and Savin, S.M. (1975) 'Oxygen and carbon isotope analyses of Cretaceous and Tertiary foraminifera from Shatsky Rise and other sites in the North Pacific Ocean', *Initial Reports of the Deep Sea Drilling Project* 32, 509–20.

Fischer, A.G. (1960) 'Latitudinal variations in organic diversity', *Evolution* 14, 64–81.

Fischer, A.G. (1984) 'The two Phanerozoic supercycles', pp. 129–50 in W.A. Berggren and J.A. van Couvering (eds) *Catastrophes and Earth History*, Princeton, N.J.: Princeton University Press.

Fischer, A.G. and Bottjer, D.J. (eds) (1991) 'Orbital forcing and sedimentary sequences', *Journal of Sedimentary Petrology* 61, 1063–252.

Hildebrand, A.R., Penfield, G.T., Kring, D.A., Pilkington, M., Camargo, Z., Jacobsen, S.B. and Boynton, W.V. (1991) 'Chicxulub crater: a possible Cretaceous/Tertiary impact crater on the Yucatán peninsula, Mexico', *Geology* 19, 867–71.

Holland, H.D. (1984) *The Chemical Evolution of the Atmosphere and Oceans*, Princeton, N.J.: Princeton University Press.

Hooghiemstra, H. (1984) *Vegetational and Climatic History of the High Plain of Bogota, Colombia: A Continuous Record of the Last 3.5 Million Years*, Vaduz: J. Cramer.

Huggett, R.J. (1990) *Catastrophism: Systems of Earth History*, London, New York, Melbourne, Auckland: Edward Arnold.

Hutchinson, G.E. (1970) 'The Biosphere', *Scientific American* 223, 44–53.

Imbrie, J. and Imbrie, K.P. (1979) *Ice Ages: Solving the Mystery*, Cambridge, Mass. and London: Harvard University Press.

Lapo, A.V. (1987) *Traces of Bygone Biospheres*, Moscow: MIR.

Lasaga, A.C., Berner, R.A. and Garrels, R.M. (1985) 'A geochemical model of atmospheric CO_2 fluctuations over the last 100 million years', pp. 397–411 in E.T. Sundquist and W.S. Broecker (eds) *The Carbon Cycle and Atmospheric CO_2: Natural Variations Archean to Present* (Geophysical Monograph 32), Washington, DC: American Geophysical Union.

Lovelock, J. (1979) *Gaia: A New Look at Life on Earth*, Oxford and New York: Oxford University Press.

McMenamin, M.A.S. and McMenamin, D.L.S. (1990) *The Emergence of Animals*, New York: Columbia University Press.

Margulis, L. and Schwartz, K. (1988) *The Five Kingdoms*, San Francisco, Calif.: W.H. Freeman.

Martin, P.S. and Klein, R.G. (1984) *Quaternary Extinctions*, Tucson: The University of Arizona Press.

Milne, D., Raup, D., Billingham, J., Niklas, K.J. and Padian, K. (eds) (1985) *The Evolution of Complex Life*, Washington, DC: National Atmospheric and Space Administration/US Government Printing Office.

Oort, A.H. (1970) 'The energy cycle of the Earth', *Scientific American* 223, 54–63.

Penman, H.L. (1970) 'The water cycle', *Scientific American* 223, 98–108.

Prinn, R.G. and Fegley, B., Jr. (1987) 'Bolide impacts, acid rain and biospheric traumas at the Cretaceous–Tertiary boundary', *Earth and Planetary Science Letters* 83, 1–15.

Raup, D. (1986) *The Nemesis Affair: A Story of the Death of Dinosaurs and the Ways of Science*, New York: W.W. Norton.

Schopf, J.W. (ed.) (1983) *Earth's Earliest Biosphere*, Princeton, N.J.: Princeton University Press.

Schopf, J.W. and Klein, C. (eds) (1992) *The Proterozoic Biosphere*, Cambridge: Cambridge University Press.

Sepkoski, J. (1994) 'Extinction and the fossil record', *Geotimes* (March), 15–17.

Singer, S.F. (1970) 'Human energy production as a process in the biosphere', *Scientific American* 223, 174–90.

Thorson, G. (1971) *Life in the Sea*, New York: McGraw-Hill World University Library.

Whittaker, R.H. (1970) *Communities and Ecosystems*, New York, Toronto, London: Macmillan.

Woodwell, G.M. (1970) 'The energy cycle of the biosphere', *Scientific American* 223, 64–74.

FURTHER READING

Ager, D.V. (1993) *The New Catastrophism: The Importance of the Rare Event in Geological History*, Cambridge: Cambridge University Press.

Lovelock, J.E. (1979) *Gaia: A New Look at Life on Earth*, Oxford and New York: Oxford University Press.

McMenamin, M.A.S. and McMenamin, D.L.S. (1990) *The Emergence of Animals*, New York: Columbia University Press.

Margulis, L. and Schwartz, K. (1988) *The Five Kingdoms*, San Francisco, Calif.: W.H. Freeman.

Raup, D.M. (1986) *The Nemesis Affair: A Story of the Death of Dinosaurs and the Ways of Science*, New York: W.W. Norton.

Schopf, J.W. and Klein, C. (eds) (1992) *The Proterozoic Biosphere*, Cambridge: Cambridge University Press.

Walter, H. (1985) *Vegetation of the Earth and Ecological Systems of the Geo-Biosphere* (3rd revised and enlarged edn, translated from the 5th revised German edn by O. Muise), Berlin: Springer.

Westbroeck, P. (1991) *Life as a Geological Force: Dynamics of the Earth*, New York and London: W.W. Norton.

Wilson, E.O. (1992) *The Diversity of Life*, Cambridge, Mass.: The Belknap Press of Harvard University Press.

4

HUMAN EVOLUTION

Bernard Wood

EVOLUTIONARY CONTEXT

The ancestors of modern humans probably lost their connection with the rest of the animal kingdom between about five and eight million years ago. There is now abundant evidence that the animals most closely related to modern humans are the two living African apes – that is, the chimpanzee (*Pan*) and the gorilla (*Gorilla*). Because both of these ape genera are non-human in their behaviour and superficial appearance, it was naturally assumed that they were more closely related to each other than either was to modern humans. However, when the proteins and the genomes are compared, there is evidence that some of the DNA in both the nucleus and the mitochondria of the cells of *Homo sapiens* and *Pan* are very similar in structure. Indeed, the DNA of *Pan* is apparently more similar to that of *Homo* than it is to *Gorilla*. While an increasing number of researchers are convinced that the similarities between *Homo* and *Pan* are significant, some of their colleagues remain sceptical and suggest that the details of the close relationships between *Homo*, *Pan* and *Gorilla* presently cannot be resolved. If, however, we accept the overwhelming evidence for a close relationship between modern humans and the African apes, how is that African ape/modern human group related to the rest of the animal kingdom?

The African ape/modern human group, together with the Asian apes, the orang-utans (*Pongo*), and the gibbons (*Hylobates*), make up the zoological superfamily called the Hominoidea, the members of which are usually referred to as 'hominoids'. The hominoids comprise one of the two superfamilies within the infraorder Catarrhini, which is one of the two major subdivisions within the order Primates. The other superfamily within the Catarrhini, the Cercopithecoidea, or the cercopithecoids, comprises the Old World monkeys; this is the group that includes the common and well-known monkeys like baboons and macaques (Figure 4.1).

Modern apes differ from modern monkeys in several ways. Apes have longer

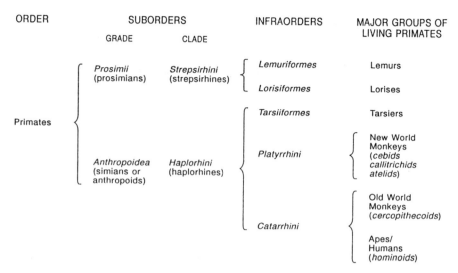

ORDER	SUBORDERS		INFRAORDERS	MAJOR GROUPS OF LIVING PRIMATES
	GRADE	CLADE		
Primates	*Prosimii* (prosimians)	*Strepsirhini* (strepsirhines)	*Lemuriformes*	Lemurs
			Lorisiformes	Lorises
	Anthropoidea (simians or anthropoids)	*Haplorhini* (haplorhines)	*Tarsiiformes*	Tarsiers
			Platyrrhini	New World Monkeys (*cebids callitrichids atelids*)
			Catarrhini	Old World Monkeys (*cercopithecoids*)
				Apes/ Humans (*hominoids*)

Figure 4.1 A classification scheme for the Order Primates showing the close relationship between modern humans and the living apes, and how the hominoid superfamily is related to other living primates.

forelimbs, a more mobile wrist that allows them to rotate the hand through 180° to grasp supports, a shoulder which allows the forelimb to be lifted above the head, a wider and shorter trunk, modifications to the teeth and a brain that, relative to body size, is larger than that of an equivalent-sized monkey.

The hominoids probably became distinct at least 25 million years ago, either at the end of the Oligocene or at the very beginning of the epoch known as the Miocene. Fossil evidence suggests that these early apes were geographically more widespread and taxonomically more diverse than their modern representatives. At this time forests were more extensive in northern latitudes than they are today. During the Miocene fossil apes were widely distributed across Europe and Asia, and they have been found in Africa as far south as Namibia.

Earlier interpretations suggested that some of these fossil apes were ancestral to living ape species, but more recent assessments suggest that most of the early and middle Miocene fossil lineages have no living representatives. They stand in relation to the living apes much as the fauna of the Burgess Shale does to the morphologically less diverse creatures that succeeded it.

APE–HUMAN SPLIT

The fossil record of the apes between about 4–12 million years ago is poor, and from this meagre evidence it is not possible to reconstruct when one group of the hominoids split off to form a separate lineage, called the 'hominids', which eventually gave rise to modern humans. It is, however, possible to use differences in the DNA, which makes up the genotype, to provide an estimate of how long a lineage has been independent. The basis of what has become known as the 'molecular clock' was the discovery that many, if not most, of the mutations that occur naturally are neutral, in the sense that they convey no particular, or discernible, advantage on the animal. If one makes the reasonable assumption that these neutral mutations have been occurring at the same rate throughout the evolution of the hominoids, then the degree of molecular difference can be used as a 'clock' to estimate the time that has elapsed since any two lineages separated. When this is done for the molecular differences between living people and the living African apes, scientists estimate that the lineage which includes humans has been separate from the rest of the hominoids for 5–8 million years.

THE SHAPE OF HOMINID EVOLUTION

For many years human evolution was likened to a ladder. The rungs of the ladder were the species within the hominid lineage. These species were thought to be 'time successive' so that an earlier, more primitive species was 'replaced' by a later, more advanced one. Modern humans were placed at the top of the ladder of ascent. However, evidence from the fossil record which has accumulated over the past two decades, and the introduction of new analytical techniques to help work out the relationships between fossil species, suggests that the ladder metaphor is no longer an appropriate one. Instead, the hominid evolutionary tree is much better likened to a bush which has multiple stems leading off from close to its base as well as higher up on the bush (Figure 4.2).

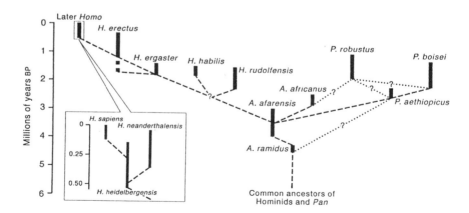

Figure 4.2 A phylogram, or family tree, showing the relationships of the early hominids for which there is fossil evidence. Note that only one hominid species, *Homo sapiens*, has survived to the present day.

All but one of the stems stop well short of the highest point of the bush. These shorter stems represent lineages of hominids which have flourished during human evolutionary history, but which have become extinct in the sense that for these 'short' lineages no direct descendant representatives exist. It seems that the hominids themselves have had their own equivalents of the Burgess Shale fauna.

The next sections will describe the major groups of these early hominids and will indicate whether they lie close to, or distant from, the lineage that was probably directly ancestral to modern humans. In reality it is very difficult to be sure that any of the early hominids can be claimed as direct human ancestors. While we know that modern humans must have had ancestors and we can be reasonably confident of tracing them back for a few hundred thousand years or so, we cannot be sure of the relationships between the creatures which are represented in the earlier parts of the hominid fossil record.

APE–HUMAN DIFFERENCES

The features which mark modern humans from the living African apes are to be found in the dentition, skull, brain, trunk and the limbs (Figure 4.3). While the apes have larger canine and incisor teeth than do modern humans, when the size of the premolar and molar teeth are related to body size, then apes and humans have chewing teeth of similar relative size. This could mean that their diets are similar, but they are not. Modern humans, and perhaps some of the early hominids too, also process their food, by cooking, before it reaches the mouth. It is probably because of this that the jaws of a modern human skull are much smaller and lighter than those of the apes.

Modern human brains are not just absolutely larger than those of the living apes, but they are also larger relative to body size. The skull is more evenly balanced on the vertebral column, so that the point where the brain attaches to the spinal cord is close to the middle of the skull in humans. In apes it is situated further towards the back of the skull.

The chest is differently shaped in modern humans and in the living apes. The thorax of modern humans does not widen towards the base as is the case in the apes. This is probably related to having to accommodate the relatively larger and longer gut in the African apes. Instead of being conical in shape the thorax of modern humans is more uniform in width from top to bottom and flatter from front to back, with the shoulder blades rotated around so that they lie closer to the vertebral column.

The legs of modern humans are longer than those of the apes and the human pelvis is arranged so that the body weight can be supported on the hind limbs alone. There are also differences in the foot, with the modern human foot making a more stable platform than does its living African ape counterpart. The human hand is more dextrous and its longer and more freely movable thumb is capable of meeting the tips of the fingers to make a precise 'pinch' grip.

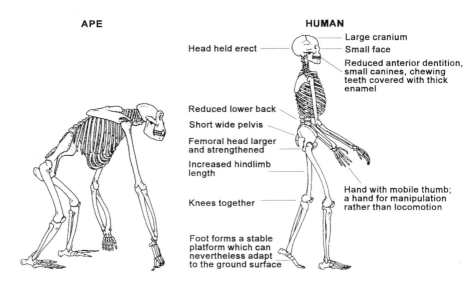

Figure 4.3 Comparisons between the skeletons of a chimpanzee and a modern human with an indication of the important modifications associated with the assumption of a bipedal gait, increased manual dexterity, and a modified diet with a reduced emphasis on chewing.

In addition to these morphological differences there are also differences in the rate that the body grows and in the order in which some structures appear. Modern humans reach maturity much more slowly than do the apes. Their teeth also erupt in a different order because the milk, or deciduous, molars wear out before the adult molars have erupted. To cope with this the adult, or permanent, premolars have to come through before the molars to replace the milk teeth. In the more rapidly growing living African apes the eruption of the premolars is delayed with respect to the permanent molars.

While this list, which is by no means a comprehensive one, provides an impressive number of contrasts between apes and humans, the differences between the earliest hominids and the apes of the late Miocene are likely to have been a good deal more subtle. What those who study human evolution have to look for in the fossil record are the first signs of the differences which are well established when contemporary modern humans and apes are compared. There may also be other differences that have not survived into the present.

AUSTRALOPITHECINES – THE EARLIEST HOMINIDS?

The first creature to show rudimentary human specializations is known as *Ardipithecus ramidus*, the first remains of which were recovered at a site called

Aramis, in Ethiopia, in late 1992 (White *et al.* 1994) (Figure 4.4). Dating evidence suggests that these remains are around 4.5 million years old. Fragments of the skull have been found, together with the remains of a lower jaw, as well as examples of the upper and lower teeth and fragments of the upper limb. The remains have some features in common with living species of *Pan*, others with the African apes in general, and some with later hominids. Because the latter features are so specialized as to make it unlikely that they would be seen in an animal other than a hominid, the discoverers of the fossils suggested that the material belongs to a hominid and not an ape species, and provisionally allocated the new species to the genus *Australopithecus* but later placed it in a new genus, *Ardipithecus*. With hindsight, the remains from Aramis are probably not the first evidence of this species to be found, for jaw bones from the sites of

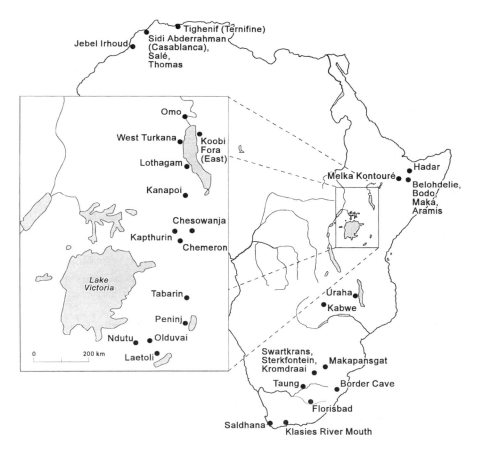

Figure 4.4 Map of Africa showing the principal sites where evidence of fossil hominids has been found.

Lothagam and Tabarin in Kenya, which both date to around 5 million years, probably also belong to *A. ramidus.*

Judging from the size of the shoulder joint this early hominid weighed about 40 kg. Its chewing teeth were relatively small and the opening transmitting the spinal cord was close to the centre of the skull, suggesting that its posture and gait were respectively more upright and bipedal than is the case in the living apes. The remains of the animals and the plants found with *A. ramidus* suggest that the bones had been buried in a location that was close to, if not actually within, woodland. As yet we have no information about the size of the brain, nor any direct evidence from the limbs about the posture and locomotion of *A. ramidus.* The thin enamel covering on the teeth suggests that its diet may have been closer to that of the chimpanzee than is the case with any of the other early hominids for which there is fossil evidence.

A little later in time, but also in East Africa, there is fossil evidence of another hominid species, which has been placed within the genus *Australopithecus*, called *Australopithecus afarensis* (Johanson and White 1979). This name was given in 1978 to hominid fossils that had been recovered from Laetoli, in Tanzania, and from the Ethiopian site of Hadar. Since then remains attributed to *A. afarensis* have been found at other localities in Ethiopia and from sites in Kenya. Nearly all of the material allocated to *A. afarensis* has been dated to between 3 and 4 million years. Thus, it seems to have lived in the same region as *A. ramidus*, but later in time.

The fossil record of *A. afarensis* is much better than that of *A. ramidus.* The collection from Hadar includes substantial fragments of several skulls, many lower jaws and sufficient limb bones to be able to estimate the body size of *A. afarensis.* It also includes a single individual, whose field number is AL-288, but which is better known as 'Lucy', which comprises just less than half of the skeleton of an adult female. The picture that emerges is of an animal which ranged in body size from about 25 kg for a small female to more than 50 kg for a large male. This is a substantial range in body size for an animal that is absolutely quite small, and this has prompted some workers to continue to speculate that the sample presently thought to be *A. afarensis* may consist of the remains of more than one species of early hominid.

The brain size of *A. afarensis* was between 400 and 500cc. This is larger than the average brain size of a chimpanzee, but, if the estimates of the body size of *A. afarensis* are anything like correct, when the size of its brain is related to estimates of its body size the brain of *A. afarensis* is relatively no larger than that of the chimpanzee. However, the chewing teeth, the premolars and molars of *A. afarensis* are larger, relative to body size, than those of the living chimpanzee. The body proportions of the 'Lucy' skeleton differ from those of a modern human of the same body size, in that the legs of AL-288 are substantially shorter than those of a modern human of similar overall stature.

Attempts to reconstruct the habitat of *A. afarensis* suggest that it was living in a more 'open' woodland habitat than has been suggested for *A. ramidus.* The appearance of the pelvis and the lower limb suggests that *A. afarensis* was capable of bipedal walking, although it was not as well adapted for long-range bipedalism as modern humans and its shoulder and arms still showed evidence of an ability to climb. This indirect evidence for the locomotion of *A. afarensis* is complemented by the discovery, at Laetoli, of several trails of fossil footprints. These provided very graphic and direct evidence that an early hominid, and we must presume that it was the only one, *A. afarensis*, for which there is fossil evidence at Laetoli, was capable of bipedal locomotion. The size of the footprints and the distance between them, the stride length, provide corroboration for the estimates of stature that had been derived from the limb bones of *A. afarensis*. These suggest that the standing height of the individuals in this early hominid species was between 1 m and 1.5 m.

RAYMOND DART'S DISCOVERIES

Prior to the announcement of *A. ramidus*, there was general agreement that *A. afarensis* was the most primitive hominid yet discovered. However, it is not the only early hominid of that general level of organization, for, nearly fifty years before the discovery of the Ethiopian remains belonging to *A. afarensis*, an early hominid child's skull had been found in 1924 in the remains of a cave at the Buxton Limeworks at Taung, in what was then Bechuanaland, in southern Africa. It was this skull which was brought to the attention of Raymond Dart, and it was Dart who proposed that it represented the remains of a hominid far more primitive than any hominid remains that had been found prior to 1925, the year that Dart's paper was published in the journal *Nature* (Dart 1925). The exact geological age of the find was not known, but animals found with it suggested that it was more ancient than any of the hominid remains that had been recovered previously in Europe, Java or China. The new hominid was thus not included in our own genus, *Homo*, but was placed in a new genus and was called *Australopithecus africanus*, which literally means the 'southern ape' of Africa.

Since the discovery of the Taung child's skull many more remains of *A. africanus* have been found at other cave sites in southern Africa, at Sterkfontein, Makapansgat and Gladysvale, but with the vast majority of the evidence coming from Sterkfontein (Figure 4.4). These australopithecine remains are not as reliably dated as those of early hominids from sites in East Africa, where the many ash layers provide material which can be dated relatively precisely. The main methods use the half-life of isotopes of potassium and argon, together with the record of the earth's magnetism, albeit a weak one, which is preserved in the sedimentary rocks. The cave sites in southern Africa can, at present, only be dated by comparing the remains of the mammals found in the caves with the mammalian fossils found at the better-dated sites in East Africa. In this way the age of the *A. africanus*-bearing rocks has been estimated to be between 2.5 and 3 million years.

Males and females differed in body size. The other animal remains found with *A. africanus* suggest that the habitat was a combination of grassland and trees, comparable to modern open 'bushland' habitat. The picture that has developed of *A. africanus* suggests that its physique was much like that of *A. afarensis* except that its chewing teeth were slightly larger, and the skull was not as ape-like in that the jaws protruded less and the underside was shorter with a more centrally located opening for the transmission of the spinal cord. Its brain was a little larger than that of *A. afarensis*, but not substantially so. Its postcranial skeleton was also arranged in a way that suggested that it was capable of, but not committed to, a bipedal gait.

The three australopithecines that have been described vary in the degree in which their morphology has been modified with respect to the living chimpanzee. Although we know that they were not descended from living species of *Pan*, one of the modern species, *Pan paniscus* (the 'pygmy' chimpanzee) is probably the closest we can get in a living animal to the likely appearance of the hypothetical common ancestor of all hominids. Although all three fossil species were initially placed in the same genus, as more evidence about *A. ramidus* accumulated, it became necessary to assign it to a new genus, to indicate that it is substantially more primitive than the two other, temporally later, hominid species.

Just as there were East African and southern African regional variants of the so-called 'gracile' australopithecines, there are also two regional variants, also allocated to different species, of another hominid genus called *Paranthropus*. They are often referred to as 'robust' australopithecines because of their massive faces and lower jaws, but because of their likely placement in a separate genus they will be referred to hereafter as 'paranthropines'.

The southern variety, *Paranthropus robustus*, postdates the known occurrences of *A. africanus*, for the cave sites where it has been found, Swartkrans and Kromdraai, are dated to between 2 and 1.5 million years. Its brain is a little larger than that of *A. africanus*, as are its face and chewing teeth. However, the incisor teeth are smaller and this had led to the suggestion that the diet of *P. robustus* differed from that of its australopithecine precursor. Little is known about its postcranial skeleton except that the shape of the hip joint is much like that of the australopithecines. The habitat of *P. robustus* is difficult to reconstruct because some of the usual clues, such as traces of fossil pollen, are unhelpful. The bones of the animals found with the hominid remains were probably accumulated by predators, and thus they provide an incomplete and perhaps biased impression of the prevailing fauna. Some workers point to differences between the hominids recovered from Swartkrans and Kromdraai, and prefer to allocate the former material to a separate species, *Paranthropus crassidens*.

The East African 'species' of *Paranthropus* probably consists of not one but two species. Both have large jaws and large chewing teeth. The earlier and more primitive of the two species is referred to as *Paranthropus aethiopicus* and is presently only known from a handful of specimens from two sites, one to the west of Lake Turkana and one to the north of the lake, along the Omo river.

Fortunately one of these specimens is a well-preserved skull. It has a smaller brain, around 400–450cc, a more ape-like face and larger anterior, incisor and canine teeth than does the later species, which is called *Paranthropus boisei*. The first evidence of *P. boisei* was discovered in 1959 by Louis and Mary Leakey while they were working in Olduvai Gorge, in Tanzania. The specimen, which is also the type specimen of *P. boisei*, is the OH 5 cranium – the 'OH' designation stands for 'Olduvai Hominid' – which was originally referred to the genus *Zinjanthropus*, and which was known as 'Zinj'. However, the largest collection of fossils belonging to *P. boisei* comes from the site of Koobi Fora which lies to the east of Lake Turkana, in Kenya.

The earlier of these two East African species of 'robust' australopithecines spans the interval between 2.5 and 2.3 million years, and the later one between 2.3 and 1.4 million years. The large chewing teeth in the two species has prompted the suggestion that the East African paranthropines were specialist feeders on a diet that required heavy mastication. Little is known about their postcranial skeleton, but what information is available suggests that it was much like that of the australopithecines. The inclusion of the three paranthropine species in the same genus, *Paranthropus*, reflects the consensus view that the three species resemble each other because they share a common ancestor from which they each inherited a basic 'body plan' which includes the combination of a large face and jaws with large-crowned chewing teeth. If future studies suggest that it is more likely that these features evolved independently, then a genus name common to all three species would no longer be appropriate.

What prompted the appearance of these large-toothed creatures in both East and southern Africa? Some palaeontologists point to analogous changes occurring in the masticatory apparatus of other large mammals, and have suggested that the common 'cause' may be a general shift to a drier climate that was occurring 2–3 million years ago. This 'climatic' explanation has been linked to a global cooling phenomenon, superimposed on 'local', but causally unrelated, elevations of the land surface in the two regions.

THE BEGINNINGS OF *HOMO*, OR ARE THEY?

Long before Louis and Mary Leakey's patient fieldwork at Olduvai Gorge had been rewarded with the discovery of the OH 5 cranium, their efforts had resulted in the discovery of stone tools, which became known as the Oldowan cultural industry. This stone tool industry was more primitive than any of the Palaeolithic handaxes which had been recovered in the previous century from European sites. It was perhaps natural to assume that OH 5 represented the species of hominid that was the maker of the stone tools. However, discoveries made in the 1960s were to challenge this assumption.

The new material included the remains of the vault, or roof, of a skull and a piece of a lower jaw, and was given the designation OH 7. Subsequently the remains of a hand, foot and lower leg were found. Although these specimens were frustratingly incomplete, the Leakeys and their colleagues were convinced that, relative to the paranthropines, the larger cranial capacity and smaller tooth size of the new material, together with what they regarded as the advanced features of the limb bones, made it a stronger contender for the role of toolmaker than *P. boisei*, the species to which OH 5 belonged. Therefore the team of researchers studying the finds proposed that the new material should be included within the genus *Homo*, as a new species, *Homo habilis* (Leakey *et al.* 1964). The name literally means 'handy man', or 'maker of the tools'.

The implication of this course of action was that the definition of *Homo* had been changed to accommodate the new material. This did not matter a great deal as long as the remains from Olduvai were generally agreed to have departed significantly from the morphology of the australopithecines and the paranthropines in the direction of later *Homo*. That consensus has now broken down in the face of convincing demonstrations that the hand and foot bones are less human-like in form and function than the original researchers had, with the limited evidence then available to them, concluded.

In the meantime, material that generally resembled the Olduvai *H. habilis* remains had been unearthed from the site of Koobi Fora in northern Kenya (Figure 4.4). As the evidence accumulated it became apparent that variation within *H. habilis* was beginning to exceed that which could reasonably be expected in a single species. In particular, the differences between crania from Koobi Fora, such as KNM ER1470 and KNM ER1813, and between mandibles such as OH 13 and KNM ER1802, were cited as being too extreme to be subsumed within the single species *H. habilis*. ('KNM-ER' is the designation for Koobi Fora. It is formed from the initials of the 'Kenya National Museum' where the fossils are kept, and East Rudolf, the 'old' name for Koobi Fora.) This led to the suggestion that there was a second species of 'early *Homo*', and this was called *Homo rudolfensis* (Wood 1993).

Present evidence suggests that *H. habilis* dates from just less than 2 million years to around 1.6 million years, and *H. rudolfensis* from around 2.5–1.8 million years. What are the differences between these two species, and in what ways do they differ from the australopithecines? The main difference between both species and the australopithecines is in the arrangement of the base of the skull. In the two 'early *Homo*' species the opening for the spinal cord is closer to the middle of the skull, and the skulls are reduced in length and increased in width, compared with the australopithecines. The two 'early *Homo*' species differ, however, in the form of the face. In *H. habilis* the face is reduced in width, and the opening of the nose is more sharply defined, than it is in *H. rudolfensis*. The brain volume of *H. habilis* is around 500–700cc, whereas the estimated brain volume for *H. rudolfensis* is in the order of 700–800cc, with an average of 750cc. However, if the admittedly crude estimates of body mass of the two species are accepted, then *H. habilis* is smaller-bodied than *H. rudolfensis*. Thus, when body size is taken into account the differences in tooth and brain size are cancelled out.

Some information about the limbs of *H. habilis* comes from a very fragmentary skeleton, OH 62, found at Olduvai Gorge. It is this skeletal evidence which suggests that the body of *H. habilis* more closely resembled that of its australopithecine forerunners than it did later species of *Homo*. Unfortunately, there is no postcranial material definitely associated with any of the cranial remains of *H. rudolfensis*. Two upper-leg bones, or femora, found at Koobi Fora in rocks that are similar in age to the KNM ER1470, may belong to the same species, but there is no certainty that they do. Thus, we must concede that presently we have no reliable information about the postcranial skeleton of *H. rudolfensis*.

Overall, the case for including both *H. habilis* and *H. rudolfensis* within our own genus, *Homo*, is weaker than it was three decades ago. However, as we shall see in the next section, by around 1.9 million years ago there is good evidence from East Africa of an early hominid that was, in terms of its physique and appearance, a much stronger candidate for inclusion in the same genus as modern humans.

HOMO EMERGES IN AFRICA

The discovery of the remains of a virtually complete skeleton of an individual has an importance which extends beyond the information contained in the specimen itself. These associated skeletons have a wider significance because they help palaeontologists to collate the remains of isolated skulls, jaws and limb bones which, even if they are well-preserved, are difficult to relate together. Such was the impact when the juvenile male skeleton, known as KNM WT15000, was painstakingly excavated from 1.5 million-year-old sediments at West Turkana (Brown *et al.* 1985).

Over the previous decade similar-looking skulls, jaws and limb bones had been recovered from the site of Koobi Fora. Although many of these had been recognized as resembling *Homo erectus* remains from Asia and elsewhere in Africa, it was not until the discovery of KNM WT15000 that their affinities with each other, and with *H. erectus* could be confirmed. Although the West Turkana skeleton was dated at around 1.5 million years, other remains attributed to the same species are as old as 1.9 million years. Some workers have given the species a new name, *H. ergaster*, while others refer to it as 'early African' *H. erectus*. What makes the inclusion of these remains into the genus *Homo* more soundly based than the similar allocation of fossils belonging to *H. habilis*, or *H. rudolfensis*?

Our own species, *H. sapiens*, differs from the australopithecines and the paranthropines in four major ways. First, our brains are larger, both in absolute size and when they are related to the size of the body. Second, our jaws and teeth are generally smaller when related to overall body size. Third, while it is apparent that the australopithecines and the paranthropines are capable of standing on their hind limbs, and of walking bipedally, the shape of their skeleton suggests that they were also still capable of climbing using all four limbs. Modern humans, on the other hand, only rest and move on all fours in childhood; adult modern humans are habitually upright in their posture and can only walk and run effectively if they are bipedal. The fourth distinction between modern humans and the early hominids, and particularly the australopithecines, is that whereas the early hominids and the apes accomplish their development relatively quickly, modern human infants are relatively helpless for longer, their locomotor skills take more time to acquire and their molar teeth erupt later. This prolonged period of dependence, together with improved dexterity, is believed to be important for the transference of the many more skills that modern humans have acquired compared to the apes. It would seem to be sensible to use the acquisition of these characteristics, or of their precursors, as criteria for including species of fossil hominids within the genus *Homo*.

When the remains of *H. ergaster* are examined for evidence of these criteria, most of them are satisfied. The lengths of the long bones of the skeleton from West Turkana confirm that this hominid species has limb proportions much like those of modern humans. It is also the first hominid we have considered thus far which has a modern human-like body shape. The jaws and teeth of *H. ergaster* are smaller than those of the australopithecines and the paranthropines; indeed, when tooth and jaw size are related to body size, *H. ergaster* is found to have teeth and jaws no larger than those of African and Australian samples of modern humans. Likewise, the timing and pattern of tooth development in *H. ergaster* is much closer to that in modern humans than is the case for any of the other hominid species that we have considered. When we turn to evidence about the brain, the affinities with modern humans are not so strong. While *H. ergaster* has a cranial capacity of between 800–900cc, thus exceeding that of *H. rudolfensis*, when this capacity is related to the larger estimated adult body size (around 60–70 kg) of *H. ergaster*, then the relative brain size of the latter is similar to that of *H. rudolfensis* and perhaps is less than that of *H. habilis*.

Thus, while details of the morphology of *H. ergaster*, such as the shape of the skull (Table 4.1), differ from those of *H. sapiens*, the overall body plan shows clear and strong links with modern humans. These resemblances are sufficient evidence to include this African material within *Homo*, making it the first and earliest species which can confidently be placed within the genus *Homo*.

HOMO MOVES OUT OF AFRICA

The first hominid species to be found outside of Africa was probably either *H. ergaster* or *Homo erectus* (Figure 4.5). It was more than a century ago that Eugene Dubois reported the discovery of a skull and femur which were collected on the bank of the Solo River, at a place called Trinil, in what was then called Java, now Indonesia. Since that time many more similar-looking remains have been located in Indonesia, mainly from a region called Sangiran where the Solo River cuts through Pliocene and Pleistocene rocks which have been thrown up into a dome. Much of modern Indonesia is intensively farmed and little soil is undisturbed.

Table 4.1 Characteristics of ten hominids

	Australopithecus afarensis	Australopithecus africanus	Paranthropus boisei	Paranthropus robustus	Homo habilis	Homo rudolfensis	Homo erectus (including Homo ergaster)	'Archaic Homo sapiens' or Homo heidelbergensis	Neanderthals or Homo neanderthalensis	'Early moderns' or Homo sapiens
Height (m)	1–1.5	1.1–1.4	1.2–1.4	1.1–1.3	1.25	?1.6	1.6–1.8	1.6–1.8	1.5–1.8	1.6–1.85
Weight (kg)	30–45	30–40	35–50	30–40	30–40	?40–50	50–70	55–65	55–65	50–80
Physique	Light build: some ape-like features (e.g. shape of thorax, long arms relative to legs, curved fingers and toes, marked to moderate sexual dimorphism)	Light build, probably relatively long arms, more 'human' features, probably less sexual dimorphism	Very heavy build, relatively long arms, marked sexual dimorphism	Heavy build, relatively long arms, moderate sexual dimorphism	Relatively long arms, moderate sexual dimorphism	No certain information about the skeleton, moderate sexual dimorphism	Robust but human skeleton, moderate sexual dimorphism	Robust but human skeleton, moderate sexual dimorphism	As 'archaic' H. sapiens' but adapted for cold, moderate sexual dimorphism	Modern skeleton, ? adapted for warmth, moderate sexual dimorphism
Brain size (ml)	350–500	400–500	410–530	approx. 525	500–650	650–850	750–1,250	1,100–1,400	1,200–1,750	1,200–1,600
Skull form	Low, flat forehead, projecting face, prominent brow ridges	Higher forehead, shorter face, brow ridges less prominent	Prominent crests on top and back of male skulls, very long, flattish face, strong facial buttressing	Crest on top of male skulls, long, broad, flattish face, moderate facial buttressing	Higher forehead, relatively small face, obvious nasal margins	Flatter forehead, large wide midface	Flat, thick skull with large occipital and brow ridges	Higher skull, face less protruding	Reduced brow ridges, thinner skull, large nose, midfacial projection	Small or no brow ridges, shorter, high skull
Jaws/teeth	Relatively large incisors and canines, moderate-sized molars	Small incisor-like canines, larger molars	Very thick jaws, small incisors and canines, large molar-like premolars, very large molars	Very thick jaws, small incisors and canines, large molar-like premolars, very large molars	Smaller jaws, narrow molars	Robust jaw, large molars, molar-like premolars	More slender jaw, smaller teeth than H. habilis, reduced size of third molar	Similar to H. erectus but teeth may be even smaller	Similar to 'archaic' H. sapiens', teeth smaller except for incisors, some chin development	Shorter jaws than Neanderthals, chin developed, teeth may be smaller
Distribution	Eastern Africa	southern Africa	Eastern Africa	southern Africa	Eastern (and southern?) Africa	Eastern Africa	Africa, Asia and Indonesia	Africa, Asia and Europe	Europe and Western Asia	Africa and Western Asia
Known date (years BP)	>4–<2.9 million	~3–<2.5 million	2.3–1.2 million	2–1.5 million	2–1.6 million	2.5–1.8 million	1.9–0.3 million	400–100 thousand	150–30 thousand	c.100 thousand to the present

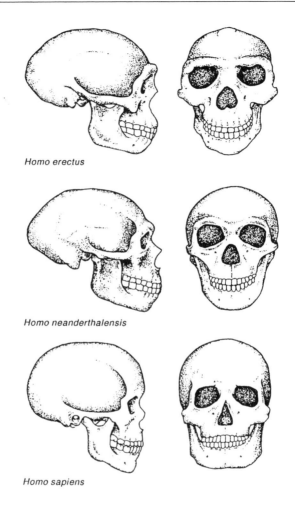

Figure 4.5 Comparisons between the cranial shape and relative jaw size of *Homo erectus*, *Homo neanderthalensis* and *Homo sapiens*. Note the taller brain case and smaller face of *Homo sapiens*.

This makes the exact location of many of the finds difficult to pinpoint and the paucity of volcanic ash and the weakness of the magnetic signals in the sediments means that the rocks are difficult to date with any accuracy. There is, however, reasonably sound evidence that some of the finds at Sangiran may date from around a million years ago, and there have been recent claims that another of the sites, and also one in China, may be in the order of 1.8 million years (Swisher *et al.* 1994, Wanpo *et al.* 1995). This would mean that the first hominids must have left Africa sometime before this.

We do not know what promoted this first exodus from Africa. The large bodies of *H. ergaster* would have provided improved tolerance to heat stress and to dehydration, and the more sophisticated stone 'tool kits' that make their first appearance in the archaeological record around 1.7 million years may have all contributed to the ability to tolerate a wider range of landscapes and climates. There is archaeological evidence that hominids had occupied Europe and the Near East by just less than one million years ago, and at the southern African site of Swartkrans there is fossil evidence of a *H. ergaster*-like hominid around 1.5 million years ago. Remains of *H. erectus* persist in the Far East, at the Chinese site of Zhoukoudian, formerly Choukoutien, near to Beijing, as late as 200,000 years BP. This means that, as we shall see below, *H. erectus* was persisting in at least one region while in other parts of the world hominids that are closer in appearance to modern humans were making their appearance.

PRECURSORS OF MODERN HUMANS

It is a paradox that of all the hominid species that we know of, the one that is the least well-defined is that to which the reader and the writer belong. Whereas other hominid species have a 'type specimen' to which the species name is irrevocably attached, there is no designated type specimen of *H. sapiens*. In addition, because it is 'polytypic' – that is, it incorporates a relatively large range of variation – a spectrum of skull shapes and limb proportions have to be taken into account when considering whether fossils can be included within the species category *H. sapiens*. However, the variation within our own species is not random, and a good deal of it is related to climate. For example, modern human populations follow Allen's Rule which stipulates that populations in warmer climes will have long extremities, while those in colder climates will have shorter, stockier, limbs. The tall, long-legged and narrow-waisted Nubian, and the short, stocky, Eskimo are much-used, but still relevant, illustrations of the effects of climate on body shape. Head shape responds in the same way, with long heads predominating in warmer climates, and people with rounder heads and smaller noses more common in colder climates. Thus, what is an example of 'anatomically modern' *H. sapiens* in one region of the world will differ from an example taken from a region with a different climate.

However, before we consider the origin of early modern humans we must consider a substantial collection of fossils which are more modern human-like than *H. erectus*, yet which are not fully modern in their appearance. Such material is known from Africa, at Ndutu and Kabwe for example; from Europe, at Mauer and Arago, and from China at Dali and at Jinnuishan.

Perhaps the best-known material in this 'archaic *Homo*' category are the remains which have been attributed to *Homo neanderthalensis*. This species has a characteristic appearance which includes a large, globular-shaped cranium, jaws and teeth which are set well forward in the face, and particularly robust limb bones with large joint surfaces (Figure 4.5).

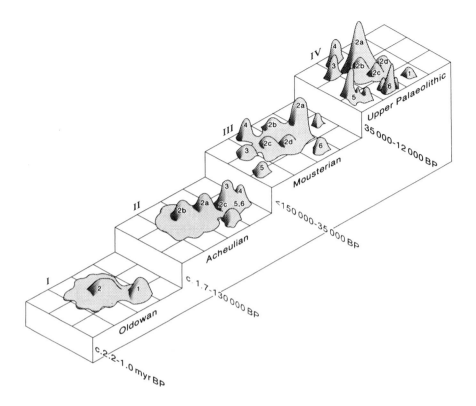

Figure 4.6 A representation of the degree of complexity and specialization of prehistoric stone-tool industries: I = Oldowan; II = Acheulian; III = Mousterian; and IV = Upper Palaeolithic. The complexity of the topology at each of the four levels is an indication of the variety of tool types, and the height of the peaks is an indication of the consistency of design in each category (after Glynn Isaac).

The earliest evidence of Neanderthals comes from Spain, and is dated to 300–400 thousand years ago, and the most recent evidence, from Western Europe, is dated to around 30,000 years. Hominids which show the characteristic Neanderthal morphology are confined to Western Europe, the Near East and adjacent parts of Asia. For much of the period when Neanderthals are present the climate was oscillating between long, cold, glacial, phases and shorter, warmer, interglacial, periods. The short stature and robust limb bones of the Neanderthals are thus just the body build that would be expected under the influence of Allen's Rule. Neanderthals are too often portrayed as primitive hominids which lacked the sophisticated technology and artistic sophistication of the populations that replaced them in Europe. This interpretation belies the fact that the tools that are associated with many, but not all, Neanderthal sites are a good deal more varied in design than the handaxes that characterize the Lower Palaeolithic (Figure 4.6). There is also good evidence that they buried their dead and used grave goods.

Populations of 'archaic *Homo sapiens*' peoples from other regions are not so characteristic in their appearance, but all have some distinguishing features. It is the extent to which those regional characteristics are continued within the regional populations that succeeded them that lies at the root of ongoing debates about the origin of early anatomically modern humans.

MODERN HUMAN ORIGINS

This is a topic which is the subject of much discussion. Two hypotheses have attracted most of the attention. One, called the 'Out of Africa' or 'Noah's Ark' hypothesis, suggests that the genetic modifications that were responsible for the shift to an anatomically modern morphology only occurred once, and in Africa. The rival 'Multiregional' hypothesis proposes that the shift to an anatomically modern human morphology was a process that occurred several times, but only once in each of the major population centres (Aiello 1993) (Figure 4.7). The

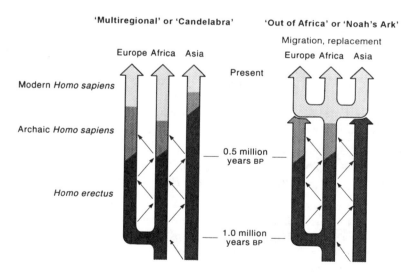

Figure 4.7 Comparison of the two competing hypotheses to explain the origin of early *Homo sapiens*. Both hypotheses stress the predominance of regional continuity in the early stages, but the proponents of the 'Out of Africa' hypothesis claim that after about 200,000 years there is a further major migration from Africa which effectively replaces all the aboriginal populations in the other regions of the Old World.

currently supported version of the hypothesis allows for gene flow between the regions, but maintains that this gene flow was not sufficient to obscure the presence of regionally distinctive morphology. The two hypotheses make different predictions about, for example, the degree to which a regionally distinct population like the Neanderthals relates to the people that succeeded it. The 'Out of Africa' hypothesis predicts that the population that succeeded the Neanderthals would have had no direct connection with them, so that it would be unlikely that any Neanderthal traits would persist in the later population. Conversely, the 'Multiregional' hypothesis would predict that there would be morphological continuity across the 'Archaic'/'Modern' boundary in each of the main regions.

Evidence about modern-human origins comes from two sources: the palaeontological record and the pattern of variation of the modern human genotype. The prediction of the 'Out of Africa' hypothesis is that any adaptively neutral variation in the genotype would be greatest in the regional modern human population which had existed for the longest time – namely, that from Africa. The first results from the analysis of variation in mitochondrial DNA suggested that this was the case. However, when the design of the analysis was more closely scrutinized it became apparent that such a result was due in large measure to the order in which the data had been analysed. When this bias was removed, the support for the 'Out of Africa' hypothesis was weakened and all but disappeared (Templeton 1993).

The earliest fossil evidence of hominids with a skull, jaws, teeth and limb bones equivalent to those of contemporary humans comes from Africa, at Klasies River Mouth on the Cape Coast, and from the Omo, in Ethiopia, and the Near East, at the Skhul and Qafzeh caves; all of these remains date from around 100,000 years. However, while this evidence does not refute the 'Out of Africa' hypothesis, it certainly does not 'prove' it. Not all other regions have fossils of this antiquity so that their 'modernity', or lack of it, cannot be judged, and the old adage 'absence of evidence is not evidence of absence' means that the case for an African origin for early modern humans is 'not proven'.

PEOPLING THE PLANET

Whatever the details of their evolutionary history, there is no doubt that anatomically modern humans had dispersed across the Old World by 25–35 thousand years ago. By that time, and perhaps as early as 50–60 thousand years BP, they had also managed to cross the barrier of water between the Asian mainland and Australasia. The periodical reductions in sea-level associated with the times of maximum glaciation would have served to shorten any sea crossings, but the colonization of Australia must have involved the ability to make a raft, or an equivalent sea-going craft.

When the polar ice-caps were at their maxima the reduction in sea-level led to a fusion of Indonesia with the Asian landmass, to form 'Sundaland', and the incorporation of New Guinea into a greater Australian landmass, called 'Sahul'. The earliest evidence of the human occupation of Sahul comes from New Guinea at around 40,000 years BP. Presently, the earliest indication of hominids in what is now Australia date from between 45–60 thousand years in Arnhem Land. However, well-dated and plentiful remains of modern humans, together with their associated archaeology and food remains, come from 30,000-year-old burials at Lake Mungo in New South Wales. More rugged-looking skulls from the Kow Swamp and Cohuna sites are more recent, around 12–15 thousand years ago. Some workers point to the morphological contrasts between these two sets of Australian remains as evidence for an earlier, Chinese-dominated wave of migration, and a later, Javan-based, wave of migrants. However, many dispassionate observers see the differences as more likely to be the result of subsequent differentiation of a single, major, wave of immigration into Australia.

Present archaeological evidence suggests that the occupation of the New World did not take place until 10–12 thousand years ago. All of the archaeological evidence thus far adduced to substantiate any earlier occupation of North America has proved to be unreliable. Modern humans would have entered the New World across the landbridge that was established between Siberia and Alaska around 20–70 thousand years ago. The earliest archaeological evidence of modern-human occupation of Siberia dates from 20–30 thousand years BP, but the landbridge would have been so barren as to provide an obstacle to migration until much later. The best evidence for any substantially earlier occupation of the New World comes from artefacts recovered in Brazil, but the dating evidence is equivocal. There are three basic language groups in the Americas: Amerindian, Na-Dene and Aleut. It has been suggested that these may correspond to three waves of immigration.

Artefacts and the remains of animals found in association with anatomically modern human remains in the Old World provide compelling evidence that by 40,000 BP, if not before, organized hunting was part of the survival strategy of these early populations. Some archaeologists place the origin of hunting much earlier, but in many cases it is possible to point to other explanations for the association of animal bones with evidence of human occupation. There is also evidence of shell middens at coastal archaeological sites by 15–20,000 BP.

Farming and permanent settlement do not become evident in the archaeo-logical record until about 12,000 years ago. While the best-known evidence comes from the 'fertile crescent' in the Near East, the discovery of drainage ditches in New Guinea and of herding in East Africa around the same time as the evidence for grass domestication along the Nile, demonstrates that changes from a hunter/gatherer lifestyle were occurring across the globe.

Thereafter, the balance between physical and cultural evolution has shifted so far towards the latter that evidence for the former all but disappears. There are changes in the nature of the inhabitants in some parts of the world, but these are the result of migrations and not of *in situ* evolution. The scene is set for the cultural changes and population expansions which were not only to shape human history, but the future of many other aspects of life on the planet.

REFERENCES

Aiello, L.C. (1993) 'The origin of modern humans: the fossil evidence', *Am. Anthrop.* 95, 73–96.

Brown, F., Harris, J., Leakey, R. and Walker, A. (1985) 'Early *Homo erectus* skeleton from west Lake Turkana, Kenya', *Nature* 316, 788–92.

Dart, R.A. (1925) '*Australopithecus africanus*, the man-ape of South Africa', *Nature* (London) 115, 195–9.

Johanson, D.C. and White, T.D. (1979) 'A systematic assessment of early African hominids', *Science* 202, 321–30.

Leakey, L.S.B., Tobias, P.V. and Napier, J.R. (1964) 'A new species of the genus *Homo* from Olduvai Gorge', *Nature* 202, 7–9.

Swisher, C.C. III, Curtis, G.H., Jacob, T., Getty, A.G., Suprijo, A. and Widiasmoro (1994) 'Age of the earliest known hominids in Java, Indonesia', *Science* 263, 1118–21.

Templeton, A.R. (1993) 'The "Eve" hypothesis: a genetic critique and reanalysis', *Am. Anthrop.* 95, 51–72.

Wanpo, H., Ciochan, R., Yumin, Gu, Larick, R., Qiren, F., Schwarcz, H., Yonge, C., de Vos, J. and Rink, W. (1995) 'Early *Homo* and associated artefacts from Asia', *Nature* 378, 275–8.

White, T.D., Suwa, G. and Pilbeam, D. (1994) '*Australopithecus ramidus*, a new species of early hominid from Aramis, Ethiopia', *Nature* 371, 306–12.

Wood, B.A. (1993) 'Early *Homo*: how many species?', pp. 485–522 in W.H. Kimbel and L.B. Martin (eds) *Species, Species Concepts and Primate Evolution*, Plenum: New York.

FURTHER READING

Brain, C.K. (ed.) (1993) 'Swartkrans. A cave's chronicle of early man', pp. 1–270 in *Transvaal Museum Monograph* No. 8, Pretoria: Transvaal Museum.

Grine, F.E. (ed.) (1988) *Evolutionary History of the 'Robust' Australopithecines*, New York: Aldine de Gruyter.

Howells, W.W. (1993) *Getting Here: The Story of Human Evolution*, Washington: Compass Press.

Jones, S., Martin, R. and Pilbeam, D. (eds) (1992) *The Cambridge Encyclopaedia of Human Evolution*, Cambridge: Cambridge University Press.

Lewin, R. (1993) *Human Evolution: An Illustrated Introduction*, Oxford: Blackwell.

Rightmire, G.P. (1990) *The Evolution of* Homo erectus, Cambridge: Cambridge University Press.

Stringer, C. and Gamble, C. (1993) *In Search of the Neanderthals*, London: Thames & Hudson.

Tattersall, I., Delson, E. and Van Couvering, J. (1988) *Encyclopaedia of Human Evolution and Prehistory*, New York: Garland.

Wood, B.A. (1992) 'Origin and evolution of the genus *Homo*', *Nature* 355, 783–90.

5

THE GEOGRAPHY OF LANGUAGE

William C. Brice

The disciplines of Geography and Linguistics evidently combine in the mapping of the present whereabouts of different languages and dialects. The next step is to consider how these patterns of distribution came to be. In this context, the study of place-names is of first concern. Other circumstances which must be taken into account are: the adaptation of speech to cultural changes; the adjustment of languages to political frontiers; and the genesis of Pidgins and Creoles under special conditions. On a more idealistic plane we must mention the several attempts over the last four centuries to construct a universal language. Finally, it seems appropriate to say something of the origin, evolution and spread of the related art of writing.

LANGUAGE GROUPINGS AND THEIR DISTRIBUTION

There was little interest in the Graeco-Roman world in the speech of 'Barbarians', so called from the perceived sound of their languages; but the understanding and recording of many remote tongues of Europe and Asia owe much to the work of Christian missionaries in translating and preaching the Gospels during the Middle Ages. This activity grew and extended widely during the Age of Discovery, and stimulated the first attempts to classify languages. Joseph Justus Scaliger (1540–1609) categorized the tongues of Europe; during the next two hundred years, partly under the stimulus of the mathematician Leibniz, several ambitious comparative accounts of languages were written, notably by Pallas (*Vocabularia Comparativa*, 1787), Hervas (1735–1809) and Adelung (*Mithridates*, 1806–17); but it was the methodical study of the Indo-European family of languages (by the grammarian Franz Bopp, 1791–1867, following the insights of the missionary Benjamin Schultze (published 1725) and of William Jones, Chief Justice of Bengal (published 1788)) which marked the beginnings of the modern science of comparative linguistics.

Affinities between languages can be recognized first by means of shared words, and second through common features of structure and usage. Of these,

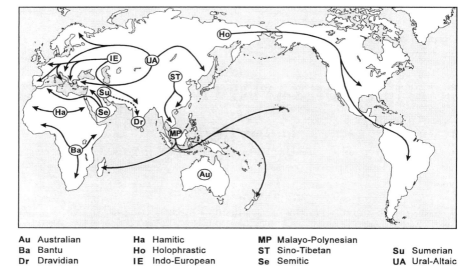

Au	Australian	Ha	Hamitic	MP	Malayo-Polynesian		
Ba	Bantu	Ho	Holophrastic	ST	Sino-Tibetan	Su	Sumerian
Dr	Dravidian	IE	Indo-European	Se	Semitic	UA	Ural-Altaic

Figure 5.1 The spread of language families.

the second criterion is evidently the more reliable, for the vocabulary or word-repertory of one language may be extended through borrowings from others that are quite unrelated. For example, both Persian and Turkish include many words, particularly in the fields of philosophy and religion, borrowed from Arabic, though all three languages belong to quite distinct families. However, word similarities in some simple and basic fields, such as those of kinship and topography, can provide significant clues to affinities of language.

Thus, the following broad categories can be distinguished (Figure 5.1):

1 *Isolating*. The *Sino-Tibetan* languages of the Far East use a system of simple syllabic sounds, whose range of expression is extended by distinctions of tone. The *Malayo-Polynesian* languages make a separate family which probably falls within the same isolating category. They include an outlier in the island of Madagascar.

2 *Agglutinating*. The Uralic language family (Finno-Ugrian, Estonian, Lappish) and the Altaic (Turkic, including Kirghiz and Manchu), though long separated, are distantly related, and may be classed together as the *Ural-Altaic* group. They are spoken over a wider area than any other of the main language groups of the Old World. Agglutination implies the elaboration of the verbal root through attachment to it of one or more of many regular adjuncts. To blend with the vowel of the root, the adjuncts (affixes) change their own vowels, a habit known as vowel harmony.

108

The other large group of agglutinating languages is the *Dravidian* of southern India (Tamil, Telugu, Canarese, Malayalam). It has been suggested that the Dravidian and Ural-Altaic groups may have had distant links, but these have not been confirmed.

Sumerian, the earliest language of Mesopotamia, and known only from literary evidence, was also of agglutinating type and may have had a remote connection with Dravidian through the lost language of the Indus civilization. Brahui, a surviving outlier of Dravidian in Baluchistan, may be a relic of such a link.

3 *Root-inflecting.* The Semitic family (Arabic, Hebrew, Ethiopic) and the Hamitic (Cushite, Berber) may be grouped together as *Semito-Hamitic*. They operate through vowel changes within trilateral verbs, on a regular system which lends itself well to rigorous grammatical analysis.

4 *Amalgamating.* This category incorporates the families of the Indian Sanskrit derivatives as well as Iranian, Greek, Romance (Latin derivatives), Teutonic, Slavonic, Celtic, Baltic, Armenian and Albanian, which together make up the *Indo-European* (earlier called Aryan) group. Amalgamation implies the use of complex and variable systems of suffixes and vowel changes to refine meanings.

5 *Classificatory.* The *Bantu* languages of Sub-Saharan Africa fall into this category, characterized by a system of prefixes which indicate the class of ideas to which a subject belongs, and which in a sentence are repeated before all the words connected with the subject.

6 *Holophrastic.* Under this general heading may be subsumed the many native *Amerindian* language families. It implies their common features of the smooth flow of expression and the absence of clear distinction between words.

In addition to these main language groups, there are several of small size ('isolates') which cannot be included therein, and which are probably local survivals of forms which have been largely overrun, replaced, or cut off very early. Such are the Australian Aboriginal dialects, and Georgian and Basque.

THE ORIGINS, SPREAD AND CHANGES OF LANGUAGES

Early ideas about a single origin or ancestry of all human speech now carry little weight, whether based on the story of the diaspora from Babel or the theory of onomatopoeic imitation. However, there has been a revival of interest in the topic, from two directions. The first, connected with the ideas of Noam Chomsky (*Syntactic Connections*, 1957) about an innate human sense of grammar, is of deep concern to linguistic theory, but in this context need not detain us.

The second, however, has more relevance here. It has arisen evidently under the influence of the burgeoning science of genetics, within which attention has been drawn to certain broad resemblances between the distribution maps of human races and languages. Separation and isolation would account, it is argued, for the parallel development of both genetic and linguistic entities; and just as modern human races may be traced back to a common ancestral group that moved out of Africa some 50,000 years ago (but see Chapter 4), so all languages should be traceable back to a common tongue, of which a partial vocabulary has been proposed – 'tik' for 'finger' or number 'one' for example.

It seems however that, if we exclude the early separation of the Australian and Amerindian stocks, the present broad pattern of languages has taken shape within the last 5,000 years at most, and that this pattern has come about largely through migrations, which often had very rapid effects.

Moreover, from archaeology and the study of place-names we have evidence of different language patterns in earlier times. For example, we know that the Americas were colonized through successive movements of peoples across the Bering Straits, beginning about 20,000 years ago. The Amerindian languages comprise several different branches or phyla (Sapir, 1884–1939, reckoned six in North America), which are so distinct that they are usually explained as having been associated with the several separate waves of immigration; but with the exception of the more recent arrivals from Siberia none can be related to any existing Old World linguistic groups.

It would appear therefore to be a formidable if not a vain undertaking to attempt to explain the present language map in terms of the early colonization from Africa of the rest of the world by the species *Homo sapiens*, whose original 'mother-language' evolved slowly in different places into distinct branches. None the less, it is reasonable to suppose that both races and languages can take on an increasingly separate identity under conditions of isolation. Sapir once observed that the longer a language is spoken the greater will be the range of its dialects. Thus there are many more English dialects in England, for example, than in Australia or North America. From this point of view, dialects are nascent languages, just as in biology varieties are incipient species. In an extreme instance, the aboriginal Australian peoples and their associated languages were cut off from South-East Asia by the rising sea-levels of the early Holocene epoch, and went their own ways in solitude.

Under conditions of less extreme isolation, we may remark the broad coincidence between the peoples of Mongoloid stock, who evolved in the cold dry steppe of Central Asia after the last Ice Age, and the isolating languages of East Asia. The clearance and settlement of the river valleys of China were made possible by the discovery of the art of cultivating rice, possibly about 5,000 years ago; thereby the isolating languages spread widely from North China through East and South-East Asia.

Likewise, the diffusion of the Bantu languages through Africa went along with the advance of the economy of shifting cultivation, made possible by the introduction of the yam from Malaysia. This forest crop, evidently brought to

110

Africa by way of Madagascar through the medium of Malay argonauts, was more productive and reliable than millet (introduced somewhat earlier from Egypt) in a regime of slash-and-burn hoe cultivation. Bantu-speaking peoples were thereby able to spread quickly from the southern Congo throughout Equatorial Africa, supplanting and restricting the early hunters who spoke Khoisan (Bushman/Hottentot) and Pygmy languages. All this has happened within the last 2,000 years.

A similar process has been hypothesized recently to account for the spread of the Indo-European languages. It is suggested that the ancestral form of all these tongues was spoken by the first Neolithic cultivators of Asia Minor some 10,000 years ago, and that it spread with them, and evolved into separate but related languages, as they made their slow way clearing and farming the ground. Westwards round the Mediterranean to Western Europe, and eastwards through Persia to India, argues Colin Renfrew (1987), they colonized, and replaced peoples speaking earlier languages, with the exception of a few chance survivors, Basque, Etruscan and Georgian, and in India the Dravidian-speakers who were pushed into the Southern Deccan.

This picture, however, runs counter to the older and more orthodox view that the Indo-European languages were carried by a dominant few who overran and imposed their own speech upon a wide variety of other peoples. With the passage of time, and through local linguistic evolution and contact, as well as further migrations, the several families and individual languages of the Indo-European stem took shape.

Only thus could be explained the common 'Indo-European' features of languages spoken by such a wide variety of peoples with different habitats and cultures. By isolating the common basic vocabulary of the Indo-European tongues, a reconstruction can be made of the domicile and lifestyle of the hypothetical speakers of the original language. They were evidently patriarchal tribesfolk, steppe-dwellers, living between the Urals and the Caspian, with domesticated animals including horses, and wheeled vehicles. They left conspicuous evidence of their presence in the form of large burial mounds ('kurgans'). Gordon Childe (1892–1957) supposed that they owed their prowess as a conquering aristocracy to their unique skills in horse-riding and wielding the double-headed battleaxe. The initial diffusion of Indo-European speech is generally thought to have occurred some 5,500 years ago, during the third millennium BC, and its subsequent further dispersion into India and the Mediterranean region about a thousand years later.

If the 'homeland' of the Indo-European languages is to be placed in the western part of the Central Asian steppes, that of the Ural-Altaic (Turkic) languages must have been in the same general area, probably somewhat further east. Again the eruption is probably to be attributed to the movement of nomadic tribes. The Lapp and Finnish peoples seem to have followed the 'circum-polar' shoreline, the Hungarians a more southerly route across the steppes, both in Bronze Age times, about 4,000 years ago. The Turkish-speaking tribes, however, newly converted to Islam, moved south and west through western Persia into Asia Minor only about 1,000 years ago. We naturally know more about this recent event, and it is remarkable that through it the language of the greater part of Anatolia was changed from Greek to Turkish within a generation, partly by displacement of population, partly through intermarriage.

Just as the Hamitic languages of North Africa are centred on the Sahara, so the Semitic belong to Arabia and its cultivated edges in Yemen and the Fertile Crescent. The latest of several eruptions from the desert, through the initial spread of Islam in the seventh and eighth centuries AD, carried the Arabic language into Egypt, North Africa and Spain, Syria and Iraq, though not into Persia or India.

The Indo-European, Turkic and Semitic languages therefore have spread through movement outwards from reservoirs of people of pastoral nomadic economy and culture, on the pattern described by Edmond Demolins in his classic work, *Comment la Route crée le type social* (1901–3). The Malayo-Polynesian languages, by contrast, diffused through movement by sea of colonists using the outrigger canoe, in the early centuries of the present era. The constituent languages of this group are remarkably close in form, from the Eastern Pacific to Madagascar.

PLACE-NAMES AND OTHER EVIDENCE FOR LANGUAGE CHANGES

The evidence of place-names is of course crucial in historical geography. If all other records were to disappear, it would be possible to write a generalized account of the colonization of North America solely from the map of modern place-names. The regions of Spanish, French and British settlement could be defined, and even the general direction of movement of the colonists. With care, traces of German, Scottish and Dutch newcomers could be found, while everywhere would be seen the evidence of a native substratum, in the Amerindian names.

Unfortunately, where there is a dearth of supporting evidence, place-names may prove difficult to interpret. It has been known for almost a century, thanks to the classic work of Paul Kretschmer (1866–1956), that before Greek speech arrived in the Aegean (most would say in the early part of the second millennium BC) there was in that region an earlier language which used place (and other) names ending in -ssos and -nthos: but this *Ursprache* has still not been identified.

The 'hill'-name *tor* is found widely in Derbyshire and Devon, and in the Middle East in the Toros (Taurus) Mountains, the Tur-el-Abdin plateau, Mount Sur (Sinai), the town of Tyre, and in the Greek word *tyrannos* ('king of the hill'). It might be a survivor from a lost language, or have belonged to the speculative group, Nostratic, a supposed common ancestor of the Indo-European and Semitic families. Another word which transcends language barriers is kayak/caïque, which means a boat across Asia from the Canadian Arctic to the Mediterranean.

The borrowing or adaptation of place-names when two cultures meet is of course only one aspect of a wider process, from which something may be inferred about the nature of the fusion. The first Turkish arrivals in Asia Minor, being nomads, seem to have limited their new coinages to descriptive topographic names (Ak Dagh – White Mountain; Yeshil Irmak – Green River): names for towns borrowed from the Greek were given a Turkish style or meaning – Tefeni for Stephanos, Ayasoluk for O Ayios Theologos. In Spain, Arabic names were often retained with only superficial transliteration – Guadalqivir for Wadi-el-Kabir. In India, British foundations often used the Mogul style – Lyallpur for Ferozepore.

In some cases, the distortion of a place-name to give some memorable shape to a meaningless sound may be all that is involved, as often where an army passes by (Hell Fire for Helfya Pass in the Western Desert). A motel in the American West was oddly named Purgatory Holiday Camp, after the original Spanish name, Purgatorio de las Animas Santas (from the ambush of some early settlers), which was doubtless considered preferable to the cowboy version, Picket-wire.

Though of less cartographic interest, a study of the borrowings and adaptations into the ordinary vocabulary of a developing language can convey much of interest regarding the evolution of a culture or economy. The English language kept the words of the Saxon peasant for his animals (cow, sheep, pig), but accepted those of the Norman aristocracy for the end-products (beef, mutton, pork). From Arabic, English took a vocabulary of trade (cheque, tariff), from Greek of science and philosophy (geography, thesis), from French and Latin of law (loyal/legal, royal/regal, treason/tradition). Hindustani borrowed English words for clothing and military drill, Arabic for business, Persian for etiquette. In the 1920s the young Turkish Republic gladly adopted European words for science and engineering needs, but endeavoured, with only partial success, to jettison loan words from Persian and Arabic in favour of Turkish equivalents.

When the romance languages were taking shape as the Roman Empire broke down, the influential forms were not the literary but the colloquial and slang. 'Equus' for horse was virtually forgotten, and 'caballus' ('nag') was used, leading to 'cheval' and its derivatives: likewise 'viaticum' for 'iter', 'mansio' for 'domus' and 'parabolare' for 'loqui' gave 'voyage', 'maison' and 'parler'. Near-'pidgin' circumlocution evolved into regular adverbial forms ('tristi mente' into 'tristement') or possessives ('de illo homine' – for 'hominis' – into 'de l'homme'). The gypsy word for a non-gypsy man, 'gadjo' (possibly from a Sanskrit form 'garha', 'house-dweller') has been adopted in North English dialects as a slang word for a stranger and may, via Spanish, be behind the Argentinian 'gaucho'.

One example from a geographical context may illustrate the subtle and unpredictable changes which such borrowings often undergo. An Early or

Middle English form 'compassus' ('equal-paced') gave rise to 'compass', meaning a pair of dividers. 'Compasso' in Renaissance Italy was a synonym for a Portolano or list of harbours in sequence along a coast. A Compass Chart was a map of the Mediterranean overdrawn – by use of dividers ('compass') – with a mesh of lines of direction emanating from the centre of a circle. The word was adopted into Arabic as simply 'kunbas' (plural 'kanabis') to mean such a chart. By the early sixteenth century, the English word 'compass' also meant a sun-dial, presumably because the divider-compass was used to lay out the hour-lines on the card. Somewhat later, the word 'compass' was applied to the mariner's magnetic compass, whose card would also be drawn by use of the divider-compass. Through a misunderstanding of the origin of the phrase 'compass-chart' for the fourteenth- and fifteenth-century sea-charts of the Mediterranean, the mistaken theory has grown up and been often repeated that these charts were so called because they were surveyed with the aid of the mariner's compass. For technical reasons this could not be so, and, in any event, as explained above, the box with magnetic needle was not at that time called a 'compass' but a 'bussola' or 'magnet'.

It remains to add the caution that bizarre cases can arise by chance convergence, where similar words can evolve even with similar meanings in quite separate languages. Thus the Sheriff of Tombstone derived his title from the English medieval office of shire-reeve, while the Sherif (Ruler) of Mecca was so called from the Arabic root 'sharafa', 'to honour'.

THE ADAPTATION OF LANGUAGE TO CULTURE

Occasionally, a virtually universal process of word invention can be observed – 'cuckoo' for example, or 'pupil' ('doll') or the like for the reflective centre of the eye; more often, vocabulary is closely related to culture. It is well known that the Eskimo have a large range of words for different kinds of snow, and the Masai for the many patterns of colouring on their cattle. Such subtleties of expression doubtless develop to meet a need, and may fade when the need is no longer there. How many would now know the difference between a piebald pony and a skewbald?

A change of lifestyle consequent on migration, or the adoption of a new religion, may call for an extension of vocabulary, through borrowing rather than invention. Both of these circumstances influenced the pastoral Turkish tribes that moved into the Middle East from Central Asia in the eleventh century AD. They changed from a tribal nomadic to a village farming economy, and about the same time embraced Islam. Their Turkish language remained structurally intact, but acquired an extended vocabulary from Greek, Arabic and Persian, and even an occasional pre-Greek 'Asianic' word, such as the toponym *ova* ('valley').

Conversion to Islam, which deeply pervades the lifestyle of its devotees and requires readings from the Arabic Koran rather than from translations, accounts for the strong infusion of Arabic vocabulary in a whole range of languages: Turkish, Persian, Urdu and Malay. The conversion of Northern Europe to Christianity in the early Middle Ages led to a widespread adoption of Latin forms into both Celtic and Teutonic speech. The place-name Eccles, for example, derives from Ecclesia.

In grammar as well as vocabulary there is evidence of the development of constructions suited to local custom. In Hindustani, causal and even double-causal verb forms can easily be constructed, usually by the addition of a vowel and semi-vowel (karna, 'to make' > karana, 'to *have* made' > karwana, 'to *cause to have* made'). The causal form in French or Latin is equally as cumbersome as in English (faire faire, faciendum curare), and the double causal would in each case call for a lengthy circumlocution. The peculiar feature of Hindu society is of course the caste system, wherein the practice of particular crafts and occupations is restricted to appropriate social strata. In these circumstances, causal and double-causal verb forms are evidently very useful, and were for this reason invented and refined in Sanskrit and its derivatives. The Turkish language has similar causal forms (yemek, 'to eat'; yedirmek, 'to feed'); certainly the reservation of crafts for particular minorities ('millets') in the Ottoman Empire would render such verb forms convenient.

Turkish has two versions of the past tense, a distinction which is certainly rare and may be unique. The verb root with the suffix -di/-du signifies that the event has happened to the full knowledge of the speaker; the verb root with -mish/-mush indicates that what has occurred is known to the speaker only through inference, or through information from a third party. Such a distinction would be convenient in the Ottoman bureaucracy, and may well have been invented to meet a need.

Benjamin Lee Whorfe, in his celebrated hypothesis (1956) based on his knowledge of Amerindian languages, viewed the phenomenon in more philo-sophical fashion. Language, he says, reflects the world-view of the speaker. Thus in Hopi units of time are expressed not in numerals but in ordinals, 'on the third day' rather than 'in three days'. This style, he claims, points to a cyclic as opposed to a linear appreciation of time.

LINGUISTIC AND POLITICAL BOUNDARIES

For various reasons, not least the invention of the printing press and the translating of the Bible into the vernacular, the nations of Europe since the Reformation have become increasingly aware of the qualities of their own languages, which they regard as crucial to their identity. In the re-partition of Europe and the Near East by the Treaties of Versailles and Lausanne after the First World War, language boundaries were used to delimit the new nation-states.

Unfortunately, language boundaries are rarely clearly defined. There are, it is true, instances where a marked geographical feature divides languages.

The terai swamps below the Central Himalayas separate the Gurkhali- and Dogra-speaking hill-folk from the Bihari- and Hindi-speakers of the plains. Between Turkey and Syria the political frontier for most of its distance follows the railway under the escarpment which sharply separates Turkish speech from Arabic.

Such cases are, however, exceptional. Usually language boundaries are wide and transitional, and within them mixed patois tend to be used, as words and expressions are borrowed in both directions. It is said that a good linguist walking across Europe would gain the impression not of regions of distinct speech, but of dialects and languages shading imperceptibly into one another. Hence the difficulty of defining satisfactory political boundaries on linguistic criteria, and the disputes that followed the Versailles settlement over the frontiers in Schleswig-Holstein, Alsace-Lorraine, the Sudetenland, the Adige, Trieste, and elsewhere.

Of course, one harsh way of making such boundaries more satisfactory is through the movement of outlying groups of speakers of a minority language to the country where their own language prevails. After the First World War, over a million Greeks from the newly defined Turkish Republic were exchanged for about half that number of Turks from the Balkans. The same two communities experienced another painful mass-exchange following the 1974 partition of Cyprus.

It has been remarked above that the central part of the southern Turkish frontier marks a neat linguistic divide: but near the Mediterranean in the Hatay (the former Sanjak of Alexandretta), and beyond the Euphrates in the Vilayet of Mosul, the situations are more complex. In both districts the grain of the relief runs north–south, and in the one case Arabic-speakers have moved northwards among Turks, and in the other Turks southwards among Arabs. The situation was complicated in both cases by the presence in large numbers of a third community – Alawites in the Hatay, Kurds in the district of Erbil. The Sanjak became part of Syria until 1939 but since then has been Turkish and renamed the Hatay. The Mosul Vilayet of course has remained part of Iraq.

At Versailles, unfortunately for them, some sizable linguistic groups were not considered large enough for political hegemony, and the resentment of the Kurds and Armenians at being denied statehood still lingers.

It is evident that since the Second World War the sense of linguistic and cultural identity within minorities, and associated demands for more independence, have generally become more vociferous and powerful, not only within Europe but also further afield. In Wales and in the Basque country in particular, demands for a greater degree of autonomy have been associated with a pride in their distinct languages. In India, after the Partition of 1947, when the country was divided into new provinces under a federal constitution, the boundaries were drawn on linguistic criteria.

PIDGINS AND CREOLES, AND OTHER MIXED FORMS

The process opposite from linguistic separation is that of linguistic fusion, where a new patois emerges from the mixing of two or more languages. This phenomenon is different from that referred to above, of the evolution of 'intermediate' dialects in the frontier zones between major language areas. Pidgins (the expression probably derives from the phrase 'business [-talk]') and Creoles are characterized by:

1 The absence of precise territorial definition, and the diffusion primarily by sea through island or coastal regions (Creoles in the West Indies, Pidgins in the East Indies, Swahili in East Africa).
2 The circumstance of the contact, usually from motives of commerce, between an intrusive higher and a more simple native culture.
3 The language of the former is drastically simplified in its grammar and vocabulary and adopts selected words from the latter. Awkward circumlocutions have often to be coined to replace abandoned grammatical forms (the suffix '-Mary', for example, to indicate the feminine).

The Creoles of the West Indies vary according to whether the main parent language is English, Spanish or French, the native element being Carib mixed with imported Bantu.

The several Pidgins of South-East Asia are generally based on a simplified English mixed with Malay and some Chinese forms, while the Swahili of the East African coast is a patois that combines local Bantu with intrusive Arabic.

There used to be a maritime lingua franca in the Mediterranean, which drew on several constituent languages and still survives to some degree in Malta: but it consisted mainly of a rich technical vocabulary to do with ships and navigation, and was too specialized to be classed as a Pidgin. Afrikaans is another mixed form (Dutch–Bantu) which grew up in a colonial time and setting, but would not fit into the strict Pidgin/Creole category, by reason of its more complex and literary structure, its inland spread, and its use in circles wider than the purely commercial.

Urdu (strictly Zaban-i-Urdu or Language of the Camp) is an artificial assemblage compiled for the mercenary armies of Mogul India, mainly from Hindi, but with vocabulary and constructions taken from Persian, Arabic and English. Hindustani, the 'civilian' version of the language (though virtually the same as the military form, Urdu) has become effectively the lingua franca of the northern part of the Indian sub-continent, as has English of the southern (Dravidian-speaking) Deccan.

Another special case is that of Yiddish, a German–Hebrew amalgam which evolved in Eastern Europe, but which now has a wider currency. In this case, as with Afrikaans and Urdu, perhaps the crucial feature which would put them in a class distinct from the Pidgins is that all three have inspired a significant body of literature.

EXPERIMENTS TOWARDS A UNIVERSAL LANGUAGE

One consequence of the decline of Latin as the vehicle of knowledge after the Reformation was the use of national languages in learning and science, which became fashionable within the academies of Europe during the seventeenth century. At the same time, there was a widespread feeling that a common means of international communication was needed to replace Latin.

Leibniz, in devising his system of mathematical notation, corresponded with Jesuit missionaries from China to find out how symbolic (logographic) writing could be used across language barriers; and he went on to speculate on the possibility of devising a universal language.

The same question concerned Descartes, Grimm and the French Encyclo-paedists. George Delgarno of Aberdeen published in 1661 his *Ars signorum*, a lexicon of ideas, as the basis for a common language. Shortly after, in 1668, Bishop Wilkins in his *Essay towards ... a Philosophical Language* devised not only a universal language but also a universal script, founded on concepts. A similar but more enduring enterprise in a less ambitious field was Linnaeus's *Systema Naturae*.

In the last century there have been several well-intentioned attempts to formulate an international language, Volapuk, Esperanto and Interlingua being the best known, but in practice C.K. Ogden's Basic English, with its 850 words, has been the most successful.

LANGUAGE AND WRITING

Writing of course is not necessarily directly related to language, and in its basic form of pictography or ideography it is a method of conveying a meaning that is quite independent of speech. The same symbol can be understood by observers who would express its significance in different languages; and this quality explains the survival of pictography for symbols intended to convey simple instructions of guidance in the fields of travel, commerce, technology and cartography. The earliest writings of this kind were devised in Egypt and Mesopotamia for the recording of taxation, trade, food distribution and landholding. In order to express personal- or place-names, for which no pictogram was available, symbols were selected not for their pictographic meaning but for the sound of their names, and were arranged in the appropriate sequence to construct the required proper names. It was quickly found that a limited repertoire of a few dozen sound–signs was adequate for this purpose, and in this way there emerged the syllabaries of the Egyptian hieroglyphic and hieratic and the Mesopotamian cuneiform scripts.

The ideographic scripts of the Far East never reached this degree of simplification, until the relatively recent invention of the Japanese syllabary. The pictographic writings of the higher civilizations of pre-Columbian America likewise remained at this undeveloped stage.

The most important advance in the history of writing occurred, probably in Sinai or nearby in the late second millennium BC, in a Semitic-speaking community who found that for their purposes it was possible to omit the vowels of the syllabic signs and thus reduce the number required to the two dozen or so of the first alphabet. This invention, which made writing more simple and flexible, swept quickly, generally as a medium of commerce, through the civilized world, in two main forms. The North Semitic letter-shapes were adapted, through the addition of vowel letters, to make the Greek and Latin and derivative European alphabets; and much later, through the spread of Islam, to write Persian, Urdu and Turkish. The South Semitic letters inspired the Indian Brahmi script, and this in turn the several alphabets of South-East Asia and that of the Old Turkish inscriptions of Mongolia.

It is manifest that the present pattern of distribution of languages has taken shape within historic times, very often through the diffusion of emigrant conquerors from areas of pastoral nomadic economy. Though languages are never static, and are subject to constant superficial modifications, basic grammar and vocabulary are remarkably conservative; even the gypsy tongue, which has been influenced by numerous local contacts through more than a thousand years, retains recognizably many of its original Sanskrit roots.

REFERENCES

Demolins, E. (1901–3) *Comment la Route crée le type social* (2 parts), Paris: Librairie de Paris.
Renfrew, C. (1987) *Archaeology and Language: The Puzzle of Indo-European Origins*, London: Jonathan Cape.

FURTHER READING

Bodmer, F. (1944) *The Loom of Language: A Guide to Foreign Languages for the Home Student*, London: Allen & Unwin.
Campbell, J. (1982) *Grammatical Man: Information, Entropy, Language, and Life*, New York: Simon & Schuster.
Gelb, I.J. (1963) *A Study of Writing: The Foundations of Grammatology* (revised edn), Chicago: Chicago University Press.
Meillet, A. and Cohen, M. (eds) (1952) *Les Langues du monde*, Paris: Centre National de la Recherche Scientifique.
Sapir, E. (1957) *Culture, Language, and Personality: Selected Essays*, Berkeley, Calif.: University of California Press.
Sapir, E. (1978) *Language: An Introduction to the Study of Speech*, London: Hart-Davis MacGibbon.
Taylor, I. (1909) *Words and Places; or, Etymological Illustrations of History, Ethnology, and Geography*, London: George Routledge.

6

RELIGION

Nature and origins

E. Geoffrey Parrinder

Religion takes the whole of life for its province and is concerned with the physical, intellectual and spiritual interests and activities of men, women and all creatures. It includes reference to the universe and to this earth, to human beings and animals, to individuals and societies, to sacred places and practices.

Definitions of religion are many, diverse and sometimes contradictory. For Karl Marx religion was an opiate, 'the sob of the oppressed creature, the heart of a heartless world', and he thought that in due course humanity would arrive at a perfect and stable condition where such a drug would no longer be required. For Sigmund Freud religion was the expression of an Oedipus Complex, a repetition of a parental fixation from which sprang not only religion but also art and morality. For other writers in the last two centuries religion was regarded as morality touched with emotion, or a personalization of society, or a worship of ancestors. But recent studies have insisted that the religions of even the most primitive peoples are complex, not reducible to animism or totemism or ancestor-worship, but inspired by apprehensions of spiritual and creative forces which are believed to have formed and to fill all life.

Religion is a universal phenomenon and no peoples have been discovered who have no expression of what may be called religion, in a broad sense, though there are infinite variations of religious concepts and practices. Even the people of Tierra del Fuego, who were thought by Charles Darwin at first to have had no religion, were later discovered to hold intricate beliefs and practices which Darwin came to acknowledge.

The origins of religion are lost in the mists of prehistory. Theories that it arose from fear of natural phenomena, or of ghosts of the departed, or from a totemic feast, or from magic, or even from language, are now discarded. For over a century Western scholars tried to reconstruct the possible origins and development of religion, linking it to theories of evolution. Obsession with the beginnings of religion took it for granted that the origins were simple or pure,

120

meaning either elementary and almost animal, or alternatively simple and perfect such as might have existed before Noble Savages were corrupted by civilization. Such theories are now obsolete, at least in scholarly circles though they still have popular currency, and they are only relevant for the history of Western speculation. It is now accepted that religion is too ancient and diverse to have had a single origin, and that even among peoples of comparatively elementary material culture there may exist varied and profound conceptions of the spiritual nature of the universe (Eliade 1973: xiv–xv).

An eminent critic of religious studies distinguishes distinct ways in which the term 'religion' is commonly understood: first a sense of personal piety, including prayer and worship; then reference to systems of beliefs, practices and values; and then religion in general, as an understanding of the many ways in which it appears in history, society and daily life (Smith 1973).

Further classifications of religions may be made according to the development of the art of writing, whether alphabetical or syllabic, which provided an invaluable tool for recording and interpreting religious belief and ritual. Major religions possessed recorded histories and sacred writings, and this often enabled them to pass from local or national concerns to international status. By contrast, among illiterate peoples, past and present, religious concepts were not expounded in writing by practitioners of the faith, and they had to be deduced and written down by external observers of ritual. Art did indeed provide some indication of religious attitudes from within, but interpretations differed widely.

With the development of written records not only were histories made available but also ritual texts, religious autobiographies and philosophical speculations came to be composed for the benefit of contemporary and future generations. Such religious recording took place chiefly in Asia and Europe and it provided a vast store of material for study: in the scriptures of the Jewish and Christian Bibles, the Islamic Koran, Persian Zoroastrian Avesta, and Buddhist and Chinese texts. An apparent exception, for over a millennium, was in the Hindu Vedas which were formulated, memorized precisely, and passed down for many generations before they were finally committed to writing. In a sense these were 'scriptures', since they were preserved and cherished texts, and they gave rise to a vast amount of commentary and development in succeeding centuries.

IN THE BEGINNING

As the earth evolved and life and people appeared, human beings came to wonder at the nature and origins of all things. Many religions record speculations, not only on the actual and material conditions of human life, but on the beginnings of this world and the universe, going on to the origins of life on earth and the appearance of people. These were formulated in hypotheses or myths, often very sophisticated, different from the material investigations of geologists, archaeologists or biologists. With all their variations, religious thinkers agreed that the world was not an accident, or thrown up by purely material forces, but they considered that it was directed or inspired by divine or spiritual intelligence. From the fact that human beings are intelligent, it has been taken for granted that they must be derived from a superhuman power that was super-intelligent. This is a fundamental assumption in all religions and distinguishes them from materialistic philosophies.

'In the beginning', said the book of Genesis, 'God created the heaven and the earth'; the book of Proverbs (8) said that wisdom was with God at creation as a master workman; and the Gospel of John wrote that 'in the beginning was the Word', and 'all things were made through him'. Zoroastrian scriptures declared that 'in the beginning' divine powers established life and non-life, while other mythologies speculated that in the beginning, when nothing else existed, power issued from the mystery of the infinite.

Diverse suggestions appear especially in basic Indian philosophical compositions, the Upanishads, from early in the first millennium BC. 'In the beginning' there was nothing, or 'in the beginning this world was soul' (*ātman*) in the form of a person. It was conscious but alone, and afraid, and it divided itself into two to make husband and wife (Brihad-aranyaka 1:4). Somewhat later in Greece Plato suggested that originally there was man-woman, 'a being which was half male and half female', and Zeus cut it in half so that men and women are always running to join together again (*Symposium* 191).

A further statement in the Upanishad was that 'in the beginning this world was Brahman', universal soul or spirit. Then developing reflection in the second Upanishad affirmed that 'in the beginning this world was just Being, only one, without a second', undifferentiated being (Chandogya 6:2). But those who said that in the beginning there was nothing, non-being alone, were contradicted. 'How could this be? How could Being be produced from non-being?' So the primacy of mind was asserted, 'in the beginning this world was just Being'.

The assertion of primary being led in India to consideration of the world. Being took thought, 'Would that I were many. Let me procreate myself', and so it emitted heat, water and food. But occasional scepticism had already appeared in the Vedas, 'whence this creation has arisen, whether he created it or not, only he who surveys it from the highest heaven, only he knows, or perhaps he does not know' (Rig Veda 10:129).

In Chinese classics the Way (*Dao*) was the unnameable from which heaven and earth sprang. The essence of the Way was to know what once there was 'in the beginning', and to master things of the present one could hold on to the Way of old, the microcosm in oneself giving knowledge of the macrocosm in the universe (Dao De Jing 14).

Myths in many religions ascribed the creation of the universe to the power of a God, gods or goddesses. One of the most ancient texts, from about the twelfth century BC in Mesopotamia, is known from its opening words, 'When on high' (*Enuma elish*). 'When on high the heaven had not been named, firm ground below had not been called by name, there was nothing but their primordial begetter, and Mother Tiamat who bore them all' (Pritchard 1950: 60–1). Like most myths, this was not just a literary composition but it had a ritual purpose and use, and it was recited solemnly on the fourth day of the New Year festival.

The Indian Vedas, perhaps about this time also, spoke of Mother Earth co-operating with Father Heaven. In Greece the Homeric Hymns sang of 'well-founded earth, mother of all, eldest of all beings'. In China the doorway of the mysterious female was said to be the base from which heaven and earth sprang.

In the Bible it was said that God created light first, then heaven, earth, seas, trees, herbs, sun, moon, stars, birds, fishes, animals, and finally humanity, both male and female made in the image of God. It is almost an evolutionary process, from seeds to animals and mankind, but all called into being by potent divine words, 'God said'. The Islamic Koran, which respected the Hebrew scriptures, followed this pattern, with God creating heaven and earth in six days and then seating himself on the throne to command all things, 'for it is his to create and to command' (7:52).

It should be noted that, contrary to some popular opinion, these ancient myths did not put mankind at the centre of the universe; for over all was God the creator, and in any case human beings were fallen creatures because of disobedience to divine command. Then there were many commentators on the ancient texts who elaborated and refined their perceptions. The Jewish philosopher Philo in the first century said that God first of all conceived the form of the universe in his mind, a form which was only perceptible to the intellect, and then the visible world was based upon the first one as a model.

HEAVEN AND THE UNIVERSE

Heaven and the universe are infinite, and so they could be used as symbols of the greatness and transcendence of a creator God, but they were not usually regarded as identical with him. Isaiah (40:12) said that the heavens have been measured with the span of the divine hand, while the seas are in the hollow of his hand and the mountains are weighed as in a balance. The nations of the world are only like the small dust that remains after weighing produce in scales, virtually worthless. The Koran declared that the throne of God extends over the heavens and the earth, while the Indian Bhagavad Gita remarked that all the universe is strung upon God like heaps of pearls upon a string. So the heavens may be compared to God in their height and mystery, but they are not God in themselves.

Nomadic or agricultural people looked up at the night sky and became filled with awe at their own insignificance, 'what is man, that thou art mindful of him?' (Psalm 8). Yet he had been created at the end of the creation process, 'thou hast made him little lower than the angels', or 'little less than God'.

Criticism has been made of religious language when it spoke of a divine body, with hands and feet, emotions and masculinity, but religious writers have long been aware of the use and abuse of symbolism. In the third century the Egyptian Christian theologian Origen declared that it would be childish to suppose that God actually planted a garden in Eden and walked therein at eventide, for these were only figurative statements. Medieval Islamic philosophers declared that the confession was that God had two hands, 'but we do not ask how this can be'. And the Anglican Book of Common Prayer affirmed that God was 'without body, parts, or passions' (Article 1).

Heaven and earth and all that lies between them are like a bellows, said the Dao De Jing (5), seemingly empty yet giving a supply that never fails. They are impartial, and the wise person will seek unity and harmony with them, so as to be able to work on all the many things in the world.

In Japan the supreme deity was the sun-goddess, Amaterasu-omi-kami, whose shrines in the ancient sacred city of Ise are still the centre of the indigenous Shinto religion, and whose lesser shrines are found throughout the country, in village sanctuaries and on modern skyscrapers. Amaterasu is the deity of the sacred Mount Fuji and the ancestress of the imperial line, so that both sacred places and the orders of society are linked to her. Traditionally this restricted the cult of the sun-goddess to the sacred land of Japan, though in recent attempts at imperial expansion the sway of Amaterasu was considered to extend over all the earth.

EARTH AND LAND

The earth has been contrasted with the sky, but beliefs in earth spirits in their own right have been potent forces in religious faith and ritual. It has been said that the Olympian gods of the invading Greeks fought, ate, drank and made love, like their worshippers, but they never did a stroke of honest work. They did not attend to the government, or promote agriculture or industries. Yet in popular Mediterranean religion there were more ancient and powerful forces, the earth-goddesses, whether they were called Artemis or Diana, Cybele or Isis. The mystery religions of ancient Greece, as far as their secrets can be deduced, for they were 'mysteries', celebrated the agricultural processes of sowing and harvest, summer and winter. In a widespread popular myth young Kore, also called Persephone or Proserpine, was daughter of Demeter, Mother Earth, also called Ceres. Kore wandered into the underworld and was held captive by its ruler, Hades or Pluto. Although the girl refused all food she did take some pomegranate seeds, and when she was allowed to return to the world with her mother she had to spend part of each year underground. This agricultural myth was played out annually, apparently, in the mystery cults of Eleusis, and similar stories were celebrated with Tammuz and Ishtar in Mesopotamia and with Isis and Osiris in Egypt.

Some theorists have assumed that such agricultural myths were universal, but in historical and literary religions Tammuz is mentioned only once in the Bible, as a foreign cult, and an earth goddess not at all. As in Islam, the Bible ascribes all the fertility of the land to the power of the supreme and celestial God, to whom alone worship is due, whether in cults with agricultural or national reference.

In Chinese Daoism the universe was seen as due to the interaction of two complementary principles, Yang and Yin. The Yang is the life-breath of heaven, corresponding to the male, strength and light. The Yin is the life-breath of earth, female, weak and dark. By the union and harmony of Yang and Yin the elements and the seasons of the year run their course, and both activity and tranquillity ensue.

In the West some modern writers (Tillich 1962: 63; Robinson 1963: 27) rejected old descriptions of God above, a heavenly father, and they tended to regard the divine as 'in the depth of our being'. It was still spatial symbolism, immanent rather than transcendent. Traditionally both types of symbol have been used; 'when I consider the heavens' brings a feeling of dependence if not insignificance, but also the divine is everywhere, 'If I make my bed in hell, behold thou art there' (Psalm 139).

PASTORAL LIFE AND ANIMALS

For pastoral and nomadic peoples in the past, and at present especially in Africa and Asia, cattle are the greatest treasure, for men, women and children. They are dependent on the milk of their herds, their bodies provide meat, and their skins furnish tools, ornaments and other domestic objects. Cattle are a means of display, valuable for obtaining wives, but also central to religious practice. When cattle are sacrificed, the most precious possessions are surrendered. Short prayers or invocations may be uttered at any time, perhaps by rubbing an ox's back; it may be dedicated to a spirit and it is regarded as potentially destined for sacrifice. In its sacrificial role the animal is considered to shield people from sin, sickness, suffering, natural or tribal disaster. It is the means by which men and women regularize their communication with God.

In ancient Israel, and down to New Testament times, animal sacrifices were abundant, so much so that prophets queried 'will the Lord be pleased with thousands of rams?', and asserted that what God required of man was 'to do justly, and to love mercy, and to walk humbly with thy God' (Micah 6:7–8).

The great Hebrew festival of Pesah (Passover) had the sacrifice of lambs as central, combined with historical reference to the Exodus, but no doubt deriving from pastoral times. Still today the Passover supper has a lamb's bone (*pesah*) as a vital component, though it is not eaten. Similarly in Islam the great Festival of Sacrifice culminates in the sacrifice of sheep, goat or camel, not only by pilgrims at Mecca but by Muslim households right across the Islamic world.

Different attitudes appear in the great cities, as in the villages, of India. There the sacredness of cows and bulls is absolute, and most people are vegetarian, though some lower and external castes may kill and eat animal flesh, and are abhorred for so doing. The inviolability of cows in India seems to have been of slow growth, since they were sacrificed in Vedic times and later, as is known from records of the emperor Ashoka in the third century BC. When penalties for killing cattle first appeared, it seems that these were for animals stolen from royal herds.

Gradually in India cows became sacrosanct; wanton killing of any animal was forbidden and that of a cow was the most serious of crimes. The 'five products of the cow' –milk, curd, butter, urine and dung – were all held to have great purifying power. Cows have been revered, not as representing any deity, but in their own right. But the bull was honoured as the mount of the god Shiva, and its image is found in most Shiva temples where it receives occasional offerings.

Cow manure is used on the land, and to decorate houses, and Indian peasants are so conservative that the methods they use can be assumed to be much like cultivation in ancient times, ploughing with shallow wooden ploughs drawn by oxen and threshing with oxen.

The numerically small but ancient Jain religion in India is the most strict against taking any form of life, even down to insects. Killing animals for sacrifice, for the sake of their hide, or flesh or blood, is strongly condemned in the extensive Jain scriptures. Some Jains strain water for fear of destroying insects, refuse to eat honey and some vegetables, and check food to take out eggs, worms or cobwebs. Jain monks wear white cloths over their mouths to prevent inhaling insects and carry small brooms to sweep their paths free. Jain temples have grills over lights to protect insects and they are renowned for their animal hospitals for birds and cattle. It followed that Jains did not engage in agriculture, but they were great traders, and much of the wealth of these small communities was devoted to building the splendid temples for which this religion is renowned. In these temples, and on cliff carvings and isolated statues are represented the *Jinas*, 'conquerors', who taught respect for all life with the doctrine of 'harmlessness' (*ahimsa*) which in modern times was adopted by Mahatma Gandhi in eclectic teachings of non-violence. It may be imagined what vegetarian Indians thought of their rulers who indulged in 'huntin', shootin' and fishin''.

Buddhism, a great international religion which began in India, also taught respect for human and animal life. The first of the Five Precepts (*pancha-shila*) incumbent on every Buddhist, lay or monk, is to 'refrain from taking life'. The Buddhist emperor Ashoka decreed that animals killed in his kitchens should first of all be limited to three kinds and eventually to all beasts and birds. Positively, he instituted medical care for human beings and animals, medicinal herbs were planted, trees were set along roads and wells dug for the use of people and animals. These decrees were inscribed on pillars and rocks, some of which remain to this day.

Avoidance of strict vegetarianism could be practised by taking the letter of the law to mean refraining from taking life oneself, but use of flesh if it had been killed by another. As Buddhism spread it mingled with other religions, Daoism and Confucianism in China and Shinto in Japan. Buddhist regulations could be disregarded by those who wanted to eat fish or meat, but they were insisted upon by ascetics, monks and nuns.

The Western Semitic religions, Judaism, Christianity and Islam, did not share the Indian aversion to meat-eating but they provided humane teachings for the care of animals, such as 'thou shalt not muzzle the ox, when he treads out the corn' (Deuteronomy 25:4). The Talmud later affirmed that preventing pain to an animal was a command of the Law. The historian Josephus expressed the shock of Jews at the Roman practices of bringing wild beasts to fight with each other and with human beings for the delight of spectators, for this was bare-faced impiety. The Deuteronomic injunction was quoted twice in the New Testament, and the care of God even for the odd and worthless sparrow out of five, was a lively example in the Gospel (Luke 12:6). The concern of God for every beast of the earth is stated in the Koran (11:8), and doctrines of providence in all three of these religions could include animals as well as human beings.

Humanity was given domination over all other living creatures, according to Genesis (1:28), though some have commented that their use for food seems first of all limited to herbs and only extended to flesh after the flood. Even then the eating of blood was forbidden and only 'clean' (*kosher*) meat was allowed, for 'the blood is the life' or soul of the animal which must not be consumed (Leviticus 17:14). The *kosher* regulations still apply, not only in Judaism but also in the much more widespread religion of Islam with its 800 million followers.

Treatment of animals in general has varied greatly. There have been many examples of their care, as with Francis of Assisi's concern for Brother Ass, Mother Earth, and all creation. But humane restrictions in Europe had to await the nineteenth century, with prohibitions of bear- and bull-baiting and cock-fighting. A sense of the unity of all life has increased with interest in Indian and other Asian religions, where even if the practice had failed yet principles of respect for all living creatures were fundamental to religion.

AGRICULTURE AND CALENDARS

With the development of agriculture major events were in sowing, first fruits and harvest, and these had religious expression. At planting, notably in parts of Africa where an earth-goddess is revered, rituals are performed before digging the ground. These may range from simple knocking on the earth and asking permission to dig, to libations of water or blood. At first fruits a common view is that token ears of corn or rice should be presented at an altar or shrine before the rest of the people cut their crops. To this day the emperor of Japan, high priest of Shinto, ritually cuts the first rice. Photographs in daily papers show him doing so and thereupon all farmers can follow his example. After the completion of the harvest, when 'all is safely gathered in' as urban congregations still sing, there are universal rejoicings. Even if it is a bad year for corn, prayers may be offered that root crops and orchards may prove more successful.

Whether agriculture was discovered by women, throwing seed into the soil while men were out hunting, cannot be proved though many have found it an attractive hypothesis. Women have often helped in sowing and reaping, and many still do. The relation of women to the land is suggested in many myths; barren women were thought to be dangerous to the seed, while pregnant women should make it grow.

Women are commonly compared to the soil. In the popular Indian epic *Ramayana* the name of the heroine Sita means 'furrow', and in the Koran (2:223) it is said that women are to men as furrows. The Hindu Laws of Manu (9:33) taught that woman may be looked upon as the soil, and the male as the seed, and the production of all life took place through their union. It has been held that some primitive peoples were ignorant of the process of reproduction, though the theory has been contested in the light of the comparison of agricultural and human production.

Even in advanced, historical and literate cultures, primitive symbols and rituals may survive. The Maypole and May tree suggest phallic symbolism, associated with May dances and May Queens with reminiscence of young couples mating ritually on ploughed earth. Whether there were witch cults or not (probably not), popular imagination supposed witches flying on phallic broomsticks to blight the earth by their harmful magic, bringing barrenness of land and people, devastation by hailstorms, milk drying up or going sour, and hens not laying.

The agricultural cycle fitted into the solar calendar, and in the historical religions adjustments were made to combine religious events with the natural year. In the Christian calendar, Christmas was fixed from a new year feast. The date of the birth of Jesus was unknown, whether in winter or summer. In the fourth century a pagan festival of the birth of the Invincible Sun on 25 December, roughly the winter solstice, was taken over by the Church in Rome for celebrating the birth of Christ, the Sun of Righteousness, and this practice rapidly spread across the Christian world. The period of Lent leading up to Easter was probably named, in English, from the 'lengthening' of the days in spring, and the feast of the Annunciation, 25 March, was at the spring equinox.

The English name 'Easter' for the great spring festival was derived, according to the Venerable Bede, from an Anglo–Saxon goddess, Eostre, dawn. The giving of Easter eggs was an ancient custom as symbols of fertility. But Easter is the Christian Passover or Pesah, and its date is calculated in the West for the Sunday after the first full moon after the spring equinox. The Christian celebration was of the death and resurrection of Christ, on Good Friday and Easter Sunday, and it illustrated the mingling of historical with natural events. Speculations that the Church feast originated from pagan cults of dying and rising gods are now generally rejected, since there is no evidence for their currency in Jewish circles where the events occurred, and the fact of the crucifixion is not doubted by serious historians.

Similarly the Jewish feast of Pentecost for the corn harvest was taken into the

Christian calendar to celebrate the descent of the Holy Spirit on the apostles on that day.

Lammas, on 1 August, was the 'loaf-mass', when harvest began in northern countries, the first ripe corn being offered in church on this day. It is still observed as a sacramental and quarter day in Scotland. All Saints and All Souls, on 1–2 November, appropriately remember the dead in the days of the dying year, with public holidays on the Continent.

The Jewish calendar is lunar, with twelve months of 354 days, but extra days added in a cycle of nineteen years to fit in with the solar year. The New Year (Rosh Hashanah) commemorates creation and is followed by the Day of Atonement (Yom Kippur), for repentance and renewal. Passover, as has been mentioned, celebrated the Exodus deliverance, the offering of lambs, and the beginning of harvest. There are many other festivals, including Booths (Sukkot) in autumn and Lights (Hanukah) in December, which are related both to nature and to past and recent history.

The Islamic calendar is lunar, though it is dated from the historical 'migration' (*hijra* or *hegira*) of Muhammad and his followers from Mecca to Medina in AD 622. The short lunar months make comparison with the Gregorian calendar difficult and special tables are compiled to aid their comparison. As lunar months decline against the solar year, Islamic festivals are affected by the climate. If the pilgrimage occurs during the summer, the heat of Arabia is intense and often fatal for older pilgrims. Similarly the fast of Ramadan, for the whole daytime of the ninth month, is long and painful if it falls during a long north European summer, but it is much more agreeable in winter.

In India Divali, 'lights', in October–November is the popular celebration of the new year, even in places that use a calendar when the new year begins in March. It is associated with Lakshmi, goddess of wealth, and traders and others pray for good fortune. As elsewhere, natural festivals may be overlaid by reference to religious events or the stories of particular deities: Krishna, Rama, Shiva, Kali, and others.

Across the Buddhist world festivities at new year open the religious calendar, which then interweaves commemorations of the birth, enlightenment and death of the Buddha with natural rituals such as retreats at the end of the rains and festivals of lights. In China a popular festival is that of hungry ghosts, like All Souls-tide, in August, while in Japan nature is revered in flower-viewing, bean-scattering and lantern festivals.

SACRED PLACES

There are countless regions throughout the world which have been regarded as holy by the people who live nearby, and to ignore or desecrate such locations has caused great turmoil in modern times, notably in Australia and North America.

For many peoples the whole landscape is alive and even small details may mean something, recalling past history, the revelation of the power of the ancestors or of divine spirits. That Australian aboriginals visit traditional sacred spots is not because of economic pressure, for they usually have modern kinds of work to provide them with livelihood, but through sacred places they maintain their union with the land and the ancestors who founded their culture.

Holy locations are not regarded as simply chosen by human beings, for they come through revelation. These rocks or trees or springs may be the media of mystical communication. A classic example is of Jacob who slept in a place of stones, perhaps an old sanctuary, and he dreamt that it was joined to heaven by steps down which angels descended. When he awoke he said, 'Surely the Lord is in this place, and I knew it not. This is none other but the house of God.' He set up a stone, pouring oil on it, and called it Beth-el, the house of God. It became a great temple, later frequented by kings and prophets (Genesis 28; Amos 7).

Sacred places have a numinous character, exercising attraction and awe from the 'tremendous and fascinating mystery', and they require both preparation and reverence. So at the burning bush Moses was told, 'put off your shoes from your feet, for the place whereon you stand is holy ground' (Exodus 3:5). Traditionally this site was Mount Sinai, to which Elijah also fled later for another revelation.

Sanctuaries were generally marked off by circles of stone, enclosures or walls, which are among the most ancient forms of temples. They were consecrated by revelation, but also by the present or past habitation of holy people, giving added history from the dwellings or graves of saints.

Sanctuaries could be changed or adapted in time, yet retain or enhance their sacredness. Bede said that in AD 601 Pope Gregory instructed Augustine on his mission to England to destroy heathen idols but their temples could be aspersed with holy water and then given altars with Christian relics, so that people would continue to attend them. To this day the altars of all Catholic churches should enclose the relics of a saint, and the same applies to Buddhist pagodas and many other holy buildings.

The world has innumerable sacred locations, surrounded by sanctified space. Hindus flock to Benares (Varanasi, Kashi), where cremations take place and where it is believed that anyone who dies within ten miles of the holy town goes straight to heaven. There are many other holy places and objects of pilgrimage for Hindus, from the Himalayas to the southern tip of the country. For Muslims, Mecca is the most holy city, the goal of the pilgrimage obligatory on all the faithful at least once in a lifetime. At ten miles from the centre of Mecca notices in Arabic and English warn that its entry is forbidden to all but Muslims, and men and women must change into pilgrimage dress.

Jerusalem is sacred to three religions. Jews pray constantly at the Western Wall (formerly called Wailing Wall), desiring the rebuilding of the ancient temple. Christians follow the Via Dolorosa in Jerusalem every Friday, in memory of the crucifixion on the first Good Friday, and they visit other locations such as the Church of the Holy Sepulchre. To Muslims Jerusalem is the third most holy city, after Mecca and Medina. The Dome of the Rock (popularly called the mosque of Omar) is a shrine on Temple Mount, the site of the legendary ascent of Muhammad to the heavens, and depressions in the slab of bare rock are still said to be his footprints. It may be that this rock was the sacrificial stone of Araunah the Jebusite from whom David bought it (2 Samuel 24), so that the holy mountain has passed through four religions and enhanced its sanctity.

Holy places all have rituals, regular and occasional, and they are supremely places of pilgrimage. This is a very ancient and important activity which has often served as the holiday of the year or lifetime, as Chaucer's *Canterbury Tales* witness, when in April 'than longen folk to goon on pilgrimages'. In Chaucer the sacred journey was to Canterbury, twice holy by its great cathedral and by the miracle-working shrine of the martyr Thomas à Becket.

In India in addition to centres holy to various deities, and rituals and pilgrimages connected to them, there are great fairs known as Kumbha Mela held every twelve years at four places in succession. The gods fought the demons for possession of a pitcher (*kumbha*) of nectar, stopping at the four sites on their way back to heaven. The Mela is a religious fair of timeless antiquity, and with modern transport many millions attend these celebrations. Ascetics emerge from the forests or mountains to bless the pilgrims, while modern teachers lecture the crowds with microphones. Rituals include bathing in the Ganges and other rivers.

Pilgrimage to Mecca is one of the five duties of all Muslims. Mecca was a pre-Islamic city and its Ka'ba, a small 'cube' shrine, was said to contain 360 idols. These were destroyed by Muhammad when he conquered the town in AD 630 (AH 8). Pilgrims had formerly circumambulated the shrine, keeping it on the fortunate right-hand side, and Muhammad retained but reversed this process by making it leftwards. He decreed the proper pilgrimage dress, probably to replace naked or careless previous attire. The 'direction' of prayer was now made towards the Ka'ba, and this is observed by Muslims throughout the world. Pilgrims circle the Ka'ba seven times and try to touch a Black Stone in its wall, said to have been sent from paradise but perhaps a meteorite whose fall here had given rise to the shrine. Legend says that the Ka'ba was first built by Adam, destroyed in the flood, rebuilt by Abraham, and enlarged in the time of Muhammad who put the Black Stone back in its place. There is a long ritual, culminating in communal confession and the 'feast of sacrifice', Id al-Adha, called Bairam in Egypt and Turkey.

For Shi'a Muslims, chiefly in Iran and Iraq, but with smaller numbers elsewhere, ten days in Muharram, the first month of the year, are dedicated to Passion Plays commemorating the death of the proto-martyr Husain, grandson of Muhammad, at Karbala in Iraq in AD 680 (AH 61). The great mosque in Karbala was damaged by Sunni Iraqis after the Gulf War in 1991.

Sacred places, with their festivals and rituals, thus can combine elements of prehistorical or legendary or historical character. The rituals continue the experience which first revealed the place to the worshipper. As specialists have commented, rituals do not only recall past events, but they 're-present' them. For the Muslim on pilgrimage, as for the Jew at Passover or the Christian in Holy Week, or for followers of other traditions, the sacred events really happen *then* before the eyes of the faithful. They feel that they are contemporary with happenings which are trans-historic. By being re-enacted the theophany, the past revelation of the divine, becomes living and actual at the present (Eliade 1958: 392).

LOCAL AND INTERNATIONAL RELIGION

Religion expresses much of the outlook of individual tribes and if they are isolated, by geography or history, the religion will tend to be chiefly of local application, though involved in tensions with neighbouring social groups. Developments can be noted in the changes from village to urban life, and from local to national and international reference, with religious beliefs and customs reflecting the changing political and commercial geography.

This process is documented in the literary religions, as in an early part of the Bible when the judge Jephthah told the Ammonites to possess the land which their god Chemosh gave them, while Israel would possess the land which their god Yahweh had allotted to them (Judges 11:24). But later the prophet Isaiah declared that Egypt and Assyria were peoples of God as well as Israel (Isaiah 19:25), and this would bring communication between these nations. No doubt most religions began as tribal ideologies, and some developed beyond that limitation while others did not.

A definition of a Hindu has been that he believes in the Veda religious texts and is born into one of the various castes. The vast complex of what is called Hinduism (a European term) has therefore virtually been confined to the Indian sub-continent, though nowadays there are travelling gurus, and organizations such as the Ramakrishna Mission which seek converts from any race or nationality throughout the world. Among the historical religions, in modern times, the most restricted in membership are the Parsis (Zoroastrians) of Iran and migrants to India, the Jews in Israel but dispersed world-wide, and the adherents of Shinto in Japan.

A major development in religious and social life occurred with the change from merely local application to inter-tribal and international concerns. The first great missionary religion was Buddhism. Originating in the fifth century BC it formally rejected the Hindu distinctions of caste, occupation, status and race. Buddhism rapidly spread throughout India, southwards into South-East Asia and then northwards to Tibet, China, Korea and Japan. It was encouraged by the Buddhist emperor Ashoka but travelled far beyond his domains. Although Buddhism developed divisions of doctrine and practice, and often took on the colours of the countries to which it went, it continued to cherish its historical founder and teachings attributed to him. At the same time it merged more or less closely with other religions that it encountered, whether with animistic customs or with older literate faiths. In China Buddhism became one of three 'ways', alongside Daoism and Confucianism, while most Japanese mingled Buddhism and Shinto. Total Buddhist numbers are difficult to calculate, in the absence of general statistics and a tendency either to count whole populations as Buddhist in South-East Asia, or as communist in China. In Japan recent statistics indicated 90 million adherents of Shinto and 89 million of Buddhism, in a total Japanese population of 123 million (*Far East and Australasia 1992*: 420).

Christianity became the next major movement towards a universal religion, of Jewish origin but using the Greek lingua franca and Roman imperial communications. The New Testament gives accounts of the admission of Gentiles into what had been a Jewish sect, with Peter accepting the Italian Cornelius and Paul insisting on commensalism of Jewish and Gentile converts (Acts 10–12; Galatians 2). Despite persecution by Roman rulers, Christianity became the official religion of the Roman Empire by the fourth century, and travelled beyond it into farther Europe, Asia and Africa in the Middle Ages. In modern times it preceded or accompanied commercial exploration into tropical Africa and the Americas. Variations of belief and practice developed, with geographical factors in distance from Rome influencing the growth of Protestantism in northern Europe and America. Today the Roman Catholic Church is the largest of any international organization, with powerful local and universal communities. Careful estimates gave the numbers of Roman Catholics as 884 million in 1985, rising to 1,170 million by the year 2000. The total adherents of all churches were calculated as 1,548 and 2,020 million respectively (Barrett 1982: 6).

In the seventh century AD the new Semitic religion of Islam broke out of its place of origin in Arabia and was rapidly diffused through the villages and cities of the Near East and North Africa, and accompanying conquering Arab armies were preachers of Islam. Islam laid great stress on the religious community (*umma*), yet while there were military and social pressures for conversion in its new domains, other religious communities were allowed temporal rights as 'people of the covenant' (*dhimmis*). Jews and Christians in particular were recognized minorities, and the Coptic church in Egypt, for example, remains one of the most ancient and strongest *dhimmis* beside the House of Islam.

Islam was the most immediately successful of the great missionary religions and by AD 732, just a century after the death of Muhammad, his religion had crossed eastwards through Persia and India and touched China, while in the West it had entered Europe from North Africa, traversed Spain and reached the heart of France. The battle of Poitiers in 732 saw the Islamic armies finally retreating before the forces of Charles Martel, though they were to remain in Spain till 1492. Numbers of Muslims world-wide were reckoned at 817 million in 1985, set to rise to 1,200 million by 2000.

Though smaller in numbers some later religious movements also claim to be international. The Sikhs, from the fifteenth century, were for long centred in the Punjab but now have important dispersions in Africa and Europe. They number about 16 million, rising to 23 million. The Baha'is, founded in Iran in the nineteenth century, also have centres in many countries. They include prophets of Judaism, Christianity and Islam among their saints, and claim millions of members world-wide.

ORGANIZATION AND INSTITUTIONS

Many people in the Western world nowadays consider religion to be a private affair, and mystics have written of it as 'the flight of the alone to the Alone'. But the social nature of religion has been dominant since earliest times and even solitary individuals follow patterns of faith and practice evolved by the community and use scriptures prepared and preserved by church, mosque or temple.

In early and tribally limited societies religion may appear to be solely a social activity to outside observers. Yet there are religious specialists, priests, prophets, witch doctors, magicians, rainmakers and the like, who practise in private as well as fulfilling functions for and on behalf of society.

The amount of organization varies from place to place, and time to time. In India the four major castes have been traditionally headed by the Brahmin priests, who preserved texts and rituals from the Vedas. Yet there was church–state rivalry, with the warrior-ruler rajahs, and it is significant that two major religions which broke away in India, Jainism and Buddhism, were both led by teachers from the rajah caste who sought followers from all castes and beyond.

In ancient Israel a literate and historical national religion was highly organized with leading priests and Levites, and minute instructions were provided for priestly and social behaviour. The destruction of the temple in Jerusalem by the Romans in AD 70 led to the decline of the priesthood, but Judaism survived under the guidance of rabbis and Pharisees.

Christianity broke through national limitations and spread to many countries with its apostles and evangelists. In time a strongly developed hierarchy and domination of Europe succeeded to the Roman Empire, the Papacy becoming the New Rome. Tensions developed between East and West, and brought a lasting separation of Western Catholics from Eastern Orthodox. Formally the schism came from difference in doctrine, but in practice it was hastened by Western military aggression. The Crusades of the eleventh to the thirteenth centuries were ostensibly against the Turks, to recover the Holy Land from Islam, but the Fourth Crusade in 1204 sacked the Orthodox Christian city of Constantinople and established temporary Latin rule and bishops there. In modern times Western European missions, Catholic and Protestant, took advantage of the changing political geography of the world to establish bases in nearly every land. But Eastern Orthodox churches generally stood aside from this imperial expansion. There are acute tensions today in the former Soviet

134

Union between Orthodox and Catholic (Uniates) who are in communion with Rome, and internally between dominant Orthodox and autocephalous Orthodox, notably in the Ukraine.

Islam also spread with its armies, and in modern times it has profited by new communications to extend its sway, especially in Africa and Indonesia but also in Europe. Different from the individualistic approach of Protestant missions, Muslims generally approached the chiefs and worked downwards, often a more successful tactic.

Religious institutions have been subject to tensions, against the state on the one hand and against internal schisms on the other. Islam, with loose organization, tolerated the emergence of mystical sects with variant practices as long as they did not challenge official doctrine. The only major division was between the majority Sunni, followers of traditional 'custom', and the Shi'a, 'followers' of Ali, cousin of Muhammad. The Shi'a differed over succession to the caliphate and divided into sub-sects.

The place of women in religious institutions has been confused and contradictory. Women usually form the major part of lay followers of religion, and as mothers they teach religious practices to their children. But women have been excluded from the higher ranks of organization. The Buddha was reluctant at first to admit women to the Sangha, the order of monks, and although he eventually gave in there are today no nuns in the Theravada Buddhism of South-East Asia. Similarly Hindu priesthood was for men, often even in the cults of powerful goddesses, yet there appeared women mystics and leaders in the modern Ramakrishna Mission. In Judaism and Christianity men have been dominant, though there were women leaders and apostles in the early church. Modern Protestant churches admit women to all ranks of ministry and leadership, but keen debate and exclusion continue in the priesthood of Orthodox, Roman Catholic, and some Anglican churches. Islam seems a great exception, in having no female object of worship (such as Mary or Kali or Kwanyin), and in excluding women even from public worship. But Muslim women pray at home, and there have been outstanding Islamic women mystics.

As they look to the future religions express doubt and hope. There has been decline in public practice and influence in Western Europe. But in America both Catholic and Protestant are powerful, and in this century there has been massive growth of organized religion in Africa and Latin America. Repression of religion was official and fierce during the seventy years of communism in the Soviet Union, but this has been halted and partly reversed since 1989 and Orthodox Christianity has assumed almost an official role. Islam has also revived in the southern Soviet states, opening new mosques and schools. Buddhism has perhaps suffered most from modern pressures, for by leaving leadership to the monastic orders their destruction in China, and repression in Tibet, left lay followers with little guidance.

Religious literature contains many prophecies of the decline of faith in the future, which seem to be borne out by the immoral or amoral activities of a secularized world. However, there are also firm beliefs in the revival of religion, at the end of the age or in the cycle of time. Some anticipate a Messianic second coming, a new Mahdi, a Buddha-to-come, and hopes will run high at the end of this present millennium. More valuable are the ideals of a better world, a kingdom of God, a wider use of the world's resources as the creation of God for the benefit of humanity. Here religion can provide the inspiration and guidance without which hopes may again be disappointed.

REFERENCES

Barrett, D.B. (ed.) (1982) *World Christian Encyclopedia*, Nairobi: Oxford University Press.
Eliade, M. (1958) *Patterns in Comparative Religion*, London: Sheed & Ward.
Eliade, M. (1973) *Australian Religions*, Ithaca: Cornell University Press.
The Far East and Australasia 1992, London: Europa Publications Limited.
Pritchard, J.B. (1950) *Ancient Near Eastern Texts* (3rd edn enlarged 1969), Princeton: Princeton University Press.
The Revised English Bible (1989) Oxford: Oxford University Press.
Robinson, J.A.T. (1963) *Honest to God*, London: Student Christian Movement Press.
Smith, W.C. (1973) *The Meaning and End of Religion*, London: SPCK.
Tillich, P. (1962) *The Shaking of the Foundations*, Harmondsworth: Penguin Books.

FURTHER READING

Arberry, A.J. (1983) *The Koran Interpreted* (tr.), Oxford: Oxford University Press.
Barker, E.V. (1989) *New Religious Movements*, Norwich: HMSO.
Bettenson, H. (1977) *Documents of the Christian Church* (tr.), London: Oxford University Press.
Boyce, M. (1987) *Zoroastrians*, London: Routledge & Kegan Paul.
Brown, A. (ed.) (1986) *Festivals in World Religions*, London: Longman.
Cole, W.O. and Sambhi, P.S. (1978) *The Sikhs*, London: Routledge & Kegan Paul.
Conze, E. (1979) *Buddhist Scriptures* (tr.), Harmondsworth: Penguin.
Evans-Pritchard, E.E. (1965) *Theories of Primitive Religion*, Oxford: Clarendon Press.
Goldberg, D.J. and Rayner, J.D. (1987) *The Jewish People*, Harmondsworth: Penguin.
Hinnells, J.R. (ed.) (1984) *A Handbook of Living Religions*, Harmondsworth: Penguin.
Hinnells, J.R. (ed.) (1984) *The Penguin Dictionary of Religions*, Harmondsworth: Penguin.
Parrinder, G. (ed.) (1990) *A Dictionary of Religious and Spiritual Quotations*, London: Routledge.
Ranke-Heinemann, U. (1990) *Eunuchs for the Kingdom of Heaven*, New York: Doubleday.
Sutherland, S., Houlden, L., Clarke, P. and Hardy, F. (1988) *The World's Religions*, London: Routledge.
Waley, A. (1977) *The Way and its Power* (Dao De Jing tr.), London: Allen & Unwin.
Zaehner, R.C. (1966) *Hindu Scriptures* (tr.), London: Dent.

7

THE MODIFICATION OF THE EARTH BY HUMANS IN PRE-INDUSTRIAL TIMES

I.G. Simmons

THE FIRES OF CULTURE

During the period from 12,000 to 200 BP, the world, including its human economies, was largely solar powered. As always, the natural ecosystems fixed the radiant energy of the sun and cascaded it through a series of organisms. Some of these organisms ate plants, others animals, some subsisted on dead organic material; a few might utilize all of these, like the chimpanzee. The systems modified by humans for their purposes were also solar powered: no matter how the flows of energy and matter were diverted in the creation of near-natural, semi-natural or cultural ecosystems, the incidence of solar energy was the limiting factor. Such confines were only broken when all these systems could be subsidized by an extra flow of energy which came from the fossil hydrocarbons (in effect stores of photosynthesis from earlier geological eras), which is the ecological basis of the Industrial Revolution irreversibly launched and well into the fairway by AD 1800 (Sieferle 1990).

It is unreasonable to suggest that there was no use of the fossil fuels before 1800, but where it occurred, it was sporadic and underlay no attempt to change the whole basis of material production: local or regional convenience was the main spur to usage (Adams 1982). Thus we find outcrops of coal in northern England being used to warm the soldiers stationed on Hadrian's Wall in the early centuries BC. In medieval times in the same region, coal was used to fire salt pans by the River Tyne, and a vigorous trade with London developed which often enough resulted in City ordinances to curtail its use because of the smoke created. Elsewhere in the world, oil seeps were used to caulk ships and provide lighting (an interesting precursor of the first reasons for the search for petroleum in the West which was to replace whale oil for lighting), and in China there are records of towns being lit at night by natural gas flares made possible by the piping of the gases from their natural seepages along the hollows of bamboos. In Burma, drilling was undertaken to get at such gas pockets (Simmons 1989). But none of this comes anywhere near the magnitudes of use prevalent during the nineteenth and, especially, the twentieth centuries. Instead, recently synthesized biological materials such as wood (both plain and as charcoal) and dung, together with edible plants and animals, were the fuels for human economies in our 12,000-year period.

One procedure which links both pre-industrial and industrial eras, however, was the key role of fire in releasing the heat from combustible materials for use in myriad purposes; it forms a further nexus between the economy of nature and that of humans with the use of fire in the landscape to manipulate vegetative and animal communities. In the time between the Upper Palaeolithic and the generation of electricity from hydropower and nuclear fission, therefore, fire has been a basic element, either directly or indirectly, in the relations of humanity and its environment. As we shall see from the material on hunter-gatherers, most terrestrial vegetation will burn at some stage in its yearly cycle, and many human communities right through to the present have taken advantage of this to produce desired plant communities or as a way of exploiting animals.

NATURAL CHANGES

It must be stressed that during this period there were natural changes with which human-induced alterations interacted (see Chapter 2). The period 9,000–2,550 BP, for example, saw the establishment of a 'hypsithermal' climate, with average summer temperatures 2°C higher than those of today in Europe, for instance (Mannion 1991: 14–22). There are findings from New Guinea which show that in 14,000 BP the tropical forests only occupied about 75 per cent of their present area. These are examples of the outcome of phenomena like the migration of plant and animal species and the consequent assembly of a variety of natural ecosystems. Sea-levels changed with the interaction of melting ice, thermal expansion of sea water and isostatic recovery, and the dislocating effects of storms, floods, earthquakes and volcanic eruptions were always possible. But in this context we need also to remind ourselves that in many parts of the world, the adjustment of natural phenomena to the changes of the postglacial period often took place during the presence of human communities so a contingent potential (though doubtless not always realized) to affect ensuing ecosystems was present.

HUNTER-GATHERERS

As we would expect from the above discussion, the energy sources of this group of people are those of recent solar origin in chemical form. This energy is not likely to be more than a few years old, even if it comes from a perennial plant or an animal of mature age. To get this energy, humans expend their own in the form of activities such as running and stalking, digging pits, setting nets, using spears, blowpipes and the bow, digging roots, collecting fruits and seeds. What is critical is that every collector is responsible for a surplus of energy, first to feed those who cannot do it for themselves like the very young and the elderly, and second to build a reserve against a season of shortage where possible. A key aid in this equation was the discovery of the control of fire, since this helped hunters to direct game towards waiting men; but beyond that it soon must have been apparent that vegetation could be manipulated by repeated firing to the point where the new plant community was more productive of human food than the old, either because the plants were different or because it harboured more and accessible game. The energy balance of the whole way of life could also be favourably affected by the use of the dog in hunting, since it could track a quarry effectively or drive beasts towards hunters yet needed only a little energy that had to be diverted from the humans. Small wonder that it became the first domesticated animal.

The distribution of hunter-gatherers occupied a gradient from high latitudes where hunting of animals was dominant, through mid-latitudes where fish were very often important, to the tropical forests where the gathering of plant food was paramount. In all groups, animals seemed to provide at least 20 per cent of the diet (Megaw 1977). Plants must therefore have been key foods in all except extreme environments like the Arctic; the hunting way of life seems to have been founded upon the consumption of as much plant food as necessary and as much meat and fish as feasible (Hayden 1981).

The interaction with nature of the pre-nineteenth-century hunter-gatherer group then involved the collection of such plant and animal material as was seasonally available within the territory of a particular group. The pulsed availability of foods, together with the kinship and other social arrangements of the people, might make it desirable or even essential for there to be a yearly movement round a territory where they had rights of access to the resources. In areas where resources are not scarce then hunter-gatherers who maintain relatively low population levels may simply live off the usufruct of nature, taking what they need without any necessity to worry that sufficient will not be there in the immediate future. Thus the !Kung bushmen of the Kalahari, studied in the 1960s (when there were perhaps 2,000 bushmen living as hunter-gatherers), appeared to be dependent upon the nut of the mongongo tree as a staple food even though their total dietary spectrum was very wide (Lee 1979). But they had no need to plant mongongo trees, or ensure their pollination or any such manipulations: in season, enough nuts could be gathered to feed everybody with a decent diet of calories and protein. Their neighbours the G/wi bushmen relied on the *tsama* melon for part of their diet and its abundance was variable: thus the G/wi had to travel further in a bad melon year to get their required quantity. Observations of the Xade San in the same region (Lee and DeVore 1976) showed that twenty-two animal groups were eaten by way of added variety, from giraffes (136 kg/yr) to termites (8 litres/ yr).

This rather romantic image of children of nature did not hold everywhere. There were environments where the people felt that nature would not always provide unless some action was taken by them. For some this was mainly of a nonmaterial kind in the shape of practices designed to propitiate the gods into maintaining food supplies; for others it was a case of the gods helping those who helped themselves, and so management of plant and animal populations was practised. In the case of plants, this might mean occasionally diverting some water over a stand of wild grasses, for example, to ensure their seed crop was heavy, or perhaps transplanting some wild yams into an easily accessible location or even one inaccessible to natural predators. Animal populations might be protected from the kill in some areas so that depleted numbers might expand again or there might be a prohibition on the killing of gravid females. In semiarid environments like part of Australia, canals might be dug in and near swamps to keep plenty of the right habitat for eels (Lourandos 1980). The fish populations of tropical lagoons and reefs were kept stable under traditional cultures by the appointment of fish wardens who controlled the volume, species and season of catch.

Not all hunter-gatherers produced ecological change, but those who did contrived to do so in both temporary and permanent ways. Impermanent alterations often centred upon the impact of a hunting group upon an animal population: this could be severe but yet not so damaging to the reproductive capacity of the beasts that their numbers could not recover. Buffalo hunting on the High Plains of the USA before the advent of Europeans is one example. Numerous archaeological excavations have shown that herds of buffalo were driven over cliffs, into box-canyons or into dune slacks, and then slaughtered wholesale (Frison 1978). Yet long-term diminution of the animals' availability was in general avoided by concentrating upon herds of males and so letting the females carry their young unmolested. Analogous practices were carried on further north, where boreal forest native people developed traditions of resting particular areas within their territories so that for example elk and beaver populations might recover from heavy hunting (Tanner 1979). The occasional use of fire as a hunting aid (i.e. to drive out animals from cover rather than to try to change the vegetation) might well not change the ecology for more than a season or two; recovery of the vegetation to the original structure would have been likely even if the species composition was in some way altered.

Of considerable interest are the cases where hunter-gatherers have affected their environments permanently. In these instances we often find that fire is one of the most important tools, and the case of the Aboriginal inhabitants of Australia exemplifies this (Latz and Griffin 1978; Head 1989). In the interior, fire was regularly used as a hunting aid since it flushed many animals out of the bush and from underground, with the result that the vegetation types encountered by Europeans were in fact fire-adapted ecosystems produced by human agency. In the north, women regularly fired the vegetation containing a particular cycad tree: it then yielded more fruit and produced them more or less simultaneously, which is always an advantage to people on the move. In North America, the burning habits of the native populations of the forest–grassland edge (again as a hunting aid, including keeping down the quantity of undershrubs that prevented a hunter getting a clear sightline for his arrows) kept a mosaic of open woodland and grassy glades; when the Indians were extirpated, the forest rapidly reclaimed the land. There is an instance of permanent change from the uplands of England and Wales during the period of the last hunter-gatherers, the later Mesolithic of about 9,000–5,500 BP. The present-day moorlands like the Pennines, Dartmoor and the North Yorkshire Moors have yielded palaeoecological evidence to show that in those times such uplands were largely forested. Yet among the woodlands there were clearings which seem to have been maintained by fire and their frequency in time and space is such that natural causes are unlikely. In some places, these openings disappeared when agriculture started, which suggests that burning was an integral part of the hunter-gatherers' way of life. But on other sites, the removal of the trees allowed the soils to become waterlogged (since the trees acted as large water-pumps) and peat grew. In favourable places, this grew to depths of 2–3 metres and formed a blanket over the land, in which condition (though often now eroding) it can still be seen today: an example of a landscape element formed by hunter-gatherers (Simmons 1990).

A further set of examples of more permanent change can often be seen when hunter-gatherers came into contact with agricultural or even industrial populations. The boreal forest again provides good examples, for many areas were almost entirely depleted of fur-bearing animals by indigenous trappers who sold their catch to the Hudson's Bay Company and similar agencies (Martin 1978). The then insatiable market for furs such as beaver wiped out many populations in spite of the Company's efforts to install rotational trapping schemes. Had fashions not changed, the beaver might have disappeared from most of North America, as it did from many areas to the south of present-day Canada.

One mediation between hunter-gatherer life and nature is through technology, which has developed a number of forms during the millennia. At the end of the Pleistocene such groups had access only to organic materials and to stone, but they quickly took up metal whenever it became available to them. In recent times, relict hunter groups have rarely been reluctant to absorb the products of the Industrial Revolution when these came their way. The traditional technology has therefore centred around wooden tools for grubbing up plants, baskets or slings in which to bring plant materials back to camp, together with hunting aids such as the spear, bow and arrow, blowpipe, slingstick, woven nets, and poisons. The rifle, the outboard motor and the snowmobile have slipped into this repertoire with ease.

AGRICULTURE

The success of agriculture has been so great that the human population grew from perhaps 170 million in the first century AD (approximately the present population of Indonesia), to 957 million in 1800, which is less than today's total for China. The energy relations of agriculture move towards a more concentrated production per unit area and per unit time than in food collecting. Hence, more people can be fed off a smaller area. A lot of human energy input is required into manifold tasks, and subsistence agriculture is often an unremitting existence. But once above a purely subsistence level it often produces surpluses that permit the differentiation of a wide variety of human occupations.

The beginnings of agriculture are still only partly known. It is likely, for example, that there existed periods of interaction between human communities and biota that are archaeologically invisible (Hillman and Davies 1990), and so only when domestication of plants and animals is fully fledged can we detect it. Such times of early husbandry of plants might involve their cultivation but not the conscious selection of seed, for instance; with animals, following a herd of wild creatures might develop into a form of loose herding with perhaps some control over the herd at one season. From there to controlling the breeding of the beasts is only a short step.

A number of places in the world were important in the emergence of different agricultural systems although they do not encompass the whole spectrum of domesticates. In the hill-lands of western Asia we know that growing cereals such as wheat and barley around permanent villages grew up in the period either side of 9,000 BP, and that these people kept domesticated animals as well: sheep, cattle, goats and pigs were almost certainly the earliest tamed animals after the dog (Zohary 1986; Harris 1989). Perhaps about 6,000 BP, nomadic pastoralism based on herds of domesticated animals became fully fledged. In South-East Asia, rice was domesticated around the same time as the western Asian cereals: millet cultivation arose in northern China in the same era. In the 6,000–4,000 BP period, New World agriculture arose on the basis of maize, potato, beans and squashes; here Mesoamerica and the Andes were the most important zones. From these origins, the various types of agricultural system spread and developed into most parts of the world, often replacing hunting and gathering on the way, though unable at that time to dispossess it from the more marginal places such as the very dry, the very cold, and the remotest tropical forests.

The reasons for the origins of the domestication of plants and animals (a phase generally labelled the 'Neolithic Revolution') are much discussed and are probably susceptible to no single answer: from place to place and time to time it is conceivable that one process was more important than the others. For some time during the twentieth century, scholars thought that climate must have been the controlling factor, with a supposed period of aridity compelling humans to move into oases where the sheer pressure of space forced the development of an intensive form of food production. Closer palaeoecological studies, however, failed to find evidence for the dry periods. 'Natural' explanations have also been put forward for animal domestication in terms of species such as the pig offering themselves, so to speak, for taming via their preference for scavenging in human settlements. This route would be favoured by a wild behavioural repertoire which included docility and curiosity; in effect the animals struck a bargain, trading security and comfort for human power over life and death (Clutton-Brock 1987). On the other hand, a largely cultural explanation for animal taming rests on the use of creatures like cattle with lunate horns in sacrifices to a moon goddess. In the case of plants, explanations tend to favour cultural practices of seed processing after harvest which bring about the selection of seeds which breed true for certain characteristics, a process which has its analogy with animals in letting only selected males breed and castrating the others. Yet another basically cultural explanation sees the whole process as an intensification of food production driven essentially by rising human populations which were not kept stable even by postulated Palaeolithic population control practices such as female infanticide. The concentration of humans is central to the thesis that hunter-gatherers may have lived in quasi-urban agglomerations and traded in commodities such as live meat (as well as consumer durables such as pigments and stone) which might then be bred much nearer the settlement, and thus domestication began.

The range of choice of hypothesis is wide and it seems unlikely that one will satisfy all conditions and be consonant with all the disparate sources of evidence. The preference of Harlan (1986: 34) for a no-model model, which 'leaves room for whole arrays of motives, actions, practices and evolutionary processes' seems at present very prudent. However, Harris (1989: 11) finds it possible to talk of 'an evolutionary continuum of people–plant interaction' in which there is increasing human energy input per unit of used land, which is paralleled by the levels of plant exploitative activity, of ecological effects, and of intensity of food-yielding system. Only after a number of early phases have been traversed is true domestication and farming established (Figure 7.1).

Though many accounts focus on the origins of domestication, it is worth repeating that it is a continuous process. Many later cultures have tried to enlarge the spectrum of tamed biota and some have succeeded: oats and rye perhaps 5,000 years later than wheat and barley, the dormouse during Roman times, and attempts to do the same to the musk ox today. The development of scientific plant and animal breeding made possible by the discovery of the nature of genetics, and today's genetic engineering, is to some extent the continuation of the Neolithic by other means. Only very recently, after all, have fisheries made any transition from hunting and herding to farming.

The biology of domestication is well documented. The intention is to replace natural selection with human-directed selection, and so the full force of culture may come into play: preference for fluffy-cooking or soggy-cooking rice, for example, or for toy poodles as against Rottweilers. In plants, one aim is usually to have all the edible and harvested parts ripen simultaneously. Once selected

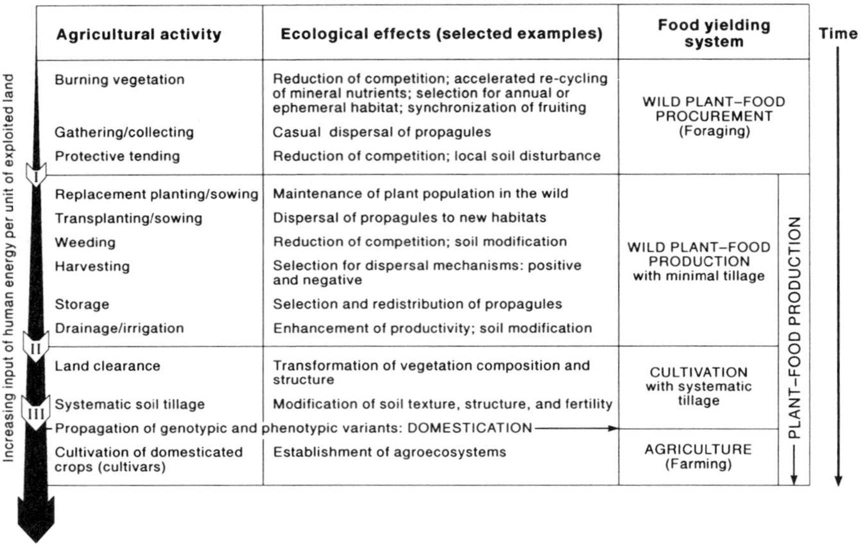

Agricultural activity	Ecological effects (selected examples)	Food yielding system	Time
Burning vegetation	Reduction of competition; accelerated re-cycling of mineral nutrients; selection for annual or ephemeral habitat; synchronization of fruiting	WILD PLANT–FOOD PROCUREMENT (Foraging)	
Gathering/collecting	Casual dispersal of propagules		
Protective tending	Reduction of competition; local soil disturbance		
Replacement planting/sowing	Maintenance of plant population in the wild		
Transplanting/sowing	Dispersal of propagules to new habitats		
Weeding	Reduction of competition; soil modification	WILD PLANT–FOOD PRODUCTION with minimal tillage	
Harvesting	Selection for dispersal mechanisms: positive and negative		
Storage	Selection and redistribution of propagules		
Drainage/irrigation	Enhancement of productivity; soil modification		
Land clearance	Transformation of vegetation composition and structure	CULTIVATION with systematic tillage	
Systematic soil tillage	Modification of soil texture, structure, and fertility		
Propagation of genotypic and phenotypic variants: DOMESTICATION ⟶			
Cultivation of domesticated crops (cultivars)	Establishment of agroecosystems	AGRICULTURE (Farming)	

Vertical left axis: Increasing input of human energy per unit of exploited land

Vertical right axis: PLANT–FOOD PRODUCTION

Figure 7.1 Stages of environmental manipulation on the route to agriculture. After Harris (1989).

seed is used, there often occurs a doubling or further multiplication of the chromosome numbers of the plant, which produces an increase in its size and robustness: the volume of the fruit of the domesticated red pepper is some 500 times that of its wild ancestors, and maize cobs are ten times the length of early domesticated varieties (Bender 1975). Defences against grazing in the wild, such as bitterness, toxicity and thorns are often lost, as is the ability to disseminate seed. In modern maize the husks make harvesting easy, but since the silks must protrude beyond them the seeds cannot be dispersed by natural means.

Many animal species, by contrast, produce smaller individuals under domestication (cattle, sheep, goat, pig). A few, like the horse, are rather bigger, though the Shetland pony shows the variety that can be achieved. Size difference between males and females is frequently reduced and adults often retain features of juvenile stages of development, a process known as neotony. It appears in the shape of the skull and jaws of dogs, pigs and cattle, the short faces and small horns of some cattle, in curly rather than straight hair and in pied rather than monotone coats (Mason 1984). Lengthy ears are common features of domestication (the horse is again an exception), and a greater proportion of fat under the skin and through the muscle has been favoured until very recently. A smaller brain is common. Castration is not of course inherited but it does produce docility, more fat and longer bones.

The period of pre-industrial agriculture was the greatest period of environmental change at human hands that the world had experienced. In the course of garnering crops from plants and animals, humans altered the pre-existing ecosystems both deliberately and accidentally, suddenly and gradually, temporarily and permanently. Agriculture (including pastoralism) is central because it feeds people. But farming-based societies also reached into their environments for a variety of other purposes: for materials to use in construction as well as in food production, in cities and as fuel. They also sought pleasure in parks and gardens, and like most humans before and since engaged in warfare which was sometimes at the expense of the environment as well as of themselves.

At the heart of this *genre de vie* was the evolution of a number of agricultural systems. Table 7.1 gives a typology which will serve for the period under discussion provided we remember that such systems are always undergoing metamorphosis. It is possible to divide them first into shifting and permanent systems, though the former may have been altered into the latter and they in turn may not have been as permanent as their operators had wished. The shifting systems more or less obliterated nature by turning a patch of forest or savannah into cultivated field, though care was taken in Africa, for instance, to 'cultivate' trees for wood, shade and possibly fruits. But when fertility declined or weeds became too pervasive, then the patch was abandoned and another took its place. The colonization by wild vegetation then to some extent restored the nutrient levels. By contrast, permanent agriculture removed the natural landscape and kept it erased, though a matrix of wilder systems might have remained. The same was true of land reclaimed from salt marsh or wetland: drainage works or sea-walls prevented (it was hoped) the re-reclamation of the land by wild vegetation or the sea, just as fences kept land sequestered from wild animals or neighbours. Terracing is a very widely spread phenomenon that may perform one or all of a number of functions: retain water (as in *padi*); retain soil (as on Mayan or Mediterranean hillslopes); and mark out ownership. Common to many systems were the environmental alterations used in pursuit of sustained nutrient cycles (Cooter 1978). Holes were dug for calcareous marls, for example: seaweed and shells gathered on coasts; twigs and leaves swept onto fields or fed into domestic beasts which were then depastured on the fields; water management systems evolved which brought nutrients to fields as in *padi* but also in water meadows and warped fields in Europe. A few agricultural practices, on the other hand, seem to have been unique to restricted areas as with the pit growth of tubers in Oceania.

One result of the era of transoceanic contacts that resulted from the European voyages of discovery and colonization was an intensification of the transfer of species into new habitats and of the extinction of rare and endemic species. The bringing of sheep and cattle to the Americas, and of species like maize, tobacco, the potato and tomato to the rest of the world, are examples. Many a rat jumped ship onto an oceanic island which was host to species of flightless birds and ground-nesting species; pigs and goats left on islands ate their way through

Table 7.1 Pre-industrial agricultural systems and their environmental impact

Type characteristics	Distribution	P/T	Crops	Environmental impact
Irrigation	Nile, Mesopotamia and other Old World rivers, mainland and insular E and SE Asia; Middle East and Med under Muslim influence	P (but occasionally abandoned)	Wheat, barley, rice, millet, sesame. Cattle often associated	Clearance of savannah, swamps, forests and conversion of grasslands. Terracing and slope conversion. Salinization and waterlogging of soils outside E and SE Asia. ± complete transformation of landscape
Dry farming	Early original type in SW Asian hill-lands. As shifting cult in savannah zone of Africa; soy and millet in China	Both	Wheat, barley, pulses, sorghum, millet, mulberry (China); pigs, cattle, sheep and goats	Clearance of oak–pistachio woods (and ? use of fire) in SW Asia. Plough comes into use in all of Old World: soil changes and soil erosion with valley aggradation and floods. Latosols prone to erosion
Mediterranean	Mediterranean littoral	P	Wheat, barley, olive, grape, fig, goats	Clearance of forest and scrub for fields. Terracing. Soil loss especially on limestones, with subsequent effects on rivers, deltas in lagoon formation at coasts. Bare limestone or semiarid steppe may result from intensive cultivation
Permanent temperate	N and W Europe, Ukraine, N America	P	Wheat, barley, rye, oats, maize, pigs, cattle, sheep, pulses	Clearance of deciduous forests for fields. Stable system depends on nutrient input from fallowing or manuring. Soil loss not absent, but soil type and climate reduces rates

Tropical vegeculture	Mainland and insular Asia, tropical Africa and S America	Both	Taro, yam, bananas, coconut, manioc, pigs, poultry	If shifting then nutrient cycles restored during fallow period. Domestic animals may be absent (e.g. in S America). Cultivation by axe and digging-stick rather than plough
Root cultivation	Various environments in Meso- and S America	Both	Manioc, yams, potato	Neither plough not domestic animals in crop system. Encroachment onto steep slopes encourages rapid soil erosion
Drainage agriculture	Typical of Maya and Aztec areas; peat and other wetlands world-wide	P	Maize, beans, cotton, tomato, squash in Americas, cereals elsewhere, perhaps grassland	Islands of drier soil and mud developed in *chinampa* system; permanent drainage may shrink soils and make drainage more difficult – need for e.g. windmill-powered pumping. Loss of wild food resources like fish and waterfowl
Pastoralism	Marginal environments ± everywhere; interstitial between croplands in e.g. India; may be seasonal (transhumance)	P	Horse, camel, yak, llama, cattle, sheep, goats (*not* pig, water buffalo, for example)	Gradual change of vegetation under grazing pressure: more xeric species and those with e.g. thorns and toxic parts survive at expense of the palatable; waterholes may have severe soil puddling and wind erosion problems. May contribute silt to valley irrigation systems

Note: P = permanent; T = temporary.

much vegetation in a series of population explosions (Crosby 1972, 1986). Much of the habitat on a protected area like the Galapagos Islands today is semi-natural because of the imported species of animals that are now naturalized there. All these processes were to intensify once again in the nineteenth century, after our chosen time period.

Agriculture (including pastoralism) is central because it feeds people. Thus most societies make attempts to extend their area, either to feed more people or the better to feed some of them. But pre-industrial societies also reached into their environments for a variety of other purposes: for materials to use in construction or in farming or in cities and as fuel, for example. They also sought pleasure in parks and gardens, and like most humans before and since engaged in warfare. These other activities can be simply classified into production and pleasure, though some, like gardens, might partake of both, and others like warfare, although productive and pleasurable for some, inflict damage on people and environments alike.

Forests have always had two main environmental roles in human societies after the dissemination of agriculture. The first is of land banks: a reservoir of fertile land which might be eventually converted to crop production. The second has been for their immediate products. Wild foods to supplement basic diets are usually present but above all there is wood. Most pre-industrial societies depend upon wood for all their fuel requirements (domestic and industrial), for much construction (shipbuilding, scaffolding, the frames of buildings) and for a host of small things like tools and animal fodder. Woods can be managed to supply these materials: for example, coppicing and pollarding provide a supply of poles (for fencing or for charcoal making); shredding is a good source of animal fodder (Rackham 1980). Deciduous trees growing in semi-open conditions may develop crooked branches which are essential for ship timbers; by contrast, dense stands of conifers produce the best masts (Albion 1926; Bamford 1956). Practices like coppicing were allied specifically to industrial processes (such as iron-making and glass manufacture) when they produced large quantities of standardized poles for conversion into charcoal, which is higher in energy content per unit weight than even dry wood. Demands for charcoal in pre-industrial Europe, for example, were so high that sustained yield practices had to be adopted over large areas.

Energy needs were also addressed through peat digging. Although peat when just dug is about 90 per cent water, when air-dried it has a high energy content. The environmental impact of the extraction of large quantities can best be seen in parts of the Netherlands like Loosdrecht and the Norfolk Broads in England, where large peat diggings combined with a high water table have left strings of lakes, their origin being given away by linear peat baulks breaking the water surface at regular intervals. Pre-nineteenth-century extraction of coal from seams near the surface may leave a legacy of dimples from the bell-shaped pits that were dug: even modern land uses may not eliminate these depressions, just as some prehistoric mining (of iron ore or flint for instance) leaves the sunken remains of shafts in among piles of waste material (Shepherd 1980).

The harnessing of energy from wind and water were pre-industrial developments; in the case of windmills the impact upon nature was very low. Waterpower, on the other hand, might necessitate the control of river flow in order to provide a good head of water even in dry weather (Jones 1979). Thus leats, sluices and weirs were all known in Europe and China in classical times. The aqueducts of the Roman Mediterranean are well-known structures devoted to bringing clean water to towns by extraction from rivers higher in their courses, and they might power a mill as well. In medieval England, tidal mills were also used to grind corn. Water control was also brought to something approaching a fine art for pleasure, as will be discussed below.

The demands for production extended also to the seas. Artisanal fisheries and whale hunting have very long histories: the bowhead whale of the Arctic, for example, is a traditional Inuit prey. In early modern times, cod fishermen ventured to Newfoundland from France, England and Portugal, and by the eighteenth century most of the world's oceans had been scoured by explorers who took a strong interest in the whale populations which at that time were more or less ubiquitous. Whaling vessels have been described as the oil tankers of the pre-petroleum world (by Jones 1981) and concentrated hunting went on in the Atlantic and the Arctic from no later than AD 1150. Processing the whales at sea from the fifteenth century brought about truly pelagic whaling. Many populations could not sustain the kill rates to which they were subject and so by 1880 most whaling regions had shrunk in size and yield even though the technology employed was pre-industrial. A few stocks, like the Atlantic right whale, were scarce as early as the seventeenth century (Cushing 1988). Later, industrialized catching has of course reduced whale populations to very small fractions of their pre-exploitation size.

If we exempt fanatics like Captain Queeg in *Moby Dick*, probably nobody has hunted whales for pleasure. Yet on land, one of the greatest pleasures for men (and some women) down the ages has been the hunting of wild birds and mammals for sport. The kitchen may have subsequently benefited as well, but the hedonism of the chase was central. Where urban populations might not themselves take part, then vicarious entertainments might cater to the same desires, as in the great circuses of Rome, with fights between wild animals as well as between humans; the wild fauna of much of North Africa and Asia Minor was ravaged to supply the coliseums of the Roman Empire: Trajan's conquest of Dacia was celebrated with the slaughter of 11,000 wild animals. Nero even exhibited polar bears catching seals (Hughes 1975).

Actual hunting, however, has usually been an aristocratic pastime. To secure it, areas of land have been set aside in many cultures to protect the hunted species, and communities have furnished officers to secure the animals against poaching by the common people or by land-use practices which might interfere with the feeding or breeding of the protected taxa. So in ancient India, in the Andes during Aztec time, in dynastic China and in medieval Europe alike, hunting parks up to 4,080 ha in area are found, separated from the land around them by various combinations of walls, ditches or fences, and with a set of laws different from the surrounding areas (Brandon 1971). The desired end was common to all: the provision of a good day's hunting to some lord or other. Where large areas could be set aside for hunting, then actual enclosure might be impossible. In that case, enactment of special laws meant that, for example, depasturing of domestic beasts was forbidden in case they competed for food resources; dog size might be controlled so that the quarry of the chase might not be hunted down illegally. The Royal Forests of England between the Norman Conquest and the Commonwealth are good examples: many were moorland and not forest, but peasants were subject to draconian laws and if they wanted to extend their cropland, or pasture cattle and sheep, they had to pay the monarch a large sum for the privilege. In almost every place, destruction of the natural predators of the favoured species was enjoined upon everybody in the neighbourhood. In Europe, at least, this must have resulted in declines in wolf numbers.

The garden, too, has been common to many cultures throughout the world, though flourishing most perhaps in Islamic cultures, in ancient Egypt and in southern and western Europe (Hyams 1971; Berrall 1978). The garden is an enclosed space which yields both pleasure and utility. The pleasure may come from trees which give shade, many species of plant with attractive flowers and scents, decorative animals (the peacock is a good instance), a congenial layout tailored to the fashion of the time, added structures like grottoes, statuary or glass-houses and, very frequently indeed, controlled water as pools, canals and fountains. The utility might combine well with the pleasure: the trees might bear fruits after the blossom; garden walls retain heat that ripens fruit in relatively cool climates and conservatories enable the cultivation of exotic flowers and fruits as well as providing conditions for year-round romances. Water led in for fountains might in due course water vegetables, and pools might harbour edible fish as well as lilies and the lotus. Small wonder that the ancient Persian word for such a garden came into English (via Greek) as 'paradise'. Opinions differ as to the apogee of this delectable combination, but somewhere near the top must be the gardens constructed by the Islamic rulers of Spain in Andalusia – above all in Granada. In terms of spatial impact, the reaction to this type of garden was perhaps greatest in the landscape gardens of eighteenth-century Britain where large parks were often laid out to look 'natural', with winding water, irregular clumps of trees and an absence of visible boundaries. These parks were in part a reaction to other massive impacts such as the very largely formal park at Versailles, where the canals, fountains and wooded compartments clearly all know their place. The exception is a mock rustic hamlet in the grounds of the Petit Trianon where Marie Antoinette, influenced by Rousseau, played at being a dairy maid.

Never far from such pleasures in past times (as in our own) was the threat of war, and this too had its environmental impacts (Cowdray 1983; Westing

1986). Forests were especially vulnerable since they could hide enemy troops or guerrillas and so fire was used to smoke out the men and simultaneously to deprive them of any future cover. After battles victors might add environmental pillage to the more usual spoilings by imitating the Romans who sowed the fields around Carthage with salt (Hughes 1975) and who in conquering Britain, says Tacitus, 'made a wasteland and called it peace'. For a grisly alternative, there is the scene of the defeat of the Teutons at Massalia where, according to Plutarch, the inhabitants fenced their fields with human bones, and the blood of the injured and slain produced 'an exceeding great harvest in after years'.

THE PRE-INDUSTRIAL CITY

In environmental terms, the city represents a comprehensive transformation of a local environment, though some pre-existing features like the shape of the land, a river and some trees may remain. Internally, even the smaller pre-industrial cities developed their own faunal and floral assemblages. Non-cultivated plants are mostly limited to those that flourish in odd corners and waste spaces and on structures such as walls, fences and roofs. Amenity planting may however increase the number of trees. Animals which flourish are scavengers like feral dogs and cats, and birds like the red kite, which were plentiful in most streets of medieval and early modern Europe. Rats, too, adapted well to such habitats and carried some of the parasites whose populations were to benefit from urbanism.

Some of the pre-industrial cities were of no mean size: in 1500, for instance, the four largest European cities had populations between 100,000 and 200,000. Such cities as always transformed land to a condition of higher energy throughput, and acted as concentrators of materials. Thus they often had industries which necessitated the gathering of fuel from a large area or the diversion of watercourses to power mills. In turn some of these manufactories produced wastes that contaminated rivers and estuaries: tanning, for example, is notorious in this respect. In ancient Rome, country-dwellers remarked upon the smoke and dust which obscured the Sun and made them pale-skinned. Such contaminations of the air were multiplied many-fold when coal was used to fire boilers or heat rooms. In the summer of 1257, Queen Eleanor was forced to cut short a visit to Nottingham because of the stench of coal smoke, and in 1285 a commission was set up to look at the soot and sulphur residues from coal which were causing problems in London. Queen Elizabeth I later prohibited the use of coal in brew-houses within a mile of the Court. However, it seems likely that the lime industry was the biggest source of smoke from coal-burning (Brimblecombe 1987). Getting rid of the contents of cesspits and privies was not always taken very seriously and so disease always had a set of ready seedbeds; many authors have commented that the one environmental feature of medieval cities that we cannot properly reconstruct is the smell. Not all was bad, of course. In the Paris of the 1870s when 96,000 horses still powered the city's transport systems, the manure fertilized 7,800 ha of market gardens where the temperature from the fermenting matter added 6 per cent to the solar flux, and Paris had a flourishing export trade in salad crops (Stanhill 1977).

The urban population produced a call for food most of which had to be met from local sources, although salted and smoked meat and fish might be traded long distances. Local land use was therefore intensified by such demands though there might be a contrary trend in the retention of pasture for horses; the richer the city, the more horses. The rich, too, inevitably wanted their country houses and estates outside but not too far from the town, as in the villas to the north of Florence or inland from Venice. Competition for the land immediately outside any walls increased when new ways of fortification required a bare area or *glacis* to give the gunners a clear fire zone and a cover-bereft area to discourage sappers. Walls and moats, naturally enough, took up space while they were needed but once outdated have provided a land bank which has furnished many a city with a ring road (as in Vienna or Avignon) or a ring of parkland, as at Lucca in Tuscany.

OVERVIEW

One fascinating question is whether in the period 14,000–200 BP, nature or humanity has been the dominating influence. This can be analysed in terms of scale. At a cosmic level, the laws of thermodynamics and gravity still determined certain aspects of all actions on this planet, as presumably they have since the universe came into being, and always will. Globally, the withdrawal of ice at the end of the Pleistocene made many habitats available for all forms of life while sea-levels were still low enough to provide bridges like the Bering land bridge and the Sahul. At a continental scale, the major changes of climate have had broad-scale influences on, for example, the immigration of plants and animals: the movement of beech species in both Europe and North America, for example, seems to be driven by climate even in the presence of human communities (Huntley and Webb 1988). Phases like the Little Ice Age of Europe (which occurred from the fifteenth century to the nineteenth century) were influential in lowering the altitudinal limits of some crops or in facilitating the adoption of other, more cold-tolerant, varieties. Locally, however, areas occupied by agricultural societies tend to be made over in the causes of settlement, production and pleasure with occasional incursions of war. Hunter-gatherers may alter their surroundings permanently but agriculturalists always do, with the possible exception of shifting cultivators. Even they, given a long enough tenure, produce a forest or savannah whose species composition is not the same as the ecosystem which they originally colonized. So by AD 1800 many species had been domesticated and many ecosystems altered or replaced. A map of the world in 1800 would however still show considerable areas where human hands had transformed the natural state to only a limited extent. Many of the tropical moist forests, for example, were inhabited by hunters and small-scale agriculturalists whose impact upon forest physiognomy and fauna was very small and virtually unnoticed by most of the travellers who recorded their visits: rubber exploitation came later (Dean 1987). Large areas of semiarid savannah still retained their full complement of mammal populations even though human-directed fire was a long-standing feature of the annual round; many temperate grasslands had an analogous ecology. Equally, tundras and boreal conifer forests were populated by hunters at low densities. By contrast, many tropical forests in monsoon zones, and many temperate deciduous forests, were vanishing or had indeed already gone in the cause of advancing agriculture to feed a rapidly growing set of world populations.

Behind these human-led changes in the natural scene were the lineaments of human culture which made it possible to transform natural ecosystems. Technology is the major mediator between humans and their surroundings and during this period most societies were not hugely different in their access to tools, as they would become after the Industrial Revolution. The possession of iron, knowledge of metals generally, the virtues of the plough, awareness of navigation techniques, the recording of information on paper, were all common to most human groups, with most groups lacking one or two of these inventions. Just as interesting, perhaps, is that the deployment of such technologies was given sanction by widely differing cultural ideologies. Thus the Chinese transformed the surface, the vegetation, soils and water regimes of much of south and central China during the ascendancy of a Taoism which preached a kind of environmental quietism (Tuan 1970; Perdue 1987). In the end, the result ecologically was not very different from the Christianized cultures of Europe where the Benedictine motto of '*labore est orare*' is symbolic of a divine approval of clearing forests and draining marshes. Nor should we forget that agricultural economies provided enough surplus energy for tangible accomplishments like the Egyptian pyramids or Chartres cathedral, or less physical achievements like the *haiku* of Basho Matsuo (1644–94) or the music of J.S. Bach (though Bach owned shares in a coalmine and so might be said to have one foot in industrialism, just as the Archbishop of Salzburg, W.A. Mozart's early patron, was very rich from salt revenues).

So the outcome of 12,000 years of human occupation is much as we might expect, given that in those years the human population grew from 4 million to 957 million. Where it was possible to produce edible materials, then they were produced in ever-increasing quantity to feed many people adequately and a few very well. At the margins, the original hunters were allowed to survive or a new breed of hunters-for-pleasure took over. Often at the junction of both, cities grew up to regulate the exchange of goods. The result by 1800 was an Earth of which large parts were physically humanized. The imminent erection of an efficient net of rapid human communication, importing more information into many cultural ecologies, was to carry on and virtually complete the process of humanizing the Earth's ecosystems so thoroughly carried forward in the period considered in this chapter.

REFERENCES

Adams, R.N. (1982) *Paradoxical Harvest: Energy and Explanation in British History 1870–1914*, Cambridge: Cambridge University Press.

Albion, R.G. (1926) *Forests and Sea Power: The Timber Problems of the Royal Navy 1652–1862*, Cambridge, Mass.: Harvard University Press.

Bamford, P.W. (1956) *Forests and French Sea Power 1660–1789*, Toronto: University of Toronto Press.

Bender, B. (1975) *Farming in Prehistory*, London: John Baker.

Berrall, J.S. (1978) *The Garden: An Illustrated History*, Harmondsworth: Penguin Books.

Brandon, M. (1971) *Hunting and Shooting, from Earliest Times to the Present Day*, London: Weidenfeld & Nicolson.

Brimblecombe, P. (1987) *The Big Smoke: A History of Air Pollution in London Since Medieval Times*, London and New York: Routledge.

Clutton-Brock, J. (1987) *A Natural History of Domesticated Mammals*, Cambridge: Cambridge University Press and British Museum (Natural History).

Cooter, W.S. (1978) 'Ecological dimensions of medieval agrarian systems', *Agricultural History* 52, 458–77.

Cowdray, A.E. (1983) 'Environments of war', *Environmental Review* 7, 155–64.

Crosby, A.W. (1972) *The Colombian Exchange: Biological and Cultural Consequences of 1492*, Westport, Ia.: Greenwood Press.

Crosby, A.W. (1986) *Ecological Imperialism: The Biological Expansion of Europe 900–1900*, Cambridge: Cambridge University Press.

Cushing, D.H. (1988) *The Provident Sea*, Cambridge: Cambridge University Press.

Dean, W. (1987) *Brazil and the Struggle for Rubber: A Study in Environmental History*, Cambridge: Cambridge University Press.

Frison, G.C. (1978) *Prehistoric Hunters of the High Plains*, New York: Academic Press.

Harlan, J.R. (1986) 'Plant domestication: diffuse origins and diffusions', pp. 21–34 in C. Barigozzi (ed.) *The Origin and Domestication of Cultivated Plants*, Amsterdam: Elsevier.

Harris, D.R. (1989) 'An evolutionary continuum of people–plant interaction', pp. 11–26 in D.R. Harris and G.C. Hillman (eds) *Foraging and Farming: The Evolution of Plant Exploitation*, London: Unwin Hyman.

Hayden, B. (1981) 'Subsistence and ecological adaptation of modern hunter/gatherers', pp. 344–421 in R.S.O. Harding and G. Teleki (eds) *Omnivorous Primates*, New York: Columbia University Press.

Head, L. (1989) 'Prehistoric Aboriginal impacts on Australian vegetation: an assessment of the evidence', *Australian Geographer* 20, 37–46.

Hillman, G.C. and Davies, M.S. (1990) 'Domestication rates in wild-type wheats and barley under primitive cultivation', *Biological Journal of the Linnean Society* 39, 39–78.

Hughes, J.D. (1975) *Ecology in Ancient Civilizations*, Albuquerque, N.Mex.: University of New Mexico Press.

Huntley, B. and Webb, T. (1988) 'Migration: species' response to climatic variations caused by changes to the Earth's orbit', *Journal of Biogeography* 16, 5–19.

Hyams, E. (1971) *A History of Gardens and Gardening*, London: Dent.

Jones, E.L. (1979) 'The environment and the economy', in B. Purke (ed.) *The New Cambridge Modern History*, vol. XIII, ch. II.

Jones, E.L. (1981) *The European Miracle: Environments, Economies and Geopolitics in the History of Europe and Asia*, Cambridge: Cambridge University Press.

Latz, P.K. and Griffin, G.F. (1978) 'Changes in aboriginal land management in relation to fire and to food plants in central Australia', pp. 77–85 in B.S. Hetzel and H.J. Frith (eds) *The Nutrition of Aborigines in Relation to the Ecosystems of Central Australia*, Melbourne: CSIRO.

Lee, R.B. (1979) *The !Kung San: Men, Women and Work in a Foraging Society*, Cambridge: Cambridge University Press.

Lee, R.B. and DeVore, I. (eds) (1976) *Kalahari Hunter-Gatherers: Studies of the !Kung San and their Neighbours*, Cambridge, Mass.: Harvard University Press.

Lourandos, H. (1980) 'Change or stability? Hydraulics, hunter-gatherers and population in temperate Australia', *World Archaeology* 11, 295–364.

Mannion, A.M. (1991) *Global Environmental Change: A Natural and Cultural Environmental History*, London: Longman.

Martin, C. (1978) *Keepers of the Game: Indian–Animal Relationships in the Fur Trade*, Berkeley and Los Angeles: University of California Press.

Mason, I. (ed.) (1984) *Evolution of Domesticated Animals*, London and New York: Longman.

Megaw, J.V.S. (ed.) (1977) *Hunters, Gatherers and First Farmers beyond Europe*, Leicester: Leicester University Press.

Perdue, P.C. (1987) *Exhausting the Earth: State and Peasant in Hunan 1500–1850*, Cambridge, Mass.: Harvard University Press.

Rackham, O. (1980) *Ancient Woodland: Its History, Vegetation and Uses in England*, London: Edward Arnold.

Shepherd, R. (1980) *Prehistoric Mining and Related Industries*, London: Academic Press.

Sieferle, R.P. (1990) 'The energy system – a basic concept of environmental history', pp. 9–20 in C. Pfister and P. Brimblecombe (eds) *The Silent Countdown: Essays in European Environmental History*, Berlin: Springer-Verlag.

Simmons, I.G. (1989) *Changing the Face of the Earth*, Oxford: Blackwell.

Simmons, I.G. (1990) 'The mid-Holocene ecological history of the moorlands of England and Wales and its relevance for conservation', *Environmental Conservation* 17, 61–9.

Stanhill, G. (1977) 'An urban ecosystem: the example of nineteenth century Paris', *AgroEcosystems* 3, 269–84.

Tanner, A. (1979) *Bringing Home Animals: Indigenous Ideology and Mode of Production of the Mistassini Cree Hunters*, New York: St Martin's Press.

Tuan, Yi Fu (1970) 'Our treatment of the environment in ideal and actuality', *American Scientist* 58, 244–9.

Westing, A.H. (ed.) (1986) *Global Resources and International Conflict: Environmental Factors in Strategic Policy and Action*, Oxford: Oxford University Press.

Zohary, D. (1986) 'The origin and early spread of agriculture in the Old World', pp. 3–20 in C. Barigozzi (ed.) *The Origin and Domestication of Cultivated Plants*, Amsterdam: Elsevier.

FURTHER READING

Crosby, A.W. (1986) *Ecological Imperialism: The Biological Expansion of Europe 900–1900*, Cambridge: Cambridge University Press.

Cushing, D.H. (1988) *The Provident Sea*, Cambridge: Cambridge University Press.

Edmonds, R.L. (1994) *Patterns of China's Lost Harmony*, London: Routledge.

Goudsblom, J. (1994) *Fire and Civilization*, London: Penguin Books.

Grove, R.H. (1995) *Green Imperialism. Colonial Expansion, Tropical Island Edens and the Origins of Environmentalism 1600–1800*, Cambridge: Cambridge University Press.

Harris, D.R. and Hillman, G.C. (eds) (1989) *Foraging and Farming. The Evolution of Plant Exploitation*, London: Unwin Hyman.

Hughes, J.D. (1994) *Pan's Travail: Environmental Problems of the Greeks and Romans*, Baltimore: Johns Hopkins Press.

Jones, E.L. (1981) *The European Miracle: Environments, Economies and Geopolitics in the History of Europe and Asia*, Cambridge: Cambridge University Press.

Mannion, A.M. (1991) *Global Environmental Change: A Natural and Cultural Environmental History*, London: Longman.

Pfister, C. and Brimblecombe, P. (eds) (1990) *The Silent Countdown. Essays in European Environmental History*, Berlin: Springer-Verlag.

Rackham, O. (1981) *Ancient Woodland: Its History, Vegetation and Uses in England*, London: Edward Arnold.

Simmons, I.G. (1993) *Environmental History. A Concise Introduction*, Oxford: Blackwell.

Turner, B.L. *et al.* (eds) (1990) *The Earth as Transformed by Human Action*, Cambridge: Cambridge University Press.

II A WORLD TRANSFORMED BY THE GROWTH OF A GLOBAL ECONOMY

INTRODUCTION

Richard Huggett and Mike Robinson

We have suggested that the first great transformation of the planet arising from the evolution of people was their capacity to adapt to different environments and to improve and consolidate their strategies for procuring food. Dependent upon these developments, but deriving from them only partially, some human societies evolved social and economic systems of relative complexity that allowed them both to consolidate within 'national' boundaries and simultaneously to enlarge political control over non-contiguous territory. In the process of enlargement cultures inevitably came into conflict and out of this conflict they began slowly to homogenize. Smaller and less well-organized cultures effectively disappeared; larger and more powerful ones steadily extended their spheres of influence and control. The focus for cultural conflict lay in a variety of apparent 'causes': religious belief; perceived population pressures; the pursuit of economic gain; and so on. For whatever reason or combination of reasons, in different places on earth, and at different historical phases, dominant cultures arose and under their influence global demography changed; biomes were transformed, and the relationships between the governed and the governing were inevitably contested.

Throughout most of this time, the energy resources of human societies came in the form of heat from the sun and from burning plants; from the motion of water and wind; and from the muscles of living creatures, especially those of people. The raw materials from which tools were fashioned, and the methods by which they were fashioned, changed little from the first millennium BC to a mere 300 years ago. The movement of both goods and people depended on vehicles and vessels built from organic materials. Economies which flourished did so on terms of trade determined more by political and military power than by competitive efficiency. By the eighteenth century, however, a new ecospheric transformation was in the offing. It depended on the combustion of fossil fuels and on the control of the heat produced to create pressures capable of translation into controlled mechanical motion. In the first instance, this transformation rested on coal. We have learned to call it 'the Industrial Revolution'.

Coal, although it had been used in some places for direct heating for two thousand years or so, was of little significance in the social energy budget of human societies before the eighteenth century. For a few people it was physically accessible and, although it was more difficult to ignite than wood, it burned for longer and gave a much greater output of heat. The application of coal to industrial processes at an altogether unprecedented scale, however, stemmed from a still-mysterious combination of conditions, some demographic, some financial, some commercial, and some intellectual and socio-political. Wedded to the development of the steam engine and employed to fire new types of furnaces for the reduction of metallic ores, the widespread availability of coal raised energy output in absolute terms while simultaneously affording increased control over energy applications. In consequence industrial production soared and innovation blossomed.

The industrial developments of the eighteenth and nineteenth centuries transformed social economies by creating new goods and also by creating new systems for their exchange and valuation, new markets for their consumption, and new means for their transport over greater distances and at greater speeds than ever before. Some of the most important of these goods were designed to increase still further the potential to produce food surplus. Others depended incestuously on the combination of new energy resources, new inventions, and a re-educated market to create an enlarged demand for different kinds of crops: 'industrial' crops, such as cotton and wool and rubber. The growing control of energy resources, first of coal and later of gas and oil, greatly accelerated the role of people in changing and creating habitats, as well as beginning to stir concern for an ecosphere that became, in parts, palpably threatened. It also accelerated the role of governments in projecting and exercising political power at every level from the local to the international. With the evolution of urban industrial cultures, and powered by differentials in industrial output, the political and economic transformation of all societies began to accelerate. Variations in the rate of change, however, exacerbated the differences between them. Countries began to separate more emphatically into those that 'had' and those that 'had not'. Europe, in particular, turned to ever-grander notions of 'Empire'.

An integral part of the industrial transformation of the earth has been its demographic transformation. By 1800 world population had probably reached 1 billion in a gradual process of enlargement that had extended over half a million years. In the course of the century which followed this number doubled. In the century which is now drawing to a close 1 billion has expanded to more than 5 billion. This unprecedented surge of people demanded a vast increase in agricultural output. Previously untouched natural ecosystems went under the plough and farming methods of increasing intensity were invented to improve the efficiency of existing agricultural ecosystems. The same surge of population also demanded housing and employment within the rapidly changing social environments that nurtured them, or in new environments purposely sought or wilfully imposed. For most of human history, the majority of people lived in small and weakly connected communities of, perhaps, fewer than 5,000. Even in 1900, in the most urbanized of societies in North America and north-western Europe, more than 70 per cent of people lived in settlements smaller than 20,000. Today, that figure has fallen dramatically to only 40 per cent or below. Growing numbers of people are now housed in cities and seek their livelihood there: perhaps more than 2 billion in total. With city growth, however, arose problems of government, of organization, of transport, of planning, of security and the like, on a scale previously unknown. Out of them, too, came new ideas of social reform, new ideas of the concept of government and competing tides of human creativity and ambition. The economically strong and powerful exploited resources on a global scale, rather than on the scale of a single region or country. At every level, the race to have more, and the struggle to have enough, was emphatically engaged in the wake of an industrializing world. By the end of the nineteenth century, and through the early years of the twentieth, the industrial nations of the earth were poised to shift global competition into the arena of global war.

8

EUROPEAN SETTLEMENT, 1450–1750

Len Guelke and Jeanne Kay

INTRODUCTION

This chapter addresses the causes, nature and legacy of European migration and colonization overseas in the period from about 1450 to 1750. It is an impossible task to condense into a single chapter a momentous episode of world history spanning 300 years and involving millions of people. The expansion of European culture overseas from its bases on the North Atlantic Ocean to encompass the whole world was a complex historical process that changed, directly or indirectly, the lives and landscapes of almost all the peoples of the earth. In this chapter we seek to highlight the nature of these changes in a general way with the aid of some concrete examples. The process of European expansion and settlement was carried out by different nations at various times and involved many different autochthonous peoples with a wide variety of social, techno-logical and economic systems. Yet notwithstanding the enormous complexity of Europe's encounters with the peoples and environments of other continents, some general patterns and consequences of those encounters can be described and analysed.

The period from 1450 to 1750 witnessed an explosive growth of European contact with lands overseas (Verlinden 1970). European mariners, traders and settlers invaded all the continents of the earth, with the exception of Antarctica, and modified and disrupted in various degrees the economies and lives of the indigenous peoples they encountered. European settlement almost everywhere took place on lands occupied by native peoples. This process ranged from relatively peaceful land purchases and intermarriage to the violent acquisition of land and the extermination or expulsion of prior inhabitants. The success of Europeans in settling and exploiting the lands they occupied depended on more than force alone. They had to devise profitable ways of using the lands and coastal waters they acquired if trade and settlements were to prosper, and in the process fashioned new systems for exploiting resources and labour that differed from those of Europe. The results of these efforts and their unintended consequences such as the spread of diseases, produced a global transformation that increasingly integrated continents and peoples.

The outcomes of colonization efforts varied, depending on the aims, culture, population and economies of both the colonizers and those colonized. In South and East Asia the principal objective of Europeans was trade: these numerous and sophisticated peoples were in a strong position on their own to supply European needs; in part to serve their own objectives. Thus, nations such as China and Japan were able to resist European colonization. In the Caribbean and Central America the initial contact populations were also very numerous but far less able to withstand the encounter with the European invaders and vulnerable to the latter's diseases (Lovell 1992; Crosby 1972). In temperate and arctic North America and in most of South America (except Peru) the number of indigenous inhabitants was small and European colonization of temperate, prime agricultural land often resulted in their expulsion or demise. They were replaced by European colonists and, in many tropical and subtropical areas, also with African slaves.

Lest Europeans be singled out as guilty of unusual greed and domination or praised for unusual heroism, it should be noted that imperial conquest, immigration, and subjugation have always been global phenomena. During the period in question, the Ottoman Turks were in control of much of the Arab World, following up on the Asian-based conquest of Eastern Europe by Ghengis Khan, and earlier records of Assyrian, Babylonian and Egyptian empires extending across much of the known world of their eras. European conquest from 1492–1750 was noteworthy for the vast distances it was able to overcome through advances in ship design and navigation, but not for the invention of imperialism and exploitation.

VALUES OF EMPIRE

The European colonizers came primarily from Spain, Portugal, Britain, France and the Netherlands (Boxer 1961, 1965; Parry 1966; Verlinden 1970). Each of these nations had its own systems of law, religion and government. Democracy as we understand it today did not exist. This was an era in Europe where peasants could be brutalized by their overlords and public displays of violence such as witch burnings, floggings and public executions were commonplace. Extremes of privilege and inequality for people of different classes, origins, religions and gender were the norm. The colonizers themselves were not representative fragments of their home societies and included disproportionate numbers of males as administrators and colonists, many from the working and poorer classes of society. These agents of European colonization took with them their values and aspirations of social advancement. They founded societies that were often as violent if not more so than the ones they had left. For example, Spanish conquistadors murdered native populations at the same time Spanish rulers implemented the Inquisition at home (Parry 1966).

During the period 1492–1750, both Europeans and indigenous people viewed their environments as accessible to human use for subsistence, wealth, and exchange; yet both had a significantly more spiritual or magical approach to their landscape than would be the case today (Merchant 1989). European farmers and gardeners planted according to phases of the moon and had a local lore about sympathetic magic associated with powers of various plant species. As Christians, they sometimes viewed the wilderness as inhabited by demons, witches or other spirits, such as voodoo spirits diffused by Africans. Native peoples of Africa, Australia and the Americas saw Nature as sentient and often animate. Each culture had its natural shrines and sacred places, yet Europeans, having left many of theirs behind, tended to view native environmental beliefs as incompatible with Christianity, and therefore superstitions to be abolished.

A critical assumption of European colonizers was that lands overseas were theirs for sharing, if not for taking. The arrogance of European peoples in this period owed much to Christian religious ethnocentrism and non-Christians were regarded as heathens or infidels. European Christians did not consider darker-skinned heathen peoples to be their equals. Although attitudes to conquered peoples varied widely, most Europeans had a sense of their own superiority on the basis of their race, religion and culture. Yet there was also a missionary zeal among some colonists to convert native peoples and slaves to Christianity. The idea that Europeans had a right to settle new lands and displace and enslave peoples who were not like themselves was generally taken for granted and provided a basis for many colonial enterprises that were brutal and destructive.

European settlers often adapted and modified the biblical Exodus legend to describe their own colonization experiences. In the biblical narrative the elect or chosen people left 'Egypt' or the Old World, following divinely gifted leaders under God's mandate. After a series of physical hardships and difficult adventures, they reached the Promised Land. The indigenous people, according to this narrative, forfeited their right to the land because they hadn't used it properly (i.e. had been hunter-gatherers rather than settled farmers) and were dangerous or morally unfit. The new colonizers then proceeded to clear the land, to build towns and cities, and generally to extend their domain. The land responded by blessing them with economic improvements according to many colonial histories.

European colonizers generally anticipated a more privileged existence in their overseas possessions than they could have sustained at home. They established colonies on principles that derived from those of their homelands, yet varied to serve their goals in a new economic situation. Slavery, for example, was a largely moribund institution in Western Europe that re-emerged with renewed vigour in many colonies. Yet Europeans of the upper and middle classes continued to limit the rights, mobility, and access to economic opportunity of servants and tenant farm workers. Most colonizers deliberately established new societies in which there would be unequal rights for people based on religion, ethnicity, class, culture and gender. The character of colonial societies changed as they grappled with defining the rights of increasing varieties of peoples that resulted from the mingling of blood and cultures and conversion of many slaves and native peoples to the religion of the colonizer (Meinig 1986).

While colonized peoples had various age/sex ratios and patterns of gender relations, the colonizers themselves were overwhelmingly young adult men. In the male-dominated hierarchically-ordered societies of Europe women experienced limited freedom to live independent lives. Colonial women usually immigrated overseas as wives, daughters, or prospective wives of European male settlers (Kolodny 1984). The number of European women who emigrated was never great, and in most colonial possessions they were greatly outnumbered by their male counterparts. This situation prompted colonial authorities in some areas to encourage immigrants to marry native or freed slave women who had adopted some European culture traits. Intermarriage and inter-racial sexual relations were common, whether defined by the practices of the Church, indigenous society, or by new syncretistic mores. New mixed-race populations emerged in many areas, such as the Métis of Canada, Louisiana Creoles, Latin American Mestizos and the Coloureds of South Africa.

The importance of religion in European overseas enterprises is well illustrated by the Portuguese, who had a long history of battling Islam in their own homeland. In 1415 the Portuguese achieved a major success by capturing the North African town of Ceuta, and retained it as a symbol of their crusading zeal. At Ceuta the Portuguese acquired further knowledge about the gold trade with West Africa, and the notion that this land could be reached by sea gained acceptance. Prince Henry of Portugal, the Navigator, initiated a well-ordered programme for exploration of the African coast. This programme would eventually culminate in Vasco da Gama's successful voyage to India (1497–9). These efforts clearly had commercial objectives, but the promotion of Christianity was an integral component in their planning and execution. One of a series of Papal Bulls issued at the request of the King of Portugal authorized him to

> attack, conquer and subdue Saracens, pagans and other unbelievers who were inimical to Christ; to capture their goods and territories; to reduce their persons to perpetual slavery, and to transfer their lands and properties to the King of Portugal and his successors.
>
> (Boxer 1961: 21)

The Dutch East India Company that ousted the Portuguese from much of their Asian Empire in the seventeenth century were also concerned to promote their reformed Christian faith. Yet they seemed more prepared to compromise their crusading zeal when commercial considerations made such action desirable and showed less enthusiasm for converting the peoples they ruled. For example, the Spanish and Portuguese traders found their missionary efforts in Japan increasingly unappreciated by the Japanese court and they were expelled in 1624 and 1638 respectively. The Dutch, on the other hand, 'studiously avoided advancing the cause of Christianity' (Israel 1989: 172) and by 1640 they were the only European nation left with business and trade connections in Japan. In areas they controlled the Dutch promoted their religion with a little more vigour, but even in these cases they did not show much missionary zeal. Although conversion of autochthonous peoples to Christianity was given as one of the explicit objectives in the colonization of the Cape of Good Hope no plan was formulated to accomplish this, and only a few of them adopted Christianity as their religion in the seventeenth and eighteenth centuries.

TECHNOLOGY AND ORGANIZATION OF EMPIRE

If the ideas of their own superiority and resource needs provided a basis for colonizing and settling the lands of other peoples, the success of the enterprise was made possible by European organization and technology (Braudel 1981). Improved sailing ships and navigation skills made highways of former ocean barriers and connected continents. Europeans took with them their ideas, diseases, pens, books, many of their animals and plants, swords and guns and much else besides. After the initial colonization ships were often converted to carry slaves to the new settlements. And ships kept many colonies connected with their home countries, allowing the frequent exchange of goods, ideas and people between them. The combination of European diseases, technological and military sophistication and organizational skills became a powerful engine for the domination or destruction of many native societies.

This engine worked differently in different places, but some of its essential features can be summarized. In Spanish strongholds of Latin America, New England towns and South African farm belts, Europeans conquered the original inhabitants and occupied land on the strength of their military superiority, which was often combined with native susceptibility to European diseases. In French Canada there was greater co-operation between Québecois fur traders, habitants and the original inhabitants (White 1991). European colonization in Asia involved conquering strategic areas, using them to dominate local peoples and control ocean trade. Although disease killed many, especially in the Americas, native peoples often survived in large numbers and became integrated into complex new colonial societies (Denevan 1992). In each of the preceding examples, mixed-blood populations emerged, in places forming their own distinctive societies.

European colonial ventures required sophisticated organization to mobilize the capital and people needed for long sea voyages to distant foreign lands. Capital was needed to build and man the ships that went overseas and organization was necessary to ensure the success of such ventures in different circumstances. These needs were often met by the creation of joint-stock companies whose purpose was to promote colonial and commercial objectives overseas. A fine example of one such company was the Dutch East India

Company. The charter of the Company, founded in 1602, gave it wide powers, including the right to wage war and conclude treaties with foreign powers. The Company was given a monopoly of all Dutch trade between the Cape of Good Hope and the Straits of Magellan (an area comprising more than half the globe). The executive committee of the Company comprised seventeen members from different areas of the Netherlands and was accountable to Company shareholders. Overseas operations were in the hands of appointed governors and councils. The India Council headed by a Governor-General conducted the company's business in South-East Asia from its headquarters in Batavia and reported directly to the company's executive committee in the Netherlands. There were smaller regional councils that reported to Batavia. Thus notwithstanding the huge distances involved the activities of the Dutch were co-ordinated in a way that sought to ensure an effective and accountable trading and commercial operation whose principal objective was to secure the maximum profits possible for its shareholders.

The Portuguese achieved domination of the Asian spice trade with a strategy that maximized their technological capabilities (Boxer 1969). The key to success was their control of the oceans by armed fleets, which were supported by a few land bases. In Asia the Portuguese fortified settlements at Goa, Hormuz and Malacca, and subsequently supplemented them with others (Subrahmanyam 1993). These settlements provided the safe harbours for Portuguese ships and were used as collection points for trade. Elsewhere Portuguese gained permission from local rulers to establish unarmed trading posts, most notably at Macao in China and, for a while, were represented in Japan. The colonial system was based almost exclusively on seapower and involved very little in the way of territorial acquisition. The whole endeavour was accomplished by about 10,000 people (Boxer 1961) and was maintained for almost a century before being challenged by the rising power of the Netherlands and England at the close of the sixteenth century.

Notwithstanding the power that a superior military machine conferred upon most European invaders of new territories, many autochthonous peoples resisted the colonizers with whatever means were available to them. Many of these efforts did not succeed in expelling the invaders, but some interim successes were achieved. For example, the Khoikhoi of South Africa acquired a reputation for fierceness when, in a violent encounter with a party of Portuguese on the shores of Table Bay, they killed fifty of the intruders and drove the rest off (Elphick 1985: 73). This action put back any thoughts Europeans might have had of colonizing this land for over a century. But resistance could also produce punitive counter reactions from the European invaders who used their military superiority to crush in a ruthless and genocidal fashion those peoples who sought to defend themselves and their ways of life. In asserting European control of conquered lands, some European officials pondered the terrible destruction their actions were wreaking on the original inhabitants of the lands they occupied. Yet in this era Eurocentricism was generally pervasive.

It is also important to realize that not all military clashes were between colonists and indigenous people. Oftentimes European agency tipped the balance of power among indigenous nations themselves by conferring technological advantages (guns, horses) to one society or band over another. European disease epidemics also forced the remnants of native ethnic groups to band with other, more powerful neighbours for protection. Truly fierce inter-tribal wars were fought among native American nations to prevent one tribe from blockading trade routes or from monopolizing the trade through obtaining middleman status between French traders and more distant peoples. In such cases there were winners as well as losers. In the western Great Lakes region of North America, the Chippewa (Ojibway) people greatly expanded their population and territory during the eighteenth century, forcing out the eastern Sioux (Dakotah) who in turn extended their territory onto the lands of some of their western neighbours.

West of Lake Michigan in North America, for example, initial direct European contact was through the French missionary-explorer Father Marquette in 1634. Yet the first records of the Huron nation trading French goods to these more western nations for beaver pelts dates from 1620. The Iroquois Wars and severe smallpox epidemics subsequently all but destroyed the Huron, whose middleman role was adopted first by the Potawatomie who expanded from the eastern shore of Lake Michigan to Green Bay, and subsequently by the Mesquakie (Fox). Although the latter tribe was relatively small, it occupied a strategic position along the Fox and Wisconsin Rivers that was the most reasonably accessible to the abundant fur-bearing regions of the upper Mississippi Valley. When the Fox nation was severely harried by the French during the 1730s and 1740s, they reformed as a confederated nation with their neighbouring allies, the Sauk. However, by 1750 the formerly rich fur-bearing streams west of Lake Michigan were over-trapped, the locus of the fur trade moved to the upper Mississippi and Missouri rivers, and the Sauk and Fox were decimated by another smallpox epidemic. Within eighty years, however, their population recouped, only to be decimated again in the nineteenth century by more disease and the incursion of white settlers.

While guns for warfare and hunting, together with metal beaver traps, were important to the redistribution of indigenous people of the western Great Lakes, the technology of simple domestic goods had an equal impact. People used to preparing firewood with stone hatchets, food with bark or pottery containers, and clothing out of skins often actively sought out traders with their European-manufactured hatchets, kettles, woollen blankets, and clothing. The principal economic value of the fur trade proved to be European goods sold to Indians (hence the contemporary expression 'Indian trade' rather than 'fur trade'). Where native people could not be counted on to supply a steady market for these goods, traders often resorted to selling or giving European manufactured alcohol to attract customers.

The role of native women in this process has often been overlooked. For the

Indian nations of this region, women were the owners of household goods – including the furs and skins they prepared for the trade and the trade goods obtained for the household.

The combined technologies offered by European traders to native people resulted in a redistribution of population throughout the study period. Populations could rebound after disease epidemics, and they often chose to locate in new areas where fur bearers were still undepleted. In the western Great Lakes/Northern Plains area, the most notable examples were the Chippewa (Ojibway) who merged with small local bands near Sault Sainte Marie, then eventually expanded as far westward as present-day North Dakota. They forced out the eastern Sioux (Dakotah) who in turn extended their own territory westward across the northern Great Plains.

USING THE LAND

European expansion overseas was fuelled primarily in the expectations of profit from such activities as trade, plunder, trapping, hunting, mining, agriculture and fishing. The economic strategy of colonization varied greatly in relation to the ability of the indigenous societies to provide the European intruders with commodities of value and the natural resources of the land bases that were occupied. European efforts to develop profitable colonies included the introduction of European crops and animals overseas and also involved experimenting with the production of local crops that appeared to have commercial value, such as tobacco in Virginia (Kulikoff 1986) or spices in South Asia. In some areas gold and silver mines, timber or furs became the basis of colonial economies, and provided markets for local produce. The economic consequences of Europe's expansion were to enlarge and rationalize the use of the earth's resources, to take advantage of new crops and animals, and to make possible the exploitation of new areas of the world for raising crops and animals not readily grown in Europe. There emerged a large-scale global economy that connected Europe with the rest of the world, and European markets drove production of minerals, crops, animals and other products.

Zelinsky (1973) proposed the 'doctrine of first effective settlement' – that the first group of immigrants or settlers to arrive in significant numbers had a disproportionately great impact on a frontier region's emerging cultural landscape. Subsequent immigrant groups might be more numerous or more proficient, yet the initial colonists normally would establish the region's vernacular architecture, land survey and tenure systems, government, legal system, etc. Thus the first effective European settlers for much of the US were English. French pockets persisted in southern Louisiana but French culture was overwhelmed by later yet more 'effective' Anglo-Americans in the former Francophone fur-trading posts of Detroit, Michigan; Green Bay, Wisconsin, and Vincennes, Indiana. The Dutch were the first effective settlers of New York's Hudson River Valley. The Spanish were the first effective settlers of the borderlands bridging from the Rio Grande of Texas to California. While Zelinsky's doctrine is debatable today over much of the American landscape due to modern technology, communications and the spread of a homogenized American culture, it worked reasonably well until 1750.

For example, Europeans took their ideas about landholding to their colonies, and created new cadastral landscapes (Conzen 1990). In New Spain and New France essentially feudal village models of landholding prevailed; while manorial plantations and yeoman freehold farms were found in the British colonies and Dutch South Africa. Regardless of the actual land system in place all these systems provided mechanisms for legitimizing European control of land at the expense of native survivors of conquest and disease. European settlers had access to the bureaucracies that confirmed land ownership. Native peoples seldom did and often had no mechanisms for entrenching their own rights in land. In this era native reservations were uncommon and native peoples who were seen as impediments to settlement were often removed forcibly to land that was not wanted, or simply left to adapt to a more precarious existence on marginal land not registered in Europeans' private hands.

There were essentially two models of European colonial commercial agriculture. Staple export crops were common in the tropical and sub-tropical plantations, such as a Louisiana or a Caribbean sugar-cane plantation (Curtin 1990). General farming was common on dispersed family farms or agricultural villages in temperate regions. Sometimes the two types were interfingered within the same region, as in the American South. The remoteness of many frontier settlements and the relative poverty of many of their inhabitants encouraged colonial agricultural families to be as self-sufficient as possible.

Most settler societies were extremely selective in their transfer of European culture overseas. New Englanders believed in the value of hard work: not simply as essential to survival in a new land, but as a Christian moral condemnation of idleness in any form (Earle 1992). Both men and women were involved in New England's distinctive form of mercantile capitalism. Europeans typically carried with them concepts of a gendered division of labour, where men managed the principal commercial crops and livestock, built ships, went fishing and whaling; and women produced woollen cloth, and tended gardens, poultry, and a dairy for household provisioning as well as for local markets (Cott 1977).

Massachusetts Puritan frontiersmen preferred to settle areas contiguous to previously settled lands, with a few exceptions along the Connecticut River (Cronon 1983). They brought with them the English idea of small nucleated settlements laid out around a village green or commons, with a larger surrounding political 'town' boundary encompassing woods and fields much like a manorial estate. Landholdings were initially allocated from several large fields in scattered parcels somewhat like the English open field system. In a landscape where trees were even more plentiful than stones, wood became the preferred building material.

The Puritan church (today, Congregationalist) was the centre of social and intellectual life. Each new settler group was a congregation headed by a minister, as well as a community. Settlement in towns was enforced partly to enhance protection on an unsafe frontier and principally to ensure moral living according to rigid codes of conduct and church attendance. However, dispersed farmsteads with consolidated landholdings became increasingly common with the passage of generations. Population growth was high, and arable land in its glaciated and often hilly terrain was scarce. The most frequent reason cited for new congregations leaving an older settlement and venturing out into the wilderness was a lack of sufficient good land to go around (Merchant 1989). Nevertheless, within a generation or two, the new towns also supported individual farmsteads located upon the original fields; and sermons denounced the unfaithful who located independently into isolated sections of the forest.

The Dutch colonizers of South Africa fashioned a new agricultural economy that incorporated new crops such as wheat and grapes (Elphick and Giliomee 1989). These crops were grown on freehold farms near Cape Town where they were cultivated in a far more extensive fashion than would have been the case in Europe. The Dutch East India Company left large areas of land between individual farms as grazing for the livestock of both settlers and Khoikhoi (Guelke 1989). This land soon proved inadequate for the growing flocks and herds of the new settlers who took advantage of Company policies to acquire large tracts of frontier land on loan. As the settler population grew, more and more people moved inland to occupy a substantial part of the southernmost tip of Africa. The pastoralism of European settlers was based on livestock traded and taken from the Khoikhoi whose animals were far better adapted to South African conditions than early European imports. However, unlike the Khoikhoi whose way of life was nomadic and communal, European settlers developed individually controlled farms, which they managed with the help of Khoikhoi and slave workers.

French interests in the St Lawrence River began with the exploratory voyages of Jacques Cartier in 1534, 1535 and 1541 (Harris 1966). He acquired knowledge of this part of North America in search of a sea route to Asia and prospects of finding gold. These efforts were a prelude to the establishment of French agricultural settlements along the St Lawrence River in the seventeenth century. The settlements were based on feudal principles and involved the allocation of large landholdings to well-connected individuals who sought to make them profitable by encouraging land-hungry French farmers to become their tenants. These tenant farmers or habitants acquired strips of land within specific seigneuries fronting on the St Lawrence River. The French brought with them to North America their crops (wheat, oats and peas) and their domesticated animals (cattle, sheep and pigs) and sought to replicate the agriculture of their homeland in a forested region of long and harsh continental winters. The new settlers adapted their agriculture to a situation in which settlers were few and land was abundant. The forest was cleared to accommodate rough fields that received little attention. There emerged a new landscape in which agriculture gradually encroached on forested land, but remained confined to the areas close to the banks of the St Lawrence.

The English settlements of Virginia were also closely tied to the sea and the waterways that give access to the land along their banks. Virginia found a native crop, tobacco, that commanded a market, and the huge profits to be derived from it spurred settlement and expansion. The cultivation of tobacco involved the exploitation of a crop previously unknown to Europeans within a system of market agriculture. The Virginia planters cleared rich forested lands to plant tobacco and moved on to new lands when old fields were exhausted. The idea that one should strive to create a sustainable system of agriculture using crop rotations and manuring was abandoned in a region where new land was readily available and profits were derived from making the most of scarce labour both free and indentured. A newly cleared field would produce tobacco profitably for three years and then needed a fallow period of twenty years before it was again ready for cultivation.

In addition to tobacco fields the Virginia settlers needed land to grow corn. They used the unimproved woodlands as pasture for livestock and as a source of timber for buildings and for the manufacture of crates ('hogsheads') that were used to ship tobacco to market. One contemporary observer estimated that a planter needed fifty acres per field hand if he or she were not to 'want for room' (Kulikoff 1986: 48). There were considerable economies of scale associated with large plantations.

LABOUR AND THE LAND

Europeans' development of profitable colonial enterprises created a demand for labour that the colonists could not meet in many areas. Indigenous peoples seemed an obvious source of labour to Europeans, and in some areas such as Indonesia and Mexico they were sufficiently numerous and capable of meeting entrepreneurial demands. In most areas of North and South America, however, the local peoples were not present in sufficient numbers to be the basis of a colonial workforce, or were hunter-gatherers or subsistence cultivators whose independent traditions made them resist the regular and sustained work demanded by plantation owners (Lamar and Thompson 1981: 33). In the British colonies of New England, levels of migration and natural increase enabled a principally British workforce to emerge. The needs of plantation agriculture were more demanding and within a few years of its establishment West Africans were captured and transported to the Americas, typically under appalling conditions. As plantation agriculture spread and flourished so did the slave trade with Africa (Curtin 1969). The relatively advanced state of West African agriculture provided the basis for a trade in people of enormous magnitude. African women were considered effective fieldworkers precisely because they had a tradition of being the principal farmers in their own cultures.

On his second voyage to the Caribbean, Columbus introduced sugar-cane to

Hispaniola with the view of using native slave labour for its cultivation. Both the Spanish and Portuguese had earlier developed sugar plantations on their islands off the coast of Africa using African slave labour. When the natives of the Caribbean succumbed to European diseases the Spanish turned to Africa to find the slave workers they needed for their colonial enterprises. The cultivation of sugar in the Caribbean proved profitable and the Spanish expanded their activities into Mexico and Peru. The Portuguese followed the Spanish example, developing sugar plantations in north-eastern Brazil. Later in the seventeenth century the British became sugar producers on their Caribbean islands of Barbados, Montserrat, Antigua and Jamaica. The cultivation, harvesting and processing of sugar were labour-intensive activities calling for a large well-organized workforce. The plantation system required both field and factory workers with different levels of skill (Mintz 1985).

In meeting these needs the planters developed a regimented and coercive form of slavery. The field hands worked in gangs under the supervision of a whip-wielding overseer. The production of sugar from the harvested cane was an industrial-type operation involving milling, boiling and curing. These activities required skilled people who had to work in exceptionally dangerous and unpleasant conditions. All these tasks involved back-breaking, unremitting toil and many slaves died prematurely or were crippled as a result of mistreatment, poor food and unsafe working and living conditions. Yet the profits to be realized in the sugar industry in the late seventeenth and eighteenth century were of such a magnitude that few plantation owners paid much attention to the human cost of their activities as they competed with each other to maximize their profits.

The relationship between labour and the family thus varied considerably from region to region: from the nuclear or extended family farms of Pennsylvania, to the slave societies of South Carolina plantations (where the mixed-race child of a white master could be sold as a slave) to the intermarried fur trade societies of northern Ontario. In New England children would be apprenticed at an early age and lived away from home; in Maryland a single male indentured servant would accept voluntary servitude with the promise of freedom and free farmland at the end of the contracted period.

COST OF EMPIRE: PEOPLE

The nature of the impact of European expansion on autochthonous peoples varied greatly depending on the social, economic and epidemiological factors. In general peoples of Europe, Africa and Asia were better able to resist each other's diseases, in contrast to the Americas where introduced diseases created great devastation. The impact in terms of population loss has appropriately been described as a 'demographic catastrophe'. In Asia the population was better able to withstand the European presence for a combination of reasons. To begin with they were very numerous and not particularly susceptible to European diseases.

They were also technologically sophisticated societies with developed economies based on the production and export of spices and other products. Many of these societies were strong enough to resist European efforts to dominate them and retained their religions, cultures and languages.

Slave traders sought to maximize their profits by minimizing the distance between African population centres and colonial plantations. In the Dutch colony of the Cape of Good Hope, for example, the enslaved peoples were seized from the Indian Ocean area where the Dutch East India Company had its monopoly. The misfortune of West Africa was its location opposite the Atlantic coasts of the Americas where the most successful plantations developed and the European demand for slave labour was correspondingly great. In the period from 1500 to 1807 hundreds of thousands of Africans were forcibly transported to the Americas often via the Caribbean (Curtin 1969).

The magnitude and duration of the Atlantic slave trade is testimony to the economic advantages of slave labour in the production of plantation crops. Notwithstanding the inhuman conditions in which many African slaves were forced to work, their labour did produce profits for those employing them. In strictly economic terms the employment of slaves reduced labour costs associated with 'reproducing' the labour force, because these costs were taken care of in Africa. As terrible as it now seems there were economic advantages for tropical planters in working people mercilessly and replacing them when they died with new ones, rather than seeking to create a self-reproducing family workforce.

In other places where Europeans established colonies on the lands occupied by hunter-gatherers, pastoralists and subsistence cultivators the impact on native peoples was generally devastating. In many areas of North and South America European diseases reduced and ravaged populations even before sustained contact with European settlers had been achieved. The native peoples suffered additional losses in bloody clashes resisting the advance of European settlers. Once settlement had occurred native peoples had insufficient land or access to key resources to sustain their old ways of life, and were denied access to the privately held lands of the invaders. In these circumstances their old societies disintegrated, leaving a few survivors to eke out a meagre subsistence as best they could.

In South Africa the original inhabitants the Dutch encountered were pastoralists and hunter-gatherers. These peoples fared very differently in their struggle to survive the occupation of their lands. The pastoralists for many decades after initial European settlement shared the land with the European newcomers. However, as the European settler population grew it acquired more and more of the best-watered land and disrupted the Khoikhoi patterns of transhumance (Guelke 1989). The settlers with access to the land office of the Company that registered their land claims were able to defend their gains legally and physically. Khoikhoi weapons were no match for horses, firepower and Eurocentric government. As the Khoikhoi lost the ability to sustain themselves as autonomous peoples many of them became workers on European farms where they mingled with African and Asian slaves losing in the process much of their distinct cultural identity. The San (Bushmen) hunter-gatherers retreated before the invader as long as they could, but eventually mounted a long and violent resistance to European encroachments on their lands. Many of them were killed and the survivors – mostly women and children – were pressed into forced service on European-owned farms.

The human cost of colonization extended beyond the disruption and destruction of autochthonous peoples in the lands conquered and invaded by Europeans. The organizations and individuals involved in transporting Africans across the Atlantic put profits ahead of human lives. Tens of thousands of people died on the slave ships crossing the Atlantic (Curtin 1969). Those who survived the journey were subjected to a life of unremitting toil on the plantations of the Americas.

The high death-rate further stimulated the slave trade to find replacements. As a general rule it seems that the conditions of slavery were a function of the profits to be made. For example, the huge profits to be made in the production of sugar in the eighteenth century were associated with harsh working conditions and high death-rates among slaves (Mintz 1985; Schwartz 1985).

The human cost of slavery was not confined to the Africans who died on the transportation ships or were enslaved in the Americas. There was also an impact on the societies of Africa. The demand for slaves on the coast of West Africa created much political and social unrest in the interior. This demand commenced shortly after 1492 and finally terminated in the mid-nineteenth century, reaching its maximum levels in the late seventeenth and eighteenth centuries.

If an advantage of slavery from the point of view of the producers of tropical crops was the acquisition of adult able-bodied workers, a negative consequence for the societies of Africa was that they had to bear the cost of raising these workers without the benefit of their labour. Many societies of Africa expended much care and energy raising children who became plantation workers for the Americas: slave-raiding thus deprived those societies of the benefits from their efforts. The economic loss was compounded by terrible social costs to the societies of Africa. The acquisition and transportation of slaves within Africa itself frequently involved much violence and loss of life (Hallett 1970). These disruptions, furthermore, often affected food production and many died of famine either at home or while being transported to the coast.

Notwithstanding these negative factors, most African societies were able to survive the worst ravages of the slave trade. One mitigating factor was the dynamic character of the internal slave trade itself which moved across vast areas of the continent hitting different societies at different times and giving many societies an opportunity to recover from their economic and human losses (Curtin 1969). Although most African societies were able to survive the centuries of the slave trade it did not provide them with the kind of connection with the outside world that did much to stimulate economic development. Africans, however, did voluntarily adopt exotic crops such as maize which gradually diffused across the continent and added a new and valuable element to many African agricultural systems.

COST OF EMPIRE: ENVIRONMENT

Michael Williams (Chapter 9) deals at length with environmental impact, but a few paragraphs here on colonial settlement and environmental change are in order. The world in 1750 had yet to experience both the Industrial Revolution and the steep upswing of exponential population growth that have since characterized human impact on the environment. European settlement overseas was still largely restricted to nodes and coastal strips. Massive global deforestation, chemical pollution, and the like were still in the future; Europeans had yet to journey into most of the interiors of Africa, North or South America, yet wildlife depletion (though not species extinctions) was widespread in some areas.

The North American fur trade was centred on beaver, whose barbed hairs were converted into high-quality felt for hats. Beaver were often trapped-out of local streams within a couple of decades of the local inhabitants' involvement in the fur trade, generally encouraging them to trap other less-valued species. In this example, native people, not Europeans, were the principal hunters and trappers, yet the mercantile system for which they produced furs and skins was centred in Paris, Montreal, London or New York.

The New England Puritans' practice of 'high grading' their forests for large white pines spurred some of the earliest forestry protection laws in the United States. Predators (bears, wolves, cougars) were deemed dangerous unwanted vermin; and many an old New England history records with pride the early dates by which they were eradicated. Settlers importing European livestock and crop seeds often inadvertently transported weed seeds mixed in with them or in ballast that, in the absence of competition from native species and in the presence of disturbed soil around the settlements, began to flourish. Certainly settlers voluntarily introduced weed species that were part of their European herbal pharmacopoeia not realizing that these might spread and prove noxious later on.

In the mid-seventeenth century the whole of southern Africa teemed with all the kinds of game that one associates with the African continent (Elphick and Giliomee 1989). The Khoisan pastoralists and hunter-gatherers shared this vast and rugged land with the wild animals in a situation in which neither people nor animals gained a decisive advantage over the other. From the early days of Dutch settlement hunting was an important activity. As the settler population expanded the numbers of wild animals, including such species as elephant, lion, leopard and hippopotamus, were greatly reduced and some species were eliminated from the more closely settled areas. The Cape Flats, near present-day

Cape Town, witnessed its last herd of wild elephant in 1702. The hippopotamus was particularly favoured for its meat and was soon hunted to extinction in the rivers of the south-west Cape. The new settler economy also put greater pressure on the land itself. The Khoikhoi pastoralists were nomadic peoples who grazed their animals in specific areas for short periods before moving on. The settlers by contrast attached themselves to specific places and pastured their animals in the same general area year in and year out. This practice had a detrimental impact on the original vegetation which did not regenerate itself as before. Hardier plants unpalatable to livestock gained hold at the expense of other species.

COLONIAL FRONTIERS

One of the most noteworthy geographical expressions of early European expansion to other continents was the presence of extensive frontiers, or ecumenes where neither the indigenous people nor the immigrants were completely in control and where they commingled. On the frontier indigenous ways rapidly changed through the diffusion of innovations, through depopulation or through military struggle. But European ways changed as well in the absence of close government control, well-developed economic links to the home country, or large numbers of their own countrymen and women. Frontiers were zones both where European 'seeds' germinated in foreign soil, and where new hybrids or mutations quickly evolved. These ranged from new populations of mixed ancestry to new laws, political institutions and government structures better suited to meet the needs of isolated settlers as the location of the frontier zone continually shifted towards the interior. Above all, frontiers were zones of transition – not only as gradients across space, but also across time and cultural norms.

For non-indigenous immigrants unused to survival in strange and therefore seemingly harsh environments, fixed absolute Eurocentric arrogance towards native ways was potentially suicidal. Although there is much interest today among social scientists in analysing the dominant and unyielding European 'gaze' on native inhabitants, such smug inflexibility was not a luxury that the majority of first-generation settlers could afford. Sale (1991: 271–3), for example, notes in his discussion of Jamestown Virginia (1607–25) that 'The most compelling fact of life in the Virginia colony ... was death', caused largely by the colonists' ignorance of food procurement techniques, adherence to English standards of gentlemanly inactivity, or, most probably, severe psychological depression brought on by an inability to escape overwhelmingly unfamiliar conditions. The first generations of frontier settlers, unless they were unusually well supported by financial backers from home, often survived their first years on the benevolence of indigenous people, and subsequent years by adopting a few of their subsistence practices. There was much vicious Eurocentrism and many brutal attacks on native people to be sure, but in the final analysis, European survival in foreign soil could not be accomplished without innovation, adaptation and flexibility in the face of unaccustomed conditions. In the American colonies many successful strategies were adopted from native people; these ranged from native crops (maize, tobacco) to methods of waging war.

The planting of European settlements in the Americas, Australia and Africa essentially created nuclei that provided the basis for their geographical expansion into sparsely populated lands. In Europe people were generally associated with specific territories and the borders between national groups were reasonably stable. Colonial possessions, in contrast, were seldom clearly demarcated and their borders were porous. The newcomers in settler colonies were able to move in and out of newly occupied lands relatively easily and they forged a variety of relationships with the peoples of the interior. The existence of an 'open' frontier was a typical colonial phenomenon, and it provided a wide variety of economic and social opportunities for European settlers. The colonial settlers took advantage of frontier opportunities to trade, hunt and plunder. These activities were largely unsupervised, because colonial authorities lacked the means or will to control the vast territories lying beyond the continuously settled areas they formally administered. Shortly after the establishment of settler beachheads on the margins of a continent, European settlers began to move inland to trade, hunt and pasture their livestock, relying on their guns, horses and occasionally legal status to protect them from the aboriginal peoples whose lands they were invading.

The French and Indian frontier in the fur trade of the St Lawrence/Great Lakes region of North America was probably one of the more benign forms of European colonization (Harris 1966). Its geographic basis was a sparse French population, a vast territory of native inhabitants, and wildlife furs and skins – notably beaver – that were in great demand in Western Europe. French ethnocentrism and disease epidemics notwithstanding, the fur trade relied on a stable, well-disposed native population. The profit motive was indeed present, but worked to discourage French traders' hostility towards their native suppliers and customers. The French 'bourgeois' (principal traders and merchants) encouraged their men to marry native women in order to cement the trade with the women's relatives, as well as to ensure their wives' various skills and services in the backcountry. Within the Indian villages and trading post communities, a new mixed-blood population – the francophone Métis – emerged.

The Portuguese plantation settlements in São Paulo produced groups of male settlers who banded together in military units called *bandeiras* to take advantage of frontier opportunities. In the late sixteenth century the interior of Brazil was inhabited by a large Indian population. The *bandeiras* initially sought to participate in the Indian trade, mingling with Indian women; in the process a large mixed-race population emerged. In the seventeenth century the *bandeiras* switched from trade to slave raiding. A vast trade in slaves developed between the interior of Brazil and the north-east sugar region. The raiders roamed far and wide (Braudel 1981) in search of slaves and penetrated Spanish-controlled areas, raiding newly settled Indian villages in Paraguay and elsewhere. It is estimated that in the seventeenth and eighteenth centuries about one-third of Brazil's slave economy was comprised of Indian workers (Curtin 1990). In the eighteenth century the demand for slaves had decreased, and the *bandeiras* again took up trading or prospecting for gold and diamonds in search of a livelihood.

THE COLONIAL LEGACY

The world was transformed by the expansion of Europe overseas in the period from 1450 to 1750 (Turner and Butzer 1992). Continents long isolated from each other were connected and peoples, plants and animals were exchanged. New peoples, new technologies, new plants, new animals and modified environments appeared in many places, yet from an environmental perspective this era was in some ways a prelude to the next. There was only so much damage that could be done to the environment by a relatively small global population a portion of which had sailing ships, animal-drawn ploughs, guns, water mills and other elements of pre-industrial technology. Much of the earth was not directly affected by the expansion of Europe overseas, and European settlements were largely confined to continental margins. Huge areas of Africa and North and South America still teemed with wildlife, and vast tracts of sparsely inhabited forest and prairie remained virtually undisturbed. A global trans-formation had however begun. If the earth as a whole was largely undamaged, notwithstanding regional and local problems, it had been changed. New plants and animals were spreading across continents, new peoples had established themselves at the expense of others and a truly global network of trade exchanges was in place. These changes provided the starting point for the accelerated transformations that would occur as the world industrialized itself and the world's population exploded in the next two centuries.

REFERENCES

Boxer, C.R. (1961) *Four Centuries of Portuguese Expansion, 1415–1825: A Succinct Survey*, Johannesburg: Witwatersrand University Press.
Boxer, C.R. (1965) *The Dutch Seaborne Empire 1600–1800*, London: Hutchinson.
Boxer, C.R. (1969) *The Portuguese Seaborne Empire, 1415–1825*, London: Hutchinson.
Braudel, Fernand (1981) *Civilization & Capitalism 15th–18th Century. Vol. III: The Perspective of the World* (trans. Sian Reynolds), London: William Collins Son & Co. Ltd.
Conzen, Michael P. (ed.) (1900) *The Making of the American Landscape*, Boston: Unwin Hyman.
Cott, Nancy F. (1977) *The Bonds of Womanhood: 'The Women's Sphere' in New England, 1780–1835*, New Haven: Yale University Press.
Cronon, William (1983) *Changes in the Land: Indians, Colonists, and the Ecology of New England*, New York: Hill & Wang.

Crosby, Alfred W. (1972) *The Columbian Exchange: Biological and Cultural Consequences of 1492*, Westport, Conn.: Greenwood Pub. Co.

Curtin, Philip D. (1969) *The Atlantic Slave Trade*, Madison: The University of Wisconsin Press.

Curtin, Philip D. (1990) *The Rise and Fall of the Plantation Complex*, Cambridge: Cambridge University Press.

Denevan, William M. (1992) 'The pristine myth: the landscape of the Americas in 1492', *Annals of the Association of American Geographers* 82, 369–85.

Earle, Carville (1992) 'Pioneers of Providence: the Anglo-American experience, 1492–1792', *Annals of the Association of American Geographers* 82, 478–99.

Elphick, Richard (1985) *Khoikhoi and the Founding of White South Africa*, Johannesburg: Ravan Press.

Elphick, Richard and Giliomee, Hermann (eds) (1989) *The Shaping of South African Society, 1652–1840* (2nd edn, 1st Wesleyan edn), Middletown, Conn.: Wesleyan University Press.

Guelke, Leonard (1989) 'Freehold farmers and frontier settlers 1657–1780', in Richard Elphick and Hermann Giliomee (eds) *The Shaping of South African Society, 1652–1840*, Middletown, Conn.: Wesleyan University Press.

Hallet, R. (1970) *Africa to 1875: A Modern History*, Ann Arbor: The University of Michigan Press.

Harris, Richard Colebrook (1966) *The Seigneurial System in Early Canada: A Geographical Study*, Madison: University of Wisconsin Press.

Israel, Jonathan I. (1989) *Dutch Primacy in World Trade 1585–1740*, New York: Oxford University Press.

Kolodny, Annette (1984) *The Land Before Her: Fantasy and Experience of the American Frontiers, 1630–1860*, Chapel Hill: University of North Carolina Press.

Kulikoff, Allan (1986) *Tobacco and Slaves: The Development of Southern Cultures in the Chesapeake, 1680–1800*, Chapel Hill: Published for the Institute of Early American History and Culture, Williamsburg, Virginia by the University of North Carolina Press.

Lamar, Howard Roberts and Thompson, Leonard Monteath (eds) (1981) *The Frontier in History: North America and Southern Africa Compared*, New Haven, Conn.: Yale University Press.

Lovell, W. George (1992) '"Heavy shadows and black night": disease and depopulation in colonial Spanish America', *Annals of the Association of American Geographers* 82, 426–43.

Meinig, D.W. (1986) *The Shaping of America: A Geographical Perspective on 500 Years of History*, New Haven, Conn.: Yale University Press.

Merchant, Carolyn (1989) *Ecological Revolutions: Nature, Gender, and Science in New England*, Chapel Hill: University of North Carolina Press.

Mintz, Sidney Wilfred (1985) *Sweetness and Power: The Place of Sugar in Modern History*, New York: Viking.

Parry, J.H. (1966) *The Spanish Seaborne Empire*, London: Hutchinson.

Pulsipher, Lydia Mihelic (1986) 'Seventeenth century Montserrat: an environmental impact statement', *Historical Geography Research Series* 17: 1–96. England: s.n.

Sale, Kirkpatrick (1991) *The Conquest of Paradise: Christopher Columbus and the Columbian Legacy*, New York: Plume.

Schwartz, Stuart B. (1985) *Sugar Plantations in the Formation of Brazilian Society*, Cambridge: Cambridge University Press.

Subrahmanyam, S. (1993) *The Portuguese Empire in Asia 1500–1700: A Political and Economic History*, Harlow: Longman.

Turner, B.L. and Butzer, Karl W. (1992) 'The Columbian encounter and environmental change', *Environment* 34(8), 16–44.

Verlinden, Charles (1970) *The Beginnings of Modern Colonization* (trans. Yvonne Freccero), Ithaca, N.Y.: Cornell University Press.

White, Richard (1991) *The Middle Ground: Indians, Empires, and Republics in the Great Lakes Region, 1650–1815*, Cambridge: Cambridge University Press.

Zelinsky, Wilbur (1973) *The Cultural Geography of the United States*, Englewood Cliffs, N.J.: Prentice-Hall.

FURTHER READING

Boxer, C.R. (1965) *The Dutch Seaborne Empire 1600–1800*, London: Hutchinson.

Boxer, C.R. (1969) *The Portuguese Seaborne Empire, 1415–1825*, London: Hutchinson.

Braudel, Fernand (1981) *Civilization & Capitalism 15th–18th Century. Vol. III: The Perspective of the World* (trans. Sian Reynolds), London: William Collins Son & Co. Ltd.

Curtin, Philip D. (1990) *The Rise and Fall of the Plantation Complex*, Cambridge: Cambridge University Press.

Elphick, Richard and Giliomee, Hermann (eds) (1989) *The Shaping of South African Society, 1652–1840* (2nd edn, 1st Wesleyan edn), Middletown, Conn.: Wesleyan University Press.

Gaspar, David Barry (1985) *Bondmen and Rebels: A Study of Master–Slave Relations in Antigua with Implications for Colonial British America*, Baltimore: Johns Hopkins University Press.

Lamar, Howard Roberts and Thompson, Leonard Monteath (eds) (1981) *The Frontier in History: North America and Southern Africa Compared*, New Haven, Conn.: Yale University Press.

Merchant, Carolyn (1989) *Ecological Revolutions: Nature, Gender, and Science in New England*, Chapel Hill: University of North Carolina Press.

Morgan, Edmund Sears (1975) *American Slavery, American Freedom: The Ordeal of Colonial Virginia*, New York: Norton.

National Geographic Society (US) Cartographic Division (1983) *Atlantic Gateways*, Washington: The Society.

Parry, J.H. (1966) *The Spanish Seaborne Empire*, London: Hutchinson.

Rego, A. Da Silva (1965) *Portuguese Colonization in the Sixteenth Century: A Study of the Royal Ordinances (Regimentos)*, Johannesburg: Witwatersrand University Press.

Sale, Kirkpatrick (1991) *The Conquest of Paradise: Christopher Columbus and the Columbian Legacy*, New York: Plume.

Sauer, Carl Ortwin (1966) *The Early Spanish Main*, Berkeley: University of California Press.

Weber, David J. (1992) *The Spanish Frontier in North America*, New Haven, Conn.: Yale University Press.

9

EUROPEAN EXPANSION AND LAND COVER TRANSFORMATION

Michael Williams

The expansion of Europe overseas 500 years ago was the prelude to what was probably the greatest change in the natural land cover of the world since the Ice Age. Events were put in train that had far-reaching consequences on the distribution of peoples, plants, animals, and diseases, all of which had knock-on effects on soils, hydrology and other biogeographical phenomena.

The story, however, is not only one of changing distributions but also one of changing technologies and changing magnitudes of scale. Humans could now intervene and alter nature on an unprecedented scale, increase the production of all commodities to unparalleled levels, and move commodities in unheard-of quantities, all to cater for accelerating human needs. There is a tendency to see territorial expansion and control, technology, and the conditions of material life, as separate, independent topics. But nothing could be further from the truth. Each was intertwined with the other to produce an ever-upwardly ascending spiral of increased production and consumption, economic change and biome modification, leading to the permanent transformation of the land use of most parts of the globe.

The 500 years from about 1500 to the present is a vast and varied canvas to look at, and it is best divided into three equal periods, 1500–1649, 1650–1799, and 1800–1950. Such a periodization emphasizes the emergence of the European core of intensive (usually urban-oriented) land use, and the creation of successive waves or frontiers of exploitation in the periphery where the land and its stored energy were extracted for the benefit of the core. Such a core/periphery conceptualization makes it clear that the exploits of the seaborne, capitalist, Western European economies of first northern Italy and then Spain and Portugal during the sixteenth century, followed by Britain, the Low Countries, and to a lesser extent France, in succeeding centuries, united the world ecumene and brought people and places into contact with each other, so laying the foundations of biome modification (Braudel 1984). But it was the new

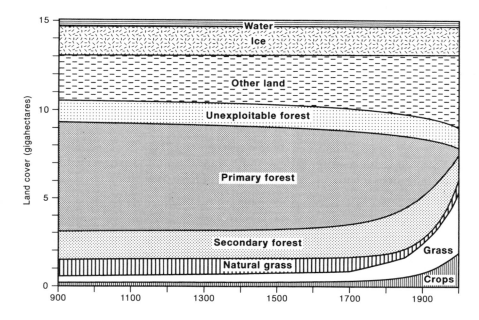

Figure 9.1 Transformation of land, 900–1977. Small differences cannot be depicted at this scale. The most important transformation took place during the last two hundred years. After Buringh and Dudal, 1987.

technology of the industrial mid-nineteenth century that facilitated and 'deepened these contacts into a constant flow of goods, people and ideas' (Headrick 1981: 176). From then on, isolated subsistence societies with limited trade relationships became part of a single world market in basic commodities that drew on natural resources, such as top soil, wood, water, minerals and wild game from throughout the world. Mechanical energy (particularly steam), medicine, and the agricultural modification of land shattered traditional relationships, and laid the foundations for a transformation of the land cover of the world (Figure 9.1).

1500–1650: BREAKING THE BOUNDS

Why Europe?

Before 1500 civilization had been essentially land-centred and contacts by sea were relatively unimportant. Therefore the continents were isolated one from another, except for a few overland trading routes which linked Europe with India and Africa. But the exploits of Columbus and others after 1492 established sea contacts between continents, and started a process of cultural and economic

183

expansion and domination by Europe (Figure 9.2). Before these voyages Europe was a peripheral appendage of the civilized world, which consisted of the land-based empires of Ming China, Ottoman Middle East and North Africa, Safavid Persia and Mogul northern India. After about 1500 Europe gradually moved to being at the centre of world innovation, trade and change. With this shift in focus the relative peace and harmonious relationship between the inhabitants and their environment in the peripheral parts of the world was shattered as Europe extracted the unused and stored-up potential of land and its vegetation for its own use.

The success of Europe after 1492 was staggering. To a detached observer of the early fifteenth century, Europe and Europeans would have seemed to have had no qualities inherently superior to those of Eastern civilizations. They were not necessarily more progressive, venturesome, achievement-oriented or more modern than societies elsewhere, and they had no advanced maritime technology. Much that was innovative in medieval society (e.g. compass, astrolabe, chainmail, crossbow, gunpowder, methods of rigging, paper making and printing, to say nothing of mathematics and trigonometry) had first been

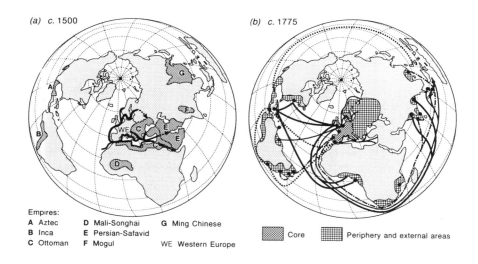

Figure 9.2 European world economies. (a) *c.*1500. Western Europe was only one of a number of land-based empires, and extra-continental trade was restricted to its coastal areas and overland routes. After Braudel (1984) and Barraclough (1978). (b) *c.*1775. By the mid-eighteenth century the core of Western Europe was firmly established, and the spectacular extension of ocean trading meant that large areas of the world constituted its periphery or external areas which were economically (and often, politically) tributary to it. After Braudel (1984) and Barraclough (1978).

received by Islam from China and had been brought back from the Levant during the Crusades (Jones 1987).

Also, long distance oceanic voyages were being undertaken by mercantile–maritime communities everywhere, from Africa to South-East Asia, from India to East Africa. But critically, Spanish and Portuguese sailors were already exploring the currents and winds down the coast of Africa and out to the Atlantic islands of the Canaries, Azores and Madeira. They were aided in their enormous leap across the Atlantic by advantages of accessibility and constant and reliable wind patterns, all of which compare very favourably with the unreliable winds elsewhere in the world. The Trade Winds *do* blow from the Canaries to the Caribbean and the return voyage *could* be made northward via the westerlies. Simply, environmental conditions aided long sea voyages from Europe (or hindered them from elsewhere). By the late fifteenth century it was Europe that leaped ahead, breaking out of the bounds of its land-based civilization and turning the continents inside out (Whittlesey 1944) by, in effect, reorienting them to face the sea. More important than the voyages, however, was the penchant of Europeans to rationalize encounters with new lands and 'to develop the resources they brought within their reach'; that really changed the existing world (Jones 1987: 80). Portents of what was to come were already in place in the fourteenth century with the vigorous and aggressive politico-commercial networks of Genoa and Venice which had reached at least a thousand miles across the Aegean to outposts in the Levant and the Black Sea, and they 'connected with all the great trade routes of western Asia' (Meinig 1969: 211).

Ecological imperialism: disease

Because of the assured contact that the currents gave, the supply of Old World biological forces of humans, plants, animals, and, above all, diseases, was constant and unrelenting, all of which reinforced the settlement process. Europeans carried with them many pathogens to which the native population had no immunity. The long isolation of the New World continents excluded exposure to a whole range of Old World epidemic diseases, many of which had evolved with domesticated animals (Crosby 1986; McNeill, 1976). With these Old World pathogens the Indians met their 'most hideous enemy' (Crosby 1986: 198). Each new European contact brought a new wave of diseases, first influenza, smallpox, measles, mumps and pneumonic plague, followed later by diphtheria, trachoma, whooping cough, chicken pox, malaria, typhoid fever, cholera, yellow fever, and scarlet fever (Chapter 8).

The evidence is fragmentary but it is abundant, and it parallels the well-documented experience of susceptibility of isolated peoples in more recent times, from the Inuit of Northern Canada, to the Aborigines of Australia and to many tribes in the depths of tropical rain forests. In the Americas each new pandemic not only displaced or eliminated hundreds of cultural groups, but the cumulative effect was a demographic collapse. Denevan (1992a, 1992b) estimates conservatively that the population for the hemisphere was 53.9 million in 1492, but only 5.6 million in 1650, an 89 per cent drop, and that in North America it fell from 3.8 million in 1492 to 1 million in 1800, a 74 per cent drop. Others would put the original population total for the hemisphere much higher at 100–13 million, which would merely make the decline even more dramatic (e.g. Dobyns 1966; Lovell 1992).

Whatever the exact numbers the fact remains that depopulation did happen on a catastrophic level. Metal weapons – to say nothing of gunpowder and horses (the Amerinds had none of these) – though decisive in campaigns, were nothing compared to the biological advantages possessed by Columbus and those who followed him. 'It was their germs, not the imperialists themselves, for all their brutality and callousness, that were chiefly responsible for sweeping aside the indigenes and opening the Neo-Europes to demographic takeover' (Crosby 1986: 196). One could say, fancifully, that they triumphed through breath alone. It was an ecological imperialism far more effective and far more terrible than an imperialism of arms.

Ecological imperialism: plants and animals

The loss of up to 90 per cent of the population led to the widespread abandonment of agricultural land, and it is possible that in places the forest actually expanded in extent and density. Much land use was dominated by shifting agriculture and it usually supported only moderate population densities, but at certain times and places it had been capable of producing vast food surpluses. In this way the cities of the Mayan civilization (c. AD 250 to 1000) had been supported, replete with their sophisticated water and waste management systems, and a monumental architecture that ranked with the greatest in the world at the time (Doolittle 1992; Turner and Whitmore 1992).

Thus, the availability of the cleared fields and their crops of abundant and nutritious plants – such as squash, tomatoes, potatoes, and maize – was a major factor in the ability of the newcomers to gain a successful toe-hold on the continent. They thought it was a God-given reward for their endeavour. Little wonder a settler in tidewater Virginia could say many years later:

> The objection that the country is overgrown by woods, and consequently not in many years to be penetrable for the plough, carried a great feebleness with it for there is an immense quantity of Indian fields cleared ready to hand by the natives, which, till we are grown over-populous, may be every way abundantly sufficient.
>
> (Quoted in Maxwell 1910: 81)

While nutritious food crops were utilized other flora and fauna disappeared and were replaced by opportunistic and aggressive Old World varieties. From the St Lawrence to the Rio de la Plata, ferns, thistles, nettles, artichokes and plantain (weeds we would call them), as well as the more common grasses and clovers such as Blue Grass and white clover, found eco-niches on the bared and eroded ground of abandoned Amerind fields or new European clearings. Plants moved with feral stock in advance of people. The story of the famed Kentucky Blue Grass country is instructive in this respect. *Poa pratensis* is not a native American grass yet when the English pioneers topped the Appalachians at the end of the eighteenth century they found vast areas of Blue Grass and English white clover awaiting them and their herds. The grasses must have entered with the French in the seventeenth, or early eighteenth, century and have been spread northwards by wild stock. In other places in the Caribbean, soils were changed as the humus layer was leached or destroyed, with detrimental long-term effects.

The ecological advantage of the Europeans over the indigenes, however, was not so much a matter of crop plants as of domesticated animals. Their multiplication was prolific as they ran wild in a world without natural predators and competitors. In particular, pigs flourished in forested environments; they are omnivorous and healthy sows can have litters of ten or more piglets. The eight pigs brought by Columbus to Hispaniola in 1492 multiplied prodigiously in the wild. Within ten years leave was given to hunt them as they were damaging crops and over-running the Island. They spread to mainland Mexico during the late 1490s and were soon said by the Spanish to be '*infinitos*'.

Cattle were said to have multiplied tenfold in three or four years. Horses adapted more slowly, as did sheep. Slower they may have been, but the rate of increase was unimpeded. These feral goats, sheep, horses, pigs, mules, and cattle were the forerunners and forebears of the millions that were to populate the land from the Prairies in the north to the Pampas in the south. Felix Azara, an unusually keen observer of natural phenomena, noted in 1700 that the numbers of cattle in the Pampas had reached 48 million, an exaggeration, perhaps, but even if he got it only half or even quarter right it was still a staggering number (Crosby 1986).

All stock had a devastating effect on the land because they put a premium on pastures and hence on the creation of wholly cleared land which had not been a part of the native economy. They also trampled and over-grazed land, causing erosion and, incidentally, aiding the spread of Old World crops and weeds.

Rats, bees, mice and rabbits; the story was the same. Old World stock seemed to do well, if not better, in their New World situations than in their homeland, and the Europeans benefited with better nutrition.

'The Great Frontier'

The social energy and surging economies of the countries of the core of Western Europe were breaking out of national confines and beginning to devour resources from the wider world. In so doing they tightened their grip on local and regional economies. In the words of Walter Prescott Webb (1952) the whole process was part of the 'Great Frontier' of European outward expansion that gave it unprecedented windfalls of land, minerals, fish, timber, furs, and crops.

Jones (1987) suggests that there were four main ecological zones of biological exploitation on the periphery of the core (to which must be added the inorganic minerals and precious metals of Latin America). All were becoming evident, though not all were fully developed, from the mid-sixteenth century onwards. Some, as of old, were on Europe's edge, most were overseas. They were:

1   The fisheries on the Grand Banks and Atlantic generally, particularly for cod and whales, which provided protein and essential oil for lamps.

2   The boreal forests of Russia and Scandinavia provided furs, and the essential timber, tar and turpentine for merchant ships. Around the Baltic fringes, especially the southern fringes, surpluses of grains were exported to Western Europe, which trade might have been as important in sustaining the Low Countries and Britain during the late sixteenth and early seventeenth centuries as bullion had been for Spain and Portugal at an earlier period (Wallerstein 1974).

3   In time the newly acquired tropical and sub-tropical regions provided a range of new industrial and food crops, such as tea, sugar, tobacco, cotton, indigo and rice, which varied diet, influenced consumption patterns and stimulated trade.

4   By the nineteenth century the forests and grasslands of the newly settled neo-European lands of North America, South America, Australia, and southern Africa and the steppes of southern Russia provided timber, grain and livestock.

As this unmatched flow of abundant and cheap resources gathered momentum 'the expansive process of land transformation in the modern world began' (Richards 1990a: 167). Space does not permit the examination of all these resource frontiers, so only those in the extra-European world are looked at.

The Caribbean and Brazilian peripheries

Whatever the primary motive of the colonizers, all wanted to achieve an economy with an assured domestic food base and some sort of export staple. Thus, sugar-cane was introduced from the Canary Islands to Hispaniola (Dominican Republic) by 1498. Fourteen years later, bananas, again from the Canaries, were introduced. In between those two dates many other Old World (specifically Spanish) crops and vegetables were imported, including wheat, barley, rice, onions, radishes, melons, grapevines, cabbage, lettuce, cauliflower, carrots, garlic, pomegranates and aubergines. Yams came from Africa with the slaves, as did castor and pigeon peas. They all flourished wonderfully.

Although not very extensive the islands of the Caribbean experienced intense change in a remarkably short period of time. They were a microcosm and example of how the new, alien, commercial objectives of the Europeans, serving a world-wide market, let loose a whole host of ecological and biogeographical changes.

By 1642 there were about 80,000 European settlers in the Caribbean (mainly in Barbados and St Kitts) which compared to only about 49,000 in New England and Maryland at this time. Many of the settlers were sponsored by trading companies that were attempting to settle the land and stimulate trade. The record of the impact on the forests is particularly clear in Barbados where after 1625 the aim was to plant sugar as a commercial crop. Initially, small patches of forest were selected and removed by either clear-cutting or ring-barking followed by the burning of the dried-out debris and trees towards the end of the dry season. Soon, more systematic clearing resulted in vast swathes being cleared. By 1647 about one-fifth of the rain forest and the secondary scrub along the coasts had been cleared and extensive sugar plantations established. Later, the large infusion of capital and the importation of black slaves from Africa ensured profitability of the plantations, and they expanded rapidly, consuming the bulk of the remainder of the forests. By 1665 the once forest-covered landscape of Barbados was almost totally open except for forest on the highest peaks, and dominated by large sugar estates.

By 1672 the same devastation and transformation had happened in the neighbouring islands of St Kitts, Nevis, and Montserrat. By 1690 the same was true of Martinique and Guadeloupe, and Jamaica was not far behind. Throughout the Caribbean the severity of clearing was highlighted by the scarcity and high price of timber for construction and fuel wood, particularly for refining the sugar, so that by 1671 it was reported from Barbados that 'all the trees are destroyed' and that coal had to be imported from England (Watts 1987).

In contrast to the relatively small area of forests in the Caribbean islands those of Brazil were amongst the most extensive in the world. In addition to the well-known Amazonian rain forest there was once a vast (approximately 78,000,000 ha) and very similar sub-tropical rain forest that stretched along the Atlantic coast from Recife (Pernambuco) in the north to Rio de Janeiro in the south, and it covered most of the eastern portions of the states of Bahia, Minas Gerais and São Paulo.

In pre-Columbian and early European times shifting, slash-and-burn cultivation was practised, and crops of manioc, maize, squash, beans, peppers and peanuts grown. When European contact intensified after 1600 the Amerind population was gradually replaced by a mixed mestizo population, but the concomitant introduction of iron axes and machetes, and the herding of pigs intensified the shifting regime. Whereas the Amerinds might have cut and burned about one hectare per family per annum, the Europeanized mestizo was capable of destroying three hectares or more with his tools of greater efficiency. The forest did not regenerate under such widespread cutting and burning, and the open ground left as a result of periodic shifting was colonized by exotic grasses, ferns and weeds (Dean 1983).

This more extensive and thorough clearing was the first step in the establishment of export-oriented plantation crops. Sugar, for example, had been introduced from Madeira in about 1560 and cultivation concentrated in a narrow strip from a little north of Pernambuco to just south of Bahia, with a minor concentration around Rio de Janeiro. Primary forest was always favoured for sugar growing, and the clearing and harvesting was done by slaves. Demands on the forest were not confined to clearing the land for cultivation alone as extreme heat was required to crystallize the juice. Thus, vast areas were cut for wood fuel. By the mid-eighteenth century the remaining forest on the coast was under immense pressure.

Wealth in nature

One could go on piling up example after example of biological change but it would be pointless. What is important, perhaps, is to recognize that the emergent commercial/social system of Europe, often typified as incipient capitalism, saw wealth in nature. It was the major purposeful driving force in the ecological transformation. Nature could be commodified and ecological changes were translated into economic advantages. The land cover of the world would never be the same again.

1650–1800: CONTINUITY AND CONSOLIDATION

Cores and peripheries

Ever since the breakdown of feudalism in the late fourteenth century, Europe had been moving towards a more entrepreneurial and capitalist system. But the transition was not clear-cut. Historians have debated long and fiercely about the onset of modernity. Was it 1510, 1600, 1650, 1750 or some other date? What we do know, however, is that a qualitatively different social system, and systems of production and distribution, emerged during the late sixteenth century when compared to anything which had existed in medieval times. Also, the role of state structures strengthened considerably (Tilley 1990). Thus, we see the emergence of truly capitalist states from about 1650, first in the Low Countries and then Britain, and the rise of Amsterdam and London as international trading and financial centres (Wallerstein 1980). Continuity with consolidation was the theme of the age.

Geographically, the boundaries of the world created by the voyages of discovery did not change significantly until after 1750. The world-system remained much the same, but the dominant role of Europe was consolidated. Specifically, the countries in the core (Britain, the Low Countries, northern France) strengthened their position while others declined. Spain and Portugal waned as did Sweden and, to a lesser extent, northern Italy, but Brandenburg-Prussia rose in importance. Uneven development was as much a feature of seventeenth-century Europe as of the modern world. Outside Europe the New England and Middle Atlantic colonies, and the Caribbean, were drawn into the core via the triangular trade in slaves, molasses, and manufactured goods. Beyond lay the new peripheries of exploitation noted already, and beyond them again lay the external areas of Asia, Africa and parts of Latin America that were tributary to the core.

Driving forces

Implicit in the above discussion is the idea that there were a number of forces that drove change – a shift in the economic system and means of production, a strengthening of state structures and control. But in addition there were a number of other factors that interacted with each other in an upwardly ascending spiral of consumption and production.

First, Europeans had the means to ship and organize this trade. It is estimated that in 1650 there were about 20,000 ships in the world carrying trade of which the Dutch – 'the carryers of the world' – owned over two-thirds. By the end of this period the British had usurped Dutch hegemony. The timber and naval stores supplies from the Baltic 'frontier' (later superseded by North America) were essential for this vast construction.

Second, just as the development of fast-moving and reliable sailing ships in large numbers was essential to the establishment of this new global trade so was the development of a financial infrastructure. Increasingly, barter as a means of exchange gave way to a money economy (boosted by the shipments of gold and silver bullion from Latin America after 1550) and the development of credit. Credit was important in this new system of global interchange. Goods took months, if not years to collect, dispatch and sell, and this was done only through the development of bills of exchange, notes of credit, and other promissory instruments and the refinement of banking. The founding of the Bank of Amsterdam (Wisselbank) in 1609, the Bank of Rotterdam in 1635, the Royal Exchange in 1576, and the Bank of England in 1694 were all institutional landmarks in the organization of money and credit (Kindleberger 1984).

Symptomatic of the new organization needed was the rise of the trading or joint-stock company, in which people ventured capital in hazardous overseas enterprises in return for monopolies guaranteed by royal charter. Such were the British East India Company (1600), the Dutch East India Company (1602), the Dutch West India Company (1621), and there were many minor ones. The companies were the powerful and well-organized spearheads of exploitation and the precursors of colonial land empires. They spread the risks and concentrated settlement, and combined the various tasks of conquest, settlement, investment and defence all in one.

Third, global population numbers certainly rose during this period, probably from about 425 million in 1500 to 679 million in 1700, which must have stimulated consumption. But rising numbers were of less significance than the rise in the purchasing power of the rapidly expanding European population. During the later sixteenth century there was a shift from the medieval preoccupation with a land-based trade in small quantities of high-value luxury goods, such as spices, perfumes, porcelain, dyestuffs, and silk clothes and rugs from Asia and Africa, to a seaborne mass trade of far bulkier commodities from the Americas, for an increasingly affluent and growing home population.

Wallerstein typifies this shift as a move from 'preciosities' from the largely socially unaffected 'external area' of contact in Africa and most of Asia, in contrast to the trade in lower-ranking goods from the 'periphery' of the Americas, and the East Indies, which trade had led to massive social change, division of labour, *and* land use in those areas (Wallerstein 1974: 301–2).

Many of these 'lower-ranking' goods were tropical products such as tea, coffee, chocolate, sugar, potatoes and cotton that related to needs of food and clothing, though some were more optional like tobacco. For example, the stimulants and tonics of tea, coffee and chocolate were introduced into Britain during the 1650s and the Low Countries a few decades earlier. Braudel (1973) attempted to trace the origin and diffusion of these commodities of material life. What is clear is that, initially, in Europe these goods were regarded as exotic products or luxuries that only the wealthy could afford. But from the mid-seventeenth century onwards, rising affluence for many, better transport, and a cheapening of production meant that they became the staples or necessities of a large part of the population. When that transition got under way the impact on the biomes of the world really began. The impact became greater and more purposeful with the settlement in the temperate lands of North America, Argentina and southern Brazil in later years.

Land cover change

The history of vegetation change is very largely the history of the expansion of arable land and the large-scale conversion of natural into managed ecosystems. Data on this expansion are confused and blurred, and the best we have is Table 9.1, which is a modelled estimate of overall changes, 1700 and 1850, with population numbers for the same years. It would be fascinating to know what had happened to land use between 1600 and 1750. Table 9.1 does suggest a significant change during the eighteenth century, especially in the core, and Figure 9.1 puts the shift into a temporal perspective. Between 1700 and 1850 world forests had declined by 250 million ha (–4 per cent), pastures by 23 million ha (–0.3 per cent), and world cropland had risen by 272 million ha (+102.6 per cent). In some cases, pastures increased as forests were cleared. The rise in cropland is important as it is the most easily traceable change. Predictably, the greatest proportionate increases were in Europe, USSR, and North America, but also occurred in China and Latin America.

Table 9.1 Possible increase in population and cropland, 1700–1850, and change in cropland, forests and woodland, grassland and pasture, 1700–1850 (land use and land use change in millions ha)

	Population (millions)		Cropland (million ha)		Cropland change 1700–1850	Forests and woodland change 1700–1850	Grassland and pasture change 1700–1850
	1700	1850	1700	1850			
Core							
Europe	92	208	67	132	+65	−25	−40
USSR	30	79	33	94	+61	−71	+10
N. America	2	25	3	50	+47	−45	−1
Total	124	312	103	276	+173	−141	−31
Periphery and external areas							
Africa	107	111	64	84	20	−26	+5
Asia	435	371	57	78	21	−19	−2
China		430	29	75	46	−39	−7
Latin America	10	34	7	18	11	−25	+13
Oceania	3	2	5	6	1	−	−1
Total	555	948	162	261	99	−109	+8
World total	679	1,260	265	537	272	−250	−23

Sources: Richards (1990); Demeny (1990).

The neo-European world

In the neo-European world the virtues of agriculture, freehold tenure, dispersed settlement, 'improvement' and personal and political freedom were extolled, which all led to a rapid and successful expansion of settlement. The United States was the classic example of land transformation through forest clearing. Before 1850 approximately 46 million ha were felled (Williams 1989a), the overwhelming bulk of it to create farmland. Such clearing was a combination of 'sweat, skill and strength' (Ellis 1946: 73), and the pioneer farmer was seen as the heroic subduer of a sullen and unyielding wilderness that needed to be tamed. It was a universal and integral part of rural life. 'Such are the means', marvelled the French traveller, the Marquis de Chastellux in 1789,

> by which North-America, which one hundred years ago was nothing but a vast forest, is peopled with three million of inhabitants ... Four years ago, one might have travelled ten miles in the woods I traversed, without seeing a single habitation.
>
> (Chastellux 1789, Vol. I: 47)

But it was not only the United States where European pioneers were hacking out a life for themselves and their families in the forest. There were other parts of the world, such as Canada, New Zealand, South Africa and Australia where the same was beginning to happen. The initial probes into the forest lands were destined to accelerate and expand rapidly during the ensuing nineteenth century.

Europe

However, in focusing on the overseas expansion of Europe one should not forget that the continent was also colonized internally during these centuries (Darby 1956), particularly eastwards into the mixed forest zone of central European Russia. The increase in the amount of land cultivated was phenomenal and rose from an estimated total of *c.* 100 million ha in 1700 to a known total of 231.9 million ha by 1860 (Table 9.2).

Much of this was a response to expanding population numbers which more than doubled during the same period. But it is likely that as much was due to the vast expansion of world trade that had occurred in the first three-quarters of the seventeenth century well before the Industrial Revolution got under way. Europe's trading connections with America and Asia were making an important contribution to its prosperity as they offered large markets for her manufactures in exchange for the exotic foods and raw materials that Europe itself could not produce. In addition, capital was invested in land improvement, particularly in the Low Countries and England, so that there was land-use intensification as well as the extension of land use. The 'frontier' was being actively colonized by

Table 9.2 Area converted to regular cropping, 1880, 1920, 1978 (million ha)

	Area cropped			Net change		
	1860	*1920*	*1978*	*1860–1919*	*1920–1978*	*1860–1978*
Core						
Europe	120.0	140.6	141.7	20.6	1.1	21.7
USSR	81.0	169.0	231.9	88.0	62.9	150.0
Canada/USA	76.0	237.2	235.7	161.2	−1.5	159.7
Total	277.0	546.8	609.3	269.8	62.5	331.4
Periphery						
Africa	63.0	79.9	164.4	16.9	89.5	106.4
Asia	211.0	302.5	445.4	91.5	142.9	234.4
Latin America	17.0	57.1	140.5	40.1	83.4	123.5
Oceania	4.0	19.1	59.1	15.1	40.0	55.1
Total	295.0	458.6	809.4	162.4	356.8	519.4
World total	572.0	1,005.4	1,418.7	433.4	418.3	850.8

Source: Richards (1986).

agriculturalists, and over 6.7 million ha of mixed forest in central European Russia was eliminated between the end of the seventeenth century and the beginning of the twentieth (French 1983).

China

These land frontiers in Europe and the neo-European lands were paralleled by the vast inland frontier of China, which was a world unto itself. The Ching Dynasty (1644–1911) was a highly centralized, bureaucratic and expansionist regime. None the less, at the local level a fairly liberal and flexible market system encouraged a merchant class, credit, and thriving market towns. The feudal system of landholding and tenure gave way to a greater commercialization, and the country undoubtedly profited from the exports of the new-found European delight – tea – which was said to be more valuable in trade than gold.

Richards (1990a: 171) calculates that the cultivated area rose 'from 33 (+/–7) million ha in 1600 under the Ming to 63 (+/–7) million ha in 1766, or nearly double . . . Over the next century, cultivated area grew more slowly, to 81 (+/–3) million ha in 1873.'

Much of this new cultivation was in the northern forests and grasslands, but it must have happened in all parts of the Chinese realm, as is shown in Perdue's (1987) detailed study of mainly seventeenth-century forest clearance and land reclamation in the south-central province of Hunan.

The tropical and sub-tropical world

Elsewhere, and particularly in the sub-tropical and tropical forests, European systems of exploitation led to the harvesting of indigenous tree crops (particularly hardwoods for domestic and naval construction), but also to the replacement of the original forest by the mono-culture of crops grown for maximum returns in relation to the labour and capital inputs. The 'plantation' system was now perfected, worked either by indentured labour but mostly by slaves, one of the new 'commodities' traded around the world. Classic examples of this were the highly profitable crops of sugar in the West Indies (Watts 1987); coffee ('green gold'), and sugar in the sub-tropical coastal forests of Brazil (Monbeig 1952; Dean 1983); cotton and tobacco in the southern United States, and spices in the East Indies.

A variation of this agricultural clearing occurred in southern Asia where peasant proprietors were drawn into the global commercial market. Throughout the Indian sub-continent there was a complex and varied expansion of all types of crops, often for cash, that led to massive forest clearing (e.g. Richards and McAlpin 1983).

In emphasizing the direct and indirect European impacts on these traditional societies one should not forget that the latter were also hierarchical, exploiting their forests vigorously in a manner no more egalitarian or caring than Europe.

There is plenty of evidence from individual studies of, for example, south-west India (Nadkarni *et al.* 1989) and Hunan province in China (Perdue 1987) from the sixteenth century onwards to show that the commercialization of the forest was not a European invention. In pre-British south-west India permanent agricultural settlement existed side by side with shifting cultivation, and village councils regulated how much forest exploitation could be undertaken by agriculturalists. The forest was not regarded as a community resource; larger land owners dominated forest use in their local areas. Scarce commodities such as sandalwood, ebony, cinnamon, and pepper were under state and/or royal monopolies. In Hunan, a highly centralized administration encouraged land clearance in order to enhance local state revenues so as to increase the tax base and support a bigger bureaucracy and militia.

Simply, natural ecosystems everywhere were being exploited and were diminishing in size as a response to the expansion of cultivation which was related to increasing population numbers and the increasing complexity of societies. The changes were just slower than those unleashed by the Europeans with their new aims and technologies, and, most importantly, their consuming market and intercontinental trade links.

1800–*c.*1950: GLOBAL CONTROL

Land control

The nineteenth century witnessed an unprecedented escalation of change in the world's vegetation as European industrial technology interacted with European imperial (territorial) ambitions. At the opening of the period, Europe occupied or controlled approximately 35 per cent of the world's land surface; by 1878 that proportion had risen to 67 per cent; and on the eve of the First World War it was 84 per cent – more than four-fifths of the globe. Nowhere was too small, too remote, or too barren not to be incorporated in the new nationalistic empires (Meinig 1969). Concomitant with territorial control went the vigorous, expansive spread of European industrial technology, with its 'omnivorous demands for materials, its need for expanding markets and its development of even more efficient tools for overcoming distance and conquering peoples' (Meinig 1969: 231). Successive inventions and discoveries made the conquest of the new territories cheaper in cost, cheaper in lives, and more effective than ever before.

The global territorial control was broadly of two kinds. First there was the continued settlement and exploitation of the neo-European temperate lands of the United States, Canada, New Zealand, Australia, southern Africa and southern Latin America. For all intents and purposes these were 'plantations' in the older sense, simplified social and political offshoots of the European mother countries that were quickly incorporated into the periphery. Some, like the United States, were independent by this time, to be sure, but most of the others were not. But independent or not, they resembled each other in that they

196

were vigorous neo-European 'settler empires' that took on a life of their own. They experimented with, and exploited their resources, particularly the land, its soil and vegetation.

But the significant diffusion of the nineteenth century compared to previous centuries was the concentration on the tropical world, particularly Africa and Asia. These were the 'external areas' of intermittent contact, into which forays had been made to create coastal way stations and trading points in the past. These contacts had been shallow in depth, temporary in duration and rarely the basis of settlement and control. They contrasted to the 'periphery' of the Americas and Caribbean (Wallerstein 1974) which had been a theatre of colonization and settlement, as well as exploitation. Now Europe attempted to create colonies in Asia and Africa that were politically submissive and economically profitable to them in crops, minerals and other raw materials.

'The tools of empire'

It would be impossible to mention all the 'tools of empire' that evolved during the nineteenth century; there was a dazzling array of interacting inventions and discoveries. Some used natural resources, some integrated economic systems, and some allowed a greater manipulation and control of lands. However, some were especially significant in the creation of the 'new' imperialism in the new, tropical areas of colonization in Africa and Asia (Headrick 1981, 1986).

Disease, which had so favoured Europe in its expansion in the New World and the Antipodes, was now an obstacle to penetration. The mortality of European settlers, missionaries, and traders was in the order of 250–750/1,000, particularly in Africa. In addition, the sheer number of people in these territories made political control difficult. But solutions were at hand. Mortality dropped dramatically to 50–100/1,000 between 1820 and 1840 with the discovery and widespread prophylactic use of quinine to counteract malaria. Rates of European mortality were still ten times higher than for similar groups in Europe, but Africa was no longer 'the white man's grave' (Headrick 1981).

Now that access was possible the penetration of the interior was facilitated in succeeding decades by the development of steamboats (often gunboats) on the navigable rivers and sheltered coastal waters of Africa and South-East Asia, China, the Persian Gulf, and India, often transported to these locations in pieces and then reassembled *in situ*. Gunboats came into their own with devastating effect in the First Opium War in China where they demonstrated their effectiveness in prompting 'political persuasion' (Headrick 1981: 54). Another key technology establishing early superiority was the development of the breach-loading gun during the 1860s, and eventually of the machine gun in the 1880s; the 'arms gap' between colonizer and colonized was awesome.

But the 'political persuaders' of gunboats and machine guns counted for nothing if the conquered territory was not connected effectively to the metropole. And that was what distinguished the 'new' imperialism (whether in the settler or nationalistic empires) from the old imperialism of the fifteenth to seventeenth centuries. Goods, information and reports were the life-blood of the new global thrust, and Europe acquired the means 'to communicate almost directly with their remotest colonies, and to engage in an extensive trade in bulky goods that would never have borne the freight costs of any previous empire' (Headrick 1981: 130).

The world was transformed to a greater extent than ever before as space and time were compressed to allow an integrated global market of communications and goods. Previously backward, isolated subsistence economies were now drawn into the commercial world market. Political control usually meant taxes, which could only be paid for by selling crops for cash and the importation of foreign expertise. The network of communications meant that goods could be transported cheaply, and information flows meant instant price fluctuation, adjustment, and fine-tuning. All meant financial infrastructures, banking, credit, and a reliable monetary system.

Railways and steamships increased cargo loads, increased reliability, made human travel comfortable, but above all, reduced time and hence costs. One example must suffice. Before the 1830s a message and its reply from London to, say, Calcutta, may have taken up to two years. By the 1850s, with steamships and trains, it could take about three months. With the opening of the Suez Canal, a letter took about thirty days. With the advent of the telegraph a few years later a message could be sent and a reply received in a day. By 1924 the submarine cable network allowed a message to circle the globe in eighty seconds (Headrick 1983). Economic networks were established and new techniques were developed in order to exploit these territories.

Global land-cover change

While it is relatively easy to demonstrate how these technical innovations bound all parts of the world together and were instrumental in facilitating land-use change, it is much more difficult to be precise about the amount and nature of change on the land itself. But there are two lines of enquiry. First, there is the expansion of the world's arable land and the large-scale conversion of natural ecosystems to managed systems. Second, there is the abundant evidence of the processes whereby expansion occurred: for example, clearing, irrigating, etc.

A growing and spreading world population has caused more land to be brought into cultivation as it searches for food. World population started its upward and accelerating trajectory during this period, rising from 957 million in 1800, to 1,650 million in 1900. But after that medicine and better hygiene and nutrition triggered the exponential growth of modern times to 2,515 million in 1950, to double again to 5,000 million in 1985. Concomitant with this growth was the spread of people; about 60 million left Europe between 1850 and 1950, primarily for the Americas, but also for Australia, New Zealand and southern Africa.

198

Table 9.3 Area of different ecosystems converted to cropping, 1860–1978 (million ha)

	Forests	Woodlands	Savannahs	Grass lands	Wetlands	Deserts	Total
Core							
Europe	7.8	0.3	1.8	8.4	3.4	–	21.7
USSR	44.0	13.6	1.7	88.4	–	3.2	150.9
Canada/USA	51.1	13.0	10.7	84.9	–	–	159.7
Total	102.9	26.9	14.2	181.7	3.4	3.2	332.3
Periphery							
Africa	18.0	28.9	23.9	32.5	–	3.1	106.4
Asia	59.2	62.8	50.0	37.4	17.7	7.3	234.4
Latin America	38.8	25.3	16.1	40.1	0.1	3.1	123.5
Oceania	23.9	12.3	13.3	5.1	–	0.5	55.1
Total	139.9	129.3	103.3	115.1	17.8	14.0	519.4
World total	242.8	156.2	117.5	296.8	21.2	17.2	851.7

Source: Revelle (1984).

Note: In Richards (1986), the conversions were revised as per Table 9.1 and the overall adjustment of 7 million ha was assigned to forest ecosystems in the above table by this author, as many of the adjustments concerned forest regrowth.

Undoubtedly, the need for subsistence has fuelled, and continues to fuel, the drive for expansion, but probably far more important during the earlier years of this period was the commercialization of agriculture and market orientation of cropping by peasant proprietors, sharecroppers and estate and plantation workers, for an integrated global economy. All the communication and fiscal tools of expansion mentioned so far played a complementary part in this expansion.

The total amount of land in regular cropping in 1860 was approaching 600 million ha (Richards 1986). During the succeeding sixty years to 1919, an additional 433.4 million ha net were added to this amount, and a further 418.3 million ha net from 1920 to 1978 (Table 9.2), so that the world's arable land increased by 851.7 million ha, to become 2.5 times the extent it had been in 1860.[1] It is also possible to calculate the approximate areas of the different natural ecosystems that were affected by the transformation of land to agriculture (Table 9.3). Approximately 242.8 million ha of forest and 156.2 million ha of woodland, or 46.9 per cent of the total global area, were converted to agricultural land. Also another 117.5 million ha of savannah were also cleared. In addition to woodlands of varying density, 296.8 million ha of grasslands were also converted.

The expansion of arable land was equally vigorous in both the New and the Old World areas alike. Everywhere the trend has been upward. Only in the more industrialized parts of North America, Europe and Eastern Asia has there been any reversion, largely through the retreat from agriculturally marginal areas. Also, the intensification of output from the most productive areas with chemical innovations such as fertilizer, biological innovations such as higher-yielding crop varieties, and mechanical innovations such as more comprehensive farm machinery, has lessened the amount of land needed (Grigg 1987).

The processes of change

It goes without saying that processes of land transformation modify, replace, or eliminate existing vegetation covers. Such processes can be divided into three broad, and at times overlapping, categories. They are:

1 *Transference*, which includes fertilizing, applying water from an external source, using machinery and hence fossil fuels.
2 *Replacement*, which includes such actions as irrigating, draining, clearing/firing/cutting, terracing, and ploughing and planting.
3 *Harvesting*, which includes timber cutting, fuelwood gathering, grazing, hunting and fishing, gathering, mining, and slash-and-burn cultivation.

It would be impossible to deal with all of these in the space available so just three of the most important in the replacement category (clearing, draining and irrigating) are singled out for treatment.

Clearing

It is possible that the cutting down of trees has been the most important factor in altering the face of the earth (Williams 1989b, 1990a). The forest has been subjected to a sustained and steady attack by humankind throughout the centuries. The fact that the forest provides wood for construction, shelter and warmth as well as the creation of 'new' land for growing crops, explains the vast devastation that has happened.

Overall figures are difficult to calculate but it is probable that during the course of human history between 750 and 806 million ha have been eliminated, over 50 per cent of that total during the years since 1850. Some of the better-documented and larger regional impacts are those in the United States where over 77 million ha were cleared for agriculture between 1850 and 1910 (Williams 1989a); in Brazil, where about 14 million ha was cleared for coffee growing by 1931 (Williams 1990a); and South and South-East Asia where forests declined from 88.8 million ha in 1880 to 67.5 million ha in 1950 to 54.2 million ha in 1980 as a result of expanding cultivation, while interrupted forests declined overall by 9.4 million ha during the same two end dates (Richards *et al.* 1987). Smaller regional stories of deforestation are too numerous to mention.

Clearing will not diminish in extent and intensity in the future. Tropical forests, in particular, are one of the world's last great frontiers in which new agricultural land can be created. But it will no longer be Europe that spearheads that transformation, but the rapidly expanding population of the developing world.

Draining

Whereas the process of forest clearing is piecemeal, individual and largely unrecorded, the process of draining land of its excess water is well planned, communal, and reasonably well documented. Moreover, draining systems have to be maintained in order to function efficiently, and that has to be an agreed co-operative project. Without vigilance, the carefully controlled agricultural landscape could soon become wetland again.

Draining has existed for probably well over two millennia as a means of transforming land and raising agricultural productivity. The knowledge and experience that periodically flooded lands often produced above-average yields when dry was a spur to draining. For over a thousand years the coasts of the North Sea, particularly in the Low Countries and in similar lowlands in England, such as the Fens, the Somerset Levels and Romney Marsh, all experienced small-scale communal reclamation often organized under the aegis of powerful ecclesiastical houses or secular landholders (Williams 1990b). Similar, and largely unrecorded, reclamation occurred in low-lying wetlands in the early civilizations of South and South-East Asia (Richards 1990b).

But the big increase in wetland draining came from the mid-nineteenth century onwards when western technology in the form of steam-powered excavators and draglines (first demonstrated successfully in the construction of the Suez Canal in 1869) lowered costs and increased returns. In addition, powerful mechanical pumps replaced windmills and animal power, cheaply manufactured earthenware pipes facilitated underdraining, and generally increased hydrological knowledge made larger and more ambitious schemes possible.

Table 9.4 Documented changes in wetland due to draining

	Approximate date of activity	Approximate area ('000 ha)
Temperate wetlands		
Fens	Post *c.* AD 800	700
Holland	Post *c.* AD 800	2,000
SE of S Australia	Post 1865	1,700
USA	Post 1780	39,000
Tropical wetlands		
Undarbans	Post 1800	800
Irrawaddy Delta	Post 1800	700
Zhuziang Delta	?	1,200

Source: Williams (1990b), pp. 181–233, and Richards (1990b), pp. 217–33.

How much land has been drained is difficult to calculate. A modest estimate is 160 million ha, nearly three-quarters of which has occurred in the temperate world. Like clearing, instances of draining are numerous and scattered and it is difficult to get an overall picture. Table 9.4 shows a number of major regional examples.

Irrigation

Like draining the process of irrigation is old, but it too has to be well-organized and co-operative because it involves the control of a highly dynamic element, water. Modern Western technology, in the form of mechanical means of excavating and the development of engineering skills to build high dams and convert falling water to electricity, have ensured the continued expansion of the process. In addition, as Richards (1986: 63) points out, it is 'a much more visible, and appealing' process than any other because 'making the deserts bloom responds to some of our deepest aesthetic and cultural instincts. The drama of towering dams, huge turbines and massive canal systems has made large irrigation systems one index of modernity.'

Since the latter part of the nineteenth century the pace of irrigation works in the world's drier regions has accelerated almost exponentially. Early twentieth-century multiple-use projects on the Colorado and Columbia in the American West are well known, and during the subsequent years these schemes were augmented and accompanied by others on the Missouri and by the withdrawal of underground waters. By 1978 there was 17.7 million ha of irrigated land in the western USA.

Less well known, but larger in size, were the massive irrigation systems initiated by the British on the waters of the Ganges, Godavari, and Krishna in Northern India from the 1850s onwards, and then on the Indus and its five territories in the dry Punjab and Sind in Pakistan during the early years of this century. By 1917 the Triple Canal project watered over 2.2 million ha of the Punjab. By 1939 approximately 11.6 million ha of land were irrigated in colonial British India by over 116,000 km of canals (Richards 1986), and by the late 1970s that figure had risen to 26.1 million ha in India and Pakistan, supplemented by 19.8 million ha of tube-well irrigation. In total, therefore, a massive 45.9 million ha came under irrigation in the sub-continent.

Similar stories could be told of Egypt, Sudan, Australia, Mexico, Iraq, South Africa, and the former Soviet Union, so that by 1979 it has been calculated that 207 million ha, or 14.3 per cent, of the world's arable land is irrigated land (Richards 1986). The point has been reached where few rivers flowing through arid or semiarid regions are left untapped.

However, not all is gain and there is much loss of the irrigated land which has been won so expensively. Over-watering causes waterlogging and salinization, erosion causes sedimentation of reservoirs – all of which may be eliminating each year as much productive arable land as is being created.

Sluggish irrigation water encourages the spread of killer diseases such as malaria and schistosomiasis. But whatever the downside, the glamour of irrigation ensures its continuation as an earth-transforming process.

CONCLUSION

The breaking down of the biological isolation of the continents that had existed for millennia, and the replacement of one vegetation cover by another, was the supreme example of what Columbus's voyages really meant. Although not in conscious control initially, the biological alterations Europeans wrought were soon augmented by many purposeful changes as large areas of the earth's surface were either cultivated or grazed more intensively.

The human control of the earth is increasing. The comprehensive global political control begun by Europe in the early nineteenth century has now been replaced by comprehensiveness of individual state land use and ecological control. This is not only a result of settlement expansion but also of tenure control and intervention to manage activities on every hectare of land in some form or another.

NOTE

1   In 1990 Richards recalculated these figures for slightly different dates. He suggested that the amount of land in regular cropping in 1850 was 537 million ha, which rose to 913 million ha in 1920, 1.17 billion ha in 1950 and 1.501 billion ha in 1980, a total increment of 964 million ha in 130 years.

REFERENCES

Barraclough, G. (ed.) (1978) *Times Atlas of World History*, London: Times Books.
Braudel, F. (1973) [1967] *Capitalism and Material Life, 1400–1800* (trans. by M. Kochan), London: Weidenfeld & Nicolson.
Braudel, F. (1984) [1981] *Civilization and Capitalism, 15th–18th Century*. Vol. III, *The Perspective of the World*, New York: Harper & Row.
Buringh, P. and Dudal, R. (1987) 'Agricultural land use in space and time', pp. 9–44 in M.G. Wolman and F.G.A. Fournier (eds) *Land Transformation in Agriculture*, SCOPE 32, Chichester: John Wiley & Sons.
Chastellux, F.J., Marquis de (1789) *Travels in North America in the Years 1780, 1781, and 1782*, 2 vols, New York: White, Gallaher & White.
Crosby, A.W. (1986) *Ecological Imperialism: The Biological Expansion of Europe, 900–1900*, New York: Cambridge University Press.
Darby, H.C. (1956) 'The clearing of woodland in Europe', pp. 183–216 in W.L. Thomas (ed.) *Man's Role in Changing the Face of the Earth*, Chicago: University of Chicago Press.
Dean, W. (1983) 'Deforestation in southeastern Brazil', pp. 50–67 in R.P. Tucker and J.F. Richards (eds) *Global Deforestation and the Nineteenth-Century World Economy*, Durham, N.C.: Duke University Press.
Demeny, P. (1990) 'Population', pp. 41–54 in B.L. Turner, II, W.C. Clark, R.W. Kates, J.F. Richards, J.T. Mathews and W.B. Meyer (eds) *The Earth as Transformed by*

Human Action, New York: Cambridge University Press.

Denevan, W.M. (ed.) (1992a) [1976] *The Native Population of the Americas in 1492*, Madison: University of Wisconsin Press.

Denevan, W.M. (1992b) 'The pristine myth: the landscape of the Americas in 1492', *Annals, Association of American Geographers* 82, 369–85.

Dobyns, H.F. (1966) 'Estimating aboriginal American populations. An appraisal of techniques with a new hemispheric estimate', *Current Anthropology* 7, 395–449.

Doolittle, W.E. (1992) 'Agriculture in North America on the eve of contact: a reassessment', *Annals, Association of American Geographers* 82, 387–401.

Ellis, D.M. (1946) *Landlords and Farmers in the Hudson–Mohawk Region, 1790–1850*, Ithaca, N.Y.: Cornell University Press.

French, R.A. (1983) 'Russians and the forest', pp. 23–44 in J.H. Bater and R.A. French (eds) *Studies in Russian Historical Geography*, Vol. 1, London: Academic Press.

Grigg, D.B. (1987) 'The industrial revolution and land transformation', pp. 79–109 in M.G. Wolman and F.G.A. Fournier (eds) *Land Transformation in Agriculture*, SCOPE 32, Chichester: John Wiley & Sons.

Headrick, D.R. (1981) *The Tools of Empire: Technology and European Imperialism in the Nineteenth Century*, Oxford: Oxford University Press.

Headrick, D.R. (1986) *The Tentacles of Progress: Technology Transfer in the Age of Imperialism, 1850–1940*, Oxford: Oxford University Press.

Jones, E.L. (1987) *The European Miracle: Environments, Economics and Geopolitics in the History of Europe and Asia*. 2nd edn. Cambridge: Cambridge University Press.

Kindleberger, C.P. (1984) *A Financial History of Western Europe*, London: George Allen & Unwin.

Lovell, W.G. (1992) '"Heavy shadows and black night": disease and depopulation in colonial Spanish America', *Annals, Association of American Geographers* 82, 426–43.

McNeill, W.H. (1976) *Plagues and People*, Garden City, N.Y.: Anchor.

Maxwell, H. (1910) 'The use and abuse of the forests by the Virginia Indians', *William and Mary College Quarterly* 19, 73–104.

Meinig, D.W. (1969) 'A macrogeography of Western imperialism: some morphologies of moving frontiers of political control' pp. 213–40, in F.G. Gale and G.H. Lawton (eds) *Settlement and Encounter: Geographical Essays Presented to Sir Grenfell Price*. Melbourne: Oxford University Press.

Monbeig, P. (1952) *Pionniers et Planteurs de São Paulo*, Paris: Librairie Armand Collin.

Nadkarni, M.V., Pasha, S.A. and Prabhakar, L.S. (1989) *The Political Economy of Forest Use and Management*, New Delhi: Sage Publications.

Perdue, P.C. (1987) *Exhausting the Earth: State and Peasants in Hunan, 1500–1850*, Cambridge, Mass.: Harvard University Press.

Revelle, R. (1984) 'The effects of population growth on renewable resources', in *Population, Resources, Environment and Development*, International Conference on Population, 1984. Department of International Economic and Social Affairs. *Population Studies* 90, 223–40, New York: United Nations.

Richards, J.F. (1986) 'World environmental history and economic development', in W.C. Clark and R.E. Munn (eds) *Sustainable Development of the Biosphere*, Cambridge: Cambridge University Press.

Richards, J.F. (1990) 'Land transformation', pp. 163–78 in B.L. Turner, II, W.C. Clark, R.W. Kates, J.F. Richards, J.T. Mathews and W.B. Meyer (eds) *The Earth as Transformed by Human Action*, New York: Cambridge University Press.

Richards, J.F. (1990b) 'Agricultural impacts in tropical wetlands: rice paddies for mangroves in South and Southeast Asia', pp. 217–34 in M. Williams (ed.) *Wetlands: A Threatened Landscape*, Oxford: Basil Blackwell.

Richards, J.R., E.S. Haynes, J.R. Hagen, E.P. Flint, J. Arlinghaus, J.B. Dillon and A.L.

Reber (1987) 'Changing Land Use in Pakistan, Northern India, Bangladesh, Burma, Malaysia and Brunei, 1880–1980'. *MS Report to US Department of Energy.*

Richards, J.F. and McAlpin, M.B. (1983) 'Cotton cultivating and land clearing in the Bombay Deccan and Karnatak, 1818–1920', pp. 68–94 in R.P. Tucker and J.F. Richards (eds) *Global Deforestation and the Nineteenth-Century World Economy*, Durham, N.C.: Duke University Press.

Tilley, C. (1990) *Coercion, Capital and European States, AD 990–1990*, Oxford: Basil Blackwell.

Wallerstein, I. (1974) *The Modern World System I: Capitalist Agriculture and the Origins of the European World-Economy in the Sixteenth Century*, London: Academic Press.

Wallerstein, I. (1980) *The Modern World System II: Mercantilism and the Consolidation of the European World Economy, 1600–1750*, New York: Academic Press.

Watts, D. (1987) *The West Indies: Patterns of Development, Culture and Environmental Change since 1492*, Cambridge: Cambridge University Press.

Webb, W.P. (1952) *The Great Frontier*, Boston: Houghton Mifflin.

Whitmore, T.M. and Turner, B.L., II (1992) 'Landscapes of cultivation in Mesoamerica on the eve of the Conquest', *Annals, Association of American Geographers* 82, 402–25.

Whittlesey, D.S. (1944) *The Earth and the State: A Study in Political Geography*, New York: Holt.

Williams, M. (1989a) *Americans and their Forests: A Historical Geography*, New York: Cambridge University Press.

Williams, M. (1989b) 'Deforestation: past and present', *Progress in Human Geography* 13, 176–208.

Williams, M. (1990a) 'Forests', pp. 179–202 in B.L. Turner, II, W.C. Clark, R.W. Kates, J.F. Richards, J.T. Mathews and W.B. Meyer (eds) *The Earth as Transformed by Human Action*, New York: Cambridge University Press.

Williams, M. (1990b) 'Agricultural impacts in temperate wetlands', pp. 181–216 in M. Williams (ed.) *Wetlands: A Threatened Landscape*, Oxford: Basil Blackwell.

10

THE ORIGINS OF THE CAPITALIST WORLD ECONOMY

John Langton

HUMAN ECOLOGY, HUMAN GEOGRAPHY AND THE CAPITALIST WORLD ECONOMY

Fewer than 10 per cent of the employed population are involved in producing food and other sources of energy in modern industrial societies, and over 98 per cent of total energy output is free for circulation, surplus to the needs of its producers (Cook 1971). People are narrowly specialized, partially producing a good or service which is consumed by many others, for each of whom it represents a very small part of total consumption. About 40 per cent of energy consumption occurs at specialist places of work outside people's homes. A further one-quarter is used in the movement of people, energy, goods and services necessitated by the complexity of interdependence (Cook 1971).

All these activities need capital: fixed capital in the form of machines, buildings and other infrastructures on and in which work is done, and variable capital comprising stocks of raw materials, finished goods, and the money used to pay for work in advance of selling products. This capital links people to nature and to each other (Smith 1984). Used as capital, money retains the worth of something previously possessed which has been exchanged for it. It can be invested to buy work, plant and materials to add value to some current possession before it is sold as a commodity: something created specifically to sell (Harvey 1982). The difference between the cost of the capital and work required to make a commodity and its sale price is profit.

> It is only in so far as the appropriation of ever more wealth in the abstract becomes a person's operations that he functions as a capitalist ... Use-values must therefore never be looked upon as the real aim of the capitalist; neither must the profit on any single transaction. The restless never-ending process of profit-making alone is what he aims at.
>
> (Marx, *Capital*, quoted in Goldsmith 1978: 64)

206

The machinery and other fixed capital through which energy, labour and materials are combined to make commodities change perpetually as money endlessly chases them endlessly to multiply itself.

Money is borrowed at interest in order to invest, land is purchased to yield rent, and the labour of workers is bought for wages. Capitalist production involves 'no less a transformation than that of the natural and human substances of society into commodities' – or rather, into fictional commodities, because 'land, labor and money are obviously *not* commodities: the postulate that anything that is bought and sold must have been produced for sale is emphatically untrue in regard to them' (Polanyi 1957: 42, 72). None the less, they must be treated as commodities for capitalist production to occur.

Commodification requires 'the Rise of the Market, the Rise of the Exchange Economy . . ., which is antecedent to [the] Rise of Capitalism' (Hicks 1969: 7), and for which money is also necessary as a measure of the relative value of other commodities. Its use allows each exchange in the web of interpersonal economic relationships to be entered into voluntarily by both parties, who can accurately evaluate it relative to every other exchange in which they are or might be involved, and terminate it completely in the act of making it. Money brings anonymity to economic exchange and allows it to be extricated from social interaction (Hart 1986).

Maximum mutual economic benefit comes from maximum freedom to enter into exchange relationships (Wallace 1990: 18). The complete freedom of buyers to pursue the cheapest price by choosing between all possible suppliers of commodities maximizes the efficiency of production. Complete freedom of sellers to seek the biggest profit by directing effort to where the greatest relative scarcities are indicated by the highest market prices maximizes the total value of what is produced, the rate of investment, and the rate of capital accumulation.

For individuals to benefit from producing according to market signals, the resources transformed in doing so must belong to them. A market economy needs private rights in property: the complete freedom of owners of commodities to alienate them at will (North and Thomas 1973: 19–24). It also requires the existence of free pools of commoditized capital and labour which producers can purchase competitively to switch between different production projects in response to price signals. These necessary freedoms are contradictory. Producers benefit from limited competition in the markets for their commodities, but maximum competition in the markets for labour, capital and raw materials, all of which are commodities sold by others. To maximize the rates of profit and capital accumulation within the system as a whole, consumption must be minimized; but this reduces markets for the commodities produced. If wages, interest rates, rents and raw material prices are low, so is the amount of commodities their recipients can buy. Owners of capital benefit from impoverishing their workers and suppliers, but only in so far as they are not their customers. In which case overaccumulation of capital results, as the capacity to produce exceeds that to consume (Harvey 1982: 192–6). Free markets are not, therefore, benign arrangements operating to everyone's benefit, but create contradictions between individuals' interests as buyers and sellers, and between the rates of profit and capital accumulation and the rate of consumption.

These contradictions are resolved by imposing controls on the freedoms of buyers or sellers, even in 'free' markets. 'The combination of freedom and unfreedom that results is the defining characteristic of a capitalist world-economy' (Wallerstein 1979: 159). States formulate and exert these controls across their territories, guaranteeing order and security in return for taxes and other dues. From this tribute (which flows outside, and therefore withdraws exchanges from, the market), states equalize minimum levels of well-being, supervise uniform rules for economic and other interpersonal relationships within their boundaries, and articulate their members' interests externally by manipulating international markets so that foreign goods are cheap and goods sold abroad are dear. All states aim to export poverty and import wealth simultaneously. Like individuals, they are inevitably in economic conflict, and succeed to the extent that they have the power to manipulate markets to their advantage.

Capital, money, commodities, markets and states are the defining elements of the capitalist world economy (Wallerstein 1979). They ensure that economic relationships are no longer embedded in environmental and social connections (Germani 1973). Where people live has little effect upon what they are able to consume. They can move as individuals between towns in the four corners of the globe without having greatly to alter their patterns of consumption, social relationships or culture (Peet 1986). Capitalism supports 'biosphere people' who, wherever they are, can draw upon a vast array of ecosystems and interact economically and socially with people scattered across the earth (Klee 1980). Nowhere do they depend on natural processes and other people in the place where they live. In the process of producing commodities, capital produces a space through which places are linked, and in which they are organized (Smith 1984). Its position in this space determines what a place is like relative to other places. In consequence, the difference between the highest and lowest national energy consumption per capita in 1980 was 1,162:1 (*Times Atlas*, 1980: xx–xxi), compared with the 20:1 difference in the natural productivity of the earth's biomes (Odum and Odum 1976). Global cities and backward provinces, cores and peripheries, First World and Third World are not inherent in nature, but produced by the organization of space by flows of capital and commodities.

The basic proposition of human ecology is obviously true: the culture and human geography of economies reflect the amounts and sources of energy used in them (White 1943): 'money can go around only if energy flows through the system to support the work that money buys' (Odum and Odum 1976: 52). The availability of mineral energy in what are effectively limitless quantities, at the cost of as little as one four-hundredth of the amount derived (Odum and Odum 1976), brought the 'ecological transition' which unbound Prometheus (Landes

1969). The diminishing returns to human effort inherent in organic economies were replaced by increasing returns to scale and subdivision of labour by the Industrial Revolution (Wrigley 1987a). The use of mineral energy in its own movement makes it extremely mobile. Even the conversion of coal to electricity and long-distance transmission to the point of consumption give a net yield of 5:1 (Odum and Odum 1976). The massive spatial concentration of consumption of things produced at widely scattered places, represented by urbanization and the global patterning of wealth, could not exist without the spatially footloose energy provided by minerals.

However, towns, money and market exchange over long distances developed in agricultural societies, and it can be argued that the ecological transition *followed* the development of a fully capitalist market economy (Wallerstein 1974).

AGRICULTURAL SOCIETY AND THE ORIGIN OF THE WORLD CAPITALIST SYSTEM

Despite the chronological and geographical span, the Nile Valley cultivator of Pharaonic Egypt and the Mecklenburg farmer in the fifteenth century would have understood each other's daily struggle to sustain family life because both were tribute producers in essentially similar tributary societies.

(Martin 1991: 92)

Commandeered tribute was prevalent in agricultural societies because dependence on a crop tied its producers to the place where it was grown, making them vulnerable to groups offering protection from natural and human threats to their harvest. Through fear (Hicks 1969), cultivators became *peasants*: people who satisfied their own needs from their crops, and produced a surplus to support others 'who [did] not carry on the productive process themselves, but [assumed] instead special administrative and executive functions, backed by the use of force' (Wolf 1966: 3).

Improved agricultural techniques increased per capita energy availability from 5,000 to 26,000 kilocalories per head per annum, and population densities from $c.0.4$ to $c.4.9$ per km^2 (Cook 1971; Deevey 1960). Increased peasant output also yielded more tribute. However, agriculture has a high ratio of energy consumed to energy produced; in rural India, 82 per cent of total energy use is directly related to food production, 55 per cent in the country as a whole (Simmons 1989). In medieval Europe, nearly 30 per cent of energy consumption was still in the form of food, and 70 per cent occurred in domestic homes (Cook 1971). From 80 to 98 per cent of the population belonged to self-provisioning peasant households (Crone 1989). Less than one-third of output was free to pass as tribute to the patriciate. A thousand years earlier the proportion must have been much smaller. In parts of early modern Europe it was over one-half (Blum 1982).

'The division of labour between producers and maintainers of order is ...

209

conducive to a highly unequal relationship' (Crone 1989: 7). Patriciates, freed from direct work on nature, accumulated and exchanged movable wealth, developed systems of writing and abstract thought, produced art, and intellectualized religious belief. Although peasants 'bore the weight of the entire social pyramid' (Blum 1978: 29), rooted to the soil, they were thought to be 'a hybrid between animal and human' (Blum 1982: 65).

Peasant economy, society and culture

Peasants were 'ecosystem people' (Klee 1980: 1–2), which was reflected in their economic practices, social relations and cultures. They were completely reliant upon their own resources, and had little use for production beyond what they consumed. Apart from giving up tribute, they remained largely autarkic economically, as they strove to minimize the risk to their subsistence rather than maximize monetary profits.

In consequence, the peasant's economy was 'submerged in his social relationships' (Polanyi 1957: 46). Family households were units of production and socialization as well as consumption (Shanin 1987). As much of total subsistence as possible was produced through polyculture from land belonging to each household. Marriage needed enough land to support a separate family. Only family labour was used; at stages in the family life-cycle when labour was insufficient, boys or girls (usually kinsfolk) surplus to their own families' labour needs were brought into the household and lived as fully-fledged members of it. Land was passed from parents to children who, in the absence of spare land, could not marry until their parents stopped farming. Neither land nor labour was commoditized, but undetachable from family households. This made peasant societies demographically homeostatic (Wrigley 1969). Their history was immobile (Le Roi Ladurie 1979).

Co-operation, reciprocity and common property with village neighbours yielded economies of scale and best guaranteed that enough would always be available (Dahlman 1980), and powerful redistributive mechanisms existed to eliminate any differentiation in wealth which did emerge between families (Foster 1965). Communalism locked them together into maximally secure systems of uncosted and unquestioned mutual aid (Redfield 1960). The peasant economy was a *moral* one: exchanges within it were mediated by mutual obligations and reciprocal duties defined according to social criteria (Scott 1976).

Peasants developed complex systems of exchange, spreading risk by linking the outputs of varied environments. But the goods involved and their prices were normally fixed by custom rather than the interaction of supply and demand. Where commoditization did develop in the exchange of surpluses (Hodges 1988), peasants' 'survival and reproduction [was] not *dependent* on the sale of their products on the market . . . they [did] not have to *compete* in terms of their productive powers' (Brenner 1977: 37, emphasis added). Because the objective of their work was not financial gain through the sale of commodities, peasants could not conceive of investing money. Their experience of markets was to their disadvantage. When there was something to sell after harvest, prices were low; when foodstuffs had to be bought in years of dearth, they were high. Rents, tithes and taxes were levied after the harvest, so that the market seemed simply to be a means of transferring part of their output into the hands of the patriciate: an instrument of exploitation. 'The net effect was ... the temporary wide diffusion of coin followed by its sudden reconcentration as the city sucked the countryside dry again' (Spufford 1988: 385). Any cash remaining in peasant hands was hoarded against future dearth, to provide daughters' dowries, or to purchase land for younger sons.

Through space, one economically autarkic cell succeeded another almost *ad infinitum*. Adjacent communities differed little, but they varied with environmental conditions, and therefore with the type of farming that would best guarantee survival. Wholly dependent upon their own work in their local environment, peasants were both conservative and conservationist (Scott 1976; Klee 1980), with a backward-referencing culture of survival (Berger 1985). They were born, lived and died within one of a myriad 'little communities' (Redfield 1960), carrying forward orally and unchanged the varied 'little cultures' appropriate to the existence of ecosystem people (Anderson 1971). 'This self-sufficient society was indestructible' (Kautsky 1976: 3). Inimical to the development of markets in commodities, land, labour or money, it 'could not provide the basis for [capitalist] economic development' (Brenner 1976: 63), but was an obstacle to it (de Vries 1975).

Patrician economies, societies and cultures

The far wealthier and more spatially uniform great cultures supported by peasant surpluses provided justice, protection, goods and services which people could not produce for themselves, and spiritual welfare (Anderson 1971). They comprised, in early modern European parlance, the three estates of lords, clergy and burgesses (the artisans, merchants, lawyers, doctors and so on of the towns). Each had its own privileges relative to the peasantry and other estates. They alone organized and were represented in government (Myers 1975).

Redistributive exchange between peasants and patricians did not simply bring net flows of goods from the former to the latter, but also between the places where they dwelled. Money, markets and towns developed to mediate them through, and therefore *organize*, space. It is to here that the origin of the capitalist world economic system is commonly traced.

The world economy, the state and colonialism

Influential theories of capitalist development attribute primary significance to the emergence of the global-scale pattern of flows which brought accumulation in a core area by draining a periphery of resources (So 1990). The establishment of this 'world system' has recently been traced back through Islamic hegemony over the Middle East, North Africa and parts of southern Europe, to Imperial Rome, China and India (Frank 1991), and beyond to the Middle East between 1700 and 500 BC (McNeil 1990). Trade between these empires brought to Europe knowledge of printing, gunpowder and the compass, the silk and sugar which first whetted its appetite for exotic goods, and the plantation, factory and slavery later imposed on its overseas colonies (McNeil 1990; Sollow 1986/7). However, whether this was exchange of a kind which represented or could lead to the emergence of capitalism *tout court* is another matter. It occurred within and between 'world empires', representing 'the extraction of tribute from otherwise locally self-administered direct producers that was passed upwards to the centre and redistributed to a network of officials' (So 1990: 177): markets were uninvolved in production and only partly in exchange.

A world system of flows articulated by market exchange and producing a pattern of core and periphery only came into being after the subdivision of labour through space was stretched across a complex of independent political units when, 'without a [single] political structure to redistribute the appropriated surplus, the surplus [could] only be redistributed via the "market"' (Wallerstein 1979: 159). The origin of the capitalist world economic system was therefore dependent upon the emergence of 'a single system of states in which change in one cell affected all the others ... [states] which were engaged in ... extending the market system, albeit for political ends' (Jones 1981: 104, 107). This happened, uniquely, in Europe in the century after 1450 (Anderson 1974).

The High Culture of medieval Europe had a veneer of homogeneity and interaction through its common inheritance from the Roman Empire and Catholicism. The re-emergence of empire was thwarted by immense topographical variety, coastal convolutions, and wide-openness to the east, which made Christendom difficult to organize and defend as a whole from Viking, barbarian and Islamic attacks round its perimeter; by natural temperateness, which rendered large-scale disasters needing widespread rescue of populations rare (Jones 1981); and by the very devolved political system represented by feudalism, which encouraged the fissipation of whatever larger units did develop. However, the introduction of pikemen, gunpowder and large infantry formations into warfare gave decisive advantages to units big enough to supply the numbers of men and amounts of money needed to deploy them (Anderson 1974). 'The organisational tasks attending the wars of the state system were so vast by the late seventeenth century that even kings were obliged to attend to them, becoming in the process more like heads of corporations rather than surrogate gods, ... A large part of the system's dynamic was an arms race' (Jones 1981: 110, 119). The erstwhile feudal nobility became the landed estate of these corporations, transformed into the holders, buyers and sellers of government offices (Anderson 1974), and the clergy became detached from Roman Christendom to justify and glorify the state and its monarch.

To finance war and the administration it required, states became territories from which to raise revenue, rather than hierarchies of personal fealties through which to summon feudal hosts of knights; and territory became a source of revenue to the state, in so far as it could be captured and defended. The bourgeois third estate benefited disproportionately from these changes. Centralized bureaucracies developed to raise money from territory through taxation and other fiscal devices, and laws were passed specifically to maximize the amount of revenue which would accrue to state governments from the areas they controlled (Jones 1981). These mercantilist policies were two-pronged. First, they laid burdens of taxation and military conscription on the peasantry, which was therefore protected as an indispensable resource of the state (Jones 1981). Second, they encouraged a positive balance of coin and bullion flows through trade or capture, which could be tapped through taxation and loans from its bourgeois organizers (Anderson 1974). England was peculiar because its government could not raise tax from land until the end of the seventeenth century. The only revenue available to the Crown without Parliament's permission was from customs and excise duties and the sale of trade monopolies (North and Thomas 1973).

Successful state revenue-raising was reflected in almost incessant wars of territorial expansion in sixteenth- and seventeenth-century Europe (de Vries 1976). Overseas campaigns to bring more territory, bullion and trade were an outgrowth of intra-continental conflicts. A trade system developed which in the broadest terms represented the import of American silver to acquire Asian goods (Tracy 1990). The wealth this brought increased the power of states: 'upon the navy depend the colonies, upon the colonies commerce, upon commerce the capacity of a state to maintain numerous armies' (Anderson 1974: 41). However, although colonial possessions were highly prized at the time, the colonial trade of Western Europe was actually of rather small significance, even in the late eighteenth and early nineteenth centuries.

> For the 1790s the geographical destination of commodity exports which crossed the boundaries of European states was: to other European states 76 per cent, to North America 10 per cent, to Latin America and the Caribbean 8 per cent, to Asia 5 per cent and to Africa 1 per cent. The 'periphery' (of Latin America, the Caribbean, Africa and Asia) purchased about 14 per cent of Europe's exports and in 1830 the same regions supplied some 27 per cent of European imports.
>
> (O'Brien 1982: 4)

England had a relatively small trade with the rest of Europe, at 68 per cent of the total in 1700, declining to 47 per cent by 1773 (Kriedte 1980).

A capitalist global economy pivoted on Europe cannot, anyway, have emerged directly from this colonial system. Market allocation of land, labour and capital in agriculture was blocked by protection of the peasantry for fiscal purposes. 'As long as labour was not separated from the social conditions of its existence to become "labour power" ... rural relations of production remained feudal' (Anderson 1974: 17). The use of slaves to produce exports from land conquered

213

overseas, in the absence of a native peasantry, brought a more extreme version of the same effect (Friedmann 1980). Most land and labour remained wholly outside the remit of market exchange, as did capital represented by bullion commandeered from colonies. The mercantilist political and legal infra-structures that produced colonialism were designed to keep them there, in rural Europe as well as abroad. It is not surprising, therefore, that an 'urban autonomy' model of capitalism proposes that its origin can be traced and understood with reference to urban history alone (Clarke 1992).

The world capitalist economy and urbanism

Money and absolute private rights in property survived from Roman times in European towns. After the Crusades reopened trade routes to Asia, cities revived to organize exchange of the silk, sugar, spices and other precious goods on which lords spent some of their peasants' tribute, for cloth and other wares produced by urban artisans. Merchants inevitably had to invest in organizing this trade, and commodity markets, and therefore the power to alienate goods at the owner's will, were necessary to circulate its fruits. Towns in northern Italy and the Low Countries, at each end of the spine along which these exchanges were organized within Europe, became immensely wealthy. It was there that the Renaissance occurred, involving the rediscovery of Roman law, which sanc-tioned individual ownership of property (including land) and the powers of sovereigns to make law (rather than interpret it), as well as art and humanist philosophies. Like the Church, these cities (and those of the Hansa which channelled northern produce to the Low Countries), gained a large measure of juridical independence from the polities in which they lay – hence, like the Church, their incorporation into them as a separate privileged estate.

Unlike the peasant economy, that of towns was completely monetized: 'money is the same as saying towns' in medieval and early modern Europe (Braudel 1973: 397). Its circulation created urban niches for artisans working up local and long-distance raw materials into commodities for merchants, the Church, landowners and each other. Large congeries of urban artisans in particular industries (especially cloth) came to depend completely on merchant capitalists as the volume of commodities they produced could only be vented by selling in markets across the whole world system (Langton and Hoppe 1983).

> The revival of long-distance trade [brought] the reemergence of a settled, free burgher class that later expanded into a broader middle class in the towns. The later rise of the absolutist state went hand in hand with the consolidation of the power of this class. Thus was ensured the continued ascendance of the towns because the political changes that occurred involved the institutionalization of the economic relation (i.e. merchant capitalism) of which the town was the spatial reflection.
>
> (Timberlake 1989: 7)

Centralized absolutist states sucked money, people and produce into the capital cities from which they operated, where conspicuous consumption symbolized state power and royal splendour. A basic aim of mercantilist policy was 'the suppression of particularistic barriers to trade within the national realm [and the creation of] a unified domestic market for commodity production' (Anderson 1974: 35). In consequence, regularly hierarchical national systems of towns developed (de Vries 1984). Through capital cities and ports they were connected to colonies where towns burgeoned to supervise the colonial ends of these links. Rio de Janeiro had 100,000 people by 1800. Lima and Mexico City were not far behind (Braudel 1967). At their home bases, trading companies set up by governments deliberately to exploit overseas possessions spawned manufacturing, banks and other funding establishments, commodity and stock exchanges, accountancy and legal business, and educational institutions. If, like Amsterdam, London, Venice, Naples, Copenhagen and Lisbon, a capital city was also its country's major port, it grew immensely. In 1500 four European cities had more than 100,000 people, in total 0.6 million, 7 per cent of the urban total. In 1750 there were fourteen cities of that size, their 2.2 million people representing 18 per cent of the urban total (Bairoch 1988). Paris contained nearly 600,000 people and London nearly 700,000 in 1750 (de Vries 1976), when no other English town was much more than 10 per cent of London's size (Langton 1978).

This evidence of growth and evolution in the urban system, as merchant capital integrated the European economy and articulated it with overseas colonies, is mesmerizing only when removed from its geographical and economic contexts. The whole system contained a very small proportion of the European economy before the nineteenth century, let alone that of other parts of the world (Tracy 1990; van Niel 1990). In 1750, only 12.2 per cent of the European population lived in towns, little more than the 10.4 per cent of 1300 (Bairoch 1988). Even in its London epicentre, '100,000 people at most could have lived on the profits of trade in 1700, [and] taken all together the profits did not add up to the civil list allocation granted to William III' (Braudel 1967: 415). Apart from a handful of Atlantic trading cities, Europe's large towns *lost* population between *c.*1600 and 1750 (Wheatley 1986/7; Wrigley 1987b).

Moreover, this aetiolated urban system was inherently and inevitably one of *merchant* capitalism; it was organized by monopoly companies, and the suppliers of goods for world trade were artisans, making commodities to sell to merchants. In so far as an urban proletariat existed, it served the artisan guilds with additional labour: a supplement, not an alternative, to a pre-capitalist mode of production which remained concerned with putting 'limits on output . . . uniting the artisans against the employer, interfering with the rate of wages, fixing the conditions of apprenticeship and limiting the hours of work' (North and Thomas 1973: 133). The 'characteristic creations [of mercantilism] were the royal manufactures and state regulated guilds in France and the chartered companies in England . . . The classical bourgeois doctrines of laissez-faire, with their rigorous separation of political and economic systems, were to be its antipode' (Anderson 1974: 36).

Neither free commodity markets nor the fictional commodities of labour and land emerged in an autonomous urban component of the early modern world system. They could not do so whilst artisanal guilds and merchant companies were able to protect and perpetuate their monopoly interests and themselves as the instruments of the third estate of mercantilist governments. Except in England, where the estates system of government decayed in the seventeenth century, and after the Revolution in France, this did not occur until the whole European *ancien régime* was swept away in the middle of the nineteenth century (Myers 1975).

The emphasis on exchange relations common to the globalization and autonomous urban models of the origins of the capitalist world system cannot account for how market relations penetrated the *production* of commodities. It might well be true that

> if we find ... that the system seems to contain wide areas of wage and non-wage labour, wide areas of commodified and non-commodified goods and wide areas of alienable and non-alienable forms of property and capital, then we should ... wonder whether this 'combination' of ... free and non-free is not itself the defining feature of capitalism as a historical system.
>
> (Wallerstein 1987: 320)

But this is to take as given exactly what needs to be explained: where did the fully capitalist economic 'freedoms' come from if they were not precipitated by the operation of the system itself?

From 80–90 per cent of labour and resources in early modern Europe were rural labour and land. It is not surprising, therefore, that rural landowners were responsible for the emergence of the notion of urban land as real estate; nor that any wage workers who did exist in towns were migrants from the countryside, where they had already been proletarianized.

THE GEOGRAPHY OF THE COMMODIFICATION OF LAND AND LABOUR IN EUROPE

The penetration of market relations into allocating labour and land in peasant farming was inevitably forced by the patriciate. It occurred in few of the many variants of peasantry created by the blending of natural environmental differences with the 'variegated fiscal pattern [which] developed in the main Western European countries' (Anderson 1974: 45). They can be classified into peasant ecotypes, each being 'a pattern of resource exploitation within a given macro-economic framework', including the structure of domination by the patriciate (Löfgren 1976: 101).

From Russia to the Atlantic and from Sweden to the Mediterranean, the peasantry was effectively freed from serfdom by the middle of the fifteenth century (North and Thomas 1973). The rents into which labour services were commuted when feudal landlords switched their demands from produce to money (to satisfy military needs and their predilection for exotic goods) were made virtually worthless by inflation during the thirteenth and fourteenth centuries. Then the huge fall in population after the Black Death in the fourteenth century brought inducements for peasants to stay on their lords' domains. This free peasantry changed in different ways in different parts of Europe as population began to grow again in the sixteenth century, and as absolutism provided 'a redeployed and recharged apparatus of feudal domination, designed to clamp the peasant masses back into their traditional social position' (Anderson 1974: 18).

Communal open field grain farming on the heavy-soiled plains north of the Alps was tailor-made for the abstraction of servile labour dues on lords' land interspersed with that of peasants. Across Eastern Europe, from the Baltic to the Balkans, serfdom was reimposed by nobles whose power remained great relative to the monarchs they elected and other estates (Kula 1976). Nobles could also fix prices to favour grain, causing towns to wither. Grain surpluses wrung from serfs were pumped towards Western Europe in return for manufactures and Oriental imports. Servile dues were wracked ever upward to chase falling terms of trade, and serfs became ever more dependent on their dwindling communal village economies (Brenner 1977). Market trade of noble and serf land was impossible, and servile labour could not be proletarianized. The first and most abject periphery of the West European economic core was Eastern Europe, where serfdom was not abolished until the late nineteenth century (Blum 1978).

In southern Europe, light soils and the tending of vines and tree crops were associated with more individualistic farming, and polities were generally fractured. Landlords (many of whom lived and made money in towns in Italy) regained power over land by buying it and leasing it out to share-croppers. In Spain, the importance of tax on sheep in state finances gave the *Mesta* shepherds' guild enormous privileges of grazing over peasant land, which obstructed the development of private agricultural property (North and Thomas 1973). In most of central and northern Western Europe where peasants remained free, a complex tangle of impositions was installed, including reversionary rights over peasant land, tolls, compulsory processing at high charges in lords' mills and bakeries, lordly monopolies of alcohol production and on hunting over peasant land, as well as increased state taxation and the monopoly prices of urban guilds (de Vries 1976). It kept peasants out of commodity markets on their own account by abstracting more of their output, distorting their terms of trade, or locking them into the archaic farming practices which allowed lords' impositions to be levied (Brenner 1977). In so far as it stopped peasants freely disposing of their land and kept common property and communal obligations in existence, private property rights in land and individual farming enterprise were prevented (North and Thomas 1973). Like share-cropping in the south, Church tithes in kind baulked new farming systems in Roman Catholic Europe, as well as depriving peasants of the fruits of their work (de Vries 1975).

An arable farming peasantry remained free of increased impositions only in the Low Countries, due largely to the power of the wealthy burghal estate (within which guilds were weak) and the concomitant small influence of landowners in government (de Vries 1975). Z goods were freely available: goods of a kind or quality which peasants could not produce for themselves, but could get in the well-developed towns of the region, which also provided outlets for peasant products (de Vries 1975). A remarkable pattern of specialized cash cropping developed (with grain imported from Poland) around what eventually became the 'Randstad' (de Vries 1975). Incorporation of peasants into commodity markets helped make the Low Countries the most prosperous part of early modern Western Europe. But farms remained small: in general, no more than 7–10 acres (North and Thomas 1973). Highly specialized in labour-intensive cash crops, they retained the peasant capacity to use only family labour on family land: it was their *surpluses* which entered the market (de Vries 1975).

In arable Europe, fully fledged agrarian capitalism emerged only in England (Brenner 1977). There, lords retained control over rents through customary three-life copyholds with variable entry fines at the succession of lives, which could be raised with inflation. Landowners were able to convert these leases to short-term tenancies, and abolish common land and farming on interspersed strips through enclosure agreements, which they could force through Parliament against their tenants' and local freeholders' wishes. Unable to tax peasant land, the English state had no concern to keep its peasantry in being. Landlords continually increased the size of tenant farms, up to about 300 acres, to accommodate evolving best-practice techniques and least-cost systems of labour organization (O'Brien 1993). Freed from the restrictive grid of common property and communal obligations, and forced to maximize their sales to pay rents beyond their control, English yeomen operated in a largely commoditized economy, buying Z goods and necessities as well as selling crops. London was the pivot of a regular von Thunen-like pattern of farming across southern and midland England by the eighteenth century (Kussmaul 1990). Dealers, craftsmen and providers of services multiplied to supply yeomen and their proletarianized labourers, many of them situated in an urban hierarchy through which there was increasing interaction with London (Mui and Mui 1989).

Tenant farms were secure sources of predictable income for landlords, who could invest in improvements and trade land unencumbered by common rights at prices reflecting current rents. Land became a commodity (Tribe 1981). Over 70 per cent of English arable farmland was owned by landlords and farmed by tenants by the late seventeenth century. The need for non-family workers increased continually, per acre and even more per farm, as labour-intensive fodder crops were incorporated into rotations and farms expanded. Without commons or smallholdings, they were completely dependent on wages. Some were paid by the task, but daily or weekly rates were more appropriate in the harvest work peak: labour *time* was being bought. The proletariat was recruited from peasants dispossessed by engrossment and smallholders who could not survive without access to commons (Snell 1985). Unrestrained by the need to provide land to children, and able to produce in large families the commodity it sold, this proletariat multiplied much faster than continental peasants, ensuring that wages did not rise parallel with the demand for labour (Wrigley 1983).

The proletarianization of labour for which demand varied seasonally and from year to year with the size of the harvest, with supply increasing continuously, was traced by the emergence and expansion of a national system of poor relief, also unique to England, 'affording the kind of provision for those in need which gave individuals a degree of protection ... that in typical peasant cultures [was] provided by kin' (Wrigley 1988: 120). Some of the work provided through this system, on road repair or in farming, was paid by the day or week.

The treatment of urban land as real estate to be improved by investment and the payment of workers for their time rather than per task performed or piece of a commodity, both entered the English urban economy through the building industry when suburbs spread to engulf agricultural estates where landowners, employers, workers and poor relief administrators already operated in agricultural land and labour markets. This represented the 'ruralization' of a hitherto artisanal urban economy (Clark 1992), in which producers possessed the means of production and the property they occupied was evaluated in terms of use rather than exchange values (Langton 1975).

Agrarian capitalism could not have developed without the enormous London food market, and in so far as this was a result of urban mercantilism, that system was complicit in its own end. At the same time, pastoral regions throughout Europe were being directly affected by merchant capital (Jones 1979). Peasants made clothes, tools and other things for their own use from farm products and resources from common land. Pastoral farmers were well-supplied with such materials; they had to work on milk to preserve it as butter or cheese, and could add value to wool, skins, bones, bristles and so on by rendering them into the forms in which they were eventually used. The labour inputs of pastoral farming are less seasonally peaked, and their variations through the year more predictable, than those of arable farming. Animal husbandry had few communal tasks, and although daily commitments were unremitting, there was usually time every day to work up products for sale. The safe delivery of young animals and the movement of flocks and herds between pastures used at different seasons needed large labour forces in each household. But even when everyone was on hand for herding, lambing, calving or farrowing, much of the time was spent walking or watching and they could spin, knit, sew, whittle, or plait while doing so.

Largely confined to land that was too rugged, wet or hungry for arable use, pastoral populations were sparse and scattered through isolated farms and hamlets. Their common pasture, woodland, reedbeds and so on on marshes, mountains or heaths were usually extensive. The large households of pastoral regions were much more disengaged from neighbours and remote from lords than those of arable regions. Their cultures were kin-based, self-reliant and anti-authoritarian, prone to heresy and the individualistic creeds of Protestantism (Langton and Hoppe 1983). Abundant (albeit poor) land allowed kin solidarity to be expressed in partible inheritance, and the opportunity of immigration to farmers' younger sons and other landless people from arable regions. The extension of arable farming chipped away at pastoral margins incessantly, and fen and coastal reclamation schemes removed huge swathes from them. As population grew and resources shrank, and as the innovation of mixed farming in arable regions worsened the terms of trade for their animal products, pastoral peasants were thrown into greater dependence on selling craft wares (Jones 1960). By the seventeenth century, and increasingly thereafter, they were travelling in their slacker seasons to sell craft goods or labour throughout Western Europe (Lucassen 1987).

Alternatively, they stayed at home and sold to merchants seeking wares for urban and colonial markets, or, especially in England, for arable regions where labour was becoming specialized in the task of farming itself, and therefore had to buy things that peasant households made for themselves (Shammas 1990). Unrestricted by guild rules, and with land and commons to provide at least partial subsistence, this rural labour force could readily be expanded or contracted in response to intermittent overseas markets or fluctuating fashions at home. As rural clothmakers and metalworkers multiplied beyond the capacity of local resources, and their holdings shrank too small to sustain credit, they became dependent on merchants for raw materials as well as markets (Langton and Hoppe 1983). 'Domestic production' was the consequence: rather than buying commodities, merchants delivered raw materials free and paid wages for work done on them. Rural artisans became rural proletarians, with the demographic effects, and therefore falling wage-rates, of agrarian proletarianization (Levine 1977). Intricate subdivisions of labour were organized over hundreds of square kilometres as the scale of production grew to accommodate narrow specialization by individual workers within the regional template of the original peasant crafts.

This process could develop fastest and furthest where internal and colonial markets were largest and least baulked by restrictions on exchange, where merchants were freest to ignore guild monopolies over craft production, where wage labour was becoming endemic to other parts of the economy, and where poor relief provided a refuge to wage-workers when demand was slack. England, therefore, was the archetypal, though by no means the only, country where the domestic system flourished (Pollard 1981). The difficulty of adequately supervising these far-flung enterprises led to the concentration of workers in large premises outside their homes. That is how factory production began, in the process of 'proto-industrialization', before being used to harness the power of water and steam during and after the late eighteenth century (Clarkson 1985).

These factories were not the first large industrial plant. Fixed capital installations employing scores of workers already existed at iron furnaces, forges and collieries. They, too, were rural in location. Iron smelting still needed charcoal and water power, and was therefore geographically dispersed. Because a colliery needed to despoil a large area of surface land and acquire an extensive subterranean property in unitary ownership, mining was inevitably rural in location (Langton 1979). By the time factory production, at first using water power, became common in English textiles manufacturing, techniques of harnessing steam were already well-advanced in mine drainage and iron forging. Like land, timber and coal were retained in the hands of English landlords: enclosure freed them from common rights and other customary restrictions on private exploitation as effectively as agricultural land, and already-proletarianized work practices moved smoothly from farms to pits and other industrial plant on landlords' estates.

These industries represented the movement of agrarian capitalism into industry, and old coalfield towns were as effectively 'ruralized' by the industrial plant around them as commercial cities were by suburban expansion. Towns were drawn into – and after the employment of steam power, produced by – the particular regional rural industrial economies surrounding them, sharing cultural as well as economic attributes which stemmed ultimately from the peasant by-employments to which local resources had given rise (Langton and Hoppe 1983).

England's pastoral ecotypes produced industrial, as its arable regions were producing agricultural capitalism in its most developed form. Mercantilism was, again, heavily involved at one remove. Even coal-mining, most 'agrarian' in its emergence, developed precociously because of the size of the London fuel market (Wrigley 1967) and the monopoly profits available to Newcastle coal dealers. As late as 1840 the Great Northern Coalfield, largely serving London, was still the most productive in Britain (Pounds 1979). Iron output and rural domestic and factory industry grew quickly to serve a commodified home economy and colonial trade which were heavily protected from overseas competition, funnelling all their multipliers back into the English economy. It was the projection of this huge demand and finance capital onto limited local resources of charcoal, water power and labour which made English industrial regions so technologically dynamic.

Like the agricultural and urban versions, English industrial capitalism developed from landlords' power over resources, and the concomitant drive to increase customs and excise revenues by a state which could not tax land. Labour, land, and the resources it yielded were, at very bottom, released into market exchange because English landlords had the power to destroy, and government had no incentive to protect, peasant and artisanal economies. They survived very much intact until the middle of the nineteenth century in most of Western Europe (Blum 1978), where market culture, especially in the allocation of labour, was slow to develop (Reddy 1984). European GNP per head was 58 per cent of Britain's in 1800. By 1860 it was 55 per cent (Berend and Rànki 1982). Britain still raised 78 per cent of Europe's coal in 1840 (Pounds 1979). It is fanciful to propose a capitalist Western European economy as the core of a world system in the early nineteenth century. Even in England, bitter conflict still raged in the eighteenth and early nineteenth centuries between communal moral and capitalist market systems of allocating land, labour, and commodities in industry and agriculture (Randall 1991; Wood 1992). As late as 1841 only one in five of England's industrial workers was in 'revolutionized industry', 'even with a deliberately exaggerated definition' of that term (Gregory 1990).

ENERGY AGAIN

A last crucial component of the capitalist system could only come *after* the heavy incursion of large-scale mechanized production into industry. As long as wages were paid per task or piece, labour was not fully commodified. These wage forms represent the purchase of 'dead labour', which has already been expended when it is paid for, not labour power (Middleton 1985). How long the job took or how efficiently it was done was of no concern to the capitalist who paid for it. Nor did recipients of piece rates 'respond to wage levels very strongly when disposing of their labor' (Reddy 1984: 5). In early factories, collieries and other large plant, where piece-rate wages survived from domestic industry and farm task work, investment in more productive techniques benefited workers by increasing the number of tasks or quantity of output they could achieve each day, rather than the capitalist who paid for it. Workers responded by reducing the hours they worked, not by increasing the amount they produced. The surest strategy for increasing profits was to pay less per task or piece, pauperizing workers who became inefficient and incapable of buying commodities made by others. This in-built checking mechanism to growth plagued domestic industrial regions throughout Britain and the rest of Europe in the late eighteenth and early nineteenth centuries (Gregory 1990).

It is only when capitalists pay for workers' time, buying live rather than dead labour, that they become concerned to maximize the efficiency with which goods are produced, and therefore invest in technological innovation. It is the market in labour power which gives capitalist economies their dynamism. Factory production allowed labour to be sufficiently closely supervised, and made the production process so highly capitalized, intricately subdivided, large scale, and continuous that workers could be treated as interchangeable hands, hired by the hour or day (Polanyi 1957). Complete commoditization of industrial labour had to await extensive large-scale factory production and the use of mineral energy. Even then, it was vigorously and sometimes successfully contested by workers (Reddy 1984; Randall 1991).

CONCLUSION

Colonial trade and urbanization were certainly involved in the emergence of the global capitalist economic system. But as soon as we look away from exchange to production, which is where wealth is created, their explanatory limitations become glaring. The economic and cultural impacts of global trade and urbanization depend on their capitalist nature; both can exist in incompletely capitalist economies without these effects. However widespread the early modern world trade system, and however large and dynamic the cities through which it was organized (and both are easily exaggerated), to discover the origins of the commodification of resources and labour we have to switch our gaze to agriculture, and narrow its focus onto human ecological relationships in very small parts of Western Europe in the seventeenth and eighteenth centuries.

In so far as global capitalism cannot exist without capitalist production, and capitalist production without commodified land and labour, even the kernel of a global capitalist system could not exist until well into the nineteenth century, after the ecological transition had been wrought by the Industrial Revolution.

REFERENCES

Anderson, P. (1974) *Lineages of the Absolutist State*, London: NLB.
Anderson, R.T. (1971) *Traditional Europe: A Study in Anthropology and History*, Belmont, Calif.: Wadsworth.
Bairoch, P. (1988) *Cities and Economic Development from the Dawn of History to the Present*, London: Mansell Publishing.
Berend I.T. and Rànki, G. (1982) *The European Periphery and Industrialization 1780–1914*, Cambridge: Cambridge University Press.
Berger, J. (1985) *Pig Earth*, London: Chatto & Windus.
Blum, J. (1978) *The End of the Old Order in Rural Europe*, Princeton, N.J.: Princeton University Press.
Blum, J. (1982) 'From servitude to freedom', in J. Blum (ed.) *Our Forgotten Past: Seven Centuries of Life on the Land*, London: Thames & Hudson.
Braudel, F. (1967) *Capitalism and Material Life 1400–1800*, London: Weidenfeld & Nicolson.
Brenner, R. (1976) 'Agrarian class structure and economic development', *Past and Present* 70, 30–75.
Brenner, R. (1977) 'The origins of capitalist development: a critique of neo-Smithian Marxism', *New Left Review* 104, 25–92.
Clarke, L. (1992) *Building Capitalism: Historical Change and the Labour Process in the Production of the Built Environment*, London: Routledge, Chapman and Hall.
Clarkson, L.A. (1985) *Proto-industrialization: The First Phase of Industrialization?*, London: Macmillan.
Cook, E. (1971) 'The flow of energy in an industrial society', *Scientific American* 225, 135–45.
Crone, P. (1989) *Pre-industrial Societies*, Oxford: Basil Blackwell.
Dahlman, C. (1980) *The Open Field System and Beyond: A Property Rights Analysis of an Economic Institution*, Cambridge: Cambridge University Press.
de Vries, J. (1975) 'Peasant demand patterns and economic development: Friesland 1550–1750', in W.N. Parker and E.L. Jones (eds) *European Peasants and their Markets: Essays in Agrarian Economic History*, Princeton, N.J.: Princeton University Press.
de Vries, J. (1984) *European Urbanization 1500–1800*, London: Methuen.
Deevey, E.S. (1960) 'The human population', *Scientific American* 203, 195–205.

Foster, G.M. (1965) 'Peasant society and the image of the limited good', *American Anthropologist* 67, 293–315.

Frank, A.G. (1991) 'A plea for world system history', *Journal of World History* 2, 1–28.

Friedmann, H. (1980) 'Economic analysis of the post-bellum South: regional economies and world markets', *Comparative Studies in Society and History* 22, 639–52.

Germani, G. (1973) 'Urbanization, social change and the great transformation', in G. Germani (ed.) *Modernization, Urbanization and the Urban Crisis*, Boston: Little, Brown.

Goldsmith, M.M. (1978) 'Mandeville and the spirit of capitalism', *Journal of British Studies* 17, 63–81.

Gregory, D. (1990) '"A new and differing face in many places": three geographies of industrialization', in R.A. Dodgshon and R.A. Butlin (eds) *An Historical Geography of England and Wales* (2nd edn), London: Academic Press.

Hart, K. (1986) 'Heads or tails? Two sides of the coin', *Man* 21, 637–56.

Harvey, D. (1982) *The Limits to Capital*, Oxford: Basil Blackwell.

Hicks, J. (1969) *A Theory of Economic History*, Oxford: Clarendon Press.

Hodges, R. (1988) *Primitive and Peasant Markets*, Oxford: Basil Blackwell.

Johnson, B.L.C. (1951) 'The charcoal iron industry in the early eighteenth century', *Geographical Journal* 118, 167–77.

Jones, E.L. (1960) 'The agricultural origins of industry', *Past and Present* 40, 58–71.

Jones, E.L. (1979) 'The environment and the economy', in P. Burke (ed.) *The New Cambridge Modern History*, Vol. 13, Cambridge: Cambridge University Press.

Jones, E.L. (1981) *The European Miracle: Environments, Economies and Geopolitics in the History of Europe and Asia*, Cambridge: Cambridge University Press.

Kautsky, K. (1976) Parts of *The Agrarian Question*, summarized by J. Banaji, *Economy and Society* 5, 2–49.

Klee, G.A. (1980) 'Introduction', in G.A. Klee (ed.) *World Systems of Traditional Resource Management*, London: Edward Arnold.

Kriedte, P. (1980) *Peasants, Landlords and Merchant Capitalists: Europe and the World Economy 1500–1800*, Leamington Spa: Berg.

Kula, W. (1976) *An Economic Theory of the Feudal System: Towards a Model of the Polish Economy 1500–1800*, London: NLB.

Kussmaul, A. (1990) *A General View of the Rural Economy of England 1538–1840*, Cambridge: Cambridge University Press.

Landes, D. (1969) *The Unbound Prometheus*, Cambridge: Cambridge University Press.

Langton, J. (1975) 'Residential patterns in pre-industrial cities: some case studies from seventeenth-century Britain', *Transactions of the Institute of British Geographers* 65, 1–28.

Langton, J. (1978) 'Industry and towns 1500–1730', in R.A. Dodgshon and R.A. Butlin (eds) *An Historical Geography of England and Wales*, London: Academic Press.

Langton, J. (1979) 'Landowners and the development of coalmining in south-west Lancashire 1590–1799', pp. 123–44 in H.S.A. Fox and R.A. Butlin (eds) *Change in the Countryside*, London: Institute of British Geographers Special Publication 10.

Langton, J. and Hoppe, G. (1983) *Town and Country in the Development of Early Modern Western Europe*, Historical Geography Research Series 11, Norwich: GeoBooks.

Le Roi Ladurie, E. (1979) 'Peasants', in P. Burke (ed.) *The New Cambridge Modern History*, Vol. 13, Cambridge: Cambridge University Press.

Levine, D. (1977) *Family Formation in an Age of Nascent Capitalism*, London: Academic Press.

Löfgren, O. (1976) 'Peasant ecotypes: problems in the comparative study of ecological adaptation', *Ethnologia Scandinavica*, 100–15.

Lucassen, J. (1987) *Migrant Labour in Europe 1600–1900*, London: Croom Helm.

McNeil, W. (1990) 'The rise of the West after twenty-five years', *Journal of World History* 1, 1–22.

Martin, D.A.L. (1991) Review of L.S. Stavrianos, *Lifelines From Our Past*, in *Journal of World History* 2, 91–5.

Middleton, C. (1985) 'Women's labour and the transition to pre-industrial capitalism', in L. Charles and L. Duffin (eds) *Women and Work in Pre-industrial England*, London: Croom Helm.

Mui, H.-C. and Mui, L.H. (1989) *Shops and Shopkeeping in Eighteenth-century England*, London: Routledge.

Myers, A.R. (1975) *Parliaments and Estates in Europe to 1789*, London: Thames & Hudson.

North, D.C. and Thomas, R.P. (1973) *The Rise of the Western World: A New Economic History*, Cambridge: Cambridge University Press.

O'Brien, P.K. (1982) 'European economic development: the contribution of the periphery', *Economic History Review*, 2nd Series 35, 1–18.

O'Brien, P.K. (1993) 'Introduction: modern conceptions of the Industrial Revolution', in P.K. O'Brien and R. Quinault, *The Industrial Revolution and British Society*, Cambridge: Cambridge University Press.

Odum, H.T. and Odum, E.L. (1976) *Energy Basis for Man and Nature*, New York: McGraw-Hill.

Peet, R. (1986) 'The destruction of regional culture', in R.J. Johnston and P. Taylor (eds) *A World in Crisis? Geographical Perspectives*, Oxford: Basil Blackwell.

Polanyi, K. (1957) *The Great Transformation*, Boston, Mass.: Beacon Press.

Pollard, S. (1981) *Peaceful Conquest: The Industrialization of Europe 1760–1970*, Oxford: Oxford University Press.

Pounds, N.J.G. (1979) *An Historical Geography of Europe 1500–1840*, Cambridge: Cambridge University Press.

Randall, A. (1991) *Before the Luddites: Custom Community and Machinery in the English Woollen Industry 1776–1809*, Cambridge: Cambridge University Press.

Reddy, W. (1984) *The Rise of Market Culture: The Textile Trade and French Society, 1750–1900*, Cambridge: Cambridge University Press.

Redfield, R. (1960) *The Little Community*, Chicago: Chicago University Press.

Scott, J.C. (1976) *The Moral Economy of the Peasant: Rebellion and Subsistence in South-East Asia*, New Haven: Yale University Press.

Shammas, C. (1990) *The Pre-industrial Consumer in England and America*, Oxford: Clarendon Press.

Shanin, T. (1987) 'Introduction: peasantry as a concept', in T. Shanin (ed.) *Peasants and Peasant Societies* (2nd edn), Oxford: Basil Blackwell.

Simmons, I.G. (1989) *Changing the Face of the Earth: Culture, Environment, History*, Oxford: Basil Blackwell.

Smith, N. (1984) *Uneven Development: Nature, Capital and the Production of Space*, Oxford: Basil Blackwell.

Snell, K.D.M. (1985) *Annals of the Labouring Poor: Social Change and Agrarian England 1660–1900*, Cambridge: Cambridge University Press.

So, A.Y. (1990) *Social Change and Development: Modernization, Dependency and World-system Theories*, Newbury Park, Calif.: Sage.

Sollow, B.L. (1986/7) 'Capitalism and slavery in the exceedingly long run', *Journal of Interdisciplinary History* 17, 711–37.

Spufford, P. (1988) *Money and its Uses in Medieval Europe*, Cambridge: Cambridge University Press.

Timberlake, M. (1985) 'The world-system perspective and urbanization', in M. Timberlake (ed.) *Urbanization in the World Economy*, London: Academic Press.

Times Atlas of the World: 6th comprehensive edition (1980), Edinburgh: John Bartholomew.

Tracy, J.D. (1990) 'Introduction', in J.D. Tracy (ed.) *The Rise of Merchant Empires: Long-distance Trade in the Early-modern World 1350–1750*, Cambridge: Cambridge University Press.

Tribe, K. (1981) 'Enclosure: reorganisation of landscape and labour', in K. Tribe, *Genealogies of Capitalism*, London: Macmillan.

van Niel, R. (1990) 'Colonialism revisited: recent historiography', *Journal of World History* 1, 109–24.

Wallace, I. (1990) *The Global Economic System*, London: Unwin Hyman.

Wallerstein, I. (1974) *The Modern World System: Capitalist Agriculture and the Origins of the World-economy in the Sixteenth Century*, New York: Academic Press.

Wallerstein, I. (1979) *The Capitalist World Economy*, Cambridge: Cambridge University Press.

Wallerstein, I. (1987) 'World-system analysis', in A. Giddens and J.H. Turner (eds) *Social Theory Today*, Stanford, Calif.: Stanford University Press.

Wheatley, P. (1986/7) 'European urbanization: origins and consummation', *Journal of Interdisciplinary History* 17, 415–30.

White, L.A. (1943) 'Energy and the evolution of culture', *American Anthropologist*, New Series 45, 335–56.

Wolf, E. (1966) *Peasants*, Englewood Cliffs, N.J.: Prentice-Hall.

Wood, E.M. (1992) 'Customs against capitalism', *New Left Review* 195, 21–8.

Wrigley, E.A. (1967) 'A simple model of London's importance in changing English society and economy, 1650–1750', *Past and Present* 37, 44–70.

Wrigley, E.A. (1969) *Population and History*, London: Weidenfeld & Nicolson.

Wrigley, E.A. (1983) 'The growth of population in eighteenth-century England: a conundrum resolved', *Past and Present* 98, 121–50.

Wrigley, E.A. (1987a) 'What was the industrial revolution?', in E.A. Wrigley, *People, Cities and Wealth*, Oxford: Basil Blackwell.

Wrigley E.A. (1987b) 'Urban growth and agricultural change: England and the continent in the early modern period', in E.A. Wrigley, *People, Cities and Wealth*, Oxford: Basil Blackwell.

Wrigley, E.A. (1988) *Continuity, Chance and Change: The Character of the Industrial Revolution in England*, Cambridge: Cambridge University Press.

FURTHER READING

Blaut, J.M. (1993) *The Colonizer's Model of the World: Geographical Diffusion and Eurocentric History*, New York: Guilford.

Brenner, R. (1977) 'The origins of capitalist development: a critique of neo-Smithian Marxism', *New Left Review* 104, 25–92.

Dodgshon, R.A. (1987) *The European Past: Social Evolution and Spatial Order*, London: Macmillan.

Hoppe, G. and Langton, J. (1994) *Peasantry to Capitalism*, Cambridge: Cambridge University Press.

Hugill, P.J. (1993) *World Trade since 1941: Geography, Technology and Capitalism*, Baltimore: Johns Hopkins University Press.

Jones, E.L. (1981) *The European Miracle: Environments, Economies and Geopolitics in the History of Europe and Asia*, Cambridge: Cambridge University Press.

Langton, J. and Hoppe, G. (1983) *Town and Country in the Development of Early Modern Western Europe*, Historical Geography Research Series 11, Norwich: GeoBooks.

North, D.C. and Thomas, R.P. (1973) *The Rise of the Western World: A New Economic*

History, Cambridge: Cambridge University Press.

Pollard, S. (1981) *Peaceful Conquest: The Industrialization of Europe 1760–1970*, Oxford: Oxford University Press.

Schwartz, H.A. (1994) *States versus Markets: History, Geography and the Development of International Political Economy*, New York: St Martin's Press.

Shanin, T. (ed.) (1987) *Peasants and Peasant Societies* (2nd edn), Oxford: Basil Blackwell.

So, A.Y. (1990) *Social Change and Development: Modernization, Dependency and World-system Theories*, Newbury Park, Calif.: Sage.

Wallerstein, I. (1974) *The Modern World System: Capitalist Agriculture and the Origins of the World-economy in the Sixteenth Century*, New York: Academic Press.

Wallerstein, I. (1980) *The Modern World-system II*, London: Academic Books

Wallerstein, I. (1989) *The Modern World-system III*, London: Academic Books

Wolf, E. *Peasants*, Englewood Cliffs, N.J.: Prentice-Hall.

Wrigley, E.A. (1969) *Population and History*, London: Weidenfeld & Nicolson.

11

INDUSTRIALIZATION AND WORLD AGRICULTURE

Brian W. Ilbery and Ian R. Bowler

INTRODUCTION

The history of world agriculture can be interpreted as the continual transformation of natural ecosystems and managed agroecosystems. To begin with, humans were integrated into the natural environment as hunters and food gatherers; but over time that environment has been increasingly managed, most recently through using agricultural technologies developed by manufacturing industry in developed market economies. These technologies are now being adopted throughout the world economy and an increasingly globalized food supply system has emerged to integrate input manufacturers, farm producers and food processors (Figure 11.1). Intervention by the state has characterized these developments, with distorting effects on the allocation of resources both in agriculture and world food trade. But in recent years international agreement has been reached on reducing state support for agriculture through the General Agreement on Tariffs and Trade (GATT) and by successive revisions of the Common Agricultural Policy (CAP) of the European Union (EU). Even so, the reduction in the international level of state support for agriculture has not addressed the problem of the damaging environmental consequences of contemporary world agriculture.

THE INDUSTRIALIZATION OF AGRICULTURE IN DEVELOPED MARKET ECONOMIES

The transformation of agriculture in developed market economies over the last fifty years has been marked by four main developmental phases: farm mechanization, chemical farming, food manufacturing, and agricultural and food biotechnology (Bowler 1992). Whilst the first two phases encouraged the

Figure 11.1 The food supply system. After Bowler (1992).

modernization of farm businesses through purchasing manufactured inputs, the third and fourth have been more concerned with the increasing links between farms and manufacturing firms involved with the processing of food. Together the four phases have brought about an increasing *industrialization* of agriculture (Healey and Ilbery 1985). Farming has become dependent on industries that supply farm inputs and process/market farm outputs. Some farming systems are now organized on scientific and business principles, whereby agribusinesses control most stages of the food chain, from supplying agricultural inputs, farming the land, to processing, storing, packaging, marketing and retailing

229

food. Such complete agricultural industrialization is found only in a few types of farming, as for example in the glasshouse sector of horticulture and intensive livestock production (e.g. pigs, poultry, feed-lot beef). Instead, integration between farming and manufacturing is usually achieved through a system of renewable contracts (contract farming), but still initiated and controlled by agribusinesses. Such processes have favoured a concentration of agricultural production in a number of specialized farming regions, each characterized by an advantageous climate, land quality or farm-size structure; less physically attractive and peripheral areas, especially uplands, have become increasingly marginalized.

The twin driving forces of agricultural industrialization have been techno-logical developments and strong state support for agriculture. Technologically, key changes have occurred in a fairly distinct temporal sequence, often initiated in the USA before spreading to other developed market economies.

Farm mechanization: 1950s

Mechanization has helped to transform agriculture by increasing crop and livestock yields, reducing the labour force, and thus producing a higher output per worker. Major changes have occurred in both arable and livestock farming, mainly in the form of new types of machinery and redesigned farm buildings. The most notable change was the substitution of the tractor for the horse. In Great Britain, for example, the number of tractors increased fourfold between 1950 and 1966; numbers have since remained fairly stable, but the machines have become more powerful. Similarly, the number of combine harvesters increased twelve times over the same time period (Figure 11.2(a)). Other major advances have included seed drills, potato and vegetable pickers, and mechanized farm buildings. Examples of the latter include silos for the storage of grain and silage, and intensive livestock units (factory farming) where pigs and poultry are

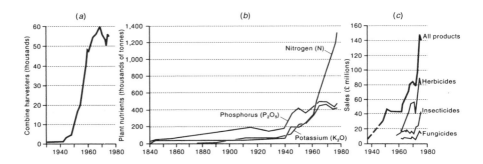

Figure 11.2 Indicators of agricultural industrialization in the United Kingdom. After Grigg (1982) and Ilbery (1992).

230

housed in computerized buildings. Mechanical milking parlours, bulk milk tanks and semi-automated dairy cow feeding systems are other examples of mechanization which have allowed herd sizes to increase considerably in the dairy sector.

Chemical farming: 1960s

Mixed farming in developed market economies was superseded in the 1960s and 1970s by an intensive, specialized agriculture. Although aided by favourable price support measures from the state, the pace of change was quickened by a rapid increase in the use of nitrogenous fertilizers, herbicides, pesticides and fungicides (agrichemicals). Together these products helped to increase crop and grass yields and control weeds, pests and diseases. By the mid-1970s, most cereal land was being treated with herbicides and pesticides; the application of agrichemicals was aided by modern spray machinery (including aircraft). Thus the consumption of chemical fertilizers increased eight times in the United Kingdom between the 1930s and 1980s (Chalmers *et al.* 1990); similar advances were recorded in other parts of the EU and the USA.

The major increase in fertilizer application since the 1960s has mainly been in the form of nitrogen (in which soils are most deficient) (Figure 11.2(b)). Application rates doubled between 1969 and 1982 in the United Kingdom, since when they have fallen back; in Germany and the USA they quadrupled between the 1950s and 1970s, and in France they increased sixfold between 1950 and 1974. These increases occurred especially in prime areas of arable farming, such as the Prairies, the Paris Basin and East Anglia, leading in later years to problems of excessive nitrate seepage into public water supplies. Concern over high nitrate pollution (where levels exceed permitted limits) has led to the designation of vulnerable zones and nitrate sensitive areas in parts of the EU. Within such zones, farmers are now compensated for reducing rates of fertilizer application.

Herbicides offer major savings in labour costs to farmers and their use on cereals, sugar-beet and potatoes in particular increased considerably in the 1960s and 1970s (Figure 11.2(c)). Pesticides and fungicides experienced less dramatic, but still significant, growth; they were not always successful in clearing weeds and pests, but contributed to the destruction of wildlife habitats. There can be no doubt that the chemical revolution has increased yields and encouraged changes in agricultural land use since the 1950s. However, it is not only arable land that has been affected; in England and Wales, for example, half of the fertilizer being used in the late-1970s went onto grassland.

Food manufacturing: 1970s

Farms are now part of a much broader food supply system, which includes the manufacture of purchased inputs as well as the processing and distribution of food (Bowler 1992; Figure 11.1). State-supported marketing boards and the traditional farmer–wholesaler–retailer chains no longer dominate. Food retailing has been revolutionized and is characterized by convenience (often frozen) processed food, supermarkets and media advertising. The marketing strategies of food processing and retailing companies reflect such factors as changing tastes and patterns of demand (caused by an increasingly urbanized and affluent population), the falling proportion of family income spent on food, higher female participation rates in the labour market, and increasing amounts of leisure time. These factors have helped to lengthen the food chain and encourage greater integration between production (farmers) and processing and marketing (food processing and retailing industries). Very little food reaches the consumer without some form of value-added processing.

One important dimension of the greater integration between farming and manufacturing has been the greater involvement in the food chain by agribusiness companies. Often of a multinational nature, they usually hold back from full integration (where they control all stages of the food chain) and instead offer renewable contracts to farmers. This is because returns to capital are higher in industry compared with farming, and agricultural land is a locationally fixed rather than mobile asset. Contracting favours certain types of farming, such as vegetables, fruit, pigs and poultry. In France, for example, 50 per cent of poultrymeat and 93 per cent of vegetables are marketed under contract; the figure for both products in the Netherlands is over 90 per cent. Contracts are rarely initiated by farmers; instead the policies of processing companies dominate and, whilst the risks of production are passed to the farmer, the companies dictate the times of sowing and harvesting, the varieties of crops to be grown, and the types and timing of chemical applications. In this respect, the farmer is relegated to the role of 'landed labourer'.

Agribusinesses often seek out the more business-oriented farmers with larger farms and modern farming practices when placing their contracts; this practice perpetuates farm–size and spatial differences in agriculture. In turn, food processing is increasingly dominated by a small number of multinational firms operating large production units. This process of concentration also character-izes food retailing which is now dominated by large multiple supermarket chains. Such large concerns are able to determine the price, quality and quantity of food purchased from manufacturing and wholesaling sectors and, increas-ingly, directly from large farms. These companies view agriculture merely as the supplier of raw materials for industrial processing and marketing. Indeed, multinational companies are beginning to substitute chemical/synthetic raw materials for 'natural' farm produce and so eliminate their dependence on agriculture even further. Examples include rayon and nylon (for wool and cotton) and different food components (starch, vegetable protein and glucose). In addition, the multinational companies seek out the lowest-cost agricultural raw materials on the international markets for their inputs, with important implications for agriculture in developing economies.

Agricultural and food biotechnology: 1980s

Improvements in plant varieties (grass, field and tree crops) and livestock breeds have been made through scientific research on experimental farms for many decades. There are numerous examples, including rising milk yields in dairy cow breeds, more efficient feed conversion ratios in pig and beef cattle breeds, disease-resistant potato and cereal varieties, and a dwarfing root-stock in apple and pear trees. Some of these developments have had controversial health implications, such as the use of growth-promoting hormones in animal feeds or through direct injection into livestock. Production patterns have also been affected by these developments. During the 1970s, for instance, maize was introduced into the southern counties of the United Kingdom and oilseed rape into the Midland counties, with the development of new varieties suited to more northerly climates (Bowler 1991). Thus cumulatively biotechnology has made a major contribution to the productivity of modern agriculture in developed market economies as well as reshaping patterns of production.

Two new developments in biotechnology have emerged since the early-1980s. First, developments in genetic engineering now offer considerable potential for raising crop and livestock yields by creating new drought- or pest-resistant crop varieties, grass and crop varieties that require lower levels of artificial fertilizer to achieve equal yields, and livestock that mature at even earlier ages or provide even higher yields of milk or meat. Genetic engineering poses similar ethical problems in agriculture as it does in the field of human medical science.

Second, the agrifood industry has developed new technologies as regards enzymes, biosensors and fermentation (Traill 1989). For example, developments in biotechnology have resulted in textured vegetable protein as a competitor for meat products, isoglucose as a competitor for sugar, and ethanol as a petrochemical substitute. To feed these new processes, agrifood firms are demanding crops with improved processing characteristics, especially as regards cereals, sugar and oils, and greater volumes of raw material with a uniform quality. Agrifood firms are prepared to search international markets for their raw materials, and their new demands are beginning to impact on agriculture throughout the world (Le Heron 1993).

State support for agriculture

These four recent developmental phases in agriculture have occurred in every developed country under active intervention by the state. A summary of the underlying reason for such intervention is as follows (Bowler 1979). Productivity gains made by advances in agricultural technology (as previously discussed) have increased production to such an extent as to oversupply the domestic market; this has exerted a downward pressure on both product prices and farm incomes, especially as there have been no compensating falls in the cost of manufactured farm inputs (the price–cost squeeze). As a result a significant movement of farm families from the land to the cities has taken place, but the rate of transfer has not been fast enough to enable those remaining in production to maintain their incomes at a level found elsewhere in the economy.

The main justification for government intervention, therefore, has been to support falling farm incomes and thereby enable the reallocation of labour resources to take place at a socially acceptable pace. There has also been political advantage to certain political parties in maintaining their 'rural vote', in ensuring a minimum of socio-political unrest in the countryside and, as regards the urban population and production costs in the wider economy, in maintaining a 'cheap food' policy. In addition, governments have been able to argue that from an economic viewpoint the orderly marketing of food, and the security of food supply, has been ensured for their populations, while a relatively efficient agricultural sector has been maintained. But intervention by the state has been costly. First, policy measures to support farm incomes through product prices have proved increasingly expensive, either directly for the taxpayer, or indirectly for the consumer. Second, in practice, intervention measures have favoured larger farm businesses while failing to deliver sufficient income support to the sector most in need – namely, smaller farms. Indeed average farm incomes have tended to remain below the level of average incomes in the whole economy in most countries. Third, domestic agricultures have been protected against lower-priced agricultural exports from other countries, especially by the CAP, with an inevitable distortion of world food trading patterns. Finally, while the main exodus of 'surplus' farm labour has been achieved in most developed countries, the 'surplus' of productive land and capital remains. Of these two, attention is at present focused on how best to utilize the land resource: options exist in the set-aside (idling) of farmland, conversion to non-food crops, afforestation, diversification into non-agricultural uses, including environmental conservation, and land abandonment. For each option there exists environmental costs and benefits.

Intervention measures by the state can be grouped under four headings: demand increasing (e.g. domestic food subsidy, intervention buying, export subsidy and food aid); supply reducing (production quota, co-responsibility levy and land set-aside); cost reducing (input subsidy, restructuring grant, farmer retirement grant and farm management grant); and income supporting (deficiency payment, headage payment and direct income support). Measures in the first group have been widely developed by governments and have become most costly. Under more recent policy revisions, especially in the EU, price levels for intervention buying and export subsidies have been substantially reduced and greater emphasis has been placed on direct income supports, land retirement (set-aside) and production quotas.

The agricultural effects of state intervention can be summarized in the following ways. Farmers have been encouraged to intensify and specialize their farming systems with marked gains in production efficiency but damaging consequences for the environment. The latter include the pollution of water and air, increased rates of soil erosion, the loss of habitat (wetlands, moorlands, woodlands), and the reduction in variety of flora and fauna. Agriculture has become polarized between a relatively few large farms producing most of the agricultural output, and a large number of small farms whose occupiers increasingly have to supplement their incomes with off-farm work. The income objective has been reached for only the occupiers of the larger farms. Food supplies have been maintained and stabilized for consumers. The farm population has been reduced with a minimum of disruption to society as a whole.

THE DIFFUSION OF AGRICULTURAL INDUSTRIALIZATION THROUGHOUT THE GLOBAL ECONOMY

Agricultural industrialization, as developed in advanced market economies, has diffused throughout the global economy as part of the process of rural and economic development. Agriculture in formerly centrally planned (socialist) economies has developed many of the features of industrialized agriculture, including large-scale farming, mechanization and a reliance on agrichemicals; there have been similar damaging consequences for the environment, especially the pollution of water resources, for example in Eastern Europe. Many developing countries have been similarly affected, although the benefits of agricultural modernization and industrialization have not always been retained but instead have 'leaked back' to developed economies.

Two major aspects of the transfer of industrialized farming systems to developing countries will be considered: transnational agribusinesses and the Green Revolution.

Transnational agribusinesses

Early attempts to transform agriculture in developing countries can be traced back to the seventeenth and eighteenth centuries when foreign controlled and operated plantations were created (Courtenay 1980). Indeed, the export of refined cane sugar to Europe led to the development of the world's first tropical plantation for overseas trade, using African slave labour, in Madeira in 1420. Plantations, therefore, were effectively created by Western capitalism to provide raw materials for industries in developed countries and to increase their security and variety of food supply. Today, they are often part of transnational corporations (TNCs) which are attracted to former plantation locations to take advantage of cheap land and labour and, in many cases, indigenous government support. However, plantations were not exclusive to the colonial period; the number of state-owned, large, centrally managed units increased after the end of the colonial period, often as newly independent governments took over previously private estates (for example in Sri Lanka and Zimbabwe). Never-theless, despite a general reduction in the number of foreign-owned plantations, they remain significant producers of a number of Third World crops.

Today plantations can be more correctly described as large-scale farms (here termed 'estates'); they are characterized by a high degree of product and labour specialization, the employment of wage labour, capital and labour intensiveness, the application of advanced technology and high land and labour productivity (Loewenson 1992). Most estates specialize in one crop, especially perennials such as tree crops or shorter-term crops such as sugar-cane and bananas. Crops produced for the world market which cannot be grown widely because of the need to meet restrictive natural environmental conditions include cocoa (Ghana, Nigeria, Trinidad), coffee (Côte d'Ivoire, Kenya, Uganda, Colombia), rubber (Liberia, Nigeria, Malaysia, Sri Lanka), palm oil (Zaïre, Nigeria, Malaysia), coconut (Philippines, Sri Lanka, Samoa), tea (East Africa, India, Sri Lanka, Assam), sugar (West Indies), and bananas (Cameroon, Central America, Philippines). Similar large-scale and centrally managed livestock farms (ranches) exist in Argentina, Uruguay, Venezuela and Brazil.

Located mainly in areas of low population density, where fertile land is available for exploitation, the estates of TNCs usually devote a relatively small proportion of their total area to the major commercial crop. However, this can still be several thousand hectares. Yet the estates often occupy a small amount of the total agricultural area of a country. Because of their relative isolation, estate managers are often forced to import labour from considerable distances and develop their own roads, services and supply and marketing systems. Overhead costs are thus higher than on traditional peasant smallholdings, but government support is often forthcoming because of the perceived benefits for local development. Greatest success is achieved where high standards of quality are paid a premium (as in rubber) and where central control and labour specialization create substantial scale economies.

The output from the estates is not for local consumption but instead is destined for Western export markets. Estates, therefore, are a main export earner; this, together with imported technology and assumed benefits for local economic development, are often enough to secure government support. National governments in developing countries have seen the commercialization of agriculture as an answer to the challenge of poverty and a rapidly expanding population.

However, the transnational corporations (sometimes referred to as agribusinesses) are benefiting most from the internationalization of food production (Loewenson 1992) and they remain a growing phenomenon in the rural economy of developing countries. Such corporations reduce their political risk by spreading operations among a number of countries: investment can be decreased in one country and increased in another; capital and technology can be made highly mobile. For example, the recent increase in pineapple production in Thailand can be attributed to a relocation of investment from Hawaii by one major agribusiness corporation in the 1970s. Economic risk can also be spread by diversifying corporate interests into new, higher value-added

Table 11.1 Multinational control of major developing world crops

Crop	Companies	Comments
Cocoa	Cadbury-Schweppes, Gill & Duffus, Rowntree (all UK), Nestlé (Swiss)	These four companies control 60–80 per cent of world cocoa sales
Tea	Brooke Bond, Unilever, Cadbury-Schweppes, Allied Lyons, Nestlé (all European); Standard Brands, Kellogg, Coca-Cola (all US)	Jointly hold about 90 per cent of tea marketed in Western Europe and North America
Coffee	Nestlé (Swiss), General Foods (US)	Jointly hold around 20 per cent of the world market
Sugar	Tate and Lyle (UK)	This company buys about 95 per cent of cane sugar imported into the EC, although the market is threatened by European beet sugar
Molasses	Tate and Lyle	The company's subsidiary, United Molasses, controls 40 per cent of world trade
Palm oil	Unilever (UK–Dutch), Lesieur (French)	Unilever dominates trade in palm oil
Tobacco	BAT (UK), R.J. Reynolds (US), Philip Morris (US), Imperial Group (UK), American Brands (US), Rothmans (UK–South African)	Together control between 89 and 95 per cent of world leaf tobacco trade
Cotton	Velkart, Cargill (US), Bunge (Dutch), Ralli Brothers (UK), Soga Shosho (Japan), Bambax, Blanchard	Are major transnationals in the cotton trade, and together with nine other multi-commodity trading groups they dominate 85–90 per cent of world cotton trade

Source: Dixon (1990: 18).

products such as winter vegetables, strawberries, peppers, melons and cut flowers.

Transnational agribusinesses often have a higher financial turnover compared with the total agriculture of the countries within which they operate. While many have been nationalized, others are able to exert increasing control over agriculture in developing countries through supplying inputs (e.g. seeds, fertilizer, equipment), directing farming practices under the terms of production contracts, providing the marketing infrastructure and processing agricultural produce (Table 11.1). For example, transnational agribusinesses and their affiliates control a large proportion of the heavily concentrated industries producing tractors, harvesters, agrichemicals and seeds. Again, between 60 and

90 per cent of the world's trade in eight leading primary food products – wheat, sugar, coffee, corn, rice, cocoa, tea and bananas – is controlled by up to fifteen transnational agribusinesses in each sector, with just three corporations accounting for the bulk of the market in most cases. Thus transnational agribusinesses have successfully linked the regional economies and crop sectors of developing countries with global systems of food production and consumption. Their power cannot be overestimated.

Criticisms of transnational agribusiness

At one level the industrialization of agriculture in developing countries may be perceived as beneficial. However, the behaviour of transnational agribusinesses has attracted a number of criticisms as regards their economic, social and environmental impacts on host countries (Espiritu 1986). These criticisms include the following:

1 Agribusinesses specializing in exports, industrial raw materials and luxury food crops, result in a sharply dichotomous pattern of income, productivity and technology compared with the sector producing domestic staples.
2 Labour-saving innovations by agribusiness both reduce the need for permanent, resident labour and exaggerate its seasonality; permanent labour is effectively replaced by casual wage workers and specialized, semi-skilled labour. The result is an increase in the landless casual rural proletariat, while those in work are often paid a low wage and live in poor conditions. Rather than reducing rural poverty, the development of agriculture by agribusiness increases the struggle for access to those resources essential for day-to-day survival.
3 The benefits of agricultural industrialization accrue disproportionately to foreign investors and, in the case of nationally owned estates, to urban-based people and local elite groups. The balance of power lies in the hands of Western companies, while agriculture has become overly export-oriented.
4 Large estates, especially those of transnational agribusinesses, often take the best land for export crops. This land absorbs the most inputs, investments and expenditure; yet it only occupies a small proportion of the total agricultural area. Nevertheless, agriculture remains the main livelihood for the majority of Third World people but, unable to compete, localized subsistence production begins to break down. Indeed, local peasant farmers can be displaced and have their land confiscated in favour of large-scale farming.
5 By importing labour from considerable distances, agribusiness estates can introduce large numbers of people of alien culture: for example, the movements of Tamils from India to Sri Lanka for coffee planting and tea picking, and into Malaysia to tap rubber. Problems of assimilation and conflict can be created.

6　Environmentally, the introduction of industrialized farming practices can lead to considerable degradation. Agribusinesses can show limited concern for the conservation of natural resources: rain forests are being cleared, soils eroded, fertility undermined and pollution from fertilizers and agrichemicals introduced. There are many recorded instances of agribusiness corporations moving on to new areas when soils have become depleted. In sum, many of the agricultural practices associated with the large-scale farming of agribusiness are unsustainable.

The Green Revolution

The Green Revolution is a package of modern agricultural technologies, drawing on scientific and industrial knowledge from developed market economies, to increase crop yields in developing countries. Central to this 'scientific farming' are high-yielding varieties (HYV) of cereal crops, especially wheat and rice. But the HYV perform more effectively when supported by the full package of technologies which includes irrigation, chemical fertilizers, and agrichemical pest and crop disease sprays. In addition, while these technologies are available to all farmers, the highest economic returns are gained when the technologies are implemented under farm mechanization and on larger landholdings: that is by industrialized agriculture.

The first HYV breeding programme started in Mexico in 1943, funded by the Rockefeller Foundation and concentrating on spring-sown wheat. The objective was to improve on native varieties of wheat which tended to be narrowly adapted to local conditions of soil, water supply, temperature and disease. The HYV breeding programme was so successful that the United Nations, through its Consultative Group on International Agricultural Research (CGIAR) and funding provided by the Ford Foundation and the Rockefeller Foundation, established the International Rice Research Institute at Los Banos in the Philippines in 1962 and the International Maize and Wheat Improvement Centre in Mexico in 1966. Other international research centres followed, for example the International Potato Centre in Peru (1972) and the International Laboratory for Research on Animal Diseases in Kenya (1973). National research centres were established in many other countries in following years. Together the research centres continue to produce an increasing range of hybrid HYV seeds with properties such as increased responsiveness to artificial fertilizers, greater disease and drought resistance, shorter and stiffer straw lengths to support a heavier head of grain, higher plant densities, smaller root systems, short upright leaves (for increased sunlight penetration), shorter growing cycle and lower sensitivity to changes in day length. While hybrid wheat, rice and maize varieties have proliferated to the greatest extent, more recently attention has turned to other food staples such as sorghum and millet.

The Green Revolution's impact on cereal production in developing countries has been rapid and dramatic. Crop yields per hectare have increased and yields

239

have become more certain. Taking wheat as an example: Mexico became self-sufficient by 1956; production in India doubled between 1966 and 1972; and production in Bangladesh increased from virtually nothing to 1.2 million tonnes by 1980. Taking all developing countries together, cereal yields have risen by 2 per cent a year during the last three decades – in line with population increase.

But the impact of the Green Revolution has been very uneven by crop type, country, region and size of farm. For crop type, the penetration of HYV has been greatest in wheat, followed by rice, and then other cereals such as maize. By country, the Green Revolution has had a relatively limited impact within Africa where problems of access to irrigation and chemical fertilizers exist for the small, family farmer, while little progress has been made in developing HYV for the main staple crops (sorghum and millet); most progress has been made in the adoption of HYV in maize production – wheat and rice are produced in a relatively limited number of African countries (Grigg 1993). Rather, the gains from the HYV have been concentrated in India, Pakistan, Sri Lanka, the central Philippines, Java, peninsular Malaysia, northern Turkey and northern Colombia. For example, over 80 per cent of the wheat area in non-communist Asian countries is in HYV, while a similar figure is achieved for rice in communist Asian countries. The penetration of HYV in maize production is similarly variable between countries, being high in communist Asian countries (over 70 per cent) but lower in African and South American countries (just over 50 per cent). There have been further spatial variations in the penetration of HYV within individual countries (Simpson 1994): favoured regions have been those with an established irrigation system (e.g. the Punjab within India), a completed distribution pattern of fertilizer centres, an efficient network of promotional demonstration farms, a larger than average farm-size structure or a pre-existing market orientation. At the individual farm level, the Green Revolution has favoured landholders with the greatest access to the means of production – especially land and water, and capital to purchase the new technologies. Since the economic results of the purchase of the new technologies remain uncertain, not least because of the vagaries of weather and crop pests, farmers with more land and capital to sustain the risk have been selectively favoured.

Criticisms of the Green Revolution

The Green Revolution, as an example of technology transfer (industrialized agriculture) between developed and developing countries, has attracted a number of criticisms. These include the following:

1   Systems of agricultural production in developed and developing countries are fundamentally different and the imported technology is often not suited to sustaining developing economies. Production gains are limited because small (peasant) farms are already efficient in the allocation of resources. What is needed is development appropriate to local farming environments rather

than an industry-led model of economic growth.

2   The new technology has been associated with rising indebtedness in rural societies: many farmers have had to borrow capital to purchase the technology of the Green Revolution. When the economic returns from HYV varieties have proved problematic, for example because of a shortage of water or fertilizer, and loans have not been repaid, either more money has to be borrowed (the debt cycle) or the money lenders acquire the forfeited land. As a result increased landlessness is recorded.

3   Agriculture in developing countries has become dependent on energy consumption (mechanization and fertilizer production), Western farming technology and the international agribusiness firms that are the suppliers of it. For example, in all developing countries fertilizer applications rose from approximately 1.5 kilogrammes per hectare of arable land in the early 1950s to 77 kilogrammes by the late 1980s (comparable figures for developed countries are 22 and 125 kilogrammes).

4   The new technologies, being most effective under farm mechanization, are labour displacing in economies which already possess a labour surplus. In particular, landless labourers are deprived of their employment.

5   A new capitalist class has penetrated rural society, purchasing farmland, consolidating the land into large holdings, and extracting the financial benefits of the new technologies. This development has widened economic and social divisions in rural societies between rich and poor farmers.

6   The new technologies have adverse effects on the environment, for example the depletion of local water resources through water abstraction for irrigation and agrichemical residues in the soil and food crops (Hoggart 1992).

Many of these criticisms are contested by research at the regional level: instances of an increased demand for labour in land preparation and fertilizer application, with rising agricultural wages, have been found (Hayami 1984); some of the changes in the structure of rural society can be attributed to factors other than the Green Revolution, for example the creation of new irrigation schemes and population change through migration; the Green Revolution brings about a needed modernization of agriculture including the development of a supporting infrastructure (banks, co-operatives, farm services), and the introduction of commercial attitudes amongst the farm population.

WORLD TRADE IN AGRICULTURAL PRODUCTS

Developed world imports and exports

Long-distance world trade, especially of food products, is of relatively recent origin (Simpson 1994): it was associated with developments in rail and sea transport during and following the Industrial Revolution in Western Europe.

Thus the increased volume and value of world agricultural trade over the last four decades, and its domination by developed market economies, continues a trend first established in the nineteenth century. Nevertheless, only a small fraction of total agricultural production is traded today.

Excluding the former centrally planned economies, developed countries account for nearly 75 per cent of the value of world food imports. The most important imports are tropical beverages (coffee, tea and cocoa), which accounted for 25 per cent of the value of developed countries' imports in 1991. These are followed by livestock products, cereals, oilseeds, vegetable oils, fruits and sugar. Since the 1950s, Eastern Europe and especially Russia have ceased to be net exporters of grain and have become importers. Russia is now the world's largest single purchaser of cereals. Japan has also increased its imports of cereals, unlike Western Europe where cereal imports have experienced considerable decline.

Exports from developed countries continue to be dominated by cereals. These originally came from the USA, Russia and Eastern Europe and more recently from Canada, Argentina and Western Europe. In 1988, four countries accounted for 73 per cent of cereal exports: the USA (36 per cent), France (16 per cent), Canada (11 per cent) and Australia (10 per cent). The developed world has increased its share of world exports (by value) from 56 per cent in 1961 to 69 per cent in 1980 and 76 per cent in 1991 (Table 11.2). Most noticeable within this trend has been the growth in the share of world exports from Western

Table 11.2 The structure of food exports by value, 1961–91

	1961 (%)	1972 (%)	1980 (%)	1991 (%)
Developed				
North America	19.7	23.1	21.4	15.6
Western Europe	17.6	26.1	36.2	47.0
Oceania	8.2	7.1	5.4	1.5
East Europe and USSR	7.9	6.0	4.2	2.1
Other developed	2.1	1.6	1.8	9.9
Total developed	55.5	63.9	69.0	76.1
Developing				
Africa	9.1	6.5	4.5	1.7
Far East	12.5	8.3	8.5	14.6
Latin America	17.1	15.4	13.8	3.7
Near East	3.7	3.2	2.2	3.8
Asia CPE*	1.8	2.4	1.8	0.1
Other developing	0.3	0.3	0.2	0.0
Total developing	44.5	36.1	31.0	23.9

*CPE: Centrally Planned Economies
Source: FAO, *Trade Yearbooks* and *The State of Food and Agriculture* (various years).

242

Europe, from 17.6 per cent in 1961 to 36.2 per cent in 1980 and 47 per cent in 1991. North America's position has remained relatively stable, whereas the positions of Oceania, Eastern Europe and Russia have declined.

Developing world imports and exports

To a considerable extent, the agricultural imports and exports of developing countries are the reciprocal of the developed world. However, one important context of such trade should be noted: while per capita food production in the developed world has been increasing, in the developing world trends have been more variable, with many individual countries suffering falling levels of per capita food production.

Developing countries account for a quarter of world food imports by value, and a significant feature of the last three decades has been an increase in those imports, especially in Africa. By value, over a third of all imports are cereals (70 per cent by weight): over a half of world cereal imports, and two-thirds of wheat imports, now move into the developing world. Those countries most dependent on cereal imports include Peru, Bolivia and Ecuador in South America, and most countries in western and southern Africa. Meat, dairy products and eggs are the other main categories of food imports.

Developing countries as a whole display a falling level of food self-sufficiency. Three factors have been influential. First, population growth has outpaced the rate of increase in agricultural production in many African countries, for example Egypt, Mauritania and Botswana; imports, especially of cereals, have been necessary just to maintain per capita food consumption levels. Second, many countries have developed their agricultural exports at the expense of their domestic food staples, for example Brazil; this has been necessary so as to earn foreign currency to repay debts incurred in the late 1970s which funded economic development programmes. Third, incomes have risen in some developing countries, for example in oil-exporting countries such as Saudi Arabia, Libya and Venezuela; the wealthier sections of the population have been able to increase their consumption of wheat, rice and livestock products (Table 11.3). Indeed nearly a half of cereal imports into such countries are fed to domestic livestock.

Agricultural exports from developing countries account for approximately a quarter of world exports and they have been increasing substantially in volume and value in recent decades. Nevertheless, as a proportion of the total export earnings (by value) of all developing countries, they have continued to fall (currently standing at 10 per cent). One outcome has been the increasing inability of many developing countries to pay for their food imports by their agricultural exports. Indeed an increasing number of countries are now net food importers (Grigg 1993).

However, the dependence of several developing countries on agricultural exports remains (e.g. in eastern Africa, central America and Argentina), while

Table 11.3 The structure of food imports by value, 1988

	Developed (%)	Developing (%)
Cereals	12.6	37.4
Meat	20.3	10.6
Dairy and eggs	13.7	19.4
Fruit and vegetables	11.7	6.5
Sugar	5.7	7.3
Coffee, cocoa, tea	13.3	5.3
Oilseeds and cake	10.3	7.3
Wine and beer	6.0	1.3
Oilseeds	6.4	4.9

Source: Based on Grigg (1993: 239).

in other countries there is dependency on just one or two products. For example, North Korea bases 70 per cent of its agricultural exports on rice and, until recently, sugar comprised over 90 per cent of Cuba's agricultural exports. In total, tropical beverages contribute over 40 per cent of agricultural exports in African countries. Product dependency by some countries can be problematic: non-food (industrial) crops tend to be replaced by artificial products (e.g. cotton by synthetic fibres); many products face inelastic demand in the markets of the developed world (e.g. beverages); weather-induced variations in production lead to fluctuations in market prices and export earnings (e.g. coffee); exports that compete with products grown in developing countries face a protected market (e.g. meat, citrus fruit); the terms of trade favour manufactured at the expense of agricultural products.

Food aid to developing countries

A considerable proportion of food exports from developed to developing countries has been in the form of food aid (Tarrant 1980). This was formalized in the USA in 1954 (Law PL480) and permitted the sale of USA grain, and later milk products, on concessionary terms to friendly and needy countries in the developing world. Throughout the 1950s and 1960s, up to one half of all cereal imports in developing countries were in the form of food aid; this figure continued to fall, however, from 38 per cent of developing countries' imports in 1970 to 18 per cent in 1977 (Tarrant 1980). By 1982, food aid accounted for just 5 per cent of international trade in cereals. Nevertheless, food aid to low-income countries remains a high, if declining, proportion of total imports, and in the period 1990–1 food aid in cereals to low-income countries accounted for 20 per cent of cereal imports.

Food aid has become increasingly concentrated in Sub-Saharan Africa since

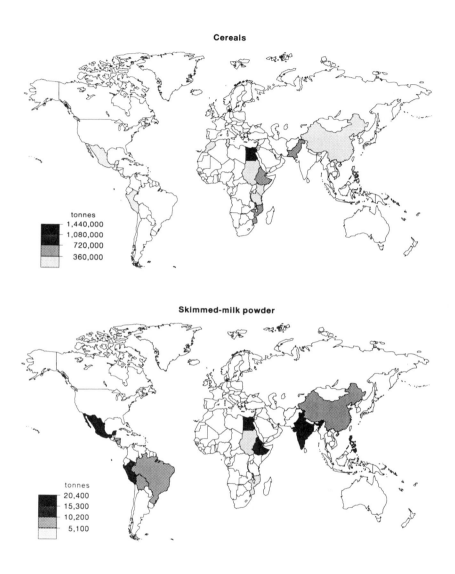

Figure 11.3 Major recipients of food aid in cereals and skimmed-milk powder, 1988–9.
Based on FAO statistics, 1989.

1980. This can be demonstrated with reference to food aid in cereals and skimmed–milk powder in the period 1988–9 (Figure 11.3). In terms of cereals, for example, the main recipients are in Africa and Asia, with the top five

245

recipients being Egypt, Bangladesh (both over 1 million tonnes, 1988–9), Ethiopia, Mozambique and Pakistan. For skimmed-milk powder, the Philippines heads the list, followed by India, Mexico, Peru and Sri Lanka.

The source of food aid has also changed over time. In 1965, the USA provided 94 per cent of all food aid; by 1976 its share had been reduced to 68 per cent. This was partly due to rising surpluses of milk products and cereals in the EU (Tarrant 1982). By the late 1970s, the USA, Canada and the EU provided four-fifths of all food aid. Early criticism over the direction of food aid, which was determined more by political and military factors than by need, led to an increasing proportion going through the World Food Programme run by the Food and Agricultural Organization (FAO). FAO receives food aid from donor countries and determines its destinations.

In theory, food aid has a number of benefits (Cathie 1982). First, it can augment available resources to help foster growth and economic development. Second, it can help to reduce inequalities in the distribution of income by generating employment for the poor (through food-for-work programmes and granting them access to food at concessional prices). Third, it can help to improve the nutrition and health of disadvantaged groups, including women and children (through supplementary feeding programmes). Fourth, it can provide food to vulnerable groups in emergency situations (emergency food aid). Finally, it can furnish multilateral lending organizations with an effective instrument (food-for-adjustment) to support economic policy reforms (structural adjustment programmes, balance of payments adjustment programmes, market restructuring programmes) and to protect the poor from hardships imposed by such reforms. In practice, however, food aid has been criticized on a number of grounds (Maxwell and Singer 1979):

1 The disincentive effect on local agriculture through the price mechanism. Food aid sold on the open market can depress market prices for local production and thus act as a disincentive to increase domestic food production. Subsidized food production in developed market economies is detrimental to Third World exports because it limits their competitiveness. Such actions have been described as the 'dumping' of food on world markets.

2 The allocation effect, whereby the distribution of food aid does not reflect criteria of need but rather the political, economic and military interests of donors and receiving governments. This use of food aid has been described as 'priming the market'.

3 The dependency effect, whereby food aid encourages greater dependence rather than greater self-reliance. In turn, this allows the penetration of recipient countries by foreign capital.

4 The inferiority effect, where food aid is 'second best' aid and dependent on surplus food products in developed countries; cash aid would be better.

There are also examples of inappropriate food aid: for example, Swiss cheese sent to Biafra.

Food aid from the developed world provides agricultural produce that could not otherwise be purchased by developing countries, and possibly the strongest argument for food aid is that it replaces commercial imports. For the totality of food trade, however, the internationalization of food production undoubtedly favours the large transnational corporations based in developed market economies. Their future competitiveness is likely to be enhanced by the liberalization of world trade following the completion of the Uruguay Round of the GATT negotiations in December 1993.

CONCLUSION

The industrialization of agriculture in the developed market economies has ensured the security of food supplies for the population, a relatively efficient and technologically advanced farm sector, and food surpluses available on the world market through trade or food aid. However, there have been considerable financial costs brought about by state intervention in the process of agricultural industrialization, together with distortion to patterns of international trade in food products and growing environmental costs resulting from the implementation of industrialized farming practices.

The industrialization of agriculture under technological transfer and large-scale agribusiness farming also poses many problems for developing countries. With the emphasis placed on export crops, the supply of staple foods has been undermined in several countries with a failure to meet local food needs. There is an increasing general dependence on food imports and, with droughts, continued desertification and military conflict in many countries, food shortages and famine have become commonplace. These trends in turn have necessitated the provision of food aid and a dependency on the farming systems and agricultural technology of the developed world.

Whether applied in developed or developing countries, the industrialization of agriculture has placed the interaction between agriculture and the natural environment under increased strain (Hoggart 1992). A growing international concern is being expressed on the long-term sustainability of this form of agricultural development. However, any resolution of the 'sustainability' problem will have to recognize two realities: the food demands of a growing world population, and the domination of the food supply system by TNCs processing and trading farm products on a global scale.

REFERENCES

Bowler, I.R. (1979) *Government and Agriculture: A Spatial Perspective*, London: Longman.
Bowler, I.R. (1991) 'The agricultural pattern', pp. 83–114 in R.J. Johnston and V.

Gardiner (eds) *The Changing Geography of the United Kingdom*, London: Routledge.

Bowler, I.R. (ed.) (1992) *The Geography of Agriculture in Developed Market Economies*, London: Longman.

Cathie, J. (1982) *The Political Economy of Food Aid*, Aldershot: Gower.

Chalmers, A., Kershaw, C. and Leech, P. (1990) 'Fertilizer use on farm crops in Great Britain', *Outlook on Agriculture*, 19, 269–78.

Courtenay, P. (1980) *Plantation Agriculture* (2nd edn), London: Bell & Hyman.

Dixon, C. (1990) *Rural Development in the Third World*, London: Routledge.

Espiritu, C. (1986) 'Transnational agribusinesses in the Third World', in D. Dembo, C. Dias, W. Morehouse and J. Paul (eds) *The International Context of Rural Poverty in the Third World*, New York: Council on International and Public Affairs.

Grigg, D. (1982) *The Dynamics of Agricultural Change*, London: Hutchinson.

Grigg, D. (1993) *The World Food Problem*, Oxford: Blackwell.

Hayami, Y. (1984) 'Assessment of the Green Revolution', in C.K. Eicher and J.M. Staatz (eds) *Agricultural Development in the Third World*, Baltimore: Johns Hopkins University Press.

Healey, M.J. and Ilbery, B.W. (eds) (1985) *The Industrialization of the Countryside*, Norwich: GeoBooks.

Hoggart, K. (ed.) (1992) *Agricultural Change, Environment and Economy*, London: Mansell.

Ilbery, B.W. (1992) *Agricultural Change in Great Britain*, Oxford: Oxford University Press.

Le Heron, R. (1993) *Globalised Agriculture: Political Choice*, Oxford: Pergamon Press.

Loewenson, R. (1992) *Modern Plantation Agriculture*, London: Zed Books.

Maxwell, S. and Singer, H.W. (1979) 'Food aid to developing countries: a survey', *World Development* 7, 225–47.

Simpson, E.S. (1994) *The Developing World: An Introduction* (2nd edn), London: Longman.

Tarrant, J.R. (1980) 'The geography of food aid', *Transactions of the Institute of British Geographers* 5, 125–40.

Tarrant, J.R. (1982) 'EEC food aid', *Applied Geography* 2, 127–41.

Traill, B. (1989) *Prospects for the European Food System*, London: Elsevier.

FURTHER READING

Bowler, I.R. (1985) *Agriculture under the Common Agricultural Policy*, Manchester: Manchester University Press.

Bowler, I.R., Bryant, C.R. and Nellis, M.D. (eds) (1992) *Contemporary Rural Systems in Transition: Agriculture and Environment*, Wallingford: CAB International.

Bryant, C.R. and Johnston, T.R. (1992) *Agriculture in the City's Countryside*, London: Belhaven Press.

Dixon, C. (1990) *Rural Development in the Third World*, London: Routledge.

Ellis, F. (1992) *Agricultural Policies in Developing Countries*, Cambridge: Cambridge University Press.

Goodman, D. and Redclift, M. (1991) *Refashioning Nature: Food, Ecology and Culture*, London: Routledge.

Grigg, D. (1982) *The Dynamics of Agricultural Change*, London: Hutchinson.

Grigg, D. (1993) *The World Food Problem*, Oxford: Blackwell.

Sarre, P. (ed.) (1991) *Environment, Population and Development*, London: Hodder & Stoughton.

Wallace, I. (1985) 'Towards a geography of agribusiness', *Progress in Human Geography* 9, 491–514.

12

CHANGES IN GLOBAL DEMOGRAPHY

John I. Clarke

WORLD POPULATION IN 1800

The population of the world at the beginning of the nineteenth century differed from that of today in almost every respect. Most notably, it totalled only about one billion, less than one-sixth of what it is projected to be in 2000 and equivalent to only 11–12 years of world population growth at the current rate in the 1990s. This estimate for 1800 is of course no more than a crude guess, because only the Scandinavian countries and America had held modern censuses of population; the first rather skeletal censuses were taken in Britain and France in 1801, and it was not until 1837 that centralized vital registration was introduced into Britain. Indeed, most of the world's population remained unenumerated during the nineteenth century and the early part of the twentieth century.

Even the concept of a world population was in the minds of very few in 1800, as most people lived in small communities, did not travel extensively, and did not think globally. Moreover, writings about population were rare, largely theoretical and necessarily speculative. Nevertheless, the seminal work by Thomas Malthus (1766–1834), *An Essay on the Principle of Population*, had just appeared in 1798, expounding the thesis that the reproductive capacity of human beings was so great that it is always capable of exceeding their ability to provide adequate subsistence, thus possibly causing starvation. Available resources were regarded as fixed, and therefore population growth created its own checks: 'positive' checks like famine, disease or war, and 'preventive' checks in the form of celibacy and marriage delay. Emphasizing the significance of food as the primary resource, the unavoidability of the law of diminishing returns, and the restricted relief provided by invention or innovation (Livi-Bacci 1992), Malthus's stark message about the dangers of unrestrained population growth, though variously criticized and attacked, was to survive the next two centuries.

Although the 'mortality revolution' had begun in the second half of the eighteenth century in Europe, human fertility and mortality were generally high everywhere on earth, though they fluctuated markedly following environmental and human crises and catastrophes, which Malthus regarded as 'natural checks'. Populations were young and life expectancy was short; the old were few. Consequently, population growth was irregular in the short term and slow in the long term.

Probably 97 per cent of the world's population lived in rural areas, deriving a living directly or indirectly from agriculture. Modern industrialization and urbanization were young seedlings in Britain, steam and the railways yet to transform transportation. Mobility was limited and most people lived in villages; urban nodality was largely restricted to the older peasant civilizations and concentrations of population in Europe and Asia, and the total number of urban dwellers on earth probably amounted to little more than the number of inhabitants in Mexico City today.

It was in such conditions that human fragmentation and ethnic differentiation had evolved, engendering a plethora of peoples – it is estimated that there are 5,000 different indigenous peoples today – speaking thousands of different languages and adhering to many different religions. In many parts of the world, notably in Africa and South America, the relative isolation of 'closed' tribal economies assisted ethnic and cultural separatism, though few parts experienced the prolonged isolation and consequent ethnic homogeneity of Japan under the Tokugawas. Populations were therefore more identifiable by ethnicity and culture than by political units, which bore little resemblance to those of today, the world having yet to be colonized extensively by European powers bearing among their baggage the concept of statehood.

Accordingly the overall distribution of humanity was also quite different two centuries ago. As now, huge 'negative' areas were uninhabited or very sparsely populated, but the inhabited areas were much more evenly peopled than now. Population densities reflected more the persistence of peasant civilizations than the sporadic distribution of natural resources, most of which were still awaiting exploration and exploitation by new technologies devised during the Industrial Revolution. Energy resources, for example, remained largely undisturbed, and were yet to have a major influence upon population patterns.

At a continental level, the imbalance in population distribution was much more marked (Table 12.1). Probably nearly 86 per cent of the world's population lived in the 'Old World' (as seen through European eyes) of Asia and Europe, and the latter alone contained more than all the Americas and Africa put together. Indeed, more lived in Britain than the whole of North America, where most European settlers were also of British descent. Asians were even more preponderant in the world than now, living especially in the great heartlands of China and the Indian sub-continent; in fact these two concentrations along with Europe comprised two-thirds of the world's population. The Americas, Africa, Australasia and Inner Asia, though inhabited by numerous indigenous peoples, were much less populous and were

Table 12.1 Percentage world population distribution by continental region, 1800–2000

	1800	*1850*	*1900*	*1950*	*2000*
Asia	64.7	65.3	54.6	54.6	59.2
Europe & USSR	21.0	23.0	26.9	22.8	13.1
Africa	10.9	6.7	9.0	8.9	13.9
North America	0.7	2.2	5.2	6.6	4.7
Latin America	2.5	2.7	4.0	6.6	8.6
Oceania	0.2	0.2	0.4	0.5	0.5

Source: UN, *World Population Prospects, 1992.*

subsequently regarded by Europeans as relatively 'empty areas' ripe for colonization.

This sketch of the patterns and dynamics of world population in 1800 has implied some stability, but it must also be emphasized that it resulted from the continual interplay of numerous natural, economic, social and political forces over many millennia, especially at local level, where there was considerable diversity reflecting local cultural conditions.

AGENTS OF CHANGE

The relative stability of world population size, distribution, composition and dynamics was transformed during the next two centuries, at first slowly but ever more rapidly over the last fifty years. Over the 130 years up to 1930 world population doubled, and reached 2.5 billion by 1950. Thereafter growth accelerated so that the 3 billion total was reached by 1960, 4 billion by 1975, 5 billion by 1987 and 6 billion before the end of the century. This phenomenal growth was brought about by demographic transition, the transfer from high birth- and death-rates to low birth- and death-rates accompanied by high natural increase during the transition as well as profound changes in the age compositions of populations and to a lesser extent in the sex compositions. Initially, transition took place slowly in the richer more developed countries, so that by the mid-twentieth century there was a marked demographic contrast between the low vital rates and low growth in those countries and the high rates in the poorer less-developed countries. Subsequently, demographic transition has progressed much more rapidly in many of the poorer countries, provoking considerably more rapid world population growth and greater differentiation in demographic rates throughout the world. We shall look at demographic transition in more detail on pp. 256–60 and 265–9.

Since 1800 world population has also undergone massive redistribution through major streams of inter-continental, international and internal migration, populating previously sparsely inhabited areas and changing the balance of continental population distribution, but also concentrating in economic core areas and urban centres. Between 1800 and 2000 the world level of urbanization rose from about 2 per cent to nearly 50 per cent.

Before examining the processes of demographic transition and population redistribution, we should emphasize the principal economic, political and social agents of change. Undoubtedly, the scientific and technological advances in Europe which instigated the agricultural and industrial revolutions (Wrigley 1988) were the main trigger. The advent of steam and other sources of energy, the introduction of manufacturing, the growth in human mobility through sea, rail, road and later air transport, and the spread and increase in trade led to rapid economic developments which had close interrelationships with mortality and fertility declines in the more-developed world, influencing their progress and affected by them. Population growth does not appear to have hindered the economic growth of the more-developed countries, where it is probable that the countries enjoying the most population growth have also enjoyed major economic growth. Unfortunately, this generalization does not hold water for the later demographic transition in the less-developed countries, where the relationship between population and economy is far less discernible, population growth frequently outstripping economic growth and reflecting social, cultural and political factors much more.

Economic changes also led to extensive diffusion of European peoples and cultures which transformed economies, polities and societies world-wide. Roberts (1985) called this process 'The Triumph of the West', whereby a host of phenomena were disseminated around the world: languages, religions, social customs, states, systems of government, methods of communication, economic activities, commercial organizations, diseases, medicine and health, styles of architecture, etc. Obviously the rates and progress of the diffusion varied greatly from one part of the world to another, leading to the contrast between more- and less-developed countries which was particularly sharp by about the mid-twentieth century and was reflected in their demographies. Most of those newly colonized lands which are numbered among the developed countries today (e.g. Canada, USA, Australia and New Zealand) were relatively sparsely inhabited by indigenous peoples, whose numbers were further drastically reduced by European contact, as in the cases of the Caribs and Tasmanians.

The various changes also had profound but differing impacts upon the patterns of human locations – the densities, nodes, networks and hierarchies – leading to marked contrasts between the human geographies of more- and less-developed countries and regions. Above all the relationship between people and land was transformed. People in industrializing and urbanizing countries became less bound to agricultural land and rural life, less dependent upon seasonal cycles and the vagaries of weather. In the nineteenth century, coal, iron ore and other mineral resources became important locational factors in Europe and North America, leading to population concentration on coalfields and in city-ports. But later, as other sources of energy evolved, the influence of natural resources upon population distribution gradually declined and was replaced by other locational criteria: transport networks, the market and labour costs. In the growing towns and cities manufacturing was complemented by service industries which continually proliferated, so that today they employ at least half of the active populations of the more-developed countries and large proportions of the urban populations of less-developed countries.

Consequently, population redistribution in Europe produced an increasingly uneven pattern of population distribution throughout the nineteenth century, and this patchiness was transferred to the 'New Europes' overseas where peripheral concentrations of population evolved even more strikingly around the city-port 'gateways', like New York, Buenos Aires, Rio de Janeiro, Sydney, Algiers and Cape Town. Linking the 'New Europes' with Europe and the growing world economy, many have become major cities and extremely dominant in their urban hierarchies. In contrast, agricultural settlement in the 'New Europes' was usually at low densities because the large areas of available land, acquired in one way or another, engendered extensive farming, using machinery, few workers and much crop specialization. The rural contrasts between the 'old' Europe and the 'New Europes' overseas became very striking, provoking considerable differences in environmental utilization and degradation, still visible today. Continents developed population density gradients which sloped downwards from pressurized coastal zones to often empty interiors, modelled by Hambloch (1966), and which currently give great concern in view of the threat of rising sea-levels resulting from the greenhouse effect.

The expansion of Europe, however, did not radically transform the overall continental balance of world population distribution. Despite the flows of migrants, the 'New Europes' overseas remained relatively feebly inhabited in contrast with the masses living in Eurasia, which by 1900 still contained well over four-fifths of humanity while less than one in ten lived in either the Americas or Africa (Table 12.1 and Figure 12.1). Even by 1950 Eurasia still contained over 77 per cent and the Americas only 13 per cent. It is since 1950 that the most striking changes have taken place in the continental balance of population distribution, with the marked decline in the proportion of the population of Europe, the great increase in the populations of Africa and the Americas, and the resurgence of Asians.

One reason for continuing continental imbalance in population distribution has been the diffusion of the European state model around the world, leading to more than two hundred countries today; an increasing number have become independent since the Second World War, particularly those created initially by colonialism. The extremely irregular political mesh superimposed upon the earth's surface has engendered states of incredibly contrasting areas, shapes and populations, which bear little relationship to the underlying physical environments or natural resources. The spectrum of population sizes and densities is such that it might almost have resulted from random doodling on the world map.

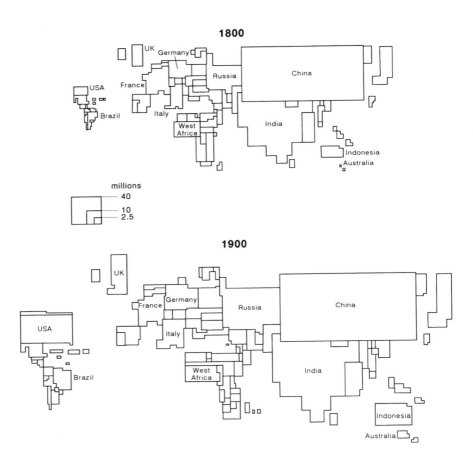

Figure 12.1 The distribution of population in 1800 and 1900. After Noin (1991).

With a markedly skewed size distribution, a few mega-states contrast greatly with a plethora of micro-states (Tables 12.2 and 12.3). Until the recent fragmentation of the USSR, the six largest states areally in the world comprised more than half of the total land area of the earth, and the six largest states in population (not the same six) comprised more than half of the total world population. In contrast, there are many micro-states with less than 1,000 square kilometres and many with less than 100,000 people. In consequence, population densities of countries range from less than 10 per square kilometre in Australia, Canada, Mongolia and Mauritania to more than 10,000 per square kilometre in some city-states such as Macao and Monaco. These political divisions have greatly affected the international migrations of peoples as well as their fertility

254

Table 12.2 Areal sizes of countries by continent, 1991

| Continent | Area (1,000 km²) | | | | | |
	More than 10,000	1,000–10,000	100–1,000	10–100	Less than 10	Total
Africa	–	12	23	12	5	52
Americas	–	8	11	9	17	45
Asia	–	7	22	12	7	48
Europe	1	–	18	13	8	40
Oceania	–	1	2	4	8	15
Total	1	28	76	50	45	200

Table 12.3 Population sizes of countries by continent, 1991

| Continent | Population (millions) | | | | | |
	More than 1,000	100–1,000	10–100	1–10	Less than 1	Total
Africa	–	–	16	28	8	52
Americas	–	2	9	15	19	45
Asia	1	5	18	18	6	48
Europe	–	1	15	14	10	40
Oceania	–	–	1	2	12	15
Total	1	8	59	77	55	200

and mortality. Moreover, through their increasingly centralized roles, governments are both explicitly and implicitly influencing not only the patterns and dynamics of their own populations but those of other countries as well (e.g. refugees, aid for family planning).

The political divisions and nationalism which characterized the period 1800–1950 encouraged little global awareness. It was only after 1950 that the rapid rise in the rate of world population growth became a common concern, at least among the educated. This global concern, fostered by the spread of education and massive improvements in communications, has gradually extended to encompass concern about the growing inequalities between rich and poor countries and peoples. In addition, the space age has encouraged greater understanding about global and local environmental changes, as well as their interrelationships with population. Similarly, it is mainly since 1950 that we have seen the widespread diffusion of two of the major social changes that had been previously experienced to a greater or lesser extent in the developed countries: the nuclear family system and changes in the status and roles of women. They have not diffused universally, and there are now strong regional variations among developing countries, as for example between China and India, as well as rural–urban contrasts, all of which cause strong social and demographic differentiation so characteristic of the world today.

DEMOGRAPHIC TRANSITION IN EUROPE AND THE MORE DEVELOPED COUNTRIES

Europe initiated the process of demographic transition during the eighteenth century. The disappearance of the ravaging plague after 1720 was an important factor, along with the gradual diminution in the number of major epidemics. Improvements in urban sanitation, especially following the construction of piped water supplies, was a contributory factor. An agricultural revolution occurred in rural areas, with the disintegration of the feudal system, scientific advances in agriculture and the acquisition of land by business men. Food shortages were also reduced to some extent by the improvements in communications. Other factors included efforts to prevent abortion and to raise the status of unmarried mothers and illegitimate children, along with the decline of celibacy. So with declining mortality and slightly rising fertility the population of Europe began to increase at a rate previously unknown. Earlier labour shortages were replaced by surpluses. Beggars and vagabonds multiplied. Of course, population growth in eighteenth-century Europe varied; rapid in England and Wales, France, Scandinavia, Germany and especially in the colonized areas of Eastern Europe and Russia, it was slower in the Mediterranean areas.

Whether rising fertility or falling mortality was the primary factor in initiating demographic transition has been a matter of considerable discussion and controversy, but in general mortality decline long preceded fertility decline leading to accelerating population growth, followed by decelerating growth as fertility later began to fall (Figure 12.2), and finally by low growth or even natural decrease, a not uncommon situation among European countries in recent decades. The classic demographic transition model (Chesnais 1986) associated these changes with development and modernization, but was never accompanied by the 'explosive' population growth rates seen in the so-called Third World today; 1 per cent was the norm rather than 2 or 3 per cent.

The process started at different times and progressed at different rates throughout Europe. It began first in Northern and Western Europe, where the period of most rapid population growth was between the 1870s (e.g. England) and 1890s (e.g. Germany) at the onset of fertility decline (Table 12.4). This period of maximum population growth was also the time when Western Europe was the keystone of the economic life of the world. France and Ireland, however, were exceptions to the general pattern of demographic transition. France experienced much slower population growth during the nineteenth century through an earlier decline in fertility from the time of the French Revolution,

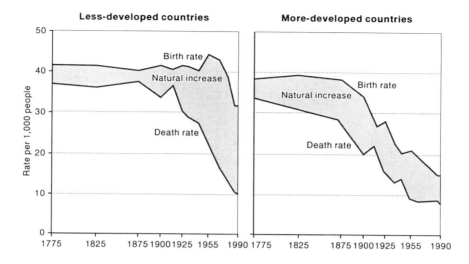

Figure 12.2 Population growth through natural increase, 1775–1990. After a diagram produced by the UN Population Division.

as well as relatively high mortality. Ireland, on the other hand, suffered population decline from the 1840s through the effects of the potato famine, emigration, delayed marriage and low fertility.

Demographic transition then diffused to Anglo-America, Australia and New Zealand, Uruguay and Argentina, and later to the countries of Southern and Eastern Europe, which were dominated by landed aristocracies, agrarian problems and absolutist monarchs for much longer. Though transition was rapid when it arrived, the eastern and southern parts of the continent did not achieve their fastest growth rates until the first three decades of the twentieth century. By the 1960s most developed countries, including a newcomer Japan, had low natural increase, fertility and mortality, a convergence later accentuated by further fertility declines during the 1970s and 1980s. Of course, these broad patterns of transition were interrupted variously by world wars, by post-war 'baby booms', by economic crises of inter-war years and by phases of early marriage. There were also marked rural–urban contrasts and socio-economic differentials which took a long time to be reduced.

If we examine the causes of mortality and fertility declines separately – and of course they are interrelated – we find a wide variety of factors. Mortality decline was influenced by different sets of factors at different times and places. Initially, it was probably assisted most by improvements in social and cultural organization which helped to reduce the number and intensity of mortality crises, especially those caused by famine and epidemics (Livi-Bacci 1992).

257

Table 12.4 Populations of selected European countries, 1800–1900 (millions)

	1800	1850	1900	Increase 1800–1900 (%)
Norway	0.9	1.4	2.2	144
Sweden	2.3	3.5	5.1	122
Finland	1.0	1.6	2.6	160
Denmark	1.0	3.5	2.5	150
Netherlands	2.2	3.1	5.2	136
Belgium	3.0	4.4	6.7	123
Great Britain	10.6	20.8	37.0	249
Ireland	5.0	6.5	4.5	−10
France	28.2	35.8	39.0	38
Germany (1871 area)	24.5	35.4	56.4	130
Switzerland	1.7	2.4	3.3	94
Hapsburg Monarchy	23.3	31.4	45.4	95
Hungary	10.0	13.3	19.3	93
Italy	18.1	24.3	32.5	80
Spain	11.5	15.0	18.6	62
Portugal	3.4	3.9	5.4	59

Source: H. Moller (ed.) *Population Movements in Modern European History*, New York, Macmillan, 1964, p. 5.

Improved methods of child care, hygiene, sanitation, clothing, housing and organization of markets all contributed, particularly in the early phase. Later, it is suggested that economic factors played a more prominent role during the nineteenth century through improving standards of living in European countries, though epidemics of cholera, diptheria, typhoid and influenza continued to exact a heavy toll. Subsequently, life expectancy has been much less affected by economic progress than medical and behavioural factors, and in fact large increases in economic development often have little impact upon life expectancy. However, these phases were not discrete, because locally diverse factors have always played a part.

The result was that life expectancy rose in France and England from 34 and 37 in 1800, to 47 and 48 in 1900, to 67 and 69 in 1950 and 77 and 75 in 1990. During much of this time the gains in life expectancy accelerated, with a growing gap between male and female life expectancy. By 1990 the life expectancy at birth of females exceeded 80 in many European countries, 7–9 years more than that of males and steadily increasing. For the developed world as a whole the life expectancy at birth of males and females was 71 and 78, a much greater gap than in the developing world where it was 60 and 62.

Fertility decline in Europe, mainly from the 1870s onwards, was associated with profound changes in Western civilization. As it results from a combination of biological and socio-economic factors, it is not easily explicable. Changes in ideas and mores were significant, associated with the waning of religious ideals

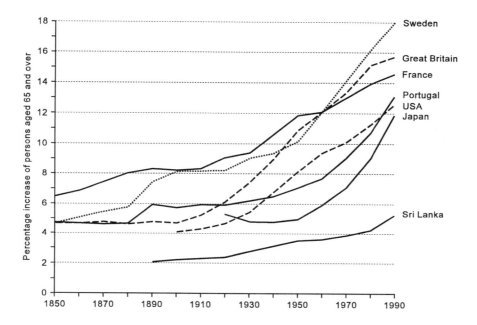

Figure 12.3 Ageing of population at the apex: increase in percentage of persons aged 65 and over in selected countries.

and the rises of rationalism and socialism; neither Marx nor Engels saw the family as the basic unit of the state. Traditional family institutions were also affected by the rising status and increased education and employment of women, the growth of divorce and the spread of more effective methods of birth control (though coitus interruptus and abortion remained the most widespread methods until the 1930s). Obviously, fertility decline was facilitated by urbanization, which created very different economic and social conditions for the upbringing (and employment) of children, who gradually left the active population and became consumers rather than producers.

Fertility decline began in France in the 1820s, but not until the last quarter of the century in Britain and other more developed regions of Western Europe, Anglo-America and Australasia (Coale and Watkins 1986). Gradually it spread towards Southern and Eastern Europe, not reaching the more peripheral and backward regions of the Mediterranean and Alpine zones until the middle of the twentieth century. Moreover, there are significant local variations where cultural factors play a more important part than economic ones, as for example in Ireland where fertility decline was much delayed, as it was in Japan where it was largely postponed until after the Second World War (Chung 1970; Noin 1983).

Fertility decline meant huge social changes; while women in Britain,

Germany and the Netherlands around 1850 had about five children, those born around 1950 had only about two children. Furthermore, those born later will have well under two, as fertility dropped sharply in the 1970s especially following the introduction of the contraceptive pill. In short, reproductivity has become much more efficient in developed countries, replacing the reproductive wastage which characterized previous human history dominated by high mortality and low life expectancy.

Demographic transition has also been accompanied by remarkable ageing of populations, by which the proportion aged less than 15 years declined from 40–50 per cent to below 20 per cent in some countries while the percentage aged 65 and over rose from less than 5 to as high as 18 (Figure 12.3). Indeed, in Sweden today the two age groups are roughly equal. The process is particularly marked in Northern and Western Europe; the proportions in Eastern and Southern Europe are more like those in North America, Australia and New Zealand. The detailed age pyramids of developed countries of course reflect their individual demographic histories, and the impact of wars, epidemics and economic and political crises can be seen in the subtle indentations in the pyramids. All countries usually have a surplus of females, mostly caused by their greater longevity; the female surplus is particularly strong among the rapidly rising numbers of people aged 75 or 85 and over, many of whom are widowed and tend to require greater community care. Only where migration of males prevails is this female preponderance upset.

OVERSEAS EMIGRATION

The demographic transition in Europe accelerated population growth beyond the needs of either improving agriculture in rural areas or nascent industrialization in towns. Coupled with improvements in transportation and the pull of relatively 'open spaces' outside of Europe and the heartlands of Asia, the nineteenth and early twentieth centuries witnessed the most spectacular trans-oceanic migrations ever known. Although exact measurement of this human current is impossible, between 1815 and 1914 probably 60 million Europeans emigrated overseas to the less peopled and less exploited parts of the world, especially the Americas (the USA alone attracted about three-fifths), Oceania and parts of Africa. In addition, some 10 million persons migrated from Russia to Siberia and Central Asia and millions of Chinese and Indian labourers migrated to work in the new colonial economies, notably in South-East Asia (Figure 12.4).

Certainly, the return movement was also large – the repatriation rate from the United States was probably about one-third, and from Argentina over one-half – and also many emigrants moved from one country to another, notably from Canada to the United States. But the main effect was to create 'New Europes' overseas, largely Anglo-Saxon, Latin or Slav, which were younger and more active demographically than Europe, and which rapidly became economic

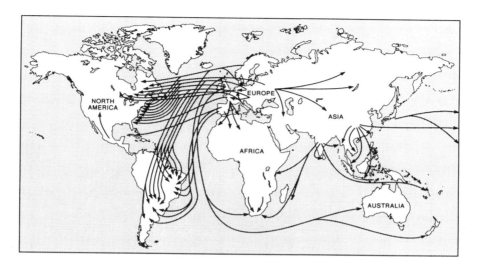

Figure 12.4 World voluntary migration, 1815–1914.

competitors, thus reducing its world dominance.

The volume of emigration overseas was far from regular, because of changing 'push' and 'pull' factors as well as the intervening obstacles. Among the 'push' factors at places of origin, apart from population growth, were political conditions, declining sizes of farms, surpluses of rural labour, fluctuations in harvests and the irregular growth of industrialization to absorb labour supply. The 'pull' factors included the abundance and cheapness of land to populate, favourability of climatic conditions and availability of new economic opportunities. Overseas emigration was also greatly facilitated, not only by tremendous improvements in transport but also by assisted passages and by the overall diffusion of an economic system and social structure which involved the development of new sources of raw materials and food, the export of capital and the growth of new markets. At the local level it also benefited from an increasing flow of information about conditions for migration and for living overseas.

Consequently, the volume of European emigration rose from 200–400 thousand annually in the period 1840–80 to over 700 thousand annually in the 1880s and, after a slight fall in the 1890s, to over one million a year during the period 1900–14, a peak never achieved before or since (Figure 12.5).

The main sources of emigrants changed considerably. In the period up to and including the 1880s the bulk of emigrants were from the industrialized countries of Britain, Germany and to a lesser extent of Scandinavia. Emigration from these countries was really a prolongation of internal migrations engendered by industrialization, and was most voluminous at times of economic crises. The

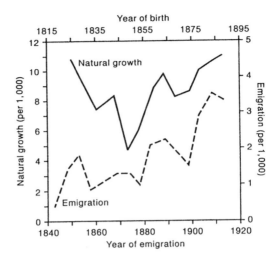

Figure 12.5 Emigration and natural growth for continental Europe. After Livi-Bacci (1992).

British formed the major contingent during the early part of the nineteenth century, although their peak departures were actually in the 1880s and in 1913, when 345,000 left. Between 1825 and 1920 about 65 per cent of British emigrants went to the United States, 15 per cent to Canada, 11 per cent to Australasia and 5 per cent to South Africa, in all of which the imprint of their language and culture has remained indelible. The 6 million German emigrants, on the other hand, mostly went to the United States, where between 1820 and 1940 they constituted the largest single group of immigrants and formed a number of 'little Germanies'.

From the 1880s onward a rather new type of overseas emigration developed, especially from countries in Southern, Central and Eastern Europe where rural overpopulation prevailed (although an earlier example was Irish emigration following the 'potato famine' of the 1840s), and soon the numbers departing from these regions greatly exceeded those emigrating from Northern and Western Europe, where only the British Isles continued to send a large stream of emigrants. This new essentially peasant emigration formed a large, unqualified, cheap, labour force for the rapidly expanding industries of the New World. Not all went to the United States: Latin America received many, especially from Southern Europe. Spaniards were preponderant in the flow to Argentina, and Portuguese in the flow to Brazil. Before 1900, however, both were outnumbered by Italians, the largest group of 'new' European overseas emigrants, though they included a high proportion of repeat and return migrants.

The impact of these massive movements upon both homelands and countries of destination was enormous. The city-ports of Europe, like Naples, Hamburg and Liverpool, expanded greatly, and many rural areas depended considerably upon remittances from the New World.

The First World War brought an abrupt halt to the powerful stream of overseas emigration, which never really recovered. Economic growth and declining natural increase meant less population pressure. Moreover, many European countries introduced unemployment benefits and health assistance, while the national socialist countries were anxious to reduce emigration, except to colonial territories (e.g. Libya), as it was regarded as a loss of human capital. At the same time, in the countries of destination there was less demand for labour and an increasing desire to safeguard their new national unity and high standards of living by quantitative and/or qualitative restrictions upon immigration. And whereas economic crises during the nineteenth century stimulated emigration, the world crisis of 1929 had the reverse effect because it was much more universal.

The massive numbers of refugees and displaced persons in Europe at the end of the Second World War helped to fuel the reopened channels of overseas emigration from Europe which between 1946 and 1964 reached an average annual gross outflow of 600,000; but there were strong return currents. The main destinations were in North America and Australasia, where prospects seemed brightest for more selected migrants. The main sources were mainly in Southern Europe, particularly Italy, but later millions of migrants from Southern Europe were to be more attracted to countries of Western Europe to fill gaps in labour supply caused by war losses and rapid and sustained post-war economic growth. Their numbers were swollen by migrants from the former colonial territories, particularly of Britain and France; West Indians, Indians, Pakistanis and North Africans, among others, increased the multi-racial character of the European countries, especially in the cities.

POPULATION MOBILITY AND REDISTRIBUTION IN MORE-DEVELOPED COUNTRIES

Overseas migrations were only one aspect of the drastic redistribution of population in Europe during the nineteenth century. There was also a considerable, though smaller, amount of international migration within Europe, especially to France and Germany, which were the only European countries with more than a million foreigners before the First World War.

A much more voluminous type of migration was the massive rural-to-urban movement, the main way which populations responded to changing economic opportunities. Declining demands for agricultural labour resulting from improved efficiency coincided with increasing demands for labour in urban-based manufacturing and tertiary industries, and consequently rural–urban migration involved a shift in employment. Initially, the burgeoning cities of Britain were unhealthy and experienced low natural increase rates, thus raising labour demand, but an important factor during later decades of the nineteenth century was the opening-up of the grain-producing lands of North America, Russia, Argentina and Australia, which enabled Europe to import cheap grain and reduce still further the demand for its own agricultural labourers. In addition, a number of socio-cultural factors later contributed to rural–urban migration, including better educational, cultural and recreational opportunities in towns and better health and public service facilities.

The growing concentration of population in towns and cities was a remarkable fact of population redistribution, and generally followed an S-shaped curve. In 1800, only about 3 per cent of the world's population lived in towns of 5,000 inhabitants or more; in 1850, 6 per cent; in 1900, 14 per cent; and in 1950, 30 per cent. Of course, the majority of these were in the more-developed countries. Indeed, in 1900 half of the world's population living in cities with 100,000 inhabitants or more were to be found in Europe, where in England and Wales 35 per cent of the total population were living in such cities. However, urbanization slowed down when 70 or more per cent of the total population were living in towns.

The growth of large cities was particularly striking. In 1850 there were only three cities in the world with more than a million inhabitants (Lowry 1991) – London, Beijing and Paris – but by 1950 there were seventy-five 'million-cities' of which fifty-one were in the more-developed world incorporating 15 per cent of its total population. In contrast, there were only twenty-four in the less-developed world with only 3 per cent of its population (UN Population Division 1992). Generally the growth of million-cities became more rapid in the New Europes than in Europe itself, but after mid-century their growth was much more rapid in the less-developed world than in the more-developed world.

Cities in more-developed countries also tended to concentrate in conurbations and clusters in economic core areas, initially influenced particularly by coalfields, industrial zones, ports and capitals. The emergence of megalopoli in the United States (Gottmann 1961) and across Europe (Hall 1977) from Britain to Italy exemplifies this polarization of population concentration. Economic forces have prevailed in this process, as they have also in the later process of urban and industrial deconcentration through changes in organizational structures, technology and communication facilities. In addition, however, direct and indirect government policies and even the nature of the state have exerted influences upon the distribution of population and settlement. In general, unitary states like France, Greece and Japan demonstrate the strength of the centripetal forces of centralization, whereas federal states like the USA, Canada and Australia exhibit more of the centrifugal forces of decentralization. Consequently, unitary state capitals tend to dominate their urban hierarchies much more than federal state capitals.

In recent decades urban deconcentration and counter-urbanization have reduced population densities in central business districts, but generally urban population densities in more-developed countries have tended to decline exponentially outwards. Trams, buses, tubes, trains, and cars enabled longer commuting distances, reduced the compactness of cities and increased the number of satellite towns, so that edge density gradients are much less steep than around large cities in less-developed countries where commuting and private transport are less.

Despite counter-urbanization, the overall settlement system began to achieve a dynamic equilibrium in which there was increasing population mobility of all sorts but low net migration and low migration effectiveness (Rowland 1979). Numerous circular and return migratory movements tend to cancel themselves out, effecting population redeployment and replacement.

DEMOGRAPHIC TRANSITION IN THE LESS-DEVELOPED COUNTRIES

By the mid-twentieth century there were striking demographic contrasts between the Europeanized world, where demographic transition and urbanization were far advanced, and the rest of the world where they were incipient or negligible. The economic dichotomy between developed and developing countries was reflected in their demographies, low fertility, mortality and natural increase in the developed countries contrasting with high fertility, mortality and natural increase in the developing countries.

Since then, rapid decline in mortality and delayed but increasing decline in fertility have led to a phase of phenomenal population growth in the developing world. While the total population of the developed world grew relatively modestly from just over 0.8 billion in 1950 to 1.2 billion in 1992, the population of the developing world escalated from nearly 1.7 billion to 4.2 billion at well over 2 per cent per annum (Figure 12.6). Inevitably, it incorporates a growing proportion of humanity; 77.4 per cent in 1992 but likely to be over 81 per cent by 2010.

However, the stark demographic contrasts between developed and developing countries have become greatly attenuated. While the vital rates of the developed countries have declined further and converged, those of the developing world have diversified, so that a wide variety of demographic conditions may be observed, from pre-transition and early transition through to almost completed transition. Some Asian countries such as China, Taiwan, South Korea, Hong Kong, Singapore, Thailand and Sri Lanka have undergone very substantial demographic transition, as have a number of Latin American and Caribbean countries like Argentina, Uruguay, Cuba and Puerto Rico. Many other countries such as Indonesia, Malaysia, Mexico and Brazil have made marked progress. On the other hand, many countries in the Middle East and North Africa have witnessed major mortality decline but much less fertility decline, and some Sub-Saharan countries have not enjoyed much decline of either.

In consequence, natural increase rates have ranged from 4 per cent per annum

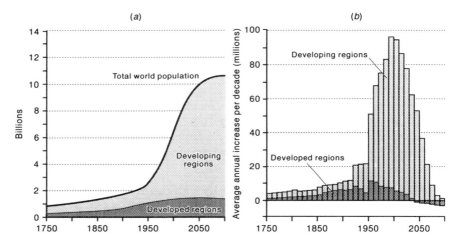

Figure 12.6 World population growth (*a*) and average annual increase in numbers per decade (*b*), 1750–2100.

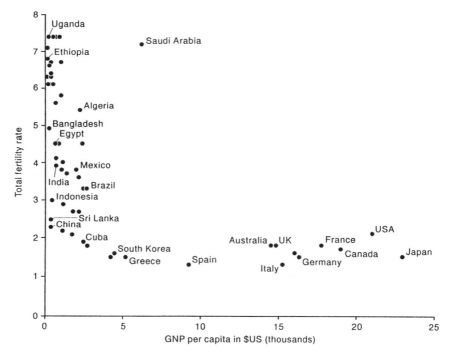

Figure 12.7 Fertility and per capita incomes in countries with more than 10 million people, 1991.

266

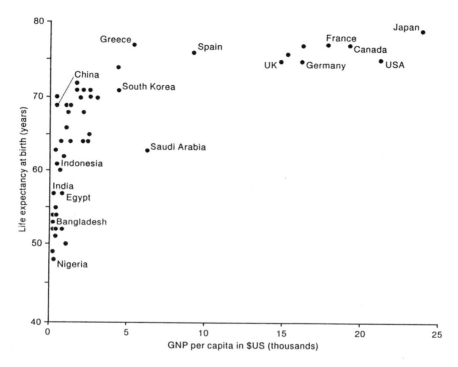

Figure 12.8 Life expectancy and per capita incomes in countries with more than 10 million people, 1991.

in parts of Africa and the Middle East to less than 1 per cent in Hong Kong and South Korea. Generally, the developing countries of East Asia, South-East Asia, South Asia and Latin America have now much lower growth rates than those of Western Asia and Africa, especially the latter.

Recent differential demographic transition in the developing countries is much less determined by economic development than was the case in the developed countries during the nineteenth and early twentieth centuries. Other factors such as the diffusion of medical and family planning services, social changes and government policies have all played important parts.

The overall decline in total fertility in the developing world from over six at mid-century to less than four in 1990 has been striking, but the rate and timing of the decline in fertility have differed greatly among the developing countries in response to levels of contraceptive prevalence, female literacy and employment, urban and rural residence, as well as age of marriage, culture and religion, and the strength of government policies. Thus a number of countries with low per capita incomes like China, Cuba and Sri Lanka have achieved total fertility rates almost as low as those of rich developed countries (Figure 12.7). Moreover,

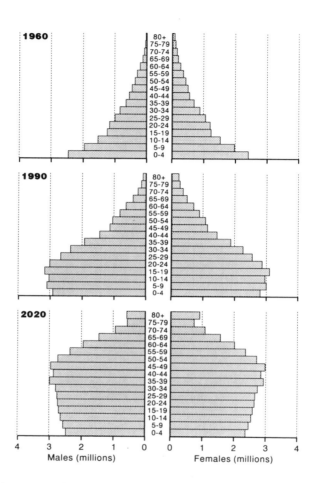

Figure 12.9 Age pyramids of Thailand, 1960, 1990, and 2020 (estimated). After US Department of Commerce.

while nearly all Chinese populations of East and South-East Asia have experienced rapid fertility decline, most Muslim populations (especially of South and South-West Asia and North Africa) have experienced very much less.

Similarly, while average life expectancy in the developing world has grown greatly from 40 years in 1950 to 63 in 1990, and with little correlation to levels of economic development (Figure 12.8) (*World Development Report 1993*), it ranges from less than 45 years in some of the poorest countries of Sub-Saharan Africa to over 75 in some East Asian countries, particularly the so-called 'dragons'. And while child mortality declined in general in the developing world

268

from 28 per cent to 10 during this period, it ranges from over 30 per cent to under 1 per cent in the same groups of countries. In too many of the poorest countries infectious and environmental diseases prevail, accentuated by malnutrition, but even in these countries some progress has been made. They have yet to achieve the 'epidemiological transition' in which the leading causes of death change from infectious and acute to chronic and degenerative. It tends to lag behind the demographic transition, as the early mortality declines usually result from reductions of infectious diseases among children. Improvements in child mortality often take place while high fertility persists leading to broadening of the base of the population pyramid, but subsequent fertility decline leads to rising median ages and to ageing populations – not so much the increase in the number of old people as in the numbers of younger adults and middle adults, posing particular problems for employment as in the case of Thailand (Figure 12.9).

Thus the demographic transition experienced by the developing countries since mid-century has been more rapid and more varied than previously experienced by the developed countries, differentiating further the demographic patterns of the world.

REDISTRIBUTION OF POPULATION IN DEVELOPING COUNTRIES

The developing countries have contemporaneously undergone massive redistribution of population since mid-century, partly generated by rapid population growth and partly by migration.

Differential demographic transition has meant that Africa is now experiencing more rapid population growth than any continental region, and its proportion of the total world population will probably rise from 8.9 per cent in 1950 to 13.9 per cent in 2000 (Table 12.1), though AIDS has yet to exert its full effect and it is not certain how much population growth rates will be lowered. In contrast with Africa, the percentage of world population living in Latin America will rise less, from 6.6 to 8.6 per cent. Moreover, the slowing down of China's population growth means that the Indian sub-continent, with more rapid growth, will soon overtake it as the largest concentration of humanity.

Since 1950 the continental balance of population has been much less affected by international migration, although tens of millions have migrated from less- to more-developed countries, including labour migrants, 'guestworkers', skilled migrants, refugees and growing numbers of illegal migrants. Apart from the long-term pulls to Western Europe, North America and to Australia, Southern and even Eastern Europe are now regions of attraction, in addition to the strong but irregular pulls of the oil-rich states of the Middle East and newly industrializing countries (NICs), which have been especially attractive to temporary migrants. All face considerable problems of immigration control and assimilation, as well as dealing with ensuing problems of cultural diversity. The

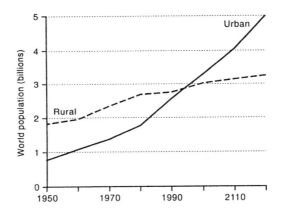

Figure 12.10 World population by urban and rural residence, 1950–2020.

high rates of unemployment in developed countries have also led to the questioning of the economic benefits of such migration.

Now the push factors in less-developed countries tend to outweigh the pull factors in more-developed countries. Apart from rapid population growth in the former, the three main triggers of international migration have been political repression, economic disparities and deprivation, and environmental deterioration, which are all closely interrelated. Many governments of developing countries have proved unable to provide political stability, democracy or the basic essentials of human life for their populations. Their economies have suffered from crippling debt burdens, shrinking currencies and dwindling foreign investment. All of these factors have contributed to widespread poverty and environmental degradation. Thus, it is increasingly realized that the way forward to reduce international migration flows from developing countries is for better government and education, slower population growth, and greater economic security and environmental responsibility, a holistic approach to Third World problems which is highly desirable but not easily achieved.

Population distribution within developing countries has tended to concentrate in economic core areas within easy access to the coast, linking up with the world economy. Peripheral population concentration is particularly strong in Latin America and Africa, and although China and the Indian sub-continent are much more evenly inhabited the densely peopled deltaic areas of Asia are very vulnerable to floods, tidal flows and potential rises of sea-level. These coastal concentrations are caused by both natural increase and net in-migration from less favourable environments. Thus uneven population distributions are widespread, as exemplified by those of Brazil, Algeria, Egypt, Iran and Indonesia.

Unevenness has been greatly intensified by urbanization. The population in

270

urban areas of the developing world grew at about 4 per cent per annum during the period 1950–90, while the rural population growth was less than half of this rate. Both are expected to decline, but urban growth will be greater than rural growth (Figure 12.10). The rate of urban population growth has been most rapid in those parts of the world which were least urbanized before; Sub-Saharan Africa rose from 12 per cent urban in 1950 to 30 per cent by 1990. Urban growth is so rapid that during the period 1990–2010 the urban populations of some developing countries will increase by more than their total populations in 1960. In 1990 the developing countries contained about 61 per cent of some 2.58 billion urban dwellers in the world; by 2025 they are expected to comprise nearly 80 per cent of over 5 billion urban dwellers.

Most striking is the growth of large cities, especially primate cities which dominate urban hierarchies, the great nodes of energy production, manufacturing, transport networks, finance, commerce, administration and all other service activities. Some primate cities, like Bangkok, Seoul, Montevideo and Buenos Aires localize well over one-third of their country's inhabitants. By 1990, 58 of the 100 cities in the world with more than 2 million inhabitants were located in developing countries, and the growth of mega-cities with multiple millions of inhabitants has become a notable phenomenon. It is expected that by 2000 two-thirds of the 24 cities with 10 million inhabitants or more will be found in developing countries. Mexico City, São Paulo, Shanghai, Seoul and Buenos Aires, among others, have joined the ranks of the world's largest cities with rapid growth rates and some of the most intransigent urban problems: slums and shanty towns; the close proximity of industries and housing; air and water pollution; waste disposal; environmental health. So far few governmental redistribution policies have had more than limited impacts in slowing down large city growth, save in a few socialist and authoritarian countries (e.g. China, Cuba, Sri Lanka). In any case, it is not always certain that the balance of benefits and disbenefits of large city growth has been adequately addressed; it certainly varies from country to country.

THE FUTURE

The last two centuries have witnessed dramatic changes in global demographic dynamics, composition and distribution, notably massive growth and urbanization, and it is certain that those processes will persist during the immediate future. As one looks forward into the twenty-first century, trends and patterns are the subject of numerous projections based upon varied assumptions, particularly concerning the possibility of fertility decline in developing countries; but all assume world population growth at least until 2050, as control of population growth is still uneven.

The broad balance of continental population distribution has seen the growing preponderance of Asians, the rapid increase in the numbers of Africans and Latin Americans and the relative decline in the numbers of Europeans and

North Americans (Table 12.1). As the future balance is likely to be influenced mainly by differential rates of natural increase rather than by massive intercontinental migrations, it will not change radically. The future of international migrations will most probably be conditioned by political insecurity and conflict as well as contrasting levels of economic development which are presently too wide.

There are also signs that the process of population concentration will also persist, especially in developing countries. While many developed countries are experiencing counter-urbanization and population deconcentration, the developing countries have more mega-cities and are seeing the relative depopulation of more hostile environments. The environmental implications of these demographic changes are profound.

REFERENCES

Chesnais, J.-C. (1986) *La transition démographique*, Paris: PUF.

Chung, R. (1970) 'Space–time diffusion of the transition model: the twentieth century patterns', pp. 220–39 in G. Demko, H. Rose and G. Schnell (eds) *Population Geography: A Reader*, New York: McGraw-Hill.

Coale, A.J. and Watkins, S.C. (eds) (1986) *The Decline of Fertility in Europe*, Princeton: Princeton University Press.

Gottmann, J. (1961) *Megalopolis: The Urbanized Northeastern Seaboard of the United States*, New York: The Twentieth Century Fund.

Hall, P. (1977) *Europe 2000*, New York: Columbia University Press.

Hambloch, H. (1966) *Der Höhengrenzsaum der Ökumene. Anthropogeographische Grenzen in dreidimensionaler Sicht*, Munster: Westf. Geogr. Studien 18.

Livi-Bacci, M. (1992) *A Concise History of World Population*, Oxford: Blackwell.

Lowry, I.S. (1991) 'World urbanization in perspective', pp. 148–76 in K. Davis and M.S. Bernstam (eds) *Resources, Environment, and Population: Present Knowledge, Future Options*, A supplement to *Population and Development Review*, Vol. 16, 1990.

Noin, D. (1983) *La transition démographique dans le monde*, Paris: PUF.

Noin, D. (1991) *Atlas de la Population Mondiale*, Montpellier-Paris: RECLUS – La Documentation Française.

Roberts, J.M. (1985) *The Triumph of the West*, London: BBC.

Rowland, D.T. (1979) *Internal Migration in Australia*, Census Monograph Series, Canberra: Bureau of Statistics.

United Nations Population Division (1992) *The World's Million-Cities, 1950–1985*, ESA/P/WP.45.

World Development Report 1993. Investing in Health, New York: Oxford University Press for the World Bank.

Wrigley, E.A. (1988) *Continuity, Chance and Change*, Cambridge: Cambridge University Press.

FURTHER READING

Chesnais, J.-C. (1986) *La transition démographique*, Paris: PUF.

Cleland, J. and Hobcraft, J. (1985) *Reproductive Change in Developing Countries*, Oxford: Oxford University Press.

Coale, A.J. and Hoover, E.M. (1958) *Population Growth and Economic Development in*

Low-Income Countries, Princeton: Princeton University Press.

Coale, A.J. and Watkins, S.C. (1986) *The Decline of Fertility in Europe*, Princeton: Princeton University Press.

Findlay, A. and Findlay, A. (1987) *Population and Development in the Third World*, London: Methuen.

Jones, H. (1990) *Population Geography*, London: Paul Chapman.

Livi-Bacci, M. (1992) *A Concise History of World Population*, Oxford: Blackwell.

Noin, D. (1991) *Atlas de la Population Mondiale*, Montpellier-Paris: RECLUS – La Documentation Française.

Population and Development Review, New York: Population Council.

UN Department of International and Social Affairs:

(various years) *Demographic Yearbooks*, New York: UN Statistical Office.

(1991) *World Urbanization Prospect, 1991*, New York: UN.

(1992) *World Population Prospects, 1992*, New York: UN.

United Nations Development Programme (UNDP) (various years) *Human Development Report*, New York: UN.

The World Bank (various years) *World Development Report*, New York: Oxford University Press.

13

ORIGINS OF MODERN ENVIRONMENTALISM

J.M. Powell

At first glance, modern anxieties over the various aspects and implications of environmental deterioration have rather shallow roots. We may be too easily persuaded to seek the origins of current concerns in the much-trumpeted coming exhaustion of non-renewable resources, intensifying fears of global overpopulation, the alarming implications in rampant pollution, recurring energy crises, and speculations on the connections between deforestation and the spectre of land degradation. The commonest inclination is to examine the environmental attitudes and behaviours of urbanized, relatively well-educated Western communities to identify the primary locus of these modern preoccupations. Anchored thus in space and time, our understanding is severely limited and ultimately unsatisfying; as a corollary, 'environmentalism' in its many guises is readily criticized for dubious intellectual and cultural foundations, and is conveniently discounted as an irritating 'peripheral' movement dependent on episodic engagements with mainstream society. That is why it needs to be asserted that environmental concern is as old as human society itself – is indeed an essential ingredient, a measure, of the human condition – and that the adoption of any constricting perspective deprives us of a rich legacy of thought and action which might prove to be our greatest resource.

STEWARDS, IMPROVERS AND 'ISSUES'

Environmental impacts attributed to early hunting and grazing cultures included the extermination of certain animal species and the extension of vast areas of grassland by the use of fire. Hindsight does not draw on a geological time-scale, and so these effects would not have been deemed 'catastrophic' or even remarkable. Because the transformations were scarcely perceptible from generation to generation, one may assume that they did not constitute 'issues' as that term is commonly conceived; yet the word is not without ambiguities, and in any case it would be foolish to underestimate the prodigious reach of oral transmission in primitive societies. Similarly, the emergence of agriculture between 10,000 and 12,000 years before the present was accompanied, necessarily, by fresh reckonings on Nature–human relationships which included the design of remedial and preventative works. For example, the great Mediterranean and Central and South American civilizations developed highly productive systems which incorporated elaborate terracing techniques to conserve water and soil resources; in a restricted sense, even the reservation of distinct enclaves as hunting preserves for the ruling elites showed a recognition of the importance of environmental protection. Again, in most pre-industrial cultures forests, water sources and selected animals were protected in varying degrees by religious myths and taboos. On the other hand, it is true to say that the rise of urban cultures over the past 6,000 years made extraordinary demands on the physical environment and widened the gap between human communities and Nature. In the Mediterranean Basin the destruction of forest cover and putatively related haemorrhaging of natural hydrological systems, soil exhaustion and widespread soil losses attracted the attention of a number of leading contemporary scholars, and so the interpretation of environmental problems entered the realm of intellectual debate.

Seen in these lights the environment has always presented 'issues', and human interest has been most enlivened by change or the prospect of change. In the days of imperial Greece and Rome, a wide range of deliberate human interventions in Nature was recorded by the classical commentators – including Cicero, Eratosthenes, Herodotus, Hippocrates, Plato, Strabo, Theophrastus and Tacitus – for whom the Mediterranean Basin offered overwhelming evidence of the transforming power of human agency. The creation of large cities and the extension of massive irrigation and drainage works, together with the selection and domestication of food grains and farm animals, suggested to philosophical minds the notion of human *stewardship* over the earth; yet speculations on deforestation–soil erosion linkages questioned idealizations of an improving 'mission', and convincing interpretations of the effects of clearing, irrigation and drainage pointed to the possibility of human-induced climatic change. The robust attraction of a mythologized 'Golden Age' when soils produced best and spontaneously, free of human interference, may have bolstered a belief that hard and intelligent work was therefore required in less fortunate eras to maintain rural productivities. In the absence of more sophisticated modes of reasoning careful observation, discussion and practical experimentation served, pre-sumably, to deepen the interest in natural processes – at least in the most demanding environmental situations. There are, in addition, intriguing anticipa-tions of twentieth-century disputes in the local and regional protests against government proposals to manipulate natural systems. In one of the more important of the latter cases, Tacitus records that major river diversion proposals were defeated by arguments calling for the protection of sacred groves and rituals, and by the contention that Nature had already made the best provision for the needs of humanity.

The ancient notion that human society possessed enormous potential for creative and destructive intervention in natural systems was revived in the high and late medieval period. This was very well shown in the work of the thirteenth-century monk Albertus Magnus, notably in his practical comments on land reclamation, soil conservation and the relationship between the removal of forest cover and changes in climate. The period between the eleventh and the thirteenth centuries has been described as Europe's great age of forest clearing. Most of the clearance was associated with the expansion of settlement frontiers and the establishment of towns and industries, but although it was encouraged by the Church it had not been universally acclaimed. On the contrary, much earlier – for example in the eighth century, during the reign of Charlemagne – efforts had been made to fashion rules which would secure a balance between the competing demands of agriculture, industry and forestry, and this advocacy of 'multiple use' principles had not been abandoned. Furthermore, Albertus Magnus was not alone in expressing an urgent interest in sustainability. By that time the need to consider more stringent forest protection had intensified with the opening-up of new regions, the expansion of towns, the use of charcoal in smelting, and the voracious demand for materials for house construction and assorted wood-working industries. And after all, the removal of vast areas of forest cover represented the most tangible of landscape modifications which must have tormented environmentally sensitive minds, both religious and lay. From the perspective of today's conservationists, however, the seventeenth century seems to offer the more recognizable antecedents of current concerns.

In Britain, John Evelyn's *Sylva: or, A Discourse of Forest Trees* (1664) made a detailed and conspicuous plea for scientific forestry, including conservation and afforestation. Evelyn was particularly anxious about the strategic dilemma raised for the Royal Navy in the relentless onslaught on Britain's traditional forest resource, and suggested that the Royal Society itself take an interest in the question of the relationship between timber and national policy. Though frequently borrowing from the classical writers, one novel contribution was his ardent advocacy of a vigorous planting programme as part of a rational approach to managing the process of continuous landscape change, rather than as an attempt to recreate some assumed Golden Age of the past. More broadly, Evelyn raised the issue of land-use planning and showed some familiarity with ancient theories on the forests–climate connection. Three years earlier, the same author's *Fumifugium* had protested against the pollution of London's air, which he decried on environmental and aesthetic grounds as well as on the basis of the undoubted effects on public health.

The pioneering French *Forest Ordinance* (1669) represented another response to a perceived shortage of timber, especially for shipping. In this case, however, the action produced a codification or synthesis of existing laws and practices, and its origins may be traced to warnings expressed at popular and official levels over the previous three centuries. Although it did not apply directly to private forests, it had the general effect of bringing greater order and regulation to all aspects of forest management and exercised an enduring influence throughout Europe. The statements of Colbert and the rest of Louis XIV's Commissioners introducing this celebrated Ordinance were just as historic as the actual codification. Glacken's authoritative account (1967) explains that a direct appeal was made on the behalf of *future generations*: close monitoring and careful regulation were essential so that 'the fruit of this shall be secured *to posterity*'. The quotation deliberately emphasizes the language of an appeal which was to echo down the centuries.

THE DISCOURSE ON NATURE AND WESTERN CULTURE

This rapid survey has transported us into the realm of practical responses to perceived resource scarcities, but it should not be allowed to obscure other motivations for concern over environmental change. Glacken and others have discussed the ramifications of a seemingly endless discourse on the relationship between Western societies and Nature – from Antiquity through the early Christian era, the Middle Ages and into the Renaissance and beyond – which tapped our deepest intellectual and spiritual currents. In so far as modern environmentalism continues that discourse (and according to Glacken, the main authority used in this section), it incorporates three fundamental and by no means unambiguous perspectives: Humanity–Nature harmony; Humanity as the modifier of Nature; and Humanity as determined by Nature.

Since the 'harmony' perspective sustains prescriptive as well as descriptive interpretations, it is readily shown that, in indicating appropriate models of environmental behaviour, its moralistic basis is timeless. The late enunciation of detailed ecological principles gave this perspective a sound scientific foundation, but for most of human time it has drawn the bulk of its considerable assurance from religious and aesthetic ideas. The origin of the interventionist or 'modifier' view has been loosely traced to mankind's special commissioning in the Book of Genesis to 'subdue' the Earth and have 'dominion' over its non-human creatures. Regularly – and all too casually – this familiar injunction has been cited as the justification for massive global transformations, including Europe's agricultural and industrial revolutions and the invasion and transformation of the New World and its native peoples. Resort to the biblical message was less frequent during the late nineteenth and early twentieth centuries when secular imperatives claimed a monopoly in introducing 'utilitarian' forms of conservationism, yet it had not been forgotten. Support for the notion of Nature as 'determinant' has ranged from enduring interpretations of climate, health and race stemming from classical Greek authors, to explorations of 'environmental determinism' in historical and geographical writings in the late nineteenth and early twentieth centuries, and some of the gloomier prophecies of our own times. Once more, the parameters were and are quite generous. They accommodate a number of apparently contradictory attitudes and aspirations, including the sanctioning of the global expansion of supposedly superior European peoples and paranoia over the forays of white settlers into tropical regions.

By some measures, the ecological crisis of the late twentieth century has its roots in the emergence of 'Enlightenment' views in the sixteenth and seventeenth centuries. Commonly, Francis Bacon is cast as the leading villain – and that is not surprising, since his influential writings assumed the pre-eminence of human beings in the Creation and argued that, without human life, all was confusion, aimless. From that position, he sketched out the ascension of science and technology and suggested that their chief role was to restore the 'dominion' which had been lost at the fall of Adam and Eve. The latter allusion is important because Bacon's pronouncements have been loosely translated into secular commands: the original emphasis was on the recovery of the divine bequest. Yet in conjuring that restoration Bacon, like several of his leading contemporaries, could not resist using revealingly robust language – 'conquest', 'mastery', 'management' and 'dominance' seemed to signal a separation from Nature; the Modern Man (*sic*) would be a technological warrior.

Such influences did not disperse, but rather coexisted with, a pervasive religiosity which continued to nurture a range of competing perspectives on the condition of the earth and human responsibility for that condition. A number of Reformation scholars subscribed to the notion that environmental deteriora-tion should be attributed to human evil. Martin Luther's interpretation, for example, argued not only that Adam had brought about the decay of Nature as well as the Fall (of 'Man'), but also that subsequent moral transgressions had ensured the continuation of the decline. This non-scientific or pre-scientific association of human guilt and impending doom remained a compelling antiphonal down to the present era: 'Repent, for the End of the World is Nigh!' There was always room for optimists, however, and they became prominent in the late seventeenth century in a series of intellectual contributions. In his best lawyer's language, Sir Matthew Hale's *The Primitive Organisation of Mankind* (1677) declared that the harmony of the whole, not its usefulness to people, reflected the wisdom of the Creator; therefore it was important to stress the adaptation of each part of the creation to every other part, *and* the special role of humans as God's 'stewards' and 'tenants'. John Ray's *The Wisdom of God Manifested in the Works of Creation* (1691) elaborated this physico-theology to expound on Nature's beauties and usefulness, human agency working through God's guidance to produce environmental changes, and the idea that such changes became integrated into a newly minted harmony. It is likely that these works are the sources of some of today's most favoured expressions and sentiments.

By the end of the seventeenth century, assisted by advances in science and technology, a characteristically *European* idea of 'progress' had emerged. It would exercise a powerful leverage during the next three centuries, but as usual the survival of contradictory viewpoints must be noted. It is the novelty and indeed the 'modernity' of the notion that should be stressed. Nostalgia for a lost Golden Age had been entertained in European antiquity and by Asian civilizations alike, and in the medieval period Christian scholars pined for the lost qualities of the Greek and Roman eras, but the gradual diffusion of the idea of progress brought increasing optimism about the *future* and the scope of human potential.

All of which is to say that, if reflections on our place in Nature constitute a crucial ingredient of any definition of our humanity, then it would be foolish indeed to expect to find the seeds of our current concern for environmental issues by thumbing quickly through the short history of half a dozen generations or so. It is tempting – and possibly ineluctably human – to seize upon particular junctures when our own preoccupations appear to have been mirrored in important events and in the activities of key individuals. The nineteenth century is especially attractive in this regard. As we have seen, however, the weight of the evidence favours a much slower evolution: certainly, it obliges us to consider the two preceding centuries and their continuity with ancient beliefs, preferences and practical experience. The task is immense. It is often useful to begin with conflicting interpretations of the tumultuous changes in the environment of the New World after the arrival of European settlers; much of the intellectual groundwork was also concerned, however, with interpretations of change in the Old World.

For the Europeans set the pace. In the late eighteenth century French naturalist (and farmer, horticulturalist, silviculturalist . . .) Georges-Louis Leclerc, the Comte de Buffon, promoted the hypothesis that Nature had been left by God for man to perfect. Buffon's schemata portrayed seven great epochs in earth history. The first six followed the biblical accounts from the Creation and the first appearance of humans and their subjection to Nature, whereas the development of agriculture in the seventh epoch shifted the balance towards humans, who had achieved an impressive degree of mastery by the eighteenth century. Buffon's distinctive contribution was earth-based. Reasoning that (what he saw to be) the disadvantages of imperceptible heat losses from the planet could be balanced by the employment of forest clearances to introduce increased solar radiation, he attributed much of the productivity of the advanced countries of his day to controlled deforestation and other types of reclamation, including the drainage of swamps. Bolstered by American speculations on the increase of local temperatures after the removal of tree cover, and by detailed empirical observations on the differences between diurnal temperature ranges and the duration of the rainy season in cleared and forested tracts in French Guiana, he concluded that Europe's climates had been greatly improved by centuries of deforestation. He was less confident of the prospects of induced change in desert regions, but used the same logic to recommend widespread tree planting.

The full hypothesis had enormous practical implications. Buffon deplored the exploitative deforestation of *developed* countries. Fulminating in Evelyn-style against overclearing in contemporary France, he advocated extensive replanting in many formerly forested regions which had been allowed to run to unproductive 'wasteland'. Similarly, his elaborations on the domestication of species warrant more attention than this short treatment offers. Briefly, his scheme sketched the making of a higher order of organized existence, of 'civilized' landscapes heavily dependent on the process of plant and animal domestication – horses, dogs, cattle, sheep, and all the carefully selected and improved farm plants, were 'secondary creations' displacing virgin Nature, and merely slaves to humans. Buffon expressed sorrow at the commensurate removal of wild species, but the main thrust of his work was an elaboration of the triumphant march of human progress.

Unfortunately – though once more by logical extension – Buffon was unimpressed with Nature in the Americas. He considered it inferior ('weaker') than in the Old World and proposed that the indigenous Americans were likewise inferior types. The New World produced fewer species, and seemed incapable of nurturing the large varieties of animals. Ergo, its primeval landscapes needed to be transformed – cleared, drained, farmed: failing to do so, the 'passive' American Indian had failed to create a high civilization. A few of Buffon's fellow-travellers took the argument further, complaining that European livestock and European immigrants actually degenerated in the baleful environments of North America.

The details of Buffon's Eurocentric theories were hotly debated in the newly independent American states. He was chastised for his factual errors, including the ludicrous opinions regarding the sizes of native animals, but above all it was asserted that the better and more authentic American challenge was to create a new civilization on moral rather than environmental bases. Thomas Jefferson, Benjamin Franklin and other American intellectuals refuted the Count's sillier findings, and indeed the New World's growing treasury of direct practical experience seemed to proclaim its peculiar relevance to the great philosophical questions of the age. French immigrant J. Hector St Jean de Crèvecoeur offered a dual image of America: an unparalleled opportunity for human industry and ingenuity in which individuals freed of the Old World's fetters could use great fertile expanses to build new communities from the wilderness; and a unique prospect of redemption from the artificiality of the Old World – another Eden, no less, in which a lost innocence might be regained.

Buffon's theorized seventh epoch seemed to have materialized in the New World conquest. He was not blind to the negative aspects of frontier expansion, as we have seen, and was more aware of the dramatic growth of the world's population than many of his contemporaries. While others thought that there had probably been a general decline since Antiquity, Buffon believed there was an identifiable upward trend: furthermore, his boundless faith in the creative, civilizing impulse persuaded him to welcome the additional numbers. His judgement on the overall trend line was enlightened, but otherwise his confidence was a trifle misplaced. Europe's population had grown from 80 million in 1500 to 140 million in 1750 (that is, during the Count's lifetime); over the next century the aggregate exploded to 266 million. In the British Isles alone, despite the tragic losses during the Irish famine, the population had trebled in the century before 1850 to 28 million. Underlying this increase were three powerful sets of processes: rising ecological pressures in town and country; a mass movement of immigrants to all parts of the New World (Australia, New Zealand and South Africa as well as the Americas) with a concomitant commercial expansion for Europe; and the metamorphosis of capitalist economies into a voracious industrialization. Adam Smith's *Wealth of Nations* (1776) had given the intellectual rationale for the capitalist system and fostered an emboldened, anthropocentric view of society–Nature relationships. In so far as the encouragement of free enterprise and the pursuit of private wealth were underlined as paramount, the land and all its resources were simply to be accounted as potentially marketable commodities. The Revd Thomas Malthus's *Essay on the Principle of Population* (1798) was much gloomier. According to this theory, human populations increased until they outstripped the available food supplies; the resultant famines restored the balance. Malthus's dire warnings were largely ignored during the nineteenth century. Instead, the idea of progress gained ground, even in popular culture, and Smith's arguments ran through the political and economic rhetoric of Western governments.

Notwithstanding the growth of intensely national interests and anxieties, in ecological terms it is vital to acknowledge the workings of a strong dynamic which was fed by rising levels of global interaction. And the agricultural and industrial 'revolutions', no less than the aggressive expansion of Europe overseas, ensured an acceleration of environmental modifications at all scales. For example, Buffon had proudly described his own country's pre-eminence in environmental management, but between 1750 and 1850 about half the forest cover had been cleared and timber shortages, floods and soil erosion had destroyed France's high reputation. Throughout north-western Europe the claims on traditional resources, especially land, continued to escalate with increasing regional specialization during the agricultural revolution. The growth of market towns and manufacturing centres aggravated these pressures. Demands for industrial charcoal, domestic fuel and construction timber, together with an expansion of farmland to supply food and fodder, put the dwindling forest cover under relentless attack. In turn, the substitution of coal for wood ensured the transformation of new resource regions. Imperialism exported more or less the same stressors around an ever-shrinking globe.

Through various forms of conquest and appropriation, the European belief in progress and the development imperative it supported were imposed on the rest of the planet. Although the role of institutionalized Christianity declined, its pivotal assumptions had already helped to forge a distinctively European world-image which was generally anthropocentric and exploitative. Indigenous peoples in the looted territories thought differently. Put simply, the traditional cultures of Africa, Asia, Australasia and the Americas were less inclined to exaggerate the worth of the human species. Of necessity, the so-called 'primitive' hunters and gatherers had developed holistic understandings of the interconnections between animals, plants and humans. Similarly, in the perspective of Jainism and Buddhism, for instance, all creatures require a release from the cycle of existence, and humans are special only in so far as they are capable of achieving enlightenment. We should therefore direct our greater intellectual and spiritual qualities to that goal and towards a compassionate view of the other creatures in a suffering world. And Zen, embodying a synthesis of Indian and Chinese cultures, stresses the importance of an aesthetic appreciation of nature; furthermore, it teaches that nature should be seen as a friend (a 'Thou', not an 'It' or object).

It was to become fashionable, in the late twentieth century, to seek productive fusions of Eastern and Western thought. Certainly there is much to be said for the cultivation of self-scrutiny – yielding a sense of humility and responsibility – in the Eastern modes, as compared with the cultivation of needs and bustling expansionism favoured in the more materialistic West. But fashion cannot compete with historical facts. One massive example that must be considered is the fate of China's forests. For the medieval and early modern eras (to use the Western chronology), it has been established that a widespread removal of forest cover brought severe economic and environmental consequences: in other words, environmental degradation was central to the history of that immense country long before the intrusion of Western values. On a related tack the development and elaboration of sophisticated water management systems, over thousands of years, point to a very high degree of sustained human intervention in natural ecosystems. The evidence of environmental calamities in ancient and pre-modern India is equally sobering: human-induced land degradation was already pronounced before the end of the seventeenth century and has been traced to much earlier periods. Ecological damage was not initiated but rather continued, and in some cases dramatically worsened, under British rule.

CONSERVATION AND ENVIRONMENTAL MANAGEMENT FROM THE LATE EIGHTEENTH CENTURY

But the most tangible connections with the present era were developed in intellectual circles in north-western Europe and the United States. In both regions, the potency of human agency was considered undeniable and engaging, and there was no consensus on the beneficial and deleterious effects. Those who belonged to or were admirers of the growing band of 'experts' employed in seemingly ubiquitous environmental change – all those dammed rivers, cleared forests, drained swamps, improved and 'acclimatized' crops and livestock – were generally optimistic about the effects of the extension of human dominion, which they usually referred to as 'progress'. On the opposing side were influential romanticists who had conjured their own revolution.

In eighteenth-century Europe a new sense of beauty accommodated the idea that Nature's own forms were aesthetically pleasing and deserved to be valued for that reason: in painting and landscape gardening the idea of the 'picturesque' gave the stamp of approval to Nature's 'irregularities'. The strong sense of rapid change, and perhaps the fear of losing cultural roots, stimulated scholars and architects to study, protect and in some cases even replicate the valued relics of previous eras. An historicist appeal to sentiment and national pride led to a minor but nevertheless significant urban preservation movement. The French government established a Department of Historic Monuments in 1830 and sponsored the compilation of a huge inventory of significant buildings. Its first official list of fifty-nine registered buildings was published in 1840. Notable edifices and precincts protected or restored through this process included Notre Dame Cathedral and the city of Carcassonne. In Britain a similar movement in the 1870s was spearheaded by William Morris's campaign for the protection of ancient buildings. Less practical romantics seldom moved beyond a vague yearning for a return to the past, but even the hardest-headed intellectuals who discovered the link with the inherent evils of runaway capitalism were unable to sketch out realistic alternatives. In this regard, however, the early communists proved the exception. A good example is a far-sighted work, *Fields, Factories and Workshops*, produced by the anarcho-communist geographer Peter Kropotkin in 1899. In this remarkable book, Kropotkin proposed a radical reshaping of the geography of Britain based upon the dispersion of small, community-owned village factories, reductions in the sizes of the cities and generally a *humanized* landscape, frequently punctuated with villages, hamlets and intensively cultivated plots. Kropotkin was far too sanguine in his expectations of sustained high yields capable of carrying a massively increased population, but the book was none the less an interesting anticipation of the appeals of some modern environmentalists for more reliance on 'appropriate' technologies and of the mystique which they have invested in China's intensively settled countryside.

These examples reinforce the observation that the great variety of appraisals of the urban condition cannot be dismissed by loose references to 'optimism' and 'pessimism', yet the terms are now well established and it must be conceded that they retain a certain appeal. Closer to the positive end of the spectrum, but with definite reasons to feel specially guarded in their enthusiasm for 'progress' and environmental change, were the groups of passionate urban reformers who crusaded for improvements in public health, safety and amenity. Some of their concern was associated with neo-classical theories positing an environmental basis for all endemic and epidemic diseases. Until bacteriological and immuno-logical research became well-founded towards the end of the nineteenth century, medicine remained an art and anxious resort to Hippocratic notions was very common. The artificial environments of the new mining and manufacturing conglomerations and the overcrowded metropolitan centres were subjected to innumerable multi-scaled critiques, and whether consciously or no, Charles Dickens and other popular novelists contributed their special skills to promote the efforts of the reformers to 'redeem' the urban population. The institution of regular national monitoring of mortality and morbidity data exposed stark differences between urban and rural milieux; this process reinforced and drew upon investigations of appalling housing conditions, and on studies of occupational health and safety in the mines and factories.

Industrialization had therefore spawned a special brand of environmental thinking. Thomas Southwood Smith founded Britain's Health of Towns Association in 1839, then served on a General Board of Health created under the path-breaking Public Health Act of 1848 to report on sanitary requirements, cholera, yellow fever and quarantine provisions. In late nineteenth-century Europe, France and Germany were probably the most advanced nations in the application of scientific methods to the treatment and control of communicable diseases, thanks to the discoveries of Pasteur, Cohn and Koch. Internationally, ordinary empirical observations had already firmly tied miners' phthisis and related illnesses to dangerous working conditions, but occupational diseases linked more insidiously with industrial chemicals required many years of research before irrefutable medical and legal arguments could be established. On the whole, it was the British approach to 'Sanitary Reform' that produced the most comprehensible environmental improvements in the field of human health – notably in the quality and efficiency of public water supplies, sewerage services and hygiene – and these emphases were duplicated quite rapidly in other advanced countries.

The inexorable concentration of population placed a premium on the skills of the new technocrats. Called upon to provide large and regular supplies of good quality water, specialized 'hydraulic engineers' developed more reliable methodologies for assessing river regimens, catchment monitoring, run-off potential, the selection of safe and adequate reservoir sites, and the measurement of evaporation rates. The proliferation of water-borne diseases and the difficulties inherent in matching supply and demand under free-wheeling urbanizing and industrializing situations guaranteed that the topic of urban water supplies would sustain many a heated public debate. Development invariably outpaced research, and when the appointed experts disagreed on matters of interpretation, and when they blamed inadequate data for unexpected visitations of drought and flood, they and their political masters were usually given short shrift. After the middle of the nineteenth century, the metropolitan and provincial press carried numerous reports of these controversies over the management of an essential resource. Far from being perceived as artificial creations removed from a primitive subservience to the physical environment, in these respects the expanding cities obviously emphasized a continuing dependence. In addition, ancient enmities between urban and rural residents were frequently revived by political and bureaucratic decisions to proceed with the construction of new reservoirs entailing the acquisition of huge areas of valued farm and forest land. In some situations, assertive water managers also succeeded in introducing 'closed catchment' policies to exclude farming, grazing, logging, housing and recreation from the run-off area, with the objective of ensuring reliable and high-quality supplies. Disputes arising from these mid- and late-nineteenth-century initiatives remain unresolved a century later.

The pace and complexity of change in the nineteenth century were nowhere better seen than in the rising scale of environmental abuse. Wildlife was hunted to extinction in some African regions, and overgrazing and soil exhaustion seemed incumbent ingredients of the empirical testing favoured by pioneers in Australia. On the Australian plains as in the interior of the United States, settlers basked for a time under the happy delusion of a potent folk-myth: supported as much by pseudo-scientific reasoning as by their own wishful thinking, they were persuaded that 'rain follows the plough' and chose to develop new country far beyond the safe environmental limits. When the dry seasons returned with a vengeance the lessons of 'marginality' seemed writ large across plundered landscapes.

There is an interesting contextual reminder here of other pre-scientific environmental ideas, including neo-classical conceptions of the nature of disease which favoured elementary climatic correlations. Before the discovery and assimilation of 'germ theory' at the end of the nineteenth century, the scourge of tuberculosis (or 'consumption') – then a major killer in the West – was imprecisely associated with urbanization and industrialization, but in terms of expert and popular responses one of the most common characteristics was a preference for dry and sunny climates.

The under-researched fear of disease was highly influential in the great migrations from Europe to the New World (and in regional migrations within North America and Australia). It is still an important ingredient in society–Nature relationships, and the lesson is this: 'vernacular' or popular conceptions and misconceptions are no less crucial in the historical record of environmental thought than economics, high science and culture.

What, then, of the fate of the New World's remaining semi-natural environments, so threatened by the march of materialistic progress? At first there was only a narrow base of protest, and it was heavily dependent upon intellectuals. It was in North America, specifically in the vigorous young democracy of the USA, that the most numerous and innovative protests were raised. In 1832, for example, the artist and author George Catlin proposed the setting aside of national reserves for the protection of wild areas and native peoples alike. A little later, ornithologist J.J. Audubon and botanist William Bartram tried to generate public enthusiasm for the conservation of native fauna and flora, and writers Ralph Waldo Emerson and H.D. Thoreau announced their regret for the rule of materialism. Transcendentalist Thoreau was particularly forthright in his appeal for the conservation of the rights of wild nature as an absolute precondition for the psychological well-being of humans. The most influential protest came, however, from the urbane man of letters George Perkins Marsh, whose *Man and Nature* (1864) synthesized historical and contemporary accounts to explain the environmental role of humans as interventionists and to challenge the conventional mystique surrounding the idea of inexhaustible resources.

Familiar with the lessons of land degradation in the ancient Mediterranean world, Marsh was also well versed in the classical literature on society–Nature relationships and had witnessed the searching impacts of deforestation in his native state of Vermont. Accordingly, *Man and Nature* drew on field and historical evidence from both hemispheres, supplemented by the author's correspondence with selected European technocrats. In lively and accessible language Marsh demonstrated that the world was being remade by human agency, that it had also been crassly managed at times, and that a radical reform was overdue. His great book was less an indictment of human culpability and a glorification of Nature's ascendancy than an affirmation of the responsibilities of a type of human stewardship which continued the subjugation of natural processes. His point was that there could be no victory over Nature without a more informed and watchful supervision of human interventions. Eschewing complicated scientific explanations, Marsh elucidated the extent and the chief causes of deforestation, disrupted river regimes, the threatening migration of the desert margins and the elimination of wildlife. Fixed, apparently, on the reshaping of Nature, society needed to be shaken into revising its goals and attitudes. Marsh wanted passionately to see an enlarged environmental awareness linked with the conscientious prediction and monitoring of change, an end to the indiscriminate destruction of wildlife, and a renewal of forest cover over a quarter of the cleared land. *Man and Nature* argued that human custodianship was an ethical or moral issue and would never be understood in purely economic terms: 'Man has forgotten that the earth was given to him for usufruct alone, not for consumption, still less for profligate waste.' Marsh looked forward to a healing of past injuries by the harnessing of modern technology, but in addition he came close to grasping the germ of modern ecological thinking when he cautioned that the interrelatedness of all life 'is too complicated a problem for human intelligence to solve, and we can never know how wide a circle of disturbance we produce in the harmonies of nature when we throw the smallest pebble into the ocean of organic life'.

The term 'ecology' was not then current, but the idea of comprehensive holism in Nature had been common enough in religious and scientific quarters since the seventeenth century. Swedish naturalist Carolus Linnaeus published a treatise entitled *The Oeconomy of Nature* in 1749 and that phrase seems the most likely precursor of 'ecology'. Certainly it was cited in late eighteenth- and early nineteenth-century appeals for the protection of rare species whose place in the 'oeconomy' had not been determined. Nash explains that such concerns touched upon the ethical idea of *community* membership and its attendant rights. Charles Darwin's *Origin of Species* appeared in 1859 and his theory of evolution serviced a marked expansion in the field sciences. Responses to this book and to the same author's *Descent of Man* (1871) excited wide controversy over the relationships between the varieties of life forms on the planet. Darwin referred to other animals as 'our fellow brethren' and in this context suggested that one criterion of true civilization was the proof of 'extended sympathies' or ethics. In fact there was some ambivalence in his position: he was not inclined to include plants, and as a witness before Britain's Royal Commission into vivisection (1875) he chose to uphold the scientific value of experimentation, provided only that excesses were checked and a degree of kindness was ensured.

Whereas *Man and Nature* made an immediate impact in Western intellectual life, Thoreau's holism – both scientific and religious in its approach to Nature's economy – was not widely appreciated until the early twentieth century. His concept of community discarded hierarchy and respected the rights of all creatures and elements: 'There is no place for man-worship,' he wrote, and 'every creature is better alive than dead, men and moose and pine trees'. Thoreau also criticized a prominent anti-slavery activist for wearing a beaver-skin coat. This bonding of abused Nature and abused people was another precursor of our most recent expressions of environmental thought. In 1866 Henry Bergh, influenced by the British precedents, founded the American Society for the Prevention of Cruelty to Animals. With slavery abolished, Bergh saw this effort as the next frontier of humanitarianism. Harriet Beecher Stowe, the author of *Uncle Tom's Cabin*, joined Bergh's movement after protesting against the treatment of Florida's birds and animals. Her brother, theologian Henry Ward Beecher, campaigned for a more commodious ethic that included the 'rights of animals'. As on the opposite side of the Atlantic, these developments were mainly founded on the narrow principle that it was morally wrong for humans to be cruel. Generations would come and go before unequivocally biocentric viewpoints established a wide constituency recognizing the existence of an entire and interdependent natural system. A start had been made, none the less.

Much the same could be said of the early conservation movement after the publication of *Man and Nature*. From the United States to Canada, Australia, New Zealand, the Russian empire and the infant Soviet Union, utilitarian anthropocentric management strategies were aimed at maximizing or sustaining resource yields. The hints about ecological integrities and broad-ranging ethical rights were only accepted as promising seeds by a few enthusiastic activists. Foremost amongst these was American poet-scientist-philosopher John Muir, who championed the cause of wilderness conservation and questioned the audacity of humans to value themselves above the rest of unified Creation. Dominant public images of wilderness had gradually changed from the initial European settlers' outright hostility and fear to an increasing show of respect during the last quarter of the nineteenth century. The transformation was based largely on aesthetic, nationalistic and scientific arguments, aided by the political influence wielded by those urban clubs and societies which had become focused on natural history and an indulgence in outdoor recreation. Wilderness addicts, Muir amongst them, were successful backers of innovative legislation for the world's first national park, Yellowstone, in 1872.

Muir's radical egalitarianism was tempered when he took issue, as a 'preservationist', with the astute conservationists whose 'progressive' notions of efficiency sat more comfortably with the politicians' and industrialists' aspirations for a new society. A leading evangelist of this 'Gospel of Efficiency' was federal forester Gifford Pinchot, who favoured 'use but wise use' policies of resource management. Pinchot's forceful representations put public forestry in the vanguard of reform, and a number of other government agencies followed the forester's lead. His framework was adapted from prior French, German and British Empire approaches, but there was no denying the special reception accorded in the USA to his proposals for 'sustained yield management'. Winning the approbation of President Theodore Roosevelt, this utilitarian line came to dominate the first modern conservation movement and set a model for other countries in the New and Old Worlds. The battle lines were drawn in 1908 over the plans of the San Franciscan authorities to dam the Hetch Hetchy Valley in the Yosemite National Park – proclaimed in 1890, it had been a personal triumph for Muir and his followers – and the controversy went unresolved for five years. Muir won the support of California's newly formed environmental activist organization, the Sierra Club; he was opposed by City Hall, Pinchot and a daunting phalanx of backers from the 'use but wise use' school. Shelving or suppressing his ecological and ethical principles to enter the political minefield, Muir adopted a more pragmatic line to emphasize anthropocentric 'preservationism' on several grounds – aesthetic and recreational (that is, with scenery conceived as a resource); spiritual nourishment (rest, regeneration, uplifting experiences); and catchment protection (the need to regulate the hydrological cycle in an insecure environment).

In this skirmish the utilitarians triumphed; but in a war which had only just begun, first the narrow preservationist stance and, ultimately, those deeper philosophical arguments which had been set aside by Muir, were to emerge as virile challengers on the national and international stage. Drawing on the very wellsprings of Western civilization, they would be nurtured by public perceptions of environmental damage and by the extraordinary popularization of a 'Gospel of Ecology' – articulated and diffused by a democratic process which is itself rooted in the legacies of ancient Greece and Rome. But the title 'Gospel of Ecology' is said to signify a fusion of aesthetic, ecological and utilitarian interests in the natural environment which derived in part from the American public's media-fed anxieties about environmental pollution. That is emphatically a very modern tendency, emerging in the 1960s – and even for that period the description inflates the degree of popular support. It is more useful to return to the early decades of the present century to identify transitional events and processes.

TWENTIETH-CENTURY TRENDS

Each of the world wars has been examined intensively from highly specialized social, economic and political viewpoints, but the immediate environmental effects – like the infinitely more elusive impacts on environmental ideas and attitudes – have been relatively neglected. Other major events and trends of the first half of the twentieth century are chronologically less discrete and their environmental history is generally better understood, yet in terms of their penetrating international influences they should be set alongside the great cataclysms. I refer to the stunning population migrations, facilitated by improvements in communication and mass transportation systems and by the very competitions between the expansionist powers which visited such brutalities on the Old World. The associated image of a 'shrinking globe' may be traced to much earlier times, but there is no doubt that it provided the West with a common focus of debate during the period 1900–45. The engrossing elaborations on the latter theme obliged discussants to explore the Malthusian legacy and they did so at the international, regional and local levels. In this regard one heavily subscribed topic centred on the earth's capacity to accommodate its increasing numbers of human inhabitants, usually entailing subsidiary disputations over ambitious contemporary designs for population redistribution and the differential values to be attached to reputedly 'under-developed' countries. The contributions made towards these debates by its leading specialist practitioners broadly enhanced geography's status as an emerging academic subject, and the best work of a number of innovative geographers drew attention to the environmental limitations on exponential growth and warned against super-confident population estimates. One of the most celebrated of these early crusaders was an Australian, Griffith Taylor. He offered a prediction of 19–20 million in 2001 for his own country when the forecasts of nationalist–imperialist optimists and opportunists were wild multiples of that total. Taylor's prediction was backed by a straightforward environmental survey which is not too badly described as a precursor of today's arguments for 'ecological sustainability'.

The orientations of modern environmentalism are at once global and local, yet it still derives much of its cutting edge from active engagement in particular political cultures. The point can be lost in digressions into broad international trends, no matter how 'anticipatory' they are judged to be. Thus, to resume our central analysis, it can be said that the most recognizable transition to the modern era was initiated in the United States before 1945, and that it was a product of the tension between three main ideas guiding policy-making. Extending the present explanatory line, these ideas may be labelled *efficiency*, *aesthetics* and *equity*. Pinchot and other exponents of the efficiency argument advocated the marshalling of technocratic expertise in the Forest Service and other big national agencies, properly aloof from the contamination of politics. These elitist groups took pride in devising elaborate planning frameworks based

289

ever more narrowly on specialized or 'expert' knowledge. Normally, developmentalism was their strongest suit. The aesthetic line drew upon romanticizations of 'wilderness' experiences, nationalistic concerns for the protection of uniquely American scenery, and a sweeping philosophical commitment to the maintenance of diversity as opposed to the crushing uniformity of modern urbanization. It was closely allied to the preservationist stand taken by the followers of John Muir and to the creation of national parks, but the other main ideas also contributed to the latter victories. Equity concerns presented the most baffling targets. They focused on the wise use of resources to foster the growth of genuinely democratic communities. Resource monopolies were therefore vigorously opposed, but by today's standards this theme was the least successful in the period under review.

In practice, the artificialities of this thematic division are quickly exposed, and it is vital to retain our wider contextual perspective. The bonding agents were science and political economy. First, Aldo Leopold and other ecologists showed how their young science could be used as a tool for a full-bored or authentic environmental *appreciation*: it could assist in the production of improved management schemes, while amplifying and refining the preservationists' programmes to give higher priority to the protection of representative biotic communities instead of relying exclusively on scenic splendours. Second, in the New Deal years under the presidency of Franklin D. Roosevelt, hostilities towards the failed, blinkered materialism of American culture ensured a growing interest in conservationism which welcomed all three perspectives. Whereas advances in federal forestry had dominated the earlier history of modern conservationism, in the 1930s both efficiency and equity were to be served by ambitious planning frameworks and spectacular federal investments in dam-building.

The publicity generated during critical episodes of land degradation, including the dramatic 'Dust Bowl' events, boosted the reformist pressures. Despite the variations in emphasis, an infectious enthusiasm for regional planning provided the preferred methodology for achieving conservationist goals. In the contemporary rhetoric, the implementation of decentralized, nation-wide plans was essential for efficiency and for ecological and social harmony. One of the earliest and arguably the best-known of the New Deal's planning models was the Tennessee Valley Authority. Unfortunately, it soon became clear that the chief business of the TVA was strictly utilitarian – that is, the production of electricity and the introduction of flood control and river improvement measures. The federal government's pioneers had sketched out a prototype for the fashioning of co-operative communities; instead, the Authority's high ideals were subordinated to the demands of regional politics and the expansion of the wider economy. While the champions of efficiency were well pleased, the TVA experience would teach conservationists of other stamps to be more suspicious about demands for regional autonomy, and preservationists would realize that the fervour for dam-building represented a huge threat to their cause. In the interim, directors of federal agencies discovered that the omission of references to equity issues would assist them to gain political support for their costly efficiency programmes.

PERSPECTIVE: ROOTS AND MIRRORS

In the popular mind, environmentalism is indeed a very modern phenomenon: its origins are interpreted as the special product of the political and intellectual ferment of the 1960s and 1970s, notably in North America and Western Europe. On the contrary, however, as this chapter has shown, the roots are deeply embedded in ancient religious and secular reflections on our humanity in its special relationship with Nature. Even the best-intentioned effort to foreshorten the history should be approached warily, for the search for meaning is itself part of that humanity: from time immemorial it has been conducted with all the arts and sciences at our disposal. Yet it is true that there was a palpable deepening of concern in the West during the nineteenth century. Anxieties over the direct and indirect impacts of urbanization and industrialization on human health and well-being, and reactions to the disturbing rapidity of landscape change – which gave, after all, such graphic evidence of the upheavals in the everyday world – seemed to ask for fresh reckonings. Inevitably, the outcomes mirrored the complexities of the human condition, and in the process proclaimed our humanity – surely a kind of 'environmentalism', unless the central part of that term is redundant. As the nineteenth century drew to a close, discomfort over environmental changes and the roles of human communities in those changes harnessed and was in turn employed in the political systems of the West. That was exemplified in the young societies of the New World, and best of all in the case of the United States, but the experiences of the European and Scandinavian countries were not dissimilar.

This linkage with political engagement should not be overdone – even in the latter stages of the period reviewed the involvement is selective and episodic. But it was there, none the less, and the strength of the observation is actually bolstered by the proven concentration of articulations of environmental stress in certain sovereign and democratic territories, as opposed to a preference for global issues transcending localized political orthodoxies. Ultimately, the narrower focus helped to build coherent 'movements' which were often characterized by a membership drawn from the scientific, technical, broadly intellectual and middle-to-upper-class elites, who were occasionally able to recruit a significant measure of popular interest. If the base broadened considerably after the Second World War so did the range of conservational thought and practice, and the scope for the expression of divergent opinions. Continuing to reflect ancient, intimate connections with human fears and aspirations, the environmentalism of today has been built somewhat precariously upon a rich legacy of ambiguities left by the conservation concerns and movements of the recent and distant past.

FURTHER READING

Crosby, A.W. (1986) *Ecological Imperialism: The Biological Expansion of Europe, 900–1900*, New York: Cambridge University Press.

Glacken, C. (1967) *Traces on the Rhodian Shore: Nature and Culture in Western Thought from Ancient Times to the End of the Eighteenth Century*, Berkeley: University of California Press.

Goudie, A. (1990) *The Human Impact on the Environment* (3rd edn) Oxford: Clarendon Press.

Grove, R.H. (1995) *Green Imperialism. Colonial Expansion, Tropical Island Edens and the Origins of Environmentalism, 1600–1860*, Cambridge: Cambridge University Press.

Hays, S.P. (1959) *Conservation and the Gospel of Efficiency: The Progressive Conservation Movement, 1890–1920*, Cambridge, Mass.: Harvard University Press.

Jones, E.L. (1981) *The European Miracle. Environments, Economics and Geopolitics in Europe and Asia*, Cambridge: Cambridge University Press.

McNeil, W.H. (1980) *The Human Condition: An Ecological Perspective*, Princeton: Princeton University Press.

Marsh, G.P. (1964) [1864] *Man and Nature, or, Physical Geography as Modified by Human Action*, Cambridge, Mass.: Harvard University Press.

Nash, R. (1982) *Wilderness and the American Mind* (3rd edn), New Haven, Conn.: Yale University Press.

Nash, R. (1989) *The Rights of Nature: A History of Environmental Ethics*, Madison: University of Wisconsin Press.

Passmore, J. (1974) *Man's Responsibility for Nature: Ecological Problems and Western Traditions*, New York: Charles Scribner's Sons.

Pepper, D. (1984) *The Roots of Modern Environmentalism*, London: Croom Helm.

Ponting, C. (1991) *A Green History of the World*, London: Penguin.

Powell, J.M. (1976) *Environmental Management in Australia, 1788–1914. Guardians, Improvers and Profit: An Introductory Survey*, Melbourne: Oxford University Press.

Powell, J.M. (1991) *An Historical Geography of Modern Australia. The Restive Fringe* (2nd edn), Cambridge: Cambridge University Press.

Seymour, J. and Girardet, H. (1986) *Far From Paradise. The Story of Man's Impact on the Environment*, London: BBC.

Thomas, K. (1983) *Man and the Natural World. A History of the Modern Sensibility*, New York: Pantheon.

Thomas, W.L., Jr (ed.) (1956) *Man's Role in Changing the Face of the Earth*, Chicago: University of Chicago Press.

Tucker, R.P. and Richards, J.F. (eds) (1983) *Global Deforestation and the Nineteenth Century Economy*, Durham, N.C.: Duke University Press.

Turner, B.L. (ed.) (1990) *The Earth as Modified by Human Action*, New York: Cambridge University Press with Clark University.

Williams, M. (1990) *Americans and their Forests. A Historical Geography*, New York: Cambridge University Press.

Worster, D. (1979) *Dust Bowl. The Southern Plains in the 1930s*, New York: Oxford University Press.

Worster, D. (ed.) (1988) *The Ends of the Earth. Perspectives on Modern Environmental History*, Cambridge: Cambridge University Press.

Worster, D. (1994) *Nature's Economy: A History of Ecological Ideas* (2nd edn), New York: Cambridge University Press.

14

THE SAVIOUR CITY

Beneficial effects of urbanization in England and Wales

Brian T. Robson

People do not necessarily live easily together. The true mark of our progress should be looked for less in great technological, scientific or medical change and more in the vast urban conglomerations where people still struggle to live and work together in harmony. That it *is* a struggle, and one whose outcome seems more uncertain as we approach the second millennium, should not cause us to join the doomsayers, but rather to see what we can learn from the age-old struggle. Obsession with the past has always been a dangerous occupation since it encourages the belief in myths. It suits the temperament of many of our current political leaders to posit the fairy tale of a past golden age. It is not quite clear when it was – forty, fifty, sixty years ago? Or perhaps, given their clearly expressed belief in free markets and liberal capitalism, they long for a return to the early and mid-nineteenth century when those doctrines swept all before them; when Manchester, the 'shock city of the age' so delighted Disraeli that he put this encomium into the mouth of one of his characters:

> Certainly Manchester is the most wonderful city of modern times. It is the philosopher alone who can conceive the grandeur of Manchester and the immensity of its future.
>
> (Disraeli 1982, Book 4: 138)

This was a rather different view from the frightening picture painted by other visitors to that archetypal industrial city – commentators such as Engels, de Tocqueville, and Faucher who looked beyond the developing factories, the overflowing warehouses and the wealthy merchants' houses to see a place where men, women and children strove to survive in unbelievably bad conditions. From their views, the aura of disease, decay and doom would cling to the city and particularly to those cities which grew as a result of industrial change.

This perception would be exacerbated as those who made money out of the industrial changes moved out of the cities into the apparently desirable countryside.

Sententious worthy historians like the Hammonds could proclaim: 'for the first half of the nineteenth century the industrial town was absorbing the English peasant used to an open-air life, learning from the landscape, in touch with nature, moving and thinking with its gentle rhythm, making rather than finding what he beheld' (Hammond and Hammond 1947: 36). In fact, for those at the bottom of the pile – the ever-present 'poor' – the industrial city probably offered more choice and opportunity for work than did the countryside which was overpopulated in terms of employment. The notion of a tranquil rural country whose Eden-like idyll was shattered by urbanization bears little resemblance to reality. Rural life may have been more pleasant from an environmental viewpoint, but fine views and fresh air were no compensation for an empty stomach and the assorted ailments associated with desperately poor living conditions. For the most part the rural poor had no voice, although occasionally their sorry state makes a brief appearance as a mute walk-on part in the main event. Gregory King (Secretary to the Commissioners for Public Accounts at the end of the seventeenth century and an amateur social statistician) suggested that in 1696, out of a population of 5.5 million, 1.3 million were cottagers and paupers, another 30,000 vagrants or gypsies, thieves, beggars, etc. and another 1.25 million were labouring people and out-servants. Leaving aside the arguments about the reliability of these or any other statistics, his perception of a considerable percentage of people being in a persistent state of poverty and unemployment was backed by other writers and observers throughout the seventeenth century. Nor was that new: the harsh laws against vagrancy and beggars in 1495, 1531 and 1536 bear testimony to the number of destitute rural dwellers. The supposedly stable feudal 'communitarian' society of medieval Britain has long since been shown to have been a period of considerable fluidity whose marked divider of natural calamity in the early fourteenth century only served to emphasize the fraught position of the labourer. At a time when population was falling and those seeking work seemed to have the upper hand, the enactment of the Statute of Labourers in 1351 clearly showed that control rested with those that 'had'. By imposing a maximum but not a minimum rate of pay; by removing freedom of choice; by fixing prices; and by forbidding the giving of alms to the able-bodied it ensured that the increasing numbers of free labourers were unable to take real advantage of the situation.

It was not that life in the cities was necessarily worse than that in rural areas, but that the problems of poverty and distress were more evident, more massed and therefore less easy to ignore within the cities; and that the real or perceived threat to social and economic stability was less readily containable when it came from the urban mass. Much of the social legislation and many of the political innovations in consequence bore an inevitable urban flavour.

PUBLIC OR PRIVATE PROVISION IN EARLY LEGISLATION

Legislation remains the favourite tool of government, whether it be of divine right, landed oligarchy or parliamentary democracy. 'There ought to be a law' is, unfortunately, all too often used as an open-ended invitation to the hasty initiation of badly made new laws. In an age of almost continuous ill-thought-out legislation it is almost too easy to see the mistakes of the past, and such hindsight aids our criticism of the actions of our ancestors. This is particularly so when, as in the case of the Navigation Acts of 1651, 1660 or of the Act for the Encouragement of Trade 1663 (the 'Staple' Act), they came in for a considerable drubbing at the behest of formative thinkers like Adam Smith (whose continuingly influential *Wealth of Nations* was published in 1776). Nevertheless they had a far-reaching and, indeed, profitable effect on the development of the British economy. The mercantilist theory of protection and its emphasis on the importance of the balance of trade ensured the advancement of British trade and commerce. Within that all-important aim and achievement, however, attention was still paid to the needs of those who were without work. 'The poor ought to be encouraged and mercifully dealt with and kindly used, until their slow hands be brought to ready working' (quoted in Wilson 1965: 232). A number of avowedly mercantilist writers, many of them successful entrepreneurs and merchants, developed the thesis that employment was all-important and to this end they encouraged the foundation of a variety of charities. Josiah Child, wealthy London merchant and governor of the East India Company, proposed a plan for an assembly to be called the 'Fathers of the Poor' which would be able to buy land, build workhouses and hospitals and enable poor people to work (Wilson 1965). He based his views on the belief that it is a 'Duty to God and Nature to provide for and employ the poor, whose condition is sad and wretched, diseased, impotent, useless'. Many of these mercantilists may well have been self-serving in that they aimed to provide cheap labour, but their overt objective could not be faulted:

> To have so many thousand Poor, who might by their Labours earn, and so eat our Provisions, and instead of sending them out, Export Manufactures, and that would bring in double to the Nation, whatever our provision doth . . . Neglect of the poor seems the greatest mistake in our government.
>
> (Brewster 1701: 52, 122)

There was a marked growth in the number of charities both in established cities like Bristol and in growing cities like Manchester. John Aikin writing of the latter city in the 1790s claimed:

> No town in England has been more exemplary in the number and variety of its charitable institutions, and the zeal by which they have been supported – a zeal in which all ranks and parties have united.
>
> (Aikin 1795: 196)

He details various bequests made during the previous century and highlights medical provision, including a Humane Society for the recovery of persons apparently dead by drowning and, more mundanely, 'a truly philanthropic society under the name of the Stranger's Friend for the purpose of relieving those poor who are not entitled to parochial assistance was formed in 1791' (Aikin 1795: 200). The clash between those who sought to provide private charity and those who felt that it should come out of the rates was further intensified by the desire of parishes to move on as many poor as they could and so diminish their rates bill. The Settlement Laws were as important to the financial well-being of local government as modern laws about the responsibility for homelessness – and equally as unkind. The milk of human kindness was as diluted by self-interest and greed in 1795 as in 1995. Indeed a writer like Aikin who draws attention to the evils of child labour and the poor housing conditions of the working man makes it very apparent that he approved of the fact that Liverpool had reduced its rate from 3s in the pound to 2s 6d. 'By great reforms in the management, though the number of poor is greater than ever, the rate has been reduced' (Aikin 1795: 351).

The argument about public or private provision for the poor was therefore not new when the vivid descriptions of the sad state of the urban poor began to proliferate. Nor was the existence of a considerable minority of poor people an urban phenomenon. What growing urbanization brought to the scene was an increased awareness of poverty, unemployment and poor housing along with their attendant companions of destitution, disease and death. A city like Manchester concentrated in one place what had been scattered and often hidden or ignored in rural England. Indeed an atlas of the care provided by different parishes and the needs of different regions up to the Poor Law Amendment 1834 would show considerable variations in time and space. Eventually the combined long-term effect of centuries of accelerating change in farming practice and patterns was allied to the short-term effect of a major war at the end of the eighteenth century resulting in considerable hardship in those areas where there was no industry to offer alternative means of subsistence. It was no accident that the much-criticized Speenhamland System (basically a system of wage supple-mentation and child allowances), introduced by the Berkshire magistrates in 1795, was more widely copied in the predominantly agricultural south than in the north where wage rates were kept up by the challenge of the new employment in the cities. The *beneficial* side of urban development is all too easily forgotten. Yet even at the beginning of the eighteenth century Defoe noted:

> here the breeders and feeders, the farmers and country people find money flowing in plenty from the manufacturers and commerce; so that at Halifax, Leeds, and other great manufacturing towns so often mentioned, and adjacent to these, for the two months of September and October a prodigious quantity of black cattle is sold.
>
> (Defoe 1968: vol. 2, 606)

It is, in fact, highly unlikely that it was concern about the growing industrial cities which brought about the implementation of the Poor Law Amendment

Act of 1834. Much more relevant was the obvious discontent of agricultural workers as expressed in the series of 'Swing' riots in 1830 which so alarmed landowners, whether of large or small acreage, that severe action was taken against the rioters with 19 men hanged, 644 imprisoned and 481 transported. The very real fear aroused by the prospect of rural revolt was coupled with growing bitterness over the rising rates bill which had been around the £2 million mark in 1775 and was about £7 million in 1831. A Royal Commission was appointed whose report (1834) was used as the basis for reform.

The Act, which was to lay the basis both for the treatment of the poor and its administration, dealt with the *symptoms* of poverty and not with the *causes*. It did, however, lay down very clear principles on which the state deigned to take some responsibility for the needs of its citizens. The kindlier, if patronizing, views of the mercantilists were now replaced by a strictly utilitarian view: it had to work and it had to be cheap. From 1834 an able-bodied man seeking help had to seek it in the workhouse; outdoor relief as practised by the Speenhamland System was to be abolished. Conditions in the workhouse were to be as rigorous as possible:

> every penny bestowed that tends to render the condition of the pauper more eligible than that of the independent labourer is a bounty on indolence and vice. As soon as the condition of the pauper is made less eligible than that of the independent labourer, then new life, new energy is infused into the constitution of the pauper; he is aroused like one from sleep, his relation with all his neighbours, high and low, is changed; he surveys his former employers with new eyes, he begs for a job, he will not take a denial – he discovers that everyone wants something to be done … let the labourer find that the parish is the hardest task master and the worst paymaster and thus induces him to make his application to the parish his last and not his first resource.
>
> (Royal Commission 1834: 227–9)

In order to facilitate these ideas – so redolent of late twentieth-century right-wing thinking – a different style of administration was introduced: a centralized board of three commissioners would appoint regional commissioners who in their turn would supervise the grouping together of parishes into unions; each union would be run by a board of guardians elected by the local ratepayers. On paper it all looked straightforward and although there was criticism, especially in the press, the second reading was passed by 319 votes to 20. Enthusiasm had cooled by the third reading which was passed by 187 to 50, and the implementation of the complete abolition of outdoor relief which was supposed to come into effect on 1 July 1835 was amended so that it was left to the commissioners to enforce through the boards of guardians.

The commissioners were guided by their secretary, the indefatigable Edwin Chadwick, whose administrative ability and eye for detail were outstanding, but whose imagination and powers of empathy were less evident, and who was determined that this act should be seen as 'the first great piece of legislation based upon scientific or economical principles' (quoted in Finer 1952: 69).

They concentrated their attention on the rural areas of the south where they considered that outdoor relief had become a wasteful scandal. It was fortunate both for them and for the agricultural workers that harvests and wages improved slightly in the two years after the Act along with an increased demand for labour in railway-building so that the harsh enforcement of the Act was not as far reaching as it might have been. Nevertheless it was unpleasant enough. The commissioners and their zealous secretary encouraged the various boards of guardians to build grim new workhouses. These latter were nicknamed 'bastilles', but it is possible that that hated symbol of the ancien régime was marginally more comfortable. Not only were families split up and all inmates were allotted monotonous, tiring work but, until 1842, a rule of silence at meal times was enforced and food was mainly bread and gruel. No extras or luxuries were allowed and even if an outsider offered to pay for a treat like a Christmas dinner it was to be refused. Chadwick even wanted to save money by denying the workhouse inmates full rites of burial and although he was overruled in that he did issue a circular to all unions cancelling the tolling of the Church bell at a pauper's funeral. This caused a public outcry, much of it no doubt hypocritical since the bulk of it came from those who hated to pay higher rates to aid the poor. This unease was further heightened by the sterling work of the editors of *The Times* and *The Lancet* as well as that of the factory reformer John Fielden MP who unearthed and publicized a number of scandals concerning the workhouses, such as that at Andover where inmates fought each other over the bits of gristle and marrow in bones they were supposed to crush. Even fervent supporters of the Act and the work of the Poor Law Commissioners had to agree to a Committee of Enquiry. The commissioners were replaced by the Poor Law Board which consisted of a president, who would be an MP, as well as important *ex officio* members like the Chancellor of the Exchequer and the Home Secretary. The boards of guardians were left in place as were the basic principles of the Act although by this time (1847) great regional differences were apparent in its application. Indeed in a review shortly before their demise, the commissioners themselves admitted that they had concentrated their attention on rural parishes where the allowance system had the most dangerous consequences and *not in the large towns of manufacturing districts.*

Such districts had shown themselves to be hostile to the Act from the start. So strong was the feeling which was manifested in riots and violence that an assistant commissioner visiting Bradford in 1837 is alleged to have said that:

> there was a very prevalent opinion in the country that the Board of Guardians had not the power to administer outdoor relief. Now he wished it to be distinctly understood that the Boards of Guardians had control of their own funds and were invested with full power to afford either outdoor or indoor relief to any extent, entirely according to their own discretion and in no instance were they dictated to by the Commissioners.

(Hammond and Hammond 1947: 101)

The industrial towns needed outdoor relief which would tide over workmen who might be temporarily laid off. Given the numbers who might be applying at any one time it might well be cheaper than trying to provide sufficient workhouse accommodation. In any case those men who had been responsible under the old Poor Law were themselves not only ratepayers but also employers and they were resentful of orders from central government. Manchester was a perfect example of what might be called the bloody minded bourgeois: the raising of the Poor Rate and care for the poor was vested in the hands of overseers and churchwardens, the former chosen by the Justices of the Peace and the latter by vestry election. So incensed were they by the 1834 Act that they applied to Parliament to have Manchester excluded. When it looked as if the commissioners were finally going to tackle the recalcitrant industrial areas they pleaded that 'the township of Manchester may be spared the calamity attendant upon the excitement which the agitation of the question might produce' (quoted in Simon 1938: 318). By 1841, Manchester, along with other growing industrial towns, had to accept the concept of unions and the board of guardians but not the abolition of outdoor relief. Instead some relief was allowed to able-bodied men in return for work such as repairing highways – thus very neatly keeping the Highway Rate down. Interestingly, Manchester accepted the format of administration of the new Act very quietly, with the Whigs and Tories issuing an agreed list of candidates to the guardians and the voters very sensibly following those recommendations. As guardians they continued to operate much as the overseers and churchwardens had done. This was hardly surprising since in some instances they were the same people and, in any case, were drawn from the same intermingled groups.

Rural unrest rather than urban need had brought about the introduction of a centralized administration for the relief of the poor. Despite that central direction however it was soon established that locally different boards of guardians 'did it their way'. Once the administrators of the Act were forced to modify the ruling about outdoor relief because of the intransigence of the urban north and to yield more power into the hands of the local boards of guardians, it became a harshly limited way of dealing with poverty and destitution. It did however conform with and exemplify the contemporary belief that people without work or who were indigent were themselves to blame for that state. Thomas Carlyle described it as 'an announcement, sufficiently distinct, that whosoever will not work ought not to live' (Carlyle 1869: 345). Yet glimmers of future light shone through the stringent provisions of the Act and the obnoxious, if self-righteous principles on which it was based: the state *was* accepting a national responsibility for its poorest citizens; the reaction of the industrial urban areas meant that the notion of an allowance, however meagre, for the unemployed man was left in place; and by its very existence the law was responsible for alerting a not-particularly-willing audience to another matter for concern – that of public health.

PUBLIC HEALTH

In 1842 the indomitably industrious Chadwick published his magisterial *Report on the Sanitary Condition of the Labouring Population of Great Britain*. The fact that it had been preceded by various alarming medical reports including those of Kay on Manchester (1832) and Kay, Southwood-Smith and Arnott on London (1838) as well as a major outbreak of Asiatic cholera during 1831–2, probably helped to make it obligatory reading amongst the close-knit influentials of the time. The underlying fear of revolution shared by all those who had wealth and power had been reactivated by the growing militancy of the urban workers who were badly affected by the depression of the late 1830s. Urbanization offered more opportunity for ordinary people to meet, talk and organize. Some movements like the Friendly Societies and Co-operatives were looked on with complacent condescension, but the trade unions and the Chartists were far more alarming. The latter grew increasingly violent and widespread and might have had more success if their initial attempt at physical force had taken place in one of the cities. The failure of the Welsh Chartist Rising in November 1839 owed more to the difficulty of the terrain than to lack of support. There were, of course, many other reasons for the eventual collapse of Chartism, but its initial threatening success helped to concentrate the minds of political leaders on the plight of working people as depicted in Chadwick's report.

The problems were nation-wide: poor housing, lack of sanitation, lack of clean water, poor diet, endemic disease; they could all be found in abundance in rural areas but not in the mass as in the urban areas. A Select Committee on the Health of Large Towns had already recommended (1840) sweeping reforms such as a general Sewage Act, a Building Act, a Board of Health in every town to look after the water supply, lodging houses, burial grounds, etc., but as is the way with governments, the Peel administration hesitated and then set up a Royal Commission into the State of Large Towns and Populous Districts which reported in 1844 and 1845. The Royal Commission, like all its splendid siblings, is an invaluable source book for contemporary ills and is full of sensible suggestions such as those noted by the Select Committee. A bill was introduced into Parliament in 1845 which aimed to co-ordinate and clarify local provision for health and sanitation and bring it under a central control which would have the power of inspection and enforcement. Naturally most local authorities were hostile; it threatened the power of many vested interests, but the Health of Towns Association formed in 1839 by the doctors Kay and Southwood-Smith, reformers like Lords Ashley and Normanby, and Messrs Tooke and Toynbee were in favour.

It came to nothing since by one of those unhappy accidents of which history is full (the 'cock-up' theory) it coincided with the Irish Famine, the Repeal of the Corn Laws and the fall of Peel's government. When the new government set to work to reintroduce the bill, they decided to model it on the 1834 Poor Law Amendment Act and have a central board of commissioners who would have equal authority with Parliament. The fact that Chadwick was to be one of them probably doomed it from the start. Possessing all the virtues of an honest public servant he managed to fall into that egregious category of being 'right but repulsive' (Sellar and Yeatman 1930: 63). In the event, the opposition was bitter and successful. Every private water company saw itself threatened; all the landowners feared for their future profits, claiming they could not afford to drain land, install sanitation, clean water; every ratepayer wondered who would foot the bill. For a while the government tried amendments to its proposed bill but eventually gave up and came back the following year (1848) with a pale shadow of the original. Nevertheless it did establish a Board of Health with some coercive powers concerning drainage and water, and by so doing invoked the wrath of those who were opposed to any form of centralized authority, no matter how worthwhile the subject matter. In cities like Manchester, Leeds and Birmingham furious many-sided battles waged between those who wanted regulation but wanted it under local control, those who did not mind who exercised the control as long as it was done and those who thought that any form of government interference was anathema. The Act was finally passed, edged on its way by another outbreak of cholera, and it set up a three-member Board of Health for five years. This Board was empowered to set up local boards although where a town was incorporated the town council could act as the local board of health. If a town refused to adopt the Act and its death-rate exceeded 23 per 1,000 or if 10 per cent of its inhabitants asked for it, the central Board might create a local health district and a local board: but having done that the central Board had no powers of ensuring that action was taken. Newcastle upon Tyne for example decided to defy the Board and did nothing; a major cholera outbreak brought the central Board to the city where it reported on the filthy condition of the town but it could do nothing to improve the situation. Birmingham, on the other hand, spent £10,000 promoting its own bill so that it did not come under the general Act, whilst Manchester basked in its own local board which pre-dated 1848, resolutely refused to accept central authority – and was left alone (Redford 1940).

In 1854 the government tried to save the situation by dropping the tactless Chadwick, but too much damage had been done and by 1858 the Board was wound up with its administrative duties being assumed by the Home Office and its medical duties by the Privy Council. Numbers of Acts were passed during the next decade including the registration of doctors and, most importantly, the creation of the Metropolitan Board of Works which gave London an efficient drainage system.

Sanitary law in the late 'sixties was scattered about in local acts, factory acts, burial acts, lodging house, vaccination alkali, smoke and food adulteration acts ... the administering authorities overlapped, yet did not cover the whole field. The sewer authority was normally the Vestry; the nuisances authority the Guardians of the Poor. Even the greatest local authorities might be apathetic.

(Clapham 1932: 423)

Recourse to that hardy annual, a Royal Commission, produced the deliberations of the Royal Sanitary Commission which lasted from 1869–71. It clearly recommended that 'all powers requisite for the health of towns and country should in every place be possessed by one responsible local authority' (Royal Commission 1871: XXXV, 3). But, of course, the good ideas so often mooted in official reports got lost in their promulgation. Not perhaps as has sometimes been suggested by argumentative academics because of any great philosophic divide, but because change is always difficult to accept and even more so when it threatens a well-established way of life or vested interests. In theory it is probable that a majority then and now would not fault John Stuart Mill's belief that:

> The authority which is most conversant with principles should be supreme over principles, while that which is most *competent* in details should have the details left to it. The principal business of the central authority should be to give instruction, of the local authority to apply it. Power may be localised, but knowledge, to be most useful, must be centralised; there must be somewhere a focus at which all its scattered rays are collected, that the broken and coloured lights which exist elsewhere may find there what is necessary to complete and purify them ... the localities may be allowed to mismanage their own interests, but not to prejudice those of others, nor violate those principles of justice between one person and another of which it is the duty of the State to maintain the rigid observance ...
>
> (Mill 1910: 357–8; emphasis added)

Put like that it all seems rather splendid and, as exemplified in the Local Government Board Act of 1871 and the codifying Public Health Act of 1875, the fine principles of all those Royal Commissions on health and on working conditions should have swept away the 'chaotic, rudimentary, corrupt system of local government' (Ensor 1936: 124). Of course, they did not. A confused and often inefficient system of central and local administration, despite considerable amendments over the years, remains to haunt us today. Yet good local government remains as desirable and as necessary an institution as when Mill published his essay in 1861.

LOCAL GOVERNANCE AND PUBLIC UTILITIES

It is ironic that some of the better efforts to achieve the goal of better local governance emerged from the hustle of commerce, the smoke and grime of industry which made the great industrial cities of Britain. Best known of all is Birmingham where, under the mayoralty (1873–6) of local businessman Joseph Chamberlain (son of a shopkeeper, himself a manufacturer of screws), the city was 'parked, paved, assized, marketed, gas-and-watered, and *improved*'. But Birmingham was not alone; what Sidney Webb called 'municipal socialism', and which is sometimes referred to as gas-and-water socialism, became widespread. Manchester, for example, had taken control of its own gas supply as far back as 1817 and it is interesting that an American engineer, reporting on the public utilities of Europe in 1907, lavished praise on the city: 'the one undertaking that comes in for practically no criticisms at the hand of engineers is Manchester – the only one which has been public from the start' (Redford 1940: vol. 2. 362).

Municipal ownership of utilities proceeded apace through the latter part of the nineteenth century. A Select Committee of both Houses of Parliament looked into the state of municipal ownership in 1900 and found that of 314 urban authorities in England and Wales, 265 were involved in some form of municipal enterprise, with ownership of the water supply being the most common (Kearns 1989). What was owned varied from city to city, but each new venture brought with it new questions and new responsibilities. If a city had invested in electric trams, as had Manchester, how long could it set its face against other forms of public transport which might be both efficient and more profitable? Indeed the profit motive loomed large in what has more accurately been described as municipal trading rather than municipal socialism. It has been suggested that Webb and the Fabians did great harm by claiming that the industrial cities and towns were instituting a form of socialism (Halévy 1961). On the other hand the Conservative government of 1895 seems to have shown itself remarkably pragmatic over encouraging or discouraging municipal power: whereas it set its face against a London County Council Water Board, insisting that the single Board which took over from eight different water companies should be elected mainly by the London boroughs, it gave way to the Midland cities who opposed a private electricity company which challenged their municipal supplies, and similarly it supported municipal competition to the National Telephone Company in 1899.

PHILANTHROPY IN THE CITIES

It was not however only a matter of public utilities and local government where the urban areas were flexing their muscles. Philanthropy, as has been noted, has a long and formidable history in Britain. Motives may not always be pure: just as the grand architectural monuments of the Victorian city fathers may well reflect a certain desire for self-glory so probably do many of the charitable institutions. That does not detract from their input, although the self-righteous moralizing apparent in most of them grates on modern susceptibilities. For example, the Manchester and District Provident Society founded in 1833 was 'for the encouragement of industry and frugality, the suppression of mendicancy and imposture and the occasional relief of sickness and misfortune'. Most organizations took a patronizing we-know-better-than-you approach and there were certainly strong elements of control in the way help was given. Nevertheless hospitals and schools owed much to private benefactors, as did housing. The latter was an urgent need for an ever-growing urban population. That is not to deny that the standard of housing was undoubtedly abysmal in rural areas. Indeed the testimony of the Royal Commissions on Agriculture of

1881, 1882, 1894 and 1897 bears eloquent witness to the misery of rural life. Mapping of the areas of greatest poverty throughout the nineteenth century show it to be worse in rural areas (Levitt 1986). However it was the more evident and dismal density of the urban areas which drew attention and prompted action. In 1841 the Metropolitan Association for Improving the Dwellings of the Industrious Classes was founded, this being followed in 1844 by the founding of the Society for Improving the Conditions of the Working Classes; but their ideas as eventually put into practice by philanthropists like George Peabody and Sydney Waterlow left a lot to be desired. The so-called model tenements certainly had better facilities but they had the same grim appearance as the hated workhouses, and if they were not well-regulated they could degenerate very rapidly. Charles Booth wrote of them that as an effort to make crowding harmless:

> the building of large blocks of dwellings, an effort to make crowding harmless may be a vast improvement but it only substitutes one sort of crowding for another. Nor have all blocks of dwellings a good character, either from a sanitary or a moral point of view: far from it.
>
> (Booth 1902: vol. 1, 32)

The essence of most of these buildings was to try to provide housing which could be let at a rent which either produced a profit or at least enough financial return to maintain the buildings. Urbanization was already producing other answers to the solution of the problem of both better housing and better conditions for working people.

If philanthropy was widespread in both rural and urban Britain – whether as a means of salving conscience, ensuring the gratitude of posterity or out of natural generosity – equally as important was the way the urban environment fostered the concept of mutual self-help. Nobody would claim that it was an invention of urban living, although the medieval guilds, archetypal self-help groups, were certainly a product of small-scale urban life. Nevertheless it is true that organizations like the Friendly Societies, which were coming into existence in the seventeenth century, became more widespread in the eighteenth *especially in towns* (Fay 1928). Most of them were small scale, like the Newcastle shoemakers who in 1719 contributed one shilling every six weeks to a sick fund and sixpence each for the funeral of a member; there was an allowance of six shillings per week for sickness for a year and thereafter three shillings and sixpence; the latter sum was also paid out as superannuation. By 1796 this Society had 160 members (Eden 1797). By 1801 Sir Frederick Eden calculated that there were 7,200 friendly societies or trade clubs in England and Wales with a membership of about 648,000 (Eden 1801). By an Act of 1793 they were given legal recognition although that appeared to be under threat when a frightened government passed the Anti-Combination Laws of 1799–1800. This did not stop their development, particularly as workers in the towns not only felt the need of potential help in times of illness or lack of work but also because many of them now earned better wages and were prepared to put them to such a use. Manchester, for example, had sixty-six lodges of the Odd Fellows and thirty-four of the Foresters in 1834, and the Odd Fellows of Lancashire numbered some 31,000 members (Clapham 1932; Kidd 1993). The parallel growth of Savings Banks, themselves a model of philanthropic endeavour, allowed Friendly Societies an outlet to deposit their small sums of money and to receive interest. By 1835 it was claimed that Friendly Societies had about one million members but official statistics collected by the Registrar-General in 1847 of enrolled societies only (enrolment was not compulsory) gave a figure of 781,722 members who had contributed £693,751 and had received £518,978 in benefits. The chief concentration of membership was in Manchester and Lancashire where the weekly subscription was fourpence per week and the initiation fee probably a guinea (Clapham 1932). Criticism was made of the societies' expenditure on idle pomp and parade, but Clapham correctly wonders why the members should not bring colour and cheer to their lives:

> There were still crowns and garters and college feasts and city dinners . . . how could Manchester or Leeds expect these adult sons of toil never to flout sound actuarial principles or play with the follies and baubles of youth, when, by their own saving, they got their chance?
>
> (Clapham 1932: vol. 1, 591)

The Friendly Societies could also be cautious when necessary; they refused to answer a questionnaire compiled by the Factory Commissioners in 1833 which was trying to probe the relationship between wages and savings, and indeed remained relatively reticent and secretive about their activities.

Similar self-help movements were found amongst the money or collecting clubs which operated rather like the modern credit unions by taking regular subscriptions from members and allowing small loans when a member had invested sufficient. Some clubs even had a lottery element whereby people balloted for the right to a loan and this love of gambling could be seen in tontine organizations. An example is the St John's Street Tontine Association formed in Swansea in 1791 which had 100 members who agreed to subscribe money for five years in order to build five houses. These houses were to be let and the surviving members divided the income until their number reduced to ten at which point the houses were to be sold and the proceeds divided amongst the survivors, or they could continue until only five members remained in which case the houses would be shared out by a lottery (Cleary 1965). Applying the idea of mutual self-help to the building of houses was developed by the terminating building societies which were less of a long-term gamble and enabled their members to subscribe for a definite goal. Basically their principle was that members should subscribe until enough money had been saved to build a certain number of houses, lots would be drawn for these and the society would continue until all members had their houses. Houses built out of this sort of funding appear marked as Club Houses in Higher Ardwick on Green's map of

Manchester in 1794. Records of a later terminating Building Society – that of the Nelson's Union Inn in Ancoats, Manchester – give the founder members of the society as two shopkeepers, a joiner, a silk manufacturer, a fustian cutter, a tailor and a milkman (Cleary 1965). Other records of early societies show that the usual monthly subscription of ten shillings per month made them more attractive to skilled working men and the lower middle classes. Eventually when Acts of Parliament (in 1836 and 1874) changed them into permanent societies, they became essentially middle-class organizations whose chief interest was to please their investors. Their influence remained no less great but shifted in emphasis.

How much influence the early building societies had on ideas about housing is difficult to say. Certainly the first societies in each region were found in the industrial cities (Robson 1973). It is possible that influential individuals within these cities who were beginning to pass by-laws about their own housing may have been influenced not only by the public health/humanitarian arguments but also by the example of the early efforts of building societies. 'By-law housing' regulations varied from city to city until central government took a hand with first the Artisans' and Labourers' Dwellings Acts of 1868 and then of 1875 which laid down model guidelines. The end-products were streets laid out on a grid-iron pattern with alley-ways between the backs of the houses to allow for the removal of refuse and night-soil where no water-closets were provided; designated amounts of space back and front; streets of more than a hundred feet in length to be at least thirty-six feet wide. Some cities started by linking these houses to mains water and drainage; others, like Manchester, dragged their heels. In fact Manchester complained bitterly about every Act which obliged it to undertake something at the behest of central government and claimed that its own policy of reconditioning houses was the best way forward. Other cities were less obstinate and set to work to build low-cost local housing units only to find, like Liverpool, that since they had to borrow money against the rates, and since the return was less than expected, the final outcome was a higher rates bill (Pooley 1989). Birmingham set to work to clear slum areas but did not build new housing, preferring instead to gain a commercial return through the development of two main shopping and commercial streets (Corporation Street and New Street).

The varied, and on the whole indifferent, response of individual cities and towns to the housing legislation led to action from central government; once more the tried and tested Victorian method of a Royal Commission was set in place. This time the Royal Commission on the Housing of the Poor was set up in 1884 and reported – in six volumes – in 1885. The outcome was another codifying Act in 1890 which still lacked the stick of compulsion or the carrot of government funds. Most cities continued to pursue a course of patch and mend whilst occasionally venturing into new build. Manchester decided to embark on the building of a 'cottage estate' in Blackley, partly because they had managed to secure land at a reasonable price (£150 an acre) and partly because they wanted to provide work for the unemployed. Like most of the industrial cities the cost of land was a problem for any council wanting to build houses, and if they were built in peripheral areas with cheaper land costs then the provision and/or cost of transport became a problem in its turn. In the survey of municipal enterprises referred to above (Kearns 1989), housing ranked the least enterprise undertaken by cities and towns; only eight authorities had invested in it between 1868 and 1898 at an average outlay of £72,000 and a rate of profit of only 1.21 per cent on capital – and even that disappeared when depreciation, costs, interest payments and the repayment of principal were taken into account. It was not until Addison's Act of 1919 that central government offered to put money into helping local authorities to provide low-cost housing.

Meanwhile the spirit of philanthropy continued to flourish: paternalistic and often authoritarian, some industrialists decided that they could ensure a well-behaved, tranquil workforce by providing reasonable quality housing. The pioneer of this had been Robert Owen with his New Lanark factory (1799) and settlement, but his vast dreams of co-operative utopias had foundered in America. Other industrialists settled for small estates of houses near their works or for the grand model village like that of Titus Salt outside Bradford, or George Cadbury in Bournville, Birmingham or the Lever brothers in Port Sunlight near Liverpool. These had their merits but, despite their attempts at better design and a more pleasant environment, were still (for the most part) yet another version of the company store and the truck system. Where, as in the case of Bournville, the estate was placed under separate management and not tied to working for the Cadburys it became very attractive to the better-off who were quick to spot the advantages. Such philanthropic industrialists were influenced in their style of building and choice of site by the a historical sentimentality of William Morris's *The Earthly Paradise*:

> Forget six counties overhung with smoke,
> Forget the snorting steam and piston stroke,
> Forget the spreading of the hideous town,
> Think rather of the pack-horse on the down,
> And dream of London, small and white and clean,
> The clear Thames bordered by its gardens green.
>
> (Morris 1911: 3)

More practical were the ideas and plans of the Garden City Movement whose chief progenitor, Ebenezer Howard, had no hatred for the city as such but only wanted to build what he saw as a city of manageable size with public buildings and houses of good design, where industry and commerce could be accommodated and the best of town and country could be combined. Apart from his own garden cities of Letchworth and Welwyn (developed before 1914) his ideas greatly influenced thinking about housing, and about urban development and planning for the next half century (Hall 1988). The Labour government of 1945 based their plans for New Towns on the garden city concept. By so doing they were tacitly admitting that if urbanization had not existed it would be necessary to invent it. It was not the *idea* of urban living that was wrong, what caused problems was the way people put it into effect.

THE MANICHAEAN DILEMMA

From their flying start in the early nineteenth century, the great industrial cities of England and Wales exemplified the two contrasting viewpoints so enthusiastically depicted respectively by Disraeli and Engels in their books published in 1844. One saw vitality, growth and enterprise; the other saw oppression, need and misery. Their strength lay in that juxtaposition which ensured an ever-changing dynamic, sometimes social, sometimes economic, but never quiescent. The pressure of the needs and the demands of the cities and *all* their citizens moved the manner and method of government, both local and central, and changed and challenged long-established ways of thinking about society. Without that pressure such rudimentary moves as have been made in Britain towards a more just society might not have taken place. The tragedy for us in the late twentieth century is not so much the Conservative-led return to liberal economics and the ruthless simplicities of the free market, which was probably inevitable in the ups-and-downs of parliamentary democracy, but the effect that has had on our towns and cities. The destruction of our manufacturing base, the growth of unemployment, the deterioration of ageing infrastructure, as well as the gimcrack nature of much of the new, have encouraged a retreat into the false anti-urban world so beloved of Morris and his ilk. Such running away from reality is no solution. Instead we should take comfort from the Manichaean nature of the city and continue the struggle towards the perfect. Now – as ever – new processes are at work, changing the nature of the urban mass, on a different scale and by different means than hitherto. The city is our past, it must be our future.

REFERENCES

Aikin, J. (1795) *A Description of the Country from 30 to 40 Miles Round Manchester*, London: Stockdale.
Booth, C. (1902) *Life and Labour*, London: Macmillan.
Brewster, F. (1702) *New Essays on Trade*, London: Walwyn.
Carlyle, T. (1869) 'Essay on Chartism', in *Critical and Miscellaneous Essays*, vol. 5, London: Chapman and Hall.
Clapham, J. (1932) *An Economic History of Modern Britain: Free Trade and Steel*, Cambridge: Cambridge University Press.
Cleary, E.J. (1965) *The Building Society Movement*, London: Elek.
Defoe, D. (1968) [1727] *A Tour Through Great Britain 1724–27*, London: Frank Cass.
Disraeli, B. (1982) [1844] *Conigsby*, Oxford: Oxford University Press.
Eden, F.M. (1966) [1797] *State of the Poor* (facsimile), London: Frank Cass.
Eden, F.M. (1801) *Observations on Friendly Societies for the Maintenance of the Industrious Classes during Sickness, Infirmity and Old Age* (in *Poor Law Tracts* vol. 49, Peyton Collection), London.

Ensor, R.C.K. (1936) *England, 1870–1914*, Oxford: Oxford University Press.

Fay, C.R. (1928) *Great Britain from Adam Smith to the Present Day*, London: Longmans, Green & Co.

Finer, S.E. (1952) *Life and Times of Sir Edwin Chadwick*, London: Methuen.

Halévy, E. (1961) *History of the English People* (trans. by E.I. Watkin), London: Ernest Benn.

Hall, P. (1988) *Cities of Tomorrow*, Oxford: Blackwell.

Hammond, J.L. and Hammond, B. (1947) [1934] *The Bleak Age*, London: Pelican.

Kearns, G. (1989) 'Zivilis or Hygaeia: urban health and the epidemiologic transition', pp. 96–124 in R. Lawton (ed.) *The Rise and Fall of Great Cities*, London: Belhaven.

Kidd, A. (1993) *Manchester*, Keele: Ryburn.

Levitt, I. (1986) 'Poor Law and pauperism', pp. 160–3 in J. Langton and R.J. Morris (eds) *Atlas of Industrializing Britain 1780–1914*, London: Methuen.

Mill, J.S. (1910) [1861] *Utilitarianism, Liberty and Representative Government*, London: Everyman.

Morris, W. (1911) 'Prologue, The Wanderers', Book 1, *The Earthly Paradise*, The Muses Library, London: Routledge.

Pooley, C.G. (1989) 'Working class housing in European cities since 1850', pp. 125–43 in R. Lawton (ed.) *The Rise and Fall of Great Cities*, London: Belhaven.

Redford, A. (1940) *History of Local Government in Manchester*, London: Longmans, Green & Co.

Robson, B.T. (1973) *Urban Growth: An Approach*, London: Methuen.

Royal Commission (1834) *Report of the Royal Commission on the Poor Laws 1832–34*, London.

Royal Commission (1871) *Royal Commission on the Sanitary Laws, 1869–71*, London.

Sellar, W.C. and Yeatman, R.J. (1930) *1066 and All That*, London: Methuen.

Simon, S. (1938) *A Century of Local Government*, London: Allen & Unwin.

Wilson, C. (1965) *England's Apprenticeship 1603–1763*, London: Longmans.

FURTHER READING

Briggs, A. (1959) *The Age of Improvement*, London: Longmans.

Briggs, A. (1963) *Victorian Cities*, London: Odhams.

Clapham, J.H. (1932) *An Economic History of Modern Britain: Free Trade and Steel*, Cambridge: Cambridge University Press.

Dennis, R. (1990) 'The social geography of towns and cities 1730–1914', in R.A. Dodgshon and R. Butlin (eds) *An Historical Geography of England and Wales* (2nd edn), London: Academic Press.

Hall, P. (1988) *Cities of Tomorrow*, Oxford: Blackwell.

Langton, J. and Morris, R.J. (eds) (1986) *Atlas of Industrializing Britain: 1780–1914*, London: Methuen.

Lawton, R. (ed.) (1989) *The Rise and Fall of Great Cities*, London: Belhaven Press.

Robson, B.T. (1973) *Urban Growth: An Approach*, London: Methuen.

Royal Commission Reports, for example: *The Poor Laws, 1832–34* (1834); *The Sanitary Laws, 1869–71* (1871), London.

15

FROM A 'CULTURAL' WORLD TO A 'POLITICAL' ONE

Paul Claval

The growth of a global economy is transforming the world. Two major technical revolutions have triggered the process. First, higher levels of energy mobilization have been reached, and this energy is increasingly available in concentrated forms. Second, more efficient ways of noting, storing, processing and circulating information have developed, which allows for increased clarity, more precise control and more sophisticated management.

As long as the exploitation of environments was limited to low-energy techniques, people had to rely on tricky and often inefficient methods in order to produce and transform food and raw materials. Exchange was limited. With the successive energy revolutions, however, methods of production have changed. They were traditionally adapted to the diversity of environments and took advantage of their specificities, but now they rely increasingly on an initial transformation of milieux, which are standardized. This allows for the use of machinery and induces economies of scale and higher returns. Transport is also made easier and cheaper because of the availability of concentrated forms of energy, but efficient exchange would have been blocked by communication problems if the new information technologies had not been available.

The technical dimension is fundamental in the growth of a global economy, but transformation had other aspects: in order to function smoothly, a society has to be organized and the adjustment of individual decisions has to be realized either through feedback mechanisms or through a political system able to enforce decisions. The relations of culture, political institutions and spatial organization have all been affected by the change of scale in human societies, but in what way and to what extent? This is a major geographical question.

It is easy to believe that traditional societies were more differentiated by their cultures than by their power organizations. Each group, after all, had to take advantage of a particular milieu without the energy inputs which could modify and standardize it. Men had to combine technique and production in a unique way if they were to survive: their existence relied on the quality of the *genres de vie* that they had been able to organize. The world of yesterday was a mosaic of groups differentiated by their ways of life. Today, by contrast, geography is increasingly characterized by the juxtaposition of standardized political constructions, the states, which have grown in importance and spread continuously over new areas from the beginning of the modernization and globalization process. Hence the idea that we have moved from a world differentiated by 'culture' to one differentiated and dominated by 'politics'.

THE MODERN STATE AND THE POLITICIZATION OF THE WORLD

The rise of the modern state

The move towards a global economy started long ago, gained speed at the time of Columbus and Vasco da Gama, in the Renaissance, and gathered its full momentum with the technical revolution of the last two centuries. By the end of the nineteenth century, the transformations of the world map were especially impressive in the political field. Within a few decades, European nations had extended their grip over Africa, Oceania, South and South-East Asia. The tearing apart of the Turkish and Chinese empires had started.

More or less nomadic societies organized along loose political models had ranged over the warm or temperate steppes outside Europe until the beginning of the nineteenth century. In the tropical rain forest and in many tropical grasslands, slash-and-burn cultivators organized into tribes or primitive kingdoms had managed to survive until then. Kingdoms or empires based on the sacredness of their sovereigns and on complex networks of feudal, caste or order relations had ruled over much of the areas where permanent agriculture had developed. Their power often looked impressive when measured by the area over which they ruled, but the control which they exercised was weak: sovereigns reigned over diverse populations and influenced only a small portion of their life.

The modern state had many forerunners from ancient times, but it developed mainly in Western Europe from the thirteenth or fourteenth century. By the end of the eighteenth century it had achieved a high measure of success in the rationalization of power. Western European economies had largely become commercial ones. Taxation and centralization were thus easy. Governments could pay for permanent armies and navies and could organize administrative bureaucracies capable of exercising a thorough control over extensive areas.

Imperialism and the world diffusion of the modern state

This type of state proved remarkably efficient. It was a product of modern economies, but predated the Industrial Revolution. From the mid-eighteenth century it gave a new impetus to the movement of discovery. It was able to co-ordinate military or naval actions in distant countries. None of the traditional forms of political organization were able to withstand the imperialist expansion of Western states. In many countries, the new balance of forces had been clearly perceived. Strategies had been devised to stop the Western expansion. The Chinese had chosen to close their harbours and boundaries: measures that did not work. Turkey and Egypt tried to mimic the Western state, but failed. Japan was the only country successfully to emulate the West.

Colonization had imposed the modern forms of state administrative control all over the world by the end of the nineteenth century. At the end of the First World War the idea of a rational world political order became popular. The only form of political organization which appeared legitimate was the Western nation-state. Hence the devolution to a Society of Nations of the resolution of international conflicts. The former Austrian Empire could easily be, and was, divided into free nation-states. The former Turkish territories in the Middle East and the former German colonies, however, were unfamiliar with modern political organization. Yet the Society of Nations was unwilling to see their return to their pre-colonial status: they could exist only as states. As a result, Britain and France were given an international mandate for introducing modern forms of political organization there in as short a period as possible.

New functions and new ambitions for the modern state

In the interwar period, the rise of the modern state continued. In the liberal economic system of the nineteenth century, states had played a decisive role, but they did not directly interfere with production and distribution. The best policy appeared to be free trade. Governments were seen to have other responsibilities: they had to create good conditions for economic life. This meant the organization of police and justice; the provision for a sound monetary system; and the defence of national firms and national interests when these were unduly attacked in or by foreign states. The economic role of the state, however, had already started to increase in Germany and the United States during the last decades of the nineteenth century. Through protectionist policies, they had built more complex economies, with higher multiplier effects and growth rates. Political structures were becoming essential for the expansion of economies.

During the interwar period, many governments increased their control on foreign exchange and tried to master more efficiently the growth of the national income. With the breaking of the world economic system during the 1930s, the former mechanisms of international economic adjustment disappeared. States had to build new systems of trade, to rely on barter for instance. They discovered, in Keynesian economics, the possibility of stimulating demand and furthering employment through direct state investment.

The Soviet experience was more ambitious. A state-controlled central planning agency had replaced the traditional market mechanisms of liberal economics. The main regulating system of social life had apparently ceased to be the economy. Instead, *everything* had become political.

Flow and ebb of the modern state

The diffusion of the modern state and the widening of its function went on in many fields after the Second World War. The former colonies were transformed into full-fledged nation-states when they became independent in the 1940s, 1950s or early 1960s. Western industrialized societies gave increasing importance to social welfare. States soon played an overwhelming role in the redistribution of incomes, and Keynesian policies were more popular than ever. In countries like France and Italy, the nationalized sector of the economy was important. Meanwhile the Soviet system expanded over Eastern and Central Europe, and eastwards into China, North Korea and Vietnam. It also gained support from Cuba and some African countries south of the Sahara. But some signs of ebb were soon present. The international economic policy launched after the Second World War was from the start based on a single belief: that the worst evils of the 1930s had grown out of the protectionist policies of developed countries. There seemed, then, to be no choice: tariffs had to be cut and free trade restored. As a consequence, the General Agreement on Tariffs and Trade, in 1948, in La Havana, created a discussion mechanism that was to prove broadly efficient in the long run and to reduce progressively the economic power that the states had built during the two world wars and in the 1930s.

From 1960, these international trade conditions opened new opportunities for the industrialization of Third World countries. South-East Asia was soon able to compete in an increasing range of activities with the older industrialized nations and, within a few years, thousands of firms disappeared in Western Europe and North America. The newly industrialized countries differed from many other poor areas in the skills and discipline of their labour force and by the strength of their states. South Korea, Taiwan and Singapore had powerful governments which strove, in many ways, and often very heavily, to develop their economies. In a way, these young economic dragons of South-East Asia were showing that growth still paid tribute to state structures, but they also provoked the ruin of state-owned companies and the demise of central planning and over-ambitious social welfare policies all over the world.

When conceived in the 1920s, the aim of the Soviet system was to overtake Germany and the United States, the most successful economies in the early twentieth century, whose industries had benefited from heavy tariffs and were relatively self-contained. Soviet leaders never grew aware of the changes occurring all over the liberal world as a consequence of lower tariffs and increased air transport and communication facilities. Until the beginning of the 1960s, it served the interests of Western economies that the locations of all plants contributing to the processing of a particular product were based in the same state: when activities were widely dispersed, the control costs were too heavy. The situation changed with the advances in air transport, the progress of telecommunications and the increased use of computers. Logistics improved so much, and control could so easily be exercised far away at low cost, that many location constraints faded away. Today, if a government tries to impose too heavy a taxation system or too strict a control of firms, they answer by locating their manufacturing capital abroad. States will not disappear, but weaknesses have become evident in many of them. They are hit by the rise of new regionalisms and new nationalisms. Our world is still a political one, but not in the same way that it was fifty years ago.

Confronting the evolution of cultures, economies and political organizations, how have geographers responded?

GEOGRAPHERS AND THE INTERPRETATION OF MODERNITY

The first evolutionist tradition and its decline

During the eighteenth century, geographers found in the ancient mythologies the idea that societies experienced, through their history, a succession of stages. They used this sequence in the new evolutionist perspective which was developing as a consequence of Enlightenment. They did not dare to speak like Hesiod of the Golden Age, the Bronze Age and the Iron Age, but they stuck to the idea of the three stages of Man (Kramer 1967). People had first been gatherers, hunters and fishers. With the domestication of cattle, sheep, goats, horses or camels, they had become nomadic herders. Then, thanks to sedentary farming, people could begin to experience what we call 'civilization'. Such a conception, of course, reduced geography to a limited role. The evolutionary sequence was similar for every group. Geographers had only to determine the stage it had reached.

This conception was criticized in the 1770s by Gottfried Herder (1774). He agreed with the idea of evolution, but he did not think that all peoples had to experience the same stages. Because France was ahead of Germany in many fields at that time, a fair proportion of German princes (including Frederick the Great) had decided that the French language had to become the support of all intellectual activities in their states. Like many of his contemporaries, Herder strongly opposed such a policy. For him, each people had a particular genius, which could only be expressed through a nation's own language. Each nation had developed an original experience of history, following a specific path. Because each group was living in a particular land, it had discovered its own road to development. Such a conception gave geography a very wide scope and limited the impact of more orthodox evolutionist interpretations until the time of Darwin.

Ratzel, *Naturvölker* and *Kulturvölker*

Even if he was not an orthodox Darwinist, Friedrich Ratzel played a central role in the introduction of new evolutionist hypotheses into the field of geography (Buttman 1977). As a result the discipline changed its main focus. Instead of dealing essentially with places, inhabited or not, its interest centred on the relations of man and environment: in Ratzel's *Anthropogeographie* human geography was born. The textbook in which Ratzel presented this new conception of geography was published in two volumes respectively issued in 1882 and 1891. For Ratzel, human evolution had been characterized by two forms of social and spatial organization: *Kulturvölker* had succeeded *Naturvölker*.

By the end of the nineteenth century there were still many peoples living in the state of nature. Ethnographers had specialized in their study, and Ratzel spent much of the 1880s writing his voluminous *Völkerkunde* (1885–8), which presented the ways of life of primitive peoples. Since their technologies were generally 'soft', they had to adapt to their environments. This did not mean, however, that primitive people had no impact on vegetation. The use of fire, for example, was the preferred tool for preparing fields or for regenerating grass for their herds. They were nevertheless unable to transform their milieux in a way which allowed for standardized and more efficient modes of exploitation. As a consequence, different primitive peoples developed distinct technologies; their cultures were specific to some ecological niches. The geographical differentiation of the earth reflected at that stage both the variety of natural environment and the diversity of skills invented by human groups to take advantage of them through a limited set of tools and without access to concentrated forms of energy.

The picture of the world of *Kulturvölker* was completely different. The techniques which they mastered buffered them, in a way, from their environment. They had no need to adapt to the minute differences existing in a milieu. They knew how to create artificial environments in which they spent a good portion of their life; and how to standardize the forests, grasslands or fields that they had chosen to exploit. Since they had developed efficient technologies of transport and communication, they had no need to produce all that they consumed in the place where they lived. In this way, too, they found a degree of protection from the environment.

Ratzel strongly emphasized the role of transport and communication in modern societies (of '*Verkehr*', which is difficult to translate into English, but is the equivalent of the French 'circulation'). He saw that social fragmentation was hampering the economic life of *Kulturvölker* when it had been harmless for *Naturvölker*. As a result, much of the energy of developed people was aimed at reducing their cultural diversity in order to facilitate all forms of relations. The variety of culture has ceased to be as striking as it was between primitive groups. And since it was through new forms of political organization that the standardization of space for more efficient economic exploitation and easier social relations was achieved, the superiority of *Kulturvölker* found a synthetic expression in their role as state builders.

Ratzel had worked through the 1880s on *Naturvölker*. He shifted to *Kulturvölker* in the third volume of *Völkerkunde*, and during the 1890s he was concerned with the analysis of their political geography (Ratzel 1897). The terminology which he used, however, is slightly misleading since the *Naturvölker* were mainly differentiated by their cultures, and *Kulturvölker* by the states in which they were living. This emphasis on the state as the major institution in the developed world had certainly something to do with the German experience. It reminds us of Hegel considering the Prussian state as the realization of reason in this world. By the end of the nineteenth century, many Germans thought, like Friedrich Ratzel, that the German state testified to the German virtues and that German civilization owed its uniqueness to the land in which it developed. The transformation of the German countries into a state, moreover, offered the German people an opportunity to consolidate (and up to a point expand) the spatial basis for its further growth.

The influence of Ratzel's views on the interpretation of civilization

The views expressed by Ratzel were readily accepted by almost all the geographers of his generation. In France, Lespagnol, who produced the first modern textbook on geography in 1907, stuck to the distinction between *Naturvölker* and *Kulturvölker* with his 'peuples primitifs, peuples civilisés'. In the United States, the work of Carl Sauer came very close to reflecting Ratzel's distinctions. He devoted all his life to the cultural geography of Indian or South American societies (Leighly 1963). For him cultural geography dealt with primitive or semi-primitive people. He had no interest in the themes which dominated the American life of his time and despised much the kind of spatial organization that American people were creating. He had more concern for the peasant societies of old Europe – and the short-lived peasantries settled in the American Middle West in the mid-nineteenth century – but he did not consider them genuine enough to deserve a lifetime's involvement. Even if he nowhere expressly endorsed Ratzel's distinction, Sauer's geography was built on the same contrast between two major levels of social and spatial organization, but with deeper views on the meaning of culture in primitive societies.

Culture in its more general sense is just the knowledge, practices and beliefs that have been transmitted to us or were added by us to the existing lore. It has to do with the ways of perceiving and conceiving the world, understanding its mechanisms and taking advantage of them, instituting social relations and giving some meaning to individual or collective life. Cultural geographers are normally mainly interested in the modalities of communication, orality or literacy, languages, practices or intellectual disciplines and so on. Carl Sauer stressed the fact that culture was also objectified in the environment. For him, groups were expressing their techniques and values through the land allotment systems which they had chosen, the plants they grew or which invaded their fields, or the building materials they extracted from their environments. Landscapes, in other words, were man-made. Moreover, as long as techniques did not allow for deep alterations of the natural environment, cultures remained place-specific. They did not master tools adapted to a diversity of environment. They offered keys, but these keys opened only a few locks. Because groups had evolved in particular environments, their cultures were unable to offer universal recipes. Hence Sauer's concern to study their material features.

Sauer was not the only geographer to accept Ratzel's view on geographical evolution, though most were more critical and only partially accepted his interpretation. Vidal de la Blache, for example, took many of his ideas from Ratzel and paid him due tribute. He thought that primitive groups were trapped in particular environments by their low mobility and their inability to promote exchange. They had to adapt to the natural conditions of the land in which they had settled and which de la Blache saw as constraining their *genres de vie*. He too stressed the role of transport and communication ('circulation') and as a consequence, he saw *genres de vie* evolving in a gradual way with the progress of transport. People could rely on non-local resources as soon as commerce developed and, in this way, environmental constraints were progressively dissolved. Instead of a two-stages vision of geographical evolution, de la Blache envisaged a multi-stages one.

PEASANTS, INTERMEDIARY SOCIETIES AND THE HISTORY OF LONG DURATION

Robert Redfield, folk societies and the dual nature of intermediary societies

During the 1930s and 1940s, more flexible interpretations of the conditions of geographical evolution were proposed by anthropologists, historians and geographers. Robert Redfield was, like Carl Sauer, a specialist in Latin American societies. The majority of his colleagues, in the 1930s, chose to study the few remaining Indian groups. They were studying societies which were so obviously far from the main centres of civilization and so jealous of their specificities that they could be considered as isolated from the outside world.

Robert Redfield turned towards peasant societies of Mexico (Redfield 1940; 1947). They were overwhelmingly of Indian stock, and some of them still used their Indian dialects – for instance in Yucatán. But they were integrated by many links into more encompassing societies. The peasants were Christian and the Roman Catholic Church was present in their everyday life and gave them outside references. Some of the farm produce was sold on the markets of the nearby cities, or to middlemen. They constituted a part of the Mexican state, even if Mexico City was far away. Redfield's fundamental interpretation was simple. The Yucatán peasants, just like peasants in many other societies, were living in big political organizations, in states or proto-states. In many respects they depended on decisions taken in the capital city. They were often exploited by landowners who were members of other ethnic groups, or had a culture deeply different from their own. In many societies, the integration of farmers into the global society was only superficial. The folk culture of peasant groups was oral. The techniques which they knew were still mainly specific to peculiar environments. They were speaking dialects which changed much over short distances, which prevented them from developing long-range relations. Their beliefs kept an important pagan component even if they had been formally converted to a monotheist religion, either Roman Catholicism in South America or Islam in Africa or the Middle East. Their social organization gave a dominant part to kinship.

The upper stratum of the society was literate. Its members spoke, or could speak, culture languages used on wide areas. Their religious practices conformed to the official norms of their faith. They were often parts of large-scale networks of social relations. Bureaucracies played a decisive role in the spatial organization of their political relations, of their churches or of their enterprises. Redfield observed that progress did not transform societies all at once. For long periods, social structures built on utterly different bases could coexist in the same region, as was the case in Latin America since the Spanish Conquest and before. They could even be present in the same city: in recent decades, for example, peasant groups have invaded the cities of Latin America bringing with them a large share of their habits, beliefs and practices.

Fernand Braudel and the slow history of intermediate societies

Societies enter history with literacy. In a way, historians never have to deal with the kind of primitive societies, trapped in their environments, which fascinated ethnographers at the end of the nineteenth century and at the beginning of the twentieth. There were states in ancient Egypt or Mesopotamia and proto-bureaucracies. Scribes established the accounts of the public treasury, or noted the feuds between farmers and temples, individuals and Pharaoh. But this alone does not mean that Middle Dynasty Egypt was a modern society.

Historians began to concentrate on the problem of the limits to history early in this century. They had explored the archives of Antiquity, the Middle Ages and modern time and knew what they could extract from them. There were still more pages to write in these fields that had been written since the birth of modern history. At the same time, it was becoming evident that many aspects of the past could not be retrieved through official archives even where the records were the best. They were speaking only of war, diplomacy and governments. Some historians had begun to explore other sources. In France, thanks to a partly geographical training, they had discovered the interest of landscapes and of documents relating to them. Legal archives were offering insights on the lives of other strata of society: instead of dealing only with the ruling classes, historians discovered that they could picture the destinies of merchants, landowners, farmers or priests rich and educated enough to have many of their important doings registered. Since the medieval and early modern societies were highly procedural ones, these private archives often shed light on the lower classes too. This was also true of police reports, which gave insights on the sub-cultures of the past, their beliefs and their rules. Last but not least, archaeology was revealing many of the aspects of the daily life of long-extinct populations: their cooking utensils, their garments, their beds, the parasites which plagued them, and so on.

In the early 1930s, progress in this new type of history was such that Lucien Febvre and Marc Bloch launched the journal *Annales* (Coutau-Bégarie 1983). Its mission was to build a coherent field out of dispersed essays. Twenty years later, Fernand Braudel summarized the main results of the *Annales* historians (Braudel 1958). He contrasted the traditional way of writing history – the *evenemential* one, with its stress on political facts, military confrontations and the biographies of kings, emperors or ministers – with the new interest in 'slow' history. Past societies were for him multi-layered realities. Traditional historians dealt with the upper level, for which plentiful evidence was available. Modern historians, though, looked deeper. They tried to discover the fate of the lower classes, the ways in which they were organized, and the techniques they used in farming or producing tools and other artefacts.

Braudel's views were the equivalent, in history, of Redfield's in anthropology. Both of them emphasized the fact that the evolution of human societies ran through a long stage of duality that has broken down only recently through the processes of modernization and industrial revolution.

Pierre Gourou and social techniques

Geographers developed similar interpretations. Pierre Gourou's first researches concerned the peasants of the Tonkin delta in Vietnam (Gourou 1936). He was fascinated by the farming techniques which allowed for an almost perfect control of water in a generally difficult environment. He reviewed the cultivated plants, stressed the significance of rice, described crop rotation and analysed the methods of irrigation. He was interested in the uses of bamboo for the fabrication of tools and as a building material. The Vietnamese house and the structure of villages were carefully described.

Gourou explored all the aspects of the material culture of the Vietnamese peasants, but was constantly aware of the significance of their thorough social organization. The material culture upon which they relied was of Chinese origin and had been developed somewhere to the north. Even if the Chinese system did not use concentrated forms of energy, it involved a high standardization of environments: the major issue was to convert the delta into patches of perfectly flat land and to control the water level in each. Some phases of cultivation had to be collectively planned in order to ensure the best use of water. Social organization, therefore, was as significant as the material aspect of culture in Vietnamese agriculture. It meant that geographers should not restrict their enquiries to sown plants, tools, crop rotations and the like because culture was made up also of social techniques ('techniques d'encadrement' as Gourou called them). For him, Sauer's approach was too restrictive. The geographically significant cultural features were *not* always directly apparent in the landscapes. There was no way to understand peasant societies like the Vietnamese if their techniques of organization (i.e. their political structures) were not explored. Later, Gourou worked mainly in Black Africa. In this part of the world, there was a juxtaposition of ethnic groups which differed both in their material culture and in their social organization. Any attempt to explain population densities, and landscapes, or whatever, without understanding their systems of societal relations was hopeless. In some cases, there was a kind of trade-off between material cultures and social ones. The less-organized groups had to live in the most difficult environments, but they had developed sophisticated farming and used manure and crop rotation in order to maintain soil fertility. Groups with a better social organization, on the other hand, could dominate more extensive areas, but they relied on cruder farming techniques, like slash and burn, which allowed for lesser labour inputs.

The lessons of intermediary societies

In the exploration of the relations between economy, culture and political organization, the study of intermediary societies has brought important results. The politicization of the world did not start with the modern industrial state. It was initiated much earlier. It relied on a change in the content of cultures. Primitive cultures were unable to organize big societies, or unwilling to do so: twenty years ago, for example, Pierre Clastres (1974) argued that they were 'societies against the State', organized on the refusal of any concentrated form of power. Their most significant dimension was then the ecological one. Because they were locally based and refused to spread, they had to rely exclusively on one milieu or a limited range of milieux. Their cultures were often so specific to these environments that they could not be translated to others. As Gallais (1967) has shown for the inner delta of the Niger there are, for instance, African regions where a dozen ethnic groups coexist peacefully in a restricted area because each of them is adapted to a specific niche.

Intermediary societies did not always master their environment more efficiently than the primitive ones. Peasant societies often used the same slash-and-burn farming techniques as tribes. But they are integrated into political organizations spread over wide areas and thus controlling different environments. The result is to allow some relaxation of environmental constraints through the medium of exchange.

The main transformation in cultures which occurred between the primitive stages of social organization and the intermediary ones concerned the attitudes relative to power. Instead of fighting any form of concentrated force and authority, people learned to appreciate them. They were ready to accept the authority of kings, for example, because of their religious status: indeed it seems likely that in many forms of early political organization the existence of authority was more important than the actual exercise of power or the particular person in whom it was vested. Thus judicial power came to be deeply rooted in this religious dimension and greatly respected, but readily vested in persons seen – for whatever reason – to possess a different or 'higher' essence.

The passage to history was in this way the result of a deep change in the content of cultures. The kind of social relations which were conducive to the building of political systems became normal and value systems integrated them so perfectly that although there were revolts against individuals in authority, there were none against authority itself. For more than three millennia, China was a semi-religious royal imperial polity. Dynasties were overthrown many times, but were always replaced by new ones. Nobody tried, before 1911, to impose another form of political organization.

The passage to intermediary societies, then, did not coincide with the end of a world differentiated principally by culture. Rather, it was marked by the integration into culture of new types of social relations which were conducive to the rise of political organizations.

The existence of a more or less formalized political organization did not lead automatically to progress. The dual societies which emerged from the institutionalization of power were stable. The peasant groups generally did not react against the elites who ruled over them. They considered that the division of labour and the separation of authority provided them with security. They recognized, or so it seems, their dependency on warrior professionals, and on the mediating power of religious or semi-religious rulers. Beyond this point, the transition to modernity would have been impossible without the existence of political organizations.

Cultures, economies and polities: the idea of world economies

Fernand Braudel devoted twenty years of his life to studying the rise of capitalism. He focused on its technical bases in the first volume of the series dealing with this issue (Braudel 1967–79). The two others were different. Their emphasis was on the geographical conditions of modernization and their main thesis was later developed by Immanuel Wallerstein (Wallerstein 1974–88).

In all civilizations, the range of economic relations has always been greater than the areas covered by political organizations or embracing the same culture. For a long time these macro-units were neglected by historians. Braudel developed his interest in them from the writing of the German historian Fritz Rörig (1933). The world had always been divided into self-enclosed macro-units which Rörig named *Weltwirtshaft*. In French, Braudel used as a translation 'economie-monde' and here we shall refer to a 'world economy'.

The structuration and evolution of these economies was set down by the relations which characterized their political and economic spheres. The classical model of a world economy was the early imperial one, epitomized, perhaps, by Rome. In it, the political and economic spheres almost coincided. Political organization was founded on a universal system of authority, and imperial ideologies were built on the idea that there were no boundaries to the exercise of the imperial power. Imperial polities had thus to be stretched to the limits of the known world since they were included in the economic relations system. Empires however generally failed in their will to extend the limits of their economic sphere because they traded with societies which could not pay for the high cost of maintaining a centralized bureaucracy and a permanent army. Thus, empires succeeded in conquering and organizing the agriculturally based core area of the economies which they were dominating, this pushing their boundaries outwards to those of barbarian peoples with low population densities: communities of hunters, gatherers, nomadic shepherds or herders. Imperial boundaries, though, were of a special kin: they were linear and dissymetrical. Their function was not to divide two different but structurally equivalent political organizations. It was to fence the politically organized world off from the raids of its unpredictable neighbours.

The world economy of today is different. For the first time in history, it is a global construction which started with the Great Age of Discoveries and has been evolving ever since. The imperial model has been explored and abandoned. The core area of the modern world economy has remained divided into competing entities, into independent states. Some of them, of course, developed colonial empires of their own. But these differed fundamentally from the old empires of the past: primitive or intermediary forms of political organization were integrated, not for security, but in order to transform and modernize them.

The modern world economy is basically economic, but its expansion and development has depended on two sets of political conditions. First, the evolution of a new type of state, rational in nature and no longer inspired by religion. It grew because of its capacity for providing citizens with a good quality of life, and the nation with power and international influence. Second, the evolution of a new kind of bureaucracy, for the new state relied more heavily than its predecessors on bureaucracies. Their role, however, was more clearly defined, and their power was consequently more efficiently controlled. The modern state was thus able to rationalize national territory and removed all the obstacles that it offered to inner trade. Meanwhile, on the international stage, it sought to support its traders and businessmen and to make sure that the world was open to their enterprises.

There were thus political preconditions to the realization of the modern world economy. It is possible to say that with it we entered a political world. But two qualifications are needed in order to avoid any misunderstanding. First, the prime characteristic of the modern world economy is to exist in a relatively unified economic space, but not by a unified political system. A large part of its efficiency lies in this feature which allows for the fierce economic competition which is its major structural advantage. When states forget this condition and try to enlarge their economic role, they impair the competitiveness of their firms: the present tendency towards less state involvement reflects this reality. Though our world has to be politically organized for the development of trade and international economic relations, it is not basically a political one. Second, there is no society without a culture, and the nature and content of cultures has evolved with modernization. In so far as cultures differ, and societies with them, this is a reflection of their structure, their evolution, and the hierarchy of values which they promote.

CONCLUSION

Primitive societies had unitary cultures. Everyone partook of the same basic set of values and had access to approximately the same lore. Beyond this, these cultures had three main features. First, they were unable to devise universally valid techniques and reflected for that reason the conditions prevailing in specified environmental niches. Because of their inability to standardize milieux and to harness the energy to allow more efficient exploitation, they were unable to revolutionize their relation to the environment. Second, they were generally very reluctant to encourage any imbalance in the distribution of power. They were 'against the State', to use Pierre Clastres's expression, and, for that reason, unpolitical or prepolitical. The joint effect of their ecological inefficiency and of their social choices made them unable to organize large groups dispersed over wide areas. Third, in all the other fields of culture there was no limit to their inventive capacities, as was shown by the study of kinship systems, for instance. Because cultural transmission was only oral, a tendency towards fragmentation was inbuilt in them.

Intermediary societies had other characteristics. To begin with, the material techniques upon which they lived were generally efficient enough to leave farm surpluses upon which their ruling classes could be fed. They could thrive only on agriculture or combinations of farming and herding. Some measure of standardization in their environments had been achieved and gave them the capacity to adapt successfully to a whole range of natural conditions. Autarky had ceased to be almost absolute for local groups, which also reduced the necessity to adapt to the natural environment. In this way, intermediate cultures were not as environmentally bounded as their forerunners. Second, the value systems of intermediary societies recognized that some institutions or personalities were rightly endowed with authority. The exercise of power was not condemned. As a consequence, these societies were politically structured and open to the formation of a wide variety of political systems. Some city-states gave their population the right to participate actively in the political life, but it was exceptional. Even in classical Athens, slaves and metics were deprived of political rights. Generally, the majority of the population accepted rule by minorities. Finally, such attitudes are more readily understood when the transmission of culture is analysed. Intermediary societies were basically dual ones. On the one hand, their ruling elite could take advantage of their literacy to dominate the other components of population and enforce their institutions and some of their values. On the other hand, the majority of the population had little access to the elite culture and relied on orality for the acquisition of its values, know-how and practices. The groups which constituted it were glad enough when they were given a right to local self-organization. Because of the dual nature of their cultures, intermediate societies were structured on a two-scale basis. Fragmentation was still evident in many aspects of their folk cultures, even if some measure of standardization had occurred in their ecological components. At the same time, a fair measure of unity was achieved by the ruling elites. The political institutions of intermediary societies paved the way for the constitution of large world economies, but they were not automatically conducive to modernization. Even if they had a fair share of inbuilt contradictions, their capacity to survive without profound change was remarkable.

Important improvements in the technical basis of material life had to be achieved in order to open the path to modernization. They were necessary, but insufficient conditions. The Han Chinese, for example, mastered an array of techniques broadly similar, except for navigation and warfare, to those practised in Europe in the seventeenth century, but nothing happened and the sociological basis of the Chinese society remained essentially a peasant one.

Modernized societies share one feature with the primitive ones: they are not split between elite and folk cultures; everyone has access to the same set of knowledge, know-how or values, even if they cannot be wholly mastered. As a result, everyone has the same right to participate in the political decision process. The modern nation-state translates this new cultural reality into institutions. Modernity is not coextensive with the state. Political organization was an essential feature of all intermediary societies, and was a precondition for the building of world economies. Nevertheless, there is something new in modern polities. The democratic participation to the political process entails the idea that social groups are free to decide their own destinies. In intermediary societies, governments had to reconcile the population with gods and to settle the domestic or international conflicts which grow out of every system of interaction. But they did not have the responsibility to remodel society itself. There was no concept of choice with respect to the nature of society. Modernized societies, however, are based on the idea that peoples are entitled to change their destiny. As a consequence, they can use of their sovereignty to launch extensive social reform or to control and plan their economies.

For modern ideologies, societies are fundamentally political since their members have the responsibility to choose and build their own future. Geography nevertheless imposes serious qualifications upon such a perspective. From the beginning of our century, many social experiments have been launched. Many nations have decided to modify their economic institutions dramatically. They have renounced free enterprise and market mechanisms and replaced them with centralized management and planning of their economic activities. Social welfare systems have completely modified the distribution systems. What we are now discovering is that this freedom of choice is not total. It has to be limited for both ecological and geographical reasons.

We have ceased to depend on local resources for our food, our shelter and our power consumption. We have standardized environments to such an extent that social life has apparently little tribute to pay to natural conditions. Some of the most dynamic and prosperous societies of today have limited farming capacities, and practically no mineral or energy resources: Japan or Taiwan are examples. However, we cannot ignore the ecological impact of such societies. Local imbalances are likely to be produced wherever human activities are concentrated. Regional and global ecological disturbances are increasingly threatening. In the political world in which we live, we are not free to set any policy. We have permanently to enhance the working of our environments.

Geography is imposing a second limit to the choices offered to our societies. The only way to reduce the inequalities and to promote development everywhere in the world is to facilitate trade, exchange and international competition. Whenever a state has chosen to protect unduly its citizens or its enterprises from the rest of the world, development has become slower, or decline has been triggered. We are certainly living in a more political world than ever, but it does not mean that we are free to do whatever seduces us. Modern culture must observe geographical and ecological concerns if we wish to avoid irresponsible political experiences.

REFERENCES

Braudel, F. (1958) 'Histoire et sciences sociales, La longue durée', *Annales ESC*, no. 4. (Reprinted in F. Braudel (1962) *Écrits sur l'histoire*, Paris: Flammarion.)

Braudel, F. (1967–79) *Civilisation matérielle et capitalisme* (3 vols), Paris: Armand Colin et Flammarion.

Buttmann, G. (1977) *Friedrich Ratzel, Leben und Werk eines deutscher Geographer*, Stuttgart: Wissenschaftliche Verlagsgesellschaft.

Clastres, P. (1974) *La Société contre l'état*, Paris: Les Éditions de Minuit.

Coutau-Bégarie, H. (1983) *Le Phénomène Nouvelle Histoire: Stratégie et Idéologie des nouveaux Historiens*, Paris: Economica.

Gallais, J. (1967) *Le Delta intérieur du Niger*, Dakar: IFAN, Étude de géographie régionale.

Gourou, P.(1936) *Les Paysans du delta tonkinois*, Paris: École Française d'Éxtreme-Orient.

Herder, J.G. (1774) *Auch eine Philosophie der Geschichte* (French trans.), Paris: Aubier-Montaigne, 1964.

Herder, J.G. (1774–91) *Ideen zur Philosophie der Menscheit* (French trans.), Paris: Aubier-Montaigne, 1962.

Kramer, F.L. (1967) 'Eduard Hahn and the end of "three stages of Man"', *Geographical Review* 57, 73–89.

Leighly, J. (ed.) (1963) *Land and Life: A Selection from the Writings of Carl Ortwin Sauer*, Berkeley, Calif.: University of California Press.

Lespagnol, G. (1907) *Géographie générale*, Collection Fallex et Mairey, Classe de Seconde, Paris: Delagrave.

Ratzel, F. (1882, 1891) *Anthropogeographie; oder, Grundzuge der Anwedung der Erdkunde auf die Geschichte* (2 vols), Stuttgart: Engleborn.

Ratzel, F. (1885–8) *Völkerkunde* (3 vols), Leipzig: Bibliographisch Institut.

Ratzel, F. (1897) *Politische Geographie*, Munich and Leipzig: R. Oldenburg.

Redfield, R. (1940) 'The folk society and culture', *American Journal of Sociology* 45, 731–40.

Redfield, R. (1947) 'The folk society', *American Journal of Sociology* 52, 293–308.

Rörig, F. (1933) *Mittelalterliche Weltwirtschaft, Blüte und Ende einer Weltwirtschafts periode*, Jena: Gustav Fischer.

Wallerstein, I. (1974–88) *The Modern World System* (3 vols), New York: Academic Press.

FURTHER READING

Badie, B. and Smouts, M.-C. (1992) *Le retournement du monde. Sociologie de la scène internationale*, Paris: Presses de la Fondation Nationale des Sciences Politiques et Dalloz.

Boulding, K.E. (1978) *Ecodynamics. A New Theory of Societal Evolution*, Beverly Hills, Calif.: Sage Publications.

Claval, P. (1995) *Géographie culturelle*, Paris: Nathan.

Crosby, A.W. (1986) *Ecological Imperialism. The Biological Expansion of Europe, 900–1900*, Cambridge: Cambridge University Press.

Dodgshon, R.A. (1987) *The European Past, Social Evolution and Spatial Order*, London: Macmillan.

Durand, M.-F., Lévy, J. and Retaillé, D. (1992) *Le Monde. Espaces et systèmes*, Paris: Presses de la Fondation Nationale des Sciences Politiques et Dalloz.

Johnston, R.J. and Taylor, P.J. (eds) (1986) *A World in Crisis. Geographical Perspectives*, Oxford: Blackwell.

Knox, P. and Agnew, J. (1989) *The Geography of the World Economy*, London: Edward Arnold.

Mann, M. (1986) *The Sources of Social Power*. Vol. 1: *A History of Power from the Beginning to AD 1760*, Cambridge: Cambridge University Press.

Vallega, A. (1994) *Geopolitica e sviluppo sostenibile. Il sistema mondo de secolo XXI*, Milan: Mursia.

III THE GLOBAL SCALE OF HABITAT MODIFICATION

INTRODUCTION

Richard Huggett and Mike Robinson

With only a brief hiatus, the 'war to end all wars' was followed by another one. The war of 1939–45, however, differed in important respects from the war of 1914–18. It signified, particularly in its concluding acts in Hiroshima and Nagasaki, a fundamental shift in perceptions of the potential of science and of its applications in the spheres of political, economic, and social activity.

Developments in engineering and technology, though inevitably displaying a timelag, gradually allowed the physical realization of scientific theory in a variety of areas which had direct bearing on the global ecosphere. Improvements in computing, in telecommunications, in radar, in remote sensing, in decoding, and in the control of nuclear energy seemed for a time to offer a future freed from the 'tyranny of distance' and the prospects of energy scarcity. The biological sciences, too, offered the hope of abundant nutrition, the effective control of insect pests, and globally improving levels of health and physical welfare.

The technologies of the post-war era proved to be less than wholly liberating. Scientific rationality, which appeared to offer so much to the recuperating world of the 1950s and 1960s, became, in the space of the decade which followed, a mistrusted implement wielded to the advantage of the few and the powerful at a cost paid by the many and the powerless. In many cases scientific applications began to reveal new environmental impacts with damaging consequences that were potentially global in their effects. Environmentalism, which had earlier been a minority concern exercised in finite spatial contexts, became a ubiquitous theme of universal significance.

In politics, too, the optimism that accompanied the creation of the supranational United Nations was severely and repeatedly tested in an atmosphere dominated by Cold War tensions, the competing ideologies of the USA and USSR, and new regional defence alignments between nations made afraid by the military history and the destructive potential that the twentieth century had revealed. The position of the states of Western Europe, meanwhile, changed drastically. Beyond rebuilding their own battered habitats, they steadily and sometimes painfully relinquished, or were relieved of, the legacies of empire. In this most radical series of political changes, new nations were created and within them new powers were contested, often on the basis of old enmities. Lip-service to notions of enlightened democracy became the stock-in-trade of the ex-colonial powers which sought to persuade younger nations to an ideology which they themselves had embraced only *in extremis*. Inevitably wars continued, but in the period after 1946 these often echoed earlier, spatially introverted conflict and only rarely extended beyond internationally recognized boundaries. The most intransigent of them reflected the lack of foresight of the boundary-makers themselves.

Political realignment in the post-war years was accompanied by economic realignment as the industrial capacity of much of the world was rebuilt following the devastation of 1939–45. Behind an 'Iron Curtain' the USSR, and the satellite countries which it controlled, sought a self-sustaining economy separated from, and wholly independent of, the influence of Western financial institutions. 'Free world' nations sought new trading alliances based on the revised geography of the post-war period. In some cases, like the establishment of the European Union, the intentions were broad-based and, in commodity terms, unspecific. In others, as in the creation of OPEC especially, producer-nations sought collective benefit from co-operation in production and marketing. Outside these nationally based blocs, however, the post-war period has been marked by the dramatic expansion of business companies which operate across national boundaries to maximize institutional efficiency and marketing potential. Their power and influence is such that they enter unavoidably into the organization of global economic activity, constraining the freedom of individual governments and posing new questions for aspirations to national autonomy and for co-operation rather than competition between nations.

The spread of industrialization in a redefined post-war world was accompanied, and to some extent dependent upon, increasing urbanization. In some countries this took the form of exaggerated growth in a primate city at a real cost to the spatial balance of national economies already threatened by volatile trading conditions in agricultural products. Hand-in-hand with rapid urban growth, the problems of urban poverty and the need for urban provisions in housing, in employment, and in services became greater and more visible. Improving the quality of urban life, however, is difficult, particularly in those countries where the need to service accumulated overseas debt resists attempts to enlarge employment opportunities and where inflation eats into the meagre material base of societies ill-equipped to negotiate a stable place in an increasingly rapacious world economy.

The end of the Second World War undoubtedly heralded a new era in the history of the earth as a habitat. It was an era that addressed itself first to the reconstruction of social and economic agendas, and then to the re-negotiation of political and geographical ones, culminating most recently in the demise of the USSR in 1989. It is a world, for some people, immeasurably better than the one it replaced. But it is a world which, for the majority, still affords little relief from an unforgiving inequality of wealth and welfare. It is a world, too, that is better understood: the inventory, as it were, is more complete; the workings of its different parts are more amenable to new generations of models and predictions. But it is not a world more easily managed under its weight of varied human occupancy and in consequence it has become a world threatened *as a world*. The urge to have more has pressured the means to survive, so that a hitherto resilient ecosphere is edged towards irretrievable depletion.

16

UNITY AND DIVISION IN GLOBAL POLITICAL GEOGRAPHY

Peter Taylor

Around the turn of the twentieth century a particular type of social theory became prominent. Relating social behaviour to the changing geography of the times, 'closed-space' thinking was all the fashion (Kearns 1984). Such theories focused on the fact that European and European-settler land grabs in the nineteenth century had reached their limit, effectively bringing to an end the era of geographical expansion: in the New World the frontier had ended; in the Old World there were no new colonies for the taking. Global closure had occurred and its social implications were just beginning to be considered. The new subdiscipline of political geography was born into this intellectual milieu. Its special contribution to understanding the new geographical circumstances was to focus on the divisions within the new world unity. Given the creation of the latter, political geographers asked what is the basic geographical patterning that constitutes the unity. This duality of unity with division defines the common thread running through the global political geographies described in this chapter.

Resolutions of the unity/division global dialectic can be reduced to just two generic types of political geography model. The geostrategic model pits sea-power against land-power in the *heartland thesis* (Figure 16.1, top). The geoeconomic model defines large economic blocs in the *theory of pan-regions* (Figure 16.1, bottom). Both models are based upon simple environmental theories of political organization: the first emphasizes absolute location and physical accessibility, the second functional location and physical resources. As such, both models had their origins in popular nineteenth-century ideas but it was political geographers who turned them into global political theories in the first half of the twentieth century. We consider this process in the first section of the chapter before dealing with the use of both models in the German geopolitics of the Nazi era. No discussion of global political geography is complete without an understanding of this notorious episode in the history of

(a) **Geostrategic model: the heartland thesis**

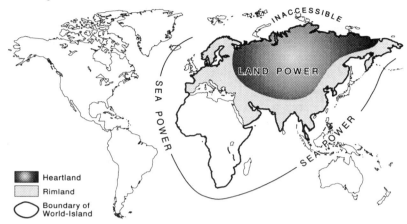

(b) **Geoeconomic model: the theory of pan-regions**

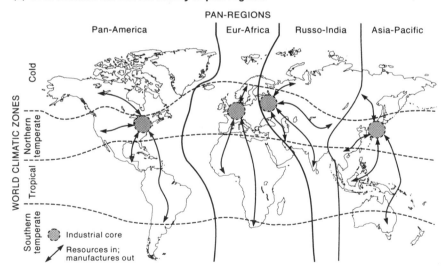

Figure 16.1 Generic types of global political model.

geography and the mark it left on subsequent political geography analysis. The main sections of the chapter then deal with this aftermath: post-Second World War global political geography from immediate post–war hopes for 'one world' to contemporary disintegration of world order. The focus is on how the heartland thesis fitted the East versus West policy needs of the Cold War and in the process marginalized the 'South' as a factor in world politics. In a brief conclusion the necessity of a return to one-world thinking is argued in a renewed concern for the role of the environment. Unlike the last *fin de siècle* when environmental factors were seen as an input to world politics, in our *fin de siècle* the environment as output – the making of global ecological fragility – is the key world political question for the future.

HEARTLAND AND PAN-REGIONS

In 1904 before an audience of the Royal Geographical Society in London, Halford Mackinder delivered his famous lecture on global political geography (Mackinder 1904). Setting political geography on the path of unity with division, his message was a dire warning of its consequences. First, he argued that global closure would lead to political instability since 'every explosion of social forces, instead of being dissipated in a surrounding circuit or unknown space, will be sharply re-echoed from the far side of the globe' (Mackinder 1904: 22). Second, the social and technological forces that had integrated the world were also responsible for accentuating the division between land- and sea-powers. In the previous four hundred years, the 'Columbian era', these forces had favoured sea-power but the new unity marked the beginning of the 'post-Columbian era' when land-powers would regain their traditional advantage over naval strength. In particular the 'heartland', the region of Eurasia beyond the geographical reach of the sea-powers, would soon re-assert itself as the natural 'pivot of history' (Figure 16.1, top). Mackinder's message was that the signs were not good either for peace in general or for the prospects, in particular, of the greatest sea-power of them all, the British Empire.

This heartland thesis was a codification of long-term British strategic concerns for the expansion of Russia into Asia (Hall 1955). Known as the 'Great Game', British imperialists had confronted the Russians from the North-West Frontier of the Indian Raj through Persia to the Ottoman Empire and the strategic 'Straits' of Istanbul in what has been latterly dubbed a 'Victorian Cold War'. Mackinder thought that modern technology in the form of the railways would enable the vast resources of the heartland to be mobilized as never before and there was little any sea-based power could do about it. Hence a land-power based upon the heartland would have the potential to become a world empire. He made this explicit fifteen years after his Royal Geographical Lecture in a book providing advice to the statesmen meeting at Versailles to reorganize Europe in the aftermath of the First World War:

> Who rules East Europe commands the Heartland:
> Who rules the Heartland commands the World-Island:
> Who rules the World-Island commands the World.
>
> (Mackinder 1919: 169)

The relevance of this famous dictum to a later era, the Cold War, is all too obvious and accounts for the heartland thesis's longevity.

The theory of pan-regions has similar nineteenth-century origins before its political geography codification. European states dominated the world in the nineteenth century but there were underlying worries for a future in which they might be displaced as world powers by large extra-European states (Bartlett 1984). The leading states of Europe were small compared with the 'continental states' of Russia, the USA and China. Hence it was just a matter of time before European states would be marginalized. The massive imperial expansion of European states in the last quarter of the nineteenth century was, in part, a response to such concerns. The end-result was that by 1900 the British Empire was the largest state in the world as measured by either land or population. But this imperial solution to the size problem generated two related problems.

First the distribution of the colonies was deemed unfair by some. It reflected to a large degree how long a country had been accruing extra-European lands and hence favoured Britain and France in particular over newly unified states such as Germany and Italy. Hence in the early years of the twentieth century the idea of a redistribution of colonies was commonplace. But there was a second problem with the European overseas empires. Unlike traditional land empires they consisted of scattered colonies with potentially impossible defence implications. The large contiguous and compact states such as Russia, China and the USA were thus strategically much more secure than Britain and France for all the size of their empires. This thinking led away from imperial redistribution as a solution for space-deprived states and resulted in the theory of pan-regions.

The feeling of space deprivation was greatly accentuated in Germany after the Treaty of Versailles stripped her of the small empire acquired before 1914. It was in this context that the German school of *Geopolitik* under the leadership of Karl Haushofer developed the concept of the pan-region (O'Loughlin and van der Wusten 1990). Viewing existing empires as haphazard, and therefore irrational, accretions of territory, the German geopoliticians proposed a complete global reorganization. Pan-regions were logical world-wide combinations of territory where each region consisted of all global environmental zones from the Arctic to the Tropics (Figure 16.1, bottom). The Americas were considered to be the clearest case of a pan-region with the US Monroe Doctrine viewed as its founding statement. This world model consisted of just three or four such pan-regions – Pan-America, Eur-Africa, Asia-Pacific and sometimes Russo-India – each of which would be economically self-sufficient (autarky) so that there would be no longer any pressures for world wars. Such a peace would be under the auspices of the pan-region organizers: the USA, Germany, Japan and perhaps Russia.

GERMAN GEOPOLITICS

There is a curious irony in the story of German geopolitics. Their heyday was from the mid-1920s to the late 1930s but when war finally came they were effectively marginalized (Stokes 1986; Heske 1987). And yet in the early 1940s, especially in the USA, geopoliticians became popularly demonized as the brains behind Hitler (Brunn and Mingst 1985; Parker 1985). Karl Haushofer, in particular, was given the role as the ruthless 'pseudo-scientist' planning the Nazi strategy for eventual world conquest. This was largely based upon what Herb Heske (1987) has called 'fairy tales', but the episode scarred the subdiscipline for the post-war period to such a degree that global political studies by geographers almost disappear.

The question of the relation of Haushofer's school of geopolitics to the Third Reich has been widely debated in recent years. The consensus is that the influence of geopolitics was much more indirect than writers during the Second World War realized (Sandner 1989). It is certainly true that in the intellectual milieux of the period geopolitics helped mould perceptions of the 'spatial injustice' of the Versailles settlement and provided concepts for reversing this political outcome (Paterson 1987). In particular the key notion of *Lebensraum* – literally 'living space', introduced into geography originally as an ecological concept – became central both to geopolitical theory and Nazi practice. It justified the expansion of the vibrant German nation into an Eastern Europe perceived as being weakly and inefficiently organized by Slav peoples. This domination of *Mitteleuropa* was to be Germany's first step in overturning its space-deficient status. The axis agreement with Japan in 1936, the Soviet–Nazi pact in 1939 and nurturing US neutrality when the Second World War began can all be interpreted as policies leading to a pan-region organization of the world. All this changed in 1941 with the German invasion of the USSR and US entry into the war. The German policy that precipitated the 'Big Three' alliance (pitting the British Empire, the USA and the Soviet Union against Germany) had no logical basis at all from a rational German geopolitical point of view.

It seems clear that the Nazi regime used and manipulated geopolitics for their own different ends. Although Haushofer had important contacts with the Nazi leadership these seem to have become less important after 1938. In fact the Nazis were always suspicious of geopolitics since its scientific materialism contradicted their romantic concern for the *Volk* and the purity of 'race' (Stokes 1986; Bassin 1987). This fundamental conflict of ideas meant that Haushofer's geopolitics could never be central to Nazi German policies whatever Allied propagandists asserted at the time. The intellectual legacy, however, is clear: geopolitics became 'notorious', a name that no geographer dared speak publicly for three decades. In a final irony, however, the failure of the German army to defeat the USSR was to give the idea of an impregnable heartland a new lease of life that was to be crucial in constructing the post-war global political geography. But that was not the initial response to the German geopolitical challenge.

GEOGRAPHIES FOR ONE WORLD

German geopolitics provided powerful world images that could be popularly understood. The problem for the Allies was that they knew what they were fighting against but had no comparable world image to guide what they were fighting for. American political geographers in particular were very ready to condemn the German geographers but seemed to have little of consequence to offer in reply (see, for instance, Bowman (1942)). An alternative was required to the geopolitics of autarkic pan-regions and aggressive heartland: for the Allies the time had come to downplay emphasis on global division and provide renewed images of global unity.

In defining a solution to the Allies' perceived deficiency of a world image, Archibald MacLeish (1943: 6) turned to the airplane to provide 'the image of victory' since he considered it 'capable of altering the geography of our world'. He argued that air travel had replaced the sea as the symbol of freedom and therefore could become an instrument for defining a new age of liberation. This 'airman's earth' was nothing less than 'the full completed globe':

> Never in all their history have men been able truly to conceive the world as one: a single sphere, a globe having qualities of a globe, a round earth in which all directions eventually meet, in which there is no center because every point, or none, is center – an equal earth which all men occupy as equals.
>
> (MacLeish 1943: 7)

But such an inspiring image still entails geographical organization within the one world. Global geographical concepts had been employed for war aims, it was now up to geographers to show how they could be pressed into the service of peace. This challenge was taken up to produce three important proposals for a future peaceful world by Halford Mackinder, now the sage of global political geography, by Nicholas Spykman, an American political scientist and by Griffith Taylor, the leading geographical determinist of the period.

In 1942 Halford Mackinder was invited to reconsider his heartland thesis by the journal *Foreign Affairs* and he responded with the appropriately titled 'The Round World and the Winning of the Peace'. He considered his heartland concept to be as relevant to policy in peacetime as in wartime (Mackinder 1943). However, it was now set in a very different context. Mackinder identified a great mantle of wilderness running through deserts, tundra and ice (encompassing both Sahara and Arctic) to encircle an inner world of heartland, Europe and North America. The problem of an aggressive Germany at the centre of this world was to be solved by co-operation between the USSR as effectively the heartland and the Western Allies as 'Midland Ocean' (North Atlantic) with the USA as resource reserve, Britain as airbase and France as bridgehead. It was crucial for Mackinder that heartland and Midland Ocean balanced one another to define a geographical stability. In these circumstances prosperity could develop with increased commerce along 'the great trunk routes for merchant aircraft between Chicago–New York and London–Moscow' (Mackinder 1943:

601). For the 'outer world' beyond the wilderness mantle, prosperity would come later in a subsequent balancing of the whole world. The end-result: 'A balanced globe of human beings. And happy, because balanced and thus free' (Mackinder 1943: 604).

Mackinder's specifically anti-German revision of his heartland thesis was effectively converted back into a general model by Spykman (1944) in his *The Geography of the Peace*. This book provided the most influential wartime global political model because it introduced the concept of rimland. Spykman interpreted the rise of Germany to indicate that the zone around the heartland was in fact much more important than Mackinder had allowed for in his original models. He renamed it the rimland and this term was to be subsequently incorporated into all post-Second World War uses of this geostrategic model. For Spykman his prescription for peace was simple: in the future the Allies had to ensure that the populous, and therefore potentially powerful, rimland was kept in check. Hence the wartime British–American–Russian alliance was an eminently sensible arrangement in this original rimland version of the heartland model and, very much like Mackinder, he argued that it should continue after the defeat of Germany. In short, continental power and sea-power had to combine to overcome the always potent power of the rimland.

As with revisions of the heartland thesis, Griffith Taylor (1946, 1951) was concerned with devising balanced political geographies for peace but his schemas drew upon the alternative tradition of pan-regions. Taylor proposed a new subdiscipline of 'geopacifics' which he promoted as 'humanized geopolitics', the 'antithesis' of warmongering geopolitics: 'Geopacifics is an attempt to base the teachings of freedom and humanity upon real geographical deductions' (Taylor 1951: 606). But this was not a pacifist approach to political geography; like Mackinder, Taylor had a keen sense of power among states based upon what he termed the 'World Plan' – the global distribution of land and resources. Hence he set himself the task of devising geoeconomic arrangements conducive to peace.

Taylor (1946: 327–8) viewed the San Francisco proposals in 1945 for setting up the UN as an improvement on the League of Nations partly because of the novel inclusion of an Economic and Social Council. If the causes of war were to be found in economic competition then peace must be promoted through fair trade. The latter was the key to permanent peace but was dependent on geographical arrangements for access to natural resources. Any state deprived of such access would ultimately have to resort to war. Hence Taylor's proposals for four 'logical "Crop-Power Blocs"' in post-war Europe (Taylor 1946: 323). Each longitudinal zone stretching from the Arctic and Baltic seas to the Mediterranean and Black seas would have its fair share of agricultural variety and fossil fuels. In this way there would be no material reason for coveting a neighbour's sovereignty – *ipso facto*, peace. Although Taylor does not extend his analysis beyond Europe, it is clear that his logic leads to world peace based upon autarkic pan-regions. Clearly Taylor was sceptical that technical attainment of one world would lead to world peace and emphasizes rational division over unity.

BI-POLAR OUTCOME: WEST VERSUS EAST

Taylor's suspicion of the one world scenario was to become quite general in the early post-war years. But the division when it came was not based on his geographical analysis. A much simpler geography prevailed and in Europe, for example, instead of his four economic regions just two security zones were created. It was Winston Churchill who set the political agenda on the road to bi-polarity in his famous 1946 speech in which he announced: 'From Stettin in the Baltic to Trieste in the Adriatic, an iron curtain has descended across the Continent' (James 1974: 2290). A year later in a speech to Congress, President Truman confirmed this division in his plea for aid to Greece and Turkey to prevent further Soviet advances in Eastern Europe. In addition this 'Truman Doctrine' expanded the bi-polarity by globalizing the stand-off in Europe. The world was described in terms of a great contest between 'alternative ways of life' in which the USA was committed to support 'free peoples' resisting totalitarianism everywhere (McCauley 1983: 121). Truman's speech is usually taken to mark the beginning of the geopolitical world order known as the Cold War. In 1949 the North Atlantic Treaty Organization (NATO) was created to consolidate the division of Europe and in 1950 the Korean War began confirming the global scope of the new world order.

The Cold War gave Mackinder's heartland thesis a vigorous new lease of life. No matter that it was the Heartland taking over Eastern Europe and not vice versa as Mackinder had assumed (Hall 1955: 120), the global contest pitted a great continental power against a great oceanic power in the manner of the original thesis. After the confusion of the immediate post-war years, the USSR as Heartland produced a strategic map that made sense. The Truman Doctrine justified a containment policy against the USSR whereby the US attempted to build a series of alliances to surround the enemy in the rimland. As well as NATO and a special defence treaty with Japan, the Central Treaty Organization (CENTO) and the South-East Asia Treaty Organization (SEATO) were formed with Asian states in the 1950s. On the other side the USSR was allied with the new communist China after 1949 and organized its Eastern European satellites into the Warsaw Pact to oppose NATO in 1955. With all this activity it is no wonder that the major conflicts through much of the Cold War era were concentrated in the rimland. From the original stimulus for the Truman Doctrine, the Greek civil war, through the Berlin crises, the Korean War, Israeli–Arab wars in the Middle East, India–Pakistan conflicts, India–China conflicts, to the Vietnam War, the active Cold War politics largely revolved around who controls parts of the rimland. Geopolitical theory and political practice have probably never been so congruent either before or since.

In this context it is hardly surprising that politicians were using Mackinder's ideas in their policy documents. Throughout all the Cold War era from 1948 with Ernest Bevin's warning as the British Foreign Secretary and architect of NATO that the Russians were aiming for 'physical control of the whole World Island' (Bartlett 1984: 274) to President Reagan's concern in 1988 for the Soviet Union's domination of 'the Eurasian land mass – that area of the globe often referred to as the world's heartland' (O'Tuathail 1992: 100), Cold Warriors have used Mackinderesque language in their arguments. The question arises, therefore, as to how instrumental Mackinder's ideas were for the production and reproduction of the Cold War Geopolitical World Order. G.R. Sloan (1988) suggests three possible relationships between geopolitical theories and practice: direct influence, whereby the theory provides the blueprint for action; indirect influence, where theories provide a general context to help mould action; and no influence, where any similarities between theory and practice are coincidental. In terms of Mackinder and the Cold War this relationship has alternated between all three. In the creation of the Cold War there is no evidence of any link between theory and practice. George Kennan, the originator of US containment thinking, has stated that he was unaware of Mackinder's ideas when warning of the post-war Soviet threat (Jones 1955: 497). In British Foreign Office documents Mackinderesque language is conspicuous by its absence in 1945 and 1946 (Taylor 1990: 129). But once the Cold War is in place it seems that the heartland thesis provided a useful *post hoc* rationalization for the bi-polar world that was created: in the post-war years the Cold War made the heartland thesis rather than vice versa. As a moulder of ideas Mackinder was influential. Walters (1974), for instance, has argued that the nuclear arms race was premised on the heartland as world fortress so that the only way to counter the Soviet Union's fundamental geopolitical advantage was to outgun her.

While the heartland thesis was enjoying this general success – Richard Hartshorne (1954) thought it geography's most famous contribution to world politics – it was relegated in status within geographical research itself in the aftermath of German geopolitics. Within political geography global strategic patterns were relegated almost to the position of an historical footnote. There was no longer a global political geography school, instead there were just a few individual scholars pursuing the Mackinder heritage. Two themes were emphasized. First there was a scepticism concerning how a model pitting land-power against sea-power could be relevant in the twentieth century where air-power was so important. This was an original criticism of the 1904 paper (Parker 1985) and seemed particularly pertinent given the rapid advances in air-power in the 1950s. In his review of global strategic views Stephen Jones (1955) introduced Seversky's 'airman's view of the world' which defined a new frontline between the superpowers laying across the Arctic. In this scenario old ideas about inaccessibility of the heartland were completely jettisoned. This was also a theme of Arthur Hall's (1955) review and culminated with William Bunge's (1966) observation that in the age of satellites the enemy was neither east nor west but 'up'. In his global strategic map all are vulnerable from inner space. However the concept of heartland could be salvaged by turning to a second property much less easily disposed of: the heartland as a great continental expanse has immense potential in terms of natural resources. Hall (1955) noted a new emphasis on resources and David Hooson (1964) identified 'a new Soviet heartland' precisely in these terms. It seemed that even among sceptical geographers the heartland would not go away.

During the Cold War, with the heartland thesis in the public mind as it were, the main contribution of geographers was to caution against overtly determinist interpretations of Mackinder's model. The idea of an absolute spatial pattern that defines a stable geographical stage upon which events are played out is anathema to modern geography. In world politics spaces are always contestable so political options are never totally closed down (Sprout and Sprout 1965). Hence global political geography models must be flexible not frozen forever even in the Cold War. Isaiah Bowman (1948: 130) cautioned against models of unchanging geography by insisting that although physical locations do not change 'the meaning of geographical conditions change'. The global modeller who took up this advice most conspicuously was Donald Meinig (1956). He proposed that 'cultural geography' would provide the antidote for the 'dangerous' illusion that 'geography' could determine politics. Focusing on Spykman's rimland he argued for a functional approach whereby countries would be allocated to either 'continental rimland' or 'maritime rimland' depending upon the political circumstances of the period under review – there was no fixed rimland character.

The most comprehensive revision of the heartland thesis in the Cold War was that of Saul Cohen (1973) in his *Geography and Politics in a World Divided*. He was particularly critical of the simplistic spatial thinking behind applications of the traditional model and emphasized division over unity in his geographically sensitive model. Using the regional tools of the geographer he defined a two-level world regionalization: geostrategic regions with a global importance and geopolitical regions with a subcontinental extent. There was a dual structure of just two geostrategic regions which were equivalent to Mackinder's original sea-versus land-power dichotomy – Cohen called them the Trade-Dependent Maritime World and the Eurasian Continental World – plus a set of ten geopolitical regions. The latter had a variety of positions; most were nested within the geostrategic regions, but others were between them as loci of competition called 'shatterbelts' and one was deemed an independent region, a potential new geostrategic region (South Asia). There was nothing inalienably fixed in this model and Cohen has modified his regionalization over time. In his 1982 version, for instance, Sub-Saharan Africa was removed from the Maritime World, its location in Cohen's initial regionalization, and was redesignated a shatterbelt (Cohen 1982).

Outside the academic confines of geography the word 'geopolitical' found its way back into the political lexicon in the 1970s via the memoirs of US Secretary of State Henry Kissinger (Hepple 1986). Using the term rather loosely to describe America's role as world political balancer in 1970s *détente*, Kissinger ironically opened the door to opponents of accommodation with the USSR to rediscover traditional geopolitics as a tool for attacking compromise. Colin Gray (1977: 8) is the person most associated with resurrecting the heartland bogeyman through a rediscovery of 'the master framework' that geopolitics was said to offer foreign policy: 'Looking at the world of the late 1970s the theories of Mackinder and Spykman yield a common logic for policy. The United States cannot afford to tolerate the effective control of Eurasia-Africa by the Soviet Union' (Gray 1977: 28). For a political geographer, reading this material is like stepping through a time warp. Nevertheless as a member of the influential pressure group the 'Committee on the Present Danger', Gray was instrumental in overturning *détente* by creating the 'second Cold War' and he was rewarded with a position in the Reagan administration in the early 1980s (Dalby 1990a, 1990b).

THE ALTERNATIVE BI-POLARITY: NORTH VERSUS SOUTH

Decolonization was undoubtedly one of the major world political movements of our times. In the period between 1945 and 1975 all the European empires were dismantled and nearly a hundred new states were created. The process peaked in 1960 when seventeen states achieved independence in the one year. This massive political upheaval was directly aided by the existence of the Cold War. The end of the Second World War marked the final demise of Western European states as major powers in world politics. Their role had finally been taken over by continental-scale countries, now dubbed superpowers, the USA and USSR. Both superpowers were explicitly anti-colonial in their foreign policy – the USA as the product of the first successful colonial revolt and the USSR as the product of twentieth-century anti-imperialism. In fact once Nazi Germany and Imperial Japan were disposed of, opposition to European empires was the only political position the superpowers continued to share. In this context decolonization was inevitable although it still had to be fought for locally in many instances.

Despite this link to the Cold War the new states did not fit easily into the basic East/West structure that was constructed (Taylor 1993a). Very early on they were viewed as a different political sector, a Third World to set against First (West) and Second (East) Worlds. Although the term 'Third World' was to be widely used as a synonym for poor countries, in its original French meaning of the early 1950s it was intended to be political as much as economic. The Third World represented a possible 'third way' separate from capitalism and communism and, like the Third Estate in the French Revolution, the new states could be a radical popular force to change the world. It is this political dimension of the Third World that is discussed here. Mabogunje (1980) has tried to relate the Third World as periphery to Mackinder's heartland thesis but in reality this model basically ignores the 'South'. In contrast the pan-region model explicitly includes countries of all latitudes as the basis of its geographical logic. Hence it would seem that although the heartland thesis can provide a way of viewing the East/West conflict, we should use the theory of pan-regions for understanding how the global economic periphery fits into the Cold War picture. Looked at from the South, the competition between the USA and USSR could easily appear as global rivalry to carve out two new pan-regions looking respectively to Washington, DC and Moscow.

Resistance to the designs of both superpowers surfaced quickly after the Second World War. In 1949 India convened the first Afro-Asian caucus at the United Nations and the third force was born. The crucial meeting that confirmed new states as a separate political group was held in Bandung (Indonesia) in 1955. Twenty-nine countries attended, ranging across the whole political spectrum from communist China to US-reconstructed Japan. The common concern to encourage decolonization did not translate into opposition to the Cold War itself. This came six years later in Belgrade with the launch of the Non-Aligned Movement. This non-alignment should not be confused with the traditional neutrality policies of small European states. Neutrality requires no entanglement in military treaties, non-alignment only targeted entanglement with the two superpowers. Hence no members of NATO, CENTO, SEATO or the Warsaw Pact were invited to Belgrade, with China and Japan similarly excluded. Sixteen of the Bandung conferencees joined the new movement along with ten new states.

In the same year the Non-Aligned Movement was formed the United Nations declared the 1960s the first 'Development Decade'. Proposed by the US this move marked a change in emphasis in world politics. Both superpowers claimed to provide the true development path for poorer countries to follow. But the change went beyond Cold War rivalries. With little remaining resistance to decolonization (with the exception of southern Africa), the economic question of development became a political issue with which the whole of the Third World could identify. Apart from Cuba, Latin American countries had not been represented at either Bandung or Belgrade, but with the focus on development they could join with the Afro-Asians as the third continent of the Third World. In the United Nations this led to a 'development bloc' of poorer countries holding a large majority in the General Assembly. From this position of power they attempted to set a new global agenda that addressed the economic disparities across the world.

In global political geography the 1970s represent a struggle between two world agendas. At the United Nations the image of one world was revived, with its original security meaning being transformed from military to economic security. The UN pushed through its new agenda in a series of 'world conferences' on global issues which were often critical of the superpowers – especially the USA. Conferences were held on population, food, human settlements, water, desertification, racism, technology transfer and agrarian reform, all issues crucial to the Third World and where programmes sponsored by the superpowers were failing. This movement for a new global agenda culminated in two events in the early 1980s. In 1980 the Brandt Commission published its report on world poverty and proposed replacing East–West politics with a new North–South politics. In a classic one-world statement the report argued that there was a basic mutality of interests across the 'North–South divide' and urgent action was required to prevent an economic catastrophe of global proportions. A year later in response to the Brandt report a meeting of world leaders met at Cancun, Mexico to discuss world economic disparities. These were interpreted as North versus South 'global negotiations' for a better world. But it was not to be. Leading the North in negotiations was a reluctant President Reagan backed by Mrs Thatcher, the iron lady herself. Reagan and Thatcher had a very different global agenda – the reactivation of the Cold War.

During the 1970s while the UN was sponsoring one world, the East–West bi-polar world was being recreated in the Third World – not through relatively benign development model competition but by superpower surrogate warfare (O'Loughlin 1989). The lesson the US learned from its Vietnam débâcle was to use local troops to do its bidding. Such 'Vietnamization' of war was cheap and incited little protest at home. From the 1970s onwards wars were sustained in the Third World, especially Africa, through the two superpowers sponsoring opposing sides. With neither superpower wishing to lose face through defeat of its ally, such wars could and did drag on for many years without any outcome except destruction of the country (Sidaway 1992). The classic case is probably Angola, where a potentially prosperous country has been destroyed after hosting South Africans and Cubans to fight the Cold War on their territory. This was the Cold War political agenda asserting itself over the UN's one-world ideal. In the 1980s with the 'Reagan doctrine' promising to confront radical regimes on all Third World continents by subsidizing right-wing opponents, the United Nations was essentially marginalized as a world institution. Arms sales replaced economic development aid as the main political link between North and South as each superpower attempted to ensure the predominance of its local ally in every regional arena. In this way the Cold War came to look less and less like a contemporary expression of the heartland thesis and more like a global disaster. In the South the hopes inspired by decolonization were crushed in the desolate 1980s.

POST-COLD WAR: WHICH WAY FROM BI-POLARITY?

The Cold War is over. Between 1989, with the collapse of the Communist regimes in Eastern Europe and the disintegration of the USSR in 1991, global political geography was turned upside down. The simple bi-polarity that had reasserted itself in the early 1980s simply disappeared from the world scene. Such dramatic turnarounds are termed 'geopolitical transitions', short periods of time when a particular world order is discarded and a successor order has to be constructed (Taylor 1993b). A key feature of such transitions is surprise

(Taylor 1992). Although in hindsight we can usually identify many symptoms of catastrophic change, for contemporaries their world order does not seem to be under threat. For instance from the time Gorbachev became the Soviet leader in 1985 it was obvious that changes were afoot, but these were interpreted as 'a new *détente*', a return to the less belligerent bi-polar relations of the 1970s (Kaldor *et al.* 1989). Hence in 1987 the Cold War seemed still to be 'an immutable fact of geography' with Europe 'divided into two blocs ... for evermore' (Thompson 1987: 14). The dismantling of the Berlin Wall in November 1989 symbolically and abruptly ended this world-image: the 1990s are a different political world.

Historically we are still very close to the 1989–91 transition so the successor world order to the Cold War is yet to be clearly identifiable. None the less, events in the last few years do suggest a pattern of change similar to the last transition of 1944–6. As we saw earlier this transition began with hopes of creating one world before the emergence of Cold War bi-polarity. A similar global political geography movement from unity to division has occurred since 1989.

With the demise of the communist half of the Cold War contest, the USA could reasonably declare victory. Even before the final demise of the USSR the communist superpower was withdrawing from its global role thus leaving the US as sole superpower. From a triumphalist perspective this meant that only the USA could reconstitute an ordered world. Hence President Bush's call for a 'new world order' under US leadership but with the United Nations called upon to legitimize it. With the USSR and then Russia no longer opposing the USA in the Security Council, the world organization could begin to take up once again its original role of representing one world. And on cue, as it were, Iraq invaded fellow UN member Kuwait to provide the perfect scenario for operating the 'new world order' according to the US. Almost unanimous condemnation of Iraq in the UN led to a task force from over thirty countries defeating the Iraqi army and re-establishing Kuwait sovereignty. Of course the task force was dominated by the US military and was under US leadership so the UN victory was clearly a US victory for its new world order. But just as in 1945, a reality of one world seems to be beyond our grasp. Victory in the Gulf War has come to be viewed both as a high point in US military achievement but also as one that masked other features of international relations that limited US power. Since the Gulf War it is the constraints on US ability to mould a new world order that have come to the fore and effectively destroyed the brief triumphalist one worldism created at the end of the Cold War.

The origins of the failure of the US new world order project can be found in the last years of the Cold War itself. The causes of the catastrophic decline of the USSR were clearly economic. In the last two decades of the Cold War the world economy experienced a period of stagnation which generated the restructuring of national economies throughout the world with serious consequences for ruling political elites everywhere. In the West most elected leaders found it almost impossible to get re-elected in the initial shock of economic stagnation in the 1970s, and in the South, especially in Latin America, modernizing military dictatorships were overturned. In the East the political stability of the communist regimes delayed their necessity for restructuring until the mid-1980s. Gorbachev came to power in the USSR precisely to overcome the inflexibility of the Soviet system and immediately instituted *perestroika* (the Russian word for restructuring). But it was too late, the communist economies were unable to cope with the changing world economy hence the political consequences of 1989–91. One factor behind the Soviet collapse can be called 'imperial overstretch' – the fact that USSR foreign commitments outstripped domestic resources. But this thesis was first proposed in 1987, not to describe the Soviet predicament but as a warning to the USA (Kennedy 1987). With the large increase in military expenditure by the Reagan administration the USA reasserted its leadership of the West in Cold War terms but at an economic price. We can be certain about who lost the Cold War – the USSR – but can we be sure who won it? Under the security umbrella provided by the US, both Japan and West Germany consolidated their economic successes in the 1980s. Perhaps they were the true winners of the Cold War. Certainly it soon dawned on US political leadership that the post-Cold War role of lone superpower, translated into the world's policeman, need not be such a good deal.

Both Japan and Germany were conspicuous by their absence from the Gulf War although, significantly, they made financial contributions. In all the triumphalist euphoria in the immediate aftermath of the end of the Cold War the changing nature of global political geography was overlooked. The emphasis on brute military capability that was the hallmark of the Cold War disappeared with that war order. The real message of the 1980s was the emergence of a geopolitical economy that replaced simple geopolitical equations as the basis of world power distribution (Agnew and Corbridge 1989). Hence the Gulf War, rather than ushering in a new world order through legitimation of US military leadership, can now be seen as part of the death throes of the old world order.

So what is replacing the Cold War? In the global pessimism that has set in since the Gulf War the common answer is 'new world disorder'. Sadly the evidence for such despondency is to be found throughout the world. Whatever our view of the Cold War we cannot but agree that it provided stability of different sorts across the world. In the West it gave a meaning for a political system under US leadership. For the East it firmly tamed an Eastern Europe that had traditionally been the most dangerous shatterbelt of the world. Its contribution to the South was much more problematic as we have seen, but the world order did provide a political space to play one superpower off against the other. With the exception of NATO all this stability has gone, resulting in turmoil in the successor states of the East and even greater instability in the South. No wonder that it all looks like geopolitical transition to world disorder.

No doubt most world orders look like disorders in their early years before the new patterns are apparent. Hence despite the political chaos, political geographers have not been slow to come forward with ideas about the future geopolitical world order (Nijman 1992). Not surprisingly, in this speculation the old global models are used prominently. For instance, in our discussion above we assumed that the end of the Cold War left Eurasia devoid of a superpower. But this is not accepted by all political geographers. Certainly there are enough nuclear weapons left in Russia to make us very concerned for possible domestic instability in that country. In what he calls a 'Mackinderian view', de Blij (1992: 16) asserts that 'the power potential of the Eurasian heartland remains' so that Cold War bi-polarity is still his image of the future. But continued belief in the Cold War is not a precondition for contemporary use of the heartland model. In a post-Cold War update of his ideas, Cohen (1991) maintains his two geostrategic regions – maritime and continental realms – but is more optimistic than de Blij. In the new Cohen model Eastern Europe becomes a 'Gateway Region', a zone that promotes accommodation between the geostrategic regions. In a final assertion of the salience of the heartland model, Cohen dismisses the alternative pan-region model by defining a 'Quartersphere of Marginality' covering Sub-Saharan Africa and Sub-Orinoco America. This large area is deemed economically and strategically outside the mainstream of the world system so neither the USA nor Europe is any longer interested in constructing 'southern peripheries'.

However, the focus on traditional geostrategic issues, epitomized by de Blij and Cohen, is no longer the central concern of global political geography. For most political geographers the geoeconomic view has eclipsed the geostrategic view, allowing the pan-region model to make a possible comeback (O'Loughlin and van der Wusten 1990). Although the economic integration of the contemporary world economy is widely emphasized as a new globalization, this is organized through three separate core regions – North America, Western Europe and East Asia which are becoming economically integrated, albeit to different degrees, through the North America Free Trade Association (NAFTA), the European Union (EU formerly EC or European Community) and the Asia Pacific Economic Co-operation (APEC). In addition one country and its currency dominates three even larger sections of the world as US dollar zone, German mark zone and Japanese yen zone. It will be appreciated immediately that these look very much like the pan-regions of another era. Whether these zones become established to produce a future economic sectionalism depends to a degree on moves to increase world trade and combat protectionism after the Uruguay Round of the General Agreement on Tariffs and Trade (GATT). Contemporary trade data suggests that if a new economic division does come to pass, however, it will be bi-polar in nature with North America and East Asia forming a Pacific economic region and Europe (both East and West) providing the counterpoise (O'Loughlin 1993). This is the 'trans-ideological' division that Wallerstein (1988) predicted before the end of the Cold War.

Global economic sectionalism will inevitably stimulate new resistances from the South. Political changes in the South may well be the most fundamental

347

processes occurring in the world economy today. Until recently political elites outside the economic core countries carried out policies that emulated states that have been economically and politically successful. This 'modernization' agenda is coming more and more under threat in the Third World. In particular the secular nationalist dimension of modernization is being challenged by reassertion of local cultural practices and hence rejection of Western influences (Corbridge 1993). Redefining 'universal' modernization as merely 'particularist' westernization is the geopolitical basis of the political revival of the Islamic world, but it can be found elsewhere, such as within Hinduism in India. The first major appearance of this new world agenda came with the overthrow of the arch-modernizer, the Shah of Iran, in the Islamic revolution of 1978. Clearly the new Islamic state of Iran fitted into neither side of the Cold War (which it viewed indifferently as godless communism versus USA, the 'Great Satan') but it was much more than simply an additional unaligned country. It pointed to a very different world of competing 'civilizations' that the rest of the world has yet to come to terms with (Wallerstein 1984; Huntington 1993). The geographical distribution of Islam straddling the world from Atlantic (Morocco) to Pacific (Indonesia) makes any future mobilization of Moslem peoples directly threatening to the North – which was one of the lessons of the Gulf War (Taylor 1992).

The key point is, of course, that we do not know what the future world order will look like. O'Loughlin (1992) provides 'ten scenarios for a "New World Order"' and we had best leave it there. There are many possible (certainly more than ten) global political geographies which is what makes new approaches to geopolitics so exciting and the old certainties of geographical determinism so staid and uninviting.

CONCLUSION: THE NECESSITY FOR ONE WORLD

The story we have told has been one of interplay between unity and division through global political geography as expressed in theory and practice. The political contest between these two tendencies has been an unequal one with idealist dreams of one world rudely interrupted by realist outcomes of division. Political idealists misunderstood the nature of power in naïve prescriptions for one world in 1919 (League of Nations), 1945 (United Nations), the 1970s (the Third World claim of mutuality) and early 1990s (New World Order). It might seem, therefore, that the ideal of one world is a chimera not worth pursuing. I will argue the exact opposite: one world is a political necessity because in the not-too-distant future a divided world will be a doomed world. The one world I am concerned for is the earth as a single ecosystem.

The environmental basis of geography has been used in political geography models as an input in the assumptions of geographical determinism implicit in both the heartland thesis and the theory of pan-regions. We have argued above that such assumptions are misleading when confronted with the complexity of actual political processes. But the environmental basis of geography remains of crucial concern as the output of political geography models. The obvious global environmental threat of the Cold War, a nuclear war (Bunge 1988; Elsom 1985), seems much less likely with the demise of that particular world order, but there is a slower more insidious threat that remains out of control. All geopolitics is predicated upon a competitive inter-state system. This competition takes numerous forms but at the heart is a materialist imperative which in this century has meant that all states pursue policies to maximize economic growth. Can you imagine a politician garnering many votes on a policy of reversing economic growth or even just lowering it? Modern states have been structured as 'economic growth machines' since the mercantilism of the seventeenth century and the earth has been able to cope reasonably well thus far. But cumulative economic growth aggregated to a global scale cannot continue indefinitely. A new geopolitics is required that takes the environment into consideration (Brown 1989). As a global political geography model this will have to emphasize one world since outputs (pollutants) are no respecters of political boundaries. Old-fashioned competitive global divisions are earth-threatening; they are a luxury we can no longer afford. The trick will be to construct a new balance between tendencies towards unity and division where the former is no longer overwhelmed and can do the job of sustaining the environment, while at the same time respecting the diversities of our social world which make our environment worth preserving.

REFERENCES

Agnew, J. and Corbridge, S. (1989) 'The new geopolitics: the dynamics of geopolitical disorder', in R.J. Johnston and P.J. Taylor (eds) *World in Crisis? Geographical Perspectives*, Oxford: Blackwell.

Bartlett, C.J. (1984) *The Global Conflict, 1880–1970*, London: Longman.

Bassin, M. (1987) 'Race contra space: the conflict between German *Geopolitik* and National Socialism', *Political Geography Quarterly* 6, 115–34.

de Blij, H. (1992) 'Political geography of the post Cold War', *Professional Geographer* 44, 16–19.

Bowman, I. (1942) 'Geography versus geopolitics', *Geographical Review* 32, 646–58.

Bowman, I. (1948) 'The geographical situation of the United States in relation to world politics', *Geographical Journal* 112, 129–42.

Brown, N. (1989) *New Strategy through Space*, London: Leicester University Press.

Brunn, S.D. and Mingst, K.A. (1985) 'Geopolitics', in M. Pacione (ed.) *Progress in Political Geography*, London: Croom Helm.

Bunge, W. (1966) *Theoretical Geography*, Lund: Gleerup.

Bunge, W. (1988) *Nuclear War Atlas*, Oxford: Blackwell.

Cohen, S. (1973) *Geography and Politics in a World Divided*, New York: Oxford University Press.

Cohen, S. (1982) 'A new map of global political equilibrium: a developmental approach', *Political Geography Quarterly* 1, 223–42.

Cohen, S. (1991) 'Global geopolitical change in the post Cold War era', *Annals, Association of American Geographers* 81, 551–80.

Corbridge, S. (1993) 'Colonialism, post-colonialism and the political geography of the Third World', in P.J. Taylor (ed.) *Political Geography of the Twentieth Century*, London: Belhaven.

Dalby, S. (1990a) *The Coming of the Second Cold War*, London: Pinter.

Dalby, S. (1990b) 'American security discourse: the persistence of geopolitics', *Political Geography Quarterly* 9, 171–88.

Elsom, D. (1985) 'Climatological effects of a nuclear exchange: a review', in D. Pepper and A. Jenkins (eds) *The Geography of Peace and War*, Oxford: Blackwell.

Gray, C.S. (1977) *The Geopolitics of the Nuclear Era: Heartlands, Rimlands and the Technological Revolution*, New York: Crane, Russak.

Hall, A.R. (1955) 'Mackinder and the course of events', *Annals, Association of American Geographers* 45, 109–26.

Hartshorne, R. (1954) 'Political geography', in P. James and C. Jones (eds) *American Geography – Inventory and Prospect*, Syracuse, N.Y.: Syracuse University Press.

Hepple, L. (1986) 'The revival of geopolitics', *Political Geography Quarterly* 5 (Supplement), 21–36.

Heske, H. (1987) 'Karl Haushofer: his role in German geopolitics and Nazi politics', *Political Geography Quarterly* 6, 135–44.

Hooson, D.J.M. (1964) *A New Soviet Heartland?*, New York: Van Nostrand.

Huntington, S.P. (1993) 'The clash of civilizations?', *Foreign Affairs* 72(2), 22–49.

James, R.R. (1974) *Winston Churchill: His Complete Speeches 1897–1963* (Volume VII, 1943–9), London: Chelsea House.

Jones, S.B. (1955) 'Global strategic views', *Geographical Review* 45, 492–508.

Kaldor, M., Holder, G. and Falk, R. (eds) (1989) *The New Detente*, London: Verso.

Kearns, G. (1984) 'Closed space and political practice: Frederick Jackson Turner and Halford Mackinder', *Environment and Planning D* 1, 23–34.

Kennedy, P. (1987) *The Rise and Fall of the Great Powers*, New York: Random House.

Mabogunje, A.L. (1980) 'The dynamics of centre periphery relations', *Transactions, Institute of British Geographers* NS5, 277–317.

McCauley, M. (1983) *The Origins of the Cold War*, London: Longman.

Mackinder, H.J. (1904) 'The geographical pivot of history', *Geographical Journal* 23, 421–42.

Mackinder, H.J. (1919) *Democratic Ideals and Reality*, London: Constable.

Mackinder, H.J. (1943) 'The round world and the winning of the peace', *Foreign Affairs* 21, 595–605.

MacLeish, A. (1943) 'The image of victory', in H.W. Weigert and V. Stefansson (eds) *Compass of the World*, London: Harrap.

Meinig, D.W. (1956) 'Heartland and rimland in Eurasian history', *Western Political Quarterly* 9, 553–69.

Nijman, J. (ed.) (1992) 'The political geography of the post Cold War world', *Professional Geographer* 44, 1–29.

O'Loughlin, J. (1989) 'World power competition and local conflicts in the Third World', in R.J. Johnston and P.J. Taylor (eds) *World in Crisis? Geographical Perspectives*, Oxford: Blackwell.

O'Loughlin, J. (1992) 'Ten scenarios for a "New World Order"', *Professional Geographer* 44, 22–8.

O'Loughlin, J. (1993) 'Fact or fiction? The evidence for the thesis of US relative decline, 1966–1991', in C.H. Williams (ed.) *The Political Geography of the New World Order*, London: Belhaven.

O'Loughlin, J. and van der Wusten, H. (1990) 'The political geography of panregions', *Geographical Review* 80, 1–20.

O'Tuathail, G. (1992) 'Putting Mackinder in his place: material transformations and myth', *Political Geography* 11, 100–18.

Parker, G. (1985) *Western Geopolitical Thought in the Twentieth Century*, London: Croom Helm.

Paterson, J.H. (1987) 'German geopolitics reassessed', *Political Geography Quarterly* 6, 107–14.

Sandner, G. (1989) 'Historical studies of German political geography', *Political Geography Quarterly* 8, 311–403.

Sidaway, J.D. (1992) 'Mozambique: destabilization, state, society and space', *Political Geography* 11, 239–58.

Sloan, G.R. (1988) *Geopolitics in United States Strategic Policy, 1890–1987*, Brighton: Wheatsheaf.

Sprout, H. and Sprout, M. (1965) *The Ecological Perspective on Human Affairs*, Princeton, N.J.: Princeton University Press.

Spykman, N. (1944) *The Geography of the Peace*, New York: Harcourt, Brace.

Stokes, G. (1986) *Hitler and the Quest for World Domination*, Lemington Spa: Berg.

Taylor, G. (1946) *Our Evolving Civilization. An Introduction to Geopacifics*, London: Oxford University Press.

Taylor, G. (1951) 'Geopolitics and geopacifics', in G. Taylor (ed.) *Geography in the Twentieth Century*, London: Methuen.

Taylor, P.J. (1990) *Britain and the Cold War: 1945 as Geopolitical Transition*, London: Pinter.

Taylor, P.J. (1992) 'Tribulations of transition', *Professional Geographer* 44, 10–13.

Taylor, P.J. (1993a) *Political Geography: World-Economy, Nation-State and Locality*, London: Longman.

Taylor, P.J. (1993b) 'Geopolitical world orders', in P.J. Taylor (ed.) *Political Geography of the Twentieth Century*, London: Belhaven.

Thompson, E.P. (1987) 'The rituals of enmity', in D. Smith and E.P. Thompson (eds) *Prospects for a Habitable Planet*, London: Penguin.

Wallerstein, I. (1984) *Politics in the World-Economy*, Cambridge: Cambridge University Press.

Wallerstein, I. (1988) 'European unity and its implications', in B. Hettne (ed.) *Europe: Dimensions of Peace*, London: Zed.

Walters, R.E. (1974) *The Nuclear Trap? An Escape Route*, London: Penguin.

FURTHER READING

Agnew, J. and Corbridge, S. (1989) 'The new geopolitics: the dynamics of geopolitical disorder', in R.J. Johnston and P.J. Taylor (eds) *World in Crisis? Geographical Perspectives*, Oxford: Blackwell.

Brown, N. (1989) *New Strategy through Space*, London: Leicester University Press.

Cohen, S. (1991) 'Global geopolitical change in the post Cold War era', *Annals, Association of American Geographers* 81, 551–80.

Corbridge, S. (1993) 'Colonialism, post-colonialism and the political geography of the Third World', in P.J. Taylor (ed.) *Political Geography of the Twentieth Century*, London: Belhaven Press.

Dalby, S. (1990) *The Coming of the Second Cold War*, London: Pinter.

Nijman, J. (ed.) (1992) 'The political geography of the post Cold War world', *Professional Geographer* 44, 1–29.

O'Loughlin, J. and van der Wusten, H. (1990) 'The political geography of panregions', *Geographical Review* 80, 1–20.

Parker, G. (1985) *Western Geopolitical Thought in the Twentieth Century*, London: Croom Helm.

Sloan, G.R. (1988) *Geopolitics in United States Strategic Policy, 1890–1987*, Brighton: Wheatsheaf.

Taylor, P.J. (1990) *Britain and the Cold War: 1945 as Geopolitical Transition*, London: Pinter.

Taylor, P.J. (1993) *Political Geography: World-Economy, Nation-State and Locality*, London: Longman.

Taylor, P.J. (1993) 'Geopolitical world orders', in P.J. Taylor (ed.) *Political Geography of the Twentieth Century*, London: Belhaven Press.

17

THE GEOGRAPHY OF CONFLICTS AND THE PROSPECTS FOR PEACE

John O'Loughlin

As we approach the millennium, it is easy to suppose that the world is a more peaceful and prosperous place than at any previous century's end. Such thoughts must have occurred in the late 1890s since large-scale war had not been seen since 1815. The second Industrial Revolution offered promises of bountiful material goods with incomes and leisure to match for the people of Europe and North America. Yet, within a quarter of a century, the world plunged into the most destructive conflict ever (over 40 million military and civilian deaths in two world wars). The expectations of stable peace may now be more realistic for residents of the richest countries but they seem as remote for much of the world as they did in 1895. The aim of this chapter is to describe and explain the current distribution of these persistent conflicts and to examine the prospects for a diffusion of peace to all parts of the world system. To accomplish that, we need first to see how conflict has changed over the course of the past half-century and to identify the various contemporary types of conflicts. Then, we can turn to global institutions and analyse their prospects of success.

One widely discussed scenario for the next century is that large-scale interstate wars will not happen. The 'end of history' thesis of Francis Fukuyama (1989, 1992) argues that with the world-wide acceptance of liberal democracy, based on the principles of individual freedom, free enterprise and political rights, the only law of international relations will come into play: democracies do not fight each other. While local nationalist and even small-scale interstate wars will probably continue in the world's poor periphery, they will not threaten the stability of the rest of the system or the prosperity of the rich core countries. Humankind, according to the Fukuyama thesis, has reached the endpoint of Hegelian ideological evolution to global acceptance of 'liberty and equality'. Only Islam and petty nationalisms continue to offer any alternative to liberal democracy, preventing its acceptance in much of the Third World. As democracy is expected to continue its global march, producing a kind of

'Commonmarketization', major war will be no more likely in all parts of the world than it is in Western Europe now.

We seem to have already reached a plateau in terms of Fukuyama's diffusion of democracy and interstate peace. Kenneth Boulding (1978) introduced the notion of 'stable peace' and demarcated regions where countries have no plans at all of going to war with their neighbours. Over the past thirty years, these zones of peace have indeed been stable but they have not expanded either. Outside the OECD bloc (Western Europe, Australasia, Japan and North America), civil and local wars (fought either between neighbours or regional powers) flare up with as much frequency as in the past and with more ferocity than ever before. While the geographic pattern of conflict is consistent, the world system is undergoing a geopolitical transition from its post-war, stable, superpower blocs as multipolarity is replacing bipolarity, states are fragmenting, state-making is incomplete and alliances shift with bewildering speed resulting in highly unlikely coalitions. Highly contingent and contextually influenced international relations are replacing rigid structures. And the United Nations, long a symbol of globally bipolar division and inaction, is reasserting its authority in peacemaking and peacekeeping, to the apparent chagrin of many of its erstwhile supporters. Geopolitical transitions are confusing, hectic and above all, surprising.

CHANGING DISTRIBUTION AND NATURE OF CONFLICTS

The year 1945 marks a watershed in the modern history of conflict. In the half-century to that date, most war casualties were a result of global conflicts; since then, most casualties have been the results of civilian and local wars. These trends are due to the changing nature of conflicts from large-scale, imperial conflicts, world-wide in scope and reach, between the great powers of the day to local and civil wars, which often involve regional or extra-regional (major) powers. As wars have changed, so have their impacts on the world community. Great power wars affect every region and corner of the world while wars such as the Iran–Iraq war (1980–8) or the long-running civil wars in Sudan, Mozambique and Ethiopia affect only populations in the most immediate surroundings. Apart from occasional television pictures of particular atrocities, Third World conflicts rarely touch the consciences of people in the core countries, unless their own country is directly involved. Thus, Vietnam, Iraq, Panama, Lebanon, Grenada, Korea, Libya, Somalia and Cambodia only became familiar to the American public because of the commitment of US troops to battles in these scattered locales in the years since the Second World War.

Any graph of the yearly distribution of battle-deaths (the usual measure of war severity) would clearly show the peaks of the late teens and forties of this century, corresponding to the two world wars, but it would also show secondary peaks corresponding to the major civil wars in Spain (1930s), China (1940s), Korea (1950s), and Vietnam (1960s). Since 1890, ten interstate war-years had more than 100,000 battle deaths (three since 1945), while there were fifteen war-years with civil war deaths exceeding that figure (nine since 1945) (O'Loughlin and van der Wusten 1993). Local and civil wars have become the stereotypical wars of the end of the twentieth century and they can be enormously destructive in large countries.

Geographically, 1945 also marks an important break. Until then, Europe was the locus of the most destructive conflicts with Russia/Soviet Union and Germany accounting for most battle-deaths in the great power world wars, accurately described as the 'European civil war'. Since 1947, except for former Yugoslavia after 1991, Europe has been peaceful, though some European countries (France, the United Kingdom and Soviet Union/Russia) have been involved in large-scale wars beyond their home regions. After 1945, the major conflicts have been in the Third World, especially in Asia, the Middle East, parts of Africa and Central America. The totals killed have fallen since the early years of the century, but this should not take our attention away from the suffering continuing in much of the Third World because of wars and related famines, refugees and social dislocations.

Many classifications of war have been produced. The author of one concludes that, over time,

> relatively abstract issues – self-determination, principles of political philosophy and ideology and sympathy for kin – have been increasingly important as sources of war while concrete issues such as territory and wealth have declined. One explanation for this pattern might lie in the ability of governments to create legal and other conflict-avoiding regimes for concrete-type issues while for abstract issues, regulation is difficult.
>
> (Holsti 1991: 321)

Between 1945 and 1989, 52 per cent of wars had territorial causes, 70 per cent had ideological overtones, 52 per cent involved nation-state building and 97 per cent generated human sympathy by one side or another. The most noticeable change from earlier periods has been the decline in territorial disputes as a cause of war as boundaries became more fixed after 1945 (Holsti 1991). But many territorial disputes remain as intractable as ever with rival claims by competing ethnic groups regenerating repetitive conflicts in the same places. Kashmir, Northern Ireland, Cyprus, Palestine, Eritrea, Bosnia, Armenia and Moldova are just some of many current examples of recurring territorial strife.

WHY WARS HAPPEN

There is no widely accepted theory of war simply because wars come in different shapes and sizes, and therefore war theories tend to be developed specifically for each different kind of conflict. For ease of discussion, wars can be classified as one of three kinds – imperialist, state-making and territorial – though the categories overlap and not all wars can be so classified. In general, decades of research in international relations have failed to find any consistent relationship between the characteristics of countries and their external behaviour, including their propensity to get involved in wars. In other words, there seems little point in searching for causes of wars in individual state attributes. Instead we need to look at more general conditions, especially the state of the international environment. Most research of this genre is devoted to building general models of conflict and, unlike historical study, it does not examine the precise origins, development and outcomes of specific conflicts.

As noted earlier, global wars are the most destructive, accounting for over four-fifths of all battle-deaths since 1500 and lasting, on average, twenty-seven years (Thompson 1988). But global wars between the great powers are really only a small, though highly destructive, set of imperialist wars that are fought by states to increase their economic, military, territorial or political position in the world system. While the First World War was the classic imperialist showdown between countries who had been preparing for it for decades (Germany, France, United Kingdom, Russia, Austria-Hungary and the United States), another kind of imperialist war occurs when an expansionist state wishes to remove local challengers to its power and presence or consolidate its historical pre-eminence. Thus, nineteenth-century colonial wars of expansion by the United Kingdom and France as well as recent American wars in Korea and Vietnam and Soviet engagement in Afghanistan all qualify for designation as imperialist contests.

Imperialist conflicts must be examined in relation to the relative power of the major players in the international community and the temporal trends in power. There is a widespread belief, though its empirical evidence is not particularly strong, that the most dangerous period for the world community is at the time of a power transition, as the competition to replace a declining global leader develops. Conversely, if there is one dominant global power or hegemon, like the US after the Second World War, conflicts will tend to remain local and not expand to hemispherical or global proportions. The superpower will operate as an organizer and mediator of possibly dangerous situations, while guarding and expanding its own national interests. Conversely, at a time of power transitions, as competition to replace a declining global leader intensifies, there is a much greater likelihood that great powers will take sides in local affairs in the belief that a gain for any opponent is a loss for itself. Local nationalism was the spark that set off the great imperialist war, 1914–18, as the competing powers rushed to take sides in the dispute between Austria-Hungary and Serbia.

The term 'historical-structural conditions' is frequently applied to such interpretations. We look to the structural conditions and changes of the world system for indications of impending war. In a sense, this becomes a search for economic rather than political insights, along the lines of John Maynard Keynes's (1936: 381) statement that 'wars have several causes . . . [Above all] are the economic causes of war, namely, the pressure of population and the competitive struggle for markets'. Economic and war conditions interact recursively, with uncertainty whether war trends lead or lag economic cycles. Lenin believed that capitalist states are more likely to engage in war when domestic economic conditions are difficult, reflected in high unemployment, declining industrial production and unrest. By looking abroad for markets and resources for growth, the capitalist state is likely to encounter other capitalist states pursuing the same remediation for general recession. The resulting competition can then rachet upwards resulting in a breakdown of order and general war.

In examining the related war-economy cycles, Goldstein (1988) found that the evidence does not support Lenin but instead shows that historically, economic upturn and growth leads war occurrences: in other words, the economy improves before the country goes to war. Goldstein interprets this as a capability problem because states cannot afford to fight wars if the economic base is not present to support it. And as Kennedy (1987), among others has noted, it is the state or coalition with the largest pool of economic resource that emerges victorious from prolonged war. The lasting importance of global war is that it reorganizes the world system with a new leader, new alliance and diplomatic patterns and an installation of the institutions that maintain the leader's grasp of power.

A second theoretical perspective also relates to a general model but has a state focus. According to Tilly (1985), war-making and state-making are inextricably linked. State functions and authority are promoted by the growth of its military apparatus. State borders are the results of wars or the spoils of treaties, victories or ceasefires. Taxes and other forms of revenues were required by the demands of the military and the almost incessant warfare that plagued medieval and early modern Europe. Even after a half-millennium, state-making is still not complete in Europe. Though the newly independent African and Asian states decided to retain the colonial boundaries in the 1960s, separatist and border conflicts indicate that state-making is far from complete there, too. As earlier in Europe, the powers departing from their overseas territories 'bequeathed to their successor states military forces drawn from and modelled on the repressive forces they had previously established to maintain their own local administration' (Tilly 1990: 199).

The European state model was replicated throughout the Third World and disproportionately large militaries became the most common form of state apparatus. The main role of these large militaries was not to protect the new state from external attack but to suppress any domestic opposition, especially from regional minorities unhappy about their incorporation within the new borders. Since 1959, the numbers killed per year in civil war have far exceeded the numbers killed in interstate disputes or post-colonial wars and their ratio is growing over time – from 16 per cent in the 1950s to 60 per cent in the 1960s and 89 per cent in the 1970s. From these developments, Tilly is able to conclude that:

The continued rise of war couples with a fixation on international boundaries. With a few significant exceptions, military conquests across borders have ended, states have ceased fighting each other over disputed territories, and border forces have shifted their efforts from defense against direct attack toward control of infiltration. Armies ... concentrate increasingly on repression of civilian populations

(Tilly 1990: 203)

Despite Tilly's dismissal of the importance of territory as a cause of war, it remains a subject of frequent dispute, both between and within states. Some of the most intractable conflicts of the 1990s involve the same territory claimed by two or more nationalist groups. Unlike the New World, where the state preceded the establishment of settlement and the indigenous populations were wiped out, in the Old World (Africa, Asia and Europe), state boundaries post-dated, and were superimposed on, pre-existing national groups in the flurry of 'nation-state' construction in the twentieth century. Large multinational empires (Russian, Ottoman, British, French, and Austro-Hungarian) were carved up into states with the intent of delimiting homogeneous national territories, but because of the spatial intermingling of the groups the effort has been foolhardy and produced dozens of ongoing conflicts over claims to territory in Bosnia, Croatia, Palestine, the Caucasus region, Central Asia, the Indian sub-continent and North-East Africa, among others.

Territorial conflict within state borders remains the most common form of conflict, though disputes between neighbours consistently occur and are usually part of a much larger set of grievances. Wealthy democratic states can resolve their territorial disputes through negotiation and treaties. Authoritarian, especially military, regimes display a far greater tendency to resort to invasion and violence, as in Iraq's invasion of Iran in 1980 and of Kuwait in 1990 and Argentina's invasion of the Falklands/Malvinas Islands in 1982. Certain regions of the world appear more likely to suffer territorial conflicts and they have been designated as 'shatterbelts' by Cohen (1982). South-East Asia, the Middle East, Sub-Saharan Africa are the usual shatterbelts, but Eastern Europe and Central America have also been mentioned as other possibilities. There is some evidence that shatterbelts are indeed more conflictual than areas dominated by super-powers. Their internal make-up (valuable resources, internecine and historical disputes, and complex cultural and social divisions) draws the attention of outside powers, both regional and global. These interested parties wish to draw parts of the shatterbelt into their geopolitical orbits by taking sides in local disputes. A clear example has been the extension of the Israeli–Arab dispute to the superpower arena after 1967.

A final overlapping kind of territorial conflict is 'recurring conflict', that remains unresolved by previous ceasefires and territorial partitions, as the examples of Ireland, Korea, Palestine, Germany and Vietnam indicate. Partition almost invites further antagonism, and the so-called 'geographical solution' is rarely permanent. And, to add to the uncertainty, there are still forty-three dependent territories in the world (Goertz and Diehl 1992) in which movements for independence and counter-movements for incorporation into the colonial power remain active. While one might be hopeful that large-scale interstate war is ebbing, one cannot ignore the endemic nature of conflict in much of the Third World and the recurrence of civil and neighbouring state violence.

THE GLOBAL ARMS TRADE

Since 1945, conflict, like disease, famine and abuse of human rights, has become a predominantly Third World phenomenon. Though not all regions of the Third World have seen significant conflict (South America is a notable exception), major casualties have occurred in South Asia, South-East Asia, East Asia, parts of Africa and the Middle East. By contrast, the previous centre of violent conflict, Europe, has been peaceful until the recent violent nationalist struggles in former Yugoslavia and in the former Soviet Union.

While war is now removed from Europe and North America, Boulding's region of stable peace, this does not mean that rich countries are not involved in war. The ex-colonial powers – the United Kingdom and France, as well as the United States – have been the three most active military powers since 1945. The US has fought major conflicts in Vietnam (1963–75), in Korea (1950–2) and in Iraq (1991) as well as engaging in approximately 260 other military actions, mostly in the 'Rimland', the countries on the perimeter of the ex-Soviet Union (O'Loughlin 1987). France and the UK fought dozens of wars in Africa and Asia as colonial status gradually gave way to independence. More recently, France and the UK have been the closest American allies in the conflagration in the Persian Gulf, 1990–3. By contrast, the other two large Western powers have remained removed from military actions and large military expenditures, due to constitutional restraints (ironically imposed by the victorious Allies of the Second World War) and by domestic public opinion. There exists some evidence for the belief that their economies have benefited as a result of the low military expenditures (Maull 1990).

Just because wars are no longer located in the core of the world economy, that is no reason for happiness. Many of the core countries are still intimately involved in the practice of war, even if they avoid the dispatch of troops overseas. The nexus linking the core to the poor countries is the arms trade, which has now become a major factor in the balance sheets of many countries, including the United States, France and Russia. The large proportion of the workforces of the richest countries engaged in the arms business makes the securing of the so-called 'peace dividend' much more difficult since no politician wants to be responsible for the unemployment of tens of thousands. Though the regional dependence on the (euphemistically called) 'defense industry' varies significantly in the United States (Crump and Archer 1993), the overall economic impact of the industry is so important and related to many other high-technology activities, that any tampering with this national dependence is perilous to any political figure (Markusen and Yudken 1992).

The global arms bazaar has been growing steadily since the wave of independence movements in the 1950s and though it has fluctuated according to the demands of war-materiel replacement cycles, the total world arms market between 1987 and 1991 was $175 billion, of which $107 billion went to the Third World (SIPRI 1992). Examples abound of the phenomenon of desperately poor countries spending unaffordable billions on arms imports to prop up shaky governments, defeat internal separatism and ward off threatening neighbours. Since much of the funding for the weapons purchases comes from Western and international aid and loan programmes, an unseemly circle of money from the West to the Third World state and back to the West for weapons purchases is completed. India ($17.6 billion), Afghanistan ($8.4 billion), Angola ($3.6 billion) and Pakistan ($2.3 billion) spent vast sums on weapons in the half-decade to 1991, while ranking close to the bottom of the UN Development Index.

The Middle East, however, is where the weapons spending is greatest and where the biggest threat to both regional and global stability is posed. As oil revenues have skyrocketed since 1973, weapons imports have marched in tandem. In the period 1982–92, the US sold $170 billion of weapons to Saudi Arabia, and, if one adds Egypt and the other US-allied Gulf states, the total US sales in the period were well over $200 billion (Beinin 1994). In the decade 1982–91, the Middle East accounted for $109 billion of the world total of $387 billion in imported arms.

The Middle East remains the site of some of the most dangerous conflicts in the world and became the scene of a proxy superpower competition after 1967. Nine of the top ten military spenders in the world (military expenditure as proportion of total government expenditure) are in the Middle East. Both superpowers and other major arms-sellers (China, France, the United Kingdom and Germany) armed both sides in the conflict-dyads and triads of Iran–Iraq, Israel–Syria, Israel–Egypt, Iraq–Saudi Arabia–Iran, and Israel–Jordan. By such common indices as imports per capita, military expenditure as proportion of government expenditure and soldiers per capita, the region is the most militarized in the world. The Gulf War of 1991 did not reduce the tensions but exacerbated them, as can be seen in the ensuing accelerated arms race. In the eighteen months after the end of the war, the US secured $35 billion of arms contracts from Saudi Arabia and its other allies. By the enormous attention attracted by sales such as these and by lavish promotion of the superiority of American high-technology weapons in the arms fairs after the war, the US has clearly consolidated its position as the leading arms exporter, as Russia, the leader through most of the 1980s, falls further back.

PROSPECTS FOR PEACE

There has been a sea-change in the concept of peace and its implementation over the past one hundred years. In the nineteenth century, as in previous centuries, it was felt that a state's security would be enhanced by alliance with other (preferably large) powers. The Concert of Europe, formed after the defeat of Napoleon's revolutionary France in 1815, kept the peace in Europe for a hundred years until it collapsed under German pressure for a redress of the British-led *status quo* in 1914. After the twin disasters of the world wars, attention turned to building international bodies (the League of Nations and the United Nations) representing all countries in order to avoid the ingroup/ outgroup antagonism that characterized earlier organizations. Even before 1914, international efforts to regulate the conduct of war in the Hague Conference (1907), culminating in the Geneva Conventions (1922), presaged the attempt to move beyond bloc formations to international norms and regulations. The weakness, internal strife and ineptitude of the larger bodies led to a return to the concept of bloc security after 1945 in the form of such military alliances as NATO (North Atlantic Treaty Organization) and the Warsaw Pact.

There are three underlying processes that help to promote peace-building efforts. First, most states are relatively isolated. Because of limited capability, most states interact with only a few neighbours; only great military or economic powers overcome the tyranny of distance. Of the tens of thousands of possible interstate interactions, only a tiny proportion are filled to any extent. This can be easily seen in trade, diplomatic, travel, or communication matrices. Even with the decline in travel and communication costs, most interaction is over a short distance and between neighbours. That is why, like murder, wars are disproportionately launched against acquaintances. There is, effectively, a built-in mechanism that reduces the frequency of war.

Second, as well as the 'natural tendency' of states to limit their war exposure to a handful of regional neighbours, some states have tried to remove themselves more completely from their regional environments. A policy of dissociation has characterized countries attempting autarky (economic self-sufficiency) and some forms of neutrality. At various points in their recent histories, Ireland and Switzerland have practised dissociative neutrality as a way of avoiding their neighbours' quarrels but more recently, during the Cold War years, Switzerland and Sweden represent the other model of neutrality. It involved acting as a meeting-place and a mediator for the disputes encompassing the superpowers in an effort to promote a more secure world for all.

The third reason why peace-building efforts seem to be more successful in the post-1945 period is the functional basis of international relations. States that have a lot in common, such as shared democratic traditions and wealth, will tend to maintain matching functional relationships. If good relations operate in one domain, like trade, these will tend to extend to other dimensions, like political relations. Eventually, it is expected that the process will result in more integrative relations, perhaps culminating in strong multinational organizations like the European Community. Functionalism seems to work best in democratic communities and it provides the basis for such beliefs that democracies do not fight each other (Lake 1992) and that the flag follows trade (Gasiorowski 1986). As the countries in the various regions of the world continue to build and strengthen trade relations, functionalism anticipates a more peaceful world as a result.

State control and international regulation

In the world politics literature, realism still retains its dominance. Though some other issue areas have been added to the traditional areas of concern, the assumptions underlying the field remain essentially unchanged. In realist terms, states will not co-operate in the international arena unless it is in their terms to do so. States are by far the most important units or actors in world affairs and both international agencies, like the UN, and non-governmental actors, both domestic and international, gain only as much attention and power as states allow (Russett and Starr 1992). Furthermore, since the international system is assumed to be anarchic, with all states trying actively to promote their self-interests, the outlook for any organization trying to regulate the conduct of international affairs, especially one as close to the core of a state's interests as war behaviour, is bleak.

In a sense, attempts at international regulation reflect the idealist efforts of an earlier age after the disaster of the First World War. Idealists tried to push for international safeguards against future war outbreaks by building on the co-operative aspects of human character, believing that human nature was inherently good. But the scattered seeds of good intentions floundered on the hard realist soil of international politics, and the post-Second World War period saw international organizations take on a different character. The United Nations, in contrast to its League of Nations predecessor, instituted a two-tier arrangement, with the victorious powers constituting themselves as the Security Council, where lay the real power of the organization.

By most standards, the United Nations, including its various specialized agencies, has been the most successful of all international efforts to build and maintain peace and international co-operation. Until about 1973, the General Assembly was dominated by the United States and its Western allies as newly independent countries of the Third World initially supported the position of the former colonial powers against the Soviet-led Eastern bloc. The reason lay in the changing composition of the UN. In 1945, the year it was founded, the General Assembly only had thirteen members from Asia and Africa in a total of fifty-one. In 1986, the total had risen to 159, and there were eighty-nine from Asia and Africa (Taylor 1989). Over time, the original post-colonial governments of the former colonies were replaced by more radical regimes, many of whom staked a more forceful position on the North–South debate, the redress of the growing income gap between the rich and poor countries. The Nonaligned Movement, eventually encompassing 112 members, was founded at Bandung, Indonesia in 1955 to press for economic development aid and though it did not take a position in the superpower contest, many of its members were clearly aligned with one of the superpowers. A crumb was thrown in the direction of Third World opinion with the acceptance of the Brandt report on North–South issues in 1980, though it was never followed by implementation.

In contrast to the General Assembly, the Security Council was immobile, due to the vote-blocking vetoes of the rival superpowers. The major military action of the UN in the 1945–89 period, in Korea, was made possible by Soviet absence from the critical Security Council vote to send US-led troops to the peninsula. After 1973, the situation changed dramatically as the US and its core NATO allies became increasingly isolated in the General Assembly. The issues that separated the North from the South were matters of economic development, the Palestinian question, tricky quasi-colonial conflicts in the remaining dependent territories, and US military operations world-wide. In most of the votes, the East (former Soviet bloc) and the South lined up on the same side. In the 1980s, even non-NATO core countries such as Sweden, Finland, Ireland, and Spain began to shift to the side of the South, and on many votes the US was joined by only a handful of loyal NATO and Third World allies. Next to Israel, the US became the most isolated state in the General Assembly and in one rancorous debate during the Reagan presidency, the US Ambassador told the Assembly that he would be happy to wave goodbye should the UN disband and leave New York.

Since the collapse of former Eastern Europe in 1989, the United Nations is hardly recognizable. Though Russia was given the Security Council seat of the former Soviet Union, the Cold War division has (at least temporarily) evaporated. The end of the Cold War, therefore, has opened up new vistas for the UN and has been characterized by a level of unprecedented activity in all parts of the world in the area of peacekeeeping. The large powers seem to have discovered some shared functional interests that motivate their co-operative actions and support for the UN interference in the heretofore sacrosanct domestic affairs of states from Bosnia to Iraq to Cambodia to Somalia to Angola. The Gulf War of 1991 probably never would have occurred under the old conditions of superpower competition but again, its original spark, the Iraqi invasion of Kuwait, also would not have happened in the years of Soviet influence in Saddam Hussein's Iraq. In the ancient struggle between states' realist interests and their interest in international co-operation, the decision is made easy by the temporary coincidence of the two for the US and its former adversary. Many Third World states view the loss of their erstwhile benefactor with dismay and some regard the UN as becoming no more than a servant at the bidding of the US (Amin 1994). So far, the mutual interests of the great powers have allowed an unprecedented level of UN military activity; how long this will continue is clearly dependent on the state of the relationship between the US and its former adversaries, Russia and China.

UN PEACEKEEPING IN THE POST-COLD WAR WORLD

Between 1956 and 1987, the UN mounted thirteen peacekeeping operations; in the period, 1988–92, it launched fourteen. Over time, the definition of peacemaking has changed and broadened. It is now appropriate to speak of four different kinds of UN operations (Russett 1994). Peacekeeping is still the most familiar form of military operation and is allowed by Article 40 of the UN charter, provided it is undertaken with the consent of the warring parties, without prejudice to either side and limited to conflicts between recognized states. Some sample actions in recent years have been monitoring ceasefire lines on the Golan Heights, Cyprus, the Kuwait–Iraq border, the Iran–Iraq border, and along the newly demarcated borders of Croatia, Bosnia and Serbia.

Increasingly, UN commitments are taking on the appearance of peacemaking and are confined to internal conflicts. Recent elections in Angola, Namibia and Haiti have been monitored and peace negotiatiors have brought an end to long-standing conflicts in El Salvador, Angola, Cambodia and Nicaragua. These long-term stationings are seen as a different element than the military focus of ceasefire monitorings by Russett (1994), as they are instituted before there is a full peace to keep and are more likely to be carried out by civilians than by military personnel. Peace-building is a third form of peacemaking and also involves the UN in the internal affairs of a state. Recently in Somalia, UN forces, led once again by the US, have effectively taken over the functions of a failed state and effectively provide the only authority in the country. The first aim of the troops was to produce, not monitor, a ceasefire. Other failed states, such as Angola, Sudan, Mozambique, Liberia, Zaïre and Georgia, loom as candidates for military and civilian intervention to end civil and ethnic unrest and to build the basis for stable government.

Most recently, the UN has embarked on a new line of peacekeeping, which can be called 'preventive stationing'. In Macedonia, the southern republic of former Yugoslavia, the UN has stationed monitors to watch for evidence of Serbian expansion. This is a large leap in UN operations since, potentially, there are dozens of such situations world-wide. It reflects a concern that all parts of the world and different issue-areas are subject to UN monitoring. By going so far, the UN is invading the sovereignty of individual states: weak states hardly have any say in the matter, in the face of a united Security Council and agreement by the great powers on the appropriate nature of the intervention.

The basis for intervention in the internal affairs of a sovereign member of the UN remains unclear. Article 2 (7) of the Charter states: 'Nothing contained in the present Charter shall authorise the United Nations to intervene in matters which are essentially within the domestic jurisdiction of any state.' However, Chapter VII allows the Security Council to take many sorts of actions, from the use of sanctions to military force, to maintain or restore international peace and order. Thus, if the domestic affairs of a state have an international dimension that threatens regional peace, such as large refugee flows from civil war, the Security Council is empowered to act. This clause allowed the Council to empower the US-led forces to intervene in Somalia and to provide protection to the Kurds in the aftermath of the Gulf War of 1991.

The role of the United Nations continues to expand and change, assumes less of a military character and takes on more political and social, especially humanitarian, dimensions. As its missions broaden and become less calculated and finite, the UN finds itself called upon to solve many of the world's most intractable problems. It seems to be less a last, and more of a first, resort for the world's trouble spots. While the horrible abuse of human rights in many countries has been known for years, it seems that there is an accelerated demand in the West for the international protection of minorities, government opponents and refugees. This has led to cries of double standards as some are given international protection (Kurds and Shiites in Iraq) while the plight of others continues (Palestinians, Bosnian and Kosovo Muslims). Rather than launching unilateral action that would have corresponded to the realist actions of a superpower, the US and its Western allies have turned to the multilateral UN route for intervention. In doing so, they increasingly run the risk that the UN will be viewed as an American surrogate.

Popular domestic support for the UN remains at an all-time high in many Western nations. Russett (1994) reports that 20–30 per cent more Americans prefer multilateral UN military actions, of which the US is part, to unilateral US interventions. However, the public is not willing to pay for increased UN activity, believing that current levels of spending are about right. Strong majorities in all Western European countries, Canada, Australia and the US were in favour of military action against Iraq in early 1991 because of its invasion of Kuwait. Even in Japan and Turkey, majorities came to support the attack on Iraqi forces in Kuwait and on Iraqi cities. By later 1991, majority support for the actions could be found in many Third World countries (Brazil, Costa Rica, Nigeria, Mexico) (Russett 1994). But this general support can be expected to ebb as the UN fails to take action in some places and its current deployments become more controversial. There is little doubt that the Western countries of the Security Council call the tune while other rich countries (Japan and Germany) pay the piper. It is easy to see why many in the Third World and in the former Eastern Europe believe that with the end of the Soviet constraints on the UN the West is controlling the UN as a private reserve as they have never been able to do since the end of the Second World War.

One proposal to increase global support for the UN is to broaden the composition of the Security Council, its most important body with veto power, to include both important Third World states, like India, Nigeria and Brazil, as well as the other two great powers of the twentieth century, Japan and Germany. The growing acceptance of the idea that the Security Council needs to be expanded is intended to generate greater economic support for the UN operations as well as to reduce the fear of many Third World leaders that the UN has grown so bold that it can ruin any weak state that crosses its path.

The future of the UN is examined in *Agenda for Peace*, an important policy document of its Secretary-General, Boutros Boutros-Ghali, published in 1992. In this paper, Dr Boutros-Ghali calls for member states to provide standing forces for peace enforcement units that could be called upon without delay by the Secretary-General. It would be expected that 50,000–100,000 troops might be needed, and France has already designated 2,000 troops for this purpose. Boutros-Ghali calls for full funding of the peacekeeping budget, now over $3 billion a year. The *Agenda for Peace* document states that 'There is an obvious connection between democratic practices – such as the rule of law and transparency in decision-making – and the achievement of true peace and security in any new and stable political order. These elements of good governance need to be promoted at all levels of international and national political communities.' Peace enforcement rules seem to be ever-loosening. The policy document recognizes that the UN commitment can range from truce monitoring all the way to complete trusteeship for states where central authority has failed. The plan has not yet been accepted by the member states. There remains the larger question of whether the current popularity of the UN will continue and be transformed into some sort of supernational police force or whether we will return to the anarchic world of unilateral state actions in pursuit of realist goals.

CONCLUSIONS

Realism is a pessimistic view of international relations because it believes that aggression is the normal human behaviour and therefore generates a like response. In contrast to the realist view, Fukuyama (1992) holds that the trend towards international co-operation belies that argument. As evidence, he offers such data as the 13 per cent average growth in international trade over the past quarter-century, which is a significant advance over the traditional 3 per cent growth. He believes that economic growth generates democratic tendencies since 'successful industrialization produces middle-class societies and that middle-class societies demand political participation and equality of rights' (Fukuyama 1992: 115). Democracies are 'remarkably unwarlike' and the thoroughly bourgeois character of large parts of the world precludes significant conflict.

Despite his optimistic visions, Fukuyama is forced to confront the nationalist challenge. In parts of the world that are relatively wealthy, conflict is expected to end because rich societies are supposed to recognize civil liberties and the rights of others. So, he is forced to conclude that if 'nationalism is to fade away as a political force, it must be made tolerant like religion before it' (Fukuyama 1992: 271). Yet nationalism shows no sign of ebbing and in Europe, where it was supposed to be on the wane in the face of increased integration, it has been transformed into a kind of hyper-nationalism. If anything, the attachment to place has become more local: region, rather than nation-state, is becoming the preferred scale of identity. In a way, the state is being squeezed by local and international forces and ceding autonomy in both directions. In the confusion of the post-Cold War world, people seem to be saying, 'in a time of great uncertainty, give me a place to stand and beliefs to hold'. Traditional loyalties, especially to the cultural group, offer an obvious attraction.

Most scenarios based on historical cycles anticipate global conflict again about 2030–40. For the past five hundred years, global conflicts have shown a remarkable regularity, and historical determinist models suggest that the next cycle will devolve to war at a time of economic crisis. These models, of course, take no account of the changed nature of warfare or the trend, noted in this chapter, towards international co-operation. In one fictional account, the global partnership of Japan, the European Community and the United States collapsed due to the increased economic rivalry in about 2006 (*Economist* 1992). With the US withdrawing to the Americas, the Muslim world united and China becoming a major player in world affairs, the way is paved for the kind of global collapse to conflict envisioned in 2030 by Wagar (1989).

It is unclear whether Fukuyama's global 'common market' or Wagar's global conflict will win out in the early twenty-first century. For the foreseeable future, the world will remain divided into zones of stable peace and zones of endemic conflict. The tendency for the zones to overlap and become fuzzy along their margins has a strong probability of increasing, as can be witnessed in south-eastern Europe at the present. Though domino theory could be easily dismissed as a predictive model of communist expansion in the post-1945 period, the spillover effects of conflict can emulate a series of falling dominoes. Contemporary evidence of this process can be found in the central Asian part of the former Soviet Union and in Sub-Saharan Africa.

To understand the specific direction of the spatial trends of conflict, one needs to return to the core methods of political geography. The special nature of our discipline allows us to examine the specific contextual nature of conflict as part of a larger global historical-structural process. By using a new geopolitics to tackle some of the most difficult political issues of the day, especially the causes of conflict, we can add a useful perspective to the aspatial character of peace and conflict studies. For geographers, the subject of peace and conflict is too important to continue to ignore.

REFERENCES

Amin, S. (1994) 'U.S. militarism and the new world order', pp. 218–35 in T. Mayer, J. O'Loughlin and E. Greenberg (eds) *War and its Consequences: Lessons from the Persian Gulf Conflict*, New York: HarperCollins.

Beinin, J. (1994) 'Arms transfers, the new structure of U.S. hegemony, and prospects for democratic developments in the Gulf', pp. 87–104 in T. Mayer, J. O'Loughlin and E. Greenberg (eds) *War and its Consequences: Lessons from the Persian Gulf Conflict*, New York: HarperCollins.

Boulding, K. (1962) *Stable Peace*, Austin, Tex.: University of Texas Press.

Cohen, S.B. (1982) 'A new map of global geopolitical equilibrium: a developmental approach', *Political Geography Quarterly*, 1, 223–42.

Crump, J. and Archer, J.C. (1993) 'Spatial and temporal variability in the geography of American defense outlays', *Political Geography* 12.

Economist (1992) 'Looking back from 2992: A world history, Chapter 13: the disastrous 21st century', 26 December, pp. 17–19.

Fukuyama, F. (1992) *The End of History and the Last Man*, New York: Free Press.

Gasiorowski, M. (1986) 'Economic interdependence and international conflict: some cross-national evidence', *International Studies Quarterly* 30, 23–38.

Goertz, G. and Diehl, P.F. (1992) *Territorial Changes and International Conflict*, New York: Routledge.

Goldstein, J.S. (1988) *Long Cycles: War and Prosperity in the Modern Age*, New Haven, Conn.: Yale University Press.

Holsti, K.J. (1991) *Peace and War: Armed Conflicts and the International Order, 1648–1989*, New York: Cambridge University Press.

Kennedy, P.M. (1987) *The Rise and Fall of the Great Powers: Economic Change and Military Conflict, 1500–2000*, New York: Random House.

Keynes, J.M. (1936) *The General Theory of Employment, Interest and Money*, New York: Harcourt, Brace, Jovanovich.

Lake, D.A. (1992) 'Powerful pacifists: democratic states and war', *American Political Science Review* 86, 24–37.

Markusen, A. and Yudken, J. (1992) *Dismantling the Cold War Economy*, New York: Basic Books.

Maull, H. (1990) 'Germany and Japan: the new civilian powers', *Foreign Affairs* 69(5), 91–106.

O'Loughlin, J. (1987) 'Superpower competition and the militarization of the Third World', *Journal of Geography* 86, 269–75.

O'Loughlin, J. and van der Wusten, H. (1993) 'The political geography of war and peace', pp. 63–113 in P.J. Taylor (ed.) *The Political Geography of the Twentieth Century*, London: Belhaven Press and New York: John Wiley.

Russett, B.M. (1994) 'The United Nations and multilateralism', pp. 185–97 in T. Mayer, J. O'Loughlin and E. Greenberg (eds) *War and its Consquences: Lessons from the Persian Gulf Conflict*, New York: HarperCollins.

Russett, B.M. and Starr, H. (1992) *World Politics: The Menu for Choice* (3rd edn), New York: W.H. Freeman.

SIPRI (Stockholm International Peace Research Institute) (1992) *Yearbook*, London: Oxford University Press.

Taylor, P.J. (1989) *A Brief Political Geography of States and Governments in the Twentieth Century*, University of Newcastle upon Tyne, Seminar Paper 56.

Thompson, W.R. (1988) *On Global War: Historical-Structural Approaches to World Politics*, Columbia, S.C.: University of South Carolina Press.

Tilly, C. (1985) 'Warmaking and statemaking as organized crime', pp. 169–91 in P. Evans, D. Rueschemeyer and T. Skocpol (eds) *Bring the State Back In*, New York: Cambridge University Press.

Tilly, C. (1990) *Coercion, Conflict and the European States, AD 990–1990*, Oxford: Basil Blackwell.

Wagar, W. (1989) *A Short History of the Future*, Chicago: University of Chicago Press.

FURTHER READING

Chase-Dunn, C. (1989) *Global Formations: Structure of the World-Economy*, Oxford: Basil Blackwell.

Dockrill, M. (1992) *Atlas of Twentieth Century World History*, New York: Harper-Collins.

Johnston, R.J. and Taylor, P.J. (eds) (1989) *A World in Crisis: Geographical Perspectives* (2nd edn), Oxford: Basil Blackwell.

Kidron, M. and Smith, D. (1992) *The New State of War and Peace*, New York: Simon & Schuster.

Kliot, N. and Waterman, S. (1991) *The Political Geography of Conflict and Peace*, London: Belhaven Press.

O'Loughlin, J. (ed.) (1994) *Dictionary of Geopolitics*, Westport, Conn.: Greenwood Press.

Taylor, P.J. (1992) *Political Geography: World-economy, Nation-state and Locality* (3rd edn), London: Longman.

Taylor, P.J. (ed.) (1993) *The Political Geography of the Twentieth Century*, London: Belhaven Press and New York: John Wiley & Sons.

Wallerstein, I. (1991) *Geopolitics and Geoculture*, New York: Cambridge University Press.

Ward, M.D. (ed.) (1992) *The New Geopolitics*, New York: Gordon & Breach.

Williams, C.H. (ed.) (1993) *The Geography of the Post Cold War World*, London: Belhaven Press.

18

A NEW 'GEO-ECONOMY'

Patterns, processes, problems

Peter Dicken

INTRODUCTION

The world economic map as we approach the new millennium is substantially different from that of one hundred years ago. Arguably, we now live in a more highly integrated global economy. In part, the difference between today's world economic map and that of the late nineteenth century reflects the gradual processes of evolutionary change. We can identify a succession of international divisions of labour – geographies of economic specialization – which, through time, overlap and interpenetrate, leaving traces of the old within the new. However, this evolutionary, incremental path of geo-economic change has been punctuated, from time to time, by revolutionary events in which either the pace or the direction of change is radically altered. The Second World War was one such dramatic event which, in a whole variety of ways – both political and technological – helped to reshape the global economic map. It was 'one of the great punctuation marks in human history' (Stubbs and Underhill 1994: 145); 'a great dividing line' (Scammell 1980: 2). The vast majority of the world's industrial capacity (outside North America) was destroyed and had to be rebuilt. At the same time, new technologies (including what were to become the information technologies) were created and many industrial technologies were refined and improved in the process of waging war. Necessity had, indeed, proved to be the mother of invention.

Hence, the world economic system that emerged after 1945 was, in many ways, a new beginning. It reflected both the new political realities of the post-war period and also the harsh economic and social experiences of the 1930s. The kinds of international economic institutions devised in the aftermath of the war grew out of both of these factors (Michie and Grieve-Smith 1995; Stubbs and

Underhill 1994). The major political division of the world after 1945 was essentially that between the West and the East (as shown in Chapter 16). Both major powers – the United States and the Soviet Union – made strenuous attempts to extend their spheres of influence, a process which had considerable implications for the subsequent pattern of global economic change. The Soviet bloc drew clear boundaries around itself and its Eastern European satellites and created its own economic system quite separate, at least initially, from the capitalist market economies of the West. In the West, the nature of the international economic order constructed after 1945 reflected the economic and political domination of the United States. Alone of the major industrial nations, the United States emerged from the war strengthened rather than weakened (Kennedy 1987). It had the economic and technological capacity and also the political will to create a new order, both politically and economically. The institutional basis of this new order came into being formally at an international conference at Bretton Woods, New Hampshire in 1944. It resulted in the creation of two major institutions: the International Monetary Fund (IMF) and the International Bank for Reconstruction and Development (later renamed the World Bank). The primary objective of the Bretton Woods system was to stabilize and regulate international financial transactions between nations on the basis of fixed currency exchange rates in which the US dollar played the central role. In this way it was hoped to provide the necessary financial lubricant for a reconstructed world economy. The other major pillar of the post-war international economic order was to be that of free trade. The view that the 'beggar-my-neighbour' protectionist policies of the 1930s should not be allowed to recur after the war was reflected in the establishment of a third international institution in 1947: the GATT (the General Agreement on Tariffs and Trade). The primary purpose of the GATT was to reduce tariff barriers and to regulate the use of non-tariff barriers and other forms of trade discrimination. Together, this triad of international bodies established in the immediate post-war period formed the international institutional and regulatory framework within which the rebuilt world economy evolved.

However, although the world economy was rebuilt anew after the devastation of the Second World War, such rebuilding did not take place on completely fresh, unbroken ground. The post-war 'economic architecture' did, indeed, point to a new world economic order but it was an order containing many traces of what had gone before. Indeed, there is a convincing argument to be made that an internationally integrated economy is far from new and is certainly not the unique product of recent, post-war, events. In fact, the half century prior to the First World War (1870–1913) was one in which there was

> unprecedented international integration. An open regulatory framework prevailed: short- and long-term capital movements were unsupervised, the transfer of profits was unhampered; the gold standard was at its height and encompassed almost all the major industrial countries by the period's close and most smaller agrarian nations . . .; citizenship was freely granted to immigrants; and direct political influence over the

allocation of resources was limited ... Under these conditions, markets linked a growing share of world resources and output; exports outgrew domestic output in the core capitalist countries ... and the migration of labour was unprecedented.

(Kozul-Wright 1995: 139–40)

In quantitative terms, therefore, the world economy was perhaps at least as integrated economically before 1913 as it is today – in some respects, even more so. But the nature of that integration was qualitatively very different.

Integration can take two main forms. 'Shallow' integration occurs largely through trade in goods and services and international movements of capital. 'Deep' integration extends to the level of the production of goods and services and, in addition, increases visible and invisible trade. Linkages between national economies are therefore increasingly influenced by the cross-border value adding activities within TNCs and within networks established by TNCs.

(UNCTAD 1993: 113)

International integration before 1913 – and, in fact, until only about three decades ago – was essentially shallow integration. Today, we live in a world in which deep integration, organized primarily by transnational corporations (TNCs), is becoming increasingly pervasive. The processes of deep integration are creating a more truly global, rather than merely an international, economy. But we should not fall into the trap which depicts globalization as a universally achieved state. Globalization is a set of complex processes which are extremely uneven in space, in time and across economic sectors (Dicken 1992). Very few industries are truly and completely global although most display some globalizing tendencies. The financial sector is probably the most global industry in which financial transactions take place within highly integrated markets across national boundaries.

Similarly, we need to beware of attributing the processes of globalization to a single dominant cause such as the transnational corporation or technological change. Although the TNC is arguably the single most important institution generating global economic integration it is neither a homogeneous form nor does it operate independently of other institutions, notably the state. Likewise, although technology – notably the technologies of time–space compression and of flexible production and organization – is undoubtedly a most important contributory influence on the internationalization and the globalization of economic activity we must beware of adopting a position of technological determinism. Technology is, first and foremost, an enabling agent: it makes possible new structures, new organizational and geographical arrangements of economic activities.

The aim of this chapter is to interpret the changing map of economic activity during the past fifty years in terms of the transition from shallow to deep integration. The next section provides a brief description of the changing geography of the global economy by comparing the 1945 'map' with that of the late 1990s. This is followed by discussions of the changing nature of transnational corporate activity and of the growing tendency towards regional integration between national states.

FROM ONE WORLD TO ANOTHER

The half century between the end of the Second World War in 1945 and the late 1990s witnessed a truly fundamental transformation in the geography of the world economy. On the eve of the war, in 1938, 71 per cent of world manufacturing production was concentrated in just four countries and almost 90 per cent in eleven countries (League of Nations 1945). Japan produced only 3.5 per cent of the world total. The United States accounted for almost one-third of the world total. After the war, the United States was even more clearly the hegemonic economic power, accounting for 40 per cent of world production in 1963. More striking, however, was the geographical structure of world trade. The international division of labour was essentially a simple core-periphery structure. The group of core industrial countries sold some two-thirds of its manufactured exports to the periphery and absorbed some four-fifths of the periphery's exports of primary products (League of Nations 1945). A substantial proportion of this trade was conducted within the colonial empires which (apart from that of Japan) still prevailed in 1945.

The international economic relationships which prevailed were those of shallow integration. As noted above, such integration is based upon trade in goods and services and international movements of capital. In the world constructed after 1945 (the Bretton Woods world) international capital movements were highly regulated, primarily through national systems of exchange controls. More generally, the international financial system was relatively stable, governed as it was by an agreed set of international regulatory mechanisms and the hegemonic power of the United States.

The world economy of the late 1990s is very different from the relatively simple world of half a century ago. The economic hegemony of the United States has disappeared. The power structure of the world economy is now triadic with three major regional foci (Figure 18.1): North America (centred around the United States); Europe (centred around the European Union) and East Asia (centred around Japan). The colonial empires have disappeared to be replaced by a proliferation of independent states including the elements of the former Soviet Union (Chapter 16). The relative stability of the Bretton Woods-regulated international financial system has also disappeared to be replaced by a system of virtually unregulated volatility in which electronically based financial transactions occur almost instantaneously on a 24-hour global basis across national boundaries (Corbridge et al. 1994). Peter Drucker (1986) has termed this new world of financial flows (which far exceed trade flows) the 'symbol economy' to distinguish it from the 'real economy' of the production and trade of goods and services. In effect, the two systems – of finance and of production – have become increasingly decoupled. Since the continuation of

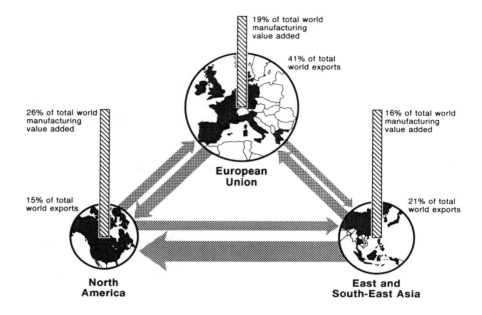

19% of total world
manufacturing
value added

41% of total
world exports

26% of total world
manufacturing
value added

16% of total world
manufacturing
value added

**European
Union**

15% of total
world exports

21% of total
world exports

**North
America**

**East and
South-East Asia**

Figure 18.1 The 'triadic' structure of the contemporary geo-economy. After Dicken (1992).

production and trade in 'real' goods and services depends fundamentally on the supply of credit, such decoupling, together with the volatility of the inter-national financial system, poses a major threat to the world economy and to people's lives and livelihoods.

The global geographical structure of the 'real economy' – the international division of labour – has also been transformed, especially since the 1960s (Dicken 1992). The simple core-periphery structure, in which manufacturing was overwhelmingly located in a small number of established industrial countries, has been fractured and reconstituted into an infinitely more complex, kaleidoscopic structure. Although world manufacturing production is still relatively concentrated geographically some entirely new centres of production have emerged, notably the so-called newly industrializing economies (NIEs) of East and South-East Asia. In fact, for the 'first wave' of Asian NIEs (South Korea, Taiwan, Singapore, Hong Kong) the term 'NIE' looks increasingly inappropriate. The production of services has also become increasingly internationalized; indeed, the simple distinction between manufacturing and services is no longer tenable. But, although some parts of the former periphery of the world economy have become virtually part of the core, large parts of the developing world (notably, but not exclusively, in Africa) remain peripheral.

The transition from one world to another over the past fifty years has been far from smooth. Time after time, the predictions of the economic forecasters have been proved wrong. During the 'golden age of growth' between the early 1950s and the early 1970s it seemed that all economic problems could be solved as the size of the cake appeared to increase inexorably. It was not to be. From around 1968, a number of important changes had been occurring which ultimately came together to throw the world economy into reverse. Commodity prices (other than oil) had been rising steeply. Labour costs in all the industrialized nations had begun to accelerate. The international financial system created at Bretton Woods eventually collapsed in 1971 when the United States moved to a floating exchange rate for the dollar and other countries followed. In 1973, the OPEC oil crisis quadrupled the price of the single most important commodity on which rapid economic growth had been based. Since 1973, economic growth has been a roller-coaster; it has also been very uneven geographically. The leading Asian NIEs, despite some minor blips, have continued to grow at rates far exceeding those of the older industrialized countries. Until the burst of its speculative bubble in the late 1980s, Japan had come to dominate several of the most important industrial sectors and, in total, to challenge the United States' economic supremacy. One major indicator of the two countries' relative standing is their strongly contrasting trade and budgetary positions: the United States has a huge deficit, Japan has a huge surplus. Although, through the auspices of the GATT, world trade in manufactures has become less and less subject to tariff barriers it has become more subject to non-tariff barriers. Trade tensions and trade frictions have intensified.

At the same time, the technological environment has become highly volatile, with both revolutionary and incremental changes in processes and products (notably more flexible technologies based on information technology) and the emergence of new generic technologies generating rapid and often unpredictable change for all participants in the global economy. Technology, then, is one of the major contributory influences reshaping the world economic map. But, as we noted earlier, we need to beware of adopting a position of technological determinism. It is all too easy to be seduced by the notion that technology makes particular outcomes inevitable or that the path of technological change is linear and sequential. What is certain, however, is that

> we have entered a new technological paradigm ... based upon information technologies ... This technological informational revolution is the backbone (although not the determinant) of all other major structural transformations:
> – It provides the basic infrastructure for the formation of a functionally inter-related world economic system.
> – It becomes a crucial factor in competitiveness and productivity for countries, regions, and companies throughout the world, ushering in a new international division of labour.
> – It allows for the simultaneous process of centralization of messages and decentralization of their reception, creating a new communication world made up at the same time of the global village and of the incommunicability of those communities

that are switched-off from the global network. Thus, an asymmetrical space of communication flows emerges.

(Castells 1992: 3–4)

In sum, the nature and the bases of international interconnectedness in the world economy have been transformed. To the long-established shallow integration based upon trade and simple capital movements has been added a more recent process of deep integration based upon production and, especially, on the strategies and structures of transnational corporations. Although it is important to emphasize that both forms of international integration continue to exist in a rather complex and geographically uneven combination, there is little doubt that the phenomenon of deep integration, generated by the spread of internationally integrated production processes, is increasingly significant. It is to these processes that we turn in the next section.

A WORLD OF TRANSNATIONAL CORPORATIONS

The business firm operating across national boundaries and controlling or co-ordinating production and distribution outside its home country is not a recent phenomenon (Dunning 1993; Jones 1993, 1994; Wilkins 1991). Its origins can be traced back to the activities of the early merchant capitalists from the fourteenth century onwards, including such trading companies as the Hanseatic League, the British and Dutch East India Companies and the Hudson's Bay Company. However, the first really major development of transnational corporation (TNC) activity occurred during the nineteenth century, with the rise of industrial capitalism. The modern TNC emerged, in particular, in the second half of the nineteenth century and, especially, after 1870. The expansion of TNC activity was part and parcel of the dramatic increase in international economic activity which occurred between 1870 and the First World War. Indeed, modern research suggests that the scale and extent of foreign direct investment (FDI) in that period were very much greater than has been assumed hitherto, with British, American, German, French and Dutch firms especially active. It was during that period that the two primary motivations for firms to extend their operation across national boundaries became apparent: to seek new markets and to acquire productive resources. Although the operations of TNCs today are infinitely more complex than those of the nineteenth and early twentieth centuries these two basic motivations still apply. However, they are manifested through increasingly intricate organizational networks, both internal and external to the firm.

Although it is now clear that the importance of TNCs in the pre-1914 period was underestimated it is equally clear that the really spectacular expansion occurred after 1945 and, especially, from the 1960s onwards (Dicken 1992). Not surprisingly, the post-war expansion was led by United States firms building upon the unprecedented strength of their domestic economy, their technological superiority and their huge reserves of investment capital. In 1960, the United

States accounted for almost 50 per cent of world foreign direct investment (the British share was 18 per cent and that of Germany and Japan a mere 1.2 per cent and 0.7 per cent respectively). By 1993, however, the geographical composition of this investment had changed dramatically (UNCTAD 1994). The United States share had declined to 25.4 per cent, Japan had become the second most important source of FDI with 12.4 per cent of the world total, followed by the United Kingdom (11.6 per cent), Germany (9.2 per cent) and France (8.6 per cent). At the same time, an increasing number of TNCs is emerging from the NICs in both Asia and Latin America.

Julius (1990) calculates that FDI growth displayed an even greater upsurge during the 1980s than in the 1960s: whereas in the 1960s FDI grew at twice the rate of GNP, in the 1980s it grew more than four times as fast (Julius 1990). After a slow-down of FDI growth during the recession of the early 1990s the upward trend has resumed. Not only has FDI been growing faster than GNP, but also it has been growing at a much faster rate than world exports, particularly since 1985. This figure alone suggests that FDI has become a more significant integrating force in the global economy than the traditional indicator of such integration: trade. Indeed, because TNCs are themselves responsible for a large proportion of international trade (much of this as intrafirm transactions), their global significance becomes even more marked.

The United Nations (UNCTAD 1994) calculates that some 37,000 parent firms controlled over 206,000 foreign affiliates in the early 1990s. But this almost certainly underestimates the true scale of TNC activity. This is because of the immense difficulty involved in identifying the bewildering variety of organizational forms involved in modern production. Simply equating TNC activity with a firm which owns overseas operations is to reveal only the tip of a very large iceberg. The activities of TNCs are expressed not only through direct ownership but also through strategic alliances and complex relationships with suppliers and customers. Entirely new organizational forms are developing at the international scale which are best conceptualized as enterprise networks. The contemporary world economy is primarily being structured and restructured through a highly intricate concatenation of production chains and organizational networks which are co-ordinated by transnational corporations operating at an increasingly international, and even a global, scale. The picture, therefore, is one of immense variety rather than of uniformity. TNCs come in a whole range of shapes and sizes. TNCs also bear the imprint of the geographical environments in which they are embedded, particularly that of their country of origin (Dicken 1994; Hu 1992; Stopford and Strange 1991).

Over the past few decades, therefore, TNCs have not only extended their operations spatially but also they have reorganized and restructured themselves into more elaborate and more flexible forms. In so doing, they have benefited from the spectacular developments in information technology. Such technologies, through their compression of time and space and through their revolutionary effects not only on how things are produced but also on how they are organized and co-ordinated, have played a critical role. Currently, TNCs are restructuring their activities in ways that involve: (1) reorganizing the co-ordination of their production chain functions in a complex realignment of internalized and externalized network relationships; (2) reorganizing the geography of their production chains internationally and, in some cases, globally; (3) transforming their relationships with their supplier firms.

Not surprisingly, therefore, the geography of transnational corporate activity is becoming increasingly complex. Certain functions tend to show a fairly predictable pattern, most notably corporate headquarters and the highest levels of decision-making which show a strong propensity to concentrate geographically into major urban centres. The most notable of these, of course, are the so-called 'global cities', of which New York, London and Tokyo are the most prominent (Friedmann 1986; Sassen 1991). Certainly these are the cities which have emerged as the major geographical control points of the international financial system. In contrast, the geographical pattern of transnational production functions is far less predictable and has been undergoing very substantial restructuring during recent years.

Figure 18.2 is a broad-brush attempt to capture some of the major alternative geographical patterns of transnational production units. Figures 18.2a and 18.2b represent opposite ends of the geographical spectrum. Globally concentrated production (Figure 18.2a), as its name suggests, is where all production of a given product or service occurs at a single location (or within a single country). Its products are then exported to overseas markets. This has been the kind of strategy followed by many Japanese companies prior to their relatively recent move towards a more dispersed production strategy. Host-market production (Figure 18.2b), conversely, is the situation where each production unit produces directly for the host nation market in which it is located. In such circumstances, the size of the production unit is limited by the size of the national market it serves.

Globally concentrated production and host-market production are the longest-established forms of geographical production orientation among TNCs. During the past three decades or so, however, a radically different form of production organization has emerged: production as part of a rationalized product or process strategy. Figures 18.2c and 18.2d show a hypothetical example of each of these. The situation in Figure 18.2c is one of product-specialization for a global or regional market. Here the units located in specific countries specialize in producing for a market which extends far beyond that of its immediate national market. The market might be global in extent (in which case the position is equivalent to that in Figure 18.2a) but, more commonly, it is regional – for example at the scale of the individual elements of the global triad discussed in the previous section. Clearly, the potential scale of production is very much greater. The trend towards the formation and strengthening of regional trading blocs, to be discussed in the next section, is a very important stimulus to this kind of production arrangement. The other type of rationalized

a. Globally concentrated
production

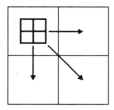

All production occurs at a
single location. Products are
exported to world markets.

b. Host-market
production

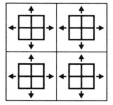

Each production unit produces
a range of products and
serves the national market in
which it is located. No sales
across national boundaries.
Individual plant size limited by
the size of the national market.

c. Product-specialization for a
global or regional market

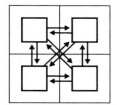

Each production unit produces
only one product for sale
throughout a regional market
of several countries. Individual
plant size very large because
of scale economies offered by
the large regional market.

d. Transnational vertical integration

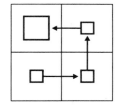

Each production unit performs
a separate part of a production
sequence. Units linked across
national boundaries in a
'chain-like' sequence – the
output of one plant is the input
of the next plant.

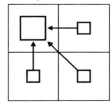

Each production unit performs
a separate operation in a
production process and ships
its output to a final assembly
plant in another country.

Figure 18.2 Some alternative ways of organizing the geography of transnational
production units. After Dicken (1992).

production strategy involves geographical specialization by process or by semi-
finished product. In Figure 18.2d this is shown as transnational vertical
integration of production. Innovations in the technology and organization of
many production processes, together with the shrinking of time–space, permit
the fragmentation of processes into separate parts and a much greater flexibility
in their geographical location at a world scale. Hence, TNCs can take advantage
of geographical variations in the costs of production, a relatively recent variant
on the resource-orientation theme (see p. 376). In this way, a genuine
internationally integrated production system becomes possible. In such a system
there is no longer necessarily any direct relationship between the location of a
TNC production unit and the domestic market in which it is located. As a result,
a highly complex geographical pattern of flows comes into being – of materials,
components, information – which is very different from what has gone before.

In effect, each unit becomes an international sourcing point for the firm as a whole.

However, the international organization of production is infinitely more complex in reality than Figure 18.2 suggests. This is not only because the diagram itself is highly simplified but also because it deals only with the organization of production within a single TNC. Each of the units themselves has to be supplied with materials and components from other independent or quasi-independent firms. One of the most significant developments of recent years has been the changing form of the relationship between firms and their suppliers. On the one hand, there has been a move towards global sourcing of some kinds of materials and components. On the other hand, there has been a move towards just-in-time sourcing in which materials and components are delivered to their point of use when they are needed rather than being held as large stocks. At first sight, this form of sourcing suggests the need for close geographical proximity between firms and their suppliers. In some cases, this is exactly what happens: geographical clusters of supplier firms oriented towards specific customers. This is the classic pattern developed within Japan. But there are dangers in assuming that such clusters are inevitable. Modern transportation and communications technologies permit considerable variation in the geographical range over which firm–supplier relationships occur. High-value, light components can be shipped by air over vast distances and still be delivered very frequently. The extreme example of such activity occurs in some of the service sectors where electronic data transmission has allowed firms to move some of their functions offshore to specialist data processing firms in low-cost countries. Although the geographical relationships between firms and suppliers are far more complex than is often suggested one tendency is clear: firms are forging much closer functional links with their major suppliers. Systems of preferred suppliers are becoming common across a very wide range of sectors and activities. Suppliers are increasingly being expected to take on far greater responsibility – and risk – in developing subsystems to be incorporated into finished products.

The international production system, therefore, is primarily articulated and co-ordinated by transnational corporations. There is still a great deal of geographical variation in how and where production takes place but the trend is towards far more complex networks of production in which firms of different kinds are co-ordinated. The networks themselves operate at different geographical scales, from the local to the global and, in so doing, integrate places into a more intricate economic structure. The scale and extent of TNC activity inevitably poses problems for states. It has often been pointed out, for example, that some TNCs are as large as, or even larger than, some national states. By extension it is then argued that TNCs are displacing states as the key economic units in the world economy. There is some truth in this. The 'global reach' of the major TNCs does, indeed, pose a threat to state autonomy simply because such firms, in effect, incorporate parts of a state's economy into their own domain. It is also true that states have become increasingly active in competing against each other to attract international investment projects to locate within their own territories. Such competitive bidding occurs at all geographical scales: global, regional, national, local. As a result, TNCs do have the capability to play off one state against another in order to win the best deal.

However, the relationship between TNCs and states is far more complex than this. The bargaining power does not inevitably lie with the TNC in all cases although it is certainly true, as Stopford and Strange (1991: 215) observe, that 'governments *as a group* have indeed lost bargaining power to the multi-nationals'. However, they also point out that 'intensifying competition among states seems to have been a more important force for weakening their bargaining power than have the changes in global competition among firms' (Stopford and Strange 1991: 215). In fact, while recognizing the undoubted power of the TNC to shape and reshape the geography of international production, it can be argued that the changing structure of the global economy is the outcome of a complex combination of processes involving both TNCs *and* states (Dicken 1992, 1994: Gordon 1988). All states attempt to regulate the economic transactions occurring across, and within, their borders. They operate regulatory structures which determine the extent to which firms may have access to markets or resources contained within state territory. They operate regulatory structures which define the rules of operation for firms operating within their particular jurisdiction. The real issue is to do with the effectiveness of such regulatory mechanisms. Here the problem is not just the extent to which TNCs have the power to circumvent state policies but also the extent to which states in general implement competitive policies between each other which undermine the effectiveness of an individual state. This is the inevitable outcome of a world in which all states have become overtly competition states.

As Taylor points out in Chapter 16, 'modern states have been structured as "economic growth machines" since the mercantilism of the seventeenth century' (p. 349). During the past few years, in fact, a 'new mercantilism' has emerged in which the leading industrial countries have become strategically competitive in a more overt way. In some respects, indeed, states have taken on some of the characteristics of business firms as they strive to develop strategies to create competitive advantage (Dicken 1994). It is no accident that one of the world's leading writers on the competitive strategies of business firms has turned his attention to 'the competitive advantage of nations' (Porter 1990). Specifically, states compete to enhance their international trading position and to capture as large a share as possible of the gains from trade. They compete to attract productive investment to build up their national production base which, in turn, enhances their competitive position. States, like firms, therefore, compete to increase material advantage *vis-à-vis* other states. Like firms – which increas-ingly engage in strategic alliances with their competitors – states also collaborate. They, too, form economic alliances with other states. As in the case of firms, inter-state collaboration can range from the simple bilateral arrangement over a single issue to the complex collaborative networks of a supranational economic bloc.

A WORLD OF REGIONAL ECONOMIC BLOCS?

Figure 18.1 portrayed an increasingly widely accepted view of the world economy: that it is crystallizing around three major regional blocs focused upon North America, Europe and East Asia. To a considerable extent, the formation of such geographical concentrations of economic activity can be explained simply in terms of the basic geographical phenomenon of proximity. Firms benefit from being close to their major markets as well as being close to their suppliers. Spatial agglomeration is a concept that can be applied at different geographical scales, not only at the micro-scale. A spatial division of labour within such a large-scale agglomeration becomes increasingly feasible and attractive to business firms, especially TNCs. In terms of the long-established principles of cumulative causation, growth begets further growth.

Thus, one of the forces leading to regional economic integration in the world economy is 'simply geographical', as a number of economists are beginning to recognize (see, for example, Thomsen 1994) as part of that discipline's recent rediscovery of economic geography (Krugman 1991; Porter 1990). It is important to make this basic statement of the significance of economic–geographical processes in the formation of regional economic blocs because, too often, they are explained solely in political terms. Too often, also, the term 'regional economic bloc' is used in a highly undiscriminating way to imply a homogeneity of form. But, as Cable (1994) points out:

> In some cases regional integration is taking place spontaneously, through market forces, impelled by the dictates of economic geography. In others, formal structures are being created in the shape of free trade areas, customs unions, common markets and various types of preferential association. These approaches – which differ considerably in their motivations and effects – have unfortunately attracted the all-purpose designation of 'trade blocs'. This term carries connotations both of confrontation and uniformity of style which may be quite wrong.
>
> (Cable 1994: 1)

Whether regional blocs are spontaneously created through the economic geography of market-driven forces (as in the Asia-Pacific) or are the deliberate creation of political agreement, their fundamental basis is that of trade. Short of complete political union, we can identify four types of politically negotiated regional economic arrangements of increasing degrees of integration:

1 *The free trade area* in which trade restrictions between member states are removed by agreement but where member states retain their individual trade policies towards non-member states.
2 *The customs union* in which member states operate both a free trade arrangement with each other and also establish a common external trade

policy (tariffs and nontariff barriers) towards non-members.

3 *The common market* in which not only are trade barriers between member states removed and a common external trade policy adopted but also the free movement of factors of production (capital, labour, etc.) between member states is permitted.

4 *The economic union* is the highest form of regional economic integration short of full-scale political union. In an economic union, not only are internal trade barriers removed, a common external tariff operated and free factor movements permitted, but also broader economic policies are harmonized and subject to supranational control.

Figure 18.3 identifies the most important regional trade groupings currently in existence. However, although many of them were initially established (at least in embryonic form) between thirty and forty years ago there was a very substantial development of such arrangements during the late 1980s and early 1990s (Cable and Henderson 1994; Gibb and Michalak 1994). In the 1980s,

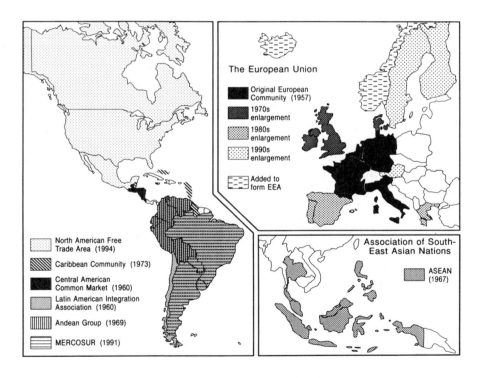

Figure 18.3 Major regional trading blocs.

membership of the European Community expanded further from nine to twelve. In 1991, an agreement was reached with all but one of the members of the European Free Trade Association (EFTA) to form a European Economic Area (EEA). Towards the end of the 1980s, the political processes aimed at the final completion of a single European Community Market by 1992 reached their climax. In 1989, the United States and Canada implemented the Canada–United States Free Trade Agreement. In 1994, the North American Free Trade Agreement came into being, whereby Mexico joined Canada and the United States in the first example of a developing country becoming fully integrated with highly developed countries. In the first few months of 1995, the regionalizing trend accelerated further. In Latin America, the MERCOSUR customs union between Argentina, Brazil, Paraguay and Uruguay was given added impetus; further development of the Andean Pact (involving Bolivia, Colombia, Ecuador, Peru and Venezuela) occurred; a free trade area involving Mexico, Colombia and Venezuela came into force. In Europe, the European Union (successor to the European Community) expanded its membership from twelve to fifteen with the incorporation of Austria, Finland and Sweden; free trade agreements were implemented between the European Union (EU) and the three Baltic republics of Estonia, Latvia and Lithuania. In the Pacific Basin – where, apart from ASEAN, there is no formal regional trade bloc – the nineteen member states of the Asia-Pacific Economic Forum agreed to remove all regional trade barriers by 2020. Other, more tentative regional trading arrangements are at least on the agenda for discussion. The United States has ambitions for a free trade area encompassing the whole of the Americas (from Anchorage to Tierra del Fuego) while it has even been proposed that there might be a Transatlantic Free Trade Agreement – a 'TAFTA' – between NAFTA and the EU.

The vast majority of the regional economic groupings shown in Figure 18.3 fall into the first two categories of the classification shown on p. 382 (the free trade area and the customs union). There is a small number of common market arrangements but only one group – the European Union – which comes close to being a true economic union. In fact, not only is there enormous variation in the scale, nature and effectiveness of these regional trade groupings but also there is, in some cases, a considerable overlap of membership of different groups, especially in Latin America. Such diversity must be borne in mind when considering the likely geographical effects of regional integration both internally (on member states and communities) and externally (on the rest of the world economy).

Regional trading blocs are essentially discriminatory in nature. Most have a strongly defensive character; they represent an attempt to gain advantages of size in trade by creating large markets for their producers and protecting them, at least in part, from outside competition. Consequently, the most important of the regional blocs – particularly the European Union and NAFTA – have a very considerable influence on patterns of world trade. They also have a major influence on the investment decisions of transnational corporations. The classic analysis of the trade effects of regional blocs (specifically of customs unions) identifies two opposing outcomes: trade creation and trade diversion:

> when trade barriers are reduced between partner countries, trade between them will increase; 'trade diversion' refers to the replacement of trade with other countries by trade with partners, while 'trade creation' refers to trade replacing home production or associated with increased consumption. There can also be 'external trade creation' if the integration process leads to a reduction in trade barriers with the rest of the world.
>
> (Smith 1994: 18)

One of the fears generated by the growth of regional trading arrangements is that each bloc will become inward-looking and will, therefore, divert trade from non-member countries. Such a fear has been articulated very strongly, for example, in the case of the Single European Market process and expressed in the popular term 'Fortress Europe'. Whether or not such fears are exaggerated there is no doubt that concern about subsequent exclusion from, or restricted access to, the European Union had a major influence on the investment decisions of transnational corporations from outside Europe. In the run-up to the target date of December 1992, there was a major upsurge in new investments in Europe by North American, Japanese and other Asian firms in particular. Some of this investment was in the form of new 'greenfield' operations, but much of it was through the acquisition of European firms as well as through the forging of strategic alliances between European and non-European firms.

Internally, the removal of trade barriers between member states of a regional bloc may facilitate the geographical reorganization and rationalization of production; the kind of situation depicted in Figure 18.2c. Again, Europe provides the most developed example of such processes although, as Thomsen (1994) argues, it was to a large extent the economic pressures created by major firms that accelerated the political process of integration. Many of the leading firms had begun to reorganize their European operations on a regional, rather than on a national, basis long before the single market process began. A good example is the American automobile manufacturer, Ford, which established a European-wide organization as early as 1967 (Dicken 1992). However, it is certainly the case that the past few years in Europe has seen a veritable tidal wave of massive organizational and geographical restructuring by both European and non-European firms. Most, if not all, European industries are becoming increasingly regionalized as firms strive to create organizational networks oriented to the European market as a whole rather than to individual national markets.

Such regionally extensive reorganization of production contributes to the redirection, and intensification, of intraregional trade. It also leads to shifting patterns of uneven development between different parts of the region. Some areas undoubtedly benefit; others undoubtedly do not. Internal regional differentiation in levels of economic and social well-being are the inevitable outcome of the processes of regional integration. These problems are intensified where there are major economic differentials between the member states of a regional bloc. Of course, in the kind of economic union exemplified by the European Union, there are redistributive mechanisms – in the form of the European Social Fund and the European Regional Development Fund – to alleviate the extremes of regional poverty and economic decline. But in the less politically developed regional blocs – including NAFTA – such redistributive mechanisms do not exist. It is not surprising, for example, that the creation of the North American Free Trade Agreement led to immense concern in both the older urban-industrial and poor rural areas of the United States over the possible flight of investment to the far lower labour cost areas of Mexico.

A final consideration in this brief discussion of the regionalization of the world economy concerns the extent to which this process is likely to continue and even intensify. Lawrence (1993: 63–4) argues that 'at least in the near future – a decade, if not longer – increased regional integration is inevitable. The critical question, though, is whether the regional arrangements will become "building blocks" in a more integrated global system or "stumbling blocks" that cause the system to fragment.' There are real concerns amongst 'outsiders' about the likely future openness of such blocs. Such concerns include fear about increased trade diversion; about possible increasing introvertedness as the regional integration process proceeds; about the possible dominance in policy-making by the more protectionist member states which might lead to new external barriers to trade and investment.

Not surprisingly, prognoses about the global implications of regional economic integration are strongly polarized. The critical issue will be the extent to which such blocs are open to external interaction. The optimistic view, as expressed by Lawrence, is that 'open regional blocs can actually promote and facilitate external liberalization, that is, trade with parties outside blocs' (Lawrence 1993: 48). He points to the history of the European Community in which

> with the noteworthy exception of agriculture ... increased regional integration among the original six members of the EC was associated with extensive participation in multilateral tariff reductions ... The European experience also demonstrated that excluded countries may have stronger incentives to liberalise in a system with emerging regional arrangements.

(Lawrence 1993: 48–9)

CONCLUSION

One hundred years ago, driven by the momentum created by the forces of industrialization, economic activity was becoming increasingly internationalized. However, it was a 'shallow' form of integration based upon trade in goods and services and international movements of capital, much of which was investment in major infrastructure projects such as railways. The dominant actor in that putatively integrating world economy was the nation–state which acted as the primary regulator of economic activity. The world economy, quite legitimately, could be regarded as a set of interlocking national economies. Trade and investment in the world economy were, literally, inter-national.

During the past fifty years, since the end of the Second World War, we have seen the beginnings of a major shift from shallow to 'deep' international economic integration; from a world economy towards a global economy. The basis of such deep integration is not merely trade but, more importantly, the geographical and functional reorganization of production, primarily through the institutional form of the transnational corporation. As a result, the spatial form of world economic activity has become far more complex. The old-established, rather simple, international division of labour of the pre-war period has been replaced by a more kaleidoscopic structure composed of extremely intricate networks of flows operating within, or co-ordinated by, transnational corporations. But this does not mean that states have lost their significance. Although their effectiveness in some areas of policy-making may have been reduced, virtually all states are increasingly behaving as competition states striving to capture a maximum share of economic welfare. One aspect of this has been the acceleration in the formation and development of regional trading blocs. Both TNCs and states, whether acting separately or in interaction with each other, operate within a world of increasing technological volatility. In particular, the information technologies have helped to transform time–space relationships although they have not, as some have argued, written the obituary of geography. Paradoxically, in a world in which the time–space relationships between places have become compressed to an unprecedented degree, the particularities and characters of individual places matter even more.

The increased complexity and volatility of the geo-economy pose serious questions about its governance (Michie and Grieve-Smith 1995). One of the major political–economic inheritances of the Second World War was the system of international financial management based upon the Bretton Woods agreement. For almost a quarter of a century, this system – founded upon the strength of the United States and the consensus of other major economic powers – provided a relatively stable environment for national and international economic activity. This was the 'golden age of growth' in the world economy; it was also the period in which the shape of the globalizing economy first began to appear. However, the collapse of Bretton Woods in the early 1970s and the subsequent series of crises, both political and economic, have replaced a relatively stable and predictable system of governance with a highly unstable and unpredictable one. This problem is especially acute in the financial system in which instantaneous switches in capital movements are now commonplace within a sector which is by far the most globally integrated of all economic sectors. But because of the centrality of finance, and especially credit, to the 'real' economy of the production of goods and services, the instability of the financial system has enormous repercussions on all other economic activities (Corbridge *et al.* 1994). In comparison, there has been rather more success in the governance of international trade although the successive 'rounds' of the GATT have been more successful in some areas than in others. For example, the general level of tariffs has fallen precipitously since 1948. On the other hand, the use of nontariff barriers to protect domestic industries and producers has proliferated. In addition, the GATT has had no influence in the areas of agriculture or the rapidly growing service industries. The latest GATT negotiations (the Uruguay Round) almost failed totally because of the problems in these areas. The new World Trade Organization (WTO), which replaces the GATT, has a very difficult task. Trade frictions remain.

In the light of these developments in the organization of economic activities, and their deep embeddedness not only in political processes but in increasingly complex geographical structures, it is far from clear what the future shape of the world economy will be. It will certainly not be the kind of homogenized, standardized world promulgated by the popular 'globalizing' writers. It will certainly continue to be a highly *uneven* world; the kind of world, indeed, that geographers are especially well equipped to understand. But geography and geographers face major challenges. Other disciplines, including the usually aloof discipline of economics, are belatedly discovering (or, in some cases, rediscovering) geography. The more significant challenge, however, is how to develop an agenda which not only enables us better to understand the rapidly changing geo-economy itself but also which makes meaningful connections with other branches of the discipline. There is, fortunately, a growing accommodation – at least in British geography – between economic, political, social and cultural geographers. So far, however, the links with physical geography and with those working on the environment remain tenuous. They need to be developed if we are fully to understand – and, therefore, to be able to solve – the fundamental problems of material existence within an increasingly complex geo-economy.

REFERENCES

Cable, V. (1994) 'The changing context of regional arrangements', in V. Cable and D. Henderson (eds) *Trade Blocs? The Future of Regional Integration*, London: Royal Institute of International Affairs.

Cable, V. and Henderson, D. (eds) (1994) *Trade Blocs? The Future of Regional Integration*, London: Royal Institute of International Affairs.

Castells, M. (1992) *European Cities, the Informational Society, and the Global Economy*, Amsterdam: Centrum voor Grootstedelijk Onderzoek.

Corbridge, S., Thrift, N. and Martin, R. (eds) (1994) *Money, Power and Space*, Oxford: Blackwell.

Dicken, P. (1992) *Global Shift: The Internationalization of Economic Activity*, London: Paul Chapman Publishing.

Dicken, P. (1994) 'Global–local tensions: firms and states in the global space-economy', *Economic Geography* 70, 101–28.

Drucker, P. (1986) 'The changed world economy', *Foreign Affairs* 64, 768–91.

Dunning, J.H. (1993) *Multinational Enterprises and the Global Economy*, Reading, Mass.: Addison-Wesley.

Friedmann, J. (1986) 'The world city hypothesis', *Development and Change* 17, 69–83.

Gibb, R. and Michalak, W. (eds) (1994) *Continental Trading Blocs: The Growth of Regionalism in the World Economy*, Chichester: John Wiley.

Gordon, D.M. (1988) 'The global economy: new edifice or crumbling foundations?', *New Left Review* 168, 24–64.

Hu, Y.-S. (1992) 'Global firms are national firms with international operations', *California Management Review* 34, 107–26.

Jones, G. (ed.) (1993) *Transnational Corporations: A Historical Perspective*, Aldershot: Avebury.

Jones, G. (ed.) (1994) *The Making of Global Enterprise*, London: Frank Cass.

Julius, DeAnne (1990) *Global Companies and Public Policy: The Growing Challenge of Foreign Direct Investment*, London: Pinter.

Kennedy, P. (1987) *The Rise and Fall of the Great Powers*, New York: Random House.

Kozul-Wright, R. (1995) 'Transnational corporations and the nation state', in J. Michie and J. Grieve-Smith (eds) *Managing the Global Economy*, Oxford: Oxford University Press.

Krugman, P. (1991) *Geography and Trade*, Cambridge, Mass.: The MIT Press.

Lawrence, R.Z. (1993) 'Futures for the world trading system and their implications for developing countries', in M. Agosin and D. Tussie (eds) *Trade and Growth: New Dilemmas for Trade Policy*, London: Macmillan.

League of Nations (1945) *Industrialization and Foreign Trade*, New York: League of Nations.

Michie, J. and Grieve-Smith, J. (eds) (1995) *Managing the Global Economy*, Oxford: Oxford University Press.

Porter, M.E. (1990) *The Competitive Advantage of Nations*, London: Macmillan.

Sassen, S. (1991) *The Global City: New York, London, Tokyo*, Princeton: Princeton University Press.

Scammell, W.M. (1980) *The International Economy Since 1945*, London: Macmillan.

Smith, A. (1994) 'The principles and practice of regional economic integration', in V. Cable and D. Henderson (eds) *Trade Blocs? The Future of Regional Integration*, London: Royal Institute of International Affairs.

Stopford, J.M. and Strange, S. (1991) *Rival States, Rival Firms: Competition for World Market Shares*, Cambridge: Cambridge University Press.

Stubbs, R. and Underhill, G.R.D. (eds) (1994) *Political Economy and the Changing Global Order*, London: Macmillan.

Thomsen, S. (1994) 'Regional integration and multinational production', in V. Cable and D. Henderson (eds) *Trade Blocs? The Future of Regional Integration*, London: Royal Institute of International Affairs.

UNCTAD (1993) *World Investment Report 1993: Transnational Corporations and Integrated International Production*, New York: United Nations.

UNCTAD (1994) *World Investment Report 1994: Employment and the Workplace*, New York: United Nations.

Wilkins, M. (ed.) (1991) *The Growth of Multinationals*, Aldershot: Avebury.

FURTHER READING

Cable, V. and Henderson, D. (eds) (1994) *Trade Blocs? The Future of Regional Integration*, London: Royal Institute of International Affairs.

Corbridge, S., Thrift, N. and Martin, R. (eds) (1994) *Money, Power and Space*, Oxford: Blackwell.

Dicken, P. (1992) *Global Shift: The Internationalization of Economic Activity*, London: Paul Chapman Publishing.

Dicken, P. (1994) 'Global–local tensions: firms and states in the global space-economy', *Economic Geography* 70, 101–28.

Dunning, J.H. (1993) *Multinational Enterprises and the Global Economy*, Reading, Mass.: Addison-Wesley.

Gibb, R. and Michalak, W. (eds) (1994) *Continental Trading Blocs: The Growth of Regionalism in the World Economy*, Chichester: John Wiley.

Michie, J. and Grieve-Smith, J. (eds) (1995) *Managing the Global Economy*, Oxford: Oxford University Press.

Stopford, J.M. and Strange, S. (1991) *Rival States, Rival Firms: Competition for World Market Shares*, Cambridge: Cambridge University Press.

UNCTAD (1993) *World Investment Report 1993: Transnational Corporations and Integrated International Production*, New York: United Nations.

19

THIRD WORLD URBANIZATION

Alan Gilbert

URBAN GROWTH IN THE THIRD WORLD

In 1950, the pattern of urbanization across the globe was rather easy to describe. Most people in developed countries lived in urban areas whereas most people in less-developed countries did not. Table 19.1 shows that in 1950 more than half of the population of the developed countries lived in urban areas but only about one in six in the so-called Third World; 160 million more urban people lived in developed countries than in the less-developed world.

By 1992, the distinction was still valid but it was much less clear-cut; the world had changed. After 1950, the pace of urban growth in most developed countries was slow; averaging only 1.6 per cent per annum. In places, improved transportation and higher incomes allowed more people to work in the city and live in the country; a process of 'counter-urbanization' had taken place (Berry 1976; Hall 1980). By contrast, the pace of urban growth in virtually every Third World country was extremely fast. The total urban population in what the World Bank calls the low- and middle-income countries, rose from 287 million in 1950 to 1.5 billion in 1992. The annual rate of urban growth for the Third World as a whole was 4.1 per cent.

As a result of these changes, the differences between developed and less-developed societies have become less marked. As Table 19.2 shows, some poorer regions show very few differences to the pattern in the advanced industrial economies. In terms of urban development, Latin America differs more from Sub-Saharan Africa than it does from the United States. Whereas the shift from rural to urban life is very advanced in Latin America, the Middle East and the Far East, most of Africa and South and South-East Asia are still far from containing an urban majority. In terms of these broad regions, South Asia is the least urbanized part of the world with only 25 per cent of the population living in towns and cities.

Clearly, there is danger in generalizing about urbanization in terms of broad

Table 19.1 Urbanization in developed and less-developed regions, 1950–90

Year	Developed countries Urban population (millions)	Per cent	Less-developed countries Urban population (millions)	Per cent
1950	447	54	287	17
1960	571	61	460	22
1970	698	67	673	25
1980	798	70	966	29
1992*	874	78	1,554	36

Sources: UNCHS (1987) and World Bank (1994).
Note: *For reasons of consistency the 1992 figures continue to include the former Soviet Union in the ranks of the developed countries. The World Bank now includes the new republics among the ranks of the less-developed countries.

Table 19.2 Urbanization by region, 1965–92

Region	Urban (%) 1965	1992	Annual growth (%) 1965–80	1980–92
Sub-Saharan Africa	14	29	5.8	5.0
East Asia	19	29	3.1	4.2
South Asia	18	25	3.9	3.5
Middle East and North Africa	35	55	4.6	4.4
Latin America	53	75	3.9	2.9
Industrial market economies	70	78	1.4	0.8
Poorer countries*	24	46	3.7	3.7
World	34	42	2.7	2.8

Sources: World Bank (1989 and 1994).
Note: *Low- and middle-income countries.

groupings of developed and less-developed countries. This is underlined by Table 19.3 which presents data on selected countries at widely different levels of development.

Table 19.3 shows that there is only a weak link between levels of per capita income and levels of urbanization. Both Venezuela and Belgium have more than 90 per cent of their people living in urban areas, Chile has a much higher level of urban development than either Ireland or Austria. In part, the expected correlation is undermined by differences in the ways in which countries measure urbanization. Few countries seem to measure urban development in the same way. In Ghana, India and the Lebanon, urban areas are localities with more than 5,000 people; in Argentina, Cuba, Denmark, Germany and Kenya they are centres with more than 2,000 people; in Peru all populated centres with 100 or

Table 19.3 Levels of urbanization, per capita income and life expectancy for selected countries, 1992

	Percentage living in urban areas	Per capita income (US$)	Life expectancy
Burundi	6	210	48
Oman	12	6,480	70
Thailand	23	1,840	69
India	26	310	61
Mozambique	30	60	44
Ireland	58	12,210	75
Austria	59	22,380	77
Peru	71	950	65
United States	76	23,240	77
Chile	85	2,730	72
Venezuela	91	2,910	70
Belgium	92	20,880	76

Source: World Bank (1994: Tables 1 and 31).

more dwellings are included; in many other places the definition is anything but clear (UNCHS 1987). In part, too, Table 19.3 provides salutary warning against putting too much belief in per capita income as a measure of development. But even when an additional indicator of the quality of life is included – life expectancy – the correlation between 'development' and urbanization does not really improve.

The differences between the countries in the so-called Third World apply not only to overall levels of urbanization but also to the pace of urban development. In certain regions, urbanization has virtually stopped whereas in others it is only now beginning to accelerate. As Table 19.2 shows, urbanization in Latin America has begun to slow right down and has slowed in most of Africa and parts of Asia. The most urbanized areas are now experiencing much slower rates of urban growth. In North Africa and the Middle East, the rate of change fell from 4.6 per cent between 1965 and 1980 to 4.4 per cent between 1980 and 1992. In Latin America and the Caribbean, the change was more dramatic, falling from 3.9 per cent to 2.9 per cent. If we exclude the highly exceptional and distorting case of China from the East Asian figures, the same trend would be true of that region. The least urbanized regions, therefore, are enduring a flood of people to the cities while the more urbanized are experiencing a slowing of rural–urban migration. In China, where migration was officially discouraged for many years, large numbers of migrants are now streaming towards the cities.

WHAT DETERMINES THE PACE AND LEVEL OF URBAN GROWTH?

The pace of urban growth in the less-developed world accelerated after 1950 because increasing numbers of people moved from the countryside to the cities. The pace of cityward migration increased for a number of reasons. First, the numbers of people born in the countryside rose rapidly as mortality rates declined throughout the Third World. Life expectancy in low-income countries jumped from 35 years in 1950 to 42 years only ten years later; by 1990, it had reached 62 years (World Bank 1981, 1992). With fertility rates remaining well above the level required for population replacement, population growth was very rapid. Between 1973 and 1980, for example, the combined population of all low- and middle-income countries increased annually by 2.3 per cent. Not surprisingly, there was considerable pressure on the available resources in rural areas. In many parts of the Third World, particularly in those areas with high population densities and with highly unequal landholding systems, it was difficult for rural populations to survive.

Second, economic growth began to create new opportunities in the urban areas. Industrialization and what we call 'modernization' led to urban productivity rising much faster than rural productivity. The result was a movement of labour from rural to urban areas. People responded in large numbers to the increasing differentials in wage levels between urban and rural areas. But it was not just wages that differed, so too did the general quality of life. Education, health care, public services and infrastructure, entertainment and consumer goods were all more freely available in the cities than in the countryside.

Third, economic growth also began to transform the rural areas. Improvements in transport and communications allowed farmers to move their produce to town much more quickly. Better transport allowed rural folk to became acquainted with modern consumer products, many soon acquired radio sets and even televisions. In the process, tastes and cultural expectations began to change. But economic growth sometimes increased levels of rural poverty. In Brazil, the rising demand for sugar, to produce gasohol which drives most Brazilian cars, led to commercial producers greatly expanding their estates and pushing tenant farmers off the land.

Differences in the level of urbanization between countries in the Third World reflect differences in the process of economic and social change. Such differences are well illustrated by the experiences of Venezuela and India. In Venezuela, the discovery of oil in the late 1920s stimulated economic growth in what had previously been an economic backwater. After the Second World War, the government managed to retain a much higher share of the country's oil revenues and began to encourage the process of industrial development. Huge sums were invested in the major cities, particularly in Caracas, in an attempt to improve living conditions and to modernize the country. With mortality rates falling, the national population was growing very rapidly. The outcome was inevitable, large numbers of people moved to the cities. During the oil boom of the 1970s, the whole process accelerated and Venezuela became a predominantly urban country.

In India, by contrast, economic and social change was very different and produced a much slower process of urbanization. First, mortality rates fell in India long after they had dropped in Venezuela. As a result, the population of India was growing at only 1.2 per cent during the 1940s compared to 3 per cent in Venezuela. Second, the industrialization process in India, while impressive in absolute terms, employed only a small proportion of the labour force. Without oil revenues to increase government revenues, there were less opportunities for urban employment than in Venezuela. The system of landholding in India was also different from that in Venezuela and was able to retain more people on the land. Consequently, the differences in the quality of life between urban and rural India were not so great as in Venezuela and consequently relatively fewer people moved from the countryside.

THE MIGRATION PROCESS

A great deal of research has investigated the nature of the migration process (Balán *et al.* 1973; Cornelius 1975; Gilbert and Gugler 1992; Portes 1972; Skeldon 1990). Much of this research was motivated by the common belief that the cities were being swamped by ill-educated migrants (Germani 1973). Such migrants supposedly moved to the cities without any clear idea of what urban conditions were like. They moved either out of a misguided belief that the city streets were paved with gold or because they were forced from the countryside by famine and pestilence. Once in the city, they were likely to be unemployed, to live in slums, and to cope only through developing some kind of 'culture of poverty' (Lewis 1959).

Research undermined this stereotype, showing that migrants were highly rational and generally made very sensible decisions about when and where to move. Most moved with full knowledge of what they were getting into. Many had previously visited the city and very few lacked contacts. By the 1960s, it was highly unusual in Latin America for someone not to have a brother or sister, uncle or aunt living in the urban areas. Relations and *paisanos* offered the potential migrant advice about whether to move or not. They also provided them with social networks and accommodation when they arrived (Gilbert and Gugler 1992; Gilbert and Ward 1986).

Perhaps the clearest indicator of the migrants' common-sense is that living conditions in the cities are generally much better than those in the rural areas. In Peru, three out of four country dwellers live in acute poverty compared to less than one quarter of urban dwellers (Fresneda *et al.* 1991). Services and infrastructure are much more abundant in the cities than in the countryside; in Colombia, only one rural family in three has access to electricity compared to almost every urban family (Gilbert 1994). In India, babies are twice as likely to

Table 19.4 Urban sex ratios in selected areas of Africa, Asia and Latin America

	*Men per 1,000 women**	*Year*
Korea	990	1985
Philippines	949	1980
Bangladesh	1,182	1981
Pakistan	1,043	1981
Egypt	1,009	1986
Turkey	1,061	1985
Ethiopia	872	1984
Sudan	1,098	1983
Zimbabwe	1,188	1982
Chile	957	1982
Colombia	931	1985

Source: Gilbert and Gugler (1992: 75–7).

Note: *The urban sex ratio has been adjusted to allow for differences in the national sex ratio. The figures listed have been divided by the national sex ratio and multiplied by one thousand. These figures are generally much less skewed than in the past. In Colombia, for example, the adjusted urban sex ratio in 1951 was 866 compared to 931 in 1985. The reason why is that increasingly the urban sex structure is determined by natural increase in the city rather than by migration.

die at birth in rural areas than in urban areas; in the Philippines urban babies have a 50 per cent higher survival rate (Pryer and Crook 1988). Ironically, more people seem to be malnourished in the Latin American countryside than in the cities (Wilkie *et al.* 1988). In sum, people move to the cities because urban living conditions are better.

Another sign of the sense and rationality of the migrants is demonstrated by figures on who actually moves to the cities. Generally, it is those in the working-age group who move, those who are better educated and trained. If the move to the cities were motivated wholly by poverty and rural push factors, then it would be the poorest of the poor who would move. Typically, however, the migrants include a majority who can read and write and who have some skill that they can use in the city. Illustrative of this rationality is the differential response by men and women to the opportunities available in the cities. In most of Africa and Asia, more men move to the cities than do women; in Latin America it tends to be women who are more likely to move (Table 19.4). This is a response to the availability of jobs. Whereas most commercial workers and many domestic servants in the Indian sub-continent are men, in Latin America most such jobs are performed by women (Brydon and Chant 1989).

A further reflection of peasant rationality has been the way in which migrants have responded to changing economic circumstances. During the 1980s, per capita income in Latin America and the Caribbean fell by more than 8 per cent. Incomes fell further in the urban areas because the cities bore the brunt of structural adjustment programmes (Gilbert 1994). Migrants responded sensibly; they stayed at home. Migration to Caracas, Mexico City, Santiago and São Paulo slowed markedly. Although there are signs that recession also reversed the direction of migration in some African countries (Mabogunje 1990), this was not true in most of the region. Drought, famine, war and civil disorder singly or in combination managed to push large numbers of people into the cities. In Africa, the decision to move was still highly rational, people just had little choice.

TO WHAT EXTENT HAS URBANIZATION BEEN A POSITIVE FORCE FOR CHANGE?

For many years the positive role of urban growth in the development process was hardly questioned. The cradles of ancient civilization lay in Athens, Babylon, Rome and urban China. More recently, the Industrial Revolution in Europe and the United States had linked economic growth with urban development. In Latin America, the most successful economies at the turn of the century had developed major urban centres. It was clear that urbanization and development were integrally linked. The countries that were most developed were generally the most highly urbanized.

This conventional wisdom began to be questioned in Latin America during the 1950s. Faced with accelerating rates of cityward migration, the urban authorities began to be alarmed about the pace of growth. Observers argued that people were moving to the cities before economic development had occurred. Their premature arrival was reflected in the proliferation of shanty towns and casual workers. Excessive numbers of migrants would eventually lead to disasters and epidemics, even to social revolution. World famous sociologists produced studies which fanned the fears. Kingsley Davis was associated with statistical analyses which seemingly proved that many less-developed countries were 'overurbanized' (Davis and Hertz 1954). Based on correlation of the relationship between per capita income, industrialization and urban development, it was clear that 'countries in the early stages of industrialization suffer an imbalance in both the size and the distribution of their urban populations implying primarily that they have a higher percentage of people living in cities than is "warranted" at their stage of economic growth' (Abu-Lughod 1965: 343).

Bert Hoselitz (1957) drew a distinction between the urbanization process as it had occurred in Europe and North America and as it was occurring in the less-developed world. Unlike the 'generative' cities of Europe which had contributed fully to economic transformation by leading the Industrial Revolution, most Third World cities were 'parasitic'. They lived off the rest of the country, sucking the resources from the countryside and producing little but ostentatious consumption and suffocating bureaucracy.

Somewhat later, Michael Lipton (1977) argued that 'urban bias' was the principal reason why 'poor people stay poor'. The process of urbanization in

Third World countries was neither efficient nor equitable. Too many resources were concentrated in urban areas and too little was invested in agriculture even though it usually employed a majority of the population. Not only was this inefficient, it was also inequitable because the rural areas contained a much higher proportion of poor people. Such a situation arose because of the political power of the urban classes and the weakness of the rural masses.

Such anti-urban views were embraced only by governments lying on the political left. In China, the Maoist vision gave much greater priority to rural development than was the case in most parts of the Third World. During the Cultural Revolution urban intellectuals were banished to the countryside to rediscover true social values. Maoist philosophy also encouraged regional self-sufficiency and rural industrialization. In Kampuchea, the Pol Pot regime took a violently anti-urban stance in 1975 and evacuated Phnom Penh. A major factor behind this high-handed approach was to increase agricultural production and to undermine the anti-revolutionary urban middle class.

Elsewhere few governments have given much credence to criticism of urban development. Indeed, their general economic policy has tended to favour urban growth. Few governments failed to adopt a policy of import-substituting industrialization (ISI), seeking to encourage indigenous development behind high tariff walls and an overvalued local exchange rate. Since most companies located in the urban areas, industrialization favoured the cities at the expense of the countryside. In addition, by distorting the exchange rate, ISI policies tended to discourage the export of agricultural products.

Generally, therefore, most governments have accepted urbanization as an inevitable accompaniment of industrialization. If urban development created problems, such as slum housing, that was a fair price to pay for the creation of a more prosperous society. Indeed, the general feeling among the majority of economists is still that because most urban-based activities are so much more productive than most kinds of agriculture, no efficient economic policy can eschew urban-oriented development policies. At times, urbanization has even been recommended as a highly effective way of accelerating the pace of economic development (Currie 1971).

Increasingly, urbanization is criticized on entirely different grounds. The left, particularly in Latin America, has suggested that urban ills are less the fault of urban development than a consequence of the distorted policies of most governments. If governments make little attempt to redistribute land and wealth and make little effort to introduce progressive systems of taxation, the urban process is bound to be distorted in favour of the rich. In so far as economic development in the Third World is dependent on decisions made in the centres of power in Washington, London, Tokyo and Frankfurt, distorted patterns of urbanization are an inevitable outcome of dependent patterns of under-development.

A not dissimilar argument is now being preached by institutions on what is sometimes regarded as the ideological right (Linn 1983; Harris 1992). The

World Bank (1991) argues that cities must be made more efficient and must contribute more to generating economic growth. Many of their faults have arisen from inappropriate macroeconomic policies. Change those policies and cities will be forced to become more efficient; an approach that has been one of the cornerstones of structural adjustment packages. In addition, higher priority needs to be given to managing urban development. Cities need to be well-planned if they are to offer rewarding lives to their inhabitants. In short, there is nothing wrong with Third World cities that some good sound economic and planning practice cannot resolve.

METROPOLITAN GROWTH

In 1950, the world had only two 'megacities' and both, London and New York, were in developed countries. By 1990, there were approximately twenty cities with more than 8 million inhabitants, of which all but six were in the Third World. Table 19.5 shows how rapidly those cities have grown, how large the biggest have become, and that most are now to be found in two of the world's poorest countries, China and India.

Not only did these cities grow to enormous size but many also became highly 'primate', dominating the national urban system and becoming many times larger than the second city in their countries. Of course, this is not something wholly unknown in the developed world, as experience in Britain and France will testify. But, in some Third World countries, the extent

Table 19.5 The growth of Third World megacities

City	Millions of inhabitants		Annual growth 1950–90
	1950	*1990*	
São Paulo	2.3	15.2	4.8
Mexico City	3.1	15.0	4.0
Shanghai	5.3	13.4	2.3
Calcutta	4.4	11.8	2.5
Bombay	2.9	11.2	3.4
Seoul	1.0	11.0	6.2
Buenos Aires	4.6	10.8	2.2
Beijing	3.9	10.8	2.6
Rio de Janeiro	2.9	9.6	3.1
Tianjin	2.9	9.4	3.5
Jakarta	2.0	9.3	3.9
Cairo	2.4	9.0	3.4
Delhi	1.4	8.8	4.7
Manila	1.5	8.5	4.4

Source: Author's calculations based on UNDIESA (1991) and Villa and Rodriguez (1994).

of distortion does seem to have become extreme. In Argentina, for example, Buenos Aires is nine times larger than Córdoba, the second city of the country; in Thailand, Bangkok is some fifty times larger than Khon Khaen; and, in Iran, Tehran has seven times the population of Esfahan (Gilbert and Gugler 1992).

Urban primacy was generally an outcome of export expansion and the channelling of most of the benefits from trade into the country's major city. The dominance of the city was influenced by the amount of foreign exchange generated and the extent to which the funds could be monopolized by the national government and elites based there. The dominance of most primate cities increased during the phase of import-substituting industrialization because most new plants were established in the largest cities. Companies were located at the centre of the largest domestic market, because infrastructure and services were generally superior to those in other parts of the country, and because the national authorities were to be found in the capital city. The latter consideration was important in so far as import substitution involved constant negotiation with the government over import licences, labour regulations and foreign exchange permission. The social interaction of leading industrialists, government officials and major politicians no doubt helped ease many kinds of commercial difficulty. As a result of import-substituting industrialization, manufacturing came to be heavily concentrated in the major cities. In Mexico, the national capital increased its share of manufacturing employment from 35 per cent in 1950 to 47 per cent in 1975.

Once underway, industrial concentration was difficult to stop. Commercial, wholesale and transport businesses developed to service manufacturing companies. More jobs in the large cities increased the level of prosperity which widened the domestic market and attracted more businesses. Increasing commercial and manufacturing activity in turn increased local tax revenues, which enabled the urban authorities to improve infrastructure and services which attracted more companies. The emergence of trade unions put pressure on national governments to improve living conditions for industrial and government workers. The introduction of food and transport subsidies, social security systems, and better health and education facilities were a direct outcome of this development. It took a real shock to the economic system to change this balance of advantage.

Such a shock occurred during the 1980s. As a result, the pace of growth of many of the Third World's largest cities began to slow. In the 1980s, Mexico City probably grew by only 3 per cent per annum (compared to 5.6 per cent between 1940 and 1980). In addition, the share of the urban population living in the largest cities of Africa, Asia and Latin America fell between 1980 and 1990 in twenty-six countries and rose in only fourteen (World Bank 1992, 1989); the comparable numbers between 1960 and 1980 were nineteen and twenty-one respectively. While this is admittedly a very crude measure, it is a tentative sign that something important has occurred in the national settlement systems of many Third World countries (Gilbert 1993).

During the 1980s, there was a major shift in the nature of the development model. Import substitution gave way to trade liberalization and an explicit strategy to develop manufacturing exports. The change removed many of the advantages of the major cities. Trade liberalization allowed imported products to undermine the previously highly protected domestic markets of companies based in the primate cities. The flood of manufactured imports unleashed by the lowering of import tariffs in many middle-income countries has sometimes devastated local industry. Many companies that boomed under the old regime have been put out of business under the new. In Mexico, it is estimated that the national capital may have lost one-quarter of all its manufacturing jobs between 1980 and 1988 (Garza 1991). In Mexico, as well as in China and Korea, the shift towards an export-oriented development model has encouraged a more deconcentrated pattern of industrial growth (Suárez-Villa and Han 1991). In Mexico, export-oriented industrialization has been concentrated along or close to the US–Mexico border (Shaiken 1990; Sklair 1992; South 1990). In China, while foreign companies can invest in almost any part of the country, most foreign investment has been concentrated in the largest cities and in the Special Economic Zones (SEZs) (Phillips and Yeh 1990).

Austerity programmes have also had their effect on many large metropolitan areas. The domestic market has shrunk as the twin forces of the recession and structural adjustment policies have taken hold. The largest cities have borne the brunt of cut-backs in government spending, programmes of privatization and reduced government intervention.

Of course, the slowing of metropolitan growth is not only a consequence of the changing economic model. It also owes something to the rate of natural increase. In Latin America and East Asia, urban growth rates have fallen in part because fertility rates have fallen substantially. In the early 1960s, the average Brazilian woman had 6.2 children during her lifetime, by the early 1980s the number of offspring had fallen to 3.8 (Merrick 1986). Fertility decline has been particularly rapid in the cities where women have easier access to modern contraceptive methods and are more likely to be engaged in paid employment. And, notwithstanding the vast numbers of women of child-bearing age living in the city, falling fertility rates are beginning to cut the urban growth rate.

In places, too, the slowing of urban growth is also the outcome of branch or assembly plants moving out of the central city and relocating in the nearby hinterland (Amsden 1991; Gwynne 1990; Richardson 1989; Storper 1991). The trend is occurring partly because urban diseconomies in the major cities have increased to such a point that it has persuaded companies to move out. Deconcentration has been encouraged by improvements in transport and communications which mean that contact with the megacity can be easily maintained. Companies have chosen to reduce the disadvantages of their site costs while retaining their locational advantages.

THE FUTURE

Levels of urbanization will continue to rise in most parts of the Third World but at very different rates. The pace of urban growth is already slowing in the more affluent countries of Latin America and Asia, partly because fertility rates are falling in those countries and partly because a majority of the national population already lives in cities (Gilbert 1994). In Latin America, natural increase has long been the most significant contributor to metropolitan growth, massive cityward migration was the critical ingredient only during the 1940s and 1950s (Merrick 1986).

In Africa and South Asia, however, the pace of urban growth will continue so long as those societies move towards urban majorities. There is so much impoverishment in the countryside and, in many parts of rural Africa, so much hunger, that millions are bound to move towards the cities. How the cities will cope is open to question, but it is difficult to believe that living conditions can get so bad that they will be worse than life in the country.

Despite the flood of people into African and South Asian cities, few governments will attempt to limit migration to the cities. In the past such controls have not been a great success (Skeldon 1990). When tried in non-authoritarian societies controls had little effect and the consequences, in the form of corruption and intimidation of minority groups, were highly undesirable (Gilbert and Gugler 1992). The three countries that were most successful in limiting cityward migration, China, Cuba and South Africa, no longer do so. Their governments have realized that under their new economic and political circumstances there is no way that such a policy can be effective. In China, the huge mass of impoverished rural people who wish to move to the cities makes rapid urban development inevitable. In South Africa, the influx controls adopted by the apartheid regime have been swept away (Smith 1992).

There is little point in governments trying to limit population movement. What is necessary is to eliminate manifest forms of 'urban bias'. Governments should remove distortions in the economic model which favour urban groups at the expense of those who live in the countryside. Farmers should receive the full market price for their produce. Governments should provide adequate infrastructure and services throughout the country. Under more balanced urban–rural regimes, migrants can then respond sensibly to real differences in the quality of urban and rural life. If distortions in social prices between urban and rural areas are removed, then the cities will grow less rapidly.

Of course, any potential improvement in Third World urban living conditions will be hampered by the tendency for many impoverished people to try to escape from the penury of the countryside. This is a particular problem in China and Sub-Saharan Africa but may even occur in relatively urbanized areas should international competition destroy domestic production systems. In Mexico, for example, imports of cheap cereals from the United States may

402

undermine the demand for the maize produced by peasant farmers (Browne 1994).

But, there is only partial reason to believe that living conditions in the cities will deteriorate. In Latin America and Asia, there are signs that governments are making sensible changes to their economic and social policies. Nevertheless, more must be done to help the poor while looking for ways to earn more money in the international marketplace. Urbanization in the Third World is potentially manageable and only in Africa does the prospect of continued urbanization really seem problematic (Stren and White 1989). Where there is economic decline, famine and war, the urban future is bound to be bleak.

As the population of Third World countries continues to grow, some very large cities will develop. The megacities of the future will include at least six in the Indian sub-continent (Bombay, Calcutta, Delhi, Dhaka, Karachi and Madras) and many more in China (Beijing, Shanghai, Tianjin, Guangzhou and Shenyang). However, the goliaths of Latin America will be much smaller than was once feared; the largest cities in that continent are today growing rather slowly.

The form of these giant cities will also change. Polynuclear growth and 'polarization reversal' seem to be producing conurbations in the place of monocentric cities (Richardson 1989; Gilbert and Gugler 1992). While urban sprawl is not without its horrors, it rarely produces urban breakdown. In economies with increasing numbers of road vehicles, the urban future lies in huge, spatially dispersed megacities.

Governments are no longer anxious to control megacity growth. Since the urban planning literature could never make an entirely convincing case either for or against very large cities, there is little reason to take direct measures against metropolitan growth (Gilbert and Gugler 1992; Henderson 1991; Richardson 1989). And, with the recent slowing of megacity growth in Latin America, no action is actually necessary; the great irony of 'polarization reversal' is that regional policy contributed very little to it. Deconcentration occurred when regional planning was at its weakest (Gilbert 1993). Few governments in the debt-ridden Third World had enough funds to invest heavily in infrastructure in the poorer regions or to offer incentives to industrialists to locate in the periphery.

The future will certainly see stronger links developing between urban and rural areas. As transportation facilities improve, the divide between urban and rural life will break down. Urbanization is increasingly penetrating deeply into the countryside of South-East Asia (McGee 1989; Brookfield et al. 1991; Goldstein 1990). This trend is likely to become more widespread and it will be increasingly necessary to question the distinction that we normally draw between town and country. Regional policies will then be more necessary than urban policies.

The critical question is whether governments will be able to manage their cities better than most have in the past. Governments need to control urban diseconomies and to improve living standards (World Bank 1991). Air and water pollution should be discouraged by the introduction of polluter-pay principles. Only if infrastructure and services can be provided for the poor, will both productivity and health standards rise (Hardoy *et al.* 1990).

At first sight, the answer to that question is unclear. At one level, there appears to be a trend towards greater democracy. More governments are freely elected in the Third World than for many a year and there is a strong tendency for more governments to engage in popular consultation. Governments are also delegating more responsibility to the local authorities; a belated reaction against the disadvantages of centralism and also a response to more forceful social and environmental activism. Unfortunately, the trend may also be a response to the fact that so many central governments lack the financial resources to implement policies from above. After all, decentralization is an excellent way of saving central government expenditure on providing local infrastructure and services. Privatization and decentralization are flavour of the decade, not just because of the past failure of alternative approaches but also because they save money!

Political decentralization and the delegation of powers to tax may improve the quality of government but few cities can generate enough taxes to make real improvements. Only the more affluent cities will be able to invest heavily in urban infrastructure. It is difficult to be optimistic about what will happen in the countryside and in those urban areas where there is limited economic activity (Stren and White 1989).

Ironically, one of the major forces towards centralization in the past was the manifest inability of local authorities to do anything properly. In Latin America, local government was renowned both for its political chicanery and its incompetence. Political games cost little and were permitted because few people complained openly. The danger is that in the future administrative decentralization will produce new forms of government incompetence. Local authorities in poor areas will have too small a tax base, too few competent administrators and too little technical know-how actually to achieve very much. If that scenario is accurate, decentralization will benefit the largest cities but not small towns and villages.

In sum, the future of Third World cities is unclear because the so-called Third World is so diverse. In places, urbanization will continue to provide people with better lives than they could conceivably have in the countryside. Elsewhere, urban life will provide a totally unsatisfactory refuge against the ravages of life in the rural areas. The outcome will vary because although the Third World contains some countries on the verge of real economic success it also has too many basket cases. The urban future will differ across the Third World because while fertility rates are declining fast in some places, many countries have still to enter the later stages of the 'demographic transition'. Optimism is possible because the Third World contains cities which are relatively well run, but pessimism is also in order because there are far too many cities that are managed appallingly. We can only hope that in a couple of decades there are more success stories than outright failures.

REFERENCES

Abu-Lughod, J.L. (1965) 'Urbanization in Egypt: present state and future prospects', *Economic Development and Cultural Change* 13, 313–43.

Amsden, A.H. (1991) 'Big business and urban congestion in Taiwan: the origins of small enterprise and regionally decentralized industry (respectively)', *World Development* 19, 1121–35.

Balán, J., Browning, H.L. and Jelín, E. (1973) *Men in a Developing Society: Geographic and Social Mobility in Monterrey, Mexico*, Austin, Tex.: University of Texas Press.

Berry, B.J.L. (ed.) (1976) *Urbanization and Counter-urbanization*, Beverly Hills, Calif.: Sage.

Brookfield, H., Hadi, A.S. and Mahmud, Z. (1991) *The City in the Village: The In-situ Urbanization of Villages, Villagers and their Land around Kuala Lumpur, Malaysia*, Singapore: Oxford University Press.

Browne, H. (1994) *For Richer, for Poorer*, London: Latin American Bureau.

Brydon, L. and Chant, S. (1989) *Women in the Third World: Gender Issues in Rural and Urban Areas*, Aldershot: Edward Elgar.

Cornelius, W.A. (1975) *Politics and the Migrant Poor in Mexico City*, Stanford, Calif.: Stanford University Press.

Currie, L.L. (1971) 'The exchange constraint of development: a partial solution to the problem', *Economic Journal* 81, 886–903.

Davis, K. and Hertz, H. (1954) 'Urbanization and the development of pre-industrial areas', *Economic Development and Cultural Change* 4, 6–26.

Fresneda, O., Sarmiento, L., Muñoz, M. and associates (1991) *Pobreza, violencia y desigualdad: retos para la nueva Colombia*, Projecto Regional para la Superación de la Pobreza, Programma de las Naciones Unidas para el Desarrollo, Bogotá.

Garza, G. (1991) 'Dinámica industrial de la ciudad de México, 1940–1988', El Colegio de México, Mimeo.

Germani, G. (1973) 'Urbanization, social change, and the great transformation', pp. 3–58 in G. Germani (ed.) *Modernization, Urbanization and the Urban Crisis*, Boston, Mass.: Little, Brown & Co.

Gilbert, A.G. (1992) 'Third World cities: housing, infrastructure and services', *Urban Studies* 29, 435–60.

Gilbert, A.G. (1993) 'Third World cities: the changing national settlement system', *Urban Studies* 30, 721–40.

Gilbert, A.G. (1994) *The Latin American City*, London and New York: Latin American Bureau and Monthly Review Press.

Gilbert, A.G. and Gugler, J. (1992) *Cities, Poverty and Development: Urbanization in the Third World* (2nd edn), Oxford: Oxford University Press.

Gilbert, A. and Ward, P. (1986) 'Latin American migrants: a tale of three cities' pp. 24–42 in F. Slater (ed.) *Peoples and Environments: Issues and Enquiries*, London: Collins Educational.

Goldstein, S. (1990) 'Urbanization in China, 1982–87: effects of migration and reclassification', *Population and Development Review* 16, 673–701.

Gwynne, R.N. (1990) *New Horizons? Third World Industrialization in an International Framework*, Harlow, Essex: Longman.

Hall, P. (1980) 'New trends in European urbanization', *Annals of the American Academy of Political and Social Science* 451, 45–51.

Hardoy, J.E., Cairncross, S. and Satterthwaite, D. (eds) (1990) *The Poor Die Young: Housing and Health in Third World Cities*, London: Earthscan.

Harris, N. (ed.) (1992) *Cities in the 1990s: The Challenge for Developing Countries*, London: UCL Press.

Henderson, J.V. (1991) *Urban Development: Theory, Fact and Illusion*, New York: Oxford University Press.

Hoselitz, B.F. (1957) 'Generative and parasitic cities', *Economic Development and Cultural Change* 3, 278–94.

Lewis, O. (1959) *Five Families: Mexican Case Studies in the Culture of Poverty*, New York: Basic Books.

Linn, J.F. (1983) *Cities in the Developing World: Policies for their Equitable and Efficient Growth*, New York: Oxford University Press.

Lipton, M. (1977) *Why Poor People Stay Poor: A Study of Urban Bias in World Development*, Cambridge, Mass.: Harvard University Press.

Mabogunje, A.L. (1990) 'Urban planning and the post-colonial state in Africa: a research overview', *African Studies Review* 33, 121–203.

McGee, T.G. (1989) 'Urbanisasi or kotadesi? Evolving patterns of urbanization in Asia', pp. 93–108 in F.J. Costa (ed.) *Urbanization in Asia*, Honolulu: University of Hawaii Press.

Merrick, T.W. (1986) *Population Pressures in Latin America*, Population Bulletin 41, no. 3, Washington, DC.

Phillips, D.R. and Yeh, A.G.-O. (1990) 'Foreign investment and trade: impact on spatial structure of the economy', pp. 224–44 in T. Cannon and A. Jenkins (eds) *The Geography of Contemporary China: The Impact of Deng Xiaoping's Decade*, London: Routledge.

Portes, A. (1972) 'Rationality in the slum: an essay in interpretive sociology', *Comparative Studies in Society and History* 14, 268–86.

Pryer, J. and Crook, N. (1988) *Cities of Hunger: Urban Malnutrition in Developing Countries*, Oxford: Oxfam.

Richardson, H.W. (1989) 'The big, bad city: mega-city myth?', *Third World Planning Review* 11, 355–72.

Shaiken, H. (1990) *Mexico in the Global Economy: High Technology and Work Organization in Export Industries*, San Diego: Center for US–Mexican Studies, University of California.

Skeldon, R. (1990) *Population Mobility in Developing Countries*, London: Belhaven Press.

Sklair, L. (1992) 'The maquilas in Mexico: a global perspective', *Bulletin of Latin American Research* 11, 91–108.

Smith, D.M. (ed.) (1992) *The Apartheid City and Beyond: Urbanization and Social Change in South Africa*, London: Routledge.

South, R.B. (1990) 'Transnational "maquiladora" location', *Annals of the Association of American Geographers* 80, 549–70.

Storper, M. (1991) *Industrialization, Economic Development and the Regional Question in the Third World*, London: Pion.

Stren, R.E. and White, R. (eds) (1989) *African Cities in Crisis: Managing Rapid Urban Growth*, Boulder, Colo.: Westview Press.

Suárez-Villa, L. and Han, P.-H. (1991) 'Organizations, space and capital in the development of Korea's electronics industry', *Regional Studies* 25, 327–43.

UNCHS (1987) *Global Report on Human Settlements 1986*, Oxford: Oxford University Press.

UNDIESA (United Nations, Department of International Economic and Social Affairs) (1991) *World Urbanization Prospects 1990*, New York: United Nations.

Villa, M. and Rodriguez, J. (1994) *Dinamica sociodemográfica de las metropolis latino-americanas*, Santiago, Chile: CELADE.

Wilkie, J., Lorey, D.E. and Ochoa, E. (1988) *Statistical Abstract of Latin America*, Los Angeles: University of California.

World Bank (1981) *World Development Report 1981*, Washington, DC: World Bank.

World Bank (1989) *World Development Report 1989*, New York: Oxford University Press.

World Bank (1991) *Urban Policy and Economic Development: An Agenda for the 1990s*, Washington, DC: World Bank.

World Bank (1992) *World Development Report 1992*, New York: Oxford University Press.

World Bank (1994) *World Development Report 1994*, New York: Oxford University Press.

FURTHER READING

Bradnock, R.W. (1984) *Urbanisation in India*, London: John Murray.

Cannon, T. and Jenkins, A. (eds) (1990) *The Geography of Contemporary China: The Impact of Deng Xiaoping's Decade*, London: Routledge.

Costello, V. (1977) *Urbanization in the Middle East*, Cambridge: Cambridge University Press.

Devas, N. and Rakodi, C. (eds) (1993) *Managing Fast Growing Cities*, Harlow: Longman.

Drakakis-Smith, D.W. (1988) *The Third World City*, London: Methuen.

Gilbert, A.G. (1994) *The Latin American City*, London and New York: Latin American Bureau and Monthly Review Press.

Gilbert, A.G. and Gugler, J. (1992) *Cities, Poverty and Development: Urbanization in the Third World* (2nd edn), Oxford: Oxford University Press.

Gugler, J. (1988) *The Urbanization of the Third World*, Oxford: Oxford University Press.

O'Connor, A. (1983) *The African City*, London: Hutchinson.

Parnwell, M. (1993) *Population Movements and the Third World*, London: Routledge.

20

FROM RICHES TO RAGS

The international debt crisis

Stuart Corbridge

INTRODUCTION

This chapter provides a straightforward guide to the debt–cum–banking–cum–development crises that have rocked so many countries over the past twenty years. It also challenges those conventional stories of 'the debt crisis' which begin their accounts with the Mexican default of August 1982 and which work their way back to the OPEC oil price rises of 1973–4. These accounts spotlight the debt crises that affected Latin America in the 1980s and 1990s (and which threatened the commercial banking system of the United States), when a debt–cum–development crisis has been at least as severe in many parts of Asia and Sub-Saharan Africa. The conventional account also concentrates attention on the destabilizing actions of a small group of countries in the 'Middle East' in the mid-1970s. As such, it fails to recognize certain changes in the international financial system that have led to a massive expansion of private credit monies over the past thirty years.

The chapter is organized with an eye to both these aims. A chronological approach to the 'debt crisis' is adopted, but the chronology charted starts before 1973–4 and plays down the importance of August 1982 as the signifier of a developing countries' debt crisis. The second section outlines the prehistory of the debt–development–banking crises which (not unreasonably) captured the popular imagination in the 1980s. Brief mention is made of the debt crises of the 1920s and 1930s and of a history of debt default in Africa in the 1970s. The OPEC oil price rises of the 1970s are then placed in the context of an unregulated expansion and transmission of private forms of international liquidity post-1960. The third section focuses on the years 1982–5, when a generalized crisis of solvency in Latin America threatened parts of the international banking system. Possible reasons for this unhappy conjuncture are reviewed – namely, poor domestic policies in the defaulting indebted countries, an unfavourable global macroeconomic environment, or both – before the narrative turns to the 'official' (non-banking) debt crises that have affected many more countries in Africa, Asia and Eastern Europe. The fourth section briefly considers how a general debt-cum-banking-cum-development crisis in the 1980s was coded as a clutch of singular debt crises and dealt with by the Bretton Woods institutions on this basis. This section also shows how a punitive policing of the 'debt crisis' was relaxed in the late 1980s and 1990s, through the Baker and Brady plans pre-eminently. This was partly in recognition of the growing stability of the US commercial banking system, and partly in response to influential groups in the West (exporters) and the South. The fifth section continues the narrative into the 1990s. It questions the claim that the debt crisis ended in 1991, when net capital flows to Latin America were positive for the first time since 1982. One problem with this claim is that attention is focused once more on Latin America (and then in very restricted terms), and not on the continuing debt-cum-development crises that affect parts of Africa and Asia. The geography of the developing countries' debt crisis changed significantly through the 1980s and 1990s and this fact is reflected upon in the conclusion to the chapter (the sixth section). The conclusion also considers what is at stake in writing about 'the debt crisis'. Discourses about debt and development strongly influence the policies that are deployed to deal with a chosen set of debt-related problems. Writing about 'debt' matters for this reason.

A PREHISTORY OF THE '1980s DEBT CRISIS'

Fernand Braudel has argued convincingly that capitalism is characterized by periods of intense capital accumulation in the 'real economy', culminating in unsustainable bouts of 'financial expansion' prior to retrenchment and devaluation (Braudel 1982). More recently, Giovanni Arrighi has suggested that this boom–bust cycle has taken on a greatly extended temporal and spatial reach as the periphery of the capitalist world economy has been more fully integrated with the core (Arrighi 1994; see also Harvey 1982). The debt crises of the 1980s (and 1990s) are thus old and new at the same time. They are old because they highlight a periodic tendency to financial excess, indebtedness and default in the capitalist world economy – as witness the bond-market crises that racked Mississippi and Louisiana in 1839, or the defaults on debt that affected many Latin American countries (along with Turkey and Egypt) in the 1870s, and again in the wake of the Great Crash of 1929 (Congdon 1988). They are new in so far as the monies owed by Latin American and other debtors today are largely owed to commercial banks or official creditors like governments, development agencies and the International Monetary Fund (IMF). The debts outstanding in the 1980s threatened the integrity of an interlinked global banking system rather than the dispersed accounts of individual and institutional bondholders.

In some African countries, too, if not always in Latin America, the total debts outstanding at various points in the 1980s and 1990s were historically high in relation to gross domestic product (GDP) or the value of exported goods and services. The debt crises of the '1980s' engulfed more people, and threatened more pivotal financial institutions, than the previous debt crises to which Congdon, Arrighi, Braudel and others have directed our attention.

But why should the debt crisis have developed in the 1980s, and did it come out of the blue as the events in Mexico in August 1982 seem to suggest? The second part of this question is easily answered. Robert Wood has noted that 'twenty Third World countries renegotiated their debt to bilateral official creditors [between 1979 and 1983] ... Indeed, up through 1982, there were considerably more Paris Club reschedulings of debt to bilateral official creditors than there were [London Club] reschedulings of debt to private creditors' (Wood 1986: 234). Zaïre defaulted on its debts as early as 1975.

The first part of the question is not so easily answered, although several factors suggest themselves as worthy of consideration. We might begin, for example, with an ideology of developmentalism which took hold in the post-colonial world (Taylor 1989), and which held out to developing countries the prospect of a rapid transition from Third World to First World status if they imitated the patterns of development mapped out by the pioneer countries of the advanced industrial world. Foreign aid would be a handmaiden to local strategies of import-substitution industrialization (and more rarely export-oriented industrialization), with extra funds coming from private institutions willing to invest in the development of the periphery. Going into debt was not something to fear, or to fear greatly. Countries had to borrow to prosper, as most white settler colonies had discovered in the nineteenth century (Lal 1983). In this lexicon credit is entered as another word for debt, and the ability to go into debt is read as a sign of a country's creditworthiness and growing maturity (Beenstock 1984).

This made sense in the 1950s and 1960s, when most of the funds that flowed from the First World to the Third World came in the form of official development assistance (ODA), military spending and foreign direct investment. The debts that were built up tended to be small in relation to a given country's GDP, and debt servicing was not usually a problem. This was also the Golden Age of Capitalism in the advanced industrial world (Marglin and Schor 1990), and the core Fordist countries provided growing, if heavily protected, markets for many of the commodities produced by the Third World. In the 1970s, however, patterns of lending to the Third World changed significantly (see Figure 20.1). The United States was now heavily embroiled in Vietnam and it faced a recurrent deficit in its balance of payments (Gilpin 1987). Two consequences flowed from this change in the geopolitical and economic position of the United States which impact on our story here. Most directly, the United States effected a change in what Wood calls the 'Bretton Woods Aid Regime' (Wood 1986), such that the private sector was encouraged to play a more central

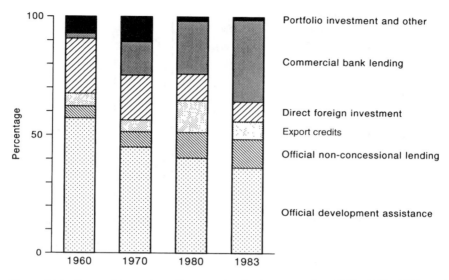

Figure 20.1 Composition of net flows to developing countries, 1960, 1970, 1980, and 1983.

role in the recycling of funds in the world economy. The main agency of this credit recycling, the Euromarkets that grew up in the 1960s and 1970s and which were and are centred on London, then saw its deposit base massively expanded by the growing balance of payments deficits of the United States. The first Nixon administration refused to pay for the war in Vietnam by raising taxes in the United States. It preferred to run up large deficits with many of its trading partners and to finance a growing debt with new issues of the trading world's main unit of account, the US dollar. Other countries were required to hold dollars as IOUs, and most did so until the late 1970s because the dollar faced no serious rival as a unit of account for transacting international trade (Corbridge 1994).

This last point might seem rather opaque, but the consequences of US exceptionalism in the 1970s and 1980s bear directly on one common account of the developing countries' debt crisis (see pp. 413–17). Most commentators are agreed that the debt crises that affected many countries in Latin America in the 1980s, and some countries in Africa and East Asia, resulted in part from the willingness of public-sector institutions in the developing world to borrow 'excessively' from the Euromarkets in the 1970s. Greed and naïvety are sometimes mentioned in this regard, as we shall see later. But the other side of this coin relates to the growth of the Euromarkets themselves, and the willingness of the United States to surrender much of the control it had once

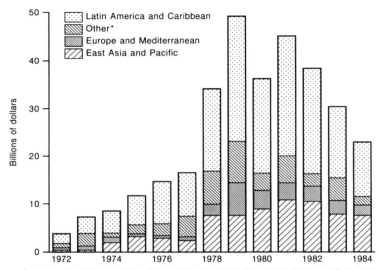

Figure 20.2 Syndicated Eurocurrency lending to developing countries, by region, 1972–84

exercised over credit creation and transmission to a small group of (mainly) US-based money-centre banks operating in and through the London Euro-markets (Wachtel 1986). The particular achievement of the commercial banks – and this is the essence of the Euromarket phenomenon – was to expand a growing deposit base not on the basis of a cash reserve requirement (as per national banking systems in the 1960s and 1970s), but according to the banks' own estimates of the creditworthiness of their potential customers (Edwards 1985). Here too stories of greed and naïvety abound. What is undeniable, however, is that syndicated Eurocurrency lending to developing countries took off in the early 1970s and expanded quickly in the years 1978 and 1979 (see Figure 20.2). The major money-centre banks now found it profitable to lend large amounts of money to so-called middle-income countries in areas like East Asia and the Pacific and Latin America (Lissakers 1991), and many bankers took the view, made public by the boss of Citibank in the 1970s, Walter Wriston, that countries could not go bust (Wriston 1986). Bankers working on a commission basis apparently fell over themselves to lend billions of dollars to countries like Brazil and Mexico at floating rates of interest and for an average of 5–7 years (Gwynne 1983).

Most accounts of the debt crisis further suggest that countries in Latin America and elsewhere went into debt in the 1970s in order to finance the threefold or fourfold increase in the price of imported oil effected by OPEC in the period 1973–4, and were able to do so because OPEC petrodollars were on offer in the Euromarkets. But this account is too simple. It clearly is the case that many developing countries were hit hard by the oil price rises of 1973–4 and again at the end of the 1970s during the Iran–Iraq war, but it is not the case that the Euromarkets were stoked up solely or mainly by the actions of OPEC members. If the reader looks again at Figure 20.2, he or she will see that syndicated Eurocurrency lending to developing countries did not exceed $16 billion in either 1974, 1975 or 1976 – the years following the first oil price rises. The real growth of this market had to await the late 1970s when Eurodollar lending grew mainly on account of a massive expansion of US IOUs in circulation as the US sought to balance its books by a reckless and inflationary resort to the printing press. Most estimates suggest that dollar deposits by the OPEC countries accounted for less than 20 per cent of the deposits taken by the Eurobanks in the 1970s. Curiously perhaps, at least in terms of conventional narrative accounts of the 1980s debt-cum-banking crises, most of the credits made over to countries in Latin America and elsewhere 'came' not from the Middle East but from North America.

Many Latin American countries were also persuaded to take out large loans from US (and other) commercial banks by development and Treasury officials in the United States itself. Although a process of 'blaming the victims' for their profligacy became common once the bank–debt crisis 'broke' during 1982–3, it was not clear to the US government or the Bretton Woods institutions or the commercial banks as late as 1980–1 that a 'debt-bomb' was in the making. In its first *World Development Report* of 1978, the World Bank suggested that the debt ratios it was predicting up to 1983 were 'not unacceptably high ... and should pose no general problem of debt management provided exports can grow at the projected rates' (World Bank 1978: 31). Less than five years later the cover of *Time Magazine* (January 1983) featured the earth as a ticking bomb, with a short fuse coming out of the belly of Latin America. Jay Palmer, the author of the magazine's leading article, warned *Time* readers that: 'Never in history have so many nations owed so much money with so little promise of repayment' (Palmer 1983: 4). As Brazil and Argentina joined Mexico in debt default, the idea was floated that the banking system of the Western world was in danger. According to Palmer, the savings of ordinary depositors in the US were at risk directly from the threat of debt-induced bank failures, and indirectly from the threat of a finance-induced slump on a par with the Depression of 1929–33. The 'debt crisis' was front-page news.

CRISIS YEARS: 1982–5

Although the Mexican default of August 1982 took many observers by surprise – and left some bankers in a state of shock (Volcker and Gyhoten 1992) – it is clear that the wider context for the Latin American defaults of the early 1980s lies with the changing international economic situation between 1978 and 1982.

413

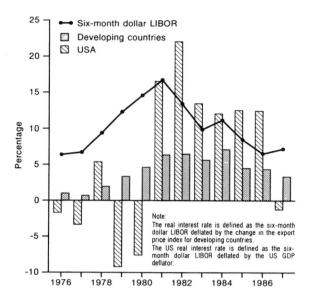

Figure 20.3 Real interest rates and LIBOR, 1976–87.

In September 1978 the US dollar came under such pressure from investors and bill-holders that President Carter was obliged to restore confidence in the greenback by ending the inflationary and devaluationist policies that America had pursued in the 1970s. As a symbol of America's new commitment to sound money policies Paul Volcker was installed in 1979 as Chairman of the Federal Reserve Board. The money supply of the United States was swiftly tightened by Volcker, and certain versions of monetarist economics were advanced by the administration to justify America's new monetary policy. By June 1982 the prime (interest) rate in the USA had reached 16.5 per cent and unemployment had risen to 9.3 per cent of the total labour force. Similar tales were to be told in Western Europe and especially in Mrs Thatcher's Britain. The New Right was in the ascendancy and a process of economic restructuring was underway.

The implications of these changes for developing countries were twofold and varied as such from country to country. On the one hand, the interest rate charged on most Eurocurrency loans (the six-month dollar London Inter-Bank Offered Rate, or LIBOR) climbed from an average of 9.2 per cent in 1978 to an average of 16.63 per cent in 1981. Real interest rates were also on the rise and they were to remain at high levels until 1986 (see Figure 20.3; they rose again in the early 1990s). William Cline estimates that 'total excess [or unanticipated] interest payments on developing country debt amounted to $41 billion in 1981–2' (Cline 1984: 12). In so far as a country's debt was denominated mainly in US dollars this unanticipated expense was compounded by the rapid appreciation of the dollar from 1980 to 1985. The other blade of an emerging scissors crisis was closed by a decline in the value of commodity exports from developing countries from 1979 to 1987. Again, the nature and impact of what really was a variegated set of commodity price movements is the subject of continuing debate. It is not unreasonable to suggest, however, that a decline in non-oil commodity prices placed many less-developed countries (LDCs) in a vulnerable position as their loan repayments fell due. Cline concludes that

> high interest rates and the global recession imposed large cumulative losses on the non-oil developing countries in 1981–82. In all, these countries lost approximately $141 billion in higher interest payments, lower export receipts, and higher import costs as the consequences of adverse international macroeconomic conditions.
>
> (Cline 1984: 13)

Whether default was inevitable in this context is a moot point. Many on the right have argued that it was not. Buiter and Srinivasan are not alone in claiming that the developing countries' debt crisis is really a debt crisis of a few high-profile countries in Latin America, and that this debt crisis was brought on by the poor economic policies of the countries concerned and the poor lending decisions of some commercial banks (Buiter and Srinivasan 1987). They note that a number of well-managed developing countries did not default on their debts in the 1980s – they point to South Korea in particular, the fourth largest developing country debtor in 1982, as well as Malaysia, Colombia, Thailand and Zimbabwe – even though they faced the same unfavourable international economic conditions as defaulting countries in Latin America (and Africa). Buiter and Srinivasan further argue that 'profligate' debtors should not be rewarded with debt relief or debt writedowns for fear of inducing moral hazard in the international monetary system (or the idea that bad behaviour is to be rewarded: Vaubel 1983). Debt relief of any kind would have the effect of rewarding the profligate and punishing the prudent (like South Korea) and poor (countries like Bangladesh that could not avail themselves of commercial bank loans in the 1970s and early 1980s).

There is something to be said for these arguments and it is significant that both India and China have opposed preferential debt relief for Latin America on just these grounds. It is also the case that some indebted countries in Latin America, Africa, Eastern Europe and elsewhere suffered from poor or ill-advised economic policies in the 1970s and early 1980s. Exchange-rates were overvalued in many Latin American countries (making export goods uncompetitive), and citizens were sometimes cushioned by fiscal policies which used foreign loans as a substitute for domestic tax increases or spending cuts. Tales also abound of bank loans being put to dubious uses, such as roads to nowhere in the Amazon, of poor project assessment procedures, or of massive capital flight to Miami, New York City and offshore financial centres in the Caribbean. Peter Bauer complains that: 'In 1987 the government of Peru had externally held reserves of about $1.5 billion, at a time when it refused to pay a few million dollars on servicing its sovereign debt' (Bauer 1991: 60).

But even if this is accepted it makes little sense to argue that the 1980s debt crises were brought about mainly by the actions of the indebted countries themselves. Carlos Diaz-Alejandro has shown that a majority of Latin American countries fell into debt default in the 1980s, even as they followed very different mixes of exchange-rate policies, fiscal policies and policies in respect of the public–private sector mix (Diaz-Alejandro 1984). Diaz-Alejandro accepts that some Latin American countries were in need of reform policies in the period 1980–1, but he maintains that 'nothing in the situation called for traumatic depression' (ibid.: 348), or the precipitate cutting off of loans by commercial banks in the period 1982–3. It was this latter action, Diaz-Alejandro charges, that sparked off the Latin American debt-cum-banking crisis. The collapse in net inflows from 1982–3, combined with a decline in the absolute dollar value of exports prior to the decline in loans, meant that a debt crisis was inevitable in Latin America in general, and regardless of domestic policy mixes. The comparison with South Korea is thus ill-judged; a more meaningful basis for comparison lies within Latin America itself.

Most commentators agree with this conclusion, which bears upon the debt crises in Africa and Asia as well as those in Latin America. In the case of Latin America and the Caribbean, it is clear that the region as a whole was damaged by three factors in the early 1980s: the increasingly short-term nature of many Latin American debt stocks, the Falklands–Malvinas conflict of 1982, and a calamitous decline in new net bank lending to the region after 1981. The South Atlantic conflict 'immediately led to a suspension of new lending by banks and to a *de facto* default by Argentina. More than anything else, the crisis in Argentina in the spring of 1982 paved the way for the really big crisis later that year – the Mexican default' (Kuczynski 1988: 81). Almost as directly, the Falklands–Malvinas war encouraged many banks to rein back on new lending to Latin American borrowers in general, to the extent that net transfers on private banking dropped to minus $3,816 million in 1982. Even in 1981 a positive net transfer was supported largely by short-term lending and borrow-ing. (In itself this is not a worry, so long as the money is used for short-term purposes such as trade financing. It is a worry when short-term loans are advanced in a market palpably not blessed with perfect information – either on an interbank basis or through the World Bank's Debt Reporting System – and where the funds are either flowing out of the country in the form of capital flight, or are being used to finance long-term public deficits.) In sum, if 'short-term borrowing . . . did not cause the debt crisis [it did enable] borrowing to continue at a time of hesitation by major commercial bank lenders' (Kuczynski 1988: 80). It is for this reason that the Latin American debt crisis did not break officially until August 1982 in Mexico (with defaults following in Brazil in November 1982 and Venezuela in February 1983). By 1984 there was a generalized debt crisis in Latin America and the Caribbean.

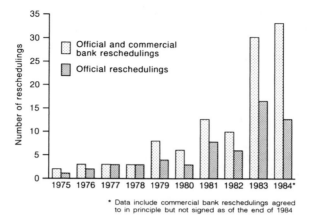

Figure 20.4 Debt reschedulings, 1975–84.

Elsewhere, a debt crisis was already apparent by 1981 and we should not forget this (see Figure 20.4 for evidence on debt reschedulings). The oil price rise of the late 1970s caused particular problems for countries in Sub-Saharan Africa and Eastern Europe, some of which were forced to take out new loans from official creditors to fund growing current account deficits. For such countries the US-inspired recession of 1979–92 was the straw that broke the camel's back. Faced by rising interest rates and shrinking commodity export markets, many of these countries were badly placed to 'adjust' their economies to the changing dictates of the world market. In the real world it is not possible to switch painlessly from the production of coffee and iron to the production of computer software and video-games. Although a wave of debt defaults ensued in Africa, the Sub-Saharan African debtors did not receive the level of attention from the media and policy-makers that was common in Latin America. In Latin America the debt crisis threatened a banking crisis in the United States and Western Europe. It had to be dealt with quickly, publicly and not ineffectively. In Africa, by contrast, where debt–GDP and debt–export ratios were often higher than in Latin America, no such threat was posed to the integrity of the international financial situation. The development crisis that was deepened in parts of Africa by enforced debt repayments to Paris Club members did not figure in the media's everyday accounts of 'the debt crisis'. The debt crisis meant Brazil and Mexico, Citicorp and Chase Manhattan; it often didn't extend in print to the indebted and impoverished countries of Sub-Saharan Africa, or indeed the Philippines and Indonesia.

417

POLICING THE DEBT CRISIS: 1982–90

Hindsight suggests that the debt crisis (or crises) has/have been policed in three main ways: by means of containment, austerity and adjustment policies (1982–5); by means of adjustment with growth (the Baker Plan years of 1985–8); and by means of the Brady initiative and various official and 'market-menu' forms of debt writedown (1988/9 onwards).

Containment, adjustment, austerity

The advanced capitalist countries were not slow to appreciate the significance of Mexico and Brazil defaulting on accumulated external debts of nearly $200 billion in 1982. When Paul Volcker addressed the Subcommittee on International Finance and Monetary Policy of the Committee on Banking, Housing and Urban Affairs, US Senate, on 17 February 1983, he warned that:

> We face extraordinary pressures in the international financial system . . . This is not an abstract, esoteric problem of marginal interest to our [US] economy. Failure to address these problems will jeopardize our jobs, our exports and our financial system. Unless it is dealt with effectively, it could undermine both our own recovery and the economies of our trading partners and friends abroad. I am confident that the situation can be managed – but it won't manage itself.
>
> (Volcker 1983: 175)

Later on, Mr Volcker confirmed that: 'Our concern for maintaining a well-functioning international financial system is rooted in our self-interest, not in altruism' (ibid.: 17).

Volcker's remarks on altruism and its absence were intended for a particular audience; in effect, he was asking Congress to approve an increase in IMF quotas as part of a broader debt management programme. The American authorities recognized that the banking/debt crises that were breaking out across Latin America had first to be dealt with on an emergency basis. To head off possible bankruptcies of major US money-centre banks it was necessary to provide countries in default on their private debts a lifeline of new funds to meet at least some interest payments. This lifeline was provided by official US credits, by the increase in IMF quotas already referred to, and by new rounds of involuntary bank lending to the indebted countries. (Such lending was involuntary because the banks had no wish to send good money after bad, but were required to make such loans in order to gain IMF support for debt rescheduling and adjustment packages.) Having put these lifelines into place, the US authorities, the IMF and various banking officials then pressed the defaulting countries to agree to adjustment – or austerity – programmes that supposedly would allow them to service their debts in the medium term. Countries were allowed to reschedule their debts at a premium (in terms of bank fees and a greater spread above LIBOR) as and when they agreed to a package of reforms that tended to take on the following broad shape: (a) rescheduling of the principal due to

418

commercial banks (often over an eight-year period); (b) maintenance of reduced trade lines and inter-bank deposits; (c) large involuntary loans; (d) significant cuts in public sector deficits of indebted countries; (e) sharp reductions in the importation of goods and services to the indebted country; and (f) attempts to increase the exports of goods and services from an indebted country (Woodward 1992).

Creditor agencies and institutions have consistently argued that indebted countries are treated on a case-by-case basis, but the evidence suggests otherwise. In the early to mid-1980s virtually all indebted countries, in Latin America and elsewhere, were encouraged to accept some version of the standard structural adjustment programme outlined above. The programme was justified intellectually on two main grounds: that public sector deficits were both bad in themselves (they betokened economic inefficiency and waste) and were linked intimately to the external debt position of a country, and that debts could only be serviced in the medium term by a country earning hard currencies from a current account surplus (Sachs 1989).

There is something to be said for these arguments, of course, but it is also clear that the debt-cum-banking-cum-development crises of the mid-1980s could have been managed otherwise had the balance of political forces been different. The banking system of the Western world was at risk in the period from 1982–5 and it is not clear that the indebted countries would have benefited from extravagant debt writedowns that would have hastened the collapse of the world's financial system (notwithstanding some rhetoric to the contrary by Castro in Cuba and Garcia in Peru). But the debt-cum-banking crisis could have been managed in a more symmetrical fashion, with taxpayers in some northern countries, and shareholders of the commercial banks at risk, bearing a fairer share of the burdens that were dumped on the indebted countries and their populations. Politically, the containment–adjustment years were a great success for the West/North, and a clear sign that the political power of the South (expressed in the 1970s in demands for a new international economic order and producer power: Hoogvelt 1982) was at an end. By coding a banking crisis as a debt crisis the creditors were able to blame the indebted countries for the problems they found themselves in, and persuaded them to embrace structural adjustment programmes that would secure bank balance sheets in the short run and hasten the privatization and outward orientation of the indebted economies in the longer term (Walton and Seddon 1994). The creditors were also careful to offer side-payments to Mexico when it seemed that the big four debtors in Latin America (Brazil, Mexico, Venezuela and Argentina) might form a debtors' cartel and threaten a collective debt repudiation. These side-payments took the form of special deals in relation to the terms of rescheduling and the details of adjustment packages (Fryer 1987).

The Baker Plan: adjustment with growth

The containment–adjustment years were clearly successful from the point of view of the banks, none of which went under in the critical years from 1982 to 1985, but the effects upon development were predictably dire. Indebted countries in Latin America and Africa (where not dissimilar programmes of structural adjustment were being imposed by the Bretton Woods institutions and Paris Club members) began to 'underdevelop' as their goods and services were sold abroad in increasing amounts and as capital disappeared to service still growing debts. In many countries matters were made worse as authoritarian governments cut welfare budgets savagely, or as populist regimes tried to minimize the political fall-out from austerity measures by borrowing money domestically at high interest rates or printing money (Dornbusch and Edwards 1991). In Brazil in 1985 the annual rate of inflation reached a disturbing, if not surprising, 242.2 per cent, while in Bolivia in the same year inflation exceeded 8,000 per cent. In country after country sustainable development was sacrificed to secure the stability of the international banking system, and people suffered accordingly.

By 1985 it was clear that the containment–adjustment policy had served its main purpose as far as the creditors were concerned, and the second Reagan administration began hesitantly to respond to a much wider political constituency; notably, to exporters in southern US states who were anxious about lost export markets in Latin America, and to governments in Latin America which the United States was keen to protect from a populist backlash that would not serve its geopolitical interests in the region. It fell to Secretary of State James Baker to propose a new 'debt-crisis initiative' at the annual joint meeting of the World Bank and the IMF held in Seoul in October 1985. After first thanking his hosts and praising South Korea, whose 'market-oriented approach and emphasis on private initiative are a lesson for us all', Secretary Baker acknowledged that the industrial countries must take up some of the burden of adjustment in the developing countries' debt crisis. More exactly, Baker acknowledged the successes of the containment strategy, and he noted that the strategy had become more flexible with the introduction of Multi-Year Rescheduling Arrangements (MYRAs) in 1984. He particularly endorsed the suggestion that MYRAs should be offered 'as rewards for countries that had made strong progress on policies to deal with their balance of payments problems' (World Bank 1989: xviii). At the same time, he proposed that:

> If the debt problem is to be solved, there must be a Program for Sustained Growth incorporating three essential and mutually reinforcing elements:
> 1   First and foremost, the adoption by principal debtor countries of comprehensive macroeconomic and structural policies, supported by the international financial institutions, to promote growth and balance of payments adjustment, and to reduce inflation.
> 2   Second, a continued central role for the IMF, in conjunction with increased and more effective structural adjustment lending from the multilateral development

420

banks, both in support of the adoption by principal debtors of market-oriented policies for growth.

3   Third, increased lending by the private banks in support of comprehensive economic adjustment programs.

(Baker 1985)

Thus was the 'Baker Plan' born and with it a new optimism about the debt crisis. The watchwords now were *adjustment with growth* and the promise seemed to be that new resources would be made available to some debtor countries (Selowsky and van der Tak 1986). In part, these resources would be delivered by a 'serious effort to develop the programs of the World Bank and the Inter-American Development Bank ... [to] increase their disbursements to principal debtors by roughly 50 percent from the current annual [1985] level of $6 billion' (Baker 1985). Other resources would flow in the form of foreign direct investment and as renewed and expanded private bank lending. At the end of his address, Secretary Baker suggested that the commitment 'by the banks to the entire group of heavily-indebted, middle-income developing countries would be new net lending in the range of $20 billion for the next three years' (Baker 1985).

From Baker to Brady: debt reductions at last

The Baker Plan was warmly welcomed by the West's financial press. In February 1985 *Fortune Magazine* had run an article by Gary Hector entitled 'Third World Debt – The Bomb is Defused', in which Hector claimed that: 'Evidence is building that the international debt crisis is over' (Hector 1985: 24). Although: 'A feeling of hard times ... still pervades the major Latin American countries ... [on balance] confidence is growing in the debt-restructuring process' and in the booming Latin American exports that complemented the 'US economic boom' engineered by President Reagan.

In practice, though, the Baker Plan continued to buy time for financial institutions rather than kick-starting economic growth in the seventeen high indebted middle-income countries it targeted. Moreover, Baker, like Hector, failed to look across the Atlantic to the deepening debt and development crises gripping large parts of Sub-Saharan Africa. In the years 1985–7, GDP per capita increased by a respectable 3.4 per cent in Latin America and the Caribbean (ECLAC 1989), but it fell by 18 per cent in Sub-Saharan Africa (IMF 1988). The Baker Plan failed to address this 'official debt crisis'. It was not a pressing geopolitical concern for the United States in the mid-1980s. Matters finally changed in March 1989 when a new Secretary of State, Nicholas Brady, reoriented US policy on debt relief to take account of market-led developments in Bolivia (where an official debt buyback scheme operated from 1988), and calls for conditional debt writedowns made at the same time by President Mitterand of France, Chancellor Lawson of the UK, the Japanese Finance Minister Miyazawa, and the Chairman of American Express, James Robinson. By this time, too, some money-centre banks were seeking to exchange existing debt for new assets in the secondary markets. Citibank had also rebuilt its capital base sufficiently to make bad debt provisions equivalent to 25 per cent of its debt exposure.

A formal change of approach was signalled by Treasury Secretary Brady in a speech to the Bretton Woods Committee on Third World Debt in March 1989. (His speech coincided with reports that: 'In Argentina and Venezuela, social unrest was evoked by austerity measures associated with the countries' external debt burden' (World Bank 1990: 17), and signs that LIBOR was beginning to rise again in the wake of the stock market crash of October 1987.) Having made a ritual declaration that 'the fundamental principles of the current [Baker] strategy remain valid', Mr Brady suggested that 'the path towards creditworthiness of severely indebted countries [including those in Africa] should involve debt and debt service reduction on a voluntary and case-by-case basis, in addition to rescheduling of principal and new money packages' (World Bank 1990: 21). The main difference between the Brady initiative and

> the *ad hoc* debt reduction that had taken place until then through [a] market-based menu strategy was its inclusion of official support for debt and debt service reduction. The World Bank and the IMF were asked to provide funds for debt and debt service reduction operations for countries with high external debt burdens and strong adjustment programmes.
>
> (World Bank 1990: 21)

The Brady proposals were directed mainly at the SIMICs, as the severely indebted middle-income countries were now labelled. The severely indebted low-income countries of Africa (SILICs) would also be offered debt writedown packages under terms agreed by Paris Club members (the Toronto Terms of 1988 and the Enhanced Toronto Terms of 1991). The Brady package was accepted by the Bretton Woods institutions in May 1989 when both the World Bank and the IMF 'adopted operational guidelines and procedures for debt and debt service' (World Bank 1990: 21). These guidelines and procedures made it possible for SIMICs which had carried out medium-term adjustment pro-grammes acceptable to the World Bank and/or the IMF to apply for resources from these two organizations to support local market-based proposals for debt and debt-service reductions. The expectation was that the IMF and the World Bank would commit resources to the programme of the order of $20–25 billion over a three-year period from 1989, and possibly beyond. Half of this lending would come from existing lending programmes, with the rest being in the form of additional lending to provide interest support on exchanged debt. Finally, it was hoped that Japan would boost the Brady proposals by committing up to $10 billion in co-financing over a period of three to four years. This was duly secured. By the early 1990s Brady Plan agreements were in place in Mexico, the Philippines, Costa Rica, Venezuela, Morocco and Uruguay. Several others have followed since. Taken in tandem with official debt writedowns in parts of Africa, sponsored by the UK and France in particular, it seemed that the debt crisis was at last being tackled in a manner that would impact positively upon development. The Bush administration maintained that the debt crisis had indeed been associated with a lost decade of development in the 1980s, but was now being turned around in time for renewed economic growth and development in the 1990s. In June 1990 President Bush announced an 'Enterprise for the Americas Initiative', which included a debt reduction programme that was taken up by Argentina, Bolivia, Chile, Colombia, El Salvador, Jamaica and Uruguay.

THE DEBT CRISIS IN THE 1990s: THE END OF THE AFFAIR?

It would be churlish to deny that the debt writedowns of the late 1980s and early 1990s found some favour with the indebted developing countries, or the indebted transitional economies of Central and Eastern Europe. Partial debt writedowns, in combination with increased current account surpluses (or reduced deficits), did allow some countries to resume a process of economic growth even as they serviced their external debts. But things need to be kept in perspective. The partial debt writedowns of the post-1988 period have affected less than 10 per cent of the debt stocks outstanding from developing countries to private and official creditors, and the much vaunted debt-for-equity swaps and debt-for-nature swaps have been little more than icing on a very small cake. Biersteker rightly concludes that: 'The Brady Plan is ... a move in the right direction, but it remains seriously underfunded [and] has thus far benefited only a small number of countries' (Biersteker 1993: 11). In 1992 total external debt as a percentage of GNP was still in excess of 100 per cent in countries as diverse as Nicaragua (823 per cent), Mozambique (584 per cent), Tanzania (268 per cent), Guinea-Bissau (290 per cent), Laos (168 per cent), Nigeria (111 per cent), Egypt (116 per cent), Jordan (178 per cent), Jamaica (178 per cent), Panama (112 per cent), and Bulgaria (111 per cent) (World Bank 1993; see Figure 20.5). Total debt service as a percentage of exports of goods and services in 1992 was 30 per cent or more in countries as diverse as Uganda, Burundi, Madagascar, Niger, India, Kenya, Nicaragua, Honduras, Indonesia, Zambia, Bolivia, Côte d'Ivoire, Ecuador, Colombia, Turkey, Algeria, Chile, Hungary, Argentina, Uruguay, Brazil and Mexico. These countries are hardly out of the woods when it comes to external debt. Moreover, the total external debt of developing countries continued to rise into the early 1990s, even though the rate of increase slowed from 1987. In 1992, the total external debt of Brazil, still the largest Third World debtor, was $121 billion (compared to $98 billion in 1983). In the same year, the total external debt of Mexico was $113 billion, Indonesia $84 billion, India $77 billion, Argentina $67 billion, China $69 billion, Turkey $55 billion and Poland $49 billion. South Korea and Venezuela dropped out of the list of 'big five' debtors in the mid-1980s and in 1991 recorded total external debts of $40.5 billion and $34.4 billion respectively. (In the case of South Korea this is offset by substantial foreign assets; the country is expected to be a net creditor by 1996: Seth and McCauley 1987.) The total external debt of the developing

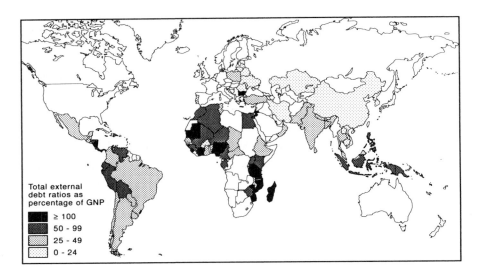

Figure 20.5 Ratios of total external debt to GNP, 1992, selected countries.

world stood at $1,662 billion in 1993, compared to $829 billion in 1982 (World Bank 1993).

Figures such as these prick the cosy optimism expressed by officials of the Bush administration. Efforts are being made to writedown the official debts of some African countries – often for geopolitical reasons (Lipietz 1989) – but debt service burdens remain an obstacle to sustainable development in large parts of the ex-colonial world. If the debt crisis seems to have disappeared from view it is for two main reasons. First, the debt crisis-as-banking crisis has largely disappeared. Most of the world's leading money-centre banks have made provisions for bad debt and have restructured their businesses to make profits from fee incomes in the derivatives markets that took off in the 1980s. Some of these banks are now exposed in First World land and property markets, but that is another story. The banks were also successful in the late 1980s in persuading governments to socialize some private debts, so that the share of official debt to total external debt in the developing world has increased since 1985 (Branford and Kucinski 1988). Second, the geography of the debt crisis continues to change. Precisely because the debt-cum-banking crises of Latin America were dealt with so adroitly by the creditors, at least from their point of view, the adverse consequences of indebtedness for development are now felt most acutely in Africa and in countries like Indonesia and the Philippines. The African countries seemingly matter very little to the Americans, or the OECD world economy more generally. In a very real sense, parts of West and East Africa do not 'count' (Agnew and Corbridge 1995). Meanwhile, India has joined the ranks of heavily indebted countries. Although debt/GDP and debt/export ratios remain quite low in India, the country was all but bankrupt in 1991 and has been following an IMF-inspired programme of economic liberalization since then (Bhagwati 1993). It should also be noted that the former Soviet Union faces a growing debt crisis, and particular problems relating to the allocation of debts between Russia and the other countries that made up the Soviet Union. The external debt of the former Soviet Union was an estimated $81.5 billion at the end of June 1993, up from about $30 billion in 1985 (World Bank 1993). In 1992 the countries that made up the former Soviet Union met less than 10 per cent of the payments due on their external debt, and Russia continues to reschedule its debt payments on a regular basis through the offices of the Paris Club (World Bank 1993). Against this background, even a seeming reversal of fortunes in parts of Latin America makes it premature, if not offensive, to talk about the 'end of the [1980s] debt crisis'.

CONCLUSION

Recent changes in the geography of indebtedness confirm how difficult it is to speak of a developing countries' debt crisis. It is now widely recognized that the United States is the world's largest debtor, and that its total external debt (of about one trillion dollars) is almost as great as the entire debt of the developing world (Corbridge and Agnew 1991). That said, the United States has no difficulty in servicing its debt – unlike many former Eastern bloc countries. There are signs too that private capital flows to parts of the developing world are beginning to take off again after a decade or more of stagnation and negative capital flows. The World Bank reports that: 'In 1992–3, for the first time in a decade, the volume of private flows has been larger than the volume of official flows, and, in real terms, private flows are now about the same as they were in the early 1980s' (World Bank 1993: 3). Most of these private flows are being made in the non-bank sector in the form of bonds, foreign direct investment (FDI), and equity portfolio investment. FDI in turn is being concentrated in services and in middle-income countries or countries that 'have either avoided the debt problem or dealt with their commercial debt overhang' (World Bank 1993: 3). Among low-income countries, China and India have also seen an upsurge in FDI.

There is room for optimism here, but little cause for rejoicing. It bears repeating that 'development' in large parts of the ex-colonial world in the 1980s (and beyond) was sacrificed to secure not just the stability of the international financial system in the short term, but also the long-term profitability of the world's major money-centre banks. This was achieved in no small measure by coding a banking crisis as a debt crisis and ignoring the fall-out for indebted developing countries that ensued from austerity programmes that supposedly were put in place to secure the long-term economic health/(adjustment) of the countries concerned. The ability to police 'the 1980s debt crisis' in this way in turn depended on portraying the indebted countries as the authors of their own misfortune, and on failing to see/acknowledge/write about the destabilizing monetary and fiscal policies pursued by successive US administrations from the late-1960s to the mid-1980s (Corbridge 1992). The creditors were largely oblivious to 'the difficulty of pursuing long-term economic reform in the midst of a major debt overhang' (Biersteker 1993: 11).

Even the World Bank's new approach to the plight of the SILICs (a deliberately anaemic acronym?) is something of a double-edged sword. The World Bank has finally recognized that: 'There is a continuing need to address the debt burden of two dozen severely indebted low-income countries (SILICs), most of which are in Sub-Saharan Africa' (World Bank 1993: 3). At the same time it accepts that, while:

> Official debt relief and new concessional financing have been the key to maintaining positive resource transfers for these countries . . . this has, inevitably, resulted in a growing debt stock. Over the past decade the debt stock of SILICs has tripled; and for most of them the capitalization of interest and the accumulation of arrears have outweighed the benefits of debt forgiveness.
>
> (World Bank 1993: 3)

What all this means in human terms is hard to say. Many countries in Africa have been battered by famines and civil wars over the past fifteen years. Indebtedness alone is not responsible for all of their woes, nor is structural adjustment (Callaghy and Ravenhill 1993). It is clear, though, that people have suffered unreasonably in many indebted countries – from rising unemployment, declining purchasing power, reduced public services, environmental degradation and so on – and far more than can be justified according to any recognized tenets of international justice (Corbridge 1993). For this reason alone it is worth ending this chapter with an account of what a 'lost decade of development' means at a household level. In her book *A Fate Worse Than Debt*, Susan George recounts the plight of a Bolivian mother, Zona San Jose Carpinteros in La Paz, amidst the debt-induced evils of rising inflation and unemployment. Says the mother:

> Since everything is so expensive, I don't give my children breakfast any more. For lunch I give them a little rice soup. I don't buy sugar now that it has gone up. To eat, I have to make do any way I can, because the children can't get along without food. Us adults, we manage without when we have to. Sometimes I say to myself, 'I'm going to give away my children to someone'. But then I think of what my parents might do to me – that's what I'm afraid of.
>
> (George 1989: 147)

Few shareholders of Citicorp and Chase Manhattan face similar problems.

CODA

1 '*Mexico: 1994 is not 1982*', leader in the *Financial Times*, London, 29 December 1994. Possibly not, but the collapse of the peso in December 1994

did signal the fragility of the orthodox economic policies pursued by Mexico since the mid-1980s. Neoliberal measures in Mexico squeezed inflation and average living standards, even as they encouraged the growth of a middle class which would lead an import-led consumer boom in the early 1990s. A trade deficit of about $5 billion in 1989 grew to an estimated $30 billion in 1994 and was largely financed by speculative flows of foreign capital from the United States. The devaluation of the peso, when it came in late 1994, encouraged massive capital flight from the country, and Mexican officials found themselves, as they had done in 1982, 'before angry lenders at a meeting at the New York Federal Reserve to explain a financial crisis' (*Financial Times*, ibid.). Mexico is now being encouraged to set 'a firm timetable for extensive privatisation, along with proposals for faster deregulation and reform of the social security system' (ibid.). Continued US financial support for Mexico is likely to be dependent upon the adoption of such measures – regardless of their consequences for the living standards and employment prospects of many Mexicans.

2   'Chemical Banking, one of the US' largest banks, has lost $70 million on an unauthorised deal in the Mexican peso following the unexpected devaluation of the currency two weeks ago. The loss stems from a position in the peso built up by one of its New York-based foreign exchange traders before the devaluation. The trader, who was not named, had deliberately hidden the size of the position, which far exceeded authorised limits, Chemical said' (*Financial Times*, London, 4 January 1995: 1).

3   '*A new year's debt resolution*', leader in the *Financial Times*, London, 3 January 1995. 'Will 1995 herald the return of the first-world debt crisis? The twin rise in rich country debt and long-term real interest rates means that the question cannot be dismissed out of hand ... Sweden and Canada will have debt/GDP ratios close to or exceeding 100 per cent in 1995, while Italy's will top 125 per cent ... if real interest rates were simply to remain at their current level, the OECD calculates that all three nations would have to tighten fiscal policy by roughly another 1 per cent of GDP to maintain their current debt-reducing plans. Add to this the possible slowing of economic growth due to higher interest rates, and, in the OECD's words "the risk of serious debt spiral in several countries becomes clear"' (ibid.).

REFERENCES

Agnew, J. and Corbridge, S. (1995) *Mastering Space: Hegemony, Territory and International Political Economy*, London: Routledge.

Arrighi, G. (1994) *The Long Twentieth Century: Money, Power, and the Origins of our Times*, London: Verso.

Baker, J. (1985) Statement before the Joint Annual Meeting of the IMF and the World Bank, Seoul, South Korea, 8 October.

Bauer, P. (1991) *The Development Frontier: Essays in Applied Economics*, Hemel Hempstead: Harvester Wheatsheaf.

Beenstock, M. (1984) *The World Economy in Transition* (2nd edn), London: George Allen & Unwin.

Bhagwati, J. (1993) *India in Transition: Freeing the Economy*, Oxford: Clarendon.

Biersteker, T. (1993) *Dealing With Debt: International Financial Negotiations and Adjustment Bargaining*, Boulder, Colo.: Westview.

Branford, S. and Kucinski, B. (1988) *The Debt Squads: The US, the Banks, and Latin America*, London: Zed.

Braudel, F. (1982) *The Wheels of Commerce*, New York: Harper & Row.

Buiter, W. and Srinivasan, T. (1987) 'Rewarding the profligate and punishing the prudent and poor: some recent proposals for debt relief', *World Development* 15, 411–17.

Callaghy, T. and Ravenhill, J. (eds) (1993) *Hemmed In: Responses to Africa's Economic Decline*, New York: Columbia University Press.

Cline, W. (1984) *International Debt: Systemic Risk and Policy Response*, Washington, DC: Institute for International Economics.

Congdon, T. (1988) *The Debt Threat*, Oxford: Blackwell.

Corbridge, S. (1992) 'Discipline and punish: the New Right and the policing of the international debt crisis', *Geoforum* 23, 285–301.

Corbridge, S. (1993) *Debt and Development*, Oxford: Blackwell.

Corbridge, S. (1994) 'Bretton Woods revisited: hegemony, stability, and territory', *Environment and Planning A* 26, 1829–59.

Corbridge, S. and Agnew, J. (1991) 'The US trade and budget deficits in global perspective: an essay in geopolitical-economy', *Society and Space* 9, 71–90.

Diaz-Alejandro, C. (1984) 'Latin American debt: I don't think we are in Kansas any more', *Brookings Papers on Economic Activity* 2, 335–89.

Dornbusch, R. and Edwards, S. (eds) (1991) *The Macroeconomics of Populism in Latin America*, Chicago: University of Chicago Press.

ECLAC (Economic Commission for Latin America and the Caribbean) (1989) *Economic Survey of Latin America and the Caribbean, 1988*, Santiago: United Nations.

Edwards, C. (1985) *The Fragmented World: Competing Perspectives on Trade, Money and Crisis*, London: Methuen.

Fryer, D. (1987) 'The political geography of international lending by private banks', *Transactions of the Institute of British Geographers* 12, 413–32.

George, S. (1989) *A Fate Worse Than Debt*, Harmondsworth, Middlesex: Penguin.

Gilpin, R. (1987) *The Political Economy of International Relations*, Princeton: University of Princeton Press.

Gwynne, S. (1983) 'Adventures in the loan trade', *Harper's*, September.

Harvey, D. (1982) *The Limits to Capital*, Oxford: Blackwell.

Hector, G. (1985) 'Third World debt: the bomb is defused', *Fortune*, 18 February, 24–9.

Hoogvelt, A. (1982) *The Third World in Global Development*, London: Macmillan.

IMF (1988) *International Financial Statistics*, Washington, DC: International Monetary Fund.

Kuczynski, P.-P. (1988) *Latin American Debt*, Baltimore: Johns Hopkins University Press.

Lal, D. (1983) 'Time to put the Third World debt threat into perspective', *The Times* (London): 6 May.

Lipietz, A. (1989) 'The debt problem, European integration and the new phase of world crisis', *New Left Review* 178, 37–56.

Lissakers, K. (1991) *Banks, Borrowers, and the Establishment: A Revisionist Account of the International Debt Crisis*, New York: Basic Books.

Marglin, S. and Schor, J. (eds) (1990) *The Golden Age of Capitalism: Re-interpreting the Post-war Experience*, Oxford: Clarendon.

Palmer, J. (1983) 'The debt-bomb threat', *Time Magazine*, 10 January.

Sachs, J. (ed.) (1989) *Developing Country Debt and the World Economy*, Chicago: University of Chicago Press/NBER.

Selowsky, M, and van der Tak, H. (1986) 'The debt problem and growth', *World Development* 14, 1107–24.

Seth, R. and McCauley, R. (1987) 'Financial consequences of new Asian surpluses', *Federal Reserve Bank of New York Quarterly Review* 12, 32–44.

Taylor, P. (1989) 'The error of developmentalism in human geography', pp. 303–19 in D. Gregory and R. Walford (eds) *Horizons in Human Geography*, London: Macmillan.

Vaubel, R. (1983) 'The moral hazard of IMF lending', *The World Economy* 6, 291–303.

Volcker, P. (1983) 'Remarks before the Subcommittee on International Finance and Monetary Policy of the Committee on Banking, Housing, and Urban Affairs, US Senate, February 17th', *Federal Reserve Bulletin* 69, 175–7.

Volcker, P. and Gyhoten, T. (1992) *Changing Fortunes: The World's Money and the Threat to Economic Leadership*, New York: Times Books.

Wachtel, H. (1986) *The Money Mandarins: The Making of a Supranational Economic Order*, New York: Pantheon Books.

Walton, J. and Seddon, D. (1994) *Free Markets and Food Riots: The Politics of Global Adjustment*, Oxford: Blackwell.

Wood, R. (1986) *From Marshall Plan to Debt Crisis: Foreign Aid and Development Choices in the World Economy*, Berkeley: University of California Press.

Woodward, D. (1992) *Debt, Adjustment and Poverty in Developing Countries* (2 vols), London: Pinter/Save the Children.

World Bank (1978) *World Development Report, 1978*, Oxford: OUP/World Bank.

World Bank (1989) *World Debt Tables, 1989/90*, Washington, DC: World Bank.

World Bank (1990) *World Debt Tables, 1990/1*, Washington, DC: World Bank.

World Bank (1993) *World Debt Tables, 1993/4*, Washington, DC: World Bank.

Wriston, W. (1986) *Risk and Other Four Letter Words*, New York: Harper & Row.

FURTHER READING

Biersteker, T. (1993) *Dealing With Debt: International Financial Negotiations and Adjustment Bargaining*, Boulder, Colo.: Westview.

Cline, W. (1984) *International Debt: Systemic Risk and Policy Response*, Washington, DC: Institute for International Economics.

Corbridge, S. (1993) *Debt and Development*, Oxford: Blackwell.

Diaz-Alejandro, C. (1984) 'Latin American debt: I don't think we are in Kansas any more', *Brookings Papers on Economic Activity* 2, 335–89.

George, S. (1989) *A Fate Worse Than Debt*, Harmondsworth, Middlesex: Penguin.

Kuczynski, P.-P. (1988) *Latin American Debt*, Baltimore: Johns Hopkins University Press.

Sachs, J. (ed.) (1989) *Developing Country Debt and the World Economy*, Chicago: University of Chicago Press/NBER.

Walton, J. and Seddon, D. (1994) *Free Markets and Food Riots: The Politics of Global Adjustment*, Oxford: Blackwell.

Woodward, D. (1992) *Debt, Adjustment and Poverty in Developing Countries* (2 vols), London: Pinter/Save the Children.

21

MONITORING, MODELLING, AND MOTHERING THE ENVIRONMENT

The impact of science and technology since the Second World War

Richard Huggett

The environmental and geographical sciences have experienced enormous and speedy changes during the last half century. Many of these changes were undoubtedly sparked off by technological developments begun during the Second World War. The war effort forced science and technology to evolve swiftly. Discoveries and inventions started before the outbreak of hostilities were turned hastily into practical applications. The applications involved new technologies, many of which were later to play a key role in environmental and geographical research. Radar, nuclear fission, ballistic missiles, electronic digital computers, and pesticide (DDT) manufacture were all to have a deep impact on environmental and geographical enquiry. Some of these developments impinged directly upon environmental and geographical thinking; others worked indirectly through later, related innovations. Digital computers had a direct, if delayed, influence that facilitated the wide application of complex mathematical models and so environmental modelling, and that made possible the establishment of geographical information systems. Ballistic missiles had an indirect influence when rockets were used to launch artificial satellites, a result of which was the rapid proliferation and development of remote sensing technology. Remote sensing and geographical information systems technologies were then combined in specialized systems for earth observation.

One of the most ironical influences of wartime technology was that of nuclear fission and DDT. Atomic power and a pesticide may seem like odd bedfellows, but there are close parallels between them. Both were initially thought of as panaceas. Nuclear fission was deemed to be an inexhaustible supply of cleanly won energy that could be provided so cheaply as to make billing unnecessary.

DDT, once its insecticidal properties had been discovered, was seen as a marvel pesticide that would guarantee the world's food supply. Both the nuclear and DDT dreams turned into nightmares, the nuclear dream because of horrendous military applications, radiation leakages, and accidents at power plants, and the DDT dream because of its unexpected retention in ecosystems and accumulation towards the top of food webs. Together, nuclear energy and DDT can be said to have bred modern environmentalism which, in its more extreme forms, adopts a motherly attitude towards Nature.

The purpose of this chapter is to explore post-war technological impacts on environmental monitoring, modelling, and mothering.

MONITORING THE ENVIRONMENT

From balloons to satellites

Airborne images of the earth were available before the invention of aircraft and satellites. The first aerial photograph was taken from a balloon by Gaspard Félix Tournachon, better known as Nadar, in 1858. His image, developed whilst he was still airborne, was of Petit Bicêtre, a village near Paris. He intended using it to make a map. An aerial view of Boston was taken in 1860, and pictures over the fortresses of Kronstadt and Petersburg, in Russia, were shot in 1886. Other airborne objects, such as kites, were used as platforms from which to take aerial photographs. There is even a report of a pigeon having had a 70 gm camera strapped to its chest! However, balloons, kites, and pigeons are not particularly navigable platforms from which to take images. Not until rockets were used as platforms, at the instigation of Albert Maul, did aerial photography make significant advances.

Aerial imaging took off with the arrival of aviation. Motion picture images were taken by Wilbur Wright on 24 April 1909 while flying his aeroplane over Centonelli, Italy. Photo-reconnaissance was used during the First World War, the first practical aerial camera being introduced by J.T.C. Moore-Brabazon of the RAF in 1915. In the interwar years, the United States Geological Survey and the Tennessee Valley Authority were the chief exponents of aerial survey techniques. The first issue of *Photogrammetric Engineering and Remote Sensing* appeared in 1934. Photo-reconnaissance and photo-interpretation procedures were greatly improved during the Second World War. They allowed British intelligence to detect the build-up for an intended invasion of the British Isles and the launch site for V-1 rockets. However, the main developments during the war were the advances made in detecting wavelengths beyond the visible regions of the spectrum, in the infra-red range and in radio detection and ranging (RADAR). A radar system was first developed in 1935 by Robert Alexander Watson-Watt and his team. In Britain, a chain of twenty radar stations, running from the Solent to Solway Firth, was built in 1937. Britain and the USA had viable systems of radar for ship and plane detection during the war. In 1945, the first radar signals were reflected from the moon. After the war, development of radar concentrated on side-looking airborne radar (SLAR), which was operational in 1954 and used by high-altitude aircraft of the US Strategic Air Command. An environmental application first arose in 1968 when US Air Force scientists showed that radar could be used to detect wind shifts and precipitation.

Space technology was adumbrated by rocket research. The first rocket test took place at Peenemünde in 1937 under the direction of a team that included Wernher von Braun. The construction of the V-2 rocket was planned in the same year. In 1938, a rocket propelled by liquid fuel was produced that travelled 18 km. The V-1 rocket, propelled by a jet engine and controlled by an autopilot, was used against the United Kingdom in early 1944, and in September of that year the V-2 rocket was used. Soon after the war, the United States established the White Sands proving ground for rocket research in New Mexico, and the first photographs from space were obtained from captured V-2 rockets. Twelve years later, in 1957, the first artificial satellite was sent into orbit about the earth – the Space Age was begun. In 1959, the first photographs of the earth were obtained from an orbiting satellite – Explorer 6. Since then, space technology has been an invaluable tool for monitoring parts of the terrestrial environment and other planets and satellites in the solar system.

A succession of satellites during the 1960s, starting with the TIROS meteorological satellites, paved the way for the 1972 launch of the first Earth Resources Technology Satellite, ERTS-1, later renamed Landsat-1. Broadly speaking, there are four kinds of satellite: weather satellites, resources satellites, geopositional satellites, and communications satellites. Technical specifications of the various series of these may be found in the *Remote Sensing Yearbook*, *Manual of Remote Sensing* (Colwell 1983), and *Space-Based Remote Sensing of the Earth* (United States National Oceanic and Atmospheric Administration 1987). Weather, resource, and geopositional satellites are a boon to environmental scientists and geographers. Communications satellites have influenced global environmental awareness by encouraging the creation of Marshall McLuhan's global village.

Weather and resource satellites carry sensors for gathering information about the earth's surface and atmosphere at some distance from them, usually by measuring radiation from the electromagnetic spectrum. The sensors may also be mounted on other platforms, as the craft carrying the sensors are called, such as aircraft and other spacecraft. Four types of imaging sensors are normally used: photographic (visible and near-infra-red), linescan (visible and infra-red), side-looking airborne radar (active microwave), and passive microwave radio-meter. Linescan sensors involve multispectral scanners (MSS) that record several spectral channels at once. Infra-red linescan (IRLS) is a system for detecting the thermal radiation emitted from the earth, and not reflected near-infra-red. Microwave radiation (in the range 1 mm to 1 m in wavelength) naturally emitted by the earth is recorded by a passive microwave system.

Microwave radiation emitted by, and sensed on returning to, a platform is an active microwave system or radar. The most common form of radar system used in earth resource surveys is side-looking airborne radar (SLAR).

Geopositional satellites were predicted in 1945 by Arthur C. Clarke, the science-fiction writer. They aid all those who need to know exactly where on earth they are. The Global Positioning System (GPS), originally developed for use by the American military, is a set of satellites and associated control systems that allow a suitable receiver to find its location anywhere on earth, day or night. GLONASS, a Russian positioning system, does the same thing.

Environmental and geographical applications

Applications of remote sensing to the environmental and geographical sciences can be considered to fall into three periods (Estes and Cosentino 1989). Before 1950, the initial applications of aerial photography were made. From 1950 to 1972 was a transitional period from photographic applications to unconventional imagery systems (such as thermal infra-red scanners and side-looking airborne radars), and from low-altitude aircraft to satellite platforms. From 1972 to the present the application of multispectral scanners and radiometer data obtained from operational satellite platforms has been predominant. A key development took place in the second of these periods. About the time that the first satellites were being launched to observe the earth, computer technology had advanced to the stage where digital processing of remotely sensed data was possible. Thus arose a new field of scientific and technological research and development – modern remote sensing – that saw a burgeoning of systems using non-photographic techniques to create digital images of reflected, emitted, and transmitted energy from objects at and near the earth's surface.

The first use of a computer in image analysis occurred in 1966. Multispectral measurements collected by an aircraft over an Indiana field were analysed by computer at the Purdue Laboratory for Agricultural Remote Sensing. By processing the image, wheat could be distinguished from oats, even oats planted within a crop of wheat, and scattered wheat along a stream from seeds transported downstream. Since its first use with remotely sensed data, the computer has proved to be an excellent means of storing, retrieving, manipulating, and displaying spatial information. Remotely sensed images are a form of spatial information, and their incorporation within geographical data handling systems is an effective way of teasing out the information that they contain. Subsequent developments in remote sensing and computer-based geographical information (integrated spatial data handling) systems have generated a new potential for conducting investigations of the ecosphere at all scales. The emphasis now is on global ecology, on studying the earth as an integrated system, on exploring what Lovelock (1989) has dubbed 'geophysiology'.

Since 1972, remote sensing, often in tandem with geographical information systems, has been applied to many subject areas (e.g. Rudd 1974; Barrett

Table 21.1 Global Positioning System fieldwork applications

Field site	Application	GPS method	Problems	Accuracy
Nepal	Sampling locations	Absolute	In steep mountain valleys, especially on valley sides	Horizontal good; vertical poor
Nepal	Sampling locations	Differential	No radio communication permitted, thus problems co-ordinating differential measurements	Horizontal ± 50 m; vertical ± 5 m
Antarctica	Site location	Absolute	None	Tens of metres
Jordan	Mapping of playas	Absolute	None	100 m latitude and longitude; altitude 300 m
Himalayas	General experimentation with equipment	Absolute	Restricted window to satellites in narrow valleys	Vertical poor
Namibia	Ground truthing and site location	Absolute	Technical fault with one receiver	Not available
Siberia	Baseline environmental database	Absolute	Within wooded valleys and forest stands	100 m latitude; 200 m longitude; altitude very poor

Source: After Cornelius *et al.* (1994).

1975; Verstappen 1977; Townshend 1981; Coulson 1990; Hobbs and Mooney 1990; Mather 1992, 1995). A major advance has been using different wavebands of the electromagnetic spectrum to gather information on various parts of the ecosphere. Within the biosphere, agricultural areas, wetlands, forest and rangeland, and global vegetation have received close attention. Measurable ecosystem properties include soil-surface moisture and temperature, terrestrial primary productivity, marine photosynthesis, leaf-area index, litter and soil organic matter decomposition, water and energy exchange, canopy bio-chemistry, and trace gas fluxes. The spatial and temporal dynamics of vegetation can be remotely sensed. Within the hydrosphere and cryosphere, remote sensing has been employed to study the land-surface water budget and ice sheets. Within the surface lithosphere and pedosphere, it is possible to map soils and surficial deposits, to measure some landscape process, and to sense biogeochemical cycles in drainage basins.

The Global Positioning System is proving to be a valuable tool in geomorphology, especially when it is linked to geographical information systems (Cornelius *et al.* 1994). It enables latitude, longitude, and altitude to be determined, though the accuracy in doing so depends on the method used to obtain a 'fix' (absolute, differential, or static). Some examples of GPS fieldwork applications are given in Table 21.1.

MODELLING THE ENVIRONMENT

The warp-speed development of the electronic computer after its invention is unparalleled in the history of technology. Mechanical calculating machines, such as those constructed by Blaise Pascal, Gottfied Wilhelm Liebniz, and Charles Babbage, ingenious contraptions though they were, had little practical value. During the 1930s, scientists investigated the possibility of using vacuum tubes (electronic devices that depend on the control of electron behaviour) to perform mathematical operations. These investigations became intensified by defence needs during the Second World War. In Germany, Konrad Zuse had devised an electromechanical coding machine to generate the Enigma code, thought to be unbreakable, for secret messages. The British response was to build a series of electronic machines, known as Colossus, to decipher German communications. The decrypting was successful and may have influenced the outcome of the war. Experience with Colossus helped engineers under the direction of Alan Turing to build Manchester University's Mark I prototype, the first electronic computer operating on a program fully stored in memory. (Incidentally, a moth is believed to have caused a malfunction in the circuitry of this machine, and all subsequent computer malfunctions have been termed 'bugs'.) The same team completed the first Automatic Computing Machine (ACE) in 1950. This was programmed using part numerical and part alphabetical code.

In the United States, John V. Atanasoff and Clifford Berry built the first operational computer using vacuum tubes in 1942. A year later work was started on an electronic computer that could keep up with the calculations of trajectories needed for gunners' firing tables. This, the ENIAC (electronic numerical integrator and computer) was completed in 1945. Its successor, EDVAC (electronic discrete variable computer), which had many features designed by mathematician John von Neumann, appeared in 1952. Unlike ENIAC, EDVAC had a central processor and a read–write memory with random access. However, the first electronic stored program computer in the USA was BINAC (binary automatic computer), which went into operation in 1949. The first electronic computer to be commercially available was UNIVAC 1. This computer stored data on magnetic tape. In 1952, the year after its introduction, it was used by CBS television to predict the results of the US presidential election, successfully as it happened. But, because its operators could not believe the predicted landslide to the right, they reprogrammed it to predict, incorrectly, a close contest. The subsequent evolution of electronic computers was spectacularly swift. Developments in the machines themselves, such as the introduction of keyboards and parallel processing (both in 1967), went hand in hand with the advent and development of high-level languages – FORTRAN (1956), COBOL (1959), BASIC (1965), PASCAL (1971), and several others. A significant development was the production in 1975 of the first personal computer available in kit form, the Altair 8800, and two years later the arrival of the Apple II, the first personal computer available in assembled form and the first to be truly successful. The monumental success of personal computers since then is well known.

The evolution of the electronic computer has had an enormous impact on the environmental and geographical sciences. At least three areas can be identified where the impact has been keenly felt: in the application of multivariate statistical methods; in the construction and solution of mathematical models; and in the storage and manipulation of data, as in the geographical information systems discussed earlier.

Multivariate statistical models

The theory of multivariate statistics was evolved at the turn of the present century by Karl Pearson and Francis Galton. It was elaborated in the 1920s, particularly by Ronald Aylmer Fisher who devised methods for studying the simultaneous operation of several variables in agricultural surveys. The multivariate character of most problems in environmental and geographical sciences was recognized well before the arrival of computers, but, without some means of taking the drudgery out of immensely time-consuming hand calculations, it was not possible to do more than indicate the complexity of most problems. Standard indications were to say that the variable studied was the most important among a host of possible controls, or else to say that variable y was a function of x, and add a *ceteris paribus* (other things being equal) clause as a rider. This univariate approach to multivariate problems at best led to gross oversimplifications in cause and effect, and at worst nurtured an extremely narrow determinism (Chorley 1965). A few pre-computer studies were made: Kendall (1939) used factor analysis to derive a productivity index for crops in England, and Hagood (1943) to define regions in the USA based partly on agriculture. But not until the appearance of electronic computers were multivariate problems widely employed in environmental and geographical research.

High-speed electronic computers have allowed conventional problems to be attacked more vigorously and have speeded up what might otherwise be impossible or unimagined (Chorley 1967). In geography, younger members of the profession started using high-speed electronic computers in the late 1950s and early 1960s (see Coppock 1962). In physical geography, Melton's (1957) precocious and penetrating investigation of factors controlling slope steepness and drainage density employed multiple regression. About this time, geologists started to take a keen interest in multivariate methods, too (e.g. Krumbein 1959). Early applications of multivariate methods in human geography include Haggett's (1964) use of multiple regression to probe the relative influence of different factors on the extent of deforestation in south-east Brazil, and Henshall and King's (1966) use of factor analysis to classify peasant agriculture in Barbados on the basis of crop–livestock combinations on 150 farms.

The fortunes of multivariate techniques have ebbed and flowed somewhat in human geography, and are at a low ebb in the present post-modern world. However, physical geographers and ecologists still use multivariate techniques, and indeed have refined them to a considerable degree. A fine example is Whittaker's (1987, 1989) study of vegetation–environment relationships on the Storbreen glacier foreland, Jotunheimen, Norway. Whittaker took a subset of 108 permanent vegetation plots, at which he measured twelve environmental variables: terrain age, altitude, soil depth, litter depth, root depth, maximum slope, frost churning, slope movement, disturbance, exposure, moisture, and snowmelt. Four vegetation gradients (ordination axes) were extracted using detrended correspondence analysis (DCA). The relations between the environmental variables and the four vegetation gradients were computed as rank correlation coefficients. A plexus diagram was then constructed from the matrix of correlation coefficients using non-metric multidimensional scaling. Two of the four vegetation axes were closely related to two environmental factor complexes: the terrain-age factor complex (comprising terrain age, frost churning, disturbance, altitude, and the three soil variables) and the exposure–moisture–snowmelt (microtopographic) factor complex. Interrelationships within the factor complexes, and in relation to the vegetation axes, were then explored using simple and partial correlation. Next, direct gradient analysis was performed using canonical correspondence analysis (CCA). The first CCA axis was closely related to variables in the terrain-age factor complex; the second axis to maximum slope angle, exposure, frost churning, litter depth, and snowmelt. Moisture was the only important variable on CCA axis III. By this lengthy but informative process of applying indirect and direct gradient analysis, Whittaker established that terrain age and associated factors (frost churning, slope movement, and the overall disturbance regime) formed the key to explaining vegetation distribution at this site. His work demonstrates the value of a multivariate approach that tackles the inherent complexity of environmental interactions head-on.

A cautionary note is in order at this point. There is evidence that the ready availability of computers and canned multivariate programs has led to much thoughtless collection of data and number-crunching in the hope that something meaningful might emerge. As one geographer put it:

> It is tempting to plug data into programs without understudying or considering the deeper implications, assumptions, and meanings of a particular statistical routine. All too often, instead of freeing the scholar from number drudgery so that more time can be devoted to creative thoughts, the computer is used only to crank out more studies of the same kind.
>
> (Terjung 1976: 205)

Many researchers may heartily agree with Terjung's cutting comments. Happily, the current literature suggests that the lesson has been learnt, and that computers are now generally used effectively and thoughtfully in statistical modelling.

Mathematical models

Mathematical models have a relatively long history within the geographical sciences, and a somewhat briefer history in the environmental sciences. In geography, mathematical ideas can be traced back at least to Johann Heinrich von Thünen's (1826) model of agricultural location, and the Reverend Osmond Fisher's (1866) paper considering the disintegration of a chalk cliff, which, he claimed, offered a 'slight contribution to the elucidation of questions of denudation, and at the same time an exemplification of the application of mathematics to a geological problem'. In ecology, a younger science than geography, mathematical models of predator–prey systems were developed by Alfred James Lotka and Vito Volterra in the 1920s. Later experimental work built on these ideas, but mathematical ecological modelling flowered with the arrival of electronic computers in the late 1950s. At this time, analogue and digital computers were used to simulate the dynamics of relatively simple ecosystems. Early workers in this field included Howard T. Odum and Jerry S. Olson. The models looked at the interactions of several species, thus extending Volterra's ideas to more complex cases, and at the passage of radionuclides, and a little later DDT, through food webs. The International Biological Programme, started in 1965, created the role of the ecological modeller. Systems models of several IBP biomes were constructed. The builders of these models, including George M. van Dyne, Robert V. O'Neill, and Bernard C. Patten, saw them as fair facsimiles of real ecosystems. Not all ecologists were so favourably disposed towards systems models, and said so in colourful terms. To an extent, the ecological systematists were probably misunderstood. It has been claimed that they were lured into using system analysis by the charm of turning ecology into a space-age science, with the attendant jargon borrowed from engineering and physics (Simberloff 1982). It matters not whether that is the case, for mathematical modelling in ecology, along with that in geography, has entered the dawn of a new era in the last few years. This era was heralded by four developments: the building of spatial models; the construction of non-linear dynamical systems theory; the accessibility of time-series from remotely sensed images; and the appearance of supercomputers and parallel processing.

Over the last decade or so, scientists have become acutely aware of the significance of the spatial dimension in comprehending the behaviour of systems generally, and of environmental systems in particular. The first signs that including spatial variables in dynamical systems models would make a big difference to the system's dynamics came in early simulations of ecological systems in the late 1950s, though spatial models in geography and fluid dynamics had existed well before that time. The addition of spatial variables to the classic Lotka–Volterra predator–prey equations produced radically different system dynamics. This finding helped to create a new field of study – population interactions in heterogeneous environments (e.g. Tilman 1994; Tilman *et al.* 1994). Spatial models have emerged, virtually independently, in geography,

Table 21.2 Some mathematical models that tackle spatial systems

Discipline and type of model	Modelling objective	Typical variables used
Geography		
Geometrical	Establish relations between form variables	Shape, size, direction, distance
Demographic	Predict movement of people between regions	Births, deaths, population density, information
Network	Minimize cost of movement between points in a network	Production, consumption, flow rates
Fluid dynamics		
Hydrodynamics	Predict velocity and mass distribution in the spatial field of a fluid	Distance, momentum, acceleration, friction
General circulation	Predict velocity, direction, and mass flows in the atmosphere–ocean system	Distance, density, pressure, temperature, turbulence, water vapour content, salinity
Ecology		
Population	Predict the size of populations	Births, deaths, immigration, emigration, population density, resources
Ecosystem	Predict resource distribution and the size of several populations	Births, deaths, immigration, emigration, population density, resources
Landscape ecology		
Stochastic	Predict state changes in a spatial system	Distance, habitat type, density, state transition rates
Deterministic (process based)	Predict storage, transfers, and transformations of matter and energy in a spatial system	Mass, density, births, deaths, movement rates, momentum, resources

Source: After Sklar and Costanza (1991).

ecology, landscape ecology, and fluid dynamics (Table 21.2). In the geographical and environmental sciences, the development of large-scale spatial models was made possible by developments in computing, geographical information systems, and remote sensing. Large-scale spatial models require the grid-based data provided by remote sensing for input and for testing, and the power of the latest generation of computers to cope with the sheer number of calculations involved within a reasonable time. The most sophisticated spatial models, apart from those used to study the general circulation of the atmosphere, are those built in landscape ecology. These models, though in the early stages of development, can predict change in entire landscapes, and try to link landscape change at local, regional, and global scales. Eventually, the hope is that they will lead to the constructing of global biosphere–geosphere models that allow a deeper appreciation of the human impact on the environment. At present, the best simulators of landscape change are the process-based landscape ecology models, such as the Coastal Ecosystem Landscape Spatial Simulation (CELSS) model built by Robert Costanza and his colleagues (1990). These models are convincingly realistic: spatial variables are included explicitly and cause and effect links are represented by a rich web of interactions. They hold out much promise as effective tools of global and regional environmental management.

Non-linear dynamics is revolutionizing the physical and space sciences. It may radically alter the current picture of geographical and environmental systems. In essence, it involves exploring mathematically the dynamics of systems away from equilibrium. The results of the exploration have been breathtaking: it is 'as if we had opened some magic casement to find, between chance and necessity, one dimension and the next, a whole new world of chaotic motions, strange attractors, and periodic windows' (Culling 1985: 70). The chief finding is that away from equilibrium the dynamic of non-linear systems is surprisingly rich and complex involving periodic and chaotic behaviour, and that non–equilibrium conditions are a source of order in systems. A system near equilibrium, being stable, can accommodate fluctuations from the mean state. When forced to move away from equilibrium, a critical point may be reached where the fluctuations can no longer be accommodated and instead are amplified to produce a new macroscopic order – complexity. Recent work with mathematical models reveals that large interactive systems, comprising millions of elements, naturally evolve towards a critical state in which a minor event leads, by way of a chain reaction, to a catastrophe affecting any number of elements in the system. This critical state appears to be poised at the edge of chaos (Kauffman 1993; Lewin 1993). Notions of self-organized criticality and systems at the edge of chaos may explain the dynamics of many phenomena, including ecosystems. Complex communities of species appear to evolve towards persistent and emergent states, the species membership of which is largely determined by happenstance. Pimm (1992) found that, in computer models of ecological communities, species-poor communities were easy to invade. Indeed, communities of up to about twelve species were easily entered by intruding species. Beyond that number, in species-rich communities, there were two results: newly established species-rich communities were more difficult to invade than species-poor communities, but long-established communities were even harder to invade. Similar mathematical experiments, conducted by Drake (1990), started with a 125-species pool of plants, herbivores, carnivores, and omnivores. Species were selected one at a time to join an assembling community. Second chances were allowed for first-time failed entrants. An extremely persistent community emerged comprising about fifteen species. However, when the model was rerun with the same species pool, an extremely persistent community again emerged, but this time with different component species than in the first community.

There was nothing special about the species in the communities: most species could become a member of a persistent community under the right circumstances. It was the dynamics of the persistent communities that was special: a persistent community of fifteen species could not be reassembled from scratch using only those fifteen species.

Supercomputers have allowed the formulation of very complex, spatially detailed models, including the later generations of general circulation model. Starting from serial machines in which a single processor interacts with a single memory, advances in computer technology have produced vector processor computers that use pipelined functional units fed from a vector register, and, most recently, vector processor computers with multiple processors. Used in conjunction with parallel processors, supercomputers greatly shorten the time needed to manipulate exceedingly large and realistic arrays of spatial data. They allow data to be integrated, the effects of competing processes to be sorted, and natural scales of distance and time to be maintained. In addition, they offer magnificent facilities for displaying spatially detailed information generated in simulation runs.

Most models, and expressly spatial models, are data-hungry. To calibrate a spatial model, information on several variables is needed for each of the many grid-cells. Obtaining this information from ground surveys is nigh on impossible, especially where the model covers inaccessible terrain. In some cases, existing maps of vegetation, soils, climatic variables, and so on have been used to calibrate spatial models. This is how the Osnabrück Biosphere Model was calibrated (Esser 1991). The process involves superimposing grids of the required density on the maps, and measuring the dominant 'state' of particular variables. A disadvantage with this practice is the need to interpolate data. For instance, climate is measured at points (meteorological stations), whereas the climatic variables needed in the calibration of a model must be fitted to all grid-cells, whether or not they contain a meteorological station. All mapped data are to a degree interpolated. They also tend to provide a composite 'snapshot', a picture of the varying conditions within the region being modelled sketched from scraps of information gathered in various surveys carried out over several years or even decades. Far more valuable are the data gleaned from remotely sensed images. These data are, normally, readily fitted to any grid size above a minimum level of spatial resolution, and each pixel in an image can be made to correspond to a grid-cell. Remotely sensed images provide complete coverage of the system's spatial domain at a particular time. In some cases, images may be available for different times of year and for different years. This source of environmental data is a boon to environmental modellers and is invaluable in calibrating climate models. The use of geographical information systems in large-scale modelling has become fairly commonplace, and has made possible the calibration and testing of complex dynamic spatial models.

It is fair to say that mathematical modelling in the environmental and geographical sciences has shown its calibre over the last couple of decades. It has shown itself to be a flexible process adaptable to a variety of geographical and environmental systems (e.g. Dendrinos and Mullally 1985; Thomas 1992; Hardisty *et al.* 1993; Kirkby *et al.* 1993). At the same time, it is firmly founded upon a base of substantial theory. In 1972, Skellam pointed out the hazards of too much theorizing in an empirical science like ecology: 'Without enlightenment and eternal vigilance on the part of both ecologists and mathematicians there always lurks the danger that mathematical ecology might enter a dark age of barren formalism, fostered by an excessive faith in the magic of mathematics, blind acceptance of methodological dogma and worship of the new electronic gods' (Skellam 1972: 28). This foreboding of bad things to come has, it can be argued, proved to be wrong. Geographical and ecological scientists with a mathematical predilection have injected new life into their subjects without losing sight of the 'real world'. Field ecologists may not find Kauffman's theoretical work with mathematical models interesting or informative, but few people could read Lewin's (1993) book on complexity without being exhilarated by the sheer daring and brilliance of the ideas that, like Kauffman's, have flowed from the exploration of non-linear dynamics.

MOTHERING THE ENVIRONMENT

The Age of Ecology opened on 16 July 1945 in the New Mexican desert 'with a dazzling fireball of light and a swelling mushroom cloud of radioactive gases' (Worster 1994: 342). Observing the scene, a phrase from the *Bhagavad-Gita* came into project leader J. Robert Oppenheimer's mind: 'I am become Death, the shatterer of worlds.' For the first time in human history, a weapon of truly awesome power existed, a weapon capable of destroying life on a planetary scale. From under the chilling black cloud of the atomic age arose a new moral concern – environmentalism – that sought to temper the modern science-based power over Nature with ecological insights into the radiation threat to the planet.

The publication of Rachel Carson's *Silent Spring* (1962), the first study to bring the insidious effects of DDT application to public notice, was more grist for the environmentalist's mill. DDT is a chlorinated hydrocarbon. It was first prepared by Othman Zeidler in 1874. Paul Müller discovered its insecticidal properties in 1939 and, for doing so, received the Nobel Prize for Physiology. Like nuclear energy, DDT was deemed to be a panacea, in this case a complete solution to pest control. By the early 1960s, its persistence in the environment and accumulation in the food chain were becoming apparent. In 1972, after years of forceful lobbying and petitioning in the United States, DDT was banned for all but emergency use by the Environmental Protection Agency. Interestingly, a mathematical model of DDT storage and transport in the atmosphere, water, and living biomass, was included as evidence in a hearing conducted by the American State Department of Natural Resources considering a petition to ban the application of DDT in Wisconsin.

Modern environmentalism has influenced some of the developments considered in this chapter. Generalizations are difficult to make since there are several brands of environmentalism, each of which promotes a different attitude towards the environment (see O'Riordan 1988). These brands tend to polarize around, at one extreme, ecocentrism (Gaianism and communalism), with its decidely motherly attitude towards the planet; and, at the other extreme, technocentrism (accommodationism and cornucopianism) with a more *laissez-faire*, exploitative attitude. The 'green' end of this spectrum was initiated by the atomic bomb and galvanized into action by pesticide abuse.

It would be wrong, if excusable, to imagine that all ecologists and environmental scientists hold ecocentric views. The relationship between ecocentric environmentalists and ecologists has not always been as cosy as might be supposed: ecologists do not always say what ecocentrics want to hear. Paul Colinvaux (1980: 105) wrote: 'If the planners really get hold of us so that they can stamp out all individual liberty and do what they like with our land, they might decide that whole counties full of inferior farms should be put back into forest.' His displeasure with land-use planning and environmentalism is evident. And later in the book (Colinvaux 1980: 119), his words smack of social Darwinism and 'Nature red in tooth and claw', when he talks of different species 'going about earning their livings as best they may, each in its own individual manner', and what 'look like community properties' being 'the summed results of all these bits of private enterprise', though elsewhere he sees peaceful coexistence, not struggle, as the outcome of natural selection, the peace being broken in upon by a deviant aggressor species such as *Homo sapiens* who seeks to encroach on another's niche (Colinvaux 1980: 131). Today, few ecologists would wholeheartedly accept Colinvaux's viewpoint, and many approach Nature green in head and heart. One of the many is Daniel B. Botkin (1990) who advocated using modern technology in a constructive and positive manner, a position that tries to bridge the middle ground between ecocentrism and technocentrism.

Technology has now reached a point where the likely consequences of implementing environmental strategies advocated by the different groups of environmentalists can be predicted with some degree of confidence. A case in point is the models used by the Intergovernmental Panel on Climate Change (IPCC) to predict climatic change over the next hundred years using four different scenarios devised by Working Group III, one of which was business-as-usual and another, in effect, a 'green' scenario. The results, not unexpectedly, show the importance of shifting speedily to renewable energy resources to minimize global warming (see Houghton *et al.* 1990).

CONCLUSION AND PROSPECT

Major technological developments during the Second World War have had a deep-rooted influence on environmental observation, on environmental modelling, and on attitudes towards the environment. Smaller-scale technological developments have also made their mark. A prime example is the slope-profile recorder, hot news for geomorphologists in the early 1970s, the design for which was based on a contrivance for surveying enemy-held beaches. There were also technological innovations, including television, that were delayed rather than accelerated by the war. It is not unreasonable to suggest that television was later to influence public perception of environmental issues and to facilitate the growth of the global village and of global thinking, and to promote what Richard Dawkins regarded as the British Broadcasting Corporation's projection of Nature as a marvellous balance and harmony. A recent creation spawned by computers and communications satellites is cyberspace, a network of computers round the world – Internet (international network). Internet was set up by the US Defense Department to allow its researchers to transfer information between their computers. It now links about 25 million computers, and has evolved its own language and 'netiquette' for easy cruising on the information superhighway. Its main use is as a channel for e-mail, though it allows technofreaks from all parts of the world to 'chat', in pairs or in groups. In addition, it opens access to information as text and images. With the right software, high-quality sound can be received and downloaded. Soon, this technology will allow 'video-on-demand' – films and television programmes selected from a menu and played on request.

It seems undeniable that novel wartime technologies, and their developments, prompted the evolution of fresh environmental theories. To an extent, the process worked in the other direction as well: some theoretical innovations in environmental disciplines drove the design of hardware based on the new technologies. This was true of film for aerial photographs. The first suggestion to provide colour processing for infra-red film, for instance, came from researchers at the Kodak Laboratories, England, who saw its potential use in detecting military camouflage. In addition, the instruments used in earth-observing systems were designed for purposes demanded by the different environmental sciences. These matters are part of a bigger issue – the relationship between technology and science. The examples discussed in this chapter suggest that, although science and technology to some extent have gone hand in hand, the impetus for a decade or so after the Second World War came from technology. Only later did environmental theory occasionally lead hardware design. To be sure, without the hardware, some theories would not have appeared. Without a computer on which to simulate the weather, Edward Lorenz is unlikely to have stumbled across the importance of initial states in deterministic systems (the Butterfly Effect) and so sparked off the growth of the new science that was later to be called chaos theory. Equally, theory has prompted the technological applications and developments. Lovelock's Gaia hypothesis led to the discovery of dimethyl sulphide and halocarbons in the atmosphere, and global change research, which seeks to understand how the earth functions as an integrated system, necessitates the development of appropriate remote sensing technology. In the field of geographical information systems, conventional systems are not efficient at handling very large spatial databases. This problem can be surmounted by changes in computer design (to make access to large databases faster and more efficient) and in user-interface design (to make interaction with the database simpler).

Evidently, as Lewis Wolpert (1992) claims, the relationship between technology and science is complex. There is not a simplistic unidirectional determinism in which technology drives science or science drives technology. None the less, the last fifty years have seen an unprecedented technological revolution that has had a substantial impact on the geographical and environmental sciences. Perhaps the most frightening impact of all is that high-speed electronic computers have enabled the human species to glimpse into the future as never before. Sophisticated computer models have provided a ghastly vision of life after a nuclear war, and predicted a hotter, more humid world next century. For the first time in human history, as the indirect result of a major military conflict, the human race has acquired a means of obtaining reliable warning of impending change. It can only be hoped that the decision-makers will have the good sense to face the challenges of global change in a responsible and humane way; that they will live up to the name given to the human species by an earlier generation – 'wise man'; and that they will think, not just of the present generation, but of generations to come.

REFERENCES

Barrett, E.C. (1975) *Climatology from Satellites*, London: Methuen.

Botkin, D.B. (1990) *Discordant Harmonies: A New Ecology for the Twenty-First Century*, New York: Oxford University Press.

Carson, R. (1962) *Silent Spring*, Boston, Mass.: Houghton Mifflin.

Chorley, R.J. (1965) 'The application of quantitative methods to geomorphology', pp. 147–63 in R.J. Chorley and P. Haggett (eds) *Frontiers in Geographical Teaching: The Madingley Lectures for 1963*, London: Methuen.

Chorley, R.J. (1967) 'Application of computer techniques in geography and geology', *Abstracts and Proceedings of the Geological Society, London* 1642, 183–6.

Colinvaux, P. (1980) *Why Big Fierce Animals Are Rare*, Harmondsworth, Middlesex: Penguin Books.

Colwell, R.N. (ed.) (1983) *Manual of Remote Sensing* (2 vols, 2nd edn), Falls Church, Va.: American Society of Photogrammetry and Remote Sensing.

Coppock, J.T. (1962) 'Electronic data processing in geographical research', *The Professional Geographer* 14, 1–14.

Cornelius, S.C., Sear, D.A., Carver, S.J. and Heywood, D.I. (1994) 'GPS, GIS and geomorphological field work', *Earth Surface Processes and Landforms* 19, 777–87.

Costanza, R., Sklar, F.H. and White, M.L. (1990) 'Modeling coastal landscape dynamics', *BioScience* 40, 91–107.

Coulson, M.G. (ed.) (1990) *Remote Sensing and Global Change. Proceedings of the 16th Annual Conference of the Remote Sensing Society*, Nottingham: The Remote Sensing Society, Department of Geography, University of Nottingham.

Culling, W.E.H. (1985) *Equifinality: Chaos, Dimension and Pattern. The Concepts of Non-linear Dynamical Systems Theory and Their Potential for Physical Geography*, Graduate

School of Geography, London School of Economics, Geography Discussion Paper, New Series No. 19.

Dendrinos, D.S. and Mullally, H. (1985) *Urban Evolution: Studies in the Mathematical Ecology of Cities*, Oxford: Oxford University Press.

Drake, J.A. (1990) 'The mechanics of community assembly and succession', *Journal of Theoretical Biology* 147, 213–33.

Esser, G. (1991) 'Osnabrück Biosphere Model: structure, construction, results', pp. 670–709 in G. Esser and D. Overdieck (eds) *Modern Ecology: Basic and Applied Aspects*, Amsterdam: Elsevier.

Estes, J.E. and Cosentino, M.J. (1989) 'Remote sensing of vegetation', pp. 75–111 in M.B. Rambler, L. Margulis and R. Fester (eds) *Global Ecology: Towards a Science of the Biosphere*, Boston: Academic Press.

Fisher, O. (1866) 'On the disintegration of a chalk cliff', *Geological Magazine* 3, 354–6.

Haggett, P. (1964) 'Regional and local components in the distribution of forested areas in south-east Brazil: a multivariate approach', *Geographical Journal* 130, 365–80.

Hagood, M.J. (1943) 'Statistical methods for delineation of regions applied to data on agriculture and population', *Social Forces* 21, 287–97.

Hardisty, J., Taylor, D.M. and Metcalfe, S.E. (1993) *Computerised Environmental Modelling: A Practical Introduction Using Excel*, Chichester: John Wiley.

Henshall, J.D. and King, L.J. (1966) 'Some structural characteristics of peasant agriculture in Barbados', *Economic Geography* 42, 74–84.

Hobbs, R.J. and Mooney, H.A. (eds) (1990) *Remote Sensing of Biosphere Function* (Ecological Studies, vol. 79), New York: Springer.

Houghton, J.T., Jenkins, G.J. and Ephraums, J.J. (eds) (1990) *Climate Change: The IPCC Assessment*, Cambridge: Cambridge University Press.

Kauffman, S.A. (1993) *The Origins of Order: Self-Organization and Selection in Evolution*, New York and Oxford: Oxford University Press.

Kendall, M.G. (1939) 'Geographical distribution of crop productivity in England', *Journal of the Royal Statistical Society* 102, 21–62.

Kirkby, M.J., Naden, P.S., Burt, T.P. and Butcher, D.P. (1993) *Computer Simulation in Physical Geography* (2nd edn), Chichester: John Wiley.

Krumbein, W.C. (1959) 'The "sorting out" of geological variables illustrated by regression analysis of factors controlling beach firmness', *Journal of Sedimentary Petrology* 29, 575–87.

Lewin, R. (1993) *Complexity: Life at the Edge of Chaos*, London: J.M. Dent.

Lovelock, J.E. (1989) 'Geophysiology', *Transactions of the Royal Society of Edinburgh: Earth Sciences* 80, 169–75.

Mather, P.M. (ed.) (1992) *TERRA-1: Understanding the Terrestrial Environment. The Role of Earth Observations from Space*, London: Taylor & Francis.

Mather, P.M. (ed.) (1995) *TERRA-2: Understanding the Terrestrial Environment. Remote Sensing Data Systems and Networks*, Chichester: John Wiley.

Melton, M.A. (1957) *An Analysis of the Relations among Elements of Climate, Surface Properties and Geomorphology* (Office of Naval Research Project NR 389-042, Technical Report 11), New York: Department of Geology, Columbia University.

O'Riordan, T.J. (1988) 'Future directions for environmental policy', pp. 168–98 in D.C. Pitt (ed.) *The Future of the Environment: The Social Dimensions of Conservation and Ecological Alternatives*, London: Routledge.

Pimm, S.L. (1992) *Balance of Nature? Ecological Issues in the Conservation of Species and Communities*, Chicago, Ill.: Chicago University Press.

Rudd, R.D. (1974) *Remote Sensing: A Better View*, North Scituate, Mass.: Duxbury Press.

Simberloff, D.S. (1982) 'A succession of paradigms in ecology: essentialism to materialism and probabilism', pp. 63–99 in E. Saarinen (ed.) *Conceptual Issues in Ecology*, Dordrecht, Holland: D. Reidel.

Skellam, J.G. (1972) 'Some philosophical aspects of mathematical modelling in empirical science with special reference to ecology', pp. 13–28 in J.N.R. Jeffers (ed.) *Mathematical Models In Ecology*, Oxford: Blackwell Scientific Publications.

Sklar, F.H. and Costanza, R. (1991) 'The development of dynamic spatial models for landscape ecology: a review and prognosis', pp. 239–88 in M.G. Turner and R.H. Gardner (eds) *Quantitative Methods in Landscape Ecology: An Analysis and Interpretation of Landscape Heterogeneity*, New York: Springer.

Terjung, W.H. (1976) 'Climatology for geographers', *Annals of the Association of American Geographers* 66, 199–222.

Thomas, R.W. (1992) *Geomedical Systems: Intervention and Control*, London: Routledge.

Tilman, D. (1994) 'Competition and biodiversity in spatially structured habitats', *Ecology* 75, 2–16.

Tilman, D., May, R.M., Lehman, C.L. and Nowak, M.A. (1994) 'Habitat destruction and the extinction debt', *Nature* 371, 65–6.

Townshend, J.R.G. (ed.) (1981) *Terrain Analysis and Remote Sensing*, London: George Allen & Unwin.

United States National Oceanic and Atmospheric Administration, United States Department of Commerce, and National Aeronautics and Space Administration (1987) *Space-Based Remote Sensing of the Earth*, Washington, DC: US Government Printing Office.

Verstappen, H.Th. (1977) *Remote Sensing in Geomorphology*, Amsterdam: Elsevier.

von Thünen, J.H. (1826) *Der Isolirte Staat in Beziehung auf Landwirtschaft und Nationalökonomie*, Hamburg: F. Perthes.

Whittaker, R.J. (1987) 'An application of detrended correspondence analysis and non-metric multidimensional scaling to the identification and analysis of environmental factor complexes and vegetation structures', *Journal of Ecology* 75, 363–76.

Whittaker, R.J. (1989) 'The vegetation of the Storbreen gletschervorfeld, Jotunheimen, Norway. III. Vegetation–environment relationships', *Journal of Biogeography* 18, 41–52.

Wolpert, L. (1992) *The Unnatural Nature of Science*, London: Faber & Faber.

Worster, D. (1994) *Nature's Economy: A History of Ecological Ideas* (2nd edn), Cambridge: Cambridge University Press.

FURTHER READING

Hardisty, J., Taylor, D.M. and Metcalfe, S.E. (1993) *Computerised Environmental Modelling: A Practical Introduction Using Excel*, Chichester: John Wiley.

Hobbs, R.J. and Mooney, H.A. (eds) (1990) *Remote Sensing of Biosphere Function* (Ecological Studies, vol. 79), New York: Springer.

Huggett, R.J. (1993) *Modelling the Human Impact on Nature: Systems Analysis of Environmental Problems*, Oxford: Oxford University Press.

Hyatt, E. (1988) *Keyguide to Information Sources in Remote Sensing*, London and New York: Mansell Publishing.

Jakeman, A.J., Beck, M.B. and McAleer, M.J. (1993) *Modelling Change in Environmental Systems*, Chichester: John Wiley.

Kirkby, M.J., Naden, P.S., Burt, T.P. and Butcher, D.P. (1993) *Computer Simulation in Physical Geography* (2nd edn), Chichester: John Wiley.

Lewin, R. (1993) *Complexity: Life at the Edge of Chaos*, London: J.M. Dent.

Mather, P.M. (ed.) (1992) *TERRA-1: Understanding the Terrestrial Environment. The Role of Earth Observations from Space*, London: Taylor & Francis.

Mather, P.M. (ed.) (1995) *TERRA-2: Understanding the Terrestrial Environment. Remote Sensing Data Systems and Networks*, Chichester: John Wiley.

Rambler, M.B., Margulis, L. and Fester, R. (eds) (1989) *Global Ecology: Towards a Science of the Biosphere*, Boston: Academic Press.

Schneider, S.H. and Boston P.J. (eds) (1993) *Scientists on Gaia*, Cambridge, Mass. and London, England: The MIT Press.

Worster, D. (1994) *Nature's Economy: A History of Ecological Ideas* (2nd edn), Cambridge: Cambridge University Press (especially Part Six).

22

ENVIRONMENTALISM ON THE MOVE

Timothy O'Riordan

CLAYOQUOT SOUND: SYMBOL OF ENVIRONMENTAL PROTEST

Clayoquot (pronounced 'Claquet') Sound may never make the environmental big time on a par with Seveso, Chernobyl or Exxon Valdez. But what is happening on the western slopes of Vancouver Island in British Columbia is a classic example of the modern environmental struggle.[1] It raises all the contemporary issues – rights of natural objects to exist, respect for the values of indigenous peoples, the scope for 'green tourism' and sustained non-consumptive economic opportunity, citizen power over land-use decisions where property rights belong to the world as a whole, not wholly to the commercial timber licence holders, loss of trust for a government that consults through due process then ignores the result, the failure of an independent mediating body to arbitrate an outcome through public consent, and the willingness of thousands of law-abiding people to go to jail, if necessary, to stop the bulldozers. Clayoquot Sound speaks for the spirit of post-Rio environmentalism.

British Columbia still depends for about 20 per cent of its economy on the forestry industry, even though the push towards more efficient wood utilization and the growing effectiveness of paper recycling campaigns are lowering the profitability and urgency of new timber removal. The Clayoquot area of the west coast of Vancouver Island is one of the last remaining reserves of old growth temperate forest in the Pacific Northwest, not otherwise protected by national park or provincial/state status. Four different bands of indigenous 'first-nation' British Columbian (BC) Indian tribes still live in the area. These are relatively marginal communities in economic terms, but proud of their traditions and their robust self-reliance. Above all, they believe that the area is an integral component of their heritage and continue to depend on the Sound's marine and coastal resources for their economic, social and cultural needs and aspirations.

Successive BC governments have tried to approach forest management in a

449

manner that academic resource analysts would generally favour. The cabinet created a Wilderness Commission in 1986 to review the wider implications for land use, employment, wildlife and regional development of logging and environmental protection. This body took hearings from all manner of interests, including the first-nation peoples. In addition the government established three mediating forums. The first, formed in 1989, was the Clayoquot Sound Sustainable Development Task Force composed mostly of local and provincial government representatives. This was widened into a 'round table' called the Clayoquot Sound Sustainable Development Strategy Steering Committee, consisting of representative opinion over aquaculture, tourism, small business, mining and environment plus provincial and federal government representatives.

Finally in 1992 the government created a Commission on Resources and the Environment (CORE), charged with looking comprehensively at a host of resource extraction and environmental protection matters. For the timber industry its remit was to examine the wider questions of timber property rights, economic and legal safeguards and 'best practice' logging. The last is a very woolly concept designed to placate moderates. It is essentially a wish to log sustainably in the narrow sense of being both efficient, in terms of minimizing waste, and environmentally protective (if not friendly) via replanting of timber on a one-for-one basis. Ecologically speaking this is no safeguard, but it was at least a sop to the old clear-cut logging practices so widely condemned in the modern age, and a recognition that timber management at least could be made to be replenishable for future generations.

By late 1992 the round table had reached an impasse. Because some interim logging had continued, the environment representative had already walked out. In any case there was no strategy. CORE should have been activated, but was not: the government never gave a good reason for this decision. So neither a consensus-based approach nor an independent arbitration mechanism was available.

The trouble lay in the failure of the BC government to read the mood of the times. Logging wild forest is no longer tolerable to a majority of BC residents, to say nothing of concerned citizens world-wide. The mood is to contract the forest industry to existing areas of cut and to insist on programme replanting, to divert some of the job specification of loggers into forest restoration, and to swing forestry economics towards green tourism and value-added services not dependent on logging at all. Residents and citizens alike want to see the BC forest industry come of sustainable age: this means ruthless protection of the pristine, and a radical new look at the stewardship role of the logging contractor. More money, over a long period of time, can be made from carefully managed tourism, whale-watching and community support services, so long as the value of these old growth forests is comprehensively calculated. Indigenous peoples can continue to run their coastal lives as they have done for years, with more control over education and economic development. This was the consensus of

450

the round table exercise, beyond the logging companies and the majority of their employees.

But the BC government pre-empted both the round table and CORE by announcing in July 1993 that it would issue forest licences for two-thirds of the million hectare forest to companies on a 'good practice' basis, and that 350,000 ha would be set aside as a BC forest heritage. MacMillan Bloedel, the US-owned but former BC company that all but ran BC governments in the past, was unable to log due to countless acts of civil disobedience by people from all over the Pacific Northwest, and from all walks of life. The vast majority had never blockaded before: to save Clayoquot was a matter of fundamental conviction and symbolic principle.

In the event over 800 people were arrested, with many facing a $1,000 fine or up to four months in prison. Many of the better off will appeal, dragging the process on for many months. The trial judge doubled the usual sentences, because, he noted, contempt of court was a serious offence and, in any case, defying court orders is not the way to work within a democracy. This in turn led five government politicians to admonish the court publicly, on the grounds that the blame for harsh treatment of protesting citizens lay with the courts, not the government. The BC Chief Justice reacted strongly: the criticism, he noted, was ill-informed. He added: 'we are entitled to expect better from politicians and lawyers' (one of the government protesters was a professor of law).

Meanwhile, the protests continued to the point of crisis. In desperation the government announced, on the recommendation of CORE, the creation of yet a new advisory forum, a scientific panel of twenty members, to advise on 'sustainable logging practices'. This panel is composed of academics and consultants with little direct experience in 'coalface' logging. It will inevitably become bogged down in definitions of sustainability, let alone sustainable logging, so will almost certainly split on key matters of principle and practice. The net result will be a fudge, with guidelines on forest management that could well prove uneconomic or unworkable.

The problem of the BC government is that by issuing the tree felling licences, it has opened itself up for a massive compensation claim should these licences be revoked. To convert the area to the Pacific Rim National Park would involve similar compensation, a figure probably equivalent to the discounted value of the net profits forgone. This would probably exceed $1 billion. However, to attach 'sustainability' conditions to 'best logging practice', even though unworkable, may well avoid the charge of compensation, while allowing the government to claim that it is in the vanguard of a 'post-Rio' forest management strategy. For the majority who want Clayoquot strictly left alone this political ruse is not only despicably devious, but wholly avoids a crucial matter of principle. For them Clayoquot should have intrinsic rights to exist: it is a biodiversity treasure, the home of peoples who have valued its existence for far longer than whites have lived in Canada.

Finally it should be noted that the BC government has reluctantly acceded to a request to re-establish sustainability panels at the local level, with the aim of reaching shared decisions (i.e. consensus). This has not worked before, mostly because representative opinion is at cross-purposes, even on a community level. Few members of the community panels command the loyalty of a large body of opinion: the majority of activists are reflecting minority interests. Meanwhile, to placate environmental opinion further, the government has recently created two new wilderness-biosphere reserves run by joint boards of officials and residents. In addition, in order to respond to the suspicion of the first-nation peoples, partly as a result of a very critical ombudsman's report, the government created a Central Region Resources Board. This is essentially a provincial indigenous tribal government task force with powers to reach consensus on logging, environmental management and appropriate economic development. This does not provide a veto power for the first-nation people, but they enter the Board with the moral authority to do so. In practice this will do nothing to overcome the deep decisions that run through all strands of consultation and decision-making in the Province. Anger, despair and resignation intermingle, and no one is any the wiser on how to manage west coast logging and environmental protection.

The Clayoquot story reveals a number of themes about modern environmentalism that will be elaborated in the text that follows.

1 Native earth now has considerable emotional attachment for people who see in its despoliation a great need to stand firm on habitats that are precious and irreplaceable. The social valuation of such areas can be estimated by clever willingness to pay studies: truly natural areas are invariably worth far more left alone than as commercial and substitutable timber. The ecological economics of biodiversity safeguard are very powerful because the valuation implies a global sharing of the protection burden. Clayoquot should not be saved simply by BC money, though as yet the mechanisms for the transfer of such resources remain poorly developed.

2 Governments have yet to grasp the significance of sustainable economics, partly because the revenue arising from sustainable resources management is not always obvious or evident, and partly because there are no readily discernible lobby groups promoting the cause of 'green budgets'. Understandably therefore, governments tend to fall into uneasy compromises so as to protect their political support from industry and trade unions, in the absence of identifiable income from green tourism and bequest economics. In short, it is one thing to calculate an economic value, quite another to obtain a revenue. Forests are still more profitably cut down than saved, particularly when the property right remains vested in a commercial tree-felling licence.

3 Consultation exercises involving many representative stakeholder groups rarely succeed unless the actual round tables are vested in arbitration powers. If these bodies remain merely as advisory talk shops, their role is

452

counterproductive. Committed people lose heart and trust as the various antagonists have no incentive to settle. Similarly any mediating body, such as CORE, is all but worthless unless its role is defined clearly and its reports are public and debated by legislative committees. Again it is false legitimization of public trust to establish such bodies simply to leave them twisting in the political wind.

4 Civil disobedience is a legitimate tactic, even if illegal, when the normal procedures of democracy fail or command no faith. People reluctantly feel they have to take matters into their own hands to pursue their civil conscience.

5 At the political margins environmental concern can make or break a government. The appropriate context for this to happen is a closely contested election where a minority of votes can propel in one party or displace the other. In the Clayoquot issue the Canadian MP for the area won against the provincial trend on the basis of seeking to convert the forests to a national park. In 1983 the Australian Labour Party gained power on the backs of a protest vote against the Franklin Dam in Tasmania, supported by the state as well as the ruling Liberals, but opposed by Labour. In the UK environmental groups are planning a campaign of targeting marginal constituency prospective parliamentary candidates on the basis of their green credentials. Given good organization and discipline this tactic might just make the difference in truly marginal constituencies, with vitally important repercussions for the main political parties.

6 Clayoquot is not a BC issue: it speaks for earth. The property right of the forest is the property right of all global citizens including those yet to be born. The internationalization of environmental issues creates a whole new perspective on the role of democracy, diplomacy and legality. No longer can democratic governments act in isolation of informed and articulate world opinion. For the majority of national government, this message remains obscure. Yet as global conventions tighten and independent surveillance becomes more codified, so the room for obscurity diminishes. We have yet to see formal sanctions applied against an environmental miscreant. But consumer boycotts are not uncommon. Witness the successful campaigns against animal fur, Norwegian whaling and do-it-yourself stores selling tropical timber.

The BC round table on Sustainable Development for Vancouver Island broke up over the Clayoquot issue. Initially the environmental group representatives walked out, but soon the forestry groups also stopped attending. Sadly this two-year exercise in participatory mediation failed because it was insufficiently supported by the main political interests. In April 1994 the Commission on Resources and the Environment reported that 12 per cent of the area should remain unlogged, that 26 per cent should be subject to sustainable logging practices based on the highest environmental safeguards, that 40 per cent should be logged commercially but with replanting, and that the remaining 22 per cent should be incorporated into integrated catchment management programmes. This meant in effect that 900 logging jobs would go, though the BC government created the post of a forest commissioner whose task it is to find alternative employment for all workers made redundant by the deal.

The reaction was very hostile on the part of the logging industry and the families of the loggers. In July 1994 over 2,000 protested against the decision on the lawns of the BC Legislature. Environmental groups and native Indian organizations looked for the fine print. The precise configuration of the unlogged areas was obviously of critical importance, as was the nature of the remaining three categories of logging practices, as well as their location. The round table has reconvened to look at all this. Ironically the future of CORE is in doubt. Nobody liked the way it handed down its recommendations, boxing in the government and apparently ignoring the views of the various key players. Arbitration is one matter, but supportive mediation is very much another. Modern environmentalism is fundamentally a participatory exercise these days: middle-managing think-tanks and advisory bodies often lose popular and political support simply because they are regarded as too aloof. The final result is probably a workable and acceptable outcome, but it would also have been arrived at more quickly and harmoniously had the round table and CORE been enabled to co-operate as one.

THE CHANGING CHARACTER OF MODERN ENVIRONMENTALISM

Environmentalism is the endless search for reconciliation. On the one hand humans seek to control their means of survival, but in so doing they endanger the very life-support systems that maintain their existence; on the other hand is their ability, through willingness to compromise on that control, to adapt their outlook and behaviour in the cause of global, and thus human, survival. From the beginning of human time, *Homo sapiens* has altered the conditions that influence its existence. In many ways proto-humanity was merely asserting its authority over environmental barriers and threats. The wholesale removal of forests by fire was in part a method of reducing the scope of marauding wild animals to inflict injury or death. Likewise the New Zealand Maori tribes burned bush to expedite the growth of the nutritious bracken fern. Ian Simmons (1993: 3–6) examines the evidence and concludes

- that hunter-gathers were ubiquitous over the globe, and very active in transforming almost all local environments;
- that their mixture of herbivorous and carnivorous eating habits dramatically influenced plant and animal populations;
- that there is no scholarly consensus on whether early societies were deliberately environmentally destructive, but that better explanations lie in

subtle changes in population numbers, and especially density, kinship rivalry, local wars caused by episodic migration of new cultures, and climatic stress.

So it seems that a degree of environmental transformation has been part of the human tradition, that this has more likely been convulsive rather than continuous, and that value change caused by interaction between new peoples or environmental conditions are important 'drivers' of the transformational process. Similar conclusions are found in Bennett (1976) Mannion (1991) and Ponting (1991).

A number of companion chapters in this volume chart the interesting but unresolved relationship between society and nature. Suffice to say here, from the burning of grasses and trees, to toxification of dryland soils to extensive soil erosion on desert margins, that even a primitive human technology caused extensive alterations in habitats and species. Environmental protection is as old as the recognition of environmental damage. In pre-mercantile days the protective element was provided by the application of inclusive common property rights. Forests, fisheries, fuelwood, water reserves were owned by tribal or village collectivities. Use was strictly controlled by a combination of reciprocal obligations both to replenishment as well as removal, and sophisticated sharing ceremonies in times of plenty. Inclusion meant exclusion of use by foreigners as well as a communal sense of obligatory survival to restrain excessive use. Individual resource managers negotiated with their neighbours as well as with the ignorance of environmental limits by responding to the vicissitudes of scarcity and of social disapproval if boundaries of replenishability were transgressed.

At its heart, environmentalism is the inner tension between the search for mastery over the earthly household and the recognition that such mastery will always be elusive. To come to terms with this sense of failure, humans turn to myth to reconcile their inherent vulnerability. This point has been well made by Burch (1972). Myth may be in the form of fable, or it may emerge as a belief in the nurturing power of 'mother earth' or, as in modern days, it may reflect a belief that the earth is a self-regulating organic whole, within which humanity must play its part to adapt and to respond to the parameters of survival. This last point is now contained in the wide-ranging discussion of the 'Gaia thesis', originally pinpointed as a scientific treatise by Lovelock (1979), and subsequently widened into an ethical imperative of stewardship, trusteeship and sustainable living (see Schneider and Boston 1991).

Modern environmentalism may be depicted as a multiple layering of this fundamental tension. A summary of the four layers – views of nature, views of intervention, views of political economy, the infusion of social movements – follows.

Views of nature

In scientific terms three views of nature can now be discerned. The first is empiricist, positivist and seeks to be objective. It is based on the post-Newtonian principles of hypothesis testing, verification, falsification, replication and modelling. The key is the representation of reality by symbols and principles, couched in data, relationships, inference and causality. This approach still remains the basic core of contemporary environmental science, since it forms the essence of modelling, predicting and evaluating. This perspective is given fuller treatment by de Groot (1992).

The second view is more interactive. It regards the earth as a set of processes that only have value in human terms. This value can be given economic expression by estimating the productive value of natural processes in assimilating, shielding, buffering and replenishing. Thus tropical forests sequester carbon at the rate of 100–300 tonnes per ha per year, or at least this is the best guess/estimate (Brown and Adger 1993). By another set of calculations one additional tonne of carbon causes $20 per year in environmental damage in the form of sea-level rise, land-use change, water-cycle disruption and temperature variation (Frankhauser 1993). Empiricist science provides these figures, admittedly with huge margins of error and uncertainty, but it is the interactive science of social valuation that allows us to estimate the 'worth' of tropical forests at $1,500–2,500 per ha per year in sequestration capacity alone (Brown and Adger 1993). This is still a very young variant of environmental science, but as the sustainability debate intensifies, of which more below, so this version of ecological economics will expand (see Barbier et al. 1994).

The third view is regarded by many as not scientific at all. It is the Gaian perspective of an earth that provides the parameters for species survival, with enormous resilience in the face of disturbance. Gaianism places no special purpose or role for humans. It provides an earth-centred perspective in which humans have to find their place in the natural order of things. Consequently the great cycles and rhythms of earth nourishment have their own intrinsic rights and purposes. Humans must respond in creative solidarity, exhilarated by the recognition of being but an infinitesimal part of a grander whole. This is the basis of concepts such as deep ecology and transpersonal philosophy, and is finding its way into new legal doctrines of natural rights. The main arguments are best summarized by Johnson (1991).

None of these 'sciences' sits easily in the proximity of the other. The scope for reconciling these views through imaginative interdisciplinarity is simply enormous, but as such has barely been addressed. Had it been, the Clayoquot débâcle would never have taken place.

Views of intervention

These three approaches to scientific interpretation of the earth give us three variants of 'greenness'. The word 'green' is a term of art to recognize some degree of empathy with the need to respond to the constraints set by natural systems. The nature of this response is very different.

Dry greens believe that there is no inherent restriction on human endeavour. Given good use of the market, scope for managerial excellence, and the guiding hand of self-discipline, there is nothing that cannot be achieved. This is generally a discredited view, but arguably it remains at the core of fiscal economics, or the economics of ever-expanding development and material acquisition. Still the most comprehensive statement of this view is found in Simon and Kahn (1984), though the standard right-of-centre think-tank discussion document is also imbued with this perspective, as are many editorial commentaries in *The Economist*.

Shallow greens are now in the centre of the frame. Here the belief is that by adjusting the traditional approaches to social and economic change, to take into account ecological economics, so humanity can flourish in an environmentally sensitive manner. This is the nub of the so-called sustainable development debate, of which more below (see pp. 463–7). Shallow greenism promotes the cause of ecologically sensitive cost–benefit analysis, policy-centred environmental impact assessment, and the burgeoning acceptance of the precautionary principle.

Deep greens reflect the Gaianist, earth-centred view already discussed. They represent solidarity with natural processes and ecosystems, including the original populations of earth-sensitive peoples who become mythologized as the symbols of sustainable survival. Pepper (1991) examines the philosophies of those living in 'new age communes' to see how far deep-green thinking pervades the attitudes and behaviour of residents. He discovered a large amount of disillusionment, brought on by lack of basic necessities, constant change of personnel and a tendency towards autocracy amongst the original leaders and followers. He concludes that, over time, most, if not all, 'communards' adopt the values of liberal orthodoxy, not uncommon in well-meaning, mainstream society.

It is important to realize that each of these dimensions of greenness are within most, if not all individuals. Each perspective will surface at various times, and in different ways in the numerous roles people play – as parents, industrialists, entrepreneurs, carers, observers and political activists. All three dimensions are apparent in the Clayoquot issue: those who oppose the logging for the most part still drive cars, demand goods that cause disturbance of ecosystems, and place their savings in pension funds that chase around the world looking for profitable investment.

Views of political economy

The three modes of greenism in turn reflect three visions of a political economy. Again the models are neither unique nor independent of each other. They are simply metaphors for a preferred social order. Dry greens seek centralized government but decentralized markets: regulation is minimal and only directed at common property regimes, where ideally a variant of market forces can be extended to allocate environmental resources fairly and efficiently.

Shallow greens prefer a federated political economy, divided into international regimes for global management, as is now appearing in the aftermath of the UN Conference on Environment and Development. This should be matched by local initiatives at the personal and household level but linked to communities of common identity, through which the striving for sustainable lifestyles can best be enforced. This vision suggests a return to the idealized interpretation of the communal reciprocity of inclusive commons that sets the framework of citizenship.

Deep greens prefer a far more decentralized version of a political economy, essentially a network of self-reliant communities interacting through a mixture of the money economy and service exchange on a reciprocal basis. Some sophisticated studies by economists and sociologists indicate that proto-barter exchange economies could become the basis of a future political economy, though few can contemplate the reality of such a radical transformation of markets.

Infusion of social movements

Environmentalism thrives by being injected with ideas and adherents of other social movements, whose aims and political purpose overlap with environmentally more stable and socially more peaceful and just futures. These include *the right to know*, by no means guaranteed, even in so-called 'sunshine law' societies. Yet arguably, information about changing environmental conditions and their human/ecological effects are of the utmost significance. A second major theme is *the right to health*, signed by the UN as a legitimate aspiration for all peoples, but of course not practised. In the developed world, health threats are an important driving force for environmental concern over air and water pollution, the reduction of toxic wastes and the clean-up of contaminated sites (see Hays 1987). But in the developing world, much commoner ailments associated with poor sanitation and inadequate prophylactic medical care dominate this debate, and environmental threats are given less prominence. The third push comes from *consumer pressure*, notably the demand for boycotting goods from countries with a poor environmental record (e.g. Norway for whaling, and McDonalds for alleged cattle ranching in Central America). The growth of ecolabelling is another aspect of this theme, though it is likely that ecolabelling will always prove unsatisfactory for effective consumer choice. The rise in *ecofeminism* is also an important development, notably in the developing world, where women's rights to credit, to their own fertility and to collective economic enterprise should have considerable social advantages for future societies. Equally important is the attempt to break down the false dichotomy between man and nature, essentially as a metaphor of man's supremacy over woman (see Mies and Shiva 1993). Finally there is the growth of *animal rights*, linked in turn to views on *intrinsic natural rights* (see Stone 1989) and to more complicated questions of direct action, of possibly a violent kind, in the name of saving the earth. At one end of the spectrum the cause of solidarity with the earth dominated the Clayoquot issue, as is evident in much of ecofeminism writings. In the middle the theme of precaution, coupled to an ecologically centred view of environmental harm, occupies the political debate (O'Riordan and Cameron 1994). At the extreme is the controversial promotion of violence in the form of civil disobedience, as now practised by active cells such as Sea Shepherd and Earth First! in the US, and the Environmental Liberation Front in the UK. Links to subversive terrorist groups are not uncommon, so it is no wonder that this outpost of environmental concern is being closely watched by the intelligence services and by Special Branch.

ENVIRONMENTAL MOVEMENTS, INTEREST GROUPS AND PUBLIC AWARENESS

The upwelling of ideas, activists and constellations of ideologies into the environmental movement for these social infusions has given environmentalism vitality and dynamism. It is the adaptability and flexibility of environmentalism in the face of changing scientific evidence and economic values that allow it to remain a potent political force. Naturally, the public interest environmental pressure groups have to adjust to these changing times as well (see Rose 1993). By and large they are doing so, across a broad range of activities measures from policy influence to scientific critique, from active and litigious campaigning to various levels of civil disobedience. The variety of policy and campaigning styles and targets is itself a powerful source of the continuing political and educational influence of these groups.

In Britain the total membership exceeds 3 million different individuals and some 5 million actual subscriptions. The total amount of money spent annually by some 2,000 groups (of which all but a handful are very small) exceeds £150 million annually. This is a most effective force in the land, easily equivalent to the media and even to the judiciary in influencing political opinion. The most important recent developments in the non-governmental environmental movement in the UK are:

1 *Corporate management* complete with experienced chief executives, sophisticated computerized membership and recruitment drives, and organized

campaign teams. This creates a sense of coherence, continuity and order to the groups, as well as a drive and a purpose.

2 *Loose coalitions with non-environmental groups*, such as development charities, social welfare organizations and Amnesty International as well as other civil rights groups. These coalitions are presently very embryonic, but are growing in importance and cohesiveness. It is a development that deserves careful scrutiny in the next few years.

3 *Industry outreach* in the form of advice, information campaigns and constructive targeting of responsive industrial directors. The links to industry, no matter how apparently 'green' they appear to be becoming, are a matter of much controversy within the environmental movement. The majority of moderate opinion see this as unavoidable, given the prominence of industry in a free trading world, now the General Agreement in Tariffs and Trade is finally signed, and the scope for industry to undertake self-policing environmental audits. These last programmes are promoted by the voluntary ecoaudit directive of the European Union, but are increasingly being regarded as essential for good management. These large corporations use the auditing exercise to become accredited for sound environmental practice. An example is the growing willingness to adopt the British Standard 7750. This is a formal management audit aimed at waste minimization, up-the-pipe pollution control, the adoption of energy- and materials–saving technology, and the creation of community advisory panels to advise and to consult on local priorities and communications issues. (The monthly magazine *ENDS*, published by Environmental Data Services, provides excellent coverage of these issues.)

4 *Public education* via campaigning on issues that attract public interest, for example dolphins caught in tuna fishing nets, tropical hardwood in do-it-yourself stores, and realistic alternatives to the car for urban and leisure transport. In addition the major NGOs are providing good copy for local sustainable development strategies, as well as greening the school curricula, and some are now actively engaged in education for the so-called sustain-ability transition.

5 *Interactive science* is a key component of most environmental campaigning. All the large organizations employ science teams and education officers whose job it is to assess the latest scientific state of the art, infuse it with the precautionary principle and seek to deploy it in the big issues of our time. These include ozone depletion, acidification, wildlife and habitat loss, climate change and the steady toxification of whole ecosystems by micro-emissions of persistent and bioaccumulative substances. Without that scientific credibility no NGO campaign is likely to be successful. Small teams can work for up to two years before a truly powerful campaign is launched.

PLANETARY CHANGE AND PUBLIC CONCERN

The advance of information technology has transformed modern science. Satellite imagery provides estimates of changes in land use, atmospheric and ocean chemistry, soil fertility and water-resource availability. Electronic tagging reveals the territorial range of fish and mammals, while regular surveys of 'indicator species', key representatives of habitat diversity and health, record changes in species distribution on comprehensive data sets. Such measurements must be backed up by field surveys and by very careful anthropological studies of the meaning of resource use for local populations. For example, soil erosion is not regarded in the same way by all societies. It depends on the remedial techniques on offer, on adaptive responses to soil and water conservation, to the amount of labour available, and to traditional rights and obligations to property rights. These points are well covered by Blaikie and Brookfield (1987) and by Chambers (1988).

The search is on for a global environmental audit, an assessment of the vulnerability and resilience of earthly systems. Vulnerability equates with thresholds, irreversible change and systems breakdown, as in the chemical transformations that give rise to eutrophication, or excessive enrichment by nutrients, or acidification, the destruction of base-poor soils and waters. Resilience is the scope for ecological or biogeochemical buffering, the capability to withstand alteration, at least for the time being.

Many natural systems are initially resilient, so it would be wise to examine how they work, through what processes they stabilize change, and what are the indicators of stress. This is the basis of *critical load* studies for acidification research. Such surveys examine the resilience and vulnerability of key species and habitats to assess how much sulphur and nitrogen compounds can be tolerated over large regions. This work is controversial because the state of knowledge and the historical record are not well known, so once again the precautionary principle has to be applied. This means, in essence, acting in advance of scientific proof, and placing in the benefits stream the possible costs of being horribly wrong. At the point of decision, therefore, costs outweigh known benefits, but society chooses to hesitate, in the name of taking care of future generations.

Any science audit is bound to be approximate. Physical processes are more variable across space and time than had previously been imagined. Methane releases from rice paddies may fluctuate by as much as a factor of 500 (Mudge and Adger 1993), the desert margin is a product of rainfall variability, grazing patterns and soil–water chemistry. Plankton on the sea surface fluctuate with ocean currents, sea surface temperature, cloudiness (which they in part create via the transmission of cloud-forming aerosols from marine phytoplankton) and winds. Projecting life-threatening futures on the basis of reporting short-term trends is hazardous. In part, science can only move forward by carefully calibrated monitoring and recording. Funding basic science capability and data collection is not very appealing to the non-scientist, but even a modest investment in instrumentation and training would pay rich dividends.

The beginnings of a scientific audit are evident, however. This formed the basis of the UN Conference on Environment and Development held in Rio de Janeiro in June 1992. Four publications capture this picture. One is the proceedings of a specially convened science congress held in Vienna in 1991 (Dooge *et al.* 1992). Another is the annual report published by the World Resources Institute (1992). A third is the annual *State of the World* essays produced by Worldwatch Institute (Brown and Wolf 1988), and the final one is the UN Environmental Data Report prepared by the UN Environment Programme (1992). In addition geographers and other scientists have assessed the transformation of the earth over the past 300 years in major analyses (Turner *et al.* 1991), and Gaia Books has recently revised its excellent *Gaia Atlas of Planet Management* (Myers 1993). Add to this the World Development Report of the World Bank (1992) and the reader should gain access to all the relevant material to form a view.

1   *Land degradation*: since 1945 1.2 billion ha or 11 per cent of the vegetated surface of the earth has been moderately or severely eroded. Of this, 300 million ha are so degraded as to be uncultivatable. In much of this area the cost of restoration is more than the cost of reclaiming new lands, so as population expands more and more of this area will become vulnerable in the absence of sustainable cultivation measures. Vulnerability is mostly induced by inappropriate cultivation techniques, usually imported via aid or commercial agencies, by the need to meet foreign revenue debt by cash cropping, and by population pressure.

2   *Water availability*: throughout the Middle East and North Africa annual rates of water withdrawal currently exceed supply. Freshwater will either have to be made via desalination, or water efficiency will have to be legislated and priced. In areas such as western North America, central Europe and South-East Asia contamination of groundwater by agricultural chemicals is beginning to affect the potability of domestic water supply.

3   *Fisheries*: in eight of the world's seventeen ocean fisheries annual catches now exceed the maximum sustainable take. The annual catch in 1990 was 84 million tonnes – an increase of 35 per cent on 1980 levels. No wonder fishing is being toughly regulated. At current rates of depletion the world's fisheries may take decades to restore.

4   *Fuelwood*: even today about 60 per cent of the developing world's peoples are dependent upon trees for their energy. Population growth and economic change are increasing this dependency by up to 4 per cent per year. Yet the scarcity of fuelwood means not only soil erosion from formerly forested soils, it also creates a longer journey for fuel. This can be dangerous in regions torn by civil strife. So animal dung can be diverted from fertilizing the soil to a fuel. Yet the dung is worth much more as a fertilizer. In Ethiopia, for

example, this practice of burning animal dung has cost the impoverished nation almost 10 per cent of its annual wealth in added soil erosion and siltation of farmland (Pearce 1993).

5 *Pollution and waste*: in most Third World economies, including recently democratized Europe, environmental pollution is so universally bad as to pose a health risk. Estimates vary, but the average life span of Russians in the twenty-three ecological disaster zones could be shortening by as much as one year per decade (World Resources Institute 1992). Waste disposal is creating environmental and economic problems, especially as liability for any adverse side effects is increasingly being placed on the disposer.

6 *Biodiversity loss* remains one of the great incalculables. The best guess is that there are 10 million species on earth, of which less than 1.5 million are identified. In the tropical forests alone some 5 million species at least may exist, with less than one-third fully recorded. Estimated guesses on the basis of current rates of removal suggest that as much as 25 per cent of all species could disappear by 2050 with three-quarters of tropical forest plants at risk.

This is a sorry tale. The consequences for humanity are simply not forecastable with present scientific knowledge. Human societies have displayed remarkable powers of adaptation and resilience in the face of appalling stress, so there is scope for a creative response. A start can be made by mobilizing public opinion. A survey of some 40,000 people in twenty-four countries as part of a health of the planet study for the Rio Conference (Dunlap *et al.* 1993) showed that surprisingly similar numbers of people from all ranges of income and social experience rated environmental problems as serious, as affecting their health, and as worthy of being stopped by higher prices and higher taxes so long as the revenue was clearly directed at environmental problem-solving. In sum the evidence is that concern over environmental futures is equally shared across the globe and that environmental degradation is almost universally regarded as a threat to material welfare and personal health. Also relevant was the pervasiveness of individual contact with disease and damage. At least the prerequisite for strong political action locally and globally is in place, though how this should be done and through what auspices remains a matter of great controversy. This takes us to the achievements of the Rio Conference and the likely future of the so-called sustainability transition.

THE SUSTAINABILITY DEBATE

The elusive goal of sustainable development dominated the UN Conference on Environment and Development held in Rio de Janeiro in June 1992. The phrase itself emerged in the mid-1960s as part of an effort by international conservation organizations to protect the habitat of emblematic game in Africa, notably the rhinoceros, the elephant and the big cats (see McCormick 1989: 43–6). At the time Africa was emerging from colonialism, with shifting administrations and the possibility of a lower profile in land, forest and water management. In addition there was a fear that unless African agriculture operated within the bounds set by natural processes, any resulting degradation of resources would push migratory populations into the prized game parks (see Hillaby (1961) and Adams (1990) for a comprehensive review).

There was an element of European self-preservation about all this. Land degradation in Africa is in part a product of the disruption of traditional property rights, land-ownership regimes and planting practices that allow flexibility against variable weather and soil conditions (see Watts 1983: 249–57). This disruption was caused in part by the introduction of cash cropping, commercial relations in commodity markets and the privatization of what was formerly common land. The game parks came to be seen as sacrosanct from human interference. So the notion of sustainability was born in an increasingly unfair resource allocation regime, imposed by external ideologies in the name of saving for the world the remaining representatives of highly adapted but environmentally vulnerable species. Big game became the metaphor for global care in the tumultuous sea of neighbouring resource mismanagement.

All this was encapsulated further in the *World Conservation Strategy* (*WCS*) published by the International Union for the Conservation of Nature in 1980. Bill Adams (1990) provides an admirable summary of this document and its inherent flaws. The *WCS* identified three cardinal areas for the maintenance of sustainable livelihoods. These were ecologically based resource management, protection of genetic diversity, and sustainable utilization of life support systems. In this fairly narrow context, sustainable development meant the use of natural resources roughly to the point of replenishability, the application of the precautionary principle where the thresholds of irreversibility could not be determined, and the progressive switch from a reliance on non-renewable resources towards the non-declining use of renewable resources.

The *WCS* ignored the underlying crisis of the sustainability debate. This lay in power structures that exploit both the poor and the natural world at the expense of the already wealthy. It also lies in the denial of civil rights to oppressed minorities, notably indigenous populations, as well as to women generally. And it is identified with the failure of both modern science and ethics to grant some sort of property right of existence in natural systems that provide the lifeblood of the earth's habitability.

This last point allows us to return to the changing notion of property rights in global commons that nurture humanity to keep us all alive. At the surface of the ocean marine plankton emit gases which regulate both the cloudiness of the oceanic atmosphere and the production of ozone forming radicals. (See Nisbet (1991: 36–45) for a good review.) Similarly, forests of all kinds sequester about one-third of the excess carbon emitted by modern humanity, though this figure is subject to a large amount of uncertainty (Brown and Adger 1993). Those regulating processes, of which there are literally many thousands for the earth as a whole, are owned by nobody yet provide inestimable value for humankind.

Lovelock (1994) suggests that the 'air conditioning' value of the tropical forest (from which latent heat of evaporation is transferred to temperate latitudes), could cost as much as the total value of global wealth to replace by artificial means. This is clearly a guestimate, but it does suggest that global life-support systems are immensely valuable if fully costed in terms of the replacement expenditures to undertake the same task.

The sustainability debate has hardly addressed the theme of investing appropriate property rights for such processes, and of ensuring some sort of legal intrinsic right to exist for key ecosystems and processes. The American environmental lawyer, Christopher Stone (1989) has begun to address this issue, as has the international lawyer Edith Brown Weiss (1989) with regard to the rights of future generations. Yet international action in this regard remains very elusive, simply because the designation of such rights would challenge the existing orders of power and wealth, and would shift the burden of proof onto those who propose change. This in turn would bestow power on potential victims of environmental calamities associated with any resource development or man-made process that unduly disturbed life-support commons regimes.

This brief discussion reveals that sustainable development endures because it is both ambiguous and elusive. Like more established precepts such as democracy and justice, sustainable development becomes a worthy goal that sets down markers for current performance and struggles for redesignated civil rights. The growing sense of unease over global change, identified above, is built upon a combination of moral guilt and pragmatic fear (O'Riordan 1993). We have started to create a democracy that is environmentally sensitized, but not at all sure of how to proceed. The sustainability notion is becoming bound up with five great liberating movements in the modern world:

1 *The growing globalization of international environmental accords* and regulatory regimes. This means that the democratic nation-state has less room for independent manoeuvre, but of course can act as a vital broker for its society that needs to shift from self-preservation to global citizenship.

2 *The rise of local environmental survival initiatives* that capture both community power and the need to identify with a larger whole through informal networks of community action. Right now this is impressively operating in many parts of the Third World where non-governmental environment and development groups are combining to create a new civic consciousness and an identifiable alternative 'government'. In turn this is bound to increase in the developed world, but the record to date is very patchy.

3 *The emergence of political and social acceptance of environmental economics,* generally and especially its derivative ecological economics. This point will be discussed below (see pp. 468–70). Here we only need mention the convergence in theory between market-oriented political ideologies and economic instruments for efficient allocation of environmental goods and tools, and the growth in valuative studies of the ecological significance of

species, habitats and life-support processes. There is a huge literature on all this, but a good summary can be found in Pearce and Turner (1990) and Turner *et al.* (1993).

4 *The equally important emergence of corporate sensitivity to environmental matters* now found widespread amongst the international business community. The initial driving force is good housekeeping – namely, the cost saving of more efficient use of energy and materials, the advantages of by-product recycling and conversion, and the overall gains on lower ultimate disposal costs of waste minimization throughout the production process. The more contentious problem lies with industry–community relations, the role of industry in revitalizing economic development in environmentally degraded areas, and the task facing well-meaning business when the good house-keeping practice is shown to be far too insufficient for a survivable earth. Right now these tricky issues are being dodged.

5 *The growth of regulation of a more common international character.* This began with environmental protection, notably pollution control and waste management. Subsequently it has extended to environmental impact analysis and eco-auditing of industry including life-cycle analysis and environmental management systems. Finally the extension of the national state of the environment reports has opened up data that were rarely made public before, and exposes governments and industry to performance targets (i.e. measurable evidence of real commitment).

So the sustainability debate has begun to get incorporated into much bigger themes, such as the environmental safeguards over the new trading order now that the General Agreement on Tariffs and Trade (GATT) has been approved, the relationship between debt and environmental destruction, and the links between aid and sustainable development. In all these areas a measure of accountability is beginning to appear. But the hurtful truth remains that on all of these key parameters the sustainability paradigm is nowhere near being addressed. The reaction is almost exclusively accommodating to existing ideologists on growth and wealth, and to current distributions of power and corruption.

The difficulty is that sustainable development needs much more than shirt-sleeve greenery and the latent guilt of the liberal-minded wealthy for its successful outcome. It confronts modern society at the very heart of its contradictions. Humanity is still a colonizing species; it has no institutional or intellectual capacity for equilibrium. At least five conditions must be met before anything like sustainable development can be attained. None of these was addressed at the UNCED Conference in Rio, nor is there any institutional mechanism in place to address these points now that sustainable development is an official international objective.

These conditions include:

1 *A form of democracy that transcends the notion of the nation-state* as the binding

entity of political responsiveness. This suggests creating a partnership between the citizen and government at all levels in a joint search for a more survivable future, where purely short-term political considerations are given less emphasis. Ultimately this means altering the social meaning of 'self-interest' and 'sacrifice' into a more enduring notion of global survival and collective endeavour. In this respect, education and social morality will also have to shift: but the rate of such metamorphosis will always be painfully slow.

2 *Guarantees of civil rights and social justice* to oppressed peoples the world over, so that they are allowed to develop culturally acceptable resource management practice that retains a dynamic equilibrium, yet fully appreciates the intrinsic rights of the global life-support processes.

3 *Commitment of real, and new, resources,* notably technology, know-how, and cash, to impoverished and environmentally vulnerable regimes, many of which are run by politically unstable and inherently corrupt governments (as indeed is the case in some rich nations).

4 *Elimination of debt* where debt is induced by unfair terms of trade and a historical legacy of exploitation. It is by no means clear how far the recent agreement on GATT will address this. Certainly it cannot remove existing debt: that will require sensitive international financial action. But even GATT as it is now emerging contains potentially serious distortions of environmental protection and the use of life-threatening chemicals, for which at present there are no agreed safeguards.

5 *Establishment of a variety of public–private, non-governmental mechanisms for delivering natural resources,* training and management to regions and communities in need, in such a way as to be socially acceptable yet improving the scope for democracy. Slowly the international aid and developmental agencies see the need to utilize the private sector as a vehicle for environmental improvement and protection utilizing the modern notion of joint implementation. This suggests a partnership between rich and poor through which the rich 'offset' possible global environmental change proposed by the poor in the name of more sustainable development. Examples include the 'purchase' of future CO_2 or SO_2 emissions or the planting of carbon sequestering vegetation as part of a deal to allow 'new' CO_2 emissions in the wealthy north (see Pearce and Bann 1993).

The sustainability debate will forever be with us. Its excitement is its scope for endless novelty and emancipation from the old order. But to examine the herculean task ahead, we must briefly look at the awkward issues facing science and economics in this transitional process, and the enormity of the task of stabilizing society–nature relationships through post-UNCED international regulatory and financing institutions.

THE CHANGING ECONOMICS OF SUSTAINABILITY

Economics is not inherently nasty: it is essentially a tool of vested interests and dominant world views. If society fails to value environmental damage, it is because it has not yet created the political will to do so. Environmental cleanup, the payment for past unwillingness to cost environmental damage as an economic loss, amounts to 2 per cent of gross domestic product for most developed countries. Yet losses to habitat and ubiquitous disagreeables such as noise or the extension of street and building lighting that blot out starry skies, are not counted in. In poorer countries, especially in Central and Eastern Europe, the cost of cleanup may exceed domestic wealth creation. In a special survey the World Resources Institute (1992) estimated that for Poland alone the costs of environmental cleanup could exceed $260 billion over the next twenty-five years. Meanwhile the Polish government has designated some twenty-seven areas of critical ecological hazard that encompass over a third of the population. Here levels of air pollution and water and land contamination exceed by up to twenty times the internationally agreed environmental health standards. This is one reason why, on average, life expectancy in these environmentally impoverished areas is five years lower than in Western Europe.

Economists define sustainability in terms of five kinds of capital, namely:

- *natural capital*, or resources such as assimilatively clean air and water, soil, forests, etc.;
- *critical natural capital*, or life-support systems whose functions are essential for maintaining habitability and biodiversity;
- *man-made capital*, or machinery, transport facilities, computers;
- *intellectual capital*, or knowledge itself including the ability to adapt and to cope with social and environmental change;
- *cultural capital*, or social values and morality generally linking individual action to acquired norms.

For the majority of economists such capital formation is substitutable, in the sense that resource scarcities usually generate innovation and improved managerial skills. For others, however, substitutability is not perfect: critical natural capital and intellectual capital may not be so readily replaced. This in turn has led to four definitions of sustainability:

1 *Very weak sustainability*. This is rooted in the substitution principle, but includes rules for ensuring that the rights of future generations are not unduly removed, and that profits from resource use are explicitly converted into substitutes or new man-made and intellectual capital. Note that there is an element of directional commitment to environmental taxation, by earmarking revenue for specific substitutable alternatives.

2 *Weak sustainability*. This applies the precautionary rule of ensuring that stocks of critical natural capital (and some historically and culturally

significant human capital) are maintained. If necessary protective buffers of added protection should be provided where there is scientific uncertainty, or where indicator species are not fully representative of critical or threshold conditions. This is one reason why genetic diversity is such a problem concept for the safeguard of key habitats and species. The ecological buffers interpreted by economists are a *safe minimum standard*, or a protective zone of non-extraction or non-sustainable use to ensure that the earth has a little breathing space.

3   *Strong sustainability*. Here aggregate natural capital is assumed to be held constant, not just critical natural capital. This position is based on the assumption that the natural world provides the basic nurturing capability for all living matter, and that criticality is frankly too narrow a concept. Also the problem of ensuring that indicators of critical natural capital are truly reflective of sustainable conditions remains under fire by some ecologists (see McGarvin 1994). Strong sustainability places great emphasis on intellectual and man-made capital to rise to the challenge of 'wise environmental usage'.

4   *Very strong sustainability*. This is a very much more contentious area. Essentially it applies to the scale of economic activity, stressing the need to decouple development from environmental systems, and generate a much more decentralized economic form of politics and society. The most penetrating statement has been made by Daly and Cobb (1989), though there are plenty of humanistic economists who advocate this approach (see for example Sachs 1993).

As these economic definitions are surreptitiously and insidiously changing the concept of nature, critical natural capital means no longer ecosystems in danger but potentially vital commercial assets that are being squandered. Cost–benefit analysis, not ethical absolutes, becomes the medium for their protection. If the poor South happens to contain more of this capital, so the North is justified, having destroyed much of its own, in investing in that stock via international agreements. Offset purchases such as debt for nature swaps or pharmaceutical royalty payments, or simple additionality grants-in-aid are being considered to reflect the added value of global survival. This is an extension of power-directed commercial capitalism of the most obvious self-interested form. Ecological economics is in danger of being hijacked as a tool for the established order, not as a new interdisciplinary science.

The sustainability debate also presumes a faith in science, at least to the point of identifying resilience and vulnerability in natural and human communities, and in a modified market that is supported by enlightened and far-seeing democracies. In fact these conditions palpably do not hold. This is why thoughtful commentators such as Lee (1993) argue for a civic science, a science of negotiation, participation and adjudication, rather than a dogmatic science of modelling and prediction set in widening bands of uncertainty.

The fact that economic dimensions of sustainability also require a sensitive knowledge of biogeochemistry and ecosystem processes, requires the emergence of a more interdisciplinary science. Interdisciplinarity is not confined to the fusion of disciplines. It also embraces awareness of the institutional biases of data collection, the political manipulation of scientific agendas, and the constant struggle between science as a liberating process and the power structures that stifle both the advancement of science itself and its effectiveness as a policy-guiding tool.

Take, for example, the ozone depletion issue. At first flush the chloro-fluorocarbons that allegedly were the prime culprits could be removed over time, because substitutes were available, and because the international chemical industry, for the most part, had made its profits on the early CFC investments. But scientific enquiry is now indicating that the HCFCs and other substitute chemicals are still ozone depleting, while new chemicals used to the point of dependence in more economically sensitive walks of life, for example the fungicide methyl bromide, became serious culprits. How is the science now treated? The significance of uncertainty of the long-term effects of ozone removal becomes more prominent in the debate: scientific enquiry is more marginalized as companies and governments struggle to set phase-out dates that meet their commercial viability. For a good review of this issue, see Rowlands (1994).

So the politics of sustainability became enmeshed in established power structures and the instinct to protect capital formation, profit-making and continued production of material wealth. This is why the aftermath of the Rio Conference is subject to so much speculative interpretation.

THE RIO EXPERIENCE

The UN Conference on Environment and Development was the most ambitious attempt yet to integrate development with environmental well-being. Critics like to dwell on its lack of specific policy measures and resource transfers, together with its enormous scope for political rhetoric, as justifications of its failure. Sachs sums up this view:

> the governments at Rio came around to recognising the decline of the environment, but insisted on relaunching development. As worn out development talk prevailed, attention centred on the South and its natural treasures, and not on the North and its industrial disorder. There were conventions on climate, biodiversity and forests, but no conventions on agri-business, automobiles or free trade. This suggests that UNCED attempted to secure the natural resources and waste sinks for economic growth in favour of the global middle class, rather than to embark on a path towards industrial self-limitation and local regeneration.
>
> (Sachs 1993: xvi)

For others, UNCED was regarded as a phase change in the political perceptions of developmental pathways and community empowerment rather than an isolated event. They point to the slow evolution of institutional arrangements both nationally and internationally as governments of every hue grapple with the obligations of global partnerships. (For a good review, see Haas *et al.* 1993.)

What were the main achievements of Rio?

1   To force 173 political leaders actually to think about the symbiosis between environment and development, many for the first time, and hence to put the sustainability theme, with all its ramifications for society and political structures, into cabinets, presidential offices and ministerial advisory networks so that key people had to address ideas about civil rights, global survival and economic salvation.

2   To create internationally binding conventions on climate change and biodiversity, with future conventions on forests, desertification and oceans in the offing, plus possibly international action on population growth, health and women's rights.

3   To establish a sustainable development Agenda 21 for all countries, each presenting annual reports on progress to a UN Commission on Sustainable Development. This in turn has the powers to review progress, to comment on the appropriateness of the strategies suggested, and to organize regular conference on key concepts such as sustainable cities, tourism and toxic waste management and disposal.

4   To incorporate nine groups of interests, or stakeholder groups, permanently into the post-Rio negotiations. These are indigenous peoples, women, youth, science, agriculture, industry, trade unions, local government and environmental/developmental/social non-governmental organizations.

5   To create formal mechanisms for regular updates on a wide variety of fronts, such as the Partnerships for Change Conference in Manchester 1993, the Global Forum also in Manchester in 1994, the opening up of local Agenda 21 initiatives at local government level, and the formation of Conferences of the Parties to run the two major conventions on climate change and biodiversity.

The success of Rio cannot be evaluated for a while. The crucial indicators are the extent to which national interests become incorporated in global obligations and commitments, the power of international governmental organizations to enforce national co-operation, and the ability of international financing institutions to ensure a real resource transfer from North to South on lines that are culturally acceptable and environmentally just. All this is difficult enough, but one also has to bear in mind the legacy of the past, and, perhaps above all, the momentous change in the mores of society from individualism, materialism, acquisitionism and short-termism, towards an affinity with all the earth and its living matter for all time. All these challenges are tough: but none is more tough than the last.

Let us look briefly at the new institutional arrangements emerging from Rio to see how far the process has evolved:

1   *The UN Commission on Sustainable Development* consists of fifty-three member states. At the time of writing it has held one meeting, in June 1993, where almost all of its attention was directed on organizational matters. It still has no effective money, but is planning a series of regional workshops to focus on specific Agenda 21 items. This will be supported by two intercessional working groups on finance and technology transfer. These are to report specific recommendations for action to the May 1994 meeting in New York.

2   *The UN Department of Policy Co-ordination and Sustainable Development* is designed to be a central focus for ensuring that sustainability principles become embedded in all UN agencies, including international financing institutions. Intriguingly, the CSD is only one agency within this mega-department. The six divisions are:
  • sustainable development, via the new Commission;
  • social development preparing for the World Conference on Social Development in 1995;
  • poverty elimination and development, incorporating poverty, malnutrition, good security;
  • inter-UN agency co-ordination via an interagency committee on sustainable development;
  • advancement of women and secretariat for the Fourth World Conference on Women in 1995;
  • co-ordination of operational development at the local level in Third World countries.

3   *The Global Environment Facility* (GEF) was not created at Rio but was formed in 1990 as a three-year pilot scheme to transfer resources, technology and know-how to the developing world (Jordan 1993). The Facility consists of three partners, the World Bank which is the main player, the UN Environment Programme and the UN Development Programme. By 1993 it had spent some $860 million on four classes of schemes: namely, transition from CFCs, international river basin management, combating global warming, and stemming the destruction of biodiversity. The bulk of the cost is hooked onto existing World Bank schemes, much to the annoyance of most Third World aid activists who see the World Bank as a US-dominated commercial enterprise. A review of its performance to date suggests that the GEF is subject to the usual biases of political patronage and fudging of project justification. Post-Rio the Facility was renewed with a grant of $3–4 billion and a promise to finance the resource transfers arising out of the climate change and biodiversity conventions. The GEF still faces serious problems of credibility with recipients, and has run into enormous trouble in trying to define 'additional' or 'incremental' costs. These are supposedly the costs that a recipient national incurs, over and above what it would spend on itself for its own good, to meet a global commitment. An example would be the safeguarding of a key biodiversity reserve on a larger scale than for

472

local use, or investing in CO_2 removal (or displacing) technology.

4   *The Conference of Parties* of the conventions on climate change and biodiversity are still in the process of formation. In principle they will consist of a small number of representative member states, possibly 20–25, who will be responsible for a vital sequence of performance indicators:

- receiving and reviewing national strategies to reduce greenhouse emissions, create new sinks, safeguard existing sinks, and develop a biodiversity plan;
- monitoring the accuracy and scientific credibility of the data sites upon which performance targets are to be set and assessed. This is no mean task as the data are based on highly politicized assumptions in the absence of a good statistical record;
- advising on strategies for implementation, including the use of regulatory and economic instruments both to control greenhouse gas emissions and to value biodiversity;
- generating scientific and technical back-up to all these points, including access to models and to predictive mechanisms;
- creating enforcement mechanisms that are credible enough to work but not so draconian as to be unacceptable to member states. This matter will be fraught with issue linkage, as enforcement is bound to be an exceptionally politically sensitive topic.

All these organizational arrangements, with the exception of the Global Environment Facility, are very new. Many have little or no money and embryonic secretariats. So far the signs are far from encouraging. The 'system' is not particularly supportive, and few member states seem to be remotely aware of their need to subsume national interest with a broader global commitment. Even the scientific data on which much of their deliberations will be made, in the words of Sachs,

> provide a knowledge which is faceless and placeless, ... [consigning] the realities of culture, power and virtue to oblivion. It offers data but no context; it shows diagrams but no actors; it gives calculations but no notions of morality; it seeks stability but disregards beauty ...
>
> (Sachs 1993: 19)

PERSPECTIVE

The new, new environmentalism is still emerging through the mists. We are all environmentalists now, so the term needs redefining in the context of power, responsibility and justice. Environmentalism is too all-inclusive a concept for the subtleties of shifts in ideology over concepts such as ecology, sustainability, criticality, vulnerability, resilience and equity that are slowly taking over the new international management speak. There is a desperate need to focus on specific programmes of salvation – of people, culture and ecology in endangered places

473

– by means which are purposefully fair, openly advocative of the interests of local communities, and trusting of the financial and social institutions that they slowly build in their search towards the painful transition to sustainability. Sadly there are still far too few such initiatives, and surprisingly little attention has been paid to measuring their effectiveness, or the success of various experimental measures to convert from the old to the new. This is the place for the future geographer, whose training should prove invaluable so long as it is peppered with plenty of social and political reality, and a thoughtful recognition that there are many, many natures and cultures, as many as there are pathways to sustainable development.

NOTE

1   There is no accessible literature on this issue. Readers interested in following up the case should write to the BC Commission on Resources and the Environment, 1802 Douglas Street, Victoria BC V8V 14.

REFERENCES

Adams, W. (1990) *Green Development: Environment and Sustainability in the Third World*, London: Routledge.

Barbier, E., Burgess, J. and Folke, C. (1994) *Paradise Lost? The Ecological Economics of Biodiversity*, London: Earthscan.

Bennett, J.W. (1976) *The Ecological Transition: Cultural Anthropology and Human Adaptation*, New York: John Wiley.

Blaikie, P. and Brookfield, H.E. (1987) *Land Degradation and Society*, London: Methuen.

Brown, K. and Adger, N. (1993) 'Forests for international offsets: economic and political issues of sequestration', *CSERGE Working Paper GEC 93–15*, Norwich: University of East Anglia.

Brown, L.C. (ed.) (1992) *State of the World 1993*, London: Earthscan.

Brown, L.C. and Wolf, E.C. (1988) 'Reclaiming the world', in L.C. Brown (ed.) *State of the World*, New York: W.W. Norton.

Brown Weiss, E. (1989) *In Fairness to Future Generations: International Law, Common Patrimony and Intergenerational Equity*, New York: Transnational Publishers and United Nations University.

Burch, W. (1972) *Daydreams and Nightmares: A Sociological Essay on the American Environment*, Englewood Cliffs, N.J.: Prentice-Hall.

Chambers, R. (1988) *Rural Development: Putting the Last First*, London: Longman.

Costanza, R. (ed.) (1993) *Ecosystem Health: New Goals for Environmental Management*, London: Earthscan.

Daly, H.E. and Cobb, J.B. (1989) *For the Common Good: Redirecting the Economy Towards Community, the Environment and a Sustainable Future*, Boston: Beacon Press.

de Groot, W. (1992) *Environmental Science Theory: Concepts and Methods in a One World, Problem Orientated Paradigm*, Dordrecht: Kluwer Academic Publishing.

Dooge, J., de la Riviere, M., Goodman, G., O'Riordan, T., Marson Lefevre, J. and Breenan, M. (eds) (1992) *An Agenda for Science for Environment and Development*, Cambridge: Cambridge University Press.

Dunlap, R.E., Gallup, G.H. and Gallup, A.M. (1993) 'Of global concern: results of the health of the planet survey', *Environment* 35, 6–15 and 33–9.

Frankhauser, S. (1993) 'Global warming economics: issues and state of the art', *CSERGE Working Paper GEC 93–28*, Norwich: University of East Anglia.

Haas, P., Keohane, R. and Levy, M. (1993) *Institutions for the Earth: Sources of Effective International Environmental Protection*, Cambridge, Mass.: MIT Press.

Hays, S.L. (1987) *Beauty, Health and Permanence: The Environmental Policies in the United States, 1955–1985*, Cambridge: Cambridge University Press.

Hillaby, J. (1961) 'Conservation in Africa: a crucial conference', *New Scientist*, 31 August, 536–8.

Johnson, L. (1991) *A Morally Deep World: An Essay on Moral Significance and Environmental Ethics*, Cambridge: Cambridge University Press.

Jordan, A. (1993) 'The international organisational machinery for sustainable development: Rio and the road beyond', *CSERGE Working Paper GEC 93–11*, Norwich: University of East Anglia.

Lee, K.N. (1993) *Compass and Gyroscope: Integrating Science and Politics to the Environment*, London: Earthscan.

Lovelock, J. (1979) *Gaia: A New Look at Life on Earth*, Oxford: Oxford University Press.

Lovelock, J. (1994) 'Taking care', in T. O'Riordan and J. Cameron (eds) *Interpreting the Precautionary Principle*, London: Cameron & May.

McCormick, J. (1989) *The Global Environmental Movement*, London: Belhaven Press.

McGarvin, M. (1994) 'The precautionary principle and biological monitoring', in T. O'Riordan and J. Cameron (eds) *Interpreting the Precautionary Principle*, London: Cameron & May.

Mannion, A. (1991) *Global Environmental Change: A Natural and Cultural History*, London: Longman.

Mies, M. and Shiva, V. (eds) (1993) *Ecofeminism*, London: Zed Books.

Mudge, F. and Adger, N. (1993) 'Methane fluxes from artificial wetlands: individual country contributions and mitigation options', Norwich: CSERGE, University of East Anglia.

Myers, N. (ed.) (1993) *Gaia Atlas of Planet Management*, Stroud: Gaia Books.

Nisbet, E.G. (1991) *Leaving Eden: To Protect and Manage the Earth*, Cambridge: Cambridge University Press.

O'Riordan, T. (1991) 'The new environmentalism and sustainable development', *The Science of the Total Environment* 108, 5–15.

O'Riordan, T. (1993) 'The politics of sustainability', pp. 37–69 in R.K. Turner (ed.) *Sustainable Environmental Economics and Management*, London: Belhaven.

O'Riordan, T. and Cameron, J. (1994) *Interpreting the Precautionary Principle*, London: Cameron & May.

Pearce, D.W. (1993) 'Sustainable development and developing country economics', pp. 70–105 in R.K. Turner (ed.) *Sustainable Environmental Economics and Management: Principles and Practice*, London: Belhaven.

Pearce, D.W. and Bann, C. (1993) 'North–South transfers and the capture of global environmental value', *CSERGE Working Paper GEC 93–24*, Norwich: University of East Anglia.

Pearce, D.W. and Turner, R.K. (1990) *Economics of Natural Resources and the Environment*, Hemel Hempstead: Harvester Wheatsheaf.

Pepper, D. (1991) *Communes and the Green Vision: Counter Cultures, Lifestyle and the New Vision*, London: Green Print.

Ponting, C. (1991) *A Green History of the World*, London: Sinclair Stevenson.

Rose, C. (1993) 'Beyond the struggle for proof: factors changing the environmental movement', *Environmental Values* 2, 285–99.

Rowlands, I. (1994) 'The Fourth Meeting of the Parties to the Montreal Protocol: report and reflection', *CSERGE Working Paper 93–18*, Norwich: University of East Anglia.

Sachs, W. (ed.) (1993) *Global Ecology: A New Arena of Political Conflict*, London: Zed Books.

Schneider, S. and Boston, P.J. (eds) (1991) *Science of Gaia*, Cambridge, Mass.: MIT Press.

Simmons, I.G. (1993) *Environmental History: A Concise Introduction*, Oxford: Basil Blackwell.

Simon, J. and Kahn, H. (1984) *The Resourceful Earth*, Oxford: Blackwell.

Stone, C. (1989) *Earth and other Ethics: The Case for Moral Pluralism*, New York: Harper & Row.

Turner, B.L., Kates, R.W., Clark, W.B. and Meyer, W. (eds) (1992) *The Earth Transformed by Human Action*, Cambridge: Cambridge University Press.

Turner, R.K., Pearce, D.W. and Bateman, I.J. (1993) *Environmental Economics: An Elementary Introduction*, Hemel Hempstead: Harvester Wheatsheaf.

UN Environment Programme (1992) *Environmental Data Report*, Oxford: Blackwell.

Watts, M. (1983) 'On poverty of theory: natural hazards research in context', pp. 231–62 in K. Hewitt (ed.) *Interpretations of Calamity*, London: Allen & Unwin.

World Bank (1992) *World Development Report*, Oxford: Oxford University Press.

World Resources Institute (1992) *Global Resources 1992/3: Towards Sustainable Development*, Oxford: Oxford University Press.

FURTHER READING

Dobson, A. (1990) *Green Political Thought*, London: Unwin Hyman.

Eckersley, R. (1993) *Environmentalism and Political Theory*, London: UCL Press.

Goodwin, R.E. (1994) *Green Political Theory*, Oxford: Polity Press.

Grubb, M. (1993) *The Earth Summit Agreements: A Guide and Assessment*, London: Earthscan.

Murphy, R. (1994) *Rationality and Nature: A Sociological Enquiry into a Changing Relationship*, Oxford: Westview Press.

O'Neill, J. (1994) *Ecology, Policy and Politics*, London: Routledge.

O'Riordan, T. (ed.) (1994) *Environmental Science for Environmental Management*, Harlow: Longman.

Sachs, W. (ed.) (1994) *Global Ecology: A New Arena of Political Conflict*, London: Zed Books.

Also *The Ecologist* and *Global Environmental Change*.

IV A WORLD
OF QUESTIONS

INTRODUCTION

Richard Huggett and Mike Robinson

The aims of science, rather like the aims of religion, are directed towards resolving uncertainty. The two operate in spheres of human curiosity and concern which are seldom the same, of course, and they employ methods which are very different and which are sometimes seen as competing. Be that as it may, they both address the unknown. Moreover, just as the human contemplation of religion has affected the human conceptualization of it, so the human exploration of science has altered the range of scientific questions which humans explore. Nowhere is this more clearly demonstrated than in those areas of concern which address the macroenvironment of the earth, for a product of the closing years of the second millennium AD has been the scientifically derived realization of impacts which stem from possibilities that were scientifically identified. They are impacts displayed on the gross physical stuff of the earth: on climates, on seas and freshwaters; on soils, on vegetation, and on a host of other socially defined 'resources'. These are significant, not in themselves, but because they stem from, and they affect, people and the possibility of peopled futures. They are aggravated, too, by the needs of people themselves and, even more, by the ambitions and competitions of human institutions.

Of all living things that inhabit the earth, people are best able to reflect and to influence, not merely the environment of life as it exists, but the environment of life as it might become. Through most of history these reflections appear to have been nested and, in the centre, to have been powerfully affected by egocentric constructions of ethics and moralities that were related in some complex ways to a need for security. The sense of what is important, therefore, has focused powerfully on the preservation, protection, or improvement of the immediate environment of living. This same sense seems to have weakened as individual reflection moved outwards in socio-economic and political space, as well as in earth space and time. A similar model applied to institutions and governments: activities were devised, rationalized and justified, on egocentric and egotemporal scales with protection and preservation at their core and improvement as their goal. It is a measure of the significance of post-war changes in human impact on the earth that the grip of egocentrism is beginning to weaken, wider concerns are beginning to find expression, and a fear of 'going too far' in earth exploitation is extending the domain of an emerging common morality and informing a new world of scientific responsibility directed to essentially 'new' scientific problems.

It is difficult to untangle the interrelationships of ecosystems. Everything, we now believe, is part of everything else. Though this idea is by no means new, in the post-war period it has acquired a new significance through newly impending threats. Climatic and atmospheric measurement, for example, has raised the spectre of a redistribution of world climatic belts as a consequence of the augmentation of natural 'greenhouse' effects. This is probably occurring in consequence of the number of people who now occupy the earth and the manner in which they live. At its simplest, the exaggeration of this effect could produce global warming on a scale likely to affect the oceans through some degree of ice cap melting. It would also affect the hydrological cycle. And in both these ways it would affect the land surfaces on which human beings live, altering their littoral shape and the qualities of land and weather which influence their capacity to produce food.

The oceans, of course, represent the last of the earth's frontier zones. They occupy most of the earth's superficial area. Yet, despite the aura of impregnability induced by the concept of 'ocean', their stability is not immutable and their living resources are increasingly threatened. This applies especially in the areas of continental shelves which are of the most immediate value to people, but which are simultaneously the most vulnerable to their actions. To an even greater extent, perhaps, the earth's freshwater environments are also threatened, not only in terms of pollution and the life-forms which they support but also in terms of their quantity and availability, especially across national boundaries. There are, of course, areas of the earth where water has always been scarce, and people have found ways to adapt to that scarcity. But the post-war period has seen the idea of scarcity redefined in consequence of competitive abstraction and user-pollution.

Changes to the macroenvironment of human habitats, we now realize, grow out of accelerated population growth and accelerated resource use. Driven by an expectation of 'growth', advanced Western economies have behaved as if the earth is not a closed system. Methods of wealth accumulation have evolved which appear divorced from the ultimate fount of their creation in earth resources. A form of 'opportunism' which, in the past, was occasionally displayed in vibrant economic competition at a much smaller scale, now dominates and drives global economic systems in which jockeying for commercial advantage in immediate wealth accumulation has become the hallmark of both institutional and political success and the focus of planning processes which are increasingly constrained by the perceived necessities of 'short-termism'. The need to re-assert control over these seemingly out-of-control systems has never been more apparent. It involves, however, a series of choices which ultimately revolve around new moral imperatives likely to tax to the limit the human capacity for co-operation, self-denial, and organizational ingenuity.

The belief that science and technology can provide answers to the threats of a physically depleting and socially polarizing earth, of course, still survives. But ranged against it are the competing scientific scenarios of an enlarging environmental lobby. It is a lobby that urges control over the exploitation of resources of all kinds, and the redistribution of wealth and welfare at all geographical scales and in all locations. Central to this perspective is an unavoidable engagement with moral and ethical questions and the realization of the critical role exercised by government, by commercial and financial institutions, and by the 'political' will of individual people.

23

CLIMATIC VARIATION AND GLOBAL CHANGE

F. Kenneth Hare

CLIMATE AND THE GLOBAL HABITAT

'Climates are still changing', announced the prospectus for this encyclopedia. Not long ago such an assertion would have been considered an oxymoron, since climate was seen as the unchanging stability behind the endless variation of daily weather. 'Weather is forever changing,' said the conventional wisdom, 'but climate goes on all the time.' To use another metaphor, weather was the fickle mood, and climate the enduring personality. I was brought up on such orthodoxies.

The prospectus was nevertheless quite right. Changes in daily weather are indeed superimposed on a more stable background. But that background itself varies, on longer time-scales. And behind these longer scales are even slower changes. In logic, there is neither end nor beginning to this progression, other than the time when the atmosphere came into being.

For practical purposes, we have decided in modern times to generalize atmospheric statistics over thirty-year periods. Doing this gives fairly stable means, moments or sums, which are misleadingly called *climatic normals*. Thirty years is a reasonable space of time for individuals who might hope to live through three such periods. The heavy tread of official climatic statistics goes forward, however, by decadal strides. Currently we use the period 1961–90 to define climatic normals (see Figure 23.1 for elucidation), replacing those for 1951–80.

When we follow this rule the difference between end-on thirty-year normals (e.g. 1931–60; 1961–90) may exceed what random noise can account for. Variability *within* each period (noise) will permit differences *between* periods that have no statistical significance (Figure 23.2). But there may be real variation between the periods, and statistical tests soon show whether or not this is so.

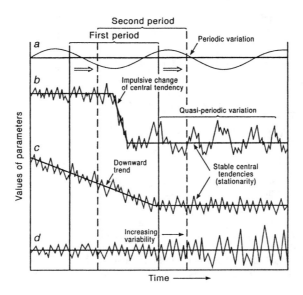

Figure 23.1 Idealized time-series (curves *a* to *d*) of a representative parameter of a climatic element that is continuous in time (such as temperature or pressure). Vertical bars indicate arbitrary averaging or integrating periods (usually thirty years) that are recalculated each decade (see dashed bars).

There may even be prolonged trends, in which change goes consistently in the same direction for several successive periods. Geological history shows that such climatic variation has been the global norm, though there were long periods of quasi-stability, such as the Cretaceous, which ended catastrophically 65 million years ago (Chapter 2).

This encyclopedia is being written at a time when global environmental change appears to have accelerated. Part III concerns itself with the impact of social, economic, political and physical processes on the human habitat. Their influence is dramatically visible in the world's landscapes, and can be detected in the oceans as well. There is a strong sense that human action is the main cause of current change. Though supply-side economists and some industrialists insist that all is well, and that the market will modulate the process, the rest of the informed world – and most of the public in Western countries – fear the worst. A sense of impending disaster hangs over the debate about global change. This was also the mood during the Depression of the 1930s, the period in which the writer was educated. In recent years geographers have had to run hard to keep up with the rest of the pack. Gloom is now fashionable.

Climate and climatic change are central to this view of the future. Is

483

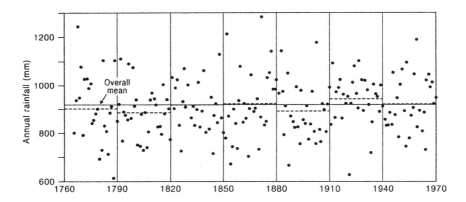

Figure 23.2 Spatially averaged and standardized annual rainfall for 1776–1970 over England and Wales, showing a typical, highly variable climatic element. No significant trend or periodicity is present, but very large differences within years are common. Differences in variability between periods are also visible. After Wigley *et al.* (1984). Thirty-year averages added by present author.

humankind damaging its own habitat by inducing climatic change? Is climate central to the productivity of the global habitat? And is human intervention really the cause of the variations already observed? These questions elude final answers, but will be discussed in what follows.

THE CLIMATIC SYSTEM: HOW DOES IT DIFFER FROM ECOSYSTEM?

In the late 1930s and 1940s a radical change affected the thinking of ecologists, who in the previous years had focused their attention on Clementsian concepts of climax and succession, and a variety of themes in distributional biogeography. Then in 1935 Arthur George Tansley coined the term 'ecosystem' to encapsulate the view that plant and animal communities could best be studied together with the physical environment that supported them; he argued, and demonstrated in *The British Islands and their Vegetation* (1949), that the biota and its physical supports constituted a natural, mutually self-sustaining system. The concept had its origins in nineteenth- and early twentieth-century Russian, German and American literature. But it was Tansley's advocacy that made us aware of the immense value of the paradigm (for such it is). Climate, in this sense, is an aspect of ecosystem; and ecosystem is the key to what we call habitat.

At the same time, G. Evelyn Hutchinson, at Yale University, and his tragically short-lived student, Raymond Lindeman, developed the notion of what Lindeman called *trophic-dynamic ecology* (Lindeman 1942). Lindeman did not originate the idea that the autotrophic producers (the green plants) fed energy and nutrients into a food web, or that consumers and then decomposers ultimately returned them to the air, soil or water from which they had come. But it was Lindeman's neat, precise definition of biological and ecosystem productivity that inserted the idea so effectively into ecological thinking – an innovation that heavily influenced the International Biological Programme of the 1960s and 1970s. Out of this work there grew the concept, still current, that the cycles of atmospheric properties and constituents – water, energy, carbon, oxygen, nitrogen and sulphur – were central ecosystem functions, and should be seen as key elements in climate.

Atmospheric scientists were slow to respond to these ideas. Their grounding in physics, chemistry and mathematics made them uneasy when confronted with biological thought, much of which was qualitative; there were no comforting differential equations in Tansley, Lindeman or Hutchinson. But naturalists and geographers had no difficulty; in many ways they had anticipated the developments. Wladimir Köppen had derived similar ideas from Augustin de Candolle almost a century ago, and his climatic classification (Köppen 1900) stressed the vegetation–climate link (though his approach was naïvely empirical). In the 1930s and 1940s C. Warren Thornthwaite (1935, 1948) used his background in the US Soil Conservation Service to develop what he called a rational, process-based classification of climates. It stressed atmosphere–biota linkages, and went well beyond Köppen's empiricism. Ironically, it was Köppen's simple scheme that survived into the modern literature of the atmospheric sciences; some of the general circulation models of the 1980s simulated Köppen's map of climatic regions effectively.

By 1970, however, key atmospheric scientists had realized that a large broadening in the concept of climate was needed if the rising tide of environmental concern was to be accommodated. The 1972 UN Conference on the Human Environment prompted an outpouring of joint effort by meteorologists, chemists, oceanographers and biologists. The Conference had before it, in particular, two large documents that contained syntheses of the climatic system (Matthews, Smith and Goldberg 1971; Matthews, Kellogg and Robinson 1971). The same idea emerged in more detail in two 1975 documents, *Understanding Climatic Change*, from the US National Academy of Sciences (1975), and *The Physical Basis of Climate and Climate Modelling* from the World Meterological Organization (1975). At the World Climate Conference in Geneva in 1979, one of the theme papers was entitled 'Global Ecology and Man'. It was delivered by Bert Bolin (1979), a distinguished Swedish meteorologist who subsequently became chairman of the Intergovernmental Panel on Climate Change (IPCC), whose reports lay before the 1992 UN Conference on Environment and Development (UNCED) at Rio de Janeiro. The gap had thus apparently been closed.

Or had it? The idea that climate pervades the global system – atmosphere, ocean, biota, ice masses and human economic activity – is now a commonplace. All scientists are aware that water, solar energy, carbon, oxygen, nitrogen and sulphur are cycled through the environment and are vital to human beings and other living species. Nevertheless, intellectual schisms still lurk in the undergrowth. In 1988, the International Council of Scientific Unions (of which the International Geographical Union is part) launched the International Geosphere–Biosphere Programme (IGBP). Its primary sponsors were the geoscientists (chiefly geologists, but with some geographers) and the biologists; the IGBP came into being to encourage creative dialogue between these groups. Atmospheric scientists tended to stand aside from the Global Change movement, of which IGBP is the organized core. They themselves had launched a World Climate Programme in 1979, with a World Climate Research Programme as its core component. The latter has insisted on maintaining its separate identity and independence of action. But behind the surface jostling there has been growing a healthy spirit of co-operation between the sciences in the understanding of global issues.

The climatic system and ecosystem are thus related ideas; they denote the same reality, which is the global interdependence of climate, life, soil, rock, freshwater, ice and ocean. But they are seen from different viewpoints. Atmospheric scientists and oceanographers are likely to talk about the climatic system, whereas biologists prefer ecosystem. Inorganic processes are central to the approach of the former, organic processes to the latter. In reality there is much common ground between them. To mark this healthy development, many other multi-member, interdisciplinary programmes have been set up in recent years, and will be stressed in what follows.

THE REALITY OF CONTEMPORARY GLOBAL CLIMATIC CHANGE: OBSERVATIONAL FACT

Is the anxiety about global climatic change empirically based? Or does it depend on the assertions of headline writers?

Earlier chapters have stressed the signs of pressure on the global habitat: the burgeoning human population; the list of biological extinctions; the cutting and burning of the rain forests; and many others. Except for population growth, and the things that follow directly from it, these signs are largely qualitative. They are real, but we have no firm estimates of their magnitude.

As regards the global climatic system we are a little better off. We have disturbing evidence of changing atmospheric composition, due largely to the release of stable gases by human agencies. We have estimates of global mean surface air temperatures for the past 150 years, and parallel data (since 1960) from the free atmosphere up to the stratopause at about 50 km above sea-level. We are less well-off as regards rainfall and snowfall. Nevertheless we are able to say with some assurance that the world climate has changed measurably since 1910.

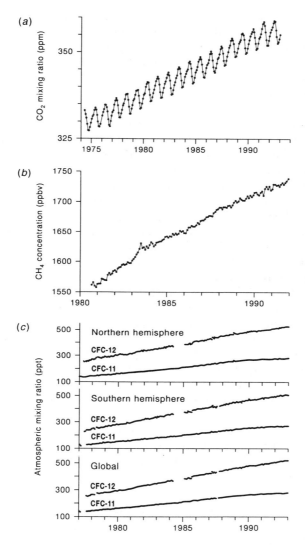

Figure 23.3 Recent growth in concentrations of (a) monthly atmospheric carbon dioxide (CO_2) mixing ratios at Mauna Loa, Hawaii; (b) global monthly concentrations of atmospheric methane (CH_4); and (c) monthly global and hemispheric mixing ratios of two chlorofluorocarbons (CFC-11 and CFC-12). Note the seasonal variations in carbon dioxide at Mauna Loa, and the tendency of growth rates to decrease, especially since 1990. After: (a) Thoning *et al.* (1994); (b) Khalil and Rasmussen (1994); (c) Elkins *et al.* (1994).

There is also a rich record of palaeoclimate deduced from a variety of proxy records dating back through the Holocene (the past 10,000 years) and much of the Pleistocene (about the past 2 million years). We can even make reasonable estimates from geological evidence of world climate back to the Proterozoic. All this is the outcome of a flood of research conducted in the past few decades, aided by a proliferation of new technologies. Previous chapters have described this haul of information.

In this chapter we deal first with recent trends in climate, and then proceed to the question of prediction. Figure 23.3 shows recent changes in global concentrations of minor but vital atmospheric constituents. As far as can be detected, the dominant gases in the atmosphere, i.e. nitrogen, oxygen, argon and a variety of minor inert gases, have remained fixed in relative concentration for millennia, though small seasonal variations in oxygen occur in connection with cyclical biological change. The variable gases include water vapour, not treated in Figure 23.3, carbon dioxide, methane, nitrous oxide and ozone (the latter two also not treated in Figure 23.3). To these must be added a growing list of synthetics, notably the chlorofluorocarbons (CFCs). Most of these gases can absorb radiation of the sort emitted by the earth's surface and atmosphere, at wavelengths in the range 4–40 µm (in which range the fixed gases like oxygen and nitrogen are largely transparent). The rises in concentration make the air more opaque (and hence resistant) to the flux of infra-red radiation from earth to space.

The dominant natural greenhouse agents are actually water vapour and clouds, and we have no data to suggest that their abundance is increasing. They and the other natural constituents raise global surface air temperatures by 33°C, to the mean annual value of 15°C (288 K). This is the natural greenhouse effect. Carbon dioxide contributes substantially to this beneficial effect – on which the habitability of land areas depends. The sensitive question is the role of the augmented effect attributable to the work of humanity. The entire current debate is about this augmented greenhouse effect.

Figure 23.4 shows the variation of annual global mean surface air temperature since 1860. This estimate was prepared, after quality control, by the Climate Research Unit of the University of East Anglia (Jones *et al.* 1994). It uses all reliable surface land data (after an attempt to remove bias due to urban heating), together with the marine data set prepared by the UK Meteorological Office and others (Folland *et al.* 1990). Comparable analyses by the Goddard Institute for Space Studies, New York (Wilson and Hansen 1994) and the State Hydrological Institute in St Petersburg (Vinnikov *et al.* 1994) yield similar results. The relevant data sources are summarized by Folland *et al.* (1990) and tabulated by Boden *et al.* (1994). The latter compilation is vital to research into climatic change.

It is apparent from Figure 23.4 that recent surface warming has affected the globe in two main phases: (a) 1910–40, of a little less than 0.4°C; and (b) 1975–present day (1994), of about 0.2°C. Between these periods there

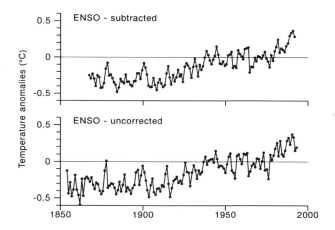

Figure 23.4 Global annual temperature anomalies, 1854–1993. These curves, prepared by the Climate Research Unit of the University of East Anglia, UK, express variations in global mean surface annual air temperature as departures from the 1951–80 means. The upper curve was prepared by eliminating the quasi-period influence of the El Niño–Southern Oscillation (ENSO) on global temperatures. After Jones *et al.* (1994).

were three and a half decades with little trend. Overall, the IPCC concluded that the real change to 1990 over the past century was $0.45° \pm 0.15°C$ (i.e. in the range 0.3 to 0.6°C). The lower figure allows both for uncertainties in the data and the opinion of sceptics who do not believe that the values given by the above groups remove all sources of bias (such as poor spatial coverage).

But Figure 23.4 also shows that large variations occur from year to year, even on the global scale. The underlying trends are masked by these short-term changes, which are highly visible to the public, and hence to the political community. The trend, which in 1994 was still upward, is a faint signal behind strident noise, the latter caused by internal changes (such as air–sea exchanges like the El Niño effect), and by external factors like volcanic eruptions. Mount Pinatubo's eruptions in 1991, for example, injected vast amounts (probably 15–30 million tonnes) of sulphur dioxide into the stratosphere, creating a world-wide veil of reflective sulphate particles that lowered global surface temperatures appreciably, an anomaly that persisted for a few years. It is hard for the casual observer to believe in a trend that can be so effectively hidden by short-term events.

A similar complication is that the observed warming has been unevenly distributed about the world. Thus the US has experienced only minor effects (Karl *et al.* 1988), whereas much of Canada has seen rises larger than the global average (Gullett and Skinner 1992).

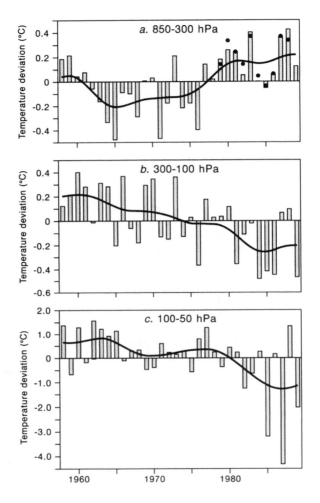

Figure 23.5 Temperature anomalies in the troposphere and lower stratosphere, 1958–89. (a) Annual global values for 850–300 hPa (dots are values from Spencer and Christy 1990); (b) 300–100 hPa; (c) annual values for Antarctica (60°S–90°S) for 100–50 hPa. After Angell (1988).

We cannot offer comparable data for precipitation, though in many land areas – the CIS, for example, and Sub-Saharan Africa – remarkable trends have been apparent in the past few decades. But we have no reliable measurements of precipitation at sea. Nor can we offer long-term analyses for the physical and chemical state of the ocean, or the response of the world's ecosystems. We are thus dependent on the global air temperature record, and on measured changes

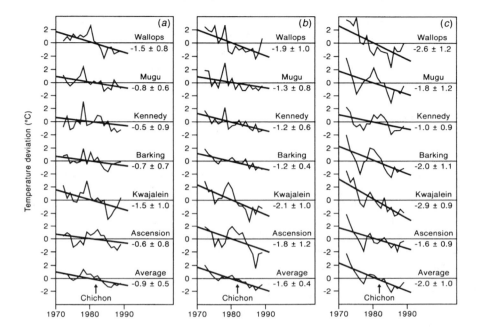

Figure 23.6 (a) Variation of annual temperature deviations for the 26–35 km layer at each of six rocketsonde stations, as well as their average deviation (at bottom). Shown at right is the temperature change (°C per decade) given by the linear regression lines and the approximate 95 per cent confidence limits as estimated from the standard error of annual values about the regression line; (b) Same criteria for 36–45 km layer; (c) Same criteria for 46–55 km layer (after Angell 1991).

in atmospheric composition, for our conviction that global changes in the climatic system are indeed in progress.

It is widely believed – but not yet accepted by some authorities – that this warming has been due to the augmented greenhouse effect, i.e. to the build-up of greenhouse gases shown in Figure 23.3. Simple prognostic models (e.g. Wigley and Raper 1992) confirm that the observed temperature record comes near the middle of the range of estimates, if the countereffects of sulphur dioxide release are considered. The models also predict that the surface warming should have been accompanied by a cooling of the stratosphere (the layer from 10/15 km to 50 km). Angell (1988, 1991, 1994) has analysed all balloon and rocket soundings since 1960 (balloon) or 1970 (rocket). Figures 23.5 and 23.6 show his results. In Figure 23.5(a) it is clear that the strong surface warming since 1970 was also present throughout the troposphere – represented by the layer from 850 hPa (~1.5 km) to 300 hPa (~10 km). Figure 23.5(b), 23.5(c) and all of Figure 23.6 show that all levels of the stratosphere underwent cooling after 1970. Qualitatively, the record therefore supports the hypothesis that the greenhouse effect has been the prime cause of the changes. But the spasmodic character of the observed warming since 1910, with a long standstill between 1940 and 1975, is *not* explained. Some commentators see this as good reason for rejecting the hypothesis of greenhouse warming.

THE PROBABILITY OF FUTURE CHANGE: CLIMATIC MODELLING

To answer the question 'will the observed warming continue, or intensify?' we have to consider climatic modelling, now a widely practised form of experiment. It is important to realize the strengths and limitations of the method.

Ideally we ought to model the entire climatic system, but that is beyond our present powers. The most recent general circulation models (GCMs) couple together the atmosphere and ocean in three dimensions, as nature does so strongly. A few attempts have been made to couple climatic and cryospheric models, and these will have to be pressed much further. What has so far eluded the research community is an equally effective coupling of the atmosphere with the biotic component of ecosystems. The same is true of coupling with the human economy. This account must accordingly deal mainly with the atmospheric and atmosphere–ocean generation of models.

Climatic modelling starts with equations from classical physics – those expressing Newtonian mechanics, bulk thermodynamics, and continuity (i.e. the conservation of mass), all developed for the rotating spheroid on which we live. This is hard science, and in principle deterministic. We then have to treat radiative energy transfer, which is more difficult, but still largely deterministic. We assume values for the solar constant, gravity, atmospheric composition and state, and define suitable boundary conditions. We end up with a system that is governed by deterministic laws, all expressed in mathematical terms. We can choose to analyse the system in zero (average for the globe), one, two or three spatial dimensions, and to integrate over time. Why do we end up with a multitude of solutions? Nature solves this problem effortlessly; since the world began it is likely that no two days have had identical weather patterns. Why do we find it so hard to predict this capricious response?

In part the answer lies in uncertainty. It is impossible to specify the initial state precisely except for the simpler zero- or one-dimensional models. The deterministic laws just enunciated are expressed by non-linear equations (if one pushes twice as hard, one is unlikely to get twice the response). Many years ago Lorenz (1963) showed that in such a medium the smallest initial errors might amplify into widely different outcomes in time. In so doing he revived interest in the mathematical concept of chaos. The atmosphere is just such a system: it is governed by deterministic laws, but its behaviour is part-chaotic. Numerical weather prediction (NWP) models, in three spatial dimensions, display this effect spectacularly. Over short periods – up to three days – they provide good hemispheric predictions. But after about five days their usefulness decays rapidly, and after 10–15 days they have no skill. Errors of observation and assumed initial state, plus the need to use a limited number of grid-points or harmonics, allow the errors to amplify, and to overwhelm the real solution – which is the global sequence of weather.

Although simple zero-, one- and two-dimensional models can be used with valuable results, notably to test hypotheses (see, for example, Wigley and Raper 1992), our picture of future climate has been derived largely from hemispheric or global three-dimensional models. When applied to the problem of climate, the NWP models can be manipulated mathematically so that they become global general circulation models (GCMs). These may be for the atmosphere alone, with surface conditions simulated as fixed or weakly interactive, or for the ocean alone. Very recently – since 1988 – the technique has been extended to coupled atmosphere–ocean models, which alone are likely to permit realistic modelling of the physical parts of the climatic system. The first requirement is that the models should simulate the mean state, statistics and diurnal or seasonal changes for the contemporary climate. Only a few of the models fulfil all these requirements, and none yet does so to a fully satisfactory level.

Difficulties also remain in other aspects of the validation and use of these predictive models. One concerns the role of clouds. In early models cloud was often specified on a climatological basis, as was humidity distribution. Cloud formation, dissipation, amount and height are important, as is microphysical state (particle size and phase). Clouds are vital to the global radiative balance, which controls temperatures. They are also an essential stage in the hydrologic cycle. A model that cannot adequately predict the present cloud regime cannot be expected to do so for future conditions.

A further weakness is that the coarse observational network, plus the modelling techniques, fail to catch many mesoscale cloud systems – massed thunderstorms, hurricanes, the moisture conveyor belts of frontal cyclones and others – that determine the details of the precipitation process. For these and other reasons, the GCMs fail to predict convincing regional detail, which is crucial to any world political response to global change. Nevertheless they are the best tools we have, and their improvement is a permanent goal.

Two families of GCM experiments can be identified:

1 *Equilibrium models*, in which the carbon dioxide equivalent concentration is instantaneously doubled, and the computation continued until apparent equilibrium is reached. Cubasch and Cess (1990) tabulated the properties of twenty-two such models introduced between 1980 and 1989. The models differed as regards the treatment of cloud, and usually confined themselves to interaction between atmosphere and the surface layer of the ocean. Some of them incorporated estimates of heat transport by ocean currents; others

assumed an immobile swamp-like surface ocean.

2   *Non-equilibrium coupled models*, which couple atmosphere and ocean together more comprehensively, thereby allowing heat exchange between the ocean's surface layer and the cold deeper layers, as well as latitudinal heat transport. It has long been surmised that this exchange would delay the observed greenhouse warming – a delay often called the transient effect. Five such models were described in Cubasch and Cess (1990). Two of these (Stouffer *et al.* 1989; Manabe *et al.* 1991; Washington and Meehl 1989) introduced the augmented greenhouse gas effect at the realistic rate of 1 per cent per annum.

Obviously the second family is closer to nature. But these models have been introduced too recently for their full implications to have been realized.

The IPCC Scientific Assessment (Houghton *et al.* 1990) adopted a variety of scenarios for the future concentrations of the greenhouse gases. Since then, revised scenarios (Houghton *et al.* 1992) have allowed more realistic modelling of probable futures (Wigley and Raper 1992). Table 23.1 lists the principal greenhouse gases (other than water vapour and ozone, which play more complex roles), and sets out the 1990 IPCC assessment of their relative ability to warm the atmosphere. The new scenarios used by Wigley and Raper (1992) are much more realistic in that they embody the latest predictions of population and economic growth, and of fossil-fuel use. They still assume, however, business-as-usual policies, i.e. that no specific anti-greenhouse measures will have been taken.

The IPCC 1990 assessment found that equilibrium-type GCMs gave for the *climatic sensitivity* (i.e. the calculated global temperature rise due to a doubling of carbon dioxide concentration) a range of warmings between 1.5°C and 4.5°C, with a best estimate of 2.5°C. Using an emissions scenario near the middle of the new range of estimates, accepting the 2.5°C value for climatic sensitivity, and incorporating the possible effects of lower strato-spheric ozone depletion and sulphate haze effects, Wigley and Raper (1992) found for the period 1990–2100 a predicted rise in global mean surface temperature of 2.5°C, and a most probable sea-level rise of 48 cm. The authors pointed out that these values were significantly smaller than those of IPCC 1990 (Houghton *et al.* 1990), which were themselves more conservative than some earlier estimates. Nevertheless, the new estimates of global warming *rates* (of the order 0.2°C per decade) are five times as great as anything observed in the past century.

The modelling effort is still in progress, and it is premature to arrive at firm conclusions. All that can be said is that the results – still forthcoming several times a year in the widely read journals – generally suggest smaller warming effects for the next few decades than did earlier models, and correspondingly lower values of sea-level rise; but they also confirm the reality of the warming.

Table 23.1 Present atmospheric concentrations, annual rates of change, and atmospheric lifetimes of greenhouse gases other than water vapour and ozone, for which comparable data cannot be provided.

Parameter	Carbon dioxide	Methane	CFC-11	CFC-12	Nitrous oxide
Pre-industrial value, 1750–1800 (parts per million by volume, ppmv)	280	0.80	0	0	0.288
Current value (1990) (ppmv)	353	1.72	0.000280	0.000484	0.310
Current annual rate of accumulation (%)	0.5	0.9	4	4	0.25
Atmospheric lifetime (years)	50–200	10	65	130	150
Relative greenhouse forcing effectiveness, per unit mass ($CO_2 = 1$)	1	58	3,970	5,750	206

Source: Houghton *et al.* (1990), Tables 1.1 and 2.3.

In spite of the great efforts being made, the models are still crude in many ways, as regards the role of the oceans, the part played by clouds (notably in the tropics), the dynamical mechanisms not yet properly incorporated (such as the Julian–Madden waves of 45–50 days period, and the El Niño–Southern Oscillation – ENSO – events on the 2–5-year scale) and the entire question of mesoscale precipitation mechanisms.

Several commentators have called attention to the apparent non-fit between the observed warming pattern in time and space, and the details of the model predictions. One such review puts the criticism in these words: 'Observed temperatures indicate that much more warming should already have taken place than predicted by earlier models in the Northern Hemisphere, and that night, rather than day, readings in that hemisphere show a relative warming' (Michaels and Stooksbury 1992: 1563). These writers (who refer to themselves in the first person singular throughout the article) assume that the prime cause of the inconsistency is the effect of anthropogenic sulphate and other aerosols in the northern troposphere. The more usual view – adopted, for example, by Wigley and Raper (1992) – is that stratospheric aerosols are likely to be more effective, because they remain there for a considerable period.

As time goes by, the tendency among atmospheric scientists is to lay less emphasis upon the augmented greenhouse warming, and to stress instead the growing capacity of the modelling community to foreshadow or predict internal fluctuations of the climatic system – the effect of the ENSO phenomenon, the behaviour of the African–Asian–Australian monsoon system, and other fluctuations on decadal to century-long scales. There is little evidence to substantiate this conclusion: but the spurt in global research made possible by funds released by fear of the greenhouse effect may yet add new predictability to the future climate of the earth – even if the greenhouse warming is smaller than originally feared.

Another cited weakness of the greenhouse warming hypothesis is that the surface temperature of the earth is sensitive to small variations in the solar constant – the supposedly invariant intensity of the incoming solar beam outside the atmosphere. Such variations are indeed observed, but are of small amplitude (of the order 0.1 to 0.2 per cent) We have detailed satellite-borne observations of the solar constant only since 1983, but these have been used to calibrate a retrospective prediction of the variation of the constant through the past century, i.e. through the period of observed warming (Foukal and Lean 1990). The current majority view is that to account for the observed warming the variations would have had to be much larger than are hindcast or observed – and that variations big enough to override the expected future warming, or to augment it significantly, are unlikely (but see Nierenberg *et al.* 1989).

In spite of these doubts, the Climate Convention signed at UNCED in 1992, and now in force after being ratified by many states, clearly accepts the probability of an accelerated warming in the next five to ten decades, and advises measures – seen as weak by many commentators – to combat the expected changes. It will still be many years before the signal of global warming emerges from the year-to-year noise with sufficient clarity to silence the doubters, and to justify the advocates of a powerful avoidance strategy for the world community.

MARINE INTERACTIONS

Climatologists have not found it easy to incorporate the oceans into their thinking about changes in the climatic system. The oceans have their own modes of behaviour, their own climates, very different time-scales for their dynamical processes, and an enormous capacity for heat storage. Like the atmosphere they are thermally stratified, but in inverse mode; the warmest waters are at the top of the column, which tends to be thermodynamically stable; convection is hence inhibited. Ocean circulation is constrained by the presence of land barriers, capable of totally prohibiting certain modes, especially in the northern hemisphere. Submerged barriers may impede deeper-level flows. The observational base is much poorer than for the atmosphere. And finally the oceans have their own body of science, and cadre of professional students, the oceanographers. Though the two scientific specialisms have been on cordial terms, and rely on the same fundamental sciences, they have not yet fully overcome the separate evolution of ideas and methods.

Nevertheless, the World Climate Research Programme includes the ambitious World Ocean Circulation Experiment (WOCE), in which a full decade of study of ocean circulation is already launched. WOCE aims at an understanding of the links between variable climate and ocean circulation, and, specifically: to measure the large-scale fluxes of heat and freshwater (i.e. rain, snow, and river discharge), with their variability; the dynamic balance of the ocean circulation, and its response to changes of air–sea fluxes; the components of ocean variability, analysed on both time- and space-scales; and the rate and nature of the formation, ventilation and circulation of water masses influencing climate, on the decade to century time-scales. WOCE has started with a five-year programme of observation, involving the research of many nations, specialized satellites and drifting buoys. The observational base thus provided will feed an intensive effort at dynamical modelling on various scales up to the ocean GCM, and fully linked atmosphere–ocean models (Canada, n.d.).

The climatic interactions involve analysis of the major oceanic gyres, which are largely wind-driven, and the interplay of surface water masses of varying salinity, temperature, and ice regimes, if any. Also vital is the development of a quantitative picture of the thermohaline circulation, the vast convective regime involving subsidence of cold salty waters in the North Atlantic and Antarctic Oceans, probably in the form of detached columns that reach into the deep waters. At these depths, the rate of exchange of mass with the surface layers is on the scale of centuries to millennia. W.S. Broecker and his colleagues among the marine geologists have given prominence to the slow inter-ocean exchange of deep water, and its potential influence on long-term climatic variation. The coldness of the deeper waters – the lowest are currently near $2°C$ – implies that the oceans are a huge potential heat sink; but their thermodynamic stability largely isolates this cold water – except in the regions and seasons of convective activity (chiefly winter), and in areas of upwelling. These processes, and the quasi-periodic ENSO phenomenon, probably play a cardinal role in climatic variation. WOCE, and preceding experiments, offer the possibility of putting numbers into models of these interactions.

Air–sea interaction is also crucial to an understanding of the carbon cycle, and hence of the global warming phenomenon. Global exchanges of carbon, mostly as the dioxide, between air and sea are of the order of 100 billion (10^9) tonnes of elemental carbon per annum. Upward and downward fluxes do not quite balance at present, and crude estimates imply a net downward flux into the ocean of about 4 billion tonnes. In the surface layers of the ocean there is a high concentration of various inorganic carbon species, and also an abundance of life, with biologically fixed carbon. The precipitation of organic detritus into the deep waters (which contain a much larger store of carbon) involves the sedimentation of fecal or skeletal material, chiefly from the plankton. But oceanographers have generally found difficulty in balancing the various carbon fluxes, and in accommodating the large influx from the atmosphere implied by terrestrial modelling and observation.

Another major world programme, this time under IGBP auspices, is in progress to put better numbers into our picture of the marine carbon cycle. The Joint Global Ocean Flux Study (JGOFS) has three themes – input of CO_2 at the sea surface, transformations and transports of carbon in the ocean, and processes leading to burial of carbon. The research vessels of several nations are at sea, and satellite inputs are in progress (JGOFS 1989).

Here again the relevance to climatic change is striking. Ice-core evidence from Antarctica and Greenland has made it obvious that atmospheric carbon content tends to march in step with temperature, warm periods having atmospheres rich in carbon dioxide and methane (though the detail of this correlation is not yet properly understood). A more sharply defined picture is needed of current carbon exchanges between atmosphere and ocean, and within the ocean.

Physical oceanographers and atmospheric scientists have thus been forced into collaboration by the logic of the natural climatic system. New technologies are making possible experiments and observations that have never before been attempted. These initiatives promise more effective atmosphere–ocean modelling, on which prediction of future climates depends.

THE HYDROLOGICAL CYCLE

The broadening perception of climate has led on land to intensive study of the hydrologic cycle by many groups – the hydrologists; the fluvial and glacial geomorphologists and glaciologists; a variety of engineers; and most recently by the climatic modellers. In a recent review of the need for an integrated approach, Chahine (1992: 379) wrote: 'Rather than fragmented studies in engineering, geography, meteorology and agricultural science, we need an integrated program of fundamental research and education in hydrological science . . . The shape of the emerging discipline is still evolving; there are difficult mental adjustments to be made, and the transition has only just begun.' Actually, the need is not for students of terrestrial water to consolidate themselves into a professional discipline. The situation calls for broadening of interactive scientific approaches, not element-by-element specialization. The hydrological cycle is an integral component of climate, and hence of ecosystem; it is no one's intellectual property.

One component of the cycle – surface water balances at specific sites, often coupled with streamflow, groundwater change and soil–water exchanges – has been extensively studied, and forms a vigorous part of what geographers like to call physical geography. It has an important part to play in many other processes, notably in geomorphic change, glacial behaviour, soil differentiation and crop productivity.

The meteorological community, however, sees this component as an input to atmospheric models. It has been difficult to incorporate adequate water-exchange processes into the GCMs; evaporation, regionally representative precipitation and run-off estimates, and available soil moisture are all needed as inputs, but are also predicted quantities requiring validation against field observations.

Out of this concern have come plans for a major international experiment, the Global Energy and Water Cycle Experiment (GEWEX), a central component of the World Climate Research Programme. GEWEX is already in progress. It is planned for at least a fifteen-year span (World Meteorological Organization 1989). Included in its early work is a series of continental-scale river-basin studies; better modelling of cloud–precipitation–radiation processes, as part of a multi-region cloud system study; a satellite-based cloud climatology; and a global precipitation climatology project.

But the most impressive part of the GEWEX plan is the development of highly evolved, satellite-borne remote sensing devices which will be used in an intensive field study of the earth's hydrological cycle in the period 1998–2003. If this survey turns out to be as successful as its proponents claim, the impact on understanding of the climatic system will be large.

ECOSYSTEM LINKAGES

If climate really plays a central role in global change, it ought to be possible to identify quantitatively its function within ecosystems; and vice versa it ought to be possible to proceed from other ecosystem characteristics to a description of climate's role. In practice, neither can yet be done to anyone's satisfaction. But major efforts are in progress.

This, of course, is an old tradition in geographical study. From von Humboldt to de Candolle and onwards, the existence of close links between climate, soil and vegetation has been asserted repeatedly. The author himself published two such statements. In 1950 I claimed that distinct, physiognomically defined zonal divisions could be identified in the circum-polar Boreal Forest (basing my ideas on the pioneer work of Ilmari Hustich) (Hare 1950). I suggested that these divisions were highly correlated with specific values of C. Warren Thornthwaite's (1948) potential evapotranspiration, which could be related in turn to net radiative energy income. In 1972, with J.C. Ritchie (Hare and Ritchie 1972), I elaborated on this correlation, using direct estimates of net radiative income. We claimed, moreover, that the spatial correlation involved a mutual interlocking; climate was a determinant of vegetational physiognomy, but physiognomy in turn influenced the surface climate (by controlling surface roughness, snowcover distribution, albedo and soil temperatures). Both studies are still quoted in the ecological literature; but neither led to practical outcomes.

The reason is apparent: we are no longer satisfied with mere spatial correlations. The scientific *zeitgeist* requires that we base ourselves on measurable, interactive processes. We have to know how things work. And we have to make measurements that enable us to confirm or refute our hypotheses.

Major studies of individual biomes require the creation of large, co-operative

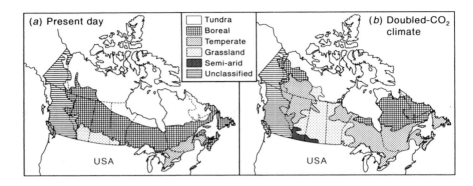

Figure 23.7 Predicted northward shift, for doubled greenhouse climate, of bioclimates corresponding to present biomes. After Canada (1993).

teams. The International Biological Programme of the 1960s and 1970s laid emphasis on biological productivity, and on the energy cycle in nature. Since that exercise there have been many attempts to focus on specific biomes in similar ways; the tropical rain forests, the arid zone, the grasslands, the temperate and boreal forests, and the Arctic tundra biomes have all been the targets of interdisciplinary attack.

Since the climatic models imply that by far the largest changes of climate will be in high northern latitudes, it is natural that close attention should be paid to the northern ecosystems. A study by Shugart *et al.* (1992) attempted nothing less than a systems modelling of the circum-polar Boreal Forest. Twenty-five investigators from six northern countries collaborated in the exercise, which involved the International Institute for Applied System Analysis; the World Meteorological Organization; the Swedish Academy of Sciences; the University of Stockholm's Institute of Meteorology; the Hungarian Academy of Sciences; and many US agencies (including the National Center for Atmospheric Research). This effort identified some of the processes that link climatic variables to the Boreal Forest, and offered a first attempt at a global numerical simulation of this remarkably uniform and homogeneous biome encircling the northern polar belt, except for the ocean barriers.

The various GCMs predict a major change of thermal climate for what is now the Boreal Forest zone, with longer and much warmer summers, and milder winters, probably with some increase in snowcover depth. Figure 23.7 shows how one such model predicts the northward displacement of warmer bioclimates into the present boreal zone of Canada. There is no firm agreement as to how the biota would respond to such a rapid northward shift of the bioclimatic zones. One line of argument stresses the comparative ability of dominant tree species to advance as climate warms (arguing from the abundant Holocene pollen record). Another considers the observed migration rate of species or communities into territories damaged by forest cutting, fire, industrial activity or storms. Most commentators recognize that the present pattern of spatial zonation is unlikely to persist during a prolonged warming, but no consensus is yet available as to the probable course of events in the next century – other than that there will be many stresses to endure before a new equilibrium is reached.

Other studies have been launched as parts of the IGBP and the Global Change programmes of Canada and the United States. The recently completed Northern Wetlands Project (NOWES), originated by the Canadian Institute for Research in Atmospheric Chemistry, and involving US and Canadian official agencies, aimed at defining the role of the boreal wetlands – huge in extent in North America and the CIS – in the carbon balance of the atmosphere (Mortsch 1990; Glooschenko *et al.* 1994). It had been widely assumed that these wetlands might serve as an important source of methane, a significant greenhouse gas, especially because measured atmospheric concentrations are high in boreal latitudes. But field measurements showed that the Hudson's Bay Lowland in Canada released only between 10 and 40 per cent of the predicted amount, and that these releases were strongly temperature-dependent. If the same regime holds in the Siberian wetlands, the net contribution of methane by the boreal zone may be only a half to a quarter the usually assumed value. The zone's role in carbon dioxide exchanges is also moot.

A more ambitious field study now under way is the BOREAS project launched in 1992 by Canada and the US (primarily the National Aeronautics and Space Administration). This includes field studies of the relevant surface–atmosphere exchange processes for energy, water, carbon and trace gases characteristic of Boreal Forest cover types (Canada 1993). At the core of the enterprise is the use of remote sensing by aircraft and satellites to complete biome-wide mappings of the relevant exchanges and surface processes, such as fire and insect infestation.

In the Boreal Forest biome there is thus in progress, or in prospect, research on air–surface exchanges that actually exceeds in committed resources all previous investigations combined. I am optimistic, as an early worker in this biome, that great progress is about to be made.

The same can probably be said for other biomes, most notably the tropical rain forests, but space prohibits a detailed review. The tropical rain forest biome consists of a series of discrete areas on several continents. Only two of these – the forests of the Congo and Amazon basins – have been thought large enough to influence climate on the global scale. Accelerated cutting of these forests, especially if followed by a grassland cover, many transfer carbon to the atmosphere, and soil degeneration after cutting may alter the dissolved and suspended loads of materials carried into the world ocean by the rivers. The latter has been thought vital in the case of the Amazon, whose large volume is unique.

Modelling of the regional climatic impact of progressive cutting in Amazonia has not, however, yielded consistent results. The effect of a removal of all the forest, with replacement by grassland, has been examined in GCM experiments by Dickinson and Henderson-Sellers (1988) and Lean and Warrilow (1989). These studies agreed on reduced evapotranspiration, increased albedo, and increased surface temperature, but differed as to precipitation and run-off changes. A third study (Shukla *et al.* 1990) showed large decreases in precipitation, evaporation and run-off (with a longer dry season than at present). The general view among global systems modellers seems to be that these regional results, if realized, would have little detectable effect on the global circulation. The preservation of the remaining tropical rain forests is desirable on many grounds. But their importance in the global climatic system has been exaggerated.

This discussion of the global climate–ecosystem linkages is obviously scrappy, because a consensus has not yet emerged; even the key questions have not been properly sorted out. A very partial list of such questions illustrates the problem:

1 Is the effect of the warming of the past eighty years yet detectable in world ecosystems? The answer may be 'yes' in relation to ocean fish stocks. But can we be equally confident about terrestrial biomes? Almost certainly not.
2 How will existing ecosystems respond to the predicted future warming? Should we attempt to answer this question by analysing the response of individual species? Or should we attempt to identify probable community responses? No clear answer is yet available.
3 Is it possible to distinguish, in the field, between effects due to natural processes from those induced by human interference? Is it, in fact, too late to ask this question? In both cases, the answers are unclear.

Failure to sort out these questions rendered the IPCC treatment of the climate–ecosystem linkage discouragingly weak (Melillo *et al.* 1990). The first and third questions were barely touched upon, and the second could be answered only in the most general terms. Without doubt this area needs to be greatly strengthened, as the results of the present wave of research become available.

CONCLUSIONS AND DISCUSSION

To the questions posed at the outset I can give only guarded answers. Science is as much concerned with asking new questions as with answering old ones. The two will be intermingled in these final paragraphs.

Is humankind damaging its own habitat by inducing climatic change? The answer must be that the danger does exist. The increase in human numbers alone guarantees the risk, because of our material demands, our need for energy, and our unquenchable outpouring of wastes. If the latter are released to the atmosphere, and are of the sort that do not easily fall out, or get washed out by rain or snow, then global climatic effects are inescapable; and some have already been detected.

Is climate central to the productivity of the global habitat? We have thought so for many decades. In plain terms the answer must surely be 'yes'. Nevertheless I am bothered by the imprecision of our understanding of the relationship. The large studies now in progress should help. But we have been too uncertain in our approach to the climate–biota link. Our knowledge of past climates depends heavily on palaeoecological data. Yet we have felt ourselves incompetent to say quantitatively how contemporary ecosystems link mechanistically with climate. By the same token, we have been unable to speak with any confidence of ecosystem response to future climatic change – other than in naïve, species-by-species terms.

Is human intervention, mainly via the augmented greenhouse effect, really the cause of climatic variations already observed? The answer in this case is still uncertain. On the global scale the rise of temperature since 1910 is consistent with the latest recalculation of the expected greenhouse warming. It is not, however, uniform in space, and is anything but smooth in time. The temperature curve is *not* parallel to the greenhouse gas build-up. And in many regions, no real warming has been observed. We shall have to wait a good many years for a firm answer.

There are, however, certain conclusions that can be firmly drawn about global change. Of these, the fact that the change is truly global needs stressing. This arises in at least two ways. One is that human folly is world-wide, its effects being detectable even at sea. Bad land use, industrial and marine pollution, and uncontrolled population growth acknowledge no geographical confines. But the second is the key factor: that the atmosphere (and the ocean, in a more confined sense) carries certain materials world-wide, and retains them for years – or in some cases almost for ever. The augmented greenhouse effect is truly and inescapably global. Only global action can contain it: hence the Climate Convention now being signed by enlightened nations everywhere.

A third firm conclusion is that the study of global change belongs to no group, no discipline, no superpower, and no special interest. It is everyone's responsibility to cope with it. Even the planet itself, as J.E. Lovelock (1988) maintains, may deliver a suitable riposte. In academic terms (though this is not an academic question), global change absolutely demands an interdisciplinary treatment. We need to group together the relevant understandings, and the competent skills, to deal with the situation that has arisen. Geography as an intellectual field is well suited to host at least part of this immense enterprise.

REFERENCES

Angell, J.K. (1988) 'Variations and trends in tropospheric and stratospheric global temperatures', *Journal of Climate* 1, 1296–313.

Angell, J.K. (1991) 'Stratospheric temperature change as a function of height and sunspot number during 1972–89 based on rocketsonde and radiosonde data', *Journal of Climate* 4, 1170–80.

Angell, J.K. (1994) 'Global, hemispheric, and zonal temperature anomalies derived from radiosonde records', pp. 636–72 in T.A. Boden *et al.* (eds) *Trends '93: A Compendium of Data on Global Change*, CRNL/CDIAC-G, Carbon Dioxide Information Analysis Center, Oak Ridge, Tenn.: Oak Ridge National Laboratory.

Boden, T.A., Kaiser, D.P., Sepanski, R.J. and Stoss, F.W. (eds) (1994) *Trends '93: A Compendium of Data on Global Change*, CRNL/CDIAC-G, Carbon Dioxide Information Analysis Center, Oak Ridge, Tenn.: Oak Ridge National Laboratory. (This is the definitive compendium of data on climatic change since 1850.)

Bolin, B. (1979) 'Global ecology and man', in *Proceedings of the World Climate Conference*, WMO-587, Geneva: World Meteorological Organization.

Canada (1993) *The Canadian Climate Program*, Downsview, Ontario: Environment Canada. (The present writer was the principal author of this overview.)

Canada (n.d.) *WOCE: The World Ocean Circulation Experiment*, WOCE Canada Secretariat.

Chahine, M.T. (1992) 'The hydrologic cycle and its influence on climate', *Nature* 359, 373–80.

Cubasch, U. and Cess, R.D. (1990) 'Processes and modelling', pp. 69–72 in J.T. Houghton, G.J. Jenkins and J.J. Ephraums (eds) *Climate Change: The IPCC Scientific Assessment*, Cambridge: Cambridge University Press.

Dickinson, R.E. and Henderson-Sellers, A. (1988) 'Modelling tropical deforestation: a study of GCM land-surface parametrizations', *Quarterly Journal of the Royal Meteorological Society* 114, 439–62.

Elkins, J.W. *et al.* (1994) 'Global and hemispheric means of CFC-11 and CFC-12 from the NOAA/CMDL flask sampling program', pp. 422–31 in T.A. Boden *et al.* (eds) *Trends '93: A Compendium of Data on Global Change*, CRNL/CDIAC-G, Carbon Dioxide Information Analysis Center, Oak Ridge, Tenn.: Oak Ridge National Laboratory.

Folland, C.K., Karl, T. and Vinnikov, K. Ya. (1990) 'Observed climate variations and change', pp. 195–238 in J.T. Houghton, G.J. Jenkins and J.J. Ephraums (eds) *Climate Change: The IPCC Scientific Assessment*, Cambridge: Cambridge University Press.

Foukal, P. and Lean, J. (1990) 'An empirical model of total solar irradiance variation between 1874 and 1988', *Science* 247, 556–8.

Glooschenko, W.A., Roulet, N.T., Barrie, L.A., Schiff, H.I. and McAdie, H.G. (1994) 'The Northern Wetlands Study (NOWES): an overview', *Journal of Geophysical Research* 99(D1), 1423–8. (This issue of *JGR*, pp. 1423–963, is entirely devoted to NOWES.)

Gullet, D.W. and Skinner, W.R. (1992) *The State of Canada's Climate*, SOE Report 92-2, Ottawa: Environment Canada, Atmospheric Environment Service.

Hare, F.K. (1950) 'Climate and zonal divisions of the Boreal Forest formation in eastern Canada', *Geographical Review* 40, 615–35.

Hare, F.K. and Ritchie, J.C. (1972) 'The boreal bioclimates', *Geographical Review* 62, 333–65.

Houghton, J.T., Jenkins, G.J. and Ephraums, J.J. (eds) (1990) *Climate Change: The IPCC Scientific Assessment*, Cambridge: Cambridge University Press.

Houghton, J.T., Callander, B.A. and Varney, S.K. (1992) *Climate Change 1992: The Supplementary Report to the IPCC Scientific Assessment*, Cambridge: Cambridge University Press.

JGOFS (Joint Global Ocean Flux Study) (1989) *Canadian National Plan*, Dartmouth, Nova Scotia: Bedford Oceanographic Institute.

Jones, P.D., Wigley, T.M.L. and Briffa, K.R. (1994) 'Global and hemispheric temperature anomalies – land and marine instrumental records', pp. 603–8 in T.A. Boden *et al.* (eds) *Trends '93: A Compendium of Data on Global Change*, CRNL/ CDIAC-G, Carbon Dioxide Information Analysis Center, Oak Ridge, Tenn.: Oak Ridge National Laboratory.

Karl, T.R., Baldwin, R.G. and Burgin, M.G. (1988) *Time Series of Regional Season Averages of Maximum, Minimum, and Average Temperature, and Diurnal Temperature Range across the United States: 1901–1984*, Asheville, N.C.: National Climate Data Center.

Khalil, M.A.K. and Rasmussen, R.A. (1994) 'Global CH_4 record derived from six globally distributed locations', pp. 268–72 in T.A. Boden *et al.* (eds) *Trends '93: A Compendium of Data on Global Change*, CRNL/CDIAC-G, Carbon Dioxide Information Analysis Center, Oak Ridge, Tenn.: Oak Ridge National Laboratory.

Köppen, W. (1900) 'Versuch einer Klassifikation der Klimate', *Geografisches Zeitschrift* 6, 593–611.

Lean, J. and Warrilow, D.A. (1989) 'Simulation of the regional climatic impact of Amazon deforestation', *Nature* 342, 411–13.

Lindeman, R.L. (1942) 'The trophic-dynamic aspect of ecology', *Ecology* 23, 399–418.

Lorenz, E. (1963) 'Deterministic nonperiodic flow', *Journal of the Atmospheric Sciences* 20, 130–41.

Lovelock, J.E. (1988) *The Ages of Gaia: A Biography of Our Living Earth*, New York and London: Norton.

Manabe, S., Stouffer, R.J., Spelman, M.J. and Bryan, K. (1991) 'Transient responses of a coupled ocean–atmosphere model to gradual changes of atmospheric CO_2, Part I: annual mean response', *Journal of Climate* 4, 785–818.

Matthews, W.H., Kellogg, W.W. and Robinson, G.D. (1971) *Man's Impact on the Climate*, Cambridge, Mass.: The MIT Press.

Matthews, W.H., Smith, F.E. and Goldberg, E.D. (1971) *Man's Impact on Terrestrial and Oceanic Ecosystems*, Cambridge, Mass.: The MIT Press.

Melillo, J.M., Callaghan, T.V., Woodward, F.I., Salati, E. and Sinha, S.K. (1990) in J.T. Houghton *et al.* (eds) *Climate Change: The IPCC Scientific Assessment*, Cambridge: Cambridge University Press.

Michaels, P.J. and Stooksbury, D.E. (1992) 'Global warming: a reduced threat?', *Bulletin of the American Meteorological Society* 73, 1563–77.

Mortsch, L. (ed.) (1990) *Eastern Canadian Boreal and Sub-Arctic Wetlands: A Resource Document*, Climatological Studies 42, Downsview, Ontario: Atmospheric Environment Service.

Nierenberg, W.A., Jastrow, R. and Seitz, F. (1989) *Scientific Perspectives on the Greenhouse Problem*, Washington, DC: The George C. Marshall Institute.

Shugart, H.H., Leemans, R. and Bonan, G.B. (eds) (1992) *A Systems Analysis of the Global Boreal Forest*, Cambridge: Cambridge University Press.

Shukla, J., Nobre, C. and Sellers, P. (1990) 'Amazon deforestation and climate change', *Science* 247, 1322–5.

Spencer, R.W. and Christy, J.R. (1990) 'Precise monitoring of global temperature trends from satellites', *Science* 247, 1558–62.

Stouffer, R.J., Manabe, S. and Bryan, K. (1989) 'Interhemispheric asymmetry in climate response to a gradual increase of atmospheric CO_2', *Science* 342, 660–2.

Tansley, A.G. (1935) 'The use and abuse of vegetational concepts and terms', *Ecology* 16, 284–307.

Tansley, A.G. (1949) *The British Islands and Their Vegetation* (2 vols), Cambridge: Cambridge University Press.

Thoning, K.W., Tans, P.P. and Waterman, L.S. (1994) 'Atmospheric CO_2 records from sites in the NOAA/CMDL continuous monitoring network', in T.A. Boden *et al.* (eds) *Trends '93: A Compendium of Data on Global Change*, CRNL/CDIAC-G, Carbon Dioxide Information Analysis Center, Oak Ridge, Tenn.: Oak Ridge National Laboratory.

Thornthwaite, C.W. (1935) 'The climates of North America according to a new classification', *Geographical Review* 21, 633–55.

Thornthwaite, C.W. (1948) 'An approach toward a rational classification of climate', *Geographical Review* 38, 55–94.

US National Academy of Sciences, US Committee for the Global Atmospheric Research Program (1975) *Understanding Climatic Change*, Washington, DC: National Academy of Sciences.

Vinnikov, K. Ya., Groisman, P. Ya. and Lugina, K.M. (1994) 'Global and hemispheric temperature anomalies from instrumental air temperature records', in T.A. Boden *et al.* (eds) *Trends '93: A Compendium of Data on Global Change*, CRNL/CDIAC-G, Carbon Dioxide Information Analysis Center, Oak Ridge, Tenn.: Oak Ridge National Laboratory.

Washington, W.T. and Meehl, G.A. (1989) 'Climate sensitivity due to increased CO_2: experiments with a coupled atmosphere and ocean general circulation model', *Climate Dynamics* 4, 1–38.

Wigley, T.M.L. and Raper, S.C.B. (1992) 'Implications for climate and sea level of revised IPCC emissions scenarios', *Nature* 35, 293–300.

Wigley, T.M.L., Lough, J.M. and Jones, P.D. (1984) 'Spatial patterns of precipitation in England and Wales and a revised homogeneous England and Wales precipitation series', *Journal of Climatology* 4, 1–26.

Wilson, H. and Hansen, J. (1994) 'Global and hemispheric temperature anomalies from instrumental surface air temperature records', pp. 609–14 in T.A. Boden *et al.* (eds) *Trends '93: A Compendium of Data on Global Change*, CRNL/CDIAC-G, Carbon Dioxide Information Analysis Center, Oak Ridge, Tenn.: Oak Ridge National Laboratory.

World Meteorological Organization, Global Atmospheric Research Programme, Joint Organizing Committee (1975) *The Physical Basis of Climate and Climate Modelling*, Geneva.

World Meteorological Organization (1989) *GEWEX: The Global Energy and Water Experiment*, Geneva.

FURTHER READING

Ambio (1994) 'Integrating earth system science: climate, biochemistry, social science, policy', *Ambio: A Journal of the Human Environment* XXIII(1), 3–103.

Coward, H. and Hurka, T. (eds) (1993) *Ethics and Climate Change: The Greenhouse Effect*, Waterloo, Ontario: The Wilfrid Laurier Press.

Environment Canada (1994) *Modelling the Global Climate System*, Special edition of *Climate Digest*, CCD94–01.

Lovelock, J. (1988) *The Ages of Gaia: A Biography of Our Living Earth*, New York and London: Norton.

Mintzer, I.M. (1992) *Confronting Climate Change: Risks, Implications and Responses*, Cambridge: Cambridge University Press.

Santer, B.D. (1994) 'The detection of greenhouse-gas-induced climate change', *DOE Research Summary* no. 29, Carbon Dioxide Information Analysis Center, Oak Ridge National Laboratory, Oak Ridge, Tennessee.

Schneider, S.H. (1994) 'Detecting climatic change signals: are there any "fingerprints"?',

Science 263, 341–7.

Stouffer, R.J., Manabe, S. and Vinnikov, K. Ya. (1994) 'Model assessment of the role of natural variability in recent global warming', *Nature* 367, 634–6.

Taylor, K.E. and Penner, J.E. (1994) 'Response of the climate system to atmospheric aerosols and greenhouse gases', *Nature* 369, 734–7.

World Meteorological Organization (1995) *WMO Statement on the Status of the Global Climate in 1994*, Geneva: World Meteorological Organization.

24

OCEAN USES, ENVIRONMENT AND MANAGEMENT

Alastair D. Couper

INTRODUCTION

Life began in the oceans some 3.5 billion years ago, possibly as single-cell microscopic bacteria and algae similar to the phytoplankton that flourish in the ocean surface layers today (see Chapter 3). It was millions of years after this that invertebrates emerged from the sea to colonize dry land. Most of the human population of the world still lives quite close to the sea. As well as deriving climatic and economic benefits from this location it seems that many people instinctively want to be near the ocean, it is literally in their blood. Professor Keikichi Kihara of Japan speaks of the 'rhythms of the sea, its tides, winds, waves and sky ... it is at the beach where human beings can best feel in the workings of their own bodies and in their breathing that they are indeed an organic part of nature' (Kihara 1989: 11). Sadly, he was referring to the closing of the windows to the open sea in Japan as a result of urban and industrial spread. It seems certain that the areas of greatest population growth in the foreseeable future will continue to be in coastal lands, many of which are already heavily overcrowded.

When we stand on the beach and look out to the wide horizon the ocean appears almost empty compared with the land; this is part of its appeal. There is of course much life and activity; a cubic metre of sea water may contain 700,000 plants (Joint 1993). However, the most visible feature from landwards, and especially from space, is the enormous expanse of water.

The world ocean covers over 70 per cent of the surface of the globe, although this is only in fact a thin veneer, averaging around 4,000 metres, when compared with the 3,963-mile radius of the earth. Nevertheless, the total quantity of water in the world is fixed at 120 million cubic miles (Agardy 1993). Approximately 97 per cent of this is in the ocean, another 2 per cent is frozen, and the remaining

1 per cent is contained in rivers, the ground, the atmosphere and in living organisms.

A relatively small amount of water is removed from the ocean by evaporation and returned to the surface as rain, sleet and snow. This hydrological cycle together with sunlight and atmospheric gases sustains life on land. The penetration of light from the sun into the upper layers of the ocean, the nutrients washed from the land and the horizontal and vertical mixing of the hydrosphere, together with gases from the atmosphere, form the basis of marine life. The ocean also stores and transports solar heat and prevents wide variations in the temperature of the global surface. Without the ocean there would be no life as we know it on the planet.

Concern today is for the well-being of ocean life and life-supporting systems. A recent report from the United Nations Environment Programme starts with the observation 'In 1989 man's fingerprint is found everywhere in the oceans. Chemical contamination and litter can be observed from the poles to the tropics and from beaches to abyssal depth' (GESAMP 1990). The ocean has for long provided a convenient place for the disposal of effluent from the land. Industrial waste, sewage, agrochemical runoff, as well as discharges from ships have been widespread.

Most of the damage being done by pollution is concentrated in the coastal zones, the bays, estuaries, the shallow waters above the continental shelves and in some areas of coastal upwelling. These are biologically the most productive zones of the ocean. It is from the continental shelves also that hydrocarbons and many other minerals are obtained, and where dredging, land reclamation and recreational and commercial sea-based activities predominate.

The living resources of the continental shelf seas are all under stress from pollution and overfishing. Even beyond the coastal zones in the wide expanse of the ocean the whale, dolphin, salmon and other highly migratory species are threatened with destruction from the activities of technologically advanced fishing fleets.

Science and technology have also revealed new resources in many frontier zones of the ocean. The abyssal depths beyond the continental shelves are being explored by remote sensing, drilling and manned and unmanned submersibles. As a result the muds of the deep-seabed, submarine volcanoes, mid-oceanic ridges, deep trenches and plate movements associated with sea-floor spreading are becoming better understood.

On the deep-sea-floor at depths beyond 6,000 m there is an unexpected range of living species, and new fish resources are being found on seamounts. But the greatest commercial interest in the deep-sea lies in the polymetallic nodules on the seabed and the mineral extrusions from hydrothermal vents. The ecological implications relating to the exploitation of these deep-sea resources are still uncertain, as are the rights and obligations associated with their acquisition. Other possible resources are the millions of shipwrecks on the seabed. These can now be reached by modern technology. The legal status of many of these ships and their contents, which range from Roman age to modern cargo vessels, is uncertain.

Much uncertainty certainly exists in establishing the full effects on the ocean of future global warming and ozone depletion. Global warming will raise the temperature of sea water, increase its volume and add massive quantities from the melting of polar ice. This would undoubtedly raise the level of the sea. When, how fast, and by how much are still the subjects of investigation and debate. Some of the consequences of major sea-level changes are predictable; including flooding of deltaic land, estuaries and other low-lying areas, inundation of coastal cities, raising of water tables, losses of mangroves, wetlands and coral reefs, and the submergence of low islands. Ozone depletion could in turn result in greater surface layer penetration of ultraviolet rays which may damage plankton and fish larvae (Hardy and Gusinski 1989).

When we consider military uses of the sea, deliberate environmental damage during conflicts (as in the Gulf War), the contamination of areas of the Arctic and other zones by disposal of radioactive substances, and likely future disputes over ocean space and resources, it is clear that a comprehensive approach to safeguarding the ocean is essential.

Many of the problems mentioned so far have been recognized for some time and much research has been conducted. It is only in recent years that attempts have been made on an intergovernmental basis to find solutions. There have been four notable events which have signalled greater global awareness of the need for co-ordinated and integrated management of the marine environment. These are:

- The Stockholm Conference on the Human Environment (1972);
- The Third United Nations Convention on the Law of the Sea (1982);
- The World Commission on Environment and Development (1986);
- The United Nations Conference on Environment and Development (1992).

The problems of ocean space, resource uses, the health of the ocean, and use management, will now be considered in some further detail.

DIVISIONS OF OCEAN SPACE

To appreciate the problems and opportunities for ocean uses it is necessary to understand some of the geographical divisions of the ocean, and who has the right to ocean space and resources. There are many possible ways of subdividing the ocean, including biological characteristics, large-scale ecosystems and regional seas designations. As far as resource exploitation is concerned divisions on the basis of geomorphology and legal entitlements are the most important.

Geomorphological divisions

The term 'coastal zone' is frequently used to describe a broad belt of land and sea. There is no precise definition of the width of a coastal zone since it depends on the nature of the coast and activities taking place. The extent of tidal influence landwards and the territorial sea seawards taken together provide one basis. The 'shoreline' or 'coastline' on the other hand may be defined as the zone where marine erosion and deposition is taking place.

Beyond the shoreline extending under the sea is the natural prolongation of the land, as continental shelf, continental slope and continental rise; these together comprise the continental margin. The continental shelf has a gentle gradient and extends to several hundred miles in some instances, although it is less than a mile in others. It terminates at a break of slope where there is a steep increase in gradient to become the continental slope; this plunges to depths of one to two miles. It is succeeded by a much gentler slope, the continental rise, which comprises partly of sediments which have rolled down from the continental shelf edge. Beyond this is the abyssal depth where the basalts of the oceanic crust are overlain by thick oozes.

The abyssal seabed beyond the continental margin is an undulating surface broken by hills, seamounts and deep trenches. Towards the centre of the ocean basin are the mountainous mid-oceanic ridges with active volcanoes and hydrothermal vents. It is in this unstable region that new seabed is being created by extrusions of magma, and where sea-floor spreading starts (Figure 24.1a).

Legal divisions

Coastal states can lay legal claims to various zones of the ocean which partly correspond to geomorphological divisions. Baselines drawn along the coastline and between some islands and promontories define the boundary between 'internal waters' (which lie within the sovereign state) and the 'territorial sea' over which the state can exercise many juridical powers (but must allow the ships of all states the right of innocent passage). The breadth of the territorial sea is nowadays accepted as extending twelve nautical miles from the baseline.

Under the 1982 United Nations Convention on the Law of the Sea it is also permissible for a coastal state to claim a 200-nautical-mile exclusive economic zone (EEZ), as measured from the baseline. In this EEZ the coastal state has sovereign rights over the resources of the water column and the seabed and subsoil. For small island states their territories were transformed. Kiribati, for example, has a total land area of about 700 square kilometres, whereas with its EEZ it has 3.6 million square kilometres of ocean (UNCTAD 1992). Beyond 200 nautical miles there are the 'high seas'. No state has dominion over the high seas or their resources.

The 1982 convention also defines legal entitlements to areas below the sea. Every coastal state has jurisdiction over its continental shelf. This does not

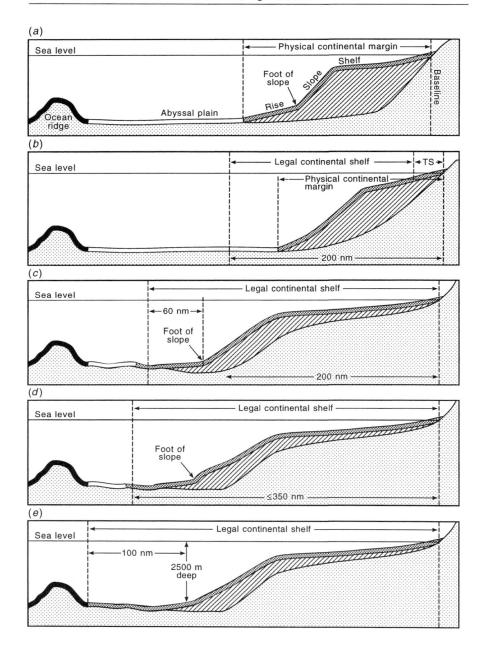

always correspond to the physical continental margin. Narrow margin states (West Coast South America) are entitled to a continental shelf of 200 nautical miles, which could therefore extend over the abyssal plain. Others, and especially broad margin states (Canada), can claim a continental shelf according to formulae laid down in the convention, including 60 nautical miles from the foot of the continental slope; but in all cases this shall not exceed 350 nautical miles from the baseline or 100 nautical miles from the 2,500-metre isobath (Figure 24.1b–e).

These divisions have now added over 35 per cent of the ocean and the related resources to coastal state jurisdiction, whereas for many hundreds of years the ocean beyond three nautical miles was regarded as free and open, belonging to no one. As a result over 90 per cent of the world's fish, and practically all of the known marine hydrocarbons, now belong to individual coastal states. This still leaves almost 50 per cent of the world ocean in the category of high seas. These divisions create problems in relation to highly migratory living resources, and fish which straddle both high seas and EEZ boundaries. The divisions present particular problems for mineral extraction from the deep-seabed which lies beyond national jurisdiction. These aspects will be dealt with under specific resource headings.

MARINE LIVING RESOURCES

Marine living resources are unevenly distributed but are found in every part of the ocean. The most productive zones are in regions of upwelling at the western margins of continents where surface currents move towards the equator. The rising replacement waters are cold and nutrient rich. In the shallow seas over continental shelves the waters are similarly rich and highly productive.

The phytoplankton which bloom in these conditions form the base of the oceanic food webs and result in concentrations of fish, including herrings, sardines, and anchovies; these constitute the highest categories of world catches. They are followed by cods, hakes and haddocks. Most fish stocks are thus concentrated within 200 miles of coasts.

The composition of plankton varies from place to place, seasonally and from year to year. In the mid-ocean beyond the highly productive zones there are less

Figure 24.1 Definitions of ocean margins. (*a*) Diagrammatic profile of the physical continental margin (PCM), not to scale. (*b*) Legal continental shelf (LCS) extends to a distance of 200 nautical miles from the baseline from which the territorial sea (TS) is measured. In narrow margin areas the LCS can overlap the abyssal plain. (*c*) Where the PCM extends beyond 200 nautical miles, the LCS can be delineated by reference to fixed points related to thickness of sediments, or to a distance no more than sixty nautical miles from the foot of the slope. (*d*) The LCS shall not exceed 350 nautical miles from the baseline. (*e*) Alternatively, the LCS should not exceed 100 nautical miles from the 2,500 m isobath. For all figures (24.1a–e), see *United Nations Convention on the Law of the Sea* (1982), Article 76, for details.

nutrients. The mid-ocean fish migrate over vast distances for feeding and breeding. These include tuna, squid, shark, swordfish and the marine mammals whales and dolphins, as well as the anadromous salmon which breeds in fresh water.

Fish are also found at every level of the ocean. Some middle-water species come to the surface at night to feed, other benthos remain feeding at great depths, including species around the hydrothermal vents of the mid-ocean ridges.

The Food and Agricultural Organization (UN) estimates that 100 million tons of fish per annum is about the limit that is likely to be taken from the ocean each year. In 1989 landings were very close to this. With demand for fish growing at about 2 per cent per annum it means a deficit of some 20 million tons by the end of the century (Garcia 1992). This signifies increased pressure in many areas on stocks, and related conflicts for the resource. It will also mean more concentration on aquaculture, rearing fish, molluscs and marine plants in the coastal zones, with the possibility of adverse environmental impacts.

In most fisheries there is overfishing, and stocks are under stress from environmental degradation. There are also losses of natural fish nurseries such as mangroves, coral reefs and estuary areas as a result of industrial and other developments.

A major problem in commercial fishing has been the common property status of fish stocks, whereby the fish belong to no one until caught. The EEZ framework has transferred fish within 200 nautical miles to national sover-eignties. This, however, has not altered the common property characteristics of the fish for most national fishermen, who compete for catches within an overall total allowable catch (TAC). This is based on an estimated maximum sustainable yield (MSY) for the national zone, or some similar system.

There are conflicts also between small-scale and large-scale fisheries in many areas. There are probably some 12 million fishermen in the world. Of these 11 million fish with small craft and simple gear. The other million operate larger vessels and land about 70 per cent of the total world catch. They fish their own EEZ, those of other states (increasingly with licence payments) and the high seas stocks. Some modern gear can be destructive, such as heavy beam trawls on the bed of continental shelves, and the unselective long monofilament drift nets in the open ocean.

Fish provide about 40 per cent of the protein intake of about two-thirds of the world population (Garcia 1992). The top fishing nations in 1989 were the USSR (11.3 million tons), China (11.2), Japan (11.2), Peru (6.5), Chile (6.5) and the USA (5.7) (FAO 1991).

ENERGY AND MINERALS

The extraction of hydrocarbons from continental shelf areas has been under-taken since the late nineteenth century. It is only since the 1960s that major activities have been conducted from drill ships and floating platforms in water depths exceeding 200 metres.

The petroleum industry in its search for oil and gas offshore first seeks the right geological features in sedimentary basins. In these basins traps are identified primarily by seismic techniques. The oil indicators are analysed and exploration drilling then takes place.

It is a high risk business, with exploration success in some regions being one out of ten holes drilled. The cost of a drill ship in the late 1980s was around $200,000 per day. Risks are therefore spread between different companies undertaking various supporting activities such as supply base operations, supply vessels, tugs, helicopters and materials.

About 25 per cent of world oil is now being supplied from beneath continental shelves. These oil and gas resources are the property of coastal states, and the seaward extensions of resource entitlements are defined from the formula in the 1982 Convention. Problems can arise when opposite and adjacent states share a continental shelf and hydrocarbon sources straddle boundaries. In these cases it is a matter of negotiation on joint development, as between the United Kingdom and Norway in the mid-North Sea. Some hydrocarbon sources are not exploited when there are disputes, as in Asian waters. International oil companies have to be assured of a valid title by the state to the resource before they will even explore.

Hydrocarbon exploitation provides income for a coastal state, employment of nationals, and the generation of related industries and services. It may however also conflict with other activities in the EEZ such as fishing and navigation, and with land availability and port space in the coastal zone. There is also always pollution from drilling and extraction operations as well as from accidents. Furthermore, the oil business is utilizing a non-renewable resource, many of its marine structures are highly mobile, and the industries can move out of a region leaving socio-economic problems.

In addition to hydrocarbons, there are renewable energy resources to be derived from the ocean. These have had limited application so far. Tidal generation of energy started in the Rance Estuary at St Malo, France during 1967. Wave energy has yet to be developed commercially, and ocean thermal energy conversion (OTEC) has not proceeded beyond successful experimental stages. The principle of OTEC has been known since the nineteenth century. It is based on vertical temperature differences in the ocean layers within tropical and sub-tropical regions. A prototype OTEC is generating energy and freshwater in Hawaii. There are several projects for OTEC floating structures to provide electricity, fresh water supplies, and fish farming facilities in the open ocean.

The mining of hard minerals from the ocean has been confined to the continental shelves. Marine aggregates of glacial origin are dredged in large volumes from the North Sea and offshore areas. Salt from sea waters is a major coastal industry, and from continental shelves placer deposits including tin, diamonds, gold and iron sands are won.

There are vast quantities of minerals on the deep ocean floor beyond national jurisdiction. The existence of seabed manganese nodules has been known since the *Challenger* expedition of the 1870s. Modern surveys have revealed these to cover thousands of square kilometres at depths of 3,000–5,000 m. Nodules contain copper, nickel, cobalt, manganese, molybdenum, vanadium, and titanium (United Nations 1984).

The deep-sea manganese nodules belong to no one. A resolution of the United Nations General Assembly declared that the area of the seabed and the resources therein are the common heritage of mankind; and accordingly the 1982 Convention stipulates that no state shall claim or exercise sovereignty or sovereign rights over any part of the area or its resources (United Nations 1982). This has not found favour with some of the developed nations where there is technical capacity for deep-sea mining and opposition to the powers of a proposed 'Seabed Authority'.

The mining of deep-sea minerals depends on the price of nickel in particular. There is a slow growth predicted for this and other metals in the future, which makes it unlikely that there will be extensive ocean mining before the end of the century.

In addition to nodules there are metalliferous deposits of hydrothermal origin in the areas where new ocean floor is being formed. These comprise copper, zinc, lead and nickel. More may lie buried, and 'huge fields with millions of metric tons of precipitated sulphide ore have been discovered on the mid Atlantic Ridge, and the North Pacific Ridge' (Edmond and Van Damm 1992: 80).

Major companies which have been concentrating on high technology for military purposes are now refocusing on the opportunities for energy and minerals in the ocean (Noland 1993).

MARINE TRANSPORT

There are some 79,000 merchant ships (above 100 gross registered tonnage) in the world. The total carrying capacity of the world fleet is 720 million deadweight tonnage (1994). Of this tonnage 40 per cent is oil tankers, mainly transporting crude oil, but also refined products, chemicals and liquid gas. The total of all cargoes transported across the ocean per year is around 4,000 million tons, 1,200 million of which is crude oil (Couper 1992a).

The world economy depends on the regularity, reliability and safety of shipping. Freedom of access and freedom of navigation for all ships on the high seas are valued and defended; as are the rights of innocent passage in territorial seas, and the even stronger rights of transit passage through straits used for international navigation. In the new 200-nautical-mile EEZs ships have retained high seas freedom of navigation.

The geographical distribution of marine transport on the world ocean shows marked concentrations off capes, straits, canals and major ports. Ships follow least distance routes in the ocean commensurate with safe navigation, but also least time tracks taking weather conditions into account and the advantages of following the axes of major ocean currents.

Marine traffic encounters other user activities in various parts of the sea. In continental shelf waters and in straits, shipping can meet with high density fishing fleets, oil rigs and platforms, aggregate dredgers and complex crossing traffic. In places such as the southern North Sea, there are many space conflicts between users. This is often made more acute by large deep-drafted ships following narrow water channels, and where they are sometimes dependent on the rise of tide for safe transit over banks. These ships have little room for manoeuvre.

In territorial seas coastal states have designated compulsory routes for various types of traffic. However, collisions and groundings still occur under conditions of bad visibility, storms, equipment failure and human error. These accidents, as well as involving loss of life, can have serious consequences for the coastal environment. The freedom of navigation is balanced therefore by the freedom of coastal states to protect themselves against the consequences of shipping off their coasts. International agreements in these respects are fundamental and are adopted at the International Maritime Organization (UN), but are implemented by the state under whose flag a ship is sailing, and increasingly by states whose ports a ship enters.

In the coastal zones, it is ports which are the most important single user of the marine environment. Breakwaters, dredged channels, dumping of dredged spoil impinge on the regimes of currents, tides and erosion and deposition, as well as on habitats for marine life. Port activities generate concentrations of marine traffic, and the accidental and deliberate discharges of materials into the marine environment.

OTHER OCEAN USES

The ocean has great biodiversity, and in the deep sea and within the ecology of mangroves and coral reefs there are as yet unidentified species which may have important significance for medicine. Indigenous societies have of course been using both marine plants and animals for medicinal purposes over many centuries. The pharmaceutical industry has made some progress in this but has possibly much more to discover about the toxic and venomous marine life as sources of antibiotics. Already algae and seaweeds are used as a basis of fertilizers, and useful compounds have been isolated from creatures such as squid and octopus (Schmidt 1989).

The recreational value of the ocean has also been widely utilized, and in recent years scuba diving has made fragile reef and other ecosystems available as a resource for the tourist industry. New technology will also make it possible for tourists to descend to vast depths, and for underwater habitats and hotels to be established. Similarly, there are millions of shipwrecks becoming accessible with advanced locational and diving technology. There are ships with cargoes of enormous value off the coast of Florida and elsewhere (Aunapu *et al.* 1993), and these are complex legal issues.

THE OCEAN AS A DUMP

In every part of the ocean there is evidence of its use for human waste disposal. Most of this is in coastal waters; the open ocean, although affected, is considered by the United Nations Environment Programme as still being relatively clean. However, with a doubling of the world population within the next seventy years there will be a corresponding increase in waste and the world disposal problem. The deep ocean may then be used for implanting nuclear waste capsules and other toxics beneath the seabed. Deep–sea mining with ship or seabed processing of ore may also take place with uncertain environment consequences from the disposal of spoil.

The origins of waste materials and their principal pathway to the ocean are from land 44 per cent, atmosphere 33 per cent, shipping 12 per cent and dumping 10 per cent (GESAMP 1990). Rivers deliver many millions of tonnes of dissolved and particulate material to the ocean. In estuaries dredgers extract and dispose transported sediments further seawards, dumping over 200 million tonnes each year. This smothers benthic communities, and it is estimated about 10 per cent of dredge spoil is contaminated (GESAMP 1990).

The runoff of agricultural fertilizers causes over–enrichment of coastal waters and eutrophication. This leads to mass mortality of fish. Sewage has a similar effect, and also contributes to gastro–intestinal infections in people as a result of eating contaminated shell fish.

The fishing fleets dump waste in the sea, including nylon nets which are not biodegradable. It is estimated that in one surveyed year 155,000 tonnes of nets were discarded or lost at sea (GESAMP 1990). This type of gear continues to ghost fish for many years after it has been discarded.

Merchant ships also discharge plastics and other material into the ocean and have been a constant cause of oil pollution. Most waste oil actually comes from land sources, but accidental and operational discharges from ships contribute almost one million tons per annum. This has been diminishing as a result of reduced carriage of crude oil in recent years and the improved operational procedures and new ship designs under the International Maritime Organization Conventions. Ship groundings such as the *Amoco Cadiz* and the *Exxon Valdez* brings large and catastrophic localized effects.

Offshore hydrocarbon activities make small but steady additions to oil pollution. Like ships, major accidents to rigs and platforms can be disastrous in a locality. The blow-out in the Ixtoc field in the Gulf of Mexico during 1979 allowed 400,000 tons of crude oil to enter Gulf waters (Clark 1989).

Other waste disposals include the dumping of weapons and the release of radionuclides from nuclear energy plants through operations and by accidents.

ENVIRONMENTAL IMPACTS OF OCEAN USES

Human uses of the ocean must clearly have had undesirable impacts on coastal areas and on specific species over a long time-scale. Some societies recognized the potential dangers of this early in their development. In the oceanic coral islands there are no continental shelves and beyond the reef's edge the slope plunges to great depths. The ocean which separates island groups is not rich in nutrients and only highly migratory fish species are found, often on a seasonal basis. Coral island soils are extremely poor and the islands are often vulnerable to droughts and hurricanes. The people of the islands were aware that the rich and diverse marine life of their homeland reefs was finite and that they were part of an isolated and almost self-contained ecosystem. As elsewhere in indigenous societies people evolved conservation rules, taboos, and sea-tenure systems, and had reverence for the marine life and habitats on which their lives depended (Couper 1993; Johannes 1978).

It took a long time for most of the inhabitants of continental coastal zones, with access to more land resources, to appreciate the need for protection of the marine environment and conservation of living resources. Another part of the problem was the difficulty in separating anthropogenic changes from natural fluctuations in fish availability, and in phenomena such as coastal erosion and flooding which could be caused by human interference or natural processes. This still presents some difficulties. Drastic reductions in, for example, catches of North Sea haddock can be attributed to the natural population volatility of that species or to over-fishing or both.

Natural causes underlie collapses of the South American anchovy fisheries, and periodically reduce world catch figures very appreciably. This is related to weakening of the trade winds over the South Pacific. These winds would normally drive surface waters westwards and generate coastal upwelling. Their reduction brings warm water from the west which stifles the upwelling of nutrient rich cold water. This regional El Niño phenomenon results from very complex interactions of ocean, atmosphere and land. The wider mechanisms may affect the Indian Monsoon, bring droughts to North and South America and storms to the Eastern Pacific (Leetmaa 1989).

Large-scale phenomena, which are only partly understood, are also responsible for changes in sea-level over geological time. These have been totally independent of human activities. In the past the human element in changes has been mainly responsive. The striking feature of the present era is that human activities are inducing enormous environment changes, but these are also superimposed on global natural fluctuations, which leads to very great uncertainties.

Global warming is the main example. It appears primarily as the result of a build-up of human-activity-generated carbon dioxide, methane and other greenhouse gases in the atmosphere. In prehistoric times forest clearing would have contributed to this. In the Industrial Revolution the use of fossil fuels

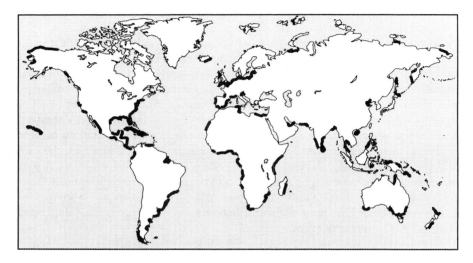

Figure 24.2 Darkened areas show some of the coastal regions vulnerable to sea-level rise. There is now a global network of about 300 sea-level monitoring stations initiated by the Intergovernmental Oceanographic Commission of UNESCO. After Peck and Williams (1992).

added enormous quantities. In the present era of population growth, megacities, cars and concentrated industry, thousands of millions of tonnes of gases are emitted.

The ocean absorbs massive quantities of CO_2. It is an important element in ocean life. It is not proven, however, whether a rise in sea temperature as a result of global warming would mean absorptions of more CO_2 by the ocean or the release of the gas to the atmosphere (Takahashi 1989). This is one of the many uncertainties in the problem of global warming.

The greatest uncertainty involves the magnitude of temperature increase and sea-level rise. Temperature increases in the range of 1.5°C to 4.5°C are postulated. This would result in a sea-level rise of 22–140 cm (GESAMP 1990). Other findings give a 21–71 cm rise in sea-level by the year 2070 (Houghton *et al.* 1990). Worst-case scenarios which include the melting of glaciers and ice caps, produce almost 2 m with accelerated temperature rises thereafter (Jacobson 1989). Figure 24.2 depicts some of the regions vulnerable to erosion and inundation with sea-level rise (Peck and Williams 1992).

Even a rise of about one metre, as has been predicted, could create major problems with the inundation of low-lying land and islands, losses of coral reefs and mangroves, bird migration sites, and saltwater intrusions further inland. A total of 5 million km² of land is at risk; this would affect one-third of coastal cropland and the homes of 1 billion people (United Nations 1991). Warmer seas will also affect the distribution of fish, although possibly with some advantages in the Arctic giving longer periods of fish growth and more inshore availability. On the other hand the breakdown of ozone layers due to CFC emissions into the atmosphere may adversely affect plankton growth in the Arctic and elsewhere by exposure to ultraviolet rays. Again there is uncertainty (Hardy and Gusinski 1989). Warmer seas may alter the paths and severities of storms, the distribution of icebergs, and changes in the Gulf Stream and Kuroshio currents, but by how much is unknown.

Characteristics of present-day anthropogenic changes on the marine environment are the great increases in volume, diversity and spatial extensions over a very brief time-scale, and the virtual independence of the actors involved. Many of the consequences of these multiple human activities are unpredictable. They may take a long time for the full effects to manifest themselves, and when they do they could be catastrophic and irreversible. It is this realization which has brought about the search for more careful and more integrated management of ocean uses. A brief outline of the trends in use management will show that it has been a slow process, but there are some achievements.

OCEAN-USE MANAGEMENT

The planning and management of multiple uses of the ocean is clearly a much more complex exercise than land-use planning. The ocean is mobile, so too are most of its living resources, as are many activities on and under the sea surface and the movement of pollutants. All of these can circulate regardless of national boundaries.

What ideally is being sought in use management are global environmental safeguards, clear entitlements, methods of harmonizing sectoral uses to reduce conflicts, and the maximizing of net returns from the optimal combinations of uses within regions, without foregoing opportunities and resources for future generations (Couper 1992b). This is far from being achieved globally, and has not been attained regionally or in any national EEZ, although the management of the Netherlands section of the southern North Sea is as close as has been achieved so far (Miles 1992).

The basic problems in achieving integrated management are that most ocean uses evolved independently of one another and under the concept of high seas freedoms. This did not present a problem while ocean space and resources were perceived as plentiful. There were no use conflicts then and little need to divide the ocean. The attempt to do so by Spain and Portugal in 1494 was premature and abortive. It was generally accepted for many centuries before and after this attempt that the ocean was free and bountiful. Such a view acquired legal legitimacy with the publication of *Mar Liberum* by the Dutch jurist Grotius in 1604. The exception to the total freedom of the ocean was coastal state jurisdiction over a belt of territorial sea. This was eventually recognized as extending up to three nautical miles from the coast.

The situation did not change appreciably until after 1945. Then President

Truman proclaimed the rights of the United States to the subsoil and seabed of the continental shelf beneath the high seas contiguous to the United States. This was not a claim to the resources of the sea surface or water column, but it precipitated such claims, as well as those to continental shelves world-wide. It was the beginning of the sea enclosure movement. Several international conferences were held to define national entitlements to ocean space, resources and navigational rights. These culminated in the third UN Convention on the Law of the Sea which met periodically from 1973 to 1982. The outcome was the largest and most comprehensive document ever produced relating to the ocean.

The 1982 Convention brought into being the 200-nautical-mile EEZ, rules for continental shelf rights, and many other measures, including the declaration that the area of the seabed and its resources lying beyond national jurisdiction was the 'common heritage of mankind'.

Many of the 320 articles in the 1982 Convention, or something very like them, were incorporated into national legislation of numerous countries. Only the seabed section with the common heritage concept remained unacceptable to the United States and several other developed countries. The Convention obtained sufficient ratifications (sixty) in November 1993 to enable it to come fully into force in November 1994. In the interim period the seabed provisions were made more acceptable to the developed states by an agreement in July 1994 recognizing market principles. The 1982 Convention thus has universal acceptance. What was never fully implemented between 1982 and 1993 however was the underlying concept of the treaty which said 'the problems and opportunities of ocean space are closely interrelated and need to be considered as a whole' (United Nations 1982).

The holistic view of the ocean had already been expressed at the 1972 Stockholm Conference on the Human Environment. It had in fact resulted in the creation of UNEP and the regional seas projects. These were based on the concept of getting integration between states that were sharing the same sea areas and common stocks of resources. The Brundtland Commission in 1987 in turn laid emphasis on the need to integrate economics and ecology for sustainable development (WCED 1987).

Only a little was achieved in practice towards these aims in most parts of the world, despite the establishment of the EEZs and other legal entitlements. In the majority of states uses of the seas were regulated, if at all, by separate ministries with few horizontal linkages between them. Over-fishing, pollution, marine accidents, unsustainable practices, and losses of coastal areas, habitats and species increased.

It was the pressure of non-governmental organizations (NGOs) and the force of scientific evidence that finally increased the awareness of governments to the likely catastrophic consequences of people-induced environmental changes – in particular, global warming, sea-level rise, and reductions in the stratospheric ozone protection shields. It was now imperative that the proposals of the previous world conferences were implemented, including integrated management and environmental strategies for ocean uses.

During July 1992 the United Nations Conference on Environment and Development (UNCED) was held in Rio de Janeiro. This was composed of 180 countries and attended by over 100 heads of state. All of the results of the Rio earth summit have a bearing on the ocean, but particularly Agenda 21, Chapter 17, entitled 'Protection of the Oceans, all kind of Seas, including Enclosed and Semi-enclosed Seas, the Coastal Areas, and the Protection, Rational Use, and Development of their Living Resources' (United Nations 1992).

The critical nature of the global situation was reiterated at UNCED and a series of principles and future actions were agreed. Important in these respects was the requirement to implement the 1982 Convention on the Law of the Sea, priority action on land-based sources of pollution, and the 'polluter pays' principle. The many uncertainties regarding cause and effect in people/environment interaction also brought the 'precautionary principle' into more prominence, whereby lack of scientific certainty should not be a reason for not taking precautionary measures. This is in line with the thinking behind the 1983 moratorium on the dumping of low-level radioactive waste in the sea, and that of 1985/6 on the catching of whales.

In summary, seven programme areas were designated at UNCED as requiring new approaches. These were to be 'integrated in content, and precautionary and anticipatory in ambit', namely:

- integrated management and sustainable development of coastal areas, including EEZs;
- marine environmental protection;
- sustainable use and conservation of marine living resources of the high seas;
- sustainable use and conservation of marine living resources under national jurisdiction;
- addressing critical uncertainties of the marine environment and climate change;
- strengthening international, including regional, co-operation and co-ordination;
- sustainable development of small islands.

The objectives of UNCED Chapter 17 on the ocean will require surveys and monitoring, ranging from global level to traditional conservation practices of indigenous maritime communities. Vast international expenditure will be required to carry through the programmes and the creation of new human resource capacities to implement them. In this latter respect UNCED emphasized the need for wide co-operation, multi-disciplinary skills, and for conceptual approaches based on an understanding of the interaction between human and physical elements.

The conference clearly recognized that human activities have become so powerful as to bring major unintended and undesirable environmental changes. It follows that social, legal, economic and institutional changes and responses must now be made to restore balances. Failure to do so means gambling with the lives of future generations. Important in trying to reach some form of balance in sustainable development will be the widespread implementation of the UNCED basic principle of 'thinking globally and acting locally', along with major ethical approaches (Cole-King 1993). This will only be achieved through education at every level. Geography has much to offer, and new academic foci such as Marine Geography, Marine Affairs, and Marine Resource Management are increasing in universities (Smith 1992). They are directed towards the skills required for the management of coastal and ocean areas and the integration of uses in a rational and sustainable way.

REFERENCES

Agardy, T. (1993) 'The beautiful and the bizarre: an unparalleled diversity', in L. Silcock (ed.) *The Oceans: A Celebration*, London: Ebury Press.

Aunapu, G., Epperson, S., Gibson, H. and Skari, T. (1993) 'The ocean gold rush', *Time* (October) 59–67.

Clark, R.B. (1989) *Marine Pollution*, Oxford: Clarendon Press.

Cole-King, A. (1993) 'Marine conservation', *Marine Policy* (May), 171–85.

Couper, A.D. (1992a) *The Shipping Revolution*, London: Conway Maritime Press.

Couper, A.D. (1992b) 'History of ocean management', in P. Fabbri (ed.) *Ocean Management in Global Change*, London and New York: Elsevier.

Couper, A.D. (1993) 'The human oceans: protectors and plunderers', in L. Silcock (ed.) *The Oceans: A Celebration*, London: Ebury Press.

Edmond, J.M. and Van Damm K.L. (1992) 'Hydrothermal activity in the deep sea', *Oceanus* 35(1), 76–81.

FAO (1991) *Fisheries Statistics* (vol. 70), Rome: FAO.

Garcia, S.M. (1992) 'Ocean fisheries management, the FAO programme', in P. Fabbri (ed.) *Ocean Management in Global Change*, London and New York: Elsevier.

GESAMP (1990) *The State of the Marine Environment*, United Nations Environment Programme, Nairobi: UNEP.

Hardy, J. and Gusinski, H. (1989) 'Stratospheric ozone depletion implications for marine ecosystems', *Oceanography* 18, 12.22.

Houghton, J.T., Jenkins, G.J. and Ephraums, J.J. (eds) (1990) *Climatic Change: The IPCS Scientific Assessment*, Cambridge: Cambridge University Press.

Jacobson, J.L. (1989) 'A really worst case scenario', *Oceanus* 32(2), 37–9.

Johannes, R.E. (1978) 'Traditional marine conservation methods in Oceania and their demise', *Annual Review of Ecology and Systematics* 9, 349–64.

Joint, I. (1993) 'The plankton powerhouse, energy capture in the sea', in L. Silcock (ed.) *The Oceans: A Celebration*, London: Ebury Press.

Kihara, K. (1989) 'People's movements on coastline conservation', *Proceedings of the Cardiff Coastal Management Seminar*, London: Centre for Economic and Environmental Development.

Leetmaa, A.L. (1989) 'The interplay of El Niño and La Niña', *Oceanus* 32(2), 30–5.

Miles, E.L. (1992) 'Future challenges in ocean management: towards integrated national ocean policy', in P. Fabbri (ed.) *Ocean Management in Global Change*, London and New York: Elsevier.

Noland, G. (1993) 'Ocean frontiers initiative in the 21st century', Law of the Sea Institute Conference, Seoul, South Korea (to be published by University of Hawaii Press).

Peck, D.L. and Williams, S.J. (1992) 'Sea level rise and its implications in coastal planning and management', in P. Fabbri (ed.) *Ocean Management in Global Change*, London and New York: Elsevier.

Schmidt, R.J. (1989) 'Drugs from the sea', in A.D. Couper (ed.) *The Atlas and Encyclopaedia of the Sea*, London: Times Books.

Smith, H.D. (1992) (ed.) *Advances in the Science and Technology of Ocean Management*, London: Routledge.

Stockholm Conference (1972) 'Declaration on the human environment', *Encyclopaedia of the United Nations and International Agreements*, London: Taylor & Francis.

Takahashi, T. (1989) 'Only half as much CO_2 as expected from industrial emission is accumulated in the atmosphere. Could the oceans be a storehouse for the missing gas?', *Oceanus* 32(2), 22–9.

UNCTAD (1992) *Group of Experts on Island Developing Countries*, Statistical Annexe, Geneva: United Nations.

United Nations (1982) *Convention on the Law of the Sea*, New York: United Nations.

United Nations (1984) *Analysis of Exploration and Mining Technology For Manganese Nodules*, London: Graham & Trotman.

United Nations (1991) *Global Ocean Observing System*, General Assembly, A/Conf.151/PC/70:16, New York: United Nations.

United Nations (1992) *Conference on Environment and Development* (UNCED), Agenda 21, Chapter 17, New York: United Nations.

World Commission on Environment and Development (WCED) (1987) *Our Common Future* (Brundtland Report), Oxford: Oxford University Press.

FURTHER READING

Barston, R.P. and Birnie, P. (1980) (eds) *The Maritime Dimension*, London: George Allen & Unwin.

Blake, G.H. (ed.) (1987) *Maritime Boundaries and Ocean Resources*, Beckenham: Croom Helm.

Coull, J.R. (1993) *World Fisheries Resources*, London: Routledge.

Couper, A.D. and Gold, E. (eds) (1992) *The Marine Environment and Sustainable Development: Law, Policy and Science*, Law of the Sea Institute, William S. Richardson School of Law, University of Hawaii, Honolulu.

Earney, F.C.F. (1990) *Marine Mineral Resources*, London: Routledge.

Glassner, M.I. (1990) *Neptunes Domain: A Political Geography of the Sea*, Boston: Unwin Hyman.

O'Neill, B. (1990) 'Cities against the sea', *New Scientist* (February), 46–9.

Peet, G. (1987) *The Status of the North Sea Environment: Reasons for Concern* (2 vols), Amsterdam: Stichting Werkgroep, Noordzee.

Prescott, J.V.R. (1985) *The Maritime Political Boundaries of the World*, London: Methuen.

Ross, D.A. (1982) *Introduction to Oceanography*, London: Prentice-Hall.

Smith, H.D. (ed.) (1991) 'The development of marine regions', *Ocean and Shoreline Management* (Special Issue) 15(4), 261–338, London: Elsevier.

Vallega, A. (1992) *Sea Management: A Theoretical Approach*, London and New York: Elsevier.

25

WATER

Confronting the critical dilemma

Peter Crabb

INTRODUCTION

Water, more than anything else, demonstrates the interdependence of the physical and human environments. No natural resource is more important than water, one writer calling it 'The first commodity' (Anon. 1992a). Because water is essential for all life and virtually every human activity, it is central to the study of the planet Earth as the habitat of all living things. It is central to geography.

Except for those who live in the world's arid regions, the very familiarity with water has contributed to many inaccurate and damaging views of and attitudes towards this ubiquitous commodity. Although renewable, 'Fresh water is a finite and vulnerable resource, essential to sustain life, development and the environment' (ICWE 1992: 4). A global water crisis, with a potential impact far greater than the 'energy crisis' of the 1970s, is looming, not so much because of any inherent supply limitations, but rather because of what human beings are doing to the earth's aquatic systems. We can no longer go on disrupting the hydrological cycle with impunity. Individual and community attitudes to water have to change. Contributing to such change is a vitally important task for geography.

THE EARTH'S WATER RESOURCES

Overall, the 'Earth's water cycle is an abundant provider' (Hammond 1990: 161). The amount of water on and close to the surface of the earth is vast (Table 25.1). However, most of it is in the oceans, while all but a small fraction of the freshwater is stored in the polar ice caps, in glaciers and in groundwater. Only a very small quantity is readily available for human use. Essentially, this is the

Table 25.1 The earth's water resources

Category	Total volume (km³)	% of total	% of fresh	Replacement period	Annual volume recycled (km³)
World ocean	1,338,000,000	96.5	–	2,650 yrs	505,000
Groundwater					
(to 2,000 m)	23,400,000	1.7		1,400 yrs	16,700
Predominantly fresh groundwater	10,530,000	0.76	30.1	–	–
Soil moisture	16,500	0.001	0.05	1 yr	16,500
Glaciers and permanent snow	24,064,100	1.74	68.7	–	–
Antarctica	21,600,000	1.56	61.7	–	–
Greenland	2,340,t00	0.17	6.68	9,700 yrs	2,477
Arctic Islands	83,500	0.006	0.24	–	–
Mountain areas	40,600	0.003	0.12	1,600 yrs	25
Ground ice					
(permafrost)	300,000	0.022	0.86	10,000 yrs	30
Lakes	176,400	0.013		17 yrs	10,376
Freshwater	91,000	0.007	0.26	–	–
Salt water	85,400	0.006		–	–
Marshes	11,470	0.0008	0.03	5 yrs	2,294
Rivers	2,120	0.0002	0.006	16 days	48,400
Biological water	1,120	0.0001	0.003	–	–
Atmospheric water	12,900	0.001	0.04	8 days	600,000
Total water	1,385,984,610	100.00*			
Freshwater	35,029,210	2.53*	100.00*		

Source: Laycock (1987).
Note: *Some duplication in subcategories and categories.

water in the world's rivers and lakes. It is only because the hydrological cycle constantly replenishes these waters that they can be used and so classified as renewable resources. Because of its relatively rapid replacement, human beings have access to much more water than what is present in the rivers and lakes at any one time. However, even this quantity needs qualification, for nowhere near all of the approximately 41,000 km³ of water that are in circulation as part of the hydrological cycle are actually available for human use and consumption. The reasons for this are many.

The limited quantity of the earth's available freshwater resources is compounded by its temporal and spatial variability. While the annual variability in the earth's total river runoff is relatively small, significant temporal variability can affect any place on the earth's surface. It can be seasonal, part of the 'normal' pattern, or it can occur from one year to another, resulting in drought or flood, such as the variations associated with the El Niño effect on the west coast of

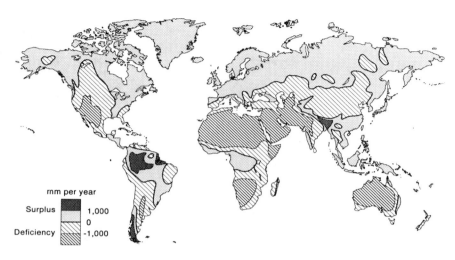

Figure 25.1 Global water surplus and deficiency: areas characterized by water surplus and water deficiency. If precipitation provides less water than needed by a well-established type of vegetation, there is water deficiency (negative values); in the opposite case there is water surplus (positive values). After Falkenmark (1977).

South America, the islands of the Pacific, and the countries of the western Pacific from the Philippines to Australia. There can also be significant seasonal variations in the availability of water from surface sources and shallow alluvial aquifers, though not from the deeper groundwater resources. Thus it is the stable runoff or dry-weather flows sufficient to meet in-stream requirements and permitted abstractions that really determine the year-round quantity of water that is available for human use.

> Of the nearly 41,000 km³ of net annual input into the continents, the stable runoff amounts to only 13,000 km³. However, since about 5,000 km³ flows through sparsely populated regions, like the Amazon Basin, the total volume readily available for human use is whittled down to about 9,000 km³.
>
> (Hinrichsen 1986: 125)

There is also considerable variation in the spatial variability of water, at global and continental scales and between and within individual countries. Figure 25.1 provides one illustration of this. The variations within individual countries may be most obvious in the larger ones, such as Australia and Canada, but they are also evident in many smaller ones, such as the United Kingdom and New Zealand.

As the remainder of this chapter will demonstrate, every aspect of the world's freshwater resources is affected by one other factor – namely, population. This is particularly evident when the resources are considered on a per capita basis (Table 25.2). The limited availability of water is most evident in the developing countries of the arid and semiarid Middle East and North Africa, where, in many cases, rapid population growth is only adding to their difficulties. Their profligate use of water for irrigation is a further complicating factor.

> But water shortages afflict other parts of the world as well. In Poland, the draining of bogs that once soaked up rainfall [and out of which it drained extremely slowly], combined with hideous pollution, has left the country as short of usable water as Bahrain.
>
> (Anon. 1992b)

WATER USE

It has been estimated that withdrawals from freshwater sources have increased tenfold this century and tripled since 1950. In the United Kingdom, for example, water use has increased by 70 per cent in thirty years, largely due to such things as dishwashers, washing machines, car washes, garden sprinklers, and industrial processes (Pearce 1992). Globally, it is estimated that approximately 3,500 km^3 of water are now withdrawn annually for human use. Of this total, some 2,100 km^3 is for consumptive use. The remainder is returned to the surface and groundwater systems, more often than not in a polluted state (Hammond 1990).

Water use is dominated by irrigation, globally and on a continental basis (Table 25.3). Long established in virtually every country, it is an extremely water intensive activity. Though accounting for a declining proportion of total withdrawals, the quantities are increasing, especially in the developing countries of Asia, Africa and Latin America. In quantitative terms, industrial water withdrawals, for cooling, processing, cleaning, and removing industrial wastes, are second only to irrigation. Major increases are taking place, especially in the developing world. World-wide, domestic and municipal uses account for about 7 per cent of all withdrawals, but in the developed world the figures are in the 13–16 per cent range. The quantities involved increase with rising living standards, from around 20 litres per capita per day to over 500. Two-thirds of the world's population, mainly in Asia and Africa, use less than 50 litres per capita per day. Industrial and domestic uses are accounting for an increasing proportion of total water withdrawals.

Such levels of water use are made possible by all kinds of water supply and storage systems. They range from the simplest individual or collective systems based on rivers or wells, to large and sophisticated ones providing water to large numbers of users for all kinds of purposes. The former meet no more than the basic needs of countless people in developing countries, though often with a very low degree of reliability in terms of quantity and quality. The latter provide a high quality and reliable water supply for most of the populations of the developed world, especially those living in the urban areas.

Table 25.2 Annual internal renewable water resources for selected countries: total and per capita, 1990

| Countries | Annual internal renewable water resources | |
	Total in km³	*Per capita, 1990 in '000 m³*
Algeria	18.90	0.75
Angola	158.00	15.77
Chad	38.40	6.76
Congo	181.00	90.77
Egypt	1.80	0.03
Ethiopia	110.00	2.35
Ghana	53.00	3.53
Kenya	14.80	0.59
Libya	0.70	0.15
Nigeria	261.00	2.31
Somalia	11.50	1.52
South Africa	50.00	1.42
Sudan	30.00	1.19
Uganda	66.00	3.58
Zimbabwe	23.00	2.37
Africa	4,184.00	6.46
Canada	2,901.00	109.37
El Salvador	18.95	3.61
Mexico	357.40	4.03
United States	2,478.00	9.94
North and Central America	6,945.00	16.26
Argentina	694.00	21.47
Brazil	5,190.00	34.52
Ecuador	314.00	29.12
Peru	40.00	1.79
Uruguay	59.00	18.86
Venezuela	856.00	43.37
South America	10,377.00	34.96
Bangladesh	1,357.00	11.74
China	2,800.00	2.47
India	1,850.00	2.17
Iraq	34.00	1.80
Israel	1.70	0.37
Japan	547.00	4.43
Jordan	0.70	0.16
Kuwait	0.00	0.00
Lebanon	4.80	1.62
Malaysia	456.00	26.30
Pakistan	298.00	2.43
Saudi Arabia	2.20	0.16
Thailand	110.00	1.97
Turkey	196.00	3.52
Vietnam	376.00	5.60
Asia	10,485.00	3.27
Belgium	8.40	0.85
Denmark	11.00	2.15

Finland	110.00	22.11
France	170.00	3.03
Germany	96.00	1.25
Greece	45.15	4.49
Hungary	6.00	0.57
Italy	179.40	3.13
Netherlands	10.00	0.68
Norway	405.00	96.15
Poland	49.40	1.29
Portugal	34.00	3.31
Romania	37.00	1.59
Spain	110.30	2.80
United Kingdom	120.00	2.11
Europe	2,321.00	4.66
Former USSR	4,384.00	15.22
Australia	343.00	20.48
New Zealand	397.00	117.49
Oceania	2,011.00	75.96
World	40,673.00	7.69

Source: Hammond (1990: 330–1).

THE CRITICAL DILEMMA

Globally and continentally, the quantities of water withdrawn for human use are significantly less than what is available. This is also true for projected withdrawals by the end of the century (Table 25.3). However, just as total river runoff misleads opinion on the water effectively available for human use, so the global figure obscures the actual water used and available at continental, national and regional levels.

However, matters of water availability and use cannot simply be considered at such macro levels. In many situations around the world, at local and regional levels, as well as the national level in some instances, there is a growing disparity between the available supplies and the ever-increasing demands for water. It is at these levels that the critical issues really exist, not only in the poor countries of the arid and semiarid zones, but even in such a large and seemingly well-endowed country as Canada (Foster and Sewell 1981). As has been indicated, demand is growing as a consequence of both increasing population and increasing per capita consumption resulting from rising living standards. It is estimated that twenty-six countries (including eleven in Africa and nine in the Middle East) already have more people than their water supplies can support (Postel 1992). As demand exceeds supply, water is a resource of increasing scarcity, in both real and economic terms, and supply deficiencies are increasing in size and number. This shortfall in supply is the critical dilemma concerning the earth's water resources and their use. Although complicated by a great many factors, many people continue to ignore the dilemma and act as though it does

Table 25.3 Global and continental withdrawals for irrigation, industry and domestic purposes

(a) In the 1980s

Continent	Irrigation	Water withdrawals, in km³ Domestic/municipal	Industry
Europe	110	48	193
Asia	1,300	88	118
Africa	120	10	6.5
North America	330	66	294
South America	70	24	30
Australia and Oceania	16	4.1	1.4
USSR	260	23	117
World	2,206	263.1	759.9

(b) Projections for the year 2000

Continent	Irrigation	Water withdrawals, in km³ Domestic/municipal	Industry
Europe	125	56	200–300
Asia	1,500	200	320–340
Africa	160	30	30–35
North America	390	90	360–370
South America	90	40	100–110
Australia and Oceania	20	5.5	3.0–3.5
USSR	300	35	140–150
World	2,585	456.5	1,531–1,308.5

Source: Hammond (1990: 172–3).

not exist. Attempts to find solutions continue to be frequently based on the belief that further supplies can be found, rather than on the more effective use of water. The following are some of the most common solutions.

Trying to increase supplies

Increases in the demand for water have long been met by increasing supplies. With over 36,000 large dams built world-wide, three-quarters of them in the last forty years, this has been the simplest solution, and in many respects it still is. Such a 'solution' is still pursued, not least in the context of many of the world's major metropolitan centres, especially in the developing world. Thus, though the numbers are significantly fewer than in previous decades, storage and reticulation schemes continue to be built, but very often regardless of the many costs involved (Goldsmith and Hildyard 1986; Anon. 1992b).

Nigeria has already built one dam, the Tiga, on the Kano River; another, at Challowa Gorge, is under construction. Both are being built in the name of greening the Sahel; both will dry out productive wetlands downstream. Once, peasants there planted rice as the river began to flood and beans as it receded, reaping two crops a year. That is no longer possible. Meanwhile, lands irrigated by the dam are already becoming too saline to use. One study concluded that the benefits provided by the wetlands in the form of crops, fishing and fuel-wood probably exceeded the gains from damming the Kano.

(Anon. 1992b; see also Madeley 1993)

Even where the water resources are available, new reservoirs are increasingly difficult and costly to construct and the unit costs of the water made available very much higher than from existing sources (i.e. the marginal costs of augmenting supplies and/or increasing reliability are often extremely large). Also, when it comes to water supply projects, often little account is taken of the availability of funds or their opportunity costs. Further, the true environmental costs are frequently ignored. For example, the siltation of reservoirs and the consequent loss of storage capacity and reductions in hydroelectric generating capacity, especially in the tropics, runs into billions of dollars (Pearce 1991a). In spite of adequate examples from within their own country, the Chinese are going ahead with a huge dam (essentially for hydroelectricity generation) on the Chang (Yangtze) River that will drown the Three Gorges area, involve the resettlement of over 1 million people, and cost over US$10 billion (Edmonds 1992). Rarely, it seems, are the costs of resettlement taken into account, often, no doubt, because they only affect minority groups of people, as with the Garrison Dam in the United States, Hydro-Québec's massive schemes in Canada, and projects on India's Narmada River.

Some measures to increase water supplies seem to hold only limited promise of success, such as cloud seeding. Desalination of sea water and brackish groundwaters is a costly option that can only be justified where water is scarce and needed for domestic or other high value purposes, as in Bahrain, Saudi Arabia and Malta. However, particularly with improvements in reverse osmosis technology, this may become a viable alternative to more traditional sources in some locations.

Other 'solutions' will likely create more problems than they solve. Of note are schemes for major inter-basin transfers of water, such as Mexico's proposal to pump water up over 1,000 m to the Valley of Mexico, and China's intention to divert 5 per cent of the Yangtze's flow to the dry northern provinces. However, these are small compared with the continental-scale schemes that have been put forward for North America, such as the North America Water and Power Alliance and the Great Recycling and Northern Development Canal concept (Bourassa 1985) – proposals that continue to arouse much opposition in both Canada and the United States (Day and Quinn 1992). Though feasible in engineering terms, the economic, environmental and social consequences of such proposals are almost totally ignored, quite apart from their political implications. For the developing countries, many of the proposed 'solutions' involve the transfer of often outdated Western technology, rather than looking at appropriate local solutions, as for example in India's Narmada Valley (Postel 1992).

Wasting scarce resources

At the same time as enormously costly attempts continue to be made to add to the supply of water, what is already available continues to be destroyed in all kinds of ways. In many parts of the world, the large quantities of water used for irrigation provide the clearest illustrations of misallocation and waste. Among the most extreme cases are where the irrigation is dependent on the 'mining' of groundwater, as such irrigation cannot be sustained. One example is the Ogallala aquifer in the United States, which underlies much of the Midwest High Plains from South Dakota to Texas (Beaumont 1985). In Saudi Arabia, excessive use could drain the country's underground water resources in 10–20 years (Hammond 1990), while in Libya, the US $25 billion 'Great Man-Made River Project' to pump massive quantities of fossil groundwater from Sub-Saharan aquifers in the southern part of the country to irrigate coastal areas in the north will destroy a non-renewable resource in 40–60 years (Pearce 1991b; McLoughlin 1991). Overuse of groundwater is occurring in other Middle East countries (Al-Ibrahim 1991; Abdel-Rahman and Abdel-Magid 1993). In many parts of the developing world groundwater is of critical importance for both rural and urban people, but overuse by large urban centres has resulted in subsidence and other problems in a number of cases (for example, Mexico City, Manila and Bangkok).

Only relatively small transfers of water away from irrigation would go a long way to meeting the increasing needs of other users in many locations. Further, as is considered below, reductions in water use would contribute significantly to reducing the environmental problems associated with irrigation – in particular, rising water tables, waterlogging and soil and water salinity.

Rampant pollution

There is no more significant waste or destruction of surface and groundwater than the widespread and serious consequences of pollution, in all parts of the world, significantly reducing available supplies of water. The dimensions of the problem are poorly understood, despite the UNEP Global Environmental Monitoring System programme. Given

> our present state of knowledge, we simply do not know the extent of contamination that has already occurred which may render some water sources unusable in the future without expensive treatment . . . it is quite possible that many sources of water that could be economically developed, may not be considered to be appropriate, especially for drinking purposes, in the future.
>
> (Biswas 1991: 221)

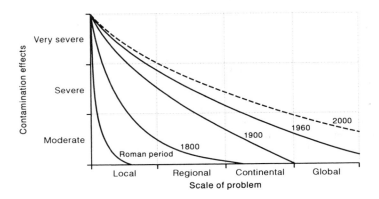

Figure 25.2 The evolution of water pollution problems. After Meybeck *et al.* (1989).

None the less, three observations can be made. First, while some improvements are evident in parts of the developed world, there is little doubt that, globally, the problems are increasing in every way: in number, scale, range, and severity (Meybeck *et al.* 1989) (Figure 25.2). As some three-quarters of industrial and domestic water withdrawals are returned to the aquatic systems and as these uses are accounting for an increasing share of total withdrawals, especially in the developing countries, it is inevitable that without a massive increase in the provision of treatment facilities, water pollution will increase. Further, the potential for dilution of wastewaters is increasingly limited. Second, the main sources and kinds of pollutants can be identified: domestic wastewater, industrial effluent, land-use runoff, leaching from mine tailings and solid and liquid waste dumps, and atmospheric sources (e.g. acid rain). Third, the nature and scale of the problems and their consequences can be clearly articulated at the local and regional levels, as the following illustrations demonstrate.

Water quality affects human health. At the start of the International Drinking Water Supply and Sanitation Decade 1981–90, it was estimated that only 40 per cent of the world's population had an adequate safe water supply and only 25 per cent a sanitation facility. Much of the progress since 1980, especially in terms of water supply in urban areas, has been more than offset by population growth. Without a major re-ordering of national and international priorities, the funds are not available to remedy such deficiencies. 'Diseases associated with filthy water kill over 5m people a year, many of them children, and incapacitate millions more' (Anon. 1992a), particularly in developing countries such as Bangladesh, India and Madagascar, where there are so many problems in providing a minimally safe and reliable water supply. There is little or no treatment of domestic and industrial wastes before they are discharged into rivers, lakes and groundwaters. As a result, many water supply sources are highly contaminated with pathogenic bacteria, viruses and many other pollutants. In the early 1980s,

> Out of India's 3,119 towns and cities, only 217 have partial (209) or full (8) sewage treatment facilities. The result is severely contaminated waters. A 48-kilometer stretch of the Yamuna River, which flows through New Delhi, contains 7,500 coliform organisms per 100 ml of water before entering the capital, but after receiving an estimated 200 million litres of untreated sewage every day, it leaves New Delhi carrying an incredible 24 million coliform organisms per 100 ml. Industry is no better. That same stretch of the Yamuna River picks up 20 million litres of industrial effluents, including about 500,000 liters of 'DDT wastes' every day.
>
> (Hinrichsen 1986: 135)

There are also serious problems in Latin America, where at least 8 per cent of rivers have faecal coliform counts of over 100,000 per 100 ml, compared with 4 per cent elsewhere: the World Health Organization (WHO) recommended level for drinking water is 0 per 100 ml. Most developing countries are a long way from meeting the UN objective of a minimum of 40 litres of clean water per person per day. Similar problems exist in both rural and urban areas, though in many respects the rural ones are the worst. On the other hand, urbanization concentrates both the demand for water and the disposal of human wastes.

Pollution problems are not confined to the developing countries. Eastern Europe provides many illustrations of severe industrial and urban pollution. In Western Europe, the collapse of outdated nineteenth-century sewers and treatment systems poses a significant pollution threat (Speed and Rouse 1980). Elsewhere in the developed world, some of the more evident manifestations of pollution may have been brought under control, though by no means completely. Even with strict controls and sophisticated treatment facilities, there

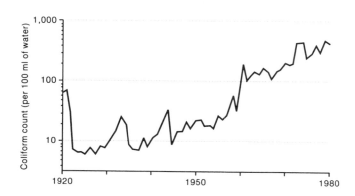

Figure 25.3 The long-term development of pathogen pollution in the River Seine. After Meybeck *et al.* (1989).

are still problems, and not just from accidents. For example, the overloading of urban sewerage systems and urban surface runoff are significant sources of river pollution in such diverse locations as Canberra and Paris (Figure 25.3). Only 57 per cent of the Canadian population is served by sewage treatment plants (Science Council of Canada 1988). Many effluents from pulp and paper, chemicals, petrochemicals and refining, metalworking, textiles, and food-processing industries are inadequately treated or controlled. Industrial effluents contain an increasing range and complexity of substances, e.g. organochlorines and other synthetic organic chemicals, heavy metals, radioactive wastes, and other toxic substances, many of them highly toxic in only minute quantities. Over 360 chemical compounds have been identified in North America's Great Lakes. Some 100,000 lakes in Canada have been damaged by acid rain and a further 600,000 are at risk (Science Council of Canada 1988). In addition, there is effluent, leaching and runoff from former and existing mining activities containing toxic substances such as mercury and cadmium. There are particular dangers in exceeding the natural absorptive and regeneration capacities of aquatic systems, especially fragile ones, when not enough is known about them, such as the Mackenzie in northern Canada.

Of increasing concern is groundwater pollution. Shallow aquifers in partic-ular are frequently damaged by numerous activities in both rural and urban areas, including agriculture, manufacturing industry, mining, septic systems, land fills, and so on. On a local scale, leaking oil storage systems have created serious problems, as has oil drilling in Alberta. More than a quarter of Canada's population depends on groundwater, but the extent of its contamination is poorly understood (Science Council of Canada 1988).

Modern agriculture is increasingly seen as a major source of water pollution. It is also a particularly difficult problem to tackle, not least because of the multiplicity and complexity of the issues involved and the diffuse or non-point sources of the pollutants. Among the problems is the contentious issue of the effects of nitrate pollution. In the United Kingdom and the USA, for example, this has resulted in unacceptable water quality in many rivers because of its believed impact on human health. For example, in thirty-eight towns in Nebraska, bottled water has to be used for babies because of the high nitrate content of local drinking waters (Biswas 1991), while the European Community has made it possible to declare nitrate-sensitive areas and impose restrictions on the application of nitrates where underlying aquifers are threatened (Conrad 1990; Evans et al. 1993). Much more serious and widespread is nutrient pollution or eutrophication. The main sources are runoff from agricultural land, which contains a wide range of chemicals from fertilizers, pesticides, etc., livestock pollution (especially feedlots), organic wastes (from processing agricul-tural commodities), and inadequately treated urban sewage and urban runoff. Such runoff and effluent contain large quantities of nutrients, especially nitrogen and phosphorus. Nutrient pollution has given rise to algal blooms in many locations, such as Venice and Lakes Balaton and Geneva. However, none

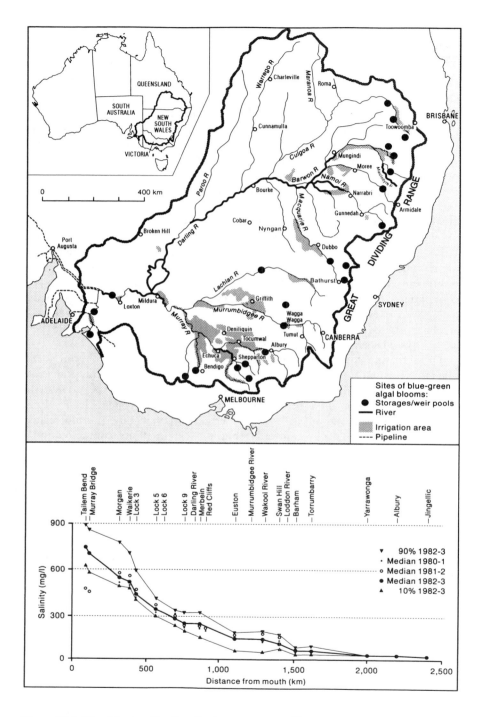

Figure 25.4 The Murray–Darling Basin: waterways affected by toxic blue-green algae; downstream increase in river water salinity levels. After Murray–Darling Basin Commission (1994) Meacham (1984).

have been on a larger scale than the toxic blue-green algal blooms (cyanobacteria) that have afflicted many parts of Australia's Murray–Darling Basin, including up to 1,000 km of the Darling River (Gutteridge, Haskins & Davey 1992) (Figure 25.4). There is perhaps no better illustration of the complexity of water quality problems, the long-term nature of the damage, the multiplicity of issues involved, and the long-term difficulties of bringing the problems under control. It is not only the result of too much nutrients, but also too little water; for the Basin as a whole, two-thirds of the water that once reached the sea is now taken out and used, very largely by irrigation.

Ecological destruction

A fourth aspect of the dilemma is the growing recognition of the environmental needs for water, needs that are essential for ecological and human survival. Over-abstraction and pollution of waterways are seriously damaging many of the world's aquatic ecosystems and affecting water users. Clearly, there has to be a reassessment of the use of water bodies for effluent disposal, not least when and where it is beyond their assimilative capacities, thus threatening and even destroying the natural ecosystems. The toxic algal blooms and salinity problems in the Murray–Darling Basin and the many problems of North America's Great Lakes provide stark illustrations of aquatic systems that are in crisis. In terms of effluent disposal, the 'sinks' are full.

But just as areas of limited water resources are under stress, so too are those where there is a seeming abundance. As was indicated earlier in this chapter, some see such water as 'surplus', available for inter-basin transfer and/or 'development'. But as Bocking (1987) observed, water is where it is needed by nature. Water has to be preserved for the aquatic systems of which it is an essential part, even where this means making less available for human use. The alternative is ecological destruction, which ultimately also means socio-economic destruction. The emergence of serious methyl mercury pollution in many of Canada's northern reservoirs serves as a warning of the potential unknown consequences of interfering with fragile aquatic systems (Hecky 1987).

TURNING THINGS AROUND

Clearly, the critical dilemma requires a new approach. In order to cope with the ever-increasing demands being placed on a finite resource, it is inevitable that we must make much better use of the water that is available. Overall, there is a need to shift the emphasis from resource-hungry, economically expensive development to sustainable management. Every effort must be made to (a) maintain water supplies, with the emphasis on integrated or total catchment management, recognizing that water management is also land management, especially in terms of quality matters, and (b) reduce and manage the demand for water. 'With technologies and methods available today, farmers could cut their water needs by 10–50 per cent, industries by 40–90 per cent and cities by a third with no sacrifice of economic output or quality of life' (Postel 1992: 23). Conservation and efficient use are the most cost-effective and environmentally sound methods of satisfying the world's water needs.

The essential transition from development or exploitation to sustainable management (Biswas 1991) must be rapid, before more irreversible damage is done. None the less, the transition will confront many difficulties and constraints – economic, political, social and institutional. Three issues of particular importance are water pricing, the regulation of water use, and the institutional arrangements for the management of water resources.

Water pricing

Central to almost every aspect of water is the relationship between the price the consumer pays for it and the real cost of its supply. Whilst water is a unique commodity, it is also a natural resource that can be bought and sold, with, ideally, the price reflecting its full costs. Rarely, however, is the real price paid for its provision and use. If it was, many of the ways in which better use can be made of the world's resources would follow. This is especially true in terms of the largest user of water, namely irrigation, in both developed and developing countries. Water does have economic value (Gibbons 1986).

The price mechanisms can be an important element in managing the demand for water. Demand is responsive to price, as, in turn, is the efficiency with which water is used. For example, in the western USA 'a fivefold increase in the price of water leads to a fiftyfold decrease in use' for thermal power-station cooling (Hinrichsen 1986: 134). Comparable evidence from the irrigation sector is also available: 'wherever (as in Punjab) farmers draw their water from privately owned tubewells and thus pay something closer to a true cost, they irrigate more efficiently than where water comes from public supplies' (Anon. 1992a). There is also growing evidence from the domestic sector. Inefficient price regimes result in the waste of water in all kinds of ways, such as the irrigation of low-value agricultural commodities and greatly increased effluent disposal. Under-charging for water and waste disposal makes a major contribution to water-related environmental problems (IC 1992). As was illustrated earlier, significant improvements in efficiency are possible now with known technology.

Increasing the price of water and the cost of effluent disposal are major incentives to increase the recycling and re-use of water. In the United States, 'water consumption by manufacturers in 2000 is expected to be only one-third of what it was in 1977, mainly because of the increased costs of disposing of wastewater that environmental legislation has imposed' (Anon. 1992a). The same point can be made with regard to finding alternative ways of disposing of waste, such as the use of sludge from Sydney's sewage systems to produce fertilizer. Such changes also make significant contributions to reducing water pollution and improving water quality. Much more contentious are proposals for charging water users for the costs they impose on others by the disposal of polluted effluent, such as from irrigation, abattoirs and feedlots.

Creating a market for water and its realistic pricing is one way of improving the allocation of water and overcoming many supply problems. Water entitlements can be transferred from one use and/or user to another, contributing to the more efficient use of water. For example,

> Los Angeles has done a deal with the farmers of the Central Valley in California: by paying for improvements to reduce wastage from irrigation channels, the city acquired more water at less than half the cost of the cheapest alternative, while farmers got cash and no reduction in their irrigation supply.
>
> (Anon. 1992a)

However, the market cannot take over completely. Allocation and reallocation have to operate within the constraints of the hydrological cycle and built reticulation systems, and also have to take account of those environmental needs that cannot pay for their water in a market place situation.

A further aspect of the under-pricing of water is that the users of the water and the consumers of the goods produced are being subsidized. The full costs of water resources schemes are frequently not paid. Thus the costs of many hydroelectric power schemes are often much more than the value of the electricity produced, especially in the developing world. The situation is even more acute in terms of irrigation, where in most countries – developed and developing – water charges do not even cover the operating costs of schemes (IC 1992).

However, regardless of how inefficient current water pricing may be, it is not something that can be changed easily or quickly (though it should not be perpetuated in new schemes). For example, what would be the consequences of charging an unsubsidized price for irrigation water in developing or even developed countries? Especially for domestic consumers, it is important to differentiate between wants and needs, to ensure that basic needs are met at a price people can afford. There are many issues involved and many potential consequences of changing current systems, quite apart from their political impact. Changes in pricing need to form part of a package of agreed measures to reduce the demand for and wastage of water. Such changes will be difficult and, in many cases, painful. But what if no changes are made?

Regulation of water use

Whilst price is probably the most important single factor, with the greatest single potential to contribute to many of the needed changes, it is by no means

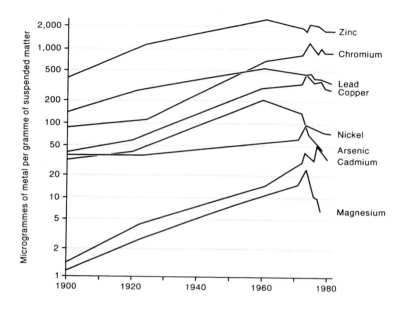

Figure 25.5 Changes in metal contamination associated with suspended matter in the River Rhine. After Meybeck *et al.* (1989).

the only one. Pricing and the market place alone are not appropriate in every location and for every issue. The market place cannot cater for essential water supplies for aquatic ecosystems and many other environmental requirements. The implications for other users of allocating scarce water resources for such purposes will have to be recognized. Further, other forms of regulated allocation may have to be used as a means of managing demand and containing conflict, such as between irrigation and ever-expanding urban areas. In other words, pricing needs to be supported and complemented by a range of other measures.

Regulations of various kinds are needed to encourage and enforce conservation through, for example, the installation of water-saving devices, the recycling of water (Postel 1985), and prohibiting some water uses at certain times. Water-quality standards will have to be set and effluent disposal restrictions enforced, regardless of an industry's ability or inability to pay for effluent disposal. Improvements are taking place, particularly in dealing with point-source pollutants, as in the River Rhine (Figure 25.5) and the Mersey Basin (Opie 1993; see Figure 25.6). Dealing with non-point or diffuse sources will continue to be much more difficult, as the Murray–Darling Basin's toxic algal outbreaks demonstrate. Water supply agencies will have to ensure that their reticulation systems have as few leakages as possible: in the United Kingdom, leakages result in losses of up to 30 per cent (Pearce 1992). In water-deficient areas, and at times of periodic scarcity in others, priorities will have to be clearly established for

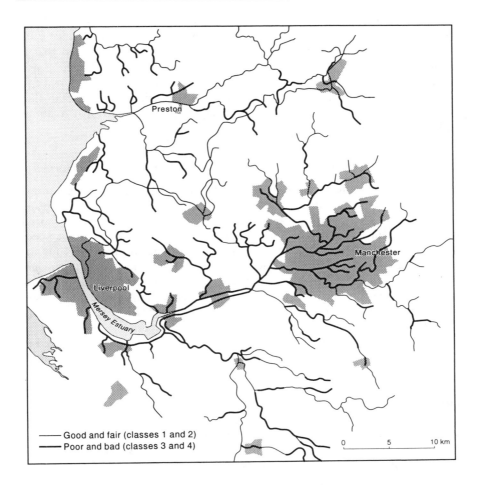

Figure 25.6 The Mersey Basin, indicating the extent of pollution in the basin's waterways, based on the National Rivers Authority classification. After Opie (1993).

water use to ensure that essential needs are met, for example, as has been done in Alberta. Apart from meeting basic human needs, such essential uses will vary from one location to another, a wetland in one place, a town or a mine in another.

Thus demand management will increasingly have to be enforced through regulation in various ways and for various reasons, because the supply is limited (in total or periodically through drought), to maintain minimum water quality standards, to ensure basic human and environmental needs are met. Increasingly, there is another reason – namely, the costs of continuing to try to satisfy increasing demand. Many of the costs of such a policy have already been mentioned; increasingly, governments and agencies are unable to finance such a policy. For example, the New South Wales government has stated that it cannot afford to build another reservoir to supply the Sydney–Wollongong area and that per capita consumption will have to be reduced. In both developed and developing countries, the demand for water and the maintenance of its quality will have to be managed and contained through a judicious mix of regulations and pricing (Crabb 1991).

Institutional arrangements for the management of water resources

Many current administration structures date from the time when their prime function was simply to satisfy demand. To cope with the radically new operating environment – for example, to ensure that a market operates efficiently, to set minimum quality standards and ensure they are met, to ensure that basic needs are satisfied – requires the creation of new institutional arrangements for the management of water resources. They will need to be strong enough to do what is needed, including providing effective interdepartmental and intergovernmental co-operation. They will have to be given a status in the government hierarchy not previously available to those concerned with water resources management.

Two aspects are particularly important. First, water cannot be considered in isolation from other natural resources, in particular land, vegetation and their uses. Natural systems have to be seen in their totality. Thus the focus and jurisdiction of the new institutions will have to be on total or integrated catchment management, certainly if sustainable management is to be achieved. This will no doubt involve many contentious issues, not least the planning and regulation of agricultural land uses: why is agriculture regarded as being beyond such regulations? Second, the institutions will have to make provision for the resolution of the many conflicts over water resources, between 'owners', users and uses. This in itself may produce bitter conflicts: 'One scientist who advocated ... very mild changes to irrigation methods has been beaten up: an official in a high position ... has heard of threats of concrete shoes waiting for anyone who interfered with water rights' (Rolls 1992). Such stories are not limited to Australia. The needed changes have many far-reaching economic and social consequences, they will create many difficulties, not least because of the long history of water 'rights' in so many countries (Teclaff 1985). But again, what are the alternatives?

TWO ISSUES THAT HIGHLIGHT THE CRITICAL DILEMMA

The following discussions are illustrative of some of the present and increasingly complex future issues involved in the management of water and related natural resources. They are also among the most contentious.

Is there a future for irrigation?

Irrigation dates from much earlier than 300 BC. It was the basis for early civilizations in the Nile, Tigris–Euphrates, and Indus river basins and in China (Teclaff 1967). Many of the irrigation systems have survived to the present century. Today, irrigation is the biggest user of water, accounting for over 70 per cent of global water use. The quantity of water involved has tripled since 1950. There are over 241 million hectares of irrigated land. One-third of the world's food is produced with irrigation and it would be very difficult to meet the world's food needs without it. In Australia, for example, irrigation production is worth over A$6 billion.

However, in almost every situation and in many ways, irrigation is a very inefficient form of agricultural production. Not only is irrigation the biggest user of water, it is the biggest waster. As one writer has noted, 'Most farmers still irrigate the way their predecessors did 5,000 years ago' (Postel 1985: 11). The efficiency of water use by irrigation is extremely low, with the world average being less than 40 per cent, largely due to over-irrigation and transport losses: even in an area such as the Colorado River Basin it is estimated that 'the efficiency of water use is below 50 percent' (Hinrichsen 1986: 131). There is considerable wastage through the reticulation systems, in getting the water from the rivers or storages to the fields, and also in the actual application of water to the plants. Thus, large quantities of water removed from aquatic systems for irrigation are never used for plant production. Instead, and in combination with inadequate or non-existent drainage systems to remove unused water, it contributes to major environmental damage, in particular waterlogging and salinization. There is a long history of salinization, and bad management continues the sterilization of large areas in all parts of the developed and developing world (Postel 1992; Madeley 1993), including locations such as parts of California's San Joaquin Valley and Australia's Murray–Darling Basin. Further, drainage from irrigation schemes contributes to the pollution of rivers and lakes, especially in terms of salinity but also through nutrients and other agricultural chemicals. Israel has demonstrated that much more efficient water use is possible, even in the most difficult environment, though it is also facing increasing salinization problems (Hinrichsen 1986).

The wastage of water takes on particularly serious forms in some locations. Among the more dramatic consequences of water extraction for irrigation are reduced water levels in the Caspian Sea and the near destruction of the Aral Sea, resulting in a very much smaller, saline and polluted lake (Perera 1993). Even more serious world-wide is irrigation that is dependent on groundwater. For example, one-fifth of irrigation in the United States depends on the Ogallala aquifer, while groundwater supports major irrigation areas in such locations as Tamil Nadu state in India, parts of northern China, and many Middle East countries. Such irrigation is little more than a 'mining' activity that cannot be sustained.

Irrigation is not only inefficient in terms of its water use; it is also inefficient in financial terms. As has been indicated, throughout the world, in developed and developing countries, the vast majority of irrigators do not pay the full cost of their water supplies. In many cases they do not even pay the operating and maintenance costs of the reticulation systems, let alone the capital costs. However, there is evidence that at least some irrigators can afford to pay more for water, as demonstrated by prices paid for the transfer of entitlements and for new entitlements at auctions. For most farmers, irrigation is only made 'viable' by the subsidized water. For example, 'A World Bank study of Cyprus argues that three-quarters of the island's crops are uneconomic and are grown only because the farmers' water is subsidised' (Anon. 1992a). In some countries, such as Saudi Arabia, additional subsidies are paid on the crops produced. It may be true that, especially in many poor countries, issues of unemployment and 'food security' cannot be overlooked, but such irrigation is not sustainable.

Other factors further complicate the price issue. Should irrigators pay for the environmental damage they cause? Many irrigation schemes are rapidly deteriorating and are now in need of major rehabilitation. Should money be spent on such works? If so, who is going to pay the cost, even assuming the funds are available? In Australia, for example, the estimated investment in irrigation schemes is of the order of A$50 billion (IC 1992). However, no water resources issue is more political than the pricing of irrigation water. As well as large numbers of individual farmers, there are whole communities that are dependent on subsidized irrigation water. Price reform is an essential part of halting the mounting financial and environmental costs associated with irrigation in many parts of the world.

The large quantitative demands of irrigation relative to all other uses is in itself a potential source of conflict. In many locations, only small reductions in irrigation use, through increased pricing, reallocation and improved efficiency of use (Postel 1985), would go a long way to easing allocation problems, making water available for other uses, not least the ecological needs of the aquatic systems themselves. They would also contribute to the reduction of many associated environmental problems. Irrigation more than anything else high-lights the links between quantitative and qualitative issues and the necessity for the integrated management of water and land. Is there a future for irrigation, especially given the economic costs and environmental costs? Especially when looked at in a historical perspective, with some 1–1.5 million hectares being salinized each year (Hinrichsen 1986), surely it has to be regarded as a non-sustainable activity.

Inter-jurisdictional conflict

Throughout history, water has been a significant source of conflict between individuals and governments. The increasing disparity between supply and demand will result in more conflict, unless acceptable conflict resolution

Table 25.4 Some of the world's major inter-jurisdictional river basins

River Basin	Countries
Amazon	Bolivia, Brazil, Colombia, Ecuador, Guyana, Peru, Venezuela
Congo	Angola, Burundi, Cameroon, Central Africa, Congo, Rwanda, Tanzania, Zaïre, Zambia
Danube	Albania, Austria, Bulgaria, Czech Republic, Germany, Hungary, Italy, Poland, Romania, Slovakia, Switzerland, former USSR, former Yugoslavia
Elbe	Austria, Czech Republic, Germany, Poland
Ganges–Brahmaputra	Bangladesh, Bhutan, China, India, Nepal
Indus	Afghanistan, China, India, Pakistan
Jordan	Israel, Jordan, Lebanon, Syria
Limpopo	Botswana, Mozambique, Rhodesia, South Africa
Mekong	Cambodia, China, Laos, Myanmar, Thailand, Vietnam
Meuse	Belgium, France, Germany, Netherlands
Niger	Algeria, Benin, Cameroon, Chad, Guinea, Ivory Coast, Mali, Niger, Nigeria, Upper Volta
Nile	Burundi, Egypt, Ethiopia, Kenya, Rwanda, Sudan, Tanzania, Uganda, Zaïre
Plata	Argentina, Bolivia, Brazil, Paraguay, Uruguay
Rhine	Austria, Belgium, France, Germany, Italy, Liechtenstein, Luxemburg, Netherlands, Switzerland
Tigris–Euphrates	Iran, Iraq, Kuwait, Turkey
Volta	Benin, Ghana, Ivory Coast, Mali, Togo, Upper Volta
Zambesi	Angola, Botswana, Malawi, Mozambique, Namibia, Rhodesia, Tanzania, Zambia

procedures and institutions can be established and accepted. This is true within individual states and more especially where resources are shared between two or more jurisdictions (FitzGibbon 1990). There are 214 river and lake basins shared by two or more countries (Table 25.4).

Nearly 47 per cent of the area of the world (excluding Antarctica) falls within shared river basins. Expressed differently, there are 44 countries (20 in Africa) where at least 80 per cent of the total area lies within the international basins.

(Biswas 1991: 223)

Without agreements to manage these shared water resources co-operatively, conflict over many of them is inevitable. The problems are particularly evident in the Middle East (Mitchell and Downey 1993; Cooley 1984; Abu-Maila 1991), where, as one writer has said, they 'will foster either an unprecedented degree of cooperation or a combustible level of conflict' (Postel 1992: 74). In 1990, the Minister of State in the Egyptian Foreign Ministry went even further, stating that 'The next war in the Middle East will be over water, not politics.' Yet while having potentially the most serious ones, the Middle East is not the only part of the world with hydropolitical problems. For example, there are the impacts on Bangladesh of activities in the Nepalese and Indian portions of the Ganges basin (Ives 1991); the Netherlands has to remove German and French pollutants from Rhine water (ICPRP 1993); the controversial diversion by Slovakia of water from the Danube for its Gabcikovo hydroelectric power-station, in spite of the opposition of other countries and the position of international law (Perczel and Libik 1989); and the fate of the Aral Sea with the break-up of the former Soviet Union (Perea 1993).

But solutions are possible, even between unfriendly neighbours, as countries realize the options are limited to co-operation or resource destruction. Basic principles for the management of international water resources have been agreed to (Crabb 1993):

1   The river basin is the basic hydrological management unit; acceptance of this principle has implications for virtually all of the other principles.
2   No state can claim exclusive sovereignty to the waters of an inter-jurisdictional river basin.
3   There should be:
    • reasonable and equitable participation in the control of the resources;
    • reasonable and equitable apportionment of the water resources of the basin;
    • protection and non-abuse of the basin's resources;
    • acceptance of the interrelationships of natural resources, especially water and land and surface and groundwater.

Perhaps even more difficult, extra-state management institutions have been established. At the international level, the Canada–United States International Joint Commission continues to be an important example. However, all inter-jurisdictional situations have much to learn from the arrangements now in place for the management of Australia's Murray–Darling Basin (Crabb 1993).

THE FUTURE

Are there any real solutions to the critical dilemma? If there are, they have to begin with the recognition that the earth's water resources – like all its resources – are finite and that 'the inhabitants of the Water Planet are in fact living at the mercy of the water cycle' (Falkenmark 1991). Subjecting waterways to over-extraction and pollution inevitably leads to the destruction of aquatic systems and everything dependent on them. Water cannot be considered in isolation, as it is at the core of all aspects of human economy and society (Edwards 1993). 'We need to switch the general approach from humid zone thinking (how much water do we need and where do we get it?) to the evident approach under water scarcity (how much water is there and how can we best benefit from that amount?)'; water is 'the ultimate constraint' (Falkenmark 1991: 239).

Water scarcity affects everything. In most cases, however, it is clear that the scarcity – in whatever form – is the result of human activity rather than nature. It is also clear that the sustainability of water resources and of all life – at global, regional, local and individual levels – is dependent upon human activity. Many current practices, such as the use of waterways for effluent disposal, are clearly not sustainable. There is more than one meaning to the slogan carried on buses in Australia's capital city, Canberra: 'Save water. Before it costs the earth'. Human beings need to value water for what it is, the most precious natural resource.

Placing a true value on water will require many changes, as has been indicated in this chapter. Demand will have to be managed, if not controlled; priorities will have to be set and allocations made, for needs rather than simply wants; the efficiency with which water is used will have to be significantly improved by all users; the real price will have to be paid. Such changes are relevant even without the complication of climatic change; they may well be even more relevant with it. Finding real solutions is central to the continuation of 'the planet Earth as the habitat of all living things'.

Water issues are inextricably linked with population. In an increasing number of countries, especially many in the developing world, water problems cannot be separated from increasing population pressures. At the global level, per capita water availability has declined by a third since 1970 solely as a consequence of population growth. Providing a potable water supply for growing numbers of people, not least the increasing urban populations, is perhaps the biggest problem. Quite apart from the actual availability of resources, it is compounded by a number of factors, especially the increased demand with rising living standards and the consequent increased effluent in situations where there are no financial resources for the needed sanitation and effluent treatment. The points that have been made in this chapter are relevant world-wide, but additional measures are needed in most of the developing world simply to meet basic human needs. Thus, given the finite nature of the earth and its water resources, we are confronted with a further dilemma. In other words, without some major global reordering of priorities, without a significant transfer of resources from the developed world, the situation can only get worse in most of the developing world (Goodland *et al.* 1991).

Unfortunately, water did not get the attention it deserved at the UN Conference on Environment and Development (UNCED) held in Rio de Janeiro in June 1992. There was little input from the International Conference on Water and the Environment held in Dublin just five months earlier (Grover and Biswas 1993). In Chapter 18 of its key report, *Agenda 21*, the UNCED gathering provided little evidence of the critical importance and urgency of water issues (Johnson 1993). More than anything, the world probably needs a 'water shock', though of a much larger and more dramatic nature than the 'oil shock' of the 1970s. It is not a question of if such a shock will occur; it is only a question of when.

Some twenty-five years ago, water provided the integrating theme for a major geographical study that sought to break down the divisions between the physical and human branches of the subject; it was a work that saw 'Water as a focus of geographical interest' (Chorley 1969: 3). Water remains the classic geographic issue, one that makes a major contribution to the discipline. In turn, the discipline of geography is making an increasingly significant contribution to the sustainable management of water, 'The first commodity'.

REFERENCES

Abdel-Rahman, H.A. and Abdel-Magid, I.M. (1993) 'Water conservation in Oman', *Water International* 18, 95–102.

Abu-Maila, Y.F. (1991) 'Water resource issues in the Gaza Strip', *Area* 23, 209–16.

Al-Ibrahim, A.A. (1991) 'Excessive use of groundwater resources in Saudi Arabia: impact and policy options', *Ambio* 20, 34–7.

Anon. (1992a) 'The first commodity', *The Economist* 322(7752), 11–12.

Anon. (1992b) 'The beautiful and the dammed', *The Economist* 322(7752), 83–4, 89.

Beaumont, P. (1985) 'Irrigated agriculture and ground-water mining on the High Plains of Texas, USA', *Environmental Conservation* 12, 119–30.

Biswas, A.K. (1991) 'Water for sustainable development in the 21st century: a global perspective', *Water International* 16, 219–24.

Bocking, R.C. (1987) 'Canadian water: a commodity for export?', in M.C. Healey and R.R. Wallace (eds) *Canadian Aquatic Resources*, Ottawa: Department of Fisheries and Oceans.

Bourassa, R. (1985) *Power from the North*, Scarborough: Prentice-Hall (Canada).

Chorley, R.J. (ed.) (1969) *Water, Earth and Man: A Synthesis of Hydrology, Geomorphology and Socio-economic Geography*, London: Methuen.

Conrad, J. (1990) *Nitrate Pollution and Politics*, Aldershot: Avebury–Gower.

Cooley, J.K. (1984) 'The war over water', *Foreign Policy* 54, 3–26.

Crabb, P. (1991) 'Paying for water: there are no free drinks!', *Australian Geographer* 22, 126–8.

Crabb, P. (1993) 'Managing Australia's major natural resource: the Murray–Darling Basin', *Canadian Water Resources Journal* 18, 67–78.

Day, J.C. and Quinn, F. (1992) *Water Diversion and Export: Learning from Canadian Experience*, Waterloo: Department of Geography, University of Waterloo.

Edmonds, R.L. (1992) 'The Sanxia (Three Gorges) Project: the environmental argument surrounding China's super dam', *Global Ecology and Biogeography Letters* 2, 105–25.

Edwards, K.A. (1993) 'Water, environment and development: a global agenda', *Natural Resources Forum* 17, 59–64.

Evans, D., Moxon, I.R. and Thomas, J.H.C. (1993) 'Groundwater nitrate concentrations from the Old Chalford Nitrate Sensitive Area, West Oxfordshire', *Journal of the Institute of Water and Environmental Management* 7, 506–12.

Falkenmark, M. (1977) 'Water and mankind: a complex system of mutual interaction', *Ambio* 6, 3–9.

Falkenmark, M. (1991) 'The Ven Te Chow Memorial Lecture: environment and development: urgent need for a water perspective', *Water International* 16, 229–40.

FitzGibbon, J.E. (ed.) (1990) *Proceedings of the Symposium on International and Transboundary Water Resources Issues*, Bethesda, Md.: American Water Resources Association.

Foster, H.D. and Sewell, W.R.D. (1981) *Water: The Emerging Crisis in Canada*, Toronto: James Lorimer.

Gibbons, D.C. (1986) *The Economic Value of Water*, Washington, DC: Resources for the Future.

Goldsmith, E. and Hildyard, N. (1986) *The Social and Economic Effects of Large Dams*, Camelford, Cornwall: Wadebridge Ecological Centre.

Goodland, R., Daly, H. and El Serafy, S. (eds) (1991) *Environmentally Sustainable Economic Development: Building on Brundtland*, Environment Working Paper No. 46, Washington, DC: Environment Department, The World Bank.

Grover, B. and Biswas, A.K. (1993) 'It's time for a world water council', *Water International* 18, 81–3.

Gutteridge, Haskins & Davey (1992) *An Investigation of Nutrient Pollution in the Murray–Darling River System*, Canberra: Murray–Darling Basin Commission.

Hammond, A.L. (ed.) (1990) *World Resources 1990–91: A Report by the World Resources Institute*, New York: Oxford University Press.

Hecky, R.E. (1987) 'Methylmercury contamination in northern Canada', *Northern Perspective* 15(3), 8–9.

Hinrichsen, D. (ed.) (1986) *World Resources 1986*, New York: Basic Books.

IC (Industry Commission) (1992) *Water Resources and Waste Disposal*, Report No. 26, Canberra: Industry Commission.

ICPRP (1993) *Ecological Master Plan for the Rhine*, Koblenz: International Commission for the Protection of the Rhine against Pollution.

ICWE (1992) *International Conference on Water and the Environment: Development Issues for the 21st Century: The Dublin Statement and Report of the Conference*, Geneva: World Meterological Organization.

Ives, J. (1991) 'Floods in Bangladesh: who is to blame?', *New Scientist* 130(1764), 30–3.

Johnson, S.P. (1993) *The Earth Summit: The United Nations Conference on Environment and Development (UNCED)*, London/Dordrecht: Graham & Trotman/Martinus Nijhoff.

Laycock, A.H. (1987) 'The amount of Canadian water and its distribution', in M.C. Healey and R.R. Wallace (eds) *Canadian Aquatic Resources*, Ottawa: Department of Fisheries and Oceans.

McLoughlin, P.F.M. (1991) 'Libya's great manmade river project: prospects and problems', *Natural Resources Forum* 15, 220–7.

Madeley, J. (1993) 'Will rice turn the Sahel to salt?', *New Scientist* 140(1894), 35–7.

MDBC (Murray–Darling Basin Commission) (1994) *Algal Management Strategy for the Murray–Darling Basin*, Canberra: Murray–Darling Basin Commission.

Meacham, I. (1984) *The River Murray Salinity Problem: A Discussion Paper*, Canberra: River Murray Commission.

Meybeck, M., Chapman, D. and Helmer, R. (eds) (1989) *Global Freshwater Quality: A First Assessment*, Oxford: Blackwell Reference.

Mitchell, B. and Downey, T.J. (eds) (1993) 'Special issue on water in the Middle East', *Water International* 18, 1–70.

Opie, R. (1993) 'Mersey mission', *Water and Environment Management* 1, 26–8.

Pearce, F. (1991a) 'A dammed fine mess', *New Scientist* 130(1767), 32–5.

Pearce, F. (1991b) 'Will Gaddafi's great river run dry?', *New Scientist* 131(1785), 5.

Pearce, H. (1992) 'In hot water', *Geographical Magazine* 64(4), 16–20.

Perczel, K. and Libik, G. (1989) 'Environmental effects of the dam system on the Danube at Bos-Nagymoros', *Ambio* 18, 247–9.

Perera, J. (1993) 'A sea turns to dust', *New Scientist* 140(1896), 24–7.

Postel, S. (1985) *Conserving Water: The Untapped Alternative*, Washington, DC: Worldwatch Institute.

Postel, S. (1992) *Last Oasis: Facing Water Scarcity*, New York: W.W. Norton.

Rolls, E. (1992) 'The river', *The Independent Monthly*, December/January, 16–21.

Science Council of Canada (1988) *Water 2020: Sustainable Use for Water in the 21st Century*, Ottawa: Science Council of Canada.

Speed, H.D.M. and Rouse, M.J. (1980) 'Renovation of water mains and sewers', *Journal of the Institution of Water Engineers and Scientists* 34, 401–24.

Teclaff, L.A. (1967) *The River Basin in History and Law*, The Hague: Martinus Nijhoff.

Teclaff, L.A. (1985) *Water Law in Historical Perspective*, Buffalo: William S. Hein.

FURTHER READING

Anon. (1993) *Sustaining Water: Population and the Future of Renewable Water Supplies*, Washington, DC: Population Action International. (See also *Sustaining Water: An Update*, published 1995.)

Ashworth, W. (1987) *The Late, Great Lakes: An Environmental History*, Detroit: Wayne State University Press.

Barrow, C. (1987) *Water Resources and Agricultural Development in the Tropics*, Harlow: Longman Scientific & Technical.

Gleick, P.H. (ed.) (1993) *Water in Crisis: A Guide to the World's Fresh Water Resources*, New York: Oxford University Press.

Hillel, D. (1994) *Rivers of Eden: The Struggle for Water and the Quest for Peace in the Middle East*, New York: Oxford University Press.

Mitchell, B. (ed.) (1990) *Integrated Water Management: International Experiences and Perspectives*, London: Belhaven Press.

Newson, M. (1992) *Land, Water and Development: River Basin Systems and their Sustainable Management*, London: Routledge.

Pearce, F. (1992) *The Dammed*, London: Bodley Head.

Pearce, F. (1994) 'Britain's other dam scandal', *New Scientist* 141(1914), 24–9.

Shrubsole, D. (ed.) (1992) *Resolving Conflicts and Uncertainty in Water Management*, Cambridge, Ontario: Canadian Water Resources Association.

26

SURFACE INSTABILITY AND HUMAN
MODIFICATION IN GEOMORPHIC
SYSTEMS

Jonathan D. Phillips and William H. Renwick

INTRODUCTION

Landforms are the stage for the drama of life. But unlike a theatrical stage, which provides only passive support for the players, landforms play an active role in biological evolution. They exist at the nexus of atmospheric, hydrospheric, lithospheric, and biospheric interactions. They influence, and are influenced by, air, water, rocks, and living organisms. Landforms provide the basic resources of soil and space, and help regulate the flow and storage of other resources, such as water and nutrients. For these reasons, and because they are quite dynamic, landforms have strong influences on human activity. Conversely, humans have important influences on landforms – both direct influences (by reshaping the earth's surface) and indirect influences (by changing climate, soils, biota, and the water cycle). In short, an understanding of human–environment relations and human roles in changing the face of the earth requires an understanding of the interaction between humans and landforms. The purpose of this chapter is to examine the stability and instability of landform (or geomorphic) systems in the context of the mutual adjustments of humans and the landscape.

Systems are sets of interrelated parts which function together as complex wholes. Geomorphic systems include interconnected landscape elements, or landforms, and related environmental factors such as geology, climate, and vegetation. Stability refers to the response of geomorphic systems to changes or disturbances. Stable geomorphic systems are able to absorb or recover from perturbations, while in unstable geomorphic systems disturbances tend to persist or even to become magnified. Two concepts of stability are relevant here: the 'restless landscape' and 'dynamical system' instability.

The restless landscape

The earth's surface is dynamic and ever-changing. It is subjected to variations in precipitation and solar energy input from above, to episodic tectonic forcings from below, and to exogenic geomorphic processes and human agency from within. These changes occur over time-scales ranging from instantaneous to millennial, and over spatial scales from micrometres to thousands of kilometres. Geological material is subject to breakdown by weathering from the moment it is exposed at the surface, and subsequently to erosion, transport, and deposition.

Even the quietest of landscapes undergoes relentless, if sometimes subtle, change. Other landscapes change rapidly: steep slopes undergo landslides and catastrophic erosion; the shifting sands of barrier islands and tidal inlets present a different landscape from day to day and season to season; rivers change course and overflow their banks; earth tremors and volcanic eruptions reshape the land overnight.

There are two basic implications of the instability of the restless landscape. First, the foundation of the built environment is often variable and impermanent. Second, the dynamic nature of the earth's surface often creates geomorphic hazards associated with catastrophic or threshold-exceeding events (Chapter 29).

Dynamical system instability

In the parlance of dynamical system theory, an asymptotically unstable system will be unable to return to its predisturbance state after a change or perturbation. Disturbances will persist, and will in fact grow exponentially in the short term. Many geomorphic systems have been shown to be unstable in this sense. Most geomorphic systems are nonlinear in that there is not a constant proportional response of outputs to inputs across the range of the inputs. Nonlinearity is common owing to the threshold-dominated nature of geomorphic system behaviour (Schumm 1979; Chappell 1983). Asymptotic instability in a nonlinear dynamical system indicates the possibility of deterministic chaos. Chaos is complex, apparently random behaviour arising from nonlinear deterministic systems, and is characterized by sensitive dependence upon initial conditions.

The dynamic instability and potential chaos which seems to be typical of many geomorphic systems means that they are vulnerable to disturbances (including those caused by human agency), that such changes may tend to persist or grow rather than to be absorbed or healed, and that consequences of landscape alterations may be unpredictable. Instability and chaos pose problems and prospects for the ability to forecast landscape evolution and geomorphic system behaviour. Thus, to understand and cope with the interactions of humans and geomorphic systems, the study of complex system behaviour is preferable to the identification of simple cause-and-effect relationships.

The scope of human agency

Several excellent reviews have been published detailing human effects on the earth (Thomas 1955; Simmons 1989; Roberts 1989; Turner 1990). But to get an idea of the global magnitude of human modification of landforms and landscapes it is useful to examine some statistics that put matters in perspective. All the figures and estimates below are taken from data in *World Resources 1992–93* (World Resources Institute 1992), a co-operative effort of the World Resources Institute and the United Nations.

The earth's land area comprises an estimated 1.313×10^{10} ha. This includes large areas of ice cap, tundra, and cold cratonic shields which are largely uninhabited. A mere 12 per cent of the total land area is considered arable. On the other hand, only about 26 per cent of the land area could be called wilderness (areas with no evidence of significant human alteration). Thus about three quarters of the subaerial surface of the earth bears clear imprints of human agency. This ranges from a low of 59 per cent in North America, mainly due to the Canadian Shield, and periglacial and ice-covered land in Canada and Alaska, to a maximum of 97 per cent in densely settled Europe.

Agriculture has well-documented impacts on erosion and deposition patterns and soil properties, among other things. Cropland accounts for an estimated 1.478×10^9 ha (11 per cent of the total), and permanent pasture for 3.323×10^9 ha (25 per cent). Deforestation and reforestation have quite dramatic impacts on landforms and surface stabilities, both on–site and downstream. In the early to mid-1980s an estimated 0.3 per cent of the total world forested area was deforested each year; a less but comparable amount was reforested (1.15×10^7 and 1.054×10^7 ha/yr, respectively, for 1981–5).

Human modification of the landscape may sometimes be benign and sometimes entirely or partly beneficial. However, soil degradation involves persistent or permanent alteration of the land surface which is both geomorphologically significant and clearly detrimental. From 1945 to the late 1980s, an estimated 1.964×10^9 ha were degraded; this is an area greater than the total amount of cropland on earth, and represents 17 per cent of all vegetated (natural and agricultural) land. Some of the degradation is chemical, but the majority is geomorphic, including water erosion (56 per cent), wind erosion (28 per cent), and physical soil degradation such as compaction (4 per cent). Thus an estimated 15 per cent of the earth's vegetated land has been geomorphologically degraded since the Second World War.

Some of the most severe geomorphic impacts, on a per–unit-area basis, are associated with urban and industrial activities. While industry accounts for only 1 per cent of soil and land degradation globally, these impacts are disproportionately concentrated in areas of high population density. Localized impacts of this nature are widespread in hinterlands as well. Road construction is well known as a major source of erosion and sedimentation problems, and roads often permanently alter flows of water and sediment. As a global average, there are 180 linear kilometres of paved and unpaved roads and railroad tracks per 1,000 km². While this form of alteration is quite low in some regions (56 km/1,000 km² in Africa), it is much higher in others (366 km/1,000 km² in North America, and 901 km/1,000 km² in Europe). Dams are another example of localized but dramatic landscape impact. As of 1986 there were 7,685 dams more than 30 m in height in the world, 36,562 more than 15 m high, and countless smaller dams. More than a thousand large dams were under construction in 1986.

Human use of water also has important geomorphic impacts on the movement and storage of sediment, and because changes in the movement and distribution of moisture affect weathering, soil formation, and many other geomorphic processes as well. The estimated annual water withdrawals for 1987 (3,240 km³) represent about 8 per cent of the earth's total renewable freshwater resource.

RESTLESS LANDSCAPES AND GEOMORPHIC HAZARDS

Rates of geomorphic processes vary in time through several orders of magnitude, with rates during relatively infrequent events being many times greater than rates during only moderately infrequent events. The long-term behaviour of the landscape is a function of a wide range of magnitudes of events, which together shape the land (Wolman and Miller 1960). In some environments relatively frequent events may dominate geomorphic landscapes, and such landscapes are relatively stable and nonhazardous. For example, the dominant channel-forming discharge for most streams is probably the bankfull discharge, or something close to it; for most alluvial channels this has a recurrence interval on the order of a few years. At the same time, there are many examples of systems in which rare events have caused changes in the landscape orders of magnitude greater than those occurring in frequent events such as bankfull flows. For most geomorphic processes the relation between applied stress and quantity of material moved is an exponential one, so that extreme weather events such as floods may cause changes well beyond those observed for more frequent events. The ability of flowing water to transport sediment is indicated by measures of applied stress such as stream power (Rhoads 1987), and this can be compared to the resistance of earth materials (Baker and Costa 1987). When applied stress exceeds thresholds of resistance the effects of floods on the landscape increase dramatically; a flood is truly catastrophic when resistance thresholds are exceeded for a large part of the area affected by the flood.

Spatial variation in geomorphic process rates is also great. Long-term average rates of change vary through several orders of magnitude between relatively dynamic environments and more stable ones. Saunders and Young (1983) compiled data on rates of denudation from a wide range of environments, showing that short- and medium-term denudation rates (averages over at least several years) vary through about three orders of magnitude, between 1 and 1,000 mm/1,000 yr, even within broad climatic and topographic groupings.

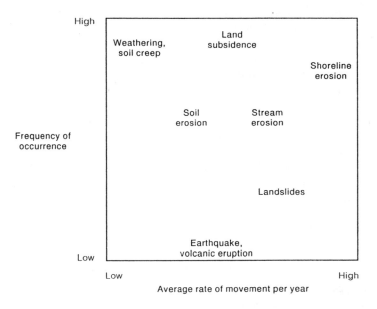

Figure 26.1 Process rates and recurrence intervals for several types of geomorphic events.

Mountain environments usually exhibit the highest denudation rates, but variability is the rule and generalization is difficult. Coastal erosion is another vivid example of spatial variability, with rates of shoreline change (averaged over periods of decades) ranging from metres per year of erosion to metres per year of accretion.

Susceptibility to rapid change during extreme events also varies within individual systems, so that a particular stress may cause rapid changes in one area and not in another (Magilligan 1992). The degree of instability in the landscape is highly variable in space and time, and the magnitude-frequency characteristics of geomorphic activity are crucial with respect to human activity. Although rates of individual process types are themselves extremely variable, a few crude generalizations are possible. Figure 26.1 depicts relative rates and frequency of occurrence of some geomorphic processes. The horizontal axis is a scale of mean rate of movement per unit time; the vertical axis is the temporal frequency of occurrence of movements. In the high frequency regions of the graph are processes which have relatively low thresholds of movement in relation to applied stresses. Some of these, such as weathering, are slow; others are quite rapid. Shoreline recession in areas of rapid beach erosion is an example of the high-rate, high-frequency end of the spectrum. Volcanic eruptions and major earthquakes are examples of phenomena with lower mean rates of change (perhaps mm/yr vertical movement when averaged over millennia), but with highly episodic movement concentrated in a few relatively rare high-magnitude events. Landslides are an example of a process that is quite active in some areas, yet occurs episodically.

The variations in rates of change create two kinds of problem for human occupancy of dynamic environments. First, the frequency of significant events may be low enough that recurrence intervals exceed either the working memory or planning horizons of decision-makers. It is common for the dominant geomorphic events in the magnitude-frequency spectrum to have recurrence intervals of decades or longer. Unless some cultural memory, written or otherwise, records the occurrences and impacts of such events, humans may not realize that they should be considered in land management. Second, as population grows and development expands to new environments, land-use practices long-established in one environment may be extended to others where they are not suitable. Because of the lack of experience in unfamiliar environments decision-makers may not be aware of relatively unusual events.

Human occupancy of hazardous areas

The spatial extent of human occupancy has increased substantially in recent decades and centuries as population has increased. Although occupancy of hazardous areas is not new, many such areas were only sparsely populated until recently. Past avoidance of geomorphologically hazardous areas may have been deliberate in some cases, and a side-effect of their relative undesirability for agriculture or construction in others. Occupation of geomorphologically hazardous areas occurs for several reasons. In some cases population pressure simply forces the use of marginal lands; as for example in Brazil where Oxisols in rain-forest environments vulnerable to degradation are increasingly utilized by peasants lacking access to better land. In other cases occupation occurs in part due to ignorance of the hazards; an example is areas prone to flood, erosion, and subsidence in south Louisiana, which were largely occupied before the nature and extent of the hazards and constraints were recognized. In still other cases development occurs in areas known to be hazardous because amenities or advantages are believed to outweigh the risks. Many beach-front property owners and occupants of the seismically active regions of California apparently hold this view. Finally, there are situations where there is a widespread and long-standing tradition or belief that geomorphic constraints can be conquered by engineering means – land reclamation and occupation of below-sea-level land in the Netherlands being the classic example.

The steep mountain area subject to landsliding and other forms of hillslope instability is an obvious example of an environment with substantial exposure to geomorphic hazards. Highland cultures have existed for millennia, and many (the Incas, for example) have thrived. But for the most part human settlement has avoided mountainous areas until recently. Today, the extent of occupation of highlands is increasing in many locations. In rapidly growing cities with limited level land, such as São Paulo and Los Angeles, residential development has crept from lowlands to adjacent mountain slopes. In some cases hazardous steeplands are occupied by poor people forced into peripheral sites; in others the mountainsides offer amenities (such as panoramic views) preferred by the wealthy. But in either case the occupancy of steep slopes increases the likelihood of slope destabilization, or exposure to the hazards of natural slope failure, or both (Alexander 1989, 1991).

Human-created hazards: slope destabilization

In addition to occupancy of hazardous sites, human occupancy may create or increase the probability of hazards by altering fundamental force–resistance relationships. A good example is slope destabilization.

Landslides and other forms of mass wasting are a classic example of a threshold-controlled geomorphic process. The threshold of stability is usually described as a factor of safety, F:

$$F = \frac{\text{Forces resisting movement}}{\text{Forces promoting movement}}$$

Soil-mechanical methods for calculating these forces are well established for a wide variety of conditions, and the material strength measurements necessary for their use are relatively straightforward. Although several different analytical methods are available, depending on the mechanics of motion and site conditions, in general three thresholds control the occurrence of landslides: slope and failure-surface geometry; material cohesion and internal friction; and pore pressure. If these factors are known, F can be calculated for any given set of conditions. Often F is estimated for conditions of maximum pore pressure (for example, complete saturation) to determine whether failure would occur under those conditions. It is thus possible to determine conditions under which a hillslope would be unstable. It is common engineering practice to design for $F > 1$ (say, 1.5) to allow for errors in measurement or estimation of strength parameters.

The major limitation to the identification of slope instability and management of human occupancy of unstable areas is therefore not the ability to predict stability, but rather in the recognition of the need for such prediction and availability of the necessary field data. Regional landslide hazard mapping has been used to provide guidance on the need for detailed site analyses. Hazard maps are usually based on a combination of topographic and geologic maps, and focus on slope steepness and length and outcrops of unstable or potentially unstable materials. Such maps can be used to identify areas where detailed measurement and analysis may be warranted prior to site modification.

Human exacerbation of landscape instability

In some cases geomorphic hazards exist independently of human agency, but are accelerated or exacerbated by human activity. In general, any environment where materials are relatively easily moved but are protected from movement by vegetation or other natural armour are especially vulnerable to accelerated rates of change due to human agency. Examples include stable vegetated dune systems destabilized by overgrazing, and accelerated soil erosion when forested areas are cleared and ploughed. Barrier islands are a particularly dramatic example of such an environment. Most barrier islands are composed of highly mobile sediments that were deposited under present environmental conditions. They are highly dynamic systems in which disturbances need not be large to cause significant sediment movement. Beaches are controlled by the balance of mass fluxes, and typically gross mass fluxes are large relative to the volume of sediment in storage at any given time. Therefore relatively small percentage changes in the rates of mass fluxes can trigger large volumetric changes in sediment storage. This in turn leads to dramatic landscape changes such as shoreline recession or accretion or tidal inlet migration. In this regard, barrier islands are highly sensitive to deliberate or inadvertent human impacts.

Examples of coastlines destabilized by human activity abound, particularly on the east coast of the United States. There the barrier island coast has mean recession rates of about 0.8 m/yr (May *et al.* 1982), with the recession driven primarily by sea-level rise. The shoreline migrates landwards by a combination of longshore transport to tidal inlets, storm overwash, offshore deposition, and aeolian redistribution. In many areas, groynes and/or seawalls have been constructed to slow longshore transport and armour the coast against wave erosion. These structures reduce the sand supply to downdrift areas, exacerbating the erosion problems there. Unprotected, sediment-starved beaches may erode at very high rates in these circumstances. On the northern coast of New Jersey, for example, the combination of groynes and seawalls resulted in local erosion rates downdrift of 20 m/yr (Nordstrom and Allen 1980).

COPING WITH GEOMORPHIC HAZARDS

Attempts to cope with geomorphic hazards can be classified into four general categories. The first category is prediction. To the extent that the nature, severity, and location of hazardous events can be forecast, such events can be avoided, mitigated, or planned for. Because the basic goal of geomorphology and any other science is prediction, and because hazardous events tend to be high-magnitude, geomorphologically significant phenomena, progress in hazard prediction will parallel progress in geomorphology in general.

A second strategy is engineered solutions. Direct physical manipulation of the environment can be used to reduce or eliminate hazards. Examples include flood control levees and dams, shoreline protection structures such as bulkheads or groynes, and stormwater detention basins. Some have been quite successful, but disadvantages often include adverse ecological impacts and the need for perpetual maintenance.

A third type of response to geomorphic hazards is land-use solutions. This approach seeks to identify locations or areas subject to hazards. Land-use planning and control methods are then used to prohibit, limit, or discourage intensive or high-risk land uses.

The final approach is referred to as adaptive solutions. Adaptation involves attempts to tailor land use to geomorphic constraints or to incorporate earth surface processes into land use systems. Examples include construction of flexible foundations to create 'earthquake proof' buildings in seismic hazard areas, the use of mobile or cheap structures on migrating barrier islands, and 'soil harvesting' whereby soil erosion from steep mountain slopes in arid areas is ignored or encouraged so as to deliver soil to valley bottoms where it can be farmed.

INSTABILITY AND CHAOS IN DYNAMIC GEOMORPHIC SYSTEMS

Instability and chaos

Deterministic chaos refers to complex, apparently random behaviour arising from completely deterministic non-linear systems. As such, chaos is independent of stochastic forcings and measurement errors and is inherent in system dynamics. Mathematically, solutions to deterministic equations are chaotic if adjacent solutions diverge exponentially in phase space, which requires that a system has a positive Lyapunov exponent (governing the rate of divergence). The diagnostic of chaotic behaviour in geomorphic systems is sensitive dependence on initial conditions. These ideas and their relevance to geography and other earth and environmental sciences are discussed in more detail elsewhere (Malanson *et al.* 1990, 1992; Phillips 1992a, 1992b, 1993a, 1993b; Turcotte 1992; Wilcox *et al.* 1991).

Classic non linear dynamical systems theory begins with a set of ordinary differential equations describing the time behaviour of the components of an earth surface system, x_i, as functions of each other:

$$\frac{dx_i}{dt} = f_i(x_1, x_2, \ldots x_n; c_1, c_2, \ldots, c_m)(i = 1, n) \tag{1}$$

The cs are parameters governing rates of processes and are omitted from subsequent equations which deal with general cases.

The co-ordinate space defined by the n components is the phase space. Such a system has n Lyapunov exponents, L, which govern the rate of convergence or divergence of initially similar starting points, i.e.,

$$\Delta(t) = \Delta(0)e^{Lt} \tag{2}$$

$\Delta(t)$ is the difference between two points at time t, and $\Delta(0)$ the initial difference. If any of the L are positive, it indicates the sensitive dependence on initial conditions and exponential divergence over time characteristic of chaos. If the average L is positive, chaotic behaviour is common.

A fortuitous property of nonlinear dynamical systems is that the stability properties of systems linearized around an equilibrium are identical to those of the original, fully nonlinear system. Thus a set of nonlinear partial differential equations describing the interactions of a nonlinear dynamical system may be transformed to a set of n ordinary differential equations in the form of Equation (1). A set of linear equations defines the growth rate of initially small perturbations, ∂x:

$$d\partial x / dt = A_{ij}\partial x_j, \quad i\, 1, 2, \ldots, n. \tag{3}$$

A_{ij} are elements of the Jacobian matrix of $f = (f_1, f_2, \ldots, f_n)$ defined by

$$A_{ij} = \partial f_i(x_1, x_2, \ldots, x_n)/\partial x_j {\mid}_{x=x0} \tag{4}$$

where x_0 is the initial equilibrium state. The complex eigenvalues, λ, of the Jacobian matrix have real parts which are the Lyapunov exponents of the system.

The Jacobian matrix is exactly equivalent to an interaction matrix, A, with elements a_{ij} signifying the positive, negative, or negligible influences of system components on each other, provided the interaction matrix derives from a nonlinear dynamical system. Linear stability analysis can be used to determine whether the interaction matrix has any positive eigenvalues and whether the average λ is positive, and thus whether there are positive Lyapunov exponents. Even when only qualitative information is known (i.e., whether each a_{ij} is positive, negative, or zero), the Routh–Hurwitz criteria can be used to determine whether there are any positive λ or L. Qualitative stability analysis is described in detail by Puccia and Levins (1985, 1991), and applications in geomorphology are given by Slingerland (1981) and Phillips (1990), among others. The stability–chaos link is established in a geomorphic context by Phillips (1992a).

Stability analysis has been used to determine that specific models, equation systems, or particular geomorphic systems are unstable in the sense above (and thus potentially chaotic). Examples include Callander (1969) on river hydraulic geometry, Loewenherz (1991) on drainage network initiation, Nelson (1990) on fluvial bedforms, and Trofimov and Moskovkin (1983) on slope stability. Qualitative stability analyses allow general conclusions to be extended to general types of geomorphic systems. These have shown instability in river hydraulic geometry (Slingerland 1981; Phillips 1990), fluvial sediment budgets (Phillips 1987), evolution of river long-profiles (Slingerland and Snow 1988), hillslope erosion and mass wasting (Trofimov and Moskovkin 1984), and generalized mass flux systems (Phillips 1991).

There have been several explicit investigations of chaos and sensitive dependence on initial conditions in geomorphic systems. Because it is impossible to know initial landscape conditions, such studies must rely on models. However, all depend on widely accepted conceptual models which reflect the state of contemporary understanding of the systems involved, and all models accurately reflect real landscapes. These studies have addressed channel network evolution (Willgoose et al. 1991; Ijjasz-Vasquez et al. 1992), evolution of river channel longitudinal profiles (Renwick 1992), hillslope evolution (Malanson et al. 1992; Phillips 1993b), hydraulic geometry of overland flow (Phillips 1992c), wetland response to sea-level rise (Phillips 1992a), soil development (Phillips 1993a), and soil moisture at seasonal time-scales (Rodriguez-Iturbe et al. 1991).

Geomorphic systems are therefore likely to be unstable, and to have a potential for deterministic chaos.

Implications of instability and chaos

Stable systems are able to recover from many disturbances or perturbations, and this property is sometimes observed in geomorphic systems. For example, erosion and sediment yields often increase dramatically after timber harvesting, but decline again to pre-logging levels within several years as the vegetation recovers (Bosch and Hewlett 1982; Morgan 1986; Patric 1976). Sandy beaches may also be stable in this sense, exhibiting cyclical behaviour as sediment eroded during storms is redeposited between storms, restoring the pre-storm sand volume and beach topography (Komar 1976; Nordstrom 1980; Wright and Short 1984). It is important to recognize, however, that even in these examples the system may be unstable when viewed from a different perspective or at a different spatial or temporal scale. The post-logging forest basin, for example, may have soil depletion and topographic changes which are semi-permanent, despite the return of sediment yields to pre-logging conditions. The beach may be unstable in response to a change in, say, sediment supply due to shore protection structures on an adjacent beach (Nordstrom and Allen 1980).

In general, geomorphic systems must be viewed as unstable or potentially unstable. This means that changes or perturbations will tend to persist and/or grow, rather than to be damped or erased. Human activity inevitably alters geomorphic systems. The ultimate implication is that human activity will generally leave a persistent (and often steadily increasing) imprint on earth surface processes and landforms.

The possibility of chaotic behaviour in geomorphic systems is relevant to the problem of prediction. Chaotic systems do not allow reliable long-term deterministic forecasting of system details. Simple cause-and-effect models of singular evolutionary pathways are largely irrelevant to chaotic systems. Efforts to forecast landscape evolution and responses to human activity must therefore focus on short-term deterministic prediction and long-term prediction in a probabilistic mode. However, chaos theory is of value in long-range deterministic modelling: an understanding of system-level dynamics via chaos theory may enhance long-range modelling of *general* system characteristics (as opposed to details) and behaviours and inform probabilistic models. Chaos analysis may enhance short-term modelling by allowing identification of a few critical variables which govern the phase space trajectories (Malanson *et al.* 1990; Tsonis and Elsner 1989).

The human role in geomorphic systems

Phillips (1991, 1992b) has analysed generalized geomorphic mass flux systems linking input, storage and output. This type of system characterizes a wide variety of real-world problems, such as inputs of sediment to channel from upland and bank erosion, sediment storage in the channel and floodplain, and output in the form of sediment yield. Runoff response of hillslopes or basins is also another example, with precipitation providing the inputs, runoff the outputs, and soil moisture and depression storage.

When only input, storage, and output and their mutual interactions are considered, such a mass flux system is asymptotically stable (Phillips 1991, 1992b). But in many real geomorphic systems, a fourth component must be considered; a threshold relationship between power (or force) and resistance. If this is exceeded, material is removed from storage and new input is transported through the system; if it is not exceeded there is a net increase in storage. In either case, the feedback links among system components are such that the system becomes unstable (and potentially chaotic) (Phillips 1992b). However, if the power–resistance threshold is not crossed (i.e. the frequency of threshold exceedances is less than the time period under consideration) stable states may be identified corresponding to conditions of degradation, aggradation, and steady state (Phillips 1992b).

In another analysis, Phillips (1991) explicitly considered human modification of mass flux systems, including humans as the fourth system component, and examining alternate assumptions regarding human goals of maximizing or minimizing system throughput. In either case, the system was found to be unstable if the human modification were persistent or recurrent. If the human perturbation is singular, or if the frequency is less than the time period considered, the system is stable (Phillips 1991). Examples regarding artificial drainage systems in wetlands and removal of large organic debris from river navigation channels have been discussed elsewhere (Phillips 1991; Phillips and Holder 1991).

The stability analyses cited above are consistent with principles of landscape sensitivity (Brunsden and Thornes 1979; Brunsden 1980). Sensitive landscapes are those where the frequency of disturbance is less than the relaxation time, implying an unstable condition. If disturbances are less frequent than relaxation times, landscapes are insensitive and stable.

If human alterations of geomorphic systems are persistent or chronic, or if episodic alterations occur more frequently than the relaxation time of the system, human agency must be considered an intrinsic component of the system rather than an external perturbation. Given the widespread, chronic nature of human agency, this suggests the need for a conceptual framework to replace or complement the existing viewpoint which emphasizes comparisons of natural and altered systems. The new framework emphasizes human agency as an inherent, intrinsic component of geomorphic systems. In particular, landscapes with frequent or chronic human agency or with long relaxation times are likely to be permanently altered (from the human temporal perspective of years to centuries). This alteration may have positive aspects, such as maintaining productive agricultural land, or negative, such as land degradation.

Example: desertification

Semiarid land degradation – persistently referred to as desertification despite some problems with the latter term – is a widespread problem in geomorphic instability. Irreversible degradation of dry and subhumid areas is estimated at 6 million ha/yr, with an additional 20 million ha/yr degraded to the point of becoming unprofitable to farm or graze (Postel 1989). Severe degradation is primarily a geomorphic problem, characterized by gully erosion, invasion of aeolian sand dunes, and exposure (by erosion) of infertile, low-permeability calcic soil horizons. The example below will examine the stability of a semiarid geomorphic system under grazing, based on the conceptual model of Schlesinger et al. (1990). This model is based on field studies in New Mexico, USA, but seems generally applicable to semiarid land degradation problems worldwide. An estimated 35 per cent of all human-induced soil degradation is attributable to overgrazing, more than to any other single cause (World Resources Institute 1992).

In the New Mexico studies it was found that 'soil heterogeneity' in the form of vegetation and organic matter removal, physical soil disturbance, erosion, and exposure of calcic horizons is the key to the semiarid land degradation process. In brief, grazing disturbance is selective, in part due to differential palatability of range vegetation. This leads to spatial heterogeneity of soils, shrub invasion, and concentration of increased runoff and erosion. This results in further resource concentration in other locations, where the plant and soil resources are more quickly removed (Schlesinger et al. 1990). The results are consistent with Thornes's (1988) theoretical analysis of erosional equilibria under grazing, which showed that any state of the erosion–vegetation cover balance between the end-points of complete vegetation cover or complete erosion is unstable.

Schlesinger et al. (1990) developed a system model of the process. Figure 26.2 shows the critical system components and their interrelationships, with two links added to the original model: a positive link reflecting wind erosion contributions to soil heterogeneity, and a positive link from climate change to

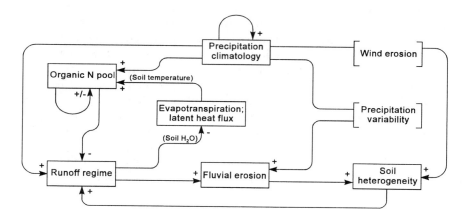

Figure 26.2 Interactions among major system components involved in desertification. Adapted from Schlesinger *et al.* (1990).

runoff regime, since a change in precipitation climatology will influence runoff response independently of changes in surface characteristics. For stability analysis purposes, components which have only one ingoing and one outgoing link (the wind erosion and precipitation variability components of Figure 26.2) can be omitted and replaced by direct links from the climate to the soil heterogeneity and fluvial erosion variables, respectively. This reduction in the number of components greatly simplifies the analysis without affecting the stability properties (Puccia and Levins 1985). Then the system may be translated into the interaction matrix of Table 26.1, where the entries represent the positive, negative, or negligible direct influences of the row on the column components.

In the interaction matrix, two self-effect loops are added. Climatological studies in deserts and semiarid areas of Africa show strong evidence of persistence in wet and dry periods, indicating internal climate self-effect feedbacks independent of climate–land surface interactions (Demareé and Nicolis 1990; Lockwood 1986; Wright 1980). Vegetation and organic matter, as reflected in the organic nitrogen component, also have self-effects independent of environmental controls, associated with direct or inverse density dependence. Note that failure to include any self-effect loops ($a_{ii} \neq 0$) or inclusion of only positive or dominantly positive entries on the matrix diagonal predetermines an outcome of instability.

Feedback at level k of a loop-model system such as this, F_k, represents the mutual influences of system components on each other at level k. Feedback at $k = 3$, for example, represents relations of the form $a_{ij}a_{jh}a_{hi}$, $a_{ii}a_{jh}a_{hj}$, and $a_{ii}a_{jj}a_{hh}$ ($i \neq j \neq h$). Defining $Z(m, k)$ as the product of m disjunct loops with k system components (disjunct loops are sequences of a_{ij} with

Table 26.1 Interaction matrix adapted from Figure 26.2

	Precipitation climatology	Organic N pool	Evapo-transpiration; latent heat flux	Runoff regime	Fluvial erosion	Soil heterogeneity
Precipation climatology	a_{11}	a_{12}	0	a_{14}	a_{15}	a_{16}
Organic N pool	0	$\pm a_{22}$	0	$-a_{24}$	0	0
Evapotranspiration; latent heat flux	0	a_{32}	0	0	0	0
Runoff regime	0	0	$-a_{43}$	0	a_{45}	0
Fluvial erosion	0	0	0	0	0	a_{56}
Soil heterogeneity	0	0	0	a_{64}	0	0

no common component i or j), feedback is computed by

$$F_k = \Sigma(-1)^{m+1} Z(m, k). \tag{5}$$

$F_0 = -1$, so the characteristic equation of the interaction matrix can be expressed in the feedback notation:

$$F_o\lambda^n + F_1\lambda^{n-1} + \ldots + F_{n-1}\lambda + F_n = 0. \tag{6}$$

The Routh–Hurwitz criteria give necessary and sufficient conditions for all real parts of the eigenvalues (and thus all Lyapunov exponents) to be negative. These are

$$F_i < 0, \text{ all } i \text{ and (for } n = 5 \text{ or } n = 6) \tag{7}$$
$$F^2_1 F_4 + F_1 F_5 - F_1 F_2 F_3 - F^2_3 > 0$$

For the semiarid land degradation system in Table 26.1, $F_1 < 0$ only if $a_{22} < 0$, and $a_{11} = 0$ or $| a_{22} | > | a_{11} |$. For this to occur the organic nitrogen pool would have to be self-limiting due to vegetation density dependence, and this relationship would have to be stronger than the self-enhancing loop of climatic persistence. However, $F_2 = 0$ unless both a_{11} and a_{22} are non-zero. Then, if the conditions for F_1 to be negative are met, F_2 must be positive. Therefore, without reference to further criteria, it can be concluded that the Routh–Hurwitz criteria are not met, all real parts of the eigenvalues are not negative, and the system is unstable.

Because the semiarid geomorphic system is inherently unstable, disturbances due to grazing, climate change, or other factors are likely to grow exponentially over the short term and to persist over the long term. Chaos is possible, and if positive self-effects dominate ($a_{11} > 0$ and/or $a_{22} > 0$, or $| + a_{11} | > | -a_{22} |$), chaos is likely (Phillips 1992a).

The management implications are that disturbance of this system should be avoided or minimized, especially during dry periods or cycles, and that long-term management must assume that geomorphic, hydrologic, and ecological changes are at least semi-permanent. However, it is instructive to include a human response in the system, in the form of erosion and runoff control. If fluvial erosion is assumed to be self-correcting ($-a_{55}$ is added to the matrix) due to implementation of soil conservation measures where erosion problems are observed, and if a negative link is added from erosion to runoff regime ($-a_{54}$) representing runoff control in response to observed erosion, a potentially stable situation is created. Feedback at levels 1 and 2 would likely be negative, as long as any positive self-effects are not too strong. Feedback at higher levels could be positive or negative, depending on the relative strength of various loops, but is more likely to be negative with the addition of the $-a_{55}$ and $-a_{54}$ links.

Intensive human agency in the form of conservation and a runoff–erosion control response at least raises the possibility of stability. Thus an apparent paradox arises – the semiarid geomorphic system is clearly unstable in response to grazing and other human disturbance, as both the model above and abundant field evidence attest; but the system is also unstable in response to other perturbations (such as climatic change), and only a form of intense, ongoing human management raises a reasonable prospect of stability.

This problem is an especially striking example of the intricate and inextricable mutual adjustments of humans and geomorphic landscapes.

Coping with instability

Instability is pervasive in geomorphic systems, and the potential for chaotic behaviour appears widespread. How can such situations be coped with? Five general principles are forwarded.

First, methods of prediction and evaluation should be based on an understanding of geomorphic systems as nonlinear dynamical systems which are often unstable and may be chaotic. This may involve direct application of nonlinear dynamical systems theory and methods, or interpretation of the results of more traditional analyses in a nonlinear dynamical system context (see Phillips 1992b; Trofimov and Phillips 1992). This will at least foster a more realistic appraisal of the reliability of forecasts and assessments.

Second, plans and policies for managing landscape resources and coping with geomorphic hazards should include considerable flexibility. Since geomorphic systems are vulnerable to change, changes are likely to persist, and predictability may be low, flexible responses or a large menu of options and responses are desirable.

Third, perturbations of geomorphic systems should be avoided or minimized, where this is feasible. Where landforms and surface processes support (or are themselves) critical resources, they should be protected from disturbances. Where land-use options which minimize or avoid interference with geomorphic processes are viable, they should be utilized.

Fourth, the timing of perturbations can be managed. The shorter the duration and the lower the frequency of disturbances, the less likely they are to result in instability and chaos, all other factors being equal. This highlights a need to know more about relaxation times, so that disturbances can be timed accordingly. This principle also implies that management options involving intense but local and short-term disturbance as opposed to less intense disturbance spread out in space and time deserve serious attention. Savory (1988), for example, has advocated this approach to avoid semiarid land degradation problems associated with livestock grazing.

Finally, there is a possibility in some cases for 'steering' inherently unstable systems. In the semarid land degradation example, it was shown that erosion and runoff control could stabilize the system. In other cases, instability can even be exploited where existing system states are unfavourable. Wetter climatic cycles or vegetation invasion, for example, could cause persistent changes in dryland environments which could be considered favourable.

CONCLUSIONS

The relationship between humans and landforms is one of mutual alteration and adjustment. The landscape is far more than a static platform on which human activity occurs; each affects, and is affected by, the other. These interactions can be viewed in two perspectives: geomorphic hazards and restless landscape instability, and dynamical system instability.

Geomorphic hazards occur wherever human populations and infrastructure are at risk from geomorphic processes or the resulting landscape instability and change. These risks may occur where humans occupy hazard zones, such as tectonically unstable areas; where human agency creates hazards, such as slope destabilization; or where natural geomorphic processes, such as barrier island drowning in response to rising sea-level, are exacerbated by human agency. Coping with geomorphic hazards involves engineered solutions, where direct landscape manipulation is aimed towards hazard mitigation; land-use solutions, where settlement and development of high-risk areas is avoided or minimized; and adaptive solutions where land use is adapted to geomorphic constraints.

Dynamical system instability implies that human (or natural) disturbances to geomorphic systems are likely to be persistent and impacts may be unpredictable. Coping with this form of instability involves methods of prediction and evaluation which recognize the dynamic, nonlinear, complex, unstable, and potentially chaotic nature of geomorphic systems, and plans and policies that incorporate flexible responses to geomorphic change. Avoiding undesirable change can be accomplished by avoiding or minimizing perturbation of critical geomorphic systems, by managing the timing of perturbations, and by 'steering' unstable systems towards particular desired system states.

REFERENCES

Alexander, D.E. (1989) 'Urban landslides', *Progress in Physical Geography* 13, 157–91.

Alexander, D.E. (1991) 'Applied geomorphology and the impact of natural hazards on the built environment', *Natural Hazards* 4, 57–80.

Baker, V.R. and Costa, J.E. (1987) 'Flood power', pp. 1–21 in L. Mayer and D. Nash (eds) *Catastrophic Flooding*, Boston: Allen & Unwin.

Bosch, J.M. and Hewlett, J.D. (1982) 'A review of catchment experiments to determine the effect of vegetation changes on water yield and evapotranspiration', *Journal of Hydrology* 55, 3–23.

Brunsden, D. (1980) 'Applicable models of long term landform evolution', *Zeitschrift für Geomorphologie Supplementband* 36, 16–26.

Brunsden, D. and Thornes, J.D. (1979) 'Landscape sensitivity and change', *Transactions of the Institute of British Geographers*, New Series 4, 463–84.

Callander, R.A. (1969) 'Instability and river channels', *Journal of Fluid Mechanics* 36, 465–80.

Chappell, J. (1983) 'Thresholds and lags in geomorphologic changes', *Australian Geographer* 15, 358–66.

Demarée, G.R. and Nicolis, C. (1990) 'Onset of Sahelian drought viewed as a fluctuation-induced transition', *Quarterly Journal of the Royal Meteorological Society* 116, 221–38.

Ijjasz-Vasquez, E.J., Bras, R.L. and Moglen, G.E. (1992) 'Sensitivity of a basin evolution model to the nature of runoff production and to initial conditions', *Water Resources Research* 28, 2733–41.

Komar, P.D. (1976) *Beach Processes and Sedimentation*, Englewood Cliffs, N.J.: Prentice-Hall.

Lockwood, J.G. (1986) 'The causes of drought, with particular reference to the Sahel', *Progress in Physical Geography* 10, 111–19.

Loewenherz, D.S. (1991) 'Stability and the initiation of channelized surface drainage: a reassessment of the short wavelength limit', *Journal of Geophysical Research* 96B, 8453–64.

Magilligan, F.J. (1992) 'Variability of flood power during extreme floods', *Geomorphology* 5, 373–90.

Malanson, G.P., Butler, D.R. and Georgakakos, R.P. (1992) 'Nonequilibrium geomorphic processes and deterministic chaos', *Geomorphology* 5, 311–22.

Malanson, G.P., Butler, D.R. and Walsh, S.J. (1990) 'Chaos in physical geography', *Physical Geography* 11, 293–304.

May, S.R., Rimball, W.H., Grandy, N. and Dolan, R. (1982) 'The coastal erosion information system', *Shore and Beach* 50, 19–26.

Morgan, R.P.C. (1986) *Soil Erosion and Conservation*, London: Longman.

Nelson, J.M. (1990) 'The initial instability and finite-amplitude stability of alternate bars in straight channels', *Earth-Science Reviews* 29, 97–115.

Nordstrom, R.F. (1980) 'Cyclic and seasonal beach response: a comparison of oceanside and bayside beaches', *Physical Geography* 1, 177–96.

Nordstrom, R.F. and Allen, J.R. (1980) 'Geomorphically compatible solutions to beach erosion', *Zeitschrift für Geomorphologie Supplementband* 34, 142–54.

Patric, J.H. (1976) 'Soil erosion in the eastern forest', *Journal of Forestry* 74, 671–7.

Phillips, J.D. (1987) 'Sediment budget stability in the Tar River basin, North Carolina', *American Journal of Science* 287, 780–94.

Phillips, J.D. (1990) 'The instability of hydraulic geometry', *Water Resources Research* 26, 739–44.

Phillips, J.D. (1991) 'The human role in Earth surface systems: some theoretical considerations', *Geographical Analysis* 23, 316–31.

Phillips, J.D. (1992a) 'Qualitative chaos in geomorphic systems, with an example from wetland response to sea level rise', *Journal of Geology* 100, 365–74.

Phillips, J.D. (1992b) 'Nonlinear dynamical systems in geomorphology: revolution or evolution?', *Geomorphology* 5, 219–29.

Phillips, J.D. (1992c) 'Chaos in surface runoff', pp. 177–97 in A.J. Parsons and A.D. Abrahams (eds) *Overland Flow*, London: UCL Press.

Phillips, J.D. (1993a) 'Stability implications of the state factor model of soils as a nonlinear dynamical system', *Geoderma* 58, 1–15.

Phillips, J.D. (1993b) 'Instability and chaos in hillslope evolution', *American Journal of Science* 293, 25–48.

Phillips, J.D. and Holder, G.R. (1991) 'Large organic debris in the lower Tar River, North Carolina, 1879–1900', *Southeastern Geographer* 31, 55–66.

Postel, S. (1989) 'Halting land degradation', in L.R. Brown (ed.) *State of the World 1989*, Washington, DC: Worldwatch Institute.

Puccia, C.J. and Levins, R. (1985) *Qualitative Modeling of Complex Systems*, Cambridge, Mass.: Harvard University Press.

Puccia, C.J. and Levins, R. (1991) 'Qualitative modeling in ecology: loop analysis, signed digraphs, and time averaging', pp. 119–43 in P.A. Fishwick and P.A. Luker (eds) *Qualitative Simulation Modeling and Analysis*, New York: Springer.

Renwick, W.H. (1992) 'Equilibrium, disequilibrium, and nonequilibrium landforms in the landscape', *Geomorphology* 5, 265–76.

Rhoads, B.L. (1987) 'Stream power terminology', *Professional Geographer* 39, 189–95.

Roberts, N. (1989) *The Holocene: An Environmental History*, Oxford: Blackwell.

Rodriguez-Iturbe, I., Entekhabi, D., Lee, J.S. and Bras, R.L. (1991) 'Nonlinear dynamics of soil moisture at climate scales: 2. Chaotic analysis', *Water Resources Research* 27, 1899–915.

Saunders, I. and Young, A. (1983) 'Rates of surface processes on slopes, slope retreat, and denudation', *Earth Surface Processes and Landforms* 8, 473–501.

Savory, A. (1988) *Holistic Resource Management*, Washington, DC: Island Press.

Schlesinger, W.H., Reynolds, J.F., Cunningham, G.L., Huenneke, L.F., Jarrell, W.M., Virginia, R.A. and Whitford, W.G. (1990) 'Biological feedbacks in global desertification', *Science* 247, 1043–8.

Schumm, S.A. (1979) 'Geomorphic thresholds: the concept and its applications', *Transactions of the Institute of British Geographers*, New Series 4, 485–515.

Simmons, I.G. (1989) *Changing the Face of the Earth: Culture, Environment, History*, Oxford: Blackwell.

Slingerland, R.L. (1981) 'Qualitative stability analysis of geologic systems with an example from river hydraulic geometry', *Geology* 9, 491–3.

Slingerland, R.L. and Snow, R.S. (1988) 'Stability analysis of a rejuvenated fluvial system', *Zeitschrift für Geomorphologie Supplementband* 67, 93–102.

Thomas, W.L. (ed.) (1955) *Man's Role in Changing the Face of the Earth*, Chicago, Ill.: Chicago University Press.

Thornes, J.B. (1988) 'Erosional equilibria under grazing', pp. 193–210 in J.L. Bintliff, D.A. Davidson and E.G. Grant (eds) *Conceptual Issues in Archeology*, Edinburgh: Edinburgh University Press.

Trofimov, A.M. and Moskovkin, V.M. (1983) 'Mathematical simulation of stable and equilibrium river bed profiles and slopes', *Earth Surface Processes and Landforms* 8, 383–90.

Trofimov, A.M. and Moskovkin, V.M. (1984) 'The dynamic models of geomorphological systems', *Zeitschrift für Geomorphologie* 28, 77–94.

Trofimov, A.M. and Phillips, J.D. (1992) 'Theoretical and methodological premises of geomorphological forecasting', *Geomorphology* 5, 203–11.

Tsonis, A.A. and Elsner, J.B. (1989) 'Chaos, strange attractors, and weather', *Bulletin of the American Meteorological Society* 70, 14–23.

Turcotte, D.L. (1992) *Chaos and Fractals in Geology and Geophysics*, New York: Cambridge.

Turner, B.L., Clark, W., Kates, R., Richards, J., Matthews, J. and Meyer, W. (eds) (1990) *The Earth as Transformed by Human Action: Global and Regional Changes in the Biosphere Over the Last 300 Years*, Cambridge: Cambridge University Press with Clark University.

Wilcox, B.P., Seyried, M.S. and Matison, T.H. (1991) 'Searching for chaotic dynamics of snowmelt runoff', *Water Resources Research* 27, 1005–10.

Willgoose, G.R., Bras, R.L. and Rodriguez-Iturbe, I. (1991) 'A coupled channel network growth and hillslope evolution model', *Water Resources Research* 27, 1671–96.

Wolman, M.G. and Miller, J.P. (1960) 'Magnitude and frequency of geomorphic processes', *Journal of Geology* 68, 54–74.

World Resources Institute (1992) *World Resources 1992–93*, New York: Oxford University Press.

Wright, L.D. and Short, A.D. (1984) 'Morphodynamic variability of surf zones and beaches: a synthesis', *Marine Geology* 56, 93–118.

Wright, P.B. (1980) 'An approach to modelling climate based on feedback relationships', *Climatic Change* 2, 283–98.

FURTHER READING

Alexander, D.E. (1991) 'Applied geomorphology and the impact of natural hazards on the built environment', *Natural Hazards* 4, 57–80.

Brunsden, D. and Thornes, J.D. (1979) 'Landscape sensitivity and change', *Transactions of the Institute of British Geographers*, New Series 4, 463–84.

Cooke, R.U. and Doornkamp, J.C. (1990) *Geomorphology in Environmental Management* (2nd edn), Oxford: Clarendon Press.

Douglas, I. (1990) 'Sediment transfer and siltation', in B.L. Turner *et al.* (eds) *The Earth as Transformed by Human Action: Global and Regional Changes in the Biosphere Over the Last 300 Years*, Cambridge: Cambridge University Press with Clark University.

Gares, P.A., Sherman, D.J., and Nordstrom, K.F. (1994) 'Geomorphology and natural hazards', *Geomorphology* 10, 1–18.

Huggett, R.J. (1985) *Earth Surface Systems*, New York: Springer.

Phillips, J.D. (1991) 'The human role in Earth surface systems: some theoretical considerations', *Geographical Analysis* 23, 316–31.

Phillips, J.D. (1993) 'Biophysical feedbacks and the risks of desertification', *Annals of the Association of American Geographers* 83, 630–40.

Scheidegger, A.E. (1983) 'Instability principle in geomorphic equilibrium', *Zeitschrift für Geomorphologie* 27, 1–19.

Thornes, J.B. (1985) 'The ecology of erosion', *Geography* 70, 222–35.

27

THE TROPICAL RAIN FOREST

John R. Flenley

DEFINITION AND DESCRIPTION

The tropical rain forest is a plant formation whose chief characteristics are that it is an evergreen, broadleaved forest very diverse in both plant life forms and plant and animal species, and found typically near the equator in hot, ever-wet climates. The name 'tropical rain forest' is unfortunate, for it is better to define vegetation by its own attributes rather than by those of its environment. This leads to ridiculous terms such as 'dry rain forest', but the term 'tropical rain forest' is probably too well established to change. It should be noted immediately that all the characteristics mentioned above may on occasion be absent, singly at least. There are tropical rain forests which are not fully evergreen, others which are not fully broadleaved, and some which are not very diverse. The distribution of the tropical rain forest extends well away from the equator and even sometimes outside the tropics (e.g. in eastern Australia). It certainly occurs in climates with up to a four-month dry season.

The structure of the forest is regarded as probably the most complex of all ecosystems (Figure 27.1). The taller trees form a canopy at about 30 m height, although emergent trees may grow to 50 m. This is well short of the tall *Sequoia* and *Eucalyptus* trees in some temperate forests, incidentally. Below the canopy there are other trees of smaller species, as well as saplings of the canopy trees. For convenience, some people group the tree crowns into three layers, but this can be shown statistically to be purely arbitrary and without objective meaning. Below the trees is a 'shrub layer' and then a 'herb layer', but these are not well developed, perhaps because very little light reaches the ground.

Certain life forms are unusually common. Palms are sometimes abundant, and some *Pandanus* species have a similar form. Lianas (woody climbers) are usually present, often climbing into the canopy and including the rattans (climbing palms). Strangler figs start life as epiphytes, but root in the ground and end up outliving their host tree. Herbaceous epiphytes are common and

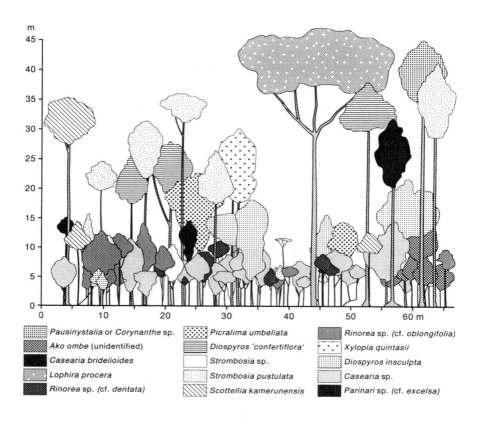

Pausinystalia or Corynanthe sp.	Picralima umbellata	Rinorea sp. (cf. oblongifolia)
Ako ombe (unidentified)	Diospyros 'confertiflora'	Xylopia quintasii
Casearia bridelioides	Strombosia sp.	Diospyros insculpta
Lophira procera	Strombosia pustulata	Casearia sp.
Rinorea sp. (cf. dentata)	Scottellia kamerunensis	Parinari sp. (cf. excelsa)

Figure 27.1 Profile diagram of primary mixed forest, Shasha Forest Reserve, Nigeria. The diagram represents a strip of forest 61 m long and 7.6 m wide. All trees 4.6 m and over are shown. After Richards (1964).

include many species of orchid and fern, as well as the ant-plants which enjoy a symbiotic relationship with ants. Total parasites occur, and include the famous *Rafflesia*, the largest single flower in the world, with a diameter up to 1 metre. Some plants are predatory on insects (e.g. *Nepenthes* has leaves modified as pitcher traps).

The trees themselves have unusual characteristics. Many have buttresses at the base. These were originally thought to have a mechanical function, but may perhaps be more important as a short cut for water travelling up the sapwood. Amazingly, over 90 per cent of trees have medium-sized (3–15 cm) entire ovate leaves with 'drip-tips'. Possibly this is the best size and shape for shedding water, or it could be enforced by the need to restrict insect feeding.

The diversity of the forest is legendary. Even if only the larger trees are counted, it is possible to find over 350 species per hectare (Figure 27.2). If small

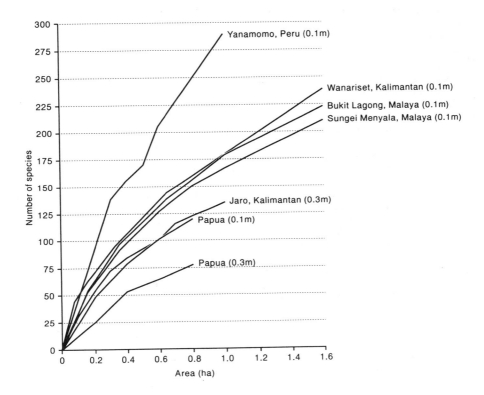

Figure 27.2 Species-area curves for tropical lowland evergreen rain forests. The Yanamomo forest is the richest yet found; every second tree on the hectare plot was a different species. These curves were mostly made by adding together the number of species found on contiguous subplots. After Whitmore (1990).

trees, epiphytes, etc. were added the number would be much higher. The total flora of the Sunda–Sahul region is *c*.30,000 species, mostly trees, many still undescribed. The orchid flora of New Guinea is perhaps 2,600 species. Mt Kinabalu in Borneo may carry 800 species of orchid.

Insects are even more numerous. The world insect fauna, concentrated in the tropical rain forest, is variously estimated at 6 million to 30 million, most still undescribed.

The causes of diversity are hotly disputed. Almost certainly the answer will be multivariate, i.e. there are many contributory causes. Popular candidates at the present time are summarized in Table 27.1, but a few comments are necessary. A hypothesis is tenable only if it applies more effectively in the tropics than elsewhere. This is one reason why the refuge hypothesis is not very likely, for temperate floras were probably more split up than tropical ones in the

Table 27.1 An outline of the basic hypotheses concerning species diversity, particularly the increased species diversity in the tropics compared with temperate and Arctic regions

---

Nonequilibrium hypothesis
Time – the tropics are older and more stable, hence tropical communities have had more time to develop.
Equilibrium hypothesis
1 Speciation rates are higher in the tropics.
   (a) Tropical populations are more sedentary, facilitating geographical isolation.
   (b) Evolution proceeds faster due to
      • a larger number of generations per year;
      • greater productivity, leading to greater turnover of populations, hence increased selection;
      • greater importance of biological factors in the tropics, thereby enhancing selection;
      • isolation in refugia by repeated Pleistocene aridity.
2 Extinction rates are lower in the tropics.
   (a) Competition is less stringent in the tropics due to
      • the presence of more resources;
      • increased spatial heterogeneity;
      • increased control over competing populations exercised by predators.
   (b) The tropics provide more stable environments, allowing smaller populations to persist, because
      • the physical environment is more constant;
      • biological communities are more completely integrated, thereby enhancing the stability of the ecosystem.

Source: Modified, after Ricklefs (1973).

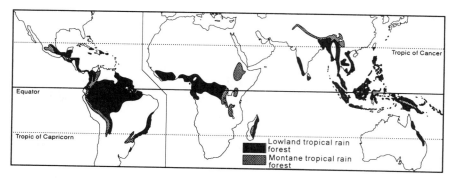

Figure 27.3 World distribution of lowland and montane tropical rain forest. After Strahler and Strahler (1994).

Pleistocene. Also the claim that concentrations of diversity today are the locations of the refugia does not stand up to examination. Some 'refugia' are artefacts of collecting density, and others did not carry rain forest in the Pleistocene at all (Flenley 1993a).

Probably we are asking the wrong questions about diversity. It is not a matter of 'why are there so many species in the rain forest?', but, 'why are there so many *rare* species in the rain forest?' It seems that rarity is actually advantageous in the forest, which suggests that a density-dependent selection mechanism is at work. One such is the 'pest pressure' hypothesis of Gillett (1962) and Janzen (1970). This mechanism exists, but may be too weak on its own to account for diversity (Ashton, pers. comm.).

DISTRIBUTION

Before recent destruction by people, the tropical rain forest existed as a broad, equatorial belt, almost continuous across the major land masses, apart from its absence in East Africa, and its replacement with montane variants in the Andes. The result is three major rain forest blocks, Amazonia, West Africa, and the Sunda-Sahul Region (Figure 27.3), with lesser concentrations in Southern India, Sri Lanka, South-east Brazil, Madagascar, New Caledonia, the West Indies, and numerous smaller islands.

ECOLOGICAL CONTROLS

There is no doubt at all that, as with most other vegetation formations, the primary control on the distribution of the tropical rain forest is climate. Most rain-forest species require hot moist conditions for survival, and these are generally provided near the inter-tropical convergence zone (ITCZ) where the north-east trade winds of the northern hemisphere and the south-east trade winds of the southern hemisphere converge, leading to rising air and heavy precipitation. Thus a typical rain forest climate has no dry season, or only a short one: four months is the absolute maximum. Temperatures are uniformly high, commonly in the 25–30°C range, both day and night, with little seasonal variation. This situation is summarized in Figure 27.4. Given the seasonal swings of the ITCZ north and south of the equator, these conditions are commonly met within 10°N and S of the equator, except where the trade winds have crossed dry land rather than ocean, or a very cool part of the ocean and have thus collected insufficient moisture. The reason for the lack of tropical rain forest in East Africa is the inadequacy of water supply carried in by the winds.

Latitudinally, the tropical rain forest is limited by the great deserts, formed where the rising air of the ITCZ descends again in a high pressure zone, forming the Hadley circulation. Faced with progressively longer dry seasons, the tropical rain forest gives way to various types of 'monsoon forest' or 'seasonal forest', in which a proportion of the trees are deciduous in the dry season, diversity is

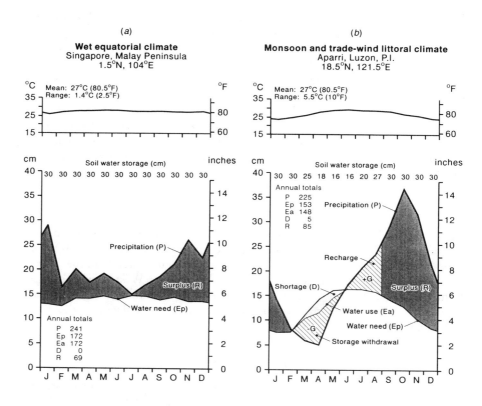

Figure 27.4 Typical climate diagrams for rain forest: (a) ever wet, Singapore; (b) short
dry season, Luzon. After Strahler and Strahler (1994).

lower and structure simplified. In Africa there is a cline between the tropical rain
forest and the Sahara, encompassing this type of forest, and then, in order,
savannah woodland, grass savannah, semi-desert and desert (Hopkins 1974).
Similar clines exist in Burma, in South and East Africa and in South America
where the shrub savannah known as *caatinga* covers enormous areas.

When the tropical rain forest encounters altitudinal variation, it is usually
temperature decline rather than lack of moisture which changes the forest.
Mountains are usually moist places on the whole, because they generate
precipitation, but tropical mountains have a peculiar temperature regime. This
has been summarized as 'summer every day and winter every night', i.e. there
is pronounced diurnal variation but little seasonal variation. The environmental
lapse rate is about 6.5°C for each 1,000 m, but there are wide variations. The
result of all this is that the typical lowland tropical rain forest gives way at about
1,000 m to a lower montane rain forest. This differs mainly by being less tall
(*c.*20 m) and slightly less diverse than lowland rain forest. Woody lianas are

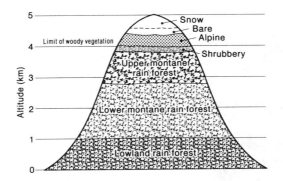

Figure 27.5 Generalized altitudinal zonation on mountains in New Guinea. After Flenley (1992).

usually less abundant, but epiphytes frequently more so. The tree taxa represented include temperate families (e.g. Fagaceae) largely absent from the lowland rain forest. Single species dominance is more common (e.g. *Nothofagus carrii* in parts of New Guinea). The trees have smaller leaves (*c.* 1–3 cm).

If the mountain is high enough, lower montane rain forest will give way to upper montane rain forest at around 2,900 m. This altitude (like all those in altitudinal zonations) varies widely, however, being much lower on isolated mountains near the sea, a phenomenon known as the Massenerhebung Effect. Upper montane rain forest is shorter still (*c.* 10 m), with even less diversity, simpler structure and tiny leaves (<2 cm). Epiphytes, especially bryophytes, may however be extremely abundant, giving rise to terms like mossy forest and elfin woodland. These forests often coincide with levels of cloud formation, and are sometimes known as cloud forests. They continue up to the forest limit at around 3,800 m (Figure 27.5).

The overall control exerted by climate must not obscure the important role of soil in distribution of the tropical rain forest. Most equatorial soils are deep latosols, developed by leaching of silica from the weathering parent materials over a very long period of time under a predominantly warm wet climate (Figure 27.6). The resulting clay soils are moisture-retaining but not inherently fertile. Most of the nutrients are held in the vegetation, and when released to the soil are rapidly taken up by feeding roots in the surface soil (or even in the litter layer in lower montane and upper montane rain forests). This is the main reason why temperate agricultural techniques (which depend on a good nutrient supply in the soil) are not very effective in the tropics. These deep clay soils provide the stored water which enables the tropical rain forest to survive during the dry season.

In marginal circumstances, soil type may become crucial. For instance in eastern Australia, under a climate almost too dry for tropical rain forest, soils on

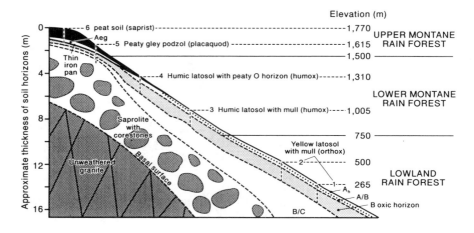

Figure 27.6 Altitudinal sequences of soils on granite in the Malayan main range near Kuala Lumpur. After Whitmore (1975).

sandstone with poor nutrient and moisture-retaining properties support sclerophyll (*Eucalyptus*) woodland, while adjacent clay soils developed on weathered basalt, and possessing good nutrient and moisture-retaining powers, support tropical rain forest.

Even in ever-wet conditions, poor sandy soils give rise to peculiar vegetation. The white sand podzols of Borneo bear the unusual *kerangas* vegetation, which is a type of tropical rain forest dominated by gymnosperms.

Under the heading of soil we may also consider swamps. Tropical swamps are naturally tree-covered, apart from the very early hydroseral stages, and include the most productive ecosystems known (Phillipson 1966). The peat-swamp forests of Borneo and Sumatra are analogous to the raised bogs of temperate regions. They are up to 30 km in diameter, very gently domed, and with concentric vegetation zones. The central zone, where the wood-peat is ever-wet and has a pH of 2, supports a dwarf vegetation of few species (Anderson 1964).

Mangrove swamps, the tropical equivalent of salt marshes, are also tree dominated. The trees develop special breathing roots (pneumatophores) to permit their survival in anaerobic estuarine mud. They, too, exhibit zonation, dependent on length of tidal inundation (Watson 1928; Thom 1967).

PEOPLE AND THE FOREST

Even in the continuous-canopy tropical rain forest there are people living. These hunter-gatherer groups include the Amerindian tribes of Amazonia, the pygmies of West Africa, and various groups in South-East Asia–New Guinea such as the Punan of Borneo (Park 1992). These people live primarily by hunting, often using poison darts projected by blowpipe, or poison arrows, as well as by fishing. They also use poison in fishing, employing natural plant products (e.g. *Derris*) for this purpose. In addition, they do a great deal of gathering of vegetable matter, fruits and insects. It is generally thought that they exist in harmony with the environment and do not have a large impact on it. This is not absolutely established, however, and it could be that such people were responsible, in the past, for the extinction of rain forest fauna such as the giant pangolin in South-East Asia. Usually, however, their population density has remained low so that the forest seems able to bear their existence unharmed. Tragically these people have suffered terribly by contact with the modern world. They often lack resistance to common diseases, and their habitat is being systematically destroyed by logging and forest clearance.

ANIMALS AND THE FOREST

The great diversity of animals, especially insects, has already been mentioned. What is especially interesting is the extent to which animal life is concentrated in the canopy. Certainly climbing forms (monkeys, apes, squirrels) are common among the mammals. Many organisms spend their whole lives there and never come to the ground. Special walkways, and the use of airships, have been necessary to enable us to uncover even the rudiments of this complex ecology (Ayensu 1980; Mitchell 1986).

The other feature only now beginning to emerge is the complexity of inter-organism relationships, many of which are only now being unravelled. For instance, it is no accident that most rain forest trees produce peculiar chemical compounds not found in temperate species. These compounds, it now appears, are chemical defences against specific insect predators. The trees can even sense when they are being devoured, and divert resources to make more of the defence compound to deter the pest.

Another example is the complexity of niche structure. For instance, there exists in Malaysia a species of pigeon which eats chiefly the fruits of fig trees. The species of fig have partitioned the year between them so that there is always a species in fruit. Thus all species have their fruit eaten and their seeds dispersed. In another example, over thirty species of animals have been found living in the tiny 'ponds' created by the bromeliad epiphytes (air plants) of tropical America. Some frogs complete their life cycle in them. In the equally small ponds provided by the leaves of the pitcher plants of South-East Asia (*Nepenthes* spp.) some species of mosquito have larvae which are resistant to the proteolytic enzyme which the plant exudes to digest them, and are thus enabled to complete their life cycle.

Another whole set of interactions is related to pollination. Many orchids have resemblances to insects which pollinate the flower while attempting to copulate. In some cases the resemblance is uncanny, the flower even producing 'legs' which move in a life-like manner.

HISTORY

There is no doubt that the tropical rain forest is of great age, and has existed as a plant formation at least throughout the Tertiary. Some elements, such as *Nothofagus* and the palm *Nypa*, date from the Upper Cretaceous. The earth appears to have been warmer then and tropical rain forests enjoyed wide distribution as far north as Britain (and possibly Greenland) and as far south as New Zealand. Plate tectonic movement accounts for some of those distributions, but there is no doubt the earth was warmer. During the Tertiary, a series of climatic declines set in, culminating in the Quaternary ice ages. Although these glaciations were more effective at the poles, they certainly affected the tropical rain forest to some extent. The main evidence for this comes from palynology (pollen analysis), and suggests that both temperature and hydrology were affected.

In the mountains, the main changes appear to have been of temperature, causing a lowering of altitudinal forest limits, and a reassortment of species in different combinations than now. Actually, since the cool periods of the Quaternary are much longer than the warm periods, it is more useful to think of the cool periods as typical, and the present conditions as a temporary aberration. In New Guinea and South America, Pleistocene forest limits were as much as 1,600 m lower than now (Figure 27.7). At present lapse rates that represents a cooling of *c.*10°C. That is at variance with the CLIMAP (1976) estimate of only 2°C lowering for tropical oceans. There are two possible ways

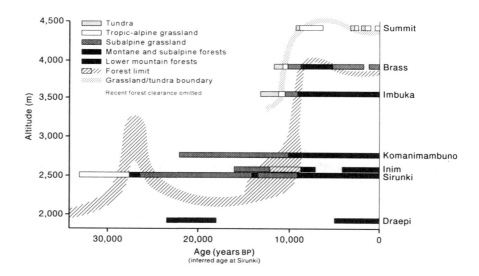

Figure 27.7 Summary diagram of Late Quaternary vegetational changes in the New Guinea Highlands. After Flenley (1979).

582

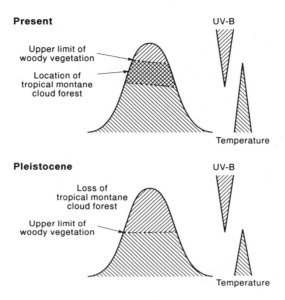

Figure 27.8 The hypothesis that lack of tropical mountain cloud forest in the Pleistocene may be explained by absence of habitat with a suitable combination of mean annual temperature and UV-B insolation. After Flenley (1993).

of reconciling the evidence. One idea is that CLIMAP have over-corrected their oxygen isotope data (Colinvaux, pers. comm.). The original data did suggest a large temperature change in the tropics to Emiliani (1955), but Shackleton (1967) explained 70 per cent of this away as due to a correction for palaeo-ice volume. The other idea is that the altitudinal zonation of tropical forests is not entirely to be attributed to temperature anyway. The tropical high mountains experience unusually high levels of ultraviolet light, and it may be that the occurrence of upper montane rain forest is related to this (Flenley 1992). This explanation has the advantage of accounting for one peculiar fact: the upper montane rain forest apparently disappeared in the Pleistocene (Figure 27.7) and this can be explained by the loss of a habitat with a suitable combination of temperature and UV-B (Figure 27.8). The two hypotheses are not mutually exclusive, and both could be partially correct. A third possibility, a much steeper environmental lapse rate in the Pleistocene (Walker and Flenley 1979), has been ruled out by climate modellers (Kutzbach and Guetter 1986; Webster and Streten 1978). A consideration of possible Pleistocene lapse rates is given by Bush (1994).

From the tropical lowlands we still have rather little evidence, but what there is usually suggests that although there was a little cooling, bringing some lower montane rain forest elements down to 600 m (Bush *et al.* 1992) or even 300 m

(Barmawidjaja *et al.* 1989), there was also in many areas a change to drier conditions. Indeed some rain forests disappeared altogether. For example, at Lake Valencia in Venezuela, forest was replaced by savannah in the Late Pleistocene. Similar evidence (not always well dated) was reported from Rondonia, Brazil (Absy and van der Hammen 1976), from Lake Bosumtwi, Ghana (Talbot and Delibrias 1977), from the Malay Peninsula (Morley and Flenley 1987), from the Misedor Borehole, Borneo (Caratini and Tissot 1985) and from the Atherton Tableland, Australia (Kershaw 1978). In many places, however, the forest also changed to include elements of drier preferences. This happened at Kau Bay, Halmahera, Indonesia (Barmawidjaja *et al.* 1989). It is therefore inappropriate at this time to speak of 'Pleistocene refuges' for rain forest taxa, and certainly those locations for them proposed on the basis of modern collection data are highly suspect (Livingstone 1982). Perhaps what happened was that in some areas tropical rain forest was restricted to 'gallery forest' along rivers. This is how it survives on the edge of its range in Australia today.

It is impossible to conclude a section on the history of the tropical rain forest without reference to the history of human impact on it. The human genus (*Homo*) probably originated in Africa, and had reached South-East Asia by 1 million years ago. *Homo sapiens* may likewise have evolved in Africa, and had reached South-East Asia, Australia and the Americas by 40,000 BP (see Chapter 4). There is little suggestion that these early people had much effect on the tropical rain forest, although they may have restricted its range by firing in Australia (Singh and Geissler 1985). Charcoal is recorded from *c.*12,000 BP in a cave in New Guinea (Hope and Hope 1976), and there is excavated evidence of some kind of disturbance in Kuk Swamp, New Guinea at 9,000 BP. By 6,000 BP the area around had been deforested, and other grasslands in New Guinea date from *c.*5,000 BP (Powell *et al.* 1975). Palynological evidence of forest decline begins in Sumatra at *c.*7,000 BP at Lake di-Atas (Newsome and Flenley 1988) and at *c.*6,000 BP near lake Toba (Maloney 1980). From Latin America there is evidence for the domestication of maize, beans, cassava and potatoes, from dates back to 5,000 BP or earlier. Pollen evidence for forest clearance coincides with the Maya and Inca civilizations. There is evidence for forest clearance in West Africa and East Africa from about 3,000 BP.

USES

The tropical rain forest has long been regarded locally as a source of timber for building and fuel, but this local use expanded into a major global trade in the latter part of the twentieth century. Tropical forests are actually not ideal for logging. The general prevalence of hardwoods is not in accordance with market demand. Certain tropical hardwoods (e.g. ramin, *Gonystylus*) have, however, taken over some of the roles of softwoods. Another difficulty is the great diversity in the forest. Loggers, therefore, have difficulty getting a tree identified

Table 27.2 Expected reduction in area of tropical moist forests, 1975–2000

	Area (10^6 ha)		Annual loss	
	1975	*2000*	*10^6 ha*	*As % 1975 area*
Pantropical total of which the main areas are:	1,120	992	5.12	0.47
West Africa	**14**	**7**	**0.28**	**2.0**
Central Africa	**170**	**166**	**0.16**	**0.09**
Eastern islands*	**172**	**149**	**0.92**	**0.53**
Asian continent	119	94	1.00	0.84
South America	**526**	**467**	**2.36**	**0.45**
Central America[†]	101	93	0.32	0.3

Source: After Whitmore (1990).
Notes: In the regions shown bold most of the moist forest is rain forest
*Malaysia, Australia, Pacific Islands; [†]Including Caribbean.

to species, and thus are unable to deliver a uniform product. For this reason, low diversity forests have been popular with loggers, e.g. the peat swamp forests of Sarawak where there are monospecific stands of *Shorea albida*.

The expansion of the log trade was dramatic. Since the Second World War the industrial countries have increased their demand for tropical hardwoods by a factor of eighteen times (Secrett 1985), and a further 50 per cent increase is expected by the year 2000. According to the Food and Agriculture Organization the rate of tropical forest clearance in 1980 was 7.5 million hectares annually (Whitmore 1990), which is about thirty acres per minute. In some regions the rate of destruction was exceptionally rapid. For instance, in Costa Rica the proportion of land which is forest covered decreased from 32 per cent to 17 per cent between 1977 and 1983. In Sumatra the decrease was about 7 per cent annually between 1980 and 1985 (Whitmore 1990). The expected further reductions in area up to the year 2000 are given in Table 27.2. More recent figures for the log trade and forest clearance are given by Park (1992).

Perhaps surprisingly, logging accounts for only 30 per cent of forest clearance in the tropics. The other 70 per cent is carried out for agriculture. It is often forgotten that vast areas were cleared for plantation agriculture in the nineteenth and twentieth centuries, the crops grown being copra, rubber, oil palm, coffee, tea, etc. This was in addition to areas cleared in prehistoric or proto-historic time, as mentioned in the previous section (see p. 584). During the post-war period, however, there was added to this an enormous rise in the requirements of small farmers, either for subsistence or for cash cropping. This was caused partly by population increases and partly by the desire for consumer goods. In all these activities, what is being used is actually the forest soil. Immediately after felling, and especially if the felled timber is dried and burnt, the soil has a high nitrogen status and yields good crops. Yields are maintained under the plantation system by weed control and the application of fertilizers. For the subsistence farmer, however, the fertilizers are generally unavailable, and weed control becomes impossible after a few years. The plot is then abandoned in favour of further forest clearance. This *swidden* or 'slash-and-burn' agriculture is sustainable indefinitely at low human populations. The disused plot reverts to forest via a series of 'secondary growth' stages, eventually (in about 50–100 years) producing a secondary forest which may closely resemble the primary forest. Unfortunately this rarely happens at the present time for two reasons. First, the rise of population in many areas has shortened the cycle, so that land is re-used before it is ready, leading to a progressive degradation in forest diversity and soil fertility. Second, in many areas with a pronounced dry season, the regrowth is never allowed to revert to forest at all because of regular burning of the regrowth. The fires may be lit accidentally, or may even be started naturally by lightning. More often they are lit deliberately as part of hunting or warfare practices, or for sheer pleasure. The result is permanent grasslands on infertile soils, of little or no agricultural value, and forest depleted annually along a 'frontier' which usually starts in the valleys and moves uphill. This situation has led to enormous grasslands, such as the *cogonales* which cover 40 per cent of the Philippines, the *kunai* of New Guinea, and possibly the *talasiga* of Fiji.

The net result of forest loss by logging and for agriculture is enormous. Often the two go hand in hand, for the farmers find it easier to clear the land once loggers have built roads and removed the large trees. Estimates of total forest loss for the whole tropical rain forest have been as high as 30 hectares per minute. Such a rate would mean the annual loss of an area the size of Great Britain or New Zealand, and if sustained would mean the complete disappearance of the tropical rain forest by the year 2015. Recently, the use of satellite photography has permitted more accurate estimates of rates of forest loss (Whitmore 1990).

A whole range of items is obtained from the forest, other than timber. Although these are sometimes listed as 'minor forest products', their importance is considerable. Principal among these, we may place medicinal drugs. The peculiar abundance of unusual chemical compounds has already been mentioned, and forest tribes have used many of these as traditional drugs. A surprising proportion of these has been found to have real effectiveness, and pharmaceutical companies now put considerable research effort into this. The classic case is quinine, the original anti-malaria drug. Although now superseded by derivatives, the original drug came from *Cinchona*, a rain-forest tree, and was well known to the Amerindians. Another well-known example is curare, an anaesthetic drug with remarkable muscle-relaxing properties, originally used as an arrow poison. More recently, a drug derived from *Rauwolfia* has shown anti-cancer properties, and it is possible the cure for AIDS is there in the tropical rain forest. The point is that the tropical rain forest has for millions of years been a biochemical laboratory, in which the ability to synthesize new compounds has been continuously evolving, and these compounds have been vigorously subjected to natural selection for biological activity. Usually, the plants were, to use teleological shorthand, just trying to protect themselves against insect pests, but the compounds often have anti-bacterial or anti-viral properties. Since valueless compounds have already been screened out, to search for drugs in the tropical rain forest is far more efficient than randomly synthesizing compounds in a biochemist's laboratory. When a drug with some effectiveness has been identified it may either be harvested from the forest, or the synthesis of derivatives may well be worth while. Actually, the value of tropical rain forest drugs is even greater than has yet been stated, for many trees rely on mammals (especially monkeys) to eat their fruits and thus disperse their seeds. They have therefore had to evolve compounds which although poisonous or distasteful to insects are harmless to mammals. It is claimed that 60 per cent of all drugs in use have their origins in the tropical rain forest (Secrett 1985).

Among other 'minor forest products' is rattan cane, which is made into cane furniture. This material is the stem of the climbing palms, and is harvested from primary forest by forest dwellers. This is sustainable as long as the rate of harvesting is not too great. Unfortunately this may not be so at the present time (Whitmore 1990). Rubber is also harvested by tapping wild trees in the forest in South America, although this is a very minor product compared with plantation rubber.

Animal products from the forest include meat, animal trophies, butterflies, skins, furs, etc. The cash value of these may not be negligible, and all are liable to over-exploitation when the demand is high, and the habitat shrinking. A good case in point is the tiger in Sumatra, whose skins are said to fetch a high price in Singapore, despite international agreements.

In recent years the value of the tropical rain forest as a sink for carbon dioxide has been realized. It has been calculated that the whole tropics removes from the atmosphere, by photosynthesis, about 23×10^9 tonnes of carbon annually (Esser 1992). Obviously in a steady state this amount would eventually be returned to the atmosphere by decay processes. Nevertheless the tropical rain forest represents a gigantic sequestering of carbon from the global budget. The present destruction of the forest, therefore, constitutes a great loss of CO_2-absorbing power, for the replacement vegetation normally photosynthesizes at a much slower rate. In addition, since the great majority of the timber is burnt (on-site in the case of agricultural clearance, or off-site later in the case of logging), there are enormous quantities of CO_2 added to the atmosphere. The atmosphere's rising CO_2 levels are probably causing global warming by the 'enhanced greenhouse effect' with potentially disastrous results in the twenty-first century for world agriculture, natural ecosystems, and sea-levels.

THE NEED FOR CONSERVATION

It is now generally accepted that there is a most urgent need for conservation of the remaining tropical rain forest. The arguments for this may be summarized as follows:

1 *Loss of diversity* (Whitmore and Sayer 1992). It has been estimated that one species of animal or plant becomes extinct every minute because of destruction of tropical rain forest. Some biologists have argued that this could lead to the collapse of the earth's ecosystem, in accordance with the Gaia hypothesis. Others simply point out that we are destroying the forest before we even know its potential to yield useful items such as drugs, genetic resources for crop plants, and agents for biological control of pests.

2 *Carbon dioxide sink.* The role of the tropical rain forest in helping to contain the rising atmospheric CO_2 levels has already been explained above.

3 *For the forest peoples.* The habitat of forest tribes is being destroyed, which amounts to a form of genocide.

4 *Runoff and flooding.* The tropical rain forests, and especially the montane rain forests, are protection forests which conserve water and control runoff, thus providing year-round flow in rivers and preventing flooding in lowlands.

5 *Soil conservation.* Forest soils do not survive well after deforestation, and some undergo irreversible degradation (e.g. laterite formation). At least some erosion following deforestation is almost universal, and according to Goudie (1986) this is the most serious problem of all in the long term. Erosion not only destroys the soil at the point of deforestation, it also pollutes water courses and lakes, damaging fishing, and causing silting up of reservoirs, estuaries and lagoons (Park 1992).

6 *Climatic reasons.* It has long been suspected that deforestation, by reducing evapotranspiration, might affect precipitation and the whole water cycle. This has now been confirmed by careful studies in Amazonia (Garstang 1991; Myers 1989; Salati *et al.* 1979).

7 *Scientific and educational reasons.* The tropical rain forest is the most complex ecosystem on earth, so should be retained for scientific research and education.

8 *Philosophical and aesthetic reasons.* The tropical rain forest is unique and it may be argued that it has a right to exist. Its aesthetic value should be preserved for posterity.

9 *Source of timber.* The tropical rain forest is a renewable resource of timber, especially hardwoods, of which the world will soon be short if it is destroyed.

10 *Tourism.* The tropical rain forest can be a source of income for tropical countries, through ecotourism.

PRESENT CONSERVATION MEASURES

Traditional conservation

The present exploitative attitude towards the rain forest must be seen against a background of long traditions, many of which were conservationist in their effect. The present Western positivist approach to nature arose only in the seventeenth/eighteenth centuries, though with roots in Greek philosophy. Earlier views in Europe often regarded the human species as a part of nature (e.g. Hildegard of Bingen in the fourteenth century). Many writers have looked to the book of Genesis in the Bible as the origin of the Western sense of domination over nature, but in the Mosaic Law the Jews were required to leave their ground fallow one year in seven, and were not even allowed to harvest permanent crops such as grapes from it in that year (Leviticus 25), so the Bible is actually quite conservationist.

The Hindu religion seems also to have fostered conservation. Certain trees (e.g. *Ficus religiosa*, the banyan) were sacred. Such trees are still usually preserved during forest clearance in Sumatra, although the Hindu religion there has been long superseded by Islam.

Many other religions and beliefs of rain-forest peoples have led to certain trees, or certain groves, being regarded as sacred or taboo (Westhoff 1983). The significance of this traditional attitude to conservation should not be under-estimated.

Selective silviculture

The procedure advocated over many years in Malaysia was selective logging of natural forest followed by a long recovery period during which the composition of the regenerating forest was managed. This was achieved by the poisoning of undesirable species, so that desirable timber species were encouraged (Whitmore 1975). The system seemed to work satisfactorily until economic pressures forced its abandonment as the logging cycles became too frequent. Whether, without this interruption, it would have been sustainable indefinitely, is untested, but it seems plausible that it would have been.

Reduction of logging

Although logging is directly responsible for only 30 per cent of forest loss, it has been indirectly responsible for far more, because the shifting cultivators tend to follow the loggers. Furthermore, it is seen as, in some ways, a less intractable problem than the small farmer problem, because it is more under the control of governments and of international companies. What has tended to happen is that conservationists from the developed countries criticize the governments of less-developed countries for selling logging rights, and the government spokespersons retort that developed countries got rich by felling their forests, and why shouldn't they do the same. This gets us nowhere. The truth seems to be slightly more complex. Developed countries have, either directly or through international bodies such as the World Bank, extended aid in the form of large loans to developing countries. The interest on these loans has to be paid in hard currencies. To obtain this the developing countries have to have exports, and timber has been one of the few items which developed countries wish to purchase from them. They therefore have no option but to permit logging. The only area where any option does exist is in the developed world. The ball is definitely in their court. If the developed world was really serious about saving the rain forest it could do it by remitting the interest payments, or by re-negotiating the terms of loans. To some extent this has been happening recently in Brazil and other countries. Ideally, tropical countries will eventually be able to reduce logging to a sustainable level, as their economies become buoyant. This is already happening in Malaysia, the economic success story of South-East Asia. Malaysia claims to be the world's largest exporter of tropical timber, and the trade is second only to petroleum in importance as an export earner for that country. In Sarawak, the main logging region of the country, output has declined from $c.24$ million m^3 in the late 1980s to 16.5 million m^3 in 1993. This has been possible partly because of a steep increase in the price of timber, but was in line with a commitment to the International Tropical Timber Organization. The country has pledged a further reduction to 9.2 million m^3 by the year 2000, at which point they claim it is sustainable (*Asahi Evening News*, 27 November 1993).

The increase in the price of timber, mentioned above, was partly the result of a deliberate restriction of output by Malaysia. Perhaps some kind of WPEC, analogous to OPEC, will develop.

Reserves

In response to the world outcry about destruction of the tropical rain forest, many governments have set up national parks, nature reserves, forest reserves, conservation areas, etc. Some are succeeding well in their aim, but two defects are common. One, and it is a major one, is that reserves tend to exist more on paper than on the ground. Many national parks are designed to have people living in them, but some other reserves are intended to have access by permit only. In practice there is often little control on access, and even the boundary of the reserve may not be marked, and certainly not fenced. Poaching of scarce resources is often rife, and reserve boundaries are frequently encroached on by agriculture. Marauding animals from the tropical rain forest are frequently the bane of the lives of nearby farmers, so that conflict results. Where politics is involved, reserves may be rescheduled for logging or clearance because of a perceived national or local need, or because of corruption.

The second problem with reserves is their size. It is a well-known fact that large reserves contain more species than small reserves, and presumably they retain these in a sustainable way. This is not very surprising. Populations of all species fluctuate, and if the initial population is small, there is a good chance of fluctuations reaching zero, i.e. local extinction. In the case of predatory animals, the top-level predators need large hunting areas in which to survive. The whole relationship between area and diversity has been given a theoretical basis as *The Theory of Island Biogeography* (Macarthur and Wilson 1967). The theory may be applied not only to islands in the ocean, but to any isolated habitats, including forest reserves in a sea of deforested land. It has thus been applied to the design of nature reserves by Diamond and May (1976). The problem is that the theory is largely empirical, and indeed it has been criticized as trivial, and explicable simply on the grounds that larger areas tend to have more different habitats, therefore they necessarily have more species. Furthermore, the exact causes of diversity are uncertain (see pp. 574–7) therefore a precise mathematical theory cannot yet be developed. The argument usually, however, simplifies down to a practical decision between a *single large or several small reserves* (the SLOSS problem). One advantage of the single large reserve is that (if it is of simple shape, i.e. nearly circular) it minimizes edge effect. This is a problem in all reserves: the marginal zone of a forest is subject to abnormal influences such as too much light and wind, incursions of fire or domestic animals, etc. Also, animals from the forest stray outside the forest and are usually then killed. Thus a small or narrow reserve may have little or no 'heartland' unaffected by these influences. On the other hand several smaller reserves, carefully chosen, can usually increase the range of habitats covered, thus probably increasing the potential diversity which can be sustained. Usually a compromise between the SL and SS positions is probably best.

Another problem highlighted in recent times is the need to accommodate environmental change. For instance, if the enhanced greenhouse effect increases world temperatures rapidly, then on a tropical mountain a few large reserves at different altitudes might be less satisfactory than a broad strip extending up the mountain. This would allow vertical migration better than separate reserves of equal area.

Compromises

Given that forest clearance and logging are bound to continue, some attempts are nevertheless being made to limit their destructiveness. For instance, there are many ways of making the logging process less wasteful. Until recently it was uneconomic to use more than 50 per cent of the timber from a logged area. There were many reasons for this. Some trees, after felling, are found to have timber defects which make them unusable. In any case the tree crown is not used at all, and usually no part above the first branch. The large buttresses of tropical rain forest trees are also wasted, and trees are often felled above these, leading to waste of a large stump. The felling process is unbelievably destructive. Logging roads must be made, which destroys much of the vegetation and often initiates soil erosion. Falling trees break a large proportion of the small trees nearby. The introduction of wood chipping has made possible the use of much of the smaller material that would otherwise be wasted. There is now a good market for wood chip products, which are often much superior to timber because of lack of warping, etc. There has been considerable experimentation with new methods of timber extraction to reduce the length of logging roads. Balloons to lift out the timber were tried but have not been successful. The use of tall steel gantries, pulling out logs with steel hawsers, has been more successful. A new development is 'heli-logging', in which large helicopters drag the steel hawsers. This practice is controversial, however. The loggers claim it is environmentally friendly, because soil erosion is reduced. Conservationists argue that it simply allows the loggers to take trees from steep terrain where road-building would be difficult, and that such terrain ought never to be logged anyway, but reserved as water catchment protection forest (*Asahi Evening News*, 27 November 1993).

A development of a totally different kind is agroforestry, in which the taking of agricultural crops is combined with the growing of timber. In a sense, this has always been practised in the tropics. Most houses are surrounded by sizeable mixed orchards growing tropical tree fruits, and it has been shown that these areas (often ignored as irrelevant by development 'experts') are not only highly productive but also retain diversity (e.g. of insects) much better than areas under herbaceous crops or single-species tree plantations such as rubber or coffee. They also require far less maintenance, and provide small timber crops (e.g. for firewood) as well. The concept of agroforestry more often advocated at present is one more akin to plantation forestry. For many years coffee has been grown interplanted with shade trees, commonly *Albizzia* sp. The coffee was thought to benefit from light shading, and the tree, being leguminous, possessed root nodules and did not exhaust the soil of nitrogen. Whether the coffee actually benefited from the shade is debatable, but certainly there was an additional product – timber – from the plantation. The extension of this idea, using species with valuable timber, is agroforestry. The underplanted crops may be small tree crops such as coffee or rubber, or herbaceous crops such as groundnuts.

Plantation forestry

In most temperate countries, when the nation's natural forest reserves are seriously depleted, plantations of fast-growing timber species are established. Most commonly these have been of softwoods such as pine (*Pinus* spp.) and spruce (*Picea* spp.), but some hardwoods such as *Eucalyptus* spp. have also been used. This is a highly successful procedure, though not without problems. For instance, some species of conifers appear to promote podzolization of the soil, which may make the use of the land for forestry non-sustainable. Also, single-species plantations have proved sensitive to attack by pests, e.g. the spruce budworm in North America.

Attempts are now being made to apply this concept to the tropics. Plantations of *Pinus kesiya* in Luzon, Philippines, and of *P. merkusii* in Sumatra, at 1,400 m altitude, were made in the 1930s and have been fairly successful. Post-war plantings of *Araucaria* spp. in the New Guinea highlands, and of *Agathis* sp. in Java have also met with success. Teak (*Tectona grandis*) has been grown in east Java, also since the 1930s and earlier. Ecologists have always been nervous about the viability of monocultures in the tropics. If the concept of pest pressure has any validity, then monocultures ought to be liable to such serious insect attack as to be impracticable. So far, provided pesticides are used, this does not seem to have been the case, and reafforestation, presumably by timber plantations, has been accepted as government policy in Malaysia (Abu Hassan, *Straits Times*, 1989). It may well be, however, that problems lie ahead. The use of chemical pesticides in the long term may prove unsustainable, and insect pests are sure to develop resistance eventually. Some time may be gained by the introduction of tree species from other regions, so that they leave their pests behind them. But they will eventually acquire pests in their new home.

Population-related matters

Since 70 per cent of forest loss comes from small farmers, mainly on subsistence agriculture, a reduction of human population pressure would be expected to control the rate of deforestation. The whole problem of human population in developing countries is beyond the scope of this chapter, but a few points may be made.

First there are the depressing statistics of the rates of population increase, which make some kind of disaster seem inevitable. It has always been argued by economists, however, that rates of increase would slow down as the economies of developing countries improved. To some extent this is already happening, the world rate of population growth having dropped significantly already. It is predicted that if this trend continues world population would stabilize about the year 2100 at a level around 10 billion, an increase of 100 per cent over today's 5 billion. Achieving this stability, however, depends on continued improvement in the economies of developing countries. Whether this can be achieved is debatable, because of resource limitations. At present one-third of the world's population, the 'developed' third, uses two-thirds of the world's resources. Clearly for all the world's population to become 'developed' in this way is impossible. The possible way out, proposed in the Bruntland Report (World Commission on Environment and Development 1987) is 'sustainable development', i.e. we must use our ingenuity to develop improved, more efficient ways of using resources so that they will be sufficient. Whether this can be achieved is the challenge of the twenty-first century. Unfortunately, the governments of the developed world have so far mostly interpreted 'sustainable development' as a licence to continue the present profligacy with resources, and to hope for brilliant discoveries which will solve the problem.

FUTURE NEEDS AND PROSPECTS

The measures currently being taken are clearly insufficient to counter the 'ecological imperialism' of the developed world. Sustainability is still a very long way off. Achieving it in tropical countries will need a broad approach, for there is no single solution.

Perhaps the following areas – efficient logging, plantations, agroforestry and population – can be singled out for special attention.

Efficient logging

As the price of timber rises, the desire to eliminate waste is strengthened. Now that woodchip materials are widely used, almost no part of the cut tree need be wasted, if it can be got to the mill. If we could eliminate the logging roads, much ecological damage would be prevented. The large Chinook helicopters can lift about 12 tonnes of timber, that is a single log 4 m long by 2 m diameter. The possibility of airships to lift larger loads, and perhaps ultimately whole trees, should be examined. If a tree could be lifted out without falling, even more damage could be prevented, and forest recovery would be more rapid.

Plantations

These will obviously be needed on an enormous scale, and they are already on the verge of economic viability. Research is needed into the possibility of mixed plantations, and into the potential for genetic engineering to introduce pest resistance into cloned individuals. Cloning could also improve growth form and growth rate, by selection of trees with good provenance, as has been done with great success in temperate forestry.

Agroforestry

As mentioned above (see p. 592), the potential for this is very large.

Population

At present the developed world has about one-third of world population, but is using two-thirds of the resources. Clearly 'sustainable development', to bring the other two-thirds of the population (plus the increase in population in the twenty-first century) to a developed state, is going to require some remarkable new efficiencies. Surely it would be simpler to accept that population limitation is absolutely essential, despite the political difficulties of such a decision.

Predicting the future is very difficult. World population could be halved by major epidemic disease. But if present trends continue, it seems likely that sustainability cannot be reached. In a major computer simulation done twenty years ago, the Club of Rome predicted an ecological disaster for the earth (Meadows *et al.* 1972). Even now that the simulation has been revised it is difficult to be confident. There remains with us the empirical test, provided by the people of the isolated world of Easter Island. They felled all their forests, and not long afterwards their civilization collapsed (Bahn and Flenley 1992): a cautionary tale for the whole earth.

If the Easter Island model is in any way reliable, the period of peak population will be accompanied by famine and warfare. I envisage a situation where the developed countries may try to take over the less-developed countries to prevent their economic advance, and to lay claim to the few remaining resources. This action would be justified spuriously. In the sixteenth to nineteenth centuries, colonialism was justified on religious grounds. In the twenty-first century it may be justified on grounds provided by the new religion – ecology.

REFERENCES

Absy, M.L. and van der Hammen, T. (1976) 'Some palaeoecological data from Rondonia, southern part of the Amazon Basin', *Acta Amazonica* 6, 293–9.

Anderson, J.A.R. (1964) 'The structure and development of the peat swamps of Sarawak and Brunei', *Journal of Tropical Geography* 18, 7–16.

Ayensu, E.S. (ed.) (1980) *Jungles*, London: Book Club Associates.

Bahn, P. and Flenley, J.R. (1992) *Easter Island, Earth Island*, London: Thames & Hudson.

Barmawidjaja, D.M., de Jong, A.F.M., van der Borg, K., van der Kaars, W.A. and Zachariasse, W.J. (1989) 'Kau Bay, Halmahera, a Late Quaternary palaeoenvironmental record of a poorly ventilated basin', *Netherlands Journal of Sea Research* 24, 591–605.

Bush, M.B. (1994) 'Amazonian speciation: a necessarily complex model', *Journal of Biogeography* 21, 5–17.

Bush, M.B., Piperno, D.R., Colinvaux, C.A., De Oliviera, P.E., Krissek, L.A., Miller, M.C. and Rowe, W.L. (1992) 'A 14,300-year palaeoecological profile of a lowland tropical lake in Panama', *Ecological Monographs* 62, 251–75.

Caratini, C. and Tissot, C. (1985) *Le Sondage Misedor: étude palynologique. Études de géographie tropicale*, No. 3, Centre d'études de géographie tropicale, Centre national de la recherche scientifique, Domaine Universitaire de Bordeaux.

CLIMAP Project Members (1976) 'The surface of the Ice-Age Earth', *Science, N.Y.* 191, 1131–7.

Diamond, J.M. and May, R.M. (1976) 'Island biogeography and the design of nature reserves', pp. 163–86 in R.M. May (ed.) *Theoretical Ecology: Principles and Applications*, Oxford: Blackwell Scientific Publications.

Emiliani, C. (1955) 'Pleistocene temperatures', *Journal of Geology* 63, 538–78.

Esser, G. (1992) 'The role of the tropics in the Global Carbon budget: impacts and possible developments', pp. 242–52 in J.G. Goldammer (ed.) (1992) *Tropical Forests in Transition*, Basle: Birkhauser.

Flenley, J.R. (1979) *The Equatorial Rain Forest: A Geological History*, London: Butterworth.

Flenley, J.R. (1992) 'UV-B insolation and the altitudinal forest limit', pp. 273–82 in P.A. Furley, J. Proctor and J.A. Ratter (eds) *Nature and Dynamics of Forest-Savanna Boundaries*, London: Chapman and Hall.

Flenley, J.R. (1993a) 'The origins of diversity in tropical rain forests', *Trends in Ecology and Evolution* 8, 119–20.

Flenley, J.R. (1993b) 'Cloud forest, the Massenerhebung effect, and ultraviolet insolation', pp. 94–6 in L.S. Hamilton, J.O. Juvik and F.M. Scatena, (eds) *Tropical Montane Cloud Forests. Proceedings of an International Symposium*, Honolulu, East-West Center, 264pp.

Garstang, M. (1991) 'Destruction of the rain forest and climate change', pp. 35–47 in T.R.R. Johnston and J.R. Flenley (eds) *Aspects of Environmental Change*, Miscellaneous Series 911, Massey University, Department of Geography.

Gillett, J.B. (1962) 'Pest pressure, an underestimated factor in evolution', *Taxonomy and Geography*, Systematics Association Publication No. 4, pp. 37–46.

Goudie, A.S. (1986) *The Human Impact on the Natural Environment*, Oxford: Blackwell.

Hope, G.S. and Hope, J.H. (1976) 'Man on Mt Jaya', pp. 225–39 in G.S. Hope, J.A. Peterson and U. Radok (eds) *The Equatorial Glaciers of New Guinea*, Rotterdam: A.A. Balkema.

Hopkins, B. (1974) *Forest and Savanna* (2nd edn), London: Heinemann.

Janzen, D.H. (1970) 'Herbivores and the number of tree species in tropical forests', *American Naturalist* 104, 501–28.

Kershaw, A.P. (1978) 'Record of the last interglacial-glacial cycle from northeastern Queensland', *Nature* (London) 212, 159–61.

Kutzbach, J.E. and Guetter, P.J. (1986) 'The influence of changing orbital parameters and surface boundary conditions on climate simulations for the past 18,000 years', *Journal of Atmospheric Science* 43, 1726–59.

Livingstone, D.A. (1982) 'Quaternary geography of Africa and the refuge theory', pp. 523–36 in G.T. Prance (ed.) *Biological Diversification in the Tropics. Proceedings of the Fifth International Symposium of the Association for Tropical Biology*, New York: Columbia University Press.

MacArthur, R.H. and Wilson, E.O. (1967) *The Theory of Island Biogeography*, Princeton: Princeton University Press.

Maloney, B.K. (1980) 'Pollen analytical evidence for early forest clearance in North Sumatra', *Nature* (London) 287, 324–6.

Meadows, D. *et al.* (1972) *The Limits to Growth*, London: Earth Island.

Mitchell, A.W. (1986) *The Enchanted Canopy: Secrets from the Rainforest Roof*, London: Collins.

Morley, R.J. and Flenley, J.R. (1987) 'Late Cainozoic vegetational and environmental changes in the Malay archipelago', pp. 50–9 in T.C. Whitmore (ed.) *Biogeographical Evolution of the Malay Archipelago*, Oxford: Clarendon Press.

Myers, N. (1989) *Deforestation Rates in Tropical Forests and their Climatic Implications*, London: Friends of the Earth.

Newsome, J. and Flenley, J.R. (1988) 'Late Quaternary vegetational history of the Central Highlands of Sumatra. II. Palaeopalynology and vegetational history', *Journal of Biogeography* 15, 555–78.

Park, C.C. (1992) *Tropical Rainforests*, London: Routledge.

Phillipson, J. (1966) *Ecological Energetics*, London: Arnold.

Poore, M.E.D. (1964) 'Integration in the plant community', *Journal of Ecology Jubilee Symposium*, pp. 213–26.

Powell, J.M., Kulunga, A., Moge, R., Pono, C., Zimike, F. and Golson, J. (1975) *Agricultural Traditions of the Mount Hagen Area*, Occasional Paper No. 12, University of Papua New Guinea, Department of Geography.

Richards, P.W. (1964) *The Tropical Rain Forest. An Ecological Study*, Cambridge: Cambridge University Press.

Ricklefs, R.E. (1973) *Ecology*, London: Nelson.

Salati, E., Dall'olio, A., Matsui, E. and Gat, J.R. (1979) 'Recycling of water in the Amazon basin: an isotopic study', *Water Resources Research* 15, 1250–8.

Secrett, C. (1985) *Rainforest: Protecting the Planet's Richest Resource*, London: Friends of the Earth.

Shackleton, N. (1967) 'Oxygen isotope analysis and Pleistocene temperatures reassessed', *Nature* (London) 215, 15–17.

Singh, G. and Geissler, E. (1985) 'Late Cainozoic history of vegetation, fire, lake levels and climate at Lake George, New South Wales, Australia', *Philosophical Transactions of the Royal Society* B(311), 379–447.

Strahler, A. H. and Strahler, A.N. (1994) *Introducing Physical Geography*, New York: John Wiley.

Talbot, M.R. and Delibrias, G. (1977) 'Holocene variations in the level of Lake Bosumtwi, Ghana', *Nature* (London) 268, 722–4.

Thom, R.B. (1967) 'Mangrove ecology and deltaic geomorphology: Tabasco, Mexico', *Journal of Ecology* 55, 301–43.

Walker, D. and Flenley, J.R. (1979) 'Late Quaternary vegetational history of the Enga District of upland Papua New Guinea', *Philosophical Transactions of the Royal Society* B(286), 265–344.

Watson, J.G. (1928) 'Mangrove forests of the Malay peninsula', *Malayan Forest Records*, no. 6.

Webster, P.J. and Streten, N.A. (1978) 'Late Quaternary Ice Age climate of tropical Australasia: interpretations and reconstructions', *Quaternary Research* 10, 279–309.

Westhoff, V. (1983) 'Man's attitude towards vegetation', pp. 7–24 in W. Holzner, M.J.A. Werger and I. Ikusima (eds) *Man's Impact on Vegetation* (Geobotany 5), The Hague: W. Junk.

Whitmore, T.C. (1975) *Tropical Rain Forests of the Far East*, Oxford: Clarendon Press.

Whitmore, T.C. (1990) *An Introduction to Tropical Rain Forests*, Oxford: Oxford University Press.

Whitmore, T.C. and Sayer, J.A. (eds) (1992) *Tropical Deforestation and Species Extinction*, London: Chapman and Hall.

World Commission on Environment and Development (1987) *Our Common Future* (The Bruntland Report), London: Oxford University Press.

FURTHER READING

Flenley, J.R. (1979) *The Equatorial Rain Forest: A Geological History*, London: Butterworth.

Grainger, A. (1993) 'Rates of deforestation in the humid tropics: estimates and measurements', *Journal of Geography* 159, 33–44.

Longman, K.A. and Jenik, J. (1987) *Tropical Forest and its Environment* (2nd edn), Longman Scientific and Technical.

Mitchell, A.W. (1986) *The Enchanted Canopy: Secrets from the Rainforest Roof*, London: Collins.

Myers, N. (1989) *Deforestation Rates in Tropical Forests and their Climatic Implications*, London: Friends of the Earth.

Park, C.C. (1992) *Tropical Rainforests*, London: Routledge.

Richards, P.W. (1964) *The Tropical Rain Forest: An Ecological Study*, Cambridge: Cambridge University Press.

Salati, E., Dall'olio, A., Matsui, E. and Gat, J.R. (1979) 'Recycling of water in the Amazon basin: an isotopic study', *Water Resources Research* 15, 1250–8.

Sandler, T. (1993) 'Tropical deforestation: markets and market failures', *Land Economics* 69, 225–33.

Whitmore, T.C. (1990) *An Introduction to Tropical Rain Forests*, Oxford: Oxford University Press.

28

HUMANITY'S RESOURCES

I.G. Simmons

Members of the human species have basic material needs. All of us must have food, shelter, clothing, and fuel. But there is a big difference between the material demands of a Palaeolithic hunter-gatherer and those of a present-day inhabitant of, say, Los Angeles. So the satisfaction of those demands is of a different economic order between the two: the hunter-gatherer would expect an area of about 26 km^2 to be sufficient to yield all the materials for survival and reproduction; a modern urbanite expects to have the kind of purchasing power to bring in the whole world's riches. The demand for materials is different ecologically as well, for the environmental impact of the hunter-gatherer is low, even if fire is used to manipulate the landscape. By contrast, the rich city-dweller not only obliterates Nature locally but causes it to be changed on the other side of the world, to provide energy, metals, baby vegetables, and roses in mid-February.

PERSONAL LEVELS OF RESOURCE USE

The advent of industrialization in the nineteenth century has meant that for members of Western economies, access to resources has expanded manifold. The rich of the world have tables loaded with 25–50 per cent more than they need for adequate nutrition, and some 30–40 per cent of that food will be in the form of fats, compared with the 15 per cent of a subsistence farmer. This means that whereas the minimal need for human calorie intake is about 9.2 MJ/day, those living in developed countries get an average of 11.0 MJ/day, with levels of 12.5–15.0 MJ/day being common.[1] Similar data can be found for energy use. In pre-industrial Europe, charcoal and wood provided about 20–40 GJ/capita/year with a 15 per cent efficiency of conversion; in less-developed countries today, access is to 25 GJ/capita/day at most, with no more than 15 per cent conversion efficiency. But in the developed countries, primary energy supply is of the order of 150 GJ/capita/ year with a 40 per cent conversion efficiency. Useful energy at the disposal

599

of developed country inhabitants is therefore immensely higher (Smil 1993).

These data for food and energy consumption are symbolic of the access to resources as a whole enjoyed by developed country populations. In the course of a seventy-year lifetime, an individual German will be responsible for the use of 460 tonnes of sand and gravel, 166 tonnes of crude oil, 39 tonnes of steel, 1.4 tonnes of aluminium and 1.0 tonnes of copper, amongst other materials. In the USA, abandoned metal- and coal-mines cover about 9 million ha, which is approximately the area of Hungary. Many of the materials used become wastes and each 'average' US citizen was putting out 864 kg/year of municipal garbage in the 1980s. These levels are much lower in less-developed countries: for example, per capita use of aluminium is twenty-five times higher in the USA than that in India; steel use per person is nine times that of China.

The totality of resources used

It is scarcely possible to list all the materials which human societies have, at one time or another, identified as resources. Even basic substances which are eaten cover an enormous range: some have eaten soils (geophagy) and others have eaten other humans. Some people feel at ease with a helping of shredded raw sea-slug but not with a portion of blue cheese. So the materials for food, liquid intake, clothing, shelter and transport, for example, are culled from a variety of both natural and human-manipulated sources; from the very wild (such as squid from the deep oceans) to the industrialized (like particle boards made from wood processing waste and sulphur from power-station emissions). When we move away from the resources deemed essential, then the outreach is even greater: consider the materials that make up a motor car: metals, plastics, rubber and even perhaps scarce timbers. Symbolic demands may have meant catching rare animals like eagles or ermine. The lesson from this plethora is, however, relatively simple: beyond the basic needs which every human must have in order to survive, grow, and reproduce, resources are culturally defined. One culture will eat raw fish, another demands leopard skins for its ceremonial occasions. There is, too, another relatively simple lesson: that every extra human on the face of the earth will add to a 'base load' of demand for resources. In addition, every extra human with the ability to accumulate 'cultural' resources will add further to those demands. Recall the current commercial energy consumption of a Nepalese at 1.0 GJ/yr (plus 10,000 MJ of biofuels) and its contrast with a Canadian's consumption of 325 GJ/yr, plus 2,500 MJ of biofuels. In these two examples is encapsulated a whole world of difference of access to resources. The remaining question is, of course, the extent to which the Nepalese aspire to be Canadians and the impact this would have on resource use and the environment.

The user population

The implication of these data is quite profound. It is that human population growth must be a vital ingredient in creating demand for the planet's materials to be used as resources. No matter how low the basic demands, they are increased with every extra person; when some at least of those extra people are rich enough to call up materials in large quantities from every corner of the globe, then the stage is set for two related concerns. The first is, put simply, are there 'enough' resources for the populations of the globe, present and future? This is often expressed primarily in terms of food and water but is often mentioned in connection with certain metals, for example. The second question considers the environmental consequences of trying to meet all the demands: will this cause environmental instability and unpredictability? Although this latter question cannot be ignored, it will not form a major focus in this essay.

The nature of human population growth is, however, central to any discussion of resource use and availability. The human population in AD 1750 was probably about 0.8 billion (1 billion = 1,000 million), of which most lived in solar-powered economies. Now (1996) the world population stands at about 5.8 billion, of whom about 75 per cent live in the less-developed countries and these nations supply about 95 per cent of the planet's population growth (Figure 28.1; see Chapter 12). Each year about 90 million people are added to the total, and a world total of 10 billion is foreseen for the year 2050 in some projections. Of that 10 billion perhaps 8.7 billion will live in the poorer countries (Findlay 1991). The big question is, 'At what level will the population stabilize?' The

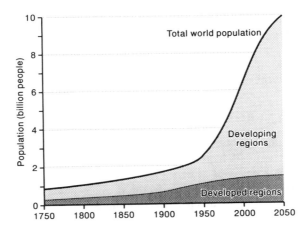

Figure 28.1 World population growth: trends and projections, 1750–2050. The darker area is for industrialized regions; the lighter shading represents the developing countries. After World Resources Institute (1994).

601

history of accuracy of population projections is rather patchy, but several estimates now suggest that a stable population of 11.5 billion in AD 2150 is a likely scenario. So at least another doubling of the present population levels, unevenly distributed between rich and poor, tropics and temperate zones, city and countryside, seems almost inevitable, though it could come sooner than 2150, as indeed it could come later (Livi-Bacci 1992).

The question in all considerations of resource use and management is, as stated above, 'Will there be enough in the future?' This is intertwined with the explicit question of who will get what, then as now: where do equity and justice lie in the matter of resource use? In some writers' thinking, the questions of justice, for example, extend even beyond the human species, exercising arguments that moral consideration should extend at the very least to animals and possibly beyond. Again, resource use is bound up with the broadest definitions of human culture. Clearly, the ways in which a resource is identified, its routes through the decisions to utilize it (which will be greatly affected by, for example, economics and politics), the ways in which its users deploy it to create their own identities, and its disposal as a waste (or its re-use) are all worthy of evaluation in the widest possible framework: any appraisal which deals with only one approach is likely to be inadequate.

Renewable or non-renewable?

It is a cliché of writing about resources that there are two main types: the renewable or flow resources, and the non-renewable or stock resources. The first category is typified by biological materials which are self-reproducing and the second by mineral sources of, for example, fuels or metals where once the coal is burned then it has gone for ever. This classification is still valid in many instances, but the intensity of use of some flow resources has in some places turned the renewable into the permanently extinct. On the other hand, technological advances have made it possible to recover and re-use many materials which hitherto would have been thrown away and buried.

ENERGY AS A MEDIATING RESOURCE

The access of human societies to energy resources is one of their fundamental material characteristics (Chapter 5). Before the nineteenth century, nearly all societies were solar-powered. Their sources of energy were plant and animal tissues which represented recently captured solar energy, wind and water which were driven by the redistribution of the Sun's energy in climatic and hydrological processes. The discovery of how to use the concentration of energy in fossil fuels is a great transformative process since many more resources now become available. Mines can be deeper, fish can be sought in far oceans and goods can be produced cheaply in machines and then transported to every niche of the earth. Applied to existing processes like

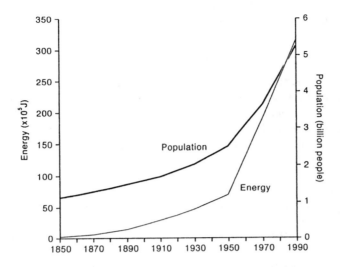

Figure 28.2 World population and commercial energy use, 1850–1990. Biofuels such as wood are excluded. After World Resources Institute (1994).

food production, fossil fuel energies allow a greater intensity of throughput with the use of more machinery, chemical fertilizers and pumped water, for instance (Figure 28.2). Above all, much of the energy of these fuels is transformed into knowledge, especially scientific and technological information, which allows more precise manipulation of productive systems. (Readers unconvinced of the last point should look at the surroundings in which they are reading and see to what extent they are a product of fossil fuel use: does the library have windows?) But the fossil hydrocarbons (coal, oil and natural gas, tar sands, and oil shales are the main sources) are non–renewables *par excellence*. Once they have been oxidized and their energy released then a few gases and perhaps some ash are left. The energy is transformed to heat in which state it can do no work; no amount of technology can reverse the second law of thermodynamics.

So a search for renewable or virtually inexhaustible energy sources has been common in the last fifty years. Some technologies were strongly developed in regions poor in oil and coal, as in pre-1960 Norwegian use of hydropower, for example. The possible running-out of hydrocarbons gave rise to the civilian nuclear power industry of the post-1945 years, and the prediction of enhanced radiative forcing in the atmosphere (the 'greenhouse effect') has allowed a limited amount of capital to be devoted to the so-called 'alternative' energy sources based on photovoltaic solar collectors, tidal power generation and wind farms, to quote a few examples. Thus the enmeshing of energy into all forms of resource use carries with it a set of attitudes towards the relation of resource

603

futures and energy supplies which transcends the simple realities of say the current price of oil on world markets (Smil 1991).

A REAL WEALTH

One argument that follows is that the only long-term wealth on which we can rely is based on renewable resources such as those of a biological nature (Eyre 1978). The key element, then, is the fixation of solar energy as chemical energy and its combination with mineral nutrients as plant tissue. The quantity of plant tissue per unit time per unit area is known as net primary productivity and is a key resource since it is the fundamental unit of all ecological systems, both wild and cultivated. What is clear from accounts of global net primary productivity is that certain natural systems are much more productive than others (Table 28.1) and that human-made systems such as agriculture trade off high levels of net primary productivity for a culturally desired product. Much of human action in recent centuries has been to reduce net primary productivity in the course of conversion of the wild to the tame: Table 28.2 shows the extent to which human societies have manipulated and appropriated net primary productivity. In resource terms, we need to think about whether the new systems we have created have the resilience and self-renewing properties of those they have replaced.

Table 28.1 Net primary productivity of major biomes, 1970s

Biome	Mean net primary productivity $(g/m^2/yr)$
Forest	1,570
Woodland, grassland, and savannah	1,408
Desert	103
Arctic–alpine	84
Cultivated land	937
Other humanized area	200
Sclerophyllous vegetation and wetland	1,783
Lake and stream	400
Marine	254
Terrestrial subtotal	899
Aquatic subtotal	255
World average	440

Source: After Vitousek *et al.* (1986).

Table 28.2 Human appropriation of net primary productivity, 1980s

Producers and human consumers in world ecosystem	Net primary productivity (Pg)
World net primary productivity:	
Terrestrial	132.1
Freshwater	0.8
Marine	91.6
Total	224.5
Net primary productivity used by humans:	
Plants eaten directly	0.8
Plants fed to domestic animals	2.2
Fish eaten by both	1.2
Wood for paper and timber	1.2
Fuel wood	1.0
Total	7.2
Net primary productivity used or diverted by humans:	
Cropland	15.0
Converted pastures	9.8
Other (cities, deforested)	17.8
Total	42.6
Net primary productivity used, diverted, or reduced:	
Net primary productivity used or diverted	42.6
Reduced by conversion	17.5
Total	60.1

Source: After Diamond (1987).

BIOLOGICAL RESOURCES

The main categories of renewable resources

These fall into three main groups, to be discussed in turn. The first are those which depend upon living organisms, such as plants, animals or microorganisms. The second is fresh water, without which humans can survive only a few days; and the third is the totality of an environment when it functions as a resource for tourism and recreation, for example, though this is not treated at any length here.

The tissues of living things appear as individual organisms of a particular species. The term biodiversity is used to mean (1) genetic diversity within species (for example, the different breeds of domestic cattle on a continent); (2) the variety of species in a given area (like the number of different moss species within a swamp forest); and (3) the number of different ecosystems within a biogeographical province, like the swamps, woodlands, savannahs and grasslands which traverse parts of eastern and central Africa.

The advent of genetic manipulation as a new element in biotechnology has meant some concentration upon genetic diversity as a focus in biodiversity research and policy-making. It is estimated that less than 1 per cent of genetic material is expressed in the form and function of the organism, so that the search for manipulable genes is always going to be costly. Nevertheless, the need is strong, for the exigencies of modern cultivation have greatly reduced the genetic base of agricultural production. Canadian wheat production, for instance (one of the world's few areas with a large surplus) relies for 75 per cent of its output on four varieties, with over one-half from a single variety. Of the 145 recognized breeds of cattle in Europe and the Mediterranean, 115 are likely to become extinct.

Perhaps 12.5 million species of plants, animals, fungi and microorganisms exist, of which about 1.4 million have been described by science (Groombridge 1992). At present, some 4 per cent per decade of these species are becoming extinct, which is estimated to be the fastest rate since the end of the Cretaceous: in absolute numbers an estimated 25,000 plant species and more than 1,000 vertebrate species and subspecies are threatened with extinction: put dramatically, the world is losing three species every hour. The tropical lowland forests which cover about 7 per cent of the land surface contain 70–90 per cent of known species.

Concern about loss of ecosystem types is bound up with changes in land cover and land use due to economic development. It is often associated with a transition from wild land inhabited with wild species to tamed land with domesticated species, and is measured by any of the data for the extension of agricultural land, the diminution of forest and grassland area since the nineteenth century, or the extension of cities and transport networks in the last hundred years. Soil erosion, too, takes the whole ecosystem (including its mineral nutrients) with it on its downward journey. The lesson of biodiversity is simple: extinction cannot be reversed and even the burgeoning biotechnology industry acknowledges that it has need of all the natural genetic diversity possible. The whole provides, though, an example of how an apparently self-renewing resource can become physically non-existent.

Above all, the focus of biological resources is on food for humans. The success of the world food system in feeding the growth of world population cannot be denied. The recent past, in particular, has been a period of rising output: in the twenty years to 1989, root crop production increased by 0.8 per cent per year, cereals by 3 per cent, milk, meat and fish by 2 per cent, and other foods by 2.5 per cent. In absolute terms, this meant for instance an increase in world production of grain from 1.0×10^9 tonnes in 1965 to 1.8×10^9 tonnes in 1989. Africa has been the only continent where production per head has in fact declined: wars, poor distribution systems and ineffective government policies have added to the uncertainties of rainfall (Table 28.3).

There is a global market in agricultural produce, with large cash crop and export trades. Several nations earn most of their foreign exchange this way:

606

Table 28.3 Indices of food production*

	Total		Per capita	
Region	*1980–2*	*1990–2*	*1980–2*	*1990–2*
Africa	102	129	99	94
North and Central America	102	112	101	96
South America	104	134	102	108
Asia	104	148	102	121
Europe	102	107	101	104
Former USSR	100	110	99	101
Oceania	97	112	96	95
World	103	127	101	105

Source: After World Resources Institute (1994).
*This set of data is for food production, not total agricultural production; it does not show whether or not the food is exported or consumed locally.

Burundi gets 93 per cent of its exchange from coffee exports, Sudan 65 per cent from cotton. Overall, Africa devotes about 13 per cent of its cropland to exports and most less-developed countries have a slightly higher proportion. The trade is to some extent dominated by North American exports of grain, which account for 87 per cent of world grain exports.

This production surge has also taken place when the quantity of cropland per head is falling. The world amount (at 0.27 ha/capita in 1995) suggests a projected decline from the 1971–5 period figure of 0.39 ha/capita to 0.25 ha/capita by 2000. In African less-developed countries the equivalent numbers are 0.62–0.32 and for South-East Asia 0.35–0.20. The global state, therefore, appears to be one of keeping pace with population growth and even of passing the threshold of minimal self-sufficiency looked at world-wide (Brown 1994). But a universal transition to food security and more varied diets has yet to be brought about, so that there are regional problems of nutrition where too many people are getting far too little. There are many views as to whether this is a systemic problem brought about by the rich, a Malthusian problem due to rapid population growth or a failure to adopt more modern technology.

Bearing in mind that organic evolution has progressed towards tree growth whenever natural conditions have permitted and that forests are both sources of useful resources and, often, reservoirs of fertile land, consideration needs to be given to them. The uses of trees, woodlands and forests are manifold. The greatest single use is still the provision of wood fuel for domestic use in the poorer nations: in 1991 the harvest was estimated to be $1.8 \times 10^9 \, m^3$ for that purpose, whereas industrial wood harvested amounted to $1.6 \times 10^9 \, m^3$. In peasant economies, the uses of wood are for fuel, construction, fencing, tools and

Table 28.4 World annual water use

Region	Total withdrawal, 1980 (km³)	Percentage of total renewable resources	Total consumption, 1980 (km³)	Percent change, 1900–80
Africa	168	3	128	126
North and Central America	663	10	224	594
South America	111	1	71	96
Asia	1,910	15	1,380	1,496
Europe	435	15	127	387
Former USSR	353	8	n/a	n/a
Oceania	29	1	15	28
World	3,320	8	1,950	2,741

Sources: After Kulshreshtha (1993), World Resources Institute (1992).

a myriad of other purposes; in industrial nations, construction and furniture are major uses but the predominant demand is for paper and paper products of all kinds (Mather 1990). The weight of these demands has led to the point where deforestation is proceeding at 10–20 times the rate of reforestation: in the last 10,000 years perhaps 33 per cent of the broad-leaved forests have gone at human hands, and 25 per cent of the savannah woodlands and sub-tropical deciduous forests. Until quite recently only 6 per cent of the tropical moist forests had been felled. Thus unless intensive production can be maintained from the current area, the global status of forests as a renewable resource is threatened.

Apart from their yields of wood, forests have other values for human societies. Unless they have no understorey and ground-layer vegetation, they can often be used to pasture domestic animals; they hold the soil and thus act as a protective cover for watersheds (a quality recognized by regulation in Tokugawa Japan, for example), and are reservoirs for wildlife. All the qualities of the forest come together in the popularity of the habitat for recreation of various kinds. Finally, forests may have regional climatic roles: in the Amazon basin, for example, much solar heat is used in evaporating the moisture produced by evapo-transpiration from the trees; without them the region would probably be much hotter. Globally, the forests are one of the sinks for carbon. Without their role in sequestering atmospheric carbon dioxide, for instance, the concentration of that gas would be rising even faster than at present, with world-wide climatic consequences. Forests, then, are an integral part of the human life-support system.

The proportion of the planet's water which is available as a human resource is small; about 0.3 per cent of the total fresh water. As a renewable resource (that is, excluding groundwater at a depth which is supplemented only on a geological time-scale), this amounts to about 41 million km³ or, in 1992, 7.42 thousand m³

Table 28.5 Uses of water

	Withdrawal (km³)		Consumption (km³)	
Use	1950	1980	1950	1980
Agriculture	1,130	2,290	859	1,730
Industry	1,768	710	14	62
Municipal	52	200	14	41
Reservoir	6	120	6	120
Total	1,370	3,320	894	1,950

Sources: After Kulshreshtha (1993), World Resources Institute (1992).

per capita, which amounts to 1,800 litres/capita/day. Such an average is exceeded by North Americans, for instance, who have access to 2,100 litres/capita/day. The use of water may be totally unconsumptive in the sense that it is not altered by uses such as power generation, flotation or recreation. On the other hand, it may be withdrawn from its original source ('intake') and only a part returned to the source in some form. The difference is called 'consumptive' use. So data presented as 'withdrawal' include those uses where some of the water is quickly returned to the river or other flow. Table 28.4 gives some of these data on a global scale.

Three main types of use can be identified: agricultural (mostly in irrigation); industrial (including food processing and the cooling of power plants); municipal (which includes domestic use and reservoir storage). As Table 28.5 shows, agriculture is by far the greatest use and thus poses the greatest problem because of the linkages to population growth and possible climatic change.

In absolute and relative numbers, the diversion of the hydrological cycle does not look very great. The highest ratio of supply to use is about 18 per cent (in Asia), with a world average of just over 8 per cent. A nation may already utilize a high proportion of its available fresh water. Examples (1990 data) include the United Kingdom at 82 per cent (the highest in Europe, with Belgium second at 73 per cent), Egypt at 111 per cent, Libya at 168 per cent, and Israel at 96 per cent. The Gulf oil states are all well in excess of 100 per cent but they can expend energy on desalination without much concern. So the nations which currently experience water stress and water scarcity (defined using criteria of supply per capita and the use–availability ratio) are the United Kingdom and Peru in the first category, and core nations of the Middle East together with Libya in the second. Using the best available assumptions about population growth, climatic change and industrial development, the more severe category might by 2025 bring in most of Europe, North Africa, the African Sahel zone and South Africa as well as the Indian sub-continent (Gleick 1993).

Water resources do not appear to be a global problem of the first order; there is untapped water and there are many possibilities for re-use of water if it is carefully managed. But that is not to ignore the fact that three-fifths of the population of the southern countries have no access to safe drinking water, let alone extra supplies for other purposes.

The living resources of the sea are without doubt the major focus of resource users. These are most common near the interfaces with the land areas and over the continental shelves, for in the open oceans the lack of mineral nutrients usually retard primary productivity by phytoplankton. Exceptions occur where cold upwellings bring mineral nutrients from the ocean floor to the surface, as with the Humboldt Current off Peru, whose capping by warmer water in some years gives rise to the El Niño phenomenon. The total carbon fixed in the oceans $(20–60 \times 10^6 \text{ g/yr})$ is about the same as that on the land; the offshore productivity is of the order of $50–170 \text{ g/m}^2/\text{yr}$ (and in upwelling zones $1,800–4,000 \text{ g/m}^2/\text{yr}$) and the open oceans no more than $100 \text{ g/m}^2/\text{yr}$, equivalent to a terrestrial desert.

Since oceanic food chains start with very small organisms, those which are actually harvestable by humans may be very high up those chains, with all that is thereby implied for energy loss from the ecosystem. A net primary productivity of $900 \text{ kcal/m}^2/\text{yr}$ of phytoplankton in the North Sea, for instance, becomes $0.6 \text{ kcal/m}^2/\text{yr}$ of adult carnivorous fish such as cod. So fish, which are the main renewable resource of the seas, may relatively easily be over-exploited, especially since they swim in shoals or lie on the bottom of the ocean bed. The global fish catch in 1950 was of the order of 20×10^6 tonnes, and in the late 1980s nearing 100×10^6 tonnes/yr; the Food and Agriculture Organization has suggested that 100 million tonnes/yr is probably the limit of sustainability. Of the current catch, about 35 per cent is converted to fish meal which is used as animal food. Fish are not the only catch: many other marine animals (molluscs, cephalopods and crustacea, mammals) and plants are also used; likewise, fishing in either artisanal or industrialized form is not the only method of exploitation, for aquaculture is widely used (and amounts to about 11 per cent of the total landings), especially in the brackish-water shore zone of Asia, often on land reclaimed from mangrove.

As an example of the key role of culture, the example of whales must be considered. There are now perhaps 200–1,100 blue whales left, whereas a century ago there were 250,000. Similar declines can be charted for all major whale groups (Cushing 1988). Nowadays, there is no need for killing whales, since their oil is not needed for lighting or for lubricants and their flesh is eaten in only a few cultures. Yet the international moratorium on whaling agreed in 1982 proved incapable of lasting for more than ten years, after which minke whales were again the target of catches by Japan, Iceland and Norway: partly to keep up employment but mainly, we might suspect, for the same reasons as others eat thick slices of beef or drive powerful motor cars.

The moral seems to be that all fisheries can be managed as renewable resources into any foreseeable future but that restraint in catch has to be exercised, otherwise fewer and fewer fish come to maturity and so the population declines (Cushing 1975). Three consequences may follow: one is that there is always a search for new stocks of hitherto unexploited animals (including squid and krill in recent years, for example); a second is more intensive production under controlled conditions of aquaculture; the third is the unplanned and querulous decline of a way of life which although dangerous had the sanctions of time and tradition.

NON-RENEWABLE RESOURCES

Apart from materials ejected into space by modern technology, the earth is a closed system. When, therefore, we talk of non-renewable resources we mean primarily inorganic materials together with hydrocarbon energy sources which are so transformed by their use that they are not immediately employable again by human societies. But in one form or another they are still present on the planet.

Major categories of resource

There appear to be three main categories: (1) those which are 'consumed' by use such as coal, oil and natural gas whose complex molecular structure is broken down into much simpler components; (2) theoretically recoverable materials such as minerals, which are technologically capable of being recovered after use; and (3) recyclable substances such as metals and glass which can be re-used without an enormous amount of reprocessing.

The general characteristics of non-renewable resources, in summary, are that they are usually products of the lithosphere, that they usually need complex processing before use (with linkages to energy consumption and the production of wastes), that they enter world trade and so are moved around the globe and have been much more important quantitatively since the nineteenth century, and that they become expended since so much of their use is a 'once-through' process. This raises the question of the optimal depletion rate: should it emphasize the perceived needs of future generations and thus conserve the material as much as possible, or will we do the best for our descendants by using as much as we wish in order to turn it into knowledge of how to do without it? The complex field of resource economics, for example, is much concerned with this last question.

How much of them is there?

In such analyses, a fundamental question about any non-renewable resource is always, 'how much of it is there?' This is not a simple question, if only for the reason that exploration of the earth's mineral resources, for example, is not complete. Indeed, the amounts available in practice at a given time depend upon five factors.

611

1 The availability of technological knowledge and equipment and their location in the right places and amounts.

2 Levels of demand, which encompass many constantly changing variables like population growth, affluence, tastes, government policies and the availability of alternatives.

3 Costs of production and processing. These reflect the nature of the material and its location, the state of the art of production as reflected in its costs, including those of energy but also capital, the rate of interest on loans, taxation, and the risks of being nationalized or terrorized.

4 End-price: this will reflect not only the factors above, but pricing policies of the producers and government subsidies or taxes.

5 The attraction and availability of substitutes, including the use of recycled products as against virgin materials.

Hence the resource is scarcely a fixed physical quantity (though this must exist) but a rather fluid economic and social construction. A common variable, for example, is price: as the price of a material increases it becomes more worth while for prospecting to take place, or for better methods of recovery to be developed. In this way, the recovery of crude oil from rock strata has risen from about 25 per cent in the 1940s to about 60 per cent in recent years.

Land resources

Each year there are land gains and losses, with the portions of the surface of the earth becoming a resource in the sense of a useful surface, or losing that status. Coastal erosion and deposition are the most obvious categories, but landslides and soil erosion are also significant. Some of these changes result from natural processes, as when cliffs of soft material are exposed to high energy seas; others result from human activity, as when for example coastal structures provide traps for silt and sand and thus build up ground above tide-levels. Occasionally, more spectacular losses occur, as when a volcano emits lava over former forests or cropland; the equivalent gains are made when a nation like the Netherlands dykes off large areas of coastal mudflat and saltmarsh for conversion to pastures and crops.

One exacerbation of land shortages is by degradation of the land surface to the point where it has little or no monetary value. Dumps of toxic waste, for example, can have no other function since they are dangerous and sterile; land prone to subsidence (due to fluid withdrawal or mining) may have a few uses but if it is unstable then it may just be left; and even in planned land-use systems, there may occur a planning blight in which land awaiting a change in function often grows nothing but weedy vegetation and discarded hypodermic needles. So although not a classic example of a non-renewable resource, land approaches the category of a material which is not being made by natural processes at anything like the same rate at which human societies are transforming it.

612

Mineral resources

These are typical non-renewable resources in the sense that the deposits from which they are taken are formed on geological time-scales of a totally different order from that of the human scales of their use. Possibly 90 per cent of the human population now depends upon minerals, not simply for industrial lifestyles but just for survival (Vanecek 1994).

About a hundred non-fuel minerals are traded and contribute about 1 per cent of world gross national product. Of these, there are twenty metals of considerable importance and eighteen non-metals of equivalent significance including aggregates, asbestos, clay, diamonds, fluorspar, graphite, phosphate, salt, cement, silica, and gemstones. Some are needed only in small, though vital, quantities like steel hardeners such as tungsten and wolfram. Some others rely for their importance upon rarity, as is the case with gemstones. Further, the use may seem humble but be very important: consider the use of metals in all phases of the food system, for example, from machinery through to processing and packing. The endpoint of this scale of values is the designation of some minerals

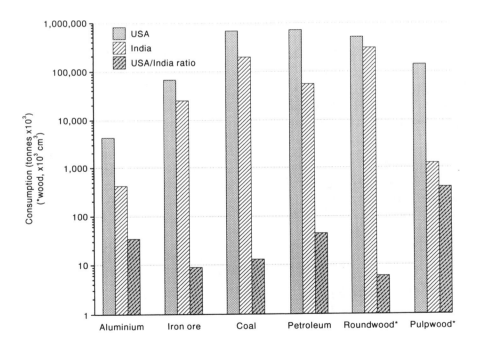

Figure 28.3 Natural resource consumption, 1991. Scale is logarithmic and is 10^3 tonnes, except wood, which is 10^3 cm^3. After data in World Resources Institute (1994).

as strategic minerals and the stockpiling of them against the kind of politically induced shortage on world markets that would impede the manufacture and use of military hardware.

In the past eighty-odd years, consumption of minerals has risen by a factor of twelve and it can be said with confidence that since about 1950 the world has consumed more minerals than in the whole previous history of humanity. Most of this has been located in North America and Western Europe (with Japan catching up fast) to the extent that the United States economy demands 20 tonnes/capita/yr of new minerals (Figure 28.3). The less-developed countries only take up about 10 per cent of the total, but their growth rates are higher at present (Blunden 1991). World trade in minerals in the past twenty years has exceeded gross national product growth by a factor of more than two and so Western Europe, Japan and the USA have come to be dependent upon imports.

The trends suggest more trade and less self-sufficiency in the consuming nations. The less-developed countries, Japan and the former Centrally Planned Economies are set to increase their consumption and in general this will lead to more large-scale production at the lowest possible cost except where environmental considerations are seriously taken into the reckoning. Exploration of less conventional sources will increase and there will be conflict over the exploitation of the resources of the ocean beds outside the Exclusive Economic Zones.

FACTORS OF EVALUATION AND CHANGES OF USE

The examples discussed above bring out certain characteristics of resource use which illustrate the renewable–non-renewable gradient. The first is substitution. In neo-classical economic theory this occurs when the consumer finds the price too high, but this assumes that there is a market in which preferences can be expressed. In the case of a species of insect not yet described by science, this is hardly likely to be the case. Some resources, such as clean air, have no effective substitute anyway. Hence, the operation of a free market mechanism is always flawed and it is here that government regulation and intergovernmental treaty become very important. Substitution may also occur 'upwards', so to speak, as when it becomes unfashionable among the smart set to go to Acapulco and they move on to Goa.

Reduce, repair, recycle

One feature of less-developed countries is that wastes are less prevalent since almost everything can find another use: picking over waste tips is very thorough. This is in contrast to the developed countries where the practices of planned obsolescence and rapid changes in fashion create 'wastes' at a rapid rate. Some revolt against this form of consumer behaviour has highlighted the desirability of reducing consumption, reusing materials and repairing rather than throwing things away. The motor car industry now claims that a very large proportion of the materials used can be recycled when the vehicle's life is over and many a car park has an overflowing waste paper skip. When one nation, such as Germany, decrees that most packaging has to be recycled, the neighbours find that it is cheaper for firms to import paper products for reprocessing than to collect them locally: the influence of traditional economics can be strong. One problem with wastes as resources is that once identified culturally as a 'waste', most households in the West prefer not to feel any responsibility of ownership: it becomes 'their' problem. This is amplified in the case of common-property resources such as the open oceans and areas of common grazing (Berkes 1989). Here, traditional pathways of allocation through folk courts, religious practices and orally transmitted mechanisms have in general not survived industrialization of the societies in which they were embedded and it became in everybody's individual interest to take a little more out of the system (a few more lobsters or graze a few more sheep) even if it undermined the biological basis of the resource. The seas have been the outstanding case, and the United Nations Law of the Sea Conference allowed nations to designate 200-mile Exclusive Economic Zones which they 'owned' and could manage more rationally.

Sustainability as shibboleth

Following the Bruntland Report (Bruntland 1987) and the Rio Conference of 1992, many national governments have committed themselves to the notion of sustainable economies (in the developed countries) and sustainable development (in the less-developed countries and those in transition to developing country levels). Sustainability is an attractive concept. It is usually based partly in a biological notion of an ecosystem which has arrived at a stable and self-renewing state so that it is resilient in the face of unpredicted change. It may also be referenced to the economic thinking in which sustainability consists of keeping up capital values, so that part of the wealth created by resource use is ploughed back into, for example, technological innovation. Thus one attitude to a newly discovered oil resource would be to use it as slowly as possible so as to eke out the material; the opposite would be to use it at a rate determined by the local markets but to make sure that some of the wealth was directed at finding out how to live without it. Using it all to fuel unemployment benefits triggered by the sudden advent of monetarist policies is neither, of course.

Sustainability is, however, a difficult concept to measure since it depends upon people's cognition of their resource situation as well as more objective measures (Redclift 1987). In particular, discussions of it often fail to mention any time dimension, as if it were an equilibrium level which, once achieved, would be in place for ever. A brief look at human history shows that in perhaps 9,000 BC, most hunter-gatherer societies would (had they formulated the concept, which is unlikely) have thought of themselves as following sustainable practices. Yet within 2,000 years, they were relegated to the margins of societies dependent upon domesticated plants and animals. Similarly, in 1750, most farmers in the world would have declared that solar-based agriculture would go on for ever. Within a very few decades, though, most parts of the world had, in some form, experienced what industrialization could do. These examples suggest that although sustainability may contain some useful short-term and localized aims, as a global target it has more value as a slogan than as an operational concept.

RESOURCE FUTURES

It is common to draw out two distinct types of resource futures (O'Riordan 1989), and these are delineated below. What is more difficult to organize into a coherent picture is the myriad of decision-making mechanisms which lead to the use or non-use of any resource. If a resource use is to be stable in the medium term, however, it ought to be capable of passing a number of decision filters.

Resource processes

The first of these is ecological – that is, the system from environment through resource use to waste disposal ought to be secure in the sense that there are no unpredictable instabilities in it. We might hope that the kind of delayed accumulation of toxins which happened in the case of chlorinated hydrocarbon pesticides or the biochemical transformation of mercury that caused Minamata disease, for example, are now foreseeable and avoidable. The second filter asks the question, is the process economic? With the breakdown of centrally planned economies, this increasingly becomes a question of a monetary profit being made at each stage. A stable system then depends upon appropriate pricing mechanisms. These must then include the whole process, including waste disposal, so that no cost-producing stage is externalized in the price structure; further, the costs must be realistic. A wood with a high amenity value, for instance, may be worth many times its timber worth, and ways have to be found of attaching suitable monetary figures to the more intangible use. The third filter could be labelled ethological. This refers to the behavioural aspects of any planned resource process: it has to be consonant with the behavioural patterns of the culture in which it is to be located. The way in which risk is dealt with can be cited as an example. How, it might be asked, is flood hazard to be dealt with if development of housing and factories takes place upon this area of low-lying but flat land? Is it acceptable in this culture that individual owners simply bear the costs in a stoic fashion? Or is it a feature of the culture that everybody must have insurance? (Lest this be thought an extreme example, recall that only about 7 per cent of private households in Kobe had earthquake insurance in 1995.) Lastly, there is the behavioural subcategory (but nevertheless often distinct in practice) of ethics. This addresses the question of what we *ought* to do. It was a feature of most cultures although overlain in the West during the technological triumphalism of the century after AD 1850. Now it is raised with considerable frequency in examples like whether we should burden successive generations with quantities of highly toxic plutonium which must be sequestered from all forms of life for at least 250,000 years, or the conditions under which calves should be raised before they are slaughtered; both of which involve the extension of moral consideration. In one case this is to the unborn; in the other to non-human animals.

Types of resource future

The first of these might be labelled 'business-as-usual' and foresees a continuation of present trends. It is essentially an optimistic trend, which regards population growth, for example, as wholly positive (1) since population pressure will force innovation and (2) people are seen as the 'ultimate resource'. The developed countries are seen as the driving forces in development from which everybody will in time benefit. Science and technology in particular are seen as the keys to better material standards, and more of these are clearly beneficial. In this scenario, the less regulation the better: free trade will allow those areas which can 'best' produce goods to do so and trade will ensure that everybody passes the economic filter. 'The news is good and it's getting better', said Julian Simon in 1994 when debating the case for this view of the world.

By contrast, there are the 'environmentalist' views which adopt an implicit model of an ecosystem with a limited carrying capacity. The environmentalists are usually neo-Malthusian and regard the levelling-off of world population growth as a first priority. Thereafter, the consumption levels of the West are seen as distorting the planet's ecology by causing the degradation of ecosystems at all scales from the very local to the global. In addition, however, they cause injustice among less-developed-country peoples by distorting the terms of trade, inducing high levels of debt and encouraging inequality of incomes when crops are grown for export rather than for subsistence. A major problem lies in the way individuals define themselves in terms of what they possess materially rather than, above a certain level of modest sufficiency, what they are.

THE CONTEXT

In the 1960s and 1970s, it was quite common to tease out these two virtually metaphysical positions and to demand adherence to one or the other. Whether in the world of the next twenty or so years such polarized positions will survive other, contextual, trends is a matter for speculation. The impact of various kinds of globalization, for instance, is unknown. Will global satellite television increase demands for material progress in less-developed countries to the point where world systems of trade and finance will have to be changed? Will the ubiquitous availability of information (said to equal power) allow less-developed countries to decide for themselves the kind of resource-use patterns they would like to have? Or will globalization finally confer upon the transnational corporations the powers formerly possessed by the nation-state, so that they determine what we eat and what we think? Is it likely that the countervailing force to globalization will produce a plethora of small ethno-cultural units whose resource demands will be different from those of the larger units which preceded them, or will they sign up to a few transnational corporations rather as a large company signs up to an outside caterer and a firm of contract cleaners?

This all raises the questions of how change in resource processes takes place, itself a subcategory of change in societies in general. There is little doubt that the environmentalist alternative has possessed many of the characteristics of a Utopia (indeed it is easy to find examples of the use of the term 'Ecotopia'); the business-as-usual scenario, on the other hand, has been much more incremental and open-ended. Bearing in mind that proponents of Utopias tend to be fanatical in their desires to implant their version of perfection, a more gradual change to implementing some of the more just and stable systems would seem helpful. What affects both positions alike are the effects of new technology and the changing attitudes of people, neither of which can be forecast in isolation let alone in synergistic interaction. Resource use in the future looks set to provide plenty of opportunity for intellectual speculation.

NOTE

1 1 MJ = 1 megajoule = 10^6 joules; 1 GJ = 1 gigajoule = 10^9 joules.

REFERENCES

Berkes, F. (1989) *Common Property Resources*, Chichester: John Wiley.

Blunden, J. (1991) 'Mineral resources', pp. 43–78 in J. Blunden and A. Reddish (eds) *Energy, Resources and Environment*, London: Hodder & Stoughton/Open University.

Brown, L.R. (1994) 'Facing food insecurity', pp. 177–97 in L.R. Brown (ed.) *State of the World, 1994*, London: Earthscan.

Bruntland, H.G. (1987) *Our Common Future*, Oxford and New York: Oxford University Press.

Cushing, D.H. (1975) *The Fisheries Resources of the Sea and their Management*, Oxford: Oxford University Press.

Cushing, D.H. (1988) *The Provident Sea*, Cambridge: Cambridge University Press.

Diamond, J.M. (1987) 'Human use of world resources', *Nature* 328, 479–80.

Eyre, S.R. (1978) *The Real Wealth of Nations*, London: Edward Arnold.

Findlay, A. (1991) 'Population and environment: reproduction and production', pp. 3–38 in P. Sarre (ed.) *Environment, Population and Development*, London: Hodder & Stoughton/Open University.

Gleick, P.H. (1993) *Water in Crisis: A Guide to the World's Fresh Water Resources*, New York: Oxford University Press.

Groombridge, B. (ed.) (1992) *Global Biodiversity: Status of the Earth's Resources*, London: Chapman and Hall.

Kulshreshtha, S.L. (1993) *World Water Resources and Regional Vulnerability: Impact of Future Changes*, Laxenburg, Austria, IIASA, RR 93–10.

Livi-Bacci, M. (1992) *A Concise History of World Population*, Oxford: Blackwell.
Mather, A.S. (1990) *Global Forest Resources*, Portland, Oreg.: Timber Press.
O'Riordan, T. (1989) 'The challenge for environmentalism', pp. 77–102 in R. Peet and N. Thrift (eds) *New Models in Geography*, Vol. 1, London: Unwin Hyman.
Redclift, M. (1987) *Sustainable Development: Exploring the Contradictions*, London and New York: Methuen.
Smil, V. (1991) *General Energetics. Energy in the Biosphere and Civilization*, New York and Chichester: John Wiley.
Smil, V. (1993) *Global Ecology: Environmental Change and Social Flexibility*, London: Routledge.
Vanecek, M. (1994) *Mineral Deposits of the World*, Amsterdam and London: Elsevier.
Vitousek, P.M., Ehrlich, P.R., Ehrlich, A.H. and Matson, P.A. (1986) 'Human appropriation of the products of photosynthesis', *BioScience* 36, 368–73.
World Resources Institute (1992) *World Resources 1992–93*, Oxford: Oxford University Press.
World Resources Institute (1994) *World Resources 1994–95*, Oxford: Oxford University Press.

FURTHER READING

Barbier, E.B. (ed.) (1993) *Economics and Ecology: New Frontiers and Sustainable Development*, London: Chapman and Hall.
Cushing, D.H. (1988) *The Provident Sea*, Cambridge: Cambridge University Press.
Gleick, P.H. (1993) *Water In Crisis: A Guide to the World's Fresh Water Resources*, New York: Oxford University Press.
Groombridge, B. (ed.) (1992) *Global Biodiversity: Status of the Earth's Living Resources*, London: Chapman and Hall.
Mather, A.S. (1990) *Global Forest Resources*, Portland, Oreg.: Timber Press.
Mather, A.S. and Chapman, K. (eds) (1995) *Environmental Resources*, London: Longman.
Myers, N. and Simon, J.L. (1994) *Scarcity or Abundance? A Debate on the Environment*, New York and London: W.W. Norton.
Rees, J. (1990) *Natural Resources: Allocation, Economics and Policy* (2nd edn), London and New York: Routledge.
Simmons, I.G. (1991) *Earth, Air and Water*, London: Edward Arnold.
Simon, J.L. and Kahn, H. (eds) (1981) *The Resourceful Earth*, Oxford: Blackwell.
Smil, V. (1993) *Global Ecology: Environmental Change and Social Flexibility*, London: Routledge.
World Resources Institute (1994) *World Resources 1994–95*, Oxford: Oxford University Press. (Available on disk.)

29

ENVIRONMENTAL HAZARDS

John Whittow

The 1990s have been declared the 'International Decade for Natural Disaster Reduction' (IDNDR) in recognition of the dramatic increase of recorded disasters resulting from floods, earthquakes, volcanoes, etc. during the last twenty years (Degg 1992). Moreover, during the same time-period human-induced catastrophes have increased in both periodicity and magnitude – for example, the Piper-Alpha oil-platform explosion, marine oil spills, the release of toxic gases at Seveso (Italy) and Bhopal (India) and the nuclear accidents of Three-Mile Island (USA) and Chernobyl (USSR), to say nothing of numerous transport disasters. If one adds to these the more insidious effects of acid rain, global warming, ozone depletion, deforestation and general resource pollution, it is little wonder that awareness of environmental hazards has never been greater (Whittow 1987a, 1988).

THE MEANING OF HAZARD

Definition of environmental hazards

Most modern geographers recognize the dichotomy which exists in our environment between the operation of physical (natural) systems and that of human systems of resource use. They also acknowledge that before there can be a better understanding of how environmental impact from hazards may be ameliorated there has to be an increased knowledge of the complex interaction between physical and human systems. Moreover, it is now understood that certain operational boundaries exist that can change systems (processes) into potential hazards which, if fulfilled, can cause disasters. It is only when society perceives the hazards as posing an unacceptable threat to life and/or property that critical behavioural responses or adjustments will be made (Whittow 1987b).

Everyone perceives hazards in different ways, and views will vary from place to place and over time. For example, so long as water is under control, in a

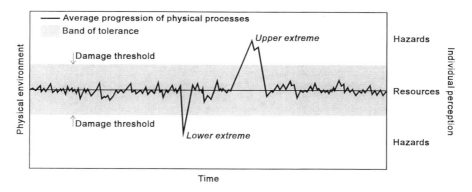

Figure 29.1 Sensitivity to environmental hazards expressed as a function of the variability of geophysical elements and the degree of socio-economic tolerance. Within the band of tolerance, events are perceived as resources; beyond the damage thresholds they are perceived as hazards. After Smith (1992).

reservoir, it will be seen as an important resource; once its volume deviates beyond the band of tolerance illustrated in Figure 29.1, it will rapidly be recognized as a flood hazard (Smith 1992).

Because of these fluctuating degrees of human perception there have been difficulties in arriving at a precise definition of environmental hazards, but the most explicit states that they are 'extreme geophysical events and major technological accidents, characterized by concentrated releases of energy or materials, which pose an unexpected threat to human life and can cause significant damage to goods and the environment' (Smith 1992: 16).

Classification of environmental hazards

There are several ways in which hazards may be classified. These can take the form of a basic division between natural hazards and human-induced hazards or a tripartite subdivision of hazards – for example, into those of endogenous origin (earthquakes, volcanoes); exogenous origin (severe weather, floods, drought) and anthropogenic origin (technological accidents). A more comprehensive classification is shown in Table 29.1 in which geophysical, biological, technological and lifestyle hazards are categorized.

Attempts have been made to compare the manifestations of certain hazards in order to show their relative magnitude and frequency at a world scale. Figure 29.2 demonstrates that because human-induced hazards occur more often they kill small numbers at fairly frequent intervals. Conversely, natural hazards result in greater death tolls but occur at less frequent intervals. What this diagram does not show is the relationship between the morbidity rate from these causes and that resulting from senility, disease or malnutrition. There is little doubt that at

Table 29.1 A classification of hazards

Geophysical		Biological		Technological		Lifestyle
Climatic and meteorological	Geological and geomorphic	Floral	Faunal	Transport	Industrial	Domestic/leisure
Blizzards and snow	Avalanches	Fungal diseases, for example:	Cancer	Air	Nuclear radiation	Fire
Droughts	Earthquakes	Athlete's foot	Bacterial and viral diseases, for example:	Marine	Fossil fuel (CO_2 release)	Smoking
Floods	Erosion (including soil erosion and shore and beach erosion)	Dutch elm	Influenza	Rail	CFC (ozone depletion)	Appliances
Fog	Landslides	Wheat stem rust	Malaria	Road	Mining/oil drilling	Gas
Frost	Shifting sand	Blister rust	Typhus	Automobile	Surface extraction	Electrical
Hailstorms	Tsunamis	Potato blight	Bubonic plague	Motorcycle	Construction industry	Mechanical
Heat waves	Volcanic eruptions	Infestations, for example:	Venereal disease	Bicycle	Plant explosion	Poisonous substances
Hurricanes	Ground surface collapse	Weeds	Rabies	Pedestrian	Structural failure	Skiing
Lightning strikes and fires		Phreatophytes	Foot and mouth		Bridge	Mountaineering
Tornadoes		Water hyacinth	AIDS		Building	Water sports
		Hay fever	Infestations, for example:		Dam	Motor sports
		Poison ivy	Rabbits		Tunnel	Aerial sports
			Termites		Fire	Contact sports
			Locusts		Toxic emissions	
			Grasshoppers		Pesticides	
			Venomous animal bites		Herbicides	
			Malnutrition		Groundwater pollution from:	
					Nitrates	
					Slurry	
					Silage	

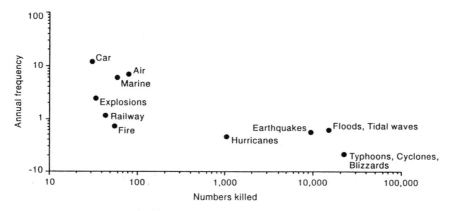

Figure 29.2 Magnitude and frequency of environmental hazards at a world scale.

a global scale the three latter causes account for by far the greater proportion of deaths but there is an interesting dichotomy between the developed and less-developed world.

More recent studies have shown that the poorer populations of the less-developed countries are becoming more disaster-prone while the developed world, although suffering increasing property and material losses, has cushioned itself against some of the worst natural hazards thereby mitigating the death tolls. North America and Europe, for example, although experiencing 44 per cent of disasters in recent decades have suffered only 3.2 per cent of the total loss of life. By contrast Asia, with 38 per cent of disasters, accounts for no less than 85.7 per cent of global death tolls. It has further been demonstrated that in the 1960–81 period each disaster event, on average, killed 3,700 people in low-income economies, 1,000 people in middle-income economies and 700 people in high-income economies (Swedish Red Cross 1984).

Calculations confirm that the world is becoming a more hazardous place, because the number of disasters reported world-wide is multiplying with each decade: 1960s (523), 1970s (767), 1980s (1,387). It must be realized, however, that the 1980s figures were inflated by the recent release of information by the former Soviet Union and China. A database, cataloguing all world-wide disasters outside the USA since 1900, is maintained by the US Office of Foreign Disaster Assistance (OFDA 1988). The criteria they use for disaster definition varies from >25 deaths for earthquakes and volcanoes, through >50 for severe weather phenomena, to >100 for human-induced accidents. Within the United States a different set of statistics has shown that in two decades (1965–85) no less than 3,726 disasters occurred which claimed 4,623 lives, although the federal government recognized a mere 531 of these as disasters and therefore gave aid to offset less than 25 per cent of the actual losses involved (Rubin *et al.* 1986).

Such data serve to highlight that the conventional techniques of measuring disasters by looking at death-toll thresholds and economic losses are very crude measures of environmental impact because they fail to take into account other less tangible effects on societies adjacent to the immediate disaster zone, the most pressing of which relates to the refugee problems. There are several studies relating to psychological stress, social disturbance and political unrest in the aftermath of disastrous events (Quarantelli 1978).

What must be emphasized is the fact that, globally, loss of life and property from hazards continues to escalate despite the phenomenal expenditure and effort that is being directed towards methods of prediction and alleviation of environmental hazards and despite the unprecedented programmes of disaster relief. Thus one must pose the question as to whether or not the task of hazardous impact reduction is insuperable. Before one is able to examine more fully the ways in which global societies attempt to cope with hazardous events it is necessary to examine the history of hazard studies.

The growth of hazard studies

The majority of early research, prior to the 1950s, was mainly concerned with descriptive accounts of catastrophic natural events with some indications of causality. Hewitt (1983) has shown how such beginnings have led to a widely held behaviouralist viewpoint that he terms the dominant view paradigm:

> The sense of causality or the direction of explanation still runs from the physical environment to its social impacts ... Few researchers would deny that social and economic factors or habitat conditions other than geophysical extremes affect risk. The direction of argument in the dominant view relegates them to a dependent position. The initiative in calamity is seen to be with nature, which decides where and what social conditions or responses will become significant.
>
> (Hewitt 1983: 5)

Hewitt believes that such a viewpoint emphasizes how risk assessment must be based primarily on geographical distributions of geophysical phenomena and their various spatial and temporal attributes. Indeed, many natural hazard management programmes, adopting the dominant view, continue to be based on the monitoring of seven major hazard variables which are capable of measurement:

1 *Magnitude* can usually be measured instrumentally and critical thresholds identified; for example, the Beaufort Scale of wind speed and the Richter Scale of earthquake magnitude.
2 *Frequency* relates to the expected recurrences of an event of a given dimension over a particular length of time; for example, the one-in-fifty-year recurrence of a damaging drought.
3 *Duration* refers to the length of time during which a hazard persists. For example, an earthquake lasts a few seconds, a blizzard for several hours, a

flood for several days, and a drought for several months or even years.

4   *Areal extent* which differs considerably, for example, between an avalanche and a hurricane.

5   *Speed of onset* relates to the time period between the first appearance of the phenomenon and the peak of the event. This variable is particularly important in planning hazard warning measures.

6   *Spatial dispersion* refers to the distribution pattern or path of the hazard. One can contrast the linear patterns of tornadoes, landslides/debris flows, etc. with the more diffuse patterns exhibited by drought, thunderstorms and by seismic/tsunami events.

7   *Temporal spacing* is the sequential spacing of events over time. Some are random (earthquakes, volcanoes), others, especially meteorological phenomena, are seasonal and/or cyclical in their occurrence.

On turning to the monitoring of technological and lifestyle hazards, it will be realized that because of the randomness of many of their temporal and spatial dimensions, a remedial programme is more difficult to erect. None the less, considerable expenditure has been directed at research into design and operational practices to lessen risk in the machine and the built environment.

In summary, the three main areas of effort that underpin the dominant view are seen as:

•   an understanding of natural processes in order to predict their likely impact;

•   a commitment to planning, managerial and technological measures to alleviate the impact by means of land-use zoning, building codes and 'fail-safe' artefacts;

•   a provision of emergency plans and relief agencies to deal with post-disaster rehabilitation programmes.

Hewitt (1983) points out that there are alternative paradigms to that of the dominant view, the most radical of which is based on a structuralist stance and is known as the theory of marginalization, in which the causal relationships between natural processes and people are fundamentally re-examined: 'It is time the popular belief in the "randomness" of disaster – the so-called "Act of God" enshrined in the law of contract and the popular mind – is questioned' (Susman *et al.* 1983: 276). Marginalization is defined as a continual process of impoverishment based on a world economy which perpetuates technological dependency and unequal exchange. Such a process is illustrated in Figure 29.3 where an underdeveloping population experiences a deteriorating physical environment and where reinforcement of the *status quo* through relief aid is seen to lead to further marginalization. Disaster mitigation which relies on the 'technological fix' is believed to exacerbate underdevelopment and it is concluded that the only way to reduce vulnerability is to place disaster planning squarely within development planning (O'Keefe and Conway 1975).

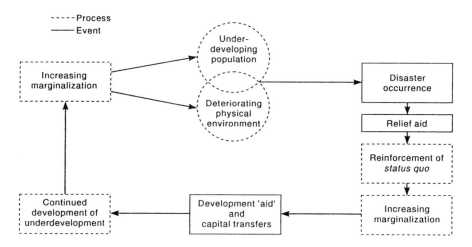

Figure 29.3 A model of the theory of marginalization, illustrating the causal relationships between physical processes and human beings. After Susman *et al.* (1983).

Some believe that the viewpoints are irreconcilable because there is still no agreement between those who believe that personal risk-taking should be tolerated in a free society and those who believe that people at risk have no real choice and are doomed to suffer from Acts of God. The majority of studies, however, agree that decision-makers and legislators (governments, planners, etc.) have a moral responsibility to reduce risks to levels that are socially acceptable. Which brings one to the difficult problem of defining risk.

The nature of risk

Because the degree of risk faced by society varies over time and from place to place it has been claimed that human response is a function of the complex relationship existing between physical exposure (i.e. to the operation of both natural and technological systems) and the amount of human vulnerability (i.e. the degree of societal tolerance). Although attempts are constantly being made to reduce the amount of risk susceptibility the inconsistency of individual attitudes towards hazards makes it difficult for acceptable safety standards to be set or for future comprehensive safety measures to be planned. 'This is because risk is not a physical quantity and what constitutes an acceptable level is a psychological problem' (Foster 1980: 16). Before a risk management programme can be implemented, therefore, the various degrees and types of risk have to be assessed.

At the outset a distinction has to be made between voluntary risks (smoking, skiing, etc.) and involuntary risks (earthquakes, floods, etc.). Figure 29.4

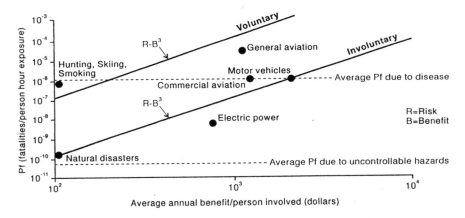

Figure 29.4 Various types of voluntary and involuntary human activities that involve exposure to hazard, showing the degree of risk (Pf) in relation to benefit (dollars) for the United States population. After Smith (1992).

illustrates how in the United States people appear to accept voluntary risks some 1,000 times greater than involuntary risks by consciously or unconsciously minimizing the risks involved in leisure activities. Despite the contention by Fischoff *et al.* (1981) that the term 'acceptable risk' is inappropriate because it really describes the least unacceptable option, four strategies to define acceptable risk levels have been suggested (Lowrance 1976). The first, termed 'risk aversion', involves a decision to achieve maximum risk reduction regardless of cost and without balancing costs and benefits. The second strategy, known as 'risk balancing', assumes that a modicum of risk is socially acceptable and is based on the concept of bounded rationality (Simon 1956) and on the belief of an individual's freedom of choice. Simon challenges the earlier economic model of profit maximization and replaces it by a decision-making based on satisfactory rather than maximized goals. Kates (1962) applied these ideas to hazard studies by showing how an individual's responses are usually based on value judgements related to experience and therefore display variously perceived alternative courses of action. In many Westernized urban societies, for example, acceptable risk standards are manifested in the amount of public and private funding already allocated to remedial measures – the revealed preference response. The third strategy is based primarily on the economic premise of cost effectiveness, suggesting that additional funding will cease when it results in significantly smaller increases in loss reduction. The final strategy is referred to as 'cost-benefit balancing' which determines both the risk level and the rigour of the safety measures by measuring the estimated social benefits against the estimated level of risk entailed by the operation of any system. Because many workers in the nuclear industry, for example, rank regular income and employment as overriding priorities they tend to balance such benefits by exposing themselves to a higher level of socially acceptable risk.

The four strategies, outlined above, have been thought to be too heavily biased towards cost–benefit analysis and have failed to place sufficient emphasis on the individual's idiosyncratic perception of personal risk acceptance (Starr and Whipple 1980). Nevertheless, most risk analysts dismiss data based on individual perceptions as being unreliable because they are too emotive, unquantifiable and less scientific. Thus, by and large, insurance company risk assessors determine their risk analysis levels by combining data culled from *probability theory* together with costings of the social consequences that resulted from the impact of past hazards. Three fundamental questions have been posed by Kates and Kasperson (1983) in an attempt to clarify and simplify risk assessment studies: first, what hazards may occur?; second, what is the probability of them occurring?; third, what losses are created by each event? Further questions have been suggested by Smith (1992): what happened after the event?; if remedial measures had been taken how effective were they?

Risk analysis is usually expressed by the formula:

$$R = p \times L$$

where risk (R) equates to the product of probability (p) and loss (L). However, as shown in the examples of hazards illustrated in Table 29.1 and Figure 29.2, hazards exhibit varying temporal and spatial dimensions, create different environmental impacts and, therefore, pose varying degrees of risk. Moreover, many less-developed countries, often the most disaster prone, lack the historical records from which both probability and loss could be calculated. If one adds to these problems that of attempting to assign a value to human life (Zeckhauser and Shepard 1984) then it will be realized how difficult it is to produce unambiguous rules on which such procedural measures as emergency preparedness and contingency planning can be based. A final complication relates to the fact that at a personal level individuals perceive risks very differently. Thus, while risk analysts often regard voluntary and involuntary risks as being of equal weight, the general public often fails to recognize voluntary risks and is concerned almost exclusively with involuntary risks. Moreover, since it has been demonstrated (Whyte and Burton 1982) that the public perceive the consequences (loss) resulting from a hazard as being of much greater significance than its probability, the standard risk analysis formula has been modified accordingly:

$$R = P \times L^x$$

where x is a power (of value greater than one) depending on a number of factors relating to individual perceptions (Table 29.2). From this table it can be assessed that risks are judged to be more serious if hazards are imminent rather than delayed and if they are spatially more concentrated rather than dispersed. It now

Table 29.2 Factors influencing an individual's perception of risk

Factors tending to increase risk perception	*Factors tending to decrease risk perception*
Involuntary hazard (radioactive fall-out)	Voluntary hazard (mountaineering)
Immediate impact (wildfire)	Delayed impact (drought)
Direct impact (earthquake)	Indirect impact (drought)
Dreaded hazard (cancer)	Common hazard (road accident)
Many fatalities per event (air crash)	Few fatalities per event (car crash)
Deaths grouped in space/time (avalanche)	Deaths random in space/time (drought)
Identifiable victims (chemical plant workers)	Statistical victims (cigarette smokers)
Processes not well understood (nuclear accident)	Processes well understood (snowstorm)
Uncontrollable hazard (tropical cyclone)	Controllable hazard (ice on highways)
Unfamiliar hazard (tsunami)	Familiar hazard (river flood)
Lack of belief in authority (private industrialist)	Belief in authority (university scientist)
Much media attention (nuclear plant)	Little media attention (chemical plant)

remains to be established to what extent an individual or a group is capable of adjusting to the implied degree of vulnerability suggested by such conclusions.

Response to environmental hazards

There is a considerable literature relating to the behavioural response to hazards and it appears that the majority of individual respondents to numerous questionnaires are either fatalistic (i.e. do nothing) or believe that it is someone else's responsibility to deal with any potential threat. But a marked distinction has to be made between those societies who are unwilling to adjust and those who are unable to adjust. In the developed world there are, in general, a greater number of available adjustments, due to greater technological expertise, greater wealth and greater mobility (due to car ownership). Contrast the case of the United States citizens along the Gulf of Mexico with those Ganges delta dwellers of Bangladesh. The former receive adequate warnings during the approach of a hurricane; they can choose to stay put or evacuate rapidly. The latter have virtually no choice in the face of a cyclone; they have only rudimentary warnings and virtually no mobility so are unable to evacuate and must bear the loss.

Many of the local or national authorities of developed nations prefer to alleviate potential loss from environmental hazards by undertaking preventive measures and/or corrective measures, depending on the degree of available funding. It is rarely possible to prevent the extreme natural event, despite experiments in seismic modification and cloud-seeding to influence severe weather. In the case of human-induced hazards, preventive measures have met with a much greater degree of success by the introduction of fail-safe mechanisms and stringent safety codes. Unfortunately, neither of these has been able to overcome human error which remains the greatest cause of accidents in the built and machine environment. Corrective measures have been widely introduced, especially by the wealthier nations who tend to become increasingly dependent on the 'technological fix' of structural adjustments in order to modify the damage susceptibility. Alternatively, there are many other ways in which the loss burden can be reduced as is illustrated in Table 29.3, which focuses on the available adjustments to the flood hazard.

In general, notwithstanding the greatest environmental disasters, world societies have survived because they have discovered ways of coping by means of adaptation and adjustment (Burton *et al.* 1978). Over thousands of years *Homo sapiens* has evolved by natural selection, a process that can be termed 'biological adaptation' to particular environments. Nevertheless, apart from inbred immunity to disease and increased adrenalin flow in times of danger, biological adaptation is far too slow to play any critical part in response to environmental hazard. Cultural adaptation, by contrast, is much speedier and can be exemplified by relocation of a settlement away from a floodplain or avalanche gully, or the change to a nomadic lifestyle to combat seasonal drought.

On turning to an examination of adjustment, a dual division can be identified between purposeful and incidental adjustments, although the number of adjustments adopted by a society may vary considerably according to the type of hazard and to the cultural stage achieved by that society. In Australia, for example, the federal government has introduced a ten-point plan to deal with the effects of the drought hazard on livestock farming (based mainly on technological innovation), whereas the Australian aboriginal will be limited to a mere handful of responses. A purposeful adjustment is one which is designed primarily to cope with a hazard, such as building above flood-level, planting drought-resistant crops or taking out insurance cover. An incidental adjustment is one that is not primarily hazard-oriented but whose spin-off will reduce potential losses. For example, an improvement of roads could increase the evacuation possibility, while the introduction of better communication systems (radio, TV, etc.) will increase the early-warning time. When the various ways of coping (biological and cultural adaptation; purposeful and incidental adjustment) are grouped together, four modes of coping can be identified, each mode separated by what Burton *et al.* (1978) term 'thresholds', respectively of awareness, action and intolerance (Figure 29.5).

The first mode of coping is loss absorption where a society absorbs the hazard impact and remains largely unaware of doing so. It has become second nature for an industrialized society, for example, to cope with traffic and factory hazards, although its ability to cope with a sudden natural hazard, such as an earthquake, may initially be slightly less than that of a poorly developed country

Table 29.3 Available adjustments to the flood hazard

Modify the flood	Modify the damage susceptibility	Modify the loss burden	Do nothing
Flood protection (channel phase)	Land-use regulation and changes	Flood insurance	Bear the loss
Dikes	Statutes	Tax write-offs	
Floodwalls	Zoning ordinances	Disaster relief	
Channel improvement	Building codes	volunteer	
Reservoirs	Urban renewal	private activities	
River diversions	Subdivision regulations	government	
Watershed treatment (land phase)	Government purchase of lands and property	Emergency measures	
Modification of cropping practices	Subsidized relocation	Removal of persons and property	
Terracing	Floodproofing	Flood fighting	
Gully control	Permanent closure of low-level windows and other openings	Rescheduling of operations	
Bank stabilization	Waterproofing interiors		
Forest-fire control	Mounting store counters on wheels		
Revegetation	Installation of removable covers		
Weather modification	Closing of sewer valves		
	Covering machinery with plastic		
	Structural change		
	Use of impervious material for basements and walls		
	Seepage control		
	Sewer adjustment		
	Anchoring machinery		
	Underpinning buildings		
	Land elevation and fill		

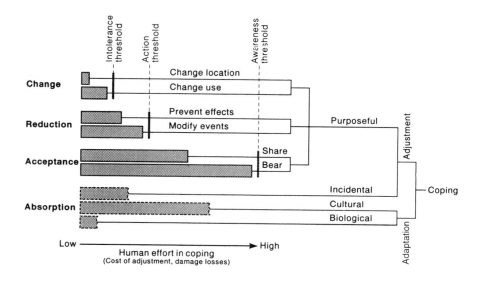

Figure 29.5 A model illustrating the various options (modes) of human adjustment to environmental hazards. After Burton *et al.* (1978).

because the former society has depended far too long on its technological 'cushion' which may have now been totally destroyed. Conversely, the wealth and expertise of an industrial society will probably offset this initial trauma and, in the long term, its absorptive capacity may well be greater than that in the less-developed nation.

The mode of loss absorption is separated from that of loss acceptance by the threshold of awareness. Once it has recognized the magnitude, periodicity or spatial aspects of the hazard a society can accept the potential loss by either bearing it or sharing it. Bearing the loss is the most common response (i.e. do nothing). However, unless the return period is very long term (e.g. catastrophic drought in Britain), societies often attempt to share the potential loss by seeking government aid, borrowing from the bank, or moving temporarily to share accommodation with family or friends.

Loss acceptance yields to loss reduction when a second threshold, that of action, is crossed, and positive measures are taken to reduce the loss. Many preventive actions are possible (see Table 29.3) but these involve significant behavioural responses; instead of taking a fatalistic stance or learning to live with the hazard positive action is taken, usually linked with varying degrees of capital expenditure. Ultimately, once the loss is thought to be of such magnitude that the threshold of intolerance is crossed, then the final mode of change may have to be adopted. This will involve such drastic decisions as permanent emigration to a different country or region, relocation of settlement, abandoning attempts

632

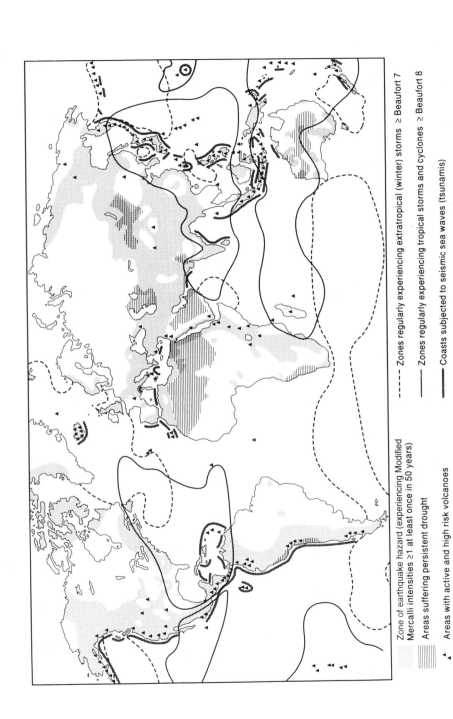

Zones regularly experiencing extratropical (winter) storms ≥ Beaufort 7

Zones regularly experiencing tropical storms and cyclones ≥ Beaufort 8

Coasts subjected to seismic sea waves (tsunamis)

Zone of earthquake hazard (experiencing Modified
Mercalli intensities ≥1 at least once in 50 years)

Areas suffering persistent drought

▲ Areas with active and high risk volcanoes

Figure 29.6 World map of major natural hazards.

to grow cash crops in an arid environment, or ceasing to abstract water from aquifers beneath such cities as Tokyo or Venice where crustal subsidence increases the flood hazard.

It will be demonstrated in the next section how different societies respond to various hazards by different approaches to loss acceptance or loss reduction measures. Figure 29.6 illustrates the world patterns relating to the major natural hazards examined below.

MITIGATION AND ALLEVIATION OF SPECIFIC HAZARDS

In the field of hazard mitigation geographers have contributed in three distinct ways: first, in order to predict the dimension of future events they have measured physical processes; second, by framing policy recommendations to decision-making bodies they have attempted to manage these processes in order to alleviate risks and/or regulate behaviour in accordance with the perceived risks; third, they have pursued studies aimed at improving disaster relief and reconstruction and rehabilitation schemes in post-disaster situations. Some of these achievements will be summarized below by examining specific hazards and suggesting how the losses may be mitigated by considering the following options: (a) modify the physical process (environmental control) largely by the application of engineering design and construction, an option which is more successful in the case of technological rather than natural hazards; (b) if it is impossible to modify the event, alleviate the impact by designing new hazard-resistant structures or modifying existing structures (retrofitting); (c) implement an emergency measures programme to minimize losses of life and property (preparedness). These will normally include a means of prediction, based on a statistical interpretation of historic records; forecasting, which is dependent on temporal monitoring of the actual event; dissemination of warnings and controlling of evacuation procedures; finally, as part of the emergency measures programme, the decision-making authority would be advised to take into account land-use zoning, to modify the degree of vulnerability in the face of environmental hazards.

Earthquakes and tsunamis

So much has been written on the spatial characteristics and mechanisms of seismic hazards that it would be impossible to even summarize the extensive literature (Bolt 1988; Smith 1992; Dudley and Lee 1988) in the present review. Instead, it will be shown the degree to which various adjustments have succeeded. Although the earthquake-prone nations of Japan, the USA and the USSR have expended substantial amounts on seismic prediction the overall success rate has remained no better than 50 per cent. Not surprisingly, because of the failed predictions credibility has been strained and many authorities suspect a growing reluctance by the public to respond to early warnings (Olson *et al.* 1989). The poor success rate is due largely to the random nature of the hazard, so far as magnitude and timing are concerned, although certain regions such as California have a plethora of sophisticated geophysical instrument arrays on all known faults. Despite the initial success of the so-called 'dilation theory' expounded by Russian scientists (in which the arrival-time pattern of seismic waves at a station gives warning of both imminence and magnitude) it has subsequently proved incapable of predicting very many earthquakes. A two-week warning period is necessary to evacuate a major city, although it has been claimed that the longer the warning period the greater the economic blight caused by the enforced closure of business and industry. Compensation for loss of output could prove financially crippling if the prediction failed, and expensive litigation would be necessary to decide where the responsibility lay. It has been argued, therefore, that seismic-prediction funding should be diverted to finance improved building design, to the drafting of accurate risk maps (Keller 1982) for land-use zoning, and to the planning of disaster-relief programmes. Although such high-risk structures as nuclear power-stations and dams have usually been designed to withstand earthquakes there have been many instances of poor construction standards leading to high death tolls from seismic shocks of relatively low magnitude, especially in less-developed countries (Whittow 1989; Degg 1992). The few expensive earthquake-proof buildings already constructed are, not surprisingly, in California and Japan (Ambrose and Vergun 1985), and it is these regions that have also experienced the most widespread retrofitting programmes (Haskell and Christiansen 1985). Engineering works to offset the impact of tsunamis (seismic-generated ocean waves) have been built on many vulnerable Pacific coastlines (Iida and Iwasaki 1983). The UNESCO guidelines for new building codes in seismic zones (e.g. replacing top-heavy stone or tile roofs with wood or corrugated iron) have largely been ignored. Moreover, the United Nations' much-heralded earthquake reconnaissance missions, meant to co-ordinate developing nations' programmes of preparedness, have achieved less than expected (UNDRO 1984). Programmes at a local level, by contrast, have achieved a much greater degree of success (Spangle 1988); likewise, the international co-operation on tsunami warning systems, based on satellite communications (Bernard *et al.* 1988), illustrates what success can be achieved at a global level.

Volcanoes

Volcanic activity, whilst creating some of the most visually spectacular of the geophysical hazards, remains one of the most complex of the physical processes because of the differences in character of the activity, the variety of the primary and secondary hazards and the disparity of the degrees of risk associated with each of these hazards. The mechanisms, distributions and phenomenology of volcanoes have been adequately discussed in Blong (1984) and Bryant (1991). This summary, therefore, will concentrate on the assessment and monitoring of the potential social and economic impact of vulcanicity and examine aspects of prediction, vulnerability and modification of risk. Despite the frequency of volcanic eruptions it has been calculated that during the twentieth century death tolls from volcanoes averaged 640 per year, although property damage during the same period was no less than $10 billion (Blong 1984). To put these figures into perspective, however, losses from the flood hazard caused some 5 million deaths between 1860 and 1960 in China alone; similarly a single 1972 flood in the United States resulted in $3 billion property damage.

A subdivision may be recognized between primary and secondary volcanic hazards: the former includes lava flows, pyroclastic flows, and tephra (ash) clouds and airborne projectiles. Secondary phenomena include volcanic gases, acid rain, *lahars* (mudflows), landslides, tsunamis and short-term climatic change due to dust veils (Lamb 1970; Stothers *et al.* 1988). All of these hazards have relatively local impact with the exception of the global climatic cooling imposed by volcanic dust circulation at stratospheric levels from such gigantic explosions as Laki in Iceland (1783), Tambora, Indonesia (1815), Krakatoa, Indonesia (1883) and Pinatubo, Philippines (1991). Because of the fertility associated with volcanic soils it is hardly surprising that settlements have frequently been located on steep volcanic slopes throughout historic time and increasingly so at present when population growth and land shortage combine to force people into hazard zones. Once a volcano passes into a dormant phase public perception of risk decreases as time elapses, although vulnerability may increase as large numbers are lured into the hazard zone. Only in a few cases, such as the volcanic islands of the West Indies Lesser Antilles (Roobol and Smith 1988) are the public kept constantly informed of volcanic risk. Paradoxically, when volcanoes remain active (e.g. Etna in Sicily) the known risks are virtually ignored by the vulnerable population, despite the high degree of risk.

Some environmental control has been attempted where property is threatened by lava flows or *lahars*, mainly in the form of diversion barriers, but other attempts to reduce the operation of the process (i.e. aerial bombing and water-jet cooling) have been largely ineffectual. Most authorities, therefore, have preferred to invest in such measures as reinforced roofs to withstand the weight of ash falls but, more particularly, by formulating programmes of community preparedness. Among the latter is a training programme for scientists of developing nations at the Centre for Study of Active Volcanoes, Hawaii, to assist them in mitigating risk by producing emergency preparedness procedures (Anderson and Decker 1992).

Great progress has been made in volcanic monitoring and prediction studies (see Latter 1989; McCall *et al.* 1992). Risk maps have been prepared for some of the most hazardous volcanoes and these have given local authorities the option of zoning land use and settlement. Evacuation procedures vary from country to country, due to the disparities in scientific monitoring and partly to lack of adequate procedural organization. Contrast the orderly evacuation of the vulnerable population around Mt St Helens in 1980 with the chaotic evacuation and instant re-opening of the hazard zone to the public in Guadeloupe in 1976, despite the prediction of a catastrophic eruption which, fortunately, did not materialize immediately.

Mass movement

The downslope movement of surface materials in response to gravity is one of the commonest of the environmental hazards and, therefore, because of its frequency, causes regular death tolls and substantial property losses every year. A broad classification can be made, according to the dominant character of the material, into rockfalls, landslides and snow avalanches. Statistics have shown that the death tolls from rockfalls and landslides are increasing because growing populations are being forced to occupy unstable slopes due to land shortages. This is particularly true in developing countries, where slope instability is exacerbated by profligate husbandry which often leads to soil erosion. Avalanches, by contrast, although less frequent in occurrence, are often triggered by the development of ski resorts and the attendant pressures of recreational pursuits (Simons 1988). None the less, it has been claimed that in several European mountain terrains some of the coniferous woodland, which serves to 'anchor' the slopes and protect the settlements, is not only being felled but is also dying from the effects of acid rain.

Rockfalls can occur on any slope steeper than 40°, but particularly where bedding planes are inclined steeply in the direction of slope, as in the case of the 1903 Turtle Mountain rockside in Alberta, Canada which buried most of a mining town. In addition to the slope gradients and the geological structures, other factors which contribute to the type of slope failure relate to the character of the vegetation cover and the pore-water pressure in the surface materials which in turn reflects the magnitude and frequency of precipitation. Moreover, the degree of seismic activity in the region is often a significant factor in the generation of slope failure, as shown in such mountainous terrains as the Himalayas, the Andes and parts of China. The world's most destructive mass movement (part avalanche and part debris and flow slide), which killed 25,000 people in the foothills of Huascaran, Peru in 1970 was triggered by an earthquake (Plafker and Ericksen 1978).

Landslides can occur on slope gradients much greater than 40°, but are commonest where slopes are steeper than 25°, especially where 'weak' or incompetent argillaceous (clay-rich) rock strata occur. Prolonged rainfall causes increased pore-water pressure which in turn can induce stress and possible slope failure, especially if there has been interference from human activity and additional loading by building construction on the upper slopes. Catastrophic landslides are common, therefore, in such heavily urbanized steep terrains as Hong Kong, Japan, Rio de Janeiro and Los Angeles. It has been pointed out, however, that death tolls cannot be used simply as a surrogate for total environmental impact. As in most hazards, economic losses due to damage and disruption are also of considerable significance (Jones 1992). In California, for example, notorious for its seismic hazard, landslides account for more than one-quarter of anticipated hazard losses, second only to those from earthquakes (Alfors *et al.* 1973).

Snow avalanches, like rockfalls and landslides, are related to steep gradients and the properties of their constituent materials, in this case the temperature and density of the snowpack (Schaerer 1981). Rises of temperature, increases of meltwater, heavy snowfall or rain can all lead to significant snowpack changes, while the aspect and shelter variables of north- and south-facing slopes can create contrasting risks even on the same mountain. Slopes above 60° are generally too steep for accumulation, those below 22° too gentle for dry-snowslope failure. Thus most avalanches occur on gradients of 30–45°, especially on *ubac* slopes (those sheltered from the sun), because snow stabilizes more slowly in lower temperatures.

On turning to an examination of hazard mitigation and risk reduction, one finds that a whole variety of technological options is available. Expensive remedial measures of slope stabilization have been widely introduced to modify rockfalls and landslides, including rock-strengthening, excavation, re-grading, filling, drainage installation and restraining structures. Some nations, notably Japan and the United States, have introduced strict land-use zoning ordinances relating to landslide susceptibility maps (Brabb 1984), the latter being based on various geotechnical, hydrological, climatological, biological and pedological factors. Similarly, snowpack can either be stabilized or artificially released into a controlled avalanche by a variety of preventive measures. Alternatively, strengthening, deflecting or retarding structures can be built to lessen the impact of the event in the vulnerable areas. In the most extreme cases tunnels and bridging structures are introduced or even a complete relocation of settlements is considered. Many national governments and local authorities have implemented forecasting and warning systems in vulnerable alpine areas, with those of Switzerland being the most sophisticated and extensive. The Swiss have also introduced an important land-use zoning system (Perla and Martinelli 1976) to mitigate avalanche risk.

Floods

Floods are the commonest of the natural hazards and in recent decades have accounted for no less than 64 per cent of all death tolls resulting from natural hazards. Moreover, in the USA during the twentieth century material losses from floods have risen at about 4 per cent per year, to say nothing of the enormous increases in the costs of the flood-control structures.

This is the price that mankind has to pay when attempting to compete with rivers for the use of their floodplains or when building on vulnerable coastlines. In earlier centuries pre-industrial societies had adapted to the flood hazard by seasonal migration and/or location of settlement above the hazard zone. This is no longer the case as world population has escalated and growing numbers have become marginalized or become dependent on artificial structures for their protection. For example, the National Rivers Authority in Britain spends £250 million each year building and maintaining its 40,000 km of flood defences. None the less, since the 1960s flood management policies are moving away from reliance on engineering structural adjustments for four main reasons: first, increased expenditure on engineering works has failed to decrease flood losses; second, over-optimistic belief in the infallibility of such technology has increased the risk by encouraging people to settle in hazardous areas; third, the costs of these structures are making them too expensive, in contrast to those of the non-structural adjustments; fourth, the improvement of flood prediction and warning, more rigorous land-use zoning and the greater availability of insurance have combined to enhance the use of alternative adjustments.

Adjustments to the flood hazard can be grouped broadly into three methods of coping: modify the flood; modify the damage susceptibility; modify the loss burden; alternatively, one can do nothing (see Table 29.3). Structural adjustments to modify the flood can themselves be classified into first 'corrective measures' in which engineering schemes modify the river course to fit the flow (embankments, channel enlargement, flood-relief channels and intercepting channels) or alternatively, regulate the flow to fit the course (reservoirs and washland storage). A second group of structural adjustments, termed 'flood abatement schemes', is dependent largely on management changes in the catchment area, including afforestation and modification of cropping practices, to delay runoff and reduce flood peaks downstream. The third group, known as 'flood-proofing measures', can be permanent or merely temporary. The former includes waterproofing cellars, raising buildings and installing pumps, while the latter depends entirely on a realistic warning time.

Of the non-structural measures, a fourfold classification can be recognized: forecasting and warning; floodplain zoning; flood insurance; public relief funding. In recent years considerable emphasis has been placed on flood-forecasting schemes (Penning-Rowsell 1986). Scientific advances have been made from initial dependence on empirical probability models (e.g. 'the 100-year event') by the introduction of satellite weather monitoring and radar scanning of precipitation. Flood forecasting is most successful on large river catchments, like that of the Mississippi, in which several weeks' warning can be given. In England and Wales, by contrast, over half the vulnerable dwellings have less than six hours' warning time. Most developed countries rely on automated telemetry for data transmission, but in countries like Bangladesh no more than twelve hours' warning is given of river flooding because of their less sophisticated radio and telephone links. Studies have shown, however, that the previously held assumption, that floodplain residents will respond rationally to a warning, cannot be justified. In Britain, for example, only about half of those given flood warnings made any positive response, due partly to previous false alarms (Penning-Rowsell *et al.* 1978).

Since it is often impractical because of costs to relocate every existing vulnerable floodplain settlement a much more acceptable alternative is that of zoning future development (Kates 1962). A tripartite division is often adopted: a prohibited zone; a restricted zone where only essential properly engineered buildings would be allowed; a warning zone where limited development could occur but where inhabitants should be given regular flood information. Certain drawbacks remain, however, partly because of the difficulties of defining accurate boundaries, partly due to the high-risk zones becoming slums, and partly due to the false sense of security engendered in the warning zone. A final problem has been highlighted by the National Rivers Authority in Britain. Because of its lack of power over local planning authorities the latter continue to give planning permission for housing in flood hazard zones.

Flood insurance simply allows the losses to be offset over a number of years, although high-risk zones will always attract high premiums. It has been claimed that in such zones compulsory insurance would ensure that only economically viable development would be built, but this would rely heavily on careful land-use zoning. Attempts are still being made to compute a reliable link between risk and insurance premiums (Arnell 1983), but many assessments are still based largely on probability theory.

Although public funding following catastrophic flooding is not uncommon there is a dichotomy between those who believe that public funds should increasingly be used and those who say that taxpayers should not be expected to underwrite losses for the uninsured who choose to live in flood-prone areas. This latter argument is not defensible, of course, when applied to marginalized communities in less-developed countries where no other funds or insurance is available. Conversely, in Western Europe for example, disaster aid is available either from short-term government loans, local charities or occasionally from the EEC Disaster Fund, in the event of a major disaster.

Drought

Droughts and associated famines are not new; there are records from all continents suggestive of periodic desiccations from long before written records began. But today some 15 per cent of the world's population live in the arid and semiarid zone which, some experts argue, is increasing annually. For example, between 1931 and 1990 records show that 63 per cent of Africa has become drier and that the arid zone has increased by 54 million ha (1.8 per cent of Africa's total area). Conversely, Africa's humid zone has decreased by 26 million ha in the same period (0.9 per cent of Africa's total area). Since 1984 starvation has threatened no less than 150 million Africans in twenty-four countries south of the Sahara, but while most would agree on the effects of aridity there is no agreement on its causes. Some scientists point to global warming, others to simple cyclic meteorological fluctuations, but there is an increasing body of opinion supporting the contention that many of the population problems in the arid zone are due to 'marginalization', inept animal and crop husbandry and political corruption. Whatever the causality, droughts often trigger catastrophic famines and today malnutrition is regarded as the world's most widespread disease (Garcia and Escudero 1982).

A classification of drought typology has been devised by Smith (1992) who recognizes four types:

1   Meteorological drought due to precipitation shortage but with no lasting ecological or economic impact and no effective behavioural response.
2   Hydrological drought which depletes water resources and leads to more rigorous management control of supply and demand.
3   Agricultural drought where widespread impact affects farming production and may lead to government compensation.
4   Famine drought, confined mainly to less-developed countries, often wipes out subsistence agriculture and leads to catastrophic mortality and massive international relief programmes.

There have been several major conferences on desertification in addition to ongoing research funded by such organizations as FAO, UNESCO, UNICEF and the World Bank. Geographers have contributed to the numerous volumes of explanation, analysis and advice, but because of the widespread nature of the hazard the practical progress on the ground has been minimal and more related to short-term crisis management inputs than to more lasting effects. Short-term adjustments include food aid, financial and medical support, and in some instances UN-backed military intervention (e.g. Somalia 1992). Long-term adjustments, by contrast, depend more on the 'technological fix', involving the expensive construction of reservoirs, pipelines, desalination plants and irrigation schemes. While all of these are favoured by the oil-rich Arab nations in the arid zone, few can be adopted by the poorer nations without considerable foreign aid. They are forced, therefore, to examine alternative, non-structural, adjustments such as growing drought-resistant crops, lowering numbers of grazing livestock, improving land-use and marketing regimes, all of which could conserve soil and groundwater resources. Low-cost technology, such as boreholes, can be more effective than costly engineering structures although care has to be taken not to decrease groundwater supplies or increase soil salinity by profligate use. Drought-stricken countries, however, will have to realize that in the main their problems arise more from a malfunctioning of their socio-political systems than from poor technology.

It has to be remembered that the first three of the drought types recognized above have also affected many of the developed nations, albeit to a lesser degree. The 1979–83 drought was the most severe of the many to have affected Australia during the last century (Gibbs 1984) despite their federal government's recent ten-point plan which combines training, technological, financial, meteorological and agronomic advice with bank loans, transport subsidies and compensation schemes. Some of the western regions of the USA are also permanent hazard zones, and major economic disasters followed the droughts of the decades starting in 1890, 1910 and 1930, the last of which were the infamous Dust Bowl years. Because of improved agricultural techniques, better marketing and substantial federal aid, the droughts of the 1950s and 1970s made a less dramatic impact, despite the massive agricultural losses – although these are now covered by insurance. Southern Britain, too, has begun to suffer periodic meteorological and hydrological droughts from 1976 onwards, though these have been largely a source of inconvenience rather than a real catastrophe (Whittow 1980).

Remote sensing from satellites has led to a remarkable advance in the monitoring and prediction of drought during the last decade. Plant and crop growth, soil moisture and groundwater deficits can now be assessed more accurately and action programmes devised more quickly. The problems remain, however, that the implementation of the schemes in the vulnerable areas is not always possible due to political pressure, distance, lack of local skills and above all prohibitive costs.

Severe wind storms

An examination of Figure 29.6 will demonstrate the extent to which large tracts of the globe are affected seasonally by tropical storms or extratropical depressions. On average 30,000 people die and some \$2.3 billion property losses are caused each year from wind-storm events (Housner 1987). Most of these severe-weather disasters are associated with atmospheric vortices and it is generally true that the magnitude of the hazard depends largely on the amount of energy released from within the revolving storm system. Thus, their kinetic energy (expressed in joules) has been calculated as follows: dust devil 4.10^7; tornado 4.10^{10}; thunderstorm 4.10^{12}; hurricane 4.10^{16}; tropical cyclone 4.10^{17}. Severe tropical storms (termed hurricanes in the western hemisphere and typhoons/cyclones in the eastern hemisphere) are associated with windspeeds of 350 kph (220 mph), rainfall approaching 80 cm (30 in.) and storm surges of 8 m (25 ft). In the vulnerable over-populated, low-lying coasts of the Bay of Bengal death tolls exceeding 300,000 are not uncommon from single storms in countries such as Bangladesh where storm surges regularly inundate vast tracts of the delta region. In Western Europe, by contrast, extratropical cyclones, though causing severe disruption and property damage, are associated with relatively low death tolls. Despite the catastrophic storm-surge losses in Britain and the Low Countries in the 1953 disaster, people have returned to the hazard zones and statistics show that human vulnerability to severe wind storms in developed countries continues to grow. In the United States, for example, more than 6 million people now live in hurricane-prone regions (Brinkman 1975), although there are more fatalities from tornadoes (113/yr) than hurricanes (75/yr). None the less, average annual US property-damage figures show a reversal, with tornadoes accounting for $75 million and hurricanes some $500 million. A contrast is again apparent in their periodicity: while only a handful of hurricane events threaten the USA each year, more than 900 tornadoes – some with windspeeds of 500 kph (310 mph) – are recorded annually.

Progress in severe-weather forecasting has been spectacular in the last few decades, partly by the development of world-wide instrumental networks and partly by the increasing sophistication of remotely sensed satellite photography. World Weather Watch (co-ordinated by WMO) has linked some 9,000 land stations, numerous weather ships and more than 140 national weather centres in recent years. Moreover, the Global Atmospheric Research Programme, by analysing satellite photography of Atlantic cloud formations, has contributed to the greater success for hurricane prediction in the Caribbean and Gulf of Mexico. Prediction of devastating storms involves forecasting at a variety of time-scales: long range (>10 days), intermediate range (3–10 days) and short range (1–3 days). Radar tracking of tropical storms and tornadoes has saved millions of lives, especially where people could take advantage of early warnings as in Galveston, USA (1969 and 1983) and Darwin, Australia (1974). The lack of such sophisticated equipment in less-developed countries has undoubtedly exacerbated the number of fatalities.

In common with other natural hazards, severe wind hazards have been the subject of much research relating to avoidance and mitigation. The former includes land-use zoning and local ordinances which restrict development on low-lying coastlines, together with improved programmes of evacuation procedures. The latter, involved with mitigation of the impact, includes improved structural design of artefacts, better building codes, construction of storm shelters, and the installation of tornado and hurricane warning watches by the local authorities. The Advisory Committee for the IDNDR in the USA has published an eleven-point plan of potential projects: improvement of global weather networks; better international co-ordination of research; post-storm assessment of building codes; mapping of hurricane windfields; improved building-design practices; improved construction practices; the role of insurance; storm-surge risk production; storm-shelter design to obviate costly evacuation; evacuation procedural planning (Housner 1987).

Temperature extremes

Excessive deviations of temperature during winter and summer seasons can cause severe disruptions to all types of social and economic activity and in many instances lead directly or indirectly to substantial losses of life (Perry 1981). Considerable falls of temperature in winter produce such hazards as blizzards, frost, fog and permafrost damage by periodic surface heave and subsidence. Large rises of temperature can cause death by heatstroke and also threats from wildfires. Most statistics related to extreme-temperature hazards fail to take into account fatalities caused by associated health hazards such as hypothermia, pneumonia and asthma (the latter associated mainly with fog and smog hazards).

Much research has focused on the economic losses resulting from frost impact on agriculture and on the possible techniques for alleviating the losses on such vulnerable assets as citrus groves, vineyards, orchards, etc. (Bush 1945). Furthermore, the increasing degree of disruption to road transport by snow and frost has led to many research projects and reports on transport hazard mitigation (Rooney 1967). The use of de-icing agents such as salt, on roads, whilst cutting down the number of accidents, has also been shown to be extremely costly, labour-intensive, operationally inefficient and counter-productive in terms of vehicle and road-structure deterioration. This is particularly true in mid-latitude maritime countries such as Britain where the average winter temperatures oscillate frequently around freezing point ($0°C$), causing many freeze–thaw cycles (Thornes 1985). In North America snow and ice control measures cost some $1.5 billion during an average winter, and salt corrosion has damaged more than 100,000 bridges (Hilberg et al. 1983).

Deaths from heatstroke have been documented, for example by Quayle and Doehring (1981), and the fatalities and destruction caused by wildfires (bush fires) have been described by Monteverdi (1973) in the United States and by Bryant (1991) in Australia; the latter require special synoptic conditions found only in certain parts of the world such as California, the Mediterranean and south-eastern Australia. Most frequently the wildfire impact is greater on the floral and faunal ecosystems than on the human population, but material fire damage and fire suppression measures in one single Australian event have been costed at A$200 million (Bardsley et al. 1983).

Lack of visibility due to fog is the factor most likely to cause transport disasters on land, at sea and in the air but, additionally, air pollution due to fog and photochemical smog can also be a killer (Whittow 1980). Moreover, loss of revenue from a single fog at a major airport can be over $500,000, while fog-induced road accidents cost the United States more than $300 million annually. Chronic smoke fogs were commonplace in large coal-burning cities throughout the nineteenth and early twentieth centuries during which layers of polluted air (hydrocarbons, CO_2 and SO_2) were often trapped for many days beneath temperature inversions. The December 1952 London fog culminated in the world's worst poor air-quality disaster when upwards of 4,000 people perished in excess of the normal mortality rate. Photochemical smog (peroxyacetyl nitrate or PAN) is created by sunlight reacting with vehicle exhaust and other air pollutants and is commonplace in many world cities with heavy traffic and high sunshine hours. Los Angeles, Tokyo, Milan and Athens have recently been joined by such cities as London and Sydney. In many countries pollution control and smokeless zones have been introduced into the worst-affected cities and in Britain the Clean Air Act of 1956 has led to better urban air quality. None the less, as will be illustrated in the next section, air pollution remains a serious environmental threat globally.

Human-induced hazards

There seems little doubt that more people are affected by technological and lifestyle hazards than by natural hazards (geophysical and biological), simply because of the contrasting incidences of natural and non-natural hazards (see Figure 29.2). What is less obvious is the degree to which human activities are increasingly influencing the magnitude and the rate at which physical processes operate, thereby turning some benign physical systems into threatening environmental hazards. Although human interference has been shown to increase the risk of soil erosion, coast erosion, avalanches and landslides the same cannot be said for the other geological and geomorphological processes (see Table 29.1). On turning to climatic and meteorological processes, however, there is a great deal of evidence that humans are seriously affecting the atmosphere (Thompson 1992). Global warming due to the release of heat-trapping gases (the greenhouse effect) poses a threat to many millions dwelling on low shorelines owing to a rising sea-level. The depletion of stratospheric ozone, associated with the use of chlorofluorocarbons (CFCs) has serious implications for skin cancer due to increasing ultraviolet solar radiation. An international agreement to control the use of CFCs is known as the Montreal Protocol of 1987.

Accelerated acidification (often termed 'acid rain') has been recognized as an environmental hazard since the 1960s, leading to the pollution of lakes and the decay of forests, and has been widely correlated with fossil-fuel consumption (Mannion 1992). For example, damage to the forests in Poland, one of Europe's highest sources of air pollution, has been calculated at 3 million m^3 per annum, with total economic losses amounting to $200 million so far (Mazurski 1990). Attempts to curb the emissions of acid-producing gases have met with varying degrees of success, with the developed world leading the way because of its wealth and more advanced technology. Flue gas desulphurization units have been fitted to many coal-fired power-stations with the possibility of more advanced biotechnology being able to desulphurize fossil fuels at source in the future. An alternative, of course, would be a gradual abandonment of fossil-fuel combustion by switching to alternative energy sources. Nuclear power, once thought to be the panacea, is now seen to be riddled with problems, not least of which is the threat of radiation hazards (Openshaw 1986). Unlike the effects of most natural hazards which manifest themselves rapidly, the impact of radiation on the environment is much less obvious and the effect of mortality levels considerably more pervasive. The effects of the Chernobyl (USSR) nuclear accident are still being discovered throughout Eastern Europe, where many billions of hectares of farming land have been seriously contaminated by radioactive fall-out.

Uncontrolled emissions of toxic gases are not uncommon events and occasionally create disasters, such as that at Bhopal, India in 1984 which led to 3,000 deaths and 200,000 injuries. Both the Bhopal and the Seveso, Italy (1976) incidents illustrate how the new and sophisticated (but often high-risk) technologies can cause stress on the socio-economic infrastructure, not only of less-developed countries but also of developed societies (Shrivastava 1992).

Marine pollution is currently regarded as one of the more serious aspects of human impact on the environment because it manifests itself in such a variety of ways. Oil-spills from supertanker accidents may be the most palpable, but the emissions of radioactive waste, sewage, pesticides. heavy metals and other toxic substances pose an even greater threat because of their effect on the oceanic food chain (Hughes and Goodall 1992). With the oceans long being regarded as a harmless repository for waste, it is now realized that the marine ecosystems of certain seas are no longer capable of neutralizing or dispersing much of the waste; the Mediterranean, the North Sea and the Black Sea are already severely polluted (Clark 1989).

Accidents at work and in the home are almost impossible to chronicle; equally, so too are accidents due to leisure activities. There is more knowledge, however, of transportation accidents, where a combination of human error and technological failure has led to many disasters throughout the centuries. The increasing incidence and magnitude of such hazards reflect both the growth of population and of technological innovation, which itself creates risks as well as benefits. It has been emphasized that most technological disasters are caused largely by: defective design and/or construction; inadequate management; sabotage or terrorism (Smith 1992). All of these reasons result from the extremely complex functioning (or malfunctioning) of human systems which themselves represent the interaction of sociological, economic and political forces. In this respect it is difficult to reach any prescriptive conclusion in respect of ameliorating the impact of human-induced hazards, no matter how desirable. World conferences on environmental pollution control take place almost annually, protocols are signed, and progress appears to be made. Increased standards of design, construction and monitoring are being slowly introduced by the developed nations. Health and Safety executives multiply, as does land-use zoning in relation to perceived risk. Forecasting and warning of human-induced hazards, however, is often not as easy as in the field of natural hazards. Thus, one must conclude that accidents will continue to happen and environmental hazards will be increasingly exacerbated by the conduct of mankind.

REFERENCES

Alfors, J.T., Burnett, J.L. and Gray, T.E. (1973) 'Urban geology masterplan for California', *California Division of Mines and Geology Bulletin* 198.
Ambrose, J. and Vergun, D. (1985) *Seismic Design of Buildings*, New York: J. Wiley.
Anderson, J.L. and Decker, R.W. (1992) *Volcanic Risk Mitigation through Training*, pp. 7–11 in G.J.H. McCall *et al.* (eds) *Geohazards: Natural and Man-made*, London: Chapman and Hall.
Arnell, N.W. (1983) *Insurance and Natural Hazards: A Review of Principles and Problems*,

Discussion Paper No. 23, Department of Geography, University of Southampton.

Bardsley, K.L., Fraser, A.S. and Heathcote, R.L. (1983) 'The second Ash Wednesday, 16 February 1983', *Australian Geographical Studies* 21, 129–41.

Bernard, E.N., Behn, R.R., Hebenstreit, G.T., Gonzales, F.I., Krumpe, P., Lander, J.F., Lorca, E., McManonon, P.M. and Milburn, H.B. (1988) 'On mitigating rapid onset natural disasters: Project THRUST', *EOS Trans. Amer. Geophys. Union* 69(24), 649–61.

Blong, R.J. (1984) *Volcanic Hazards: A Sourcebook on the Effects of Eruptions*, London: Academic Press.

Bolt, B.A. (1988) *Earthquakes*, New York: W.H. Freeman.

Brabb, E.E. (1984) 'Innovative approaches to landslide hazard and risk mapping', *Proc. of the 4th International Symposium on Landslides* (Toronto) 1, 307–24.

Brinkman, W.A.R. (1975) *Hurricane Hazard in the United States: A Research Assessment*, Monograph No. NSF-RA-E-75-007, University of Colorado, Boulder: Inst. Behavioural Science.

Bryant, E.A. (1991) *Natural Hazards*, Cambridge: Cambridge University Press.

Burton, I., Kates, R.W. and White, G.F. (1978) *The Environment as Hazard*, New York: Oxford University Press.

Bush, R. (1945) *Frost and the Fruit Grower*, London: Cassell.

Clark, R.B. (1989) *Marine Pollution*, Oxford: Clarendon Press.

Degg, M.R. (1992) 'Some implications of the 1985 Mexican earthquake for hazard assessment', pp. 105–14 in G.H.J. McCall *et al.* (eds) *Geohazards: Natural and Manmade*, London: Chapman and Hall.

Dudley, W.C. and Lee, M. (1988) *Tsunami*, Hawaii: University of Hawaii Press.

Fischoff, B., Lichtenstein, S., Slovic, P., Derby, S.L. and Keeney, R.L. (1981) *Acceptable Risk*, Cambridge: Cambridge University Press.

Foster, H.D. (1980) *Disaster Planning: The Preservation of Life and Property*, New York: Springer-Verlag.

Garcia, R.V. and Escudero, J.C. (1982) *Drought and Man. Vol. 2. The Constant Catastrophe: Malnutrition, Famines and Drought*, Oxford: Pergamon Press.

Gibbs, W.J. (1984) 'The great Australian drought: 1982–1983', *Disasters* 8, 89–104.

Haskell, R.C. and Christiansen, J.R. (1985) 'Seismic bracing of equipment', *J. Environmental Sciences* 9, 67–70.

Hewitt, K. (ed.) (1983) *Interpretation of Calamity*, London: Allen & Unwin.

Hilberg, S.D., Vinzani, P.G. and Changnon, S.A. (1983) *The Severe Winter of 1981–82 in Illinois*, Report of Investigation No. 104, Urbana, Ill.: Illinois State Water Survey.

Housner, G.W. (1987) *Confronting Natural Disasters: An International Decade for Natural Hazard Reduction*, Washington DC: National Academy Press.

Hughes,, J.M.R. and Goodall, B. (1992) 'Marine pollution', pp. 97–114 in A.M. Mannion and S.R. Bowlby (eds) *Environmental Issues in the 1990s*, Chichester: J. Wiley.

Iida, K. and Iwasaki, T. (eds) (1983) *Tsunamis: Their Science and Engineering*, Boston: D. Reidel.

Jones, D.K.C. (1992) 'Landslide hazard assessment in the context of development', pp. 117–41 in G.H.J. McCall *et al.* (eds) *Geohazards: Natural and Man-made*, London: Chapman and Hall.

Kates, R.W. (1962) *Hazard and Choice Perception in Flood Plain Management*, Paper No. 78, Department of Geography, University of Chicago.

Kates, R.W. and Kasperson, J.X. (1983) 'Comparative risk analysis of technological hazards: a review', *Proc. National Academy of Science, U.S.A.* 80, 7027–38.

Keller, E.A. (1982) *Environmental Geology* (3rd edn), Columbus, Oh.: Merrill.

Lamb, H.H. (1970) 'Volcanic dust in the atmosphere with a chronology and assessment of its meteorological significance', *Phil. Trans. Roy. Soc. Lond.* (Ser. A.) 266, 425–533.

Latter, J.H. (ed.) (1989) *Volcanic Hazards: Assessment and Monitoring*, Berlin: Springer-Verlag.

Lowrance, W.W. (1976) *Of Acceptable Risk*, Los Altos, Calif.: Kaufman.

McCall, G.J.H., Laming, D.J.C. and Scott, S.C. (1992) *Geohazards: Natural and Man-made*, London: Chapman and Hall.

Mannion, A.M. (1992) 'Acidification and eutrophication', pp. 177–95 in A.M. Mannion and S.R. Bowlby (eds) *Environmental Issues in the 1990s*, Chichester: J. Wiley.

Mazurski, K.R. (1990) 'Industrial pollution: the threat to Polish forests', *Ambio* 19, 7–74.

Monteverdi, J.P. (1973) 'The Santa Ana weather type and extreme fire hazard in the Oakland-Berkeley Hills', *Weatherwise* 26(3), 118–21.

OFDA (1988) *Disaster History: Significant Data on Major Disasters Worldwide, 1900–Present (May 1988)*, Washington, DC: Agency for International Development.

O'Keefe, P. and Conway, C. (1975) *A Survey of Natural Hazards in the Windward Islands*, Paper No. 14, Disaster Research Unit, University of Bradford.

Olson, R.S., Podesta, B. and Nigg, J.M. (1989) *The Politics of Earthquake Prediction*, Princeton: University of Princeton Press.

Openshaw, S. (1986) *Nuclear Power: Siting and Safety*, London: Routledge & Kegan Paul.

Penning-Rowsell, E.C. (1986) 'The development of integrated flood warning systems', in D.I. Smith and J.W. Handmer (eds) *Flood Warning in Australia*, Canberra: Centre for Resource and Environmental Studies, Australia Nat. University.

Penning-Rowsell, E.C., Chatterton, B.J. and Parker, D.J. (1978) *The Effect of Flood Warning on Flood Damage Reduction*, Report for Central Water Planning Unit, Middlesex Polytechnic, London.

Perla, R.I. and Martinelli, M., Jr (1976) *Avalanche Handbook*, Agriculture Handbook 489, Washington, DC: US Department of Agriculture (Forest Service).

Perry, A.H. (1981) *Environmental Hazards in the British Isles*, London: Allen & Unwin.

Plafker, G. and Ericksen, G.E. (1978) 'Nevados Huascaran avalanches, Peru', pp. 277–314 in B. Voight (ed.) *Rockslides and Avalanches I: Natural Phenomena*, Amsterdam: Elsevier.

Quarantelli, E.L. (1978) *Disasters: Theory and Research*, London: Sage.

Quayle, R. and Doehring, F. (1981) 'Heat stress: a comparison of indices', *Weatherwise* 34, 1210–24.

Roobol, M.J. and Smith, A.L. (1988) 'Volcanic and associated hazards in the Lesser Antilles', in J.H. Latter (ed.) *Volcanic Hazards: Assessment and Monitoring*, Berlin: Springer-Verlag.

Rooney, J.F. (1967) 'The urban snow hazard in the United States: an appraisal of disruption', *Geographical Review* 57, 538–59.

Rubin, C., Yezer, A.M., Hussain, Q. and Webb, A. (1986) *Summary of Major Natural Disaster Incidents in the U.S. 1965–1985*, Special Publication 17, University of Colorado, Boulder: Inst. of Behavioural Science.

Schaerer, P.A. (1981) 'Avalanches', in D.M. Gray and D.H. Male (eds) *Handbook of Snow*, Toronto: Pergamon.

Shrivastava, P. (1992) *Bhopal: Anatomy of a Crisis*, London: Chapman and Hall.

Simon, H.A. (1956) 'Rational choice and the structure of the environment', *Psychological Review* 63, 129–38.

Simons, P. (1988) 'Après ski le déluge', *New Scientist* 1595, 49–52.

Smith, K. (1992) *Environmental Hazards: Assessing Risk and Reducing Disaster*, London: Routledge.

Spangle, W. and Associates, Inc. (1988) *California at Risk: Steps to Earthquake Safety for Local Government*, Sacramento: California Seismic Safety Commission.

Starr, C. and Whipple, C. (1980) 'Risk of risk decisions', *Science* 208, 1114–19.

Stothers, R.B., Rampino, M.R., Self, S. and Wolff, J.A. (1988) 'Volcanic winter? Climatic effects of the largest volcanic eruptions', pp. 3–9 in J.H. Latter (ed.) *Volcanic Hazards: Assessment and Monitoring*, Berlin: Springer-Verlag.

Susman, P., O'Keefe, P. and Wisner, B. (1983) 'Global disasters, a radical interpretation', in K. Hewitt (ed.) *Interpretations of Calamity*, London: Allen & Unwin.

Swedish Red Cross (1984) *Prevention Better than Cure*, Stockholm.

Thompson, R.D. (1992) 'The changing atmosphere and its impact on Planet Earth', pp. 61–78 in A.M. Mannion and S.R. Bowlby (eds) *Environmental Issues in the 1990s*, Chichester: J. Wiley.

Thornes, J.E. (1985) 'Thermal mapping, road danger warning and ice on roads', *Proc. Second Internat. Road Weather Conference*, Danish Ministry of Transport, Copenhagen.

UNDRO (1984) *Disaster Prevention and Mitigation, Vol. II: Disaster Preparedness Aspects*, New York: Office of the Disaster Relief Coordinator, United Nations.

Whittow, J.B. (1980) *Disasters: The Anatomy of Environmental Hazards*, Harmondsworth: Allen Lane.

Whittow, J.B. (1987a) 'Environmental problems', *Progress in Human Geography*, 11, 417–24.

Whittow, J.B. (1987b) 'Natural hazards: adjustment and mitigation', pp. 307–21 in M.J. Clark, K.J. Gregory and A.M. Gurnell (eds) *Horizons in Physical Geography*, Basingstoke: Macmillan Education.

Whittow, J.B. (1988) 'Environmental concerns', *Progress in Human Geography* 12, 451–8.

Whittow, J.B. (1989) 'Earthquakes and building design: an overview', *Disaster Management* 2, 77–82.

Whyte, A.V. and Burton, I. (1982) 'Perception of risk in Canada', in I. Burton, C.D. Fowle and R.S. McCullough (eds) *Living with Risk*, Environmental Monograph 3, Inst. Env. Studies, Univ. of Toronto.

Zeckhauser, R. and Shepard, D.S. (1984) 'Principles for saving and valuing lives', in P.F. Ricci, L.A. Sagan and C.G. Whipple (eds) *Technological Risk Assessment*, NATO Advanced Science Institute Series, The Hague: Martinus Nijhoff.

FURTHER READING

Alexander, D.E. (1991) 'Applied geomorphology and the impact of natural hazards on the built environment', *Natural Hazards* 4(1), 57–80.

Alexander, D.E. (1993) *Natural Disasters*, London: UCL Press.

Blaikie P., Cannon, T., Davis, I. and Wisner, B. (1994) *At Risk: Natural Hazards, People's Vulnerability, and Disasters*, London: Routledge.

Chester, D.K. (1993) *Volcanoes and Society*, London: Arnold.

Coburn, A. and Spence, R. (1992) *Earthquake Protection*, Chichester: John Wiley.

Degg, M.R. (1992) 'Natural disasters: recent trends and future prospects', *Geography* 77(3), 198–209.

Drabek, T.E. (1986) *Human Systems Responses to Disaster: An Inventory of Sociological Findings*, New York: Springer-Verlag.

Green, C.H., Parker, D.J. and Penning-Rowsell, E.C. (1990) 'Lessons for hazard management for United Kingdom floods', *Disaster Management* 3(2), 63–73.

Hodgkinson, P.E. and Stewart, M. (1991) *Coping with Catastrophe: A Handbook of Disaster Management*, London: Routledge.

The Institute of Civil Engineers (1995) *Megacities: Reducing Vulnerability to Natural Disasters*, London: T. Telford.

Merriman, P.A. and Browitt, C.W.A. (eds) (1993) *Natural Disasters: Protecting Vulnerable Communities*, London: T. Telford.

Mitchell, J.K., Devine, N. and Jagger, K. (1989) 'A contextual model of natural hazard', *The Geographical Review* 79(4), 391–409.

Tiedemann, H. (1992) *Earthquakes and Volcanic Eruptions: A Handbook of Risk Assessment*, Zurich: Swiss Reinsurance Company.

Whittow, J.B. (ed.) (1995) 'Hazards in the Built Environment', *Built Environment*, 21, 77–193.

30

THE SUSTAINABILITY OF SUSTENANCE

Land and agricultural production in the Third World

William C. Clarke

Land, considered as planetary surface, is essential to the sustenance of life, for it is earth's surface that receives the solar radiation that energizes photosynthesis, which in turn provides food for plants and animals, including human beings. Each individual person (along with individuals of other species) can survive only by having access directly or indirectly to the area of land surface required to produce enough photosynthetic product to nourish that individual, either immediately or by nourishing the plants or animals upon which it feeds. Because the planet's surface is of a fixed size, the area available to feed each person is limited and, in effect, is becoming smaller and smaller as the human population continues to grow, as the productive capacity of land is degraded, and as land surface is diverted to uses other than pasture and food-crop agriculture – the methods whereby human beings now gain most of their sustenance.

The purpose of this chapter is to explore the growing pressures on land in relation to the sustainability of food production in the Third World. Is there land enough to feed adequately the rapidly growing populations of many Third World countries? For how long? Is land shortage the cause of the widespread hunger already existing in the Third World? Does population growth stimulate increased food production? How can the supply of food from Third World agricultural systems be increased? What constrains increased production? How much degradation is taking place in the agricultural environment? What are the causes of the degradation and what are its effects on agricultural production? Is sustainable agricultural development possible? To such questions, few clear-cut, direct answers can be provided, but it is widely agreed that the issues the questions raise require prompt and effective attention if the welfare of many Third World peoples is not to diminish.

THE ADEQUACY OF AGRICULTURAL LAND IN THE THIRD WORLD

Even though we can say with assurance that land as earth surface is fixed – within the slightly fluctuating range caused by changing tides or by coastal accretion and erosion – we must be much less assured if asked to say how many hectares of land are presently in use for agriculture or are potentially available for such use. As we further pursue such questions, we find that too much concern with the fixity of land area or with the precise measurement of land use or land availability diverts us from other, probably more significant, questions about the earth's ability to provide sufficient food for the present global population or for future generations. None the less, an initial consideration of the basic spatial aspects of agriculture is essential to our discussion.

The land area of our planet is 149 million km^2, which is 29 per cent of the earth's surface, 71 per cent being water. As 14 million km^2 of the land area is permanently covered by ice, some 134.4 million km^2 remain available to be used as arable land, grassland, forests and woodlands, or else to be categorized as 'other land'. Land is also used for housing, urban services, industry, mining, reservoirs, waste disposal, transportation, recreation, and other purposes. All these non-agricultural categories of use continue to expand in order to meet the needs of the growing number of persons who additionally inhabit the earth each year (some 100 million). As each of these persons also requires food, it is evident that agricultural land is under even more pressure to provide humanity's sustenance. Further, whatever few natural landscapes or comparatively undisturbed landscapes remain on planet earth are being subjected to ever more profound modifications to serve human purposes (Buringh and Dudal 1987; Kates *et al.* 1990).

As this transformation of landscapes by human beings began with fire even before agriculture was invented some 10,000 years ago, when global population was a tiny fraction of what it is today, it is necessary always to frame questions about the adequacy of land in time as well as in space. It also needs always to be kept in mind that the figures given for types of land or land uses, for yields of crop per unit of land, or for production of a single crop or aggregate production of all crops are only approximations or estimations because even the best available statistical data may be inaccurate, whether because of errors of measurement, inadequacy of the survey, lack of understanding of the significance of non-commercial production, questionable methods of averaging, or for political purposes such as an underestimation of food availability so as to gain more foreign aid for food relief. Agriculture exists, in other words, in a realm where information is often uncertain and in a state where transformation is continuous (Buringh and Dudal 1987; Simmons 1987; Turner *et al.* 1990).

Table 30.1 Generalized land use of the world

Category	1987 (million ha)	Change since 1977 (%)
Arable land	1,478	+2.2
Grassland	3,323	+0.1
Forest	4,095	−1.8
Other land*	4,233	+1.0
Total	13,129†	

Source: World Resources Institute (1992).
Notes: *'Other land' includes land in polar regions, desert land, stony and rocky land in mountains; †the difference of 2 per cent between this figure and the previously mentioned figure of 134 million km² reflects the usual sort of discrepancy between different measurements of global land area and land use.

Arable land is defined as land that is cultivated and used to grow annual and perennial crops, mainly food crops, with cereal grains (wheat, rice, maize, barley, sorghum, and several others) of particular importance in global terms – although the legumes, or pulse crops (the many species and varieties of beans, mung, gram, peas, pigeon pea, lentils, and so forth), and several so-called root crops (the true yams, the aroids such as taro, sweet potato, cassava, and others) are of great importance in many Third World tropical countries. If the area of earth's arable land – taken to be about 1,478 million ha (Table 30.1) – is arithmetically divided among the world's present population of about 5.5 thousand million people, then each person has, on average, about 0.27 ha (0.66 acre, or an area 52 m × 52 m) of land surface for his or her food production. Is this area sufficient to meet the food needs of one human being? If a wholly theoretical approach could be taken to this question, the short answer that would be given by most agricultural scientists or farmers would be 'yes' or even 'yes, easily'. But the real fields and food gardens where actual crops grow in specific countries are not theoretical; nor are they average. They contain an immense possibility for mixes of different crops, each crop having its own set of requirements and tolerances; they are subject to an immense range of climatic and weather conditions; their topographies and soils vary greatly, often over very short distances; they are situated differently with regard to markets and transport facilities; they exist under different densities of population where people with different goals practise different farming systems in different sorts of political economies; and they are subject to an immense range of labour inputs, technologies, knowledge, and management skills. Further, the present division of the world into nation-states results in a markedly unequal distribution of land resources in relation to the distribution of population so that ratios of agricultural land to people vary widely from nation to nation – and from region to region within nations. Consequently, any particular nation's attempt to achieve national self-provision of food must be undertaken under a unique set of circumstances. Once analysis begins to take into account the necessary provisos regarding these varying conditions found among fields, regions, and countries there can be no easy and only a few undisputed answers to questions about the adequacy of land to feed earth's people.

APPROACHES TO THE CAUSES OF HUNGER IN THE THIRD WORLD

The widespread, sometimes severe hunger occurring in the Third World is one reason that the issue of land adequacy is continually raised. Clear explanations that might suggest workable solutions to the problem of large-scale hunger are sought. Land shortage, which can be inverted to mean an over-abundance of people, is one such explanation; and it can be highly pertinent. Informed opinion today, however, more commonly explains hunger in terms of political, economic, and social factors such as low income and lack of entitlement because it is obvious that if global food were shared evenly, everyone in today's world would have enough to eat – or considerably more than enough if the inhabitants of the industrialized world cut down slightly on their consumption of meat, the production of which requires a much larger production of vegetable food.

The argument for the socio-economic causation of hunger is clearly expressed by the World Commission on Environment and Development (1987) in a publication on global policies for sustainable agriculture. The Commission points to the paradox of a problem with food surpluses in the industrialized market economies while at the same time nearly one-fifth of the world's people

> live in absolute poverty, in conditions of unacceptable deprivation, squalor and misery. Struggle though they do, they lack entitlements and therefore cannot produce or purchase the food and basic goods they need. They are caught in a poverty trap from which they cannot escape and suffer a hunger they cannot satisfy. Every year, 13 to 18 million people die from hunger and hunger-related diseases. Of these, about 15 million are small children (18 children under five years of age die every minute). Over 500 million people are chronically hungry.
>
> (World Commission on Environment and Development 1987: 1)

Other sources, such as Barraclough (1991), believe 500 million is too conservative a number for the world's hungry and cite evidence for double that number of undernourished people. No matter what the exact number, which is indeterminable in any case, the World Commission on Environment and Development (1987) and many other sources (e.g. Barraclough 1991) argue that world hunger is not caused by global food production being outstripped by population growth. On the contrary, between 1950 and 1983 an unprecedented progress in world agriculture accompanied the also unprecedented growth in world population from 2.51 thousand million to 4.66 thousand million, so that during the period agricultural production increased from 248 kg to 310 kg per capita. The Commission (1987) further note that 50 per cent of the world's hungry people live in just five countries, four of which are in Asia where the production increases of the Green Revolution have most famously taken place and even in one country where national surpluses have been recorded.

Other modern authorities see a more direct causal relationship between population and hunger and are more concerned with environmental constraints on agricultural production than with issues of the global distribution of food, power and wealth. Among the most meticulous expressions of an environmentalist concern with agriculture are those of Brown (1988, 1993), who points out that the period of rapid growth in agricultural production has ended, at least for the time being. Using data from various published and unpublished sources of the UN Food and Agriculture Organization (FAO) and the US Department of Agriculture, Brown (1993) discusses the recent slowdown in global agricultural production:

> Of the major economic sectors, the one most vulnerable to environmental degradation is agriculture, simply because it is so directly dependent on natural systems and resources. Environmental degradation, along with emerging agronomic constraints, is slowing the growth in world food output.
>
> The production of grain, which dominates human diets, expanded from 1950 until 1984, when per capita output peaked at 344 kilograms. From then until 1992, it grew less than 1 percent annually, scarcely half the rate of population. For soybeans, the world's leading protein crop, growth averaged 5 percent a year from 1950 to 1980. Over the next 12 years, it averaged 2 percent annually. Slower growth of grain and soybean production, both of which are used as feed, helps explain the slowdown in meat production from 3.4 percent a year between 1950 and 1986 to 2 percent annually during the following six years.
>
> This slower growth has several causes, but two stand out. One is that the growth in the use of key inputs – cropland, irrigation water, and fertilizer – has slowed dramatically. And two, the many forms of environmental degradation – soil erosion, aquifer depletion, air pollution, ozone depletion, and hotter summers – are taking a toll on agricultural output.
>
> (Brown 1993: 11–12)

A momentous feature of the great growth in food production since 1950 – whether the rate of growth has now slowed down or not – is that the increase came largely from raising yields on existing agricultural land rather than from expanding the area of land under crops. While world population climbed rapidly after 1950, total cereal production more than kept pace; but for the first time in global agricultural history, increases in population were not more or less matched by increases in the total cultivated area. Rather, the graphical representation of change for the two shows an ever more marked divergence after 1950, with the increase in cultivated area slowing to a crawl and even slightly declining in the mid-1980s before recovering somewhat after the United States returned to cultivation land previously idled under programmes intended to manage production of agricultural surpluses (Evans 1993; Brown and Young 1990).

Almost everywhere in the world today, the lands of greatest fertility are already in use; the lands remaining in reserve are marginal, suffering from one or more of several constraints: too cold, too dry, too steep, too wet, or with soils that are too nutrient-poor, too shallow, too salty, too acid, or too toxic (Buringh and Dudal 1987). Further, each year millions of hectares of what was good cropland are lost either because the land is so severely eroded that it is not worth ploughing any more or because new homes, factories and highways are built on it. Losses of this sort are now most pronounced in the densely populated, rapidly industrializing countries of East Asia, including China, Japan, South Korea, and Taiwan, where nonfarm uses claim roughly a half-million hectares of cropland each year (Brown and Young 1990). Mexico and India also suffer significant losses of cropland, and the United States is pulling back from rapidly eroding land. In what was the Soviet Union, eroded land has also been abandoned, and an increase in alternate-year fallowing to maintain yields has reduced the amount of cropland under production annually. World-wide, the potential for profitably expanding cultivated area is limited. A few countries, such as Brazil, will be able to add new cropland – though many forms of agricultural production on land opened by deforestation in Amazonia have not been successful, and the soils above the restricted zone of river flooding are known to be poor (Salati *et al.* 1990). In most of the world, during the 1990s, gains and losses are likely to offset each other. As a result, the figure for the average area of arable land per capita on earth is certain to continue its decline (Brown and Young 1990). For how much longer can population grow so much faster than the area of cropland? Are those parts of the world with a rapid population increase finally facing an imminent Malthusian dilemma?

LAND, DENSITY OF POPULATION, AND AGRICULTURAL PRODUCTION: SOME THEORY

Almost two centuries have passed since 1798 when Thomas Malthus published his influential volume *An Essay on the Principle of Population*, in which he forcefully rejected the idea of the inevitability of human progress and attempted to show the environmental limitations that human beings must understand and accept. His doctrine is based on two general ideas: the fecundity of life and the forces in nature that are constantly at work to control life's relentless expansiveness. As Clarence Glacken (1967) noted, in his lucid exposition of the Malthusian principle, Malthus likened the earth variously to a closed room, to an island, and to a reservoir. The closed-room metaphor was intended to show the irrelevancy to the principle of population of the argument that there was no population problem as long as large areas of the world were still uninhabited and great tracts of land in the inhabited world were still available for use. As Malthus said: 'A man who is locked up in a room may be fairly said to be confined by the walls of it, though he may never touch them.' The moral of the reservoir comparison is that human beings are more skilled in the utilization than in the creation of resources, for if a people's territory of fertile land is compared to a reservoir fed by a small stream, it is seen that as the population using the reservoir grows, an increasing quantity of water will be taken each year until the reservoir be exhausted, and only the small stream remain.

Malthus's arguments have often been belittled because the limitations he proposed did not quickly become manifest. Rather, the British population rose from less than 10 million in 1800 to about 41 million in 1900 while agricultural output rose fourteenfold, and life expectancy doubled (Kennedy 1993). World population is now close to six times larger than when Malthus wrote, and food output has kept pace with or surpassed population increase rather than being outstripped, as Malthus argued it must be. But instead of being a refutation of Malthus, this seeming contradiction can be reconciled with the Malthusian principle. During the nineteenth century, 20 million people emigrated from the British Isles to relatively empty lands in North America and elsewhere. Large amounts of cheap grain became more and more easily available from the vast new cultivated areas established on fertile lands in the prairies of the United States and Canada, the Argentine pampas, and the wheatlands of the Ukraine and Australia. Now that these fertile frontiers are closed, only poorer soils remain to be newly exploited, and there are no more pioneer realms open for massive migration (Kennedy 1993). As Malthus held, soils as the machines of food production vary from very good to very poor. Since the best soils cannot alone provide for an increased population, the poorer ones must be cultivated, with more and more labour being applied less and less efficiently. The conclusion is inescapable that, with population growth, it becomes more and more costly in money and human effort to obtain subsistence (Glacken 1967). Over the past several decades, neo-Malthusians have believed this process to be underway and again see the human population approaching stringent environmental limits – the walls of Malthus's locked room. But still the truly massive famines that have been predicted, for example by Borgstrom (1965), Dumont and Rosier (1969), or Paddock and Paddock (1968) have not yet occurred.

Easy though it is, with hindsight, to say that history disproves Malthus, it is as feasible to argue that his principle contains a basic truth, the dire consequences of which have not been cancelled but only postponed by the adaptive ingenuity of human beings. We can see now, however, that Malthus did hold a mistaken view about land in that he considered it as merely the unchanging container of agriculture, as a static entity. It is with regard to this misunderstanding that another major, and much more recent, theory of the interrelations of population and production needs to be considered – that of Ester Boserup, first expressed in *The Conditions of Agricultural Growth* (1965). Boserup argued that Malthus saw population growth as a dependent variable, which could increase only to the extent that an increase in food production allowed. Boserup assumed, in contrast, that the line of causation is in the opposite direction: population growth is an independent variable acting as a major factor determining agricultural production. Boserup believed that land responds more generously to additional inputs of labour and other factors of production than had been assumed. That is, output, or yield, from a given area of land is quite flexible and can be greatly increased through human effort. In her classification of farming systems of different intensities, Boserup focused especially on frequency of cropping, running through a sequence of increasing intensity from forest-fallow (where the frequency of cropping is less than 10 per cent) to bush-fallow (10 to 40 per cent) to short-fallow (40 to 80 per cent) to annual cropping (80 to 100 per cent) to multi-cropping, where the frequency is 200 to 300 per cent because two or more crops are grown in the same fields each year without any fallow. If under pressure of population growth, farmers move their land use through this sequence, the output per unit of land over time will greatly increase, lifting the Malthusian ceiling. Something like this process can be seen to have happened in many traditional farming systems around the world (Turner *et al.* 1977). The deliberate and often successful efforts made by agricultural scientists and policy-makers over many years to increase output also support her thesis. On the other hand, the rapid swallowing-up of the gains in food production by population growth in developing countries illustrates again the Malthusian view that population growth can be a dependent variable, which is able to continue growing so long as sufficient – even if no more than barely sufficient – food supplies remain available. Although Boserup's process of intensification under the push of population growth meets the needs generated by that growth, the process is not necessarily beneficial economically, socially, or environmentally. With each move to a more intensive stage, labour productivity declines (i.e. people with non-industrial farming systems have to work harder to produce the same amount of food as in the previous stage), competition for access to land may increase, land degradation of the fields may worsen, and, especially in high-technology farming, the surrounding ecosystems may be harmed.

THE AGRONOMY OF INCREASING PRODUCTION

Most developing countries seek to increase agricultural yield, whether for national or regional self-provisioning or to produce cash crops or both. There are an enormous number of agronomic techniques whereby aggregate production or yield per unit of land can be increased. Which particular techniques are used depends on a complex of economic, social, political, physical, and historical factors. No attempt can be made here to detail specific techniques. Instead, a few basic points will be made about general approaches to increasing production from an agronomic point of view.

The four main components of increased crop production are:

1 increase in area of land under cultivation;
2 increase in yield per hectare per crop;
3 increase in the number of crops per hectare per year; and
4 displacement of lower-yielding crops by higher-yielding ones.

(Evans 1993: 49)

The reduction of pre-harvest losses from pests and diseases is included under 2 above, but net crop production can also be raised by reducing post-harvest losses, which have been estimated to range as high as 50 per cent in some developing countries. Besides the four sources of greater food production listed above, two other possibilities have been suggested. One is the replacement of cash crops by high-yielding food crops, which is not likely inasmuch as the generation of income is a strong goal of national governments as well as of local farmers. The other possibility is to replace intensive livestock production with an extensive grazing system, thus freeing land now used to produce feed grains to change to producing food directly for human consumption. This also seems unlikely unless the yield of grasslands and forages can be greatly increased. In any case, although food supplies might increase, crop production would not – it might even decline if feed crops such as maize and sorghum were replaced by lower-yielding crops (Evans 1993).

Increase in area of land under cultivation

As noted above, throughout history until the mid-twentieth century the growth of the human population has been marked by a roughly comparable expansion in the cultivation of land, until today 11 per cent of the land not permanently covered by ice is under crops, 24 per cent is under grassland, and 31 per cent is under forest; but these proportions vary greatly between continents. Almost 30 per cent of Europe is cultivated, and more than 28 per cent of tropical Asia, this proportion rising to 55 per cent in India and 69 per cent in Bangladesh. By contrast, in South America and Africa, the proportion is only 5–6 per cent.

In pace with the extremely rapid population growth over the past century, half the world's 1.4 billion ha of currently arable land has been brought into cultivation over the same period, at the expense of forests and grasslands (Evans 1993). The recent marked slowdown in expansion of cultivated land is expected to continue, with growth reaching only 0.15 per cent per year by the end of the century, mostly from clearing of forests (as in the Amazon basin), from new irrigation projects in areas too arid for dryland cropping, and from the drainage of wetlands (Evans 1993). This small rate of increase is now offset, or more than offset, by land degradation and non-agricultural development on arable land.

Various, sometimes sharply differing, estimates have been made of the potentially cultivable area of the world, with a recent estimate being 3.03 billion ha, which is slightly more than twice the current arable area. Most of the potential for expansion is in the developing countries. In South America the arable area could increase by 560 per cent, and in Africa by 370 per cent, whereas in Asia, it can increase by less than 10 per cent, even on an optimistic assessment. As almost 60 per cent of the world's population lives in Asia, further increases in food production in that region must come mostly from land already in use. In Africa and South America, a large spatial expansion of agriculture is possible but it would be on marginal land, it would threaten conservation values, and it would cause further land degradation. A widespread conclusion among students of global land matters is that it will be better to intensify agricultural production on the presently cultivated land than to reclaim more land (Evans 1993; Buringh and Dudal 1987).

Greater yields per crop

Increase in yield per crop has been going on slowly for millennia, as rural folk have empirically developed more productive ways of farming. But widespread 'take-off' rises are a recent phenomenon, beginning in the nineteenth century with rice in Japan and wheat in several European countries, and since the 1960s for the staple crops of many developing countries as aid-funded international agricultural institutions spread the results of their scientifically based innovations – the Green Revolution package of plant breeding, herbicide, pesticide, fertilizer, and other high-technology inputs. Continued increases in yield are possible but always at the cost of greater inputs, whether labour or the capital investment required for Green Revolution inputs. Two further points need to be kept in mind. First, changes in yield over time are likely to follow the characteristic curve of growth – that is, the approximately S-shaped (sigmoid) form. After an early period of slow change, there is acceleration followed by a temporary or more prolonged slowing down, or yields may even fall. Second, yields in the farmers' fields never reach the levels possible under experiment-station conditions. As an example, the highest yield of wet rice ever obtained on the International Rice Research Institute (IRRI) farm in the Philippines is about 11 tonnes per hectare, four times higher than the average national yield in recent years. But agronomists warn that such comparison should not be taken to mean that there is ample head-room for further increase in the national yield per hectare (Evans 1993).

Greater frequency of cropping

In terms of frequency, the most intensive agricultural systems are annual cropping (in which a crop is sown or planted each year, leaving only a few months for fallow), and multi-cropping (in which the same piece of land bears more than one crop each year without ever being left in fallow). Because human effort must somehow make up for the lack of a restorative fallow, these high-frequency cropping systems tend to be practised only in regions or countries with high population densities, where the need for production spurs the required efforts (Boserup 1981). The difference between high-frequency systems and the shifting cultivation at the lower end of Boserup's sequence is elegantly described in Geertz's (1963) description of agricultural involution in Indonesia. As described by Geertz, wet-rice cultivation in densely peopled areas of Java brought a total transformation of the landscape and an even greater requirement for labour to maintain the system at or beyond the edge of diminishing returns. This contrasts strongly with the sparsely inhabited areas of outer Indonesia, where shifting cultivators enjoyed low labour requirements in a much less humanized environment. In densely populated countries with a high level of technology, industrial inputs can be substituted for labour in high-frequency cropping systems. Fertilizer and herbicides carry out the functions of the missing fallow.

Double cropping of rice has been known in China since the twelfth century, and indices of cropping intensity in the twentieth century show that this form of intensification is still spreading in many Asian countries. There is scope for further intensification of cropping frequency at low latitudes, not only with successive rice crops in order to maximize yield per hectare per day, but also with more diverse cropping systems such as the rice–wheat system, which now occupies 10 million hectares in South Asia. In the tropics, where high temperatures tend to cut short the grain-growth stage of cereals, thereby limiting yield per crop, increasing cropping frequency could contribute substantially to greater food production (Evans 1993).

Displacement of lower-yielding crops by higher-yielding ones

Prehistoric peoples and the traditional folk now vanishing from today's world had diets different from those in any modern community. Food plants once common such as *Chenopodium* in Europe or amaranths in tropical America are no longer grown on a significant scale, having lost out in the displacements that have gone on from the beginnings of agriculture. Since the fifteenth century, the process has been particularly dramatic, as many staple crops were successfully introduced from one region of the world to others, with, for instance, the native American crops of potato, sweet potato, maize, and cassava gaining rapid ground over local crops in many other parts of the world. The process continues today, as pressure on arable land increases. In India, for instance, the highly cherished nitrogen-fixing legume crops such as chickpeas, which were once grown widely in crop rotations to restore soil fertility, are being replaced by high-yielding varieties of staple cereals, which receive their nitrogen from commercial fertilizers. Although yields may rise because of this process, an unfortunate aspect is the loss of genetic diversity in agriculture as a few high-yielding varieties displace a wider range of varieties within a single species and as whole species disappear from the fields and food gardens (Evans 1993).

SOME ASPECTS OF THE SOCIAL ECOLOGY OF AGRICULTURAL PRODUCTION

In the enormous literature that exists on how human beings relate to agricultural land by means of social, economic, and political arrangement, there can be found studies on all parts of the world, at scales from fields, farms, and the smallest village through nations, empires, continents, to the entire world; temporally, concern ranges from the future to humankind's distant past. This profusion of studies demonstrates that agriculture is always embedded in the social world beyond its literal roots in the natural world. Agriculture is a social process. Human beings possess the agronomic knowledge to improve yields to some extent in any agricultural system anywhere; they also generally know how to sustain yields at a reasonable level for a long time; but human behaviour takes place in a social as well as agronomic world (Blaikie and Brookfield 1987). An anthropologist and a forester put it this way: 'In all parts of the world, purely technical solutions to resource management problems have rarely proved effective in achieving their stated objectives' (Rambo and Hamilton 1991: 124). Or the geographer Wilken (1991) commented in a USA-government document intended to place sustainable agriculture more firmly on national and international agendas:

> Solutions to technical problems of sustainability, such as reducing environmentally harmful inputs or managing fragile lands, are available or can be developed. However, as long as nonsustainable practices are rewarded economically or politically, nonsustainable agriculture is inevitable. Thus, achieving sustainability depends on social and economic factors.

(Wilken 1991: 25)

Among the social and economic factors that affect agricultural production, local and national food supply, and agricultural sustainability are:

1 *Land tenure and land reform.* In the past, the aim of land reform has often been less for purposes of agricultural efficiency than to attain social justice through a change in rural power structures and the redistribution of status and wealth. But the pattern of land tenure inevitably affects the productivity of labour and capital and the land's aggregate production. It is difficult, however, to make simple generalizations about the optimum size of farms in relation to agricultural production. Studies in several countries indicate that farmers with small holdings produce more per hectare than those with large holdings, in part because small farmers can use more labour per hectare, thus raising land productivity (Brown 1988). On the other hand, small farmers are less able to afford high-technology inputs. Among the continents, undoubtedly the greatest inequalities in land exist in Latin America, where large landholdings (*latifundia*) are often underused. In most African countries and the Pacific Islands, communal land ownership by local groups coupled with individual rights to use is the norm, and arrangements are generally egalitarian. In most Asian countries, small farms predominate, though inequalities in land ownership can be large, and the percentage of rural people who are landless is rising (Hrabovszky 1985; World Commission on Environment and Development 1987).

2 *Security of the rewards to labour and investment.* Aside from social and political arrangements intended to establish tenure systems that ensure the optimum size of landholdings, security of tenure crucially affects yield as well as the preservation of the land in good health. Tenant farmers, lessees, or squatters who are not assured of a secure future on a tract of land may not try to produce as much as farmers with security and certainly will not invest in conservationist measures. As R.G. Ward wrote in his classic study of land–

population relations in Fiji, where much of the sugar-cane crop is produced on leasehold land:

> lack of security results in poor farming, as the tenant is reluctant to dig or maintain drains, construct bunds or contour terraces, use fertilizer or carry out any improvement which might benefit the landlord or result in demands for higher rent.
>
> (Ward 1965: 123)

3 *International trade, subsidies and price controls.* Estimates of the economic effects on the Third World of the liberalization of international trade vary greatly. It is argued, for instance, that the gains following the acceptance of the Uruguay Round of negotiations on the General Agreement on Tariffs and Trade (GATT) will be greater than the benefits of development aid because, with national protectionist measures stripped away, markets will open in North and South alike, and the resulting free trade will allow each country to maximize its comparative advantage in production. This will boost world trade and increase global income in a way that will benefit all. The complex counter-argument asserts that *who* benefits from trade is determined now by economic-ally powerful governments and their political allies, notably the transnational companies that control 80 per cent of world trade. Further, the trading regime that results from this arrangement, which protects corporate interests, will pauperize millions in the South and increase environmental degradation (Anon. 1993; Coote 1992). But trade in subsidized food can be seen as no more beneficial. Boserup (1981) spells out both the advantages and the disadvantages for lower-technology countries of importing cheap food surpluses from higher-technology countries. Seeming advantages include (1) food prices can be kept lower in urban areas; (2) with urban areas supplied by imported food, rural development can be neglected while funds are channelled into industrial or other forms of desired development; (3) by accepting food surpluses, importing countries hope to be granted other types of aid more easily; (4) imports of food allow a shift of agricultural resources to export crops, thus providing govern-ments with much more revenue than can be gained from food production for national consumption. The disadvantages, which are slower to appear, include such low prices for locally produced food that farmers cannot gain any profit, lack of rural development, consequent rural-to-urban migration, and increasing food dependency. Price controls on food prices, which have been established in many Third World countries to quell urban discontent and to constrain inflation, also have a distorting effect, discouraging local farmers from production.

4 *Information dispersal, education, extension services, the use of indigenous knowledge.* Knowledge about agriculture is often categorized as either scientific or else traditional (or indigenous). Although some distinction between the two is warranted, both ultimately rest on empirical testing in the farmer's field. If application of the knowledge increases yields or enhances sustainability, the question then is how to disperse the relevant information so that others can take

up the technique. In traditional agricultural communities this social process takes place when such knowledge is passed from father to son and mother to daughter or when marriageable girls move to other communities as wives. In modernizing communities, formal education might fulfil the task except that agricultural training has low status compared with education for white-collar pursuits. Agricultural extension services are intended to meet the need for spreading information directly to working farmers, but are almost always underfunded in Third World countries and lack personnel and adequate transport facilities. Further, when extension officers do reach the farmers, they may not understand the local people's cognitive map, and the information the officers spread usually emphasizes the expansion of cash crops rather than the improvement of subsistence food crops or the implementation of conservationist measures for agricultural sustainability. Recently, however, there has been a growing appreciation of the value of indigenous knowledge in agriculture. Rather than introduce wholly alien techniques and crops, governments and development agencies are coming to see that the generations-old local knowledge, perhaps with slight modifications based on scientific agronomy, is already well suited to local environments and societies (Chambers 1983; Durning 1993; Richards 1985; Warren *et al.* 1993; Wilken 1987).

5 *The purposes of agriculture.* Since agriculture is a social process, farmers have goals beyond sheer production that maximizes output while minimizing input. Rather than the highest yields, farmers may, for instance, seek risk avoidance:

> Living and working in an environment of risk and uncertainty in which it is not possible to predict with confidence that such-and-such a set of inputs will yield such-and-such a return, farmers are reluctant to embark on new practices which might increase risk.

(Blaikie and Brookfield 1987: 35)

Or, social requirements for prestation and prestige may dominate the production of certain agricultural goods, a process that usually involves high labour costs, which translate into the high value of the goods. Classic examples of such goods are the labour-intensive ceremonial yams and the pigs grown for feasts and prestations in the South Pacific (Brookfield 1972; Blaikie and Brookfield 1987). In terms of agronomic efficiency or the logic of intensification, the great effort given to the production of these goods makes no sense in the absence of an increased pressure of population; in terms of social life, the intensification is a wholly reasonable way to behave in the pursuit of valued social goals.

LAND DEGRADATION AND OTHER CONSTRAINTS TO SUSTAINABLE AGRICULTURAL DEVELOPMENT

As the twenty-first century approaches, agriculture is buffeted by a convergence of deleterious trends. The frontiers of good land have closed.

Land degradation is widespread. Many important aquifers and the irrigation systems they water are in jeopardy (Postel 1993). Currently, 500 million to 1 billion people go hungry each day, while each year the multitude of mouths added to the earth's population is greater than ever before in human history, with the population anticipated to reach over 8 billion (3 billion or so larger than at present) by the year 2025 (Demeny 1990; Norse 1992). In the two regions with the fastest population growth, Latin America and Africa, per capita grain production is falling (Brown 1988). Global growth in food output is slowing (Brown 1993). Urbanization continues to consume excellent arable land. Spatial expansion of agriculture threatens tropical forests, the major source of global biological diversity (Wilson 1992). A complex of human-induced climatic-change processes may have profoundly disruptive effects on agriculture, although by no means will all of these be permanently detrimental everywhere – nor have scientists reached any consensus as to what the effects will be (Nordhaus 1994). To these threats and uncertainties can be added unresolved socio-economic issues such as poverty, inefficacious land tenure, imbalances in trade, landlessness, inadequate entitlement to food, and inequitable control of the agricultural resource base (Barraclough 1991; World Commission on Environment and Development 1987). Some of these issues have already been discussed earlier in this chapter; here, only land degradation will be considered, while bearing in mind that land degradation, as a constraint which itself consists of many parts, is only one part in a suite of constraints, all of which affect agriculture as well as affecting each other.

Less dramatic than 'global warming', or 'the ozone hole', land degradation, with its insidious, incremental deterioration of a resource basic to the human life-support system, is crucial to the future of humankind. To describe fully any of the many types of land degradation would require a separate chapter for each type, but all share a common effect on productivity:

> When land becomes degraded, its productivity declines unless steps are taken to restore that productivity and check further losses. In either case, the yield of labour in terms of production is adversely affected. Land degradation, therefore, directly consumes the product of labour, and also consumes capital inputs into production. Other things being equal, the product of work on degraded land is less than that on the same land without degradation.
>
> (Blaikie and Brookfield 1987: 1)

Land degradation includes such processes as salinization, soil erosion, soil compaction, waterlogging, decline in soil nutrients and humus, lessened nutrient-exchange capacity, desertification, acidification, and accumulation of toxic substances in the soil. That there is a wide awareness of degradation is shown by the ever-growing literature on the subject, such as two recent major studies by geographers (Blaikie and Brookfield 1987; Turner et al. 1990); yet there remain woeful gaps in the data about degradation, with much of the information, especially at the global scale, being the product of estimation rather than the result of actual measurement (Rozanov et al. 1990). Although the cause of degradation is often attributed to the direct harsh pressure of population on resources, actual local cases are rarely so simple, with the causes almost always tied to varied mixes of physical, social, and economic conditions. Nor is there any simple correlation of population density with degradation (Blaikie and Brookfield 1987). For example, a few pioneers moving into forested mountains can cause severe degradation that may later be remedied by the greater labour power that becomes available to a denser population, who might construct irrigated terraces and so rehabilitate the land and create a sustainable agricultural system.

Frequently, and ironically, agricultural development projects intended to improve people's lives in the Third World cause degradation, as is well exemplified by a recent study of a large Australian-funded rural development project in Kenya (Porter *et al.* 1991) or the irrigated rice growing sponsored by the World Bank and the International Monetary Fund (IMF) in the Sahel, where the development had the aim of bringing structural adjustment to Sahelian economies but reportedly had results that included profiteering in land, salinization, alkalinization, and an increase in the disease schistosomiasis (Madeley 1993). Although large-scale agricultural interventions for the sake of raising production do often show benefits, they also often have environmental costs sooner or later, as, for instance, waterlogging and salinity by irrigation (Shanan 1987), erosion and damage to soil structure by mechanization (Spoor *et al.* 1987), subsidence and toxic concentrations of aluminium and iron in the soil by reclamation of wetlands (Ruddle 1987), or eutrophication of surface water and groundwater by the long-term application of fertilizers (Olson 1987).

Slowing or halting degradation is seldom easy even if a fairly simple technical solution exists, for just as the cause of degradation is not solely technical, so the solution must deal with social and economic issues as well as physical ones. On the other hand, a heartening aspect of land degradation is that it is not a one-way street. Human beings can exercise restorative management and 'the land itself also has its own means of repair: new soil is formed, gullies grass over and become graded; nutrient status is restored under rest' (Blaikie and Brookfield 1987: 7).

PROSPECTS FOR THE SUSTAINABILITY OF SUSTENANCE

In 1970, an orthodox economist with an extensive background in agricultural issues argued that the efficient use of existing technology on the agricultural resource base could provide plenty of food for around 35 billion people (Clark 1970). In 1992 an authoritative joint statement was issued by the US Academy of Sciences and the UK Royal Society that,

If current predictions of population growth prove accurate and patterns of human activity on the planet remain unchanged, science and technology may not be able to prevent either irreversible degradation of the environment or continued poverty for much of the world.

(Norse 1992: 6)

The discrepancy between these two views – and many others at both the pessimistic and optimistic extremes of opinion – reflects the contrast between Boserup's postulate of a continuing intensified production in response to population pressure and the Malthusian view of inevitable limits to production, with the current issues of environmental degradation and disruptive global climatic change appended. Sheer physical limits indicate that population growth must cease eventually, but most authorities see that cessation happening long before any such ultimate spatial limit, most probably some time later in the twenty-first century. Before then, the inexorable decline in the amount of arable land per person and the likelihood of significant though uncertain climatic changes suggest to many that the future holds an emerging calamity with regard to population growth being balanced by food supply. Others see innovation and intensification as still offering opportunities. Certainly, as Brookfield puts it:

Change is a natural condition, and change at the hand of human beings is inevitable ... The role of human activity in relation to nature is to modify ecosystems and accelerate change in natural systems, while at the same time further opening systems by introducing new elements and shifting others from place to place.

(Brookfield 1991: 49)

The forms of intensification and innovation that humans have brought to agriculture over human history have resulted in an evolution of ever-changing techniques and systems, always sufficient to meet increasing needs as older methods suffered diminishing returns (Figure 30.1). But, as in the past, so today, many authorities see an inevitable levelling off in the output of food, an aggregate curve of diminishing returns.

If we move from the global scale, it becomes obvious that some particular parts of the world are worse off than others. On a continental scale, Africa is the leading example of food insecurity in the world. By the year 2000 there are likely to be 130 million hungry people in the Sahel countries of Africa as a result of drought and desertification. Whereas Asia has the largest absolute number of hungry people, Africa stands out in terms of the percentage of people affected – 25 per cent of the population. Per capita food production has plummeted 20 per cent from its peak in 1967. The World Bank warns that 'Africa's food situation is not only serious, it is deteriorating.' A downward agricultural spiral in food production is also evident in Latin America despite its copious lands. In 1987, Brazil had to approach the World Food Programme for emergency aid to feed 2.5 million peasants in its north-eastern region, partly because of unprecedented drought linked to El Niño processes. But blame for hunger in Latin America also falls on rapid deforestation, a decline in food crops at the expense of export crops and livestock, and a contraction in the role of small farmers in a region where less than 8 per cent of the population control the

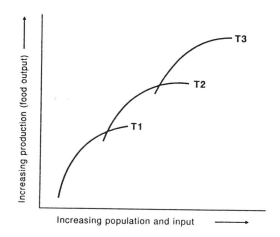

Figure 30.1 T1, T2, and T3 represent an onward progression (T4, T5, T6, and so forth) of technical innovations and continuing intensification in agriculture. With intensification, and for each innovation, there are usually increased inputs of labour, fertilizer, chemicals, water, energy and so on. During each stage there comes a point where productivity is at an optimum and beyond which returns diminish. Population-induced and economically induced demands push innovation and intensification to ever higher levels. Some forms of intensification and some new technologies conserve the agricultural resource base; others degrade it. All of them face the eventual limits in terms of the costs of inputs (and the degradation or pollution of the environment) in relation to yield. Maximum yield is not the same as sustained yield. Are there limits to the progression?

product of 80 per cent of the land (Megalli 1992).

Undoubtedly, innovation and intensification, applied in a context of management for sustainability, could maintain higher yields in Africa and Latin America if the political economy allowed. In Asia, where there are many areas of very high population density with fields already very efficiently farmed, significant increases in yields are now less likely. In a recent detailed study of one of the most densely settled rural areas in the world, the Red River Delta of northern Vietnam, it was argued that current yields are close to their upper limit biologically, with the outputs of the wet-rice fields already extremely high compared with any other known agroecosystem and already eight times more efficient at energy capture than the average of all ecosystems in the biosphere (Rambo and Le Trong Cuc 1993). It is unlikely, therefore, that yields can be significantly increased. Even maintaining current production levels in future years is problematic. The Delta's agriculture is now highly dependent on imported technology in the form of seed, fertilizer, pesticide, and fuel for irrigation pumps. So far, development and propagation of high-yielding varieties of rice have just about kept up with the ability of pests to adapt to them.

A change in pest variety, a run of bad weather over several years, or a shortfall in the supply of agricultural imports could cause production to fall steeply, perhaps even to collapse into a Malthusian disaster. There is also evidence from elsewhere in South-East Asia that land as intensively cropped as some of that in the Delta gradually loses its capability to sustain high yields without carefully managed fallows, rotation of crops, and inputs of organic fertilizers. Technological innovation and involutional intensification (Geertz 1963) have so far managed to keep production up with population growth in the Delta, but much of that production is now purchased at the cost of ever-greater human inputs (Rambo and Le Trong Cuc 1993). Another cost of the creation of an almost completely monocultural agroecosystem has been the loss of bio-diversity, which is low at all biological levels (populations, species, ecosystems).

On a global scale it has been noted that response to fertilizer is levelling off after a 3 per cent annual increase between 1950 and 1984, which boosted output 2.6 times. There is also a steady loss of irrigated land, which also grew rapidly for three decades following 1950, but these areas are now being lost because of salinization and the fall of water tables because of overpumping (Megalli 1992). Another constraint, focused on by Vaclav Smil (1991), is the availability of nitrogen fertilizer produced by industrial synthesis. Unlike many other elements, there is no substitute for the nitrogen found in food proteins derived from plants, and no way to make do with less protein in building human tissues. The lack of nitrogen in the soil in compounds available to plants was the historically most intractable constraint to the expansion of population until synthetic nitrogen compounds became available and 'allowed the expansion of populations well beyond the natural capacity delimited by the intensive manipulation of waste and plant nitrogen flows' (Smil 1991: 572). Synthetic nitrogen now supplies about half the nutrient available to the world's annual and permanent crops, the rest being supplied by natural processes of lightning and the formation of ammonia by bacteria in association with leguminous plants. There is, therefore, an enormous dependence in the world on synthetic nitrogen, but this is not to say that half the world's population would die without the synthetic input because Western nations use roughly two-thirds of their grain output for feeding animals and could easily reduce their dependence on synthetic nitrogen by lowering their high consumption of meat and other animal foods (Smil 1991). In contrast, 80 per cent of China's protein supply is derived from crops, and roughly half of all nitrogen in China's food supply comes from inorganic fertilizer, meaning that nearly 500 million people depend on its application, with very little of the potential for increased food production found in Western countries (Smil 1991). Figure 30.2 further illustrates the ever-growing dependence on synthetic nitrogen as cultivated land per capita declines. It is, of course, the increase in yield brought by the nitrogen that allows a lessening amount of land per person to continue producing food for a growing population. As synthetic nitrogen's production requires high energy inputs, the unexpected image arises that for their food the Asian peasant working synthetic

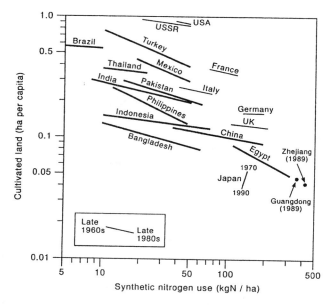

Figure 30.2 Recent trends in cultivated land per capita and the use of synthetic nitrogen per hectare of cultivated land. Largest developing and developed countries shown, late 1960s–late 1980s (logarithmic scale). Reprinted with the permission of the Population Council, from Vaclav Smil, 'Population growth and nitrogen: an exploration of a critical existential link', *Population and Development Review* 17(4), December 1991, p. 5.

fertilizer into wet rice fields with a hoe is potentially more dependent on fossil fuel than the urban North American. The only conceivable major reduction in the total use of synthetic nitrogen could come from substantially lower meat consumption in the rich countries, a shift dispensing with the cultivation of feedgrains now dominating Western farming (Smil 1991).

As the chemical- and energy-rich intensification associated with the Green Revolution increasingly turns out to have ecologic flaws, social costs, and static yields for the past ten or twenty years (Norse 1992), attention is directed more towards biological techniques of increasing crop production and achieving sustainability. This biologically focused technology is now generally known as biotechnology (Ghatak 1988). Biotechnology includes a mixture of approaches, including:

* genetic engineering in an effort to improve crops (as by improving disease resistance or salt tolerance or by introducing nitrogen-fixing genes into cereal crops, thus making them less dependent on chemical fertilizers, usually at the cost of a lowered yield);

670

- more emphasis on biological (as opposed to engineering) methods of weed control and of soil and moisture conservation, such as maintaining continuous ground cover with live mulches;
- biological inputs to integrated pest management systems to minimize pesticide application (for example, more appropriate timing of pesticide application, the use of natural enemies of pests, the use of hormones to affect breeding, or the sterile-male method);
- biological sources of nitrogen to replace or, in most instances, to complement application of fertilizer (as by the greater use of the *Azolla* fern in wet-rice fields, where it enhances biological nitrogen fixing).

All of these approaches are useful in preventing further soil degradation and would help to reduce, to some extent, the use of agricultural chemicals and, hence, the cost of off-farm inputs. But few would claim that biotechnology is a panacea that will end hunger (Brown 1988; Norse 1992):

> Unfortunately, there are no identifiable technologies waiting in the wings that will lead to the quantum jumps in world food output such as those associated with the spread of hybrid corn, the ninefold increase in fertilizer use between 1950 and 1984, the near tripling of irrigated area during the same period, or the relatively recent spread of the high-yielding dwarf wheats and rices in Third World countries. The contribution of these technologies is playing out in some situations and there are no major new technologies emerging to take their place.
>
> (Brown 1988: 37)

To some extent, it is grounds for optimism that belief in scientific-technological quick-fix solutions is waning, even in the large development agencies. This recognition is leading to more concern not just to make agriculture sustainable but, more to the point, to pay attention to the health and viability of the agricultural resource base, the land itself. Under high-technology, large-scale agriculture, land came to be little more than space in which to locate the industrial inputs that would generate high crop yields. Now, a growing concern for the old concept of 'carrying capacity', even though its estimation is notoriously uncertain and impermanent, makes us think more about limits and the inherent capability of land, which can be subjected to management that either enhances or degrades that capability (Blaikie and Brookfield 1987; Brookfield 1991; Norse 1992). Land becomes not just a factor of production along with labour, capital, and management but a capital stock itself; and if sustainability is to have operational meaning, the value of that capital stock needs to be maintained through time, or even augmented by management (Pearce *et al.* 1990).

Aside from a change of attitude towards land, there are intimations that an emerging strategy for feeding the world would also pay more attention to the social and cultural determinants of sustainability, build on indigenous knowledge, and focus on low-risk, low-cost stability of yield for small farmers, not just maximum aggregate yield (Norse 1992; Porter *et al.* 1991). Large farms, however, will continue to play a critical role in producing food for urban populations and other net food buyers and as major suppliers of export crops, but in many cases with increasingly better waste-supply management, integrated pest management, and other approaches to lessening the degradation commonly associated with large-scale, intensive farming operations (Norse 1992).

Aside from the greater attention to indigenous knowledge, which has led to the empirical development of local agroecosystems that can make maximum use of locally available resources (Gliessman 1990), there is also movement towards greater local control of agricultural resources, rather than the imposition of development from above. It must be acknowledged, however, that there is also a strong tendency in the other direction, whereby,

> through the pervasive influence of the commercial economy, decision-making is no longer localised, and a multitude of decisions with impact on resources is taken in total or partial ignorance of that impact, or in spite of knowledge that damage – somewhere – is severe.
>
> (Brookfield 1991: 52)

In this milieu, aid agencies and governments will need to take positive action if they want to increase local involvement and incorporate indigenous knowledge into agricultural projects in the hope of making those projects less likely to fail to deliver what they promised (Gliessman 1990; Porter *et al.* 1991).

The many contradictions and converging pressures that surround agriculture also suggest strongly that a pluralist approach is needed. The problem of agricultural development

> is a moving, evolving multi-faceted thing, and if it was possible to offer an answer today, it would be inappropriate tomorrow . . . Successful development practice must be able to grasp the rich and diverse forms of social life and survival which typify rural people in the Third World.
>
> (Porter *et al.* 1991: 203, 213)

This concern with local land managers, the people on the ground, is also strongly expressed by Blaikie and Brookfield (1987) and Chambers (1983), for it is where human beings interact with the land that sustainable or unsustainable practices take place, where agricultural development – if it is to occur – happens. Or perhaps 'adaptation' is a better word than 'development' for describing whatever processes may lead to providing enough food for Third World peoples (Brookfield 1991; Common 1988). To say that as long as human beings can adapt they can survive is an evolutionary truism; but that adaptation may not always run parallel with progress. On the other hand, the expanding pressure of population on agricultural resources need not necessarily bring further land degradation or even more hunger than presently exists, either in proportion to population size or absolutely, especially if relief can be found from distortions brought by international trade or from the pressure to grow export crops brought by the Third World's debt obligations or from the insecurity of tenure and income suffered by many small farmers. But population's continued growth will make the task of adaptation ever more exacting and precarious.

REFERENCES

Anon. (1993) 'Cakes and caviar? The Dunkel draft and Third World agriculture', *Ecologist* 23(6), 219–22.

Barraclough, S.L. (1991) *An End to Hunger? The Social Origins of Food Strategies*, London: Zed Books.

Blaikie, P. and Brookfield, H.C. (1987) *Land Degradation and Society*, London: Methuen.

Borgstrom, G. (1965) *The Hungry Planet: The Modern World at the Edge of Famine*, New York: Macmillan.

Boserup, E. (1965) *The Conditions of Agricultural Growth*, Chicago: Aldine.

Boserup, E. (1981) *Population and Technology*, Oxford: Basil Blackwell.

Brookfield, H.C. (1972) 'Intensification and disintensification in Pacific agriculture: a theoretical approach', *Pacific Viewpoint* 13, 30–48.

Brookfield, H.C. (1991) 'Environmental sustainability with development: what prospects for a research agenda', *European Journal of Development Research* 3(1), 42–66.

Brown, L.R. (1988) *The Changing World Food Prospect: The Nineties and Beyond*, Worldwatch Paper 85, Washington, DC: Worldwatch Institute.

Brown, L.R. (1993) 'A new era unfolds', in L. Starke (ed.) *State of the World Report 1993*, New York: W.W. Norton.

Brown, L.R. and Young, J.E. (1990) 'Feeding the world in the nineties', in L. Starke (ed.) *State of the World Report 1990*, New York: W.W. Norton.

Buringh, P. and Dudal, R. (1987) 'Agricultural land use in space and time', in M.G. Wolman and F.G.A. Fournier (eds) *Land Transformation in Agriculture*, Chichester: John Wiley.

Chambers, R. (1983) *Rural Development: Putting the Last First*, Harlow: Longman Scientific & Technical.

Clark, C. (1970) *Starvation or Plenty?*, New York: Taplinger Publishing.

Common, M. (1988) '"Progress and Poverty" revisited', pp. 15–39 in D. Collard, D. Pearce, and D. Ulph (eds) *Economics, Growth and Sustainable Environments*, New York: St Martin's Press.

Coote, B. (1992) *The Trade Trap: Poverty and the Global Commodity Markets*, Oxford: Oxfam Publications.

Demeny, P. (1990) 'Population', in B.L. Turner II, W. Clark, R. Kates, J. Richards, J. Mathews, and W. Meyer (eds) (1990) *The Earth as Transformed by Human Action: Global and Regional Changes in the Biosphere over the Past 300 Years*, Cambridge: Cambridge University Press.

Dumont. R. and Rosier, B. (1969) *The Hungry Future* (translated by Rosamund Linell and R.B. Sutcliffe), London: Methuen.

Durning, A.T. (1993) 'Supporting indigenous peoples', in L. Brown, C. Flavin, and S. Postel (eds) *State of the World 1993*, New York: W.W. Norton.

Evans, L.T. (1993) *Crop Evolution, Adaptation and Yield*, Cambridge: Cambridge University Press.

Geertz, C. (1963) *Agricultural Involution: The Process of Ecological Change in Indonesia*, Berkeley: University of California Press.

Ghatak, G. (1988) 'Towards a second green revolution in the tropics: from chemicals to new biological techniques for sustained economic development', in R.K. Turner (ed.)

Sustainable Environmental Management: Principles and Practice, London: Belhaven Press.

Glacken, C. (1967) *Traces on the Rhodian Shore: Nature and Culture in Western Thought from Ancient Times to the End of the Eighteenth Century*, Berkeley: University of California Press.

Gliessman, S.R. (1990) 'Applied ecology and agroecology: their role in the design of agricultural projects for the humid tropics', in R. Goodland (ed.) *Race to Save the Tropics: Ecology and Economics for a Sustainable Future*, Washington, DC: Island Press.

Hrabovszky, J. (1985) 'Agriculture: the land base', in R. Repetto (ed.) *The Global Possible: Resources, Development, and the New Century*, New Haven: Yale University Press.

Kates, R.W., Turner, B.L. II and Clark, W.C. (1990) 'The great transformation', in B.L. Turner II, W. Clark, R. Kates, J. Richards, J. Mathews, and W. Meyer (eds) (1990) *The Earth as Transformed by Human Action: Global and Regional Changes in the Biosphere over the Past 300 Years*, Cambridge: Cambridge University Press.

Kennedy, P. (1993) *Preparing for the Twenty-first Century*, New York: Random House.

Madeley, J. (1993) 'Will rice turn the Sahel to salt?', *New Scientist* 140(1894), 35–7.

Megalli, N. (1992) 'Hunger versus the environment: a recipe for global suicide', *Our Planet* 4(6), 4–7.

Nordhaus, W.D. (1994) 'Expert opinion on climatic change', *American Scientist* 82(1), 45–51.

Norse, D (1992) 'A new strategy for feeding a crowded planet', *Environment* 34(5), 6–11, 32–9.

Olson, R.A. (1987) 'The use of fertilizers and soil amendments', in M.G. Wolman and F.G.A. Fournier (eds) *Land Transformation in Agriculture*, Chichester: John Wiley.

Paddock, W. and Paddock, P. (1968) *Famine — 1975!*, London: Weidenfeld & Nicolson.

Pearce, D.W., Barbier, E.B. and Markandya, A. (1990) *Sustainable Development: Economics and Environment in the Third World*, Aldershot: Edward Elgar.

Porter, D., Allen, B. and Thompson, G. (1991) *Development in Practice: Paved with Good Intentions*, London: Routledge.

Postel, S. (1993) 'Facing water scarcity', in L. Starke (ed.) *State of the World Report 1993*, New York: W.W. Norton.

Rambo, A.T. and Hamilton, L.S. (1991) 'Social trends affecting natural resources management in upland areas of Asia and the Pacific', in S. Naya and S. Browne (eds) *Development Challenges in Asia and the Pacific in the 1990s*, Honolulu: United Nations Development Program/East–West Center.

Rambo, A.T. and Le Trong Cuc (1993) 'Prospects for sustainable development in the villages of the Red River Delta', in Le Trong Cuc and A.T. Rambo (eds) *Too Many People, Too Little Land: The Human Ecology of a Wet Rice-Growing Village in the Red River Delta of Vietnam*, Honolulu: East–West Center, Program on Environment, Occasional Paper No. 15.

Richards, P. (1985) *Indigenous Agricultural Revolution: Ecology and Food Production in West Africa*, London: Hutchinson.

Rozanov, B., Targulian, V. and Orlov, D. (1990) 'Soils', in B.L. Turner II, W. Clark, R. Kates, J. Richards, J. Mathews, and W. Meyer (eds) *The Earth as Transformed by Human Action: Global and Regional Changes in the Biosphere over the Past 300 Years*, Cambridge: Cambridge University Press.

Ruddle, K. (1987) 'The impact of wetland reclamation', in M.G. Wolman and F.G.A. Fournier (eds) *Land Transformation in Agriculture*, Chichester: John Wiley.

Salati, E., Dourojeanni, M.J., Novacs, F.C., de Oliveira, A.E., Perritt, R.W., Schubart,

H.O.R. and Umana, J.C. (1990) 'Amazonia', in B.L. Turner II, W. Clark, R. Kates, J. Richards, J. Mathews, and W. Meyer (eds) *The Earth as Transformed by Human Action: Global and Regional Changes in the Biosphere over the Past 300 Years*, Cambridge: Cambridge University Press.

Shanan, L. (1987) 'The impact of irrigation', in M.G. Wolman and F.G.A. Fournier (eds) *Land Transformation in Agriculture*, Chichester: John Wiley.

Simmons, I.G. (1987) 'Transformation of the land – in pre-industrial times', in M.G. Wolman and F.G.A. Fournier (eds) *Land Transformation in Agriculture*, Chichester: John Wiley.

Smil, V. (1991) 'The critical link between population growth and nitrogen', *Population and Development Review* 17(4), 569–601.

Spoor, G., Carillon, R., Bournas, L. and Brown, E.H. (1987) 'The impact of mechanization', in M.G. Wolman and F.G.A. Fournier (eds) *Land Transformation in Agriculture*, Chichester: John Wiley.

Turner, B.L. II, Clark, W., Kates, R., Richards, J., Mathews, J. and Meyer, W. (eds) (1990) *The Earth as Transformed by Human Action: Global and Regional Changes in the Biosphere over the Past 300 Years*, Cambridge: Cambridge University Press.

Turner, B.L. II, Hanham, R.Q. and Portararo, A.V. (1977) 'Population pressure and agricultural intensity', *Annals of the Association of American Geographers* 67(3), 384–96.

Ward, R.G. (1965) *Land Use and Population in Fiji: A Geographical Study*, London: HMSO.

Warren, D.M., Slikkerveer, L. Jan, and Brokensha, D. (eds) (1993) *Indigenous Knowledge Systems: The Cultural Dimension of Development* (Vol. 1, The International Library of Development and Indigenous Knowledge), London: Kegan Paul.

Wilken, G.C. (1987) *Good Farmers: Traditional Agricultural Resource Management in Mexico and Central America*, Berkeley: University of California Press.

Wilken, G.C. (1991) *Sustainable Agriculture is the Solution, But What is the Problem?*, Board for International Food and Agricultural Development and Economic Cooperation, Occasional Paper No. 14, Washington, DC: Agency for International Development.

Wilson, E.O. (1992) *The Diversity of Life*, Cambridge, Mass.: Belknap Press of Harvard University Press.

World Commission on Environment and Development (1987) *Food 2000: Global Policies for Sustainable Agriculture*, A Report of the Advisory Panel on Food Security, Agriculture, Forestry and Environment to the World Commission on Environment and Development, London: Zed Books.

World Resources Institute (1992) *World Resources 1992–93*, A Report by The World Resources Institute in collaboration with The United Nations Environment Programme and The United Nations Development Programme, New York: Oxford University Press.

FURTHER READING

Adams, W.M. (1990) *Green Development: Environment and Sustainability in the Third World*, London: Routledge.

Bayliss-Smith, T.P. (1982) *The Ecology of Agricultural Systems*, Cambridge: Cambridge University Press.

Brown, L.R. (1994) 'Facing food insecurity', in L. Starke (ed.) *State of the World Report 1994*, New York: W.W. Norton.

Byé, P. and Fonte, M. (1993) *Biotechnologies and Agriculture: Technical Evolution or Revolution*, Single-topic issue of *Agriculture and Human Values* 10(2).

Food and Agriculture Organization of the United Nations (FAO) (annually) *The State of Food and Agriculture*, Rome, FAO.

Food and Agriculture Organization of the United Nations (FAO) (1981) *Agriculture: Toward 2000*, Rome: FAO.

Food and Agriculture Organization of the United Nations (FAO), UN Fund for Population Activities, and the International Institute for Applied Systems Analysis (1982) *Potential Population Supporting Capacities of Lands in the Developing World*, Rome: FAO.

Kendall, H.W. and Pimentel, D. (1994) 'Constraints on the expansion of global food supply', *Ambio* 23(3), 198–205.

McCann, S. and Mock, G. (1994) 'Food and agriculture', in *World Resources 1994–95*, A Report by The World Resources Institute in collaboration with The United Nations Environment Programme and The United Nations Development Programme, New York: Oxford University Press.

Netting, R. (1993) *Smallholders, Householders: Farm Families and the Ecology of Intensive, Sustainable Agriculture*. Stanford: Stanford University Press.

Pearce, D.W., Markandya, A. and Barbier, E.B. (1989) *Blueprint for a Green Economy*, London: Earthscan.

Postel, S. (1994) 'Carrying capacity: earth's bottom line', in L. Starke (ed.) *State of the World Report 1994*, New York: W.W. Norton.

Sachs, I. and Silk, D. (1990) *Food and Energy: Strategies for Sustainable Development*, Tokyo: United Nations University Press.

Turner, B.L. II and Brush, S.B. (eds) (1987) *Comparative Farming Systems*, New York: Guilford Press.

Turner, B.L. II, Hyden, G. and Kates, R. (eds) (1993) *Population Growth and Agricultural Changes in Africa*, Gainesville: University Press of Florida.

Vink, A.P.A. (1975) *Land Use in Advancing Agriculture*, New York: Springer-Verlag.

Warnock, J.W. (1987) *The Politics of Hunger*, London: Methuen.

Wolf, E.C. (1987) 'Raising agricultural productivity', in L. Starke (ed.) *State of the World 1987 – A Worldwatch Institute Report on Progress Toward a Sustainable Society*, New York: W.W. Norton.

World Resources Institute (1992) 'Food and agriculture', Chapter 7 in *World Resources 1992–93*, A Report by The World Resources Institute in collaboration with The United Nations Environment Programme and The United Nations Development Programme, New York: Oxford University Press.

31

FAMINES AND SURPLUS IN WORLD FOOD PRODUCTION

David Grigg

Some years ago Professor D.G. Johnson wrote a book entitled *World Agriculture in Disarray* (Johnson 1973). Twenty years on it is still an apt description of world food production systems. Nor is knowledge of this disarray confined to the academic world. A glance at a newspaper or television confirms that the problems of world agriculture are now widely discussed. The problems, however, differ from one part of the world to another. In the developing countries of Afro-Asia and Latin America, where food production still employs much of the working population and accounts for a substantial proportion of national income, famines still occur, malnutrition is widespread, and much farming is technologically backward, with productivity of both land and labour below that in the developed countries. Furthermore, population growth continues to be rapid, and numerous writers have warned that population growth is, or will, outrun food production. Yet nor are developed countries without problems. In both North America and the European Community food surpluses mounted in the 1970s and 1980s, whilst there was increasing criticism of the elaborate systems of agricultural protection that were costly to the consumer and damaging to other food exporters. Modern agricultural methods have also attracted much criticism – beginning with Rachel Carson's celebrated book *Silent Spring* (Carson 1970) – because of their effect upon the environment.

The treatment of farm animals in modern factory-like conditions has attracted almost universal condemnation, whilst fears about the quality of food have been traced to the use of pesticides, nitrate fertilizers and growth hormones. In spite of technological advance and price protection, farmers in the European Community in the 1980s suffered declining real income, and in North America and Australia thousands suffered bankruptcy. This chapter describes and tries to explain these problems.

FOOD SURPLUSES

In the 1970s and 1980s food surpluses accumulated in both North America and the European Community. This was the result of three factors.

First was the remarkable technological advance within farming in both areas in the period after 1950, which has increased output per capita and output per hectare, and so the supply of food products. The increase in crop yield has been due to the greater use of chemical fertilizers, herbicides, pesticides and the adoption of improved seed; the latter alone accounted for half the wheat yield increases in the United States and the United Kingdom between the 1930s and 1970s (Silvey 1978; Jensen 1978). Advances in livestock productivity have been equally marked (Table 31.1). Selective breeding helped by the spread of artificial insemination, better feeding and management, and advances in disease control by antibiotics and vaccines, combined with greater livestock densities, have increased milk, meat and egg output.

The great increase in output since 1950 has thus been largely a result of increased crop yields (Table 31.2). In Europe there has been everywhere a decline in the area in arable land, the net result of the gain from the reclamation of new land and the greater loss to urban expansion and afforestation. In North America and Australia area expansion has contributed to higher output, but the dominant source has been higher yields. However, in tropical Africa and Latin America the colonization of new land and the reduction of fallowing has made a substantial contribution to the extra output (Table 31.2).

A second reason for the creation of food surpluses in the West has been the slower growth of home markets for food. The high rates of population growth that had characterized the nineteenth century declined in the 1920s and 1930s, and after the immediate post-war baby boom have remained low in Europe. More important were changes in income. As late as the 1930s malnutrition, due to poverty, was commonplace in much of Europe and North America. For

Table 31.1 Increase in yields in selected developed countries, 1948–50 to 1989–90

	Wheat (kg/ha)		Milk (kg/cow) per annum	
	1948–50	*1989–90*	*1948–50*	*1989–90*
Netherlands	3,530	7,412	3,657	6,028
Denmark	3,587	7,440	3,117	6,244
United Kingdom	2,687	6,820	2,673	5,111
West Germany	2,470	6,412	2,173	4,881
France	1,830	6,418	1,730	2,822
Belgium	3,143	6,708	3,367	4,188
United States	1,120	2,429	–	6,586

Sources: Food and Agriculture Organization, *Production Yearbook 1964*, Vol. 18, Rome 1965; *Production Yearbook 1990*, Vol. 44, Rome 1991; United Nations, *Economic Bulletin for Europe*, Vol. 35, 1983, pp. 167–8.

Table 31.2 Increase in output of cereals, 1961–90

	Percentage of increase	
	Attributable to increases in area	*Attributable to increased average yields*
Sub-Saharan Africa	47	52
Latin America	30	71
Middle East and North Africa	23	77
South Asia	14	86
East Asia	6	94
Eastern Europe and USSR	−13	113
High-income countries	2	98
World	8	92

Source: World Bank (1992) *World Development Report 1992*, Oxford: Oxford University Press, p. 135.

many, incomes were too low to allow the purchase of an adequate diet. Economic growth after the end of the Second World War saw real incomes increase rapidly. By the 1960s incomes were high enough for most of the population of North America and Europe to obtain an adequate diet, but consequently most further increases in income were spent not upon food but upon leisure and consumer durables. The growth of incomes, which in the nineteenth century had provided a steady expansion of the market for food no longer did so. By the 1970s food output in the European Community was expanding at 2 per cent per annum but consumption at only 0.5 per cent. Not surprisingly farming in Europe and North America began to produce surpluses.

In a free market overproduction would have led to falling prices and farmers would have cut output. However, virtually every industrial country protects its farmers in some way. The reasons for this are various. In some countries there is still a belief that a government should aim at ensuring self-sufficiency in basic foodstuffs, although a repetition of the circumstances of the First and Second World Wars seems unlikely (Sturgess 1992). Perhaps more important has been what American economists in the 1930s called the farm problem; in nearly all industrial countries average farm incomes were – and still are – significantly below average non-farm incomes, and this has been thought to be inequitable. The power of farm lobbies has perpetuated support for policies designed to reduce this gap, whilst some governments, notably in the USA and France, have felt that the preservation of family farming is important. A further justification for farm support is to avoid the instability of prices that has often characterized agricultural production in the past.

Not only do most countries subsidize agricultural production in some way, but everywhere the cost of protection soared in the 1970s and 1980s. In 1990 the total cost of protection – in supporting prices, subsidizing exports and financing

Table 31.3 Nominal assistance coefficients*

| | For producers | | For consumers | |
	1979–86	*1991*	*1979–86*	*1991*
Japan	2.58	2.67	1.83	1.96
Other Europe[†]	2.49	3.73	1.92	2.97
European Community	1.57	1.94	1.46	1.76
Canada	1.42	1.69	1.29	1.45
United States	1.35	1.39	1.23	1.24
Australia	1.12	1.16	1.08	1.09
New Zealand	1.28	1.04	1.11	1.04
Average OECD	1.53	1.78	1.41	1.62

Source: Organization for Economic Co-operation and Development (1992) *Agricultural Policies, Markets and Trade. Monitoring and Outlook 1992*, Paris: OECD, p. 13.
Notes: *NAC is the domestic price divided by border price for the weighted average of wheat, coarse grains, rice, beef, lamb, pork and poultry, dairy products and sugar.
[†]Switzerland, Austria, Finland, Norway and Sweden unweighted average.

farm improvements – was 44 per cent of the value of crop and livestock production in the OECD countries (Corbet 1991). However, protection was far higher in Japan and Europe (Table 31.3) than in North America or Australasia. In the early 1980s protection was greatest for sugar, dairy products and beef (Figure 31.1).

Agricultural protection was introduced in the United States as a result of the depression of the 1920s and 1930s. Under the Agricultural Adjustment Act of 1933, the United States government established parity prices for selected agricultural products, covering by 1989 70 per cent of all farm crops. If prices fell below this predetermined price the government made up the difference between the parity price and the price at which the farmer sold. An attempt was made to restrict output by only paying parity prices on specific acreage quotas, but the great advances in agricultural technology in the 1940s eventually led to surpluses which the government had to store. There have been a number of attempts to control surplus production. First was the reduction of guaranteed prices; the first substantial reduction was in the early 1970s, and then in the 1980s support prices were cut by 40 per cent (Buttel 1989; Clunies-Ross and Hildyard 1992). Second was the attempt to reduce the acreage in production. In the 1930s farmers were encouraged to adopt soil conservation practices and withdraw eroded land from cultivation; by the 1980s farmers were paid not to grow crops – set-aside. In 1983 77 million acres were withdrawn (World Bank 1986a). Third has been the encouragement of exports, beginning with the bill PL480 of 1954 which subsidized the export of wheat as food aid. United States exports rose from 10 per cent of all production in the 1950s to over 30 per cent in the late 1970s (Bohland 1988). However, a series of events made exports more difficult; after the ban by the United States on exports of wheat to the USSR

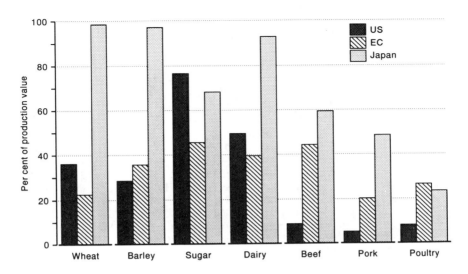

Figure 31.1 Producer subsidy equivalent by commodity, 1982–6. Producer subsidy includes all forms of subsidy and is expressed as a percentage of the value of output. After Lingard and Hubbard (1991).

in 1979 that country found other suppliers, whilst a strong dollar reduced exports to the Third World in the 1980s. Futhermore, the European Community subsidized exports of cereals and soybeans. United States carry-over stocks of cereals rose from 30 million tons in 1975 to 203 million tons in 1987; however, price cuts and set-aside reduced this to less than 70 million tons in the late 1980s (Buttel 1989; FAO 1991a). The cost of government support to agriculture soared in the early 1980s, and convinced the United States of the need for an international reduction of agricultural protection.

European agriculture protection has a much longer history than in the United States, and the Common Agricultural Policy of the European Community which came into force in 1968 was partly a result of a need to harmonize the existing policies of the six initial member states. Under the Common Agricultural Policy a target price is set annually, which in the past allowed the least-efficient farmers to make a living. If prices fall to about nine-tenths of the target price, farmers can sell at the intervention price to the state, which is obliged to buy and store and then seeks to dispose of this surplus production. Farmers are protected from outside competition by the imposition of levies upon imports. Much smaller amounts are spent upon subsidizing farm improvements, farm amalgamations, and supporting farmers in problem areas.

Not surprisingly food surpluses accumulated in the 1970s, and the European Community resorted to a variety of methods of reducing these surpluses. In the first place target prices were increased at less than inflation, or in the late 1980s

Table 31.4 Commodities in public storage in the European Community, 1983–93

	1983	1984*	1985*	1986*	1987†	1988†	1989†	1990†	1993†
	Thousand tonnes								
Cereals	9,541	9,394	18,642	14,714	10,513	9,939	8,608	14,378	24,205
Skimmed milk	957	773	513	847	593	10	22	333	37
Butter	686	972	1,018	1,297	888	101	5	251	160
Beef	390	595	803	671	753	557	158	524	719
	Million ECU								
Cereals	1,853	1,901	3,821	2,977	2,314	1,158	561	832	1,497
Skimmed milk	1,457	1,242	867	1,593	1,136	17	23	262	28
Butter	2,474	3,536	3,416	4,254	2,942	233	5	257	129
Beef	–	1,732	2,270	1,995	–	1,082	219	662	424
Other	208	339	205	541	730	813	224	80	266
Total	–	7,035	10,579	11,360	9,367	3,303	1,032	2,158	2,344

Source: Commission of the European Communities, *The Agricultural Situation in the Community*, Brussels, various years.
Notes: *European ten; †European twelve.

not increased at all; between the early 1970s and 1989 the real price of cereals for farmers halved (Ackrill 1992). Second, target prices were reduced on cereal output above a predetermined quantity. Third, in 1984 milk output was reduced by allocating specific quantities to each member state, and subsequently reducing this. Fourth, attempts were made to lower output by paying farmers to use less chemical fertilizers and pesticide in Environmentally Sensitive Areas. Fifth, the American practice of set-aside was adopted and farmers were paid to withdraw 20 per cent of their supported crop acreage from production, although this had initially a limited impact.

The most effective method of disposing of home surpluses, however, has been by exports. The prices of most products in the Community have been well above world prices and so exporting has only been possible by providing subsidies. It is this that had such adverse effects upon other exporters, particularly in North America and Australia. However it has reduced the European Community's expenditure upon the amounts in public storage (Table 31.4).

Although food surpluses in the developed countries have aroused much indignation – either because of their cost, or because of the contrast with malnutrition in developing countries – by the end of the 1980s the quantities in store in the European Community were not large; and as a proportion of world output or world exports they were not of much significance (Table 31.5). United States carry-over stocks, however, were more substantial. In the period 1988–90 they were equivalent to only one-twentieth of world output, but 40 per cent of United States annual output and nearly half of annual world exports (FAO 1991a, 1991b, 1991c). It is, however, not so much the surpluses in the West that

Table 31.5 Quantities in public storage

	As a percentage of		
	EC output 1988–89	World output 1988–90	World exports 1988–90
Cereals	5.8	0.1	4.7
Butter	5.9	0.5	2.6
Beef	9.4	0.3	3.2

Sources: Commission of the European Communities (1991) *The Agricultural Situation in the European Community 1990*, Brussels; Food and Agriculture Organization (1991) *Production Yearbook 1990*, Rome; *Trade Yearbook 1990*, Rome.

require reform but the policy of supporting agriculture which gives rise to them; the system of protection has not protected small farmers, has enriched large producers, raised prices to consumers, interrupted world trade and diverted funds from more productive sectors of the economy (Carr 1992).

FAMINE

Since the end of the Second World War numerous writers have predicted coming world-wide famine and the press have described actual famines in many parts of the world. Yet paradoxically there is little agreement as to what a famine is, and how it may be distinguished from the more widespread and persistent undernutrition and malnutrition. There is even less agreement upon the causes of famine. If famine is defined as a protracted shortage of food in a restricted geographical area that causes disease and death from starvation (Dando 1980), then there have been a number of major famines in the developing world since 1945. In China 30 million died between 1958 and 1961, and several hundred thousand in Bangladesh in 1974. Both countries had a long history of famine before then, as had Russia, with particularly severe occurrences in the period 1891–2, 1921 and 1932–4 (Arnold 1988). But since 1974 major famines have been confined to tropical Africa, notably in the Sahel in the 1970s, Ethiopia in 1984–5 and Somalia in the early 1990s.

Europe had many famines in the past, although historians have doubts as to the precise extent and nature of most of these. However major famines disappear from the historical record in England after the 1620s, in Germany and Scandinavia after the 1730s, and France after the 1790s. The last great West European famine was in Ireland between 1845 and 1849. However, whilst the incidence of large numbers of deaths due to the absence of food – or an inability to buy it – disappeared from Western Europe after the 1840s, the occurrence of undernutrition and malnutrition did not. In the 1930s malnutrition was still common in Western Europe and the United States (Grigg 1993c). However

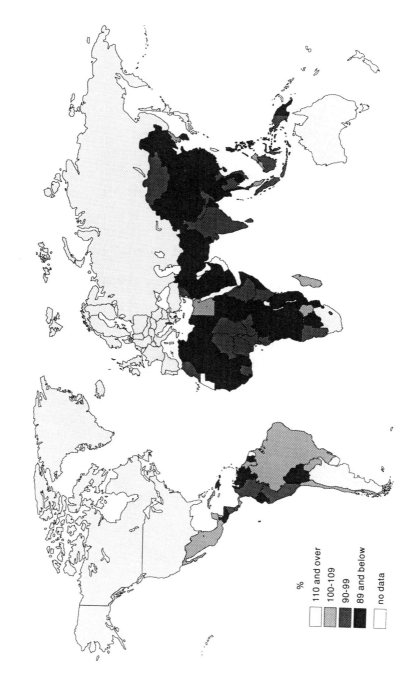

Figure 31.2 Daily calorie supply per capita per day as a percentage of national requirement, 1962–3. After FAO (1977).

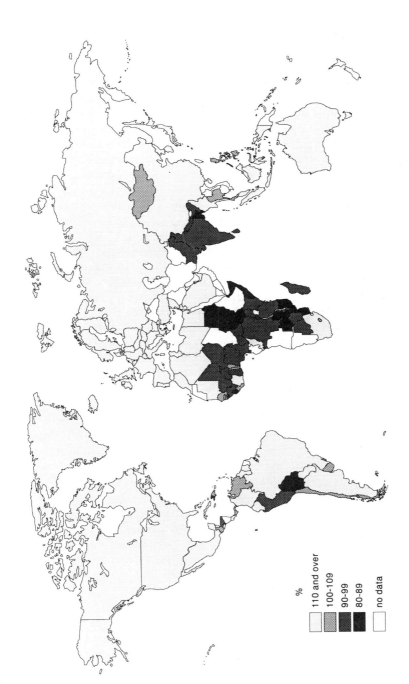

Figure 31.3 Daily calorie supply per capita per day as a percentage of national requirement, 1987–89. After FAO (1991b).

since the 1960s neither have been of much significance in the developed countries; it is the diseases which result from excess rather than deficiency that now prevail.

It is not so in the developing world, where undernutrition and malnutrition still afflict substantial numbers. Malnutrition has been defined as the inadequate intake of protein and vitamins that gives rise to diseases with specific clinical symptoms; thus the once widely prevalent scurvy was due to a lack of Vitamin C, xerophthalmia or blindness to a lack of Vitamin A, and kwashiorkor to a lack of protein. Undernutrition, on the other hand, is due to an inadequate intake of calories, giving rise to stunted growth and a reduced ability to carry out physical tasks.

The extent of undernutrition and malnutrition is difficult if not impossible to measure, and there have been many conflicting estimates of its extent. One preliminary measure is to compare the calorie supplies available in a country with the number of calories required to avoid undernutrition, providing calories are distributed according to biological need (FAO 1977). In the period 1962–3 few countries in the developing world had sufficient available food supplies – which included imports – to provide an adequate diet for their populations, even assuming it was distributed according to need (Figure 31.2). By 1987–9 far fewer countries fell into this category; these were mainly in tropical Africa and South Asia (Figure 31.3). But whilst this approach indicates those countries which are likely to have large numbers of undernourished inhabitants, it is imperfect; there are certainly countries with apparently adequate national supplies that have large numbers of malnourished people. Both the World Bank and the Food and Agriculture Organization have attempted to estimate the number of people undernourished; they have done this first by estimating available supplies, and comparing this with a minimum requirement. But of course food is not distributed according to need; income is the most important determinant of food consumption and the poor are also the hungry. Income distribution figures were applied to data on food availability and requirements in a rather complex way (FAO 1987). The FAO provides two estimates of the number of people in the developing countries: those receiving less than 1.2 and also less than 1.4 times the basic metabolic rate (BMR) for four different periods, but excluding the centrally planned economies. In the period 1983–5 over 500 million received less than 1.4 BMR, mainly in Africa and South and South-East Asia; they constituted one-third of the population in Africa, one-fifth in Asia (Table 31.6). The total number increased by 11 per cent between 1969–71 and 1983–5, but the undernourished as a proportion of the total population of the developing countries fell from 27 per cent to 21.5 per cent. However, these figures exclude China, where in the mid-1980s about 90 million received less than 1.4 BMR (FAO 1988a). The World Bank used a different method of calculating the numbers undernourished in 1980 (World Bank 1986b), but like the FAO offered two estimates: of those receiving below 80 per cent, or below 90 per cent of a minimum requirement (Table 31.7); 730 million received less than 90 per cent

Table 31.6 Estimates of undernutrition in developing market economies

	1969–71	1979–81 (% of the population)	1983–5	No. of persons (millions)		
		Below 1.2 BMR				
Africa	23.5	21.9	26.0	63	78	105
Near East and North Africa	15.7	6.7	5.6	28	16	15
Asia	19.5	15.6	14.3	190	191	191
Latin America	12.7	9.8	9.5	35	35	37
All developing countries	18.6	14.7	14.6	316	320	348
		Below 1.4 BMR				
Africa	32.6	30.6	35.2	86	110	142
Near East and North Africa	22.5	10.8	9.1	41	25	24
Asia	28.7	23.5	21.8	281	288	291
Latin America	18.5	14.6	14.2	51	52	55
All developing countries	27.0	21.8	21.5	460	475	512

Source: N. Alexandratos (ed.) (1988) World Agriculture: Towards 2000: An FAO Study, London: Belhaven Press, p. 66.

Table 31.7 Prevalence of energy-deficient diets in developing countries, 1980

	Below 90% of requirement		Below 80% of requirement	
	% of population	Population (millions)	% of population	Population (millions)
Sub-Saharan Africa	44	150	25	90
East Asia*	14	40	7	20
South Asia	50	470	21	200
Middle East and North Africa	10	20	4	10
Latin America	13	50	6	20
All developing countries	34	730	16	340

Source: World Bank (1986) *Poverty and Hunger: Issues and Options for Food Security in Developing Countries*, Washington, DC: World Bank, p. 17.
Note: *Excluding China.

of minimum requirements; this however omits China. Hence it is likely that in the mid-1980s somewhere between 400 and 820 million people were undernourished in the developing countries, between 8 and 17 per cent of the world's population, but between 11 and 22 per cent of the population of the developing countries. Undernutrition is of least importance in Latin America, North Africa and the Middle East; it is of greatest significance, in absolute and proportionate terms, in tropical Africa and Asia, particularly in South Asia.

THE CAUSES OF UNDERNUTRITION

There is now a marked spatial difference in the extent of undernutrition. It is largely absent in the developed countries, but still widespread in much of the developing world. The experience of the countries of Western Europe may shed some light upon the present. Prior to the nineteenth century most countries in Europe had available food supplies as low as those now found in tropical Africa and South Asia: as in these countries today two-thirds or more of all calories were derived from cereals and roots, and livestock products – because of their greater cost per calorie – were a small part of the diet (Grigg 1993a). However, famines had largely been overcome by the early nineteenth century because of a fall in the cost and speed of transport, and by a growing awareness by national governments of the need to provide famine relief. In the nineteenth century population grew at quite unprecedented rates, but improvements in agricultural productivity increased available supplies per capita; home production was supplemented by imports of cereals and livestock products from North America and Australasia. Industrialization led, after 1850 if not before, to higher real wages. This allowed not only the purchase of more cereals and potatoes, but

Table 31.8 Food supplies, calories per capita per day, developing regions 1950 to 1987–9

	1950	*1961–3*	*1969–71*	*1979–81*	*1987–9*	*Percentage change 1950 to 1987–9*
Africa	2,020	2,031	2,087	2,199	2,218	9.8
Latin America	2,376	2,366	2,506	2,702	2,724	14.6
Near East	1,924	2,254	2,437	2,867	2,984	32.4*
Far East		1,824	2,016	2,250	2,433	33.4*
All developing	1,977	1,930	2,103	2,333	2,474	25.1
All developed	2,878	3,060	3,224	3,326	3,415	18.7
World	2,253	2,289	2,431	2,596	2,703	20.0

Sources: Food and Agriculture Organization, *The Second World Food Survey* (Rome, 1952); *Production Yearbook*, Vol. 44 (Rome, 1990).
Note: *1961–3 to 1987–9.

greater amounts of sugar, vegetable oils, fruits and vegetables, and above all more livestock products. From the later nineteenth century livestock products rose as a proportion of all calories, and the importance of the starchy staples declined. Between 1800 and 1900 available calories rose from $c.2,000$ to $c.3,000$ per capita per day yet malnutrition was by no means eliminated; records of the height of young males indicate that West Europeans in the 1870s and 1880s were below the average height today, and contemporary social surveys indicate that the poor could not buy an adequate diet. In the 1930s many writers drew attention to defective diets. It was not until the return of full employment and rapidly rising real wages after 1945 that the poor could afford a diet that eliminated malnutrition (Fogel *et al.* 1982; Boyd Orr 1937; League of Nations 1936).

Until comparatively recently the problems of malnutrition in the developing countries were blamed upon population growth. There is no doubt about the rapidity of population growth in these regions since 1945, but world food production per capita has risen substantially over the last fifty years; furthermore it has risen in nearly all parts of the developing world. The notable exception has been tropical Africa where food production per capita has fallen since the 1960s, due to a combination of continued high rates of population growth and much slower increases in agricultural output than in other regions. These increases in food production, combined with imports, have meant that available supplies have risen in all parts of the developing world (Table 31.8). Admittedly supplies are still low in Africa and South Asia, but clearly factors other than availability are important. Over the last decade little attention has been paid by writers on hunger to rapid population growth as a cause of undernutrition, and income distribution rather than inadequate production has been seen as the prime cause. There is an abundance of information that shows that calorie intake is linked to income, as is also the composition of diet. As incomes increase so too does the total number of calories per capita, and also the amount and proportion of livestock products (FAO 1988b). Whilst it is now agreed that animal protein is not essential for a healthy diet, animal foods are still the best source of protein and vitamins. In much of South Asia and tropical Africa further increases in food output are an essential prerequisite to the elimination of undernutrition. But there and elsewhere it is also essential that incomes increase if undernutrition is to be reduced; this is a function of the general level of economic development and not simply of agricultural growth. But as population growth will continue at a high level well into the twenty-first century, the problems of further increasing agricultural output cannot be neglected.

PRODUCTION AND PRODUCTIVITY IN WORLD AGRICULTURE

A principal cause of undernutrition is poverty; thus there are major differences in the extent of undernutrition between the developed and developing regions, and between the richer developing areas (such as the Middle East) and the poorer (such as tropical Africa). But differences in the supply of agricultural products are also crucial. The agricultural systems of the developed world are far more productive than those of the developing countries; as a result the real cost of food in the industrial nations has been falling for most of this century.

Most of the developing countries still have a substantial proportion of their labour forces employed in agriculture, and agriculture is generally the chief contributor to their gross domestic product. In contrast agriculture employs a very small proportion of the labour force and contributes very little to the national income. Yet the developed countries, with only one-quarter of the world's population, produce over half the value of world agricultural output, using 41 per cent of the world's agricultural area and only 5 per cent of the workforce (Table 31.9). Not surprisingly the value of agricultural production per worker in the developed countries greatly exceeds that in the developing countries; even the value of output per hectare is much higher in the developed countries, although in parts of Asia it exceeds all but Western Europe (Table 31.9). Not surprisingly of the ten leading agricultural producers, seven are industrial nations (Grigg 1992).

Differences in agricultural productivity are shown in Figure 31.4. Only Europe has output per hectare and output per capita above the world median; its output per hectare is well above that in the other developed regions, but output per capita falls well short of North America and Australasia. Throughout much of Asia output per hectare is comparable with the industrial regions, but output per capita is very low. Tropical Africa is the only region where both output per capita and output per hectare are below the world median (Figure 31.4 and Table 31.9).

690

Table 31.9 The world distribution of the agricultural gross domestic product

Region	% of world population	AGDP US$ millions	AGDP % of world total	Agricultural area Million hectares	Agricultural area % of world total	AGDP per hectare (US$)	Agricultural workforce Millions	Agricultural workforce % of world total	AGDP per worker (US$)
Africa	9.7	55,227	5.5	798	17.1	69	136	12.5	406
Near East and North Africa	5.2	42,946	4.3	349	7.5	123	35	3.2	1,227
Far East	29.0	160,432	16.1	307	6.6	522	339	31.2	473
Asian CPEs*	23.5	125,402	12.5	553	11.8	226	477	44.0	262
Latin America	8.4	85,100	8.4	751	16.0	113	41	3.8	2,076
Other developing	0.1	1,196	0.1	2	–	598	1.5	0.1	797
All developing	75.9	470,303	46.9	2,760	59.0	170	1,029.5	95.0	457
North America	5.3	114,380	11.4	508	10.8	226	3.5	0.32	32,680
Western Europe	7.4	193,824	19.4	162	3.5	1,196	13.0	1.2	14,910
Australasia	0.4	13,818	1.4	485	10.4	28	0.55	–	25,124
Other developed	3.2	90,921	9.1	100	2.1	909	6.33	0.58	14,432
Eastern Europe and USSR	7.8	119,186	11.8	664	14.2	179	31.0	2.9	3,844
All developed	24.1	532,129	53.1	1,919	41.0	277	54.4	5.0	9,782
World total	100.0	1,002,432	100.0	4,679	100.0	214	1,083.9	100.0	924

Source: D.B. Grigg (1992) 'World agriculture: production and productivity in the late 1980s', Geography 77, pp. 97–108.
Note: *Asian Centrally Planned Economies.

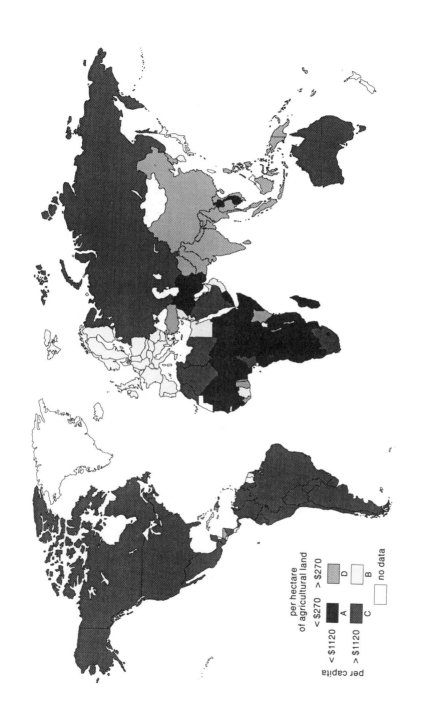

Figure 31.4 Productivity classes, 1988. After Grigg (1992).

THE CAUSES OF DIFFERENCES IN AGRICULTURAL PRODUCTIVITY

There are a great many reasons for international differences in agricultural productivity. One generally neglected cause is the environment. Most of the developing countries lie within the tropics, and these regions present great challenges to farmers. Much of Africa, the Middle East, and South Asia have large areas of low rainfall and high evapotranspiration rates, and in much of this zone agriculture is impossible without irrigation. On the margins of the deserts low and highly variable rainfall gives a meagre natural vegetation, which can be used only by nomadic pastoralists. In the wetter semiarid zones – such as the Sahel or north-west India – low rainfall greatly limits the range of crops that can be grown; crop yields are low, highly variable, and with frequent harvest failures (Swindale and Virmani 1981). In the humid tropics net photosynthesis is lower, for an eight-month growing season, than in the subtropics or temperate regions (Chang 1968); high temperatures and humidities give a greater number of plant and animal diseases than in temperate regions, and when the natural vegetation is cleared the soil is highly susceptible to erosion.

But differences in environment are not the most important reason for differences in agricultural productivity. Agricultural output per capita is highly correlated with gross domestic product per capita; agricultural productivity growth is related to the general level of economic development. There are a number of reasons for this. First is the dependence of modern agriculture upon inputs from industry. Traditional farmers – which includes most farmers in the developing countries today, and most in the now industrial countries before this century – get their inputs from their own farms; thus seed comes from the preceding harvest, power from human and animal muscle, plant nutrients from animal manure, whilst disease is controlled by rotations, fallowing and intercropping. In Sweden in the 1860s purchased inputs were only 5 per cent of the gross value of agricultural production; it is now 40 per cent and in many developed countries exceeds 50 per cent (Simantov 1967). Modern inputs are the product of a well-developed industrial economy; power comes from electricity and petroleum, the latter driving tractors and self-propelled machinery, fertilizers and pesticides from the chemical industries.

Second is the importance of research and education in modern agriculture. Modern varieties of seed are the products of research in plant-breeding institutes, pesticides and fertilizers of work undertaken in the chemical industry. Not surprisingly the state has supported agricultural research in the developed countries; the recent advances in biotechnology, however, have come increasingly from private companies. The dissemination and adoption of innovations in agriculture are also related to the level of economic development, for the provision of education – at all levels – is related to the wealth of nations.

Third has been the decline in the labour force. In most developed countries incomes in industry and services have been higher than those in agriculture for

Table 31.10 Agricultural labour productivity and inputs, 1988

| | AGDP per capita of the agricultural labour force (US$) | Agricultural labour force per 1,000 ha of agricultural land | Tractors per | | % of power on farms derived from machinery 1980 |
			1,000 ha of arable	1,000 agricultural workers	
North America	32,680	1	23	1,550	–
Australasia	25,124	1	8	750	–
Western Europe	14,910	80	91	660	–
Eastern Europe and USSR	3,844	47	16	142	–
Latin America	2,076	50	7	33	19
Near East and North Africa	1,227	100	12	28	12
Far East	473	1,100	4	4	2
Africa	406	170	1	2	2
Asian Centrally Planned Economies	262	863	9	2	–

Source: D.B. Grigg, 'World agriculture: production and productivity in the late 1980s', Geography 77, pp. 97–108.

Table 31.11 Agricultural land productivity and inputs

	AGDP per ha of agricultural land 1988 ($)	AGDP per ha of arable land 1988 ($)	Yields of all cereals kg/ha 1987–9	Kg of fertilizer per ha of arable 1984–6	% of all cereals sown with modern varieties 1982–3	% of all arable irrigated 1988	% of all agricultural land in arable 1988
Western Europe	1,196	2,061	4,433	249	–	12	58
Asian Centrally Planned Economies	226	1,140	3,860	164	56	44	20
Far East	522	592	2,034	156	40	28	88
North America	226	487	3,706	86	–	8	46
Eastern Europe and USSR	179	428	2,177	128	–	9	42
Near East and North Africa	123	523	1,604	52	19	24	24
Latin America	113	475	2,079	41	27	9	24
Africa	69	354	1,052	11	10	4	20
Australasia	28	294	1,607	32	–	4	10

Source: D.B. Grigg (1992) 'World agriculture: production and productivity in the late 1980s', Geography 77, pp. 97–108.

the last century, and this has attracted labour from farming to the towns. Indeed there has been a continuous decline in the agricultural labour force in the British Isles since the 1850s and in much of Western Europe since the 1890s, although the decline has been most dramatic since 1950. As a result farmers have been forced to substitute capital – machines and inorganic sources of power – for human labour.

There are of course many other reasons for international differences in productivity. But differences in the use of capital inputs are of prime importance and they reflect international differences in economic development. Thus in terms of labour productivity there is a close inverse relationship between output per capita and labour density, and a positive relationship with the number of tractors per agricultural worker; virtually all the power in industrial countries is derived from machinery, but draught animals and human labour provide the great majority in developing countries (Table 31.10).

Inputs that increase output per hectare are less closely related to the level of economic development; fertilizer usage, for example, is higher in China than in North America, and cereal yields in the Far East exceed those in Australasia. But virtually all the seed sown in the developed countries are modern varieties, whereas they only occupy over half the sown area in China and substantially less elsewhere in the developing countries (Table 31.11).

The rapid adoption of industrial inputs has transformed Western agriculture since 1950 and given animal and crop yields well above most parts of the developing world. Farm output in the 1950s, 1960s and 1970s increased at unprecedented rates, and indeed gave rise to the surpluses noted earlier. But it should not be thought the food output has stagnated in the developing countries; indeed it has risen more rapidly than in the developed countries since 1950, but of course increases in population growth have meant that increases in output per capita have been less impressive (Table 31.12). Thus differences in agricultural capacity remain a powerful factor in explaining the distribution of under-nutrition. The modernization of agriculture has brought great increases in productivity in the developed countries, and together with rising incomes has banished undernutrition. Neither process has made so much progress in the developing world.

SUSTAINABLE AGRICULTURE

In the post-war period farming in the developed countries, particularly in North America, Australasia and Western Europe, has experienced a remarkable increase in productivity and output, providing an abundant supply of food at falling real prices. Yet there has been a growing chorus of critics. Some have focused upon the economic consequences for those who work in agriculture, lamenting the decline of the numbers who can find employment on the land, regretting the demise of small farms, and pointing to the rise of a dual economy: a small proportion of all farms produce an increasing percentage of total output,

Table 31.12 Rates of increase in food output and output per capita (per cent per annum)

	Rates of increase in food output			Rates of increase in food output per capita		
	1952–4 to 1959–61	*1961–70*	*1971–80*	*1952–4 to 1959–61*	*1961–70*	*1971–80*
Developing countries	3.1	3.1	3.3	0.7	0.7	1.0
Developed countries	3.0	2.4	1.9	1.7	1.4	1.1
World	3.1	2.7	2.5	1.1	0.8	0.6

Sources: Food and Agriculture Organization (1970) *World Agriculture: The Last Quarter Century*, Rome: FAO, p. 9; Food and Agriculture Organization (1982) *The State of Food and Agriculture 1981*, Rome: FAO, pp. 5–6; Food and Agriculture Organization (1977) *The Fourth World Food Survey*, Rome: FAO, p. 4.

and maintain high rates of profitability (and gaining most of the benefits of protection); a large proportion of farms are small, unprofitable and contribute little to total output. Other writers have regretted the rise of agribusiness; modern agriculture depends upon inputs from industry, particularly the chemical industry, and it sells a large proportion of farm output to the food processing sector, and so the farmer, it has been argued, is becoming no more than a provider of raw materials. As such he has little control over the price of his inputs, and an increasing proportion of the sale price of food in retail outlets goes to the processors and distributors. As yet, however, there is little vertical integration of the food production system in Europe, although a large proportion of United States output of some commodities such as broilers and lettuce are controlled by food processors.

But probably the fiercest criticisms of industrialized agriculture have come because of the effect of new methods upon the environment. The first signs of alarm came in the 1950s when the impact of the use of pesticides upon flora and fauna became apparent. Herbicides were developed to kill weeds, insecticides and fungicides to control a wide range of pests. As these are delivered by sprays, a low proportion of the spray reaches the pest, much goes onto other flora and fauna. Sprays, and also seed dressings, have killed birds and thinned eggshells. Sprays may also have reduced the flora – particularly in old meadows (Conway and Pretty 1991). Nitrate fertilizers have also had adverse consequences. Since the 1960s the application of nitrate fertilizer has increased remarkably in the arable areas of North America and Western Europe; this has increased crop yields, but also increased the nitrates in surface water and groundwater supplies. Excessive nitrate content in surface water leads to rapid plant growth which clogs rivers, encourages algae growth that excludes sunlight and may thus kill fish by reducing the oxygen supply. The European Community has laid down safety limits for the content of nitrates in water supplies, which are exceeded in parts of eastern England; there are fears that a high level of nitrates may be associated with stomach cancer and the blue baby syndrome. The nitrate level may also have been increased by leakage from tanks used to store animal waste (Addiscott et al. 1991).

The use of machinery is also said to have had adverse effects upon the environment in many of the older settled areas of Europe, notably northern France and southern Britain, where fields were small and enclosed by hedges. This greatly reduces the efficiency with which machinery can be used; field enlargement has led to a serious reduction in hedges, whose length in England has fallen by over one-fifth since 1945; this has reduced the habitat of birds, animals and insects, and reduced protection against wind erosion. The use of heavy machinery has compacted soils, and so increased the risk of soil erosion.

Modern farming techniques are thought to have increased the rate of soil erosion, although erosion has of course occurred in the past and is currently widespread in much of the developing world where modern farming methods are not used. The climate of Western Europe makes the soil less susceptible to soil erosion; even so, since the 1960s the intensification of farming in lowland Britain has increased the risk of soil erosion. The high price of cereals has led to the cultivation of high-angle slopes, and farmers have ploughed up and down slopes instead of along contours; the creation of finer seed beds has increased the risk of wind erosion. The decline of livestock farming in arable areas has reduced the amount of organic matter in the soil, as has the ploughing of grassland, whilst the use of heavy machinery and 'tramlining' by chemical sprayers has made sheet and rill erosion more likely. In the United States, with its more extreme climates, soil erosion has become a major problem; the United States Department of Agriculture states that soil erosion is significant if the annual loss of soil exceeds 2 tonnes per hectare on deep soils, 0.4 tonnes on shallow soils; currently erosion exceeds 2 tonnes on 27 per cent of the total cropland (Hodges and Arden-Clarke 1986).

Not surprisingly there has been a call for a change in the methods of farming in the developed countries. This was first prompted by the rapid rise in oil prices in the early 1970s, when it seemed that modern mechanized farming would become uneconomic; but the subsequent decline in the real price of petroleum has lessened the force of this argument. A more cogent critique has suggested that farmers should reduce their inputs of fertilizers and pesticides; such an approach has been variously described as sustainable agriculture, low-input farming, alternative agriculture, radical agriculture, and in Britain, organic farming. It is in effect a return to the practices found in Western Europe before the 1950s. Most farms then grew crops and also kept livestock. Animals provided dung which, mixed with the straw from cereal crops, provided farmyard manure, then the source of most plant nutrients. Most farms had land under grass and under crops, the former improving soil structure and organic content, whilst crops were grown in rotation, providing some control of disease. Row crops – such as potatoes, turnips and sugar-beet – allowed hoeing during crop growth, and kept weeds down. A majority of the inputs were produced on the farm, soil fertility was maintained, and soil erosion minimized. Efforts have been made to encourage a return to such a style of farming, but there are serious doubts about its viability under present economic conditions; nor is it clear from where the much greater labour force it would require would come.

CONCLUSIONS

The second half of the twentieth century has seen unprecedented rates of technological change in farming in the developed countries, and a great debate upon the problems of protection and environmental degradation. In the developing countries attention has been focused upon the need to increase food output and reduce the extent of malnutrition. In both cases the future of farming lies not in the hands of farmers, but in the actions of politicians. To end – or at least substantially reduce – the degree of protection in developed countries will need a degree of unity and purpose not so far displayed. In the developing countries politicians will have to ensure both a growth of national income and its wider distribution if malnutrition is to decline.

REFERENCES

Ackrill, R. (1992) 'The Common Agricultural Policy; its operation and future', *Economics* 28, 5–11.

Addiscott, T.M., Whitmore, A.P. and Powlson, D.S. (1991) *Farming, Fertilisers and the Nitrate Problem*, Wallingford: CAB International.

Arnold, D. (1988) *Famine, Social Crises and Historical Change*, Oxford: Basil Blackwell.

Bohland, J.R. (1988) 'Rural America', in P.L. Knox, E.H. Bartels, B. Holcomb, J.R. Bohland and R.J. Johnston, *The United States: A Contemporary Human Geography*, London: Longman.

Boyd Orr, Sir John (1937) *Food, Health and Income*, London: Macmillan.

Buttel, F.H. (1989) 'The U.S. farm crisis and the restructuring of American agriculture; domestic and international dimensions', pp. 46–83 in D. Goodman and M. Redclift (eds) *The International Farm Crisis*, London: Macmillan.

Carr, E. (1992) 'Agriculture; grotesque', *The Economist* 325(7789), 1–18.

Carson, R. (1970) *Silent Spring*, London: H. Hamilton.

Chang, J. (1968) 'The agricultural potential of the humid tropics', *Geographical Review* 58, 333–61.

Clunies-Ross, T. and Hildyard, N. (1992) *The Politics of Industrial Agriculture*, London: Earthscan.

Conway, G.R. and Pretty, J.N. (1991) *Unwelcome Harvest: Agriculture and Pollution*, London: Earthscan.

Corbet, H. (1991) 'Agricultural issues at the heart of the Uruguay Round', *National Westminster Quarterly Review* (August), 2–19.

Dando, W.A. (1980) *The Geography of Famine*, London: Edward Arnold.

Fogel, R.W., Engerman, S.L. and Trussell, J. (1982) 'Exploring the use of data on height: the analysis of long term trends in nutrition, labour welfare and labour productivity', *Social Science History* 6, 401–21.

Food and Agriculture Organization (1977) *Fourth World Food Survey*, Rome: FAO.

Food and Agriculture Organization (1987) *The Fifth World Food Survey*, Rome: FAO.

Food and Agriculture Organization (1988a) *The State of Food and Agriculture 1987–88*, Rome: FAO.

Food and Agriculture Organization (1988b) *Review of Food Consumption Surveys 1988*, Food and Nutrition Paper 44, Rome: FAO.

Food and Agriculture Organization (1991a) *The State of Food and Agriculture 1990*, Rome: FAO.

Food and Agriculture Organization (1991b) *Production Yearbook 1990*, Rome: FAO.

Food and Agriculture Organization (1991c) *Trade Yearbook 1990*, Rome: FAO.

Grigg, D.B. (1992) 'World agriculture: production and productivity in the late 1980s', *Geography* 77, 97–108.

Grigg, D.B. (1993a) 'International variations in food consumption in the 1990s', *Geography* 78, 251–66.

Grigg, D.B. (1993b) 'Income, industrialization and food consumption', *Tijdschrift voor Economische en Sociale Geografie* 85, 3–14.

Grigg, D.B. (1993c) *The World Food Problem*, Oxford: Basil Blackwell.

Hodges, R.D. and Arden-Clarke, C. (1986) *Soil Erosion in Britain, Levels of Soil Damage and their Relation to Farming Practices*, Bristol: Soil Association.

Jensen, N.F. (1978) 'Limits to growth in world food production', *Science* 201, 317–20.

Johnson, D. Gale (1973) *World Agriculture in Disarray*, London: Macmillan.

League of Nations (1936) *The Problem of Nutrition*, Vol. 3: *Nutrition in Various Countries*, Geneva: League of Nations.

Lingard, J. and Hubbard, L. (1991) 'The CAP and its effects on developing countries', pp. 241–57 in C. Ritson and D.R. Harvey (eds) *The Common Agricultural Policy and the World Economy*, Wallingford: CAB International.

Silvey, V. (1978) 'The contribution of new varieties to increasing cereal yield in England and Wales', *Journal of the Institute of Agricultural Botany* 14, 367–84.

Simantov, A. (1967) 'The dynamics of growth and agriculture', *Zeitschrift fur Nationalökonomie* 27, 328–51.

Sturgess, I.M. (1992) 'Self sufficiency and food security in the UK and EC', *Journal of Agricultural Economics* 43, 311–26.

Swindale, L.D. and Virmani, S.M. (1981) 'Climatic variability and crop yields in the semi-arid tropics', pp. 139–66 in W. Bach, J. Pankrath and S. Schneider (eds) *Food–Climate Interactions*, London: Reidel.

World Bank (1986a) *World Development Report 1986*, New York: Oxford University Press.

World Bank (1986b) *Poverty and Hunger: Issues and Options for Food Security in Developing Countries*, Washington, DC: World Bank.

FURTHER READING

Body, R. (1982) *Agriculture, the Triumph and the Shame*, London: Temple Smith.

Edwards, C.A. (1990) *Sustainable Agricultural Systems*, Ankeny, Ia.: Soil and Water Conservation Society.

Goodman, D. and Redclift, M. (1989) *The International Farm Crisis*, London: Macmillan.

Goudie, A. (1990) *The Human Impact on the Natural Environment*, Oxford: Blackwell.

Grigg, D.B. (1993) *The World Food Problem* (2nd edn), Oxford: Blackwell.

Hayami, Y. and Ruttan, V.W. (1985) *Agricultural Development: An International Perspective*, London: The Johns Hopkins University Press.

Johnston, B.F. and Kilby, P. (1975) *Agriculture and Structural Transformation. Economic Strategies in Late-developing Countries*, London: Oxford University Press.

Sen, A. (1981) *Poverty and Famines. An Essay on Entitlement and Deprivation*, Oxford: Clarendon Press.

Tracy, M. (1989) *Government and Agriculture in Western Europe, 1880–1988*, London: Harvester Wheatsheaf.

32

THE NATURE OF THIRD WORLD CITIES

David Drakakis-Smith

INTRODUCTION

The popular image of cities in developing countries tends to be dominated by the acute contrasts afforded by modern high-rise commercial buildings, replete with the premises of multinational firms, and squalid, chaotic squatter settlements. Not infrequently these occur side by side, which is very convenient for the ten-second background shot of a television reporter describing another ecodisaster or reactions to it. The Rio and Cairo meetings of the early 1990s provided many such visual 'bites', as did the Cancun summit a decade earlier.

Are these instant images a fair reflection of life in the cities of the Third World? Certainly for most countries increasing proportions of their populations are facing the prospect of a permanent life in the city, as Alan Gilbert has elaborated in Chapter 19; but considerable variation exists between and within cities in relation to the conditions with which urban populations will have to cope. Clearly much will depend on the economic and political circumstances of the national or urban context, but there are a whole range of other considerations that must be taken into account in assessing the situation – the circumstances of the individual household, the class structure of the city, the level of political commitment to social welfare schemes, the political structure of local and state government, and many more.

Any attempt, therefore, to summarize the features of the Third World city must admit to a level of generalization which clearly will not exist in reality (Gilbert 1992). At best this discussion will seek to identify the common denominators that have shaped, and continue to shape, such cities. However, I will not attempt to move into the realms of theory and produce a model of the Third World city. Such a level of abstraction, given the huge variation in the history and contemporary situation of urbanization in the Third World, would be even less useful than analogous models produced for Western cities. This is not to say that schematic representations of this kind have no use, but they are perhaps more fruitfully discussed in the context of particular regions, such as South-East Asia (McGee 1967) or Southern Africa (Simon 1993). Thus, the concern of this chapter will be with the processes of change as much as with its end-product, whether this be a housing deficit, political instability, traffic congestion or pollution. Finally, we must remind ourselves that cities are not merely collections of buildings or functions, they are primarily concentrations of people – an axiom which must not elude our analysis; indeed, it has to remain central to it.

Given the complexity of cities and their multiplicity of functions, it is necessary to identify some sort of structure within which the discussions of this chapter can be located. Any brief review, such as this, must be selective and yet that selection must be informed by some conceptual framework. In this context, the growing interest in sustainable urban development perhaps provides the most useful guidelines. Sustainable development *per se* has had a growing influence on both the rhetoric and reality of the development process throughout the 1990s, but the specific application of this concept to urbanization has been relatively limited. Indeed, urbanization has been blamed for many of the problems associated with unsustainable development, particularly through its demand for resources and its production of wastes.

However, urbanization itself must be considered not only as a means to sustain growth by economic surplus, but also as a component of sustainable development at global, national and local levels. But before we can hope to structure and manage urbanization so that it brings maximum benefits to the greatest number, whilst retaining environmental awareness and integrity across space and through time, we need to be able to identify the various elements or components that comprise sustainable urban development and appreciate how they integrate with one another. In this context, I would identify five broad areas which together shape the nature of the Third World city, and format the basis of any attempt to address sustainability (see Drakakis-Smith 1995). These five areas are:

- Demographic features: these would include, for example, the relative importance of natural and migrational growth, and the degree of ethnic or cultural homogeneity.
- Economic features: these include not only the production and consumption roles of the city but how these relate to employment, incomes and poverty at the household level.
- Social issues: for example, the provision of basic needs and human rights, gender matters and the like.
- Environmental matters: this comprises the so-called 'brown agenda' associated with the role of the city in the consumption of renewable and non-renewable resources and the associated production of wastes.

703

- Political concerns: this broad-ranging section includes issues related to democracy and decentralization, both of which influence direct management of the city and the nature of the planning process.

Clearly, any of the matters raised above overlap the broad categories identified. Human rights, for example, are both a social and political issue. Moreover, the discussion of many of these issues has already been covered in Alan Gilbert's chapter in this volume. However, these criticisms serve only to underline the integrative nature of the development process as it relates both to cities and sustainability. What the remainder of this chapter will do, therefore, is to review those issues more specifically related to the city *per se*, rather than the urbanization process as a whole, whilst at the same time recognizing the fact that urbanization as a physical, economic and social process extends beyond the formal boundaries of the city into what is increasingly being referred to as the extended metropolitan region (see McGee and Greenberg 1992). The issues covered, therefore, encompass urban management, the urban economy, environmental matters and social welfare concerns.

Finally, in this introduction, a brief comment about the time-scale of the discussion which follows. Although the urbanization process has accelerated at a phenomenal rate during the second half of this century, many cities in developing countries are the product not only of their recent, often traumatic past, but also of their more distant past. This can encompass the vicissitudes of pre-colonial periods but for the purpose of this discussion primarily covers the legacies of colonialism. It is appropriate that we begin, therefore, with a brief historical overview.

THE LEGACIES OF COLONIALISM AND BEYOND

Despite the importance attached by many theories of development to the integration of contemporary social formations with the world economy, the literature on the pre-independence or colonial phase of urban development is still disappointingly sparse. Those discussions do, however, constitute a rich vein of valuable material which this section will seek to mine (see especially King 1976, 1990; Taylor 1985; Dixon and Heffernan 1991; McGee 1967; Lowder 1986; Telkamp 1978; Drakakis-Smith 1991).

Many discussions of the colonial city must immediately confront a contradiction in terminology since the processes which most authors discuss relate to imperialism which Taylor (1991: 6) describes as 'a structural relation of the capitalist world economy'. This changing relationship covers a long period of some four hundred years since capitalism first began to emerge from Europe. However, over this period it has overlapped with an accelerated colonialism *per se* (the establishment of a geographically extended political unit which is politically and economically subordinate to a metropolitan power), a phenomenon closely tied to the impetus of capital to reproduce itself.

704

The resultant processes varied in their spatial and temporal extent. In Latin America intense imperialism sustained through extensive colonialism was over by the mid-nineteenth century, but in Asia and Africa the period from 1500 to 1800 was marked for the most part more by mercantile colonialism, exploitation of natural resources structured through relatively limited metropolitan settlement. Between the mid-nineteenth and mid-twentieth centuries, however, the imperialism of industrial capital combined with aggressive territorial colonialism resulted in the structured exploitation that is now associated with colonialism.

However, even within the relatively short period of intensive industrial colonialism we must be aware of the tremendous variation that existed in terms of its impact on urbanization. In some areas colonial urbanization occurred in regions with an already lengthy urban history, such as in North Africa or South Asia; in others entirely new urban systems were created. The process can be examined fruitfully through the analysis of the combination of the economic role the colony and its cities played in metropolitan expansion, the political power structure which evolved within the colony, the technological changes that were introduced and the cultural contrasts that existed.

In real terms this produced a wide range of legacies from the colonial city, some in terms of building fabric, others related to land-use structures or to social relations. For example, many of the old pre-colonial urban areas were ignored by the colonial population who used their new technologies to create broad-boulevarded, geometrically influenced 'European' zones, geared to metropolitan transport needs and building regulations. In contrast the other 'indigenous' parts of the city were neglected and allowed to fall into disrepair (Lowder 1986).

As new planning principles were introduced into the colonies (and it must be appreciated in this context that there was often more freedom to experiment outside the metropolitan centre), so the nature of the European enclaves were expanded and transformed, replete with massive administrative or cultural manifestations of power within the built environment, such as huge city halls, libraries or 'clubs'. Often such areas were surrounded by extensive open spaces for the exclusive use of Europeans. These polo fields, parks and, later, golf courses, also served as *cordons sanitaires* separating the colonizers from the colonized. Such areas still exist in many Third World cities, often with remarkably similar spatial patterning, and have experienced varying fates. In some, the open spaces are now available for public use; indeed, some have been invaded by squatters seeking space for housing, while others retain a degree of their exclusiveness through the preservation of colonial planning values. This is particularly true of privately owned spaces, such as school playing fields, or golf courses. Meanwhile, the old indigenous quarters of many former colonial cities remain hugely overcrowded and pose massive problems for contemporary government, for example in Old Delhi.

But the legacies of colonialism for the cities of the Third World are not confined to its built environment; they relate also to the regulation of social space. Many of the social and political problems experienced by contemporary

705

Third World cities are also deeply rooted in their colonial past. Much of this is linked to the common practice in many colonies of introducing or permitting an alien ethnic group to assume responsibility for urban petty commerce. Most of the cities of South-East Asia, for example, have substantial Indian or Chinese business communities, as do many East African states. In the period immediately following independence in the 1950s and 1960s, the resentment of indigenous populations who felt themselves excluded from profitable urban businesses not infrequently spilled over in bloody ethnic conflicts, resulting in voluntary or forced outmigration by minority groups.

It is true that many cities were riven by class, ethnic and cultural tensions before the Europeans arrived but, as O'Connor has remarked

> Immigration, pluralism, (primacy) and an ethnic division of labour were not new . . . Colonial cities merely magnified and ossified these long-standing patterns. Whatever else changed, the city remained the centre of wealth, power and prestige . . . What changed was not this role and its meanings, but the outside world.
>
> (O'Connor 1983: 14)

The world has continued to change, at an ever-increasing pace, but we must bear in mind that there is a degree of continuity in the urbanization process as it affects the structure of the cities themselves. The next section will reveal this continuity only too clearly.

MANAGING THE THIRD WORLD CITY

Before any detailed discussion of the economic and social features of Third World cities, it is necessary to investigate their management and administrative structures. In this context, the most important point to appreciate at the outset is that most cities, even capital cities, are not necessarily managed by or for those who live in them. This contention is not intended to refer only to the mass of urban poor who are excluded from democratic political participation (where this exists) by virtue of franchise disentitlement; it is also intended to encompass the fact that cities are not managed solely by their elected or appointed politicians, bureaucrats or planners.

Any recent discussion of the urban crisis in Africa, Asia or Latin America will endorse this assertion by the very range of actors and agencies that are drawn into the debate (see Yeung 1990; Stren and White 1989; Gilbert 1992; Devas and Rakodi 1993). Although the 'urban chaos' of disruption in public utilities, traffic congestion or housing deficiencies is often attributed to the inefficiency and corruption of bloated urban bureaucracies (Wallis 1989), we must also incorporate into the discussion references to National Housing Banks, intervention by senior politicians, funds and 'experts' from the World Bank, instructions and guidance from the International Monetary Fund, and the growing presence of various national and international non-governmental organizations (NGOs) (see Douglass 1992). The individuals and agencies involved in this plethora of advice or instruction hold varied and often conflicting loyalties; some need to justify their policies to superiors, some to national constituencies, others to international bankers, many are simply self-seeking opportunists. Few are directly responsible to the urban residents who are theoretically the beneficiaries of their efforts.

Not all of these influences are indirect, many capital cities have had their planning processes usurped directly by national governments, and the chief executive of the city is often an eminent national political figure (see Drakakis-Smith and Rimmer 1982). For example, the ultimate authority of the Federal Capital Territory in Malaysia is the Prime Minister; in the Philippines, Metro Manila was for many years governed by Imelda Marcos who used this position to plan the city for her own self-aggrandizement. Thus urban investment was channelled into conference centres and luxury hotels, in which events such as the Miss World Pageant or world championship boxing were held (remember 'the thriller in Manila'!). Needless to say, few of the ordinary citizens of Manila benefited directly from such decisions on urban management; indeed, the parallel policies of squatter eviction ensured the opposite effect (see Bello 1992).

Although this is an extreme example of the manipulation of the urban planning and management process, it is far from unique. Capital cities in particular are vulnerable to government desires to create a showpiece of modernity; a built environment that will, parts of it at least, impress visiting dignitaries and attract potential investors. But direct intervention is not the only way in which urban management is distorted. The fact that funds for so many basic services are sectorialized and centralized means that, for example, the housing programmes in most cities are more likely to be determined by priorities decided in the capital. Apart from shortages at the centre, eventual funding may well be determined, therefore, by the political relationships between the urban authorities and the centre. In this context, the greater clout of larger or more economically important cities drastically curtails the funds available for those further down the urban hierarchy (see Lowder 1991).

However, influences on urban management need not be limited to the public sector; financial institutions of varying kinds have increasingly imposed their own conditions on hard-pressed urban authorities deprived of a substantial local fiscal resource base. Multinational firms have been able to negotiate virtually as equal partners for land and infrastructural provision with urban governments desperate to attract investment. Whilst international financial institutions, both private and multilateral, have dictated the terms, conditions and even the purpose of loans, the driving force behind sector involvement in urban management is the fact that rapid urban growth offers enormous opportunities for commercial interests to make massive profits through redevelopment projects. Many of these projects, designed to appeal to image-conscious administrations, are huge in scale and are designed and built by Western firms. Such projects include the contentious £10 billion Check Lap Kok airport in Hong Kong and the £4 billion redevelopment of the Pudong area adjacent to

Shanghai. However, even the more everyday reshaping of the built environment offers massive opportunities for profit whilst at the same time wrenching cities away from their local cultural values and pushing them towards an international norm. As Cook notes, even

> the back street *hutongs* of Beijing are being overwhelmed by the invariable growth of mid-rise and high-rise blocks ... in the last decade a total of over 2,000 residential quarters, each with a floor space of more than 50,000 square metres, have been built ... Chinese urbanists justify this policy because of the poor provision of public utilities in the *hutongs*, an argument which to me has uncomfortable echoes of the justification for British inner-city redevelopments of the 1960s.

> (Cook 1993: 5)

Of primary importance in this process has been the World Bank whose own developmental priorities, whether for aided self-help housing or six-lane highways, have determined what urban authorities in large and small cities alike may do. As the enthusiasms for certain schemes have changed, so Third World cities are frequently forced to live with the often contradictory consequences. Rahim (1988) clearly illustrates such confusion in his account of transport investments in Kuala Lumpur, which in the space of a few short years oscillated in favour of multilane highways, public maxi-bus services and small-scale paratransit. Erratic swings in the nature of national and international intervention and influence have perversely been the one reliable constant in the whole equation. In addition to the fluctuations in policy fashions and the contradictions between institutions in their developmental priorities, as noted above, there have also been major waves of advance and retreat in views on the extent to which the state or the private sector should be involved in urban management.

The current trend, in the light of right-wing policy shifts in the West, has been towards 'rolling back the state'. This is dignified by some as giving back power to civil society (de Soto, as reported in Gilbert 1992). But to whom? By others it is seen as 'anti-public and pro-private enterprise' (Wallis 1989: 30). The starting point for much of this change in attitude was, of course, the debt crises and the corresponding structural adjustment programmes which ensued. The political roll-back in state involvement in social welfare programmes, together with the decline in the availability of funds due to the global recession itself, has been particularly felt in the cities where most of the limited activity of this nature had occurred (Cohen 1990; Friedmann 1992). For some, the new policies are justified by the 'awakening of self-reliance in the urban poor' (*Time*, 11 January 1993: 33). Of course such self-reliance has always existed, it has simply been 'rediscovered' on useful occasions, such as the present. Deregulation of the informal sector, it can be argued, has drawn such activities even more closely under the scrutiny and control of the state, reliant as it is on the self-help activities of the poor to compensate for the withdrawal of welfare subsidies.

Of course, the impact of all these changes has been extremely variable across the urban Third World, and has been further affected by a wide variety of local factors such as the legacies of colonialism like ethnic or cultural antagonisms (see

Ward 1989; Brown 1994), or the extent of individual corruption which cut across national or international trends. In short, there has been little spatial or chronological synchronization in either cause or effect. Many cities are still in the grip of 1960s modernization strategy and prestige projects continue to take precedence over the needs of low-income or even middle-income populations; others, such as Singapore, are already privatizing public enterprises as rapidly as they can. Indeed, it is not uncommon to find such contradictions and inconsistencies in the management of single cities. Thus, in Harare, whilst the national government advocates (at least in its rhetoric) an enhanced role for the informal sector in the provision of basic jobs and services, the city council continues to demolish squatter settlements and persecute those in informal activities (Drakakis-Smith 1993).

Amidst all this welter of change and confusion in policy, urban planning departments continue to struggle in practice with fluctuating objectives and funds, often trying to reconcile conflicting instructions from above with planning systems, procedures and regulations which are hopelessly out of date, elitist and Western in character. Moreover, in many cases their remit does not extend to the immensely complex areas outside formal urban boundaries. In this context, it is not surprising that increasing interest is being shown in the concept of the extended metropolitan region (EMR) (McGee and Greenberg 1992; McGee and Yeung 1993) which is said to differ from conventional urban structure in terms of the way in which advances in transport have rapidly increased the flow of people and commodities, breaking down any clear distinction between rural and urban areas and incorporating rural population into the urban region *in situ*.

EMPLOYMENT IN THE CITY

There is an enormous literature on economic growth and industrialization in the Third World (see, for example, Knox and Agnew 1989; Chandra 1992; Dicken 1991; Dixon and Drakakis-Smith 1993), and it is not the intention of this section to review or examine this general topic. However, it is a strong feature of economic development in the Third World that investment policy favours urban-based industrial growth. For the last two or three decades Third World governments have sought to encourage the growth not only of import-substitution industries but also, and more particularly, export-oriented manufacturing. The goal is to boost economic earnings by producing goods cheaply through the use of abundant local resources. For the most part this means human resources and again focuses attention on the ever-growing pools of labour available in cities of all sizes. Thus, for most national governments the goal is sustained urban growth rather than sustainable urban development.

For many governments the preference for urban-based industrialization is reinforced by the positive correlation between the level of urbanization and GNP in the Third World. However, quite a different and much less positive

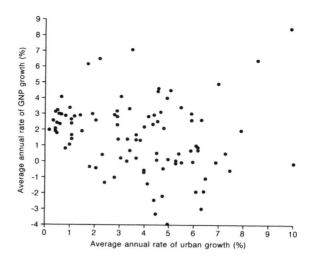

Figure 32.1 Urban and economic growth, 1980–90. After World Bank (1994).

relationship emerges if economic and urban growth are compared (Figure 32.1) since urban expansion is often associated with near zero economic growth, usually in countries where investment anywhere is minimal and where a massive shift of rural poor into the cities continues to take place. Such data reveals the selectivity of industrialization within developing countries and, indeed, until very recently there were only a handful of rapidly industrializing economies (RIEs), with most of these being concentrated in Pacific Asia. In 1990 over 60 per cent of the manufacturing exports of the Third World originated in the 'four dragons' of South Korea, Hong Kong, Taiwan and Singapore. The reasons for such selectivity are not difficult to discern. Manufacturers require more than cheap labour and investors prefer locations with an educated, trainable labour force, reliable infrastructure, good communications and the like. Relatively few cities offer such necessities to a high standard, although many Third World governments offer additional inducements through financial packages, such as duty reductions on imported and exported goods or cheap factory units. Often such packages are spatially concentrated in Free Trade or Export Processing Zones which reinforce existing tendencies by being located in the largest cities.

Of course not all investment in Third World industrial development is from Western multinationals. Several of the RIEs benefited from substantial domestic investment. Increasingly RIE capital itself has become more mobile and a complex pattern of movement to other industrializing countries, as well as to Europe, has evolved (Dicken 1991). As with most international capital movement, this has been heavily concentrated in capital cities and Dixon (1993) has estimated that almost 50 per cent of all inflowing foreign funds to Thailand are invested in the capital Bangkok.

But perhaps, as far as sustainability is concerned, the most important feature of urban industrial investment in the Third World is the extent to which it provides employment for the rapidly accelerating labour force, many of whom were drawn into the city by the prospect of such work. In this context, we must remember that export-oriented industries, whether domestic or foreign, are producing for a discerning world market, usually in developed countries. Most emphasis is, therefore, on high-quality production and high productivity per capita. Both of these mean capital investment, rather than employment of a huge, cheap but unskilled workforce. In many countries, usually those whose economies are expanding most rapidly, the proportional contribution of manufacturing to economic production far outweighs its contribution to employment (Figure 32.2). Indeed, in some countries growth in manufacturing output has even been accompanied by a decline in manufacturing employment (Todaro 1989). So in most cities the majority of employment is not in large firms but in medium to small operations, usually with fewer than ten employees. Industrial employment is, therefore, often equally important throughout the urban hierarchy in terms of relative proportions, although clearly absolute figures emphasize the concentration in the largest cities.

These structural characteristics of urban industry in the Third World lead on to several further points worthy of note. First, the built environment of industry is very varied. Although many cities have concentrations of large factories, often

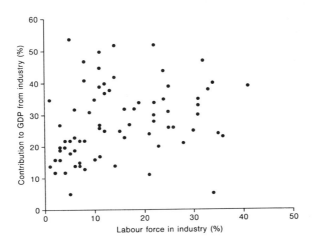

Figure 32.2 The contribution of manufacturing to employment and production. After World Bank (1994).

711

foreign-owned, in Export Processing or Free Trade Zones, much industrial employment continues to be located in small-scale businesses scattered throughout the urban area, frequently in the older central areas where they blend almost imperceptibly with the less-well-recorded activities of the informal sector. Such activities must not be underestimated since they offer considerable flexibility of operations and are said to underpin much of the success of Hong Kong and Taiwan in particular (Shibusawa *et al.* 1992).

Given the relatively limited employment opportunities created within the manufacturing sector, where is the remainder of the workforce employed? Certainly in almost all cities there are large proportions of the urban labour force found in the tertiary or service sector. In many capital cities, this would include a substantial number employed in government or government-related jobs. Indeed, in some countries, such as India, employment has been deliberately inflated in order to absorb the burgeoning secondary and tertiary graduates who, if left idle, may well foment unrest. But the service sector category can disguise quite substantial variations in character. In a few cities it would primarily encompass work in modern retail or office establishments, reflecting a rapid shift to producer services; for example, in Singapore, Seoul or São Paulo. In most cities, however, the service sector would also include rapid growth in domestic service or in more traditional retailing occupations, such as markets. The bottom line, however, is that even an inflated service sector cannot provide enough employment opportunities for the massively expanding workforces of most Third World cities. Large numbers are, therefore, forced to seek work in what is variously known as the informal or petty commodity sector.

Despite the fact that the informal sector has been closely studied for some twenty years, few have established a satisfactorily compact definition for it. This is perhaps not surprising when the range of activities which can be incorporated into the sector is considered. These stretch from the construction of squatter settlements or prostitution to illegal cultivation of food. Part of the problem in the past was that attempts to 'define' the informal sector simply reversed the features of formal employment in the city, implying also that the sectors were distinct and contrasted (see Sethuraman 1981). However, subsequent investigation has revealed that there is a considerable overlap between formal and informal sector activities (see Bromley and Gerry 1979): individuals with formal sector employment, even with the government, may live in squatter settlements; domestic sewing activities may be on contract to a local factory; the trishaw pedalled around the streets may be leased by its driver from a teacher. All these linkages imply a flow of people, commodities, services and funds between the two 'sectors'.

An informal activity, therefore, may conform to any number of features allegedly characteristic of that sector. Illegality or semi-illegality is certainly true of many activities, land is occupied without the owner's permission, premises are constructed which flaunt building laws, commodities are made or sold in defiance of health or safety regulations. In addition, informal sector activities tend to be small in scale, offer low irregular incomes, employ unpaid family labour, and be traditional in their technology (see Friedmann 1992 for a useful list). However, they may conform to almost none of these features. For example, many 'squatter settlements' in Ankara were originally designated as illegal by virtue of there being no cadastral maps against which plots could be claimed or bought. The houses themselves are often substantial stone and brick constructions (Payne 1984).

Because of the difficulties of defining the informal sector and also because many of its activities, by virtue of their semi-legal nature, go unrecorded, it is extremely difficult to draw estimates as to its statistical extent. For example, Todaro (1989) cites figures as varied and as suspicious as 25 per cent (Singapore) to 69 per cent (Pakistan). What is more certain, however, is the value of the sector – not only in creating employment, and therefore income, but also in the positive role it plays in the life of the city as a whole. It may be true that the informal sector is essentially run by the poor for the poor, but the goods and services produced through petty commodity activities benefit most urban residents, rich and poor alike. Fresh fruit and vegetables are usually moved around the city by informal suppliers and retailers (Young 1990), the better-off are usually the patrons of the trishaw operators (Forbes 1981), whilst domestic outworkers, usually female, are often essential to small businesses in absorbing fluctuations in supply and demand.

Despite this valuable role and the fact that the net movement of created wealth is from the informal to the formal sector via unequal exchange (see Forbes 1981), most urban authorities have a range of laws available which they can and do use to proscribe and control the activities of the informal sector. For many administrators the informal sector gets in the way and spoils the image of the modern city that they have worked so hard to create (see Bibangambah 1992). Roads blocked with hawkers' carts or trishaws, luxury hotels that overlook squatter huts, airport highways whose median strips are full of maize or vegetables are all anathema to most urban managers; and yet they too benefit from the operations of the informal sector.

In the 1970s, in the wake of the academic 'discovery' of the informal sector, it became the conventional wisdom of development policy to seek ways of supporting informal activities. Such encouragement came much more from the international agencies than from domestic sources, being seen in part as a way in which the energies of the poor could be combined with minimal investment to ease their frustrations and defuse potential political problems. 'Political stability through defensive modernization' as the World Bank's motives were described by Ayres (in Friedmann 1992: 59). The basic needs approach which evolved from such thinking became increasingly popular as the 'unconventional wisdom' of development strategy during the 1970s and 1980s (see Streeton and Burki 1978; Richards and Thomson 1984). It was principally structured around meeting social and physical needs rather than directly concerned with employment creation, and its features in this context will be addressed below. Certainly the overall approach has been the subject of heated and informed debate (see Streeton 1984; Wisner 1988; Hettne 1990; Grindle 1992).

Increasingly, however, the debate has taken on a rather hollow ring as a new development philosophy has emerged from the World Bank in the wake of the debt crises of the 1980s. The Structural Adjustment Programmes (SAP) imposed, in particular, on African debtor nations are essentially the product of the neo-liberal (i.e. market-oriented) responses to a growing global recession. As far as the recipient nations are concerned SAPs have several factors in common – namely, increased private sector investment, reduction of inflation, liberalization and expansion of trade, reduction of government expenditure and deregulation of the economy in general (see Riddell 1992). One of the principal consequences of these programmes has been increased unemployment as reductions in government expenditure filter through to the labour market. Despite an attempt by the World Bank to encourage 'adjustment with a human face', the real incomes of the poor, and the urban poor in particular, have fallen dramatically across the Third World (see Gilbert 1992). The knock-on effects on basic needs will be discussed below, but as far as employment is concerned the prospects for the urban poor are bleak, with the World Bank encouraging those countries struggling under SAPs to think positively about the role the informal sector can play in absorbing surplus labour. This has allowed the urban poor to be redefined from households or individuals with incomes below a certain level to those with limited job access (Urban Foundation 1993), thus permitting the state to identify increasing labour productivity as a means to alleviate poverty and to justify the withdrawal of the state from other forms of welfare subsidy. In many countries, therefore, the most rapidly expanding component of the informal sector is scavenging (Furedy 1992; Huysman 1994).

THE URBAN ENVIRONMENT

Rapid urban population growth, combined with the expenditure of limited state resources on economic development or prestige projects, has also resulted in the accumulation of considerable environmental problems in Third World cities: the so-called 'brown agenda'. These vary in their nature but the 'brown agenda' has at its core the fact that individual households cannot or should not be held responsible for the resolution of many of the problems that currently exist. As Peil (1994: 173) has noted, 'poor people get on with their lives and leave the environment to look after itself'. The problems involved range from contaminated water supply and inadequate waste disposal facilities, through air pollution from uncontrolled factory and vehicle emissions, to more spectacular disasters resulting from inadequately controlled building programmes. The one factor common to many of these urban environmental problems is the abrogation of responsibilities by the urban authorities, whether consciously or not. Many governments do lack detailed knowledge on environmental deterioration but many more see environmental concerns as being of secondary importance compared to economic growth or urban politics (Pernia 1992).

Detailed analysis by Degg (1994) has indicated that a high proportion of the major cities of the Third World are located in hazard-vulnerable zones. Add to this the rapid and uncontrolled population expansion into marginal areas within those cities and the risks of large and small environmental disasters become apparent. Major catastrophes such as the Mexico City earthquake of 1984 attract the headlines, but in many cities there are daily deaths from the collapse of unstable shops or poorly constructed buildings. On such occasions, sheer population growth and irresponsible urban management frequently exacerbate inherent vulnerability; but in many other instances the environmental problems are primarily man-made (Main and Williams 1994). One of the most infamous recent events of this type was the leak of poisonous gases from the Union Carbide Plant in Bhopal in 1984 when 3,300 people were killed and almost 150,000 seriously injured (Walker 1994). As Gupta (1988) observed, the impact was intensified by the fact that many of the areas in the proximity of the plant comprised densely crowded low-income housing. Main and Williams (1994) point out that in many other cities, too, the association between poverty and

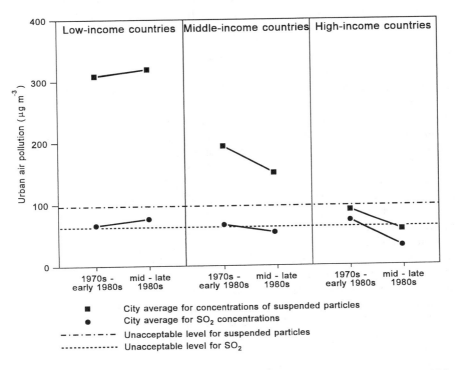

Figure 32.3 Urbanization and pollution: some recent trends. After World Bank (1992).

715

what they term 'negative externalities', whether natural or man-made, is widespread and portends many disasters to come.

The 1992 World Development Report, which focuses on development and the environment, presages many of these problems, drawing attention to a deteriorating situation in the cities of the Third World. At least 170 million urban dwellers still lack access to potable water *near* (not *in*) their homes (World Bank 1992). Between 1980 and 1990 the number of urban residents who were without access to adequate sanitation increased by almost 25 per cent to 400 million; whilst levels of air pollution also worsened in contrast to the improvements noted elsewhere in the world (Figure 32.3) (Haughton and Hunter 1994). Indeed, not only do Third World cities face rapidly worsening environmental problems, many difficulties are primarily the creation of the urbanization process itself. Foremost amongst these is the inadequate disposal of liquid and solid wastes from both residential and industrial sources. Ho Chi Minh City, for example, has no water treatment plant at all despite the fact that it produces 60 per cent of Vietnam's industrial output (Lam 1994). Moreover, the impact of such disposal problems extends far beyond the urban area once seepage of pollutants into water systems occurs (Douglas 1983).

In this depressing context, economic success seems to be no guarantee that such problems have or will be tackled. Bello (1992: 18) alleges that 'the environment has been traditionally sacrificed to the demands of export-oriented planning' and cites examples from South Korea and Taiwan to support his argument. In similar vein, Lau (1990) claims that the Hong Kong government's anti-pollution measures over the last fifteen years have been 'timid and ineffective' (see also Drakakis-Smith 1992). Little wonder that the new wave of industrializing economies see no reason to increase production costs by introducing expensive pollution control regulations. Bangkok, in particular, is testimony to the philosophy of 'grow now and clean up later', and is said to be on the brink of environmental disaster (Tasker 1990; Pernia 1992). Amongst many other problems, and in common with Mexico City, Bangkok has pumped so much artesian water out of its sub-strata and permitted so much new construction, that the city is sinking (Yeung 1990; Komin 1991). Each major storm brings even more widespread flooding, when most of the city's waste returns to threaten its residents (Douglas 1983; Gupta 1988).

Combating environmental problems of this nature requires collective action and responsible urban management. However, as the threatening instances cited above reveal, this has been slow to emerge. Traditionally, municipal governments have been responsible for urban infrastructure, but without sufficient financial resources many authorities have simply turned their backs on the problem. It has been estimated that in Bangkok, for example, 'construction of a city-wide sewerage system would cost [the] equivalent [of] 33 percent of total government expenditure' (Rigg 1991: 149). Some of the more manageable and potentially profitable services have been successfully privatized in a few cities and Gilbert (1992) presents optimistic accounts of water provision and transportation services in Latin America. However, the removal of solid and liquid wastes present less attractive possibilities for privatization.

In the mid-1980s, at least in East Asia, Yeung and McGee (1986) could proffer a reasonably positive scenario for the development of community participation in the delivery of urban services. It is doubtful whether such optimism will characterize the 1990s. The rapid pace of urban growth has itself eroded many of the traditional values on which community participation is predicated. Moreover, governments have been able to use participatory planning as a harmless cul-de-sac into which frustrations can be channelled but from which little real policy change emerges. Little wonder, in such circumstances, that urban social movements have emerged as a more effective force for change (Schuurman and Van Naerssen 1989). The 1992 *World Development Report* (World Bank 1992) does not make for encouraging reading. With sheer numbers and poverty both increasing, and urban managers feeling the pinch of SAPs, urban environmental problems are likely to worsen considerably in the future, giving the UN declaration of the 1990s as 'International Decade for Natural Disaster Reduction' a somewhat shaky foundation as far as the Third World cities are concerned.

MEETING BASIC NEEDS

The situation with regard to the meeting of basic needs, such as housing, food and health care, is almost as bad as it is for the broader urban environment. The state, in most cities, evaluates social welfare programmes in a short-term perspective as resource-absorbing rather than resource-generating, and prefers to invest in the built environment of production rather than consumption. Clearly there are short-term costs, but these can be compensated for by longer-term gains in the well-being of the population and the workforce. However, such arguments find little favour with most governments, although there are clearly exceptions – Singapore houses almost 80 per cent of its population in government-built accommodation and until recently Zimbabwe invested almost 12 per cent of its GNP in social welfare programmes for the poor (UNDP 1991).

State involvement in the provision of basic needs to urban populations probably exhibits more than any other government policy, the vicissitudes of the complex local, national and international forces that impact on the development process. Unlike basic infrastructure, such as sewerage systems, water reticulation or road provision, individual households themselves can and must compensate for lack of government action, so that meeting basic needs in most Third World cities becomes an extremely complex and varied process. Rather than list each in turn, this section will review the situation relating to just three of the principal basic needs. These are shelter, about which there is a voluminous literature; food, which, despite being the most basic of basic needs, is poorly researched in the urban context; and human rights.

717

The standard response of many governments to the ever-growing housing needs of the urban poor (noted by Dwyer in 1975) is still one of apathy, but in those countries where investment in public housing has occurred the 1970s saw a fundamental shift in the nature of the housing programme (Drakakis-Smith 1988; Gilbert and Gugler 1991; Ward and Macoloo 1992). Prior to the work of researchers such as John Turner (1969) or William Mangin (1967), public housing programmes had been structured around expensive Western designs and planning, often in the form of high-rise blocks that were unpopular and too expensive. Such costly facilities only served to discourage further investment.

In the wake of the recognition that the poor have energies and ambitions that could be harnessed to improve their shelter, a range of aided self-help programmes were developed, the principal feature of which was some form of collaboration between the state and the community. The schemes themselves varied considerably in terms of the inputs of the state. Some involved on-site infrastructural improvements with no transfer of land tenure; others encompassed the provision of serviced plots, usually in the urban periphery, on which tenants were to build their own homes within a specified time period; some of these site-and-service schemes provided small core-housing units as a starting point for such construction.

Encouraged by World Bank lending policy, within which the new approaches became the conventional wisdom from the mid-1970s onwards, aided self-help programmes rapidly became the most common form of government involvement in shelter provision. This is not to say that all governments adopted such a philosophy wholeheartedly. Many did not, or else continued to pursue quite unsympathetic policies in parallel with their externally funded aided self-help schemes. Zimbabwe, for example, has supported a range of self-help schemes in Harare since independence in 1980, and yet in 1991 could still demolish a squatter settlement in the centre of the capital, prior to the Commonwealth Heads of Government Meeting, and bus the people involved to a rudimentary camp 25 kilometres outside the city (Drakakis-Smith 1993).

Notwithstanding such inconsistencies, the aided self-help housing philosophy spread rapidly across cities in the developing world from the mid-1970s onwards and has been extremely well reviewed and critiqued (see Amis and Lloyd 1990; Payne 1984; Hardoy and Satterthwaite 1989; Gilbert and Ward 1985; Drakakis-Smith 1988). There have been almost as many opinions as writers. These range from the dismissive (Burgess 1985), on the grounds that aided self-help simply eases a few of the frustrations of the poor whilst facilitating the penetration of capitalism into the petty-commodity sector, to the more realistic (Payne 1984; Ward 1982), including a welcome critique of the relationship between gender and housing (Moser and Peake 1987).

By the mid-1980s these realistic appraisals were beginning to identify areas for further attention, in particular the bottlenecks provided by restrictive land markets or inadequate supplies of building materials. In addition, it was also becoming clear that for various reasons, the really poor were not being helped

718

even by aided self-help programmes and that in some cities considerable upward filtration of the units constructed in such schemes had occurred. At the same time the broader critique of the basic needs approach as a whole was also being extended to housing programmes, particularly the observation that they perpetuated top-down planning and failed to involve the poor other than as cheap labour (Wisner 1988).

However, many now look back at the mid-1980s almost with nostalgia, for despite the valid criticisms that were made of aided self-help programmes, at least funds were becoming more widely available and many had been assisted. In contrast, the situation today is one in which external support for, and state involvement in, construction programmes have been drastically reduced. Gilbert (1992) suggests that, at least in part, this was due to disillusionment on the part of the World Bank with the effectiveness of aided self-help. But the debate by the mid-1980s seemed to have reached a generally positive conclusion about such programmes and a constructive dialogue for change had evolved. It seems more likely that World Bank enthusiasm for social programmes in general was undermined by a more fundamental change of policy within the Bank itself. This shift to a neo-liberal approach, in which 'special' social development programmes were rejected in favour of improving welfare through an unfettered market system, has been associated by Gibbon (1992) with the replacement of Robert McNamara by A.W. Clauson as President – a politician by a banker.

In terms of its impact on housing programmes *per se*, there have been several important repercussions. The first was to tighten up the fiscal terms of World Bank loans for aided self-help schemes, with the consequence that tenants were asked to construct their homes or repay their loans within increasingly shorter periods, even as low as eighteen months in some instances (Teedon 1990). The consequence of these trends was increased defaulting and an acceleration of upward filtration of the serviced plots. As far as the state itself is concerned, the tendency has been to withdraw from involvement in the construction process *per se* and concentrate more on the provision of serviced land for the self-help process (aided or not). This is not a universal trend, but a simpler level of involvement is clearly discernible. Gilbert (1992) and also Ward and Macoloo (1992), for example, have noted that as land has become an increasingly scarce commodity, so its commodification or commercialization has ensued. The state has played a major role, either directly or indirectly, in this commodification process by such varied measures as the production of cadastral maps, making subdivisions available, restructuring the ownership of customary land. The consequence in many cities has been that it has become more difficult, more time-consuming and more costly for the poor to acquire land. This seems to be particularly true where the private sector becomes involved in legal or illegal subdivision of plots. In contrast, middle-income householders are finding that, with more funds available for mortgage loans under the new market-oriented regimes, greater prospects have opened up for them.

The new tighter controls over funding and land have drastically slowed down the opportunities available to the urban poor to improve their housing conditions, and yet urban populations continue to rise relentlessly. One consequence may have been an acceleration of sub-letting through renting and lodging. It is difficult to estimate the growth of sub-tenancies because it was so thinly documented in the past. Of course, renting has always been a feature of those cities with older quarters which have become subdivided over time into smaller units – the tenements of Asia, the *vecindades* of Latin America or the *medinas* of the Middle East. With the restrictions on alternative forms of low-cost accommodation, the role of sub-letting has become more apparent and a series of studies have produced more useful information on this important source of shelter for the poor (Varley 1993; Gilbert and Varley 1991; Teedon 1990; Bryant 1986; Angel and Amtapunth 1989). From these reports it is clear that sub-letting has spread from the old parts of the city, where it has often been displaced by commercial redevelopments, to more peripheral housing areas including many of the aided self-help schemes. Indeed, raising shacks on garden space seems to be the most profitable form of cultivation in some of the existing low-income areas. The corollary of such changes, as Gilbert and Varley (1991) point out, is that much renting is small-scale with reasonable relations existing between tenant and landlord. Of course, not all situations conform to this pattern but in general large-scale exploitation appears to be the exception rather than the rule.

The increase in sub-letting is, once again, a solution to the problem of shelter provision which has emerged from the poor themselves. It cannot, however, compensate fully for the retreat of state involvement in low-cost housing construction. The increase in sharing (as distinct from renting), much of which occurs involuntarily within families, is likely to create tensions which could escalate to unpredictable levels in the near future. The more so as the situation outlined above in relation to shelter is mirrored in the supply of other basic needs.

With regard to urban food systems, fundamentally the same processes have operated in terms affecting its supply to the poor (see Drakakis-Smith 1993 and Atkinson 1991 for more detailed discussion). An inexorable Westernization of the urban food system has occurred on a widespread scale – 'the industrialization of the food system' as McGee and MacLeod (1990) have termed it. Westernization of diets, increasing imports of food, rapid expansion of supermarket outlets and the like have all been encouraged and facilitated by the state, and have provided yet another avenue for the penetration of Western multinationals into domestic markets of the Third World. At the moment, for example, the huge Indonesian urban food market is an economic battleground between Proctor and Gamble and Unilever, with the expansion of associated supermarket chains as the crucial tactic in this war.

The consequences of such changes for the urban poor are many, but as far as food is concerned the overall effect is to restrict access. Increased food imports, food packaging and food processing must push up costs. Given the declining real incomes of the poor this must restrict their ability to buy the food they need to keep up their calorie intake. Malnutrition is not a predominantly rural phenomenon, the persistent food deprivation of urban households with low and erratic incomes has been an unheralded phenomenon of the last two decades. Significantly, of all the major health problems affecting the urban poor, malnutrition has proved to be the most intractable. Even in Zimbabwe, where expenditure on social welfare programmes has been exceptionally high as a proportion of GDP, 100,000 people per year were being brought into the hospitals of the capital Harare suffering from undernutrition, even before the recent drought.

This worsening and largely unappreciated situation has clearly put enormous pressure on poor families to maintain their access to food. As with housing, an informal supply sector has emerged to meet such needs and the poor patronize hawkers of both fresh and cooked foods extensively. Such street foods must not be underestimated in the role they play in feeding the poor. Research in South Asia suggests that the poor receive up to one quarter of their nutritional intake this way (Atkinson 1991). Nor should such outlets be criticized for being potential health risks and purveyors of less nutritious food. Hygiene conditions on the streets are often no worse than in many homes; moreover, street food usually contains a much wider range of items, and therefore nutrients, than do basic home diets.

Despite the important role that the informal food sector plays throughout the urban Third World, its activities have almost inevitably been subject to legislative proscription by the urban authorities (May and Rogerson 1995). Once again the principal objection is that the operations of the informal food sector get in the way, blocking pavements, littering streets and generally spoiling the image of the modern Western city that urban managers want to create. Of course, proscriptive legislation is not always applied. Many authorities have sought to control and organize food retailing into designated selling places. But the legislation remains active and from time to time when conditions dictate, period purges of the informal sector occur and the hawkers are swept from the street – at least for a time.

But in the worsening economic climate of the 1980s and 1990s it is not only the informal retailing sector which has expanded to meet the increasingly desperate food needs of the urban poor. Urban agriculture has also begun to revive and expand (May and Rogerson 1995). Of course, not all cities provide suitable opportunities for urban agriculture, environmental factors such as soils or water supply may be inadequate, land itself may be too scarce or have more valuable uses, the poor may simply not have time to cultivate or may not have the necessary knowledge or skills (the common view of the urban poor as recent peasants is nowadays largely a myth). Nevertheless, where circumstances permit, urban agriculture has increased either within household gardens or, more usually, on vacant land within or around the city. Ironically, not all of this increased cultivation is by the poor. Control of and access to basic inputs such as land, labour or capital (for seeds, fertilizers, etc.) is often the prerogative of the not-so-poor, so that a considerable amount of urban agriculture in the Third World is a middle-class activity (see Gefu 1992; Drakakis-Smith 1993).

Whoever is involved, urban agriculture has increased rapidly during the 1980s as a response to an increasingly costly urban food system and deteriorating real wages (Maxwell and Zziwa 1992). Unfortunately, as urban food systems have been so poorly researched and understood, they were never incorporated into the basic-needs approach in any effective way. No equivalent of aided self-help housing programmes emerged in the 1970s and 1980s. Whilst governments were winding down Westernized state housing programmes in favour of site and service schemes, they were continuing to encourage Westernization of the food supply system and periodically sweeping the streets of food hawkers. Until the recent deterioration in food supply systems, urban agriculture continued to be discouraged and crops destroyed. In Harare, even as the effects of the recent drought were being increasingly felt, the city council was sending its employees to slash and burn extensive areas of illegally planted maize and vegetables (Drakakis-Smith et al. 1995). Clearly there must be more constructive ways of reacting to the desperate actions of people trying to do no more than provide their families with the most basic need of all.

If meeting basic needs in the city has proved to be an erratic and difficult process, then the support of human rights has in general been even more disappointing (Pinkney 1993). Basic rights are difficult to define satisfactorily on an international scale because they vary with cultural context. In most Western countries human rights are vested in the individual, whereas in other countries they may take on a communal dimension (Chua 1994) linked to societal obligations; for example, in Islamic states or in the old USSR where 'rights and liberties [were] stated to be inseparable from the performance by citizens of their duties' (Grindle 1992: 68).

Although basic human and physical needs should be indivisible, this complementarity does not always occur (Vincent 1986). In some countries the basic-needs record might be good but that of human rights poor (China in the late 1980s is often cited as an example), and vice versa. Nor does the support of human rights correlate well with economic growth in developing countries, perhaps because there is still disagreement on which influences which. Certainly, the RIEs do not all show up well on ranking lists of human rights provision (Devan and Heng 1994). Indeed, a recent UNDP (1991) report shows Singapore, one of the most efficient and successful of the RIEs, to be just above the lowest category on the ranking table.

What relevance does this brief discussion of human rights have for the Third World city? The short answer is that it has enormous relevance since it is in the concentrated demographic cauldrons of the cities of the Third World that the greatest threat to human rights occurs, explicitly or implicitly. Continued exploitation of workers, particularly women and children, in appalling working conditions, denial of the right to organize or protest, violations against individuals by the state, are all still part of the urban scene in too many developing countries (as they are in parts of the West too). Unfortunately, as Grindle (1992: 71) acidly observes, the United Nations record on pursuing the cause of human rights is poor and it 'preaches a theory that it does not practise itself'. Little wonder that the urban poor turn in upon themselves and their own communities to seek support and succour. But for how long?

CONCLUSION

The conclusions that are to be drawn from the above discussion must be brief and blunt. Despite the best efforts of some governments, national and municipal, of some international agencies and some NGOs, and despite the heroic efforts made on their own behalf by most of their inhabitants, cities in developing countries are facing a deteriorating situation – one which is already bad and is going to get worse.

Urban populations continue to grow rapidly. The latest estimate is that the Third World will be more urban than rural by the year 2000. The present world situation is such that economic growth in most Third World countries is going to be very limited and the creation of new employment even more so. As economic prospects diminish so do the chances that the urban poor will be able to improve their living conditions. Hopes of a growing partnership between the state and the poor to help meet basic needs have been dashed by the new conventional wisdom of structural readjustment and the perceived need to roll back state investment and involvement in social welfare programmes. Prospects for improved human rights look even more bleak; as protests grow about the deteriorating situation in cities so the state will have less need for democracy and more for oppression. Significantly, police and armed forces have been far less affected by retrenchment than other areas of government employment.

In the 1960s when similar crisis conditions were recognized after independence, the first acceleration of urban growth and the failures of 'modernization' to trickle down benefits to the poor, both left- and right-wing observers anticipated widespread urban protests, even insurrection. But this did not occur. What was not appreciated until later was the cushioning effect of the informal sector and the coping mechanisms it offered to new migrants with hopes for the future, at least for their children.

In the 1990s, despite the apparent advance of democratic government throughout the Third World (Gilbert 1992), a new unease is beginning to grow amongst many observers. As a recent review recently admitted 'with urban areas producing about half of the world's income and governments nervous about restive urban populations, agencies such as the World Bank have begun to focus more on cities' (*Time*, 11 January 1993: 34). In contrast to the 1960s, however, the actors have changed. Most of the urban poor are no longer recent migrants from the countryside reluctant to rock the boat, they are young people born and raised in poverty in cities where inequalities are becoming more marked and less tolerable. Disillusion with the effectiveness of the more formal politics of poverty, such as trade unionism or populist political parties, has resulted in the rise of self-help politics or urban social movements (Corbridge 1993). Whilst these appear to be more a feature of Latin America (Slater 1991; Schuurman and Van Naerssen 1989; Parkes 1989), the potentialities elsewhere for such specifically focused protests seems immense (Narman 1993).

Of course, there is also a growing and equally frustrated middle class in many Third World cities. Although still small it is influential and aware of the opportunities and means to extend its own participation in the management of Third World cities (Robison 1994). Already its collective voice has been heard. The overthrow of the Marcos regime in the Philippines owed much to the mobilization of middle-class support for Aquino; the pro-democracy movement in Hong Kong is driven by a middle class that is anxious to preserve its position; elsewhere urban demonstrations are often instigated by already privileged students concerned over their future prospects. The key question is whether such bourgeois agitation for greater democracy will be extended to incorporate rights for the urban poor too. Shakur (1991) is sceptical of the possibilities of such inter-class collaboration, and if it fails then the poor must surely begin to associate their deteriorating position with the apparent democratization of their governments – the fall of the dictators in Latin America, the advent of multipartyism in Africa, the retreat of socialism.

The calamitous impact of structural adjustment on the urban poor is rapidly giving democracy a bad name; a reconceptualization of democracy along local lines is urgently needed and the Vienna Conference on Human Rights has accelerated this debate. For some, however, the scenario for the 1990s and beyond has already been set and it is not optimistic. Shakur's despairing comment on the situation in Dhaka sums it up:

> although this research [on squatters] started off with enthusiasm and optimism . . . pessimism has taken hold because under the present circumstances the problem seems unsurmountable [sic] . . . unless there is a revolutionary change in societal structure.
>
> (Shakur 1991: 14)

In particular, what is needed are ways of bridging the widening gap between a rapidly growing and frustrated group of have-nots and state administrations that seem to care more for sustained urban growth rather than sustainable urban development. If this contradiction is not addressed, the urban insurrections feared in the 1960s may well eventually materialize in the 1990s.

REFERENCES

Amis, P. and Lloyd, P. (eds) (1990) *Housing Africa's Urban Poor*, Manchester: Manchester University Press.

Angel, S. and Amtapunth, P. (1989) 'The low-cost rental housing market in Bangkok', *Habitat International* 13, 173–85.

Atkinson, S. (1991) *Nutrition in Urban Programmes and Planning*, Urban Health Programme, London School of Hygiene and Tropical Medicine, London University.

Bello, W. (1982) *Development Debacle: The World Bank in the Philippines*, Manila: IFDP.

Bello, W. (1992) 'Export led development in East Asia: a flawed model', pp. 11–27 in *Trocaire Development Review*, Dublin: Trocaire.

Bibangambah, J.R. (1992) 'Macro-level constraints and the growth of the informal sector in Uganda', pp. 303–13 in J. Baker and P.O. Pedersen, *The Rural–Urban Interface in Africa*, Uppsala: Nordiska Afrikainstitutet.

Bromley, R. and Gerry, C. (eds) (1979) *Casual Work and Poverty in Third World Cities*, Chichester: Wiley.

Brown, D. (1994) *The State and Ethnic Politics in Southeast Asia*, London: Routledge.

Bryant, J.J. (1986) 'The acceptable face of self-help housing: subletting in Fiji squatter settlements', pp. 171–95 in D. Drakakis-Smith (ed.) *Economic Growth and Urbanization in Developing Areas*, London: Routledge.

Burgess, R. (1985) 'The limits to state self-help programmes', *Development and Change* 16, 271–312.

Chandra, R. (1992) *Industrialization and Development in the Third World*, London: Routledge.

Chua, B.H. (1994) 'For a communitarian democracy in Singapore', pp. 27–31 in D. da Cunha (ed.) *Debating Singapore*, Singapore: ISEAS.

Cohen, M. (1990) 'Macroeconomic adjustment and the city', *Cities* 7, 49–59.

Cook, I. (1993) 'Urban issues in the West Pacific Rim', Paper presented to the British Pacific Rim Research Group, John Moores University, Liverpool.

Corbridge, S. (1993) 'Colonialism, post-colonialism and the Third World', pp. 173–205 in P. Taylor (ed.) *Political Geography of the Twentieth Century*, London: Belhaven.

Degg, M. (1994) 'Perspectives on vulnerability to earthquake hazard in the Third World', pp. 29–48 in H. Main and S.W. Williams (eds) *Environment and Housing in Third World Cities*, London: Wiley.

Devan, J. and Heng, G. (1994) 'A minimum working hypothesis of democracy for Singapore', pp. 22–6 in D. da Cunha (ed.) *Debating Singapore*, Singapore: ISEAS.

Devas, N. and Rakodi, C. (1993) *Managing Fast Growing Cities*, Cambridge: Cambridge University Press.

Dicken, P. (1991) *Global Shift*, London: Paul Chapman.

Dixon, C.J. (1993) 'The impact of structural adjustment on the Thai economy', in J. Dahl, D. Drakakis-Smith and A. Narman (eds) *Land, Food and Basic Needs in Developing Countries*, No. 83, Series B, No. 1. Kulturgeografiska Institutionen, Gothenburg University.

Dixon, C.J. and Drakakis-Smith, D.W. (eds) (1993) *Economic and Social Development in Pacific Asia*, London: Routledge.

Dixon, C.J. and Heffernan, M. (eds) (1991) *Colonialism and Development in the Contemporary World*, London: Mansell.

Douglas, I. (1983) *The Urban Environment*, Baltimore, Md.: Edward Arnold.

Douglass, M. (1992) 'The political economy of urban poverty and environmental management in Asia', *Environment and Urbanisation* 4(2), 9–32.

Drakakis-Smith, D.W. (1988) 'Housing', pp. 148–98 in M. Pacione (ed.) *The Geography of the Third World*, London: Routledge.

Drakakis-Smith, D.W. (1991) 'Colonial urbanization in Africa and Asia: a structural review', *Cambria* 16, 123–50.

Drakakis-Smith, D.W. (1992) *Pacific Asia*, London: Routledge.

Drakakis-Smith, D.W. (1993) 'Food security and food policy for the urban poor', in J. Dahl, D. Drakakis-Smith and A. Narman (eds) *Land, Food and Basic Needs in*

Developing Countries, No. 83, Series B, No. 1, Kulturgeografiska Institutionen, Gothenburg University.

Drakakis-Smith, D.W. (1995) 'Third World cities: sustainable urban development I', *Urban Studies* 32(4–5), 659–77.

Drakakis-Smith, D.W., Bowyer-Bower, T. and Tevera, D. (1995) 'Urban poverty and urban agriculture: an overview of the linkages in Harare', *Habitat International* 19(2), 183–93.

Drakakis-Smith, D.W. and Rimmer, P. (1982) 'Taming the wild city: urban management in Southeast Asia's capital cities', *Asian Geographer* 1(1), 17–34.

Dwyer, D.J. (1975) *People and Housing in Third World Cities*, London: Longman.

Forbes, D. (1981) 'Petty commodity production and underdevelopment: pedlars and trishaw riders in Ujung Pandong, Indonesia', *Progress in Planning* 16, 105–78.

Friedmann, J. (1992) *Empowerment: The Politics of Alternative Development*, Oxford: Blackwell.

Furedy, C. (1992) 'Garbage: exploring the non-conventional options to Asian cities', *Environment and Urbanisation* 4(2), 42–53.

Gefu, J. (1992) 'Part-time farming as an urban survival strategy: a Nigerian case study', pp. 295–302 in J. Baker and P.O. Pedersen (eds) *The Rural–Urban Interface in Africa*, Uppsala: Nordiska Afrikainstitutet.

Gibbon, P. (1992) 'The World Bank and African poverty 1973–1991', *Journal of Modern African Studies* 30(2), 193–220.

Gilbert, A. (1992) 'Housing, infrastructure and servicing in Third World cities', *Urban Studies* 29(4–5), 435–60.

Gilbert, A. and Gugler, J. (1991) *Cities, Poverty and Development: Urbanization in the Third World*, Oxford: Oxford University Press.

Gilbert, A. and Varley, A. (1991) *Landlord and Tenant: Housing the Poor in Urban Mexico*, London: Routledge.

Gilbert, A. and Ward, P. (1985) *Housing, the State and the Poor*, Cambridge: Cambridge University Press.

Grindle, J. (1992) *Bread and Freedom*, Dublin: Trocaire.

Gupta, A. (1988) *Ecology and Development in the Third World*, London: Methuen.

Hardoy, J. and Satterthwaite, D. (1989) *Squatter Citizen*, London: Earthscan.

Haughton, G. and Hunter, C. (1994) *Sustainable Cities*, London: Regional Studies Association.

Hettne, B. (1990) *Development Theory and the Three Worlds*, London: Longman.

Huysman, M. (1994) 'Waste picking as a survival strategy for women in Indian cities', *Environment and Urbanisation* 6(2), 157–74.

King, A. (1976) *Colonial Urban Development*, London: Routledge.

King, A. (1990) *Urbanism, Colonialism and the World Economy*, London: Routledge.

Knox, P. and Agnew, J. (1989) *The Geography of the World Economy*, London: Routledge.

Komin, S. (1991) 'Social dimensions of industrialisation in Thailand', *Regional Development Dialogue* 12(1), 115–37.

Lam, M.T. (1994) 'Water pollution in Ho Chi Minh city', in A. Awang, M. Salim, and J.F. Halldane (eds) *Environment and Urban Management in Southeast Asia*, Johor Bahru, Malaysia: Institute Sultan Iskandar, Universiti Teknologi.

Lau, E. (1990) 'A license to pollute', *Far Eastern Economic Review* 148(19), 23–5.

Lowder, S. (1986) *Inside Third World Cities*, London: Croom Helm.

Lowder, S. (1991) 'The context of urban planning in secondary cities', *Cities* 8, 54–65.

McGee, T.G. (1967) *Southeast Asian City*, London: Bell.

McGee, T.G. and Greenberg, L. (1992) 'The emergence of extended metropolitan regions in ASEAN', *ASEAN Economic Bulletin* 1(6), 5–12.

McGee, T.G. and MacLeod, S. (1990) 'The internationalization of the food distribution system in Hong Kong', pp. 304–35 in D.W. Drakakis-Smith (ed.) *Economic Growth and Urbanization in Developing Countries*, London: Routledge.

McGee, T.G. and Yeung, Y.M. (1993) 'Urban futures for Pacific Asia: towards the 21st century', in Y.M. Yeung (ed.) *Pacific Asia in the 21st Century*, Hong Kong: Chinese University Press.

Main, H. and Williams, S.W. (1994) 'Marginal environments as havens for low-income housing: Third World regional comparisons', pp. 151–70 in H. Main and S.W. Williams (eds) *Environment and Housing in Third World Cities*, London: Wiley.

Mangin, W. (1967) 'Latin American squatter settlements: a problem and a solution', *Latin American Research Review* 2(3), 65–98.

Maxwell, D. and Zziwa, S. (1992) *Urban Farming in Africa: The Case of Kampala, Uganda*, Nairobi: African Centre for Technology Studies.

May, J. and Rogerson, C. (1995) 'Poverty and sustainable cities in South Africa: the role of urban cultivation', *Habitat International* 19(2), 165–82.

Moser, C.O. and Peake, L. (eds) (1987) *Women, Human Settlements and Housing*, London: Tavistock.

Narman, A. (1993) *Om Folkrorelser i Afrika*, Stockholm: SVS.

O'Connor, R.A. (1983) *A Theory of Indigenous Southeast Asian Urbanism*, Discussion Paper No. 38, Institute of Southeast Asian Studies, Singapore.

Parkes, A. (1989) 'Latin American urbanization during the year of the crisis', *Latin American Research Review* 25, 7–44.

Payne, G. (1984) *Low Income Housing in the Developing World*, Chichester: Wiley.

Peil, M. (1994) 'Urban housing and services in Anglophone West Africa', pp. 173–90 in H. Main and S.W. Williams (eds) *Environment and Housing in Third World Cities*, London: Wiley.

Pernia, E. (1992) 'Southeast Asia', in R. Stren, R. White and J. Whitney (eds) *Sustainable Cities: Urbanisation and the Environment in International Perspective*.

Pinkney, R. (1993) *Democracy in the Third World*, Buckingham: Open University Press.

Rahim, M.N.A. (1988) *Public Transport Planning in Malaysia*, Occasional Paper No. 14, Department of Geography, Keele University.

Richards, P.J. and Thomson, A. (1984) *Basic Needs and the Urban Poor*, London: Croom Helm.

Riddell, J.B. (1992) 'Things fall apart again: structural readjustment programmes in Sub-Saharan Africa', *Journal of Modern African Studies* 30(1), 53–68.

Rigg, J. (1991) *Southeast Asia: A Region in Transition*, London: Unwin Hyman.

Robison, R. (1994) 'The emergence of the middle class in Southeast Asia', Paper presented at a conference on Emerging Classes and Growing Inequalities in Southeast Asia, Nordic Association for Southeast Asian Studies, Aalborg University.

Schuurman, F. and van Naerssen, T. (1989) *Urban Social Movements in the Third World*, London: Routledge.

Sethuraman, S. (ed.) (1981) *The Urban Informal Sector in Developing Countries*, Geneva: ILO.

Shakur, T. (1991) 'Policy recommendations towards sheltering the squatters in the Third World: the case of Dhaka, Bangladesh', Paper presented at a workshop on Low-Income Housing and the Environment, Staffordshire University.

Shibusawa, M., Ahmad, Z. and Bridges, B. (1992) *Pacific Asia in the 1990s*, London: Routledge.

Simon, D. (1993) 'State intervention in Africa's urban land and housing markets', in J. Dahl, D.W. Drakakis-Smith and A. Narman (eds) *Land, Food and Basic Needs in Developing Countries*, No. 83, Series B, Kulturgeografiska Institutionen, Gothenburg University.

Slater, D. (1991) 'New social movements and old political questions: rethinking state–society relations in Latin American development', *International Journal of Political Economy* 21, 32–65.

Streeton, P. (1984) 'Basic needs: some unsettled questions', *World Development* 12(9), 973–8.

Streeton, P. and Burki, S. (1978) 'Basic needs: some issues', *World Development* 6(3), 212–19.

Stren, R. and White, R. (eds) (1989) *African Cities in Crisis*, Boulder, Colo.: Westview Press.

Tasker, R. (1990) 'Bangkok on the brink', *Far Eastern Economic Review*, 29 November, 52–3.

Taylor, P. (1985) *Political Geography*, London: Longman.

Taylor, P. (1991) 'The legacy of imperialism', in C.J. Dixon and M. Heffernan (eds) *Colonialism and Development in the Contemporary World*, London: Mansell.

Teedon, P. (1990) 'An analysis of aided self-help housing schemes in Harare, Zimbabwe', Ph.D. thesis, Department of Geography, Keele University.

Telkamp, G. (1978) 'Urban history and European expansion', *Intercontinental* 1, Leiden: University of Leiden.

Todaro, M. (1989) *Economic Development in the Third World*, London: Longman.

Turner, J. (1969) 'Architecture that works', *Ekistics* 27(158), 40–4.

UNDP (1991) *Human Development Report*, New York.

Urban Foundation (1993) *Managing Urban Growth: The International Experience*, Johannesburg: Urban Foundation.

Varley, A. (1993) 'Gender and housing; the provision of accommodation for young adults in three Mexican cities', *Habitat International* 17(4), 13–30.

Vincent, R.J. (1986) *Human Rights and International Relations*, Cambridge: Cambridge University Press.

Walker, G. (1994) 'Industrial disasters, vulnerability and planning in Third World cities', pp. 49–64 in H. Main and S.W. Williams (eds) *Environment and Housing in Third World Cities*, London: Wiley.

Wallis, M. (1989) *Bureaucracy: Its Role in Third World Development*, London: Macmillan.

Ward, P. (ed.) (1982) *Self-Help Housing: A Critique*, London: Mansell.

Ward, P. (ed.) (1989) *Corruption, Development and Inequality*, London: Routledge.

Ward, P. and Macoloo, C. (1992) 'Articulation theory and self-help housing practice in the 1990s', *International Journal of Urban and Regional Research* 16(1), 60–79.

Wisner, B. (1988) *Power and Need in Africa*, London: Earthscan.

World Bank (1992) *World Development Report*, Washington, DC.

World Bank (1994) *World Development Report*, Washington, DC.

Yeung, Y.-M. (1990) *Changing Cities of Pacific Asia: A Scholarly Interpretation*, Hong Kong: Chinese University Press.

Yeung, Y.-M. and McGee, T.G. (1986) *Community Participation in Delivering Urban Services in Asia*, Ottawa: IDRC.

FURTHER READING

Allen, T. and Thomas A. (eds) (1992) *Poverty and Development in the 1990s*, Oxford: Oxford University Press.

Boyden, J. (1991) *Children of the Cities*, London: Zed Books.

Devas, N. and Rakodi, C. (1993) *Managing Fast Growing Cities*, Cambridge: Cambridge University Press.

Drakakis-Smith, D.W. (1987) *The Third World City*, London: Routledge.

Drakakis-Smith, D.W. (1995) 'Third World cities: sustainable urban development I', *Urban Studies* 32(4–5), 659–77.

Gilbert, A. and Gugler, J. (1991) *Cities, Poverty and Development: Urbanization in the Third World*, Oxford: Oxford University Press.

Hardoy, J. and Satterthwaite, D. (1989) *Squatter Citizen*, London: Earthscan.

Haughton, G. and Hunter, C. (1994) *Sustainable Cities*, London: Regional Studies Association.

Main, H. and Williams, S.W. (eds) (1994) *Environment and Housing in Third World Cities*, London: Wiley.

Moser, C.O. and Peake, L. (eds) (1987) *Women, Human Settlements and Housing*, London: Tavistock.

Schell, L., Smith, M. and Bilsborough, A. (1993) *Urban Ecology and Health in the Third World*, Cambridge: Cambridge University Press.

Schuurman, F. and Van Naerssen, T. (1989) *Urban Social Movements in the Third World*, London: Routledge.

Simon, D. (1992) *Cities, Capital and Development*, London: Belhaven Press.

UNDP (1991) *Cities, People and Poverty*, Strategy Paper, New York.

World Bank (1991) *Urban Policy and Economic Development: An Agenda for the 1990s*, Policy Research Report, Washington, DC.

33

WESTERN CITIES AND THEIR PROBLEMS

David T. Herbert

As with most other generic terms, that of the 'western city' disguises a diversity of forms. Undoubtedly there are common strands as most Western societies achieve an advanced stage of post-industrial urbanization but the variety is still there; a continuum of urban forms and situations can be identified which relate to societal processes and change. West European societies, notably the United Kingdom, were the first to experience industrial urbanization and as America urbanized it did so in even more dramatic ways; some forms of the metropolitan urbanization found their earliest and most emphatic expressions in the United States. As the isolationism of the former socialist societies of Eastern Europe breaks down in the last part of the twentieth century, their cities begin to merge into the Western urban system and both the cultural differences and the 'time-warps' are thrown into sharp focus. South African cities with extreme forms of racial segregation in residential areas have begun, in the new political order, to experience changes which have major social–geographical consequences.

The common problems relate to changing economic bases, to ageing urban infrastructures and a lack of investment; from these emerge the issues of social disorder and pockets of disadvantage which have become the hallmarks of the later-twentieth-century Western city.

GROWTH AND STRUCTURE OF WESTERN CITIES

Western cities have been typified by rapid growth over a long period of time. After an initial phase of concentration and centralization which produced the compact, high-density, congested industrial cities of the later nineteenth century, the process of decentralization took over and new urban forms, variously referred to as conurbation, metropolis, functional urban region and megalopolis, were direct outcomes of that physical and functional outward spread of the city. When Colin Clark (1951) coined the phrase 'transport is the maker and breaker' of cities, he was summarizing the main features of these processes. Initial concentration and centralization arose from the focusing of transport and movement upon the central nodes of cities; this process 'made' the industrial city. As forms of transport became more flexible with the rise of the internal combustion engine in particular, the supremacy of this central space was progressively diminished and compact cities were 'broken up' as they spread over larger amounts of space. In the 1970s and 1980s the term 'counter-urbanization' was used to describe the latter stages of this process as both people and activities appeared to favour location in small towns and rural areas within a widely dispersed urban system. Information technology, certainly in theory and increasingly in reality, provides further release from spatial constraints and will help form the future 'city'.

North American transport systems have always developed more quickly than those of European countries, especially in terms of a reliance upon the automobile. Availability of space, higher levels of affluence and cultural expectations underpinned the new urban forms of metropolitan America. There are, of course, significant regional variations within the United States and whereas less than 10 per cent of commuters might use private cars in older metropolitan areas such as New York, a comparable figure for 'new' cities of the West, such as Denver and Los Angeles, would be in excess of 60 per cent. In many European cities public transport systems are much more fully developed and maintained and this has its effect on urban form. At the end of the twentieth century, the need to control traffic has become increasingly apparent and measures already in place in some cities, such as Singapore, are advocated. Hart (1992) has predicted a decline in road traffic and a shift to public transport. There will be dramatic changes for many Western cities which have grown at the behest of the motor car, but the massive investments in new public transit systems are seen as essential solutions to this form of urban problem. Another general problem of Western cities which is being tackled in a variety of ways is exemplified in Table 33.1.

Western cities have problems of government which arise from the frag-mentation of authority and funding. In the United States the problems are

Table 33.1 The fragmented government of cities

Metropolitan areas	Population (000s)	Area (sq. miles)	Units of government*
Chicago	8,181	5,660	267
Boston	4,110	2,429	105
San Francisco	6,042	7,403	110
Houston	3,642	7,151	100
Minneapolis-St Paul	2,388	5,049	205

Source: Rothblatt (1992).
Note: *Does not include school and special districts.

731

compounded by the legal autonomy of wealthy suburbs and the depth of disparities between different parts of the metropolitan area. In the United Kingdom the removal of the urban regional scale unit of government, notably the Greater London Council, has led to the loss of overall strategic policy and planning. Different municipalities have different burdens and different fund-raising abilities. Rothblatt (1992) reported that in Chicago there was rivalry among neighbourhoods for jobs, housing and public services, and fierce competition between the city of Chicago and suburban municipalities for economic development, tax revenues and public resources. Where local government is underfunded, reliance on national government for funding is high. At the heart of the problems for Western cities are the economic bases upon which they rest, the infrastructures and, in particular, the transport systems which allow them to function, and the forms of government with which they are managed. The many other problems have some roots in this basic trilogy.

EARLY EXPRESSIONS OF SOCIAL PROBLEMS

Western cities have never been without problems. During the nineteenth century, as the industrial city grew as the flagship of a wealth-creating society, it contained large areas of tenements, courtyards and terraces, ill-served by sanitation and water supply and racked with ill-health and crime. Western cities have always mirrored the gulfs in Western societies between the wealthy and the poor. As these societies created the new class of entrepreneurs, businessmen and professionals, they also created the industrial proletariat whose labour was the essential component of economic growth. Poverty and deprivation found its expression in nineteenth-century cities, its characteristics were described in academic texts (Jones 1971), in contemporary social surveys (Booth 1892), and in the writings of authors such as Dickens who used their literary skills to portray contemporary urban life through the experiences of their imaginary characters.

> A dirtier or more wretched place he had never seen. The street was very narrow and muddy, the place was impregnated with filthy odours. There were ... heaps of children ... crawling in and out of doors ... covered ways and yards ... disclosed little knots of houses, where drunken men and women were positively wallowing in filth.
>
> (Dickens 1982: 45)

Similar contemporary descriptions can be found for American cities such as New York and Chicago:

> They were on a street that seemed to run forever ... one uninterrupted row of wretched little two-storey, frame buildings ... here and there would be a great factory ... and immense volumes of smoke pouring out from the chimneys, darkening the air above and making filthy the earth beneath.
>
> (Sinclair 1972: 24–5)

The themes for these authors were similar. As the industrial city grew it fed upon imported labour, largely from rural areas in Europe and from immigrants in the United States. Workers were plentiful and cheap and were housed inadequately; poverty and deprivation became the hallmarks of the cores of Western cities well before the end of the nineteenth century. Bad housing and overcrowding had its repercussions. Infectious diseases were endemic, outbreaks of cholera punctuated the century and devastated parts of the inner city; Mayhew (1862) described the 'rookeries' of Victorian London as places where children were born and bred to the business of crime. The social problems flowed from uneven economic distribution, bad housing and limited opportunities. Problems became 'localized' within cities, there were 'pockets' of deprivation and this geography of urban problems has proved a durable feature.

INNER-CITY/OUTER-CITY AND THE GROWTH OF DISPARITIES

The problems of the modern Western city are in some ways residual and can be traced to the disparities and disadvantages which have been inherent to the process of urbanization for a long period of time. There is an inner city–outer city schism which reaches its most dramatic expressions in the United States. As the bonds of the industrial city were loosened with improving transport, greater availability of land and an emerging housing market, it was the wealthy who first began the outward drift to the periphery and allowed suburbs to grow. The residual areas of the inner city became occupied by recent immigrants some of whom were never to achieve their place in the suburban exodus. As the suburban municipalities achieved the status of separate jurisdictions, they practised exclusionary zoning to protect their character. American inner cities became typically black or Hispanic and poor; suburbs were white and rich. This process has never really abated. Mayor Koch of New York stated in 1979 that the large majority of those who moved in were black, Hispanic and poor. This inner city/outer city dichotomy was evident though never quite as starkly expressed in other Western societies.

CITY AND SOCIETY: THE PROCESSES OF CHANGE

Economic change and de-industrialization

At the end of the twentieth century, the Western city, despite its diverse forms, has a number of general features which underpin its problems. Although it is now a truism to state that cities reflect the societies in which they are placed, the ways in which fundamental social changes find expression in cities are becoming more obvious. Terms such as post-industrial are used to describe Western urban societies which no longer rest upon an industrial or manufacturing base. As Western societies have de-industrialized the impact upon urban economies and upon the geography of the city has been profound. This process has been accompanied by huge losses of employment which have fallen disproportionately upon the blue-collar workforce and by a restructuring which has changed the nature of the urban economy.

An analysis of the countries of northern Europe (Cheshire and Hay 1989) showed losses of 7 per cent of industrial employment between 1974 and 1980 and of a further 12.8 per cent between 1980 and 1984; by 1984 manufacturing accounted for less than 30 per cent of total employment. In the same year Beeson and Bryan (1986) calculated that manufacturing in the United States made up only 21 per cent of total employment. Bovaird (1992) showed that for almost all European regions there has been a loss of manufacturing jobs from within urban cores. Between 1951 and 1981 the metropolitan inner city in the United Kingdom lost a million manufacturing jobs, the outer metropolitan areas lost a further million; 3 million manufacturing jobs (or 36 per cent of manufacturing in the UK) were lost between 1971 and 1988. De-industrialization is counterbalanced by restructuring and by 1984 the services provided 61 per cent of employment in northern Europe and 72 per cent in the United States. Disinvestment followed de-industrialization as the cities and the districts within cities in which manufacturing was in sharp decline were not those in which new employment opportunities were being created. Worse affected at an inter-urban scale were the older industrial cities with less advantageous locations. Cheshire and Hay (1989) described de-industrialization, decentralization and increasing peripherality as the triad of disadvantages affecting some European cities. Barcelona lost 88,037 manufacturing jobs (12.4 per cent of its total) in the 1970s; Liverpool lost 87,934 (36.2 per cent) and Amsterdam 46,800 (29.3 per cent).

Some cities show gains in Europe and evidence from North America suggests a similar pattern, with cities based on older manufacturing activities such as Cleveland, Detroit and Buffalo suffering radical losses of manufacturing employment and limited compensating growth in services. These are all 'frost-belt' cities and contrast with 'sun-belt' cities like Houston, Dallas and Miami where there is new industry and rapid growth of services.

During the 1980s there were signs of selective revival among major cities. Pittsburgh and St Louis, which showed all the classic symptoms of urban decline, have attracted new investments (Cheshire and Hay 1989); the latter experienced over a billion dollars worth of new investment in its core. There are regional effects to 'urban survival' and Champion and Green (1991) concluded that the North–South divide was widening in the United Kingdom at the beginning of the 1990s. There are new investments in some cities which promise to provide service employment and to generate income, not least the heritage-tourism boom which earns £3.8 billion each year for central London, but these are selective in their efforts. 'World cities' (Hall 1977) have advantages, and Frost and Spence (1991) have provided a detailed analysis of employment change in central London. Throughout the 1980s it was the producer services, particularly banking, insurance and finance, and other sources, which fuelled growth. By 1981 this sector accounted for 650,000 jobs compared with 125,000 in manufacturing. During the 1980s, central London gained 120,000 jobs in producer services as defined, though this was offset by losses in other sectors. By the end of the 1980s some of the operational difficulties facing central London were becoming more apparent: one-third of all firms involved in business and financial services reported difficulties in recruiting professional and managerial staff; office rents were almost double those of Paris; in a city where two-thirds of the workforce relies on public transport, the system was badly in need of major investment.

Although compensating growth to balance loss of the manufacturing base can be identified in some cities, in others it is not occurring. Even where the services have grown, the negative impact of change is still acute for large sections of the urban populations. The new jobs, whether in producer services, tourism or high technology, are not accessible to a discarded industrial workforce. Rising levels of unemployment are paralleled by the diminution of employment opportunities. Bovaird (1992) stated that in many parts of large industrial cities in the United Kingdom outside southern England, over half the adults were out of work; within some inner-city areas over 80 per cent of youths were unemployed. Dawson (1992) reported unemployment rates of 32 per cent in Seville in 1986, 24 per cent in Naples in 1987 and 19 per cent in Dublin in 1987. Again, unemployment is focused, and the Ballymun estate in Dublin had unemployment rates in excess of 50 per cent. A new Mazda car factory near Detroit in 1987 attracted 100,000 applicants for 2,000 jobs (Morris 1992).

Demographic change

The residual and marginalized nature of the unemployed is emphasized by the ongoing processes of demographic change in large urban areas. Trends of selective outmigration have been established for much of the twentieth century and earlier. Over the period 1951 to 1981, the six largest urban areas in the United Kingdom lost over one-third of their inner-city populations; inner Paris lost 300,000 between 1968 and 1975 and Rome 200,000 in a similar period. Preliminary results from the 1991 British census suggest a continuing urban to rural shift of population with highest rates of loss being experienced by London's inner boroughs (−5.89 per cent) and metropolitan principal cities (−8.6 per cent).

From the late 1960s through to the early 1980s a process of counter-urbanization appeared to be generally prevalent in Western societies. This refers to a pattern of population redistribution in which larger urban areas experienced decline or at least no-growth due to net migration losses, whilst smaller towns and rural areas showed significant gains. Explanations of counter-urbanization were couched in terms such as under-bounding of urban areas, anti-urbanism, fewer disparities in terms of quality of services, and new factors in economic location which allowed preferences for 'prestige environments' to come to the fore (Fielding 1990). Counter-urbanization was always an ambiguous concept. There were regional effects such as the sun-belt/frost-belt in the United States and a problem in disentangling product dispersal, involving the relocation of economic activities, from a continuing suburbanization process which reached well beyond urban fringes (Gordon 1989). Whatever its detailed mechanisms, counter-urbanization still contributed to urban problems by continuing the removal of both employment opportunities and specific population groups from the inner urban areas. Since the early 1980s, counter-urbanization seems at least to have lost its spatial generality, though whether this ever truly existed is questionable. It has become region-specific and net gains in rural areas are limited to favoured regions such as southern England. Champion (1992) has identified these shifts during the 1980s and has found that particularly in the late 1980s the relative growth of non-metropolitan areas has slowed down significantly. As Table 33.2 shows, the experience of the 1970s, with its rapid growth outside the large metropolitan areas of the United States, has not been repeated in the 1980s. Similarly in Europe, the later 1980s showed a swing backwards towards population concentration and a recovery in national core regions. Although there is continuing counter-urbanization in France, Paris recovered its losses of 1975 to 1982. The 1991 United Kingdom census indicates recovery for London and much of southern England. An understanding of these shifts remains speculative, but Champion (1992) considers the effects of a slowdown in fertility among 'dispersing' groups and the upsurge of unskilled migration into the central city. The flood of Hispanic migrants into some American cities, and of refugees, the low-skilled and others into European cities, may be symptomatic of this trend. An additional component of positive migration into the inner city is that involving gentrification. Again this is particular to some cities and is a minority phenomenon, but it is a concomitant of some of the massive urban renewal schemes such as London Docklands and its North American counterparts. Ley (1992) found that in Canadian cities such as Vancouver and Montreal in the 1980s the impetus for gentrification had been maintained throughout the severe downturn in the economy. Allied changes

Table 33.2 Population change in the United States

Period	Large metropolitan areas	Other metropolitan areas	Non-metropolitan areas
1960–9	18.5	14.5	2.2
1970–9	8.1	15.5	14.3
1980–9	12.1	10.8	3.9
1980–4	6.0	6.1	3.6
1985–9	5.8	4.4	0.3

Source: Champion (1992).
Note: All figures are positive percentage change.

in parts of the central city designed to cater for the demand for heritage tourism have assisted this process.

The positive trends clearly help the overall urban economy but do little for the basic problems of unemployment and limited job opportunities amongst traditional inner-city populations. Where such people could reasonably aspire to become factory operatives and craftworkers a couple of decades ago, their prospects are low-paid services employment, cleaners and janitors now (Wilson 1991). Such prospects may provide more opportunities for women, but the men directly affected by the disappearing industrial base have fewer options. Many of the new jobs in producer services attract more educated, professionally trained commuters and where such professionals – 'the yuppies' – move into refurbished inner-city areas the social divides are wide and obvious. White (1984) showed that throughout continental Western Europe there was evidence of inner-city population polarized into highest and lowest levels of the social hierarchy; in many of these cities however the polarization arises as much from the traditional preferences of middle-class people for an inner urban residence as from recent gentrification. As the disadvantaged recent arrivals move into inner-city areas to join the 'trapped' and residual groups already there, the inner cities have become what Champion *et al.* (1987) termed the 'repositories' of people with the least economic 'clout' in society, undermining life there by creating a situation of multiple deprivation. Similarly, Friedmann and Wolff (1982) characterized the disadvantaged groups as an 'underclass' marooned in the inner city and condemned to a life of economic and social disadvantage.

Changes in the built environment

A third process relates to the built urban environment in general and the 'affordable' housing market in particular. Older components of urban fabric, both buildings and the whole infrastructure of utilities and facilities and routeways, are concentrated in the inner city. Urban renewal has cleared the worst examples of nineteenth-century slums, but the rows and tenements which remain are often blighted by a lack of maintenance and investment. As property filters down the housing market it does not keep pace with affordable housing demand and standards continue to decline. Urban renewal has not always dealt sympathetically with such areas; where improved transport is the priority, new roads have cut swathes through the inner city and have disrupted residential communities. Where older housing has been replaced *in situ* the emphasis has been on cheapness rather than quality. High-rise solutions of the 1950s and 1960s, motivated more by low cost and expediency rather than quality planning principles, created more problems than they solved; the demolition of such high-rise projects, perhaps beginning with the demolition of Pruitt-Igoe Towers in St Louis in 1972, has become commonplace. In many American cities the stark and ill-conceived high-rise projects (Newman 1972) became afflicted by vandalism and crime to the extent that many were unmanageable; in Paris, some of the 'grands-ensembles' stand as planning disasters as does Amsterdam's Bijlmermeer; in British cities such as Liverpool the experience is the same, and in London Coleman's (1985) study of such housing in inner boroughs documented the design disadvantages from which it suffers.

Loss of community

Pressures on the inner city are perhaps best exemplified by the impact of community change or the diminution of a sense of place in inner-city areas. Traditionally, the inner city was the locale for strong communities of various kinds such as those recorded by Gans (1962) in his study of urban villages in Boston, by Jacobs (1961) in her portrayal of New York's Greenwich village, and by Young and Willmott (1957) in their studies of London's East End. As these studies and many others have documented, community in the traditional sense has long been under attack. There have been many pressures on inner-city communities from initial expansion of the commercial area to later congestion and overcrowding, the impact of urban renewal and influxes of new migrants. It now appears to be devastating impacts of economic decline coupled with lack of investment which is the final blow. Detroit with its black inner city and collapsed motor manufacturing industry symbolizes the worst problems of American inner cities: 'These are streets glittering with broken glass, dotted with hulks of abandoned cars and lined with thousands of abandoned buildings that attract drug-users, homeless people and arsonists' (Morris 1992: 33).

By the 1990s other American cities show the same signs of neglect and decay, shells of buildings set in a wasteland of dereliction, violence and crime, vividly portrayed in films such as *Fort Bronx*. These are environments in which scores of communities have little chance (see also Wilson 1987). In a reflection on London, Widgery (1991) presents a similar image of a radically changed inner city: 'What is being lost is something infinitely precious, that sense of neighbourhood, community and mutual solidarity with has given London its special character' (cited in *The Independent*, 1 November 1992, p. 24).

This inner city is commonly problematic in Western cities but it has no monopoly of urban problems. The geography of social problems reflects the extent to which the market forces have been allowed free play. In North America where this has been largely unfettered, the inner city areas stand out as the condemnations of disparities and neglect. In the United Kingdom and other European cities where governments have intervened in housing markets in substantial ways, the concentration of problems is not confined to the inner city. Some time ago the Liverpool Inner Area study concluded that concentrations of people at risk were found in many outer municipal estates such as Speke and Cantril Farm; deprivation was also recognized as a growing problem on Glasgow's peripheral estates. Cater and Jones (1989) identified as a category of problem areas, the undesirable council property typified by overcrowding, lack of amenities, disrepair and poor service provision. These are the rehousing schemes for the urban disadvantaged which at times have led to the 'difficult-to-let' or 'problem estates'. They are the stigmatized parts of municipal provision, labelled by social agencies and tenanted by a combination of self-selection or lack of choice. Barke and Turnbull (1992) described the South Meadowell estate in Tyneside. In the 1940s it was being regarded as an estate on which to locate tenants 'unsuitable in standard for new housing' (ibid. 54); in the 1960s the name was changed from 'The Ridges' in recognition of the fact that part of the area's problems was due to labelling and stigma; in the 1980s problems were heightened as the homeless were moved in, as crime and unemployment, vandalism and a sense of hopelessness rose. These areas, the product of twentieth-century planning, are as much locales of social problems as are parts of the inner city or the high-rise disasters: 'many parts of the inner city and peripheral council estates of urban Britain have become almost "cashless communities" of permanently unemployed residents dependent upon social welfare payments' (Pacione 1990: 200). This statement has generality in the Western city of the 1990s; we can now examine the theories which have been put forward to explain why these dire situations have arisen.

EXPLANATIONS OF SOCIAL PROBLEMS IN THE CITY

There have always been 'grand' theories to account for social problems. Those explanations derived from Marxist theory belong to this group which views social problems as the inevitable outcomes of a system which concentrates power and wealth in a limited number of hands and creates a society of haves and have-nots. The have-nots are not only the inevitable outcomes, they are an essential part of the grand plan to perpetuate such a society. Grand theories of this kind have depth and scholarship and provide a valid context against which to understand social problems, but they cannot account for the diversity of their expression and run the danger of operating with a deficient account of human agency. One feature of grand theories which is indisputable is the assumption that social problems in the city are in effect those which find their origins in society but find their fullest expression in the city. Herbert and Johnston (1976) made the distinction between problems in the city which can be described in those terms and those of the city. The latter form a minority which could be seen as specific products of the urban environment such as overcrowding, built environmental or neighbourhood effects. These are more difficult to sustain as independent explanations but the urban environment does present social problems in their most dramatic form.

Extant theories can be summarized in three broad types, though these are not mutually exclusive (see Figure 33.1). They range from those theories which seek explanation at a locality scale within the city and those which rest entirely upon the 'structural' origins of social problems. The 'culture of poverty' theory belongs primarily at the level of locality. It regards poverty as self-induced, a product of the kinds of people afflicted. Terms such as 'fatalism' and 'apathy'

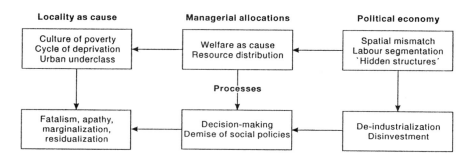

Figure 33.1 Theoretical perspectives on social problems.

were used to indicate an acceptance of poverty for many for whom obsessive consumption and present-mindedness are major barriers to economic mobility. Although largely discounted since its original formulation (Lewis 1968) elements such as fatalism and hopelessness tend to recur with studies of 'slums of despair'. The 'cycle of deprivation' idea is a related concept which has been more rigorously tested in recent years (Rutter and Madge 1976), though the evidence remains inconclusive. It argues that children born into deprived households, and typically deprived areas, have limited access through school and job opportunities to break out of that condition and are kept within the cycle. They are 'trapped' less by fatalism than by being denied access to sufficient means to achieve social mobility. The 'urban underclass' theory, most recently articulated by Wilson (1987), recognizes the ways in which a section of society becomes marginalized and residualized and again trapped in poverty. Wilson argues that this 'truly disadvantaged' group remained after those able to achieve social and geographical mobility had abandoned the ghettos. Such areas are denuded of the organizers and leaders by this process; the social institutions and networks of community disintegrate and unemployment and hopelessness become a way of life.

Some theories relate explicitly to the process of economic change. 'Spatial mismatch' recognizes the geographical implications of the fact that urban centres are undergoing an irreversible structural transformation from places of production and distribution to places of administration, information and high order services. There is a mismatch between the kinds of jobs available and the skills of the resident workforce; inner-city jobs are increasingly white-collar jobs. 'Labour segmentation' relates to the new workforce status of the most disadvantaged groups. If they are employed they are relegated to low-paid segments of the labour market which cannot be staffed in any other way. In this bottom segment are migrant workers, often lacking work permits, and women single parents who have reduced labour ability to compete in the labour market because of their roles as sole parents. The number of women in this position in

Canada increased from 291,900 in 1931 to 853,600 in 1986. Aponte (1991) relates this to the male marriageable pool index which suggests that the rise of female single parents is linked to the rise in joblessness among men. 'Welfare as cause' is a theory which has been promoted in various ways but was recently re-articulated by Murray (1984) who argued that the liberalization of welfare in the United States, by offering rewards to those who stay poor, had made both work and marriage less attractive. Similar sentiments have been expressed in the United Kingdom for many years, and a determination to 'crack down on welfare spongers' recurs as an explicit part of government policy. Whereas access to welfare clearly helps people to remain in specific circumstances with often limited incentive to work harder, its status as a causal theory is limited.

Resource allocation theories highlight the role of managers or decision-makers in the urban system. Sometimes called the gatekeepers, they have the power to allocate resources and to determine who gets jobs, housing finance, social housing allocations, etc. The managerialist thesis is criticized on the grounds that many managers merely relay established rules and guidelines; they are the intermediaries between the 'system' and the consumers. However there are levels of discretion which can be exercised in face-to-face relationships. Understanding the procedures and practices of resource allocation, and in particular its practitioners and the way they work, does have a contribution to make in understanding the incidence of deprivation and social problems in cities. Finally, there are structural or political economy theories of social problems which stress their roots in the kind of society which exists. A society which rewards differentially creates relative disadvantage and needs a class of disadvantaged population to perpetuate itself. The strength of structural explanations lies in its theoretical cohesion, the weakness in its tendency to offer mechanical explanation and to founder on the diversities of empirical experience. Whereas the relevance of many structural concepts is clear, many things do filter downwards from the 'imperatives' of society and impact upon localities; nevertheless, people are not merely the passive recipients of structural conditions, they are the creators of meaning which is the well-spring of human action and historical change.

The theoretical explanations of social problems in the Western city offer a range of options. Each of these has been subject to close scrutiny and empirical verification and none emerge unscathed. What is on offer is a 'package' of explanatory approaches which cannot be used as a kind of 'bran-tub' of ideas but which do have something to offer when used in careful and responsible ways. The theories offer a frame of reference for policy-makers, though of more immediate concern are the specific ways in which urban problems find their expression. Multiple deprivation is a common but not inevitable fact of urban problems. There are concentrations of multiple deprivation in geographical space, but a great deal of research has revealed the weaknesses as well as the strengths of area policies. There are bad areas within any major city but they will not capture all of the problems nor necessarily a majority of those in need, neither will they be uniformly disadvantaged. There is an aggregate, scale or 'ecological' effect which needs to be recognized. Some specific problems can be discussed with these caveats in mind.

POVERTY AND DEPRIVATION

Whereas the inner cities have no monopoly of poverty and deprivation, they do give these problems their sharpest expression, especially in American cities. As Aponte (1991) argued, a substantial inner-city minority in the American city is 'mired in poverty'. Wilson (1987) emphasized the problems of the black population trapped in these conditions, but other ethnic minorities such as Hispanics and Chinese have similar experiences. A central theme in Wilson's thesis (1987, 1991) was that changes were occurring to create an underclass of truly disadvantaged urban residents. As black mobility denudes the historic ghetto of any emerging middle-class and stable blue-collar families, the structure of community collapses. New ethnographic studies (Marks 1991) identify the extremity of problems in these areas – unemployment is very high, job prospects are few, traditional family life collapses, and crime and drugs become hallmarks of the 'community'. Ethnic minorities exemplify the extreme examples of disadvantage. In 1987 the incidence of poverty among American black people was 33 per cent compared to 28 per cent of Hispanics; the fact of increasing poverty was exemplified by the Hispanic rate of 16.7 per cent in 1977 (Aponte 1991; see also Knox 1989). The breakdown in traditional family life in the ghetto could be shown by the rise from 17.9 per cent in 1940 to 41.9 per cent in 1985 of the number of female single-parent families.

If ethnic minorities form one group at risk, the elderly are another. Although the group is diverse, for many people old age means a sharp decline in material standards of living and an increase in health and morale problems associated with the ageing process. Duncan and Smith (1989) showed the percentage of old people below the poverty line in the United States as 6.9 per cent among white males, 13.3 per cent among white females, 24.2 per cent among black males, 35.5 per cent among black females, 18.8 per cent among Hispanic males and 25.5 per cent among Hispanic females. Although there is nothing which resembles the 'geriatric ghetto' in Western cities, many old people occupy the same homes where they spent the latter part of their working lives. It is those trapped in the inner city or on problem estates who face the biggest problems. Their disadvantages become manifold when relative poverty is added to by loss of partners, decreased mobility and increasing dependence. Many old people have the means and fortitude to cope, but for those who do not the circumstances in which they live are an indictment upon the societies of which they are part.

During the 1980s there was evidence that the welfare dimension to Western society was proving to be increasingly inadequate. Cuts in public expenditure, underfunding of facilities and measures designed to reduce the burden of care all had their impact upon the dependent population. One measure of failure in the welfare provision system has been the growth of the homeless; Wolch (1991) argued that the withdrawal of federal welfare support systems and the shrinkage of economic opportunities had forced individual survival strategies and local responses. In the United States the problem has been exacerbated by the demise of single-room-occupancy hotels (SRO) – New York lost 30,385 such units in 160 buildings as a result of clearance schemes (60 per cent of the total) between 1975 and 1981 (Hoch 1991). Chicago's Skid-Row was devastated by renewal between 1960 and 1980; the West Madison district lost 92 per cent of its group quarters; near North side lost 84 per cent of its single-room units. Estimates of the homeless in American cities vary from 250,000 to 3 million (Rossi 1989). In Britain the number of homeless households doubled during the 1980s to an estimated 128,000, which has been described as a totally inadequate measure of the true need (Daly 1991). Even in Canada, where severe winters make homelessness a dire condition, the problems are rising. One indicator of urban poverty, if not directly of homelessness, is the rise of 'food-banks' designed to provide free food for impoverished urban populations. Following the first food-bank in Edmonton in 1981, by 1991 there were almost three hundred of these voluntary welfare agencies in Canadian cities serving around 2 million people. Economic or, more accurately, employment decline underpins the homeless problem; it is compounded by the retreat of government from welfare programmes and is most acute where refugees or illegal immigrants add to inner-city populations.

HEALTH CARE

The problems of health-care systems epitomize the difficulties facing Western cities. An inverse-care law has for some time been recognized as a succinct commentary upon a health-care system which is least effective in areas of greater need. In the early 1970s, de Vise (1973) showed the rapid decline in numbers of physicians in inner Chicago from 475 to 76 between 1950 and 1971. His conclusion was that 'if you are average poor, under 65, female, black or live in a low-income household ... you are a disenfranchised citizen as far as health-care rights go' (de Vise 1973: 1). Whiteis and Salmon (1987) noted some 150 community hospital failures in the United States from 1980 to 1984. Many poor neighbourhoods in cities such as Chicago, Philadelphia, St Louis and Atlanta were denied access to welfare hospitals. Knox (1987) described a situation in Glasgow where residual, marginalized people with high levels of illness were being cared for by elderly doctors in ill-equipped, single-handed practices. Whereas the community health movement grew in the 1980s, Farrant (1991) argued that it was severely underfunded and the health-care issue had reached the highest level of concern in Western cities by the 1990s.

It was in the area of mental health care that policy failures and the inabilities of communities to cope have been revealed most clearly. At the heart of the present problem is the policy of deinstitutionalization which was initiated in the early 1960s. Designed to reduce the ever-rising costs of health care and to turn to a 'community' solution, the number of mentally ill patients in long-stay hospitals in the United States was reduced from 560,000 in 1955 to 125,000 in 1981 (Shadish *et al.* 1989). In England and Wales a 28 per cent reduction in in-patients was achieved between 1951 and 1970. For many the idea of community care turned out to be a myth. Warner (1989) estimated that only half of the non-institutionalized mentally ill had reached anything like a domestic environment, the rest were homeless, in jail, in bad housing, rooms, nursing homes or sheltered care. Community care only works where kin and, more rarely, close friends are prepared to accept the responsibility; that may only occur in 10 per cent of cases. The problem of mental illness has certainly not diminished and out-patient care has risen dramatically, but the policy of deinstitutionalization has direct links with homelessness, alcoholism and desperate urban poverty.

SOCIAL DISORDER AND URBAN RIOTS

Crime and social disorder have become endemic in disadvantaged parts of the Western city. Crime areas or the 'pocketing' of crime and criminals, have been known for some time. Mayhew (1862) described the 'rookeries' of nineteenth-century London; Shaw and McKay's (1942) depiction of 'delinquency areas' in Chicago was a vivid portrayal of an urban problem. Little has changed. Wilson (1987) described the Wentworth Avenue district of south Chicago, which was the site of eighty-one murders and 1,691 aggravated assaults in 1983 – up to 13 per cent of a city's crime in an area which held 3 per cent of its population. As significant as the incidence of crime is the fear of victimization. Smith (1989), using figures from the British Crime Survey, showed that 34 per cent of those interviewed felt unsafe walking alone after dark; the Islington Crime Survey (Jones *et al.* 1986) revealed that one in two households could expect to fall victim to crime. Schuerman and Kobrin (1986) found that rising crime rates were linked to indicators of community deterioration such as more multifamily dwellings, broken homes and ethnic change. Incivilities appear such as graffiti, vandalism and abandoned cars which are symptomatic of the downward spiral of neighbourhood decline. Out of this mix of rising crime, heightened fear, and the rising drug cultures of the worst areas came the new phenomenon of urban riots and inter-group conflict.

In his description of urban life in Los Angeles, Davis (1992) painted a vivid and detailed picture of urban problems. A police estimate suggested 230 black and Latin gangs in Los Angeles, plus a further eighty Asian gangs in the late 1980s. Much of the inner city had become a battleground with black gangs federated into two broad groups, the Crips and the Bloods (see Figure 33.2). The epidemic of youth violence had its roots in exploding youth poverty and unemployment; the latter reaching 45 per cent among black youths in Los Angeles county. Poverty reached down to the young, affecting 23 per cent of

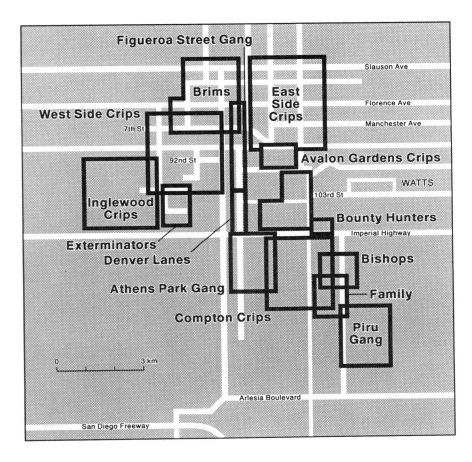

Figure 33.2 Los Angeles gangs. After Davis (1992).

children. Drop-out rates in high-school reached 30–50 per cent; in the public housing project of Nickerson Gardens there were 120 wage earners in 1,060 households. The influx of cocaine and 'crack' have made the problems of youths more deadly, and the police react in a way which criminalizes large sections of the community: 'as a result of the war on drugs, every non-Anglo teenager in southern California is now a prisoner of gang paranoia and its associated demonology' (Davis 1992: 284).

The reaction of the affluent is to protect themselves, and the divisions between haves and have-nots take new and tangible forms. Davis (1992) spoke of 'fortress L.A.' a city divided into the 'fortified cells' of the affluent and the 'places of terror' where police battle the criminalized poor. Using income

745

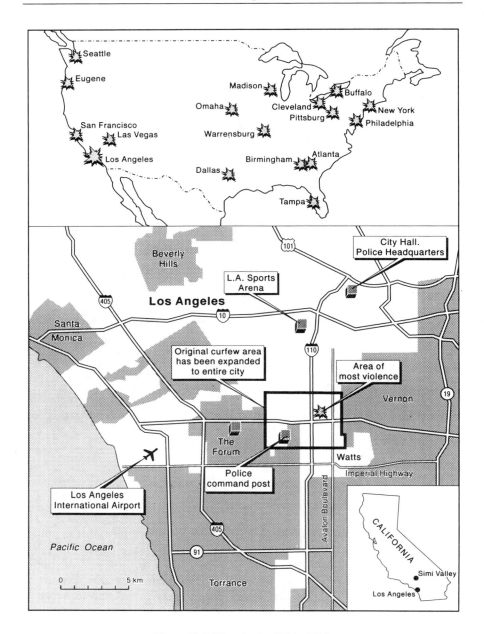

Figure 33.3 Riots in the USA, 1992.

thresholds of $50,000 and $15,000, Davis showed that in the 1980s as the affluent increased from 9–26 per cent and the poor from 30–40 per cent, social polarities became more sharply stated. Worst inner-city areas and public housing projects became powder kegs waiting to be ignited, and the severe riots which have punctuated the last part of the twentieth century are the result. In the late 1960s riots in cities like Detroit and Los Angeles were focused on the ghettos. Thirty-four died in the Watts district of Los Angeles. In April and May 1992 a new wave of violence erupted in Los Angeles and spread to other cities (see Figure 33.3). Most riots have had a racial flashpoint and involve conflict with police; in Los Angeles, forty-four were killed, 5,500 arrested and over 2,000 fires were started. Bloods and Crips announced a truce to concentrate fire on the police. 'The protests turned violent and the dreadful cycle began of looting and burning that became indiscriminate … and then came the shooting. Blacks attacking Korean shopkeepers, Hispanics attacking the Jewish shops in Hancock Park and luxury stores in Beverly Hills' (*The Guardian* 1992: 10).

There was also the telling comment that most of the violence was have-nots against haves in a city where the contrasts between the vastly rich and the really poor are as great as in any Third World city. Racial tensions gave spark to the riots but the bedrock reason was poverty and disenchantment as black household incomes fell by 44 per cent in real terms and Hispanic by 35 per cent between 1973 and 1990.

European cities have experienced riots in the 1980s. In the United Kingdom the causes again appear to be disadvantage and bad police relations in ethnic areas. Riots swept many British cities in the early 1980s and were followed by the Scarman inquiry. In Bristol in the summer of 1992 the Hartcliffe estate, with high unemployment and bad facilities, erupted when two local youths died in a collision between their stolen motor-cycle and a police car. In German cities in October 1992, out of work youths with overt racist attitudes attacked refugee hostels. Perhaps as ominous as the sporadic violence is the growing tendency for disenfranchised youths to seek confrontation with police and others in episodes which range from assault to car theft to vandalism.

POLICY INITIATIVES

Government at all levels has been involved in the attempts to ameliorate urban problems. The broadest traditional pattern has been one in which central government sets the broad agenda and provides the base funding and local government initiates and carries out specific projects. Major programmes of urban renewal and investment in transport infrastructure have followed this model:

> City planning in Western Europe has been part of a growing trend towards intervention in almost all European countries in the period since 1945 … direct

intervention by central government is rare . . . but various forms of economic policy
. . . have been incorporated into city policies.

(Burtonshaw *et al.* 1991: 222)

An additional component in the European Community has been the availability of EC funds for urban projects in similar ways to federal schemes in North America. The system, though never perfect, revealed its inadequacies in the 1980s. In the United Kingdom a gulf opened between central and local government as central policies, such as the sale of public housing, were imposed, but funding was diminished in key areas of welfare. Eyles (1989) argued that much of United States federal urban policy amounted to no more than rhetoric and had little or no real impact on cities. In the 1980s both American and British governments focused on economic recovery and private investment; social goals and public spending were lesser priorities. In order to achieve these aims in Britain, new autonomous agencies were established such as the Urban Development Corporations with the brief of achieving local area regeneration. At times in the 1980s, some of these initiatives, notably London Docklands Development Corporation, achieved success and the face of the inner city did change. There were always qualifications. Sharp variations were evident regionally, much more public money was needed outside southern England, social tensions arose between the wealthy newcomers and the older residents in London's docklands. By the end of the 1980s even the success stories were wavering. Infrastructure and transport costs soared, the bubble burst on property markets and the Canadian-based Olympia and York, with huge investments in Canary Wharf, was in serious trouble. Expectations that the 'market' would rejuvenate the inner city were foundering.

Where major social disorder occurred, government money was forthcoming. Inner areas schemes and priority estates were in a sense responsive policies. By the 1990s this was being recognized at all levels of society. As a resident of Bristol's Hartcliffe estate stated after a night of riots in 1992: 'St. Paul's [another Bristol district] got plenty of money when they rioted. Let's see what we get now the trouble has started here. They will start throwing money at the place' (Prestage 1992: 2).

This is a sad comment on the form urban policies have taken in the last part of the twentieth century. As the problems have grown the policies have become more fragmentary and niggardly as a reliance on voluntary agencies and private sector investment replaces a genuine social strategy. Urban policies respond to social problems but are in themselves on occasions the creators of problems. The record is long from the policy of 'dumping' problem households to create a problem estate, to the misconceived high-rise towers of the 1960s, to the neglect of a consistent social welfare dimension in the 1980s. In the private sector, as housing managers 'red-lined' areas to be starved of mortgage funds, they were condemning a place and its people to a downward spiral of decline. The whole experience of 'labelling' and stigmatizing sections of society does nothing but perpetuate urban problems. Western cities are areas of wealth with increasingly evident pockets of poverty and disorder; urban policy is ineffective so long as it allows these disparities to grow.

REFERENCES

Aponte, R. (1991) 'Urban Hispanic poverty: disaggregations and explanations', *Social Problems* 38(4), 516–28.

Barke, M. and Turnbull, G. (1992) *Meadowell: The Biography of an Estate with Problems*, Aldershot: Avebury.

Beeson, P.E. and Bryan, M.F. (1986) *The Emerging Service Economy*, Cleveland: Federal Reserve Bank.

Booth, C. (1892) *Life and Labour of the People*, London: Macmillan.

Bovaird, T. (1992) 'Local economic development and the city', *Urban Studies* 29(3/4), 343–68.

Burtonshaw, D., Bateman, M. and Ashworth, G.J. (1991) *The European City: A Western Perspective*, London: David Fulton.

Cater, J. and Jones, T. (1989) *Social Geography: An Introduction to Contemporary Issues*, London: Edward Arnold.

Champion, A.G. (1992) 'Urban and regional demographic trends in the developed world', *Urban Studies* 29(3/4), 461–82.

Champion, A.G. and Green, A.E. (1991) 'Britain's economic recovery and the north–south divide', *Geography* 76(13), 249–54.

Champion, A.G., Green, A.E., Owen, D.W., Ellin, D.J. and Coombes, M.G. (1987) *Changing Places: Britain's Demographic, Economic and Social Complexion*, London: Edward Arnold.

Cheshire, P.C. and Hay, D.G. (1989) *Urban Problems in Western Europe*, London: Unwin Hyman.

Clark, C. (1951) 'Urban population densities', *Journal of the Royal Statistical Society* (A) 114, 490–6.

Coleman, A. (1985) *Utopia on Trial*, London: Hilary Shipman.

Daly, G. (1991) 'Local programs designed to address the homelessness crisis; a comparative assessment of the United States, Canada and Britain', *Urban Geography* 12(2), 177–93.

Davis, M. (1992) *City of Quartz, Excavating the Future in Los Angeles*, London: Vintage Books.

Dawson, J. (1992) 'Peripheral cities in the European Community: challenges, strategies and prospects', *Public and Policy Administration* 7(1), 9–20.

Dickens, C. (1982) [1838] *Oliver Twist* (K. Tillotson, ed.), Oxford: Oxford University Press.

Duncan, G.J. and Smith, K.R. (1989) 'The rising affluence of the elderly: how far, how fair and how frail?' *Annual Review of Sociology* 15, 261–89.

Eyles, J. (1989) 'Urban policy? What urban policy? Urban intervention in the 1980s', pp. 370–86 in D.T. Herbert and D.M. Smith (eds) *Social Problems and the City; New Perspectives*, Oxford: Oxford University Press.

Farrant, W. (1991) 'Addressing contradiction, health provision and community health in the United Kingdom', *International Journal of Health Services* 21, 423–39.

Fielding, A.J. (1990) 'Counterurbanisation: threat or blessing?', pp. 226–39 in D. Pinder (ed.) *Western Europe: Challenge and Change*, London: Belhaven.

Friedmann, J. and Wolff, G. (1982) 'World city formation; an agenda for research and action', *International Journal of Urban and Regional Research* 6, 309–44.

Frost, M.E. and Spence, N.A. (1991) 'Employment changes in central London in the

1980s', *Geographical Journal* 157(1), 1–12; 157(2), 125–35.

Gans, H. (1962) *The Urban Villages*, New York: Free Press.

Gordon, I.R. (1989) 'Urban employment', pp. 232–46 in D.T. Herbert and D.M. Smith (eds) *Social Problems and the City: New Perspectives*, Oxford: Oxford University Press.

Hall, P. (1977) *World Cities*, London: Weidenfeld & Nicolson.

Hart, T. (1992) 'Transport, the urban pattern and regional change, 1960–2010', *Urban Studies* 29(3/4), 483–503.

Herbert, D.T. and Johnston, R.J. (1976) *Social Areas in Cities: Spatial Processes and Form*, Chichester: John Wiley.

Hoch, C. (1991) 'The spatial organization of the urban homeless: a case study of Chicago', *Urban Geography* 12(1), 137–54.

Jacobs, J. (1961) *Death and Life of Great American Cities*, New York: Random House.

Jones, G.S. (1971) *Outcast, London*, Oxford: Oxford University Press.

Jones, T., McLean, B. and Young J. (1986) *The Islington Survey*, Aldershot: Gower.

Knox, P.L. (1987) *Urban Social Geography* (2nd edn), London: Longman.

Knox, P.L. (1989) 'The vulnerable, the disadvantaged, and the victimised: who they are and where they live', in D.T. Herbert and D.M. Smith (eds) *Social Problems and the City: New Perspectives*, Oxford: Oxford University Press.

Lewis, O. (1968) 'The culture of poverty', pp. 87–220 in D.P. Moynihan (ed.) *On Understanding Poverty: Perspectives from the Social Sciences*, New York: Basic Books.

Ley, D. (1992) 'Gentrification in recession: social change in six Canadian inner cities, 1981–1986', *Urban Geography* 13(3), 230–56.

Marks, C. (1991) 'The urban underclass', *Annual Review of Sociology* 17, 445–66.

Mayhew, H. (1862) *London Labour and London Poor*, London: Griffin–Bohn.

Morris, W. (1992) *Biography of a Buick*, Granta Books (cited in *Independent Magazine*, 8 August: p. 33).

Murray, C. (1984) *Losing Ground: American Social Policy 1950–1980*, New York: Basic Books.

Newman, O. (1972) *Defensible Space*, New York: Macmillan.

Pacione, M. (1990) 'What about people? A critical appraisal of urban policy in the United Kingdom', *Geography* 75(3), 193–202.

Prestage, M. (1992) 'Riots mark of respect to dead local men', *The Independent*, 18 July: p. 2.

Rossi, P. (1989) *Homeless in America*, Chicago: University of Chicago Press.

Rothblatt, D.N. (1992) 'Swimming against the tide: metropolitan planning and management in the United States', *Planning Practice and Research* 7(1), 4–8.

Rutter, M. and Madge, N. (1976) *Cycles of Disadvantage*, London: Heinemann.

Schuerman, L. and Kobrin, S. (1986) 'Community careers in crime', pp. 67–100 in A.J. Reiss and M. Tonry (eds) *Communities and Crime*, Chicago: University of Chicago Press.

Shadish, W.R., Lurigio, A.J. and Lewis, D.A. (1989) 'After de-institutionalisation: the present and future of mental health long term care policy', *Journal of Social Issues* 45(3), 1–15.

Shaw, C.R. and McKay, H.D. (1942) *Juvenile Delinquency and Urban Areas*, Chicago: University of Chicago Press (revised edition, 1969).

Sinclair, U. (1972) [1905] *The Jungle*, Cambridge, Mass.: Robert Bentley Inc.

Smith, S.J. (1989) 'Social relations, neighbourhood structure and fear of crime in Britain', in D.J. Evans and D.T. Herbert (eds) *The Geography of Crime*, London: Routledge.

de Vise, P. (1973) 'Misused and misplaced hospitals and doctors: a locational analysis of the urban health-care crisis', *Association of American Geographers Resource Paper* 22.

Walker, M. (1992) 'Dark past ambushes America's city of the future', *The Guardian*, 2 May: p. 10.

Warner, R. (1989) 'De-institutionalisation: how did we get where we are?', *Journal of Social Issues* 45(3), 17–30.

White, P. (1984) *The West European City: A Social Geography*, London: Longman.

Whiteis, D. and Salmon, J.W. (1987) 'The proprietorization of health care and the underdevelopment of the public sector', *International Journal of Health Services* 17, 47–64.

Widgery, D. (1991) *Some Lives*, London: Sinclair Stevenson.

Wilson, W.J. (1987) *The Truly Disadvantaged: The Inner City, the Underclass and Public Policy*, Chicago: University of Chicago Press.

Wilson, W.J. (1991) 'Studying inner-city social dislocations: the challenge of public agenda research', *American Sociological Review* 56, 1–14.

Wolch, J. (1991) 'Urban homelessness: an agenda for research', *Urban Geography* 12(2), 99–104.

Young, M. and Willmott, P. (1957) *Family and Kinship in East London*, London: Routledge & Kegan Paul.

FURTHER READING

Burtenshaw, D., Bateman, M. and Ashworth, G.J. (1991) *The European City: A Western Perspective*, London: Fulton.

Champion, A.G. and Townsend, A.R. (1990) *Contemporary Britain: A Geographical Perspective*, London: Edward Arnold.

Davies, W.K.D. and Herbert, D.T. (1993) *Communities within Cities: An Urban Social Geography*, London: Belhaven.

Davis, M. (1992) *City of Quartz: Excavating the Future in Los Angeles*, London: Vintage.

Hall, P. (1984) *The World Cities*, London: Wiedenfeld & Nicolson.

Hambleton, R. and Taylor, M. (eds) (1993) *People in Cities*, Bristol: School of Advanced Urban Studies (SAUS).

Herbert, D.T. and Smith, D.M. (eds) (1989) *Social Problems and the City: New Perspectives*, Oxford: Oxford University Press.

Jones, E. (1990) *Metropolis: The World's Greatest Cities*, Oxford: Oxford University Press.

Paddison, R., Lever, B. and Money, J. (eds) (1993) *International Perspectives in Urban Studies*, London: Jessica Kingsley.

Robson, B.T. (1988) *Those Inner Cities: Reconciling the Social and Economic Aims of Urban Policy*, Oxford: Clarendon Press.

34

CHANGING COUNTRYSIDES

Hugh Clout

TOWN AND COUNTRY

The evolving relationship between towns and their countrysides lies at the heart of many vital issues that have enthused generations of geographers and still offer formidable challenges. Agricultural systems and their markets, flows of migrants and trade, regional formations, models of land use, settlement patterns, transport networks, and systems of land-use planning all manifest aspects of this tantalizing juxtaposition (Cronon 1991). To separate 'rural' from 'urban' uses of land is still a relatively straightforward task, once critical categories and thresholds have been decided, but to distinguish 'rural' and 'urban' people has become increasingly difficult in the largely urbanized and industrial – or, more accurately, post-industrial – economies of Europe, North America and Australasia (Best 1981). Self-contained 'rural communities' no longer exist, as urban people, national lifestyles and international information in one form or another penetrate the most tranquil rustic backwaters (Kayser 1990).

In fact, as far as Britain is concerned the notion of the stable, self-contained, healthy and harmonious rural community was something of a myth during the past three centuries (Short 1992). In the eighteenth century London was attracting young, landless migrants from a vast hinterland that covered lowland England, while, across the Channel, Paris attracted young people from the northern third of the country. Different farm goods commanded distinctive supply areas (as von Thünen was to demonstrate in 1826), and villagers were not spared their share of exploitation, poverty, disease and ill-will from their neighbours. The coming of the railways encouraged greater numbers of rural folk to leave their long-settled farms and villages in Europe in search of a better life in towns and cities. Rural crafts and services collapsed because of competition from urban factories, and the countryside housed increasingly fewer people but employed a higher proportion of them in farm-related work. The onset of depopulation came later in the newly settled farmlands of North

America and Australasia, but by the early years of the twentieth century it was widespread in areas that had been occupied by Europeans only a few decades earlier (Robinson 1990).

RURAL DEPOPULATION AND AGRICULTURAL CHANGE

The fundamental equation of opportunity seemed fairly straightforward. The peasant plots of France or Germany, the family farms of the American midwest, and the great estates of the Italian Mezzogiorno or the American Deep South spelled out poverty and restricted life-chances; towns and cities offered the promise of self-improvement. The financial crash of 1929 and the depressed 1930s clouded that optimistic image and sent some unemployed cityfolk back to the apparent security of their parents' farms, both in Europe and North America. But conscription and profound economic upheaval during the Second World War attracted even more people away from farms towards urban factories which were producing goods as part of the war effort. The cityward shift of population and the sectoral shift of national economies from agriculture towards manufacturing and then services continued unabated in the post-war years.

Rural areas shed more and more of their inhabitants; remaining farms were enlarged; agricultural productivity rose to feed the growing industrial work-forces of the so-called developed economies. To be sure the pace of urbanization and industrialization varied by country and by region, affecting Britain, Belgium and Germany rather sooner than France or Italy, and considerably earlier than Spain or Portugal. Similarly the 1950s and especially the 1960s witnessed the emergence of daily commuting hinterlands. More people had become sufficiently affluent to purchase their own cars and also had leisure time to devote to day excursions or holidaymaking in the countryside. None the less, depopulation continued apace, with the farmlands of Mediterranean Europe and North America losing more workers.

Agricultural policies and the rural environment

At the same time the volume of farm produce became even greater, thanks to chemicals, fertilizers, mechanization and official farm policies which ensured that the citizens of Western Europe, North America and Australasia would not go hungry, through their own farmers' efforts or those of their trading partners. The Common Agricultural Policy (CAP) of the European Community inflated consumer prices and guaranteed high incomes to farmers producing cereals and dairy goods (Tracy 1989). Larger farmers in northern countries of the EC fared substantially better than their southern counterparts. As a result, commodity surpluses began to rise, with 'mountains' of grain and butter and 'lakes' of table wine becoming costly to store and difficult to dispose of both technically and politically. Their existence demonstrated that the CAP was, in some respects, too successful. Policy mechanisms varied in North America, Australasia and the

European countries outside the EC but the net results were similar: fewer and fewer people were producing ever-increasing quantities of farm goods (Cloke 1988, 1989).

In other respects the CAP was not successful. Intensification, strong reliance on chemicals, farm enlargement to enable large machines to operate, and the removal of wetlands, copses, hedges and other 'negative' features from farmed landscapes were erasing traditional countryside features which other elements in society wished to conserve (Lowe and Goyder 1983). Opposition came from two directions, sometimes working separately, sometimes together. Conservationists, often hailing from middle-class suburbs, sought to halt the destruction of aesthetically pleasing 'heritage' features; while ecologists marshalled scientific arguments to protect vulnerable habitats, endangered plant and animal species, and scenery that was under threat (Lowe and Cox 1986). National and regional mechanisms varied and operated at differing speeds, but the net result was the passage of protective legislation and designation of hundreds of national and regional nature parks and tens of thousands of sites of special ecological importance.

Table 34.1 Basic structures of rural Europe, 1988

	(a)	(b)	(c)	(d)	(e)	(f)	(g)	(h)	(i)	(j)
Belgium	93	17.3	36	2.7	2.2	45.7	53	44	3	20.2
Denmark	88	32.5	44	6.3	4.0	65.4	92	8	0	11.4
France	982	30.7	48	6.8	3.5	57.1	57	38	5	26.8
Germany (W)	705	17.6	32	5.2	1.5	47.9	61	37	2	29.6
Greece	953	5.3	53	27.0	15.6	43.5	32	57	11	43.6
Ireland	217	22.7	50	15.4	10.3	81.1	18	82	0	4.7
Italy	2,784	7.7	56	9.9	4.5	57.8	52	28	20	21.0
Luxemburg	4	33.2	43	3.4	2.4	49.1	44	55	1	34.3
Netherlands	136	17.2	40	4.7	4.1	50.7	44	54	2	8.3
Portugal	769	8.3	50	20.7	6.4	49.2	64	17	19	32.2
Spain	1,792	16.0	50	14.4	5.2	53.7	57	25	18	24.8
UK	258	68.9	46	2.2	1.7	75.8	37	63	0	9.4
EC 12	8,780	16.5	50	7.7	3.2	57.1	53	38	9	23.8

Source: European Commission (1990) *The Agricultural Situation in the Community, 1989 Report*, CEE, Brussels and Luxemburg.

Key: (a) Number of farm holdings, thousands
   (b) Average farm size, ha.
   (c) Percentage of farmers over 55 years of age
   (d) Agricultural employment as percentage of all employment
   (e) Percentage of GDP derived from agriculture
   (f) Agricultural Utilized Area (excluding woodland, buildings, yards, etc.) as percentage of total land surface
   (g) Arable as percentage of Agricultural Utilized Area
   (h) Permanent grass as percentage of Agricultural Utilized Area
   (i) Orchard, vines and olives as percentage of Agricultural Utilized Area
   (j) Woodland as percentage of total land use

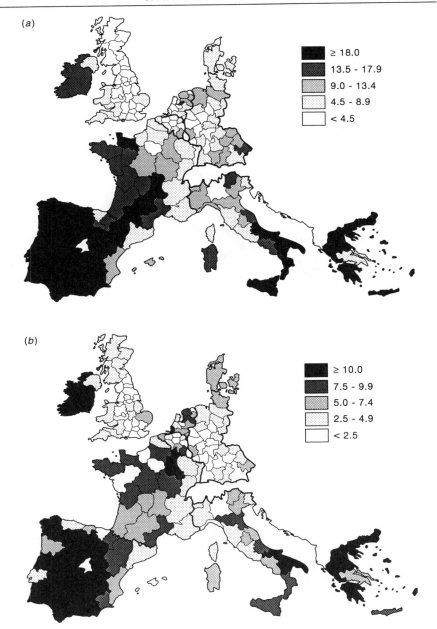

Figure 34.1 Agricultural employment is concentrated along the Atlantic and Mediterranean peripheries of the European Community, where farming still accounts for an important share in the local economy. (a) Agricultural workforce as a percentage of total, 1988; (b) share of agriculture in value added, 1985.

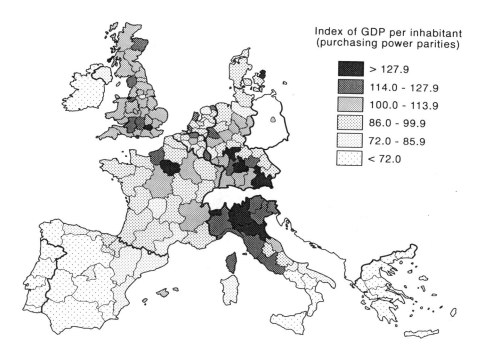

Index of GDP per inhabitant
(purchasing power parities)

> 127.9

114.0 - 127.9

100.0 - 113.9

86.0 - 99.9

72.0 - 85.9

< 72.0

Figure 34.2 Rural poverty characterizes the agricultural peripheries of the European Community. Regional inequalities expressed as GDP per inhabitant in PPPs (purchasing power parities), 1988 (EC average = 100).

Despite the rise of environmental awareness and the emergence of green policies and politics, especially in Germany, Scandinavia, California and New England, intensive agricultural output continued to rise (especially from the most productive farming areas) and far fewer people worked in agriculture. Many members of farming households, and even farmers themselves, held off-farm jobs or were engaged in non-agricultural forms of income generation on their holdings. By the 1970s to distinguish agriculture from other sectors of the economy was becoming increasingly difficult, not only in areas of well-established rural tourism (such as the Austrian Tyrol or south-west England) but also in more ordinary stretches of apparently agricultural countryside.

At the same time the distinction between rural and urban people was becoming more blurred. To be sure the so-called agricultural workforce continued to contract; thus the twelve countries of the EC contained only 9.5 million workers in farming and forestry on 8.5 million holdings in 1990 compared with over 30 million workers on 20 million holdings in 1950 (Table 34.1; Figures 34.1, 34.2). Even more dramatically, farm population in the USA fell from 24.3 million in 1945 to fewer than 5 million in 1987. However, depopulation seemed to be overtaken by a new set of repopulation processes in many parts of the so-called urban, industrial world. Depending on the strength of local planning systems during the 1950s and 1960s, cities either grew in logical order or sprawled chaotically into the surrounding countryside, which rising rates of car ownership made even more accessible and hence desirable as places to live. By contrast, city centres lost residents as more and more prime space was redeveloped for offices. All cities were suburbanizing, but in the USA it was as if cities were turning themselves inside out. Innovative growth was taking place in the outer suburbs, with their shopping malls, business parks and freeways, while old industries, long-established offices and many affluent residents were quitting the inner city. Within a few years signs of the inner-city decay that characterized New York, Chicago and Los Angeles were showing up in London, Paris and other major European cities.

RURAL REPOPULATION

As these trends were underway, so new features of population change were appearing in the countryside. Using the results of the 1970 census, American researchers noted that during the 1960s population growth had been faster in non-metropolitan counties than in metropolitan ones – in other words outside the nation's major cities (Champion 1989). Further enquiries showed that small towns and even stretches of farming country were no longer experiencing depopulation but were housing increasing numbers of people. A similar trend occurred during the 1970s with a net inmigration of about 4 million urban Americans to non-metropolitan counties. For the first time since the census began in 1790 rural areas and small towns recorded higher rates of population growth than metropolitan centres. This trend occurred in all major regions of the nation. In statistical terms, non-metropolitan counties adjacent to metropolitan counties grew by an average of 17.4 per cent (1970–80) and more distant non-metropolitan counties averaged a growth rate of 14 per cent (see also Figure 34.3). At the same time Standard Metropolitan Statistical Areas grew by 10 per cent but central cities recorded zero growth. During the 1970s suburban areas grew by 18 per cent and the population of the whole of the USA by 11.4 per cent (Lapping *et al.* 1989).

Counterurbanization

This 'population turnaround' suggested a rural renaissance or revival, and some believed heralded a 'clean break with the past'. The word 'counterurbanization' was introduced to describe the change from increasing concentration of people in urban areas to increasing dispersal of residents over wider stretches of territory. Other census results showed a broad switch from depopulation to repopulation in many rural parts of Western Europe, Australasia and Canada;

however, the population turnaround varied in degree and intensity between regions (Champion 1989). After depressing depopulation, counterurbanization was an attractive phenomenon and a seductive concept, but it was not immediately clear what its contributory mechanisms were or what its fundamental causes might be. Did the clean break just evoke the reversal of decades of rural depopulation or did it mean that incoming counterurbanites had to live completely non-urban lifestyles and no longer drive to cities to work, shop or find entertainment? Or might population growth associated with long-distance commuting be included as part of the rural renaissance? How much of the revival was truly rural? How much of it due to city people invading the countryside and using its resources in new ways? Was counterurbanization simply the most recent phase in the complex process of urbanization? Was it occurring as part of a wider shift from industrial to post-industrial society? These and many other questions forced geographers to look very carefully at repopulation in visually rural districts in order to isolate the mechanisms and causes at work. Some even rejected the term 'counterurbanization', arguing that it was too vague and represented the outcome of too many contributory processes to be helpful (Brunet 1991a).

Despite the intricacies of scholarly debate it was clear that many stretches of countryside in Europe, North America and Australasia were housing more people and new kinds of economic activity during the 1960s and 1970s than immediately before (Robinson 1990). By virtue of prolonged depopulation such areas contained important stocks of empty housing which might be purchased, restored and occupied by incomers escaping from congested cities. The 1960s certainly represented a time of increased well-being in many industrial societies, with rising wages, greater disposable income to be used for non-essential items, longer holidays with pay, securer pensions, and rising rates of car ownership and holiday taking. More and more citizens in the industrial nations had, indeed, never had it so good. After reading about the virtues of country life in magazines and seeing such pleasures on films and television, they were venturing out into the countryside in their own cars and taking holidays in more distant areas – often abroad in the case of West Europeans. More of them had purchased their own homes and were enjoying better health and living longer lives than their parents.

Far-flung commuterdom

The underused housing stock of many rural areas enabled recently implanted dreams of a place in the country to come true. By comparison with urban or suburban prices, housing in depopulated rural areas was still a bargain when restoration costs had been included in the calculation. A second option was to purchase land at low (agricultural) prices and build a new home on the plot. Low petrol costs before 1973–4, and the improved quality of highways, meant that driving to the nearest commuter train station or even right into the city was a feasible means of travelling to work each day. Two-car households became common among newcomers to the countryside enabling partners to take children to school, to shop beyond the immediate locality, or to drive themselves to work. Obtaining a place in the country was not just the preserve of the growing middle class, attracted by the appeal of fresh air, the mythical village community, and a pleasant environment in which to live and raise children, as well as the opportunity of obtaining a better home than in the suburbs (Connell 1978). Less affluent 'reluctant commuters' followed a similar route, spurred on mainly by the search for cheaper housing. By virtue of their financial circumstances, two-car households were rare among poorer incomers who were far more constrained spatially than their middle-class neighbours (Pahl 1975).

This kind of socially diverse, far-flung commuterdom invaded many country areas around major cities throughout the industrial (and increasingly post-industrial) world. In Britain the existence of 'green belts', in which house building was not encouraged, expelled this new generation of commuters even further into rural areas which had particularly attractive housing (e.g. the Cotswolds) or where old cottages could be found or new housing constructed at tolerably cheap prices (e.g. East Anglia or the East Midlands) (Herington 1984). Of course, the invasion soon enhanced the desirability of these rural areas and growing demand pushed up prices, thereby forcing would-be incomers living on low wages to look even further afield, and also creating severe problems for young couples in search of a home of their own (Dunn et al. 1981).

Retirement migration

Retirement migration was a second cause of rural repopulation, whereby retirees sold their urban (often suburban) home, bought a cheaper house in the countryside, and invested the remaining capital as a nest egg to be drawn upon in the future (Warnes 1982). Unlike commuters, retirees did not have to concern themselves with the journey to work each day and hence had a wider spatial field to search. The presence of good highways or train stations was less important to them, but attractions such as climate, landscape and village atmosphere were more so. Thus, small settlements in south-west England, with their mild climate, and in the picturesque Yorkshire Dales, the Lake District or in rural New England appealed to large numbers of retired people. As in the case of long-distance commuters, the presence of retirees inflated local house prices beyond the reach of young local people and contributed to a housing cleavage between more or less affluent groups in the repopulating countryside (Shuck-smith 1990).

In France, Italy and other European countries, where the generation gap between urban residents and their agricultural forebears was less wide than in Britain, many couples inherited farmhouses which they decided to use in retirement or else might choose to buy or build a retirement house in the same district as their ancestral home, where they might have relatives still working the land. Regardless of the process involved, the presence of considerable numbers of elderly retired migrants, far from their children and other immediate sources of help, soon gave rise to social and medical problems. Two-storeyed cottages with large gardens, located some distance away from villages or small towns, which had seemed so attractive but had required a good deal of energy to maintain when both partners were in good health, became even more burdensome with increasing age, failing health and infirmity. Death of the car-driver (usually the male) often left the remaining partner isolated, far from children or long-established friends, and literally stranded in a beautiful countryside that is poorly served with medical, shopping or care facilities and might be totally devoid of public transport.

'Urban refugees'

At another band along the age spectrum, the 1960s and 1970s saw the arrival of young adults in the countryside in search of peace and quiet and an alternative lifestyle from that of the city. This movement emerged from the widespread social protest on university campuses throughout the industrial world and took on many forms, ranging from groups of people setting up rural 'communes', to couples simply opting for an existence away from urban pressures (Chevalier 1981). Earning their living from organic farming, making pottery or practising many other crafts, they have raised children in long-depopulated areas such as the Cévennes, the Pyrenees, central Wales and Vermont. Their experience fits the anti-urban connotation of counterurbanization in its pure sense.

Without being former hippies, many newcomers to remote rural areas have forsaken daily travel to metropolitan centres for purposes of work, but maintain frequent contact in other ways. Electricity, the FAX machine and the personal computer enable 'urban refugees' to become 'telecommuters', undertaking city work in the deepest countryside and transmitting it instantly to clients in a nearby city or, indeed, anywhere in the world (Clark and Unwin 1981). Likewise the now ubiquitous deep freeze, stocked up by occasional visits to a freezer centre, can offer a range of food in the heart of the countryside, and during every season of the year, that would previously have been available only near a major wholesale market, and even then only on a seasonal basis.

Rural industrialization

Rural repopulation emerges from all these mechanisms and has been further boosted by the decision of many industrial and commercial firms to leave the congestion and high costs of metropolitan centres and to relocate in small towns or even in the open countryside, where land and running costs are lower and part of the workforce may be reinstalled in attractive surrounding villages (Heale and Ilbery 1985). After appropriate training, the remainder may be recruited from the local labour pool. New job opportunities of this kind have given rise to commuting zones around many small towns throughout the industrial world and have inserted a new kind of leader in local society. For example, recently arrived managerial staff have introduced a new sense of dynamism to attractive but isolated villages in western Ireland, and in some instances have assumed leadership roles that were traditionally performed by parish priests or local schoolteachers (Breathnach and Cawley 1986).

Second homes in the countryside have become increasingly popular in Europe and North America over recent decades (Coppock 1977). They are no longer the preserve of the very rich and their acquisition is partly explained by the growing affluence of a wider section of urban society during the 1960s and in subsequent years. Obtaining a property to use as a second home fits in with the kind of spatial searching done by long-distance commuters, retirees and even by urban refugees (Rambaud 1969). It may also be conditioned by the desire to participate in winter sports (e.g. in the mountains of central Colorado), sailing or some other specific recreational activity. In other instances, farm buildings have been inherited from rural relatives and are subsequently used by urban dwellers at weekends or in long vacations. Whatever the method of acquisition, large concentrations of second homes serve to inflate property prices and thereby set housing beyond the reach of poorly paid rural residents. Acquiring a second home may well be a critical step towards becoming a long-distance commuter (but perhaps not every day of the week) or a retired migrant to the countryside. For these reasons, second homes need to be mentioned alongside other mechanisms that contributed to the repopulation of many rural areas over the past quarter century.

ARENAS OF UNEQUAL OPPORTUNITY

As a result of all these changes, villages and small towns across the countrysides of modern industrial countries no longer form rural communities in the traditional sense of the term, with connotations of working the land, self-help, self-sufficiency and socio-economic stability. Repopulated villages involve groups of people of markedly different background and financial command. They find themselves as neighbours but have come to the countryside for different reasons, such as finding a suitable house for commuting or retirement, or simply living where their parents did or where work is located. Some incomers to the countryside are affluent, personally mobile, young (or middle-aged) and healthy; others are less wealthy, less mobile, elderly, and perhaps in poor health. To think of the rural population as the average of these conditions is misleading, since members of the two groups have quite different needs and are not equally able to resolve them. For example, closure of a village store or withdrawal of a bus service may be disastrous for the daily life of a car-less pensioner somewhere in rural Europe but may pass unnoticed by a three-car household of commuters who shop at a distant hypermarket (Archbishops' Commission on Rural Areas 1990). The absence of public transport throughout rural North America makes the ability to drive a car absolutely essential.

Rationalization and decline of rural services

Against a background of long-standing depopulation, rural services declined as the local clientele became less numerous. Grocery stores went bankrupt, cafés and village pubs closed, congregations diminished and churches closed, and village schools and local buses ceased to operate. The smaller and more isolated the settlement, the more likely that its services would cease (Shaw 1979). The story could be repeated with varying degrees of intensity across the countrysides of rural Europe and North America prior to the 1970s. Part of the reason was of course to do with market forces, since too few people remained for services to be financially viable. The remainder of the explanation had to do with cuts in government subsidies and operation of official policies to 'rationalize' service provision at selected small towns or 'key villages' which would be accessible to surrounding countryfolk travelling by car (Cloke 1979, 1983). But for those with no cars or in poor health obtaining services became a matter of asking neighbours for lifts or to bring in shopping or hoping that local people might organize a minibus or some other form of voluntary transport scheme (Moseley 1979).

Even though repopulation occurred in many areas over the past quarter century, the rationalization policies were not reversed and in some countries they have been implemented more vigorously. For example, deregulation in the US transport industry permitted companies to set higher freight rates in rural areas which often resulted in the abandonment of services to small towns and rural settlements. Cuts in spending programmes in the 1980s have channelled less federal money to rural America (Clark 1991). The trend towards privatization of service provision that has been so important in Britain since 1979 has had its impact on what public transport remained in the countryside and has led to further cuts or total closure of little-used services (Bell and Cloke 1989). Thus, espousal of market principles rather than provision of welfare support has provoked closure of small shops, schools and transport services in the Dutch countryside in recent years (Clark *et al.* 1989). The presence of affluent commuters or mobile retirees does little or nothing to strengthen demand for local facilities since services can be obtained elsewhere; but as retirees age and become less mobile, new demands for social and medical services arise. These are expensive to supply since there are few economies of scale involved in serving a relatively dispersed rural population. By virtue of the increase in house prices following the arrival of newcomers, young local couples may have to leave their home area, hence they cannot be counted upon to boost demand for services.

Rural deprivation

Country areas in industrialized countries do, indeed, represent arenas of unequal opportunity (Clout 1986). Beneath the comforting statistics which indicate repopulation profound tensions exist between those with few resources (and hence strong local needs) and those with considerable resources (and few, if any, local needs). Even in rich post-industrial countries rural deprivation remains a serious but often overlooked phenomenon which adversely affects people's mobility, access to services, and ultimately life chances (Lowe *et al.* 1986). Part of the explanation lies with the low density and thin distribution of needy people across wide stretches of countryside; the remainder is to do with the working of market principles in capitalist economies. The market philosophy prevalent in the UK, the USA, the Netherlands and many other countries offers no hope of local rural services being subsidized or restored; voluntary self-help among villagers seems to be the only workable solution. Other European countries, notably France and Italy, offer special grants and subsidies to keep village shops and post offices open, to maintain village schools in areas of harsh climate or with difficult terrain, and to encourage unconventional forms of community transport and service provision (Clout 1984). A change in government philosophy could, of course, bring such an approach to an end.

Ongoing depopulation

Repopulation unquestionably formed an interesting but perhaps deceptive phase in recent rural history. Beneath trends of net inmigration and overall population growth in many countrysides during the 1960s and 1970s, the agricultural population continued to decline and former farmworkers had to seek jobs and housing wherever they might be found (Weekley 1988). 'Gentrification' of attractive areas – as a result of commuting, retirement migration or second homes – meant that there was little chance of finding affordable housing nearby. During the 1980s the sale of council housing to sitting tenants compounded the problems for young country people in Britain. In short, counterurbanization was not ubiquitous, with many rural areas undergoing depopulation as they had for decades. Such places enjoyed no respite from rural decline. For example, the 'fragile crescent' of inner France, extending from the Ardennes, through eastern Champagne and the Massif Central to the Pyrenees, displayed population loss, increased ageing of those who remained, and ongoing withdrawal of local services (Béteille 1981; Estienne 1988).

Changes in international politics and trade relations, as well as the high value of the dollar, had a major impact on the US farm economy in the late 1970s and early 1980s. Important overseas markets were lost as former trading partners increased their own food production and the Soviet embargo of 1979–81 removed a major outlet for American grain. Similarly, decisions taken by far-away multinationals led to the closure of many rural mining operations in the USA (Fuguitt *et al.* 1989). Collapsing crop prices in the US farm economy in the early 1980s and the ensuing farm crisis, which forced many farmers out of business, fuelled a more general rural decline across many parts of the nation

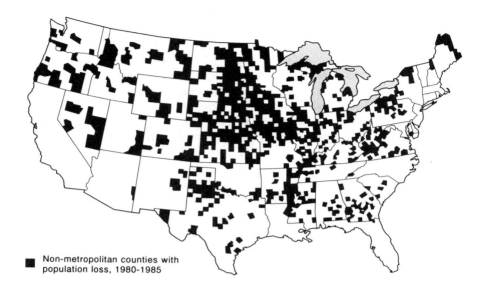

Non-metropolitan counties with population loss, 1980-1985

Figure 34.3 Non-metropolitan counties of the United States recording population loss between 1980 and 1985 (shown in black). Note the widespread occurrence of population loss during that period.

(Figure 34.3). For example, substantial population losses were recorded in the six Great Plains states (Minnesota, Iowa, Kansas, Nebraska, and North and South Dakota). Over half of the counties in those states are farm-based and those areas lost an average of 6 per cent of their population between 1969 and 1986. In many places farmhouses and barns now stand empty; stores and banks in small towns are boarded up. The rural renaissance that had become so evident in New England and parts of California avoided the deep rural interior of the continent. By contrast urban growth in the Great Plains states continued apace, with their total population increasing by 8 per cent over that same period.

That experience was paralleled in many parts of the USA during the 1980s and urban areas nation-wide started to grow again and at a faster rate than rural ones (Lapping *et al.* 1989). Likewise living standards in rural America started to fall relative to urban areas after a period of convergence in the 1970s (Fuguitt 1991). In a similar fashion, since 1980 repopulation has become less significant in remoter rural areas of Western Europe and has focused once again in countrysides within easy access of metropolitan areas. Reasons for this 'turnaround of the turnaround' are far from simple. They include dissatisfaction with life in the deep countryside among some incomers, realization of the high costs involved in driving long distances to obtain every kind of service, collapse of some mining concerns and manufacturing plant located in rural areas, and many other factors. In short, sparse and uneven data suggest that the balance seems to be tipping away from repopulation in remote rural areas of the post-industrial world, but it has to be acknowledged that there are many local deviations from that broad trend (Champion and Illeris 1990; Keddie and Joseph 1991; Walsh 1991; Stillwell *et al.* 1992).

Admittedly in very different environments from the farmlands of the USA, the family farms and estates of Western Europe's farm economy have undergone profound change in recent years and are set for even more in the immediate future. The twelve countries of the EC nominally had 9.5 million people in farming and forestry work in 1990 but many of them have additional jobs outside the primary sector (Arkleton Research 1988, 1991; Fuller 1990; Gasson 1988; Shucksmith 1989). There are about 8.5 million 'farm holdings', many of which are too small to provide sufficient work for one person or adequate income to support a family, hence many farms are just part-time operations and many people who live on farms hold non-farm jobs. The number of holdings will continue to decline since many elderly farmers have no successors to take over the land. For decades farmland has been abandoned to scrub or rough woodland in the depopulated uplands of the Massif Central, Italy and Iberia. Now between 10 and 20 per cent of farmland in the four Mediterranean countries of the EC lies abandoned, by comparison with less than 4 per cent forty years ago.

THE QUEST FOR DIVERSIFICATION

After more than two decades of a costly Common Agricultural Policy which generated unsaleable food surpluses, a new farm policy was devised in the 1980s (Tracy 1989). It changed the broad objective from increasing output to bringing European farming into line with market trends. Efforts would have to be made to curb production by operating quotas on certain commodities, reducing guaranteed levels of price support to farmers, directing them away from surplus-generating routes, encouraging some to quit farming completely, and promoting a modest withdrawal of land from productive use. However unpalatable, none of these ideas was new: Agriculture Commissioner Sicco Mansholt had voiced them in the late 1960s and efforts were made to rein in the American farm economy during the 1980s. As a result of the new CAP, farm incomes have fallen in Western Europe and the number of farming bankruptcies has increased. Exactly the same mismatch between supply and demand, policy objectives, and socio-economic consequences for farmers affects the USA, where it seems that there is just too much good agricultural land in productive use (Hart 1991). Throughout the post-industrial world, many farmers are desperately seeking new ways to use sections of their land and spare buildings in order to generate a reasonable income.

Just as the rural landscapes of the EC are changing through 'set-aside' (whereby stretches of land are withdrawn from agricultural production for specified periods) and extensification (whereby farmers are paid to practise low-intensity, ecologically sensitive agriculture), so farm-based economies are being

765

increasingly diversified (Clout 1991; Potter 1991). Depending on where the farm is located, the quality and size of its buildings, the financial footing (or creditworthiness) of the household, its stage in the family cycle, and whether the farm wife is willing to tolerate a much greater burden of new work, so there are many new opportunities to be grasped (Brunet 1991b). These include not only bed and breakfast and farm-based camping or caravanning, but also running farm-shops and restaurants, setting out farm trails or golf courses, and enabling visitors to pick their own produce. In addition, redundant buildings might be converted into craft centres, workshops for clean, high-tech industry, or even for holiday accommodation. Redundant farmland might be used for hunting or converted into woodland. Many farmers in more accessible countryside and in attractive areas that receive large numbers of visitors have already introduced enterprises of this kind; others operating in more ordinary landscapes will surely try to follow their example in the years ahead.

THE KALEIDOSCOPE OF RURAL CHANGE

As the twentieth century draws to a close, it is clear that most rural areas and most dwellers in the countryside are closely linked to trends in urban life. Ease of personal travel and telecommunication have brought both sectors of geographical space into increasingly close contact. However, alongside the affluence of many newcomers, disturbing deprivation is still to be found (Champion and Watkins 1991). Rural land is being used to accommodate a multitude of functions, of which farming is but one. Some holdings remain strongly geared to highly productive, commercial farming, others operate at a lower intensity and place more emphasis on conserving the rural environment of hedgerows, copses, historic cottages and barns. There is no single model for the future of agriculture, just as the pattern of land-use intensities has become more varied than before (Lockhart and Ilbery 1987).

This mosaic of rural life is subject to constant change, whereby spatial patterns become rearranged in response to variations in natural endowment, capital availability, human ingenuity and myriad other factors. As every observant traveller knows, the *pays* of Europe and the 'natural regions' of North America and Australasia, which geographers recognized early in the twentieth century, can still be identified but their characteristics are in a state of flux. Wheatlands are replaced by woodland, wetlands are no longer problem patches but have become nature reserves, fieldbarns are now second homes or youth hostels, and sleepy farm settlements have been transformed into up-market commuter villages or desirable retirement homes.

There is, of course, a geographical logic to this kaleidoscope, conditioned not only by the facts of terrain, climate or distance and accessibility of particular places from large urban centres, but also by how those facts are perceived by different sections of society (Lewis and Maund 1976). The diversification of economic activity and social life in the countryside has much to do with efficient responses to new demands and the creation of new 'products' (such as farm-based holidays or organically grown foodstuffs) for which a market niche has to be created through calculated promotion. The influences that work to change rural life and transform the geography of the countryside operate at numerous scales, from the energy of the individual, through the resources of the household and the enterprise of the local 'community', to the policies of individual nations and supranational organizations, and the operation of mega-processes, such as fluctuating oil prices.

FUTURE COUNTRYSIDES

In the future, countrysides will perform a wide variety of functions which will be juxtaposed in differing ways from place to place. Together with the diversity of the physical environment, this complexity of human conditions helps explain the great diversity of rural conditions in the intricately fashioned, long-settled countrysides of Europe and across the less densely woven landscapes of North America and Australasia. At the risk of oversimplifying, it is possible to envisage four 'types' of countryside in the future:

1 Pressure on rural land will continue to grow close to major cities and tourism resorts in order to accommodate suburban expansion and new facilities (such as airports, golf courses and theme parks) to serve metropolitan populations (Bryant *et al.* 1983). Surrounding settlements will function as commuter villages. Values of housing and land with obvious development potential may well continue to rise, with negative implications for less-privileged local people. Just how far commuterdom may spread will depend on many factors, including oil prices and the degree to which urban regeneration and gentrification take place thereby making inner-city districts (indeed, cities as a whole) more desirable places in which to live. Some stretches of farmland will continue to be used intensively and will yield high-value commercial crops for urban markets as they always have done. Others will assume new functions, including ecological conservation and farm-based recreation for short-stay visitors. The degree to which farmland is consumed for building purposes will depend on population growth, the health of national economies, and the relative strength of national, regional and local planning regulations.

2 At somewhat greater distance will be stretches of farmland which made major contributions to the massive increase in food production since mid-century, through application of chemicals, drainage of wetlands, reorganization of farm structures, and great injections of capital. As a result such farming landscapes have become rather bland and standardized and hold little immediate appeal to the visitor. By virtue of their high productivity, these areas may well be targeted for set-aside and other mechanisms for controlling food surpluses, hence their farm landscapes will gradually become more varied. Some stretches will continue to be used intensively but others will be given over to conservation, woodland and low-intensity farm production. Such areas enjoy the advantage of

reasonable proximity to metropolitan centres, but this is partly offset by the ordinariness of their landscapes. In the decades of high guaranteed prices their farmers had little or no need to look beyond commercial agriculture so they have little experience of diversification. In fact, the challenge of recomposing the countryside may well be greatest and most difficult to implement in these 'agricultural' countrysides.

3 Hilly and middle-mountain areas experienced depopulation until recently and some continue to do so. Their family farms were not involved in the kind of agricultural transformation that has affected the lowlands since mid-century and their incomes remained modest, hence alternative sources of income had to be sought. Many hill farms command varied, traditional landscapes and contain a range of old farm buildings. For all these reasons some, but by no means all, hill-farming areas have experienced a considerable influx of retired newcomers and city folk who have acquired second homes (Mendras 1979). Part-time farming, rural tourism and other forms of diversification are established facts of life on some hill farms, and this kind of enterprise will be exploited further in the future as more efforts will have to be made to encourage visitors to invest in the local economy. Designation of national parks and other forms of heritage area will endow some locations with particular appeal among outsiders. Demand for housing and landed property will increase and prices will rise – to the detriment of young local people and many other less affluent residents. The costly service needs of elderly retirees will contrast with the requirements of tourists and short-stay visitors. Paradoxically, pressures will mount to conserve the rural environment (for aesthetic appeal and selling power) but also to exploit it (in order to draw more finance into the local economy). Some 'farmers' will be paid to use their land in traditional, environmentally friendly ways and thereby act as landscape gardeners. But some expanses will no longer be cultivated or grazed and their appearance will change dramatically as they become colonized with scrubby woodland or are covered with rationally planted blocks of trees. Although far from major cities, visual changes in these hilly areas will give rise to great outcry from the conservation lobby.

4 Population densities in high mountain areas and very remote countrysides will remain very low and depopulation may well continue as it has done for a century or more. The few residents who remain will continue to experience problems of service provision. However, these areas will be perceived even more strongly as largely unspoiled wilderness country, whose plants, animals and habitats are of great scientific interest in our urbanizing, polluted world, and whose very existence is of importance in terms of national and international heritage. National park or equivalent status, coupled with sheer remoteness, can afford a measure of protection but much will depend on serious implementation of legislation and rigorous land management. However, having special status and being shown on television wildlife programmes serves to heighten public awareness of wilderness and increases the demand to visit. Zoning of facilities, containment of visitors in defined areas, charging for entry, and closing access

roads when agreed capacity is reached offer ways of controlling visitor pressure and ensuring that truly remote areas are left to hardy hikers and scientists as well as wildlife.

REFERENCES

Archbishops' Commission on Rural Areas (1990) *Faith in the Countryside*, Worthing: Churchman Publishing.

Arkleton Research (1988) *Rural Change in Europe: Montpellier Colloquium*, Oxford: Arkleton Trust.

Arkleton Research (1991) *Rural Change in Europe: Braemar Colloquium*, Oxford: Arkleton Trust.

Bell, P. and Cloke, P.J. (1989) 'The changing relationship between the private and public sectors: privatization in rural Britain', *Journal of Rural Studies* 5, 1–15.

Best, R.H. (1981) *Land Use and Living Space*, London: Methuen.

Béteille, R. (1981) *La France du Vide*, Paris: Librairies Techniques.

Breathnach, P. and Cawley, M.E. (eds) (1986) *Change and Development in Rural Ireland*, Maynooth: Geographical Society of Ireland.

Brunet, P. (ed.) (1991a) *Le Développement Régional Rural en Europe*, Caen: Centre de Publications de l'Université de Caen.

Brunet, P. (ed.) (1991b) *France et Grande-Bretagne Rurales*, Caen: Centre de Publications de l'Université de Caen.

Bryant, C., Russwurm, L.H. and McLellan, A.G. (1983) *The City's Countryside: Land and its Management on the Rural–Urban Fringe*, London: Longman.

Champion, A.G. (ed.) (1989) *Counterurbanization: The Changing Pace and Nature of Population Decentralization*, London: Edward Arnold.

Champion, A.G. and Illeris, S. (1990) 'Population redistribution in Western Europe', pp. 236–53 in M. Hebbert and J.C. Hansen (eds) *Unfamiliar Territory: the Reshaping of European Geography*, Aldershot: Avebury.

Champion, A.G. and Watkins, C. (eds) (1991) *People in the Countryside: Studies of Social Change in Rural Britain*, London: Paul Chapman.

Chevalier, M. (1981) 'Les phénomènes néo-ruraux', *L'Espace Géographique* 10, 33–47.

Clark, D. and Unwin, K.I. (1981) 'Telecommunications and travel: potential impact in rural areas', *Regional Studies*, 15, 47–56.

Clark, G., Huigen, P. and Thissen, F. (eds) (1989) 'Planning and the future of the countryside: Great Britain and the Netherlands', *Netherlands Geographical Studies* 92, 1–234.

Clark, T.A. (1991) 'Capital constraints on non-metropolitan accumulation: rural process in the USA since the 1960s', *Journal of Rural Studies* 7, 169–90.

Cloke, P.J. (1979) *Key Settlements in Rural Areas*, London: Methuen.

Cloke, P.J. (1983) *An Introduction to Rural Settlement Planning*, London: Methuen.

Cloke, P.J. (ed.) (1988) *Policies and Plans for Rural People*, London: Unwin Hyman.

Cloke, P.J. (ed.) (1989) *Rural Land-use Planning in Developed Nations*, London: Unwin Hyman.

Clout, H.D. (1984) *A Rural Policy for the EEC?*, London: Methuen.

Clout, H.D. (1986) Population changes in rural Britain', *Espace, Populations, Sociétés* 1986(3), 19–32.

Clout, H.D. (1991) 'The recomposition of rural Europe', *Annales de Géographie* 100, 714–29.

Connell, J.H. (1978) *The End of Tradition: Country Life in Central Surrey*, London: Routledge & Kegan Paul.

Coppock, J.T. (ed.) (1977) *Second Homes: Curse or Blessing?*, Oxford: Pergamon.

Cronon, W. (1991) *Nature's Metropolis*, New York: Norton.

Dunn, M.C., Rawson, M. and Rogers, A. (1981) *Rural Housing: Competition and Choice*, London: Allen & Unwin.

Estienne, P. (1988) *Terres d'Abandon? La Population des Montagnes françaises*, Clermont Ferrand: Institut d'Études du Massif Central, Université de Clermont Ferrand II.

Fuguitt, G.V. (1991) 'Commuting and the rural–urban hierarchy', *Journal of Rural Studies* 7, 459–66.

Fuguitt, G.V., Brown, D.L. and Beale, C.L. (1989) *Rural and Small Town America*, New York: Russell Sage Foundation.

Fuller, A.M. (1990) 'From part-time farming to pluriactivity', *Journal of Rural Studies* 6, 361–73.

Gasson, R. (1988) *The Economics of Part-Time Farming*, London: Longman.

Hart, J.F. (1991) *The Land That Feeds Us*, New York: Norton.

Healey, M. and Ilbery, B. (eds) (1985) *The Industrialization of the Countryside*, Norwich: GeoBooks.

Herington, J. (1984) *The Outer City*, London: Harper & Row.

Kayser, B. (1990) *La Renaissance Rurale: Sociologie des Campagnes du Monde Occidental*, Paris: Armand Colin.

Keddie, P.D. and Joseph, A.E. (1991) 'The turnaround of the turnaround? Rural population change in Canada 1976 to 1986', *Canadian Geographer* 35, 367–79.

Lapping, M.B., Daniels, T.L. and Keller, J.W. (1989) *Rural Planning and Development in the United States*, New York: Guilford Press.

Lewis, G.J. and Maund, D.J. (1976) 'The urbanization of the countryside: a framework for analysis', *Geografiska Annaler* 58B, 17–27.

Lockhart, D.G. and Ilbery, B. (eds) (1987) *The Future of the British Landscape*, Norwich: GeoBooks.

Lowe, P. and Cox, G. (1986) *Countryside Conflicts: The Politics of Farming, Forestry and Conservation*, London: Temple Smith.

Lowe, P. and Goyder, J. (1983) *Environmental Groups in Politics*, London: Allen & Unwin.

Lowe, P., Bradley, T. and Wright, S. (eds) (1986) *Deprivation and Welfare in Rural Areas*, Norwich: GeoBooks.

Mendras, H. (1979) *Voyage au Pays de l'Utopie Rustique*, Le Paradou: Actes Sud.

Moseley, M.J. (1979) *Accessibility: The Rural Challenge*, London: Methuen.

Pahl, R.E. (1975) *Whose City?*, Harmondsworth: Penguin.

Potter, C. (1991) *The Diversion of Land: Conservation in a Period of Farming Contraction*, London: Routledge.

Rambaud, P. (1969) *Société Rurale et Urbanisation*, Paris: Seuil.

Robinson, G.M. (1990) *Conflict and Change in the Countryside*, London: Belhaven.

Shaw, J.M. (ed.) (1979) *Rural Deprivation and Planning*, Norwich: GeoBooks.

Short, B. (ed.) (1992) *The English Rural Community: Image and Analysis*, Cambridge: Cambridge University Press.

Shucksmith, M. (1989) 'Pluriactivity, farm structures and rural change', *Journal of Agricultural Economics* 40, 345–60.

Shucksmith, M. (1990) *Housebuilding in Britain's Countryside*, London: Routledge.

Stillwell, J., Rees, P. and Boden, P. (eds) (1992) *Migration Processes and Patterns. Vol. 2: Population Redistribution in the United Kingdom*, London: Belhaven.

Tracy, M. (1989) *Government and Agriculture in Western Europe 1880–1988* (3rd edn), Hemel Hempstead: Harvester Wheatsheaf.

Walsh, J.A. (1991) 'The turnaround of the turnaround in the population of the Republic of Ireland', *Irish Geography* 24, 117–25.

Warnes, A.M. (ed.) (1982) *Geographical Perspectives on the Elderly*, Chichester: Wiley.
Weekley, I. (1988) 'Rural depopulation and counterurbanization', *Area* 20, 127–34.

FURTHER READING

Bolsius, E.C.A., Clark, G. and Groenendik, J.G. (eds) (1993) *The Retreat. Rural Land Use and European Agriculture*, Amsterdam: Royal Netherlands Geographical Society.
Bowler, I.R., Bryant, C.R. and Nellis, M.D. (eds) (1992) *Contemporary Rural Systems in Transition* (2 vols), Wallingford: CAB.
Buller, H. and Hoggart, K. (1994) *International Counterurbanization. British Migrants in Rural France*, Aldershot: Avebury.
Champion, A.G. (ed.) (1989) *Counterurbanization: The Changing Pace and Nature of Population Decentralization*, London: Edward Arnold.
Champion, A.G. and Watkins, C. (eds) (1991) *People in the Countryside: Studies of Social Change in Rural Britain*, London: Paul Chapman.
Cloke, P.J. (ed.) (1988) *Policies and Plans for Rural People*, London: Unwin Hyman.
Cloke, P.J. (ed.) (1989) *Rural Land-use Planning in Developed Nations*, London: Unwin Hyman.
Clout, H.D. (1984) *A Rural Policy for the EEC?*, London: Methuen.
Coppock, J.T. (ed.) (1977) *Second Homes: Curse or Blessing?*, Oxford: Pergamon.
Gasson, R. (1988) *The Economics of Part-Time Farming*, London: Longman.
Huigen, P., Paul, L. and Volkers, K. (eds) (1992) *The Changing Function and Position of Rural Areas in Europe*, Utrecht: Royal Netherlands Geographical Society.
Kayser, B. (1990) *La Renaissance Rurale: Sociologie des Campagnes du Monde Occidental*, Paris: Armand Colin.
Lapping, M.B., Daniels, T.L. and Keller, J.W. (1989) *Rural Planning and Development in the United States*, New York: Guilford Press.
Marsden, T., Murdoch, J., Lowe, P., Munton, R. and Flynn, A. (1993) *Constructing the Countryside*, London: UCL Press.
Robinson, G.M. (1990) *Conflict and Change in the Countryside*, London: Belhaven.
Tracy, M. (1989) *Government and Agriculture in Western Europe 1880–1988* (3rd edn), Hemel Hempstead: Harvester Wheatsheaf.
Warnes, A.M. (ed.) (1982) *Geographical Perspectives on the Elderly*, Chichester: Wiley.

35

THE QUALITY OF LIFE
Human welfare and social justice
David M. Smith

Geographical engagement with the 'quality of life' and related concepts dates back to the late 1960s. This is not to say that human geography was entirely oblivious to qualitative aspects of life before then, simply that the traditional preoccupation with natural resources, production and population characteristics tended to dominate any concern with consumption in its broadest sense. That different 'ways of life' existed in different places was central to the geographer's view of the world, but explicit qualitative comparisons tended to be avoided. What was new about the self-styled geography of social concern (or 'radical geography') which began to take shape in the latter part of the 1960s was consideration of such hitherto neglected topics as poverty, health, hunger, crime and environmental pollution, and their contribution to the general quality of people's lives as a spatially variable condition (Smith 1973).

The focus of human geography subsequently shifted. The empirical identification of spatial variations in qualitative aspects of life, in the pattern-recognition tradition, became eclipsed by a structural perspective concerned with the processes behind spatial manifestation of inequality or uneven development. More recently a revitalized cultural geography has helped to draw attention to the particular experience of deprived or marginalized groups in society, identified by such characteristics as race, ethnicity and gender. But there remains an underlying concern about social justice, or the morality of unequal life chances, which is being asserted in the changing world of the 1990s as various societies grapple with the new institutional arrangements for the distribution of scarce goods and services (Smith 1994).

This chapter is not so much concerned with actual patterns and contrasts in life quality (addressed in Chapter 19) as with the broader conceptual and technical problems of making such comparisons. The discussion proceeds from some general issues, through demonstrations of how spatial variations in life quality may be identified, to the question of social justice. The main purpose is to illustrate the criteria and methods of analysis which have been adopted in attempts to give the quality of life and related concepts empirical content, and to consider where such exercises lead the practice of geography. Particular cases are introduced to elucidate specific problems. But first it is necessary to clarify terminology.

The quality of life is an elusive concept, the meaning of which is very much dependent on the context within which it is used. Unlike physical phenomena such as temperature, climate or people's size, quality of life is not directly observable and measurable by generally agreed criteria. It is an intellectual abstraction devised to facilitate consideration of how people live, in contexts where judgements as to what may be better or worse arise. It is therefore similar to such concepts as development, welfare and well-being. In fact, all these terms refer to much the same thing. Some may have assumed rather specific and different meanings in a particular literature; for example 'welfare' in economics, which incorporates the distribution of the entire range of goods (and bads) among a population, and 'development' which has more specific if varying definitions in the literature on that subject. And it is sometimes claimed that 'quality of life' itself refers to more intangible, personal or individual attributes. However, it is a mistake to seek subtle distinctions between what must be understood as human constructs subject to malleable and often contested definition, rather than as empirically observable phenomena.

COMPARING LIVES

If Jack says that he lives better than Jill, and Jill that she lives better than Jack, it is extraordinarily difficult to adjudicate between such competing claims. What it means to live well is an intensely personal experience. To some people the good life may be manifest in material possessions, to others it may be relations with family or friends. Some people prefer to go to concerts or fancy restaurants, others to football matches and pubs. Different groups of people have different sets of values, making particular forms of recreation or religious expression, for example, more or less important to their members. And to some people, in some places, the food, clothing and shelter required for physical survival is more significant than choice of menu, wardrobe or type of house. Such common-sense observations underline the problem faced by the friend Jack and Jill may have enlisted to settle their difference.

Progress can be made, however, if people are required to be specific about what aspects of life really matter most to them. They will find that the existence of differences does not mean that they have nothing in common. They are likely to agree that they share certain material needs (e.g. for food, clothing and shelter of some kind), even if the manner of their actual satisfaction is to some extent a matter of personal preference. They may even agree that particular properties, such as daily calorific intake, possession of an overcoat or square metres of floor space, may be acceptable measures of how well fed, clothed or housed they are. They may spend their leisure time in quite different ways, but agree that if Jack has more leisure time than Jill he is better off in this respect. But the relative importance of food, clothing, shelter, leisure and so on to their overall life quality may still be a matter of contention and honest difference.

It is not therefore surprising that there are influential bodies of opinion that consider comparisons between individuals or groups with respect to quality of life ill-advised or impossible. For example, in neo-classical economics what is good for people is held to be very much a matter of their own personal preference. Individuals or households maximize their satisfaction, or 'utility', by spending their money as they wish, and no one else knows better what they should consume – or so the more dogmatic versions of this doctrine assert. Any specific statement of what goods and services are required, and in what quantities or proportions, to maximize individual utility or the welfare of society at large involves value judgements, which it is not the prerogative of economists to make. Nor is it their business to make moral judgements on the nature of the good life, or on who gets more or less of what there is to consume. Thus interpersonal or between-group comparisons of life quality are beyond the scope of a supposedly value-free social science. Free markets working perfectly will aggregate all individual preferences in deciding what is produced and for whom, and collective welfare will be maximized without recourse to any specific universal criteria.

While this utilitarian perspective is usually associated with the political right, a similar form of relativism can be found in other parts of the ideological spectrum. Marxists stress the historical specificity of human needs, and accuse some groups in society of attempting to impose their own preferences on others. What markets provide at any time (and place) reflect the wishes of those with the money, power or whatever else it takes to determine market outcomes; in a world dominated by big business, sovereignty is more plausibly exercised by the producer rather than the neo-classical consumer. Another contemporary position is that different peoples, at different times and in different places, hold different views as to what matters most in life, including different codes of morality, and that those who claim to know better are guilty of 'cultural imperialism'. For example, Western industrialized societies proscribing what is best for the 'Third World' may be the subject of such an accusation. The post-modern emphasis on difference and disdain for universals provides further reinforcement for a relativistic view of the quality of human life.

The alternative argument is that a universal conception of human need is possible, and indeed necessary if anything is to be said by way of comparison between people's lives. As with Jack and Jill, Britons and Americans, Jews and Arabs, men and women can have their differences without this entailing that they have nothing in common. In other words, it is possible to think in terms of some universal human needs without denying the authenticity of the individual, national or cultural basis of how they may be satisfied. We must all eat to live, but may choose different kinds of food. The extent to which universally recognized needs are actually satisfied, by measurable criteria such as calorific intake or housing floor space, then provides a basis for comparisons of life quality. The crucial point is that there must be some frame of reference external to the individual, group, place or historical epoch which provides at least an entry into the question of what quality of life is dependent on. And if all human beings have at least their humanity in common, then the wherewithal at least to preserve this (for example by a minimum consumption of food, clothing and shelter) could be regarded as basic and universal human needs.

This approach can be exemplified by an attempt by Doyal and Gough (1991) to set out a theory of human need. Against the various forms of relativism, they argue that all people share one obvious need: to avoid serious harm. This goes beyond the obviously harmful failure to survive in a physical sense, to incorporate impaired social participation or pursuit of objectives deemed valuable by the individual in a specific social milieu which can vary with culture, place and so on. This universal goal generates two basic needs. One is for the physical health to continue living and able to function effectively; the other is for the personal autonomy or ability necessary to make informed choices about what should be done and how to go about doing it in a given societal context. These basic needs in turn require the satisfaction of certain intermediate needs, specified as follows:

- adequate nutritional food and water;
- adequate protective housing;
- a non-hazardous work environment;
- a non-hazardous physical environment;
- appropriate health care;
- security in childhood;
- significant primary relationships (i.e. with other people);
- physical security;
- economic security;
- safe birth control and child-bearing;
- basic education.

That all people share these needs is hard to contradict.

The extent to which these needs are satisfied can be measured by appropriate indicators. For example, Doyal and Gough (1991) propose that satisfaction of the need for adequate housing could be measured by proportion of people homeless, proportion in structures that do not protect against normal weather, proportion lacking safe sanitation facilities, and proportion living above a specified ratio of persons per room. Comparisons can then be made, without denying people the freedom to express individual or cultural preferences in the construction, appearance, layout and adornment of their homes.

This approach typifies most practical attempts to make geographical comparisons of the quality of life. The concept of quality of life (development, standard of living or whatever the term adopted), is disaggregated into contributory elements or component parts in greater or lesser detail, then specific indicators are derived with respect to measurable conditions which capture as faithfully as possible the meaning of such broader conditions as adequacy of housing. The extent to which such exercises can effectively withstand charges of cultural imperialism depends on how far those responsible can distance themselves from the (usually) Anglo-American or Eurocentric predispositions which can otherwise colour what are claimed to be universal conceptions of the quality of life.

'SUBJECTIVE' SOCIAL INDICATORS

Before looking at some actual attempts at geographical comparison, an alternative approach must be considered briefly. If sceptics still insist on rejecting the notion of universal criteria, they might argue that how people report their own quality of life may provide some basis for comparison – providing of course that the findings are not so clearly inconsistent as those attributed above to Jack and Jill. For example, if a sample of people in different countries are invited to describe the quality of their own lives on a semantic differential scale (such as 'very good', 'good', 'indifferent', 'poor', 'very poor', valued at 5, 4, 3, 2, 1 respectively), then numerical comparisons of the national averages are possible. But what does this actually mean? The judgements elicited are bound to be very much a reflection of people's limited and geographically specific experience, their frames of reference being friends, neighbours or their own past, rather than knowledge of how people actually live in other countries. That a greater proportion of Americans than Britons may describe their lives as 'very good' is thus hardly a satisfactory basis for concluding that life really is better in one country than the other.

More might be learned from how people in different places view their lives in relation to their own previous circumstances, suggesting perhaps that people in some places see more improvement than in others. While less strictly geographical, comparisons between the views of different groups in the same society or nation may also be revealing. Other applications of what are usually referred to as subjective social indicators include surveys of opinions as to the relative importance of different 'domains' of life, as they vary among population groups or the inhabitants of different places.

However, all such exercises are subject to problems common to survey research using formal questionnaires, including the possible imposition of the understandings of the researcher on the researched in the form of questions used, compounding the intrinsic difficulty of investigating people's thoughts on something as personal as quality of life. An old story carries a potent warning: the experience of the researcher who asked his first respondent what mattered most in her life, only to be told, 'my privacy', before the door was shut in his face. There is also the argument that self-reported status is no less a universal

776

Table 35.1 Perceived quality of life in South Africa by race group, 1983–8

	People 'satisfied' or 'very satisfied' (%)							
	Whites		Indians		Coloureds		Africans	
Aspects of life	1983	1988	1983	1988	1983	1988	1983	1988
Own health	91	88	90	82	92	83	67	51
Dwelling	93	92	82	71	73	68	60	45
Public services	80	73	68	54	55	51	39	33
Family happiness	93	91	94	98	92	84	83	76
Own education	71	74	65	60	52	64	39	26
Job opportunities	66	73	37	35	47	46	19	17
Own wages/salary	70	59	55	44	57	47	26	15
Food	94	95	96	89	94	89	67	59
Voting rights	90	93	31	48	20	44	27	19
Life overall	89	82	89	77	81	77	48	32

Source: Smith (1990), p. 9, following Moller (1989), p. 44, based on surveys of over 4,000 people.

criterion for measuring life quality than more 'objective' assessment.

One illustration of subjective social indicators must suffice. Table 35.1 shows the results of a survey of perceived quality of life in South Africa by race group (as officially defined at the time), indicating changes in the proportion of respondents 'satisfied' or 'very satisfied' with various aspects of life in 1983 and 1988. There was little difference in reported levels of satisfaction between those classified as Whites, Indians and Coloureds, on most criteria, voting rights and job opportunities being the main exceptions – underlining White privilege. Satisfaction among Africans (or Blacks) tended to be lower, however. Changes over the five years suggest that perceived quality of life decreased for all South Africans during these turbulent times. Differences between the Whites, Indians and Coloureds appeared to have been reduced, but the cleavage between these groups and the Africans had increased. The reduction in overall life satisfaction among Africans compared with the other groups is striking. Such surveys as this can reveal sources of social tension, as well as how people actually respond to unequal life chances revealed by more objective data.

'OBJECTIVE' COMPARISONS

To juxtapose subjective and objective approaches to quality of life comparisons is, of course, an oversimplification. There is a subjective or value-laden content to any attempt at definition beyond recognition of basic needs, as should be clear from the previous discussion. Ultimately, the good life is a question of ethics, and what can be said about it is a subject of inconclusive philosophical debate. What follows illustrates how various attempts have been made to compare population aggregates on some numerical scale based on what those responsible might claim to be objective criteria relevant to the quality of life; the intention is not to prescribe any particular definition.

Interest in national comparisons of life quality (or level of development) began to attract serious attention in the 1960s. Academics, politicians and some popular writers came to question the prevailing preoccupation with material prosperity and economic growth as the be-all and end-all of life, raising issues concerning the social and (later) environmental costs (e.g. Mishan 1969). Specifically, the adequacy of such pecuniary measures of national prosperity and progress as gross national product (GNP) or per capita income was challenged. Calls for more social content to systems of national accounting generated what became known as the social indicators movement (Smith 1973), which gathered strength in the United States and soon influenced Western European thinking. A prototype *Toward a Social Report* was published by the United States government, and in Britain the annual *Social Trends* was initiated. The former provided a frequently cited definition of a social indicator, as:

> a statistic of direct normative interest which facilitated concise, comprehensive and balanced judgement about conditions of major aspects of a society. It is in all cases a direct measure of welfare and is subject to the interpretation that, if it changes in the 'right' direction, while other things remain equal, things have gotten [*sic*] better, or people are 'better off'.
>
> (US Department of Health, Education and Welfare 1969: 97)

The breadth of possible conditions amenable to social indicators was clearly expected to be considerable.

The first impetus for the compilation of sets of national indicators came from the United Nations. Concern about international disparities in what was usually termed development, accompanied by recognition that money isn't all that matters, stimulated the search for measures to augment GNP. An early example is provided by the work of the UN Research Institute for Social Development. They assembled data on forty-two variables for 115 nations, the aim being a balance between social and economic indicators and between structural (causal) and development (outcome) indicators. The list was then reduced to eighteen 'core indicators' on the basis of their strength of association with all the others (Table 35.2).

These indicators are typical of the kind of conditions chosen for international comparisons in the 1960s and 1970s (see Smith 1977, ch. 8; 1979, ch. 2). The best-known contribution by a geographer from this period (Berry 1960) grouped forty-three variables under the headings of transportation, energy, agricultural yields, communications etc., GNP, trade, demographic and 'other'. His emphasis was very much on the technical preconditions for economic development. More social content could be found in some other studies, for example the stress on 'socio-cultural' development by Adelman and Morris (1967). The prevailing methodology was to subject large sets of data to multivariate statistical analysis, so as to derive a single index which would incorporate as much information as possible from the range of conditions originally selected.

Table 35.2 Core indicators used in a United Nations development index

Indicator	Average correlation (r) *with all other indicators*
1 Expectation of life at birth	0.744
2 Population in localities of 20,000 and over (% of total)	0.730
3 Consumption of animal protein per capita per day	0.791
4 Primary and secondary school enrolment (% aged 5–19)	0.777
5 Vocational education enrolment (% aged 5–19)	0.788
6 Average number of persons per room	0.783
7 Newspaper circulation per 1,000 population	0.823
8 Telephones per 100,000 population	0.762
9 Radio receivers per 1,000 population	0.737
10 Economically active population in utilities, transport, etc. (%)	0.769
11 Agricultural production per male agriculture worker ($)	0.839
12 Adult male labour in agriculture (% of total)	0.809
13 Electricity consumption per capita (kwh)	0.687
14 Steel consumption per capita (kg)	0.769
15 Energy consumption per capita (kg of coal equivalent)	0.760
16 GDP derived from manufacturing (% of total)	0.532
17 Foreign trade (sum of imports and exports) per capita ($)	0.737
18 Salaried and wage earners (% of economically active)	0.750

Source: McGranahan *et al.* (1970).

Comparisons among nations would then be made.

Attractive though this approach proved to be, as a means of measuring something as broad in scope as development or the quality of life, it has some obvious difficulties. Many individual indicators are highly correlated with one another (as shown in Table 35.2); that is, they tell basically the same story about variations among nations. In fact, GNP per capita (deliberately excluded from the UN exercise) was found to have a correlation coefficient as high as 0.89 with the composite development index derived from data for the eighteen core indicators. Composite indicators can overcome problems of data overlaps and redundancy, but the statistical techniques required to derive them are incomprehensible to most politicians and ordinary people, which tends to obscure the meaning of the results. Added to this is the questionable accuracy of numerical data on many conditions compiled in countries lacking reliable methods of counting and compilation, along with the possible instability of definitions from one country to another. Thus subsequent refinements of the UN development index eventually yielded to the need for greater simplicity and less data, without merely reverting to a single pecuniary measure.

One simple device which has featured prominently in the literature on development is the Physical Quality of Life Index (PQLI) (Morris 1979). This is the average of just three indicators, representing infant mortality, life

779

Table 35.3 The Human Development Index (HDI) for a sample of nations

Nation	Life expectancy (years) 1987	Adult literacy (%) 1985	GDP/head ($) 1987	HDI	Rank by HDI	Rank by GDP/head
Niger	45	14	452	0.116	1	20
Benin	47	27	665	0.224	10	28
Zaïre	53	62	220	0.294	20	5
Ghana	55	54	481	0.360	30	37
Kampuchea	49	75	1,000	0.471	40	2
Burma	61	79	752	0.561	50	11
Tunisia	66	55	2,741	0.657	60	70
Mongolia	64	90	2,000	0.737	70	57
Brazil	65	78	4,307	0.782	80	85
Romania	71	96	3,000	0.863	90	84
Yugoslavia	72	92	5,000	0.913	100	90
East Germany	74	99	8,000	0.953	110	115
Finland	75	99	12,795	0.967	120	121
Japan	78	99	13,135	0.996	130	126

Source: United Nations Development Programme (from data reproduced in *The Economist*, 26 May 1990, p. 111).

expectation and basic literacy. More recently the United Nations Development Programme (UNDP 1990) has introduced a Human Development Index (HDI) of similar simplicity. It combines life expectancy at birth (in years) and the adult literacy rate (per cent) with gross domestic product per head ($) adjusted for purchasing power parity. For each of these indicators a minimum and 'desirable' value are specified: the minimum is the lowest national value actually observed, while the desirable levels were (then) the maximum of 78 for life expectancy (as in Japan), 100 per cent for adult literacy, and for GDP $4,861 which was the average official poverty line for nine industrialized countries in 1987. Logarithms rather than the absolute dollar values are used for GDP, to reflect 'diminishing returns'; that is, the higher the purchasing power the less needs are fulfilled by each additional dollar spent. The minimum and desirable values then set the end point on scales from zero to one for each of the three indicators, and their average for each nation give its HDI score.

Table 35.3 shows the relevant data for a sample of nations, from the lowest on the HDI scale (Niger) to the tenth lowest, and thereafter every tenth nation by rank on the index to the highest (Japan) of the 130 for which calculations have been made. The first three columns list the basic data, the next the index, then the nation's ranking on HDI followed by ranking on GDP per capita for comparative purposes (note that rank 1 indicates the lowest score and 130 the highest). While the index is broadly reflective of GDP, there are some substantial differences. It is beyond the scope of this treatment to pursue the reasons for these discrepancies; it is sufficient to demonstrate that the HDI does rank nations differently from the more conventional pecuniary measure.

Like all other such attempts to compare the life of nations on some qualitative scale, the HDI has come in for criticism. This includes the arbitrariness of the equal weighting of the three indicators and the effective topping off of the purchasing power (GDP) scale by a poverty level standard – which implies that income above this contributes nothing to human development. One commentator (Nasar 1992) has shown that subtle changes in measurement can make quite a difference to the ranking of individual nations. For example, the 1990 HDI ranks Japan first and the United States seventh, but if this index is adjusted to take into account inequality in income distribution within nations Japan is first again and the United States ninth. An adjustment for gender inequality places Finland first, with the United States down to tenth.

So, even the simplest attempts at international comparisons involving multiple criteria fail to yield unambiguous answers to the question of where life is best, better or worse. Some respond by arguing for one single indicator. Income or gross product (per capita) is obviously very important, as indicative of access to direct sources of individual need satisfaction and of society's capacity to provide public services, especially to the poor. Such measures are good predictors of many other qualitative aspects of national life. But not all that matters is reflected by money. And it is commonly observed that as national wealth increases the standard of living or level of development more broadly defined does not improve in the same proportion (see, for example, Smith 1979: 73); as at the individual level, there appear to be diminishing returns to increasing wealth. Life expectancy is sometimes proposed as an alternative, but long lives can be miserable just as short ones may greatly satisfy those who know no other. The conclusion must therefore be that geographical comparison of quality of life (or whatever it is called), if undertaken at all, must come to terms with the conceptual and practical problems inherent in the use of multiple criteria.

Studies of variations in the quality of life within single nations are relieved of some of the problems of international comparison. For example, numerical data for different parts of one country are more likely to have been compiled using the same definitions and with similar degrees of accuracy than in the case for different nations. There may also be more of a national consensus on the ingredients of the good life than among nations of contrasting cultural traditions.

Most intra-national studies have been conducted in the United States. These range from academic research involving sophisticated techniques and often an explicit theoretical framework to guide choice of criteria, to journalistic exercises in 'rating places' which may produce well-publicized but deeply flawed results. Cutter (1985) has provided a comprehensive review. The earliest such studies date from the 1930s, notably those of Angoff and Menken (reviewed and revised in Cutter 1985: 39–43) at the state level, and Thorndike by cities (summarized in Smith 1973: 32–7). Subsequent attempts include major investigations by Liu (1973, 1975) at both scales.

The criteria adopted by Liu are worth listing (major categories only), as representative of the kind of conditions incorporated into such studies. At the state level they were:

- individual status (e.g. opportunities for self-support);
- individual equality (by race, gender);
- living conditions (economic, social, environmental);
- agriculture (incomes, mechanization, value of product, etc.);
- technology (e.g. scientific manpower);
- economic status (income, employment, etc.);
- education (school enrolment, attainment, resources);
- health and welfare (medical services, levels of health, etc.);
- state and local government (citizen participation, professionalism, etc.).

By cities the following main components of quality of life were recognized:

- economic (individual well-being and community health);
- political (individual activities and local government);
- environmental (individual, institutional and natural);
- health and education (individual and community);
- social (individual development, equality and community conditions).

As at the international scale, different approaches generated different results, though with similar broad geographical patterns (see Cutter 1985; Smith 1973; Smith 1979, chs 3 and 4).

A number of individual cities have attracted quality of life investigations, breaking ground overlooked by traditional studies of urban morphology and the factorial ecologies of the so-called quantitative revolution in human geography. Early contributions in the United States include Dickinson et al. (1972) on Gainesville, Florida, and Bederman (1974) on Atlanta, Georgia.

Rather than going further into case studies, this section will conclude with a summary of the main problems of territorial quality-of-life indicator construction. The catch-phrase of 'who gets what where' (Smith 1974) focuses attention on the three initial decisions, as to: (1) whose experience is involved, e.g. the entire population in aggregate, or subdivisions by race, gender and so on; (2) what they get, i.e. the conditions to be measured and incorporated into some composite index; (3) how territory (the world, nation or city) is to be subdivided, i.e. national entities, regional delimitations, urban neighbourhoods.

In actual practice the 'who' is usually taken to be the entire population, in the absence of a strong rationale (or the data) to do otherwise. However, the different, and unequal, experience of people defined by gender or race may be attempted. For the 'what', conditions relating to living standards or the quality of life are selected. Each of these is measured by numerical indicators, carefully

782

chosen to maintain the validity of the condition in question, and limited in number to avoid the complexity of a large data set. The data should be the latest available, unless a historical or time-sequential study is involved. The indicators will each have a normative interpretation, such that high values are judged to be 'good' and low ones 'bad' or the reverse. The territorial subdivision may be derived from some prior analysis such as regionalization or city-tract grouping, but is often purely a matter of convenience or necessity with respect to data availability.

It is important to recognize that contestable judgements are involved at every stage in this process. For example, each condition could be measured by alternative indicators – for example, health by rates of mortality, morbidity or incapacity. Some of the normative interpretations may be open to question; for example, high numbers of hospital beds in relation to population may indicate a well-provided health service, or a population highly prone to illness, or even that people are kept in hospital rather than being cured and sent home.

Data measuring different conditions can be directly compared only if they are in the same units and on the same scale, instead of in £, $, percentages or as ratios to population. Transforming data on different conditions into a comparable form is also a necessary step in combining them into a composite indicator (as in the HDI, explained above.) The easiest method of transformation is ranking, but this loses information; that is, on the intervals between observations. A frequently adopted transformation is to calculate standard scores, in which the mean of each variable is set at zero (0) and the standard deviation at unity (1.0), thus equalizing both the average and the spread of the data. A composite indicator can then be derived by summing the scores for each territorial unit.

This derivation of a composite indicator implicitly assumes that each of the individual conditions are of the same importance. But they can be weighted differentially, according to some measure of their relative significance as contributors to life quality. Unfortunately, there is no grand theory or Delphic oracle which can be consulted to obtain such weightings, a fact which itself reflects the absence of any scientific or popular consensus on what is meant by the quality of life. One possible approach is to conduct opinion surveys; for example, asking people to rank each condition and then working out the average. To build weights into a composite indicator it is necessary to multiply each original indicator value by the measure of relative importance derived from the survey. Examples of this kind of exercise can be found in Smith (1979: 122–31; 1987: 13–19). Given the expense of surveys, it is unusual for differential weighting based on popular opinion to be used in practice. If adopted at all, weightings are likely to be derived from supposed expert opinion which invites the charge of intellectual imperialism, or from some data compression technique like factor analysis (a form of technical imperialism).

FROM PATTERN TO PROCESS

Geographical exercises of the kind reviewed above are prone to a unidimensional view of the quality of life, which lacks depth and a sense of process. To generate a column of figures which can be mapped, and the technique involved, risks becoming an end in itself. The limitations of this kind of study were quickly recognized, as the empiricism of the discipline's early engagement with social relevance gave way to a more structural concern with process informed by the resurgence of Marxism in social science. It is beyond the scope of this chapter to explore this development; the discussion here will be confined to some perspectives on process which have arisen within the mainstream discourse on quality of life.

The quality of life is both individual and social, mutually interdependent. We may try to measure individual levels of satisfaction, but how needs are satisfied can seldom if ever be a purely personal matter pursued in isolation. Even Robinson Crusoe had help from Man Friday. No man (or woman) is an island, as the poet John Donne pointed out; we are all bound together, in families and communities, taking sustenance from them and making our own contribution to their lives. The internationalization of economic relations and the global nature of environmental threats reinforce a sense, and reality, of interdependence on a world scale.

There have nevertheless been attempts to interpret human need satisfaction in individualistic terms. Neo-classical economics is a case in point, with its emphasis on individual utility maximization and tendency to play down the highly variable constraints imposed on people's consumption possibilities. More relevant to understanding how people actually behave is the notion of a needs hierarchy, as originally developed by Maslow (1954). The argument is that 'higher' needs emerge as successively 'lower' needs are satisfied. Maslow's progression is from survival or the struggle to maintain life, to security from physical danger, to belonging and love, to esteem or prestige, and finally to self-actualization or the need for personal fulfilment. Some people, in some places, will be preoccupied with survival (e.g. in the famine belt of central Africa); others will have the freedom from more basic needs to seek status in particular kinds of consumption or behaviour. However, it would be a mistake to conclude that individual-need satisfaction is always ordered with such implicit rationality. The poor family may choose to spend an unexpected windfall on one extravagant meal or a gigantic television set, rather than acting in a more conventionally prudent fashion. People's personal priorities may be subject to certain imperatives, the most obvious being survival, but even this can be transcended by the deliberate sacrifice of life to a 'higher' purpose.

Broader social conceptions of how the quality of life arises also involve some prioritization of needs. This is an important step beyond undifferentiated lists of criteria. An example can be taken from the early research stimulated by the United Nations engagement with the meaning of development. Drewnowski

(1974, summarized in Smith 1987: 35–9) made a distinction between a population's 'state of welfare' and its 'level of living', proposing that the former depended on the latter. He defined welfare in terms of people's somatic status (physical development), educational status (mental development) and social status (social integration and participation). Level of living involves nutrition, clothing, shelter, health, education, leisure, security, social environment and physical environment. Just as people's stock of pecuniary wealth is maintained (or otherwise) by flows of income, so their state of welfare stems from their level of living. This is very similar to the link between basic and intermediate needs (e.g. in Doyal and Gough 1991; see p. 775 above). Such notions provide the basis for a possible general process model of human-need satisfaction.

There are horrendous difficulties involved in the construction of such a model. Central to these is the fact that the cause and effect relationships are poorly understood. For example, it is far from clear how a nation's physical or mental status, for example, can best be enhanced by a given level of additional expenditure on health or educational services. Defining the outcomes, in terms of physical and mental status, will be a subject of contention; for example, whether health is best measured by death rates or by the ability of the living to function effectively. Then it will not be clear whether health (however defined) is best improved by spending more on health services (and within them on hospitals, doctors or nurses, reactive or preventive care), or by improving people's housing, safety at work or vulnerability to unemployment – all of which have a bearing on their physical status.

At the root of this problem is a belief, held in some quarters, that human-need satisfaction, in the broad sense of how the quality of life is determined, is analogous to manufacturing. How to produce the most shirts or washing machines from a given investment is a fairly simple exercise in optimal resource allocation, involving well-understood technology. The question of how to get the most out of expenditure on social services is far more complicated, not least because the output is less clearly defined than that of a factory. In limited spheres of activity there is evidence that certain practices are more cost-effective than others. But there is no objective or scientific means of demonstrating how best to improve a nation's health, nor that the government can improve the lives of the governed more by favouring health, education or crime protection in the next round of public spending deliberations. And it is doubtful whether the contemporary cult of numerical performance assessment within simulated market mechanisms will make decisions on resource allocation much more convincing. It may merely focus attention on what can most easily be measured. What it means to improve people's bodies, minds and lives will remain matters of contention, for such uncertainty and ambiguity is part of what it means to be human.

It is in the political arena, of course, that so many of the decisions relating to the process of human-need satisfaction are made. People presumably have some regard for political party policy when they elect their representatives, being more or less persuaded by those who promise to prioritize the health service or the police force, for example. There will be competing claims as to the relative effectiveness of public agencies and private enterprise as means of service provision, reflecting ideological conviction rather than the evidence of alternative experience. The actual outcomes may well reflect the allocation of power within a society, and the structures from which this derives, rather than (literally) measured consideration of the desired quality of life. For the geographer to attempt a research contribution to this process involves an act of faith as well as of science.

INEQUALITY, WELFARE AND SOCIAL JUSTICE

It remains to round off the discussion with some broader issues which have been muted thus far. As soon as we move from the traditional geographical curiosity over areal differentiation to a concern with spatial inequality, we engage morality. To observe that people live differently in different places need carry no moral connotations, but judging the quality of those lives inevitably raises questions of good or bad, better or worse.

That human life chances are unequally distributed, according to who and where people are, is a commonplace observation. Geographical patterns of development or territorial social indicators reveal this clearly, and the degree of inequality can be measured in various ways. But there is more to inequality than this, and more to quality of life than some property which is unequally distributed. Inequality should be regarded as an intrinsic feature of any conception of the quality of life. At the individual level, few people are indifferent to the condition of others – to the degree of inequality manifest in the homeless sleeping on London streets or starving children on the television screen. To cite John Donne again, anyone's death diminishes me, because I am involved in humankind.

The existence of severely deprived groups within society can impact on others more directly. They exact a cost, such as unemployment and social security benefits, and may threaten the 'social fabric' when civil disorder arises. Inequality has in fact been incorporated in various ways into some attempts to derive social indicators, as shown above (see p. 781). But it requires greater emphasis in a world where the gap between the 'haves' and 'have-nots' appears to be widening both among and within nations.

The concept of welfare in neo-classical economics, while the subject of much justified criticism, is helpful in crystallizing the content of a comprehensive view of the quality of life which includes distribution. In the traditional formulation, the welfare of society is maximized when the optimal (best possible) combination of goods (in the broadest sense) is distributed in the best possible way among the population. Nothing can be done to make anyone (anywhere) better off without someone else (somewhere else) being made worse off – the criterion of Pareto optimality. The practical exploration and improvement of quality of life requires not only information on the actual structure and distribution of society's product in the most general sense, but also asking who *should* get what where, and how, for the 'hidden hand' of the supposedly impartial market may be more like that which consciously guides the movements of a marionette.

Engagement with the quality of life therefore leads the geographer inevitably to the question of social justice. This was a subject of considerable interest in the early years of the social relevance movement (see Harvey 1973), but subsequently became more subdued. However, there is now a resurgence of interest in social justice, linked to practical issues relating to social change in various parts of the world (e.g. Eastern Europe and South Africa; see Smith 1992, 1994), as well as to some contemporary theoretical problems (e.g. post-modernism and the politics of difference; see Harvey 1992). A special issue of the journal *Urban Geography* (Laws 1994) exemplifies this trend. Outside geography the literature on social justice continues to grow (see the review in Kymlicka 1990), an important feature being renewed interest in the work of Rawls (1971; see Barry 1989; Peffer 1990; Doyal and Gough 1991).

What constitutes a just distribution of society's benefits and burdens has been the subject of over two thousand years of inconclusive debate. The issue is as indeterminate as the meaning of life itself. However, there is one central component of Rawls's theory which seems particularly pertinent to the contemporary world. This is his proposal, simply expressed, that social arrangements should be judged according to their effect on society's poorest members. In other words, measures that benefit the poor should be preferred to those that benefit the rich. The complexities of this principle in both theory and practice need not detain us here; it is sufficient to underline its resonance with political rhetoric in many parts of the world, which advocates narrowing the gap between rich and poor rather than the reverse. Indeed, justice as equalization is an arguable universal principle (see Smith 1994: ch. 5). Yet vast disparities continue to exist at all spatial scales: among nations, among regions and cities within nations, between cities and rural areas, and among different parts of individual cities (see Chapter 19).

Inequality is very much taken for granted. Indeed, its perpetuation and exacerbation is encouraged, not only by those economic institutions that depend for their survival and profits on selling ever more sophisticated gadgets to the already well endowed, but also by ourselves as consumers duped into believing that the quality of life is manifest in material possessions. Capitalism has been conspicuously successful in meeting the ever more frivolous needs of the wealthy and in securing comfortable living standards by many criteria for the mass of the working class. But this has been at the cost of marginalizing significant numbers of people from the mass-consumption mainstream even in the wealthiest countries, while poverty and starvation pile up in the so-called Third World. Socialism, as actually practised, failed to live up to its equalitarian expectations to the point of self-destruction, despite the notable achievements of some states in the provision of housing, health and education services for previously deprived populations.

The emerging hegemony of capitalism makes it more important than ever to question the degree of inequality in human life chances with which this system is associated. Rather than the inevitable outcome of anonymous market forces or exploitative economic relations, it should be understood as capable of some amelioration, via such means as public policy innovations, local community struggle and individual behaviour. New forms of human practice are required, to forge a new understanding of what it means to be human in a strongly interdependent world, in which consumption choices in one place can have implications for people across the globe. Rather than taking our good (or ill) fortune for granted, we need to work towards a theory of (territorial) social justice which respects human differences but recognizes the universality of a wide range of needs relevant to the quality of life. In so many aspects of life it is inequality not equal treatment which requires justification: 'the central issue in any theory of justice is the defensibility of unequal relations between people' (Barry 1989: 3).

The process-oriented human geography which has dominated the field for the past quarter of a century has forged important links with central issues in social theory. Now, renewed concern with the quality of life requires an even broader perspective, involving ethics and moral philosophy. Indeed, it requires a return to some of the classical concerns with the meaning of a good life. And it is not just a question of how we view life in general, but how we ourselves live it, and relate to others, in personal and professional practice. As Williams (1985: 119) puts it, 'The most urgent requirements of humanity are, as they have always been, that we should assemble as many resources as we can to help us to respect it.' These resources include those of geographical inquiry.

REFERENCES

Adelman, I. and Morris, C.T. (1967) *Society, Politics and Economic Development*, Baltimore, Md.: Johns Hopkins University Press.

Barry, B. (1989) *Theories of Justice*, London: Harvester-Wheatsheaf.

Bederman, S.H. (1974) 'The stratification of "quality of life" in the black community of Atlanta, Georgia', *Southeastern Geographer* 14(3), 378–86.

Berry, B.J.L. (1960) 'An inductive approach to the regionalization of economic development', in N. Ginsburg (ed.) *Essays on Geography and Economic Development*, Research Paper 62, Chicago: Department of Geography, University of Chicago.

Cutter, S. (1985) *Rating Places: A Geographer's View on Quality of Life*, Washington, DC: Association of American Geographers.

Dickinson, J.C., Gray, R.J. and Smith, D.M. (1972) 'The quality of life in Gainesville, Florida: an application of territorial social indicators', *Southeastern Geographer* 12, 121–32.

Doyal, L. and Gough, I. (1991) *A Theory of Human Need*, London: Macmillan.

Harvey, D. (1973) *Social Justice and the City*, London: Edward Arnold.

Harvey, D. (1992) 'Social justice, postmodernism and the city', *International Journal of*

Urban and Regional Research 16, 588–601.

Kymlicka, W. (1990) *Contemporary Political Philosophy: An Introduction*, Oxford: Clarendon.

Laws, G. (1994) *Urban Geography*, Special Issue, 15(7).

Liu, B.-C. (1973) *The Quality of Life in the United States 1970: Index, Rating and Statistics*, Kansas City: Midwest Research Institute.

Liu, B.-C. (1975) *Quality of Life Indicators in the U.S. Metropolitan Areas, 1970*, Kansas City: Midwest Research Institute.

McGranahan, D.V., Richard-Proust, C., Sovani, N.V. and Subramanian, M. (1970) *Content and Measurement of Socio-Economic Development: An Empirical Enquiry*, Geneva: UN Research Institute for Social Development.

Maslow, A.H. (1954) *Motivation and Personality*, New York: Harper.

Mishan, E.J. (1969) *The Costs of Economic Growth*, Harmondsworth: Penguin.

Moller, V. (1989) 'Can't get no satisfaction: quality of life in the 1980s', *Indicator South Africa* 7(1), 43–6.

Morris, M.D. (1979) *Measuring the Condition of the World's Poor: The Physical Quality of Life Index*, Oxford: Pergamon Press.

Nasar, S. (1992) 'Why international statistical comparison don't work', *The New York Times*, 8 March.

Peffer, R.G. (1990) *Marxism, Morality and Social Justice*, Princeton: Princeton University Press.

Rawls, J. (1971) *A Theory of Justice*, Cambridge, Mass.: Harvard University Press.

Smith, D.M. (1973) *The Geography of Social Well-being in the United States: An Introduction to Territorial Social Indicators*, New York: McGraw-Hill.

Smith, D.M. (1974) 'Who gets what *where* and how: a welfare focus for human geography', *Geography* 59, 289–97.

Smith, D.M. (1977) *Human Geography: A Welfare Approach*, London: Edward Arnold.

Smith, D.M. (1979) *Where the Grass is Greener: Living in an Unequal World*, Harmondsworth: Penguin.

Smith, D.M. (1987) *Geography, Inequality and Society*, Cambridge: Cambridge University Press.

Smith, D.M. (1990) *Apartheid in South Africa* (3rd edn), Cambridge: Cambridge University Press (UpDate series).

Smith, D.M. (1992) 'Geography and social justice: some reflections on social change in Eastern Europe', *Geography Research Forum* 12, 1–15.

Smith, D.M. (1994) *Geography and Social Justice*, Oxford: Basil Blackwell.

UNDP (1990) *Human Development Report 1990*, Oxford: Oxford University Press.

US Department of Health, Education and Welfare (1969) *Toward a Social Report*, Washington, DC: US Government Printing Office.

Williams, B. (1985) *Ethics and the Limits of Philosophy*, London: Fontana Press.

FURTHER READING

Attfield, R. and Wilkins, B. (eds) (1992) *International Justice and the Third World*, London: Routledge.

Baldwin, S., Godfrey, C. and Propper, C. (eds) (1990) *Quality of Life: Perspectives and Policies*, London: Routledge.

Corbridge, S. (1993) 'Marxisms, modernities, and moralities: development praxis and the claims of distant strangers', *Environment and Planning D: Society and Space* 11, 449–72.

Cutter, S. (1985) *Rating Places: A Geographer's View on Quality of Life*, Washington, DC: Association of American Geographers.

Doyal, L. and Gough, I. (1991) *A Theory of Human Need*, London: Macmillan.

Kymlicka, W. (1990) *Contemporary Political Philosophy: An Introduction*, Oxford: Clarendon Press.

Kymlicka, W. (ed.) (1992) *Justice in Political Philosophy* (2 vols), Cheltenham: Edward Elgar.

Smith, D.M. (1977) *Human Geography: A Welfare Approach*, London: Edward Arnold.

Smith, D.M. (1987) *Geography, Inequality and Society*, Cambridge: Cambridge University Press.

Smith, D.M. (1994) *Geography and Social Justice*, Oxford: Basil Blackwell.

V CHANGING WORLDS, CHANGING GEOGRAPHIES

INTRODUCTION

Richard Huggett and Mike Robinson

Since 1945, approaches to geography have been led, perhaps more than at any time before, by changes in society. The hegemony of positivism, fostered by the enormous successes and promise of science and technology in the immediate post-war years, was broken during the 1960s. The ebbing of positivism started in the USA. Social unrest, environmental degradation, and involvement in the Vietnam war all helped to promote a loss of faith in science. Poverty, race, and sex were issues that needed addressing in more humane terms than the clinical language of science offered. Science could not feed the world, despite developments in the green revolution. The rights of human races – black and white, red and yellow – required an approach beyond the reach of homologous models derived from science. How were people to be fed and housed? Where were responsibilities for welfare to be located and how were they to be distributed? How far should the views of people themselves enter into those questions which affected the conditions of their lives? To a profession dominated by white Western men, how much effort should be devoted to the needs of a world peopled mostly by different cultures? Moreover, what recognition should be given to the needs and rights of women of all cultures and minorities within every culture? A greater humanity was encouraged in a discipline which has been, perhaps more than any other, open to a full range of explanatory modes.

A more humane geography involves different ideas about how the world's ecological and social problems can be conceptualized and solved. Some people favour Marxist dialectic or theories of structuration. Some look to the abandonment of metatheory and the acceptance of a heterogeneous and contradictory world full of ambiguity and defying ubiquitous resolution. Most, however, work within a pragmatic framework derived from personal biography and experience, and constrained by resources and data availability. In this sense, as well as in many others, geography has changed very little.

36

THE EXPANSION AND FRAGMENTATION OF GEOGRAPHY IN HIGHER EDUCATION

R.J. Johnston

The academic discipline of geography has changed very substantially in recent decades, in both substance and method. It has also expanded very rapidly, with more geography students in higher education institutions, and more staff both teaching them and undertaking original research in the discipline. This chapter looks at those two phenomena, with particular reference to the situation in the United Kingdom (on the necessity of which, given the relatively short length of the treatment, see Livingstone 1995), and argues that they are linked in the growth of specialisms within the discipline. Geography is now a very fragmented discipline. Indeed, its fragmentation is the dominant theme of this chapter: within less than a century, and mainly over the last forty years, we have witnessed within the discipline the demise of the 'all-round geographer' and the rise to pre-eminence of the topical specialist.

EXPANSION

Although it is easy to make the broad generalization that the academic discipline of geography has expanded substantially in recent decades, it is very difficult to sustain that with hard data, and certainly so if you want to make international comparisons. Even in the United Kingdom, data on the discipline's size within the country's higher education sector are only available since the 1960s, and they only refer to the 'old' universities in the country – that is, those established with the title 'university' prior to 1992. The 'facts' of expansion have largely to be taken for granted, therefore; a series of general claims is presented here, with some supporting data. The only 'hard data' available to me refer to the UK, and they are cited as exemplars only. (For more detail, see Johnston 1995a.)

Origins

Universities were established to provide vocational training for the professions – notably the Church, the law, and medicine. By the late nineteenth century, the civil service had been added to this list; it required university graduates to staff the rapidly growing state apparatus in most countries of the Western world plus, as in the British case, its large colonies. Those universities were elite institutions, admitting only a very small proportion of school-leavers in their late teenage years: there was little state support for students, and so only the wealthy, plus a few who were able to obtain scholarships or sponsorship, were able to attend.

The demand for university graduates grew in the late nineteenth and early twentieth centuries, concurrent with expansion of the number of professions for which a degree qualification was considered desirable, if not mandatory. Schoolteaching, especially secondary schoolteaching, was one of those professions; the introduction of compulsory education for all and the gradual increase in the school-leaving age stimulated an increased demand for university-trained teachers.[1]

Geography came relatively late to these institutions, since it had no link to a profession other than schoolteaching. As the imperial powers extended their influence over larger portions of the earth's surface, however, so the need arose for systematically organized knowledge about those places – what became known as 'commercial geography'. Furthermore, many states wished their children to learn about the world, and especially about their nation's place in it. This stimulated demands for geography to be taught in schools, and hence for trained geography teachers.

The first university Departments of Geography were established in Germany in the late nineteenth century (Taylor 1985); they served three main markets – commercial, military, and educational. Other countries followed suit. In the UK, the Royal Geographical Society put strong pressure on the government and the university authorities to found departments at Oxford and Cambridge (Stoddart 1986). The first honours degree course in geography in a British university was established in neither of those establishments, however, but at the University of Liverpool in 1917 (Johnston and Gregory 1984). By the end of the Second World War there were twenty-five Departments of Geography – one in nearly every university and university college in the country.

Those departments in place in 1945 were all small, with only a few staff members each and probably no more than twenty students graduating annually with an honours degree in geography. The great majority of the graduates were destined to become schoolteachers, who needed a broad education as the foundation for their later career. Many of the people who taught them were generalists too, covering a wide range of subject matter in their varied lecture courses.

Rapid growth

British universities grew quite substantially during the 1950s, although the percentage of 18-year-old school-leavers admitted to degree courses was still extremely small and the universities remained elite institutions to which only the most able could aspire – even though grants for maintenance and fees were available through competitive state and county examinations. Geography departments participated in that growth, which continued through the 1960s. Few of the new universities founded during that decade included Departments of Geography within their academic profile, however. The discipline consolidated its position in the older establishments but was apparently not welcomed in the new ones – perhaps because it had a relatively low profile as a discipline which focused on the preparation of potential teachers and because it was not then associated with the 'scientific and technological revolutions' of the times.

The increased numbers of geography students were no longer predominantly destined for careers in schoolteaching, however. Other opportunities beckoned, not least in the burgeoning field of town and country planning, in which many geographers found employment that called for their particular skills. But from the late 1960s on, as universities expanded further – albeit admitting only about one-seventh of the 18-year-olds until the late 1980s – so the career opportunities for all graduates widened. Outside the professions, many employers indicated that what they were looking for in graduates was not so much subject knowledge and disciplinary understanding as general (now often termed 'transferable') skills which could be applied in a variety of occupations for which specialist postgraduate training was provided (some of it on-the-job). Geography graduates – many of them having been introduced to both the scientific and the social scientific aspects of their discipline – were well placed to move into various careers, therefore, especially in management and the non-graduate professions such as accountancy.

Although the proportion of geography graduates who became schoolteachers declined rapidly from about 1960 on, the discipline remained extremely strong in schools. Its protagonists – notably in the Geographical Association (Balchin 1993) – were very active in advancing the cause of geography in national curriculum discussions in the late 1980s, for example. Geography remained one of the largest subjects taken by students in their school-leaving examinations at both 16 and 18, thus providing a very substantial number from which university departments could attract students. The rapid growth of the discipline within higher education was sustained by this large pool of potential students (though on the relatively small flow from that pool see Lee 1985); geography may have fared less well in the universities were it not for this strength in the country's schools and the national examination boards.[2]

British higher education stagnated during most of the 1980s because of government reluctance to expand universities.[3] At the end of the decade, however, politicians realized that the UK had a very low percentage of its 18-year-olds entering higher education, according to international comparisons. Critics argued that this was threatening the country's economy, because a large, well-trained workforce is necessary for competition in a service-industry-driven world economy in which possession of and ability to use information is a key component of economic success. Hence the decision to expand university education very substantially so that about one-third of the eligible population would be entering universities by the end of the century (see also Johnston 1995b).

Geography participated in this recent expansion of undergraduate education, although lagging the trend in total numbers slightly (Johnston 1995a). By the beginning of the 1990s, the 'old' universities (the only ones for which we have data) had nearly 8,000 undergraduates reading for geography degrees (and more were studying joint degrees including geography), which was more than double the number in the mid-1960s (the first date for which comparable data are available).

This student expansion was accompanied by a growth in staff complement, at least until the early 1980s. Their numbers have stabilized since, largely as a consequence of government insistence that expansion be achieved more 'efficiently'. The resources available per student fell very substantially (over 40 per cent in no more than five years according to Jenkins and Smith's 1993 data on student:staff ratios), and the number of academic geography staff members has remained fairly constant at around 600 in the 'old' universities for the last decade.

One indicative source on the number of academic staff is the membership of the Institute of British Geographers, the UK's leading learned society for teachers and researchers of the discipline in higher education. Not all university teachers and researchers join by any means, and there are some overseas members, but the general trend is indicative (pre-1983 data are from Steel 1983). Steady growth from the end of the Second World War (when membership was about a hundred) until the mid-1960s (about 800) was followed by even more rapid growth until the end of the 1970s (nearly 2,000). There was then a substantial fall in membership, with the figure at the end of the 1980s (at under 1,300) less than two-thirds that of ten years previously. This was followed by a rapid recovery, and the 1993 membership was little more than a hundred short of the maximum in the late 1970s.

The growth in academic staff numbers over the period encouraged expansion of postgraduate study in geography, since possession of a research degree increasingly became a necessary condition for entry to the academic career system. As with undergraduates, postgraduate numbers have approximately doubled in geography over the last three decades. This is not in line with the recent trends in all disciplines, however, for geography numbers in 1991 were only the same as those a decade earlier whereas the total for all subjects was 45 per cent up (Johnston 1995a).

Within British universities, therefore, geography as a discipline has expanded very substantially since the end of the Second World War. What are now known as the 'old' universities – that is, those which achieved the designation prior to the 1992 decision to allow the former polytechnics and the larger colleges of higher education to become universities – have about 600 full-time academic staff members (people employed as both teachers and researchers); there are probably as many as 300 more in the new institutions.

Organizational structures

The organizational focus of teaching and research within most universities is the academic department, but for each staff member his or her department is only one of the contexts within which professional life is structured. Indeed, for many geographers (and for members of many other academic disciplines too) the 'home department' is at best a very partial context, because few, if any, of their colleagues will have more than a general passing interest in their specific research and teaching subjects. Interaction with like-minded others occurs in wider communities. At the formal level, these are best represented by the learned societies, the organizations which arrange meetings and publish journals, thereby facilitating interaction among research workers.

In the UK there are three main learned societies for geographers. The oldest is the London-based Royal Geographical Society (RGS), which is now more than 150 years old (Brown 1980). It has a large membership, most of whom have no degree qualification in the discipline, and have become Fellows because of their general interest in geography, especially those aspects of it concerned with exotic environments. But it does have a large academic component to its work, and considerable strengths in certain areas of the subject.

The second most important society is the Geographical Association (GA), which celebrated its centenary in 1993 (Balchin 1993). Its prospectus is very largely concerned with geographical education, and, although some of its work is concerned with universities, most of its attention is focused on geography in the country's school curricula. It is a very effective pressure group for geography as a school discipline.

The final and youngest society of the three is the Institute of British Geographers (IBG), founded in 1933 largely by a breakaway group from the RGS which they perceived as unsympathetic to their interests, especially in human geography. Its focus is very much on research within the discipline, and its membership consists almost entirely of university staff and postgraduate students. Like similar societies in other disciplines, it organizes conferences and publishes a major international journal. Since the 1960s, much of its detailed work has been devolved to its separate study groups which also organize meetings and act as smaller-scale contexts within which geographers with similar interests can interact.[4] (The IBG and the RGS merged in January 1995 to form the Royal Geographical Society (with the Institute of British Geographers).)

798

Learned societies provide the major formal institutional contexts within which many geographers interact with their peers. But there are also less formal organizations – sometimes termed 'invisible communities' – through which groups maintain contacts and circulate their ideas. More formally, published journals – many of them emanating from commercial organizations rather than learned societies – are also very important components of the process of information flow within a discipline, or at least of its component parts (Johnston 1994c). As will be argued in more detail below, the pattern of research publication activities is an important indicator of the structure of a discipline.

SPECIALIZATION

The growth of the academic discipline and the increase in the number of academic staff has been associated with several major changes in the nature and practice of geography during the period since the Second World War. Whether cause-and-effect has been involved must remain open to doubt, because we have no counterfactual situation against which to compare what happened, but there is strong circumstantial evidence that the growth strongly encouraged, if not generated, specialization and fragmentation within the discipline.

Growth was probably a necessary if not a sufficient condition for the increasing specialization of staff research and teaching interests which has marked this period. In a small university Department of Geography with only a few staff members, each person had to teach a range of courses – both substantive and regional in many cases in the 1950s and early 1960s. As more staff were appointed to teach the increased student population, so a broad range of teaching activities became less necessary for any one person. Individuals could specialize in a few related courses only, although they may have been expected to give general support to others in tutorial and field teaching. This ability to concentrate one's teaching attention was strongly linked to the growth of research – increasingly, staff only taught other than introductory courses in the subjects they were actively researching.

Universities are now widely recognized as places where research and teaching go together.[5] In most institutions, those appointed to permanent teaching positions attain that status largely on the basis of their research experience and promise, and they are expected to fulfil that promise in their careers: career advancement is very largely based on research achievements in most universities, despite the (ritual?) statements that teaching and, to a lesser extent, administrative performance are important too (Johnston 1991b, 1995b). The majority of university staff – particularly those in the 'old' universities which have the strongest research traditions – are on 'dual career tracks', having been appointed as both teachers and researchers. This is a relatively recent phenomenon, however, and until a few decades ago the pressure to do research and to publish its findings was much less than it is today.

That pressure is now institutionalized in the United Kingdom with the sequence of regular Research Assessment Exercises (1986, 1989, 1992, 1996) in which the work of the 'research-active' staff in all departments is graded according to its perceived quality by a panel of experts – from grade 5 (the best) to grade 1 (see Thorne 1993). Funding for university research is now almost entirely associated with the grades attained, thereby promoting selective allocation of funds to the perceived highest quality departments. Since much of the decision-making about research excellence involves evaluation of staff research publications, the pressure to produce substantial volumes of high-quality research output annually has become intense in British universities, with consequences for the quantity of published material, if not the quality.[6] Research is a time-consuming activity. It is much more easily carried out by those university staff with relatively small and specialized teaching loads, who do not have to spend most of their time preparing for and delivering courses and assessing student work. Similarly, staff whose teaching is concentrated into only a small part of the wider discipline are much better placed to undertake research than are those who have to occupy a broad canvas and spend much time 'keeping up to date' with a large and diverse literature.

Specialization has thus aided the growth of research activity. This has made the profession attractive to new staff appointed from the 1960s on, who were entering institutions with a very different ethos from that of their predecessors. A research-based culture rapidly came to dominate the universities, and geographers were as keen as others to participate in it (Johnston 1994d) – not only for the inherent intellectual attractions of research and the pleasure that success at doing original work brings, but also because it aided their career progress.

Specialization was almost inevitable, given the general growth not only in the number of practising academic geographers but also in their much increased research productivity. Whereas only a few decades ago additions to the discipline's literature were relatively few each year, and individuals could expect to 'keep up' with the main trends of work throughout the subject, this is not the case now. The geographical literature has expanded exponentially (Stoddart 1986), and individuals can only expect to keep track of a small part of it – especially if they are also interested in keeping pace with developments in other disciplines too. Together, size and activity rates have pushed geographers to become specialists within their discipline, a trend which has been exacerbated by developments within those separate specialisms.

University teachers of geography increasingly became specialized in their teaching and research activities; therefore, instead of being geographers who covered a very broad field, they concentrated their attention on a particular corner of it only. But that specialization led to fragmentation and tensions.

FRAGMENTATION

Specialization in research interests characterizes the staff of all academic disciplines in British and other universities, so geography is not alone in its recent developments. What may be more peculiar to geography, however, is its fragmentation – although even with that there is evidence of a similar situation in other social sciences (Johnston 1991a).

A dominant feature of geography's fragmentation is that the specialization which accompanied the development of research and teaching fields within the discipline has been centrifugal in its orientation. To enhance their research, very many geographers have closely allied their interests with those of other subjects. Again, this is not unique to geographers, but what distinguishes them relative to members of several other disciplines is the lack of any clear centripetal force holding them together, any dominant idea or concept which defines their discipline and ensures continued adherence to it.

For many geographers, adherence to the discipline is inertial rather than intellectual; they remain attached to it and to the structures within which they were socialized as academics and which gives them status, although substantial numbers of one-time geographers have migrated to other disciplines.[7] That attachment is also political. For most, their employment within universities depends on the continued flow of students to separate geography departments. Whatever their research specialisms and links, therefore, academic geographers are constrained to a continued commitment to the discipline which, in effect, supports their livelihood. Although a few (some examples are in Johnston 1985) may promote alternative academic divisions of labour, most pay allegiance to the discipline and its organizational structures. This creates a situation in which geographers have strong external links but often feel the need to enquire about their internal connections, to seek some intellectual *raison d'être* which sustains their continued separate existence.

THE CAUSES OF FRAGMENTATION: EXTERNAL PULLS AND INTERNAL PUSHES

The work of most geographers incorporates (if often only implicitly) one or more of three key concepts – space, place and environment. The discipline's *raison d'être* has been defended in the past because its studies integrated work on all three, focusing on how and why places differ; geography was defined as studying areal differentiation in human occupation of physical environments.

Early in the present century this concern with areal differentiation was structured within the framework of *environmental determinism*, which contended that human responses were conditioned by physical milieux. Such simplistic causal reasoning was soon cast aside to be replaced by a largely atheoretical approach – *regional geography* – in which the goal was to describe areal differentiation in terms of the nature of 'regions', areas of the earth's surface which shared common characteristics and which were different from the others on one or more criteria. Many regional geographies were built on spatial variations in the physical environment and so continued the deterministic mode of thinking, albeit in more muted ways.

Regional geography came increasingly under attack in the 1950s because of its perceived shallow intellectual foundations. Although they recognized that part of their educational task was to synthesize knowledge about different parts of the world, geographers found this approach less and less satisfactory as their research orientation. Science was coming to dominate the universities, and geographers wanted to be part of that programme, hoping to gain respectability by adopting its canons of inquiry to their tasks. Regional geography did not offer a viable framework for doing that, and so they switched their attention to other ways of 'doing geography'.

This response to the perceived failings of their disciplinary orientation was partly conditioned by what geographers saw happening in other disciplines, not all of which were also changing in the immediate post-war decades. It was also a response to wider demands within society, where science and technology were becoming not only increasingly important but also greatly respected. Although it had involved massive loss of life and its outcome was substantially influenced by political considerations and strategic and tactical decisions, the Second World War was the first in which mastery of technology rather than human sacrifice was the dominant influence on the outcome. Science and technology were seen as providing the 'bright new dawn' which would ensure material plenty for all. A perceived intellectually more rigorous social science was also gaining in respectability. Geographers were aware of these trends – both within and beyond their own universities – and many, especially but not only members of the profession's younger generation, were encouraged to follow them. As geographers themselves increasingly came to recognize two or three decades later, context is extremely important as an influence on who studies what, where, and why (see Johnston 1994e).

The segmentation of physical geography

Physical geographers made the first shift, as they realized the poverty of their understanding of how the earth's environment worked. Their sketchy appreciation had been based on general reasoning only, with very little experimental material on which conclusions were based: geomorphologists, for example, observed the earth's surface and the processes operating on it, and from those observations speculated on what had happened in the past to produce the present physical landscapes. This was to be replaced, according to the proponents for a new physical geography, by the adoption of 'scientific methods' and the associated quantitative procedures. The method of 'denudation chronology' was strongly criticized for its weak procedures (Chorley 1965), and geomorphology shifted from studies of regional landforms to investigations of landforming processes (Tinkler 1985). Gregory's (1985) chapters dealing with these two eras in his historical survey of *The Nature of Physical Geography* are entitled 'Chronology Continuing' and 'Processes Prevailing'; in the latter, he approvingly quotes Strahler's (1952) view that geomorphology was 'already a half-century behind developments in chemistry, physics and the biological sciences' (Gregory 1985: 88).

This major shift of emphasis required not only adopting new ways of working (involving the lexicon of models, hypotheses and laws) but also strengthening links to other physical sciences. Understanding how environmental processes operate requires an appreciation of the relevant laws of physics and chemistry: geomorphologists, for example, looked to other disciplines, such as hydrology, where those laws were also being studied. Just as geomorphologists were no longer satisfied with describing landform morphology and then speculating on its origins, so climatologists became disenchanted with generalized descriptions of climates and weather: they turned to studies of the physics of the atmosphere as undertaken by meteorologists. Biogeographers, too, decided to associate their work more closely with that of ecologists, and students of the geography of soils developed closer links with pedologists.

Over the last four decades, then, physical geography has fragmented as its practitioners have developed particular research interests which are linked more closely to those of adjacent disciplines than they are to those of other physical geographers (on which see Stoddart 1987). Their academic literature reflects this fragmentation; UK geomorphologists, for example, see *Earth Surface Processes and Landforms* as their leading journal; biogeographers focus on the *Journal of Biogeography*; and climatologists have their own *Journal of Climatology*. Very little physical geography appeared in the pages of the *Transactions, Institute of British Geographers* in the 1980s as a consequence. Institutionally, too, they have fragmented organizations. Among the IBG study groups, there is a Geomorphological Research Group (a section of the larger British Geomorphological Research Group, whose membership includes many geologists and engineers as well as geographers) and a Biogeography Study Group, which has strong links to the British Ecological Society. There is no climatology group: instead the Association of British Climatologists, whose members include virtually all geographers interested in the study of climate, has remained outside the IBG altogether, and is linked instead to the British Meteorological Society.

Attempts have been made to counter this fragmentation by bringing the study of the various parts of the environment together into holistic analyses of systems (as promoted, for example, by Bennett and Chorley 1978; see also Gregory 1985). The interrelationships among the various components are generally appreciated, but the demands for successful integration – especially if it is to be based on mathematical modelling – are substantial and rarely met. Thus whereas numerous undergraduate texts have promoted an integrated approach, especially for the beginning years of degree programmes, research has tended to lag behind.

A major stimulus to attempts at integrating the various parts of physical geography has come with increased public and scientific realization that the earth's environment is under very substantial strain as a result of the demands made on it by both the increased numbers of humans and their desires for ever-increasing material living standards. Appreciation of what has happened, what is happening, and what could well happen in the near future if these trends continue, has stimulated much work aimed at modelling whole environmental systems so as to simulate the likely outcomes of changes in some of their component parts. Studies of, for example, global warming and its possible impact on changing sea-levels have generated interest not only in the likely coastal changes but also in such issues as the size of the polar ice caps and the potential impact of their alteration on global and local climates, and thence on sea-levels. The call for reintegration of physical geography, with a bringing-together of the separate parts – which will undoubtedly remain separate in the pursuit of detailed research agenda – is being pressed as an urgent social concern.

Segmenting human geography: two-dimensional tensions

Human and physical geography drifted apart as the decline of regional geography removed a major bond between them. Regional geography was increasingly attacked as 'tired description'; physical geographers were no longer called upon to teach general courses which required some appreciation of human geography and its concerns, and so shifted their attention to substantive interests in physical geography alone. Some human geographers, especially those trained pre- and immediately post-Second World War, retained a commitment to regional description, including description of the physical landscape, but their concerns did not extend to the detailed scientific work being conducted by physical geographers; they needed to know about the landscapes, but not about the physics and chemistry involved in their production.

New generations of human geographers also attacked the traditional format of regional geography, and promoted a reorientation of their subdiscipline. Their goal was very similar to that of their physical geography colleagues: they wanted to understand the patterns of human occupance of the earth's surface. Furthermore, many of them believed, with the physical geographers, that those patterns were the outcomes of general laws of human spatial behaviour which could be uncovered through application of what was widely perceived at the time as '*the* scientific method'. Hence there were attempts, as in the joint work by Chorley and Haggett (1969) on networks, to integrate physical and human geography through a shared methodology. This did not last for long, however, and whereas the fragmentation of physical geography has been one-dimensional only – with the subdiscipline divided according to its practitioners' substantive interests and with a largely common adherence to methodological protocols – human geography since the 1960s has been fragmented in two ways, substantively and epistemologically.

The substantive fragmentation of human geography proceeded in similar fashion to that of physical geography. As the opportunities for greater specialization made available by expansion were grasped, so individual geographers concentrated their attention on particular topics only. Much of that concentration involved divisions which were linked to the subject structure of the social sciences, a major area within the academic division of labour with which geographers increasingly became associated: economic geography is concerned with aspects of the subject matter of economics, for example; social geographers share interests with sociologists; political geographers have links with political scientists; cultural geographers overlap with anthropologists; and historical geographers have concerns in common with historians. A number of those sections of geography are further subdivided: within economic geography, for example, it is usual to identify industrial and agricultural geographers separately, whereas students of the geography of service industries tend to be distinguished from the former and transport geographers consider themselves a separate group again. (Interestingly, the links between economic geographers and economists are probably weaker than those involving geographers with sociology and political science, perhaps because economics is more mathematically rigorous; certainly few geographers publish in economics journals, relative to those – still small in number – who publish in sociological and political science outlets.)

In each of these subfields (and also in some more specialized parts such as medical geography and population geography) geographers have been concerned to identify and promulgate a particular focus, thereby earning themselves niches in the academic division of labour. Space and place were commonly called upon as defining characteristics of the discipline in the 1960s and 1970s: geographers study those distributional aspects of economic, social, political, etc. phenomena which are generally given low priority by members of the 'parent disciplines'. Thus geography was presented as the social science discipline which focuses on 'where' questions, leading some to adopt the alternative descriptive term 'spatial science' for their work (see, for example, Agnew 1987).

As with the trends in physical geography, publishing activities in human geography became fragmented (Johnston 1994c). Each subgroup developed its own journal(s), some of them only relatively recently: *Economic Geography* (founded in 1927), *Political Geography*, *Journal of Historical Geography*, and *Ecumene: Environment, Place and Culture* characterize this division, as do even more specialist journals such as the *Journal of Transport Geography*. Interestingly, however, human geographers continued to publish some of their major papers in the general journals of their learned societies, such as the *Transactions, Institute of British Geographers*, but physical geographers very largely abandoned use of such outlets.

The division of human geography according to substantive interest has been paralleled in part by a spatial division, despite the decline of interest in regional geography defined as the study of specified (usually continental scale) areas of the earth's surface. Most prominent of these divisions has been urban geography, which was one of the largest in the 1960s and 1970s – as illustrated by membership of the IBG Urban Geography Study Group. Its complement, rural geography, has attracted many fewer adherents, perhaps not surprisingly given the dominance of cities in the contemporary world; even so, rural geographers are usually distinguished from agricultural geographers.

One difference between human and physical geographers is that the division within the former has not been as clear-cut as that among the latter, and some integration of the work of the various human geography branches has been attempted. Three features have probably contributed to this. First, there has been a continued realization that the various parts of human geography are not readily separated into academically impervious containers: this argument has been most strongly made by scholars such as David Harvey (e.g. Harvey 1982, 1989) and Doreen Massey (e.g. Massey 1984) who, working on Marxian foundations, have stressed the interrelationships of economic, political, social and cultural processes with special reference to their uneven spatial development. Second, at least until the 1970s there was a common quantitative methodology employed by many of the adherents to most of the individual branches. Third, quite a lot of the work undertaken had strong applied implications, and geographers of various persuasions were brought together by a commitment to such applications, especially in the burgeoning fields of town and country and of regional planning.

Integration of the various subfields of human geography was linked in the late 1950s and the 1960s with the work of other quantitatively inclined social scientists interested in spatial aspects of economic, social and political phenomena. The work of the (US-based) Regional Science Association, especially the proceedings of its conferences (*Proceedings, Regional Science Association*) and its *Journal of Regional Science*, were major sources of cross-disciplinary ideas for adherents to the 'new human geography' then. That stimulus rapidly declined in its intensity, however, and attention turned to broader constellations of interests, as reflected in the British Regional Studies Association and its journal *Regional Studies*, and in the inter-disciplinary journal *Environment and Planning* (later *Environment and Planning A*; see Johnston and Thrift 1993).

An epistemological fragmentation of human geography has been stimulated by debates throughout several of the social sciences (economics was the main exception) concerning approaches to the study of human behaviour and its artefacts. The so-called 'new human geography' of the 1960s and 1970s was based on similar conceptions of the nature of a scientific discipline to those adopted in physical geography. The goal of a science was to develop laws, through explicit procedures that were replicable and which thus established the generality of the conclusions; those procedures were invariably quantitative in their nature. Once laws were established, they could form the basis for engineering, or applied science. Thus within social science human geographers were seeking the laws of spatial behaviour and spatial arrangements which could be applied in planning for better, more efficient, societies.

As suggested above, this was a period of shared methodological and technical commitments among human and physical geographers, despite the differences in subject matter and the substantive sources of inspiration in the social and physical sciences respectively. Both moved from maps to statistics as their basic tools, and thence (somewhat more so in physical than in human geography) to mathematics. (Both human and physical geographers publish in the journal launched in 1969 to represent this approach – *Geographical Analysis*.) As a consequence, the links between cartography and geography became weaker; the former became an almost entirely separate discipline (but lacking formal departmental status in universities), though some geographers retain an active interest in the processes of map construction and interpretation.

Maps were always means and not ends for the majority of geographers, and so the need for a separate discipline dealing with the technical aspects of map production was not surprising. Nor, perhaps, was it surprising that the development of two new technical areas with strong academic bases within geography also became semi-independent disciplines. The first of these was the field of remote sensing, involving the production, processing and interpretation of imagery from remote platforms – notably earth-orbiting satellites. This imagery provides vast amounts of valuable information for many aspects of work in physical geography and for some fields within human geography, but much effort since the 1970s has focused on technical aspects of data capture and use. For this, geographers who have adopted remote sensing as a specialism have combined with physicists, applied mathematicians and other scientists in separate learned societies with their own journals, such as the Remote Sensing Society and the *International Journal of Remote Sensing* (Chapter 21).

Just as with remote sensing from the 1970s on, so the development of Geographical Information Systems (GIS) – integrated computer hardware systems and software packages for the storage, synthesis, display and analysis of spatially referenced data – has stimulated a specialism practised by some geographers in closer collaboration with other disciplines than with other members of their own. Again, there are separate learned societies and publications, and although GIS is presented as an important technical tool for much work in both human and physical geography (see Fotheringham 1993) it is but weakly integrated into the discipline as a whole (a problem exacerbated by the claims of some GIS proponents; see Taylor and Johnston 1995). Some of the important links are with applied disciplines such as planning and surveying.

By the 1970s quantitatively oriented spatial science, with its emphasis on the search for laws of spatial patterns, was under attack for its implicit model of human decision-making. Spatial science assumed that all decision-making was driven by a single set of goals encapsulated by the concepts of profit maximization for entrepreneurs and cost minimization for their customers; in both cases, since crossing space involves costs, such behaviour involves distance minimization. Some geographers, including Pred (1967), pointed to the impossibility of such decision-making in practice, and suggested somewhat weaker fundamental assumptions. Others attacked even those, however, arguing that reducing decision-making to a small number of materialist criteria very badly misrepresented the nature of humans; treating them virtually as programmed respondents to material stimuli denied them free will and denigrated the nature of humanness.

This critique of spatial science was accompanied by a series of explorations of alternative approaches to study within the social sciences, which are frequently grouped together under the rubric of humanistic geography. These, such as idealism and pragmatism, phenomenology and existentialism, share a common belief in the importance of the individual as decision-maker (Jackson and Smith 1984). Thus to appreciate what people do requires understanding what they believe, and why; to appreciate the worlds that they create involves understanding their goals and aspirations, and how these are inscribed in landscapes. The entire task was hermeneutic, therefore: geographers had to appreciate the thought behind action, and then had to transmit that appreciation to their audiences – those reading the results of geographical research.

A second major set of critiques was also launched against the spatial science orientation of much human geography in the 1970s. It too saw the implicit determinism as unrealistic, but more importantly it criticized the absence of any deep appreciation of the wider social, economic and political context within which spatial behaviour occurred and built environments were created. It was also critical of the voluntarism characteristic of work in humanistic geography, which (at least implicitly) presented individuals as free agents. Inspired in particular by the renewed interest in Marxian theory (Harvey 1973), this third group of workers emphasized the constraints of the political economy within which people acted. The requirements of the capitalist mode of production mean that spatially uneven development is a necessary outcome, for example (Harvey 1982; Smith 1984). Furthermore, economy, society and polity cannot be separated analytically: each is interdependent with the others and only a holistic approach (albeit with one part as the focus of the particular investigation) can lead to full appreciation.

This third set of approaches has also involved a great deal of experimentation with detailed methods of investigation, and has attracted nomenclature such as the political economy approach, the realist approach, the structuralist approach, and also the radical as well as the Marxian (Peet and Thrift 1989). With time, there has been something of a *rapprochement* between much of this work and that in the humanistic tradition, recognizing that the operations of a mode of production such as capitalism are both constraining and enabling – constraining because they restrict what can be done but also enabling because they provide a wide range of optional outcomes; decision-making is constrained but not determined by the context.

Development of this 'humanistic political economy' approach within human geography has advanced considerably within recent years with the realization that people are socialized into the beliefs that underpin their actions in places (see Kobayashi and Mackenzie 1989). Society is not homogeneous; it is spatially differentiated in a large variety of ways. The complex, ever-changing cultural mosaic provides a vast range of stores of knowledge on which people draw, and to which they contribute every time that they act. This has provided the context for a 'new regional geography' which emphasizes the distinctiveness of places within global society, and which is informed by arguments from postmodernism that general theory is impossible – if only because we have no generally agreed language.

Central to this *rapprochement*, therefore, has been a realization that people act in society from a particular 'position'. That realization has been very substantially promoted by the arguments of feminist geographers, who have provided convincing evidence regarding the male bias in the structuring of the discipline (Rose 1993). Women, including women academic geographers, occupy particular positions within society, which are reflected in their contributions to the discipline. Similarly, much of the interpretation of the world and its many cultures reflects the hegemony of Western imperialism, an argument strongly put by Said (1978) whose work on orientalism has stimulated geographers to appreciate the ethnocentrism inherent in much of their writing. Much of this work has thus been linked to developments in several humanities disciplines, as well as other social sciences, in what has been termed the 'cultural turn'; as Gregory (1994: x) has expressed it, 'I have started to understand my own situatedness and to think about its implications in a discipline that has had as one of its central concerns an understanding of other people and other places'.

Just as the substantive division of human geography has produced separate communities and literatures so too has the epistemological division. For example, the orientation of the journal *Antipode*, which was launched in 1969, is indicated by its subtitle, *A Radical Journal of Geography*. Just over a decade later, publication of *Society and Space: Environment and Planning D* provided an outlet for those aligned to the 'cultural turn' and to the 'humanistic political economy' approaches, whereas 1994 saw *Gender, Place and Culture* added to the list. These journals – and associated more general social science journals such as the *International Journal of Urban and Regional Research* – cover material which others might separately identify as economic geography, social geography, political geography, etc.: their coverage reflects common epistemological concerns rather than common subject matter.

Human geography today is therefore characterized by a series of cross-cutting cleavages, both substantive and epistemological. Similar topics can be approached in very different ways – compare, for example, Sheppard and Barnes (1990) and Sayer and Walker (1992) on the changing spatial division of labour, or Dicken (1992), Knox and Agnew (1989), Peet (1987) and Wallace (1990) on the world economy. The various epistemological strands to the subdiscipline all remain strong, with very substantial output of research findings and vigorous debate about approaches to the discipline. There is little prospect of a coming-together to produce a singular view of what the discipline is and how its work is done. Although some new ideas – such as many of those regarding uneven development introduced by Harvey (1982) and other Marxian scholars – may be generally incorporated into much of the discipline, continued search for the new opens up further debates, as with those currently raging regarding post-modernism (Harvey 1989).

These cross-cutting cleavages are reflected in several aspects of the academic division of labour, as well as by the separate journals which serve the different points of view. The IBG's study groups reflect them in part, too. Most of the groups for human geographers reflect the substantive cleavage – with separate groups for historical, industrial activity, medical, political, population, rural, transport and urban geographers. The Quantitative Methods Study Group clearly promotes one epistemological approach, however, whereas, less obviously from its title, the Social and Cultural Geography Study Group has largely embraced the 'cultural turn' approach.

This presentation of a divided discipline, or a 'discipline without a core' as I have termed it elsewhere (Johnston 1991a), should not be taken to imply a totally incoherent academic structure to most university Departments of Geography. Their members have many things which draw them together – not least a realization of their political strength in difficult economic times if they present a united front to the 'outside world' – and most produce degree curricula which seek to present the diversity of modern geography within a coherent framework, even if the syllabuses of most individual course modules, especially those in the later years of the degree scheme, focus on one of the cross-cutting cleavages only. But there is a great diversity within that apparent unity.

PUTTING IT ALL TOGETHER AGAIN?

To some extent, the description of the contemporary academic discipline of geography presented here could be likened to that of Humpty Dumpty after he fell from the wall: it comprises a great number of parts from a whole, but can those parts be reunited? Some argue that it can, usually around one or both of two main themes.

The first themes can be characterized as *environmental management*. The last few decades have seen enhanced realization that the earth's environment is being placed under very substantial threat by human use and abuse of many of its parts. There is considerable fear that human life cannot be sustained on earth for much longer, certainly not at current material levels, if existing rates of use and abuse continue. This leads to calls for new strategies of sustainable development, which will reinstate harmony between society and nature and ensure the long-term survival of human societies (Johnston *et al.* 1995).

A number of geographers have advanced the cause of their discipline as a leading player in the search for understanding the interrelationship between society and nature, because of its roots in both the natural and the social sciences. The role of the discipline as an 'integrating science' was promoted in the 1940s and 1950s, but largely ignored and discarded thereafter because of the emptiness of those claims at all but the most superficial level and the increased interest of most geographers in their substantive research specialisms. Arguments for such an integration focus were revisited in the 1980s (Johnston 1983) and there was increased recognition that, despite the incompatibility of some of the research perspectives of human and physical geographers (Johnston 1989), there was much to be gained from work in a discipline which introduced students both to how nature works and to how people organize their societies in the contexts of their understanding of nature.

Work on environmental issues has thus engendered some *rapprochement* both among physical geographers and between physical and human geographers, therefore, and has proved popular with students. A variety of perspectives is on offer (compare, for example, Rees 1985; Goudie 1986; Fernie and Pitkethly 1985; Blaikie 1985; Adams 1990; Simmons 1989, 1991; and Pepper 1984) and an Environmental Study Group of the IBG has been created to 'bring the two sides together'.

The other bringing-together strand is the promotion of *applied geography*. Geographers have often bemoaned their lack of influence on policy-making and practice (see Johnston 1991b), and there have been many calls for them to make their discipline relevant to their societies' needs.[8] Taylor (1985) argued that such calls are usually more strident from within the discipline, reflecting pressures from outside, in times of economic recession than in periods of relative plenty. Certainly the depressed state of the British economy throughout the 1980s produced much government pressure for academics to orient their work towards 'the nation's needs', and this was matched by calls from geographers for the discipline to become more involved in, and focused on, the needs of the 'enterprise society' (see, for example, several of the essays in Macmillan 1989, and the critique in Johnston 1993a).

Continued pressures within the UK university system and its funding methods are pressing all departments to restructure their approach to research, with an emphasis on 'team research' oriented towards a nationally defined goal of wealth creation. General funding for research within universities is being increasingly focused on a small number of departments whose work is rated as excellent in the (now quadrennial) Research Assessment Exercises. This will have substantial impacts on research-focused careers for academics, and for the type of work they are able to undertake; the practices of geography and geographers are likely to change very substantially in the next few years as a consequence (Johnston 1995b).

811

SO WHY?

This chapter has illustrated and examined the fragmentation of geography in recent decades. The facts of that fragmentation, plus the additional cross-cutting cleavages in human if not in physical geography, have been outlined and exemplified with reference very largely to developments in the United Kingdom. Very similar trends can be seen in North America, and although it would be unwise to generalize what I have described as a global phenomenon, it almost certainly is.

This fragmentation has occurred concurrently with the growth of geography as a discipline within universities, and growth has been presented here as a necessary though not sufficient condition for the onset of fragmentation. Without the rapid growth in the 1960s and 1970s, the number of academic geographers would have been insufficient to allow the development of so many specialisms, and we would almost certainly have been denied the current vibrant, if somewhat incoherent, contemporary discipline.

Specialism within a discipline has certainly not been peculiar to geography over recent decades; as I have discussed in more detail elsewhere (Johnston 1991a, ch. 1), it is certainly characteristic of both sociology and political science, and casual observation suggests that it has been typical of most other disciplines too (and certainly so in the social sciences and humanities). Expansion of all university subjects has almost certainly fuelled this increased specialization: as the literature becomes larger and more diffuse and as the methodologies employed become more sophisticated, so the vast majority of individual scholars have specialized more and more in certain parts of their discipline only, in order to maintain full appreciation of what is being done and to contribute original work to it. The beginning of the present century saw the end of the polymath; its conclusion is witnessing the end of the 'all-round scholar' within an individual discipline.

Geography did not differ in the growth of specialization, therefore, but not all other disciplines also became as fragmented. As described here, most of the specialisms within geography have strong centrifugal components; they are more closely linked with allied disciplines than they are with other geographic specialisms. This may be because geography cannot be a 'coherent discipline', as the term is generally understood, but merely a collection of specialisms with very loose links in the study of variations over space: a number of analysts suggest that geographical variations are contingent outcomes of general processes, so that space cannot be theorized (see, for example, Sayer 1992, 1994; Johnston 1993b). Alternatively, it may be that geographers themselves have failed their discipline by not sustaining and enhancing the centripetal forces which formerly held them together through adherence to a common core project.

Whatever the reason for the fragmentation, there can be little doubt about the strength of the centrifugal tendencies – although they are stronger in some parts of the discipline than others. Within human geography, for example, many of the changes in approach during the last four decades have been stimulated by contact with external literature rather than by 'revolutions' within the discipline itself. Peter Haggett's (1965) pathbreaking introduction to spatial science, *Locational Analysis in Human Geography*, was very largely based on successful explorations of literature presenting geographers with the spatial aspects of other disciplines to illustrate his original synthesis of spatial structures, for example, and many of the critics of spatial science drew on other disciplines for their proposed philosophically-based alternatives. In some senses, therefore, geography has been a 'magpie discipline': its practitioners have referenced other literatures very much more than scholars in other disciplines have referenced the works of geographers.

This is not to imply that there has been no important work within geography, which has also been influential beyond the discipline's boundaries. As spatial science developed, so the technical issues of analysing spatial data were realized, and geographers such as Andy Cliff made important contributions to the emerging subfield of spatial statistics (Cliff and Ord 1981); this type of contribution has been continued with the work on remote sensing and on GIS. The 'political economy' approach, too, has seen geographers making seminal contributions to the application of Marxian ideas for the understanding of the contemporary world – not least in Harvey's important corpus of books (for example, Harvey 1973, 1982, 1989). But analysis of the intellectual imports and exports suggests a balance of trade firmly against geography within the social sciences.[9]

IN SUMMARY

The trajectory of geography as an academic discipline since the Second World War has much in common with that of other disciplines, therefore, but it has some features which appear to be more peculiar to it alone. Specialization, in research and associated teaching, is typical of all areas of scholarly endeavour, and geography, like other subjects, comprises groups of linked workers with some common ground but little grand coherence. Fragmentation is not so typical, however: most geographers have developed research and teaching specialisms which link them much more closely to parts of other disciplines, from which they gain much of their stimulus, than to other members of their own discipline.

It is difficult to get a synoptic view of geography as a discipline, therefore. It is held together by attachment to a largely defunct core which nevertheless provides it with a political strength within many university systems – assisted in some countries by a strong representation in school curricula – and by a weak commitment to both an interest in spatial distributions and the integration of studies of the natural environment with those of human societies occupying them. But it is more outward-looking than many other disciplines, which may be both its strength and its weakness.

NOTES

1 Training of teachers did not occur in universities in all countries: in the USA, for example, much of it took place in what were called Normal Schools – comparable to the teacher-training colleges in the UK which did not offer degree qualifications and were largely used for training those who would not be qualified to teach at grammar schools, where almost all of the potential university students were pupils.

2 The UK differs from the USA in this regard, where geography is very poorly represented in the school curricula in most states. American university departments thus have to compete for undergraduate students on the basis of the attractiveness of their course offerings (US undergraduate degree programmes are much less highly structured than those in UK universities, and students have a much greater choice of modules).

3 The polytechnics – several of which had geography departments – were allowed to expand somewhat because they were perceived to be more 'efficient' (i.e. they cost less per student per year), in part because they were not funded for research as well as teaching (in the universities about one-third of the funds are provided for research rather than teaching, which means that staff there have fewer teaching hours per week), and in part because they were perceived as being more attuned to the demands of a modern enterprise society.

4 The study groups are identified on p. 803. The Association of American Geographers has a similar, though more recently initiated, structure.

5 The strong research–teaching ethos, and the wide belief that good teaching can only be provided in universities if it is 'research-driven' by individuals who are all also active researchers at the frontiers of their discipline, is relatively new to the universities; in Eastern European countries the two tasks were separated (teaching in the universities, research in the academies) during the Soviet period.

6 So far, the UK is the only country to have this regular external evaluation of all research work – and now of university teaching too (Johnston 1994a, 1994b); for the Australian procedures see Jones (1994). The 1996 Research Assessment Exercises will not collect data on the quantity of publications, thus focusing attention very much on quality; it will also have more grade points – grade 3 is to be divided into 3a and 3b, and a new grade 5* will be added above grade 5.

7 Probably more geographers have left their discipline and become members of another than has occurred in reverse, although there is no evidence to sustain this assertion.

8 In the former USSR and in Eastern Europe between 1945 and 1989, state control over academic life ensured that all work was applied work.

9 The same is probably true of physical geography too, but I am less well placed to make that claim.

REFERENCES

Adams, W.M. (1990) *Green Development: Environment and Sustainability in the Third World*, Routledge: London.

Agnew, J. (1987) *Place and Politics: The Geographical Mediation of State and Society*, Allen & Unwin: Boston.

Balchin, W.G.V. (1993) *The Geographical Association: The First Hundred Years 1893–1993*, Sheffield: The Geographical Association.

Bennett, R.J. and Chorley, R.J. (1978) *Environmental Systems: Philosophy, Analysis and Control*, London: Methuen.

Blaikie, P.M. (1985) *The Political Economy of Soil Erosion*, London: Longman.

Brown, E.H. (ed.) (1980) *Geography, Yesterday and Tomorrow*, Oxford: Oxford University Press.

Chorley, R.J. (1965) 'The application of quantitative methods to geomorphology', pp. 147–63 in R.J. Chorley and P. Haggett (eds) *Frontiers in Geographical Teaching*, Methuen: London.

Chorley, R.J. and Haggett, P. (1969) *Network Analysis in Geography*, London: Edward Arnold.

Cliff, A.D. and Ord, J.K. (1981) *Spatial Processes*, London: Pion.

Dicken, P. (1992) *Global Shift*, London: Paul Chapman.

Fernie, J. and Pitkethly, A.S. (1985) *Resources, Environment and Policy*, London: Harper & Row.

Fotheringham, A.S. (1993) 'On the future of spatial analysis: the role of GIS', *Environment and Planning A*, Anniversary Issue, 30–4.

Goudie, A.S. (1986) *The Human Impact on the Natural Environment*, Oxford: Blackwell.

Gregory, D. (1994) *Geographical Imaginations*, Oxford: Blackwell.

Gregory, K.J. (1985) *The Nature of Physical Geography*, London: Edward Arnold.

Haggett, P. (1965) *Locational Analysis in Human Geography*, London: Edward Arnold.

Harvey, D. (1973) *Social Justice and the City*, London: Edward Arnold.

Harvey, D. (1982) *The Limits to Capital*, Oxford: Blackwell.

Harvey, D. (1989) *The Condition of Postmodernity*, Oxford: Blackwell.

Jackson, P. and Smith, S.J. (1984) *Exploring Social Geography*, London: George Allen & Unwin.

Jenkins, A. and Smith, P. (1993) 'Expansion, efficiency and teaching quality: the changing experience of British geography departments 1986–91', *Transactions, Institute of British Geographers* NS18, 500–15.

Johnston, R.J. (1983) 'Resource analysis, resource management, and the integration of physical and human geography', *Progress in Physical Geography* 7, 127–46.

Johnston, R.J. (ed.) (1985) *The Future of Geography*, London: Methuen.

Johnston, R.J. (1989) *Environmental Problems: Nature, Economy and State*, London: Belhaven Press.

Johnston, R.J. (1991a) *A Question of Place: Exploring the Practice of Human Geography*, Oxford: Blackwell.

Johnston, R.J. (1991b) *Geography and Geographers: Anglo-American Human Geography since 1945* (4th edn), London: Edward Arnold.

Johnston, R.J. (1993a) 'Meet the challenge, make the change', pp. 151–80 in R.J. Johnston (ed.) *The Challenge for Geography – A Changing World: A Changing Discipline*, Oxford: Blackwell.

Johnston, R.J. (1993b) '"Real" political geography: some comments engendered by a review of Andrew Sayer's *Method in Social Science* (1992)', *Political Geography* 12, 473–80.

Johnston, R.J. (1994a) 'The "quality industry" in British higher education and the Association's publications', *The Professional Geographer* 46, 491–7.

Johnston, R.J. (1994b) 'Quality assessment of teaching: inputs, processes and outputs', *Journal of Geography in Higher Education* 18, 184–93.

Johnston, R.J. (1994c) 'Geographical journals for political scientists', *Political Studies* 42, 310–17.

Johnston, R.J. (1994d) 'Department size, institutional culture and research grade', *Area* 26, 327–42.

Johnston, R.J. (1994e) 'On spatial analysis, place and realism', *Urban Geography* 15, 290–5.

Johnston, R.J. (1995a) 'The business of British geography', pp. 317–41 in A.D. Cliff,

P.R. Gould, A.G. Hoare and N.J. Thrift (eds) *Diffusing Geography*, Oxford: Blackwell.

Johnston, R.J. (1995b) 'Geographical research, geography and geographers in the changing British University system', *Progress in Human Geography* 19, 355–71.

Johnston, R.J. and Gregory, S. (1984) 'The United Kingdom', pp. 106–31 in R.J. Johnston and P. Claval (eds) *Geography since the Second World War: An International Survey*, London: Croom Helm.

Johnston, R.J., Taylor, P.J. and Watts, M.J. (eds) (1995) *Geographies of Global Change*, Oxford: Blackwell.

Johnston, R.J. and Thrift, N.J. (1993) 'Ringing the changes: the intellectual history of *Environment and Planning A*', *Environment and Planning A*, Anniversary Issue, 14–21.

Jones, R. (1994) 'The underside of quality: an Australian viewpoint', *Journal of Geography in Higher Education* 18, 373–8.

Knox, P.L. and Agnew, J.A. (1989) *The Geography of the World Economy*, London: Edward Arnold.

Kobayashi, A. and Mackenzie, S. (eds) (1989) *Remaking Human Geography*, Boston: Unwin Hyman.

Lee, R. (1985) 'Where have all the geographers gone?', *Geography* 70, 45–59.

Livingstone, D.N. (1995) 'The spaces of knowledge: contributions towards a historical geography of science', *Environment and Planning D: Society and Space* 13, 5–34.

Macmillan, B. (ed.) (1989) *Remodelling Geography*, Oxford: Blackwell.

Massey, D. (1984) *Spatial Divisions of Labour: Social Structures and the Geography of Production*, London: Macmillan.

Peet, J.R. (1987) *Global Capitalism: Theories of Social Development*, London: Routledge.

Peet, J.R. and Thrift, N.J. (1989) 'Political economy and human geography', pp. 3–21 in J.R. Peet and N.J. Thrift (eds) *New Models in Geography*, vol. 1, London: Unwin Hyman.

Pepper, D. (1984) *The Roots of Modern Environmentalism*, London: Croom Helm.

Pred, A.R. (1967) *Behavior and Location: Foundations for a Geographic and Dynamic Location Theory Part I*, Lund: C.W.K. Gleerup.

Rees, J.A. (1985) *Natural Resources: Economics, Allocation and Policy*, London: Methuen.

Rose, G. (1993) *Feminism and Geography*, Cambridge: Polity Press.

Said, E. (1978) *Orientalism*, New York: Pantheon Books.

Sayer, A. (1992) *Method in Social Science: A Realist Approach* (2nd edn), London: Routledge.

Sayer, A. (1994) 'Comment: realism and space: a reply to Ron Johnston', *Political Geography* 13, 107–10.

Sayer, A. and Walker, R.A. (1992) *The New Social Economy: Reworking the Division of Labour*, Oxford: Blackwell.

Sheppard, E.S. and Barnes, T.J. (1990) *The Capitalist Space Economy*, Boston: Unwin Hyman.

Simmons, I.G. (1989) *Changing the Face of the Earth: Culture, Environment, History*, Oxford: Blackwell.

Simmons, I.G. (1991) *Earth, Air and Water: Resources and Environment in the late 20th Century*, London: Edward Arnold.

Smith, N. (1984) *Uneven Development: Nature, Capital and the Production of Space*, Oxford: Blackwell.

Steel, R.W. (1983) *The Institute of British Geographers: The First Fifty Years*, London: Institute of British Geographers.

Stoddart, D.R. (1986) *On Geography, and its History*, Oxford: Blackwell.

Stoddart, D.R. (1987) 'To claim the high ground: geography for the end of the century', *Transactions, Institute of British Geographers* NS12, 327–36.

Strahler, A.N. (1952) 'Dynamic basis of geomorphology', *Bulletin Geological Society of America* 63, 923–37.

Taylor, P.J. (1985) 'The value of a geographical perspective', pp. 92–110 in R.J. Johnston (ed.) *The Future of Geography*, London: Methuen.

Taylor, P.J. and Johnston, R.J. (1995) 'GIS and geography', pp. 51–67 in J. Pickles (ed.) *Ground Truth*, New York: Guilford Publications.

Thorne, C.R. (ed.) (1993) 'Arena symposium: University Funding Council Research Selectivity Exercise, 1992: implications for higher education in geography', *Journal of Geography in Higher Education* 17, 167–99.

Tinkler, K.J. (1985) *A Short History of Geomorphology*, London: Croom Helm.

Wallace, I. (1990) *The Global Economic System*, Boston: Unwin Hyman.

37

ACHIEVEMENTS OF SPATIAL SCIENCE

Arild Holt-Jensen

THE REGIONAL APPROACH AND ITS LIMITATIONS

Schmithüsen (1976: 22) describes the methodological peculiarity of geography, with which we try to understand the complex features of particular unities like landscapes and regions, as 'total-special reasoning'. A geographer seeks to describe, understand and explain the complexities that exist together and interact in a part of the earth, creating its particular regional character. This is the basis of geography as an independent science. No other discipline concerns itself with the earth's surface and its special parts in their totality, with the connections between different features within the same area. The central geographical question is 'Why is it like this here?'

The study of phenomena that are found to be interconnected in a place or region, termed the 'chorological approach', established regional geography as the central task for geographic research. The work of Paul Vidal de la Blache in France, Alfred Hettner in Germany, H.J. Mackinder and A.J. Herbertson in the UK and Carl Sauer and Richard Hartshorne in the USA established the central importance of regional geography in many universities and maintained it until the 1960s. Many MA and PhD students were expected to present a regional monograph of a tiny part of the world or to present and explain particular geographic features of an area in their work.

In my student days Hartshorne (1939, 1959) and Hettner (1927) provided the philosophical underpinning of studies in geography. They both made a distinction between systematic geography, which seeks to formulate empirical generalizations or laws, and the study of the unique in regional geography, whereby generalizations are tested so that subsequent theories may be improved. Regional geography was regarded as the core of the discipline or 'the highest form of the geographer's art'.

This view seems related to neo-Kantian arguments for idiographic rather than nomothetic approaches, which justify the scientific character of studies of the individual case. 'The regional geographer sought a causal, yet non-lawful, understanding of the relationships of humans to their environment' (Entrikin 1991: 146). Harvey (1969) argued that the concept of geography as a *chorological science* is built on the assumption of *absolute space*. The concept of space as absolute means that space is independent of empirical objects. Space is something in itself, it exists independent of its content. Thus we delimit a spatial section of the earth first, say the Manchester region, and then start to examine its content. The space is in this way treated as a container.

The regional school of geography offered a methodology that seemed to avoid environmental determinism in so far as it concentrated on concrete relations between people and their environment in particular places, rather than looking for general laws governing human adaptation to the environment. The methodology of the French school of regional geography was widely regarded as an exemplar, i.e. a model approach for successive scientists (Wrigley 1965). Vidal de la Blache and other leading members of this school formulated an inductive and historical method by which distinctive regional characteristics were shown as having developed through a long history of interaction between man and nature. *Vertical connections*, man's dependence upon local natural resources, were the basis for this type of regional study.

But already by the end of the nineteenth century the traditional, self-sufficient economy had given way to an international market economy in Europe, and the value of this type of regional study was reduced. *Horizontal connections*, state and international policies, market forces, the interplay between regions, cities and countries, became more important for local development than the local connections between man and land.

These changes led to greater interest in horizontal, spatial structures. Hettner (1927) and Hartshorne (1939, 1959) both regarded the regional synthesis as central to geography, but discouraged historical methods of analysis, basing themselves on Kant's view of geography as a contemporary, chorological science. However, they failed to establish a convincing model for such a non-historic study of regions to replace Vidal's historical model.

Some progress between the wars was, however, made within landscape geography, particularly in the field of *landscape morphology*. Landscape morphology seeks to describe and explain the form and spatial structures created by the visible phenomena on the surface of the earth. The landscape geographer sees the complex features formed by mountains, rivers, pastures, forests, roads, canals, gardens, fields, villages and towns as a unity, investigating it on a map, an aerial photo or from an outlook point. This unity forms a 'whole', a synthesis of different features contained within the absolute space being studied. But as landscape geography developed, increasing specialization led many researchers to concentrate on the study of particular morphological features in the landscape, such as village forms, land use, road networks, land forms or vegetational patterns.

Topical specialization, doubt as to the value of the established regional methods and calls for a more 'scientific' geography were emerging in the 1950s. My own master's thesis (Holt-Jensen 1963) was at the outset intended as a regional monograph of the Norwegian mountain farm municipality Rauland. However, at an international student geography seminar in Saarbrücken in 1962, a British student suggested that I should begin by considering whether the general hypothesis that mountain farm regions would suffer industrial contraction and depopulation would hold in my region. I decided to try this out by adding historical data to the extensive contemporary material I had already collected and by using elementary hypothetic-deductive reasoning. The hypothesis was proved wrong in this case: population growth, not depopulation, had occurred. So I had to use my contemporary data to explain why. I was lucky enough to get my data punched on punch cards and tables calculated on the mechanical data machines that were available at that time. A new age was coming: a fellow MA student had written a rather quantitative thesis in transport geography, which our very able but humanistic-oriented professor had difficulty in assessing.

THE GROWTH OF SPATIAL SCIENCE

During the 1950s and 1960s, geographers and their employers became increasingly dissatisfied with a geography focused upon the study of distinct sections of absolute space, partly because horizontal connections became more important in everyday life than the vertical man–land connections. In addition even Hartshorne (1939: 281) had stated 'that it is not possible to define sections of the earth surface as regions that form units in reality, that we cannot correctly consider them as concrete individual objects'.

The claim that the regional synthesis constituted geography's essential identity as well 'gave the subject a dilettantish image among the practitioners of ever more specializing sciences. As early as 1948, James Conant, president of Harvard University, had reportedly come to the view that "geography is not a university subject"' (Livingstone 1992: 311). The Department of Geography at Harvard was closed down soon after. In order to survive, many geographers saw their professional future in specialization in some systematic field as suggested by Ackerman (1945).

This was partly due to the fact that in the USA and many other countries (apart from the UK) geography enjoyed only a minor role and status in the school system. For this reason geographers were concerned to find new career outlets for their graduates in applied research. Guelke (1978: 45) points out that after the Second World War the North American 'universities were expected to produce problem-solvers or social technologists in order to run increasingly complex economies, and the geographers were not slow in adopting' theory

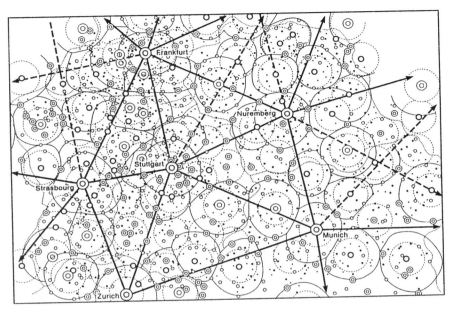

Figure 37.1 The geometrical hexagonal landscape of towns in southern Germany from Walter Christaller's classic study of central places made in the 1930s.

building and modelling methods which might promote the status of their science in these fields. Also, inspiration for a new approach was gained from earlier theoretical works that so far had been almost overlooked.

A theme in landscape morphology – the morphological network of central places in Southern Germany, as seen on the topographic map – was the starting point for Walter Christaller (1893–1969) when he developed the now famous *central place theory* (1933). As Christaller (1968: 95) explained later, he started to 'play with the maps' connecting towns of the same size with straight lines until his maps were filled with triangles (Figure 37.1). There seemed to be some regularities here. If the region had been a flat plain with uniform rural population densities, it would seem that the morphological features could be idealized in a hexagonal, hierarchical structure of urban places (Figure 37.2). Christaller, who had studied economics, used economic theory to explain this morphological structure.

About twenty years later Christaller's studies were taken up, notably in North America and Sweden, where it was realized that his theories could be applied to the planning of new central places and service establishments (exemplified in Figure 37.3), as well as to the delimitation of administrative units.

The acceleration of theoretical work was especially marked in institutions led

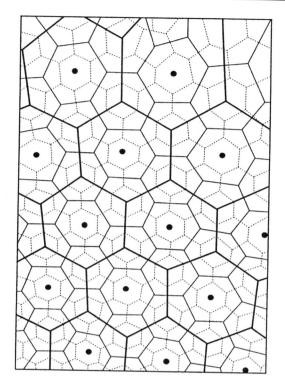

Figure 37.2 The idealized Christaller model on a flat plain with uniform population densities. When population densities are uneven, the lattice of central places adjusts to the changes, closing up in densely settled areas and opening out in sparsely settled areas.

by geographers who had studied the natural sciences, especially physics and statistics, and/or where there were good contacts with developments in theoretical economic literature. The frontier between economics and geography became very productive in new ideas and techniques during the 1950s at several North American universities, notably the geography departments at Washington, Wisconsin and Iowa. In addition the 'outsiders' J.Q. Stewart and W. Warntz developed a 'social physics' school, drawing their inspiration from physics rather than economics and introducing, for example, gravity models to measure spatial interactions.

The work of Christaller, August Lösch and others was introduced into Sweden by Edgar Kant, an Estonian geographer who had tested their theories in his homeland before taking refuge in Lund after the Second World War. His research assistant from 1945–6 was the brilliant young Swedish geographer Torsten Hägerstrand. Inspired by Swedish ethnological studies of innovation and tradition in rural areas, Hägerstrand became interested in the possibilities

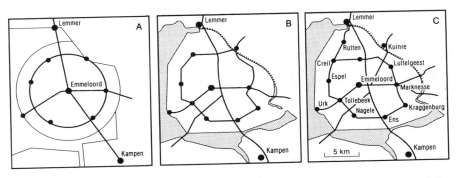

Figure 37.3 Walter Christaller's central place theory applied in the planning of the settlements in the North-East Polder in the Netherlands. (a) Geometrical diagram of the proposed settlement pattern. (b) Plan of five new villages around Emmeloord. (c) The revised plan as executed. After Meijer (1981).

of investigating the process of innovation with the aid of mathematical and statistical methods. The aim was to develop a general model of diffusion. Hägerstrand broke with the current regional tradition by focusing on process and general models. In the opening paragraph of his thesis, Hägerstrand (1953) bluntly stated that although the data used to throw light on the process of diffusion relate to a single area, this should be regarded as a regrettable necessity, rather than as a methodological subtlety.

The search for and use of models in British geography was launched by then junior geographers like Chorley and Haggett in the period 1964–5. The initial aim of the movement in Britain was to influence the teaching of geography. 'Our most clearly definable aim was to make geography at all levels a more intellectually attractive and relevant subject' (Chorley and Haggett 1989: xvii). The first result of this effort was *Frontiers in Geographical Teaching* (1965), followed by the more ambitious volume *Models in Geography* (1967). Two years later the first volume of *Progress in Geography* appeared, to be replaced by the quarterly journals *Progress in Human Geography* and *Progress in Physical Geography* in 1976. In the USA the absence of publication outlets for the new generation of quantitative geographers led to the establishment of a theoretically oriented journal, *Geographical Analysis.*

Peter Haggett's book *Locational Analysis in Human Geography* (1965) was rather influential in demonstrating the substantial amount of theoretical work already carried out by geographers. In the Geography Department at Bergen, for example, it became impossible to continue to teach geography in the same manner as before after this book appeared. Our young staff decided we had to introduce a 'staff course' in quantitative methods and modelling.

THE PRIMARY MOTIVE FOR EMBRACING THE 'SCIENTIFIC PARADIGM': A CHANGE OF ASPIRATION

New political and social trends favoured the development of quantification. Livingstone (1992: 324) demonstrates how 'the early history of quantification [thus] reveals that social and political interests evidently underlay the enterprise. This of course should not be taken to mean that quantification was *just* ideological crystallization.' Harvey (1984) explains the advance of quantification in American geography as a strategic move to escape the political suspicion falling on social science during the McCarthy era. A technocratic spatial science, crystallized in the application of control engineering to geographical systems, may even be said to have bolstered Western capitalism and imperialism and demoted value judgements.

But this is hardly the whole story. History can become a 'trick we play on the dead' in the sense that we interpret the past in terms of our own concerns and interests and even caricature former times in order to support our current viewpoints. While agreeing with Livingstone (1992) that the spatial science school could have been used to support political conservatism, in fact its most ardent advocates were actively left-wing oriented. In the context of the debates about *positivism*, which only really began in 1968, they saw no connection between spatial analysis, model building, quantification and a conservative political ideology.

For example, at my university we held a whole 'critical week' of student debates in 1968. As a young lecturer I was asked to present geography as a social science at a meeting where the well-known political scientist Stein Rokkan also was present. Not having read Marcuse or Habermas, I was unprepared for my first encounter with the concept of positivism. My enthusiastic presentation of the intriguing development of spatial science within geography was discarded as hopeless by the critical social science students. I was astonished to hear a young, female sociology student making a frontal attack on Rokkan for his models, his use of the hypothetic-deductive method and his positivism. When I realized how little I knew about the philosophy and theory of science, I prepared a second-year course on the history and philosophy of geography, and later published books on the subject (Holt-Jensen 1981, 1988).

To understand the primary motives for embracing the 'scientific paradigm' we need somehow to forget the debate over positivism. Most geographers hoped that by adopting the 'new geography' they could advance both the status of the discipline and their own academic standing. The situation was more critical in the United States than in Britain. Because the discipline seemed to be bound up with plain description and enumeration it was no wonder that geography was being eased out of the more prestigious Ivy League universities like Harvard and Yale, noted Morrill (1984: 59). The only alternative was to seek a scientific standing and develop a theoretical, model-building, quantitative geography. 'The need to develop theory precedes the quantitative revolution', stated Burton (1963: 157), 'but quantification adds point to the need, and offers a technique whereby theory may be developed and improved.' Gould (1979: 140) recalls how the new generation of geographers were sick and ashamed of 'the bumbling amateurism and antiquarianism that had spent nearly half a century of opportunity in the university piling up a tip-heap of unstructured factual accounts'. Morrill (1984: 64) claimed that the vision of the young generation, although it might seem radical to those satisfied with an inferior status for the discipline, was in fact conservative in the sense that 'we wanted to save geography as a field of study and to join the mainstream of science'.

A geography based upon spatial analysis which stressed the geographic arrangement and patterns of phenomena seemed far more rewarding than a chorological geography, emphasizing the special character of places and regions. As long as space was regarded as a container it could not in itself have any explanatory value. By introducing the concept of relative space, spatial relations and distance measured in different ways were given explanatory power. Distance could be measured in transport costs, travel time, mileage through a transport network and even as perceived distance. Spatial analysis showed how the interaction of geographical patterns of settlement, land use, diffusion processes, etc. depend on their relative position in space. As transport networks and new means of communications are developed, their relative positions change. Relative space 'shrinks', the world is getting smaller. Forer (1978) used the term 'plastic space' for a space that is continuously changing its size and form.

Geography, as presented by Hettner and Hartshorne, sought to describe the world as it is. But in a world changing at an increasingly rapid pace, simple factual accounts of a region or a country seemed to have little lasting value. An ambitious research student would not find such essentially descriptive work satisfactory. A study of past, present and future processes might give more depth, and longer standing, to a piece of research and add to the understanding of some general trends. Another line of research was to look for spatial patterns that could be explained by some general theory. In my own MA thesis I found some solution by adding the study of settlement processes to my regional account. Torsten Hägerstrand's presentation of *Innovation Diffusion as a Spatial Process* (1953) in a guest lecture in 1960 was enthusiastically greeted by the handful of master students in Norway at that time as a lasting scientific contribution. Here a general spatial theory was combined with the study of processes. Spatial analysis offered new possibilities for more significant, lasting research contributions, and was attractive for this reason.

In geography departments all around the world a leading role in promoting spatial analysis was taken by MA and PhD students. They foresaw that geography through its theoretical development would achieve the level of respect in government and business that economics had held since the 1930s. And to a large extent they proved right. The use of quantitative methods and spatial models soon persuaded many public and private enterprises to employ geographers in analysis and problem–solving. Location theory, central place theory, diffusion theory and interaction theory gave the discipline a stronger scientific image as well as promoting a market in the field of applied science.

SECONDARY MOTIVES FOR CHANGE: A REDEFINITION OF THE SUBJECT MATTER OF GEOGRAPHICAL INQUIRY BASED ON THE THEORY AND PHILOSOPHY OF SCIENCE

Many geographers who worked hard to develop geography as a 'real science' did not appreciate the philosophy of logical positivism which underlay their goal. Positivism has neatly been described as the 'hidden philosophy' of geography as a spatial science. Even the foremost methodological text of the school, David Harvey's *Explanation in Geography* (1969), said virtually nothing about positivism as philosophy.

A small minority of geographers, however, believed that the scientific ideals of logical positivism were necessary for a scientific geography. This view was most clearly expressed by Fred Schaefer in his article 'Exceptionalism in Geography' (1953). Schaefer attacked the 'exceptionalist' view of the Kant–Hettner–Hartshorne tradition; the view that geography is different from other sciences, methodologically unique because it studies unique phenomena (regions), and therefore is an idiographic rather than a nomothetic discipline. 'For Kant geography is description; for Hartshorne it is "naive science" or, if we accept the meaning of science, naive description' (Schaefer 1953: 239). Schaefer maintained that objects in geography are not more unique than objects in other disciplines and that a science must search for laws. Geography should establish itself as a scientific law-seeking discipline and use the 'scientific' (hypothetic-deductive) method.

Since Schaefer had been trained in the Vienna Circle of logical positivists, his arguments were more incisive than those of most advocates of spatial science. Schaefer, like most members of the Vienna Circle, was a political radical and had had to flee from Nazi persecution. Schaefer died before the article was printed, and the galley proofs were read by his friend Gustav Bergmann, a mathematical philosopher at Iowa, who was a major proponent of logical positivism. Bergmann had associated himself with the geographers, sociologists and economists at Iowa (Livingstone 1992), and his books *The Philosophy of Science* and *The Metaphysics of Logical Positivism* became well known to the geographic community. The department of geography at the University of Iowa became one of the early centres for quantitative geography in the USA.

There was a direct connection between Schaefer/Bergmann and William Bunge, who expressed indebtedness to Bergmann for his philosophical assistance in his *Theoretical Geography* (1962). Like Schaefer, Bunge was an ardent socialist and felt himself hounded by the establishment in which he sought an academic career. He openly acknowledged his enmity with Hartshorne, whom he accused as being responsible for his failure to pass his preliminary PhD exams at Wisconsin (Unwin 1992). Bunge had his thesis (1962) published in Lund,

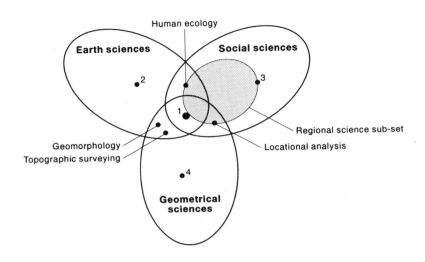

Figure 37.4 Geography and its associated subjects. After Haggett (1965).

Sweden, and dedicated it to Christaller. Bunge sought to establish geography as a strict science and extended his arguments to the effect that geography is the science of spatial relations and interrelations. Geometry, as the mathematics of space, is the language of geography. A main part of the book focused on abstract mathematics, considering theories of movements, central place and location, the meaning of spatial relations and patterns of location.

Bunge's arguments were taken up by the British geographers Peter Haggett, Richard Chorley and David Harvey who made considerable progress towards a unifying methodological and philosophical basis for the spatial science school. Haggett (1965) used a Venn diagram (Figure 37.4) to illustrate the argument that there are three traditional subject associations in geography: with the earth sciences (geology and biology), with the social sciences, and with the geometrical sciences. The geometrical tradition, the ancient base of the subject, is now probably the weakest of the three he maintained. 'Much of the most exciting geographical work in the 1960s is emerging from applications of higher order geometrics' (Haggett 1965: 15–16). Geography should be regarded as a science of distance. The study of spatial arrangements may be summarized in Haggett's diagram of spatial structures (Figure 37.5) (Haggett, Cliff and Frey 1977). Logical positivism was, however, implicitly accepted rather than actively advocated in the works of Chorley, Haggett and Harvey.

The paradigm concept was more directly used. Thomas S. Kuhn, a leading theoretician of science, introduced the concept in his book *The Structure of Scientific Revolutions* (1962). The book was widely read and provided timely arguments in support of a change of emphasis in geography. Kuhn argued that

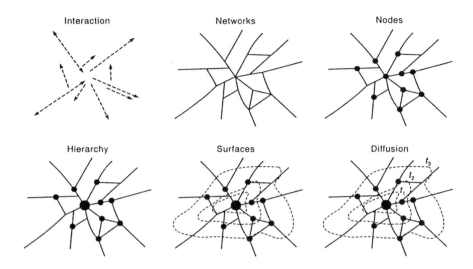

Figure 37.5 The basic elements in Haggett's model of the study of spatial systems.

academic disciplines are subject to paradigm phases, crises and revolutions as one set of basic assumptions and objectives are replaced by another. Chorley and Haggett (1967) identified the paradigm of traditional geography as classificatory and suggested that it was already under severe stress. By suggesting that geography should adopt an alternative model-based paradigm, they were advocating a fundamental change in the basic assumptions and objectives of the discipline, not merely a 'new methodology'. Each geographer was given the choice between the traditional and the new model-based paradigm. Model building was set up as the aim of geographical investigation, a task to be performed with the aid of quantitative methods and the use of computers to handle data. But here we should add that the paradigm-shift argument primarily added ammunition to the propaganda for a shift that the adherents believed would bring scientific advancement and esteem. The motives were pragmatic.

THE ACHIEVEMENTS OF SPATIAL SCIENCE

The spatial science school could be described as having to do with spatial generalization, quantitative methods and paradigm shift within the discipline. But it was much more than this, it also threw open the windows of a hitherto introvert discipline, with its major links to history and geology. Disciplinary boundaries became much more open; methods and theories were openly borrowed from geometry, physics and social sciences as geographers became involved in multidisciplinary research projects. The 1960s was an optimistic period for geographical innovators. Student numbers grew rapidly and career opportunities expanded considerably.

The redevelopment of geography as a social science raised the self-esteem of many geographers. Attempts were made to apply some of the new models in public administration and planning, and some leading geographers were called in to take part in the preparation of government reports. This was a rather important achievement as it opened up a new job market for candidates within planning and administration. The 'new' geography provided an advanced training in data handling and statistics that in any case came in handy on the job market. Generalists with technical and statistical knowledge proved to be better adapted to the job market than candidates with rather narrow specializations.

Haggett (1990: 6) argues for practical and pragmatic approaches in geography. If 'science is the art of the soluble, then much geography is the art of the mappable'. 'Thinking geographically', liking maps and thinking by means of them is, I agree with Haggett, intrinsically linked to geography. More than any other natural or social science, geography is a visual science with similarities in this respect to architecture and the history of art. Like Alice in Wonderland we may stand on tiptoe in order to get an overview, a grand survey of the geographical patterns in front of us. We try to describe and explain the world as we perceive it.

Description and mapping were also central to the traditional schools of geography, but the spatial science school developed more refined methods that made spatial correlations and statistical tests possible. The most recognizable shift was, however, the downgrading of cognitive description and development of sophisticated methods and models in morpho-metric analysis. Most models created were simplifications of spatial morphological patterns based on empirical data. Christaller's central place theory is an example of this. When trying to achieve a general, theoretical explanation of the patterns, theory was imported from other sciences. In many cases this was economic theory.

The main achievement of the spatial science school was, in my opinion, the development of sophisticated methods for the detection of spatial patterns. Many of the models, including such a simple one as 'the gravity model', are good devices to compare data and thus to describe geographical differences. These approaches have given valuable insights into the geographical patterns which form the bases for our analyses or the results of our decisions. But it might be argued that spatial science research developed greater refinement of description rather than explanation, and that explanation is the main task of any science.

This has led many students in human geography, reacting against 'the simplicities of the spatial science school', to discard mapping and description altogether and to start their investigation with some social science hypothesis which they want to test. In many cases this will lead their research into the realm of general social science, and away from the mental structuring that should be the trademark of geographic research.

Specifically geographical hypotheses may, however, be formulated which require the investigation of geographical patterns. Another approach, suggested by Haggett (1983), is to begin with a specific geographical pattern and ask: why is this pattern like this? This approach follows the 'traditional' working sequence in geographical research: localization – geographical pattern – explanation, an approach that could be used both as a simple educational device and in a more advanced research process. For example, a simple explanation of the geographical pattern of desertification in the Sahel would cite and discuss well-known theories. A research work on the same theme would be more refined, setting out to test a particular theory in a particular field situation.

Haggett (1990) points out several research areas in which geographers ought to be active if they are to advance geographical description and analysis. First, mapping is changing from a ROM ('read-only memory') format to a RAM ('random-access memory') format, through which map area, resolution, scale and contents could be printed out on the specifications given by the user on his own computer. This opens up a new era of map comparisons and analysis. Second, the completion and maintenance of a global land-use inventory, using data obtained from remote sensing and an international network of benchmark sites, would serve as a marker for measuring both future and past changes. Third, a space-proofed set of statistical tests is being established which can give bias-free estimates when analysing geographical data.

In a frenetic search for grand explanations, some geographers have also forgotten the value of descriptions which enlighten us. Orderly description is certainly a legitimate scientific endeavour as long as it creates new knowledge. Spatial analysis provided better tools for such descriptions.

CRITICISMS OF MODELLING AND THE SPATIAL SCIENCE SCHOOL

The modelling enterprise has been criticized for being essentially a design-based, deductive activity which precludes the analysis of the qualitative aspects of natural and human systems (see Chapter 38). There certainly has been a preoccupation with things that can be quantified.

Some models were rather unrealistically based upon the concept of economic man as a decision-maker blessed with perfect predictive ability and knowledge of all cost factors. Thus it became necessary to derive alternative theories to those based on economic man and to investigate the perceptions and behaviour of the decision-makers.

Sack (1972) raises the fundamental criticism that geometry alone cannot explain geographic patterns. The laws of geometry are static – they have no reference to time. Time, space and matter cannot be separated in a science concerned with explanations. Gatrell (1985) argues that a prime task for geographers is to detect spatial patterns and, if possible, to try to illuminate an observed spatial distribution through correlations with other distributions. But it is doubtful whether this could be called explanation.

Sayer (1985) points out that the regular behaviour of two or more variables tells us nothing about the causes of this regularity. Causation is not the matter of regularities, but rather of the mechanism which produces them. We need structuration theory (Chapter 39) to understand how 'real' or deep structures influence the empirical outcomes or events. The outcomes are the results of mechanisms, sometimes decisions of individual agents interpreting the constraints of the real level as well as the limitations of the empirical world. The real level may be structures of social relations – for example, the capitalist market economy. The agents or mechanisms could be called systems of social practices.

The *realist position*, as advocated by Sayer, calls for a more intensive kind of research in which a limited number of events are studied in depth so as to get a deeper understanding of the relations between structures, mechanisms and events. This calls for more qualitative study and limits the use of quantitative techniques, but does not rule out spatial science approaches; indeed it rather adds another dimension to them.

But the realist position poses a dilemma for those who still see a unity in the study of the earth as the home of man (Haines-Young 1989). The natural sciences study the relationships between empirical occurrences and permanent, general laws of nature. In the social sciences mechanisms are changing through the historical development in an interaction between all three levels: the real, the actual and the empirical. This leads Johnston (1986) to claim that natural and social sciences cannot be integrated because they have different epistemologies, are different forms of science. I agree that there is a fair gap between certain positivistic natural science methods and structurationalist social science methods, but I would argue that there is a range of choices in between. Further, I would not regard it as utterly wrong to attempt natural science approaches within human geography. In economics there never was such a bias against natural science methods. However, they have their limitations and should be balanced with other methods. Such pluralist points of view may be more appropriate in the current *post-modernist* period than they were in the past.

Quantitative geography has been criticized as being necessarily positivist. While positivism has been regarded as the 'hidden philosophy' of the spatial science school, many of the theories and models are essentially normative. The theoretical construct of, say, central place theory is not intended to show how the world is actually organized, but to demonstrate the patterns that would occur if reality were rational. But who is to judge what is rational? Rationality is a normative or political concept. Many normative theories used by planners were largely based upon the classical economics of capitalism. For instance, Berry (1973) points out that urbanization cannot be considered as a single process, but should be treated as several fundamentally different processes which have arisen out of differences in culture and time. This is yet another argument for a realist approach to human geography.

831

THE CONTRIBUTION OF SPATIAL SCIENCE TO THE CONTEMPORARY STUDY OF THE EARTH AS A HABITAT

Haines-Young (1989: 31) points out that 'the new information-based technologies' provide techniques, notably expert systems, which enable us to carry the problem of modelling geographical knowledge to a deeper level than has been possible so far. For example, a system designed to predict fire risk in the Kakadu National Park in Australia uses a geographical database together with information supplied by the users of the park.

Advanced systems analysis has proved its usefulness in physical geography and ecogeography – the study of man's role in changing the face of the earth. Goudie (1990), Huggett (1993) and Gregory (1985) have provided many examples of these developments. Researches based on the models and methods of the spatial science school have been undertaken on changes in Quaternary climates and on plate tectonics. The development of remote sensing, digital presentation of topography, real-time monitoring of surface processes, and the use of vastly improved computers in the next century may well revolutionize environmental studies. Geography has traditionally been the one discipline to bridge the division between social and natural sciences. It would be wrong to abandon this line of advance today when there is an urgent public need for both educational efforts and research to analyse natural and social factors affecting the relationship between man and his environment. Environmental studies or *ecogeography* using natural science methods, systems analyses, or further developed spatial science approaches have an important role to play here. It would be foolish to allow reservations concerning the spatial science approach in a social science context to inhibit its use in physical geography and ecogeography.

In the social science context spatial science models have had important impacts in planning. They have been used in a direct normative way as planning devices, but as planning is a normative activity in any case this does not discredit the models as such. It is, however, important to clarify the normative basis for the models. To give but one example: the central place theory was used in the development of settlement in the reclaimed North-East Polder of the Ijsselmeer (Netherlands) after the Second World War (Figure 37.3). An 'ideal' model was drawn up for the location of a town and a number of villages on the new land, planning from 'zero point' (a planner's dream!). It was believed that central place theory could be used to create an ideal, smoothly functioning society. The planners had learned from the mistakes made in the pre-war settling of Wieringermeer Polder, where the location of service centres had not initially been thought of (Constandse 1988).

With hindsight, the North-East Polder settlement plan failed to take into account the possibility of technical and societal change. The number and locations of the service centres were based on bicycle or horse and carriage as means of transport and on a much higher workforce on the farms than is feasible today. The small

832

and closely spaced villages of the plan are no longer functionally necessary, but exist due to geographical inertia. But there is no need to abandon the model or the search for more appropriate models on that account. The planning of the more recently reclaimed Flevoland Polders has been based on newer concepts both in central-place and town-planning theory. Naturally, post-modernist ideas in architecture have strongly influenced the planning of the newest Flevoland urban structure, Almere. Almere is planned as a polynuclear structure with 5–6 settlements and no clear functional division between dwelling and service/shopping areas. Further development might well involve more direct citizen participation in the whole planning process. Still, whenever a new environment has to be created, definite and concrete plans must be established, enabling current theories to be tested in practice and their utility improved for further use. In the real world an inadequate theory is better than no theory at all.

WHAT ARE THE IMPLICATIONS OF POST-MODERNISM FOR THE FUTURE OF SCIENTIFIC GEOGRAPHY?

Post-modernism has brought a renewed interest in 'place' and 'locality'. Gregory (1989: 92) has stated that 'we need, in part, to go back to the question of areal differentiation, but armed with a new theoretical sensitivity towards the world in which we live and to the ways in which we represent it'. The eclipse of Marxism is creating opportunities for a more open-minded radical geography. Reflecting upon the radical critique of the 1970s, Massey (1985) argues that 'geography' was underestimated; it was underestimated as distance, and it was underestimated in terms of local variations and uniqueness. We have entered a multi-paradigmatic world, hopefully one in which geographers of all stands and research traditions are respected for what they achieve in the way of widening and strengthening the status of geography, and not one in which some geographers stigmatize those outside their school of thought. Ideas from Marxism, structuration theories, realism and post-modernism have broadened the research basis of geography. Scientific geography developed tools and models which will continue to be important in research and practical work.

In this evolving climate of pluralism, we might disagree with Taylor (1985) when he envisages that a third pure geography will emerge. Some of his arguments are, however, relevant. Taylor identifies the first pure geography as the regional school of traditional geography; the second as the spatial school. The spatial school has left us with a very useful technical kit for description and analysis of the empirical world, which is of lasting value even if the philosophy and much of the theory does not survive. Taylor (1985) maintains that we cannot return to the optimism of positivism; neither can we expect a rebirth of regional synthesis. But the holistic approach implicit in the earlier regional geography is curiously closer to the humanistic and radical positions that each will have an impact on the 'third pure geography'. He maintains that the attractions of specialization are diminishing in the intellectual division of labour. Geography is in a unique position to prosper from such a trend as an intellectually respectable alternative to *ad hoc* interdisciplinary arrangements.

The spatial science school presented the world as a series of examples of a few simple geographical laws. The implicit message that everyone is like us and is governed by similar structures of social relations, was, however, soon discarded.

The positivist assumption that all is general is not a sound basis for future geographical research. We should not fall into the *generalization* trap. Should we then assume that nothing is general, that every region is an individual phenomenon, which can only be understood as such? Johnston (1985) points out that this is the *singularity* trap, which also must be avoided. On the empirical level every place and region may be seen as unique, but not singular, as the uniqueness is the result of the response of agents to general processes.

To conclude:

1 The approach of the spatial school is inadequate as far as a geometry of patterns is thought to provide some sort of explanation. But the spatial school provided the empirical research with advanced tools of description which are far superior to those used by the traditional regional school. These tools we need to learn and develop further, also because they are unrivalled in other sciences.

2 A realist approach is needed to structure an explanation of our empirical descriptions. Focus should be on the agents that individually and collectively take action on the basis of their interpretations of possibilities and limitations set by the empirical environment in which they live, and by the rules and structures of social relations surrounding them.

3 The critique of positivism, and especially its transfer of natural science methods to the study of social relations, should not lead us to think that nature does not matter, and that an ecological understanding of the physical world around us is not needed in geography.

Geography is an environmental and a social study, not merely a social science. Places matter, as is also clear from the analysis of time geography. Because of a combination of capability, available transportation, coupling and authority constraints, people's lives are packed in time–space prisms. Places matter because for most of the time most of us are confined to particular 'locales' – places where we have our roots and from where we interpret the world.

NOTE

I wish to thank Brian Fullerton for his collaboration in getting the manuscript and my English into a good and readable form.

REFERENCES

Ackerman, E.A. (1945) 'Geographical training, wartime research and immediate professional objectives', *Annals of the Association of American Geographers* 35, 121–43.

Berry, B. (1973) *The Human Consequences of Urbanization*, London: Macmillan.

Bunge, W. (1962) *Theoretical Geography* (2nd edn 1966), Lund Studies in Geography, Ser. C: 1, Lund: Gleerup.

Burton, I. (1963) 'The quantitative revolution and theoretical geography', *Canadian Geographical Journal* 7, 151–62.

Chorley, R. and Haggett, P. (eds) (1965) *Frontiers in Geographical Teaching*, London: Methuen.

Chorley, R. and Haggett, P. (eds) (1967) *Models in Geography*, London: Methuen.

Chorley, R. and Haggett, P. (1989) 'From Madingley to Oxford: a foreword to *Remodelling Geography*', pp. xv–xx in B. Macmillan (ed.) *Remodelling Geography*, Oxford: Basil Blackwell.

Christaller, W. (1933) *Die zentralen Orte in Suddeutschland*, Jena. (English translation C.W. Baskin, 1966, *Central Places in Southern Germany*, Englewood Cliffs, N.J: Prentice-Hall.)

Christaller, W. (1968) 'Wie ich zur der Theorie der zentralen Orte gekommen bin', *Geographische Zeitschrift* 56, 88–101.

Constandse, A.K. (1988) *Planning and Creation of an Environment*, Lelystad: Directorate Flevoland.

Entrikin, N. (1991) *The Betweenness of Place: Towards a Geography of Modernity*, Basingstoke: Macmillan.

Forer, P. (1978) 'A place for plastic space', *Progress in Human Geography* 2, 230–67.

Gatrell, A. (1985) 'Any space for spatial analysis?', pp. 190–208 in R.J. Johnston (ed.) *The Future of Geography*, London and New York: Methuen.

Goudie, A. (1990) *The Human Impact on the Natural Environment* (3rd edn), Oxford: Basil Blackwell.

Gould, P. (1979) 'Geography 1957–1977: the Augean period', *Annals of the Association of American Geographers* 69, 139–51.

Gregory, D. (1989) 'Areal differentiation and post-modern human geography', pp. 67–96 in D. Gregory and R. Walford (eds) *Horizons in Human Geography*, London: Macmillan.

Gregory, K.J. (1985) *The Nature of Physical Geography*, London: Arnold.

Guelke, L. (1978) 'Geography and logical positivism', pp. 35–61 in D.T. Herbert and R.J. Johnston (eds) *Geography and the Human Environment. Progress in Research and Applications*, vol. 1, London: John Wiley.

Hägerstrand, T. (1953) *Innovationsforloppet ur korologisk synpunkt*, Meddelanden från Lunds Universitet Geografiska Institutionen, Avhandling nr. 25. (English translation by A. Pred (1967) *Innovation Diffusion as a Spatial Process*, Chicago: University of Chicago Press.)

Haggett, P. (1965) *Locational Analysis in Human Geography*, London: Arnold; 2nd edition, 1977.

Haggett, P. (1983) *Geography: A Modern Synthesis* (3rd edn), New York and London: Harper & Row.

Haggett, P. (1990) *The Geographer's Art*, Oxford: Basil Blackwell.

Haggett, P., Cliff, A.D. and Frey, A. (1977) *Locational Analysis in Human Geography* (2nd edn), London: Edward Arnold.

Haines-Young, R. (1989) 'Modelling geographical knowledge', pp. 22–39 in B. Macmillan (ed.): *Remodelling Geography*, Oxford: Basil Blackwell.

Hartshorne, R. (1939) *The Nature of Geography: A Critical Survey of Current Thought in the Light of the Past*, Lancaster, Pa.: Association of American Geographers.

Hartshorne, R. (1959) *Perspectives on the Nature of Geography*, Chicago: Rand-McNally.

Harvey, D. (1969) *Explanation in Geography*, London: Edward Arnold.

Harvey, D. (1984) 'On the history and present condition of geography: an historical materialist manifesto', *Professional Geographer* 36, 1–11.

Hettner, A. (1927) *Die Geographie, ihre Geschichte, ihr Wesen und ihre Methoden*, Breslau: Ferdinand Hirt.

Holt-Jensen, A. (1963) 'Rauland – forsøk på en regional–geografisk beskrivelse av en fjellbygd', Unpublished MA thesis, University of Oslo: Department of Geography. (Partly published in 1968 as *Fjellbygda Rauland*, Ad Novas – Norwegian Geographical Studies No. 6, Oslo: Universitetsforlaget.)

Holt-Jensen, A. (1981) *Geography, its History and Concepts*, London: Harper & Row.

Holt-Jensen, A. (1988) *Geography, History and Concepts* (2nd edn), London: Paul Chapman.

Huggett, R.J. (1993) *Modelling the Human Impact in Nature: Systems Analysis of Environmental Problems*, Oxford: Oxford University Press.

Johnston, R.J. (1985) 'To the ends of the world', pp. 326–38 in R.J. Johnston (ed.) *The Future of Geography*, London: Methuen.

Johnston, R.J. (1986) 'Four fixations and the quest for unity in geography', *Transactions of the Institute of British Geographers*, New Series 11, 449–53.

Kuhn, T.S. (1962) *The Structure of Scientific Revolutions*, Chicago: University of Chicago Press.

Livingstone, D.N. (1992) *The Geographical Tradition*, Oxford: Blackwell.

Massey, D. (1985) 'New directions in space', pp. 9–19 in D. Gregory and J. Urry (eds) *Social Relations and Spatial Structures*, London: Macmillan.

Meijer, H. (1981) *Zuyder Zee – Lake Ijssel*, The Hague: IDG.

Morrill, R. (1984) 'Recollection of the "quantitative revolution's" early years: the University of Washington 1955–65', pp. 57–72 in M. Billinge and R. Martin (eds) *Recollection of a Revolution*, London: Macmillan.

Sack, R.D. (1972) 'Geography, geometry and explanation', *Annals of the Association of American Geographers* 62, 61–78.

Sayer, A. (1985) 'Realism and geography', pp. 159–73 in R.J. Johnston (ed.) *The Future of Geography*, London: Methuen.

Schaefer, F. (1953) 'Exceptionalism in geography', *Annals of the Association of American Geographers* 43, 226–49.

Schmithüsen, J. (1976) *Allgemeine Synergetik*, Berlin: de Gruyter.

Taylor, P. (1985) 'The value of a geographical perspective', pp. 92–110 in R.J. Johnston (ed.) *The Future of Geography*, London: Methuen.

Unwin, T. (1992) *The Place of Geography*, Harlow: Longman.

Wrigley, E.A. (1965) 'Changes in the philosophy of geography', pp. 3–20 in R.J. Chorley and P. Haggett (eds) *Frontiers in Geographical Thinking*, London: Methuen.

FURTHER READING

Annals of the Association of American Geographers (1979) vol. 69, no. 1 (Diamond Anniversary Issue: Seventy-Five Years of American Geography).

Billinge, M., Gregory, D. and Martin, R. (eds) (1984) *Recollection of a Revolution*, London: Macmillan.

Cloke, P., Philo, C. and Sadler, D. (1991) *Approaching Human Geography. An Introduction to Contemporary Theoretical Debates*, London: Paul Chapman.

Gould, P. (1985) *The Geographer at Work*, London: Routledge & Kegan Paul.

Haggett, P. (1990) *The Geographer's Art*, Oxford: Blackwell.

Harvey, D. (1969) *Explanation in Geography*, London: Arnold.

Johnston, R.J. (ed.) (1985) *The Future of Geography*, London: Methuen.

Macmillan, B. (ed.) (1989) *Remodelling Geography*, London: Blackwell.

38

GEOGRAPHY AND HUMANISM IN THE LATE TWENTIETH CENTURY

Anne Buttimer

INTRODUCTION

Two centuries ago a multi-volume *Encyclopédie* crowned the efforts of French humanists, an enterprise which aimed to reveal all secrets of knowledge and technology, liberating minds from fear, ignorance and prejudice, and prodding imaginations towards the role of science and technology in promoting civilization and a universal brotherhood of humankind. Since then there has been enormous progress in all fields of knowledge, each becoming so specialized that the worlds of science and humanities drifted apart. The latter years of the twentieth century again witness a *rapprochement* among fields of knowledge, the central focus now shifting from humanity to the earth itself. While the marvels of science and technology are still applauded there is a pervasive doubt over humanity's capacity to use them wisely. Quite in contrast with the eighteenth-century Promethean image of humanity as creator and master, there is today a pervasive image of humanity as Narcissus, pondering over the many contradictions between ideal and reality of its diverse achievements, seeking to remember forgotten features of human nature and rediscover wiser ways of dwelling.

The story of geography during the latter half of the twentieth century has mirrored, in microcosm, the general pattern of the sciences over the past two hundred years. As the universalist aims of Enlightenment science were opened to scrutiny and critique by scholars of eighteenth-century and early nineteenth-century Romanticism, so too was the exuberant triumph of mid-twentieth century 'new geography' followed by a variety of alternative approaches. Variously labelled as 'post-positivist', 'radical', 'humanist' and even now 'postmodernist', these diverse currents pointed to weaknesses in the inherited lore, and have opened up new worlds of exploration and reflection. Each has expanded the range of contacts with scholars in other fields, and most have served to revivify interest in the history and philosophy of geography. The Socratic formula 'Know Thyself' has assumed a fresh meaning for geographers today.

Geographical thought and practice is perhaps best understood as historically situated discourse (Livingstone 1992). Only in a historical context might one interpret the rationale and prospects facing change at any particular period. Breakthrough moments within intellectual and political life have been preceded by periods of critique and uncertainty over the *status quo*. Even the institutional separation of geography's physical and human branches, bemoaned today by advocates of environmental sensitivity, was itself the product of an emancipatory project launched by individuals who entered the profession during the interwar period. They sought freedom from the parental bonds of geology on the one hand and from history on the other. Fleeing, too, from ghosts of environmental determinism, they argued for a substantive focus on space rather than environment, methodological procedures inspired by positivism, and livelier interaction with other scientists. After the Second World War human geographers would proclaim themselves as social scientists; history and the humanities were to become the favoured pursuits of only a few. Yet it was from the rediscovery of late nineteenth- and early twentieth-century works of geographers attuned to the humanities (e.g. Marsh, Vidal de la Blache, Braudel, Wright and Dardel), and from the heightened awareness of cultural differences in environmental perception, that much of the enthusiasm for a 'humanistic' movement emerged in the 1960s and 1970s. Today one begins to acknowledge how context-bound all our histories and geographies have been.

GEOGRAPHY AND HUMANISM

Words such as 'geography' and 'humanism' reveal as much about their definers and their worlds as they do about any perennial truths. Proclamations about the essence of humanness, be it *animal rationale*, *Homo sapiens* or *demens*, *zoon politikon*, *homo faber*, or *homo ludens*, each claiming generality transcending cultures, history, and environments, reveal quite as much about the authors of such propositions as they do about human nature. Humanism can scarcely be regarded as an autonomous field of knowledge enquiry. Rather it is a stance on life and world shared by people of diverse walks of life, including geographers (von Wright 1978). Geography, as scholarly tradition, ranks among the oldest in the Western world (Wright 1966; Glacken 1967). Its aim is to describe the earth, *Geo-* (*Gaia*, meaning earth), and *-graphy* (*graphein*, to describe). To make sense of the great Gaian drama we dismember the whole into component parts which are more amenable to analytical scrutiny. Human geographers focus on the anthroposphere – *Gaia*'s human envelope – each subfield focusing on a distinct yet inseparable part of this complex drama. The distinctive mark of the geographic approach is its analytical emphasis on space and time, pattern elucidated in terms of process. Most distinctive are its synthetic aims, always eager to integrate insights from diverse sources in elucidating relationships between humanity and earth, from local townland and parish to global *oecumene*.

Humanists down the centuries have explored the nature of humanity, its passions and powers, while geographers have studied the earth where humans, among many other life forms, make a terrestrial home. For each facet of humanness – rationality or irrationality, faith, emotion, artistic genius or political prowess – there is a geography; for each geographical interpretation of the inhabited earth there are implicit assumptions about the nature of humanness. To slogans such as Protagoras' 'man is the measure of all things' and Alexander Pope's 'the proper study of mankind is man', geographers would note that *Homo sapiens* is a terrestrial species, its projects and plans played out in particular environments, even if its diverse cultures have created and been inspired by myths and symbols. *Humanus*, literally, means 'earth dweller'.

Interpreting the 'humanist' turn in geography

The essential quality of Western humanism, I have argued elsewhere, is a plea for freedom: freedom *from* bonds of ignorance, fear, oppression or oblivion, freedom *to* create and to explore new horizons in thought and life (Buttimer 1990, 1993). The emergence of a 'humanist' turn within geography during the latter half of the twentieth century might be described as a liberation cry for aspects of thought and experience which had been ignored or suppressed within the practices of previous generations. Too much emphasis may indeed have been placed on elements *from* which one sought to be free; to date, the horizons *towards* which a humanist spirit would aspire remain somewhat unclear. Yet herein lies a host of fundamental problems associated with inherited models of scientific and humanist modes of knowing. Whereas the former would define clear-cut analytical procedures for the analysis and representation of research results, the latter invites discovery and understanding, objectives which not only vary from one individual to another, but also imply elements of emotion, aesthetics, and moral judgement which, strictly speaking, were to be eliminated from the (positive) scientist's agenda.

The trend away from positivist hegemony took different forms throughout the various contexts in which geography was practised in Euro-American institutions. The impact of humanism might best be evaluated in terms of the observable changes in the nature and status of different practices in the field: educational (*paideia*), analytical (*logos*), applied (*ergon*), and critically reflective (*poesis*). The 'New Geography' of the 1950s and 1960s highlighted *logos* and, in some schools, there was a commitment to *ergon*. A spate of new textbooks promoted the 'modernization' of geography teaching; yet despite a marked increase in student numbers, *paideia* rarely assumed importance for rank and tenure within the profession. The undeniable change apparent already in the 1980s was the re-affirmation of *poesis*, the critical emancipatory element which remained marginal and even trivialized during the 1950s and 1960s (Gould and Olsson 1982; Harvey 1989).

A focus on these four practices could serve to elucidate the many-faceted nature of humanist concern throughout Western history and also the interplay of individual and societal concern in the career journeys of geographers themselves (Buttimer 1983, 1990, 1993). *Poesis* has involved ontological questions about the nature of reality and especially the nature of humanness (*humanitas*): diverse and recurrent themes about individuality and sociality, freedom and responsibility, rationality and hedonism, conservatism and creativity down the centuries. In terms of *logos*, there have been claims about humanist modes of knowing, which oppose scientific reductionism, seek to elucidate rather than to explain, and emphasize the subjectivity of consciousness and the intersubjective nature of scholarly discourse. Educationally (*paideia*) there is the tradition of the humanities, fields of learning deemed appropriate for the cultivation of the arts, classical literature, and the nurturing of civic virtue. Finally, there has been an enduring concern about the human condition (*ergon*), e.g. in humanitarianism, which has sought to encourage social responsibility and liberal politics. In historical perspective, these different strains intermingle, and one of the biggest challenges facing scholars today is to rediscover the bonds among them in the everyday practice of geography.

All four strands are detectable in the literature commonly labelled 'humanistic geography'. Connotations of the term are many (Tuan 1976; Ley 1981, 1983; Daniels 1985; Rowntree 1986, 1988). What was subsumed under this rubric varied from one country or language tradition to another, the terms 'social', 'radical', 'critical' and 'humanistic' were often seen as virtually interchangeable (Racine 1977; Relph 1981; Claval 1984; Ballasteros 1992; Pellegrini 1992). In terms of *paideia*, 'humanistic' has become the prefix for new orientations within such well-established subfields as historical, political, and cultural geography (Daniels 1985; Brunn and Yanarella 1987; Rowntree 1988). In terms of *logos*, humanism for some has implied a kind of mission to restore human subjectivity to a field where scientific objectivism has been so dominant (Ley and Samuels 1978; Mackenzie 1986). Some have emphasized human attitudes and values, others cultural patrimony; some have focused on the aesthetics of landscape and architecture, others on the emotional significance of place in human identity (Bowden and Lowenthal 1975; Meinig 1979; Seamon and Mugerauer 1985; Pocock 1981; Bunksé 1981). There was a humanitarian motif also in terms of *ergon*, a substantial number advocating compassion and engagement in the resolution of social or environmental problems – reminders about twentieth-century pilgrims of peace, and early Cassandra voices on environmental destruction (Thomas 1956; Buchanan 1968; Bunge 1973; Santos 1975; Guelke 1985).

From whatsoever ideological stance it has emerged, the case for humanism has usually included a call for *poesis*: for critical reflections on the nature of reality and conventional ways of interpreting it. And at the forefront of contemporary *poesis* is a distinctively 'contextual' turn, a concern about the inextricable connections between thought and context, experience and expertise, and the adequacy of all our practices for the elucidation of contemporary reality (Cosgrove 1984; Gould 1985; Pellegrini 1992).

Observation–participation–interpretation

In retrospect, it seems, the stirrings of contextual awareness in geography have echoed that profound transformation in twentieth-century approaches to knowledge generally: first a movement away from 'spectator' to 'participant' stances, followed by a 'hermeneutic' or interpretative stance (Rorty 1979; Toulmin 1983). Down the centuries the predominant image of the geographer is that of 'observer', aiming to achieve objectivity in their representations and explanations of phenomena. By the 1960s, there was a growing awareness of how human perceptions of reality had been filtered by different cultural groups, by different research instruments and by different practical agenda (Wright 1966; Lowenthal 1961). Following the epochal *prise de conscience* which followed Kuhn's *Theory of Scientific Revolutions* (1962) and the strident calls for attention to *épistemologie* in the human sciences (Foucault 1966), some geographers became aware of the values implicit in their taken-for-granted modes of practice. The social construction of knowledge was to become a burning curiosity; queries arose about what or whose interests were actually being addressed in applied geography: managerial or popular, professional or client, elite or folk? (Häger-strand 1970; Zelinsky 1970; Buttimer 1974). This led to debates about power, language, and conflicts of interest between 'insiders' and 'outsiders' in particular situations (Buttimer and Seamon 1980; *Antipode* 1985). Some pointed towards ways in which positivism had led disciplinary practice in directions which served to foster managerial interests and pointed towards alternative approaches (Samuels 1971; Seamon 1980; Karjalainen 1986). Following upon decades of *ceteris paribus* orthodoxy, there were strong pleas from existentialist, phenom-enological, pragmatist and other sources that *ceteris*, for all practical purposes, were never *paribus* (Hägerstrand 1970; Buttimer 1974). As in other fields, *epistemological* questions yielded place or became transposed to *sociological* ones; the 'foundational' concerns of the earlier (observation) phase yielded to 'dialectical' ones (Stoddart 1981; Ley 1983).

Many geographers, by the late 1970s, became aware of the fact that we were all participants as well as observers, all to some degree 'insiders' or 'outsiders' in different settings (Geertz 1983; Buttimer 1983). Awareness also grew about the myriad ways in which conventional thought and practice had been filtered through the complex orchestra of Western social experience. And how might our (Western) language and inherited distinctions (e.g. between subjective and objective, insider and outsider, humanist and scientist), help to bridge worlds of knowledge and understanding? For a few, for example Berque (1982), Bonne-maison (1984), Chapman (1985), such issues of cross-cultural dialogue became important. Preoccupations of the majority who adopted a hermeneutic stance on their work continued to focus on their own societal experience (Gregory and Walford 1989). Reality would become construed, in an era of post-modernism, as an arena of events, of mirrors and masks, of texts reflecting contexts: a theatre in which the antinomies of subjective and objective, normative and descriptive, internalist and externalist interpretations of science would be deemed anachronistic (Gregory 1981; Sugiura 1983; Pickles 1985). Some expressed the need for languages and symbols which could facilitate dialogue among different civilizations, while others polished sophisticated arguments about the futility of such a dream.

Already during the 1980s, much of the zest and optimism which accompanied the humanistic turn had faded. Some regarded it as a kind of amnesia, a turning away from problems and retreat into *esoterica*, or simply as a critique of the *status quo* (Entrikin 1976; Smith 1979; MacLaughlin 1986). Beyond the doctrinaire anti-humanism of structuralism, 'realists' found the prose of an earlier generation to be quaintly idealistic. Many, disillusioned about all the contradictions and tragedies of the Western legacy, condemned humanism as archculprit, wellspring of all hubris and vaunting ambition, a myth overdue for dismissal (Ehrenfeld 1978; Relph 1981). One prop for this judgement was the conventional practice of identifying the origin of Western humanism with the early fifteenth-century Renaissance in Italy: a movement which also bore the seeds of modernism. Humanism was thus found 'guilty by association' for the Promethean excesses of Western humanity and hence shared its condemnation. Materialist manifestos continued to highlight the enduring propensity of capitalism to produce and reproduce those power-structures which held sway in the global economy (Smith 1979; Harvey 1984; Santos 1975; MacLaughlin 1986). Post-modernist scenarios of texts reproducing texts in a labyrinth of self-reflecting mirrors would preclude any assumptions about human intentionality or meaning in lived reality. A sense of imprisonment in one's own cultural world, of claustrophobia and nihilism, of 'hitting one's head against the ceiling of language' (Olsson 1979; Dematteis 1985) characterized some critical thought within geography during the 1980s. One became aware of the many ways in which the entire intellectual heritage of the West had mirrored and been mirrored in the peculiar social history of Western humanity.

While hermeneutic concern continues to attract a growing number of younger scholars, the 1990s witnessed a return to observation and a heightened humanitarian concern about ethics and intervention in the course of public life. Global horizons beckoned in the widespread concern about humanity and earth, the shocking record of environmental destruction and radical transformations in culture and politics (White 1985; Johnston and Taylor 1986). The enormous scale and urgency of environmental issues and global change, dramatic realignments of demographic and political patterns, again challenged the geographer to more field-oriented observation. Rapid advances in analytical technology, computer-based information systems and satellite images beckoned.

As the philosophical strains of realism and pragmatism again recaptured audiences, disciplinary orthodoxy became less important than theoretical salience (Rowntree 1988; Ley 1989) and eventually problem resolution (Kates 1987).

Even over this relatively short stretch of time, one could discern some resonance of characteristic processes in the social construction of knowledge fields in the Western world. The 1960s were years of innovation and 'discovery' for many. By the decade's end and throughout the 1970s, there was a flurry of excitement as new subdisciplines, societies, and specialty groups emerged. The late 1970s and early 1980s revealed many contradictions between dream and reality, *ethos* and structure, and a critically reflective mood set in. Some became nostalgic for the past, some became fascinated by critique on the *status quo*, and some began to envisage prospects for a new dawn (Rowntree 1988; Ley 1989; Folch-Serra 1989).

GAIA-GRAPHEIN IN THE LATE TWENTIETH CENTURY

Today geography is identifiable not only as a professional field of expertise, an academic discipline, but also, a reality of everyday lived experience. To date, geographers have attuned their expertise to managerial rather than popular interests in the organization and appropriation of space. Existentialist voices have consistently championed the cause of everyday geographical experience, not only for individual human persons, but also for the diverse cultural worlds whose home territories are increasingly invaded by globally organized techno-logical and commercial systems. What humanists would advocate is co-responsible engagement in matters of territorial and social discretion in humanity's modes of dwelling. No matter how fascinating the horizons of intellectual curiosity now explored by 'humanistic geographers' are, the ultimate criteria on which this movement will be evaluated are those of enhanced ability to render disciplinary expertise more relevant to the elucidation of lived experience. Illustrations of contemporary themes for research and reflection include (1) dwelling, (2) nature and culture, (3) mindscape and landscape, and (4) geography itself as historically situated discourse.

Dwelling: essential quality of humanness

Geography is the stuff of everyday experience for all forms of terrestrial life. Mountain and plain, river and lake, woodland and wildlife may be 'explainable' in the categories of natural and social science, but in reality each cultural group has understood nature, space, and time through its own special filters. To survive on planet Earth, every creature has to develop a geographical sense, a sense of place, space, time, and movement. How to negotiate our diverse geographies is surely one of the enduring challenges of existence, from the politics of empire to the arrangement of one's kitchen, office and garden. A geographer tells stories of dwelling, and the interplay of mindscapes and landscapes, in terrestrial environments.

The progressive 'humanization' of the earth over the past half-million years, offers fascinating ground for study and reflection. Humanity has assumed a dominant role within the Gaian orchestra, shaping the surface landscape, harvesting resources, tipping the natural checks and balances among other life forms, in culturally and technologically diverse ways of life. Each cultural group has devised its own variety of ways in which Nature was understood and how humans should relate to it. Some, as in the Judaeo-Christian and Hellenic traditions, have placed humanity *above* all other life forms and have regarded intellect as the highest manifestation of human spirit. In Hindu and Buddhist traditions, humanity has been regarded as an integral part of cosmic order, recycling, like other life forms, in a continuous process of reincarnation. If one seeks a definition of what is quintessentially human, it is surely the propensity to make symbols: to re-present reality in symbolic forms in art and artefact, monument and metaphor, science, technology, and national flags (Cassirer 1944). Humans everywhere have looked to Nature for insight into the secrets of creation, birth and death, meaning and purpose in everyday life. Each innovation (discovery) in the domestication of animals, cultivation of plants, in technology, agriculture and commerce has been construed in terms of cosmology and values. Since time immemorial, humans have inscribed their cosmologies in the lived landscape.

Dwelling, according to Heidegger, is the essential quality of humanness. Dwelling involves a 'gathering of the fourfold ... earth, sky, mortals, and divinities' (Heidegger 1947). 'Worlding earth and earthing world', Heideggerians would summarize it. The orchestration of economy and ecology, place and space, home and horizon, past and future, is how human geographers might express it (Relph 1974; Karjalainen 1986). Whenever philosophers utter generalizations about 'universal' features of humanness, the geographer would note the dramatic differences in the interplay of culture and environment within particular places.

Cultural historians claim, for example, that civilizations might be best understood in terms of their overriding preoccupations: whereas East Asian societies were chiefly concerned about the control of collective living (sociality), and India with the control of consciousness (mind), the West has shown a preoccupation with the conquest of nature, internal and external (Nakamura 1980). It has also been claimed that ideologies based on the Judaeo-Christian heritage have been the source of exploitative attitudes towards the physical environment and natural resources – one fundamental pillar in the anti-humanist rhetoric of the late twentieth century (Leiss 1974; White 1967; see, however, Doughty 1981 and Kay 1989). Humanity, for many a late twentieth-century eco-catastrophist, is the villain of world geography, Gaia's cancer (Dobell 1990). Latecomers anyhow to earth history, the argument goes, humans may be already on a road towards self-destruction. Geographers today would look askance at such universal judgements. Probing the diversity of landscapes and lifeways *within* any cultural realm, they document evidence of human ingenuity in adapting lifeways to environment throughout history, and reiterate the enduring challenge of environment and ecology for global humanity today.

Nature and culture

Throughout human history Nature has constituted the source of inspiration and foundation for cosmology and human identity (Eliade 1961; Tuan 1978). Australian Aborigines revere rocks as dreaming centres, and song-lines evoke a sense of ongoing creation. The *Upanishads* celebrate the eternal Asvatta (fig-tree) and banyan tree, while North American Sioux build their entire cosmology around the sacred cottonwood tree. For East Asian societies, life was seen to emerge from the sacred lotus; devout Hindus bathe in the sacred waters of the Ganges. Artefactual and cartographic evidence suggests that most cultures have imaged their own place as the centre of the world, with horizons varying in shape and extent depending on their livelihoods and travel. The cargo cults of Polynesian islanders re-enact the stories of colonial occupation; in Vanuatu, individual and group identity revolves around the metaphors of tree and canoe (Bonnemaison 1985). Ojibway and Cree have detailed mental maps of trapping lines across the North American Plains – territory viewed by Euro-settlers as a vast arena for conquest by railway, lumber company, or settler (Turner 1980; Brody 1981).

For the 350 million so-called indigenous populations of the earth today, nature and human identity are inseparable. Multinational companies, often in league with national governments, find themselves confronted with a resounding *cri de coeur* from autochthonous populations eager to protect the integrity of their home environments and maintain discretion over minerals, forest, fish and wildlife in their ancestral lands. Chiloe islanders have halted the exploitation of their forests by giant lumber companies; women in Ladakh have defied Western rationality in ecologically sensitive cultivation of their own hillsides and valleys. The dramatic clash of human interests which characterizes the use of earth resources has at last found its way to public consciousness. The languages within which such clashes are overtly negotiated are characteristically those of national governments and multinational companies; the voices of dwelling, by people in place, are rarely included.

Human geographers would seek alternative bases on which human discourse about earth resources might be negotiated. They would seek improved self-understanding and mutual understanding among the people of the earth, highlighting fundamental considerations of *noösphere* and *biosphere*. The human body itself, Yi-fu Tuan observed, has been a universal catalyst for symbol-making in many civilizations: its anatomy suggesting a view of the world as mosaic of forms, its physiology symbolizing the world as mechanical system, its uniqueness and individuality illustrating the contingencies of context (Tuan

845

1971). Most common of all, however, has been the organic analogy inspired by the coherence and unity of a human body; for example, in the favoured (Renaissance) notion of the human person as a whole (body, soul, personality) as microcosm of the universe.

In most civilizations there has been an attempt to identify basic elements such as fire, air, earth, and water; stories of creation woven around the interactions of these elements. The I-Ching, one of the earliest recorded cosmologies identified five elements: earth, fire, wood, water, and metal. Sixth-century Chinese world views are anchored in four cardinal directions corresponding with the elements, seasons, sensory mood and colour; world horizons depicted in orthogonal zones stretching outwards from civilization to barbarism. While Western cosmologies tell of an original emergence of Cosmos from Chaos through the agency of a transcendent God, Indian cosmologies see a continuous process of ongoing creation, divinity immanent in all aspects of nature and environment. Creation myths are all enacted in geographic terms, the story usually involves an original creation, a fall, followed by a flood, and then regeneration. Noah built an ark into which pairs of living creatures were rescued until the waters subsided; on the Ganges, Vishnu gathered living creatures into the flowing water. An enduring difference between Western and other cosmologies is epitomized in our inherited landscapes and politics. While the West has sought to contain the sacred in cathedrals and churches, power within territorially circumscribed domains, and land-use planning in zones, our Indian colleagues persist in maintaining a more fluid attitude towards life: holiness through ritual bathing in flowing water, intermingling of diverse elements, in the ongoing process of life and death. Land-locked geopolitics rather than hydrological cycle? (Yi-fu Tuan 1968; Singh and Singh 1984; Buttimer 1984).

How relevant could such exploration be, a realist might question, in understanding actual experience today? While pragmatic relevance might not hold priority for many a humanist scholar, the salience of symbolism and belief in shaping contemporary political and economic life could scarcely be questioned. Besides, the humanist would argue, the journey towards understanding the Other is a vital step in achieving critical self-understanding. Ultimately without an appreciation of differences in the cultural evaluation of nature, space, and time, the geographer could scarcely elucidate *landscape*, one of the central *explicanda* facing the discipline.

Mindscape and landscape

'The humanised landscape', Vidal de la Blache wrote, 'is a medal struck in the image of a civilisation.' Goethe referred to architecture as 'frozen music'. The visible forms of landscape are the *texts* which geographers read and seek to interpret.

To interpret any living landscape geographers seek to understand its underlying processes. General laws of geomorphology and biogeography help explain why certain parts of the earth are habitable or inhospitable. The actual map of population, land uses, and lifestyles, however, defies any suggestion that the bio-physical environment always determines ways of life. To understand the humanized earth one needs to understand cultural diversity among societies: not only their economies and traditional skills, but also the sense of sacred and profane in their perceptions of nature, space, and time. Successive groups have migrated over time from one area to another bringing their talents, images, hopes and prejudices with them. Characteristically, they try to re-create the new landscape in the image of the past, usually an idealized version of it (Buttimer 1985). Contemporary landscapes have all been formed through the sedimentation of many layers of human occupance. Under the geographer's gaze, there is nothing static on the surface of the earth.

Far more influential than soil, climate, or culture, are the systems of economic and political organization which hold sway within the human world. Most places are connected to wider-scale interests: markets, administrative divisions of territory, communication and transport systems linking places into networks which transcend local and national boundaries. Virtually anywhere one travels today there is landscape evidence of the 'global economy' – from airports and TV towers, satellites and stock exchange, to computer terminals, Coca-Cola and *pommes frites*.

Idealistic approaches to the study of landscape, such as those inspired by Collingwood and others, attempt to reconstruct the geographies of past times, seeing landscape through the eyes of those who lived in them (Guelke 1985). 'Every landscape is a reflection of the society which first brought it into being and continues to inhabit it' (Jackson 1952: 5). Why not, then, 'read' every landscape as text to be decoded in terms of the vernacular, the environments of the workaday world? Among the strongest voices favouring the 'humanistic' turn in geography were those of scholars who already acknowledged the hermeneutic value of landscape interpretation (Meinig 1979; Rose 1981). 'Our human landscape', Pierce Lewis remarked, 'is our unwitting autobiography, and all our cultural warts and blemishes, our ordinary day-to-day qualities, are there for anybody who knows how to look for them' (Lewis 1985: 13). During the 1970s this *genre* was indeed revitalized (Salter 1978; Meinig 1979; Rose 1981; Sugiura 1983; Lewis 1985; Rowntree 1986). Landscape seen as the 'sedimentation' of diverse forms of discourse allowed room for interpretations of various kinds (Olwig 1981; Cosgrove 1984; Daniels 1985). It invited enquiry into culturally varying modes of symbolic transformation such as those inspired by Vico or Herder (Dainville 1964; Mills 1982), as well as semiotic enquiry into landscape texts and signs as products in their own right (Choay 1981; Marchand 1982; Duncan 1987).

The communication of insight on the relationships between mindscape and landscape, however, poses the critical issue of language. As Goethe described architecture as 'frozen music', so Emerson described language as 'frozen poetry'. The challenge is one which geographers now share with colleagues in literature, philosophy and art. It has surely been on questions of language and power, semiotics and symbolism, that some geographers found the most challenging moment of interaction with structuralists (Gale and Olsson 1979; Gregory 1981; Dematteis 1985; Rose 1981). The deconstruction of inherited meanings, as suggested in the work of Derrida, deliberately sought to reveal the multiple and conflicting readings that could be made on particular texts (Derrida 1972). In the 1980s, Gunnar Olsson pointed towards modernity's malaise in human geography: 'Thing yields to process, stability to change, certainty to ambiguity, noun to verb, being to becoming' (Olsson 1984: 73). In sharp contrast with the previous generation's deeply held convictions about epistemological certainty and images of the world as complex mechanism, this voice sought to unmask the extent to which *la condition humaine* 'is one of the predicaments lived behind prison walls' (Olsson 1984: 84).

Walls on the lived landscape remain, for the geographer, one basic focus for examining tensions among conflicting human interests. Landscape provides the visible text in which the competing demands of social, economic, and ecological processes are inscribed. Beyond its literal (bio-physical and artefactual) content, landscape reveals important symbolic meanings in lived experience. The humanist is therefore keen to understand those 'invisible' processes underlying conflicts between the 'frozen' landscapes of dole and conservation as well as the technologically transformed and homogenized landscapes of production and consumerism.

Geography as historically situated discourse

As with the earth's landscapes, the mindscapes of professional geographers have reflected the cultures and world-views of its sponsors and audiences. Its status as academic discipline has mirrored the fluctuating fortunes of nations and empires, flowering at moments of societal self-confidence, altruism, or expansionary challenge; relaxing in routine-operational tasks of housekeeping and inventory, textbook writing, and functionally specialized research in times of stability or depression. Active participants in a global community of scholars too, geographers have also negotiated their practices within the reigning scientific paradigms of particular periods. In historical perspective, the changing trends in political and intellectual life have only rarely been synchronous; their complex interplay amply revealed within the career trajectories of individual scholars and texts produced within various schools (Buttimer 1983, 1993).

One of the most important foci for critical discussions among geographers today has been the history and philosophy of the discipline itself. Far from an exercise in memorizing names and dates of the alleged ancestors, this field has become both 'contextual' and 'critical' (Berdoulay 1982; Smith 1979; Buttimer 1983, 1993). The humanist perspective would focus on ideas in practice, interplay of creativity and context, through various periods, and seek to evaluate the implications of such practices for everyday life and landscape.

848

While academic geography in Europe can be traced to Greek and Roman sources, its institutional origins during the modern period are due to the work of geographical societies, particularly those of the nineteenth century. The establishment of geography as university department, however, marked explicit political choices made within national Ministries of Education. The discipline's agenda would henceforth be harmonized with that of its major sponsor, the nation-state. To this day, with some notable exceptions, the practice of geography everywhere has borne a distinctively national character. The most widely circulated texts on world geography were produced by the major imperial and colonizing nations, projecting the world-views and global ambitions of the respective metropole. For a strident nation on the march, it was exceedingly important that the young be socialized in the ideology and expansionary hubris of the home nation, and that some be trained for the ranks of its military, mercantile, or missionary bands eager to conquer and 'civilize' the rest of benighted humanity. The geographer was quite useful for colonial conquest in mapping potential scenarios for the administrative regionalization of newly occupied territory. Early colonizers deemed it wise to pack bibles and catechisms as well as geography textbooks and guidelines for survey and inventory of resources. Later imperialists brought the catechisms of capitalistic resource exploitation.

From its beginnings as an academic discipline in modern times, geography has been valued as a training ground for the exploration and conquest of space and resources; for the imposition of order deemed rational by the conqueror; for the delivery of information on areas, distances, flora and fauna, peoples and cultures, in language categories and narrative frames which could be understandable and usable 'back home'. So effective indeed has this convergence of geography and national power interests been that its practice in post-colonial lands still bears the stamp of its original preachers. Throughout the Island Pacific, Canada, Africa, and South America, the practice of geography still reflects the national (mostly European) origins of its founders. Many of the would-be transnational paradigm shifts of recent years – those inspired by structuralism, Marxism, existentialism, positivism, or even *laissez-faire* Friedmannesque capitalism – have all been refracted through the filters of the former colonial language and pedagogical orthodoxy.

In historical and cross-cultural perspective, geography reveals two contrasting images of humanity and earth. In its interpretations of human experience, there is the record of culture, creativity, communication in the gradual appropriation and cultivation of landscapes and life; in terms of marketable expertise, there is the record of national capitols, international *Kapital*, and conquest of space. At one end of the spectrum one notes the gradual emergence of technology in the harvesting of natural resources and the social organization of space and time, a growing self-confidence in the control of natural risks and overcoming the constraints of distance; at the other end, one also notes those humanly created technologies and economic systems which have become virtually divorced from the realities of place and culture. Geographical experience and geographical expertise both are manifestations of humanity's *noösphere*: patterns of thought, belief and values which have sought to negotiate the wrapping of the earth, from local to global scales.

Western scientists describe Gaia as a vast interlocking synergy of bio-physical and economic systems. Yet the lived reality for most humans might better be described as a mosaic of culture-bound worlds, each endeavouring to negotiate values of social and ecological integrity with those of global integration. Given the urgency of global environmental challenges, many geographers have called for critical reflection on taken-for-granted ways of thought and practice, their intellectual and moral foundations, and the strengths and limitations of their cognitive claims. If such critical reflection could lead to a mutually respectful dialogue of cultures perhaps humanity could discover wiser ways of dwelling.

REFLECTIONS ON HUMAN GEOGRAPHY IN THE LATE TWENTIETH CENTURY

The impact of humanism on the practice of geography in the latter decades of the twentieth century has been felt not only in terms of cognitive styles and choice of substantive foci of research, but also in modes of vocational meaning. One of its major results has been a reaffirmation of a critically reflective element, *poesis*, long overlooked in a profession which unequivocally favoured scientifically based approaches to the sharpening of analytical skills and the marketability of geographical expertise. Ontological issues are again discussable; modes of discourse themselves have become matters for critical reflection. Substantively, there has been a renewal of research interest in relationships between humanity and environment, the meanings of place and landscape for human creativity and health. Flowing from this, environments themselves re-assume significance as contexts for life, cultural identity and historical legacy: a long overdue rediscovery after a few generations which focused on space as a *tabula rasa* on which diverse models of economic or technological rationality could be tested. Within a few decades, even mainstream approaches to *logos* and *ergon* have changed. Naïve confusions of descriptive and normative discourse which characterized some post-war ventures in applied geography are now less palatable to audience or sponsor.

Challenges for humanism

It is perhaps in the practices of *logos* and *ergon* that the most important challenges remain for the humanist. First, how is one to construe the Protagorean motto 'The proper study of mankind is man'? How is one to bridge insight on the experiences of individual persons and/or places with that of the more general experiences of regions and cultures? Are the aims of *Wissen* and *Verstehen* forever irreconcilable (Smith 1984; Guelke 1985)? And might the emphasis on 'social construction' eliminate rigorous epistemological reflection?

Methodological challenges also abound, e.g. how to relate narratives on 'events-in-context' or on 'landscapes as texts' to the wider issues facing humanity and world. Among the advocates of contextual approaches, a distinction could be made between those who emphasize *events* and study contexts in terms relevant to those events, versus those who pay primary attention to *contexts* (physical, ecological, functional) and seek a potentially universal grid or explanatory theory to account for all possible events. On one side stand the would-be guardians of local cultures and the integrity of places who abhor any prospect of universal laws of context. On the other side, there are would-be managers of universal systems to guide and plan human activities on a planetary scale.

At what scale, in space and time, might the humanist approach be applied? Contextual approaches have yielded their best results when the focus of inquiry rested on particular events or periods. One gains much better insight on particular 'nows', but one simultaneously risks losing the threads of historical flow or possibilities for cross-cultural comparisons and contrasts. While the humanist spirit would rejoice at this growing contextual sensitivity, it would also seek to affirm the values of cross-cultural understanding and historical depth.

Finally, human geographers have also tended to define 'context' in anthropocentric terms, focusing on the humanly constructed reference frames of laws, structures, and artefacts surrounding events. How to now incorporate nature and the bio-physical environment in descriptions of context, yet avoid resuscitating the ghosts of environmental determinism, remains one of the most demanding challenges for geographers today.

Geographers in the 1990s approach Gaia with better awareness of their intellectual heritage, of the strengths and limitations of models we teach and apply. Whether this new-found technical competence and vastly increased volume of information will lead to global understanding will depend, to a large extent, on the wisdom we can develop in its interpretation. Inherited biases for models of 'spatial efficiency' and political expediency have served the interests of national capitols, international capital, and territorial conquest; how to harmonize these with the equally important realities of culture, creativity, and communication remains the challenge for the future.

Conclusions

Each discovery in humanity's geographical understanding of the world has heralded changes in its self-images, hopes, and fears. Each was preceded by a period of malaise over the *status quo* before the emancipatory *élan*, often aiming at deliverance from confining orthodoxies or structures of thought, politics, or material conditions, at times uttering a prophetic note about possibilities. Each, too, witnessed a will to build structures and institutions, bequeathing its own legacy of unresolved tensions to its offspring. And in the subsequent attempt to understand and transcend those structures, there was the ardent longing of a Narcissus.

The late twentieth century has shown a radically revised relationship of mindscape and landscape. Geographers have dared to look inwards, exploring cultural differences in environmental perceptions and experience, while colleagues in the humanities dare to look outwards, freely exploring ground traditionally trod only by natural science and divinities. The occasion for joint exploration into the mystery of human creativity and wisdom in modes of dwelling has come. 'New alliances' between physical and biological sciences, and between them and the human sciences are taking shape within the academic world (Prigogine and Stengers 1984). One can scarcely again consider 'matter' as dead, or Nature as a complex of blind forces; rather one is discovering the complex and dynamic wisdom written into the nature of the universe, and the basic bonds between humans and fellow living creatures.

Meanwhile the actual map of humanity shows that Euro–America and its legacy of Hellenic and Mediterranean models of *humanitas* is but one corner in an evolving *noösphere*. Today's world map of power and politics has taken on radically new dimensions due to transformations in technology and trade. The explosion of information and trans-continental circulation of people, commodities, and ideas raise exciting new possibilities for thought and life. It was just such transformations which leant wing to the imaginations of Renaissance humanists, many of whom were jack-of-all-trades poetic types, and thereby eager to stay abreast of new developments in all realms of human becoming. What the retrospective glance deems as 'revolutionary' could scarcely have been possible if these diverse aspects of humanness had not been capable of mutually enriching one another, either in dialogue among human individuals, or within the individual person.

Implications for the practice of geography today are obvious. While each of our practices demands its own level of specialization, so too does it need to be orchestrated with the others. Inherited structures and externally imposed priorities may shed light on how our present fragmentation of effort has come to be, but to suggest that only structural changes are needed to clear the way for tomorrow's styles of practice would betray a lack of imagination. The humanist's emancipatory hope would be that as a scholarly community we begin at home, as it were, critically evaluate the priorities which have been inherited, and bravely experiment with new ways to orchestrate our energies.

Communication and mutual understanding, the tissue of the *noösphere*, is one of the greatest challenges facing geographers today. And this involves emotion and will, quite as much as cognitive brilliance, technical ingenuity or the design of media. Ultimately it redounds to an expansion of heart and spirit to embrace the twofold challenge facing humanity's dwelling in the late twentieth century: *sociality* and *ecology*. Sociality, one of the highest art forms of *zoon politikon*, faces the challenge offered by the spatial juxtaposition of culturally diverse peoples: to let these often involuntary movements and convergence of humans become the springboard for new creativity in politics and social life, rather than problems to be solved by social engineering. And ecology, in the original meaning of the term, calls for a sense of how life as a whole functions. Today's protests against environmental destruction have indeed been heard, but those very ways in which problems are being addressed often serve more to fragment rather than to unify efforts towards ameliorative action. Geographers have much to offer by way of cross-cultural and historical evidence from mankind's experiences in sociality and ecology. Attuned to the emancipatory role which humanist thought has played historically, too, they could reiterate at least the essential message that human reason cannot function without hope. Gaia's human envelope, the anthroposphere, needs to be understood as a drama more complex than simply as battleground of ecological versus economic rationality, but rather as *oecumene*, potential home for mankind, a species which urgently needs to rediscover the art of dwelling.

Much rhetoric has been aired on the Western world's anthropocentric perspectives, critics usually implying alternatives such as the deification of Nature, of the social collective, or the reification of ideologies playing out their purposes with humanity and Nature. Perhaps the West, far from overestimating *humanus*, has grossly underestimated it. Recovery of the human subject, and recognition of human agency as an integral part of the lived world, the creativity again apparent on peripheral and previously marginalized regions, all betoken fresh potential for human geography today. One is challenged to not only regard humanity and the earth in global terms, but also to understand the ecological and social implications of a world humanity now 'planetized'. The need for cross-cultural and comparative research also implies an extension of time horizons on the history of the earth and human occupants to date, and serious reflection on the potential future of Gaia.

And 'humanistic geography'? As a stance on life, humanism is one which welcomes the challenge of discerning the creative potential of individuals and groups to deal with the surface of the earth in responsible and co-responsible ways. Nor is human creativity confined to the intellectual sphere; it involves emotion, aesthetics, memory, faith and will. The humanist turn in geography has delivered inspiration to practitioners of physical, economic, cultural or social geography, and its aims are not well served by investing energy on staking claims for becoming a special branch of the field. Humanism should thus more appropriately be considered as a leaven in the dough rather than as a separate loaf in the *smörgåsbord* of geographic endeavour. The emancipatory *élan* recuperable even from our Western traditions could enable geography itself to perform as leaven in the mass of contemporary science and humanities. The Renaissance of humanism calls for an ecumenical rather than a separatist spirit; it calls for excellence in special fields as well as a concern for the whole picture. From our Western corner, perhaps the greatest Promethean challenge is that of bridging worlds through dialogue and mutual understanding. The history of geography is a history of discovery. The practice of geography at each moment of history inevitably reflects ideas inherited from previous generations; the very landscapes and spaces we study are themselves the product of culturally diverse geographical ideas. Discovering hitherto unknown lands invites deeper explorations into the taken-for-granted knowledge of familiar lands, constantly posing the challenge of re-examining one's understanding of the world. The geographer is *homo viator*, explorer, as T.S. Eliot (*Four Quartets*) claimed:

> We shall not cease from exploration
> And the end of all our exploring
> Will be to arrive where we started
> and know the place for the first time.

REFERENCES

Antipode (1985) 'The Best of *Antipode* 1969–1985', *Antipode* 17.

Ballasteros, A. Garcia (ed.) (1992) *Geografía y humanismo*, Barcelona: Oikos-Tau.

Berdoulay, V. (1982) 'La métaphore organiciste. Contribution à l'étude du langage métaphorique en géographie', *Annals de Géographie* 507, 573–86.

Berdoulay, V. (1988) *Des mots et des lieux. La dynamique du discours géographique*, Paris: CNRS, Mémoires et documents de géographie.

Berque, A. (1982) *Vivre l'espace au Japon*, Paris: Presses Universitaires de France.

Bonnemaison, J. (1985) 'Les fondements d'une identité: territoire, histoire, et société dans l'Archipel du Vanuatu (Melanésic)', Thèse pour le Docteur ès Lettres et Sciences Humaines, Université de Paris IV.

Bowden, M. and Lowenthal, D. (eds) (1975) *Geographies of the Mind*, New York and London: Oxford University Press.

Brody, H. (1981) *Maps and Dreams. Indians and the British Columbia Frontier*, Toronto: Doublas & McIntyre.

Brunn, S. and Yanarella, E. (1987) 'Towards a humanistic political geography', *Studies in Comparative International Development* 22, 3–72.

Buchanan, K. (1968) *Out of Asia*, Sydney: Sydney University Press.

Bunge, W. (1973) 'The geography of human survival', *Annals of the Association of American Geographers* 63, 275–95.

Bunksé, E.V. (1981) 'Humboldt and an aesthetic tradition in geography', *The Geographical Review* 71(2), 127–46.

Buttimer, A. (1974) *Values in Geography*, Commission on College Geography Research Report No. 24, Washington, DC.

Buttimer, A. (1983) *The Practice of Geography*, London: Longmans.

Buttimer, A. (1984) 'Water symbolism and the human quest for wholeness', pp. 57–91 in R. Castensson (ed.) *Vattnet bär livet – funktioner, föreställningar och symbolik*, Linköping, Sweden: University of Linköping Tema V Report 6. (Abridged version pp. 159–90 in D. Seamon and R. Mugerauer (eds) (1985) *Dwelling, Place and Environment*, Dordrecht: Kluwer.)

Buttimer, A. (1985) 'Farmers, fishermen, gypsies, guests: who identifies?', *Pacific Viewpoint* 26(1), 280–315.

Buttimer, A. (1990) 'Geography, humanism and global change', *Annals of the Association of American Geographers* 80, 1–33.

Buttimer, A. (1993) *Geography and the Human Spirit*, Baltimore, Md.: The Johns Hopkins University Press.

Buttimer, A. and Seamon, D. (eds) (1980) *The Human Experience of Space and Place*, London: Croom Helm.

Cassirer, E. (1944) *An Essay on Man*, New Haven, Conn.: Yale University Press.

Chapman, M. (ed.) (1985) *Mobility and Identity in the Island Pacific*, Special Issue of *Pacific Viewpoint* 26(1).

Choay, F. (1981) *Le règle et le modèle*, Paris: Le Seuil.

Claval, P. (1984) *Géographie humaine et économique contemporaine*, Paris: Presses Universitaires de France.

Cosgrove, D. (1984) *Social Formation and Symbolic Landscape*, Totowa, N.J.: Barnes & Noble.

Cosgrove, D. (1994) Editorial in *Ecumene* I(1), 1–6.

Dainville, F. de (1964) *Le langage des geographes*, Paris: Picard.

Daniels, S. (1985) 'Arguments for a humanistic geography', pp. 143–58 in R.J. Johnston (ed.) *The Future of Geography*, New York: Methuen.

Dardel, E. (1952) *L'homme et la terre. Nature de la réalité géographique*, Paris: Presses Universitaires de France.

Dematteis, G. (1985) *Le metafore della Terra. La geografia umana tra mira e scienza*, Genoa: Gianfracomo Feltrinelli.

Derrida, J. (1972) *Marges de la philosophie*, Paris: Editions du Minuit.

Dobell, R. (1990) 'Gaia's cancer: the human dimensions of global change', *DELTA*. Newsletter of the Canadian Global Change Program, vol. 1, 1, p. 7.

Doughty, R. (1981) 'Environmental theology', *Progress in Human Geography* 51, 234–48.

Duncan, J.S. (1987) 'Review of urban imagery: urban semiotics', *Urban Geography* 8, 473–83.

Ehrenfeld, D. (1978) *The Arrogance of Humanism*, New York: Oxford University Press.

Eliade, M. (1961) *The Sacred and the Profane*, New York: Harper.

Entrikin, N. (1976) 'Contemporary humanism in geography', *Annals of the Association of American Geographers* 66, 615–32.

Folch-Serra, M. (1989) 'Geography and post-modernism: linking humanism and development studies', *The Canadian Geographer* 33(1), 66–75.

Foucault, M. (1966) *Les mots et les choses*, Paris: Gallimard.

Gale, S. and Olsson, G. (eds) (1979) *Philosophy in Geography*, Dordrecht: Reidel.

Geertz, C. (1983) 'The way we think now: toward an ethnography of modern thought', pp. 147–66 in *Local Knowledge. Further Essays in Interpretive Anthropology*, New York: Basic Books.

Glacken, Clarence J. (1967) *Traces on the Rhodean Shore*, Berkeley, Calif.: University of California Press.

Gould, P. (1985) *Geographers at Work*, London: Routledge & Kegan Paul.

Gould, P. and Olsson, G. (eds) (1982) *A Search for Common Ground*, London: Pion Ltd.

Gregory, D. (1981) 'Human agency and human geography', *Transactions, Institute of British Geographers*, New Series, vol. 6, 1–18.

Gregory, D. and Walford, R. (eds) (1989) *Horizons in Human Geography*, London: Macmillan Press.

Guelke, L. (ed.) (1985) *Geography and Humanistic Knowledge*, Waterloo Lectures in Geography, Vol. 2, Department of Geography, University of Waterloo, Ontario.

Hägerstrand, T. (1970) 'What about people in regional science?', *Papers of the Regional Science Association* 24, 7–21.

Harris, C. (1978) 'The historical mind and the practice of geography', pp. 123–37 in D. Ley and M. Samuels (eds) *Humanistic Geography: Prospects and Problems*, Chicago, Ill.: Maroufa Press.

Harvey, D. (1984) 'On the history and present condition of geography: an historical materialist manifesto', *The Professional Geographer* 36(1), 1–10.

Harvey, D. (1989) *The Condition of Postmodernity*, Oxford: Blackwell.

Heidegger, M. (1947) 'Platons Lehre von der Wahrheit. Mir einen Brief über den "Humanismus"', Berne: A. Francke, pp. 53–119. (Trans. by E. Lohner as 'Letter on Humanism', pp. 204–48 in N. Langiulli (ed.) (1971) *The Existentialist Tradition. Selected Writings*, New York: Anchor Books.)

Jackson, J.B. (ed.) (1952) 'Human, all too human geography', *Landscape* 2, 5–7.

Johnston, R.J. and Taylor, P. (1986) *A World in Crisis*, Oxford: Basil Blackwell.

Karjalainen, P.T. (1986) *Geodiversity as a Lived World: On the Geography of Existence*, University of Joensuu Publications in Social Sciences 7, Joensuu, Finland.

Kates, R.W. (1987) 'The human environment: the road not taken, the road still beckoning', *Annals of the Association of American Geographers* 77, 525–34.

Kay, J. (1989) 'Human dominion over nature in the Hebrew bible', *Annals of the Association of American Geographers* 79, 214–32.

Kirk, G.S. and Raven, J.E. (1962) *The Presocratic Philosophers*, Cambridge: Cambridge University Press.

Leighly, J. (1983) 'Memory as mirror', pp. 80–90 in A. Buttimer, *The Practice of Geography*, London: Longmans.

Leiss, W. (1974) *The Domination of Nature*, Boston: Beacon Press.

Lewis, P. (1985) 'Beyond description', *Annals of the Association of American Geographers* 75, 465–77.

Ley, D. (1981) 'Cultural/humanistic geography', *Progress in Human Geography* 5(2), 249–57.

Ley, D. (1983) 'Cultural/humanistic geography', *Progress in Human Geography* 7(2), 267–75.

Ley, D. (1989) 'Fragmentation, coherence, and limits to theory in human geography', pp. 223–44 in A. Kobayasghi and S. MacKenzie (eds) *Remaking Human Geography*, London: Allen & Unwin.

Ley, D. and Samuels, M. (eds) (1978) *Humanistic Geography: Prospects and Problems*, Chicago, Ill.: Maroufa Press.

Livingstone, D. (1992) *The Geographical Tradition*, Oxford: Blackwell.

Lovelock, J.E. (1979) *Gaia. A New Look at Life on Earth*, New York and London: Oxford University Press.

Lowenthal, D. (1961) 'Geography, experience and imagination: towards a geographic epistemology', *Annals of the Association of American Geographers* 51, 241–60.

Mackenzie, S. (ed.) (1986) *Humanism and Geography*, Carleton Geography Discussion Papers, Ottawa, Canada: Carleton University.

MacLaughlin, J. (1986) 'State centered social science and the anarchist critique: ideology in political geography', *Antipode* 18, 11–38.

Marchand, B. (1982) 'Dialectical analysis of value: the example of Los Angeles', pp. 232–51 in P. Gould and G. Olsson (eds) *A Search for Common Ground*, London: Pion Ltd.

Meinig, D. (ed.) (1979) *The Interpretation of Ordinary Landscapes. Geographical Essays*, New York and Oxford: Oxford University Press.

Mills, W.J. (1982) 'Metaphorical vision: changes in Western attitudes toward the environment', *Annals of the Association of American Geographers* 72, 237–53.

Nakamura, H. (1980) 'The idea of Nature, East and West', *Encyclopedia Britannica*, Inc. *The Great Ideas Today*, pp. 234–304.

Olsson, G. (1979) 'Social science and human action or on hitting your head against the ceiling of language', pp. 287–308 in S. Gale and G. Olsson (eds) *Philosophy in Geography*, Dordrecht: Reidel.

Olsson, G. (1984) 'Toward a sermon on modernity', pp. 73–85 in M. Billinge, M.D. Gregory and R. Martin (eds) *Recollections of a Revolution*, London: Macmillan Press.

Olwig, K.R. (1981) 'Literature and reality: the transformation of the Jutland heath', pp. 47–65 in D.C.D. Pocock (ed.) *Humanistic Geography and Literature*, London: Croom Helm.

Pellegrini, G. Corna (ed.) (1992) *Humanistic and Behavioural Geography in Italy*, Pisa: Pacini Editore.

Pelt, J.-M. (1977) *L'homme re-naturé. Vers la société écologique*, Paris: Seuil.

Pickles, J. (1985) *Phenomenology, Science and Geography: Spatiality and the Human Sciences*, Cambridge: Cambridge University Press.

Pocock, D.C.D. (ed.) (1981) *Humanistic Geography and Literature. Essays on the Experience of Place*, London: Croom Helm.

Prigogine, I. and Stengers, I. (1984) *Order out of Chaos. Man's New Dialogue with Nature*, Boulder, Colo. and London: Shambhala New Science Library.

Racine, J.-B. (1977) 'Discours géographique et discours idéologique: perspectives épistemologiques et critiques', *Hérodote*, no. 6, 109–58.

Racine, J.-B. (1981) 'Problématique et méthodologie: de l'implicite à l'explicite', pp. 85–162 in H. Isnard *et al.*, *Problématiques de la géographie*, Paris: Presses Universitaires de France.

Relph, E. (1974) *Place and Placelessness*, London: Pion.

Relph, E. (1981) *Rational Landscapes and Humanistic Geography*, London: Croom Helm.

Rorty, R. (1979) *Philosophy and the Mirror of Nature*, Princeton, N.J.: Princeton University Press.

Rose, C. (1981) 'Wilhelm Dilthey's philosophy of historical understanding', pp. 99–133 in D. Stoddart (ed.) *Geography, Ideology, and Social Concern*, Oxford: Basil Blackwell.

Rowntree, L. (1986) 'Cultural/humanistic geography', *Progress in Human Geography* 10, 580–6.

Rowntree, L. (1988) 'Orthodoxy and new directions: cultural/humanistic geography', *Progress in Human Geography* 12, 575–86.

Salter, C.L. (1978) 'Signatures and settings: an approach to landscape in literature', pp. 69–83 in K.W. Butzer (ed.) *Dimensions of Human Geography*, Chicago: University of Chicago Press.

Samuels, M. (1971) 'Science and geography: an existential appraisal', Seattle, Washington, Department of Geography Ph.D. dissertation.

Santos, M. (1975) *L'espace partagé*, Paris: Editions M.-Th. Génin (trans. (1985) *The Shared Space*, New York: Methuen).

Seamon, D. (1980) *A Geography of the Lifeworld. Movement, Rest, Encounter*, London: Croom Helm.

Seamon, D. and Mugerauer, R. (eds) (1985) *Dwelling, Place and Environment*, Dordrecht: Kluwer Academic Publishers.

Singh, R.I. and Singh, R.P.B. (1984) 'Lifeworld and lifecycle in India: a search in geographical understanding', *The National Geographical Journal of India* 30, 207–22.

Smith, N. (1979) 'Geography, science, and post-positivist modes of explanation', *Progress in Human Geography* 3(3), 356–83.

Smith, S. (1984) 'Practicing humanistic geography', *Annals of the Association of American Geographers* 74, 353–74.

Stoddart, D. (ed.) (1981) *Geography, Ideology, and Social Concern*, Oxford: Basil Blackwell.

Sugiura, N. (1983) 'Rhetoric and geographer's worlds: the case of spatial analysis in human geography', Ph.D. dissertation, Pennsylvania State University Department of Geography.

Thomas, R.L. (ed.) (1956) *Man's Role in Changing the Face of the Earth*, Chicago, Ill.: University of Chicago Press.

Toulmin, S. (1983) *The Return to Cosmology*, Berkeley, Calif.: University of California Press.

Tuan, Yi-fu (1968) *The Hydrological Cycle and the Wisdom of God: A Theme in Geoteleology*, Toronto: University of Toronto, Department of Geography.

Tuan, Yi-fu (1971) 'Geography, phenomenology and the study of human nature', *The Canadian Geographer* 15, 181–92.

Tuan, Yi-fu (1976) 'Humanistic geography', *Annals of the Association of American Geographers* 66, 266–76.

Tuan, Yi-fu (1978) 'Sign and metaphor', *Annals of the Association of American Geographers* 68, 363–72.

Tuan, Yi-fu (1982) *Segmented Worlds and the Self. Group Life and Individual Consciousness*, Minneapolis, Minn.: University of Minnesota Press.

Turner, F. (1980) *Beyond Geography. The Western Spirit against the Wilderness*, New York: Viking Press.

White, G. (1985) 'Geographers in a perilously changing world', *Annals of the Association of American Geographers* 75(1), 1–10.

White, L. (1967) 'The historical roots of our ecological crisis', *Science* 155, 1203–7.

Wooldridge, S.W. and East, W.G. (1951) *The Spirit and Purpose of Geography*, London: Hutchinson University Library.

Wright, G.H. von (1978) *Humanismen som livshållning och andra essayer*, Stockholm: Rabén & Sjögren.

Wright, J.K. (1966) *Human Nature in Geography*, Cambridge, Mass.: Harvard University Press.

Zelinsky, W. (1970) 'Beyond the exponentials', *Economic Geography* 46, 498–535.

FURTHER READING

Berdoulay, V. (1988) *Des mots et des lieux. La dynamique du discours géographique*, Paris: CNRS: Mémoires et documents de géographie.

Berque, A. (1982) *Vivre l'espace au Japon*, Paris: Presses Universitaires de France.

Buttimer, A. (1993) *Geography and the Human Spirit*, Baltimore, Md.: The Johns Hopkins University Press.

Cosgrove, D. (1984) *Social Formation and Symbolic Landscape*, Totowa, N.J.: Barnes & Noble.

Dematteis, G. (1985) *Le metafore della Terra. La geografia umana tra mira e scienza*, Genoa: Gianfracomo Feltrinelli.

Glacken, Clarence J. (1967) *Traces on the Rhodean Shore*, Berkeley, Calif.: University of California Press.

Harvey, D. (1989) *The Condition of Postmodernity*, Oxford: Blackwell.

Ley, D. and Samuels, M. (eds) (1978) *Humanistic Geography: Prospects and Problems*, Chicago, Ill.: Maroufa Press.

Livingstone, D. (1992) *The Geographical Tradition*, Oxford: Blackwell.

Meinig, D. (ed.) (1979) *The Interpretation of Ordinary Landscapes. Geographical Essays*, New York and Oxford: Oxford University Press.

Pellegrini, G. Corna (ed.) (1992) *Humanistic and Behavioural Geography in Italy*, Pisa: Pacini Editore.

Relph, E. (1974) *Place and Placelessness*, London: Pion.

Santos, M. (1975) *L'espace partagé*, Paris: Editions M.-Th. Génin (trans. (1985) *The Shared Space*, New York: Methuen).

Seamon, D. and Mugerauer, R. (eds) (1985) *Dwelling, Place and Environment*, Dordrecht: Kluwer Academic Publishers.

Thomas, R.L. (ed.) (1956) *Man's Role in Changing the Face of the Earth*, Chicago, Ill.: University of Chicago Press.

Tuan, Yi-fu (1982) *Segmented Worlds and the Self. Group Life and Individual Consciousness*, Minneapolis, Minn.: University of Minnesota Press.

Wright, J.K. (1966) *Human Nature in Geography*, Cambridge, Mass.: Harvard University Press.

39

STRUCTURAL THEMES IN GEOGRAPHICAL DISCOURSE

Richard Peet

Reflecting its origins in the natural and social sciences, geography studies the interrelations between humans and the natural environment. In this definition, the term 'environment' is broadly understood to include natural processes, conditions and materials such as climate, landforms, soils and resources from and within which people continually make their livelihoods; territory or landscape as the 'stage' on which life literally 'takes place'; and earth distance or space which, whether through friction or content, shapes every movement and is socialized by human activity. In early-modern geography relations with environment were interpreted as determinism – the creation by nature of what were assumed to be innate human characteristics, and the effects of varying mental abilities on the differential evolution of civilizations (Ratzel 1896; Semple 1903; Huntington 1935). The overriding theoretical deficiency of environmental determinism consisted in its positing direct, causal connections between natural processes and racial (natural) characteristics, without adequately conceptualizing the social structures mediating between natural and socialized human being. A theory with this kind and level of deficiency could survive and prosper only because it served ideological purposes in an age of Euro-American imperialism (Peet 1986). The eventual disciplinary reaction against environmental determinism, which gained strength as overt imperial competition declined in relative significance in the inter-war years, led to the reorientation of geography – first towards a modified determinism in the regional geography of the 1930s, 1940s and early 1950s, and then towards the spatial dimension of environment in the location theories of the later 1950s and 1960s. Here space was drained of its natural qualities (in idealized landscapes typically called 'isotropic plains') while spatial phenomena were 'explained' via purely spatial causes in a movement (the 'quantitative revolution') later to be diagnosed as having a bad case of 'spatial fetishism'. Hence, the late 1960s and 1970s saw an increasing disciplinary emphasis on social structures, both as mediating between nature and human characteristics, and as causes of spatial phenomena. The argument was that in reproducing their existence people combine in social structures, accumulating experience (cultures) and systems of social relations and power (class and gender systems, states, etc.) within which they create their personalities, socialize nature and produce social forms of space.

To understand social structures geographers in the radical 1960s and 1970s turned to Marxist, structuralist and structurationist philosophies (Peet and Thrift 1989). These formed bridges between the particular interests of geography and the broader sweep of generalized knowledge summarized by philosophy and social theory. These bridges, indeed, were instrumental in the re-awakening of geography; that is, its more effective realization of the potential inherent in a unique position at the interface between the study of nature and that of society. Strengthening the connections with philosophy and politics also meant that geography was subject to the transformations in social and political theory which came with the turn from modern to post-modern philosophy, literature and culture in the 1970s and 1980s (Best and Kellner 1991). From this kaleidoscope of ideas, tendencies and fashions, this chapter outlines a quasi-coherent line of structuralist thought – first, by briefly outlining Marxist and structuralist approaches to society and space developed in the 1960s and 1970s; second, by introducing structuration theories which, in part, arose as a critical modification of the structural movement in the late 1970s and 1980s; and third, concluding with an argument stressing the continuation of structural themes even in the post-structural/post-modern approaches typical of the 1980s and 1990s.

MARXISM, STRUCTURALISM, GEOGRAPHY

In idealist and religious forms of understanding, the earth was created by the will of spirit and historical events are materializations of prior spiritual intent. Materialist philosophy, by contrast, argues that the earth originates in physical processes alone, while history has purely natural or social causes: even consciousness, a piece of God's mind lodged in the human brain for idealists, has a natural origin for materialists.

Marxism

Marx's historical materialism suggests, therefore, that social analyses begin with the provable, observable position that human beings, human activity and social relations have purely natural, material origins. In acting on nature humans produce the material basis of their continued existence. Reproduction employs the forces of production, including human labour power itself (the human's capacity to work) and the various productive means which enhance labour power (tools, machines, computers, infrastructures). In some versions of Marxism the development of rationality expressed in the growth of the forces of production leads the evolution of societies and precipitates historical transformation (see, for example, Cohen 1978). But productive forces do not grow in vacuums, nor humans act on nature as isolated atoms. Rather, individuals engage in relations with each other in the *social* reproduction of their existence. Indeed, Marx can be alternatively interpreted as arguing that social relations determine the growth (or stagnation, or retrogression) of the productive forces, the forms of their exertion on nature, the transformation of the natural environment during the productive process, and thus the directions taken by socio-natural development in general. Social relations thus become the key category in Marxian analysis (Marx 1976).

Marx elaborates a particular version of this analytical category: social relations involve systems of ownership and control of productive property, and power over people rests, most essentially, on control over the production of their existence. In the early forms of human society, Marx (1973) argues, nature and social means of production were communal property in egalitarian hunter-gatherer societies (Leacock and Lee 1982). But with the domestication of plants and animals, and with agriculture and permanent surpluses (production above that necessary for the simple maintenance of existence), class and gender inequalities were generated during social struggles over the control of reproduction, at first within kinship groups, then in the form of the hegemonic state, and finally as private ownership of the means of production by the capitalist class. In all class societies surplus produced by human labour power is taken through various forms of 'exploitation' and used in the accumulation and perpetuation of elite power (Marx 1976). Exploited people, be they peasants paying taxes to the state, women forced by 'custom' to work for male elders or husbands, factory labourers or office workers turning out a greater value of product or service than they receive back as wages, resist exploitation and challenge elite control over the conditions under which labour and working lifetime occur. Furthermore, from a *geo*-materialist perspective, the exploitation of labour leads to the exploitation of nature: contradictory social relations direct socio-economic development in contradiction to nature. Modern life therefore unfolds as a series of escalating and interlocked social and ecological crises. Hence Marxists analyse history in terms of the dynamics of class, gender, and similar forms of social struggle. They see socialism as eliminating elite power through the direct control of social reproduction by everyone involved in its institutions and processes.

Existentialism and structuralism

As intimated above, Marxism is subject to numerous interpretations. In the twentieth century, Marxism – with other social theories – has been influenced by notions stemming most importantly from existential phenomenology and structuralism. Returning to the notion of 'humans making themselves', it is clear that two quite different positions may be taken. Either humans are free to create themselves through subjective means or they are continually re-created by the structures they form in relation with nature. At their polar extremes, these competing positions stress either the human being's freedom and responsibility for its existence (existentialism) or the irrelevance of human intention in the light of structural necessity (structuralism).

Existentialism is a philosophy of the existence of the human subject. For Kierkegaard (1936), the first modern existentialist, 'existence' means the unique, concrete being of the individual which refuses any system constructed by rational thought. Sartre (1956) asserts that 'existence precedes essence', defining existence to mean concrete individual being, the conscious subject, free to choose its essence, yet paradoxically finding in freedom a lack of being. Existential phenomenology gives accounts of the active subject; humans *act*, develop their own characters, and interpret the world. To reveal the character of the subject, and the meaning of the world, existential phenomenology describes ordinary human activities. The task of existential phenomenology is to describe the characters of existent objects and active subjects, given their interaction (Hammond *et al.* 1991).

This humanist, existentialist and phenomenological position came to the peak of its prominence in the works of Sartre, Merleau-Ponty and others in post-war France, and continues to influence humanistic social science today (Poster 1975). Writing in the late 1950s, Sartre (1968, 1976) attempted to synthesize existential phenomenology with humanist Marxism in a kind of existential anthropology. Sartre accepted Marx's dictum that the mode of production of material life determines the social, political and intellectual life process in general. Marx, he says, tried to generate knowledge dialectically, rising progressively from the broadest determinations to the most precise, from the abstract to the concrete. For the existential Sartre, individual human beings *are* the most concrete entities; yet Marxism lacks a hierarchy of mediations allowing it to grasp the process which produces a person in a class, within a society, at a given historical moment. Existentialism, he said,

> intends, without being unfaithful to Marxist principles, to find mediations which allow the individual concrete – the particular life, the real and dated conflict, the person – to emerge from the background of the general contradictions of productive forces and relations of production.
>
> (Sartre 1968: 57)

However, exactly as Sartre formed this synthetic project, Marxism, along with other modes of thought, moved in the opposite direction. As opposed to the humanist, existential notion that history is created by human agency, structuralism argues that social structures have no agents, and cannot be understood by studying individual human beings. Indeed, the father of contemporary structuralism, the French anthropologist Lévi-Strauss (1966: 62), found existentialism the exact opposite of true thought by reason of its 'indulgent attitude toward the illusions of subjectivity'. While Sartre opposed reductionism, for Lévi-Strauss the very purpose of science was reduction: in anthropology, ethnographic analysis (observing and analysing discrete human groups) tries to arrive at the invariants of a general humanity, from which an attempt is made at reintegrating culture into nature and, beyond that, life within its physico-chemical origins.

Lévi-Strauss reached this position through a study of structural linguistics (Saussure 1959). Here the emphasis is placed on the supra-individual and social character of language systems, the codes linking communicators, the language system (*langue*) rather than the particular utterances of speakers (*parole*). This approach is thought to be more generally applicable to all social and cultural phenomena, interpreting these as though they were linguistic signs (Eco 1973). Thus in semiotics, the study of signs, the meanings of cultural phenomena are conditioned by the underlying social system of conventions. Space too may be considered a cultural product resulting from signifying or semiotic production (Gottdiener and Lagopoulos 1986). In this sense cultures and spaces are symbolic orders structured by social relations (Lyons 1973; Culler 1973).

Structural Marxism

Influenced by structural linguistics, anthropology and semiotics, the French philosopher Althusser (Althusser 1969; Althusser and Balibar 1968) reconceptualized Marx's notion of social structure. In Althusser's reconception the forces and relations of production determine the character of the societal totality by determining which of its aspects or 'instances' (economy, politics, culture, ideology, etc.) is *dominant* at any time. Social totalities are thus 'structures in dominance' with the various instances of society having hierarchical order determined finally by the economic. Yet also, like words in speech patterns, each societal instance is relatively autonomous. Far from being epiphenomena determined unilaterally by economy, the cultural and political instances are the economy's conditions of existence. Indeed, for Althusser (1969: 43) pure determination by the economic, even in the last instance, never happens: 'From the first moment to the last, the lonely hour of the "last instance" never comes.'

In the complex, structured unity of the social totality the way different social (and we might add, socio-natural) contradictions are articulated with each other determines the direction taken by the entity as a whole. The whole is a unity of instances developing unevenly at different speeds, the term 'conjuncture' expressing the specific complex of unevenness at any time. Events in any conjuncture are multiply determined ('overdetermined') by factors which these events influence in turn. This conception of dynamic totalities is greatly different from earlier versions in Hegelian and mechanistic (Soviet) Marxism: often misunderstood, Althusser's concept of over-determination opposes reductionism, whether to spiritual essence, or to economy (Callinicos 1976).

STRUCTURAL MARXIST GEOGRAPHY

Virtually all structural geography theorizes the social structuring of space, with 'space' being narrowly interpreted as geometrical arrangement. The deeper notion of structural relations between society and nature, relations which would include space as merely one aspect, is approached indirectly via discussions of space. Marxist geography has thus realized only a fragment of its structural potential.

Structural urban geography

While there were earlier contacts between Marxism and geography (for example, Wittfogel 1929) most connections between the two were made as part of the 'radical geography' movement of the 1960s and 1970s (Peet 1977). The first modern work in English consciously employing Marxian analysis, Harvey's influential *Social Justice and the City* (1973), used versions of structuralism developed by Piaget (1970) and Ollman (1971). Piaget argues that totalities are structured via the elaboration of the relationships within them; relationships between elements within the structure express 'transformation rules' through which the totality is changed. As Ollman alternatively puts it, the totality shapes the parts to function in preserving the existence of the whole and each element internally reflects all the characteristics of the whole. For Harvey, likewise, reality is a totality of related parts, or a structure of elements. As opposed to structural functionalism (Parsons 1949) the related elements are frequently in contradiction, from which come conflicts:

> Transformations occur through the resolution of these conflicts and with each transformation the totality is restructured and this restructuring in turn alters the definition, meaning and function of the elements and relationships within the whole. New conflicts and contradictions emerge to replace the old.
>
> (Harvey 1973: 289)

For Harvey, dialectical materialism is a way of elaborating the transformation rules by which society is constantly restructured.

Harvey uses this structural reading of Marx to re-examine the controversy over urbanism and urban origins. Following Marx he reasons that urbanism originated as a resolution to contradictions in pre-urban society, while urbanism itself generated new contradictions, like antagonisms between town and country. This he says is a quite different approach from explanations provided by non-Marxists, who explain urban origins through lists of 'causal factors' or in terms of 'emergent totalities' (that is, totalities which exist and change independently of their parts). Drawing critically on the work of the French urban sociologist Lefebvre (1970, 1972), Harvey argues that urbanism is a set of social relationships structured by relations in the society as a whole. Yet while urbanism is moulded from basic principles of social organization, it has a relative autonomy influencing, in turn, the development of social relations and the

865

organization of production. For Harvey urbanism is channelled and constrained by forces in the economic base of the society, ultimately being related to the production and reproduction of material existence.

While we almost have to read structuralism into Harvey's book, French work on cities and urban planning was more obviously influenced by structural Marxism. The French work began to appear in English in the middle to late 1970s (Preteceille 1976; Pickvance 1976; Castells 1977) while Latin American geographers attuned to theoretical developments in France (Santos 1977; Corragio 1977) also spread structural notions. Castells (1977) closely follows Althusser's reasoning in arguing that space is a concrete expression of the historical ensemble in which society is specified. As an expression of social structure space is shaped by elements of the economic, political and ideological instances, by their combinations, and by the social practices derived from them. In effect, then, space may be 'read' according to the different systemic logics which form it. An economic reading concentrates on the spatial expressions, or 'realizations', of production. In a society where the capitalist mode of production is dominant, the economic system dominates the social structure (cf. Althusser) and is the basic force organizing space. This does not mean that industry simply and directly shapes space according to the logic of profit maximization, for there are a number of tendencies within production besides profit-making, while production interacts with other instances of the society in shaping space. Hence, the politico-juridical (state) system structures institutional space according to political processes of integration, repression, domination and regulation. Likewise, space is charged with meaning under the ideological system, its forms and arrangements being articulated one with another in symbolic structures. For Castells (1977), the social organization of space in any conjuncture can be understood in terms of determination by the currently hegemonic mode of production (in the ways described above), by the persistence of spatial forms created by earlier social structures (articulated with the new forms in evermore specific concrete situations) and in terms of the specific actions of individuals, social groups, etc. in interaction with their environments.

Social process and space

These early versions of the socio-structuring of space, simple yet vague though they were, represented an advance over earlier, incestual notions of the spatial formation of space. Commenting on this, Corragio (1977: 14–15) found Castells's structural approach to space preferable to the 'spatialist approach' of positivist geography in which '"spatial structures" are produced by "spatial processes"'. But Corragio was interested in a far broader theory of the structuring of material space by social structures and processes, and in turn the effect of space on social process.

Corragio (1977) returns to Marx's 'economic metabolism' between society and nature, emphasizing in particular the 'spatial crystallizations' of the various aspects of this metabolic process. Economic metabolism in capitalist societies with advanced divisions of labour, Corragio argues, may be visualized as a chain of the production, circulation and consumption of commodities, which require fixed locations (for means of production, etc.) and flow systems (for the organic interchange of materials). These activities crystallize in space in the form of factory buildings, transport networks, storage operations, housing, etc. Systems of social regulation and control also spatially crystallize in nucleations (offices, banks) and linear traces (telephone lines, etc.). Natural resources form bases for the regionally differing operation of metabolic processes. More generally, the natural environment can be viewed 'as a (natural) metabolism, with its own processes and timings, to which the processes of social metabolism are connected' (Corragio 1977: 17). Economic landscapes form as one aspect of this dual metabolism.

For Corragio the human agents involved with the metabolic process make 'locational decisions' about the productive or consumptive apparatuses and 'coupling decisions' dealing with flows between locations. In discussing this, neo-classical (positivist) location theories stress the intentionality of the agent (entrepreneurs, companies, etc.). By comparison, for Corragio (1977: 19), 'the real regulation of spatial configurations is determined by processes that cannot be reduced to an interaction between unvarying social atoms'. The (locational) success or failure of a particular agent (company) depends on the objective structural forces and processes. Thus for Corragio (1977: 20), in a typical structuralist statement, 'observable behavioral patterns are not causes, but convey objective forces which produce spatial configurations'. In this view socialism would consist of democratic control over the 'objective' forces determining livelihood.

Similar arguments were made by Santos (1977), who finds that modes of production (combinations of forces and relations of production with elements of politics and ideology) become concrete on historically determined territorial bases: spatial forms thus constitute a 'language' of modes of production. But for Santos mode of production analysis is not fully adequate to the intricate specifics of space, for which the intermediate term 'social formation' is more fully applicable:

> The relations between space and social formation are of a different order altogether, since they are formed in a particular space, not space "in general" like the modes of production. Modes of production write history in time; social formations write it in space.
>
> (Santos 1977: 5)

Modes of production have effects on natural environments yielding a series of social formations or, rather, socio-spatial formations. Like Corragio, Santos finds that while space is matter shaped by the totality of social life, space also shapes human activity. Only 'theoretical lag', he concludes, can explain why the ideas of space and social formation have not previously been united in a single concept.

Spatial dialectics

Issues of the relations between society and space were further elaborated via discussions of the dialectics of space and the dialectics of space and society. Peet (1978, 1981) proposed moving beyond Castells's 'created space' by extending Althusser's relative autonomy to spatial relations in what would remain a structural version of Marxist geography. For Peet, differing natural conditions influence the (internal) form and development of regional social formations, while (external) spatial relations between formations modify, and even transform, internal contents and processes. Thus the geography of capitalism is composed of differentially developed spatial instances (social formations) with historical change unfolding unevenly in each through sequences of interlinked contradictions. As social contradictions build internally in a social formation (for example, economic crises in late nineteenth-century Europe) the nature of external socio-spatial relations changes (a renewed tendency for imperialism) with effects being transmitted elsewhere (the destruction of the 'natural' economies of Third World countries) or antidotal 'solutions' imported (additional markets for European manufactures, cheaper sources of raw materials, etc.). This may slow down or redirect the build-up of contradiction in one social formation (central capitalism moves into a new phase) but may qualitatively change regional social formations elsewhere (peripheral societies), where the interaction of external influences with local processes brings new hybrid social forms into existence (colonial and then post-colonial social formations). The idea essentially is to show that 'structural' contradictions are comprised of interlinked, but still particular, regional contradictions and that crisis has a geography of intensities – hot and cool spots which interchange over time. The complex interplay across space between the social formations of a global system may be termed 'spatial dialectics':

> World history is understandable in terms of the dialectics of the whole, the geographical instances of the whole, and the relations across space between these instances. The essential contribution of Marxist geography in this formulation is an understanding of the uneven development of contradictions in space, the forms in which contradictions appear as complexes of crises, and the (spatial) relations between these complexes.

> (Peet 1981: 109)

Unresolved in such discussions is the initial separation between structure and space. For Soja (1980) the problem comes from not appreciating the *dialectical* character of socio-spatial relations:

> space is not a separate structure with its own autonomous laws of construction and transformation, nor is it simply an expression of the class structure emerging from the social (i.e. aspatial) relations of production. It represents, instead, a dialectically

868

defined component of the general relations of production, relations which are simultaneously social and spatial.

(Soja 1980: 208)

For Soja social and spatial relations are not only homologous (i.e. correspond), in that both arise from the same origin in the mode of production, but are also dialectically intertwined 'in that each shapes and is simultaneously shaped by the other in a complex interrelationship which may vary in different social formations and at different historical conjunctures' (Soja 1980: 225). Space itself may originally be primordially given but, as with 'second nature' (that is, environments altered by human practice), its organization, use and meaning are products of social transformation. Once it is recognized that space is socially organized there can no longer be a question of its being a separate structure with independent rules of transformation. The dialectical relationships between created space and society become the important moments. The social–spatial dialectic calls for the inclusion of socially produced space in Marxist analysis as something more than an epiphenomenon. For Soja the interplay between the social and the spatial should be a central issue in social theory (see also Soja 1989).

The regulation school

Criticism from sources within Marxism (for example, Thompson 1978) and outside it (for example, Baudrillard 1975) led French Marxists of the 'regulation school' to develop structural concepts in new directions using novel terms. They retained the notion of society as a coherent structure of elements: as with the term 'mode of production', their analysis of society begins with the labour process and builds outwards into consumption, culture and politics. Thus, the regulationist concept 'regime of accumulation' refers to

> the stabilization over a long period of the allocation of the net [social] product between consumption and accumulation; it implies some correspondence between the transformation of both the conditions of production and the conditions of the reproduction of wage earners. It also implies some form of linkage between capitalism and other modes of production.

(Lippietz 1986: 19)

Such regimes do not materialize automatically, but need means of coercion and institutions of persuasion to assure the cohesion of people's strategies and expectations. The body of norms, habits, laws and regulating networks which ensure that individual behaviour is consistent with a regime of accumulation is called a 'mode of regulation'. Regimes and modes succeed in practice because they ensure regularity and permanence in social reproduction. However, bearing in mind criticisms of the functionalism of structuralist thinking (for example, Giddens 1979a), it is argued that modes of regulation do not simply have the 'function' of making regimes of accumulation work. 'Rather, a regime of accumulation and forms of regulation get stabilized together because they

869

ensure the crisis-free reproduction of social relations over a certain period of time' (Lippietz 1986: 20). External relations also play a regulating role for capitalism of the global centre: imperialism, for example, temporarily resolves contradictions of central capitalism. But again bearing in mind that congruent human intentions are not automatically generated by structural necessities, in Lippietz's (1986: 20–1) view imperialism was not created specifically (intentionally) to resolve contradiction, and other solutions could have been found – 'we should not confuse "results" with "causes"' and must remember 'the openendedness of history, the class struggle, and capitalist competition' as well as 'the autonomy of national social formations and the sovereignty of states'.

This proved an attractive set of notions, especially for geographers. The French regulation school divides the history and space of capitalism into regimes based essentially on the prevailing labour process: Fordism, beginning just after the turn of the century and dominant from 1940 to the late 1970s; and neo- or post-Fordism beginning with the economic crises of the 1970s and rapidly expanding in the late twentieth century (Dunford and Perrons 1983). Within each regime attention is also paid to types and levels of consumption, related to phases in the labour process, and related also to popular culture, aesthetics, and the media. In Aglietta's (1979) version of regulation theory, consumption is conceptualized as an organized set of processes which both reconstitutes the energies expended in social activities (especially production) and preserves those abilities and attitudes of the socialized individual implied by hegemonic social relations. Under Fordism, for the first time workers are paid enough to consume the sophisticated products they increasingly make. In particular, Fordism involves individual ownership by the working class of commodities like automobiles and standardized housing which permit 'the most effective recuperation from physical and nervous fatigue in a compact space of time within the day, and at a single place, the home' (Aglietta 1979: 159). But also, because they are bought on credit, these forms of consumption serve to regulate the behaviour and beliefs of consumers. Fordist consumption entails also the creation of a functional aesthetic, the calculation of consumption habits, and their social control via media, advertising, entertainment, etc. (Peet 1989a). The resulting 'consumer society' appears to resolve a number of contradictions: the problem of underconsumption in an age of mass, mechanized production; the problem of worker alienation on assembly lines (alienation at work being alleviated by consumption at home); the problem of a lack of democracy in the workplace, choices being endless in the consumption sphere. But in the 1970s, so the argument goes, Fordism entered a period of crisis, for instance as the limitations of the intensification of labour were reached in the limited physical and mental capabilities of the worker, and as workers somehow persisted in resisting through organizing the strikes which escalated in the late 1960s. Neo- or post-Fordism, based in automatic production, responds to the crisis of productivity by introducing a new flexibility of production design and location. Generally the response to the intensification of contradiction in Fordist locations is to move elsewhere and/or to discipline populations (e.g. through automation) expressing critical reactions to the prevailing production system. Hence for Aglietta (1979: 385–6) post-Fordism means 'massive socialization of the conditions of life' and 'a strong totalitarian tendency under the ideological cover of liberalism'. Additionally, in post-Fordist societies critical reactions against the system are integrated into it as sources of growth, hence the wave of environmentally sensitive and health-conscious products. The impression is of a global system spinning out of rational control and subject therefore to increasing ideological manipulation.

The appeal to geography of these notions lies in their incorporation of the dimension of space. Each regime is associated with a distribution in space of its economic and social components, together with the politico-cultural regulation mechanisms tied to them. Fordism is characterized by a division of labour, including a spatial division of labour, between conception, skilled manufacturing, and unskilled assembly work, and multinational corporations can spread these across different labour pools (Massey 1984). Hence the development of specialized production regions with different class structures and the related development of centres and peripheries in the production and distribution of popular cultures associated with different levels and styles of consumption. Recent work developing regulation school ideas has tried to capture the nature of the emerging post-Fordist regime using the term 'flexible production system' (Piore and Sabel 1984). Scott (1988) argues that the typically rigid mass production processes of Fordism are giving way to changeable, computer-enhanced processes, situated within systems of malleable external linkages and labour market relations. Storper and Scott (1989) find the turn towards flexibility marked by re-agglomeration of production within a wider international division of labour. The older foci of Fordist mass production, unionized with rigid labour relations and with governmental restrictions on producers, are avoided by the newly flexible production systems – for example, high technology industries located in the suburbs of large metropolitan areas and in previously unindustrialized communities and countries. The socio-economic contradiction inherent in post-Fordist societies is the lack of high-paying jobs, the basis of a high unemployment–alienation–social problems syndrome, although the globalization of environmental crisis is an even more fundamental contradiction, with increasing interpenetrations and interactions occurring between the social and environmental aspects of a single contradictory complex. The phases of capitalist development are also spread unevenly across space, so that post-Fordism may typify Los Angeles, Tokyo, and London, a decrepit Fordism may be the case in Detroit, Birmingham and the Ruhr, and peripheral Fordism may just be developing in Latin America and East Asia.

Largely unexplored in much of the regulationalist literature are the connections between differences in levels of consumption, degrees of resource exploitation, and thus pressures on natural environments in pre-Fordist, Fordist and post-Fordist zones of the global capitalist system. Different regulation systems may also imply varying attitudes towards environment – for example, Fordist consumption associated with increased objectification (Disneyization) of nature. Fordist development is predicated on the intensification of resource use in central regions, but more significantly the expansion of exploitation into non-Fordist peripheries, in massive processes of resource transfer which parallel flows of economic value, capital and political power.

The major differences between what might be termed a 'neo-structuralist' regulation theory and the earlier, more obviously structural work are the more fluid relations between economy and social, political and cultural characteristics, the breaking of modes of production into phases (Fordism, post-Fordism, etc.), and in general an attempt to move away from what is seen as an overly deterministic, even functional, form of structural Marxism. Even so the regulation school is a continuation of the structural notion which, in Harvey's (1989: 121) account, provides a way of representing the social, cultural and political changes of the post-modern era 'which does not lose sight of the fact that the basic rules of a capitalist mode of production continue to operate as invariant shaping forces in historical-geographical development'.

STRUCTURATION THEORY

Ideas like these allowed geographical conceptions of causality to be extended from spatial causes of spatial phenomena to social causes of spatial phenomena or, more complexly, to the reintegration of the spatial into the social in a formulation in which each aspect of the interdependency is relatively autonomous and has causal powers. But structural versions of social causality of all types soon became the targets of vigorous criticism, from which emerged a somewhat different, structurationist approach to a new regional geography.

Geographical critiques of structuralism

Recollecting earlier differences between existential and structural philosophies, structural geography was criticized from the perspective of humanistic geography for its impoverished conception of human agency. In this critique, structural Marxist geography is portrayed as a form of holistic explanation in which large-scale events occur regardless of the conscious actions of their human participants. This, it is said, leads structural geography into reifying analytical categories like mode of production, while neglecting social-psychological and other human factors (Duncan and Ley 1982).

In turn, humanistic geography (Ley and Samuels 1978) was itself criticized for its failure to clarify the relation between human agency, structure and structural transformation, a relation which for Gregory (1981: 5) ought to lie at the heart of human geography: 'any such geography must restore human beings to their worlds in such a way that they can take part in the collective transformation of their own human geographies'. For Gregory the problem for humanistic geography lies in finding a model which allows autonomy to social consciousness within a context determined, in the final analysis, by social being, one in which history is neither willed nor fortuitous, and neither lawed nor illogical. Gregory proposes the theory of structuration, most fully developed by the sociologist Giddens, as a way of resolving the differences between the two positions.

Taking a similar intermediate but critical position Thrift (1983) found human geography caught between the determinism of structural Marxism, which tries to 'read off' the specifics of place from the general laws of capitalism, and the voluntarism of most humanistic geography, in which events result from purely individual intention. Like Gregory he also sees the tension between these positions leading human geographers towards a resolution in structuration theory (Bourdieu 1977; Giddens 1979b, 1981; Bhaskar 1979). The qualities Thrift finds most attractive in this approach are: its critique of functionalist explanation as evasive; its critique of both structural-determinist and voluntarist approaches and recognition of the duality of social structure; its critique of social theory's lack of an adequate notion of the acting subject, or theory of practice, and recognition of the need for an explicit theory of practical reason in human intentionality; and its view that time and space are central to social interaction and social theory. For Thrift (1983) these common elements provide means of progressing towards a Marxism excised of its functionalist, evolutionist and essentialist excesses. With this many human geographers turned their attention from structures to more humanly proactive structuration processes, particularly as envisioned by the sociologist Giddens.

Giddens's structuration theory

Giddens reconsiders the division within social theory between functionalism and structuralism on the one side and hermeneutics, existentialism and other forms of 'interpretative sociology' on the other. Functionalism and structuralism express a naturalistic standpoint emphasizing the pre-eminence of the social whole over individual parts (for example, the constituent actors). Hermeneutics, by contrast, presents subjectivity as the preconstituted centre of experience and thus the basic foundation of human science. Giddens's structuration theory aims at ending the 'empire-building endeavors' of the imperialisms of object (society) and subject (the individual) by proposing instead that the social sciences study social practices ordered across space and time.

At the core of Giddens's theory of structuration lie the concepts 'structure', 'system' and 'duality of structure'. In structuration theory, structures are sets of rules (constraints) and resources (capacities or possibilities). 'Structure' in this sense has only a 'virtual existence' in the reproduced social practices of social systems and the human memory traces orienting social conduct. By contrast 'systems' have concrete existence as 'interconnected or articulated series of institutionalized modes of interaction reproduced in spatially distinct social settings over a determinate period of history' (Cohen 1989: 89). By the 'duality of structure' structuration theory links structuralism to hermeneutics by proposing that recursive social practices both draw on structural rules and resources and reconstitute them: 'Structure enters simultaneously into the constitution of the agent and social practices, and "exists" as the generating moments of this constitution' (Giddens 1979b: 5). Structure gives systemic form to social practices across time and space.

For Giddens, institutions are clusters of regularized practices, structured by rules and resources, which are 'deeply layered' in time and space (that is, 'stretch' through time and space). Systemic change is conceptualized in a similar manner to Marx, in that structural principles operating in relation with one another also contravene one another. Primary contradictions are oppositions or disjunctions of structural principles fundamentally involved in the system of reproduction of a society; secondary contradictions are brought about as a result of primary contradictions; and conflict is linked to contradiction, with an area of contingency between the two (Giddens 1979b). However, contrary to an evolutionary reading of Marx, contradiction can also stimulate retrograde movements of historical change.

As well as structural contradiction there is in Giddens a conception of existential contradiction in the generic relations between humans and their organic conditions. There is

> an antagonism of opposites at the very heart of the human condition, in the sense that life is predicated upon nature, yet is not of nature and is set off against it. Human beings emerge from the 'nothingness' of inorganic nature and disappear back into that alien state of the inorganic.
>
> (Giddens 1984: 193)

Giddens (1979b: 202) argues that 'most forms of social theory have failed to take seriously enough not only the temporality of social conduct but also its spatial attributes', whereas the theory of structuration portrays time–space relations as constitutive features of social systems. In this Giddens (1984; see also Giddens 1985) sees geography taking a significant position in social thought. Thus the Swedish geographer Hägerstrand's (1975) concept of 'time-geography' stresses the routinized character of daily life connected with the basic features of the human body, its mobility and means of communication, path through the life-cycle, and therefore the human's biographical project. For Giddens, developing time-geography's ideas as part of structuration theory involves reconceptualizing the notion of 'place' to mean more than point in space. Giddens (1984: 118) therefore uses the term 'locale' to refer to the 'use of space to provide the settings of interaction', these being essential to specifying the 'contextuality' of interaction and the 'fixity' underlying institutions. Ranging in size from a room to the territory of a nation-state, locales are internally regionalized, with regions being critically important as contexts of interaction.

Giddens's more innovative work on space involves connecting the time–space constitution of social systems with the question of power. Certain types of locale, he argues, permit a concentration of allocative and authoritative resources which generate power; as opposed to Marx, Giddens emphasizes the concentration of 'authoritative', rather than 'allocative', resources: these include possibilities for surveillance allowed by various settings; the possibility of assembling large numbers of people not directly involved in material production; the facilitation of the scope and intensity of sanctions (at first in the city, later in the nation-state); and the creation of conditions influencing the formation of ideology (Giddens 1985: 13–17). In Giddens (1981: 5) 'storage capacity' is a fundamental element in the generation of power through 'time–space distanciation'; that is, through the 'stretching' of societies across time and space. Information-storage devices, like writing, simultaneously are parts of the generation of power and allow the extension of social control in time–space, an extension which involves increasingly elaborate devices of societal integration. In essence this amounts to a power–space theory of history.

THE NEW REGIONAL GEOGRAPHY

Giddens's work on structuration, power, locale and space coincided with, yet also helped generate, a renewed interest in regional geography. In the new version 'regions are defined through historical material processes in which spatial structures initially are produced and, by their reproduction and transformation, become constitutive of material processes' (Pudup 1988: 380). The idea is to link regions with structuration processes in a theory at once social and spatial, a theory of generalities capable of dealing with specifics.

Social action in space

Thus for Thrift (1983) structuration processes should be specified as they occur for particular individuals or groups in particular localities. Such a historically specific, contextual theory of human action, he thinks, should stress practical reason and concrete interaction in time and space. This gives renewed emphasis to a reconstructed regional geography, which builds on the traditional but is distinct in its more theoretical orientation and emancipatory aims. The social activities in a region take place as a continuous 'discourse' rooted in shared material situations, with cultures having both limiting and activating capabilities. A reconstituted regional geography would start conventionally with a compositional account of the regional setting, followed by the organization of production, class formation, the sexual division of labour, and the local form of the state. However, since concern lies with developing a theory of social action, it is necessary to constitute the region as setting for interaction, or in Giddens's terms, a locale; that is, region provides the constraints and opportunities for action, the base for what is known about the world, and the material for changing it. In a particular landscape, certain locales (home-reproduction or work-production) are dominant, structuring people's life-paths in space and time, constraining interaction possibilities, providing the main arenas of interaction (and thus being the sites of conflict), providing the activity structures of daily routines, and forming therefore the major sites of socialization processes. For Thrift, the analysis of social actions as a discourse through and in a region is a goal rather than an achievement, one needing a concentrated programme of theoretically informed empirical research to back it up.

Place, practice and structure

A basic problem with structurationist approaches is that complex terminology and abstract theorization tend to preclude empirical research. One of the few pieces of research to have actually used structuration theory in empirical practice is Pred's (1986) study of southern Sweden (but see also Gregory 1982). Pred contrasts structuration with traditional regional geography, which he finds portraying places and regions as little more than frozen local scenes of human activity. For Pred, by comparison, place involves an appropriation of space and nature inseparable from the reproduction and transformation of society as a whole. Place is not only locale, in Giddens's sense of the term, but also what *takes* place, 'what contributes to history in a specific context through the creation and utilization of a physical setting' (Pred 1986: 6). This initiates a theory of place as a historically contingent process emphasizing both individual practices and structural features using, as linking devices, the theory of structuration, the language of time-geography, and a version of traditional human geography which, however, stresses power relations binding individuals, society and nature in place-particular practices. The theory attempts to account for the material continuities of people and the natural and human-made objects employed in time–space-specific practices. Participating individuals are regarded as integrated human beings, at once objects and subjects, whose thoughts and actions, experiences and ascriptions of meaning, are constantly 'becoming' through involvement in the workings of society.

For Pred, social structure is the generative rules and power relations built into specific historical and geographical situations and social systems. But nobody identifiable with the structuration perspective has succeeded in conceptualizing the means by which the everyday reproductive shaping of self and society come to be expressed as practices locationally specific in time and space. Pred (1981a, 1981b, 1983) contends that this limitation can be overcome through integrating structuration theory with time-geography, as with the concepts of 'path' and 'project'. According to the 'path' concept each action and event making up the existence of the individual has temporal and spatial attributes so that a person's biography is a continuous path through time and space subject to various types of constraint. A 'project' consists of the series of tasks necessary for completing any intentional or goal-oriented behaviour. Each sequential task in a project involves the coupling in time and space of people's paths and tangible inputs or resources, the task couplings having a certain logical, consecutive order and duration. Structuration processes, then, are spelled out by particular paths and projects, while places are inseparable from the everyday unfolding and interpretation of specific structuration processes.

As place-specific biographies are formed through social reproduction the physical environment is constantly transformed. Any project using objects directly or indirectly transforms nature, although this link is often opaque when non-local resources are employed; through engagement in the becoming of place, human nature is also internally transformed. The transformation of body and nature can be expressed in terms of path convergence–path divergence, creation–destruction, or presence–absence dialectics rooted in the time-geography perspective. For Pred, structuration processes also entail power relations as the structural cement holding the individual, society and nature together in time–space-specific practices. Power is the capacity to define precisely the content, or control the tasks, of a specific project involving others, while power relations involve the capacity to prevent the participation of others in a particular project. Power relations underlying local practices are themselves reproduced and transformed by these practices, with transformations brought about through conflict and involving the modification of project definitions and rules, or even the elimination of some projects. The most potent aspect of power relations lies in affecting what people know and are able to say.

For Pred, three empirical foci are suggested by this theory of place as a historically contingent process: the place-specific impact of dominant institutional projects on the daily and life-paths of participants, on the landscape, and on power relations; the formation of particular biographies as a reflection of elements of structuring processes in place; and the sense of place as part of the becoming of individual consciousness and biography formation. He exemplifies these topics in a study of changes in the agricultural villages of southern Sweden during land enclosure in the late eighteenth and early nineteenth centuries. His case study shows that the 'becoming of place' and 'formation of biographies' are intermeshed with more geographically extensive historical processes. Negotiated agreements to institute enclosures emerging from changes at the global, regional and local levels resulted in the radical restructuring of local space, new daily paths, altered power relations, and changed farming practices. Most crucially for locality study, Pred (1986: 197) finds that 'despite a shared macro-structural context, the unfolding of structuration processes in each village had some unique attributes ... local influence was exercised by intergenerationally sedimented predispositions, human creativity and agency'. This reinforces geography's move towards the local and the particular.

But these are still limited findings. Pred's empirical work is strongest in elucidating the multiple agencies mediating between large-scale causal processes and local outcomes. The return flow of influences, from the local to the global, the grassroots practices (that is, 'what contributes to history') by which people make not only village life but constitute global processes, is barely mentioned, so that in effect only one geographical aspect of the spatial dialectics of reproduction is explored. Parallel problems have been experienced in similar projects in ethnography (Marcus 1986). Structuration theories should allow the full analysis of mediations between the local and the global – a geographical version of Sartre's 'project'. Yet global patterns are still analysed as resulting from meta-level or at best meso-level processes rather than as the result of combinations, at ever-increasing scales, of local, regional and *then* global processes. Thus, in practice, the local and the particular remain 'victims' of transcendent global and general processes, rather than being seen as constitutive, active elements; structures from nowhere retain primal causality.

THE SOCIAL IMAGINARY

Structuration theory, and the attempts to found a new regional geography on it, moves social theories away from the notion of structures as *pre*constituted objects, in the direction of systems which are *re*constituted by daily practice. While this leads in the geographically useful direction of the study of localities it does not say much about purpose, imagination, and meaning, being overly preoccupied with the practical reasoning of everyday reproduction. To a degree these broader issues of the connections between everyday reproductive purpose and action and the overall social imagination have been more successfully broached by Castoriadis (1991) and Touraine (1988) in styles which draw from phenomenology and existentialism yet remain compatible with the structure/ structuration project.

The self-deployment of society

As with Marx, Castoriadis begins with the physical environment, the biological properties of human beings, and the necessity of material and sexual reproduction, for which coherent fragments of ensidic logic and applied knowledge must be created. But he claims this would be as true for apes as it is for human beings. For Castoriadis

> The construction of its own world by each and every society is, in essence, the creation of a world of meanings, its social imaginary significations, which organize the (pre-social, 'biologically given') natural world, institute a social world proper to each society (with its articulations, rules, purposes, etc.), establish the ways in which socialized and humanized individuals are to be fabricated, and insaturate the motives, values, and hierarchies of social (human) life. Society *leans upon* the first natural stratum, but only to erect a fantastically complex (and amazingly coherent) edifice of significations which vest any and every thing with *meaning* . . .
>
> (Castoriadis 1991: 41)

In socializing 'the wild, raw, antifunctionally mad psyche of the newborn' society must 'in exchange' provide the psyche with meaning. From the standpoint of the psyche, the process by which it (never fully) abandons its initial ways and cathects socially meaningful ways of behaving, motives and objects is *sublimation*; from the view of society it is *social fabrication*. The task of 'knowing' a society therefore consists in reconstituting the world of its social imaginary significations.

Contradicting the spectrum of existing theories of societal dynamics, history as the will of God, history as the action of (natural or historical) laws, as a subjectless or, for that matter, random process, Castoriadis (1991: 34) argues: 'History does not happen to society: history is the self-deployment of society.' His notion is that the elements of social-historical life are created each time (in terms of relevancy, meaning, connections, etc.) in and through the particular institution of society to which they 'belong'. Thus each social-historical instance has an essential singularity: phenomenologically specific in the social forms and individuals it creates; ontologically specific in that it can put itself into question, explicitly altering itself through self-reflective activity.

Social fields of action

Similarly Touraine replaces the construct of society as a system driven by an inner logic with society as a 'field of action'. His stress lies on the social praxis involved in the genesis of norms, and the conflicts over their interpretations. Whereas in Marxism classes are defined structurally by positions in the production process, for Touraine they are defined purely in terms of social action. Touraine also distinguishes himself from what he regards as the main message of structural/post-structural social theory. From Marcuse to Althusser to Foucault and Bourdieu the claim is that social life is nothing more than 'the system of signs of an unrelenting domination' (Touraine 1988). But the necessary decomposition of society, the passage from one cultural and societal field to another, makes possible the role of social movements in historical change. At the core of his analysis lie the 'cultural orientations common to actors who are in conflict over the management of these orientations, for the benefit of either an innovative ruling class or, on the contrary, those who are subordinated to its domination' (Touraine 1988: 155). Most significantly, for Touraine, class struggles and social movements express conscious contestation over the self-production of society, by which he means the work society performs on itself by inventing its norms, institutions and practices. Struggles over historicity lie at the centre of the functioning of society and of the process by which society is created (Touraine 1985, 1988).

In this view, then, geography would consist of the investigation of social and cultural fields of meaning and action. The idea would be to retain notions of reproductive structuring, but emphasize struggles by numerous social movements over the continually transforming social imaginary process, particularly the concept of what society ought to be. This challenges structuralism by contesting the automatic creation of intention; it challenges structuration through its integration of practical reasoning into the social imaginary.

STRUCTURALISM RECONSIDERED

The movement from theories of the general structures of existence towards the structuration processes specific to instances and localities was part of a more general tendency, the post-structural and post-modern criticism of the certainties of modernist understanding. Post-structural theorists tend to phrase their ideas in terms of critiques of universal rationality and the order this imposes on what is, for them, a more chaotic, diverse, differentiated, even anarchic world. Post-modern philosophy rejects unifying or 'totalizing' theories as rationalist myths of the Enlightenment and the modern world initiated by it. From such a perspective any theory that supposes structural coherence is criticized as reductionist, obscuring of difference, suppressive of plurality in favour of conformity. Post-modern thinkers draw on Nietzsche's attack on modern Western philosophy, and Heidegger's radical rejection of modernity, to postulate instead the importance of differences over unities and particularities over structures (Best and Kellner 1991).

Post-structural space

Post-structural theorists therefore see space fractured into micro-zones as small as that occupied or dominated by an individual human body. While networks of such individual spaces might build up, such networks would be unstable, constantly breaking down and reforming. Post-modern theory is suspicious of general tendencies in spatial relations, permanent or even semi-permanent characterizations of spatial entities, because each space and its people is constantly separating and going its own way. Lyotard (1984) would explain this in terms of 'petite-narratives' localized in distinct spaces and the lack of a space-transcending metanarrative which coheres particular localities through communication between them. Post-colonial writers and literary theorists (Ashcroft *et al.* 1989), post-modern ethnographers (Clifford and Marcus 1986) and some post-modern geographers (Barnes and Duncan 1992) would similarly see each place containing its own interpretive community, speaking its own local voice, different and experienced as 'other' by the rest of the world. The appeal, as with deconstruction (Culler 1982), would be to texts interpreted differently in different places and at different times. From this view spatial generalization is a contradiction in terms (cf. Philo 1992).

The persistence of structure

Even so, in what is supposedly *post*-structural work we find what might be called a quasi-structural position centred on discourse. A 'discourse' is a particular area of language use related to a certain set of institutions and a particular standpoint. Concerned with a given range of objects, it emphasizes some concepts at the expense of others. Most importantly, significations and meanings are integral parts of discourses just as, for example, the meanings of words depend on where the statement containing them is made (Macdonnell 1986). Hence for Barnes and Duncan (1992: 8) discourses are 'frameworks that embrace particular combinations of narratives, concepts, ideologies and signifying practices, each relevant to a particular realm of social action'.

In Foucault's *The Archaeology of Knowledge* (1972) discourses are treated as autonomous, rule-governed systems which unify systems of practices. Discursive unity makes social, political, economic, technological and psychological factors function coherently, rather than the other way round; that is, coherent institutional practices creating unified discursive practices (Dreyfus and Rabinow 1982). As Megill (1985: 238) points out, Foucault views the world 'as if it were discourse' subject to battles between discourses. Discourse, then, occupies a similar position to structure; that is, as a primordial causal force (see also Foucault 1973, 1980).

Cultural geographers borrowing the terms 'discourse' and 'discursive field' from Foucault and Althusser, 'attempt to escape the more deterministic, more structural implications of these terms whereby subjects are "produced" by an autonomous discourse which produces the "illusion" within them that they are agents' (Duncan 1990: 16). The idea is to show how discourses create landscapes which in turn may be read as texts. Yet the relevant landscape-shaping discourses have their origins, in Duncan's work, only in other discourses. The social contexts of the origins of these discourses are not pursued, nor the material and class circumstances under which the discourses change, nor those of their adoption. Discourses originate, change, diffuse, interact and are adopted with minimal reference to their social contexts in an account which, in practice, comes perilously close to the autonomy of discourse (Peet 1993). Post-structural philosophy and geography thus retain remnants of the notion of an underlying, controlling force structuring social and cultural life and forming space, landscapes, and geographies; in this case instead of material structures the emphasis lies on discursive structures. Structure, then, has been *the* theme of recent social and geographical thought. Structur*al* conceptions have prevailed even in the works of thinkers opposed to structural*ism*. Why might this be the case?

Geostructuralism and human experience

Structuralism responds to two characteristics of the human experience. Life consists of unique events (in the sense that no event is identical with another) which, however, recur in patterns which have order in time and space. More basically, life has the characteristic of effort against constraint, externality, that which already exists. Indeed the notion of complete freedom to make life at will is both difficult to imagine and, somehow, not practically interesting. On reflection, order and constraint turn out to be much the same thing – structuralism attempts to capture this order in theoretical terms.

In geography, statements about the constraints, the contexts, the regularities which recur over spaces widely separated might be termed *geo*structuralism (Peet 1989b, 1991). As intimated throughout this chapter, the geographical conception of order rightly has a natural basis, although this has been neglected during an overemphasis on geometric space. Human beings are natural creatures bound into relations with the earth that originated them, cohabiting the environment with other organisms, dependent on the world's resources for the very possibility of continued existence. Natural environments may be regarded as huge complexes of resources which condition reproductive possibilities. Specific kinds of reproductive practice develop as organized efforts against the constraints of different kinds of environments. Life is earned by transforming a limited array of natural resources, and all social activity occurs in a place and is influenced by the frictions of distance. Space intervenes to limit human interaction so that for virtually all of history people have reproduced themselves within local systems, each with its culture, language, traditions and loyalties. The thousands of local societies which environment and space structurally condition are forgotten only at great expense, as recent experience in Eastern Europe reminds.

But human beings are not solely local and are hardly automatic products of nature and natural forces. They intercede with nature by constructing social (reproductive) institutions, such as the workplace, the school, university, family and community, locales where life is made collectively and people must exist together. It is indeed this basic, structuring quality of social reproduction, its position next to, yet countering, the structuring of nature, that connects environment and locale with all aspects of existence in multiple, usually fundamental ways. Structural geography has done truly useful work here in showing how reproductive practices develop in patterns conditioned by social relations and the level and type of force exerted on nature; there remains a need to extend this work into the full range of reproductive institutions, for example how families of different kinds are arranged in space, alter environments, create places. Structuration theory continues this work into the micro worlds in which life is actually reproduced and the structures of existence continually replaced, calibrated and reconstructed. The stress here lies on practical, rather than grand, structural reason, and with Giddens and Pred links are made with the question of power and power relations in specific localities. Social field theory gives the potential of reconnecting everyday purpose with social imagining. This entire process of social reproduction can be seen as transforming natural space into landscapes, creating socialized environments in which people are increasingly made by the structures they themselves have implanted into nature. Thereby the conditions and social relations under which socialized landscapes are reproduced become increasingly significant, and structural/structuration theories assume greater intellectual importance. For while people make history they do so increasingly under conditions of their own making but not entirely of their own choosing. It is the political purpose of socialism to increase the range of choice and to democratize the entire process by which reproduction occurs.

Structuralism and freedom

By transforming nature through labour, human beings produce the very possibility of their existence, and by relating to persons of the opposite sex, and raising children, humans reproduce the members of their species. While events have unique aspects, they are also parts of these ongoing processes of the reproduction of existence and are determined by their place in the structures formed by the reproductive relations and practices. This geographical form of structuralism, re-emphasizes the necessities inherent in the social relation to nature. Geo-structuralism particularly emphasizes constraints of environment and place on the freedom of human thought and action – indeed, the task of geography in the academic division of labour involves stressing the natural determination of human events and return effects of social action on nature. But these dimensions of necessity, constraint and context should not be regarded merely as a set of negative forces preventing what would otherwise be an unbounded human freedom. In contradistinction to the early (existential) Sartre and more in accord with his later (Marxist) phase, humans discover freedom, that is *learn* how to be free, as they exert themselves against constraint. Here intention and interpretation become supremely important, and notions of the social imaginary need linking with those of structure and structuration. Theorizing geo-structuring processes therefore does not at all mean 'reducing' the complexity of social life to its basic natural-material content. Rather it means finding a *guide* to social activity in the practices and relations by which humans make and express their lives, in the socio-natural relations.

This finding of freedom is beset with ideological struggles over the interpretation of experience and control over social imaginaries. The present need is to reinterpret past and present experiences to derive a liberative value structure. In overcoming constraint in the satisfaction of need human beings find the social values which they use to guide their further action. At a time of environmental crisis, relations with nature (that is, most significantly, the way human landscape interacts with natural processes) must be re-examined to derive new value models of social relations, institutions and behaviours. The making of a social reproductive system guided by a liberative imaginary in sustainable relation with earth is a project geography cannot avoid. That is why theories of structure recur in ever-changing guises.

REFERENCES

Aglietta, M. (1979) *A Theory of Capitalist Regulation: The U.S. Experience* (trans. David Fernbach), London: New Left Books.

Althusser, L. (1969) *For Marx* (trans. Ben Brewster), Harmondsworth: Penguin.
Althusser, L. and Balibar, E. (1968) *Reading Capital*, London: New Left Books.
Ashcroft, B., Griffiths, G. and Tiffin, H. (1989) *The Empire Writes Back, Theory and Practice in Post-Colonial Literature*, London: Routledge.
Barnes, T. and Duncan, J. (eds) (1992) *Writing Worlds: Discourse, Text and Metaphor in the Representation of Landscape*, London: Routledge.
Baudrillard, J. (1975) *The Mirror of Production* (trans. Mark Poster), St Louis, Mo.: Telos Press.
Best, S. and Kellner, D. (1991) *Postmodern Theory: Critical Interrogations*, New York: The Guilford Press.
Bhaskar, R. (1979) *The Possibility of Naturalism*, Hassocks, Sussex: Harvester Press.
Bourdieu, P. (1977) *Outline for a Theory of Practice*, Cambridge: Cambridge University Press.
Callinicos, A. (1976) *Althusser's Marxism*, London: Pluto Press.
Castells, M. (1977) *The Urban Question: A Marxist Approach* (trans. Alan Sheridan), Cambridge, Mass.: The MIT Press.
Castoriadis, C. (1991) 'The social historical: mode of being, problems of knowledge', pp. 33–46 in C. Castoriadis, *Philosophy, Politics, Autonomy*, New York: Oxford University Press.
Clifford, J. and Marcus, G.E. (eds) (1986) *Writing Culture: The Poetics and Politics of Ethnography*, Berkeley: University of California Press.
Cohen, G.A. (1975) *Karl Marx's Theory of History: A Defense*, Princeton: Princeton University Press.
Cohen, I.J. (1989) *Structuration Theory: Anthony Giddens and the Constitution of Social Life*, New York: St Martin's Press.
Corragio, J.L. (1977) 'Social forms of space organization and their trends in Latin America', *Antipode*, no. 911, 14–28.
Culler, J. (1973) 'The linguistic basis of structuralism', pp. 20–36 in D. Robey (ed.) *Structuralism: An Introduction*, Oxford: Clarendon Press.
Culler, J. (1982) *On Deconstruction: Theory and Criticism after Structuralism*, Ithaca, N.Y.: Cornell University Press.
Dreyfus, H.L. and Rabinow, P. (1982) *Michel Foucault: Beyond Structuralism and Hermeneutics*, Chicago, Ill.: University of Chicago Press.
Duncan, J. (1990) *The City as Text: The Politics of Landscape Interpretation in the Kandyan Kingdom*, Cambridge: Cambridge University Press.
Duncan, J. and Ley, D. (1982) 'Structural Marxism and human geography: a critical assessment', *Annals of the Association of American Geographers* 72(1), 30–59.
Dunford, M. and Perrons, D. (1983) *The Arena of Capital*, New York: St Martin's Press.
Eco, U. (1973) 'Social life as a sign system', pp. 57–72 in D. Robey (ed.) *Structuralism: An Introduction*, Oxford: Clarendon Press.
Foucault, M. (1972) *The Archaeology of Knowledge*, New York: Pantheon Books.
Foucault, M. (1973) *The Order of Things*, New York: Vintage Books.
Foucault, M. (1980) *Power/Knowledge*, New York: Pantheon Books.
Giddens, A. (1979a) 'Structuralism and the theory of the subject', pp. 9–48 in A. Giddens, *Central Problems in Social Theory*, Berkeley: University of California Press.
Giddens, A. (1979b) *Central Problems in Social Theory*, Berkeley: University of California Press.
Giddens, A. (1981) *A Contemporary Critique of Historical Materialism*, Berkeley: University of California Press.
Giddens, A. (1984) *The Constitution of Society; Outline of a Theory of Structuration*,

Berkeley: University of California Press.

Giddens, A. (1985) 'Time, space and regionalisation', pp. 265–95 in D. Gregory and J. Urry (eds) *Social Relations and Spatial Structures*, New York: St Martin's Press.

Gottdiener, M. and Lagopoulos, A. (1986) *The City and the Sign: An Introduction*, New York: Columbia University Press.

Gregory, D. (1981) 'Human agency and human geography', *Transactions of the Institute of British Geographers*, New Series 6, 1–18.

Gregory, D. (1982) *Regional Transformation and Industrial Revolution: A Geography of the Yorkshire Woollen Industry*, Minneapolis: University of Minnesota Press.

Hägerstrand, T. (1975) 'Space, time and human conditions', in A. Karlquist, *Dynamic Allocation of Urban Space*, Farnborough: Saxon House.

Hammond, M., Howarth, J. and Keats, R. (1991) *Understanding Phenomenology*, Oxford: Basil Blackwell.

Harvey, D. (1973) *Social Justice and the City*, Baltimore, Md.: The Johns Hopkins University Press.

Harvey, D. (1989) *The Condition of Postmodernity: An Enquiry into the Origins of Cultural Change*, Oxford: Basil Blackwell.

Huntington, E. (1935) *Civilization and Climate*, New Haven, Conn.: Yale University Press.

Kierkegaard, S. (1936) *Philosophical Fragments*, Princeton: Princeton University Press.

Leacock, E. and Lee, R. (1982) *Politics and History in Band Societies*, Cambridge: Cambridge University Press.

Lefebvre, H. (1970) *La Revolution Urbaine*, Paris: Gallimard.

Lefebvre, H. (1972) *La Pensée Marxiste et La Ville*, Paris: Casterman.

Lévi-Strauss, C. (1966) *The Savage Mind*, Chicago, Ill.: University of Chicago Press.

Ley, D. and Samuels, M. (eds) (1978) *Humanistic Geography*, Chicago, Ill.: Maroufa Press.

Lippietz, A. (1986) 'New tendencies in the international division of labor: regimes of accumulation and modes of regulation', pp. 16–40 in A. J. Scott and M. Storper (eds) *Production, Work, Territory*, Winchester, Mass.: Allen & Unwin.

Lyons, J. (1973) 'Structuralism and linguistics', pp. 5–19 in D. Robey (ed.) *Structuralism: An Introduction*, Oxford: Clarendon Press.

Lyotard, J.-F. (1984) *The Postmodern Condition*, Minneapolis: University of Minnesota Press.

MacDonnell, D. (1986) *Theories of Discourse: An Introduction*, Oxford: Basil Blackwell.

Marcus, G.E. (1986) 'Contemporary problems of ethnography in the modern world system', pp. 165–93 in J. Clifford and G.F. Marcus (eds) *Writing Culture: The Poetics and Politics of Ethnography*, Berkeley: University of California Press.

Marx, K. (ed.) (1973) *Grundrisse: Foundations of the Critique of Political Economy* (trans. Martin Nicolaus), Harmondsworth: Penguin.

Marx, K. (ed.) (1976) *Capital: A Critique of Political Economy*, Vol. 1 (trans. Ben Fowkes), Harmondsworth: Penguin.

Massey, D. (1984) *Spatial Divisions of Labor: Social Structures and the Geography of Production*, New York: Methuen.

Megill, A. (1985) *Prophets of Extremity: Nietzsche, Heidegger, Foucault, Derrida*, Berkeley: University of California Press.

Ollman, B. (1971) *Alienation: Marx's Conception of Man in Capitalist Society*, Cambridge: Cambridge University Press.

Parsons, T. (1949) *The Structure of Social Action*, Glencoe: The Free Press.

Peet, R. (1977) *Radical Geography*, Chicago, Ill.: Maroufa Press.

Peet, R. (1978) 'Materialism, social formation, and socio-spatial relations: an essay in Marxist geography', *Cahiers de Géographie du Quebec* 22, 147–57.

885

Peet, R. (1981) 'Spatial dialectics and Marxist geography', *Progress in Human Geography* 5, 105–10.

Peet, R. (1986) 'The social origins of environmental determinism', *Annals of the Association of American Geographers* 75(3), 309–33.

Peet, R. (1989a) 'World capitalism and the destruction of regional cultures', pp. 175–99 in R.J. Johnston and P.J. Taylor (eds) *A World in Crisis? Geographical Perspectives* (2nd edn), Oxford: Basil Blackwell.

Peet, R. (1989b) 'Conceptual problems in neo Marxist industrial geography', *Antipode* 21(1), 35–50.

Peet, R. (1991) *Global Capitalism: Theories of Societal Development*, London: Routledge.

Peet, R. (1993) 'Review of James Duncan, *The City as Text*', *Annals of the Association of American Geographers* 83(1), 184–7.

Peet, R. and Thrift, N. (eds) (1989) *New Models in Geography: The Political Economy Approach* (2 vols), London: Unwin Hyman.

Philo, C. (1992) 'Foucault's geography', *Society and Space* 10, 137–61.

Piaget, J. (1971) *Structuralism*, London: Routledge & Kegan Paul.

Pickvance, C. (1976) 'Housing, reproduction of capital and reproduction of labour power: some recent French work', *Antipode* 8(1), 58–68.

Piore, M.J. and Sable, C.F. (1984) *The Second Industrial Divide: Possibilities for Prosperity*, New York: Basic Books.

Poster, M. (1975) *Existential Marxism in Post War France: From Sartre to Althusser*, Princeton: Princeton University Press.

Pred, A. (1981a) 'Social reproduction and the time geography of every day life', *Geografiska Annaler* 63B, 5–22.

Pred, A. (1981b) 'Of paths and projects: individual behavior and its societal context', pp. 231–55 in R. Golledge and K. Cox (eds) *Behavioral Geography Revisited*, London: Methuen.

Pred, A. (1983) 'Structuration and place: on the becoming of sense of place and structure of feeling', *Journal for the Theory of Social Behavior* 13, 157–86.

Pred, A. (1986) *Place, Practice and Structure: Social and Spatial Transformation in Southern Sweden, 1750–1850*, Totowa, N.J.: Barnes & Noble.

Preteceille, E. (1976) 'Urban planning: the contradiction of capitalist urbanization', *Antipode* 8(1), 69–76.

Pudup, M.B. (1988) 'Arguments within regional geography', *Progress in Human Geography* 12, 369–90.

Ratzel, F. (1896) *History of Mankind* (trans. A.J. Butler), London: Macmillan.

Santos, M. (1977) 'Society and space: social formation as theory and method', *Antipode* 9(1), 3–13.

Sartre, J.-P. (1956) *Being and Nothingness: An Essay on Phenomenological Ontology* (trans. Hazel E. Barnes) New York: Philosophical Library.

Sartre, J.-P. (1968) *Search for a Method* (trans. Hazel E. Barnes), New York: Vintage Books.

Sartre, J.-P. (1976) *Critique of Dialectical Reason*, Atlantic Highlands, N.J.: Humanities Press.

de Saussure, F. (1959) *Course in General Linguistics* (trans. Wade Baskin), New York: Philosophical Library.

Scott, A. (1988) 'Flexible production systems and regional development: the rise of new industrial spaces in North America and Western Europe', *International Journal of Urban and Regional Research* 12, 171–86.

Semple, E.C. (1903) *American History and Its Geographical Conditions*, Boston, Mass.: Houghton-Mifflin.

Soja, E. (1980) 'The socio-spatial dialectic', *Annals of the Association of American*

Geographers 70, 207–25.

Soja, E. (1989) *Postmodern Geographies: The Reassertion of Space in Critical Social Theory*, London: Verso.

Storper, M. and Scott, A. (1989) 'The geographical foundations and social regulation of flexible production complexes', pp. 21–40 in J. Wolch and M. Dear (eds) *The Power of Geography: How Territory Shapes Social Life*, Boston, Mass.: Unwin Hyman.

Thrift, N. (1983) 'On the determination of social action in space and time', *Society and Space* 1, 23–57.

Touraine, A. (1985) 'An introduction to the study of social movements', *Social Research* 52(4), 749–87.

Touraine A. (1988) *Return of the Actor: Social Theory in Post Industrial Society*, Minneapolis: University of Minnesota Press.

Wittfogel, K. (1929) 'Geopolitik, Geographischer Materialismus', *Unter den Banner des Marxismus* 3(1), 4, 5 (translated by G.L. Ulmen as 'Geopolitics, geo-graphical materialism and Marxism', *Antipode* 71(1) (1985), 21–72).

FURTHER READING

Benton, T. (1984) *The Rise and Fall of Structural Marxism: Althusser and his Influence*, New York: St Martin's Press.

Cloke, P., Philo, C. and Sadler, D. (1991) *Approaching Human Geography: An Introduction to Contemporary Debates*, New York: The Guilford Press.

Giddens, A. (1981) *A Contemporary Critique of Historical Materialism. Vol. 1: Power, Property and the State*, Berkeley: University of California Press.

Gregory, D. and Urry, J. (eds) (1985) *Social Relations and Spatial Structures*, New York: St Martin's Press.

Harvey, D. (1989) *The Condition of Postmodernity: An Enquiry into the Origins of Cultural Change*, Oxford: Basil Blackwell.

Peet, R. and Thrift, N. (eds) (1989) *New Models in Geography: The Political Economy Approach* (2 vols), London: Unwin Hyman.

Soja, E. (1989) *Postmodern Geographies: The Reassertion of Space in Critical Social Theory*, London: Verso.

40

CHALLENGING THE BOUNDARIES

Survival and change in a gendered world

Janice Monk

As an academic expression of a social movement, feminist geography is self-consciously political, challenging the gender equity of the existing social order and disciplinary knowledge and practice. Though its visions may be contested, marginalized and, at times, wavering, over the two decades of its existence it has become international in scope and constructed more complex conceptualizations. It incorporates diverse theoretical positions and methodologies while continuing to investigate the relationships between space, place and gender relations, women's world views and constructions of identity, and their behaviour as social agents.

In this chapter I will first comment on the origins of feminist geography. I will next provide examples of the research in order to illustrate important concepts, methodological concerns and empirical themes, first focusing on the cultural and symbolic significance of gender and then on the material world. In the final section I will discuss contemporary challenges for feminist geography – theoretically and methodologically as it confronts the promises and dilemmas of post-modernist scholarship, empirically in a changing global context, and politically, if it is to continue striving towards its idealistic goals.

THE DEVELOPMENT OF FEMINIST GEOGRAPHY

Feminist geography emerged in the 1970s as scholars realized that the academic implications of the women's rights movements of the late 1960s included not only issues about the representation of women as geographers (Berman 1977; McDowell 1979; Momsen 1980; Zelinsky 1973a, 1973b) but also required attention to the invisibility of women's experiences within the substance of the discipline. They began to incorporate gender as a variable into studies of such themes as daily travel, migration and employment in order to demonstrate that women's spatial behaviour differed from men's and to introduce new research topics especially salient for women's lives, such as access to child-care services. Most of this early research was published in the United States and Britain and has been reviewed by Zelinsky *et al.* (1982). Less widely known is the more radical and theoretical work of Dutch feminist geographers that also emerged in the late 1970s stimulated by student critiques of Marxist writing (Droogleever Fortuijn 1993).

Drawing on the ideas of feminists in other fields, by the early 1980s geographers were arguing for transformation of the discipline to address sexist biases in its theories, content, methods and purposes (Monk and Hanson 1982). In this critique they posited that knowledge is socially constructed, that *who* creates it shapes the selection of problems and the interpretations of data, and that explanations are partial (in both senses of the word) and temporary. During the 1980s, these contentions have been more fully developed by feminist scholars in other disciplines who have invoked geographic language in writing of the 'positionality' of the speaker and of a 'situated', grounded knowledge as an alternative to relativism or to false presumptions of a universal vision (Haraway 1988). Sharing this perspective and recognizing the international character of contemporary feminist geography, I will not suggest a linear development or a unified approach within the field, even though I consider commonalities to exist in goals and desires for solidarity within diversity. Instead, I will begin by suggesting how feminist geography varies in its perspectives and emphases across selected countries, a theme I have explored more fully elsewhere (Monk 1994).

Best known and most extensive is the research by English-speaking geographers in Britain and the United States, together with that of their colleagues in Australia and Canada. This literature exhibits many connections but I think it is fair to say that British geographers have emphasized theory and particular theoretical approaches, whereas geographers in the United States have been more empirically oriented and have adopted diverse epistemologies. With an overriding interest in questions of oppression, British geographers turned early to socialist feminism, especially in their work on urban Britain (Bowlby *et al.* 1989). Their identification with locality studies has also been important (McDowell and Massey 1984). Recently, British feminist geographers have become strongly interested in post-modernist theories, though have found these to be a source of dissension among themselves (Penrose *et al.* 1992).

By contrast, work in the United States, reflecting diverse philosophies, has continued to adopt positivist and humanist approaches as well as socialist and post-modernist and has long concerned itself with questions of expression as well as of oppression. Scholars have challenged the invisibility of women (Monk and Hanson 1982), documented spatial patterns of gender differences and inequalities in the United States and elsewhere (for example, Lee and Schultz 1982; Seager and Olsen 1986), examined behavioural choices (such as the journey to work) within constrained environments (for example, Hanson and

Johnston 1985), and given voice to women's responses to place and environment (for example, Wilkinson 1979; Norwood and Monk 1987). Questions about differences among women on bases of class, race, ethnicity, sexuality, and life stage are also now generating considerable theoretical, methodological, and empirical attention (for example, England 1994; Gilbert 1994; Katz and Monk 1993; McLafferty and Preston 1991; Sanders 1990).

Though feminist geographers in the Netherlands initially focused on theory, they soon took up applied research reflecting their close links with planners and state support for feminist concerns. This led to work on housing, for example, a theme not widely pursued in other contexts (Karsten 1989). The original dominant themes – women's labour problems and the relationships between home and family life – have persisted, but with a Dutch flavour that reflects the national paradox between a liberal, progressive ideology and a traditional praxis with respect to women's family responsibilities (Droogleever Fortuijn 1993).

Whereas much of the research on Britain, the United States, and the Netherlands deals with urban women, Spanish scholars, reflecting national geographic traditions (Albet et al. 1992), have paid more attention to rural women (Sabaté and Tulla 1992). They have implemented alternative approaches to measuring women's participation in agricultural work (Canoves 1989; García-Ramon et al. 1990), compared the gender division of labour on family farms in different regions (García-Ramon et al. 1991), evaluated the implications of state welfare policies for women agricultural day labourers (García-Ramon et al. 1992), and shown how economic restructuring is promoting the development of industrial homeworking by women in rural communities (Sabaté et al. 1991).

Women's roles in agriculture and problems of development dominate research on and by geographers in Sub-Saharan Africa and South Asia, particularly in the studies supported by international agencies and by foreign scholars working in Africa, Asia, and Latin America (Momsen and Townsend 1987; Momsen and Kinnaird 1993). Environmental issues, such as women's management of fuel resources, are of concern in Africa (Ardayfio-Schandorf 1993; Hyma and Nyamwange 1993) and Asia (Ulluwishewa 1993b; Wickrama-singhe 1992). So too are questions of health and family planning (Iyun and Oke 1993; Samarasinghe et al. 1990) and women's survival strategies, particularly their informal sector work and their responses to economic crises (Rasanayagam 1992a, 1992b; Okpala 1992; Owei and Jev 1992).

SPACE, PLACE AND THE CULTURAL CONSTRUCTION OF GENDER

In discussing the cultural and symbolic separately from the material strands of feminist geography I am not implying that these arenas of life are isolated from one another, rather I am seeking a way to organize the research so as to introduce important concepts while cutting across different geographic contexts and scales. Indeed, an important contribution of feminist geography has been to

890

show how gender ideologies shape material experience and how material social practices are involved in both the maintenance of such ideologies and their change.

For feminist scholars gender is a central category of analysis. They use the concept to recognize the culturally and historically varying transformations of male and female bodies into masculine and feminine gendered identities. Important related concepts are the gender roles into which people are socialized and the power inequalities of gender relations. Feminist geographers ask how space and place are implicated in the creation and expression of gender roles and relations and vice versa. In this section of the chapter, I will examine examples of the symbolic functioning of gender roles in political and environmental spheres, emphasizing ideas about the gendered meanings of 'public' and 'private' spaces, and linked notions about rationality, spirituality, morality, domesticity, mothering and caring.

As Sallie Marston (1990) has pointed out, ideologies about the natural roles of men and women have been fundamental in Western formulations of the nation and state, undergirding, for example, the concept of citizenship on which the American republic was founded. Within this conceptualization, men were associated with the reason deemed necessary for participation in public society, whereas women were identified with a natural emotionality and spirituality fitting them for the domestic sphere. This construction gave men not only personal rights in the public world but also the right to represent and control the political and economic interests of the private household. Thus in law and fact, 'citizen and wife' were one person, with women's legal and property rights ceded to their husbands (Marston 1990: 453); women's moral and emotional realm was only part of the private sphere.

In the 1970s and 1980s the gender associations of public and private spaces have been both invoked and challenged in Latin American political struggles. Scarpaci and Frazier (1993) write of the 'gendering of landscapes' in examining how male state aggressors in the Southern Cone have entered private spaces to abduct state 'enemies' while the Madres of the Plaza de Mayo have publicly manipulated the gender ideologies that sanctify their private roles as 'mothers' in order to protest state terror. Similarly, the CoMadres of El Salvador invoke their roles as mothers in public protest against state violence. They transgress in public spaces by seeking out clandestine cemeteries, taking and publishing photographs of the tortured and killed and visiting prisons. Yet they also incorporate traditional private sphere caring responsibilities of women into their public projects – organizing health and child-care services and providing food and clothing for refugees (Schirmer 1993).

In her study of Palestinian women under Israeli occupation, Tamar Mayer (1994) has found utility in the concepts of public and private space. She argues that women's resistance and sense of nationalism have been sharpened by the occupation's invasions of the private spaces of the home where Palestinian women were traditionally largely secluded. Forced to confront soldiers in their

891

homes or to defend their children in the streets, women have given new political meanings to 'motherhood'. Additionally, harassment and rape have violated private sexuality and family honour, yet Mayer (1994) reports these acts have also intensified women's public political commitments.

The political relations and manipulations of public and private gender roles are not simple functions of gender, however. Sarah Radcliffe (1993) has demonstrated the interplay between gender and ethnicity in describing the efforts of indigenous *campesinas* in Peru to resist the state's attempts to impose Hispanic culture models that associate women with the private sphere. They have employed long-standing indigenous concepts of male–female partnership in their participation in peasant unions, though with a gradually heightening consciousness of gender distinctions. While the collaborative pattern is widespread, however, Radcliffe notes that expressions of independent femininities have arisen in some regions where women's political consciousness has been shaped by local religious and ethnic identities.

In addition to examining the gendered meanings of public and private spaces, feminists have been interested in the implications of the nature–culture dichotomy in Western thought wherein woman is symbolically identified with nature and man with culture. Caroline Merchant (1980) has described the historical transitions in European thought from an organic world-view in which human, natural, and spiritual were blended to a modern, mechanistic world-view in which nature is conceptualized as an object to be controlled by humans. In this formulation, man is symbolically associated with rationality, technology and domination of a female earth and, by extension, with power over woman. Merchant considers it important to address the material as well as symbolic aspects of nature–culture and gender relations but many women who identify as ecofeminists stress the positive spiritual and symbolic power they derive from linking woman with nature, expressing visions of nature as goddess and mother. Geographers have only recently begun to write about ecofeminism (Monk 1992a; Nesmith and Radcliffe 1993; Seager 1993) though find it problematic because of its essentializing biological foundation and its apolitical, ahistorical stance. Seager questions its value as an individualistic and spiritual approach for dealing with environmental deterioration that has a basis in the acts of governments, militaries, and businesses (Seager 1993).

Nevertheless, some women have been motivated to environmental activism through their symbolic identifications as mothers and keepers of the home. Their discourse in contemporary green consumer movements, campaigns against toxic wastes, and animal rights organizations reveals these gender values, as did the involvement of early twentieth-century women in the urban environmental and forest conservation movements in the United States (Gittel and Shtob 1981; Merchant 1984; Seager 1993). Women's environmental activism may even concretely incorporate the domestic realm. Bru describes the tactics of housewives protesting toxic waste dumps in Gibraleón, Spain who nightly banged their *cacerolas* (saucepans) to signal their opposition (Bru 1993).

These studies, Seager's (1993) reflections on the gendered discourse within male-dominated 'mainstream' environmental organizations and the deep ecology movement, and her recognition that women environmental activists are commonly dismissed as 'hysterical housewives', together demonstrate the value of feminist analyses of the part played by gender symbolism within environmental politics.

Some women find personal and cultural identification in 'female' landscapes. In contrast with Anglo-American men who have variously seen the desert as a female land to be protected, exploited or conquered, American Indian, Mexican American and Anglo-American women writers and artists in the south-western United States forcefully express images of the desert as a strong and powerful woman who cannot be mastered, as a teacher, *curandera* (healer) and wise woman from whom they draw strength. Interpretation of these visions reveals the salience of gender, but also the ways in which it is inflected by ethnicity and specific historical and spatial contexts (Norwood and Monk 1987). Fay Gale's (1990) analysis of Aboriginal Australian women's political activism illustrates other aspects of the gender dimensions of spiritual relationships with the land. Because white male scholars and bureaucrats assumed the primacy of men in Aboriginal political leadership and religious life, women were not consulted in early cases to establish Aboriginal land claims. Yet women have their own sacred sites about which Aboriginal men lack knowledge. Gale describes how Central Australian Aboriginal women organized protests in order to protect their own sites.

Feminist geographers, planners, and urban historians have examined many other examples of the ways in which gender is symbolized in space and place, particularly in domestic architecture and in the meanings attached to the concepts of city and suburb throughout much of the twentieth century (for example, Hayden 1984; McDowell 1983; Monk 1922a; Wilson 1991). I will not elaborate on this literature but simply note that it reveals a complex picture of the historically and spatially varying ways in which cultural symbols of space and place can be used to impose, accept and resist oppression and to conceive of social and environmental order.

PRODUCTION AND REPRODUCTION: GENDER IN THE MATERIAL WORLD

Feminist geographers recognized early that economic geography rendered much of women's work invisible (Monk and Hanson 1982; Tivers 1978). They advocated examination of unpaid reproductive work in the household and community, women's labour in subsistence production and informal sectors, and the nature of gender-based occupational segregation in part-time and full-time employment. Such an endeavour presented new theoretical and methodological challenges. Theoretically, important contributions have been made by socialist feminist scholars, who, concerning themselves especially with urban, industrial societies, have identified how capitalism and patriarchy together supported and benefited from the spatial separation of paid employment by men and unpaid household work by women. Studies have revealed how these forces have operated at both intraurban and regional scales (Mackenzie and Rose 1983; McDowell and Massey 1984). They identified critical intersections between social and economic life, thus challenging a discipline which had largely treated these areas separately.

Methodologically, a continuing issue has been how to measure the work of women. Studies on rural women in Spain indicate the complexities of the task. Analysis of census data reveals regional differences but underestimates women's work; in-depth interviews identify gender divisions of labour within productive and reproductive work and the multifaceted aspects of women's activities but require scholars to grapple with problems such as coding women's simultaneous use of time in productive and reproductive tasks (Baylina *et al.* 1991; Canoves 1989; García Ramon *et al.* 1990). Many of the projects on women in Third World countries have also measured the specifics of women's work contributions, often using detailed records of daily time expenditures. They show women's exceedingly heavy work loads in production and reproduction together with variations across agricultural systems and throughout the year (Meertens 1993; Mwaka 1993; Oughton 1993; Raghuram 1993).

The research on Third World women demonstrates how systems of production and reproduction are interdependent while varying over space and time. Sylvia Chant's (1991) study of women's survival strategies in low-income households of three Mexican cities shows the importance of attending to variations in household structures as an aspect of labour supply as well as to differences in labour demand resulting from different local economic bases. Rohana Ulluwishewa's (1993a) research on women's roles as mothers, house-wives, and farm workers in traditional agricultural villages and in settlements within the Mahaweli water management development project illustrates contrasting, interconnected differences between the two contexts in family structures, community relations, the extent of children's and husbands' participation in reproductive work, gender divisions of agricultural labour, women's access to income-generating opportunities, and their roles in decision-making about productive tasks and household expenditures. In general, he documents the adverse effects of the development project on women whose complex relationships to work were not adequately understood by planners. Fiona Mackenzie's study (1986) of national land policies and rural development in Kenya highlights how external forces interact with local circumstances to alter the interactions between production and reproduction. Land redistribution programmes that awarded individualized titles to men, new emphases on export production and rising needs for cash income affected labour demands and access to resources. Out-migration by men has led to their becoming non-producer owners and women non-owner producers. To meet their cash needs, these women have used traditional savings and credit groups to open shops, sell firewood and charcoal and make cement blocks. In the past, such organizations mainly funded communal welfare needs; now support for those activities has been lost as women draw on the groups to finance the production that will provide household income.

A number of studies address the varying ways in which Western post-industrial states shape relations between production and reproduction through policies on dependent care and women's employment. Thus the recent growth in the employment of nannies by middle-class dual-earner households in Britain (where the state has largely avoided support for extended child care) (Gregson and Lowe 1994) contrasts with the effects of French policies on maternity leave, public child care and educational structures; together these enhance both maternity and women's full-time employment (Fagnani 1993). In Australia, policies facilitate full-time employment of 'Anglo' middle-class mothers of healthy pre-school children to a greater extent than they do the employment of mothers of disabled children, ethnic minority women or middle-aged women caring for older dependants. The policies also have spatially variant effects at the intraurban scale (Fincher 1993).

As these examples suggest, understanding the interdependency of production and reproduction requires geographic analysis at a variety of scales, from the personal and household to the community, state, and international. Maureen Hays-Mitchell's (1993) gender-sensitive research on *ambulantes* (street vendors) in Peru, in which she examines linkages between formal and informal sectors in political (institutional), social, and economic contexts, demonstrates historically and culturally specific connections across scales. She shows how production and reproduction are intertwined and also points out the fluidity of boundaries of the related categories 'formal' and 'informal'.

Challenging the definitions and boundaries of taken-for-granted categories is an important aspect of the research by Susan Hanson and Geraldine Pratt on Worcester, Massachusetts. They have suggested that gender-based class differences often exist within dual-earner households and within neighbour-hoods between those households that have dual earners and those that have single earners (Pratt and Hanson 1988). Additionally, by working at a fine scale and examining male and female occupational segregation within a local labour market, they question the more usual definition of labour markets at metropoli-tan and regional scales (Hanson 1992). Their research identifies ways in which employers construct labour forces and labour practices in relation to class, gender and ethnic geographies within Worcester (Hanson and Pratt 1992). Hanson and Pratt argue that 'home' and 'workplace' are fluid concepts, evidenced by practices whereby family and neighbourhood contacts link people to jobs, intergenerational transfers of residential property replace 'rational' decision-making about the location of workplace and residences, and working-class parents arrange sequential scheduling in shift work in order to enable one of them to be at home at all times to care for their children (Hanson 1992). Other studies of women workers meshing their employment and family histories and new life goals or adapting to the exigencies of economic restructuring show them converting home spaces into income-generating workplaces in locations as diverse as middle-class suburbs in rural Appalachia in the United States (Christensen 1993; Oberhauser 1993), a declining resource-based town in western Canada (Mackenzie 1987), a working-class suburb of Athens (Vaiou 1992), the rural periphery of Madrid (Sabaté *et al.* 1991), low-income areas of Colombo, Sri Lanka (Rasanayagam 1992a) and professional-class households in Port Harcourt, Nigeria (Owei and Jev 1992). Likewise, Gibson (1992) describes how reorganization of the workplace, with the institution of a schedule that has become known as 'the divorce roster', has affected domestic relations in coal-mining communities in Australia. Collectively, these cases challenge the 'rigidity of dualist classifications' (Vaiou 1992: 247) while demonstrating the locally specific, contingent expressions of the interactions between production and reproduction.

Feminist research on environmental degradation and politics is also extending conventional categories. Cross-culturally, women's gender roles prescribe their management of health care, child care, food preparation, and the home. Scholars adopting a materialist perspective see this social location as the basis for women's disproportionately high participation in grassroots environmental activism world-wide (Seager 1993). For women activists, the concept of a healthy environment extends from the scale of the body to scales of the home, community, and globe. In industrial societies, their activism may begin with consumer movements around issues of food safety and quality or in local campaigns against toxic environmental conditions that they have first identified through threats to their family's and neighbours' health (Bru 1993; Seager 1993). In Third World settings, concerns about fuel, food and water have engendered women's activism. Well-known examples are the Chipko forestry protection movement in India and the Green Belt Movement tree-planting campaign in Kenya organized by the National Women's Council, but many others exist (Dankelman and Davidson 1988; Seager 1993). In the Kenyan case, women favour planting of indigenous species (as opposed to export-valued alien species) because of their multipurpose utility for fuel, fodder, food, medicine and soil and water conservation. Numerous studies document the unequal gender burdens of environmental degradation, especially of the fuelwood crisis (e.g. Ardayfio-Schandorf 1993; Hyma and Nyamwange 1993; Ulluwishewa 1993b; Wickramasinghe 1992). At the international scale, women organizers clearly link environmental and social justice, connecting the environmental burdens of indigenous and minority communities as well as of women, threats to women's reproductive rights and issues of sustainable economic development to the environmental health of the planet (Seager 1993; Women's Environment and Development Organization 1992).

Feminist geographers are only beginning to analyse women's environmental activism and to frame approaches that combine the symbolic dimensions of women's relation to the environment with their material condition. Dianne

Rocheleau (personal communication 1993) is developing a framework for a feminist political ecology; Cathy Nesmith (personal communication 1993) is undertaking research on gendered responses to the environmental movement in British Columbia, drawing on the cultural ideological perspectives of ecofeminism, the materialist position that focuses on the gender division of labour, and gendered theories of political activism. Rebecca Roberts (personal communication 1994), in investigating the choices of farm families in Iowa between sustainable and high-technology strategies, is exploring how the values, characteristics and private and public roles of farm women shape these decisions. Josepa Bru favours the political ecology approach while emphasizing the importance of local conditions for three Spanish communities in understanding the commonalities and differences among women activists, their tactics and goals (Bru 1993).

REDEFINING THE BOUNDARIES: CHALLENGES TO AND FOR FEMINIST GEOGRAPHY

The many examples I have cited reveal that for feminist geographers gender is a central category of analysis, whether they are investigating structural inequities or women's visions and actions. These examples also show how gender is inflected by other social categories such as class and ethnicity, and the contextually specific ways in which it functions across time and place. They also point to the boundary problems and tensions within feminist geography and in its relations to other areas of the discipline that have arisen with the post-modernist questioning of 'metanarratives' and 'totalizing' theories. Minority and (especially expatriate) post-colonial feminist scholars have criticized white feminists' portrayals of 'woman' for their failure to represent experiences of the 'other'. Within this discussion of differences, issues of social identity predominate (Bondi 1990, 1992, 1993; McDowell 1992, 1993a, 1993b). While race, ethnicity and class are most frequently identified as sources of difference among women, sexuality, age and nationality are also recognized.

The challenges of post-modernism are several. First, feminists are concerned that dominant cultural (white, male) groups are promoting the notion that *no* category is central at an historical moment when their hegemony is being assailed by 'others' (Christopherson 1989). Second is the dilemma of so essentializing categories such as 'race', or of falling into such fragmentation that scholarship does not provide support for political commitment and action (Kobayashi 1994). Too great a stress on diversity can devolve 'into an over-reliance on individual solutions to collective social problems; conversely, seeing only the similarities in women's experiences can lead to overgeneralized and inappropriate policy decisions' (Katz and Monk 1993: 277). Third is the possibility that feminist geographers, in drawing theoretical insights from other disciplines about 'differences' among women, will lose sight of place and space as sources of similarity and difference. Recent reviews by feminist geographers on theory, difference and 'multiple voices' pay great attention to social categories and often by-pass or have difficulty dealing with the differences demonstrated in the (often positivist) research on women's lives being produced by European and Third World feminist geographers (Bondi 1990, 1992, 1993). We need ways to encompass both social differences and context to bring empirical research to bear on more complex theorizing (Peake 1993; Pratt 1993). We need to recognize that categories can remain vital even if their boundaries are fluid.

Current methodological discussions also focus on differences and on the boundaries between researcher and 'subjects'. Feminists espouse a research ethic that attempts to diminish hierarchical power relationships and to let women speak for themselves. This leads to an emphasis on qualitative, local-scale research, raises questions about gender and power relations within the interviewing process and about problems of representation in written texts, particularly when the research engages women of another sexual, racial/ethnic, or cultural identity than the scholar's (Dyck 1993; England 1994; Gilbert 1994; Herod 1993; Miles and Crush 1993). One challenge is to avoid paralysis and excessive introversion while maintaining the insights of reflexive research that acknowledge the position of the scholar yet support social change (Katz 1992). Another is to find appropriate methodologies for dealing with women's lives beyond the local scale since the interactions across global, national and local scales are vital in shaping women's experiences and necessary to address in working for social and environmental change.

The question of scale is also important in identifying empirical research questions in a changing global environment. Feminists have extended the boundaries of geography by introducing the scale of the body in work ranging from studies of ideologies to research on the experience of sexualities (Valentine 1993), girls' and women's fear in public spaces (Katz 1993; Pain 1991; Valentine 1989), and of domestic violence (Townsend et al. 1995). They have begun to address the ways in which sexuality is involved in the international movements/traffic in women, examining individual behaviour and the complicity of states and private enterprise in this 'industry' (Humbeck 1991; Meyer-Hanschen (noted in International Geographical Union Commission on Gender and Geography 1993: 7); Timar n.d.). Women's physical and mental health also offers much scope for research, as do the early and later stages of women's lives (Katz and Monk 1993). Many new questions also need to be addressed at national and international scales because of the effects of political restructuring on women's status as workers, citizens, immigrants and refugees. Interest in them is indicated by the emergence of research on women within Eastern Europe (Ciechocinska 1993; Timar n.d.) and by the current agenda of the International Geographical Union Commission on Gender and Geography (IGU Commission on Gender and Geography 1993).

The research on women and environment suggests the widening scope of feminist geography which for most of its short history has embraced urban social contexts, other than in the studies of Third World women. Silence still largely exists around the gendered construction of geographical research on the biophysical environment, though German-speaking feminist geographers have critiqued masculinism within physical geography (IGU Commission of Gender and Geography 1993). Gender divisions in the creation of geographic knowledge have also been noted, such as the concentration of women members of the Association of American Geographers within 'socially conscious' areas of human geography and in biogeography more than other subfields of physical geography (Goodchild and Janelle 1988). More attention needs to be paid to the diverse ways in which social categories construct all of geographical knowledge and professional practice (Domosh 1991; Monk 1992b).

At the level of action, other boundary challenges exist. Most effort has been made on the front of geographic education in developing feminist courses and teaching materials (for example, Monk 1988; Peake *et al.* 1989; Women and Geography Study Group 1984) and in attempting to integrate gender themes into 'mainstream' courses (Matthews 1993). International collaborative educational efforts have been organized with the support of the European Erasmus programme bringing students and faculty from several countries together since 1990. But textbooks have been slow to incorporate feminist perspectives, and systematic attempts to introduce teaching practices more likely to motivate interest and achievement in geography by girls of diverse ethnic backgrounds are only now being proposed. Multicultural, gender-sensitive research on children's geographic learning in natural settings is also sparse (Caballer-Arce and Breitbart 1993; Katz 1993).

The role of feminism in 'applied' geography is also limited. Dutch feminist geographers are linked with governmental agencies and planners (Karsten 1989; Droogleever Fortuijn 1993); Canadian scholars have worked on interdisciplinary urban projects (Andrew and Milroy 1988) and such projects have also been carried out in the United States (Breitbart 1990; Pader and Breitbart 1993); policy work has been done with women immigrants in Australia (R. Fincher, personal communication 1991). More common is engagement of both First and Third World geographers in consultancies and training projects on women and development in Asia and Africa (for example, A. Buang, J. Momsen, S. Raju, personal communications 1993; Iyun and Oke 1993). But the blossoming field of 'applied' geography, which emphasizes technical skills, has not addressed such feminist concerns as gender bias in secondary databases or the ethics of contract work (for example, in location studies) that sustain the exploitation of women in low-wage work.

Feminist geography challenges boundaries on many fronts. It calls for a rethinking of traditional categories of analysis, especially of the dualisms that have been pervasive in Western thought, while grappling with ways to incorporate the multiplicity of women's lives into its own dualist category of masculine and feminine. It is alert to the interplay between symbolic

899

constructions of the world and material experiences. It believes in the importance of ideas in the political struggle to change gender inequities and in efforts to present women's voices and visions of a more just and sustainable social and environmental order. Many challenges remain, new ones will arise, but the productivity, international expansion, and growth in sophistication of this field in two decades has been remarkable.

REFERENCES

Albet, A., García-Ramon, M.D. and Nogué-Font, J. (1992) 'Fifty years of geography in Spain: a review based on the analysis of academic journals', in J. Bosque-Maurel *et al.* (eds) *Geography in Spain (1970–1990): Spanish Contribution to the 27th International Geographical Congress (IGU)*, Madrid: The Royal Geographical Society and the Association of Spanish Geographers.

Andrew, C. and Milroy, B. Moore (1988) *Life Spaces: Gender, Household, Employment*, Vancouver: University of British Columbia Press.

Ardayfio-Schandorf, E. (1993) 'Household energy supply and women's work in Ghana', in J.H. Momsen and V. Kinnaird (eds) *Different Places, Different Voices: Gender and Development in Africa, Asia and Latin America*, London: Routledge.

Baylina, M., Canoves, G., García-Ramon, M.D. and Vilariño, M. (1991) 'La entrevista en profundidad como metodo de analisis en geografia rural: mujeres agricultoras y relaciones de genero en la costa gallega', *VI Colloquio de Grografia Rural AGE Madrid*, pp. 12–19.

Berman, M. (1977) 'Facts and attitudes on discrimination as perceived by AAG members', *The Professional Geographer* 29, 70–6.

Bondi, L. (1990) 'Progress in gender and geography: feminism and difference', *Progress in Human Geography* 14, 438–45.

Bondi, L. (1992) 'Gender and dichotomy', *Progress in Human Geography* 16, 98–104.

Bondi, L. (1993) 'Gender and geography: crossing boundaries', *Progress in Human Geography* 17, 241–6.

Bowlby, S., Lewis J., McDowell, L. and Foord, J. (1989) 'The geography of gender', in R. Peet and N. Thrift (eds) *New Models in Geography*, Vol. 1, London: Unwin Hyman.

Breitbart, M.M. (1990) 'Quality housing for women and children', *Canadian Woman Studies / Les cahiers de la femme* 11(2), 19–24.

Bru, J. (1993) 'Genero y percepcion de riesgos ambientales: el papel de las mujeres en la defensa de la salud y el medio ambiente', Unpublished research report submitted to the Instituto de la Mujer, Departament de Geografia, Universitat de Lleida.

Caballer-Arce, G. and Breitbart, M.M. (1993) *Facing Education / Enfrentando la Educacion* (copyright Michael Jacobsen-Hardy). Funded by the Massachusetts Foundation for the Humanities.

Canoves, G. (1989) 'La actividad de la mujer en la explotacion agraria familiar: una primera aproximacion en las comarcas de Osona y del Girones', *Documents d'Anàlisi Geogràfica* 14, 73–88.

Christensen, K. (1993) 'Eliminating the journey to work: home based work across the life course of women in the United States', in C. Katz and J. Monk (eds) *Full Circles: Geographies of Women over the Life Course*, London: Routledge.

Christopherson, S. (1989) 'On being outside "the project"', *Antipode* 21, 83–9.

Ciechocinska, M. (1993) 'Gender aspects of dismantling the command economy in Eastern Europe', *Geoforum* 24, 31–44.

Dankelman, J. and Davidson, J. (1988) *Women and Environment in the Third World*,

London: Earthscan and International Union for the Conservation of Nature.

Domosh, M. (1991) 'Toward a feminist historiography of geography', *Transactions of the Institute of British Geographers* 16: 95–104.

Droogleever Fortuijn, J. (1993) 'The Netherlands', in J. Monk, J. Droogleever Fortuijn, H.U. Rii, L. McDowell and M. Gilbert, *Contextualizing Feminist Geography: International Perspectives*, International Geographical Union Commission on Gender and Geography Working Paper No. 27.

Dyck, I. (1993) 'Ethnography: a feminist method?', *The Canadian Geographer* 37, 52–7.

England, K.V.L. (1994) 'Getting personal: reflexivity, biography, and feminist research', *The Professional Geographer* 46, 80–90.

Fagnani, J. (1993) 'Female activity, fertility and family policy in France', Paper presented at the Erasmus Intensive Course on Geography and Gender, Autonomous University of Barcelona, Spain.

Fincher, R. (1993) 'Women, the state and the life course in urban Australia', in C. Katz and J. Monk (eds) *Full Circles: Geographies of Women over the Life Course*, London: Routledge.

Gale, F. (1990) 'The participation of Australian Aboriginal women in a changing political environment', *Political Geography Quarterly* 9, 381–95.

García-Ramon, M.D., Cruz, J. and Baylina, M. (1992) 'Female day labourers, sexual division of labour and welfare state in Spain: the case of Andalusia', Paper presented at the pre-Congress symposium of the International Geographical Union Study Group on Geography and Gender, Rutgers University, New Brunswick, N.J.

García-Ramon, M.D., Cruz Villalon, J., Salamaña, I., Valdovinos, N. and Vilariño, M. (1991) 'Women and farm households: regional variations in gender roles and relations in Spain', *Iberian Studies* 20, 81–112.

García-Ramon, M.D., Solsona, M. and Valdovinos, N. (1990) 'The changing role of women in Spanish agriculture: analyses from the agricultural censuses', *Journal of Women and Gender Studies* 1, 135–61.

Gibson, K. (1992) 'Hewers of cake and drawers of tea: women, industrial restructuring and class processes on the coalfields of central Queensland', *Rethinking Marxism* 5, 29–56.

Gilbert, M. (1994) 'The politics of location: doing feminist research at home', *The Professional Geographer* 46, 90–6.

Gittell, M. and Shtob, T. (1981) 'Changing women's roles in political volunteerism and reform of the city', in C.R. Stimpson, E. Dixler, M. Nelson and Y.B. Yatrakis (eds) *Women in the American City*, Chicago, Ill.: University of Chicago Press.

Goodchild, M. and Janelle, D. (1988) 'Specialization in the structure and organization of geography', *Annals of the Association of American Geographers* 78, 1–28.

Gregson, N. and Lowe, M. (1994) *Servicing the Middle Classes*, London: Routledge.

Hanson, S. (1992) 'Geography and feminism: worlds in collision?', *Annals of the Association of American Geographers* 82, 569–86.

Hanson, S. and Johnston, I. (1985) 'Gender differences in work trip lengths: explanations and implications', *Urban Geography* 6, 193–219.

Hanson, S. and Pratt, G. (1992) 'Dynamic dependencies: a geographic investigation of local labor markets', *Economic Geography* 68, 373–405.

Haraway, D. (1988) 'Situated knowledges: the science question in feminism and the privilege of partial perspective', *Feminist Studies* 14, 575–99.

Hayden, D. (1984) *Redesigning the American Dream: The Future of Housing, Work, and Family Life*, New York: W.W. Norton.

Hays-Mitchell, M. (1993) 'The ties that bind. Informal and formal sector linkages in streetvending: the case of Peru's *ambulantes*', *Environment and Planning A* 25, 1085–102.

901

Herod, A. (1993) 'Gender issues in the use of interviewing as a research method', *The Professional Geographer* 45, 305–17.

Humbeck, E. (1991) 'Acculturation aspects of Thai women in West Germany', *Association of American Geographers Annual Meeting Abstracts*, p. 91.

Hyma, B. and Nyamwange, P. (1993) 'Women's role and participation in farm and community tree-growing activities in Kiambu District, Kenya', in J.H. Momsen and V. Kinnaird (eds) *Different Places, Different Voices: Gender and Development in Asia, Africa and Latin America*, London: Routledge.

International Geographical Union Commission on Gender and Geography (1993) *Newsletter No. 10*.

Iyun, B.F. and Oke, E.A. (1993) 'The impact of contraceptive use among urban traders in Nigeria: Ibadan traders and modernisation', in J.H. Momsen and V. Kinnaird (eds) *Different Places, Different Voices: Gender and Development in Asia, Africa and Latin America*, London: Routledge.

Karsten, L. (1989) 'Feminist geography in the Netherlands', *Journal of Geography in Higher Education* 13, 104–6.

Katz, C. (1992) 'All the world is staged: intellectuals and the projects of ethnography', *Environment and Planning D: Society and Space* 10, 495–510.

Katz, C. (1993) 'Growing girls/closing circles: limits on the spaces of knowing in rural Sudan and US cities', in C. Katz and J. Monk (eds) *Full Circles: Geographies of Women over the Life Course*, London: Routledge.

Katz, C. and Monk, J. (eds) (1993) *Full Circles: Geographies of Women over the Life Course*, London: Routledge.

Kobayashi, A. (1994) 'Coloring the field: gender "race" and the politics of fieldwork', *The Professional Geographer* 46, 73–80.

Lee, D.R. and Schultz, R. (1982) 'Regional patterns of female status in the United States', *The Professional Geographer* 34, 32–41.

McDowell, L. (1979) 'Women in British geography', *Area* 11, 151–4.

McDowell, L. (1983) 'Towards an understanding of the gender division of urban space', *Environment and Planning D: Society and Space* 1, 59–72.

McDowell, L. and Massey, D. (1984) 'A woman's place?', in D. Massey and J. Allen (eds) *Geography Matters*, Cambridge: Cambridge University Press.

Mackenzie, F. (1986) 'Local initiatives and national policy: gender and agricultural change in the Murang'a District, Kenya', *Canadian Journal of African Studies* 20, 377–401.

Mackenzie, S. (1987) 'Neglected spaces in peripheral places: homeworkers and the creation of a new economic centre', *Cahiers de Geographie du Quebec* 31, 247–60.

Mackenzie, S. and Rose, D. (1983) 'Industrial change, the domestic economy and home life', in J. Anderson, S. Duncan and R. Hudson (eds) *Redundant Spaces: Social Change and Industrial Decline in Cities and Regions*, London: Academic Press.

McLafferty, S. and Preston, V. (1991) 'Gender, race, and commuting distance among service sector workers', *The Professional Geographer* 43, 1–15.

Marston, S. (1990) 'Who are "the people"?; gender, citizenship, and the making of the American nation', *Environment and Planning D: Society and Space* 8, 449–58.

Matthews, S.G. (1993) 'Curriculum redevelopment: medical geography and women's health', *Journal of Geography in Higher Education* 17: 91–102.

Mayer, T. (1994) 'Heightened Palestinian nationalism: military occupation, repression, difference and gender', in T. Mayer (ed.) *Women and the Israeli Occupation: The Politics of Change*, London: Routledge.

Meertens, D. (1993) 'Women's roles in colonization: a Colombian case study', in J.H. Momsen and V. Kinnaird (eds) *Different Places, Different Voices: Gender and Development in Asia, Africa and Latin America*, London: Routledge.

Merchant, C. (1980) *The Death of Nature: Women, Ecology, and the Scientific Revolution*, San Francisco: Harper & Row.

Merchant, C. (1984) 'Women and the progressive conservation movement', *Environmental Review* 8(1), 57–85.

Miles, M. and Crush, J. (1993) 'Personal narratives as interactive texts: collecting and interpreting migrant life histories', *The Professional Geographer* 45, 84–94.

Momsen, J. (1980) 'Women in Canadian geography,' *Canadian Geographer* 24, 177–83.

Momsen, J.H. and Kinnaird V. (eds) (1993) *Different Places, Different Voices: Gender and Development in Africa, Asia and Latin America*, London: Routledge.

Momsen, J. and Townsend, J. (eds) (1987) *Geography of Gender in the Third World*, London: Hutchinson Educational.

Monk, J. (1988) 'Engendering a new geographic vision', in J. Fien and R. Gerber (eds) *Teaching Geography for a Better World*, Edinburgh: Oliver & Boyd.

Monk, J. (1992a) 'Gender in the landscape: expressions of power and meaning', in K. Anderson and F. Gale (eds) *Inventing Places: Studies in Cultural Geography*, Melbourne: Longman Cheshire.

Monk, J. (1992b) 'The occupational segregation of women geographers in the United States, 1900–1950', *Abstracts, 27th International Geographical Congress*, pp. 429–30.

Monk, J. (1994) 'Place matters: comparative international perspectives on feminist geography', *The Professional Geographer* 46, 277–88.

Monk, J. and Hanson, S. (1982) 'On not excluding half of the human in human geography', *The Professional Geographer* 34, 11–23.

Mwaka, V.M. (1993) 'Agricultural production and women's time budgets in Uganda', in J.H. Momsen and V. Kinnaird (eds) *Different Places, Different Voices: Gender and Development in Africa, Asia and Latin America*, London: Routledge.

Nesmith, C. and Radcliffe, S.A. (1993) '(Re)mapping Mother Earth: a geographical perspective on environmental feminisms', *Environment and Planning D: Society and Space* 11, 379–94.

Norwood, V. and Monk, J. (eds) (1987) *The Desert Is No Lady: Southwestern Landscapes in Women's Writing and Art*, New Haven: Yale University Press.

Oberhauser, A. (1993) 'Industrial restructuring and women's homeworking in Appalachia: lessons from West Virginia', *Southeastern Geographer* 33, 23–43.

Okpala, J. (1992) 'A comparative study of teachers' and rural women's survival strategies under the Structural Adjustment Programme (SAP) in Nigeria', *Working Paper No. 20*, International Geographical Union Study Group on Geography and Gender.

Oughton, E. (1993) 'Seasonality, wage labour and women's contribution to household income in western India', in J.H. Momsen and V. Kinnaird (eds) *Different Places, Different Voices: Gender and Development in Africa, Asia and Latin America*, London: Routledge.

Owei, O.B. and Jev, M.O. (1992) 'Coping with structural adjustment in Nigeria: professional women and contingent work in Port Harcourt', *Working Paper No. 21*, International Geographical Union Study Group on Geography and Gender.

Pader, E.-J. and Breitbart, M.M. (1993) 'Transforming public housing: conflicting visions of Harbor Point', *Places: A Quarterly Journal of Environmental Design* 8(4), 34–41.

Pain, R. (1991) 'Space, sexual violence and social control: integrating geographic and feminist analyses of women's fear of crime', *Progress in Human Geography* 15, 415–31.

Peake, L. (1993) '"Race" and sexuality: challenging the patriarchal structuring of urban social space', *Environment and Planning D: Society and Space* 11: 415–32.

Peake, L. *et al.* (1989) 'The challenge of feminist geography', *Journal of Geography in Higher Education* 13: 85–121.

Penrose, J., Bondi, L., McDowell, L., Rose, G. and Whatmore, S. (1992) 'Feminists and feminism in the academy: Women and Geography Study Group meeting, University College London, 20 September, 1990', *Antipode* 24, 218–37.

Pratt, G. (1993) 'Reflections on poststructuralism and feminist empirics, theory, and practice', *Antipode* 25, 51–63.

Pratt, G. and Hanson, S. (1988) 'Gender, class and space', *Environment and Planning D: Society and Space* 6, 79–88.

Radcliffe, S.A. (1993) '"People have to rise up – like the great women fighters": the state and peasant women in Peru', in S.A. Radcliffe and S. Westwood (eds) *Viva: Women and Popular Protest in Latin America*, London: Routledge.

Raghuram, P. (1993) 'Invisible female agricultural labour in India', in J.H. Momsen and V. Kinnaird (eds) *Different Places, Different Voices: Gender and Development in Africa, Asia and Latin America*, London: Routledge.

Rasanayagam, Y. (1992a) 'Women as food sellers in the informal sector of the city of Colombo, Sri Lanka', *Working Paper No. 24*, International Geographical Union Commission on Gender and Geography.

Rasanayagam, Y. (1992b) 'Survival strategies and female headed households in Sri Lanka', *Abstracts, 27th International Geographical Congress*, pp. 524–5.

Sabaté, A. and Tulla, A. (1992) 'The geography of gender: the state of the art', in J. Bosque-Maurel *et al.* (eds) *Geography in Spain (1970–1990): Spanish Contribution to the 27th International Geographical Congress (IGU)*, Madrid: The Royal Geographical Society and the Association of Spanish Geographers.

Sabaté, A., Martin-Caro Hernández, J.L., Martin Gil, F. and Rodríguez Moya, J. (1991) 'Economic restructuring and the gender division of labour: the clothing industry in the rural areas of the Autonomous Community of Madrid', *Iberian Studies* 20, 135–54.

Samarasinghe, V., Kiribamnua, S. and Jayatilake, W. (1990) *Maternal Nutrition and Health Status of Indian Tamil Female Tea Workers in Sri Lanka*, Research Monograph No. 8, Washington, DC: International Center for Research on Women.

Sanders, R. (1990) 'Integrating race and ethnicity into geographic gender studies', *The Professional Geographer* 42, 228–31.

Scarpaci, J. and Frazier, L.J. (1993) 'State terror: ideology, protest and the gendering of landscapes', *Progress in Human Geography* 17, 1–21.

Schirmer, J. (1993) 'The seeking of truth and the gendering of consciousness: the CoMadres of El Salvador and the CONAVIGUA widows of Guatemala', in S.A. Radcliffe and S. Westwood (eds) *Viva: Women and Protest in Latin America*, London: Routledge.

Seager, J. (1993) *Earth Follies: Coming to Feminist Terms with the Global Environmental Crisis*, New York: Routledge.

Seager, J. and Olsen, A. (1986) *Women in the World: An International Atlas*, London: Pluto Press.

Timar, J. (n.d.) 'Feminist prospects: uneven development of Hungarian geography', Unpublished manuscript.

Tivers, J. (1978) 'How the other half lives: the geographical study of women', *Area* 10, 302–6.

Townsend, J.G. (1995) (in collaboration with U. Arrewillaga, J. Bain, S. Cancino, S.F. Frant, S. Pacheco and E. Pérez) *Women's Voices from the Rainforest*, London: Routledge.

Ulluwishewa, R. (1993a) 'Development planning and gender inequality: a case study in the Mahaweli Development Project, Sri Lanka', *Working Paper No. 26*, International Geographical Union Commission on Gender and Geography.

Ulluwishewa, R. (1993b) 'Development planning, environmental degradation, and women's fuelwood crisis: a Sri Lankan case study', *Working Paper No. 28*, International Geographical Union Commission on Gender and Geography.

Vaiou, D. (1992) 'Gender divisions in urban space beyond the rigidity of dualist classifications', *Antipode* 24, 2.

Valentine, G. (1989) 'The geography of women's fear', *Area* 21, 85–90.

Valentine, G. (1993) '(Hetero)sexing space: lesbian perceptions and experiences of everyday spaces', *Environment and Planning D: Society and Space* 11, 395–403.

Wickramasinghe, A. (1992) 'Women, equity, and natural resources management', *Occasional Working Papers* 1(5), Vancouver: The Centre for Research in Women's Studies and Gender Relations, University of British Columbia.

Wilkinson, N.L. (1979) 'Women on the Oregon Trail', *Landscape* 23, 42–7.

Wilson, E. (1991) *The Sphinx in the City: Urban Life, the Control of Disorder, and Women*, Berkeley: University of California Press.

Women and Geography Study Group of the IBG (1984) *Gender and Geography: An Introduction to Feminist Geography*, London: Hutchinson (in association with the Explorations in Feminism Collective).

Women's Environment and Development Organization (1992) *Official Report, Women's World Congress for a Healthy Planet*, New York: World Women's Congress Secretariat, Women's Environment and Development Organization.

Zelinsky, W. (1973a) 'The strange case of the missing female geographer', *The Professional Geographer* 25, 101–6.

Zelinsky, W. (1973b) 'Women in geography: a brief factual account', *The Professional Geographer* 25, 151–65.

Zelinsky, W., Monk, J. and Hanson, S. (1982) 'Women and geography: a review and prospectus', *Progress in Human Geography* 6, 317–66.

FURTHER READING

Hanson, S. (1992) 'Geography and feminism: worlds in collision?', *Annals of the Association of American Geographers* 82, 569–86.

Hanson, S. and Pratt, G. (1995) *Gender, Work and Space*, London: Routledge.

Katz, C. and Monk, J. (eds) (1993) *Full Circles: Geographies of Women over the Life Course*, London: Routledge.

McDowell, L. (1991) 'The baby and the bathwater: diversity, deconstruction and feminist theory in geography', *Geoforum* 22, 123–33.

McDowell, L. and Massey, D. (1984) 'A woman's place?', in D. Massey and J. Allen (eds) *Geography Matters*, Cambridge: Cambridge University Press.

Mayer, T. (1989) 'Consensus and invisibility: the representation of women in human geography textbooks', *The Professional Geographer* 41, 397–409.

Momsen, J.H. and Kinnaird, V. (eds) (1993) *Different Places, Different Voices: Gender and Development in Asia, Africa and Latin America*, London: Routledge.

Monk, J. (1988) 'Engendering a new geographic vision', in J. Fien and R. Gerber (eds) *Teaching Geography for a Better World*, Edinburgh: Oliver & Boyd.

Monk, J. (1994) 'Place matters: comparative international perspectives on feminist geography', *The Professional Geographer* 46, 277–88.

Seager, J. (1993) *Earth Follies: Coming to Feminist Terms with the Global Environmental Crisis*, New York: Routledge.

Seager, J. and Olsen, A. (1986) *Women in the World: An International Atlas*, London: Pluto Press.

41

PLACE

Edward Relph

PLACE AS AN ASPECT OF ENVIRONMENTAL EXPERIENCE AND GEOGRAPHY

Place is a simple enough concept until you begin to think about how places are experienced, and then there is very little about it which seems entirely clear or unambiguous. As a geographical concept it refers to named localities, and it is this meaning which stands behind the frequently used definition of geography as the study of places, a definition with a robust life that has endured throughout the two-thousand-year history of the discipline. Until recently it was considered unnecessary to elaborate upon the meaning of the word 'place'; it was straightforward and self-evident.

Since about 1960, and simultaneously in geography and other academic fields such as architecture and psychology, a more complex notion of place has begun to emerge. It is not yet possible to be sure whether this is merely an academic elaboration of ideas previously taken for granted, or whether it reflects some fundamental cultural change in the way people relate to environments, though I suspect the latter is at least partly responsible. What is clear is that places have come increasingly to be understood as experiential and social phenomena consisting of territories of meanings and subject to all the inconsistencies of everyday life. And, as if this change in understanding were not difficult enough, it appears that there has been a parallel shift in the character of actual places as their interconnections and identities have been dramatically alerted by a combination of electronic communications, mass travel, and the growth of global business.

Philosophically and chronologically the changes in interpretations of place since 1960 reflect a trend from phenomenological approaches to the critical analyses of political economy. The following discussion follows the general sequence of this trend in order to summarize the most important arguments which have been made about place and to demonstrate how these arguments have responded to shifts in the character of places. If it has a central message it is that place is a microcosm of geography – it is varied, changeful, eludes simple definition, is open to a variety of interpretations, does not respond well to reduction into simple categories, and to understand it requires keen powers of observation combined with flexible thinking.

CLASSICAL NOTIONS AND DEFINITIONS OF PLACE

In ordinary language the word 'place' serves as a nebulous catch-all which refers to where something is regardless of scale or type of environment. The Australian desert is a place, so is Melbourne, and Monash University, and a house in the suburbs. Even this short list suggests why Aristotle declared that 'The question, what is Place? presents many difficulties for analysis.' He devoted a section of his *Physics* (Book 4, 209a–212b) to these difficulties, and resolved most of them to his satisfaction by concluding that place refers to the precise dimensions of the space which contains something – thus the place of a book on a shelf is the space which is exactly occupied by that book, and the place of a city is the area containing its buildings and roads. This interpretation stood behind the old geographical idea of place as whatever occupies a location, culminating perhaps in central place theory in which places have spatial attributes and no particular content.

The Aristotelian view suggests that a place-container and what is in it can easily be separated. This might seem like an innocent and accurate observation, but Eric Walter (1988), a social psychiatrist who has written an erudite account of place as a fundamental aspect of a mentally healthy life, suggests that this Aristotelian view is the philosophical foundation for policies of dislocation and uprooting, such as urban renewal.

Walter is obviously no admirer of Aristotle, and argues persuasively that the notion of place as a detachable container is a gross reduction of Plato's earlier view, expressed in the *Timaeus*, that place is one of the great modes of being in the universe, 'as it were, the nurse of all becoming', and the receptacle of forms, powers and feelings (Walter 1988: 120–6; Plato, *Timaeus*, 49a–52). This is an inestimably richer notion than the Aristotelian one, for it suggests that place is an interactive environment which influences and responds to whatever is within it. Plato's argument led Walter (1988: 215) to derive his own definition of place as the 'location of experience; the container of shapes, powers, feelings and meanings'. This emphasis on meaning and experience indicates that there is a deep connection between a place and those who occupy it; the two cannot be separated without radically changing both of them.

The definition of places as territories of meaning rather than containers of things is more commonly argued from phenomenology than from Plato. Phenomenology is a philosophical perspective which considers the world as it is directly experienced, so a phenomenological understanding regards places as tightly interconnected assemblages of buildings, landscapes, communities, activities, and meanings which are constituted in the diverse experiences of their inhabitants and visitors. From this perspective place is an existential phenomenon and places are not just geographical objects to be studied academically; they are where we live.

In some degree this recognition informs almost all the recent discussions of place, though different disciplines do bring their academic perspectives to bear on it. For example, the book *People Places* by landscape architects Clare Cooper-Marcus and Wendy Francis (1990) discusses the qualities of urban squares and public spaces in terms of their built forms and the ways in which people use and experience these. In contrast, Sharon Zukin (1992: 12), a sociologist with an orientation to economics, defines place as 'a territory ... a concentration of people and economic activity ... a cultural artifact of social conflict and cohesion'. And John Logan and Harvey Molotch argue in *Urban Fortunes: The Political Economy of a Place* (1987) that the attributes of a place result from social action rather than the qualities in a piece of land. Geographical approaches tend to be more inclusive and to understand places as combining landscapes, social and economic activities, and meanings, though the relative importance accorded to each of these elements can vary tremendously depending on the character of the place and the bias of the geographer.

HOME

A difficulty with definitions which stress meaning is that they expose accounts of place to the accusation of being subjective and having no broad relevance. This is a false charge. Place experiences and meanings are not locked up in the minds of individuals, rather they must be considered to be intersubjective – in other words, shared, because they can be communicated and make clear sense to others. The most pervasive expression of the intersubjective character of place experiences is probably found in the sense of home, which appears to be almost universally felt. Wherever it may be, home is a centre of meaning, a familiar setting in an uncertain world, it is the place where one belongs and is best known. Homelessness, in contrast, describes both a socially unacceptable condition and the loss of a fundamental aspect of human existence.

Discussions of home and sense of home are often conducted in the language of plants, especially in terms of roots. To have a strong sense of home and belonging is to have roots; to be forced to move is to be uprooted. This organic language is scarcely incidental. It implies that to have a home place is natural; it is metaphorically to belong to the earth. This is the meaning which the philosopher Martin Heidegger and other existential writers have chosen to emphasize. For Heidegger place was to be understood in terms of 'dwelling', which is a fundamental connection between human beings and the earth, and a manifestation of the very essence of existence (Heidegger 1966, 1971; see also Kolb 1992: 149–54). To dwell in a place is to be in a world complete in itself; it is both to exist and to take responsibility for the existence of other beings; it is to be at home.

This interpretation has to be qualified. If you happen to live in an anonymous apartment slab in suburban Moscow, or a squalid *favela* threatened by mud slides in São Paulo, such notions about home and place and the meaning of existence will be radically truncated. One's home may still be a familiar shelter in an alien world, but it will not be cosy and nice. Existence is not without its burdens and home, indeed all types of places, can be constraining and tedious. Homes have to be maintained, and there is considerable drudgery in doing that, usually falling upon women. Small towns and villages can be prisons to their younger inhabitants who wish for escape to the anonymous freedoms of big cities. In short, experiences of home, as those of most places, are ambivalent. They involve a fluctuating balance of feelings of attachment and of entrapment, though the former sentiment is perhaps the prevailing one.

SPIRIT OF PLACE OR *GENIUS LOCI*

In less agnostic cultures than those which now prevail in the urbanized world it was, and in some areas still is, believed that localities were occupied by spirits or gods who served both as their guardians and as a source of their identity. Mount Olympus was the home of Zeus, and every mountain top, grove and spring was the home of some lesser deity who had to be acknowledged and propitiated. The idea of spirit of place, often referred to by its Latin name as '*genius loci*', has its origins in this polytheistic sense of environments as consisting of diverse sites, each with its guardian spirit.

In its relatively secularized modern meaning, spirit of place refers simply to the inherent and unique qualities of somewhere. It is this idea which is explored by the architect Christian Norberg-Schulz in his book *Genius Loci: Towards a Phenomenology of Architecture* (1984). He argues from Heidegger's philosophy that 'place is evidently an integral part of existence' which can be best understood phenomenologically, and using this approach he examines several distinctive landscapes including Prague, Khartoum and Rome. These are locations with a strong spirit of place because they have strong visual properties which may reveal a sense of mystery about natural forces, or manifest rational order, or express some equilibrium of these. Norberg-Schulz (1984) concludes with a familiar refrain – that modern architecture and town planning are deficient in these properties of distinctiveness, they are monotonous and demonstrate a loss of spirit of place.

As a secular concept *genius loci* has a great deal to do with aesthetic qualities; it is, in effect, a way of considering places as works of art. Adele Chatfield-Taylor, writing in Lipske's book *Places as Art* (1985: 8), has written that places 'can satisfy our desire for beauty, stir our deepest feelings, link us to our history'. This type of thinking appeals to architects and planners who want to do more than design skyscrapers and plan subdivisions. So Kevin Lynch, in his textbook *Site Planning* (1962: 225) proposed that places, and he clearly means places as objects of design, 'should have a clear perceptual identity: recognizable, memorable, vivid, engaging of attention, differentiated from other locations'. Since perceptual identity in urban settings is largely a matter of distinctively built forms and well-constructed spaces, this suggests that the spirit of place is something which can be designed. And perhaps it can, given suitable social conditions and creative architects such as those of the old parts of Rome and Prague. But the evidence is that, divine or secular, *genius loci* is elusive. Even though we may recognize that somewhere has a powerful personality it is invariably difficult to identify how this is constituted and even more difficult to reproduce it. This is why new developments so often seem utterly out of context. This is fortunate. Humanity has sufficient powers of control without adding to them the ability to create the lesser deities of place. What can be done is to protect distinctive places which now exist and then perhaps to find ways to create the conditions which will, in time, allow *genius loci* to emerge.

SENSE OF PLACE

The term 'sense of place' is often used to mean the same as 'spirit of place'. This is confusing. It is more appropriate to understand 'sense of place' as the *awareness* of spirit of place, and as a faculty which individuals possess rather than a property of environments. Like a sense of judgement or a sense of responsibility, it is a synthetic faculty which embraces and extends the various senses of perception. Kevin Lynch (1962: 9) explained that 'a place affects us directly through our senses – by sight, hearing, touch and smell'. To this list should be added imagination, memory and purpose.

Sense of place is not a mandatory requirement for survival, so there are many who pay scant attention to the world around them. Indeed, a detailed survey made by the geographer John Eyles (1985: 123–4) of residents' attitudes to their small English town led him to identify four different senses of place, one of which he labelled 'apathetic' since those individuals had little interest in their surroundings. Another he called 'a social sense of place' because for many people places are defined chiefly by where family and friends are. The other two pay more attention to environments. Eyles calls them 'instrumental', an attitude which regards place primarily as a resource providing goods and opportunities, and 'nostalgic', which stresses heritage and old buildings.

Eyles's study indicates clearly that sense of place cannot be considered as a simple, undifferentiated attitude towards environments. There are considerable variations both in type and intensity of sense of place, depending on such things as familiarity, detachment, social status, gender and self-consciousness. With wealth, for example, comes the freedom to choose places to live. In contrast, the very poor are trapped in places they can afford, or in whatever is provided for them, so that their geographical experiences are constrained and their sense of place is relatively limited.

The degree to which gender effects sense of place is not entirely clear.

Daphne Spain (1992) argues that architectural and geographical space are differentiated by gender in most cultures and times, and some of her conclusions must apply to place, although she does not consider this explicitly. Eyles (1985) comments briefly that women identify places more with community than men, who are either apathetic or see them as built forms. This is reinforced by Dolores Hayden's research (1984) into the history of planning which indicates that women, when they have infrequently had the opportunity, tend to design places which encourage communal activities. From a different perspective, it is the case that women's experience of cities, especially at night, is much more constrained than that of men because of the threat of personal violence. It should follow that female experience and perception of environments differs substantially from that of men, and to that extent women's sense of place must also differ.

Degree of familiarity is a particularly important influence on sense of place, and makes it possible to distinguish what might be called an insider's from an outsider's experience, a distinction which cuts across gender and social status. Insideness is an aspect of the sense of place which comes with knowing and being known somewhere, and is mostly unselfconscious. It can be such a key component of someone's personality that they effectively identify themselves with their place and declare they can live nowhere else. Vestiges of such attachment to place can be glimpsed in homesickness, in characterizations of individuals by their home town ('She's a New Yorker') or in the local fan loyalty evoked by sports teams (even though these are mostly corporate ventures employing player-mercenaries hired from elsewhere).

An outsider's sense of place is relatively detached, and regards places chiefly in terms of ostensible and superficial characteristics. In this, as with most aspects of place, there are considerable variations. Different types of outsideness are found in the packaged experiences of mass tourism, in the standardized conveniences of international business travel which reduce the diverse identities of localities to comfortable familiarity, and in much professional expertise. The latter is mostly based on general and abstract knowledge, so it is assumed that it must have relevance anywhere. When this expertise is brought to bear on places their universal properties are stressed and specificities ignored, often with unfortunate consequences. *Planned to Death: The Annihilation of a Place called Howdendyke* is the pointed title Douglas Porteous (1989) gave to his study of the redevelopment of the village on Humberside where he had grown up. A combination of external economic pressures and planning based on theories about efficient new communities were responsible for remaking Howdendyke into somewhere which bears little relationship with its past.

Outsideness need not be destructive. A self-conscious sense of place can be cultivated and refined by improving powers of observation through the open-minded exploration of environments, and by making imaginative attempts to understand what it is like to live in a place which is not one's own. Through such means it is possible to enter empathetically into situations where one is otherwise an outsider, and to understand them almost as their inhabitants do.

Practising a self-conscious sense of place is an essential skill for geographers and anyone who cares about the quality of environments. It is sometimes instinctive, but for most of us it requires, as Eric Walter (1988) observes astutely, a continual effort to exercise a subtle balance of intellect, common sense, feelings and imagination.

TOPOPHILIA AND TOPOPHOBIA

Topophilia is defined by Yi-fu Tuan (1974: 4), the geographer who gave the term a wide currency, as 'the affective bond between people and place or setting'. It is the 'human love of place . . . diffuse as a concept, vivid and concrete as personal experience' (Tuan 1974: 92).

Topophilia is mostly a gentle human emotion induced by positive attitudes or by pleasant landscapes. Occasionally, however, when circumstances of both person and place are in a positive conjunction, topophilia can be a powerful and ecstatic experience, one which promotes great insights. If we are to believe autobiographical accounts, mountain tops are particularly conducive to such formative moments; Wordsworth climbed Snowdon at night, emerging through cloud to brilliant moonlight, and there 'beheld the emblem of mind / That feeds upon infinity' (*The Prelude*, Book XIV, 'Conclusion', ll. 70–1). Such experiences are not confined to poets; there is evidence that many people have intense topophilic encounters which provide touchstones of meaning by which much of the rest their lives are judged.

In *Topophilia* Tuan chooses not to stress such formative experiences. He writes instead about encounters with landscapes which provide a muted and sustained sense of pleasure, and the factors which influence these. On the human side good health, familiarity, culture, mythology and ideals can play significant roles in making places appear full of light and life. On the natural side there are 'environments of persistent appeal' (Tuan 1974: 114), such as seashores, valleys, islands, and the middle landscapes of carefully tended countryside, which can promote topophilia.

Yi-fu Tuan's preference is to discuss felicitous environments and experiences. However, our environmental experiences are not all pleasant. Even landscapes of persistent appeal can be the source of ugly and disturbing events. The mountains which for Wordsworth were so uplifting can quickly turn frightening; there is little joy in being lost on a mountain-top in an unexpected storm. Such unpleasant experiences of places, in which the overwhelming desire is to be somewhere else that is safe and secure, can appropriately be described as 'topophobia' – repulsion by place.

Topophobia, like topophilia, involves an emotional bond between person and place – but one in which the relationship is essentially negative. The reasons for this can reside with the person's attitude and social context, such as depression, ill health, aesthetic repulsion, unfamiliarity, insecurity, or situation of despair. Alternatively, the reasons might reside in the character of the landscape.

Environments of persistent repulsion are well known to dramatists and the writers of horror stories – they include barren heaths, dereliction, storms, slums, and anywhere dark, dank, polluted, uncared for, and otherwise threatening.

A THEORETICAL INTERJECTION

Topophilia and topophobia, like belonging and entrapment associated with home, can be understood as good and bad aspects of place experience. It is, I think, inappropriate to see them as independent for they are really two facets of a single phenomenon – experience of place – in which sometimes the positive aspects are dominant and sometimes the negative, but both are always present. The patterns within each of them, and the relationships between them, are complex and subtle, so there is little about the experience of place which is entirely unambiguous and predictable, though not everyone acknowledges this. For example, Winifred Gallagher (1993), in *The Power of Place: How our Surroundings Shape our Thoughts, Emotions, and Actions*, offers a simple deterministic account completely explained in the title.

A far more sophisticated view is the one argued by Nicholas Entrikin in *The Betweenness of Place* (1991) that place is a concept which does not fit into standard methodological and epistemological categories. Conventional thinking which separates objective and subjective approaches, and which divides the particular from the general, does not apply. Place, he suggests, has to be viewed both with regard to the objective characteristics of location and in terms of subjective experiences. Writing about place should consider both the particular features of localities and the generality of the idea.

Entrikin's argument is a sensible one. It reveals in an abstract way the tension which exists in all places between particular (or local) features, and general (or global) processes. In any environment there are things which are locally specific, such as festivals, building styles and historical events. There are also manifestations of non-local fashions and influences, such as gothic revival architecture, fast-food franchises and globally diffused pollutants. If the local and specific aspects of an environment are those which enable somewhere to be discussed as a 'place', then the non-local, international and general influences can appropriately be referred to as 'placelessness'. It is, I believe, misguided to treat these as two separate phenomena, or even to see them necessarily as being in conflict. Rather, they are each implicated in the other, the local in the non-local, the general in the particular. In some contexts, such as old villages with traditional cultures, the particular qualities of a locality dominate and placelessness is subservient; in other cases, such as airports, standardized design prevails and the specifics of the place are scarcely discernible. Whatever is local contributes to distinctiveness; whatever is placeless helps to make places comprehensible to outsiders. In some balance, then, the particular and the general in places always occur together, and always need to occur together.

This is not how most writing about place considers the matter. The overwhelming tendency is to present the issue as one of confrontation. Occasionally, as in Le Corbusier's (1929) modernist polemics about cities, it is maintained that standardized international design will bring enlightenment from parochialism and the burdens of local history. Most recent arguments reverse this and claim that the bad forces of uniformity are destroying the good qualities of places.

PLACELESSNESS

Tony Hiss (1990: xv) writes in *The Experience of Place* that 'the fading and discoloration of places has been going on for years'. And Michael Hough (1990: 2) complains in *Out of Place* that 'The influences that at one time gave uniqueness to place – the response of built form to climate, local building materials, and craftsmanship, for instance – are today becoming obscured ...' The clear message is that the ability to construct places rich in local identity and meaning has been lost.

The best evidence for the loss of this ability is placelessness, or the proliferation of modern landscapes which look alike. It is easy to see examples – suburbs, shopping malls, airports, corporate skyscrapers, international franchises, modernist housing projects, and so on. Perhaps more important than similarity of appearance, however, is the levelling of experience and meaning which placelessness apparently involves. It is quite possible for placeless environments to have distinctive appearances; for example, theme parks are imagineered to be 'unique', each with its own arrangements of quaint buildings, pretend mountains, fake lakes, roller-coasters and fantastic images, but they are all predictably similar fabrications. The real issue is not that they look alike, but that they feel so much the same, that there seems to be nothing truly distinctive about them.

Of course, something similar can be said about Greek temples, Gothic cathedrals, and Mogul forts, because they also employed standardized architectural elements. And in some respects these are placeless; however, each temple or fort was adapted to its site and situation, used local construction materials and was the product of intense craftsmanship, so it is now usually seen as emphasizing place rather than diminishing it.

Since the Enlightenment the adaptation of standardized techniques to locality appears to have decreased, with the consequence that placelessness has increased. There are several connected reasons for this. One is philosophical – the widespread adoption of scientific methods which stress the general and measurable characteristics of culture and nature over what is particular. A second is technological – the development of technologies such as those of concrete and air travel which have progressively overcome the constraints of locality. A third is social and economic – mass production and consumption, global trade, and the promotion of international design fashions. In the mid-twentieth century these effectively coalesced into processes such as urban renewal, suburban development and new town planning, with the result that many localities were profoundly transformed. These processes have subsequently intensified with the emergence of what might be called a global mass culture in which electronic information, images and goods are disseminated to many places simultaneously, with scant regard for cultural history or political boundaries and permitting no time for local adaptation. Under such an onslaught perhaps place is destined to disappear altogether. This is presumably what Joshua Meyerowitz (1985) anticipates because he called his book on modern electronic culture *No Sense of Place*. In it he observes that electronic communication goes from nowhere in particular to anywhere in general, and that firsthand place experience is being substituted by vicarious television experience in which locality and variations in geography have little intrinsic importance.

Arguments of place decline have a powerful emotional charge. We almost want them to be true because they reinforce nostalgic concerns about the disappearance of a world of attractive villages and urban neighbourhoods where everyone knew and was known by everyone else. Eric Walter (1988: 2) writes that we have now reached a point where 'for the first time in human history people are systematically building meaningless places', and this seems to confirm our topophobic reactions to ugly shopping malls and hideous housing projects. Be this as it may, I believe he is being too extreme. However deeply we may dislike certain modern landscapes, the very act of construction gives them some significance. And are we to dismiss as meaningless the lives of those who live and work in these 'meaningless places'?

While I accept that placelessness must have serious consequences, because our individual, social and political identities are inexorably tied up with the places where we live and work, I do not think that categorical representations of placelessness as some massive, ahuman force of uniformity, or dismissals of it as meaningless, will be helpful in redressing these consequences. A more reasoned assessment is that increasing placelessness indicates a substantial tipping of the balance from what is particular in place towards what is general, and it has to be acknowledged that this has had the benefit of democratizing geography by making remote places more accessible. Relatively placeless suburbs and tourist resorts reflect great improvements in overall standards of living, and travel to international academic conferences would be far more difficult without the familiar conveniences of standardized airport terminals and hotel chains. In short, both place and placelessness are subtle and ambiguous phenomena, and just as attachment to place has negative aspects, so placelessness has positive aspects. Ignoring such subtleties will make it more difficult to grasp and redress the deepening imbalances between the local and the non-local because we will be deluded into expecting the obvious. And what now seems to be happening to local geographies is anything but obvious.

PLACE EXPLOITATION

The idea of placelessness as a monolithic modernist uniformity invading landscapes, slowly obliterating everything distinctive, is not consistent with the subtle social and economic processes of the late twentieth century. There has, in fact, been a marked revival of interest in the overt qualities of place. In architecture and planning this has something to do with a post-modernist interest in historical and regional context, but in simple economic terms what seems to have happened is that the value of distinctive places has increased as they have become more scarce. Much of the impetus for this reawakening of interest in place identity comes from outside, and the primary motive is not so much to maintain the integrity of a place as to turn it into an attractive opportunity for money-making, an opportunity often realized through post-modern design and heritage planning. It has, in short, become worth while to invest in local identities. This is placelessness, but in a particular and most subtle guise.

Sharon Zukin (1992: 15) describes one reason for this shift in attitude to place as being 'a simple imbalance between investment and employment: capital moves, the community doesn't'. As capital has become flexible, moving through the abstract electronic networks of financial markets, it seeks out the best locations for returns on investments. In these circumstances, David Harvey (1989: 295) points out, the 'qualities of place stand ... to be emphasized in the midst of the increasing abstractions of space. The active production of places with special qualities becomes an important stake in spatial competition between localities, cities, regions and nations.' What he means by 'qualities of places' could include an educated labour force, proximity to an international airport, an attractive landscape, an interesting history, or anything which makes a location stand ahead of its competitors.

Communities in economic decline because they have been bypassed by flexible capital have to do whatever they can to revitalize themselves, including the exploitation of their own place identity, commonly by reworking local heritage into a tourist attraction. This has been done most spectacularly with festival marketplaces, such as South Street Seaport on the waterfront in New York City, or Covent Garden in London, where former working environments have been remodelled into boutiques, outdoor cafés and cobbled streets with licensed street entertainers. Smaller-scale examples of this process include the industrial history centre at Wigan Pier in Lancashire, complete with The Orwell Restaurant, and the former sardine fishing town of Monterey in California, with Cannery Row architecture and Steinbeck's Lady Boutique. This sort of place resurrection may bring money into the local economy, but it also involves a radical metamorphosis. Formerly grimy and unpretentious working-class settings are turned into thoroughly sanitized and attractive settings for informative family outings and school trips. Destroyed buildings may be rebuilt, artefacts imported, dead traditions revived, and the otherwise unemployed of the twentieth century dressed in costumes and employed to represent the employed of the past. Industrial production has been replaced by heritage consumption. And, with remarkable irony given that each of these reconstituted places is a deliberate attempt to foster local individuality, the design and management of exploited places is often done by outside development corporations which employ common design elements and images so that they all have a similar ambience. Manufactured place identity is placeless.

TOPOMORPHIC CHANGES AND DISLOCATED GEOGRAPHY

Eric Walter (1988: 23) suggests that place identities sometimes undergo what he calls 'topomorphic revolutions' – fundamental changes in the structure of their internal relations which occur in association with social transformations. He notes, for instance, that industrial tenements and slums of the nineteenth century had no precedent as elements of urban form, and that the ways which their residents found to relate to them necessarily had to be innovative. The same could equally well be claimed for twentieth-century automobile suburbs. In other words place has to be understood as contingent upon social and historical circumstances and not a geographical constant.

I suspect that a topomorphic revolution on a grand scale is now underway in the geographies associated with global electronic culture, and that placelessness – and its transformation into place exploitation – is a large part of this. It is difficult to gain a firm perspective on what is happening, partly because we are in the midst of the process and partly because so many of the changes fit into old structures rather than creating original forms. Unlike steel frames which imposed rectangularity on skyscrapers, or railways which were forced through landscapes to link cities, the recent technologies of polymers and electronics and gene splicing are malleable and mostly invisible. Plastic oak beams, genetically designed laboratory mice, and electronic marketplaces represent fundamental changes in the nature of things, yet leave them looking much the same. They confound assumptions about what is real or fake, about what is natural or artificial, about where is here and whether geography has any relevance.

It used to be that places were associated with a local environment, economy and culture, and the connections between these, though they may have been complex in practice, were in principle clear and direct. The identity of a place was something made locally, by the people who lived there, perhaps using elements from outside but always bending these to local needs. This is no longer the case. 'To a degree never known before,' Logan and Molotch (1987: 249) observe, 'local interests in place are being shaped by the changing order of international spatial relations.' This is a radical transformation. In the geography of global culture it involves a sort of space–time–culture compression, the global village, in which a diverse mixture of international practices and tastes is being made more or less equally available everywhere. From this mixture of possibilities individuals or groups can select as they need or wish. The consequence is that any spatial variety of place, and indeed placelessness, is now being supplanted by a locally constituted, social variety based on such things as ethnicity, gender preferences, reproduced best bits of other townscapes, or even tastes in music and clothes. So the process of forming place identity increasingly consists not in local development within a geographical context, but in the ways many similar fragments of global culture are combined somewhere. Distinctiveness is given either by the self-consciously preserved or reconstructed fragments of old landscapes, by an emphasis on a selective social activity, or by particular combinations of uprooted and transported global fragments. I am struck, for instance, by the gondolas which ply the modernist waterfront in Toronto, and Texas-style restaurants decorated with American antiques in the Victorian centre of Glasgow.

In the new geographical logic any place distinctiveness which is not inherited is largely an illusion because there are now few necessities about why anything has to be somewhere specific or have a particular appearance. Investment capital can easily move elsewhere and identities are mostly detachable images, to be contrived in any one of countless ways. In the strategies of global marketing the aim is to make money rather than to sustain local integrity. In this process place has apparently been divorced from context. It is as though the previously fixed points of reference of geography have been uprooted and can now be exchanged at will. Geography itself has literally been dislocated.

THE RECOVERY AND DESIGN OF PLACE

Placelessness and place exploitation have met with three types of resistance. There have been countless local political reactions, often taking the form of neighbourhood protests against the threat of potentially place-destroying intrusions, such as new highways or corporate developments or even single franchise outlets. Such exclusionary protests have protected particular places, and the lessons learned can be used elsewhere, but these are essentially isolated actions which do little to address the larger issues of topomorphic change. Of more fundamental importance are those forms of resistance which either stress the need for personal sensitivity to places as a foundation for recovering something that is disappearing from the world, or emphasize the need to find technical ways to design and maintain distinctive places.

Yi-fu Tuan (1977: 203) concludes his book on *Space and Place* with the remark that its 'ultimate ambition' is 'to increase the burden of awareness'. By making ourselves more aware of the subtleties and ambiguities of our experiences of place, and then raising questions about conventional planning practices, Tuan suggests that we can open up issues which planners and professionals have found it convenient to forget. Eric Walter (1988: 213) similarly suggests that to redress the recent loss of ability to make meaningful places 'we can start by rebuilding ourselves. The archaic way of seeing, thinking and caring is not lost. We can bring it to the surface and change its position in

918

the structure of experience.' When this is done 'we may begin to enlarge public sensibility and to rediscover the expressive intelligibility of human locations'.

The main aim of Tuan and Walter is to show that place is a phenomenon of human environmental experience which should not be taken for granted. They also believe that solutions to the problem of a decline in place identity must lie first with select individuals, who by their sensitivity and their strength of reasoning will be able to convey the message that place matters. Presumably their hope is that this will result in a situation in which places will once again come to be made with all the distinctive qualities still apparent in the remnants of pre-modern landscapes.

This is undeniably a worthwhile hope. It is, however, far from clear how individual sensitivity and reason will combat processes of placelessness and the marketing strategies of place exploitation.

In *The Experience of Place*, Tony Hiss (1990: xv–xvi) connects personal sensitivity with what he calls a 'brand-new science of place'. First he recommends the development of 'simultaneous perception' – a sort of self-consciously diffuse experiential watchfulness – so that 'we can salvage experiences of place'. He then links this to the work of various landscape architects, environmental designers and regional planners who have developed scientific techniques for simulating, redesigning, and managing places. Through the methods of these place scientists we are, so Hiss claims (1990: 100), 'finally in a position to get on with the job of making sure that all places are worth experiencing'. David Canter (1977: 157), an environmental psychologist who once proposed the slogan 'The Goal of Environmental Design is the Creation of Places', might well agree. He would like to bring all the techniques of behavioural and environmental psychology to bear on this goal. Michael Hough (1990: 2) begins his book *Out of Place* by suggesting that the question of regional character has become a question of choice, an argument which would seem to be consistent with what is happening in place exploitation, and therefore is a matter of design rather than of necessity. For him place design would use the methods of landscape architecture to build upon the local character and processes of the natural environment.

These are clarion calls to action. It should be noted, however, that they require considerable shifts in perspective for design and planning professionals. Planning and landscape architecture, for example, are mostly concerned with tangible, material things such as street networks or trees. But a place is not just a material entity or a container of manipulable bits. It is also a location of experiences and meanings. To design a place is therefore to try to design meaning and value. For this there are no firm assumptions, no clear guidelines, and no body of established practice. And the science of place is useless because scientific methods cannot resolve issues of value.

A further, very important caution is also warranted. It is probable that any techniques developed for the scientific design of meaningful places will be equally useful for fabricating place identities for economic exploitation.

These concerns about placemaking have not gone unnoticed. Some designers maintain that it is impossible for professionals to design meaningful places for others, and that to do so would be an act of imposition which suppresses rather than generates significance. To minimize this possibility they emphasize self-help, usually by devising methods which allow communities in effect to create their own places. An excellent example of this is the work of Randy Hester, a landscape architect, who was asked by the residents of the island of Manteo in the Outer Banks of North Carolina to prepare a plan for economic development which would not undermine the most valued aspects of their daily lives and local landscapes. By participating in the community and talking with the residents he was able to identify a 'sacred structure' of valued places, which includes a local marsh, the drugstore, the Duchess Restaurant where locals gather for morning coffee, a statue of Sir Walter Raleigh and the town cemetery (Hester 1993). The plan for the town which Hester devised keeps tourists and related development away from these special places, so that their meaning for residents can be preserved.

Self-help design approaches to placemaking are political rather than merely technical in that they promote local empowerment. Their initial hope is that the processes of active involvement in design will lead to a revived sense of local responsibility. The larger hope is that they could lead to what Kenneth Frampton calls 'critical regionalism' (cited in Kolb 1992: 165–6, 180). Unlike a strategy of simple place protection, which attempts to ignore or resist global forces of change and capital, often by resorting to nostalgia and heritage preservation, critical regionalism calls upon local resources to shape and mollify the impact of outside forces.

We can only speculate about how critical regionalism might apply in practice. It might lead to something like the 'place utopia' imagined by Kevin Lynch (1981) in which there would be an 'urban countryside' comprised of a patchwork of regions, each with its own style of living, yet linked by electronic communications systems. At a less idealistic level it has to involve a careful attempt to discover what might be called a geographically responsible way of doing things, in which global processes and fashions would cease to be imperatives and would be used only when refracted through the lens of locality and implemented in a locally responsible way.

The balanced approach that is required by critical regionalism will not be easy to establish. First of all, there is the difficulty of finding local solutions for problems caused by processes which are not local. How, for example, is it possible to modify transmissions from direct broadcast satellites so that these respect local customs and accord with local interests? It seems that non-local process must always have an advantage because whatever is local operates at a smaller scale and is necessarily fragmented.

There is another, opposite, and no less difficult, problem to resolve. The forceful promotion of place carries with it the possibility of a descent into parochialism and sectarian politics. At its worst the unbalanced assertion of the importance of place can serve as the foundation for strategies of regional purification and ethnic cleansing. In such cases, David Harvey (1989: 351) notes quietly, 'respect for others gets mutilated in the fires between the fragments'. For all its merits, place is not an untarnished concept, and a strong attachment to place does not invariably have pleasant consequences.

CONCLUSION

In the increasingly dislocated and fabricated geographies of the late twentieth century, places have taken on unprecedented forms and appearances. Whether we like these or not, our individual and social identities are still implicated in them, and we have no choice but to live and work or be unemployed in them. Given such inevitabilities I believe it is essential to redress the imbalance which has developed in favour of remote agencies and abstract universal processes, and to try to find methods to re-establish local integrity and responsibility. This is unlikely to be achieved by the uncritical use of a baggage of concepts such as *genius loci* and topophilia which assume that everything to do with place is good and placelessness is always bad. That will lead to crass simplifications which will ignore the subtlety and originality of the processes which are transforming the geographies in which we live. What is required is a broad-ranging and critical approach which acknowledges that place, like geography, is diverse, continually changing, ambiguous, and inclined to take on new forms as soon as the current ones have been explained.

REFERENCES

Aristotle, *Physics*, in R. McKeon (ed.) (1941) *Aristotle: Basic Works*, New York: Random House.
Canter, D. (1977) *The Psychology of Place*, London: The Architectural Press.
Cooper-Marcus, C. and Francis, W. (1990) *People Places*, New York: Van Nostrand Reinhold.
Entrikin, N. (1991) *The Betweenness of Place*, Baltimore, Md.: The Johns Hopkins University Press.
Eyles, J. (1985) *Senses of Place*, Warrington, England: Silverbrook Press.
Gallagher, W. (1993) *The Power of Place: How our Surroundings Shape our Thoughts, Emotions and Actions*, New York: Poseidon Press.
Harvey, D. (1989) *The Condition of Postmodernity*, Oxford: Basil Blackwell.
Hayden, D. (1984) *Redesigning the American Dream: The Future of Housing, Work and Family Life*, New York: W.W. Norton.
Heidegger, M. (1966) 'Memorial address', in *Discourse on Thinking* (trans. by J. Anderson and E.H. Freund), New York: Harper & Row.
Heidegger, M. (1971) 'Building dwelling thinking', in *Poetry Language Thought* (trans. by A. Hofstadter), New York: Harper & Row.
Hester, R.T. (1993) 'Sacred structures and everyday life: a return to Manteo, North Carolina', pp. 271–97 in D. Seamon (ed.), *Dwelling, Seeing and Designing*, Albany: State University of New York Press.
Hiss, T. (1990) *The Experience of Place*, New York: Alfred A. Knopf.

Hough, M. (1990) *Out of Place*, New Haven, Conn.: Yale University Press.

Kolb, D. (1992) *Postmodern Sophistications: Philosophy, Architecture and Tradition*, Chicago, Ill.: University of Chicago Press.

Le Corbusier (Charles Édouard Jeanneret) (1929) *The City of Tomorrow*, Cambridge, Mass.: MIT Press.

Logan, J. and Molotch, H.L. (1987) *Urban Fortunes: The Political Economy of Place*, Berkeley, Calif.: University of California Press.

Lynch, K. (1962) *Site Planning*, Cambridge, Mass.: MIT Press.

Lynch, K. (1981) *A Theory of Good City Form*, Cambridge, Mass.: MIT Press.

Meyerowitz, J. (1985) *No Sense of Place*, New York: Oxford University Press.

Norberg-Schulz, C. (1984) *Genius Loci: Towards a Phenomenology of Architecture*, New York: Rizzoli.

Plato, *Timaeus*, in F.M. Cornford (1937) *Plato's Cosmology*, London: Routledge & Kegan Paul.

Porteous, J.D. (1989) *Planned to Death: The Annihilation of a Place called Howdendyke*, Toronto: University of Toronto Press.

Spain, D. (1992) *Gendered Spaces*, Chapel Hill, N.C.: University of North Carolina Press.

Tuan, Yi-Fu (1974) *Topophilia*, Englewood Cliffs, N.J.: Prentice-Hall.

Tuan, Yi-Fu (1977) *Space and Place*, Minneapolis, Minn.: University of Minnesota Press.

Walter, E. (1988) *Placeways*, Chapel Hill, N.C.: University of North Carolina Press.

Zukin, S. (1992) *Landscapes of Power*, Berkeley, Calif.: University of California Press.

FURTHER READING

Entrikin, N. (1991) *The Betweenness of Place*, Baltimore, Md.: The Johns Hopkins University Press.

Hewitt, K. (1983) 'Place annihilation: area bombing and the fate of urban places', *Annals of the Association of American Geographers* 73, 257–84.

Hiss, T. (1990) *The Experience of Place*, New York: Alfred A. Knopf.

Hough, M. (1990) *Out of Place*, New Haven, Conn.: Yale University Press.

Lipske, M. (1985) *Places as Art*, New York: Publishing Center for Cultural Resources.

Lynch, K. (1960) *The Image of the City*, Cambridge, Mass.: MIT Press.

Lynch, K. (1972) *What Time is this Place?*, Cambridge, Mass.: MIT Press.

Lynch, K. (1981) *A Theory of Good City Form*, Cambridge, Mass.: MIT Press.

Meyerowitz, J. (1985) *No Sense of Place*, New York: Oxford University Press.

Norberg-Schulz, C. (1984) *Genius Loci: Towards a Phenomenology of Architecture*, New York: Rizzoli.

Relph, E. (1976) *Place and Placelessness*, London: Pion.

Relph, E. (1989) *The Modern Urban Landscape*, Baltimore, Md.: The Johns Hopkins University Press.

Seamon, D. (ed.) (1993) *Dwelling, Seeing and Designing: Toward a Phenomenological Ecology*, Albany: State University of New York Press.

Tuan, Yi-Fu (1974) *Topophilia*, Englewood Cliffs, N.J.: Prentice-Hall.

Tuan, Yi-Fu (1977) *Space and Place*, Minneapolis, Minn.: University of Minnesota Press.

Walter, E. (1988) *Placeways*, Chapel Hill, N.C.: University of North Carolina Press.

VI GEOGRAPHICAL FUTURES

INTRODUCTION

Richard Huggett and Mike Robinson

Beyond determining the basic approach and the overall structure of this encyclopedia, the editors have interfered as little as possible with the content of individual chapters and the freedom of individual authors. Some people have chosen to treat their themes at considerable length; others have opted for brevity: some have made extensive use of illustration; others have not. Most have reflected our vision of the many different issues more closely than we could have hoped; a few have not. The encyclopedia, we feel, would be incomplete without some consideration of the future directions which might be taken by the discipline, and for this reason we have invited four distinguished geographers from different backgrounds and traditions to express their thoughts and aspirations for the discipline. No constraints of any kind were placed upon them, and it would be presumptuous of any editorial team to look for substantive changes in the texts that they have offered. The final words in this encyclopedia, therefore, go to them.

42

CONCERN FOR GEOGRAPHY

A case for equal emphasis of the geographical traditions

Adetoye Faniran

INTRODUCTION

Ordinarily, the word 'concern' connotes the idea of such fundamental needs of man as food, shelter, security, etc., while the United Nations University, one of the authorities on the theme/topic of human concerns, recently categorized them in terms of peace, poverty, the economy, the environment, and appropriate technology (UNU 1989).

The issues being raised in this chapter for the discipline of geography are similar in that they have a lot to do with the demands on geographers to provide for those ingredients that will make their subject matter continue to live and prosper, just as food, shelter, security, peace, etc. determine man's continued existence and prosperity.

With respect to geography, Faniran (1990) distinguished between two major types of concern. He distinguished the category of concerns 'detectable from the various attempts to define the nature, philosophy, scope, content, rationale, approach, methodology, and utility value of the discipline' (Faniran 1990: 306), from those which relate to contributions in the form of research and publication.

One analogy often drawn to illustrate the difference between the two types of concern is the way Jesus Christ is presented in the four Gospels of Matthew, Mark, Luke and John. It has been argued that whereas the first three books, often referred to as the synoptic gospels, present the historical Jesus, the book of John in particular, and the other books written by this beloved disciple of Jesus Christ, present the real (spiritual, divine) Jesus, calling him the Word of God, the Son of God, the Saviour of Mankind, etc.

THE DISCIPLINE OF GEOGRAPHY

This is not the place to start another discussion of what is and what is not geography. Among other things the theme of the 27th International Geographical Congress, 'Geography is Discovery', would seem to have laid to rest any such fruitless argument. Geography, as a point of view and a way of life, has been firmly established in the literature as:

- an important component of basic education and the communication of ideas; and
- a way of knowing and understanding the universe, the home of man, and as a means of earning a living through its practice.

However, and before we go on to discuss these broad areas, it is necessary to raise the rather vexing and recurring issue of the unity of the discipline of geography. It is discussed here as a concern in itself as well as a way of introducing the other more substantive concerns, namely: geographic education, information, and communication, and the relevance of geography.

THE UNITY OF GEOGRAPHY

There is substantial evidence in the literature to show that geographers have shown varying levels of concern for the discipline as a whole or for its parts thereof. This is not surprising since geography itself is a way of expressing humanity's feelings about, as well as concern for, total environment, including physical and human environments and, especially, the interactions, interdependence, and interrelationships among them.

Basically, and possibly on account of the accident of history, the concern of geographers for the environment has assumed a dichotomous posture. The so-called physical geographers show concern essentially for the naturally occurring aspects of man's environment (cf. Faniran and Ojo 1980; Goudie 1984), while the human geographers concern themselves with the man-made, artificial or cultural aspects. The most extreme cases in this dichotomy are the universities and other institutions where the two exist as separate departments (for example, at the University of Reading, UK); fortunately, however, and at the global level, the strains and stresses that attend this division have not succeeded in splitting the discipline apart. On the contrary, the unity of the discipline persists, made possible, among others, by mutual adjustment among the concerned adherents of the two groups of geographers, especially those who see humanity as a focal point of geographical study. This position is in sharp contrast with that which tries either to see the two sides as two separate coins rather than two sides of the same coin, or to subsume one under the other – for example, the view that treats physical geography as an appendage as it were (for instance, as the physical basis of geography), or even fails to recognize physical geography at all. Straw

927

(1973), in disagreeing with the separation of physical from other geographies, observed as follows:

> Physical Geography can now be seen to interact with human or cultural geography at all levels ... the Physical Geographer now appreciates the controls exerted by man on coastlines, rivers, soils, vegetation and microclimates; the human geographer increasingly acknowledges that relief, climate and soil still influence individual and collective human activity.
>
> (Straw 1973: 8)

This is the beauty and advantage which geography as an integrative discipline provides and which makes it relevant to the ongoing debate on environment and development (Crump 1991; UNDP 1991; Faniran 1993).

Attempts to unify geography via the adoption of a focus for geographical work and research have been made on several occasions. In a brief review, Faniran (1989a) identified two foci among others, namely: (a) paradigms, techniques, skills, and approaches; and (b) topics, themes, and objects or materials studied by geographers. Among the paradigms already proposed are locational analysis and the inculcation of skills such as literacy, numeracy and cartography. The topics that have been proposed include the earth as a whole or parts such as water (Chorley 1969), humanity (Faniran and Ojo 1980), and energy (Gregory 1987). Although no consensus has been reached on any of these, geography largely remains a single discipline with legs fully established in both the sciences and humanities.

The success achieved to maintain the unity of geography has yielded tremendous benefits. Again, as observed by Straw (1973), both sides of the discipline, in spite of the temptations to separate, have come to accept that the traditional or intrinsic dichotomy within the discipline, rather than being the cause for dismay and pessimism, has turned out to be a source of dialogue, challenge, stimulus, and comparison, and has created an internal variability that has generated new ideas and techniques. Consequently, both sides have benefited from mutual interaction: the development of spatial analysis in human geography has influenced the researches of physical geographers, while the study of connectivity of transportation systems, as well as flows of people, goods, and services has benefited from the study of natural networks such as rivers, trees, and other branching systems in the natural sciences.

Geographical education, information and communication

The idea contained here is perhaps best expressed by two of the several statements made at the opening ceremony of the 27th International Geographical Congress referred to above. In one of these statements, President George Bush of the United States of America, in a letter sent to the Congress, observed as follows:

This Administration [US government] has made the study of geography a priority of 'America 2000' the strategy by which we will achieve our National Education Goals and revitalize our schools. Without a knowledge of Geography, the events of world history would not speak to us with the same urgency, the realignment of national and political borders would possess no significance, and the diversity of our earth's terrain and the need for conserving its precious resources would remain unknown.

(27th *IGC News* 2: 1)

President Bush added:

Greater understanding of our common heritage is not only fitting at a time we celebrate the 500th Anniversary of Columbus' first journey to the Americas, it is crucial to meeting the global challenges that we face. That is why a mastery of geography is so important.

(27th *IGC News* 2: 1)

In the same vein Robert E. Peary, in his opening address to the Congress, observed:

Today the people of all nations and all walks of life need Geography as never before ... As a special citizen of the planet Earth – and, indeed, you are special with your geography skills – you have the opportunity, the power, the obligation to nurture planet Earth for the 21st Century.

(27th *IGC News* 2: 1)

All these are among the latest efforts to spell out the purpose and place of geography not only in education but also in national life.

However, it is pertinent to refer to the work of Ojo (1981) in Nigeria, who, writing under the general title of 'the purpose of geography today', listed the objectives of school, especially university, education to include:

- the provision of a sound knowledge of the principles of spatial organization of natural and human phenomena on the earth's surface;
- the creation of an awareness or appreciation of the nature and distribution of natural resources and the impact of these on human activities and problems; and
- the inculcation of suitable analytical and technical skills required for tackling the problems of spatial planning and development.

It is this third objective that Balchin (1972) has described in terms of development of human avenues of intelligence, or modes of communication, namely: verbal, social, visual, and numerical. These modes are the near equivalents of the traditional skills of literacy, cartography and numeracy which have featured prominently at one time or the other in the development of the discipline of geography (Crist 1969). It has, therefore, been suggested that what is probably needed most, in addition to the spirit of enquiry/discovery, and which should constitute the greatest concern of geographers, is effort devoted to the application, in as equal and balanced a manner as possible, of all the modes or traditions of communication at both individual and group levels. That is to say that whatever type of geographical enquiry we engage in we should emphasize all the modes to ensure that we are in the position, not only to master our subject matter but also to communicate our findings effectively among geographers and non-geographers alike.

Unfortunately, as James (1972) has observed:

> ever since Pythagoras there has been a mathematical [numerical] tradition in geography; and ever since Anaximander, there has been a cartographic [graphical] tradition; and ever since Hecateus there has been a literary tradition. Although all these traditions can be traced to the early days of Geography, not all of them were usually dominant at all times ... particular periods have been characterised by particular traditions.
>
> (James 1972: 26)

Crist (1969) likened this phenomenon to a lengthy play with many scenes.

Furthermore, the phenomenon of information explosion as well as the process of specialization in geography in recent years has led to a situation whereby geographers have not only developed expertise in one or the other of the modes or traditions but have also argued for their own area of expertise in preference to the others. A good example of this position is provided in some of Barry Floyd's publications on the virtues of the literary tradition (Floyd 1961, 1966, 1976, 1978, 1980), while the so-called 'quantitative' revolution of the 1960s and beyond at some stage argued that the numerical/mathematical or quantitative tradition in geography is superior to the others. Cartography has virtually severed links with geography and has formed its own professional body, while many geographers have completely discarded maps and diagrams from their work. The result is marked deficiency, not only in the development of the major avenues of human intelligence but also a minus in the utilitarian value of geographic education.

It is necessary to enter some caveats here. First, other subjects or disciplines also develop one or the other of these skills or modes. Some people will, in fact, argue that some of them perform better than geography in such areas as language. Similarly, numeracy is tremendously enhanced by mathematical training. Perhaps these are parts of the reasons why these two subjects among others are compulsory requirements at various stages of higher education.

Second, it has been argued that these skills or modes as well as paradigms are only ways designed for the promotion of the study and the reporting and application of the knowledge of the objects which properly constitute the subject matter of geography. The point, however, is that the specialization of skills, which has taken place at the expense of an all-round acquisition of skills, has made it difficult for geographers effectively to communicate their ideas to a wider audience (cf. Emberger and Hall 1955; Balchin 1955, 1976; Freeman 1971; Floyd 1980; Faniran 1984, 1990). Unfortunately, this is precisely what the craze about specialization and the production, whether through the school system or working life, of narrow-minded specialists is seriously endangering. There is, therefore, the need for geography to rediscover its roots as 'the mother of the sciences', not necessarily in the sense of a 'prolific mother' (Ojo 1981) but as a discipline which makes serious, persistent, and deliberate efforts to develop the in-born abilities of people to discover the secrets of nature to the benefit of humankind. Faniran (1990) puts it thus:

> Geography should not . . . be allowed to degenerate into such a state of specialization that it cannot offer . . . the opportunity to fulfil the purposes originally intended. Geographical education [and practice] . . . offers among the best opportunities for the flowering and fruition of man's inborn abilities and the realization, by its beneficiaries, of full-fledged and balanced education.
>
> (Faniran 1990: 309)

Similar pleas for geography to 'return to its roots' have been made in recent years. In what Jeje (1990: 302–3) described as 'reassertion of the geographic core', a strong plea was entered, supported among others by the works of Sack (1980), Bennett (1985), and Orme (1985), for geographers to adopt an integrated (human–ecology) approach to spatial analysis in the human and physical environments, using the most appropriate methodologies. This is with a view to reaffirming the position of the early geographers that by investigating the earth's surface phenomena, in an integrated and comprehensive manner, we are contributing in no small measure to the solution of the problems confronting man in both the natural and the human sciences, especially at their interface; problems which are 'often created by misunderstanding of nature and . . . defiant attempts to engineer solutions' (Jeje 1990: 302). The goal is the creation of a harmonious relationship between people and nature in order to ensure the continued existence of both (Faniran 1989b).

GEOGRAPHY AND RELEVANCE

By far the greatest concerns among geographers today, as reflected in the geographical literature, relate to the relevance of their subject matter generally and the topic(s) of their research in particular. The examples of both the themes of national and international conferences and the contents of the presidential addresses at such conferences clearly illustrate this point. Some Nigerian examples of presidential addresses delivered at the Nigerian Geographical Association's conferences over the years are used to illustrate. As contained in Ayeni and Faniran (1990a) the following are some of the titles:

- Geography and National Development;
- Geography and National Reconstruction;
- Geography in Contemporary Society;
- Geography in the Nigerian Environment;
- Geography and Planning in Nigeria;
- Geography and the Environmental Challenge;
- Jobs for Geographers: Career Outlets for Geography Graduates in Nigeria;
- The issue of Relevance in Geographical Research with Reference to Nigeria;
- Whither Geography?

931

Commenting on the content of these and other addresses, Ayeni and Faniran (1990b) observed thus:

> It is clear ... that much concern has been shown for such issues as the nature of the subject, its relevance to the identification of societal problems and concerns for utilizing the subject by policy makers and by government.
>
> (Ayeni and Faniran 1990b: 327)

They then concluded that the collection of these addresses represents a 'compendium' on applied geography (Ayeni and Faniran 1990a: 14).

However, while it is possible to read from these topics the desire to make geography relevant, the topics and their contents show less concern for geography than for development. Nevertheless, after over twenty-five years of campaign, development remains mainly within the purview of economists, engineers, architects, and the other more traditional professionals, while the contributions of geography and geographers remain at best subsidiary and personal. Nigerian geographers continue to be involved in development issues more in their personal than in their professional capacities (Faniran 1982a, 1982b).

This attitude has also led to certain undesirable consequences (Faniran 1989b). Many geography departments have changed their names and course contents. The most popular examples are those which have added 'planning' or 'regional planning', to the traditional name of geography. Also many geographers who have acquired additional qualifications and experience in other fields prefer to be known by their newly found interest as planners, hydrologists, economists, etc.

Yet another noticeable undesirable consequence of the craze for relevance for its own sake is detectable in the attitude of some geography students and researchers. Students at both undergraduate and postgraduate levels tend to rush to recommend solutions to generally ill-diagnosed and poorly understood problems. They are neither interested in the phenomena that make geography what it is, nor are they enthusiastic about mastering, not to talk of applying, the traditional skills of the discipline. They neither write good prose (literary presentation), discuss intelligently (articulatory skill), illustrate effectively (graphical or cartographical mode), nor provide concrete, objective, and quantitative proofs of their presentations in figures and other numerical values (numerical/mathematical skill), to aid clarity.

Moreover, most final year and postgraduate students prefer options which they see as having direct relevance to some known professions (for example, hydrology, land-resource evaluation, urban geography, transportation geography, etc.), or those labelled 'applied' (for example, applied climatology, applied geomorphology, etc.). The situation appears to have degenerated most among some human geographers who seem to have forgotten everything about the 'geo' in their discipline but feel more at home among economists, sociologists, mathematicians, etc., than among their colleagues in the department of geography. The result is the doubt currently being raised as to the status of geography as a discipline (Elliot-Hurst 1985).

The issue of relevance has been very well discussed by, among others, Thornes (1978), Walker (1978), Brunsden (1985), and Faniran (1990). Walker (1978) put it thus:

> as applied research is aimed at solving specific problems, usually those with pressing economic significance, it is easily and often emphasized at the expense of basic research. Such emphasis will almost certainly lead to fewer fundamental discoveries that actually make applied research meaningful.
>
> (Walker 1978: 203)

Fortunately, however, and as observed by Brunsden (1985) among others, basic and applied research need not be approached differently. In fact, as shown by examples from geomorphology, properly conducted pragmatic (applied) research has contributed significantly to knowledge, including the development of theory and fundamental discoveries. The case being made, therefore, is not against relevance but against shoddy or deficient geographical teaching, research and practice in the name of relevance. Indeed, geographers have every right to be accorded the status of professionals in their own right (Ajaegbu 1979; Faniran 1982a, 1982b), and quite a number of colleagues now indicate their profession as 'geographer' on official forms. This is, however, not enough; we need to work to justify that appellation. Professionalism is not earned by campaign in the process of filling forms or on pages of books, etc.; it is earned through clear evidence of expertise and contributions to societal advancement.

The simple dictionary meaning of the word 'profession' is an occupation requiring advanced education and special training, both of which geography gives. What other recognized professions have that geography seems to lack is a sufficient degree of specificity in terms of what they have to offer. Geography, by and large, is still perceived as a 'jack-of-all-trades' subject, without any recognizable focus (Faniran 1991: viii).

This situation has, however, not always been so; geographers in several countries and at different times have earned recognition on account of the information and service they provided. Perhaps the best examples are the United States of America and the former Soviet Union where the position of Government Geographer, or its equivalent, was established at Cabinet level (Faniran 1982b) and based on the expertise and proficiency shown both in research and, perhaps more importantly, in the manner of presentation of research findings for public consumption, as well as contributions to solving developmental problems. The example of landform study or geomorphology can be used to illustrate this point.

Geomorphology, like geography, shares many of our concerns for the physical environment, with particular reference to the materials, processes and forms of the earth's surface features and their relationship to people (Faniran *et al.* forthcoming). On all these, geomorphologists have accumulated vast amounts of information, a lot of which has made minimum impact on decision-making. As observed among others by Faniran (1984) decision-makers hardly have time to go through large volumes of lengthy reports: rather what they need is the presentation, in concise, manageable, and authoritative form, of the findings and recommendations of their research. This will require expert and balanced application of the geographical traditions of discovery, literacy, graphicacy, and numeracy, among others. Thus, among the geomorphologists who have received the widest recognitions are those who have imbibed these traditions, especially those applied in landform/geomorphological mapping, the tradition which brought geomorphology into the limelight, especially in Europe and the former Soviet Union. Practical work in geography is, however, one of the most neglected areas, despite the efforts made by concerned geographers to reverse the trend (cf. Ajaegbu and Faniran 1973; Faniran and Okunrotifa 1981).

It is therefore not difficult to see the link between the neglect of geographical traditions and the decline in esteem of the discipline in the public view. The example of the United States of America is again apposite in this connection. It is therefore necessary to ensure, for example, that the revitalization mentioned by President Bush will also include the reactivation of all the traditions to the teaching and practice of the discipline, as much as possible, on an equal basis.

It is not so much that practical work is not being taught as part of geography at the different levels. Fieldwork, surveying, statistics, mathematics, map reading, and map-making, etc., are still being taught. However, the questions that arise include the following: How much of these are reflected in our students' final year projects or subsequent research outputs – dissertations, theses, and other publications? How many of our geography graduates and even lecturers/ professors see practical work as part and parcel of as well as aids to the mastery of their respective areas of specialization? How many of them are capable of presenting the information they collect on an area or theme in concrete, concise, and manageable forms, especially on maps? How many are even capable of designing appropriate maps and diagrams to illustrate their work? Geography and maps go together, and as Faniran (1990) observed, the projects which have been given to the Nigerian Geographical Association by government so far concern mapping of communities. Several others in this and related areas are begging to be attended to. We will be able to do something concrete about them and earn recognition for geography when we have taken positive steps to correct the mistakes in our training programmes and when we have incorporated sufficient doses of practical work and expertise into our products.

The seriousness of this problem varies from country to country; there is, therefore, no universal solution to be prescribed. Instead, an appeal is being made to national associations and other bodies of geographers to assess the situation locally and find out what has gone wrong, where, and since when. They will then be in the position to work out a corrective programme or set of programmes. Such programmes should include a deliberate crusade to improve both the training of geographers and the practice of geography, with special reference to the avenues or modes of learning and communication – particularly discovery, literacy, graphicacy and numeracy.

SUMMARY AND CONCLUSIONS

Although part of the title of this chapter has been borrowed from that of a book published to mark the occasion of the appointment of Professor W.G.V. Balchin as Emeritus Professor of the University of Wales, Swansea, as well as from the author's presidential address in 1990 to the Nigerian Geographical Association (Faniran 1990), the treatment is different in varying degrees from both. The University of Wales's book reproduces Professor Balchin's major works, while the presidential address focused on the Nigerian situation. This chapter, by contrast, while addressing the same concerns for, of, and about the discipline of geography as the lecture did, attempts to broaden the scope of the discussion by addressing the international or global community of geographers.

A case is made for the rearticulation or reintroduction of the established avenues of intelligence in the training of geographers – namely, fieldwork and discovery, literacy, graphicacy and numeracy, with a view to strengthening the unity of the discipline and to enhance its professional status. The point is also made that the strong links between these modes and the so-called geographical traditions indicate that the discipline of geography, as originally conceived, should be placed on a very high pedestal when compared to other disciplines and professions.

Unfortunately the present image of geography in many countries of the world does not fully reflect this position because of the erstwhile partial and differential development and application (if at all) of the avenues of intelligence and modes of communication in time and space, as well as their almost total neglect in recent years. The case is therefore made that if these deficiencies can be corrected, and if the facilities offered by modern technology can be utilized, then geography will shed most of its inferiority complex and take its rightful place among the disciplines and professions which are directly relevant to our problems and welfare. This will happen when we have succeeded in combining expertise in basic research with the professional competency of practical work and effective reporting, similar to what obtains presently in the traditional professions such as architecture, engineering, accountancy, and so on. Recent developments in computer-based geographical information systems (GIS) if widely accepted will go a long way in achieving this purpose.

REFERENCES

Ajaegbu, H.I. (1979) *Geography for Development, The Jos Experiment Inaugural Lecture*, Mimeo, Jos, Nigeria: University of Jos.

Ajaegbu, H.I. and Faniran, A. (1973) *A New Approach to Practical Work in Geography*, London: Heinemann.

Ayeni, B. (1981) 'Techniques in human geography', pp. 99–106 in A. Faniran and P.O. Okunrotifa (eds) *A Handbook of Geography Teaching for Schools and Colleges*, Ibadan: Heinemann.

Ayeni, B. and Faniran, A. (1990a) 'Introduction', pp. 7–15 in B. Ayeni and A. Faniran (eds) *Geographical Perspectives on Nigeria's Development*, Nigerian Geographical Association, University of Ibadan.

Ayeni, B. and Faniran, A. (1990b) 'Epilogue', pp. 327–32 in B. Ayeni and A. Faniran (eds) *Geographical Perspectives on Nigeria's Development*, Nigerian Geographical Association, University of Ibadan.

Balchin, W.G.V. (1972) 'Graphicacy', *Geography*, 57, 185–95.

Balchin, W.G.V. (1976) 'Graphicacy', *American Cartographer* 3(1)

Balchin, W.G.V. (1981) *Concern for Geography: A Selection of the Works of Professor W.G.V. Balchin*, Swansea: Dept. of Geography, University College of Swansea.

Bennett, R.V. (1985) 'Quantification and relevance', in R.J. Johnston (ed.) *The Future of Geography*, London: Methuen.

Brunsden, D. (1985) 'Geomorphology in the service of society', pp. 225–57 in R.J. Johnston (ed.) *The Future of Geography*, London: Methuen.

Chorley, R.J. (1969) (ed.) *Water, Earth and Man*, London: Methuen.

Crist, H.P. (1969) 'Geography', *The Professional Geographer* 21(5), 305–7.

Crump, A. (1991) *Dictionary of Environment and Development*, London: Earthscan Publications.

Elliot-Hurst, M.E. (1985) 'Geography has neither existence nor future', pp. 59–91 in R.J. Johnston (ed.) *The Future of Geography*, London: Methuen.

Emberger, M.R. and Hall, M.R. (1955) *Scientific Writing*, New York: Harcourt Brace.

Faniran, A. (1982a) 'The geographer in government', Address delivered, Students' Geography Week, University of Ilorin, Ilorin, June 1982 (mimeo).

Faniran, A. (1982b) 'The role of geographer in nation building', Address delivered, Geography Week, University of Ife (now Obafemi Awolowo University), June 1982 (mimeo).

Faniran, A. (1984) 'Landform and man: a geomorphologist's viewpoint on the man and nature controversy', Inaugural Lecture, University of Ibadan.

Faniran, A. (1989a) 'Quest for a unified approach to the study of geography: a review of energy and energetics in physical geography', Department of Geography, University of Ibadan, Staff/Postgraduate Students Seminar paper (mimeo).

Faniran, A. (1989b) 'Reality harmony and environmental planning in developing countries: a review of recent developments in Nigeria', *Third World Planning Review* 11(2), 175–88.

Faniran, A. (1989c) 'The development of geography in Nigeria', *Transactions*, Second International Geomorphology Conference, Frankfurt, W. Germany, September 1989, Japanese Geomorphological Union.

Faniran, A. (1990) 'Concern for geography', pp. 306–26 in B. Ayeni and A. Faniran (eds) *Geographical Perspectives on Nigeria's Development*, Nigerian Geographical Association, University of Ibadan.

Faniran, A. (1991) *Water Resources Development in Nigeria*, University Lecture, University of Ibadan: Ibadan University Press.

Faniran, A. (1993) 'A framework for the implementation of long-time environmental planning and management in Nigeria', *Zeitschrift für Geomorphologie, Supplementband* 87, 41–9.

Faniran, A., Jeje, L.K. and Ebisemiju, F.S. (forthcoming) *General Geomorphology*, Ibadan: Ibadan University Press.

Faniran, A. and Ojo, O. (1980) *Man's Physical Environment*, London: Heinemann.

Faniran, A. and Okunrotifa, P.O. (eds) (1981) *A Handbook of Geography Teaching in*

Schools and Colleges, Ibadan: Heinemann.

Floyd, B. (1961) 'Towards a more literary geography', *The Professional Geographer* XIII(4), 7–11.

Floyd, B. (1966) 'Some comments on the scope and objectives of geography as polite literature', *Nigerian Geographical Journal* 10(2), 133–6.

Floyd, B. (1976) 'Whither geography, a cautionary tale from economics', *Area* VIII(1), 15–23.

Floyd, B. (1978) 'Style in professional writing', *Area* IX(2), 90–2.

Floyd, B. (1980) 'Scientific writing and the Nigerian geographer', Contribution to Symposium on the Old and the New in Nigerian Geography, 23rd Annual NGA Conference, University of Calabar (mimeo).

Freeman, T. (1971) *The Writing of Geography*, Manchester: Manchester University Press.

Goudie, A. (1984) *The Nature of Physical Geography*, London: Edward Arnold.

Gregory, K.J. (1987) *Energetics in Physical Geography*, Chichester: John Wiley.

Guildford, J.F.C. (1989) 'Three faces of intellect', *American Psychologist* 14, p. 459.

James, P.L.C. (1972) *All Possible Works: A History of Geographical Ideas*, New York: Odyssey.

Jeje, L.K. (1990) 'Whither geography', pp. 290–365 in B. Ayeni and A. Faniran (eds) *Geographical Perspectives on Nigeria's Development*, Nigerian Geographical Association, University of Ibadan.

Ojo, G.J.A. (1981) 'Geography today: its purpose, content and methods', pp. 17–28 in A. Faniran and P.O. Okunrotifa (eds) *A Handbook of Geography Teaching for Schools and Colleges*, Ibadan: Heinemann.

Orme, A. (1985) 'Understanding and predicting the physical world', in R.J. Johnston (ed.) *The Future of Geography*, London: Methuen.

Sack, R.D. (1980) *Conception of Space in Social Thought*, London: Macmillan.

Straw, A. (1973) *Concerning Physical Geography*, Inaugural lecture, University of Exeter.

Thornes, J.D. (1978) 'The character and problems of theory in contemporary geomorphology', in C. Embleton, D. Brunsden and D.K.C. Jones (eds) *Geomorphology: Present Problems and Future Prospects*, London: Oxford University Press.

United Nations Development Programme (UNDP) (1991) *The Challenge of the Environment, 1991 Annual Report*, New York: UN Plaza.

United Nations University (UNU) (1989) *Work in Progress* 12(1), 1.

Walker, H.J. (1978) 'Research in coastal geomorphology: basic and applied', pp. 203–23 in C. Embleton, D. Brunsden and D.K.C. Jones (eds) *Geomorphology: Present Problems and Future Prospects*, London: Oxford University Press.

FURTHER READING

Adalemo, I.A. (1990) 'Geography and the environmental challenge', pp. 180–96 in B. Ayeni and A. Faniran (eds) *Geographical Perspectives on Nigeria's Development*, Nigerian Geographical Association, University of Ibadan.

Alao, N. (1990) 'Spatial dynamics and the dynamics of geography', in B. Ayeni and A. Faniran (eds) *Geographical Perspectives on Nigeria's Development*, Nigerian Geographical Association, University of Ibadan.

Amedeo, D. and Colledge, R.D. (1975) *An Introduction to Scientific Reasoning in Geography*, New York: John Wiley.

Barbour, K.M. (ed.) (1972) *Planning for Nigeria: The Geographical Approach*, Ibadan: Ibadan University Press.

937

Faniran, A. (1990) *Geomorphology*, University of Ibadan External Degree Programme (Monograph), Department of Adult Education.

Hartshorne, R. (1959) *Perspectives on the Nature of Geography*, Chicago, Ill.: Rand-McNally.

Harvey, D. (1974) 'What kind of geography and what kind of public policy?', *Transactions, Institute of British Geographers* 63, 18–24.

Johnston, R.J. (1979) *Geography and Geographers*, London: Edward Arnold.

Johnston, R.J. (1983) *Philosophy and Human Geography*, London: Edward Arnold.

Natoli, S. (1980) *Geography: Tomorrow's Career*, Washington, DC: Association of American Geographers.

Ritter, D.F. (1988) 'Landscape analysis and the search for geomorphic unity', *Geological Society of America, Bulletin* 100, 160–71.

Yatsu, E. (1992) 'To make Geomorphology more scientific', *Transactions, Japanese Geomorphological Union* 13(2), 87–124.

43

HOME AND WORLD, COSMOPOLITANISM AND ETHNICITY

Key concepts in contemporary human geography

Yi-Fu Tuan

One way to describe progress in human-cultural geography in this century is to say that the concept 'human' has undergone progressive elaboration. 'Man as a geological agent' was an early formulation. Then came the notion of culture – that human beings are cultural beings. From the 1960s onwards a psychological dimension was added. And now, increasingly, human beings are seen as moral beings as well, subject to moral demands and dilemmas, and endowed with moral aspirations, which may be manifest as the yearning for the good life. As geographers become more fully aware of the complexity of human nature, they find simultaneously a need to re-examine certain keywords or concepts – traditional words in their subfield such as 'home' and 'world', and newer words, thrust upon them by the events of the day, such as 'ethnicity' (or 'ethnicism') and 'cosmopolitanism'. How do we geographers understand these words? Will the attempt to articulate the meaning of these words give us a handle to understand real-life situations better at both the intimate personal and the impersonal political levels?

HOME AND WORLD

Geography is the study of the earth as the home of human beings; alternatively, it is the study of how human beings have transformed 'environments' into 'worlds'. The two definitions of geography are much alike, except that the second definition puts more emphasis on process than does the first. Both contain a key word – 'home' in the first case and 'world' in the second – that defines what it is to be human. Human beings have homes and they have worlds. 'Home' may be the earth itself; at the other extreme it is the porch on which we have placed our favourite rocking chair. To understand what it is to be human is to understand, intimately, the variant meanings of home, and with it, the variant meanings of 'not feeling at home' – of estrangement or alienation. The word 'world', whose root meaning is 'man' (*wer*), shows how intimate is the bond between a person and his circumambient reality. Not to have a world is to be merely a body in space, an object in an environment. The difference between 'having a world' and 'being a body in space' can be illustrated with numerous commonplace examples from daily life. Suppose you take a walk in the woods after dark. Upon your return, you see that you have left the light on in your bedroom and reflect that before long you will be asleep in your bed – a body in space, a worldless unconscious object that you now look upon, in imagination, with a mixture of protective fondness and pity. Upright and alert, you command a world, which may include, strange as it sounds, your own future self out there asleep.

What is the difference between 'home' and 'world'? One difference is scale. Although we speak of the whole earth as our home, normally that term calls to mind something smaller and more intimate – homeland, neighbourhood, house, or, in imagination, a 'secreted island', utopia. By contrast, world implies something large, open, and broadly shared. Wordsworth hints at the difference in the following passage:

> Not in Utopia – subterranean fields, –
> Or some secreted island, Heaven knows where!
> But in the very world, which is the world
> Of all of us, – the place where, in the end,
> We find our happiness, or not at all!
>
> (Wordsworth, *The Prelude*, XI, 140)

Home, because of its smaller scale, is accessible to the proximate senses of touch, smell and hearing. The house we live in, for example, is known to us in an intimate, multisensorial way and for this reason it is felt to have a core – a centre, and experience of home is a 'centred' experience. By contrast, the world is 'out there' – a visual-conceptual horizon spread before a person as in the expression, 'the world at one's feet'. In America, many suburban houses have picture windows, and it is a common practice for the guest to enter a friend's home and then move to the picture window to admire the view or prospect. Prospect, which means both expansive space and future time, is the world out there. How does one know another person? One might visit his home, the core of his being and the locus of intimate experiences, and then examine his prospect – his world.

Home evokes a greater sense of materiality than does world. The earth, our home, is a material object and so, of course, is our town, neighbourhood, house. World, by contrast, is more abstract – a generalized image and concept rather than a specific composition of tangible particulars. Yet 'world' is often contrasted with 'spirit'. A worldly person is someone whose values are anchored in the things of this life. Geographers, in their concern with home and world, thus reveal a bias towards the material and the tangible, a bias that they share with the bulk of humanity. On the other hand, home and world are obviously far more than just material objects. They are human and social ideals; both are soaked in human values and aspirations. Moreover, the condition of 'not feeling at home' in the world and of otherworldliness can also be human values.

PEOPLE: KINSFOLK, NEIGHBOURS, AND STRANGERS

Home may well be another person. 'We make a home for each other, my grandfather and I. I don't mean a regular home ... a building, a house ... of wood, bricks, stone. I think of home as being a thing that two people have between them in which each can ... well, nest–rest–live in, emotionally speaking.' This sentiment from a play by Tennessee Williams (1962, Act 3) is easy enough to appreciate. What makes a home 'home' is its human quality – other people, the family. When children fly the nest, as the common saying goes, the nest feels empty. This is true of larger scales too: a neighbourhood is made up of people who behave in a neighbourly fashion. It is this neighbourly behaviour that makes a neighbourhood. At a still larger scale is the city, but as Shakespeare put it in his play *Coriolanus*, 'What is the city but the People? True, the people are the city.'

Home, then, is people. As for world, I have suggested that it can be pictured as open, hopeful space 'out there'. If home is a social bond and inward directed, world is an individual's command of space – his or her prospect. Yet, even this conception of world is not necessarily individualistic – the project of a particular individual. Hope may well be the gift of another person rather than an attitude that can be generated solely by oneself through an exercise of will. We need friends to open up the world – to see mere space as prospect. A friend may be defined as someone who creates resonant space for us – an inviting world into which we would wish to enter (Barthes 1978). Indifference or hostility on the part of another, by contrast, shrinks our world and makes space feel 'dead'. As the world shrinks, so does our sense of self, since the self is so intimately bound to world.

There is, however, another sense of world – world as society. The emphasis here is on people, not so much the kinsfolk and neighbours we know well, but the friends, acquaintances, and strangers whom we meet irregularly on occasions which are often casual but which can also be formal to a degree beyond affairs at home or in the neighbourhood. To be a man or woman of the world implies freedom, opportunity, and a certain social standing. Society at large, with its wealth of human types and individuals, resources and institutions, provides such freedom and opportunity. On the other hand, just as space 'out there' can seem hostile, a threat to self-identity nurtured in home or neighbourhood, so can the world and society at large. Moreover, society can be hostile to its marginalized members, in fact rather than merely in perception. It can put pressure on their world, diminishing its scope to that of a besieged ghetto. Society can threaten an individual as well as a group. The consequences, however, differ. An individual whose world has been pressed back to the boundary of his body by society or simply by another individual is likely to suffer personality disintegration, whereas a group whose world has withdrawn to the walls of its ghetto under pressure from the larger society may well develop a strong sense of identity, though such identity – it may be argued – is distorted by having developed under oppression.

COSMOPOLITANISM

Home and world bespeak a difference of scale: the one intimate and warm, the other impersonal and cool. If we focus on people, then again 'family and neighbour' differ from 'acquaintance and stranger' in degree of warmth and involvement. The small, moreover, precedes the large in time; that is, one establishes a home first before moving on to the larger world, and one secures relationship with family and neighbour before moving on to cultivate acquaintance and stranger. At first glance, ethnicity and cosmopolitanism would seem to be another such pair. There is, however, an interesting difference: in the last named pair, the small does not precede the large. Indeed, it is the other way round: first cosmopolitanism, then ethnicity. Ethnicity, as a self-conscious stance and ideology, is a reaction against cosmopolitanism. Likewise, movements such as localism, regionalism, and nationalism are reactions against the universalist imperium.

What, then, is cosmopolitanism? At a very general level, the word implies something universal, inclusive, and orderly. If we attend to this general meaning, then all premodern communities, to the extent that they existed with little conscious awareness of the world beyond them, were cosmopolitan in outlook. Even the smallest hunting-and-fishing village considered itself the geographical, population, and cultural centre of the world. Villagers put themselves – the true people – at the heart of a cosmos which also included the heavenly bodies, the earth and its topographic features, plants, animals, and spirits. In no sense did they see themselves as marginal, or even local, merely one group, with its cultural traits, among other groups, with their different cultural traits. Strange as it might sound, the peoples that were the subject of so many ethnographic studies – Eskimos, Ainus, the Semang, Australian Aborigines, etc. – were all originally 'cosmopolites' rather than 'natives' tied to place, or, more recently, 'ethnics' at the margins of a larger sociopolitical entity.

I am clearly using 'cosmopolitan' here in a specialized, yet (I believe) nonarbitrary sense. Its reasonableness will be more apparent if we think not of (what seems to us) a small village or a band of hunter-gatherers, but of an imposing Mayan ceremonial centre, with its monumental architecture, or of a

Chinese capital city, with its monumental walls oriented to the cardinal points. The cosmic aspiration of these places is apparent; their scale implies the existence of complex sociopolitical organizations with the power to command space and resources, including human resources. We can readily see how the ancient Mayans and Chinese could take themselves to be world citizens – inhabitants of a reality that, by its scale and splendour, put all other possible realities into shade.

A ceremonial complex or capital city that succeeds in drawing a multitude of peoples of diverse background and occupations to its compound introduces another conception of cosmopolitanism, at odds with the first. Whereas the first conception emphasizes order – cosmic and social harmony – the second conception emphasizes heterogeneity. Cosmic harmony implies one people and one culture. This harmony is threatened by a diversity of peoples and cultures. In China, since the cosmic world-view was agricultural in nature and inspiration, traders and merchants, and even artisans could not be comfortably accommodated within its compass. And this was true even if they were ethnically Chinese, and often they were not. Students, who might be rowdy, were sometimes suspect; entertainers far more often so, for the authorities tended to associate them with the criminal elements of society. In Ch'ang-an, capital of T'ang China (618–907), special quarters were set aside for commerce and related activities, so that an official eye could be kept on them (Schafer 1963).

Yet the authorities had the wisdom to see that they could not, and moreover it was not in their interest to, impose one encompassing way of life. T'ang Ch'ang-an was in fact a highly cosmopolitan city in the second sense. The official religion was Confucianism, but religious tolerance prevailed. Foreigners – Syrians, Arabs, Persians, Tartars, Tibetans, Koreans, Japanese, Tonkinese, and many other groups – were conspicuous in the city. They spoke different tongues and practised different faiths in close proximity to one another, in remarkable freedom from the ferocious religious and racial strife that afflicted Europe at this time. The capital's National Academy attracted some 8,000 students, of whom perhaps half were Chinese; the other half consisted of Koreans, Japanese, Tibetans and students from Central Asia (Goodrich 1959).

Here, then, we have two conceptions of cosmopolitanism that, despite their oppositional character, lived synergistically with each other during a glorious period in the T'ang dynasty. The official cosmos was strong enough to be enriched, rather than threatened, by accommodating alien peoples and cultures in its interstitial space. Law and order, emanating from the cosmos and its traditional values, coexisted with the spontaneity and heterogeneity of commercial life. Foreigners and Chinese knew that they benefited from one another's presence: there was competition but no deep antagonism.

Nevertheless, traders and merchants, despite the wealth they brought to the city and empire, were always somewhat suspect in China because they were viewed by the governing class and perhaps by farmers too as rootless. They seemed ever ready to pick up their chattels and make a more profitable living elsewhere. They were not diligent in worshipping the household and local gods. They readily socialized with strangers and acquaintances. They were, in short, 'cosmopolites'. This meaning of cosmopolite as someone who can make a home for himself anywhere in the world adds another dimension to 'cosmopolitanism'. However, it is obvious that traders and merchants did not embrace rootlessness – or, to put it more positively, the ability to be at home anywhere in the world – as a desirable value in itself. In this absence of an ideological bent, they differed from practitioners of universal religion and philosophy. To practitioners of a universal religion, intense loyalty to place, kinsfolk, and local spirits, could easily turn into idolatry – to kneeling before gods that were too small. Their God, by contrast, was all-encompassing and impartial; and for that reason he could seem impersonal and remote from ordinary human passions and prejudices. The following well-known passage from the Gospel according to Matthew puts the position well:

> You have learned that they were told, 'Love your neighbour, hate your enemy.' But what I tell you is this: Love your enemies and pray for your persecutors; only so can you be children of your heavenly Father, who makes his sun rise on good and bad alike, and sends the rain on the honest and the dishonest. If you love only those who love you, what reward can you expect? Surely the tax-gatherers do as much as that. And if you greet only your brothers, what is there extraordinary about that? Even the heathen do as much.

> (Matthew 5: 43–8)

Philosophers were, by the nature of their calling, homeless. So, at least, Aristotle believed. The same idea, more positively expressed, would be: because philosophers lived in realms of thought, their material habitat and location mattered little. They were – or should be – citizens of the cosmos rather than of a particular city. They moved to wherever their calling found an opportunity to flourish. By the fourth and third centuries BC, the men who dominated the intellectual life of Athens were all foreigners – Aristotle, Chrysippus, Cleanthes, Theophrastus and Zeno. Even Epicurus, who came of Athenian stock, was a colonial by birth (Dodds 1951). Stoic philosophers were particularly vehement in declaring themselves citizens of the world. Thus Seneca (1969: 76): 'We should live with the conviction, "I wasn't born for one particular corner; the whole world's my home country".' Thus Marcus Aurelius (1956: 73–4): 'My city and country, so far as I am Antoninus, is Rome, but so far as I am a man, it is the world.' Dante and Pico della Mirandola, during the Italian Renaissance, said much the same thing. To the poet and the humanist, God created man to know the laws of the universe, its greatness and beauty. Such a creature could not be bound to any fixed place (Burckhardt 1951).

ENLIGHTENMENT, CIVILIZATION, IMPERIALISM

The meaning of cosmopolitanism is enriched by association with three closely related terms – enlightenment, civilization and imperialism. Practitioners of universal religion and philosophy decried local cults and superstitions. Buddhism, as articulated by its founder, had everything to do with enlightenment and truth, and nothing to do with the worship of gods and goddesses; philosophical Taoism constantly struggled against its tendency to degenerate into magic; Confucius preferred to talk about society and people, which he knew, than about spirits and unusual events, which he could not know; Christianity was intent, in its early phase, to eradicate the worship of local spirits of the earth, which it considered pagan or heathen. As for enlightenment in ancient Greece, its brightness can still astonish us today. Hecateus found Greek mythology 'funny'. When Heraclitus declared 'dead is nastier than dung', he deeply offended Greek sentiment, for in these few words he dismissed 'all the pother about burial rites' (Dodds 1951: 181–2). Critical ideas flourished during the Renaissance and during the eighteenth century, when the *philosophes* condemned tyranny and superstition in their own world and, at the same time, praised the rationality and benign government in China.

The concept 'civilization' emerged to denote a refined way of life, removed as far as possible from the crude needs and passions of the body and from barbarism. To China's mandarinate, the very hallmark of a civilized people was the proper observance of ritual and, in its more mundane form, social etiquette (Wechsler 1985). In the West, too, civilization originally meant courtly manners – the refined life-way of the upper class, more and more self-consciously cultivated in such matters as dress and food, during the eighteenth century, when the word itself was coined. Civilization at this time also came to be associated with commerce – with *doux commerce*, an activity of exchange, almost as gentle as the exchange of ideas (the commerce of polite conversation), in contrast to the older passions of the aristocracy, which led to the glory and debauchery of war (Hirschman 1977). Civilization, as conceived in nineteenth-century Europe, was contrasted with not only barbarism but primitivity: 'primitive' lay at the bottom and 'civilized' at the top of a ladder of progress. In the course of time, every primitive group could in theory rise to become civilized. Primitivity was diverse – every group had its own peculiar and barbarous customs – whereas civilization was one and universal (Barnard 1973).

The word 'empire' and its variants (emperor, imperial, etc.) are derived from the Latin *imperium*, which means power over a family, a great household, a state – supreme power. Power is ultimately armed or military power. Empires command armies, and it is a characteristic of imperial and nation-state armies, both in the past and in the present, to transcend tribal-regional affiliations and loyalties. The army, in the interest of efficiency, demands strict equality within each rank; it promotes not only uniformity but impersonality, as the slogan 'salute the uniform, not the man' implies. An imperial elite creates engineering and architectural monuments, from roads and waterworks to palaces and gardens, to maintain its own economic and military power, to promote a general state of prosperity, and to project an image of imperial glory that puts local cultural achievements in the shade. An empire encourages both uniformity and heterogeneity. In imperial China, a high degree of architectural and ritual uniformity is evident: the rectangular walled city and its attendant rites are standard throughout the sinicized empire, imposed from above or willingly adopted for the prestige they carry (Rudolph 1987); and in the Roman Empire one can probably find bathhouse, theatre, and forum – the same basic architectural set pieces – in Britain as in North Africa. Yet, an empire also promotes cosmopolitanism and colourful heterogeneity, as peoples from different corners of a vast domain converge on the imperial capital, seeking economic opportunity, official emolument, and the sheer excitement of seeing a larger and more splendid world.

ETHNICISM AND NATIONALISM: A NEGATIVE VIEW

In the absence of a larger political order, peoples of different ethnic background lived in peace, exchanging goods as need and opportunity arose. But such separate communities could also live in a state of tension and antagonism that periodically erupted into raid or war. Military conflicts and slave raids were known in Africa before the European imperium (Goody 1980); in the ancient Greek world, wars between city-states occurred so frequently that Thucydides treated them as the normal state of affairs (Knox 1990). One might justifiably apply the words 'ethnicism' and 'nationalism' to the self-awareness and awareness of difference in such communities and states. However, I should like to examine ethnicism and nationalism as attitudes and ideologies that have developed under the umbrella of a larger order – empire, nation-state, *cosmopolis*, civilization.

First, let me cast ethnicism and nationalism in a rather negative light as driven by resentment and envy. Needless to say, there is ample ground for resentment, even hatred. An empire creates centres of power and glory at the expense of the provinces – and note that the literal meaning of 'province' is 'vanquished region' (Foucault 1980: 69). People who once could consider themselves *the* people at the centre of their world if not at the centre of *the* world find themselves displaced to the margins; once 'cosmopolites' in their own eyes, they have become provincials, their customs and beliefs treated by the metropolitan sophisticates as quaintnesses and superstitions – that is, with either patronizing condescension or outright contempt. The provincial or ethnic at the capital suffers from the contumely of officials and clerks, but such rudeness does not pose an insurmountable threat, for it shows that urbanites, for all their pretensions, are morally underdeveloped. An offense more injurious to self-esteem is the undeniable wealth, power and beauty of metropolitan life. Confronted by metropolitan cultural superiority, the provincial or ethnic can either deny it against all the evidence of his senses, or try to imitate it with the risks of failure and with the knowledge that even the successes will be judged mere imitations.

Perhaps the deepest threat to a people rooted in place and its emotionally charged customs is rationality. All human beings use reason, of course, to conduct personal and social life. In cosmopolitan society, however, reason is more deliberatively and systematically applied to nature and society, and over greater expanses of space and time, than has ever been ventured in small, nonliterate communities. Such application creates wealth and power, great technological achievement, new types of human relationships based on shared temporally bounded goals and interests, while at the same time it corrodes kinship ties and enduring commitments to sanctioned ways of life. And then there are the thinkers, both religious and nonreligious, who carry their rationality to an extreme that threatens fundamental human institutions. The Hindu and Buddhist conception of space and time is so vast and impersonal that, like conceptions in modern astronomy, it reduces human striving and achievements to utter insignificance. The human body itself is worthless, 'just an old bag with orifices on top and bottom, and stuffed full of groceries', according to one Buddhist doctrine (Allen 1959: 162–3). In the family–centred Roman world, blood ties are subverted by Christ's famous question, 'Who is my mother? Who are my brothers? . . . Whoever does the will of God is my brother, my sister, my mother' (Mark 3: 31–5). Nonreligious thinkers are, expectedly, even more radical in their challenge to folk beliefs and practices. We have noted that, 2,500 years ago, Hecateus already thought mythology funny; Protagoras wanted to eliminate 'barbarian silliness' from traditions; Epicureans and Sceptics wanted to banish the passions from human life, and urged that contemplation of the starry heavens be substituted for earthbound cults.

Greek enlightenment (as Socrates' detractors could have told him) led to the decline of religion and, soon after, public life, since the two were closely linked. It did not lead to an Age of Reason. Whereas a talented few might be drawn to contemplate the stars and live in the cosmos (that is, nowhere in particular), most people found little appeal in such placeless, thin-blooded activity. Sustained thinking is a sort of death, for it can only occur at the expense of sensual engagements with particular individuals and localities. Freedom from economic constraint is desired by all, but the radical freedom demanded by thought is likely to provoke horror rather than delight in all but the supremely confident few. Not surprisingly, from 300 BC onwards, reason underwent a slow decline. Superstitious beliefs rose in its place, egregiously astrology, which promised people the security of astral determinism in the disorienting megacities of the Hellenistic empire (Dodds 1951).

The Greek story has parallels in other parts of the world and in other times. Universal religions such as Buddhism and Christianity, for example, quickly acquired vast pantheons of place-bound deities, spirits, and saints, with satisfyingly emotional modes of worship in small, tight-knit communities. The modern secular world does not offer its own universal philosophy or religion, but it does offer instrumental rationality and the freedom to choose, or even invent, goals of one's own making that are corrosive of local communal values and customs. Reason, as in ancient Greece, has succeeded in weakening

947

established religions that have close ties with public life and the state. In their place, a plethora of new and old beliefs has emerged: all promise emotional balm and many are communally based. Astrology itself has seen a revival. But among the most powerful and widely adopted answers to the impersonality of the modern secular-global order is not New Age religion, astrology, or religious fundamentalism, but ethnicism and nationalism – the warm glow of the local and the specific.

ETHNICISM AND NATIONALISM: A POSITIVE VIEW

Ethnicism and nationalism take on a positive aura – a warm glow – if we see them against the sombre backcloth of universalist compulsions – not cosmopolitanism as such – and ideologies such as imperialism, scientism, and global economism. Imperialism, whether economic–political or ideological (as in communism), carries the dark images of military conquest, ruthless exploitation, and the emasculation of lively regional cultures so that they become merely colourful folk dances for the entertainment of metropolitan elites. Imperialism makes for peace – a peace imposed from above – and it makes for uniformity, though this tendency to impose a common sociopolitical style is hidden by the colourful congregation at the capital of subject peoples from the outposts of empire. Scientism is the passionate belief that science has only one language, one narrowly conceived methodology that all seekers after real knowledge (as distinct from pseudo-knowledge) should accept. Scientism is austere, reductionist, and universalist. It promotes uniformity, but again this uniformity is hidden by local flashes of colour such as the international conferences, where scientists from different nations, wearing suits of different cut and speaking in varieties of faintly accented English, congregate. Lastly, there is the phenomenon of a global economy. It promotes uniformity, but within that general uniformity are countless islands of heterogeneity. Even a modest-sized town offers a cosmopolitan array of consumer goods (Sack 1992).

The global economy has produced a wealth of goods that people from all parts of the world appear to want. Yet the good life, even when material affluence reaches a historically unprecedented level, remains elusive. Consumer reality seems shallow, all glitter and fluff, lacking in substance and authenticity, ultimately boring for its want of genuine difference, deficient in passion, seriousness, and virtue. In reaction, people seek emotionally richer, better grounded, more distinctive ways of life. The answer to the bland, amoral universalism of consumer society is commitment to a specific people – one's own people – and to place, not only its sights, but also its elusive odors, myriad sounds and tactile qualities. The answer, in short, is ethnicism and nationalism.

This answer is not new, as scholars (Berlin 1991; Penrose and May 1991) have recently shown. In the seventeenth and eighteenth centuries, French culture dominated Europe such that its culture – not only its science and philosophy but its language, arts, and even its etiquette – came to be widely accepted as civilization, the acme of human achievement. Achievement invited imitation. So the Germans, relatively backward compared with not only the French but the British and the Dutch, imitated. Imitation could lead to a sense of inferiority and resentment. The Germans grew resentful of French dominance and the French intellectual elite's easy assumption of superiority – their arrogance. Increasingly, German thinkers saw French culture as empty, materialistic, and superficial, compared with which their own was spiritual and deeply moral, concerned with the inner life, man's relation to man, and man's relation to God, family and happiness rather than the *salon* and vainglory. At a deeper and more philosophical level, they challenged the concept of universalism – of elevating the ways of one people, however admirable, to a model for all humanity. The outstanding spokesman for the ethnic-national point of view was Johann Gottfried von Herder. He denied that values could be universal and maintained that every human society, every people, possessed its own unique tradition and standards – its own genius embodied in language and nurtured by soil.

POSSIBLE RESOLUTIONS

A human being, to develop fully, needs both home and world. Women suffered in the past because men tried to confine them to just one sphere – the home. In classical antiquity, the men arrogated the public realm to themselves. They operated in the larger sphere of the world; they aspired to the exhilarating expansiveness of universalist-cosmopolitan views. But they also recognized the essential role of home to the development of their full human stature. Home was not only the biological and economic base of life; it was also a place of withdrawal, a sheltered reality of inchoate feelings, thoughts, and experiences, a dark subconscious reality, out of which new understandings, perspectives' and ideas could emerge. What would be the point of meeting in the public realm – on the world stage – if one had nothing distinctive or fresh to bring to it (Arendt 1958)?

Home. That may be the key justification of ethnicism and nationalism. Home as a particular people and place – a unique agglomeration of qualities that can be sustained and evolve in accordance with its own genius only if home (ethnic group or nation) is somewhat isolated, not constantly buffeted by exotic customs and views. On the other hand, without any buffeting wind from the outside, or, to change the metaphor, without germinating ideas and values from an external source, home becomes ingrown and sterile – it becomes homely, with its negative sense of dullness and inconsequentiality.

Cosmopolitanism and ethnicism depend on each other for vitality. The one provides for openness, the other for boundedness, and it is the creative calibration of the two that keeps both alive and vital. Still, there are important differences between the two that resist easy accommodation; for instance, the difference between, on the cosmopolitan side, the universality of science and of certain conceptions of good and evil, and on the ethnic side, particularisms in individuals and groups that defy translation and generalization. Yet, whatever the conceptual difficulties, an individual or a group can embrace the two extremes – the universal and the particular – without feeling undue intellectual stress. It is possible for people to retain a profound attachment to local colours, smells, ways of dressing and cooking, literary and musical styles, etc. *and* a passion for 'light and truth', science, and universal human rights. But certain middle-level beliefs and habits do have to give way, such as magic, witch-hunt, slavery, wife sacrificing herself on the funeral pyre, etc., for while these practices are compatible with ethnicism or nationalism, they cannot be reconciled to science and the concept of universal human rights.

HUMAN GEOGRAPHY AS AN APPLIED DISCIPLINE

Human beings have to cope with nature and with each other to survive. Beyond mere survival, they wish for the good life. Planning is necessary to survival; and planning, with the boost of imagination, is necessary to attaining a better life. Geographers who study people creating 'worlds' out of 'environment' may feel that they are simply noting what has occurred. But I don't think that is all that geographers do – not if they are deeply engaged in their subject. Inevitably, as they look at a particular situation from a broader perspective than can the struggling actors themselves, geographers are able to offer certain clarifications, certain possibilities for adjustment and change – often implied rather than explicitly stated – that are at the germinal core of planning and politics.

REFERENCES

Allen, G.F. (1959) *The Buddha's Philosophy: Selections from the Pali Canon and An Introductory Essay*, New York: Macmillan.
Arendt, H. (1958) *The Human Condition: A Study of the Central Dilemmas Facing Modern Man*, Garden City, New York: Doubleday Anchor Books.
Barnard, F. (1973) 'Culture and civilization in modern times', *Dictionary of the History of Ideas* I: 613–21.
Barthes, R. (1978) *A Lover's Discourse: Fragments*, New York: Hill & Wang.
Berlin, I. (1991) *The Crooked Timber of Humanity: Chapters in the History of Ideas*, New York: Knopf.
Burckhardt, J. (1951) *The Civilisation of the Renaissance in Italy*, London: Phaidon Press.
Dodds, E.R. (1951) *The Greeks and the Irrational*, Berkeley: University of California Press.
Foucault, M. (1980) *Power/Knowledge: Selected Interviews and Other Writings 1972–1977*, New York: Pantheon Books.
Goodrich, L.C. (1959) *A Short History of the Chinese People*, New York: Harper & Brothers.
Goody, J. (1980) *Technology, Tradition, and the State in Africa*, Cambridge: Cambridge University Press.
Hirschman, A.O. (1977) *The Passions and the Interests*, Princeton: Princeton University Press.

Knox, J. (1990) 'Thucydides and the Peloponnesian War: politics and power', in *Essays Ancient and Modern*, Baltimore, Md.: Johns Hopkins University Press.

Marcus Aurelius (1956) *Meditations*, Chicago: Henry Regnery.

Penrose, J. and May, J. (1991) 'Herder's concept of nation and its relevance to contemporary ethnic nationalism', *Canadian Review of Studies in Nationalism* 18, 165–78.

Rudolph, S.H. (1987) 'Presidential address: state formation in Asia – prolegomenon to a comparative study', *Journal of Asian Studies* 46, 731–46.

Sack, R.D. (1992) *Place, Modernity and the Consumer's World: A Relational Framework for Geographical Analysis*, Baltimore, Md.: Johns Hopkins University Press.

Schafer, E.H. (1963) 'The last years of Ch'ang-an', *Oriens Extremus* 10, 133–79.

Seneca (1969) *Letters from a Stoic*, Harmondsworth, Middlesex: Penguin Books.

Wechsler, H.J. (1985) *Offerings of Jade and Silk: Ritual and Symbol in the Legislation of the T'ang Dynasty*, New Haven, Conn.: Yale University Press.

Williams, T. (1962) *The Night of the Iguana*, New York: New Directions.

FURTHER READING

Foster, R.J. (1991) 'Making national cultures in the global ecumene', *Annual Review of Anthropology* 20, 235–60.

Gellner, E. (1992) *Reason and Culture*, Oxford: Blackwell.

Highham, J. (1993) 'Multiculturalism and universalism: a history and critique', *American Quarterly* 45, 195–219.

Jackson, J.B. (1994) *A Sense of Place, A Sense of Time*, New Haven, Conn.: Yale University Press.

Kolakowski, L. (1990) *Modernity on Endless Trial*, Chicago, Ill.: University of Chicago Press.

Radhakrishnan, R. (1994) 'Postmodernism and the rest of the world', *Organization* 1, 305–40.

Schlereth, T.J. (1977) *The Cosmopolitan Ideal in Enlightenment Thought*, Notre Dame, Ind.: University of Notre Dame Press.

Singh, Rana P.B. (ed.) (1994) *The Spirit and Power of Place: Human Environment and Sacrality*, National Geography Society of India Research Publication 41.

Starobinski, J. (1993) 'The word civilization', pp. 1–35 in *Blessings in Disguise; or, The Morality of Evil*, Cambridge, Mass.: Harvard University Press.

Tuan, Y.-F. (1989) 'Cultural pluralism and technology', *Geographical Review* 79, 269–79.

44

PALAEOENVIRONMENTAL NARRATIVE AND SCENARIO SCIENCE

Frank Oldfield

Way back in 1975, feeling giddy and disoriented on the cusp between Marcuse and the managers, as well as fearful of the demise of empiricism in a world of model formulation, I wrote a piece called 'Model mocking birds and empirical eggshells'. Part of it, reproduced below and minimally revised, forms my starting point now in the mid-1990s.

'There was once a society of people who, after many long years spent learning what they could about their past, enjoying their present life and believing in progress (hence giving only fitful thought to their future) began to worry more and more about what lay around the next corner.

'They found many new and interesting ways of preconceiving this future and some of the more creative among them began to see it lurching towards them, vivid and impractical like a misshapen giant, limbs unmatched but mouth agape, voracious and compelling. This scared all those who possessed either refined sensibilities or the illusion of control over their destiny, and it prompted some to seek more soothing metaphors.

'One group began to see the future as an egg laid miraculously by an ever-present but invisible bird and they began to worry a lot about how the system of perpetuation worked. They decided to study it by trying to build models of the bird and by coaxing each of them to lay a range of hypothetical eggs. These could then be looked into to discover which, if any, were both consistent with the eggs of the past and contained the capacity for self-renewal so obviously desirable for the future.

'The business of building the bird models was truly confusing. Few indeed were capable of taking part in this task, and not many more (and perhaps even fewer) could really understand what it was all about. Reasonably enough, it was contended that any bird models made should be constructed in accordance with the best available information about eggs. Since the present and indeed the past (of which there was a lot more) had once been the future, the egg fragments that

952

remained should tell all that was necessary to know about the nature of desirably procreative birds. The task of making the model birds began. Oddly enough, some nasty, impeding paradoxes cropped up almost immediately. For whereas future eggs (of the hypothetical kind) could be produced with great fluency, nobody could fully perceive the egg of contemporary reality (since they were living in it) and, sad to say, the numberless eggshells of futures past lay in fragments, each piece so worn and small as to be hard to match up securely with any other.

'Even so, the bird-building went ahead, for it was an invigorating task for the builders, and perhaps even a sort of therapy when the egg of the present seemed so hard to comprehend. Moreover, it proved possible to arrange and rearrange the egg shards of the past in ways which generated powerful illusions about the egg–bird–egg system. The illusions could be quantified and birds simulated by the manipulated interactions of the variously enumerated illusions. And the birds laid eggs and the eggs became the business of serious academic debate (for whereas the future was manifestly not an academic matter, the eggs were, and so could become the subject of positivist evaluation).

'Now different people used the simulated birds and their hypothetical eggs for different purposes and this affected the way they put the eroded shell fragments from the past together in order to derive structures and functions for their models. Some picked the pieces they liked and when they saw that they could use them to produce future eggs which appeared to give them more power, pleasure or control, they thought it would be nice to establish their own egg-reality and so began to persuade the rest of the population accordingly. This was made easier by the way in which it was often possible to limit discussion to the technicalities of bird building, to confuse people with new languages and to convince people with the general argument, "If the best model bird we can build lays that sort of hypothetical egg, you'd better get ready to live in it."

'Eventually a new group began to emerge from this band of self-confident craftsmen. Suspicious of unconstrained abstractions, they began to look again at the litter of shell fragments accumulated from the past and to try to piece them together in more honest ways. Gradually they began to suspect that some of the more important pieces were in fact the dull or even smelly ones which the first groups had tended to ignore. When at last they managed to construct procreative birds, their eggs cracked open very quickly and were not only bad, but smelled even worse than the most disreputable fragments of old shells from which the bird models had been derived. These new egg-bird breeders were sure that it was a terrible warning. The problem of avoiding such bad smells from hypothetical future eggs became a subject of fierce contention (and funding).

'Thus it was that those who were not builders of model birds became deeply perplexed. They perceived that too often the bird builders, having arranged their empirical eggshell fragments of the past into their own designs and then seen the way in which the hypothetical eggs of the future were fulfilments of their preconceived ideas, simply rejoiced in their own perspicacity. Too seldom did the builders realize that the birds they created could at best mimic and more often mock them.

'*Moral:* The future grows from our efforts to know better and to reshape the real egg of the present from the inside, not from our attempts to build mocking birds.'

My anxious allegory predated the shifts in agenda setting in palaeoenvironmental studies through deductive to what I have elsewhere termed 'projective' modes (Oldfield 1993), of which more anon. More generally, it also predated my awareness of any formulation of post-modernist attitudes to science. Yet, filtering through the Luddite paranoia of the innumerate, dirty booted empiricist I was and still am, there is a sense of concern with the nature of validation, with the relationships between environmental science and 'reality', with the mismatch between our appraisal of the past and our sense of the future, and with the increasingly relativistic nature of 'scenario science'. These concerns have, I believe, been justified by trends over the last twenty years.

Are we now, as our preoccupation with *future* environmental change strengthens to the point of dominating our thinking, moving towards a deeper understanding of *environmental* processes as they generate change on *human* time-scales, or are we in that pre-dawn condition identified by T.S. Eliot 'when the past is all deception, the future futureless'? Put less poetically the issues are about the extent to which we can reconstruct past environments with the accuracy, precision and depth of understanding that allows us to shed light on the near future; the extent to which the models, projections and scenarios upon which the growing scientific concerns about future environmental conditions depend are realistic as guides to action; and the extent to which the modes of research, the agenda setting that constrains them and the external influences that form their cultural, socio-economic and political context are functional or not in relation to the task in hand. I find these issues impossible to address either at an academically detached level, or from the point of view of a particular discipline. The best I can do is address them at a purely personal level, and this requires me to preface my views with a brief confessional digression noting some of the relevant curiosities of personal experience and attitude that colour my subsequent evaluation.

My training up to first degree level was almost exclusively in the 'Arts' and in Human Geography. From this base I worked successively in the Cambridge Botany School and the Muséum National d'Histoire Naturelle in Paris before 'returning' to Geography, from which, subsequently, I have made excursions into a Museum of Anthropology, a Department of Environmental Science, a School of Biological and Environmental Studies and a School of Independent Studies located in a Sociology Department. Yet, here I am since 1975 not only happily resettled in Geography but keen to defend my preference for it as a context for my work. I see no epistemological rationale for the subject, nor for any past or current definition or circumscription of it. I am unable to apply even *post hoc* criteria to define the 'geographical', save in terms of individuals and their affiliations. I see (and seek) no escape from the need for 'physical' geographers to submit their work to and accept the judgement (and sometimes the more narrowly prejudiced priorities) of fellow scientists in neighbouring disciplines. This implies aspiring to meet their criteria, often in addition to those consciously drawn from, or unconsciously endowed by a different background and training. For me, these observations, which could militate against finding my home in Geography, are all outweighed by two related characteristics which Geography, as historically and pragmatically embodied in UK universities, tends to promote. First, there is the opportunity to share, or at least share disputation between, quite disparate frames of reference. This offers to the relatively open minded, a chance to grow out of the disciplinary xenophobia that scars so many academics. Second, springing from this, there can grow a willingness to dig for, examine, challenge, exchange and perhaps eventually share fundamental assumptions, both those which form part of the tenet of defined disciplines or methodologies, and those which often fall unevaluated down the conceptual cracks between them. To quote: 'Each method is a language and reality replies in the language of the question. Only a constellation of methods can capture the silence between each language asking questions' (de Sousa Santos 1992: 39); and from the same source: 'If it is true that each method clarifies only that which is convenient for it, and when it does clarify anything, does not allow any major surprises, scientific innovation consists in inventing persuasive contexts that allow the application of methods outside their natural habitat' (de Sousa Santos 1992: 39). This makes for improved science since more depends on the formulation of questions and the articulation of assumptions than on the proliferation of answers. Forgive me if, in the light of the above, I fail to speak for, or even from within, 'established' Geography and lapse into eclectic heresy. One aspect of this (though quite a congenial aspect for a 'geographer'), is my reluctance to accept fully any of the commonly held dichotomous stereotypes of 'science' versus 'arts'. They tend to emphasize differences of language and culture which coexist alongside important parallels in structure of reasoning and development. Two schema (Figures 44.1 and 44.2) and their annotations help to make this point – the first in terms of deductive models of science; the second in terms of parallels between 'environmental' and 'documentary' history. In the first instance, the strongest and most crucial similarity between this model of the scientific reasoning and the creative 'Arts' is at the stage of the initial conception of a scientific problem or concern. The second scheme shows how, even during the later, 'productive', stages of empirical research, there can be structural and conceptual parallels between science-based and literature-based study. The big difference between science and the creative arts (for me) lies in the nature of the criteria and the procedures used in evaluation. As an example, at the extremes, it is just as pointless for the performer or critic to evaluate works of art in terms of their repeatability and consistency with existing *oeuvres* as it would be for scientists to value in their sphere, the inextricably related richness and ambiguity immanent in great texts

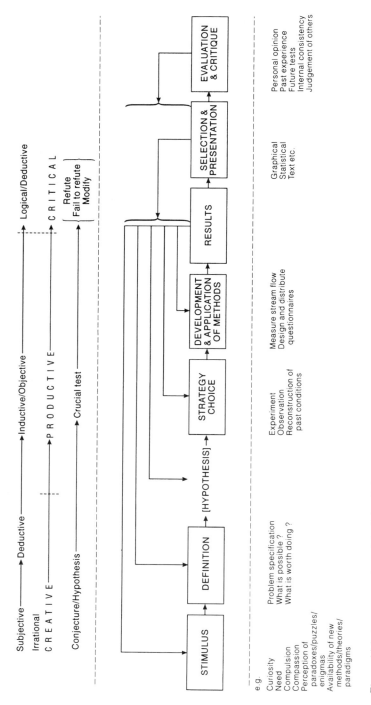

Figure 44.1 A conceptual model of the deductive approach to research. The flow diagram in the middle section sets out possible stages in a deductive research strategy and illustrates the way in which results and their evaluation feed back into the earlier stages as new ideas and knowledge serve to redefine the problem. The 'stage' to which feedbacks occur will depend on the nature of the new knowledge. For example, poor quality results may force a change in methods within the same overall deductive framework; incomplete data relative to the problem posed may lead to a change in strategy; refutation or modification may lead to a redefinition of the hypothesis to be tested; evaluation of the research strategy as a whole may redefine the scope of the research and, finally, the results and their evaluation may provide stimuli for new lines of deductive reasoning.

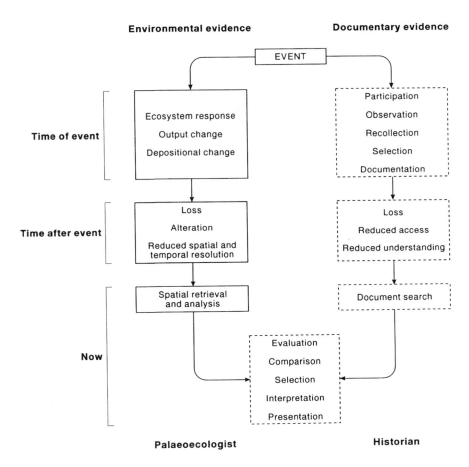

Figure 44.2 A comparison of the 'filters' that environmental and documentary evidence go through between the event they record and the interpretation by palaeoecologists or historians. The diagram is divided into three sections that show processes at work at the time of creation and preservation of the evidence through the period of alteration or destruction afterwards, to present-day methodological limitations. Boxes bound by dashed lines indicate processes that are solely dependent on humans; those bounded by solid lines are affected by environmental processes or human activities. After Oldfield and Clark (1990).

or pictures. Yet inspiration, honesty, self-criticism, technique and application are just as important to both. Moreover, just as fashions, schools of thought, frames of reference and models of performance (or research) change in the arts and humanities, so paradigms in science shift, as Kuhn's well-known book (1974) reminds us. By way of illustration, it is suitably chastening for someone whose career began in the 1950s and who shared and can remember well the scepticism and even scorn heaped on Wegener (1912), Milankovitch (1930) and, at a more modest, but now rather significant level, Schove (1955), to see their basic propositions of, respectively, 'continental drift', astronomically controlled long-term climatic change and global synchroneity of short-term climatic secular variation, part of the current canon. Does this move us beyond Kuhn to an essentially post-modernist view of environmental science, and specifically environmental history, as providing no more than a range of alternative and equally valid narratives?; and science as a whole, as providing merely a range of versions of the world all neither more nor less interesting, valid or valuable than those arising from other strands of thought? Not for me. Interestingly, it is Bill Cronon (1992), the historian, who helps me out of this dilemma. He proposes three criteria which narratives of environmental history must meet. First, they should not contradict known facts about the past. Second, they must make ecological sense – and we can expand this criterion and express it as an embargo on contravening our current understanding of environmental and ecological processes (in essence, a somewhat Huttonian point). Finally, the work of environmental historians reflects their affiliation to the communities, cultural and academic, to which they belong. The network of overlapping peer groups which this statement implies ensures critique and imposes evaluative constraints. This brings us to the question of what the current academic and cultural context of historical aspects of environmental science, and of 'physical geography' implies.

Part of this context for palaeoenvironmental research arises from changes over the last thirty years or so in the dominant inferential structures within which problems are defined and research pursued. My contention is that a largely inductive mode of research dominated the field until the late 1960s, after which an increasingly important contribution was made by deductive approaches, culminating in major studies like those reconstructing the histories of eutrophication and, more recently, surface-water acidification, with a view to testing competing hypotheses about their origins. This in turn is being superseded, at least at the policy level of prioritizing programmes and setting the agenda for major funded research, by what I term a 'projective' mode. This arises from apprehension about the course and consequences of future environmental change and the realization that improved knowledge about past environmental change may contribute to our insight into and evaluation of potential future changes. Fears about the consequence of this new type of research mode are the subject of fuller discussion in Oldfield (1993). They hinge essentially on the problem of reinforcement (Watkins 1971) and the difficulties of falsification (Popper 1963), both of which are raised metaphorically in the introductory allegory with which this chapter begins. There is a sense in which a culture preoccupied with a growing concern about its *environmental* future places the environmental scientist alongside the economist as a creator of explanatory models designed to help us to adapt to and optimize future quality of life. The lead and lag times are generally longer for the environmental

958

scientist, hence the likelihood of experiential refutation during the currency of any given projection is reduced. Moreover, the greater degree of determinism inherent in *most* of the explanations formulated in environmental science lend more credence to the potential value of models in that sphere than to those in economics which tend to be all too easily discredited by reality during their currency. To my mind, both these observations *heighten* the danger for environmental scientists placed under pressure to establish the course and consequences of future environmental change. Questioning the operational limits of deterministic models in environmental systems and expressing worries about the potential loosening of inferential frameworks are, nevertheless, a very long way from opting for a post-modernist view of environmental science in either its 'affirmative' or 'sceptical' forms (Rosenau 1992).

At the heart of the problem for an empiricist of my generation lies the relationship between empirical research and theory. In the Popperian realm, it seems clear enough. Hypotheses are valued for their ability to generate falsifiable corollaries (preferably of low probability, hence high value). Testing may fail to refute them or it rejects or redefines them. Theories unify insights derived in this way and in their turn generate new hypotheses for testing. Indeed the whole process hinges on that upward-spiralling cycle of renewal. Within this realm, pieces of empirical research serve to test each other's outcomes. Hypothesis-testing brings each piece of significant research to an ever-changing starting line in a never-ending race, rich in participants but almost entirely lacking in permanent winners or losers.

At least from outside physics, it seems as though experimentalists in that subject live comfortably and work interactively not just with theory but with theory articulated as quantitative model. I infer that the comfort springs from the more controlled nature of the experimental environment, from the dominantly mechanistic nature of the frame of reference for most practical purposes, and from the statistical rigour achievable in the relatively simple and coherent systems under consideration; simple, that is, when compared with the much more complex environmental systems on a global scale. There appears to be no essential conflict between this 'simpler' sort of science and the Popperian model. In environmental science, and more specifically with that part of it focused on the reconstruction of palaeoenvironmental conditions, experiments are *post-hoc* in so far as we can, in Ed Deevey's words (1969) 'coax' history to conduct them. They are therefore rarely complete or fully controlled. The processes involved are never entirely mechanistic and have become less and less so as the impact of human activity on environmental change has increased. Moreover the models with which palaeoenvironmental reconstructions interact are not so much an articulation of theory as a symphony of possibilities within which the number and diversity of themes, and the dynamics of their non-linear interaction are unimaginably complex and only imperfectly quantified. In the physicist's experiment–model dichotomy, Heisenbergian uncertainty can largely be encompassed within the framework of statistical probability. In the palaeoenvironmental context, the 'uncertainties' are different. They arise from uncontrolled error, loss of resolution and information as we dip into the real past, ignorance of crucial aspects of contemporary global function, non-linear responses and so forth. All this alters the quality of the relationship between empirical research and model building. It lengthens the pathway along which both must proceed from an early stage where 'tests of mutual consistency between model simulations and reconstruction derived from proxy records often lack recourse to any verifiable reality that might independently validate the one before it, in turn, has to do service in validating the other' (Oldfield 1993), to a later stage where each may at least realistically constrain if never validate the other.

Who controls the pathway and who the pace of this interaction between model building and palaeoenvironmental reconstruction? My contention is that for the reasons outlined above and others outlined in Oldfield (1993), the researcher in 'environmental change' finds the 'empirical–modelling' framework not only different from the classical deductive framework of Popperian science, but sufficiently different from the 'experimental–theoretical' framework of the physicist to make that also an imperfect paradigm for understanding the activity or optimizing research within it. Nevertheless, the 'empirical–modelling' framework appears to be the only one available for environmental science in the 'projective' mode; and the projective mode is not going to go away. The notion of a range of alternative palaeoenvironmental 'narratives' serving to constrain a myriad of future scenarios and vice versa must surely prompt the hollow chuckles of the post-modern sceptic. All the more reason for learning how to operate effectively and rigorously within the new framework (Figure 44.3) as a matter of urgency; but can this be reconciled with responding to those external pressures which require operational answers to practical questions on prescribed time-scales? This issue places environmental scientists, and among them palaeoenvironmentalists, in the forefront of the dialectic between internal, academic agenda setting and external agenda imposition. During the recent past, academics have experienced a period when their own capacity to set the agenda, the pace and the priorities has been respected and the internal criteria of the academic community left relatively undisturbed. Now, the pendulum has swung to the opposite extreme to a degree which changes almost out of all recognition the context in which projective science must learn its business.

In its inductive stage, palaeoenvironmental research was largely driven by perceptions and definitions of problems by and for the academic practitioners. Under these circumstances it is natural (though not satisfactory in the society within which the researcher operates) to prioritize entirely what, in a similar scientific context, Passmore (1974) calls 'problems in ecology' as distinct from 'ecological problems'. The former are academic problems defined within a research field and only of direct relevance and interest to those within the field; the latter are ecological problems which, because of their impact on human activities, are also inevitably, in part, technological, economic, cultural and

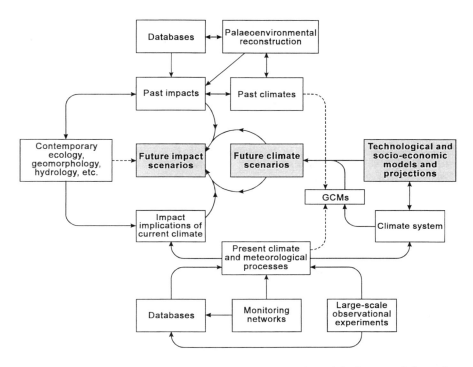

Figure 44.3 A tentative and much simplified conceptual model of some of the major components and inferential linkages involved in 'scenario science' as considered in the text. The pecked lines indicate opportunities for the partial validation or constraint of models and scenarios.

political. These kinds of problems became established as key parts of the palaeoenvironmental research agenda in the deductive mode from the late 1960s onwards and have by no means reduced the quality of the scientific research achieved. It is not surprising that they have grown to dominate the overall formulation of environmental research policy in the current projective mode.

Without a doubt, 'applied', deductive-mode, hypothesis-testing research into problems like the origin and history of lake acidification has involved first-rate scientific research, but it has depended in part on, and 'mined' an immense store of earlier, 'pure', inductive research into, for example, diatom taxonomy and ecology. The 'within-discipline' questions raised by the acid waters research (e.g. on diatom physiology) received less funding and inevitably less attention than the pragmatic, applied aspects of the programme. In financial terms it is tantamount to saying that there was a greater emphasis on yield than on growth or reinvestment. This is the nub of our dilemma, for, within the projective mode, as outlined above and elsewhere (Oldfield 1993), it is even more difficult

961

to define and pursue those fundamental aspects of research upon which improved reconstructions, models and insight into environmental processes and ecological responses depend. All that is certain is that the new fundamental questions raised by the research agenda of the future will require basic and often unglamorous research to keep pace with them as they arise. Much of this is likely to be less attractive to external funding agents than the research which appears to address more directly pragmatic questions of future management. Yet without it, the quality of the answers to these questions will be severely limited. At present, the implicit judgement by which much funding is guided, is that it is more important to make what we already know more coherent and accessible – hence databases, GIS-based syntheses, etc. This sort of activity seems wise and timely, provided it neither replaces nor entirely defines all the fundamental research required to underpin more applied sciences. Fundamental research is defined more by problems in understanding than by 'gaps' in coverage.

Of course these are not black and white issues, they are ones of degree. I have simplified and possibly exaggerated the antithesis, but an important general point must be made and perhaps dramatized somewhat in the interest of a better contribution from environmental science to a future research agenda so critical to understanding how the environment may change, at what pace and with what consequences. My contention is that scientific research in those areas with which I have been associated for over thirty-five years, has, like so many aspects of higher education and academic life, lost rather too much autonomy and been forced to yield too much control to externally directed managerialism. It fares little better under the sway of growing pressure for an external accountability rooted in limited comprehension than do the arts under totalitarianism. In both cases, self-defined, within-community-acknowledged excellence is prone to suffer. I am not making a plea for a return to the past, or for an Olympian academic freedom beyond the limits of acceptability in a responsible, well-educated, sympathetic democracy. I am simply saying that if environmental scientists in general and, closer to my own sphere of concern, palaeoenvironmentalists within the broader environmental community, are to play the best role they can, there needs to be stronger support for the fundamental research which underpins model development and testing, and a greater proportion of science funding going into some of the less glamorous, basic work as a strategic investment against future needs. Equally, it seems to me counterproductive to force so much of the planning and progress of research into the 'business project' mode where stress is laid on time-scales, goods and 'deliverables'. The only totally secure way to meet projected targets in environmental science is to take no risks and to do nothing of real originality. That way, programmes of research can be designed at the outset and, given persistence and good management, their outcomes can be predicted. My experience of working on the fringe of large research groups of North America has convinced me that this approach to complex, inter-disciplinary science, however competent and well managed, is no substitute for the less predictable approach in which each new step grows out of the interaction between hypothesis formulation and empirical testing, and each new turning point is defined by an interrogative approach to all the ideas and data available at that stage. I have memories of just such an approach in the seminal work of the much respected Hubbard Brook Research Group. Alas, all the changes I am pleading for here seem unlikely without a change in the nature of the dialogue, the pattern of responsibility and the balance of power between researchers and those who now provide the context of management and control within which we work.

Some little time ago I was fortunate enough to receive a volume of manuscripts from the Monastery at Brno. It contained a transcript of one of the Abbé Mendel's appraisal interviews. It starts off pleasantly enough: 'Well Brother Gregor, how has your time in our gardens helped you to enrich the quality of your spiritual life and to contribute to the welfare of our community?' After a few lines, it is possible to detect a widening rift between the perceptions of the senior monk and Mendel's ideas and activities: Brother Gregor gets on to the subject of pea plants. Not all the peas went into the pot . . .

REFERENCES

Cronon, W. (1992) 'A place for stories: nature, history and narrative', *The Journal of American History* 78(4), 1347–76.

Deevey, E.S. (1969) 'Coaxing history to conduct experiments', *Bioscience* 19(1), 40–3.

Kuhn, T. (1984) *The Structure of Scientific Revolutions*, Chicago, Ill.: Chicago University Press.

Milankovitch, M. (1930) 'Mathematische Klimalehre und astronomische Theorie der Klimaschwankungen', in W. Koppen and R. Geiger (eds) *Handbuch der Klimatologie* 1, Teil A, Berlin: Borntraeger.

Oldfield, F. (1993) 'Forward to the past: changing approaches to Quaternary palaeoecology', in F.M. Chambers (ed.) *Climate Change and Human Impact on the Landscape*, London: Chapman Hall.

Oldfield, F. and Clark, R.L. (1990) 'Environmental history – the environmental evidence', pp. 137–61 in P. Brimblecombe and C. Pfister (eds) *The Silent Countdown*, Berlin: Springer-Verlag.

Passmore, J. (1974) *Man's Responsibility for Nature*, London: Duckworth.

Popper, K.R. (1963) *Conjectures and Refutations*, London: Routledge & Kegan Paul.

Rosenau, P. (1992) 'Modern and post-modern science', *Review* xv(1), 49–89.

Schove, D.J. (1955) 'The sunspot cycle', 649 B.C. to A.D. 2000', *Journal of Geophysical Research* 60, 127–46.

de Sousa Santos, B. (1992) 'A discourse on the sciences', *Review* xv(1), 9–48.

Watkins, N.D. (1971) 'Polarity events and the problem of the reinforcement syndrome', *Comments on Earth Sciences: Geophysics* 2, 23–42.

Wegener, A. (1912) *Die Entstehung der Kontinente und Ozeane*, Braunschweig: Vieweg.

FURTHER READING

Battarbee, R.W., Mason, J., Renberg, I. and Talling, J.F. (eds) (1990) *Palaeoliminology and Lake Acidification*. London, The Royal Society.

COHMAP Project Members (1988) 'Climatic changes of the last 18,000 years:

observations and model simulations', *Science* 141, 1043–51.

Eddy, J.A. and Oeschger, H. (1993) *Global Changes in the Perspective of the Past*, New York, John Wiley & Sons.

Kutzbach, J.E. and Webb, J.T. III (1991) 'Late quaternary climatic and vegetational change in eastern North America: concepts, models and data, in L.C.K. Shane and E.J. Cushing (eds) *Quaternary Landscapes*, London, Bellhaven Press.

PAGES Core Project Report (1995) *Palaeoclimates of the Northern and Southern Hemispheres*, PAGES Report Series 95-1. Bern, Switzerland.

Popper, K.R. (1963) *Scientific Knowledge*, London: Routledge & Kegan Paul.

Roberts, N. (ed.) (1994) *The Changing Global Environment*. Oxford, Basil Blackwell Ltd.

Street-Perrott, F.A. (1991) 'General circulation (GCM) modelling of palaeoclimates: a critique', *The Holocene* 1, 74–80.

Troen, I. (ed.) (1993) *Global Change: Climate Change and Climate Change Impacts*, European Commission Science Research Development EUR 15921 EN, Brussels.

USGCRP (1995) *Forum on Global Change Modeling*, US Global Change Research Program, USGCRP Report, 95-02, Washington, DC.

45

GEOGRAPHICAL FUTURES

Some personal speculations

Peter Haggett

An open-ended invitation to speculate about the future of geography is an irresistible temptation. It allows speculation full range yet, if one is already in one's sixties, more or less guarantees that one will not be around to face one's critics if the predictions all fail.

For on looking around at previous predictions by geographers, failure is the most likely outcome (Haggett 1994). This can occur in two ways. Under-prediction or failure to predict the scale and nature of change (what the statistician calls Type I errors) and overprediction or the prediction of events that failed to occur (Type II errors). An earlier review of geographical predictions (Haggett 1972) suggested Type II errors were rare, implying that our past attempts to look forward have been too cautious.

So here I will try and err towards being incautious and spell out my personal guesses about future change in two arenas. The first is world geography itself. For if geography early in the next century follows the past record of the discipline then it will continue to be concerned to describe 'the earth as the home of the human population'. And as our planet's geography changes so will the discipline of geography (literally 'writing about the earth') change in sympathy with it.

The second arena is changes within the discipline of geography. Here the questions relate to how geographers as an academic and scholarly group will go about the business of describing these world changes.

ARENA I: CHANGES IN WORLD GEOGRAPHY

Forecasts and projections cannot be separated from the time-period to which they relate. So before we begin to look forward we have to decide on a useful target date at which to aim. How do we decide on this? We know that, as we

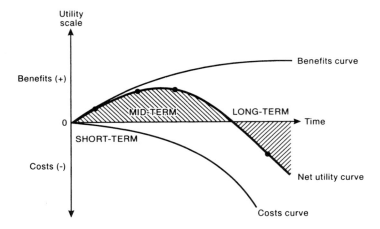

Figure 45.1 The changing balance of benefits and costs over time in forecasting. After Haggett (1990).

try to project further ahead from the present, so the band of probable errors flares out to give an ever-wider cone of uncertainty. That realization would tend to suggest we should make only very short-term projections. But forecasts of the immediate future may have limited value since they are likely to be so like the present. Thus it would be helpful to have guidance somewhat further into the future but not too far, for – in terms of the span of human life – very long-term forecasts have only academic or philosophical interest. Figure 45.1 shows how the balance of benefits and costs described by the net utility curve reaches a peak in the medium term. In practical terms that means we might usefully try to look a decade ahead (say to AD 2005) but that it would be foolish to try and look more than a generation ahead (say to AD 2020). This is the band within which I will try to stay.

I isolate here just seven trends in world geography. They represent a personal selection only and other geographers would choose other lists. They range from trends which are highly likely, to those which are more speculative.

Change 1: Increases in world population

The most important of changes in world geography is also the one that one can predict with most confidence. Although the world rate of annual increase is now declining (it peaked at around 2 per cent in the late 1960s) the slowdown is very gentle and the momentum of increase will take world population to over 6.5 billion by 2005 and to 8.5 billion by 2025 (United Nations 1991). The population is not only changing in overall number but is changing in three other important respects. First, the age structure is edging towards an older pattern with the over-65 group overtaking the under-15 group around 2075. This has implications for the patterns of health and disease and social support: by our first target date China's elderly population alone will have passed 100 million. Second, the greatest proportion of new growth will come in what are now developing countries: in 1950 one-third of the world's population was located in industrialized countries but by 2025 that fraction will have been halved to 16 per cent. Third, the population is becoming more urbanized. 'Urban' can be defined in so many ways that absolute levels are misleading, but the upward trend in the share of the world's population that lives in cities continues to grow apace. By 2005 more than thirty world cities are expected to have populations of more than 10 million inhabitants and over 500 to be beyond the one million mark. Again, most of these will be in what are now developing countries.

Change 2: Increased resource consumption

Along with the increase in total population the multiplier of resource use per capita will have increased. Per capita energy consumption in OECD countries has remained rather steady for the last twenty-five years but that in the newly industrialized countries has doubled. Since the rest of the world's per capita energy use is still only one-sixth of that in industrialized countries, the potential impact on resource demand over the target period if current growth rate is simply maintained is very great. We can expect even modest per capita changes in the developing world to push up global resource demand threefold by 2025.

Change 3: Increased environmental pressures at both local and global levels

The combined effects of the first two changes is to produce a greatly increased pressure on the fragile environment within which the human population lives (United Nations Environment Programme 1992). Every part of that environment – the lithosphere, the hydrosphere, the atmosphere, the pedosphere and the biosphere – will be under increasing pressure. What distinguishes the pressures of the past from those of the future is that the scale of the assault will be increasingly global rather than local or regional.

Change 4: Further collapse of long-distance space

The notion that the world is becoming smaller is a commonplace. What is not often appreciated is the exponential scale of that collapse. Figure 45.2 shows the average travel within France over a two-hundred-year period. If the exponential increase for France is maintained until 2025 the present average of 50 km will have increased to over 150 km. We can illustrate that change dramatically for Australia: in the period since the end of the Second World War the Australian population has doubled (from 8 to 16 million) but the flux of population moving

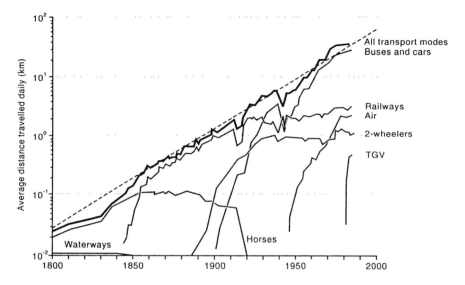

Figure 45.2 Increased spatial mobility of the French population over a two-hundred-year period, 1800–2000. Note that the vertical scale is logarithmic so that increases in average travel distance increase over time. After Grubler and Nakicenovic (1991).

into and out of that country through travel has increased a hundredfold (from 40,000 to 4 million travellers). The rapid world-wide spread of new infectious diseases is partly linked to this hugely increased mobility.

Change 5: Further switch of resource development into offshore areas

The main increase in hydrocarbon extraction in the second half of the present century has come from offshore rather than onshore areas. A review of the potential areas for new oil fields suggests this trend will be reinforced in the next two decades. With such a switch the areas of future international dispute and conflict are likely to come from overlapping claims in offshore areas. Unlikely flashpoints like the Spratly Islands in the South China Seas will become the foci of international tension. Geographers have for too long neglected the 70 per cent of the earth's surface that is water covered ('terrographers' might be a more accurate description of our profession) so a change in concern towards the oceans has long been overdue.

Change 6: Trend away from hierarchically organized structures

The electronic revolution is only at its beginning. The trend in computing from major central computers to widely distributed systems linked into global, regional and local networks is a parable of what is happening more generally in much human organization. Power and influence is likely to come through access to networks so that long-established hierarchic structures are likely to be increasingly replaced by ring structures. I expect one implication of this trend is that more countries will short-circuit the conventional development paths, missing out on one or more of the stages of economic growth foreseen by Rostow over thirty years ago (Rostow 1960).

Change 7: Instability in major geopolitical hegemonies

The implications of the weakening of hierarchic structures has implications for geopolitical power. Each of the last four Kondratieff waves has focused on a different geographical power centre. In the last decade we have already seen the collapse of one superpower (the Soviet Union) and the global position of the United States at the end of this century is now relatively much weaker than that at mid-century. The sharp rise of Japan is a phenomenon of the last thirty years (though its roots lie much deeper), but its position seems considerably less secure in the time-frame towards which we are now looking. The rise of other smaller Pacific Rim states such as South Korea and Taiwan is a foretaste of what may transpire when the Chinese economy begins to draw away. By AD 2025 the world power map may look a very different one than at present with China and possibly India occupying much more dominant positions than at the present day.

In looking back over the seven changes, it is important to remember that these are reasoned guesses based on trends which are now discernible. That for population (Change 1) is far more certain than geopolitics (Change 7). Each will affect the kind of geography which our children and grandchildren will study in their ever-changing world.

ARENA II: CHANGES IN GEOGRAPHY AS A DISCIPLINE

Both in general parlance and in this chapter the term 'geography' is used in two different ways. One is as a shorthand for the world we inhabit and the other is as the name of the discipline which studies that world. So in the second arena I try to isolate changes which I expect to see occurring within the academic discipline (Haggett 1990). Here I am being speculative and take just seven of those where I hope significant advances will be made, which will either involve geographers or affect geography, or both. The reader will need to allow for an optimistic temperament.

Change 1: ROMs to RAMs in mapping

Mapping is likely to continue to change from a ROM to a RAM format (Monmonier 1985). Map production has conventionally been geared to a final printed sheet which displays information in a fixed format; in that jargon of computer technology a map library with all its printed sheets might be likened to a 'read-only memory' (ROM). But the need for maps is often for selective, personalized data that may be needed only for a brief period to answer a specific question. When we sketch on a memo pad a map for a newcomer to find his way to our home for supper, the resultant 'map' is cheap, quickly produced, single-purpose, area-specific and disposable (indeed we would be embarrassed were it to be kept!). Huge changes in computer storage already allow mapping to change to that interactive, on-demand form within the next decade. In computer language the 'random-access map' (RAM) is increasingly likely to replace the fixed ROMs. Selection of area, resolution or scale, and content to be displayed are already available on military systems, and the main barrier to wider dissemination is likely to be cost. The fall in the cost of both storage and purchasing should ensure that this barrier is steadily reduced.

Change 2: Synthesis in climatic modelling

A second breakthrough could come from a general paradigm for climatic evolution developed to bridge the gap between detailed meteorological studies (at the micro-level), geophysical models of the earth's atmosphere, and Pleistocene history (both at the macro-level). This should allow some return to an integrated systematic picture of the earth's surface, welding both biological history and palaeoclimatic history in a new synthesis for our understanding of man's physical environment.

Change 3: Analytical regional geography

I expect a new regional geography to evolve which will supplement traditional skills by integrating quantitative models from economic geography, regional economics and regional science (Braat and van Lierop 1987). I have argued elsewhere that one of the main reasons for the quantitative revolution of the 1960s centring on systematic fields was that the methods then available were unable to handle regional complexity (Haggett 1990). Although the problem remains formidable, the development of spatial impact and regional interaction models allows a start to be made on designing a new generation of regional geographies with some additional capacity for forecasting regional change.

Change 4: Demographic history

The gap between micro- and macro-studies of the historical evolutions of the population of developed countries may be closed (Denecke and Shaw 1988). Geographers have conventionally been dogged by this problem of linking their findings at different scales, and nowhere has this been more evident than in population geography. Work during the last two decades on individual migration pathways at one scale and on migration modelling at another promises a fuller and more personalized view of migration.

Change 5: Integration of geographical thinking into the history of science more broadly

A philosophical basis for geography will slowly be worked out to encompass the existing ideas of the nature of geography within the broader base of a man–land ethic (Mitchell and Draper 1982). The questions of how man relates to the natural world and the cosmic order have been asked by philosophers and theologians down the centuries. The impact of the environmental crisis has brought a new interest by philosophers in geography and by geographers in philosophy. This convergence is leading to a re-evaluation of the historical evidence showing how people of different faiths (and therefore of different conservation ethics) have used resources in past periods.

Change 6: Establishment of global benchmarks for measuring change

A global land-use inventory will be completed and maintained using remote-sensing data and an international network of benchmark sites (Lindgren 1984; Denegre 1988). This will serve as a marker for measuring both future and past change. One of the duller but necessary jobs for geographers to finish is the establishment of international measurement standards for metropolitan areas.

Change 7: A coming of age for spatial analysis

A space-proofed set of statistical tests is being established which can give bias-free estimates when analysing geographical data (Cliff and Ord 1975). Analysis of time-series on a secure basis was begun in the 1920s but, for spatial series, serious problems of bias and error remain to be cleared up.

Other geographers would make other predictions. But all would agree that there is no lack of major geographical problems that we can expect to be solved given a reasonable input of resources. Most of the ways forward depend critically on advances being made in other areas. For geography to be able to put its piece in the jigsaw puzzle it will need other pieces to be placed in position by others. Most of the advances expected depend on research activity in microcomputers and information processing, in genetics and biochemistry, in remote sensing and space exploration. Within the next decade, entirely new possibilities may have revealed themselves which will allow new phases of geographic work, not yet foreseen, to go forward.

THE SEARCH FOR IDENTITY

One of the structural problems which faces geography early in the next century stems from its success. Even half a century ago geographers could work away at their problems and see only distantly on the horizon the far-off smoke from other academic neighbours. Today the same academic space is more heavily occupied and the campfires of near neighbours – in environmental chemistry, in engineering hydrology, in conservation economics, in environmental history, in area studies and the like – burn brightly around us. Many of the topics which once formed the basis of our own undergraduate courses are now matters of grave public concern; funds are there for their study if we bid for them, and some countries have whole government institutes devoted to them. Our favourite wild picnic spots are now attracting many other visitors, and partnerships have to be forged with our new and closer academic and institutional neighbours.

In this context of interchange it is tempting to argue for the sweeping away of the old disciplinary boundaries and the merging of geography with its neighbours. The argument comes from two directions. First, from those who see the intellectual ideas sweeping major areas of study (e.g. the role of social theory within the social sciences or molecular biology within the life sciences) as so powerful as to warrant major restructuring of conventional subjects. Second, from those who find the whole university system archaic, intellectually superfluous, self-indulgent and morally intolerable (Hurst 1985). The first works to dismantle subjects and the second to demythologize experts. With all this noise we might well expect the Walls of Jericho to come tumbling down and much broader and more open fields of study to be left.

I am sympathetic to these changes in some fields but hostile to their application to geography. I hope this inconsistency is not just local conservatism or hardening academic arteries. To see why it may not be we need to look at the 'map' of disciplines surrounding geography. With which neighbours should these alliances be made? The greatest pressure for merger has come from human geographers within the social sciences, but a merger here would leave both physical geography stranded and cartography adrift, just at the time when an increasing number of our graduates are finding job opportunities in such areas as Environmental Impact Analysis or in Geographical Information Systems (GIS). To conceive of geography in terms of human geography only (important though that is) is, I think, profoundly misguided. To do so is to miss one of the essential points of the subject – whether taught at school or college, used in employment or for public education.

Geography's relations with other subjects form a complex series of inter-secting planes. Our contributions lie at the interface of several scientific traditions and not within the bounds of one. That remains a central core in our contribution, and as Sauer argued, 'If we shrink the limits of geography, the greater field will still exist: it will be only our awareness that is diminished' (Sauer 1956).

REFERENCES

Braat, L.C. and van Lierop, W.F. (1987) *Economic–Ecologic Modelling*, Amsterdam: North Holland.

Cliff, A.D. and Ord, J.K. (1975) 'The comparison of means when samples consist of spatially autocorrelated observations', *Environment and Planning A* 7, 725–34.

Denecke, D. and Shaw, G. (1988) *Urban Historical Geography: Recent Progress in Britain and Germany*, Cambridge: Cambridge University Press.

Denegre, J. (ed.) (1988) *Thematic Mapping from Satellite Imagery*, London: International Cartographic Association.

Grubler, A. and Nakicenovic, N. (1991) *Evolution of Transport Systems*, Vienna: IIASA Laxenburg.

Haggett, P. (1972) 'Forecasting alternative spatial, ecological and regional futures: possibilities and limitations', pp. 217–36 in R.J. Chorley (ed.) *Directions in Geography*, London: Methuen.

Haggett, P. (1990) *The Geographer's Art*, Oxford: Blackwell.

Haggett, P. (1994) 'Prediction and predictability of geographical systems', *Transactions, Institute of British Geographers*, New Series 19, 6–20.

Hurst, M.E. (1985) 'Geography has neither existence nor future', in R.J. Johnston (ed.) *The Future of Geography*, London: Methuen.

Lindgren, D.T. (1984) *Land Use Planning and Remote Sensing*, Dordrecht: Martinus Nijhoff.

Mitchell, B. and Draper, D.L. (1982) *Relevance and Ethics in Geography*, London: Longman.

Monmonier, M.S. (1985) *Technological Transition in Cartography*, Minneapolis, Minn.: University of Minnesota Press.

Rostow, W.W. (1960) *The Stages of Economic Growth*, Cambridge: Cambridge University Press.

Sauer, C.O. (1956) 'The education of a geographer', *Annals of the Association of American Geographers* 46, 289–99.

United Nations (1991) *World Population Prospects, 1990*, New York: Population Division, United Nations.

United Nations Environment Programme (UNEP) (1992) *State of the Environment, 1972–1992: A Report*, Nairobi: United Nations.

FURTHER READING

Haggett, P. (1990) *The Geographer's Art*, Oxford: Blackwell.

Johnston, R.J. (ed.) (1985) *The Future of Geography*, London: Methuen.

Johnston, R.J., Taylor, P.J. and Watts, M.J. (eds) (1995) *Geographies of Global Change: Remapping the World in the Late Twentieth Century*, Oxford: Blackwell.

Macmillan, W. (ed.) (1989) *Remodelling Geography*, Oxford: Blackwell.

Stoddart, D.R. (1986) *On Geography and its History*, Oxford: Blackwell.

Unwin, T. (1992) *The Place of Geography*, Harlow: Longmans.

World Resources Institute (1992) *World Resources, 1992–93*, New York: Oxford University Press.

INDEX

acid rain 82, 537, 645
acidification 461, 645, 961
aerosols 495
Afghanistan 356, 360
Africa 12, 191
    animals 74, 176–7, 179, 463–4
    arid zone 640–1
    coast explored 165, 185
    and colonialism 705
    debts 408, 410, 416, 417, 420, 421,
        422, 424, 426, 714
    development 343
    economies 374
    erosion 31
    ethnic groups 320
    famine 683
    farm land 659
    farming 146, 147, 188, 240, 243, 244,
        464, 690
    food supply 244–6, 606, 667
    forests 54, 56, 188
    fossil hominids 91–7, 101, 103–4, 584
    game parks 464
    immigration 260
    languages 109, 110–11, 112, 117
    landscape 34
    migration to cities 396, 397, 402
    population 250, 253, 265, 267, 268,
        269, 270, 271, 665
    rift valley lakes 56
    settlement 105, 178, 196, 197
    and slave trade 175
    southern 188, 196, 343
    states 357, 358
    Sub-Saharan 347, 358, 367, 421

    trade 213
    undernutrition 402, 640, 667, 686, 688
    urbanization 391, 393, 403
    wars 355, 359
    see also East Africa; South Africa;
        West Africa
Afrikaans 117
agribusiness 229, 247, 698
    and food 233
    multinational 232, 237
    nationalized 237
    transnational 235–9, 247
agrichemicals 231, 239, 241, 669, 670,
        671
    and pollution 462, 470, 537
agriculture 11, 81, 183
    adaptation 672
    arable 217, 219, 220, 230
    in colonies 164, 169, 170, 171–2, 193,
        213–14
    commercial 170, 199, 241–4
    constraints 669
    contract 230, 232
    cycle 128, 219
    in developing countries 235, 238, 239,
        240–1, 663–4, 672, 689–90
    diversification 765–6, 767, 768
    and energy 142, 209
    and environment 145, 146–7, 148,
        152–3, 228, 234–5, 239, 241, 247,
        555, 655, 693, 698, 754, 766
    evolution 13, 78, 105, 128, 142, 143,
        223, 275, 277
    extensive 253, 765, 767
    farm incomes 234, 235, 677, 679, 753,

756, 765
and genetics 606
harvest 127, 129
hill farms 768
increased output 160, 233, 688–9, 690, 696
individualistic 217
industrialization 229, 230, 232, 235, 238, 239, 247, 616, 661, 696, 753
innovation 667, 668, 693
intensification 160, 233, 234, 659, 667, 668, 671, 754, 767
livestock 142, 230, 233, 699
local knowledge 663–4, 672
markets 606–7
and manufacturing 230, 231, 232, 752, 753
mechanization 200, 228, 230–1, 240, 241, 693, 696, 698–9
mixed 220, 231, 699
modernization 229, 241, 256, 281, 696, 753
Neolithic 111, 142, 143
organic 699
part-time 765, 768
pastoral 219–20
permanent 145, 146, 196, 314
plantation 172–3, 174, 190, 585, 586
planting 127, 164
polarized 235
pre-industrial 145
production 233, 654, 655, 657, 658–9, 677, 688, 690–6, 753, 767
protection 677, 679–80, 681, 683, 699
research 239, 693
rituals 127, 128
shifting 111, 145, 186, 189, 196, 311, 585–6, 660–1
'slash and burn' see shifting
social aspects 661–4
spread 142, 153
state support 228, 230, 233–4, 247, 679–80, 681, 683
subsistence 142, 585–6
sustainable 247, 276, 654, 662, 668, 671–2, 699
systems 145, 146–7, 658
threats 665
traditional 693
urban 721–2
agroforestry 592, 594
AIDS 269

Aikin, John 295–6
air power 337, 341
air–sea interaction 489, 492, 493–4, 497
air–surface interaction 501
albedo 61
Albertus Magnus 276
Aleut languages 105
algae 40, 70, 77, 517
    blooms 437, 439, 542
Algeria 270
Allen's Rule 101, 102
Althusser, Louis 864, 866, 881
altiplano 33
altithermal 58
Amazonia 501–2, 608, 656
Americas 105, 110
    see also Central America; Latin America; North America; South America; United States of America
Amerindian languages 105, 109, 110, 115
Amerindians 140, 185–6
ammonites 77, 80
amphibians 75, 76
Amsterdam 190, 191, 215
Anatolia 112
Andean Pact 384
Andes Mountains 33, 142
andesite 27
angiosperms 77
Anglican Church 124, 135
Angola 344, 360, 363, 364
animals 67, 71
    as agents of plants 77
    breeding 143, 144
    castration 144
    in caves 75
    depleted stocks 176
    diseases 239
    domestication 14, 78, 139, 142–3, 144, 187, 275, 280, 314
    dung 462–3, 699
    exploited 138
    exterminated 274
    extinction 145, 606
    fights organized 149
    as food 13, 125, 126–7, 138, 140, 143, 677
    forest 153
    fur-bearing 141
    hunted 140
    husbandry 219
    introduced 169, 186, 187

land 74
limnic 75
managed 140
marine 71, 73, 517
migration 73, 75, 138, 145, 152, 179
native 186, 280, 285
naturalized 148
predatory 176
protection 127, 275, 285, 287
rain forest 153, 581, 590, 591
rights 287, 459
sacrificed 125
savannah 153
scavengers 151
treatment 127
see also the names of various groups
animism 120
Annales 319
Antarctica 35
Antigua 173
APEF *see* Asia Pacific Economic
Forum
apes 77, 86–7
African 86, 88, 89, 91
bipedal 11, 77, 90
fossil 87
and humans 87, 89–90
Appalachian Mountains 34
apprentices 173
aquaculture 514, 610, 611
aquifers 534, 537, 665
Arabia 112
Arabic (language) 108, 112, 114, 115, 116
Aral Sea 53, 545, 548
Araucaria 29, 77, 593
Archaebacteria 69
Argentina
cities 400
debt 413, 416, 419, 422
immigration 260, 262
Pampas 187
population 265
settlement 192
urbanization 392
war 358
argon 37
arid zone 640, 641
Armenians 116
arms race 212, 340
arms trade 344, 359–60
art 78, 210
and religion 121

Stone Age 77
arthropods 75, 76
artisans 221
guilds 216, 220
rural 220
urban 214, 215
Asia 104, 191
central steppes 111
colonization 166, 167, 173–4, 705
debts 416
development 343
east 347, 373, 374
farm land 659
farming 146, 240, 668, 690
fertility rates 401, 402
and food aid 245–6
forests cleared 200
High 61
languages 110, 111
migration 260, 396
population 173, 250, 253, 267, 268,
270, 271
settlements 142, 167
south 391
states 357, 358
strong societies 174
trade 163, 166, 167, 194, 213, 214
undernutrition 654, 667, 686, 688
urban growth 391
wars 355, 359
see also South-East Asia
Asia Minor 111, 112, 113
Asia-Pacific 335
Asia-Pacific Economic Forum 347, 384
asteroids 80
asthenosphere 21
astrology 947
Atlantic Ocean 23, 24, 25, 30, 48–9, 337
crossing 185
fisheries 188
islands 185
atmosphere 12, 15, 21
change 486, 487, 488
composition 37–8, 41
modelling 492
origin 36–7
of planets 37, 41
radiation 60–1
atomic power *see* energy: nuclear
Audubon, J. J. 285
Australasia 12, 104
immigration 262, 263

population 259
Australia 35, 74
  Aboriginals 109, 110, 129, 130, 140–1,
    893
  animals 74
  Bitter Springs Chert 40–1
  child care 895
  colonization 178, 188, 193, 196
  drought 285, 630, 641
  erosion 32
  fossil hominids 104–5
  immigration 105, 269, 281
  irrigation 545, 546
  landscape 34
  mountains 33
  Murray–Darling Basin 539, 542, 548
  population 289, 967–8
  wildfires 644
Australopithecines 90–5, 96, 97, 98
Australopithecus afarensis 92–3, 99
Australopithecus africanus 93–4, 99
Australopithecus ramidus 90–1, 94
Austrian Empire 312
avalanches 637, 638
Avesta 121
Avignon 152
Azara, Felix 187

baboon 86
Bach, J. S. 153
Bacon, Sir Francis 278
bacteria 40, 68, 69, 71, 76
  limnic 75
  marine 72, 73, 508
  parasitic 70
  subterranean 75
  symbiotic 71
Baha'i religion 134
Baikal, Lake 75
Baker, James 420, 421
Baker Plan 409, 418, 420–1
ballistic missiles 430
Baltic region 188, 191
banana 188, 236, 238
Bandung conference 343
Bangkok 271, 716
Bangladesh
  cyclones 629, 642
  debt 415
  famine 683
  floods 639, 642
banking 191, 198

crises 413, 418, 419, 424, 425
  global 409, 417, 420
  involuntary loans 418
  private lending 420, 421, 424
  profitability 425
  stability 420, 425
Bantu languages 109, 110–11
Barbados 173, 189, 436
barley 142, 143
Bartram, William 285
basalt 17, 18, 20, 25, 27, 30, 36
Basho Matsuo 153
Basque language 109, 111, 116
Batavia 167
beaches 563
beans 142
beaver 141, 168, 176, 178
beech 51, 57, 152
Beecher, Henry Ward 287
Beijing 264, 708
belemnites 80
Belgium 392
Benares 130
Benioff Zone 27
Bergh, Henry 287
Bergmann, Gustav 826
Bergson, Henri 3
Berlin 339, 345
Bertalanffy, Ludwig von 5
Bevin, Ernest 340
Bhagavad Gita 123
Bhopal 715
Bible 121, 123, 124–5, 127, 926
  and attitudes to nature 277, 278, 589
  as model for colonization 164
  quoted 122, 123, 125, 130, 132, 944,
    947
  translations 107, 115
bioclimate 500–1
biodiversity 473, 605, 606
  genetic 606
  loss 463, 669
  marine 517
  protection 452, 469
biogeochemical cycles 434
biogeography 29, 484, 803, 899
biomes 182, 192, 499, 503
biosphere 6, 845
  duration 83
  evolution 78
  fluctuating 78
  investigation 434

limnic 74, 75
modelling 441
realms 20, 67, 72
subterranean 75
terrestrial 74
see also oceans
biotechnology 228, 233, 605, 670–1, 693
biotic communities 71, 72, 83
crises *see* extinction
degradation 81
diversity 72, 73, 78
marine 74
migration 81, 82
modelling 440–1
restricted 81
richness 72
terrestrial 71
tropical 72
birch 74
birds 77
affected by pesticides 698
extinction 81
flightless 145
ground-nesting 145
models 952–4
scavengers 151
Birmingham 301, 302, 306, 307
birth control 255, 259, 260
bivalves 75, 77
Black Sea 73
blizzards 643
Bloch, Marc 319
bogs 51
bolide 80
Bolivia
debt 421, 426
inflation 420
Bonneville, Lake 53
Book of Common Prayer 124
Booth, Charles 304
Bopp, Franz 107
Bosnia 363, 364
Bosumtwi, Lake 55, 56
bourgeoisie 212, 213, 214
brachiopods 76, 77
Brady, Nicholas 421, 422
Brady Plan 409, 418, 421–3
Brandenburg *see* Prussia
Brandt Commission 344, 363
Braudel, Fernand 319, 321, 409, 838
Brazil 105, 173, 178–9, 189–90
debt 413, 416, 417, 418, 419

deforestation 436
farming 243, 656
fertility rates 401
food supply 667
forests cleared 200, 656, 667
immigration 262
inflation 420
loans 412
marginal land 558
population 265, 270
settlement 192
sugar production 190, 195, 394
Bretton Woods system 371, 373, 375,
387, 409, 410, 422
Bristol 295, 748
Britain
arms sales 360
child care 895
cities 263, 303, 308, 734, 735, 743, 748
Clean Air Act 644
climate 45, 50
development 190
drought 642
emigration 261, 262, 657
environmental groups 459–60
farming 698, 699
flood defences 639, 640
foreign investment 377
forests 276
and geopolitics 337
government 308
gross national product 221
higher education 794, 796–7
historic buildings 283
in ice ages 48, 50
immigration 263
industrialization 221, 294, 730
latitude 45
local government 302
manufacturing 734
medieval 294
North–South divide 734
philanthropy 303, 304
population 250, 259, 281, 294, 657
public health 284, 300–1
and Romans 151
universities 795–8, 811
urbanization 730
wars 355, 356, 359
water use 529
see also England
British Columbia

civil disobedience 451, 453
Commission on Resources and the
    Environment 450, 453, 454
forestry 449, 450–1
indigenous people 449, 450, 452, 454
logging 450–1, 452, 453–4
British East India Company 191, 376
British Empire 163, 170, 182, 334, 335
Bruhnes, Jean 3
Bruntland Commission 522, 593, 615
bryozoans 76, 77
Buddhism 121, 126–7, 282, 945, 947
    festivals 129
    pagodas 130
    repressed 135
    spread 132–3, 134
Buenos Aires 271
buffalo 140
Buffon, G. L. Leclerc, Comte de 279–80,
    281
building societies 305–6
built environment 554
    colonial 705
    industrial 711–12
    materials 145, 148
    preservation 283
    urban 737–8
Bunge, William 826–7
bureaucracy 318, 322
Burgess Shale 87, 89
Burma 138
Bush, George 345, 928–9
Bushmen 139–40, 175

caatinga 578
Cainozoic see Cenozoic
calcite 15
calcium carbonate 15, 79
calendar 127, 128, 129
California 288, 558, 638
Cambodia 354, 363, 364
Cambrian period 76
Canada
    cities 736, 743
    debt 427
    emigration 260
    and food aid 246
    French 166, 168, 171–2
    immigration 262
    pollution 537
    settlement 193, 196
    warming 489, 500, 501

water supply 533
wheat production 606
see also British Columbia
Canada–United States Free Trade
    Agreement 384
Cancun meeting 344
Canterbury 131
Cape of Good Hope 166, 174
capital 208
    accumulation 207–8, 409, 869
    for colonies 166
    commoditized 207
    cycle 409
    exchange controls 373
    fixed 206, 207
    flight 415
    flows 208, 214, 425
    intellectual 468, 469
    international movement 373, 386, 710
    man-made 468, 469
    natural 468–9
    variable 206
capitalism 208, 223, 321, 409, 808, 870
    agrarian 218, 219, 221
    development 190, 207, 211, 212, 222,
        281, 871
    divisions 870–1
    evils 283
    'golden age' 410
    industrial 221, 221–2, 376
    mercantile 170, 214–15, 216, 219, 221,
        295, 381
    spreading 704–5, 787, 788
carbon 15, 20, 37, 78, 79, 456
    atmospheric 498
    cycle 497
    and life 41, 67, 68
    in oceans 497–8
    reserves 41
carbon dioxide 15, 37, 40, 41, 78
    as greenhouse gas 63–4, 82, 488, 498
    in ocean 498, 520
    and rain forests 587, 588
    variation 487, 488
    and warmth 45
carbon monoxide 37
carbonate 40, 77, 78
Carboniferous 76
Caribbean
    colonization 163, 173, 188–9, 191
    debt 416, 421
    trade 213

urbanization 393, 396
Carlyle, Thomas 299
Carson, Rachel 442, 677
Cartier, Jacques 171
cartography 807, 928, 929, 930, 970
Caspian Sea 53
caste system 115, 132, 134
cat, feral 151
Catlin, George 285
cattle 125, 126, 142
   breeding 144
   colonial 171
   sacrificed 125, 126, 143
   transfer 145, 187
   varieties 606
caves 75
Cenozoic 19, 77
   climatic decline 44–5, 47, 61
CENTO *see* Central Treaty Organization
Central America 142, 147
   and colonization 163
   wars 355, 358
Central Treaty Organization 339
cephalopods 76
cereals 142
   exported 242–3
   as food aid 244, 245, 246
   imported 242, 243, 688
   prices 682, 699
   stocks 681
   yields 239, 240, 661
Ceuta (North Africa) 165
CFCs *see* chlorofluorocarbons
Chad, Lake 53, 56
Chadwick, Edwin 297, 298, 300, 301
Chamberlain, Joseph 302
chaos theory 441, 444,
   and geomorphic systems 554, 561–4
chaparral 74
charcoal 148, 281, 584
charities 295, 296, 303
Chartists 300
Chelford interstadial 50
Chicago 732, 743, 744
Child, Josiah 295
Childe, Gordon 111
children 259
Chile 392
chimpanzee 86, 90, 91, 94, 137
China 50, 184, 195, 283, 311, 335, 969
   arms sales 360
   cities 399, 403, 943, 946

civil war 354
crops 142, 146, 195
customs 945
emigration 260
empire 946
famine 683
farming 146, 153, 195, 196
festivals 129
food supply 669
foreign investment 401, 425
forests 282
fossil hominids 101
importance 367
and India 339
languages 110, 943
land loss 656
migration to cities 393, 402
Opium Wars 197
political organization 321, 324
population 251, 265, 267, 269, 270
religions 122, 123, 133, 153, 943
resists colonization 163, 312
rural development 398
Three Gorges dam 533
town lighting 137–8
undernutrition 686, 688
urbanization 393
and USSR 339
water supply 533
world-view 943
chlorine 38
chlorofluorocarbons 63, 470, 488, 521, 645
chocolate 192
cholera 300, 301
Chomsky, Noam 109
Chorley, Richard 4, 823, 827, 828
Choukoutien *see* Zhoukoudian
Christaller, Walter 821, 822, 823, 827, 829
Christianity 127, 944, 947
   calendar 128–9
   churches 130
   and colonization 165, 166
   Crusades 134
   defended 212
   ethnocentrism 164, 281
   holy places 130
   missions 107, 130, 134, 164, 165
   and nature 164
   rituals 132
   spread 115, 133, 134

and women 135
chromosomes 70
Churchill, Winston 339
cities 78, 145, 148, 941
   capital 214, 215, 264, 706, 707,
      710–11, 943
   colonial 704–6
   commercial 221, 223, 705
   deconcentration 264–5, 403, 404, 730
   in developing countries 271
   dominant 253, 264, 271
   efficiency 399
   employment 734, 735, 740
   European 215, 397, 730, 734, 735,
      736, 747–8
   funding 707, 732, 734
   garden 307
   global 208, 378
   growth 161, 214, 215, 253, 263, 264,
      271, 275, 400, 401, 402, 403, 703,
      710, 730, 757
   immigration 736
   independent 214
   industrial 283, 293–4, 296, 302, 303,
      306, 307, 308, 730–1, 732, 734, 735
   infrastructure 308, 400, 401, 403, 404,
      707–8, 716, 732, 747, 748
   inner 733, 735, 737, 738, 742, 748
   living conditions 395–6, 401–2, 403,
      732–3
   location 715
   management 704, 706, 709, 715,
      731–2, 748
   marginal areas 715, 757
   Mayan 186, 942–3
   and migration 395–7, 706, 736
   overcrowding 705, 733
   peripheral estates 738–9
   as places of exchange 153, 214
   polarization reversal 403
   polynuclear 403
   as ports 215, 252, 253, 263, 517
   pre-colonial 705
   pre-industrial 151–2
   primate 330, 399–400
   redevelopment 707–8, 709, 736
   riots 724, 744, 746–7
   services 716–17
   size 151, 215, 264, 271, 283, 399–400
   social polarities 747, 748
   streets 306
   suburbs 732, 733, 736, 757

   tensions 705, 706, 745–7
   Third World 397, 399, 400–1, 702–24
   vulnerability 715–16
   walls 152
   Western 730–49
city-states 323
civil disobedience 453, 459
civil rights 467
civilization 316, 944, 945
Clarke, Arthur C. 433
Clayoquot Sound, British Columbia
   449–54
Clements, F. E. 3–4
clergy 211, 212
climate 41
   arid 142–3
   change 44, 46, 56, 58, 60–1, 62, 64,
      74–5, 79, 152, 275, 276, 473, 480,
      483–4, 486, 498, 502–3, 665, 958
   classification 485
   concept 485, 498
   cooling 44–5, 47, 60
   and ecosystems 484, 485, 486,
      499–502, 503
   fluctuation 58, 61
   global 485–6, 502–3
   greenhouse 500, 503
   and human activity 63, 64, 502–3
   and human anatomy 101
   hypsithermal 138
   instability 35, 48
   modelling 492–5, 498, 970
   modified 61
   regions 485
   systems 485, 486
   variation 45, 63, 482–3
   and vegetation 2, 3, 56, 485
   warming 56, 58, 60
   zones 2
climatic normals 482–3
climatic optimum 58
climatology 803
clones 69
cloth 214, 221
clouds 488, 493, 495
clover 187
coal 41, 137, 159–60, 252, 602
   imported 189
   and industry 160, 281
   mining 148, 221
   monopoly 221
   ownership 221

and pollution 151, 645
  trade 137
cocoa 236, 237, 238, 242, 244
coconut 236
cod 188
coffee 192, 200, 236, 237, 238
  exports 607
  imported 242, 244
  interplanting 592
Cohen, Saul 341, 347
Cold War 329, 339, 340, 341, 342
  aftermath 345, 346, 347, 363, 366
  opposition 343, 348
  revival 344, 345
  and UN 363
Colombia 395, 415
colonialism *see* colonization
colonists 163
  European 163–4, 178, 189, 193, 250
  first generation 177–8
  on frontiers 177, 178
  and labour 170
  mortality 197
  survival 178
colonization 176, 213, 332
  and Christianity 165–6
  distribution 335
  economics 169, 214
  effects 163, 164, 252, 281, 312
  'first effective settlement' 169–70
  future 595
  industrial 705
  and land systems 170
  legacy 705–6, 709
  legendary 164
  organization 166, 169
  resisted 163, 167, 312
  tropical 197, 278
  and urbanization 705
Columbus, Christopher 172, 183, 186,
  187, 203
comets 80, 81
commensalism 71
commerce 213, 215, 337, 945
commodification 207
commodities 206
  flow 208
  markets 214, 216, 217, 218
  prices 375, 415
Common Agricultural Policy *see*
  European Union: CAP
communication 153, 310, 315, 317

of geographical work 929–30, 934–5
  and imperialism 198
  improvement 313, 314, 329, 375–6
  satellite 432
  speed 198
communism 283, 344, 345, 367
communities
  co-operative 290
  disrupted 738, 742
  participation 717
  peasant 210, 211
  revitalized 916–17
  rural 752, 761, 762, 763
  self-help 763
  settled 13
  small 161, 762
  see also biotic communities;
    settlement
commuting 265, 731, 753, 767
  long-distance 758–9
  reluctant 759
  telecommuting 760
companies
  joint-stock 166, 215, 216
  multinational 330, 707, 845
  see also corporations
compass 212
computers 314, 968–9
  development 435–6, 445
  digital 430
  electronic 435–6
  in image analysis 433
  in mapping 969–70
  in modelling 438, 444–5
  personal 435–6
  precursors 435
Concert of Europe 361
conflict 353
  change 353, 354
  colonial 363
  endemic 359, 367
  ethnic 706
  global 367
  models 356
  pattern 354
  post-independence 706
  recurring 353, 358, 367
  territorial 358
  in Third World 354, 359, 706
Confucianism 127, 133, 943
Congregationalist Church 170–1
conservation 589

development 288, 291
  of historic buildings 283
  of nature 285, 286, 287–8
  of resources 288
  'utilitarian' 277
  of wilderness 287–8, 290
Constantinople 134
consumer organisms 71, 72, 73
consumer pressure 458
consumption 207, 208, 214–15
containment 340
continental drift 22–4, 29–30, 35, 61, 958
continental rise 511
continental shelf 30, 509, 511–13, 522
  resources 509, 511
continents 21, 30
  age 30
  form 45, 74
  isolated 183
  linked by sea routes 183–4
  margins 27, 30, 73
  and oceans 30, 31
conurbations 264, 730
convection currents 27
co-operative societies 300
Copenhagen 215
coral islands 519
coral reefs 73, 76, 82, 83, 517
corn see maize
corporations 381
  transnational 372, 373, 376–81, 385,
    387, 618
cosmology 846
cosmopolitanism 939, 942–3, 944, 946,
  947, 949–50
cotton 160, 192, 195, 237, 244, 607
counter-urbanization 264, 265, 272, 391,
  398, 731, 735–6, 757–8, 760, 763
countryside
  diversification 766–7
  explored 758
  future 767
  green belts 759
  as refuge 760
  repopulated 758, 762
  services 760, 762, 768
  and towns 752, 759, 764, 766
cows 125–6, 231, 233
craftsmen 219, 220
creation myths 846
credit 191, 198
  control 411–12

and debt 410
  official 410
  private 408, 410–11
  recycling 411
  short-term 416
  supply 374, 387
  unions 305
Creoles (languages) 107, 117
Cretaceous 35, 77, 483
Crèvecoeur, J. H. St Jean de 280
crime 733, 744, 745
Croatia 364
crops 146–7, 653
  areas used 192, 193, 194, 199, 238,
    555, 607, 653
  cash 218
  commercial 169, 169, 188, 195
  dependency 244
  distribution 152
  exotic 176, 188
  export 170, 188, 236, 238, 247, 672
  fodder 148, 218, 243, 655, 659, 669,
    670
  frequency 660–1
  higher-yielding 200, 239–40, 658, 659,
    661, 678
  improvement 670
  industrial 160, 232, 244
  intensive 669
  introduced 169, 171, 187, 188, 661
  local 169, 186
  Old World 188
  protection 209
  rotation 699
  specialization 253
  staple 170, 247
  Third World 236, 237, 653
  tropical 693
  underplanting 592
  varieties 233
  yields 655, 657, 658–9, 660, 668–9
Crusades 214
cryosphere 434
Cuba 343
  exports 244
  population 265, 267
  urbanization 392, 402
culture 3, 12, 939
  conflict 159
  diversity 269, 315, 847
  electronic 917
  elite 324

and environment 153, 275, 281–2, 316–17
expansion 159
exported 170
folk 324
global 917, 918
and language 116
as man 892
and nature 844–6, 892
oral transmission 323, 324
primitive 320
and society 310, 311, 315, 316, 323
unitary 323
urban 275
customs union 382–3, 384
cyanobacteria 40
cycads 77, 141
cyclones 629, 642
cymatogeny 33
Cyperaceae 56
Cyprus 116

dams 290, 532–3, 556
Danube basin 548
Daoism 122, 123, 124, 125, 126, 133
Dart, Raymond 93
Darwin, Charles 69, 120, 286–7
Dawkins, Richard 444
DDT 430
    banned 442
    disadvantages 431, 442
    in food web 438
Dead Sea 53
debt
    adjustment 419, 420, 421, 708, 714, 724
    buy back 421
    crises 408–9, 410, 411, 412–13, 415, 416–17, 419, 425, 708
    default 413, 415, 416
    in developing countries 241, 330, 409, 410, 414–15, 422–3
    First World 427
    and GDP 410, 427
    and GNP 423–4
    and growth 420–1
    policing 409, 418, 425–6
    reduction 422, 423
    relief 415, 421
    rescheduling 410, 417, 418–19, 420
    servicing 410, 418, 422, 423
    Third World 421, 422

writedown 418, 421, 422, 423, 424, 467
decolonization 342, 343, 344, 373
decomposer organisms 68, 71, 72
defence alignments 329
Defoe, Daniel 296
deforestation 81, 286, 555
    and climate change 279, 588
    global 176, 608
    and soil erosion 275, 588
de-industrialization 733–4
Delgano, George 118
democracy 466–7
    and peace-keeping 361
    spread 353–4, 366, 404
    in Third World 353, 404, 723, 724
demography see population
Demolins, Edmond 112
Denmark 392
depopulation, rural 752–3, 757, 763, 768
deprivation 740
    rural 762–3, 766
    urban 732, 733, 738, 741, 742
Descartes, René 118
desertification 565–8, 641, 830
deserts 48, 74
    expansion 74
    as female 893
    margins 52, 693
    sand 48
determinism 959
    environmental 801, 838, 860
Detroit 738, 747
development 325, 343, 363, 773, 932
    appropriate 240
    assistance 362, 410
    and the environment 716
    indigenous 398
    level 778, 779, 780
    'lost decade' 426
    post-colonial 410
    sacrificed 420, 425
    strategy 713, 714
    sustainable 420, 457, 460, 463, 464, 465–70, 472, 593, 594–5, 615–16, 703–4, 810
    uneven 810
    and urbanization 393, 397, 398, 703, 706–7
Devonian 76
Dickens, Charles 283

diffusion models 823
Dimlington stadial 51
dinosaurs 35, 77, 80
disasters 620
  see also environmental hazards
disease 185
  in cities 151, 283, 284, 300, 733
  control 284
  of deficiency 686
  epidemics 185–6, 256, 258, 284
  of excess 686
  in European colonies 163, 166, 168,
    173, 174, 185–6, 197
  feared 285
  and mortality 269
  occupational 284
  spread 968
  in Third World 359
  water-borne 284
Disraeli, Benjamin 293, 308
dog 139, 143, 144, 151
Dominican Republic *see* Hispaniola
domino theory 367
dormouse 143
Dravidian languages 109, 111
drought 44, 630, 640–2
drugs: rain-forest sources 586–7
Dryer, C. D. 3
Dubois, Eugene 98
dunes 30, 65
  desert 45, 52–3
  destabilized 560
  fields 55
  fossil 55
dust 62, 80
dust storms 60, 65
Dutch East India Company 165, 166–7,
  171, 191, 376
Dutch West India Company 191
dwelling 844, 908–9

Earth 3, 22, 67
  area 555
  axis 48, 79
  bombarded 35, 80, 81
  as closed system 611
  condition 278
  core 20, 21
  crust 15, 16, 17, 20–1, 25
  evolution 17, 42
  expanding 25, 29
  as female 892

  as habitat 1, 6, 7, 179, 330, 456, 832,
    939
  heat 21–2, 35, 83
  and Heaven 123, 124
  homeostasis 35, 41
  humanized 153
  as integrated system 5, 15, 444
  as magnet 25
  management 462
  mantle 20, 22
  materials 11, 15–17
  as mother 123, 124, 455, 892
  movement 32
  observation 430
  orbit 61–2, 79
  origin 35, 40, 121–2
  productivity 208
  stewardship 275, 278, 443
  structure 20–2
  surface 3, 6, 20, 60–1
  temperature 49
  tenants 278
  topography 30
earthquakes 24, 557–8, 634–5
  distribution 25, 27
  insurance 616
  prediction 634–5
East Africa 56, 92, 105, 117
East India Company *see* British East
  India Company; Dutch East India
  Company
East Indies 195
Easter Island 595
EC *see* European Community
echinoderms 76
ecofeminism 892, 897
ecogeography 832
ecolabelling 458–9
ecology 3, 286, 443, 803
  crisis 278, 595
  and geography 832, 853
  global 433, 485
  human 208, 931
  imbalance 325
  mathematical 442
  modelling 438–41, 442
  predator–prey systems 438
  pressures 281, 334, 461
  principles 277, 286, 288, 290
  problems 960–1
  protected sites 754
  rain forest 581

sustainable 289
tolerance 41, 461
trophic-dynamic 484–5
economics 806
ecological 465–6
and geography 805, 822, 830, 831
global 480
international 313, 344, 370–1, 386–7
Keynesian 312
liberal 312–13
neo-classical 774, 784
Soviet 313
sustainable 452, 468–70
and war 356–7, 359, 370
economy 4
alliances 381–2
capitalist 209, 211, 213–14
centrally planned 235, 312, 313, 325, 330
colonial 169
competition 323, 338, 367
development 252, 312
disparity 344
European 182, 330
global 169, 208, 310, 311, 347, 370, 371, 387, 480, 847, 948
growth 252, 349, 375, 427, 710
integration 347, 371–3, 376, 382, 386, 387
nationalized 313
peasant 210–11
post-war 370, 371
'real' 373, 374, 387, 409
realignment 330, 373–4
regional blocs 347, 373, 382–6
regulation 381
restructuring 414, 418–19, 425–6
stagnation 345–6
sustainable 615
'symbol' 373
and technology 375–6
unions 383, 386
urban 214–15, 219
world 321–3, 324, 345, 346, 347, 373, 376, 387, 388
ecosphere 6, 11
human impact 12, 13, 140
investigation 433, 434
threatened 160, 331
ecosystems 137, 348, 484
and agriculture 228, 606
carrying capacity 617

and climate 486, 499–502
degradation 617
exploited 196, 199
fire-adapted 140–1
human impact 145, 152–3
interrelationship 480
lost 606
managed 192, 198, 199
measurable properties 434
modelling 438, 440
natural 138, 192, 485
northern 500
productivity 485
transition 606
education 693
EEA see European Economic Area
Eemian interglacial 51
EFTA see European Free Trade Association
Egypt 270, 312, 318
El Niño 489, 519, 527
El Niño–Southern Oscillation 489, 495, 497
El Salvador 364
electricity 138
elephant 176, 177
elm 51, 57
Emerson, Ralph Waldo 285, 847
empires 322, 335, 945–6
divided 358, 373, 946
empiricism 952, 959
employment
female 737
inflated 712
informal 712–13, 714
losses 734
low-paid 740
manufacturing 711, 734
of the poor 295
rural 760–1
in services 712, 734–5, 737, 760
Third World 711
urban 330, 711, 734, 735, 737, 740
see also labour
energy 21
concentrated 310, 315
consumption 208, 599–600, 967
control 160
distribution 250
and economies 208
for life 68, 72, 83, 138–9
mechanical 183

mineral 208–9
mobilization 310
nuclear 329, 430, 431, 442, 603, 645
renewable 515, 603
as resource 602–3
solar 79, 137, 138, 159, 602
sources 206, 252, 603
steam 252
surplus 139, 153, 206
thermal 515
tidal 515
use 603
from water 149, 159
from waves 515, 603
from wind 149, 159, 603
Engels, Friedrich 259, 293, 308
England 50
   cities 264, 308
   farming 218, 698
   gardens 150
   government 216
   industrialization 221–2
   local government 303
   lowlands 201
   moors 141
   Norfolk Broads 148
   northern 137
   overseas settlements 169
   poor relief 219, 220, 295–6, 297–9
   rainfall 484
   revenue 213, 221
   Royal Forests 150
   trade 213
   urbanization 293–4
   see also Britain
English language 113, 114, 115, 118
enlightenment 944, 945, 947
ENSO see El Niño–Southern Oscillation
entropy 41
environment 275, 801, 860
   appreciation 290
   artificial 283, 315
   audits 460, 461–2
   carrying capacity 12
   change 11, 56, 78, 275, 282, 291, 483,
     954
   cleanup 468
   and colonization 164, 176, 179, 190,
     253
   critical load 461, 971
   degradation 253, 282, 286, 655
   deterioration 274, 278, 288, 455, 468

   and fire 140–1
   future 954, 959–60
   and human activities 56, 60, 81,
     140–1, 148, 151, 152, 239, 274–6,
     278, 281, 285–6, 443, 454–5,
     479–80, 519, 521, 845, 939, 972
   improvement 479
   limitations 289
   management 228, 275, 277, 290, 440,
     453–4, 458, 465–6, 768, 810
   marine 519–21, 523
   modelling 430, 431, 438–42, 959–60,
     961
   monitoring 431, 461–2
   past see palaeoenvironment
   and politics 334, 458, 463
   and population 255, 272, 601
   preservation 349, 479, 971
   pressure 967
   problems 463
   protection 455, 467, 479, 754, 845
   regulation 466, 467
   research 430, 456, 962–3
   as resource 605
   rural 766, 768
   standardized 325
   strategies 443, 465–6
   systems 438, 804
   theories 444
   Third World 714–16
   threatened 804
   tropical 693
   urban 714–16
   utilization 253, 278, 281, 479–80
   and war 289
Environmental Data Services 460
environmental hazards 620–1, 624–5,
   715–16
   adjustment 629, 630–2
   aftermath 624, 629
   causality 624, 626
   classification 621–2
   death tolls 621, 623
   distribution 625, 633
   duration 624
   frequency 623, 624, 625
   human-induced 621, 645–6, 714, 715
   loss absorption 630, 632
   magnitude 623, 624
   marginalization 625, 626
   mitigation 634, 638, 717
   monitoring 624, 625

natural 621
   prediction 624, 625, 629, 634
   prevention 629–30, 632
   relief 624, 625, 632
   risk analysis 624, 626–9, 638
   Third World 715
environmentalism 274, 329, 431, 454
   development 279, 288, 289–90, 291,
      442–3, 455, 473–4
   and education 460
   feminist 892–3, 896–7
   flexibility 459
   international 465–6, 481
   issues 449, 452–3, 458–9, 481
   and politics 453, 458
   and posterity 277
   precursors 277–9
   pressure groups 459–60
   themes 452–3, 455, 458–9, 617
   varieties 443
epiphytes 573–4, 579, 581
epistemology 809, 810, 841
Ericaceae 51
erosion 30–2, 34, 38, 81, 82
   see also soil: erosion
escarpments 34
ethics 616–17
Ethiopia
   civil war 354
   famine 683
   fossil hominids 91, 92, 101, 102, 104
   fuel 462–3
ethnicity 897, 939, 942, 946, 948, 949,
   950
ethnocentrism 809
ethnography 315
Eubacteria 68, 69
eucaryotes 40, 70, 76
Eur-Africa 335
Eurasia 74, 334, 340, 347
Eurocentrism 177
Europe
   Central 50
   cities 264, 397
   colonies 162–3, 164, 173, 187–8, 213,
      215, 252, 279
   conservation 291
   core 183, 184, 190
   'Crop-Power Blocs' 338
   development 184, 190–1, 194
   divided 345
   dominance 335

Eastern 217, 257, 334, 339, 346, 347,
   358
   economy 347, 373
   eighteenth-century 282–3
   emigration 162, 198, 260–3, 285
   empires 160, 162, 163–5, 166, 169,
      170, 179, 182, 185, 196, 330, 342
   expansion overseas 187–8, 190, 196,
      252, 253, 260–1, 278, 281, 311,
      312, 332, 335
   famine 683, 688
   farming 146, 153, 194–5, 218, 219,
      680, 690, 698, 699, 765
   food production 209, 690
   forests cleared 276
   fossil hominids 101, 102
   government 216
   gross national product 221
   immigration 263, 269
   malnutrition 683
   medieval 212–13, 276
   Mediterranean 12, 256
   migration within Europe 263
   multi-racial 263
   nationalism 366
   nineteenth-century 283
   north-western 161, 281
   partition 115
   periphery 184, 213, 217
   population 250, 253, 256, 257, 258,
      259, 264, 271, 280–1, 678
   rural areas 764–5
   security zones 339
   southern 217, 257, 263
   states 357, 358
   storms 642
   trade 213, 215, 385
   undernutrition 688–9
   vegetation 51, 56–7, 276–7
   wars 355, 359
   Western 183, 184, 213, 256, 311, 329,
      342, 347
European Community 347, 361, 367, 384,
   386
   and agriculture 754, 755, 756, 765
   Common Agricultural Policy 681–2,
      753–4, 765
   food surpluses 679, 681–2, 753
   limits on pesticides 698
European Economic Area 384
European Free Trade Association 384
European Union 330, 347, 373, 384

Common Agricultural Policy 228, 234
and food aid 246
food surpluses 681–2
redistribution 386
single market 385
surplus products 246
trade 385
eutrophication 461
Evelyn, John 276
evolution 11, 69, 286, 314
of life 20
in tropics 72
existentialism 808, 862, 863
exports
agricultural 243–4
European 213
manufactured 373, 401
Third World 246
extinction 76–7, 80, 81

Fabian Society 303
factories 220, 222, 283, 284
factory farming 230, 677
Falkland Islands 358, 416
family 255, 159
planning see birth control
famine 13, 247, 281, 641, 683
causes 683
and drought 640
predicted 657, 683
relief 688
in Third World 359, 677
and war 355
see also malnutrition; undernutrition
farming see agriculture
faulting 31
fauna see animals
Febvre, Lucien 3, 319
feminism 809
and ecology 459
post-modern 889
socialist 889
Fertile Crescent see Near East
fertility 265
decline 252, 257, 258–60, 265, 267–8,
271, 401
fluctuations 250, 269
and income 266
among migrants 736
rates 257, 401
rise 256
Third World 394

fertilizers 145, 152, 200
chemical 231, 698
in developing countries 240, 241
mineral 81–2, 145
response 669
feudalism 212
Fielden, John 298
figs 581
Fiji 663
finance
exchange rates 371, 375, 415
flows 373, 387
global 372
international 371, 373–4, 375, 378,
408, 418
unregulated 373
Finland 363
fire 77, 138
control 139
and environment 140–1, 153, 315
and hunting 139, 140, 141
First World (West) 208, 342
First World War 329, 353, 355, 356
aftermath 312, 334, 361
causes 361
Fisher, Osmond 438
Fisher, R. A. 436
fisheries 143
artisanal 149, 514, 581, 610
Atlantic 188
commercial 514, 610
overfishing 509, 514, 519, 610
regulation 462, 514, 610–11
stocks 502, 513, 514, 519, 610–11
fishes 73, 75, 76
distribution 513–14, 520–1
farmed 140
as food 127, 139, 140, 514
global catch 610
meal 610
ownership 513
tropical 140
floods 519, 520–1, 556, 616, 621, 638–40
adjustment 631, 639, 640
forecasting 639–40
defences 639
losses 636, 638
flora see vegetation
Florence 152
fluorocarbons 82
fog 643, 644
food 188

banks 743
choice 137, 139
consumption 599, 600
distribution 686
'dumping' 246
exotic 194
exports 242–3, 244
and festivals 125
fish 127, 139, 140, 152
imports 243–4, 247, 663
local 663
manufacturing 228, 231, 232
meat 125–6, 127, 139, 140, 143, 152,
  654, 669, 670, 690
prices 234, 663, 681–2, 690
processing 228, 232, 233
quality 677
salted 152
staple 243, 247
subsidized 663
surpluses 186, 246, 247, 654, 655, 663,
  678, 681–2, 753
food aid 244–7
Food and Agriculture Organization 246,
  514, 686
food chain 82, 229, 232, 610
food production 143, 159, 206, 607
in developed world 243
in developing countries 243–4, 246,
  607, 651, 669–70, 689
and energy use 209
global 238, 247, 665, 667
growth 606, 655, 657
integration 698
limits 667
problems 677
self-sufficiency 243, 679
and technology 233, 666–7
transnational control 238
food supply 13, 105, 139, 140, 148, 247
costs 720
in developed world 599, 688–9
global 228, 607, 654
increasing 153, 689
industrialization 720
informal 721
and population growth 249, 256, 281,
  651, 658, 689
retailing 232, 234, 720, 721
rural 760
sustainable 671–2
systems 229–30, 243, 720

Third World 651, 672, 686, 689,
  720–2
urban 152, 232, 234, 300, 717, 720–2
food webs 438, 485, 513
forestry 276, 288, 290, 592–3, 594
forests 76, 179
African 54, 56
as air conditioners 464–5, 608
American 56, 57, 58
area 659
boreal 50, 65, 188, 499, 500, 501
clearance 64, 141, 153, 171–2, 189–90,
  193–4, 195–6, 199, 200, 239, 275,
  276, 279, 281, 282, 585
cloud 579
commercialization 196, 452
composition 152
coniferous 153, 637
conservation 276
deciduous 51, 74, 153
decline 192, 193
expansion 186
harvesting 563, 607–8
Jurassic 77
as land banks 148, 200, 608
limit 579
management 277
margins 141, 591
planting 276
products 148, 188, 200, 276, 607–8
protection 176, 275, 276–7
reclamation 141
replanting 280, 286
reserves 591
Royal 150
temperate 51, 74
Tertiary 44
tropical see rain forest
in war 151
Forster, J. R. 2
fossil fuels 41, 64, 75, 82, 137, 159,
  519–20, 602–3, 645
fossils 20, 40, 67, 73, 75–7, 87, 93
Foucault, Michel 881
France 190
archives 318–19
arms sales 359, 360
child care 895
Department of Historic Monuments
  283
Empire 163, 168, 182, 335
farming 698

foreign investment 377
forestry 276–7, 280, 281
and geopolitics 337
government 216
immigration 263
influence 948–9
nationalization 313
overseas settlements 169, 171–2, 178
population 256–7, 259
travel 967, 968
and UN 366
wars 355, 356, 359
Francis, St 127
Franklin, Benjamin 280
French language 114, 314
freshwater 497, 527
Freud, Sigmund 120
friendly societies 300, 304–5
frontiers 177, 178
frost 643
fuel 138, 145, 148 *see also* fossil fuels; oil; wood
Fukuyama, Francis 353–4, 366, 367
fungi 70, 71, 75
fungicides 231
fur trade 141, 166, 168, 169, 173, 176, 178, 188

Gaia 5–6, 443, 444, 455, 456, 457, 462, 838, 844, 850
Galápagos Islands 148
Galton, Francis 436
Gama, Vasco da 165
Gandhi, Mahatma 126
Ganges basin 548
gardens 145, 148, 150, 283
gases
atmospheric 486, 487, 488, 495, 645
control 160
greenhouse 63–4, 473, 487, 488, 494, 495, 498, 519, 645
natural 138, 515
toxic 645–6, 715
gastropods 77
GATT *see* General Agreement on Tariffs and Trade
GCMs *see* general circulation models
gender 809, 900
and class 895–6, 897
and employment 893–4, 895–6
and ethnicity 897
inequality 781, 898

and post-modernism 897
roles 891, 896
and sense of place 910–11
significance 888, 893
symbolism 892, 893
General Agreement on Tariffs and Trade 228, 375, 388, 460, 467
established 313, 371
Uruguay Round 247, 347, 388, 663
general circulation models 492, 493–4, 498, 502
genetic code 69, 70
genetics 110, 143
and biodiversity 606
engineering 233, 670
genius loci see place: spirit of
Genoa 185
gentrification 736–7, 763
geochemical cycle 20
geo-economics 338, 347
change 370, 372–3, 387, 388
regions 373, 374, 382–5, 386
Geographical Association 796, 798
Geographical Information Systems 430, 433, 434, 807, 813, 972
geography 1, 3, 4–5, 7, 813, 829, 843, 849–50, 927
applied 806, 811, 820, 825–6, 829, 850, 899, 931–3, 950
chorological 818, 819
'closed-space' 332, 334
commercial 795
cultural 3, 316–17, 341, 772, 805, 809, 881, 939
as discourse 838, 847, 848, 880–1
dislocated 917–18, 921
economic 382, 805, 809, 822
education 794, 795–8, 823, 839, 849, 899, 928–9, 934–5, 972
and environment 348, 349, 927
evolutionist 314–15, 317
expansion 794–8, 972
feminist 809, 888–99
fragmentation 794, 799, 803, 812, 813, 843
future concerns 926, 929–31, 935, 965–6, 969–72
global 337–8, 347, 812, 842, 849, 852
historical 112, 805, 838, 848
human 1, 4, 7, 315, 436, 772, 788, 798, 804–7, 809–10, 811, 813, 838, 845, 851, 899, 927, 928, 972

humanistic 793, 808, 809, 837, 838–40, 842, 843, 844, 848, 850–1, 853, 872
integration 803–4, 806, 811, 927–8, 931
landscape 819–20
marine 524
physical 1, 2, 4, 7, 436, 802–4, 807, 811, 832, 838, 899, 927–8, 958, 972
political 316, 332–4, 340, 341, 347, 348–9, 805, 809, 813, 849
post-modern 4, 809, 837
practical work 934
processes 825, 955
quantitative 4, 802, 806, 820, 823, 824–5, 826, 828, 829, 830, 831, 930
radical 772, 808, 809, 833, 837, 865
regional 3, 801–2, 804, 805–6, 809, 818–19, 820, 833–4, 875–8, 970
research 430, 799–800, 811, 820–1, 822, 932–3, 956–8, 962
research publications 799, 800, 803, 805, 806, 823, 933, 935
rural 806
and science 802, 803, 804, 820, 822, 824–5, 826, 827–8, 831, 833, 852, 954–5, 971
social 809, 822, 824, 829, 842, 860–1
societies 798–9, 803, 805, 807, 849
spatial models 439–40, 772, 823, 824, 825, 827, 828, 829, 832, 860
specialization 799, 800–1, 805, 812, 813, 820, 834, 852, 930–1
structuralist 772, 808, 831, 833, 842, 861, 865–6, 872, 881–2, 883
systems 4–5, 439, 440, 444–5
university departments 795–8, 799, 800, 801, 810, 820, 824, 825, 826, 849, 972
urban 806, 865–6
geological cycle 17–18, 78, 81
geological time 11, 19, 68
geology 17
geomagnetic polarity 25, 26
geomorphic systems see landforms
geomorphology 802–3, 933–4
and prediction 560, 563–4
geopacifics 338
geophysiology 433
geopolitics 332, 336, 338, 367
and environment 349
German school 335–7, 340

instability 969
in practice 340, 341–2, 348
regions 341
transitions 344–5, 346, 348, 354
Georgia 364
Georgian language 109, 111
geostrategy 341, 347
model 332, 333, 338
geostructuralism 881–2, 883
Germany 316
arms sales 360
colonies 312, 335
culture 949
currency 347
economy 312, 346
emigration 261, 262
foreign investment 377
geopolitics 332, 335, 336–7
immigration 263
and Japan 336
lack of space 335, 336
Nazi regime 336
and Soviet Union 336
Third Reich 336
and UN 365
urban areas 392
in World Wars 336, 355
GEWEX see Global Energy and Water Cycle Experiment
Ghana 55, 56
urban areas 392
gibbon 86
Giddens, Anthony 873–5
glacials 48, 50, 56, 61
glaciation 35, 49, 58, 78
glaciers 48, 51, 58, 520
glacio-isostasy 50
glass-making 148
Global Atmosphere Research Programme 643
Global Energy and Water Cycle Experiment 499
Global Environment Facility 472, 473
Global Positioning System 433, 434
global warming 44, 63, 82, 488–9, 491–2, 495–6, 503, 519–20, 587, 645
impact 804
and oceans 510, 520
Goa 167
goat 142, 144, 145, 187
gods and goddesses 122–5, 127, 129, 136
Goethe, J. W. von 2, 846, 847

gold 165, 169, 188, 191
Golding, William 5
Gondwanaland 24
Gorbachev, Mikhail 345, 346
gorilla 86
Gourou, Pierre 319
governments
    central 308, 311
    local 308, 707, 732
    mercantilist 216
    officials 212, 213
    representation 211
    responsibilities 312
    systems 215, 216
    and urban problems 398, 707
GPS *see* Global Positioning System
grain 275
    as animal food 670
    farming 217
    in food aid 244
    imports 264
    production 606, 655, 657, 665
    surpluses 217, 606
    trade 188, 217, 264
granite 15, 17, 18, 20, 21, 27, 30, 36, 41
grapes 171, 217
grasses 12, 56, 77, 187
grasslands 51, 74, 77, 187
    areas 192, 193, 659
    borders 141
    clearance 199
    colonial 188
    ecology 153
    extended 274, 659
    and fertilizer 231
    ploughing 699
    replacing forests 501, 502, 586
    tropical 311
gravity sliding 33
Gray, Colin 342
'Great Game' 334
Great Lakes region 75, 178, 539
Great Transformation, The 11, 12
Greece 123, 944, 947
    aid 339
    civil war 339
    and nature 275
    religion 124, 945, 947
Greek language 112, 116
green politics 452, 756
Green Revolution 239–41, 654, 660,
    670

greenhouse effect 45, 63–4, 73, 75, 78,
    480, 645
    augmented 488, 491, 492, 494, 495–6,
      500, 503, 587
    natural 488
Greenland Ice Core Project 51
Greens 457, 458
greenstone 36
Grenada 354
Grimm, Jakob 118
gross domestic product 410, 427, 779,
    780–1
gross national product 423–4, 779
groundwater 526, 534
    and irrigation 545
    mining 534
    pollution 537
growth 420, 421, 423, 710
Guadeloupe 189
guilds, urban 217, 218, 304
Gulf Stream 49, 521
Gulf War 345, 346, 348, 359, 360, 363,
    510
gunboats 197
Gunflint Chert 76
gunpowder 186, 212
Guyot, A. H. 3
gymnosperms 76, 580
gypsies: language 113, 119

habitat 484
    affected by pollution 231, 235
    isolated 591
    marine 517
    saving 452
Hadar (Ethiopia) 92
Hadean stage 35, 38
Hägerstrand, Torsten 822–3, 825, 874
Haggett, Peter 4, 813, 823, 827, 828, 829,
    830
Haiti 364
Hale, Matthew 278
halites 40
Hanseatic League 376
Harare 709
Hartshorne, Richard 818, 819, 826
Harvey, David 824, 826, 827, 865–6, 916,
    921
Haushofer, Karl 335, 336
hazel 51, 57
health 458
    care 717, 743–4

insurance 263
  mental 743–4
  and water 535
heartland theory 332–5, 336, 337–8, 339, 340, 341, 342–3, 347
heaths 51
hedges 698
Hegel, G. W. F. 2, 316
Heidegger, Martin 908
helium 37
Henry, Prince of Portugal, the Navigator 165
Herbertson, A. J. 3, 818
herbicides 231
Herder, Johann Gottfried von 314, 949
Hettner, Alfred 818, 819, 826
Himalayas 116
hippopotamus 176, 177
Hindu religion 121, 122, 126, 128
  and conservation 589
  and creation 846
  holy places 130, 131
  festivals 129, 131
  revival 348
  society 114, 132
Hindustani language 113, 114, 117
Hispaniola 173, 187, 188
holism 1–2, 3, 5, 6, 286, 287, 803
Holocene 48, 51, 56, 58, 63
Holy Roman Empire 134
home 908–9, 939–41, 942, 949
homelessness 743, 908
Homeric hymns 123
hominids 11, 12, 77, 87–8, 99
  bipedal 90, 93
  evolution 88–9, 90–5
  spread 100–1, 103–4
hominoids 86, 87, 88
Homo 77, 86, 101
  and apes 86
  origins 96, 97, 98, 103–4
  spread 98, 100–1, 103–4
Homo erectus 97, 98, 99, 100, 101
Homo ergaster 97, 98, 99, 101
Homo habilis 96, 97, 98
Homo heidelbergensis 99
Homo neanderthalensis 99, 100, 101–2, 104
Homo rudolfensis 96, 97, 98, 99
Homo sapiens 77, 83, 88, 97–8, 99, 100, 101, 110, 443, 445
  adaptation 630

archaic 101, 103
  spread 584
Hong Kong 724
  airport 707
  industrialization 374, 710
  pollution 716
  population 265, 267
Hormuz 167
hornbeam 51, 57
horse 186, 187
  breeding 144
  manure 152
  and transport 151–2
horticulture 230
housing 284, 717, 775
  affordable 763
  cleavage 759
  company 307
  in country 152, 758, 759–60, 767
  and gender 718, 890
  high-rise 737–8
  low-cost 306, 307, 720, 737, 759
  low-income 715–16, 718, 719
  public 306–7, 718, 738–9, 748
  regulations 306
  rented 720
  rural 303–4, 758–9, 762
  second homes 761
  self-help 718, 719
  sharing 720
  slums 398
  urban 303, 305–6, 307, 330, 707, 717–20, 733, 737–8
  for working men 296, 300, 304
Howard, Ebenezer 307
Hudson's Bay Company 141, 376
Human Development Index 779, 780–1
human rights 704
  abuse 359, 365
  in Third World 722–3
humanism 808, 838–40, 842, 850, 853
humans 11, 67, 939, 940
  adaptation 159
  anatomy 89–90, 97, 98, 101
  ancestors 86, 89, 96, 97–8, 110
  and animals 287
  and apes 86, 87, 88, 89
  basic needs 599, 713, 717, 722, 773, 774–5, 784, 785
  brain 89, 97, 98
  and environment 81, 82, 83, 138, 140, 145, 148, 151, 152, 182, 274–6,

277, 286, 291, 443, 454–5, 479–80,
519, 521, 844
evolution 12, 13, 86–104, 159
origin 122, 123, 584
place in nature 278, 279, 282, 844,
845, 853
precursors 101, 103
spread 12, 100-1, 104–5, 110
as symbols 845–6
teeth 89, 90, 97, 98
urban 599
· variation 104
Humboldt, Alexander von 2–3
hunger *see* malnutrition; undernutrition
hunter-gatherers 12, 13, 105, 138–41,
152, 228, 282, 314
and environment 454–5, 599
Pleistocene 13
Pliocene 77
in rain forests 580–1
replaced 142, 164, 615
hunting 81, 105, 153
in Africa 176–7, 285
for food 139, 140, 141, 153, 177
for fur 141, 176
laws 150
for pleasure 149–50, 153, 176–7, 217,
275
regulated 140
hurricanes 65, 629, 642, 643
Hutchinson, G. E. 484
Hutton, James 2
hydrocarbons 968
fossil 137, 603
inorganic 41
marine 513, 514–16, 968
hydroelectricity 202, 533, 603
hydrogen 37, 67, 68
hydrogeology 2
hydrological cycle 493, 498–9, 509,
526–7, 609
hydrology 803
hydrosphere 15, 21, 36, 38, 434
hypsithermal 58

ice ages 35, 44, 46, 48, 62
end 152
in tropics 52–3, 582
ice caps 50, 51
melting 65, 480, 520, 804
ice sheets 48, 50, 51, 74, 78, 79
icebergs 521

Iceland 25
ichthyosaurs 77
idealism 808
IGBP *see* International
Geosphere–Biosphere Programme
IMF *see* International Monetary Fund
imperialism 163, 704–5, 870, 944, 945–6
ecological 186–7, 190, 594
European 166, 186, 196, 197–8,
311–12, 809, 948
and military force 167, 197, 311, 312,
356, 945
tools 197–8, 849, 860
values 163, 166
imports
agricultural 243, 264
European 213
manufactured 401
Oriental 217
substitution 398, 400, 401
incomes
in developing countries 243, 714
and diet 686, 688, 689–90
of ethnic minorities 747
of farmers 677, 679
inequality 781
low 714, 747
national 312
per capita 266, 267
redistribution 313
rising 678–9
and urbanization 391, 396
India 846
agriculture 126, 195–6, 656
caste system 115, 125
and China 339
cities 399, 403
colonization 113
debt 424–5
divided 116
ecological damage 282
emigration 260
food production 209
foreign investment 425
and Hinduism 348
infant mortality 395–6
irrigation schemes 202
landholding 395
languages 109, 111, 113, 115, 116, 117
Mogul 184
north-west 56, 334
and Pakistan 339

population 251, 269, 270, 395
religions 122, 125–6, 131, 132
sanitation 536
southern 111
urbanization 392, 395
water 534, 536
weapons bought 360
indigenous peoples 250, 316, 942–3
cooperation 168
displaced 164, 166, 170, 174, 189
and environment 164, 315, 845
and European colonization 162, 163,
164, 168–9, 174–5, 185–6, 195,
252, 277, 281–2, 706
and European diseases 185–6
excluded 706
female 165, 168–9, 178
in frontier regions 177, 178–9, 449
integrated 317
inter-tribal wars 168
killed 163, 175, 358
as labour 172, 175
loss of land 170, 174
and marine life 517, 519
and nature 845
numbers reduced 252
in rain forests 580–1, 588
redistributed 168, 169
religion 164, 165, 166
reservations 170, 285
resistance 167, 168, 174
trade goods 168, 169
Indo-European languages 107, 109, 111,
112
Indonesia 98, 104
debt 417, 424
farming 660–1
food supply 720
fossil hominids 98, 100
population 265, 270
Industrial Revolution 137, 159, 277, 353
ecological transition 223, 250, 252
and urban development 397
industrialization 161, 179, 250
and agriculture 229, 230
and capitalism 221, 281
and cities 398
competition 313
and environment 284
export-oriented 401, 410, 709
factories 220
impact 291

import-substituting 398, 400, 401,
410, 709
investment 710–11
and migration 261
newly industrializing countries 269,
313
newly industrializing economies 374,
375, 710
and resources 599
rural 760–1
spread 330
Third World 313, 397, 710–11
and urbanization 397, 398, 400,
709–12, 753
industry 78
capacity 370
in cities 151, 221, 400, 401, 710–11
domestic 221
and effluents 537
and environment 460, 466
factory 220, 221, 711–12
global 372
innovation 160
and land degradation 555
production 160
reorganization 385
rural 221, 760–1
service 253, 374
waste material 151
water use 529
inequality 786, 787–8
infanticide, female 143
inflation 330
information management 310, 617, 930
information technology 370, 375, 832
effects 377–8, 387, 461, 731
insects 74, 75, 76, 575
and pollination 77, 581
rain forest 581
symbiotic 71
Institute of British Geographers 797, 798,
803, 806
study groups 810
interest rates 414, 590
London Inter-Bank Offered Rate 414,
422
interglacials 48, 51, 75
Intergovernmental Panel on Climate
Change 443, 489, 494, 501
intermarriage 165
International Biological Programme 438,
485

International Geosphere–Biosphere
   Programme 6–7, 486, 501
International Maritime Organization 517
International Monetary Fund
   and debt crises 418, 420, 422
   and development 706
   established 371
international relations 361–2, 366
Internet 444
interpluvials 52, 55, 56
interstadials 50–1
investment
   domestic 710
   foreign direct 376–7, 385, 410, 421,
      425, 710
   productive 381
IPCC *see* Intergovernmental Panel on
   Climate Change
Iran
   cities 400
   population 270
   religion 132
   revolution 348
   war 354, 358, 364
Iraq 116, 345
   wars 354, 358, 363, 364, 365
Ireland
   emigration 262
   famine 281, 683
   neutrality 361
   population 256, 257, 259, 281, 761
   and UN 363
iridium 80
iron 12, 37–8, 153, 221
   forging 220, 221
   making 148
   mining 148–9
   ores 41, 252
   smelting 220–1
Iron Curtain 330, 339
irrigation 146, 200, 202–3, 545–6
   cost 540, 546
   effects 275, 545
   in developing countries 241, 693
   problems 534, 666
   and water use 529, 534, 541, 545–6,
      665
Islam 112, 124, 125, 127, 353
   area of influence 212, 367
   calendar 129
   and diet 127
   distribution 348

fast 129
festivals 125, 129
holy places 130–1
revival 135, 348
rituals 131, 132
Shi'a 131, 135
spread 112, 114–15, 133, 135
Sunni 131, 135
and women 135
islands 523
   barrier 560, 569
   oceanic 145, 148
Israel (ancient) 125, 132, 134
Israel (modern) 132
   and UN 363
   wars 339, 358
Italy 190
   colonies 335
   debt 427
   emigration 262
   landlords 217
   nationalization 313
   northern 214
   and overseas 182

Jain religion 126, 134, 282
Jamaica 174, 189
Japan 312, 969
   currency 347
   economy 346, 373, 375
   farming 680
   festivals 129
   foreign investment 377, 385
   and Germany 336
   homogeneity 250
   land loss 656
   manufacturing 373, 380
   missionaries 165
   in pan-region 335
   population 257, 259, 508
   religions 124, 132, 133
   resists colonization 163, 165–6
   trade 166
   and UN 365
   and USA 339, 367, 375
Java Man *see Homo erectus*
Jefferson, Thomas 280
Jerusalem 130–1
joint-stock companies 191, 216
Jones, William 107
Judaism 125, 127, 128–9, 132, 135
Julian–Madden waves 495

Jurassic 77

Kalahari 53
Kampuchea 398
Kant, Edgar 822
Kant, Immanuel 2, 819, 826
Kennan, George 340
Kenya
  development 666, 894–5
  fossil hominids 91–2, 95, 96, 97
  urban areas 392
King, Gregory 294
kingdoms 311
Kiribati 511
Kissinger, Henry 341–2
Köppen, Wladimir 485
Koran 114, 121, 123, 128
Korean War 339, 354, 356, 363
Kretschmer, Paul 112
Kropotkin, Peter 283
Kuala Lumpur 708
Kuhn, Thomas S. 827, 841, 957, 958
Kulturvölker 315–16
Kurds 116, 364, 365
Kuroshio current 521
Kuwait 345, 358, 363, 364, 365

labour
  in agriculture 234, 235, 238, 241, 263,
    264, 690, 693, 696, 752, 753, 755,
    756–7, 763, 765, 890
  blue-collar 734
  child 296, 722
  commodification 215, 221, 222, 223
  costs 375
  'dead' 222
  demand in colonies 172, 213–14, 263
  division 209–10, 212, 373, 374, 387
  family 173, 210, 218
  female 232, 722, 740, 890, 893–4,
    895–6
  and gender 170, 893
  industrial 732, 734
  'live' 222
  manufacturing 263, 294, 734
  market 207, 213, 216, 218, 219
  migrant 238, 269, 733, 740
  power 222
  proletarianized 216, 218–19, 221, 733
  rural 216, 219, 220, 294
  segmentation 740
  as service 216, 217

slave 171, 173, 174, 213
Third World 712, 722–3
urban 300, 304–5, 394, 709, 712
wage 216, 218, 220, 222
lakes 38, 58, 75
  duration 75
  fossil 55
  pluvial 53
  spread 74–5
Lamarck, J.-B. P. A. de Monet,
  Chevalier de 2
land 651
  abandoned 186, 234, 765
  allocation 213, 214, 656
  arable 198, 199–200, 202, 220, 231,
    555, 653, 656, 659, 665
  area 652
  banks 148
  bridges 152
  for building 767
  clearance 187, 200, 281
  colonized 12, 164, 171
  commodification 215, 217, 218, 221,
    223, 281, 671, 719
  common 218
  configuration 61
  cultivated 14, 659
  degradation 462, 555, 565, 612, 659,
    665–6
  distribution 61, 338, 659
  drainage 145, 147, 200, 201–2, 275
  enclosure 218, 221
  exploitation 182, 187, 203
  harvesting 200
  for housing 719
  improvement 194
  irrigated 202
  loss 612, 656
  management 671
  marginal 558, 655, 659
  and marriage 210
  over-grazed 187
  ownership 662
  pastoral 219, 220
  pressures 651, 656
  purchased 207
  reclamation 145, 201, 220, 276, 558
  replacement 200
  resources 612
  rural 216, 219, 766, 767
  semi-arid 565, 567, 568, 569
  set-aside 234, 655, 682, 765, 767

shortage 651, 654
suburban 767
surplus 234
tax 213
terracing 145, 275
transference 200
unstable 559
urban 216, 219
wilderness 337, 338, 555, 768–9
land use 652, 653
  adaptive 561, 569
  agricultural 651, 652, 655, 767
  in colonies 170, 171
  extended 194, 203
  in hazardous areas 558, 559, 560, 561
  intensive 182, 194, 203
  inventory 830, 971
  managed 568–9
  planning 276
  rural 752
  transformed 182, 183, 188, 190, 192, 193, 198
  and urbanization 665, 752, 971
  zoning 638, 640
landforms 32, 553, 933
  dynamical systems 553, 554, 562, 568, 569
  glacial 50
  and hazards 560–1, 569
  instability 557, 562, 563, 564, 568
  mass flux 560, 564
  rate of change 556, 557, 560
  systems 553, 554, 569
landholding 662, 667–8
  in colonies 170, 171
  communal 662
  leased 217, 218
  peasant 210, 217
  reform 662
  size 662
  tenant 218, 662–3
landlords 217, 218, 297
  monopolies 217, 221
land-power 332, 334, 340
landscape 30
  age 34
  architecture 919
  change 440, 554, 556, 560
  and colonization 164, 280
  evolution 31, 34
  farmed 754, 767–8
  and fire 138

  gendered 891, 893
  humanized 283, 555, 563, 564–5, 846, 847, 852
  and ice sheets 50
  instability 557, 559, 560
  interpretation 846–8
  lacustrine 50
  management 568
  man-made 317, 553
  models 439–40
  modified 276, 280
  morphology 819–20, 821
  old 34–5
  painting 283
  palaeotropicoid 48
  restless 553–4, 569
  sacred 129–30
  sensitive 564
  stable 564
  symbolic 848
  urban 908
landslides 558, 559, 637–8
language 11, 12, 250, 864
  agglutinating 108–9
  amalgamating 109
  American 105, 110
  boundaries 115–16
  classification 107, 108–9
  classificatory 109
  and culture 114–15, 847–8
  dialects 110, 116, 117
  distribution 107, 111–12, 116, 119
  European 107, 115
  fusion 117
  holophrastic 109
  international 118
  isolating 108
  national 314
  origin 110
  and power 848
  root-inflecting 109
  spread 111–12
  universal 107, 118
  vocabulary 108, 112–15
  see also Creoles; Pidgins
Latin America 191, 317
  debt crises 408, 409, 410, 411, 413, 415, 416, 417, 420, 421
  development 343
  dictatorships 346
  exports 421
  fertility rates 401, 402

food supply 667
immigration 262
and imperialism 705
land tenure 662
local government 404
migration to cities 395, 396–7
population 267, 269, 270, 271, 665
religions 135
resources 188, 191
social structure 318, 724
southern 196
Spanish colonies 166
trade 213, 384
urbanization 391, 393, 397, 398
water pollution 536
women in public life 891, 892
Latin language 114, 118
Laurasia 24
Laurentian Shield 50
lava plains 30
League of Nations 348, 361, 362
Leakey, Louis 95
Leakey, Mary 95
Lebanon 354, 394
Lebensraum 336
Leeds 301
Leibniz, Gottfried von 118
Lenin, V. I. 357
Lespagnol, G. 316
Lévi-Strauss, Claude 863–4
lianas 573, 578–9
Liberia 364
Libya 354, 534
lichens 71
life 67
  aerobic 69
  anaerobic 41
  and environment 11, 12, 42, 83
  evolution 20, 41, 69
  nature 67
  origin 11, 37, 40–1, 67–8, 72, 121–2
  protected 126
  quality *see* quality of life
  unity 127
life expectancy 393, 780, 781
  in developing countries 268–9, 394
  female 258
  and income 267
  male 258
  rising 258, 268, 657
  short 250
life-forms 11, 68, 70, 74

Lima 215
lime (tree) 57
lime-burning 151
limestone 20, 36, 41, 75, 76
Lindeman, Raymond 484–5
linguistics, comparative 107
Linnaeus, Carolus 118, 286
Lisbon 215
literacy 318, 780, 928, 929
lithosphere 21, 434
Little Ice Age, the 44, 58, 60, 152
Liverpool 296, 306, 307
livestock
  breeds 233
  food 243, 659
  production 230, 236, 243, 659, 678
  products 688, 689, 690
local government 302-3, 404
localism 942
locality 833, 914
locational analysis 4, 823, 826, 860, 867, 928
Loch Lomond Readvance 51
loess 50
logging 589–90
  efficient 594
  rain forest 584–5, 589, 591–2
  selective 589
London
  banking 191, 411–12
  crime 744
  Docklands 736, 748
  employment 734–5
  Euromarkets 411–12, 413
  and farming 218
  fog 644
  food market 219
  fuel market 221
  immigration 752
  inner city 738
  management 732
  population 735, 736
  public health 301
  size 215, 264
  smoke 137, 151, 276
  tourism 734
  as trading centre 190, 215, 218
  transport 735
  water supply 303
Lorenz, Edward 444
Los Angeles 744–7
Lösch, August 4, 822

Lothagam (Kenya) 92
Lotka, Alfred 5
Louisiana 169, 558
Lovelock, James 5–6, 433, 444, 455, 503
Low Countries *see* Netherlands
Lucca 152
'Lucy' *see Australopithecus afarensis*
Luther, Martin 278
lycopods 76

Macao 167
macaque 86
Macedonia 364
Mackinder, Halford 334, 337–8, 339–40, 341, 342, 347, 818
MacLeish, Archibald 337
Madagascar 108
maize 142, 144, 145, 176, 178, 186, 238, 240
Malacca 167
malaria 197, 203
Malayo–Polynesian languages 108, 112
Malaysia
    debt 415
    Federal Capital Territory 707
    logging 589, 590
    population 265
malnutrition 677, 678–9, 683
    defined 686
    in developed world 689
    in developing countries 689–90, 699, 700, 721
    extent 686
    see also undernutrition
Malta 117
Malthus, Thomas 249, 250, 281, 656–7
Malvinas *see* Falkland Islands
mammals 74, 75, 77
    climbing 581
    extinction 75, 81
    marine 514
    rain forest 581
man *see Homo*; humans
Manchester 293, 296
    building societies 306
    charities 295–6, 303
    friendly societies 305
    gas supply 302–3
    housing 306
    poor relief 299
    public health 301
    transport 303

manganese 516
mangroves 580
Manila 707
manufacturing
    base 308
    concentrated 373, 374, 400
    decline 734
    dispersed 314, 374, 401
    export-oriented 709, 711
    and farming 230
    growth 711
    introduction 252
    Pacific Asian 710
    small-scale 712
    suppliers 380
    Western European 217
maps 807, 829, 830, 834, 969–70
maquis 74
marble 15
Margulis, Lynn 5, 6
markets 160, 169, 208, 774
    colonial 220, 221
    common 383, 384
    competition for 356–7
    development 211, 221, 222
    domestic 215, 220, 221, 401
    as exchange 207, 209, 212, 216
    as exploitation 211
    'free' 207–8
    global 198, 216
    international 208, 215
    prices 207
    protected 221, 244, 410
Mars 37
Marsh, George Perkins 285–6, 838
marshes 153
Martinique 189
Marx, Karl 120, 259, 862, 863
Marxian theory 808, 813, 865
Marxism 774, 833, 861–2
    humanist 863
    regulation school 869–72
    structural 864, 865–72
Maryland 189
Massachusetts 170–1
mathematical models 430, 438–42
Maul, Albert 431
Mayans 186, 942–3
Mecca 130
Mediterranean 117, 146, 275
megacities 399, 403
megafauna 13

megalopolis 264
Mendel, Gregor 963
merchants 215, 943–4
MERCOSUR 384
Mersey basin 542, 543
mesas 36
Mesoamerica *see* Central America
mesosphere 21
Mesozoic 19
metals 153, 602, 613
    precious 169, 188
Metaphyta 70, 71, 74, 76
Metazoa 70, 71, 74, 75, 76, 78
meteorites 35, 80
meteorology 2, 803
methane 37, 38, 63, 461, 488, 498, 501
metropolitan regions 709, 730
Mexico 172, 173, 187, 384
    crop research 239
    debt crises 408, 410, 413, 416, 417,
        418, 419, 426–7
    development 401
    farm land 656
    loans 412
    peasant societies 317–18
    population 265
Mexico City 215, 271, 400, 401
    earthquake 715
Michigan, Lake 168
microbes 76
Middle East 212
    conflict 358, 547
    farming 146
    Israeli–Arab wars 339, 358
    population 265, 267, 269
    urbanization 393
    wars 355, 359, 364
    water 547
    weapons imported 360
Middle Ocean 337
migrants
    and business 706
    Chinese 706
    to cities 395–7, 706, 733, 736, 752
    female 396
    Indian 706
    male 260, 396
    retired 759–60
    unskilled 736
migration 105, 251–2, 289, 847
    of animals 73, 75, 138, 145, 152
    assimilation 269

assisted 261
causes 261, 263, 270, 285, 396–7
from cities 735, 736, 757, 759, 760–1
to cities 736, 752
from colonies 263, 706
and colonization 253, 263, 706
control 269
from developing countries 270
from Europe 260–3, 281, 285, 657
illegal 269
internal 261
international 254, 263, 269, 270, 272
and language 110, 111–12, 113, 114
modelling 971
overseas 260–3
pace 397
peasant 262, 396–7
of plants 138, 145, 152
reduced 263, 263, 397
in retirement 759–60
return 265
sources 261–2
temporary 269
to towns 216, 263–4, 394, 395–7, 402
volume 261
Milankovitch hypothesis 62, 958
milk 682
    in food aid 244, 246
    products 244, 245, 246
Mill, John Stuart 302
millet 111, 142, 239
minerals 15, 16, 17, 188, 613
    consumption 614
    marine 509, 515–16
    polymetallic 509
    production 169
    sequences 40
    sources 614
    strategic 614
    trade 613–14
mining 169
    centres 283
    coal 148, 221
    iron ore 148, 220, 221
    rural 763
    safety 284
minorities: protection 365
Miocene 77, 87
missionaries 107
Mississippi valley 50
Mitteleuropa 336
Moho *see* Mohorovicic Discontinuity

Mohorovicic Discontinuity 21
molluscs 75
Monera 40, 69, 70, 71, 76
money 191, 198
    borrowed 207
    as capital 206
    circulation 208, 209, 213, 214
    and towns 214
    development 211
    see also credit
monkeys 77, 86
monks 126, 127, 135
monoculture 195, 593, 669
monsoon 45, 519
Montevideo 271
Montreal Protocol 645
Montserrat 173, 189
Moore-Brabazon, J. T. C. 431
moorlands 141
moraines 30
Morris, William 283, 307, 308
mortality
    in Africa 197
    causes 269
    child 268–9
    decline 252, 256, 257–8, 265
    fluctuation 249–50
    infant 395–6, 779
    low 257
    rural 284
    transition in rates 251, 256, 265,
        269
    Third World 394
    urban 284
mountains 27, 30
    formation 31–3, 61
    slope destabilization 559, 569, 637
    slopes inhabited 558–9
    tropical 74, 578, 591
Mozambique 354, 364
Muir, John 287–8, 290
Mumford, Lewis 3
Mungo, Lake 105
musk ox 143
mutations 69, 88
myths 121–5, 846

Nadar 431
Na-Dene languages 105
NAFTA see North America Free Trade
    Association
Namibia 364

Naples 215
national parks 288, 290, 768
    rain forest 590
nationalism 255, 353, 942, 946, 948
    persistence 366, 950
    regional 366, 942, 949
    secular 348
    and war 356
nation-states 312, 946
    boundaries 358
    building 355, 358
    from former colonies 312, 313, 330,
        358
    and geography 849
    individuality 314
    and language 115–16, 314
native peoples see indigenous peoples
NATO see North Atlantic Treaty
    Organization
natural selection 69–70
Nature 811
    attitudes to 443, 456, 459, 589, 844
    commodified 190
    'determinant' 277
    and culture 277, 281, 282, 589, 844–6,
        892
    as female 892
    and human intervention 275, 277,
        278, 279, 286, 291, 455
    picturesque 283
    protected 285–6
    reserves 591
    sentient 164
    and society 810, 811, 844, 947
    unity 2
    as wealth 190
Naturvölker 315, 316
navigation 153, 184
Neanderthal Man see Homo
    neanderthalensis
Near East 112
    fossil hominids 101, 102, 104
    partition 115
    settlements 105
neighbourhood 941
neoglaciation 58, 62, 63
Neolithic Revolution 142
Netherlands
    development 190
    Empire 163, 165–6, 182
    farming 218
    lakes 148

land reclamation 558
overseas settlements 169
peasantry 218
settlement patterns 823, 832–3
trade 214
neutrality 361
Nevis 189
New England 166, 170, 191
plants and animals 176
settlers 189
workforce 172, 173
New France 170
New Guinea 105, 138
New Mexico 565
New Spain 170
New World *see* North America; South
America
New York 733, 743
New Zealand 48, 50, 193, 196
Newcastle 221, 301, 334
NICs *see* industrialization: newly
industrializing countries
Nicaragua 364
nickel 516
NIEs *see* industrialization: newly
industrializing economies
Nigeria 533
nitrates 698
pollution 231, 537, 698
safety limits 698
nitrogen 37, 76
biological 669, 671
fertilizer 231, 669–70
salts 82
synthetic 669–70
nitrous oxide 63, 82, 488
nomads 113
Non-Aligned Movement 343, 362
non-government organizations 706–7
noösphere 845, 850, 852
North America 12, 27, 28, 48
animals 74, 176, 179
colonization 112, 163, 174, 178, 188,
192, 277, 280, 332
eastern 57
economy 347, 373
environment 279, 280, 285
exploration 168, 178
farming 146, 280
food surpluses 679
foreign investment 385
Great Lakes 168, 169

immigration 105, 263, 269, 280, 281,
285
labour supply 172
native Americans 129, 168–9, 185–6,
280, 358
population 186, 250, 253, 272
products 191
religions 135
states 358
trade 194, 347
urbanized 161
vegetation 57–8
wars 359
water supply 609
North American Free Trade Association
347, 384, 386
North Atlantic Current *see* Gulf Stream
North Atlantic region 44, 48
North Atlantic Treaty Organization 339,
346, 361, 363
North Korea 244
North Sea 515, 521
Northern Hemisphere 48
Nottingham 151
nuclear power *see* energy: nuclear
numeracy 928, 929, 930
nuns 127, 135

oak 51, 56–7
oats 143
Oceania 260
oceanic gyres 497
oceans 12, 21, 30, 75, 480, 968
abyssal regions 509, 511
advance 50
aeration 73, 76
age 30
anaerobic 68, 73
biodiversity 517
and carbon dioxide 497–8, 520
circulation 73–4, 496–7
and climate 493–4, 495, 496–8
coastal zones 509, 511, 513, 519, 523
components 38
contamination 510
and continents 30–1, 511, 513
convection 497
crust 22
currents 45, 61, 185
deep layer 72, 73, 497
distribution 61
divisions 510–13

and dumping 518
and energy 515
exchange 497
extent 508–9
form 45, 72
functions 509
and global warming 510, 519–20
and ice 48–9, 480
'internal' 511
Jurassic 77
law 510, 511–13, 515, 521, 522, 523
and life 72, 509
linking continents 183
management 510, 521–4
margins 511–13
mixed layer 72, 73
origin 36
pollution 509, 518, 646
productivity 72–3, 509, 513
protection 519, 523, 524
resources 509, 510, 511, 513–16, 610,
    615
ridges 25, 27, 509, 511
salt 38
shallow 38
spreading 25
temperature 73, 510, 520–1
territorial 511, 521
thermocline 73, 79
and tourism 517
trenches 509
upwelling 73, 497, 513, 610
voyages 183, 184, 185, 516–17
in war 510
OECD *see* Organization for Economic
    Co-operation and Development
Ogden, C. K. 118
oil 41
    control 160
    and development 394, 395
    drilling 515
    marine sources 515, 968
    pollution 518, 646
    price rises 375, 408, 412–13, 417, 699
    search 515
    seeps 137
    trading alliance 330
    transport 516
oil shale 41
Olduvai Gorge (Tanzania) 95–7
Olsson, Gunnar 848
OPEC 330, 375, 412–13

ophiolites 22
Oppenheimer, J. Robert 442
orang-utan 86
orchids 574, 575, 581
Ordovician 76
Organization for Economic Co-operation
    and Development 354, 427
Origen 124
Orthodox Churches 134–5
    and women 135
Ottoman Empire 115, 163, 184, 311, 312
outmigration 735
overgrazing 565
Owen, Robert 307
oxen *see* cattle
oxygen 15, 37–8, 40, 41
    and life 67, 69, 76
    in ocean 73
Oyo (Sahara) 58, 59
ozone 82, 488
    depletion 470, 510, 521
    shield 38, 82

Pacific Ocean 25, 27, 29, 30
Pacific region 347, 384, 969
Pakistan 339, 360
palaeoclimate 488
palaeoecology 957
palaeoenvironment 954, 957, 958, 960–2
Palaeogene 45
Palestine 363, 891–2
palm oil 236, 237
palms 573
Panama 354
Pan-America 335
Pandanus 573
Pangaea 24, 74, 78
pan-regions 332, 333, 335, 337, 338,
    342–3, 347
paper 608, 615
Papua New Guinea 34
paranthropines 94, 95, 97
Paranthropus aethiopicus 94
Paranthropus boisei 95, 99
Paranthropus crassidens 94
Paranthropus robustus 94, 99
parasites 70, 71, 151, 574
Paris
    immigration 752
    population 735, 736
    size 215, 264
    transport 151–2

Paris Club 422, 425
Park, Robert 3
parks 145, 148
  for hunting 149–50, 275
Parsis *see* Zoroastrian religion
pastoralism 145, 147
  colonial 171, 174, 177
  nomadic 142, 176, 177
pastoralists 219–20
pasture, permanent 555
Patagonia 48
patricians 210
  feudal 212, 217
  and peasants 211, 216, 217
peace 334, 367
  building 364
  diffusion 353
  dividend 359
  global plan 337, 338
  implementation 360, 363–6
  interstate 353, 354
  stable 353, 354, 359, 367
  zones 354
Pearson, Karl 436
peasant proprietors 195, 199
peasants 209, 210–11, 219, 221, 250, 316
  craftsmen 219, 220
  free 216–17
  Mexican 317–18
  military conscription 213
  pastoral 219, 220
  taxation 213
  Vietnamese 319–20
  yeomen 218
peat 51, 141, 148, 580
pedology 803
pedosphere 434
peneplain 31
peppers 144
permafrost 50, 64, 643
Permian 76
Persia 184
Persian (language) 108
Peru 163, 173
  debt 416
  reserves 415–16
  urban areas 392–3, 395
  women in public life 892
pest pressure 593
pesticides 231, 430, 671
  abuse 443
  adverse effects 698

chemical 593, 698
  see also DDT
petroleum
  industry 515
  for lighting 137
  see also oil
Phanerozoic 40, 41
phenomenology 808, 863, 907
philanthropy 303, 304, 307
Philippines 396, 417, 424, 724
philosophers 944, 947
photography
  aerial 431, 433, 444
  from space 432, 433
photo-interpretation 431
photo-reconnaissance 431
photosynthesis 37, 41, 68, 69, 71, 651
Physical Quality of Life Index 779–80
phytoplankton 71, 73, 508, 513, 610
Pidgins (languages) 107, 117
pig 142, 143, 144, 145, 187
pilgrimage 131
  Hindu 130, 131, 132
  Islamic 129, 130–1
Pinatubo, Mount 489
Pinchot, Gifford 288, 289
pine 56, 57, 176, 593
pineapple 236
pitcher plants 581
place 801, 805, 833, 834, 921
  central place theory 821, 822, 823,
    829, 831, 832–3
  definitions 906–8, 913
  design 918–19, 920
  exploitation 916–17, 918, 919
  identity 917–18, 919
  as process 876, 877
  sense of 738, 910–12, 918–19
  spirit of 909–10, 921
  topomorphic change 917, 918
placelessness 913, 914–15, 916, 917, 918
place-names 107, 110, 112–13
plankton 77, 79, 461, 464, 497, 508,
  513–14
planning 290, 806, 807, 821, 829, 832,
  919, 932, 950
  urban 399, 402–3, 705, 709, 747–8,
    806, 866, 910, 911, 915, 920
plant communities 74, 77, 138, 139
plantations 172, 195
  American 172, 174, 190
  foreign-owned 235–6

forestry 592–3, 594
  output exported 236
  specialized 236
  state-owned 235, 236
plants 68, 151, 485
  affected by pesticides 698
  ancestors 40
  breeding 143–4
  dispersal 77
  domestication 14, 142–4, 280, 584
  extinction 145, 606
  flowering *see* angiosperms
  as food 13, 127, 138, 139
  herbaceous 51, 56, 57
  higher 70
  hybridization 75
  introduced 176, 186, 187, 189
  land 79
  leguminous 71
  and magic 164
  managed 140
  marine 517
  migration 138, 145, 152, 176, 179
  native 176, 186, 187, 280, 285
  pioneer 51
  productivity 604–5
  as sources of drugs 586–7
  symbiotic 71
  threatened 463
  varieties 233
  see also crops; weeds
plateaux 30, 61
plate tectonics 24–5
  activity 32, 35, 74
  cycle 27
  gravity 34
  uplift, 31–3
Plato 1–2, 122, 275, 907
Pleistocene epoch 44, 48, 62
  forests 582–3, 584
  refuges 584
plesiosaurs 77
Pliocene 48
plough 153
pluvials 52, 56, 58
Poa pratensis 187
podzols 51
Poland 645
polar regions 73, 74
politics
  authority 321, 323
  and economics 312, 347–8

and environmentalism 453
and geography 849
organization 321, 324, 332
power 160, 255, 323
realignment 330
systems 291, 310, 311, 321
see also geopolitics
pollen 77
  analysis 582
  arboreal 55, 56, 57
  records 56, 57, 58, 59
pollination 581
pollution 646
  air 151, 235, 276, 404, 644
  chemical 176, 239, 645–6
  control 460, 644, 716
  environmental 288, 460, 463
  marine 509, 518, 523, 646
  nitrate 231
  oil 518, 646
  penalties 404, 523
  urban 714–16
  water 231, 234, 235, 404, 462, 480,
    534–9
Poor Law 296, 297, 298, 299
poor relief 219, 295–6, 297–9
population 438
  adaptation 82, 105–6
  ageing 260, 268–9, 966–7
  censuses 249
  'checks' 249, 250
  coastal 508
  composition 251, 271
  concentration 264, 270–1, 272, 284,
    736
  demands 83
  demographic transition 251, 256, 257,
    260, 265, 267–9, 404, 735
  density 209, 250, 253, 254, 264–5
  density gradients 253, 265
  distribution 5, 250, 251–2, 253, 254–5,
    264, 271, 967
  employed 206
  and environment 255, 289
  ethnic minorities 742
  fall 217
  female 260
  and food supply 243, 247, 249, 654,
    657, 665, 667, 672–3, 677, 689–90
  fragmentation 250
  future growth 594, 601–2, 665, 667
  government influence 255

growth 13, 81, 106, 142, 143, 153, 160,
  176, 179, 191, 194, 198, 217, 250,
  252, 266, 271, 280–1, 289, 486,
  503, 593, 601, 617, 651, 966
history 249–51, 970
inner-city 737, 742–3
Islamic 268
limitation 594, 657, 667
location 252–3, 508
low growth 256
middle-class 724, 737
mixed-race 165, 178, 189
mobility 252, 265, 967–8
natural decrease 256
natural increase 257, 263, 265–7, 272
Old World 250
polarized 737
pressure 486, 558, 593, 617, 658, 666,
  667
purchasing power 192
rate of growth 255, 256, 260, 265, 266,
  517, 593, 665, 966
redistribution 251–2, 253, 264,
  269–71, 289, 735
reduced by disease 173, 186, 217
restricted 83, 143, 249
retired 742, 762
rural 161, 250, 256, 262, 270, 271,
  284, 303–4, 394, 709, 752, 753,
  756–7, 761
sex ratio 396
spread 198
stability 593
Third World 256, 265, 391, 394, 651,
  714, 723
urban 161, 215, 250, 253, 256, 264,
  270, 271, 284, 303, 304, 391, 396,
  397, 400, 714, 723, 967
Portugal 190
  emigration 262
  Empire 163, 165, 167, 173, 178–9, 182
  explorers 185
positivism 793, 824, 826, 827, 831, 833,
  834, 841
post-modernism 774, 809, 810, 833, 842,
  880
  and gender 897–8
post-structuralism 880
potato 142, 145, 186, 192, 239
poverty 472
  culture 739–40
  and environmental hazards 715–16
  inner-city 742, 744–7
  line 780, 781
  rural 238, 294, 296–7, 298, 304, 756
  and self-reliance 708
  trap 654
  and undernutrition 690, 721
  urban 294, 296, 298–9, 304, 330, 706,
    715–16, 723–4, 732, 743
  youth 744–5, 747
pragmatism 808
prairie 57, 58, 179
Precambrian 19, 20, 40, 41
precipitation see rainfall
predation 81
predators 150
priests 134, 135
Prigogine, Ilya 5
Primates 77, 86, 87
printing 115, 212
producer organisms 71, 72
production 182
  capitalist 207, 216, 222, 223, 870
  chains 378
  colonial 169
  concentrated 378
  dispersed 313–14, 374, 378
  'domestic' 220
  efficiency 207
  and finance 373–4
  host-market 378
  international organization 378–80
  means 219, 310
  mechanized 222
  phases 871–2
  pre-capitalist 215
  reorganization 385
  and reproduction 894, 895, 896
  rural 219
  strategy 378–9
  surplus 210, 211, 212
  systems developed 190, 191, 215, 378
profit 206, 207, 208, 222
progress: European view 278–9, 281, 282,
  283
prokaryotes 40, 41
proletarianization 220
proletariat 215, 218–19, 220
property
  private 214, 217
  rights 464, 465
Proterozoic time 76
Protestant Churches 133, 135

missions 134, 166
rural 220
and women 135
Protoctista 70, 71, 73
proto-industrialization 220
protozoa 70, 71
Prussia 190
pterosaurs 77, 80
public health 283, 284
in Britain 300–1, 302
occupational 284
Puerto Rico 265

quality of life 772, 773, 784, 788
comparisons 774, 775–8, 781
criteria 782
index 779–80
indicators 775, 776–7, 778–9, 781,
782–3
process model 785
satisfaction of needs 784, 785
unequal 786, 787
quartz 15, 80
Quaternary period 35, 48, 51, 582
Quebec 166, 171

race 793, 897
radar 430, 431–2, 433
radiation 645
and climate 61
infra-red 63
solar 48, 60, 69, 78
Rafflesia 574
railways 198, 334, 556, 752
rain forest, tropical 582–4
as air conditioner 464–5, 587, 588,
608
clearance 153, 189–90, 200, 501–2,
584, 585
and climate 501–2, 577, 582–3, 584,
608
conservation 588–9
decline 584, 588, 590
distribution 573, 576, 577, 582, 583–4
diversity 573, 574–5, 577, 588, 606
drugs from plants 586–7
fires 586
as food source 139, 580–1
inhabitants 580–1, 584, 588
loss 586
lowland 575, 583, 606
monsoon 577

montane 578–9, 582–3
products 586–7
regrowth 586
reserves 590–1
seasonal 577
structure 573, 574, 583–4
sub-tropical 189
threatened 665
timber 584–5
tropical 56, 74, 83, 138, 152–3, 189,
311, 456, 573, 573–7
worth 456
rainfall 58, 490
annual 484
patterns 60, 64
Ramayana 128
rat 145, 151
rattans 573, 587
Ratzel, Friedrich 3, 315–16, 317
raw materials 194, 235
Ray, John 278
Reagan, Ronald 340, 342, 344
Redfield, Robert 317–18, 319
redwoods 77
reforestation 555
refugees 255, 263, 355, 365
regionalism 920–1, 942
relativism 774
religion 11, 12, 78, 120–36, 250, 479
and art 121
Asian 127, 943
and colonization 165
conversion 165
decline 135, 947, 948
definitions 120, 121, 134
and diet 125, 126, 127
festivals 125, 128
future 135–6
intellectualized 210
international 132–4
and language 114–15
local 132
missions 132, 133, 134–5
'mystery' 124
national 124, 125, 132
of native peoples 164, 318
origins 120–1
repression 135
revival 136
rituals 131–2
sects 135
Semitic 127

social 134
solitary 134
symbolism 124, 125
varieties 120, 944, 945, 947
views of nature 278
and writing 121
see also the names of individual
    religions
remote sensing 430, 432, 434, 441, 642,
    807, 813
Renaissance 214
rents 216–17, 218
repatriation 260
repopulation, rural 757, 759, 760–1, 762,
    763
reproduction 68, 69, 76, 128
reptiles 75, 76, 77
reservoirs 284, 533
resources
    access 338, 600, 602
    allocation 464, 593, 594, 741
    biological 604, 605–10
    of colonies 169, 187
    consumption 612, 967
    demand 600
    and developed world 593, 594
    distribution 338, 341, 467
    economics 611
    exploitation 161, 183, 185, 187–8, 281,
        287, 480
    future 616–17
    local 325
    management 288, 289–90, 467
    marine 509, 510, 513–16, 523, 968
    mineral 252
    natural 183, 252, 287, 341
    non-renewable 602, 611–12, 614
    ownership 207, 221, 290
    and population 601–2
    price 612, 614, 616
    processes 616–17, 618
    recycling 611
    renewable 602, 614
    satellites 432
    substitution 614
    sustained yield 288
    transfer 471, 472
retirement 759–60
Rhine, River 542
rice 110, 127, 142, 143, 188, 238
    exported 244
    yields 239, 240, 660, 661

rimland 338, 339, 341, 359
Rio Conference *see* United Nations:
    Conference on Environment and
    Development
Rio de Janeiro 215
Ritter, Karl 3
rivers 58, 75
    age 34
    basins 547–8
    diversion 275, 533
    modified 639
    pollution 536, 537
    as source of water 527
roads 152
    construction 555–6, 737
    de-icing 644
rockets 431, 432
rockfalls 637, 638
rocks 15, 16, 17
    basement 41
    continental 30
    cycle 17–18, 20, 36
    dating 20, 22
    erosion 30
    folding 33, 34
    magnetism 25
    oceanic 30
    sedimentary 20, 36, 38
Roman Catholic Church 128, 133, 134
    in America 135
    churches 130
    independence 214
    influence 212
    missions 134, 165
    in South America 317, 318
    tithes 217
    and women 135
Roman Empire 151, 322
    aqueducts 149
    circuses 149
    influence 212, 946
    religion 133, 134
Romance languages 113
romanticism 282, 283
Rome 149, 151, 275, 735
Rorig, Fritz 321
Royal Geographical Society 798
rubber 153, 160, 236, 238, 587
rural areas *see* countryside
Russia 335
    arms sales 359, 360
    debt 425

ecological disaster zones 463
expansion in Asia 334
emigration 260
famine 683
forests 188, 194
instability 347
in pan-region 335
and UN 363
and USA 363
wars 355
see also Soviet Union
Russo-India 335
rye 143

sacred places 124, 129–32, 164
Sahara 53, 58, 59, 74
Sahel
    development 666
    drought 44, 60, 667
    famine 683
    hunger 667
    limit 58, 59
Sahul 105
St Kitts 189
saints 130
salinization 202, 545
salt 38, 45, 137, 515
sanctuaries 130
sand 20
sand-dunes *see* dunes
sand seas 52
sandstone 20, 36
Sangiran (Indonesia) 98, 100
sanitation 535
    urban 256, 284, 300, 301–2, 716
Sanskrit 109, 115, 119
São Paulo 271
Sapir, Edward 110
Sartre, Jean-Paul 863
satellites 432, 461
    artificial 430, 432, 433, 807
    communications 432
    geopositional 432, 433
    meteorological 432
Saudi Arabia 360, 534
Sauer, Carl O. 3, 316–17, 818, 972
savannah 58, 77, 152, 199, 578
savings 305
Scaliger, Joseph Justus 107
Scandinavia 48, 50
    conservation 291
    emigration 261

forests 188
Schaefer, Fred 826
Schelling, F. W. J. 2
Schultze, Benjamin 107
science 78, 479, 806
    applications 329, 617, 802, 806
    biological 329
    civic 469
    development 430
    distrusted 793
    environmental 430, 456, 460, 461–2, 479
    and geography 430, 802, 803, 804
    Enlightenment 837
    interdisciplinary 470
    Islamic 185
    languages 118
    medieval 184
    methods 804, 831, 955–7
    paradigm 824, 827–8, 957
    physical 803, 822
    post-modern 958
    potential 329
    purpose 278
    and technology 444, 445, 461
scientism 948
Scotland 34
scrublands 51
sea-floor 21, 22
    age 27
    animals 509
    spreading 25, 27, 29, 30, 509
sea-levels 32, 33, 78
    change 138, 804
    fall 50, 104
    rise 65, 253, 494, 510, 519, 520
sea-power 167, 332, 334, 340
seas *see* oceans
SEATO *see* South-East Asia Treaty Organization
Second World (East) 342
Second World War 329, 336, 353, 355
    aftermath 313, 330, 333, 342, 361, 370
    axis 336
    'Big Three' alliance 336
    science and technology 430, 431, 443–4, 802
sedimentation 202
sediments 17, 18, 20, 40, 560
semiotics 864
Semitic-Hamitic languages 109, 112

Seoul 271
Serbia 364
serfdom 216, 217
servants 173
settlements 13
    attractive 759
    colonial 169–70, 178
    commuter 767
    fortified 167
    nucleated 170
    patterns 821, 822, 823
    permanent 105, 142, 143
    Puritan 170–1
    small 161, 759
    squatter 713
    system 265
    see also communities
sewage 151, 306, 536–7
    treatment 536, 541
shale 20, 36, 73
Shanghai 271, 707–8
sharecroppers 199, 217
shatterbelts 341, 358
sheep 142
    breeding 144
    sacrifices 125
    tax 217
    transfer 145, 187
Shinto religion 124, 127, 132, 133
ship-building 148, 188, 191
shipping 516–17
ships
    and colonization 166
    owners 191
    for slaves 166
    and trade 191, 516
shipwrecks 509–10, 517–18
shorelines 30, 511
    destabilized 560
    recession 557
shrines 124
sial 21, 30
Siberia 41, 105
Sikhs 134
silica 27
silicon 15
silicon dioxide 15
silk 212
Silurian 76
silver 169, 191
sima 21, 30
Singapore 313

housing 717
human rights 722
industrialization 374, 710
population 265
privatization 709
Sino-Tibetan languages 108
slavery 195
    abolished 287
    in European colonies 164, 172–3, 174,
        178–9, 212, 213
    and sugar production 173, 175, 179,
        189, 190, 191, 195, 235
slaves
    African 163, 172, 173, 174, 175, 189
    Amerindian 178–9
    death rates 175
    female 165, 172
    Indian Ocean 174
    mixed-race 173
    religion 165
    transport 166, 175
Smith, Adam 281, 295
Smith, Thomas Southwood 284, 300
smog 643, 644
Smuts, J. C. 3
snow see avalanches
social disorder 724, 730, 744–7, 748, 786
social indicators see quality of life:
    indicators
social justice 772, 787
social problems 738, 739–41
social sciences 805, 806, 808, 813, 824,
    829–30
social welfare 313, 786–7
    access 741
    components 785
    inadequacy 742–3
    state involvement 708, 714, 717, 743
    subsidy 714
    voluntary 743
socialism 787–8, 862
    municipal 302, 303
sociality 852
society 159, 941
    African 175
    agricultural 209, 241, 311
    civil 708
    colonial 164–5, 166, 170, 705–6
    deployment 878–9
    differentiation 809
    European 163, 165
    evolution 319

fields of action 879
intermediary 320–1, 323–4
marginal 942
medieval 319
mixed-race 166
modernized 315–16, 319, 324
native 166, 316, 317
nomadic 311
organization 310, 317
patrician 210
peasant 316, 317–18, 319
post-industrial 733, 758
'primitive' 315, 316, 317, 320, 323, 324
rural 236
survival 810
traditional 310–11
tributary 209
sociology 3, 805
sodium 20, 38
soils 803
conservation 275, 276, 568, 588, 671
degeneration 501
degradation 555
denudation 556–7
desert 45
deteriorating 51
erosion 235, 239, 275, 461, 555, 557, 560, 588, 606, 698–9
as female 128
fertile 51
mapped 434
mineral 51
poor 657
tropical 579–80
volcanic 636
waterlogged 141
solar constant 496
Somalia 354, 363, 364, 683
Somerville, Mary 3
sorghum 239, 240
South Africa 170
animals 176–7
cities 730
colonization 167, 174–5, 193
farm belts 166
farming 171, 240
immigration 262, 281
migration to cities 402
quality of life 777
South America 12, 74, 359
animals 179

colonization 163, 174, 178, 185, 188, 277, 332
environment 279
farm land 659
farming 147
immigration 260, 281
labour supply 172, 175
population 186, 189, 250, 253
rain forest 56
states 358
sub-Orinoco 347
trade 194
South Carolina 173
South-East Asia 12
colonization 167
conflict 358
crops 142
'dragons' 313, 374
economies 374
forests cleared 200
immigration 260
industrialization 313, 374, 375
Pidgin languages 117
trade 167
undernutrition 686
urbanization 391
wars 359
South-East Asia Treaty Organization 339
South Korea 313, 969
debt 415, 420, 424
industrialization 374, 710
land loss 656
population 265, 267
Soviet Union (former) 329
alliances 339, 363
Asian part 367
and China 339
decline 345
debt 425
dissolution 330, 344, 346, 373, 969
economy 312, 313, 345, 346, 371
foreign policy 342, 346
and geopolitics 337, 339, 340, 342
religions 134–5
satellite countries 339, 371
sphere of influence 313, 330, 339, 340, 363, 371
and Third World 363
and UN 363, 365
and USA 342, 343, 345
wars 336, 355, 356, 359
space 432, 801, 805, 819

plastic 825
private 891
public 891, 941
and society 866–7, 868–9, 871
structured 864, 866
Spain 190
civil war 354
emigration 262
Empire 163, 173, 182
explorers 185
finance 217
fossil hominids 101
Muslim 133, 150
overseas settlements 169–70
and UN 363
spatial mismatch 740
spatial science 805, 807–8, 813, 824, 829,
834
analysis 825, 827, 830, 833, 928, 971
criticism 830–1
dialectics 868, 871
feminist 889–90
multidisciplinary 828–9
patterns 830, 831, 834, 867
spatial statistics 813
speciation 74, 75
species
diversity 576
endemic 75
indicator 461
tropical 576
vanishing 463, 606
Speenhamland System 296, 297
Spencer, Herbert 3
spice trade 167, 169, 195
sponges 76
spruce 57, 58, 74, 593
Spykman, Nicholas 337, 338, 342
squashes 142, 186
squids 73
Sri Lanka 265, 267
stadials 50, 51
states 208, 311
alliances 354, 361
boundaries 357
capitalist 357
coastal 511, 513, 521–2
and colonization 253
competition 349, 381
concept 250, 312
control 208, 214
co-operation 362

development 190, 191, 212, 316
economic role 312, 349, 387
European model 253, 312, 347, 357
federal 264
intervention 228
isolation 361
making 357
military power 357
modernized 311–13, 322, 348
new 342, 357, 373
power 213, 215
revenue 213
'rolling back' 708
security 361
size 254
Third World 347, 357
unitary 264
at war 212, 213
weaknesses 314
statistics 436–7, 441
steam power 220, 221, 252
steamships 197, 198
steppe 74
Sterkfontein (South Africa) 93
Stoddart, David 4
storms 642–3
Stowe, Harriet Beecher 287
strangler figs 573
strata 17, 20, 34
stratigraphy 20, 75–6
streams see rivers
stromatolites 76
structuralism 808, 842, 861, 862, 863–4,
881
and freedom 883
Marxist 865–9
reconsidered 880–1
regulation school 869–72
structuration 831, 833, 873–5, 876, 878,
882
subduction 27, 29, 33
suburbs 219, 221
succession 4
Sudan 354, 364
Suez Canal 201
sugar 170, 188, 192, 212, 236, 238
in Africa 173
in Caribbean 172–3, 188, 189, 190,
191
exported 244
for gasohol 394
imported 242

production 173, 190, 195, 235, 237, 394
sulphates 41
sulphur cycle 82
Sumerian language 109
Sun 21
Sundaland 104
supercomputers 438, 441
supercontinents 34
superpowers 342, 345, 346, 363
    allies 362, 363
    frontline 340
    influence 371
    surrogates 344
    and war 344, 356
surveying 431, 807
sustainability *see* development: sustainable
Swahili 117
swamps 580
Swartkrans (South Africa) 94
Sweden 34, 190, 361
    debt 427
    farming 693
    and UN 363
'Swing' riots 297
Switzerland 361
symbiosis 70, 71
symbolism 124, 125, 128
Syria 116
systems theory 5

Tacitus 275–6
TAFTA *see* Trans-Atlantic Free Trade Agreement
taiga 74
Taiwan 313, 969
    industrialization 374, 710
    land loss 656
    population 265
Tajikistan 50
Tanganyika, Lake 75
tanning 151
Tansley, A. G. 484
Tanzania: fossil hominids 92, 95–7
Taung skull *see Australopithecus africanus*
taxation 198, 213, 217
Taylor, Griffith 289, 337, 338–9
tea 188, 192, 195, 236, 237, 238
    imported 242, 244
teak 593
technology 141, 153, 182

agricultural 228, 230–1, 239
appropriate 283
development 430
for drainage 201
dynamic 221
enabling 372, 617
European 166, 179, 186
new 375, 444
nineteenth-century 183, 196, 201
post-war 329, 370, 444, 802
'primitive' 315
purpose 278
and science 444, 445
transfer 240, 241
see also information technology
tectonics *see* plate tectonics
tektites 80
television 444
temperature 60
    changes 46, 48, 49, 58, 486
    decline 44, 47, 48, 50, 56
    extreme 643–4
    rise 64, 82, 138, 488–9, 494, 520
    and tree cover 279
    variation 489, 490, 491
Tennessee Valley Authority 290, 431
terranes 27, 28
Tertiary era 19, 44, 45, 77
    climatic decline 582
    warmth 48
tetrapods 76
Thailand
    cities 400
    debt 415
    population 265, 268, 269
Thatcher, Margaret 344
thermoregulation 77
Third World 208, 342
    arms imports 360
    cities 397, 399, 400–1, 404, 702–24
    composition 343
    conflicts 359
    debts 410, 421, 422, 714
    and democracy 353, 723
    environment 714–16
    farming 666
    food supply 651, 653, 663
    industrialization 313, 394, 709
    modernization 394, 723
    population 394, 651, 723
    rural areas 394, 398
    states 357

and superpowers 344, 363
trade 663
and UN 365
urbanization 391–3, 394, 397–8, 402,
403, 404, 702, 703, 705, 723
wars 344
Thoreau, H. D. 285, 287
Thünen, J. H. von 4, 438
Tierra del Fuego 120
tiger, Sumatran 587
Tilly, Charles 357–8
timber *see* wood
tithes 217
TNCs *see* corporations, transnational
tobacco 145, 169, 172, 178, 188, 192, 195,
237
Tocqueville, Alexis de 293
tomato 145, 186
tools 12, 153, 159
metal 141, 168, 189
stone 77, 95, 102–3, 168
wooden 141
topography 30
topophilia 912, 913, 921
topophobia 912–13
tornadoes 642–3
totemism 120
tourism
heritage 734, 737, 916–17
marine 517
rain forest 588
rural 756, 767, 768
towns
colonial 171, 215
and country 752, 767, 768
development 209, 211, 214, 215,
264
hierarchy 215
industrial 221, 253, 281, 303
market 281
new 307
satellite 265
size 215, 264
see also cities
toxins, synthetic 82
trade 159, 179
agricultural products 243–4
alliances 330, 347
in arms 359–60
barriers 375, 385, 388
and colonies 163, 165–6, 167, 169,
189, 198, 213, 215, 222, 373

commodities 143, 169, 182, 183, 213,
214
creation 385
development 190, 194, 212
distorted terms 617
distribution 373
diversion 385, 386
and finance 373–4
in food 241–4, 247
free 312, 313, 338, 371, 382, 384, 617
friction 375, 388
global 183, 184, 189, 191, 194, 211–12,
214–15, 223, 241, 663
growth 252, 313, 347, 366
international 322, 323, 325, 361–2,
366, 377, 663
land-based 183, 184, 192, 214
liberalization 247, 401
manufactured products 244, 375
profits 215
regional blocs 347, 378, 382–6
sea-borne 183, 184, 185, 191, 192
systems 312
terms 467
within colonies 168
trade unions 300, 400
trading-posts 167
traffic control 731
Transatlantic Free Trade Agreement 384
transport 159, 160, 250, 310, 315, 317
accidents 646
air 313, 314
car 731, 760, 761
marine 516–17
networks 825
public 731, 761, 762
road 644, 731
rural 762
urban 265, 403, 708, 709, 731, 732
travel 337, 967–8
tree ferns 76
tree lines 74
trees 56
coniferous 148, 592, 593
as crops 195, 592–3
cultivated 145, 592–3
deciduous 148
distribution 56–7, 75
dominant species 500
indigenous 195
migration 501
planting 151, 279, 280

rain forest 574, 579, 581, 591
removal 141, 200, 591–2, 594
sacred 589
selection 594
Triassic 35, 77
tribute 209, 210, 212
trilobites 76, 77
Trinil (Indonesia) 98
tropics 52, 72
Truman, Harry 339
tsunamis 635
tuberculosis 285
tundra 50, 64, 74
and hunting 153
replaced 51
Turing, Alan 435
Turkana, Lake 94, 95
Turkey 116, 312, 339
see also Ottoman Empire
Turkish (language) 108, 112, 116
vocabulary 113, 114–15
typhoons 642

underclass 737, 740, 742
underdevelopment 420
undernutrition 13, 665, 683
causes 654, 688–9, 690
defined 686
in developed countries 688–9
in developing countries 651, 654, 686, 688, 721
distribution 686–8, 696
unemployment
benefit 263, 299, 615
growth 308, 714
high rates 270, 735
inner-city 737, 740–1
male 741
urban 734–5, 735
United Kingdom *see* Britain
United Nations 329, 348
Afro-Asian caucus 343
Commission on Sustainable Development 472
Conference on Environment and Development 458, 462, 463, 466, 470–3, 485, 510, 523–4, 549, 615
Conference on the Human Environment 485, 510
Consultative Group on International Agricultural Research 239
Convention on the Law of the Sea 510, 511, 516, 522
and development 343, 472, 778–9
Environment Programme 462, 472, 518, 523, 534
established 338, 361
future 365–6
General Assembly 362, 363
and global issues 343, 344
and human rights 723
marginalized 344
membership 362
military operations 345, 363–4, 365–6
monitoring 364
peace-building 364
peacekeeping 354, 363–6
power 362
precursors 312, 338, 348, 361
'preventive stationing' 364
protection of minorities 365
role 364–5
Security Council 362, 363, 364, 365
support 365
and Third World 362, 365
and world order 345, 348
United States of America 34, 191, 329, 335
agricultural protection 680–1
alliances 339, 359, 362, 363
arms sales 359, 360
banking 408, 409, 412, 413, 418
and China 363
cities 264, 731–2, 733, 734, 736, 738, 742, 743, 744
coastlines 560
colonization 166, 169–70, 193, 196
conservation 285, 290, 291
continental shelf 522
currency 347, 371, 375, 411, 413, 763
deficit 410–11, 425
drought 642
Dust Bowl 290, 642
English settlers 169
economy 312, 346, 371, 373, 375, 420, 421–2
ethnic minorities 742
exploration 168
farm crisis 763–4
farm land 655, 656
Federal Reserve Board 414
and food aid 244, 246, 680–1
food surpluses 682
foreign investment 376–7, 386

foreign policy 342
forestry 288, 289–90
forests cleared 200
French settlers 169
frost belt 734, 736
and geopolitics 337, 339, 342, 969
hurricanes 642–3
immigration 260, 262, 733, 736
irrigation schemes 202
and Japan 339, 367, 375
malnutrition 683
manufacturing 373, 734
military operations overseas 345, 346,
    354, 363
money supply 414
national parks 288, 290
New Deal 290
Office of Foreign Disaster Assistance
    623
old people 742
overseas aid 339
in pan-region 335
plains 140, 185, 764, 845
population 736, 757, 763–4
quality of life 781–2
rainfall 60
religions 135
reserves 285
rural areas 762, 763–4
social problems 738, 742, 743
soil erosion 699
south-west 53, 56, 893
Spanish settlers 169–70
sphere of influence 371
sun belt 734, 736
and Third World wars 344
tornadoes 642–3
urbanization 730, 764
and UN 343, 362, 363, 364, 365
and USSR 342, 343, 345, 363
warming 489
wars 336, 356, 359, 410
water 533, 534, 541
western 61, 202, 642
wildfires 644
withdrawal 367
United States Geological Survey 431
universe 121–3
Unstead, J. F. 3
Upanishads 122
Ural-Altaic languages 108, 109, 111–12
uraninites 37, 41

urban bias 402
urbanism 865–6
urbanization 11, 214, 222, 223, 264, 265,
    831
    and colonization 705
    and consumption 209
    criticism 398
    in developed countries 391, 392
    development 307–8
    effects 259, 270–1, 291, 296, 300, 304
    in England 219, 250, 264, 296
    increasing 330
    and industrialization 398
    and land use 665
    level 252, 271, 392, 393, 394, 402
    measurement 392–3
    pace 393, 394, 402, 704
    patterns 398
    planning 709
    and pollution 715–17
    problems 271, 296, 397
    resisted 397–8
    selective 710
    spreading 704
    sustainable 724
    Third World 391–3, 394, 397–8, 402,
        403, 404, 702, 703, 704, 705, 709,
        716, 723
Urdu 117
Uruguay 265
USSR see Soviet Union

Vail curves 31, 33
valleys 30
Vedas 121, 122, 123, 132
vegetarianism 125, 126
vegetation 12
    affected by pesticides 698
    burning 138, 139, 315
    of dry areas 74
    change 192, 203
    and climate 485, 499, 500
    colonization 145
    dynamics 434
    European 51
    forest 44
    gradients 74, 437
    Holocene 57–8
    North American 57
    phases 51
    regeneration 177
    temperate 48

tropical 56
types 3–4
zones 56
Venezuela 36, 392
   debt 419, 422, 424
   oil boom 394, 395
   urbanization 394–5
Venice 152, 185, 215
Venus 37
Versailles 150
   conference 334, 335, 336
Vidal de la Blache, Paul, 3, 317, 818, 819,
   838, 846
Vienna 152
Vietnam 319–20
   farming 668–9
   pollution 716
   war 339, 344, 354, 356, 410
violence 163
Virginia 169, 172, 177, 186
vivisection 287
volcanoes 20, 27, 635–7
   distribution 25
   dormant 636
   dust 62, 636
   eruptions 63, 489, 557–8
   monitoring 636–7
   outgassing 36, 37, 38, 636
   prediction 636
   under-sea 509
Volcker, Paul 414, 418
von Neumann, John 435

wages 305
   rural 394
   rates 296
   rising 688, 689
   urban 394
Wales 116
   Chartist rising 300
   cities 264, 308
   local government 303
   moorlands 141
   rainfall 484
Wallerstein, Immanuel 321, 347
Walter, Eric 907, 917, 918
war 145, 148, 150–1, 161, 330, 945
   battle-deaths 354–5, 356
   casualties 354
   causes 338, 355–8
   civil 354, 355, 357–8
   classification 355

   and colonization 166, 213
   conduct 361
   and economics 356–7, 359
   effects 289, 357
   feudal 212
   finance 213, 357
   global 354, 356, 357
   ideological 355
   impact 354
   imperialist 355, 356
   interstate 353, 358
   invasion 358
   large-scale 353, 354
   length 356
   local 354, 355
   nationalist 353
   nuclear 349
   and oceans 510
   post-colonial 357–8, 359
   regulated 361
   small-scale 353
   state-making 355, 357
   territorial 213, 355, 358
   in Third World 344, 354, 355
   see also First World War; Second
      World War
Warsaw Pact 339, 361
waste 614–15
   disposal 463, 502–3, 518, 540, 716
   management 186
   radioactive 518, 523
   recycling 614–15
   toxic 518, 612
water 15, 526, 548
   availability 462, 527, 528, 529, 531,
      549
   conflict 546–8
   conservation 275, 540, 549
   control 149, 319–20
   costs 533
   crisis 526, 531, 549
   demand 531, 536, 539, 540, 544, 549
   desalination 533, 609
   distribution 30, 527
   exchange 498–9
   in gardens 150
   leakage 542
   and life 67, 72, 605
   management 145, 186, 282, 284,
      539–40, 542–4, 546, 549, 550
   origin 38
   and population growth 529, 549

polluted 231, 234, 235, 462, 534–9, 542
power 149, 202, 220, 221
pricing 540–1, 542
properties 38, 39
quality 535, 539, 542, 544
recycling 540, 609–10
resources 527–9, 530–1, 549
scarcity 480, 529, 531–2, 542, 548–9, 609
sluggish 203
supply 149, 256, 284–5, 300, 301, 303, 462, 529, 531, 532–3, 535, 536, 608–10, 716
transfer 50, 149, 533-4, 539
treatment 535, 536–7
use 529, 531, 532, 556, 608–9
vapour 488
wasted 534, 540, 545
waterlogging 202
watermills 149
wealth
    accumulation 480–1
    inequality 331
    movable 210
    pattern 209
    private 281
    and state power 213, 481
weapons 168, 197
    for hunting 141
    metal 186
    nuclear 347
    trade 359–60
weather
    change 480, 482
    forecasting 643
    prediction 492–3
    satellites 432
    severe 643
    variety 492
weathering 34, 82, 554, 557
Webb, Beatrice 303
Webb, Sidney 302, 303
Weber, Alfred 4
weeds 176, 186, 671
Wegener, Alfred 22, 958
welfare see social welfare
West Africa 165
    and slave trade 172–3, 174, 175
West Indies 195
    Creole languages 117
wetlands

boreal 501
drained 201–2
reclamation 666
whales 610
    hunting 149, 523, 610
    oil 137, 188, 610
wheat 142, 143, 171, 238
    exported 680–1
    as food aid 680–1
    imported 243
    research 239
    varieties 606
    yields 239, 240, 660, 661
Wheeler, W. M. 3
White, Gilbert 2
Whitehead, A. N. 3
Whittlesey, Derwent 4
Whorfe, Benjamin Lee 115
wildfires 643, 644
Wilkins, John 118
willow 74
Wilson cycle 78, 79
windmills 149
winds 185
    storms 642–3
    trade 185, 519
wolf 150
Wolpert, Lewis 445
women 793
    and agriculture 128, 890
    change in status 255–6, 472
    as colonists 165, 170
    differences 890
    education 259
    emigrants 165
    employment 170, 259, 740–1, 893–4
    environmental activism 896–7, 898–9
    experiences 888, 898
    indigenous 165, 168–9, 178
    as mothers 891–2, 895
    political activism 892–3, 896–7
    and religion 135
    responsibilities 891, 894–5, 896
    rights 889, 891
    rural 890
    sexist bias 889
    as single parents 740–1
    slave 165
    in society 809, 889–90, 949
    Third World 894
    urban 890, 893–4
wood

chips 592
for construction 148, 170, 189, 195, 200, 276, 281
for export 590
as fuel 138, 148, 189, 200, 462, 607, 896
ownership 221
products 607–8
from rain forests 584–5, 586, 588
for ship-building 188, 191, 195, 276
shortage 276
trade 169, 195, 584, 585, 590
tropical 584, 588
woodlands 51, 74, 141, 148, 199, 276
wool 160
WOCE *see* World Ocean Circulation Experiment
Wordsworth, William 912, 940
workhouses 297, 298
world 939–41
as society 941–2
World Bank 371, 413, 420, 421, 422, 425
and debt relief 426
and development 706, 708, 713, 714
estimates of undernutrition 686
Global Environment Facility 472
loans 590, 719
social programmes 718, 719
World Development Report 462, 716, 717
World Commission on Environment and Development 510, 654

World Conservation Strategy 464
World Ocean Circulation Experiment 496–7
world order 344
future 348
geopolitical 340, 347, 348
new 339, 343–4, 345, 346, 348
World Trade Organization 388
World Weather Watch 643
world-island 334, 340
Wright, Wilbur 431, 838
writing 107, 118–19, 153
alphabet 119
archives 318–19
development 210, 318
ideographic 118–19
pictographic 118–19
and religion 121
WTO *see* World Trade Organization

yam 110–11, 188
Yiddish language 117–18
Younger Dryas stadial 51, 63
Yucatán 80
Yugoslavia (former) 355, 359, 364

Zaïre 364, 410
Zelinsky, Wilbur 3
Zen 282
Zhoukoudian (China) 101
Zimbabwe 415, 717, 718, 721
Zinjanthropus 95
Zoroastrian religion 121, 122, 132